Handbook of Superconductivity

Handbook of Superconductivity

Fundamentals and Materials, Volume One

Second Edition

Edited by

David A. Cardwell

David C. Larbalestier

Aleksander I. Braginski

CRC Press
Taylor & Francis Group
Boca Raton London New York

CRC Press is an imprint of the
Taylor & Francis Group, an **informa** business

The cover figure shows a timeline of the discovery of new families of superconductors, courtesy of Pia Jensen Ray. Copyright © Pia Jensen Ray (www.pjray.dk)

Second Edition published 2023
by CRC Press
6000 Broken Sound Parkway NW, Suite 300, Boca Raton, FL 33487-2742

and by CRC Press
2 Park Square, Milton Park, Abingdon, Oxon, OX14 4RN

First Edition published by IOP Publishing 2003

CRC Press is an imprint of Taylor & Francis Group, LLC

ISBN: 978-1-4398-1732-2 (hbk)
ISBN: 978-0-367-68767-0 (pbk)
ISBN: 978-0-429-17918-1 (ebk)

DOI: 10.1201/9780429179181

Typeset in Minion Pro
by KnowledgeWorks Global Ltd.

Contents

Foreword

It is a pleasure to introduce the second edition of the *Handbook of Superconductivity*, now with the subtitle *Theory, Materials, Processing, Characterization and Applications*. In combination, the enlarged title expresses the very broad scope of this publication. It is a mark of the ongoing vigour of the field of superconductivity that, in the 15 years or so since the first edition, tremendous progress has been made in theory, materials discovery and applications. Completely new topics have emerged, including topological superconductors, single-atomic-layer superconductivity and twistronics. New superconductors have been predicted and demonstrated, most notably the clathrate superhydride LaH_{10}, which superconducts close to room temperature (though at several megabars). New applied technologies, such as flux pumps, have been demonstrated in motors, generators and MRI. This new edition is therefore timely in presenting the broad vista of our current knowledge in basic and applied superconductivity.

Superconductivity is one of the most remarkable physical states yet discovered. More than any other known effect, superconductivity brings quantum mechanics to the scale of the everyday world where a single, coherent quantum state may extend over a distance of metres – or even kilometres – depending on the size of a coil or length of superconducting wire. And perhaps less well known, the underlying physics extends in scale to the very size of the universe, for the Higgs mechanism, by which all matter particles acquire their mass, has the same symmetry-breaking origins as superconductivity.

There is something in this mysterious state that never fails to enchant its beholders: researcher, student or lay-person alike. The allure is to be found not only in the demonstrations of infinite conductivity and levitation, not only in the ongoing intellectual puzzle of its root physics, but also in the bold technologies that superconductivity enables.

The puzzle of superconductivity resisted explanation for a very long time. The eventual breakthrough with the Bardeen–Cooper–Schrieffer theory, 46 years after their discovery, actually preceded, by about a decade, any significant practical and commercial development of these remarkable materials. Today, half a century later still, the progress of science and technology is greatly accelerated. And yet a theory of the cuprate high temperature superconductors (HTS), discovered 33 years ago, still remains elusive. For these materials the tables are turned – a range of applications are now nudging their way onto the market even though we do not really understand them.

That, of course, overstates the matter. We know that the supercarriers in HTS are Cooper pairs and their symmetry in reciprocal space is predominantly *d*-wave. We quantitatively understand many properties of HTS materials such as the effect of impurities in suppressing superconductivity and the temperature dependence of the specific heat, thermal expansion and superfluid density. However, this description is primarily based on thermodynamics and the observed *d*-wave symmetry. What we lack is a clear understanding of the mechanism that binds the pairs in the first place and a relationship between the magnitude of this interaction and the energy scale of the superconductivity set by the maximum *d*-wave gap amplitude. The difficulty here is the strongly interacting electronic system and HTS cuprates, which, in this sense, are just part of a much wider problem of strongly correlated transition metal oxides that incorporate manganites, ruthenates, cuprates, vanadates and tungstates to name just a few, not to mention hybrid materials, such as the ruthenocuprates, in which magnetism and superconductivity coexist.

Such materials not only formally defy a suitable perturbation treatment, but they exhibit many different types of ground-state correlation that compete with each other. Thus, in the HTS cuprates, we currently struggle with the issues of charge ordering, spin ordering, nematicity and superconductivity and the question as to whether these are intimately linked or, in fact, compete. What if we could methodically remove each of these competing states one by one? Would superconductivity remain at all as the ground state? Would it fade away along with its supporters? Or might it even be enhanced? What is the importance of fluctuations or of short-range versus long-range correlations? A current challenge to the community is to develop the tools to do precisely these kinds of elucidating experiments.

The central approach then to these issues is systematic measurement in high-quality materials. With the combined improvement in quality of single crystals as well as resolution in low-energy experimental techniques, much progress has been made in the last several decades leading to many surprising new results. By "systematic" I mean variation in properties with carrier concentration, temperature, magnetic field, ion size, pressure and disorder. It is really only recently that the effect of small increments in carefully controlled doping levels have been employed across a variety of spectroscopies and these studies have uncovered abrupt changes in physical properties, such as a ground-state metal insulator transitions and the possibility in many cases of a quantum critical point driving the essential physics and phase behaviour.

Materials issues also lie at the heart of commercial application. Superconducting technologies, such as magnets, motors, power cables, transformers, flux pumps, NMR, MRI, telecommunications and computing, all push the present horizons of physical performance and their improvement, not to mention ultimate commercial success, predicated on ongoing materials development.

If any relatively uncharted territory were to be identified it might, even today, be high pressure. Not a few pressure-dependent studies have been reported and, somewhat recently, elemental superconductivity has been discovered at high pressure in iron, sulphur and lithium. Nonetheless, high pressures allow one to significantly modify the magnetic exchange interaction in oxides and much more could be done to explore the links between magnetism and other competing correlations through the combined investigation of pressure and doping-dependent systematics.

This might have been the situation up until just a few years ago. But beginning in 2015, dramatic developments in high pressure superconductivity have altered the superconductivity landscape forever. Demonstration of transition temperatures as high as 203 K were first reported for sulphur hydride at 155 GPa. Then, following what proved to be a remarkably accurate theoretical prediction, superconductivity was reported in LaH_{10} at 280 K at pressures around 200 GPa. The result is groundbreaking. It effectively achieves the century-old dream of room-temperature superconductivity – quantum mechanics is now brought to everyday temperatures. The only remaining barrier, and it is no small challenge, is to reduce these enormous operational pressures to ambient conditions – everyday pressures.

Such ongoing achievements continue to enliven the subject. But the point must be emphasized that in all these experiments pressure is used only to *enable* superconductivity. The challenge to the community is to utilize pressure as a tool to probe the essential physics by tuning spin or charge interactions to investigate their role in pairing. This entails the difficult task of incorporating spectroscopic studies in diamond-anvil experiments.

So, while much has been done in this remarkable field, there is much yet to be done. These intellectual and engineering challenges form part of the ongoing *puzzle and promise* of superconductivity. The present handbook provides a snap-shot of our current knowledge of the field. It is necessarily in introductory form, but its sheer scope illustrates just how broad and multidisciplinary this field is. More than that, the wide range of authors underscores a further critical element of science in any field, namely the organic human element. Those of us who have spent a good few years in the field have collaborated, debated and otherwise interacted with many scientists around the globe and, in the process, formed lasting friendships that otherwise would never have eventuated. Sadly, a number of members of this research community have passed away since the first edition of this handbook. I would note just two, as representative of the many others, Professors John Clem and Koichi Kitazawa who, with me, wrote prefaces to that first edition. Each played a critical role in the development of our subject, each were passionate advocates and each were gentlemen in the finest sense of the word. It is our task to continue the legacy of these and so many others who have built the field to its current highly complex and impactful status.

In the present edition, it is especially pleasing to see the names of so many research friends and collaborators as authors of the various sections of this handbook. Collectively, we trust that this volume will not only provide a comprehensive information base for the physics and phenomenology of superconductors, but will also communicate something of the puzzle and promise of superconductivity that continue to fascinate and motivate us as the years roll by.

Jeffery L. Tallon

Preface

This *Handbook of Superconductivity; Theory, Materials, Processing, Characterization and Applications* is the second, much expanded edition of the *Handbook of Superconducting Materials*, edited by David A. Cardwell and David Ginley and published originally by IOP in 2003. That large encyclopedic publication had quite favorable reviews, was rather quickly sold out to major distributors and is no longer readily available in retail. This situation alone would have justified the preparation of the 2nd edition. The second edition also became necessary for the reasons outlined below.

The past nearly 20 years have seen rapid and dynamic progress in superconducting materials, with many new compounds and entire isostructural families discovered and eventually synthesized into usable wires, thin film structures and other forms suitable for applications. Of the newly discovered materials, some exhibited the conventional, phonon-mediated Cooper pairing, while others, notably the Fe-based superconductors, still present a formidable challenge to be understood theoretically. The discovery of these new materials is another reason why the 2nd edition is overdue. Progress has been less dynamic in the development of theory and the elucidation of unconventional superconductivity mechanisms, including those of high critical temperature, T_c, cuprates or iron-containing pnictides. As a result, it has been possible to reprint in the 2nd edition a number of rather fundamental chapters published originally in the 1st edition, without or with only minor updates. In a few cases, such reproductions are also *in memoriam* of 1st and 2nd edition authors and section editors who have passed away over the past two decades. Brian Pippard, E. Helmut Brandt, Heinz Chaloupka, Harry Jones and, the most recently departed, Archie M. Campbell are so honored. The inclusion of reprinted chapters has resulted in some heterogeneity in the *Handbook*'s style, such as in referencing and citing references within the text.

Viable new applications of superconductivity have emerged in the interim time period of nearly 20 years, and, in some cases, have even became state-of-the art dominant technology, such as, for example, various types of novel highly sensitive radiation and particle detectors, which enabled astronomers to discover many distant galaxies and even distant exoplanets. Both established and novel superconductors have been useful in this particular application. Another example of a currently dominant technology in a completely novel field is that of superconducting qubits and circuits for quantum computing, which most recently reached the milestone of first "quantum supremacy" demonstration. Such new and dominant superconducting technologies have necessitated the addition of several new *Handbook* chapters and justified changing its title to the present one.

Although we have broadened and expanded this *Handbook*, we do not claim it covers the field of superconductivity in its entirety. Nevertheless, we hope it presents a relatively detailed and up-to-date introduction into that field, suitable for both graduate students and practitioners in experimental physics and multiple engineering disciplines: electronic and electrical, chemical, mechanical, metallurgy and others. Finally, the ultimate phase of our editorial activity coincided with the rapid progression of the Coronavirus pandemic in early 2020, which presented a particular challenge to the timely completion of this work. As a result, the Handbook contains some nonessential defects of presentation, such as occasional use of cgs units system rather than the SI units. We considered it tolerable given the need to publish the work as soon as possible.

David A. Cardwell
David C. Larbalestier
Aleksander I. Braginski

Acknowledgements

The Editors-in-Chief would like to acknowledge the contribution of a number of individuals and institutions who have played a major role in the development of the *Handbook of Superconductivity* in its evolution from the first edition to its production in a new and substantially revised format. The *Handbook* would not have been possible without their input.

We are particularly grateful to Lara Spieker, the senior editorial assistant at CRC Press/Taylor & Francis Group, for steering the *Handbook* skilfully from the author submission to the production stage. Her constructive and supportive approach throughout critical phases of the *Handbook* has been pivotal to its publication.

Jeffery L. Tallon, Peter B. Littlewood, Peter J. Lee, and Jianyi Jiang each made a contribution to the *Handbook* that went way beyond their responsibilities as section editors, invariably at critical times in the publication process, from advising on structure and content to managing the glossary and general format of the *Handbook*. All four managed to retain their enthusiasm for this exceptionally demanding project over an extended period and provided continuous support to the Editors-in-Chief. Doug Bennett provided helpful advice on the contents of Volume 3.

We acknowledge the contribution of Archie M. Campbell as a section editor and note with great sadness his death in November 2019. Archie made a fundamental contribution to both editions of the *Handbook* and he will be greatly missed as a friend, colleague and collaborator. We also note the great contributions of Harry Jones to the first edition of the *Handbook*. He was a section editor of the second edition but passed away in 2015 before being able to complete his efforts for this edition.

Finally, we would like to acknowledge the support of our home institutions: the Department of Engineering, University of Cambridge; the National High Field Magnet Laboratory at the University of Florida; and Forschungszentrum Jülich.

David A. Cardwell
David C. Larbalestier
Aleksander I. Braginski

Editors-in-Chief

David A. Cardwell
David C. Larbalestier
Aleksander I. Braginski

Section Editors and Advisory Board

David C. Larbalestier (A1, E3, G3, 2015-2019)
Archie M. Campbell (A2, G2, 2015-2019 †)
Alexander V. Gurevich (A2, 2017-2020)
Alexander A. Golubov (A2, 2019-2020)
David A. Cardwell (A3)
Peter J. Lee (B)
Jeffery L. Tallon (C)
Peter B. Littlewood (D)
Kazumasa Iida (E1, E2, E5)
Jianyi Jiang (E3, 2019-2020)
Michael Lorenz (E4, 2015-2018)
François Weiss (E3, E4, 2019-2020)
Ray Radebaugh (F1)
Lance D. Cooley (E3, G1)
Fedor Gömöry (G2, 2015-2019)
Antony Carrington (G3, 2019-2020)
Harry Jones (H1, † 2015)
John H. Durrell (H1, 2016-2020)
Mark Ainslie (H1, 2019-2020)
Aleksander I. Braginski (E4, H3, 2017-2020)
Horst Rogalla (H2, 2019-2020)

Contributors

Mark D. Ainslie (H1, H1.10, H1.11)
Department of Engineering
University of Cambridge
Cambridge, United Kingdom

Naoyuki Amemiya (H1.5)
Kyoto University
Kyoto, Japan

Tabea Arndt (H1.7)
Karlsruhe Institute of Technology
Institute for Technical Physics
Eggenstein-Leopoldshafen
and
Siemens AG
Corporate Technology (until September 2019)
Erlangen, Germany

Amalia Ballarino (E2.7, H1.4)
CERN, TE-MSC
Meyrin/Geneva, Switzerland

Nobuya Banno (E3.9)
Low-Temperature Superconducting Wire Group
National Institute for Materials Science
Ibaraki, Japan

Kees van der Beek (A3.2)
Centre National de la Recherche Scientifique
Université Paris-Saclay
Palaiseau, Paris
France

Peter A. Beharrell (H1.13)
Department of Physics and Astronomy
University of Southampton
Southampton, United Kingdom

Kamran Behnia (G3.2)
Laboratoire Physique et Étude de Matériaux (CNRS-UPMC)
Ecole Supérieure de Physique et de Chimie Industrielles
Paris, France

Victor Belitsky (H4.6)
Department of Earth and Space Sciences
Chalmers University of Technology
Gothenburg, Sweden

Emilio Bellingeri (C3)
Consiglio Nazionale delle Ricerche
Genova, Italy

Simon Bending (G2.10)
Department of Physics
University of Bath
Claverton Down
Bath, United Kingdom

Douglas A. Bennett (H4.2)
Quantum Sensors Group
National Institute of Standards and Technology
Boulder, California

Luca Bottura (E2.7)
CERN, TE-MSC
Geneva, Switzerland

Aleksander I. Braginski (E4.6, H3)
Forschungszentrum Jülich (FZJ), retired
Jülich, Germany

Marcel ter Brake (F1.4)
Faculty of Science and Technology
University of Twente
NB Enschede, the Netherlands

E. Helmut Brandt (A3.1) † 2011
Max-Plank-Institut für Metallforschung
Stuttgart, Germany

Archie M. Campbell (A2.1, G2.1) † 2019
Department of Engineering
University of Cambridge
Cambridge, United Kingdom

Ahmet Cansiz (H1.8)
Istanbul Technical University
Sarıyer/Istanbul, Turkey

Haishan Cao (F1.4)
Department of Energy and Power Engineering
Tsinghua University
Beijing, China

David A. Cardwell (A1, A3, E2.2)
Department of Engineering
University of Cambridge
Cambridge, United Kingdom

Antony Carrington (G3, G3.1)
H.H. Wills Laboratory of Physics
University of Bristol
Bristol, England, United Kingdom

Francesco Cerutti (E2.7)
CERN, EN-STI
Geneva, Switzerland

Heinz J. Chaloupka (H2.3) † 2014
Department of High-frequency Technology and Communication
University of Wuppertal
Wuppertal, Germany

Serguei Cherednichenko (H4.6)
Department of Microtechnology and Nanoscience
Chalmers University of Technology
Gothenburg, Sweden

Ching-Wu Chu (D4)
Department of Physics and Texas Center for Superconductivity
University of Houston
Houston, Texans

Gianluigi Ciovati (H1.14)
Thomas Jefferson National Accelerator Facility
Newport News
Virginia

Edward W. Collings (G2.7)
Center for Superconducting and Magnetic Materials
Materials Science Department
The Ohio State University
Columbus, Ohio

Lance D. Cooley (E3.7, G1, G1.1)
Applied Superconductivity Center
National High Magnetic Field Laboratory
Florida State University
Tallahassee, Florida

John R. Cooper (G2.11)
Cavendish Laboratory
University of Cambridge
Cambridge, United Kingdom

Timothy Davies (E5.3)
Department of Materials
University of Oxford
Oxford, United Kingdom

Nuria Del-Valle (G2.6)
Departament de Fisica
Universitat Autonoma de Barcelona
Bellaterra, Barcelona
Catalonia, Spain

Gianluca De Marzi (B1)
Superconductivity Section
Fusion and Technology for Nuclear Safety and Security Department
ENEA, Frascati Research Centre
Frascati RM, Italy

Liangzi Deng (D4)
Department of Physics and Texas Center for Superconductivity
University of Houston
Houston, Texans

Marc Dhallé (G2.3)
University of Twente
NB Enschede, The Netherlands

Pavel Diko (G1.5)
Institute of Experimental Physics
Slovak Academy of Sciences
Košice, Slovak Republic

John H. Durrell (H1)
Department of Engineering
University of Cambridge
Cambridge, United Kingdom

Michael Eisterer (G2.5)
Institute of Atomic and Subatomic Physics
Technische Universität Wien
Vienna, Austria

Jack W. Ekin (E5.1)
National Institute of Standards and Technology (NIST), retired
Boulder, Colorado

Christian Enss (H4.4)
Kirchhoff Institute of Physics (KIP)
University of Heidelberg
INF 227
Heidelberg, Germany

Andreas Fleischmann (H4.4)
Kirchhoff Institute of Physics (KIP)
University of Heidelberg,
INF 227
Heidelberg, Germany

Jaap Flokstra (H3.2)
Faculty of Science and Technology, Emeritus
University of Twente
NB Enschede, The Netherlands

René Flükiger (C3, E2.5, E2.7, E3.6, E3.11)
Department of Quantum Matter Physics
University of Geneva
Geneva, Switzerland

Stephan Friedrich (H4.1)
Lawrence Livermore National Laboratory
Livermore, California

John Gallop (H2, H3, H4.7)
National Physical Laboratory, Emeritus
Teddington, Middlesex, United Kingdom

Loredana Gastaldo (H4.4)
Kirchhoff Institute of Physics (KIP)
University of Heidelberg,
INF 227
Heidelberg, Germany

Wilfried Goldacker (G3.4)
Institute of Technical Physics
Karlsruhe Institute of Technology
Eggenstein-Leopoldshafen, Germany

Alexander A. Golubov (A2.9)
Institute for Nanotechnology
University of Twente
NB Enschede, The Netherlands

Fedor Gömöry (G2)
Department of Superconductors
Institute of Electrical Engineering
Slovak Academy of Sciences
Bratislava, Slovakia

Colin Gough (A2.7)
Department of Physics and Astronomy, retired
University of Birmingham
Birmingham, United Kingdom

Francesco Grilli (G2.2)
Karlsruhe Institute of Technology
Institute for Technical Physics
Hermann-von-Helmholtz-Platz 1
Eggenstein-Leopoldshafen, Germany

Chris Grovenor (E5.3)
Department of Materials
University of Oxford
Oxford, United Kingdom

Alexander V. Gurevich (A2)
Department of Physics
Old Dominion University Norfolk, Virginia

Damian P. Hampshire (B3, G2.4)
Department of Physics
Durham University
Durham, United Kingdom

Ling Hao (H4.7)
National Physical Laboratory
Teddington, Middlesex, United Kingdom

Eric E. Hellstrom (E3.1 and E3.12)
Applied Superconductivity Center
National High Magnetic Field Laboratory
Florida State University
Tallahassee, Florida

Bernhard Holzapfel (E4.1)
Institute for Technical Physics
Karlsruhe Institute of Technology
Eggenstein-Leopoldshafen, Germany

Hideo Hosono (C5)
Laboratory for Materials and Structures
Tokyo Institute of Technology
Yokohama, Japan

Rudolf P. Huebener (A2.4, A2.5)
Experimentalphysik II, retired
Universität Tubingen
Tubingen, Germany

Kazumasa Iida (E1, E2, E5)
Department of Materials Physics
Nagoya University
Nagoya, Japan

Yasuo Iijima (E3.9)
Low-Temperature Superconducting Wire Group
National Institute for Materials Science
Ibaraki, Japan

Yoshihiro Iwasa (D3)
Department of Applied Physics and Quantum-Phase Electronics Center
The University of Tokyo
Tokyo, Japan

Jianyi Jiang (E3, E3.1)
Applied Superconductivity Center
National High Magnetic Field Laboratory
Florida State University
Tallahassee, Florida

Carmen Jimenez (E4.3)
Université Grenoble Alpes, CNRS, Grenoble INP*, LMGP
Institute of Engineering Univ. Grenoble Alpes
Grenoble, France

Harry Jones (H1.1) † 2015
Clarendon Laboratory
University of Oxford
Oxford, United Kingdom

Stephen R. Julian (D1)
Department of Physics
University of Toronto
Ontario, Canada

Alan M. Kadin (H3.6)
Consultant
Princeton Junction
Princeton, New Jersey

Debra L. Kaiser (E2.4)
Office of Data and Informatics, MML, NIST
Gaithersburg, Maryland

Fumitake Kametani (G1.3)
Department of Mechanical Engineering
FAMU-FSU College of Engineering
Florida State University
Tallahassee, Florida

Sebastian Kempf (H4.4)
Kirchhoff Institute of Physics (KIP)
University of Heidelberg, INF 227
Heidelberg, Germany

Peter H. Kes (A3.2)
Kammerlingh-Onnes Laboratory
University of Leiden
Leiden, The Netherlands

Simon A. Keys (G2.4)
Department of Physics
Durham University
Durham, United Kingdom

Akihiro Kikuchi (E3.9)
Low-Temperature Superconducting Wire Group
National Institute for Materials Science
Ibaraki, Japan

Caroline A. Kilbourne (H4)
NASA/Goddard Space Flight Center
Greenbelt, Maryland

Johannes Kohlmann (E4.5)
Department Quantum Electronics
Physikalisch-Technische Bundesanstalt (PTB)
Braunschweig, Germany

Victor K. Kornev (H2.3)
Department of Physical Electronics
Lomonosov Moscow State University
Moscow, Russia

Panagiotis Kotetes (D7)
CAS Key Laboratory of Theoretical Physics
Institute of Theoretical Physics
Chinese Academy of Sciences
Beijing, China

Hans-Joachim Krause (H3.4)
Research Center Jülich
(Forschungszentrum Jülich, FZJ)
Jülich, Germany

M'hamed Lakrimi (H1.1.2)
Siemens HC Ltd
Wharf Road
Oxford, United Kingdom

David C. Larbalestier (A1, E3.7, G3.5)
National High Magnetic Field Laboratory
Florida State University
Tallahassee, Florida

Peter J. Lee (B, E3.7, G1.4)
The Applied Superconductivity Center
National High Magnetic Field Laboratory
Florida State University
Tallahassee, Florida

Steve Lee (G2.10)
School of Physics and Astronomy, SUPA
University of St. Andrews
St. Andrews, United Kingdom

Anthony J. Leggett (A2.2)
Department of Physics, Emeritus
University of Illinois at Urbana-Champaign
Urbana, Illinois

Peter B. Littlewood (D)
James Franck Institute and Department of Physics
University of Chicago
Chicago, Illinois

Nicholas J. Long (H1.9)
Robinson Research Institute
Victoria University of Wellington
Lower Hutt/Wellington, New Zealand

Ralph Longsworth (F1.3)
Engineering Department
Sumitomo (SHI) Cryogenics of America, Inc.
Pennsylvania

Michael Lorenz (E4)
Felix Bloch Institute for Solid State Physics
Universität Leipzig
Leipzi, Germany

Bing Lv (D4)
Department of Physics and Texas Center for Superconductivity
University of Houston
Houston, Texans

Judith L. MacManus-Driscoll (E3.4)
Department of Materials Science and Metallurgy
University of Cambridge
Cambridge, United Kingdom

Milan Majoros (G2.7)
Center for Superconducting and Magnetic Materials
Materials Science Department
The Ohio State University
Columbus, Ohio

Arkadiy Matsekh (H1.12)
Foucault Dynamics
Gold Coast, Australia

Benjamin A. Mazin (H4.3)
Department of Physics
University of California
Santa Barbara, California

Christoph Meingast (G3.3)
Institute for Quantum Materials and Technologies
Karlsruhe Institute of Technology
Karlsruhe, Germany

Brian H. Moeckly (E4.6)
Commonwealth Fusion Systems
Milpitas, California

Antonio Morandi (H1.6)
University of Bologna
Bologna, Italy

Michael Mück (H3.4)
Institut für Angewandte Physik
Justus-Liebig-Universität Gießen
Gießen, Germany

Oleg A. Mukhanov (H3.5, H3.6)
Seeqc, Inc.
Elmsford, New York

K. Alex Müller (A1.3)
Department of Physics
University of Zurich
Zurich, Switzerland

Luigi Muzzi (B1)
Superconductivity Section
Fusion and Technology for Nuclear Safety and Security Department
ENEA, Frascati Research Centre
Frascati RM, Italy

Carles Navau (G2.6)
Departament de Fisica
Universitat Autonoma de Barcelona
Bellaterra, Barcelona
Catalonia, Spain

Daniel E. Oates (H2.1)
Quantum Information and Integrated Nanosystems Group MIT Lincoln Laboratory
Massachusetts Institute of Technology
Lexington, Massachusetts

Terry P. Orlando (A1.2)
Department of Electrical Engineering and Computer Science
Massachusetts Institute of Technology (MIT)
Cambridge, Massachusetts

Michael Parizh (H1.3)
GE Research
Niskayuna, New York

John M. Pfotenhauer (F1.2, F1.5)
Department of Mechanical Engineering
University of Wisconsin – Madison
Madison, Wisconsin

Brian Pippard (A1.1) † 2008
Cavendish Laboratory
University of Cambridge
Cambridge, United Kingdom

Britton Plourde (H3.7)
Department of Physics
Syracuse University
Syracuse, New York

Anatolii A. Polyanskii (G3.5)
Applied Superconductivity Center
National High Magnetic Field Laboratory
Florida State University
Tallahassee, Florida

Ian Pong (E3.8)
Accelerator Technology and Applied Physics Division
Lawrence Berkeley National Laboratory
Berkeley, California

Adrian Porch (A2.6, G2.9)
School of Engineering
University of Cardiff
Cardiff, United Kingdom

Kosmas Prassides (D3)
Soft Materials Group
AIMR
Sendai, Japan

Ray Radebaugh (F1, F1.1)
Applied Chemicals and Materials Division
National Institute of Standards and Technology (NIST), retired
Boulder, Colorado

Mark J. Raine (G2.4)
Department of Physics
Durham University
Durham, United Kingdom

Mark O. Rikel (E2.1)
Nexans SuperConductors GmbH
Hannover, Germany

Horst Rogalla (H2)
Electrical, Computer and Energy Engineering Department (ECEE)
University of Colorado
Boulder, Colorado

Alain Rüfenacht (H4.7)
Superconducting Electronics Group
National Institute of Standards and Technology (NIST)
Boulder, Colorado

Athena Safa Sefat (E3.3)
Materials Science and Technology Division
Oak Ridge National Laboratory
Oak Ridge, Tennessee

Gunzi Saito (D2)
Toyota Physical and Chemical Research Institute
Nagakute
Aichi, Japan

Charlie Sanabria (G1.4)
Commonwealth Fusion Systems
Cambridge, Massachusetts

Alvaro Sanchez (G2.6)
Departament de Fisica
Universitat Autonoma de Barcelona
Bellaterra, Barcelona
Catalonia, Spain

Tilmann H. Sander Thoemmes (H3.3)
Department of Biosignals
PTB – The National Metrology Institute of Germany
Berlin, Germany

Kenichi Sato (E3.2)
Cryogenics and Superconductivity Society of Japan
Tokyo, Japan

Christian Scheuerlein (E5.2, G1.2)
TE Department – Magnets, Superconductors and Cryostats (MSC)
European Organization for Nuclear Research (CERN)
Geneva, Switzerland

Jörg Schmalian (D6)
Institute for Theoretical Condensed Matter Physics
Karlsruher Institut für Technologie (KIT)
Karlsruhe, Germany

Lynn F. Schneemeyer (E2.4)
Montclair State University
Montclair, New Jersey

Thomas Schurig (E4.5)
Kryophysik und Spektrometrie
Physikalisch-Technische Bundesanstalt (PTB), retired
Berlin, Germany

Bernd Seeber (E3.10)
scMetrology SARL
Geneva, Switzerland

Paul Seidel (E4.4, H3.2)
Institute for Solid State Physics
Faculty of Physics and Astronomy
Friedrich Schiller University of Jena
Jena, Germany

Yunhua Shi (E2.2)
Department of Engineering
University of Cambridge
Cambridge, United Kingdom

Jun-ichi Shimoyama (C2, E2.3)
Department of Physics and Mathematics
Aoyama Gakuin University
Tokyo, Japan

Theo Siegrist (G1.1)
Department of Chemical and Biomedical Engineering
FAMU-FSU College of Engineering
Florida State University
Tallahassee, Florida

Alejandro V. Silhanek (G2.10)
Experimental Physics of Nanostructured Materials
Q-MAT, CESAM
Université de Liège
Liège, Belgium

Enrico Silva (A2.6)
Department of Engineering
Università Degli Studi Roma Tre
Rome, Italy

Roman Sobolewski (H4.5)
Department of Electrical and Computer Engineering
University of Rochester
Rochester, New York

Susie Speller (E5.3)
Department of Materials
University of Oxford
Oxford, United Kingdom

Tiziana Spina (E2.7)
Applied Physics and Superconducting Technology Division
Fermi National Accelerator Laboratory
Batavia, Illinois

Wolfgang Stautner (H1.3)
GE Research
Niskayuna, New York

Michael D. Sumption (G2.7)
Center for Superconducting and Magnetic Materials
Materials Science Department
The Ohio State University
Columbus, Ohio

Francesco Tafuri (A2.9, H3.1)
Department of Physics
University of Napoli Federico II
Napoli, Italy

Takao Takeuchi (E3.9)
Human Resources Division
National Institute for Materials Science
Ibaraki, Japan

Jeffery L. Tallon (Foreword, C, C1, C6)
Robinson Research Institute
Victoria University of Wellington
Lower Hutt, New Zealand

Saburo Tanaka (H3.4)
Toyohashi University of Technology, Toyohashi, Japan

Chiara Tarantini (B2)
The Applied Superconductivity Center
National High Magnetic Field Laboratory
Florida State University
Tallahassee, Florida

Edward J. Tarte (A2.8)
Department of Electronic, Electrical and Systems Engineering
University of Birmingham
Birmingham, United Kingdom

Sergey K. Tolpygo (E4.5)
Lincoln Laboratory
Massachusetts Institute of Technology (MIT)
Lexington, Massachusetts

Volker Tympel (E4.4)
Helmholtz Institute Jena
Helmholtzweg 5
Jena, Germany

Ruggero Vaglio (A2.6)
Department of Physics
University of Napoli Federico II
Napoli, Italy

Philippe Vanderbemden (G2.8)
Department of Electrical Engineering and Computer Science
University of Liège
Liège, Belgium

Orest G. Vendik (H2.2)
Department of Physical Electronics and Technology
St. Petersburg Electrotechnical University
St. Petersburg, Russia

William F. "Joe" Vinen (A1.2), retired
Department of Physics and Astronomy
University of Birmingham
Birmingham, United Kingdom

Ming-Jye Wang (D5)
Institute of Astronomy and Astrophysics, Academia Sinica
Taipei, Taiwan

James H. P. Watson (H1.13)
Department of Physics and Astronomy
University of Southampton
Southampton, United Kingdom

Harald W. Weber (E2.6)
TU Wien – Atominstitut
Wien, Austria

François Weiss (E4, E4.3)
Université Grenoble Alpes, CNRS, Grenoble INP*, LMGP
Institute of Engineering Univ. Grenoble Alpes
Grenoble, France

Jeremy D. Weiss (E3.12)
Advanced Conductor Technologies LLC
Boulder, Colorado

David Welch (A2.3)
Brookhaven National Laboratory
New York, New York

Frank N. Werfel (E2.1)
Adelwitz Technologiezentrum GmbH (ATZ)
Torgau, Germany

Jonathan White (G1.6)
Laboratory for Neutron Scattering and Imaging
Paul Scherrer Institute
Villigen, Switzerland

Jörg Wiesmann (E4.1)
Incoatec GmbH
Geesthacht, Germany

Frank K. Wilhelm-Mauch (H3.7)
Institute for Quantum Computing Analytics
Research Center Jülich
Jülich, Germany

Dag Winkler (H4.6)
Department of Microtechnology and Nanoscience
Chalmers University of Technology
Gothenburg, Sweden

Roger Wördenweber (E4.2)
Forschungszentrum Jülich (FZJ)
Jülich, Germany

Judy Z. Wu (C4)
Department of Physics and astronomy
University of Kansas
Kansas City
Kansas

Maw-Kuen Wu (D5)
Institute of Physics, Academia Sinica
Taipei, Taiwan

Phillip M. Wu (D5)
Solarcity Inc.
San Mateo, California

Mingyao Xu (F1.3)
Engineering Department
Sumitomo (SHI) Cryogenics of America, Inc.
Pennsylvania

Akiyasu Yamamoto (E3.11)
Department of Applied Physics
Tokyo University of Agriculture and Technology
Tokyo, Japan

Ayako Yamamoto (E3.5)
Shibaura Institute of Technology
Graduate School of Engineering and Science
Tokyo, Japan

Yukihiro Yoshida (D2)
Department of Agriculture
Meijo University
Nagoya, Japan

Xiaoqin Zhi (F1.2)
College of Energy Engineering
Zhejiang University
Zhejiang, China

Part A

Fundamentals of Superconductivity

A1

Introduction to Section A1: History, Mechanisms and Materials

David A. Cardwell and David C. Larbalestier

Section A1 is the cornerstone of the handbook, acting not only as a general introduction to superconductivity, but providing significant insight into its overall organization and motivation. Chapter A1.1 is an excellent account of the history of the subject, followed in Chapter A1.2 by a basic introduction into the nature of superconductivity. Finally, Chapter A1.3 presents a detailed and up-to-date description of the evidence, both experimental and theoretical, that the basis for superconductivity in the high-temperature superconducting cuprates is polaronic.

Chapter A1.1 delves into the history of superconductivity via a series of key events in the field that defined the subject. This begins with Kamerlingh Onnes' preconceptions of the low-temperature behaviour of the resistivity of metals and the shock of observing a sudden loss of resistance to the flow of DC current in mercury at 4.2 K. This led to a number of early theories based on conjected mechanisms for the zero-resistance state and, notably, the failure to manufacture a high-field superconducting solenoid. With the subsequent discovery of the Meissner effect, this led to the analysis of superconducting phenomenon as a thermodynamic phase change and, eventually, to the London equations. These led, in turn, to the development of a two-fluid model of charge flow and the concepts of magnetic penetration depth, coherence length and the existence of non-local effects. The early work culminated with the development of the Ginzburg–Landau equations, which resulted in the identification and initial understanding of a number of important superconducting phenomena, including the importance of the sign of the surface energy at the S–N interface to classify types I and II superconductors, Abrikosov vortices, flux quantization and the role of non-superconducting impurities in flux pinning. The intense interest of leading theorists such as Heisenberg and Frőhlich in developing theories of the superconducting state is described. Finally, there came the watershed of the Bardeen, Cooper and Schrieffer (BCS) theory, which seemingly laid to rest most of the then unanswered theoretical questions. The advent of the high T_c cuprates was subsequently to revitalize interest in superconducting theory more than 30 years after the BCS Nobel prize-winning work.

Chapter A1.2 presents a detailed discussion of the nature of the superconducting state. This begins with the observation that superconductivity derives from an ordering transition, which leads to the occurrence of zero resistance and the Meissner effect. The Ginzburg–Landau and BCS theories describe this ordered state from thermodynamic and physical principles. The chapter builds on fundamental theory to discuss persistent currents and trapped flux, which leads to a discussion of the mixed state and the nature of types I and II superconductors. The chapter concludes with a description of unconventional superconductors and briefly introduces the high T_c cuprates.

Chapter A1.3 provides a detailed insight into the search for superconductivity in complex materials with possible Jahn–Teller polarons, written by Alex Muller, who co-discovered the first HTS materials with Georg Bednorz in 1986. His conviction that stronger interactions were possible with the electron-phonon interaction of metallic superconductors is what motivated his search for superconductivity in nickelates and cuprates. The world took intense notice, especially in 1987 when the transition temperature soared above the boiling point of liquid nitrogen with the discovery of $YBa_2Cu_3O_{7-\delta}$ with T_c of 92 K. The detailed description of the experimental evidence for polaronic superconductivity is especially interesting given the mention by Pippard in Chapter A1.1 of Froehlich's conjecture that attractive interactions were possible between electrons when there was a polaronic distortion in an ionic lattice.

Overall, Section A1 provides a context for the history of the subject, the underlying physics of superconductivity and the form and nature of high-temperature superconducting materials required to realize practical applications.

A key attribute of superconductivity that has kept us both in its orbit through most of our long careers has been its continuous surprise, its deep physical complexity and the very wide range of material types, which support the superconducting state, and, after 1960, the realization of superconducting devices with unique and valuable capabilities. It is useful, thus, to read the article by Brian Pippard who grew up in the still very poorly formed state of the field in the late 1940s, when none of the applications that started to emerge in the 1960s could be considered. The challenges described by Pippard were then still largely intellectual, above all to understand the physics behind the "strange" mechanism that allowed perfect conductivity and bulk diamagnetism. Reading it now reminds us of the scientific freshness of a poorly understood field with many dead-ends, even when some of the best minds were cracking their heads against the very tough and stubborn problem of the superconducting mechanism. Also clear in Pippard's article is the great localization of science into many national stovepipes with only imperfect contact between them. In particular, he speaks of the failure of Western scientists to appreciate the Russian literature. But it is also true that the Russians themselves did not fully appreciate all that had gone on within their own borders, especially the work of Shubnikov in Kharkov in 1934–1936 when unambiguous evidence of the transition from type I to type II superconductivity was seen. Landau who had been in the same institute as Shubnikov at the same time did not take Shubnikov's experiments into account when he and Ginzburg developed their foundation theory in 1950. When his student Abrikosov considered the case of a high kappa superconductor a couple of years later, Landau discouraged his publication of the vortex state in a high kappa superconductor with negative, not positive surface energy. Only after the discovery of vortices in Helium did Landau relent, but this still had no effect on applications until very late 1960 when Kunzler and colleagues, unguided by theory but curious, showed that a primitive Nb_3Sn wire could carry more than 1000 Amm^{-2} at almost 9 T. Within 12 months, multiple groups had made superconducting solenoids out of Nb_3Sn and Nb-Ti with fields above 6 T, fields never before achieved except by brute force and very high-power dissipation in copper solenoids. We now have superconducting magnets with magnetic fields above 30 T made out of the polaronic cuprates described by Müller in Chapter A1.3.

A1.1

Historical Development of Superconductivity

Brian Pippard

The story of superconductivity falls into rather well-marked periods. From the discovery in 1911 until 1933, exploration was largely confined to Leiden, Toronto and Berlin, almost the only laboratories with liquid helium. A great change came with the discovery of the Meissner effect in 1933, which stimulated widespread interest in superconductivity at the same time as the exodus of Jews from Germany was causing a diffusion of low-temperature physics. After 1945, the post-war reconstruction of European research also saw the emergence of the United States as a leading participant (greatly helped by the Collins liquefier and unprecedented government funding). With the revelation of the BCS theory in 1957 and the introduction of superconducting solenoids at about the same time, a new era opened; we shall take this to define the end of our historical period.

Early in 1911, Kamerlingh Onnes measured the resistance of a thin thread of mercury and was not surprised that it fell to an imperceptibly low value as the temperature was reduced below 4.2 K [1] (see Figure A1.1.1). He had, after all, chosen mercury precisely because it could be made pure enough to show clearly the steady fall of resistivity to zero that he expected in an ideal metal. Only when he and Holst repeated the measurement with greater sensitivity did they experience the shock of seeing an abrupt disappearance just below the normal boiling point of helium. Within months, superconductivity* had been found in lead and tin, and the now voluminous catalogue of superconducting elements, alloys and compounds began to grow. By observing the persistence for hours of a current induced in a lead ring, Onnes convinced himself that below a well-defined transition temperature the resistivity was, in Casimir's words, 'the zeroest quantity we know' [2, 3]. By 1914, it had been found that a moderate magnetic field sufficed to restore the resistance, and that too large a current had the same effect (and could explode the wire); the connection between the two effects was noted in 1916 by Silsbee [4]. It was thereafter taken for granted that solenoids wound with superconducting wires were useless for strong fields, until about 1954 when Yntema achieved 7 kG with a niobium-wound coil [5]; to be sure, it had an iron yoke, but this probably marks the moment when hope was renewed. Within a few years high-field solenoids began to appear, as will be mentioned later.

Once superconductivity had been found in several metals, it was an open season for the theorists, and in the next 20 years, almost as many hopeful theories were proposed and forgotten. All started, naturally enough, from the premise of perfect conductivity for which some mechanism had to be found that would allow unopposed movement of the electrons; condensation into a rigid lattice was a favourite. Since during most of this period there was no quantum mechanics, let alone a coherent picture of metallic conduction, there was unlimited scope for guesswork. With regret, we pass over this phase to take notice of the radical change that began in 1933 with the experiments of Meissner and Ochsenfeld, in Berlin, on the spontaneous expulsion of flux in the transition to superconductivity with a magnetic field present [6].

It should not be supposed that this important result made the truth immediately plain to all; it was not just a matter of cooling an ideal sample in a magnetic field and observing the complete expulsion of flux at the transition. The use of both solid and hollow cylinders confused the issue, and even with solid cylinders, expulsion was far from complete. It is not surprising that behaviour so unexpected should take time to be grasped, and the now-accepted interpretation, that $B = 0$ in a superconductor came not from Berlin but from Gorter in Haarlem [7]. He had been impressed by the success of thermodynamics in accounting for the temperature variation of the critical magnetic field, and puzzled because the argument was invalid if the transition at H_c was not reversible. The new observation convinced him (though Meissner was doubtful at first) that indeed it was, and that superconductivity was something more than perfect conductivity. It was not long before repetitions of the experiment, with variations, set doubts at rest, and in 1935, F and H London published their phenomenological theory that took Gorter's interpretation as the starting-point [8, 9]. They accepted the earlier idea that a

* Onnes's original name, whose form survives in the German Supraleitung, was gradually replaced by superconductivity as English became the preferred language during the 1930s.

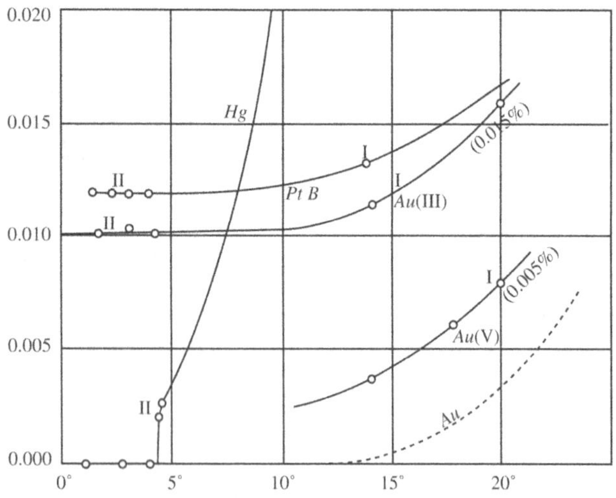

FIGURE A1.1.1 At the first Solvay conference in November 1911, Kamerlingh Onnes presented an account of the extraordinary resistive behaviour of mercury at about 4 K. The resistance of gold and platinum wires drops steadily to a constant value as the temperature is lowered; in mercury, there is a sudden drop to an imperceptibly low value.

supercurrent was accelerated by an electric field, $\Lambda\dot{J} = E$, but were dissatisfied because it implied too many solutions – it was not enough for them that the internal magnetic field should be unchangeable, it must be zero†. This they achieved with the postulate $\Lambda \boldsymbol{J} + \boldsymbol{A} = 0$. Their first paper noted that such a relation would follow from the quantum-mechanical expression for $\boldsymbol{J}$ if only it could be assumed that a moderate magnetic field was powerless to perturb the electronic wave functions. This is not so in a normal metal, and they suggested that interactions between the electrons might confer the required rigidity. It is not clear how much notice was taken of this proposal at the time, but in the following years, the latest round of tentative (and still abortive) molecular theories tended to seek an energy gap separating condensed and excited electron states. This might indeed make the condensed electron assembly unresponsive to a magnetic field, as in an insulator, but as yet no theories met the stiffer requirement of an energy gap that was compatible with a conduction current.

Meanwhile the new phenomenologies, thermodynamic (Gorter) and electromagnetic (London), encouraged attack from several quarters on the different problem of phase stability [10]. Penetration of magnetic field into a thin superconducting lamina allows it to survive in fields greater than H_c; why then is the transition of a pure metal at H_c so abrupt, yet in an alloy so gradual, with persistence of perfect conductivity even after complete field penetration? Mendelssohn was most concerned with alloys and gave inhomogeneity as the reason – a network (sponge) of thin superconducting filaments of different composition from the rest could survive to higher fields.

Gorter was of the opinion that in a pure metal, there was a minimum size for a superconducting domain, so as to preclude a fine mixture of phases, stable in high fields. Heinz London postulated a surface energy at a phase boundary that would have the same discouraging effect. Gorter [11] admitted that a minimum size for normal domains was equally necessary, but any worries this may have caused him were small compared to what troubled both him and London – why should alloys be more tolerant than pure metals of finely divided phases? Little light was cast until 1950, but the likely existence of an interface energy remained in the minds of those concerned with a related problem, the structure of the intermediate state. In essence, the problem is that the magnetic field can be uniform only at the surface of a long and thin superconductor. With any other shape, some parts should reach H_c before others, and it was clear that then the superconducting body must become partially normal. Only phase division into thin laminae can prevent the field, H_c at each interface, being smaller somewhere in the normal phase. Landau in 1937 made the first theoretical attempt to resolve the difficulty, and there have been many delicate probings of the phase configuration at the surface, and progressively more sophisticated attempts to explain the observed patterns. But this is a study in itself, and we must be content to quote a source of references and return to the development of phenomenological models, especially those that came to be known, by analogy with similar ideas about superfluid helium, as two-fluid models [12, 13].

The idea of a two-fluid model, with coexistent 'superelectrons' and normal electrons, was probably not new when Heinz London wrote his doctoral thesis in 1933, but it seems that the first published mention was by Casimir and Gorter in 1934. London had searched unsuccessfully for resistive loss when the presumed normal electrons are stimulated by a high-frequency electric field in the penetration layer. His frequency of 40 MHz was far too low, but he succeeded in 1940 with 1500 MHz [14]; by then the two-fluid model was well established as a theoretical concept. In the explicit form of Casimir and Gorter, it reproduced the thermodynamic behaviour, including the second-order critical behaviour in zero magnetic field – a recent and controversial invention by Ehrenfest, which Laue and others (mistakenly) attacked as thermodynamically impossible. Casimir and Gorter postulated a steady diminution of the normal fraction, x, as the temperature fell below T_c and a corresponding increase in the superconducting fraction $(1 - x)$. The superelectrons, carrying no entropy, contributed free energy proportional to $-(1 - x)$, while the normal electrons contributed not the expected $-\gamma T^2 x$ but $-\gamma T^2 x^{1/2}$. When x took the value t^4, i.e. $(T/T_c)^4$, that minimized the free energy, the resulting critical magnetic field had the observed parabolic temperature variation. The model, an interesting artificial construction, received little attention until it proved relevant to measurements of the penetration depth [15, 16].

It was inherent in the London phenomenological theory, as in earlier acceleration theories, that the currents screening magnetic fields from the interior should flow in a surface layer

† They were not sure that L had to be interpreted as m/ne2 as in earlier acceleration theories.

of effective thickness λ, i.e. $(\Lambda|\mu_0)^{1/2}$, in modern notation. The first direct studies of field penetration were made in 1940 by Shoenberg using pharmaceutical 'grey powder', finely ground mercury in chalk with droplets little more than 10^{-6} cm in diameter. This remains one of the few determinations of relative values, λ/λ_0, at different temperatures, and only in 1948 was it shown to agree well with the Casimir–Gorter model which predicted $\lambda/\lambda_0 = (1 - t^4)^{-1/2}$. Most measurements of penetration depth have used samples much larger than λ, and have determined $\lambda - \lambda_0$; on the assumption that the Casimir–Gorter model is generally valid, it is possible to derive λ_0 [17, 18]. The few values obtained in the early post-war years were two or more times greater than expected from a naive acceleration theory. It also became clear that the high-frequency resistive losses were not readily explained, except in general terms, on the basis A3.1 of the two-fluid model, but these deficiencies of the phenomenological models were not seen as grave, and the proposals for modifying the London picture had other origins; they came in 1950 and shortly after from two independent sources, Cambridge (England) and Moscow. Since the political situation at the time practically excluded mutual discussion between the two countries‡, it is convenient to treat them separately.

The Cambridge side of the story began with measurements that showed little variation of λ as the magnetic field was increased to H_c, despite the thermodynamic requirement of an entropy increase which, if confined to the penetration layer, would imply a very significant change to the number of D2.1 superelectrons [19]. It seemed likely, then, that the entropy change and disturbance to the electron assembly were spread to a depth perhaps 20 times greater than λ, as if the superconducting phase could tolerate only gradual changes. This view was supported by the sharpness of the second-order transition and the absence of any foreshadowing of superconductivity above T_c, as if fluctuations in very small regions were not permitted. Gorter's 1934 suggestion of a minimum size had been forgotten, but the concept of coherence, as it came to be called, served to explain the origin of London's interface energy. If only gradual change was possible, the extended transition from superconducting to normal would exclude the magnetic field from a layer where the full energy of condensation was not available [20]. A little evidence that the penetration depth in an alloy was more susceptible to change by a magnetic field gave support to the view that the range of coherence was limited by electron scattering, so that the interface energy might be less in an alloy than in the pure metal, and might even become negative.

Substantial support for this interpretation came in 1953 when the penetration depth was found to be considerably larger in a tin–indium alloy than in pure tin, although the thermodynamic parameters were hardly affected. To incorporate the idea of coherence into the equation for the supercurrent, the local relation of the London theory was replaced by a non-local relation; J was now to be determined by an average of A over a range of about the coherence length, and therefore affected by impurity content. The integral equation, with solutions agreeing well with the limited data available, was inspired by the already-discarded Heisenberg–Koppe theory of superconductivity, but was soon more soundly derived by Bardeen, even before the BCS theory gave it refined form [21, 22]. In 1950, to which we return for the Russian side of the question, such refinements did not concern Ginsburg and Landau; they adhered to the London local model until 1957 and the advent of BCS (or Bogoliubov who seized and developed the idea of Cooper pairs independently).

The Ginsburg–Landau (GL) theory was a full-dress professional performance in comparison with the naive coherence idea [23]. It allowed arbitrary spatial variations of an order-parameter Ψ (which determined Λ) but at the cost of an extra energy of the form $|\text{grad } \Psi|^2$, with a multiplicative constant that was in no way arbitrary but fixed by the penetration depth and other parameters. When it was discovered that impurities could change λ without affecting the transition temperature, they maintained this rigid connection at the price of credibility. The re-examination by Gorkov of the GL equation in the light of BCS theory consolidated its status, and in its revised form, it remains a pillar of the theoretical structure [24]. In their original treatment, Ginsburg and Landau explained the interface energy in the same way (but in quantitative detail) as the coherence model. Strangely, they declined to consider the possibility that the parameters might take such values as would lead to a negative interface energy, and when Abrikosov took this step in 1952, they did their best to discourage him, so that he did not publish for some years [25]. He then pointed out that a negative interface energy would considerably change the shape of the magnetization curve [26]. Even in a long rod the field would begin to penetrate, before H_c was reached, to form a mixed phase which retained enough of superconductivity to conduct without resistance, and this could persist until the field was much greater than H_c. He drew attention to the pre-war experiments of Schubnikow and his co-workers, who had taken trouble to homogenize alloys of indium and thallium, and measured magnetization curves like what he expected when the interface energy was negative. He distinguished between superconductors of type I (typically pure metals with positive interface energy) and those of type II (in ideal form homogeneous alloys with negative interface energy), as illustrated in Figure A1.1.2. Until then, Schubnikow's work had been little regarded, partly because the Meissner effect had overshadowed it and partly because it was taken for granted that alloys were inhomogeneous and their

‡ There were as yet no journals devoted to systematic translation of Russian papers, and the Russian publication, *Journal of Physics*, had been discontinued. Except for Cambridge, which enjoyed in David Shoenberg a fluent Russian-speaking physicist, there was great ignorance in the West of Russian work. Many Moscow physicists, on the other hand, spoke English well.

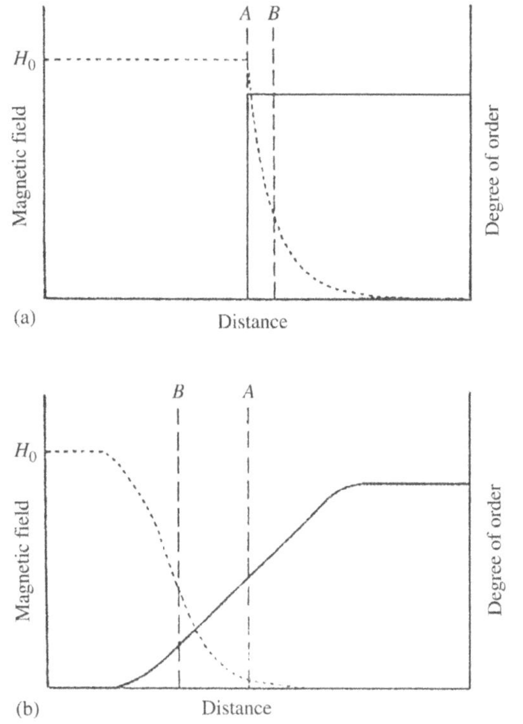

FIGURE A1.1.2 Diagrams to illustrate the origin of the interface surface energy [15]. In the upper diagram (a), the interface is sharp (superconducting phase on the right), and field penetration puts the effective boundary of the magnetic field at B, within the superconductor; this leads to negative interface energy. In the lower diagram (b), the transition is broadened, and B lies outside the effective phase boundary A, to give positive surface energy.

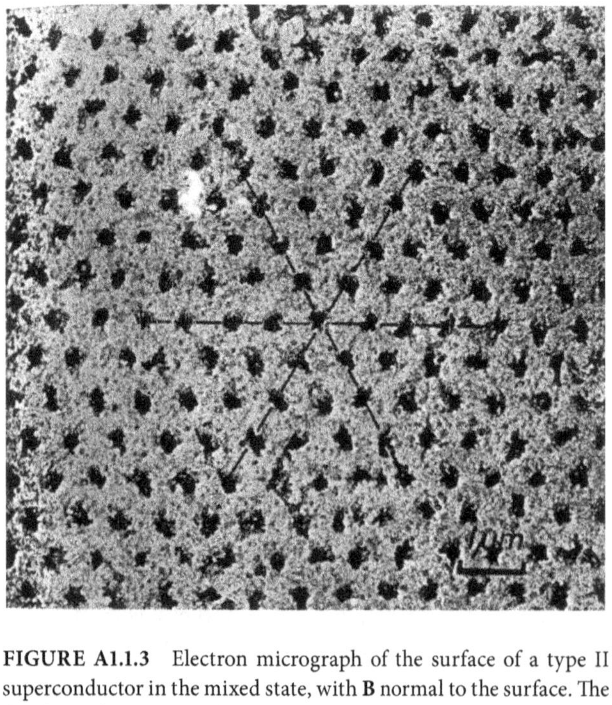

FIGURE A1.1.3 Electron micrograph of the surface of a type II superconductor in the mixed state, with **B** normal to the surface. The flux lines, decorated with iron powder, have taken up a regular triangular lattice configuration [35].

magnetic behaviour explicable in terms of the Mendelssohn sponge. It was easy in those days to follow mandarins like Pauli in their scorn of 'dirt effects'.

The details of Abrikosov's analysis went much further than the categorization of different types of superconductor. In the Ginsburg–Landau theory, the order parameter Ψ had the character of a wave function for the superelectrons, and was complex with a phase that depended on A. Abrikosov realized that continuity of Ψ meant that the magnetic flux through a normal region within a superconductor a2.6 could not take an arbitrary value, but only a multiple of a flux quantum h/e. This quantum was foreseen for similar reasons by Fritz London in 1948 [27] but was not observed in a macroscopic ring until 1961 (and then with half London's value). In a superconductor with negative interface energy, the smallest normal domain was what would become known as a flux line, with a core parallel to H on which Ψ dropped to zero, and around it a sheath where Ψ rose to its full value; one flux quantum was trapped in the sheath. The flux lines remained hypothetical, though not doubted, until 1966 when electron microscopy of a surface decorated with magnetic powder revealed a triangular array, as illustrated spectacularly in Figure A1.1.3, hardly different from the square array Abrikosov had analysed [28]. The existence of flux lines justified a more precise picture of the migration of flux within the mixed state of superconductors, as well as the influence of inhomogeneities and defects in pinning flux lines, and the origin of electrical resistance accompanying flux flow.

The consequences were not realized until the first high-field superconducting solenoids became commercially available. They followed the initial exploration by Hulm and Matthias of new alloys and a1.2 compounds with high transition temperature [29], which, as extreme forms of type II superconductors such as Nb_3Sn and NbTi, could retain flux lines in very strong fields and, being microscopically inhomogeneous, inhibited flux flow. It was only in 1962 that Goodman [30] pointed out the significance of Abrikosov's work and set the technological development on a sound basis. By this time, the BCS theory had radically changed the physical understanding of superconductivity, and its applications were at long last being recognized as worthy of governmental and industrial support, such as has continued and enlarged with the discovery of cuprate superconductivity.

Here we may stop, except for brief remarks about the origin and immediate consequences of BCS a1.2 theory. The turning point for theory was Fröhlich's appreciation (1950) that polaronic distortion of anionic lattice could lead to an attractive force between electrons and create an energy gap at the Fermi surface – the conditions for realizing the Londons' dream of 1935 [31]. Simultaneously, it was found that the transition temperature depended on isotopic mass in the way predicted by Fröhlich [32], so that his mechanism was immediately accepted even though his detailed theory of the ground state was seriously flawed. Several years elapsed before Bardeen, Cooper and Schrieffer overcame the enormous difficulties of a many-body theory in which electron pairs of opposite momentum and spin, Cooper's seminal inspiration, were coupled in a ground state [33]. The excited states had properties reminiscent of those postulated in the two-fluid model but, because of entirely novel symmetry properties, they were sufficiently different to succeed where the other had failed. For a while, there was resistance from experienced theorists who were only well aware of the pitfalls besetting a path that they themselves had failed to discern. But experimenters received the theory enthusiastically and made such advances as convinced the most sceptical. Giaever's tunnelling that revealed the energy gap and the enhanced density of excited states above it; Josephson tunnelling with all its complexities and applications; the Knight shift in superconductors and the attenuation of ultrasonics; Andreev reflection and its consequences; the halving of the flux quantum to $h/2e$ as a direct consequence of electron pairing in the ground state; these are items in an unprecedented catalogue of success, and all within a year of two – witnesses to one of the scientific triumphs of the century.

Yet not, it seems the complete answer. There were critics, from the beginning, who agreed the theory described the phenomenon as no other had done, but were dissatisfied because it had little power to predict which metals would be superconductors, or what their transition temperature would be. With further theoretical development, they had fallen silent before the events beginning in 1986, when the new class of cuprate superconductors was discovered and made clear that BCS might not be a universal theory [34]. Despite valiant efforts, the high-temperature superconductors remain fundamentally mysterious. But those with transitions below about 25 K, and well described by BCS, have acquired and seem likely to retain the classical status of 'conventional superconductors'.

References

[1] Dahl P F 1992 *Superconductivity: Its Historical Roots and Development from Mercury to the Ceramic Oxides* (New York: American Institute of Physics) Chapter 3

[2] Dahl P F 1992 *Superconductivity: Its Historical Roots and Development from Mercury to the Ceramic Oxides* (New York: American Institute of Physics) p 83

[3] Casimir H B G 1983 *Haphazard Reality* (New York: Harper and Row) p 339

[4] Dahl P F 1992 *Superconductivity: Its Historical Roots and Development from Mercury to the Ceramic Oxides* (New York: American Institute of Physics) Chapter 4 and p 98

[5] Yntema G B 1995 *Phys. Rev.* **98** 1197

[6] Dahl P F 1992 *Superconductivity: Its Historical Roots and Development from Mercury to the Ceramic Oxides* (New York: American Institute of Physics) Chapter 9

[7] Dahl P F 1992 *Superconductivity: Its Historical Roots and Development from Mercury to the Ceramic Oxides* (New York: American Institute of Physics) Chapter 10

[8] Shoenberg D 1952 *Superconductivity* (Cambridge: Cambridge University Press) p 180

[9] Bardeen J 1956 *Encyclopedia of Physics* **Vol 15** ed S Flügge (Berlin: Springer) p 284

[10] Dahl P F 1992 *Superconductivity: Its Historical Roots and Development from Mercury to the Ceramic Oxides* (New York: American Institute of Physics) Chapter 11

[11] Gorter C J 1935 *Physica* **2** 449

[12] Shoenberg D 1952 *Superconductivity* (Cambridge: Cambridge University Press) Chapter 4

[13] Faber T E 1958 *Proc. Roy. Soc. A* **248** 460

[14] London H 1940 *Proc. Roy. Soc. A* **176** 522

[15] Shoenberg D 1952 *Superconductivity* (Cambridge: Cambridge University Press) p 194

[16] Dahl P F 1992 *Superconductivity: Its Historical Roots and Development from Mercury to the Ceramic Oxides* (New York: American Institute of Physics) p 280

[17] Shoenberg D 1952 *Superconductivity* (Cambridge: Cambridge University Press) Chapter 5

[18] Bardeen J 1956 *Encyclopedia of Physics* **Vol 15** ed S Flügge (Berlin: Springer) p 244

[19] Pippard A B 1950 *Proc. Roy. Soc. A* **203** 210

[20] Pippard A B 1951 *Proc. Camb. Phil. Soc.* **47** 617

[21] Dahl P F 1992 *Superconductivity: Its Historical Roots and Development from Mercury to the Ceramic Oxides* (New York: American Institute of Physics) p 245

[22] Bardeen J 1956 *Encyclopedia of Physics* **Vol 15** ed S Flügge (Berlin: Springer) p 299

[23] Bardeen J 1956 *Encyclopedia of Physics* **Vol 15** ed S Flügge (Berlin: Springer) p 324

[24] Waldram J R 1996 *Superconductivity of Metals and Cuprates* (Bristol: Institute of Physics Publishing) p 176

[25] Dahl P F 1992 *Superconductivity: Its Historical Roots and Development from Mercury to the Ceramic Oxides* (New York: American Institute of Physics) p 250

[26] Abrikosov A A 1957 *J. Phys. Chem. Solids* **2** 199

[27] London F 1948 *Phys. Rev.* **74** 562

[28] Waldram J R 1996 *Superconductivity of Metals and Cuprates* (Bristol: Institute of Physics Publishing) p 72

[29] Dahl P F 1992 *Superconductivity: Its Historical Roots and Development from Mercury to the Ceramic Oxides* (New York: American Institute of Physics) p 256

[30] Goodman B B 1962 *IBM J. Res. Dev.* **6** 63

[31] Bardeen J 1956 *Encyclopedia of Physics* **Vol 15** ed S Flügge (Berlin: Springer) p 359

[32] Dahl P F 1992 *Superconductivity: Its Historical Roots and Development from Mercury to the Ceramic Oxides* (New York: American Institute of Physics) p 252

[33] Bardeen J, Cooper L N and Schrieffer J R 1957 *Phys. Rev.* **108** 1175

[34] Bednorz J G and Müller K A 1986 *Z. Phys.* **B 64** 189

[35] Träuble H and Essman U 1968 *J. Appl. Phys.* **39** 4052

A1.2

An Introduction to Superconductivity

William F. "Joe" Vinen* and Terry P. Orlando

A1.2.1 Superconductivity as an Ordering Transition

Any material becomes more ordered as it cools, its entropy decreasing, provided it remains in thermodynamic equilibrium. As the material approaches the absolute zero of temperature, its entropy tends to vanish, so that it becomes completely ordered. Often the ordering proceeds through one or more discrete phase transitions. Perhaps the most remarkable of these phase transitions, and surely the most unexpected, are those leading to superfluid or superconducting phases, typically at temperatures of a few kelvin. At a superficial level, these phases can be seen to exhibit frictionless flow, but they have other striking macroscopic quantum properties that are a reflection of the unique and special type of quantum ordering that takes place within them.

In the case of superconducting metals, the phase transition involves the electron fluid that is responsible for electrical conductivity in the normal (high-temperature) phase. The frictionless flow is seen as a loss of electrical resistivity, the material often showing a resistivity that is unmeasurably small. Electrical resistivity is due to the scattering of the conduction electrons by imperfections in the crystal lattice in which they are moving, so one might be tempted to think that in a superconductor these scattering processes are mysteriously turned off. As explained in Pippard's historical introduction, this view is quite misleading. All superconductors exhibit the Meissner effect: in sufficiently small applied magnetic fields, they behave as perfectly diamagnetic materials, in the sense that the magnetic flux is excluded in a reversible manner; they behave just like conventional diamagnetic materials, except that they have a much larger diamagnetic susceptibility. The required screening current is maintained as part of the equilibrium state of the system just as it is in the diamagnetic screening current in a diamagnetic atom or molecule. Scattering processes, far from having disappeared, are helping to maintain this equilibrium, as we shall see a little later.

The superconducting state is a new thermodynamic phase of the metal, distinct from the normal phase. The form of the heat capacity of the metal in the neighbourhood of the transition is very similar to that found in ordering transitions of the type seen in, for example, a paramagnetic material when it transforms into a ferromagnet. In the superconducting phase, the conduction electron fluid is in a more strongly ordered state than it is in the normal phase, and the diamagnetism observed in the Meissner effect is an equilibrium property of this ordered state.

Distinct phases can be conveniently described in an appropriate phase diagram: that for most pure metals (type I) is shown schematically in Figure A1.2.1, where the material has a shape with zero demagnetizing factor. The transition between the normal and superconducting phases in zero field is second order (no latent heat); in a finite field it is first order. The transition from superconducting to normal phase, as the field is increased at fixed temperature, takes place when the free energy associated with the induced magnetic moment exceeds the free energy difference ($F_s - F_n$) between the two phases in zero field, which leads to the following relationship for the critical field.

$$F_n - F_s = B_c^2/2\mu_0 \qquad \text{(A1.2.1)}$$

Other and more complicated phase diagrams are also possible, as we shall see later.

A1.2.2 The Meissner Effect and the Nature of the Ordering in a Superconductor

The simplest form of diamagnetism in a single atom arises when, as is often the case, the atomic electron wavefunctions ψ are not significantly perturbed by the applied magnetic field.

* This version has been reviewed by Terry P. Orlando (MIT), with some minor suggested updates.

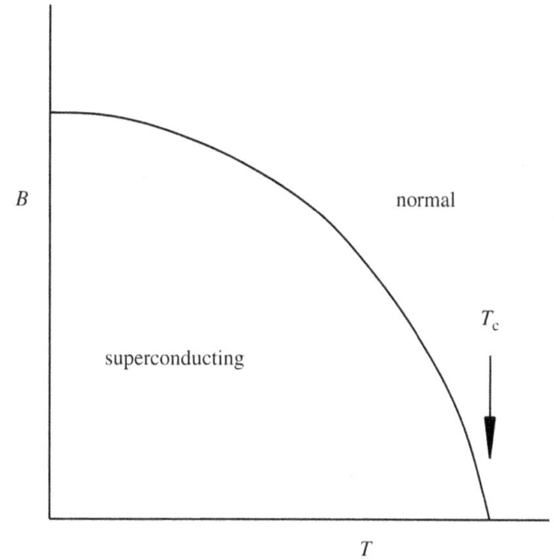

FIGURE A1.2.1 Schematic phase diagram for a type I superconductor: applied magnetic flux density (B) plotted against temperature (T).

The current electric density in the atom given in terms of the vector potential $\boldsymbol{A}$ for each electron acting independently by

$$\boldsymbol{J}(\boldsymbol{r}) = \frac{ie\hbar}{2m}(\psi^*\nabla\psi - \psi\nabla\psi^*) - \frac{e^2}{m}\psi\psi^*\boldsymbol{A}(\boldsymbol{r}) \quad \text{(A1.2.2)}$$

then reduces to

$$\boldsymbol{J}(\boldsymbol{r}) = -\frac{e^2}{m}\psi\psi^*\boldsymbol{A}(\boldsymbol{r}) \quad \text{(A1.2.3)}$$

since the unperturbed wavefunctions make no contribution. The resulting diamagnetic moment is very small, because the atom is very small. Fritz London noticed that, if Equation (A1.2.3) was to apply to each conduction electron in a whole metal, the resulting diamagnetism would be very large, as in a superconductor, because the electron wavefunctions ψ extends over the very large volume of the whole metal. This does not happen in a normal metal because the conduction electron wavefunctions are not unperturbed; each electron wavefunction is strongly perturbed and describes a quantized cyclotron orbit. The two terms in Equation (A1.2.2) almost cancel, so that there remains only the very weak 'Landau' diamagnetism. But if the conduction electron wavefunctions were not modified by the field, so that Equation (A1.2.3) were to apply to a macroscopic number density (n_s) of electrons in the whole superconducting metal, the Meissner effect would be more or less correctly described. Equation (A1.2.3) would then imply that the total current in the superconductor would be given by

$$\boldsymbol{J} = -n_s\frac{e^2}{m}\boldsymbol{A} \quad \text{(A1.2.4)}$$

from which we obtain, by taking the curl,

$$\text{curl}\,\boldsymbol{J} = -\frac{n_s e^2}{m}\boldsymbol{B} \quad \text{(A1.2.5)}$$

where $\boldsymbol{B}$ is the magnetic induction. Combining Equation (A1.2.5) with the Maxwell equation curl $\boldsymbol{B} = \mu_0 \boldsymbol{J}$, we obtain

$$\nabla^2\boldsymbol{B} = \frac{1}{\lambda_L^2}\boldsymbol{B} \quad \text{(A1.2.6)}$$

showing that the field tends to zero within the superconductor, with a penetration depth given by

$$\lambda_L^2 = \frac{m}{\mu_0 n_s e^2}. \quad \text{(A1.2.7)}$$

If n_s is of the order of the total number of conduction electrons per unit volume in the metal, λ_L is very small (of order 100 nm) so that flux exclusion is almost complete, as observed. Careful experiments show that a finite penetration depth does exist, with the order of magnitude given by Equation (A1.2.7), so we seem to have a good description of the superconducting behaviour.

The 'rigidity' in the wavefunction of the electrons in a superconductor that leads to the Meissner effect must presumably be a result of the ordering process that marks the onset of superconductivity. The nature of this ordering process became clear only when Bardeen, Cooper and Schrieffer had developed their theory, although the suggestion given much earlier by Fritz London that superfluidity in liquid helium and superconductivity in metals might have a common origin can now be seen to have pointed the way. Superfluidity is associated with Bose condensation: an ordering described by the accumulation of the helium atoms in a single quantum state. The BCS theory indicated, perhaps somewhat surprisingly, that a similar process is occurring in a superconductor. Since electrons are fermions, the individual electrons themselves cannot exhibit Bose condensation, but BCS told us that an attractive interaction between electrons, due to local distortion of the lattice (phonon exchange), leads to the formation of electron pairs (Cooper pairs), which can and do undergo a form of Bose condensation. A satisfactory description of this condensation is not straightforward, since each electron pair occupies a large volume (connected with the coherence length ξ_0 to which we refer later), so that there is massive overlap between the pairs. Bose condensation leads to long-range order in a particular type of correlation function, which, in the case of liquid helium, is the single-particle density matrix. In the superconducting electrons, it is the two-particle density matrix that exhibits the same long-range behaviour, as can be shown directly from the BCS wavefunction.

It is the formation of the electron-pair condensate that gives rise to the rigidity of the superconducting wavefunction. Small perturbations to the wavefunction can be achieved only by mixing in states in which electron pairs have been removed from the condensate, and it turns out that this requires a minimum energy, Δ. This minimum energy is that required to produce thermal excitation of the system. In a normal metal, the excited states of the system are obtained by taking an electron from below the Fermi surface and placing it above, thus creating an electron excitation and a hole excitation. In the superconductor, the excitations involve the breaking of electron pairs, and they correspond to linear combinations of electron-like states and hole-like states in the normal metal. Each excitation has an energy equal to $(\Delta^2+\varepsilon_k^2)^{1/2}$, where ε_k is the energy of the corresponding excitation in the normal state. As the temperature of the superconductor is raised above absolute zero, more and more excitations are produced, and the energy gap Δ falls. Eventually, at the critical temperature, T_c, Δ vanishes and the superconductor becomes a normal metal. In the BCS theory, T_c is related to the energy gap at $T = 0$ by the relation $\Delta(0) = 1.76k_BT_c$. However, it should be added that the existence of an energy gap is not essential for superconductivity; for example, superconductors containing magnetic impurities may be gapless; and superfluid helium is also gapless. The condensate in liquid helium owes its 'rigidity' to the form of the excitation spectrum, which is itself determined by the existence of the condensate.

Only the condensed electrons contribute to the Meissner effect, and the value of n_s in Equation (A1.2.4) is the effective number of such electrons. The excitations behave to some extent like electrons in a normal metal. We are led, therefore, to a two-fluid model, in which n_s electrons behave as superconducting and the rest as normal. The falling value of n_s with increasing temperature is reflected in an increasing penetration depth ($\lambda_L \to\infty$ as $T \to T_c$).

A1.2.3 The Ginsburg–Landau Equations; the Pippard and Ginsburg–Landau Coherence Lengths

It is convenient in developing an understanding of superconductivity to introduce the 'condensate wavefunction' or Ginsburg–Landau wavefunction. This can be formally defined in terms of the correlation function to which we have already referred, but in essence, it is a complex function, ψ, the phase of which is the phase of the wavefunction of the condensed pairs and the amplitude of which is proportional to the local concentration of condensed pairs. The function ψ obeys the Ginsburg–Landau equations when the temperature is near T_c; at lower temperatures, these equations cease to be strictly correct, but they still provide a correct qualitative description. One of the G–L equations gives the supercurrent density.

$$J(r)=\frac{ie\hbar}{2m}(\psi^*\nabla\psi-\psi\nabla\psi^*)-\frac{2e^2}{m}\psi\psi^*A(r) \quad \text{(A1.2.8)}$$

(cf Equation [A1.2.2]), while the other, which has the form of the nonlinear Schrodinger equation, describes in essence how the amplitude of ψ varies with position, such as might occur near a normal-superconducting boundary. It is found that ψ cannot change abruptly with position, but only gradually over a characteristic distance, ξ_{GL}, called the Ginsburg–Landau coherence length. We note the presence of the first term on the right-hand side of Equation (A1.2.8), which implies that current-carrying states of the superconductor are possible independently of the magnetic field. We discuss such states in a moment. Since the modulus of ψ is a measure of the density of condensed electrons, it also is a measure of the extent of the superconducting ordering. Indeed Ginsburg and Landau originally introduced ψ as a (complex) 'order parameter,' and it is frequently referred to in this way. Often it is referred to in other ways: 'the gap parameter,' since it is proportional to the energy gap in a spatially homogeneous situation; or the 'pair potential,' since it appears as a self-consistent potential in some formulations of the theory of superconductivity.

The superconducting wavefunction is not completely rigid in its response to a perturbation. It is rigid only for perturbations that vary slowly with position. The relevant characteristic length scale is the size of a Cooper pair ξ_0, often called the Pippard coherence length. In practice, ξ_0 is often larger than λ_L, so that our derivation of λ_L as the penetration depth, which relied on the assumption of a rigid wavefunction, breaks down. In Equation (A1.2.4), the relation between the current density and the vector potential has to be replaced by a nonlocal relation, involving a kernel with range ξ_0, and the expression for the penetration depth becomes more complicated ($\lambda \neq \lambda_L$). For a pure metal, the coherence lengths ξ_{GL} and ξ_0 are equal at low temperatures and equal to $\pi\Delta(0)/\hbar v_F$, where v_F is the Fermi velocity in the normal metal, but at higher temperatures, they behave differently, ξ_0 remaining constant, but ξ_{GL} diverging to infinity at T_c.

A1.2.4 Persistent Currents and the Quantization of Trapped Flux

Suppose that the superconductor has the form of a long hollow cylinder, with radius R and wall thickness t. If t is large compared with the penetration depth, and if an external magnetic field is applied parallel to the axis of the cylinder, no magnetic flux penetrates to the inside of the cylinder. Suppose, however, that the cylinder is cooled into the superconducting phase while exposed to the external magnetic field. The state of lowest energy must surely then be one in which magnetic flux is trapped inside the cylinder. In such a state, the Ginsburg–Landau wavefunction must, as it turns out, have the form

$$\psi=\psi_0\exp[iS(r)] \quad \text{(A1.2.9)}$$

where the ψ_0 is the unperturbed wavefunction, and the phase S varies round the ring. The corresponding current density, obtained from Equation (A1.2.8), is

$$J = \frac{e\hbar\psi_0^2}{m}\nabla S - \frac{2e^2\psi_0^2}{m}A. \qquad \text{(A1.2.10)}$$

Taking the curl of this equation we recover Equations (A1.2.5) and (A1.2.6), so that the magnetic field is still screened from the bulk of the superconductor. However, there is now a magnetic field within the cylinder, as we see by taking the line integral of Equation (A1.2.10) round a circuit C enclosing the hole in the superconductor and lying at a distance much greater than the penetration depth from the surface (Figure A1.2.2, where we have assumed that $\lambda \ll t$). $J = 0$ on this circuit, and therefore, the flux within the circuit, equal to the line integral of **A** round the circuit, is given by

$$\Phi = \frac{\hbar}{2e}\oint \nabla S \cdot dr. \qquad \text{(A1.2.11)}$$

The wavefunction ψ must be single-valued, and therefore, the line integral in Equation (A1.2.11) must be equal to $2\pi q$, where q is an integer. Therefore, the trapped flux Φ can be nonzero, implying the existence of a magnetic field inside the cylinder, but the magnitude of the trapped flux is quantized in units of $\phi_0 = 2\pi\hbar/2e$. The factor $2e$ here has its origin in the fact that the condensate is composed of electron pairs. This quantization of trapped flux is rather directly related to the existence of the condensate wavefunction.

If the external field is removed from the superconducting ring, the trapped flux remains. The superconductor is not then in a state of minimum possible free energy. It is, however, in metastable equilibrium, with an associated local minimum in the free energy. A transition to the state of absolutely minimum free energy would require that the flux passes through the superconductor and therefore penetrates it, at least transiently, by much more than the penetration depth, which is energetically very unfavourable (a consequence of the Meissner effect). More striking is the situation when the thickness, t, of the ring is much less than the penetration depth. A persistent current is still possible, although an extension of the argument in the preceding paragraph shows that the trapped flux is quantized in units less than $2\pi\hbar/2e$. The metastability of this persistent current does not follow simply from the Meissner effect. It is necessary to note that a sudden loss of the trapped flux, or equivalently, loss of the persistent current, could occur only if all the Cooper pairs in the condensate were simultaneously to undergo a transition between two states (Equation [A1.2.9]) with different $S(\mathbf{r})$, which has negligible probability. Alternatively, the persistent current would disappear if it exceeds a critical value, equal to roughly Δ/p_F, at which excitations are produced in such large numbers that superconductivity is suppressed (p_F is the Fermi momentum); below this critical current, extra excitations may still be produced, the excitations being in a state of thermal equilibrium maintained by scattering, but the effect is only to reduce n_s without destroying it altogether. Under certain circumstances, loss of a persistent current can occur with nonzero probability through the important process of gradual 'phase slippage,' which we shall discuss later. The persistence of the current then depends on phase slippage having a sufficiently small probability. We emphasize that the persistent current is an equilibrium state of the system (a minimum in the free energy, maintained by collisions), subject only to the constraint that the phase of C does not change.

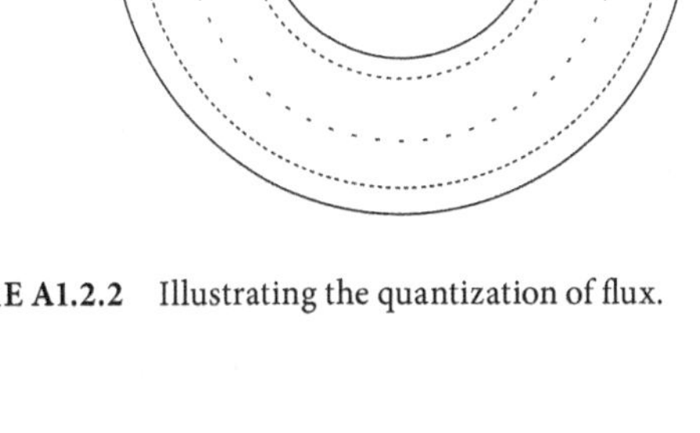

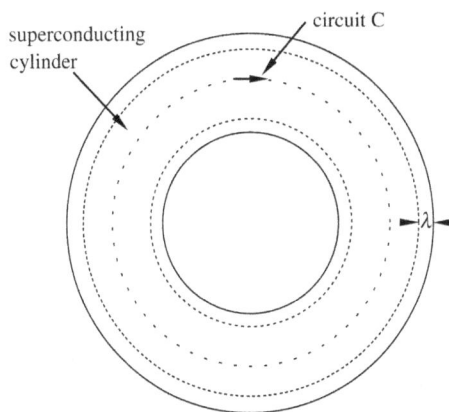

FIGURE A1.2.2 Illustrating the quantization of flux.

A1.2.5 Flux Lines, the Mixed State, and Type I and Type II Superconductors

We might ask whether a single quantum of flux, for example, could be trapped inside a bulk sample of superconductor. It can be easily shown from Equation (A1.2.10) that in such a case the current would diverge to infinity along a line in the superconductor. In the presence of a very large current, it is energetically favourable for the Cooper pairs to break up, so the material would become effectively normal along this line. The situation can be described by an appropriate solution of the Ginsburg–Landau equations. The amplitude of the Ginsburg–Landau wavefunction vanishes along the line, and it rises to its normal value over a distance from the line of order the Ginsburg–Landau coherence length ξ_{GL}. The phase $S(\mathbf{r})$ changes by 2π as the line is encircled, so that one flux quantum is associated with the line. The magnetic field penetrates from the line to a distance of order of the penetration depth λ (Figure A1.2.3). The resulting structure is called a 'flux line' or 'vortex,' the latter name originating from the fact that close to the line the electron velocity is similar to that in a hydrodynamic vortex (proportional to $1/r$). The ratio λ/ξ_{GL} is called κ. This ratio turns out to be approximately independent of temperature.

Whether it is energetically favourable for such a structure to exist depends on its energy, ε, per unit length. Suppose that

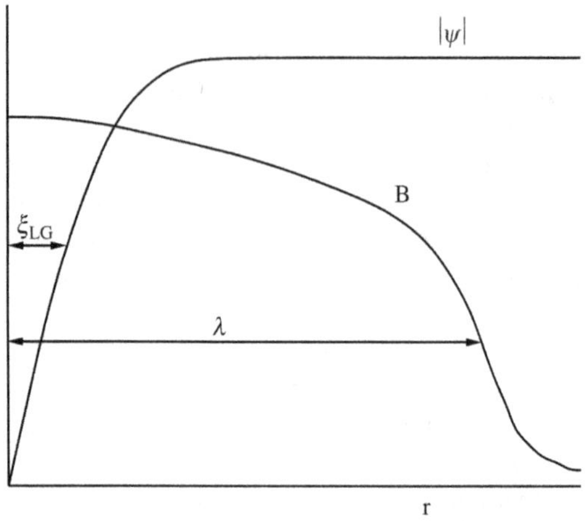

FIGURE A1.2.3 The modulus of the order parameter (|ψ|) and the magnetic flux density (*B*) plotted against radial distance (*r*) from the centre of a flux line (schematic).

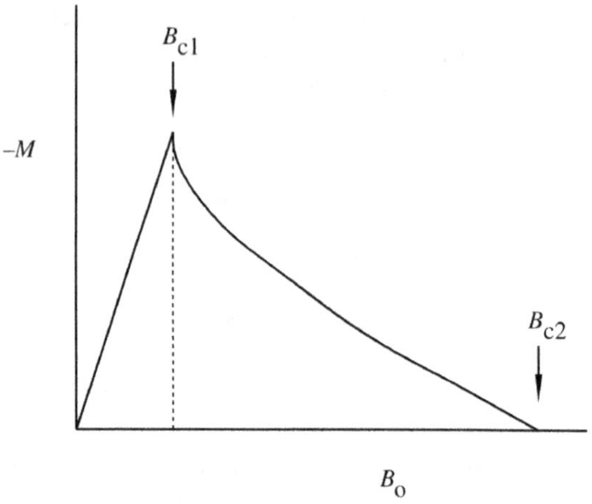

FIGURE A1.2.4 Magnetization curve for a type II superconductor. Magnetization (*M*) plotted against applied magnetic flux density (B_0) (schematic).

a superconductor in the form of a long thin solid cylinder (radius $\gg \lambda$) is exposed to a magnetic field, *B*, directed along its length. It can be shown that it is energetically favourable for a flux line to exist in the superconductor if B exceeds the value given by $B = B_{c1} = \varepsilon/\phi_0$. We recall that if B exceeds the value given by Equation (A1.2.1) (the 'thermodynamic critical field') superconductivity is destroyed. Therefore, flux lines can be formed only if ε is sufficiently small, so that $B_{c1} < B_c$. It turns out that this condition is equivalent to the condition that $\kappa > 1\sqrt{2}$.

If $\kappa > 1\sqrt{2}$ the superconducting cylinder passes directly from the 'Meissner state' (no field penetration except in the penetration depth) to the normal state when *B* exceeds B_c. A superconductor of this type is called a type I superconductor; its phase diagram was shown in Figure A1.2.1. If $\kappa > 1\sqrt{2}$ flux lines can penetrate the superconducting cylinder at fields greater than B_{c1} ($< B_c$), the diamagnetic moment of the cylinder being reduced. In fact more and more flux lines will penetrate until the (repulsive) interaction between them causes it to be no longer energetically favourable for them to form. As the applied field is increased, the density of flux lines increases, the diamagnetic moment falling, until the 'cores' of the flux lines (the regions of size ξ_{LG} where the superconductivity is suppressed) overlap, when the material becomes normal; the transition to the normal state occurs when $B > \sim \phi_0 / \xi_{LG}^2$ which is equivalent to $B > B_{c2} = \sqrt{2}\kappa B_c$. A superconductor with $\kappa > 1\sqrt{2}$ is called a 'type II superconductor.' When a type II superconductor contains an array of flux lines, it is said to be in the mixed state. The magnetization curves for a type II superconductor is shown schematically in Figure A1.2.4, and the phase diagram for a type II superconductor is shown in Figure A1.2.5. (It is necessary to consider a long thin cylinder here because otherwise demagnetizing effects become important and result in a different type of behaviour. Even in type I superconductors, flux penetration can then occur through the formation of the intermediate state in which relatively large areas of normal state are embedded in the superconductor.)

Type II superconductors are of great practical importance. Values of B_c are typically quite small (of order or less than 0.1 T), so that type I superconductors are useless in applications involving high magnetic fields. In contrast, values of B_{c2} can be very large, allowing, for example, the construction of superconducting coils for the generation of high magnetic fields. However, no pure single-element metal has a large value of κ, and indeed only niobium has a value of κ large enough (but only just) to make it type II. We are led, therefore, to look at other types of superconducting material, which we shall do in later sections.

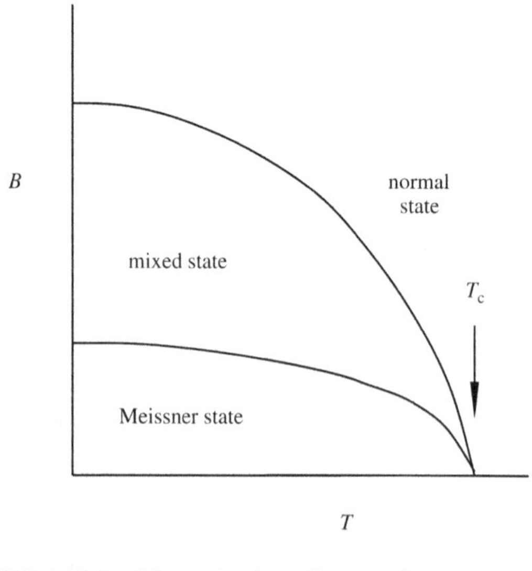

FIGURE A1.2.5 Schematic phase diagram for a type II superconductor: applied magnetic flux density (*B*) plotted against temperature (*T*).

A1.2.6 Phase Slip

The existence of flux lines allows us to see how the 'persistent currents' we described earlier can decay by 'phase slippage.' If a single flux line were to pass through the walls of the cylinder described in Section A1.2.4, the phase change round a circuit enclosing the hole in the cylinder would change by 2π. If the change is a decrease, the effect is a decrease in the persistent current. This phase slippage, which involves only a localized perturbation to ψ, can take place much more easily than any change involving simultaneously all the superconducting electrons. Creation of the flux line and its passage across the superconductor still generally encounters an energy barrier (some of it arising from an interaction between the flux line and the boundary of the superconductor). However, in the presence of a large enough current (the critical current), the barrier may be eliminated or reduced enough for thermally activated phase slippage to occur; the supercurrent can then decay in a relatively short time.

A1.2.7 Weak Links and Quantum Interferometers

This type of phase slippage often occurs in narrow constrictions in the superconductor, and we then talk of a weak link. A particularly simple type of weak link is the Josephson tunnel junction, formed by connecting two bulk volumes of superconductor through a thin insulating layer through which electrons can tunnel. The Cooper pairs can also tunnel, so that a supercurrent can pass across the junction. A simple analysis shows that for a weak junction the supercurrent is related to the difference in phase, ΔS, of the superconducting wavefunction across the junction by the relation

$$I = I_0 \sin \Delta S. \tag{A1.2.12}$$

The critical current is I_0, and we have described the dc Josephson effect. If I exceeds I_0, a potential difference, V, appears across the junction. A Cooper pair on one side of the junction then differs in energy by $2eV$ from one on the other side. Like an ordinary wavefunction, the superconducting wavefunction, ψ, has a time dependence $\exp(-\mathrm{i}Et/\hbar)$, so that the potential difference V gives rise to a continual slippage of the phase on one side of the junction relative to the other at a rate given by

$$\frac{\mathrm{d}\Delta S}{\mathrm{d}t} = \frac{2eV}{\hbar}. \tag{A1.2.13}$$

The junction therefore carries an oscillating supercurrent of angular frequency $2eV/\hbar$ (the ac Josephson effect). The phase slippage can be viewed as due to the steady flow of flux lines across the junction, each flux line causing a phase slip of 2π. The ac Josephson effect has been important in defining the volt in terms of frequency and the fundamental constants e and $\hbar$.

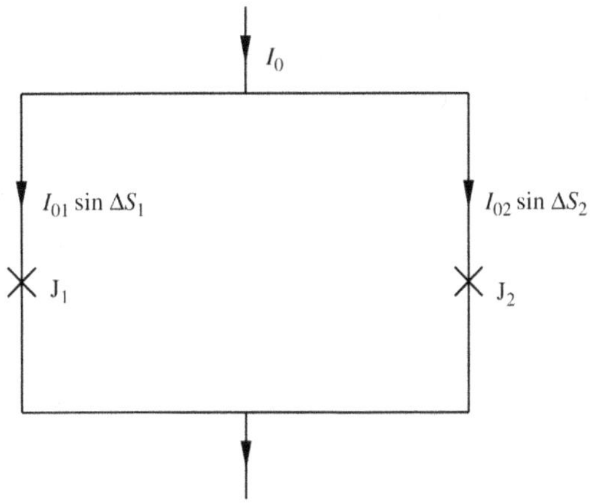

FIGURE A1.2.6 Schematic quantum interferometer.

Josephson junctions can be used to form quantum interferometers. Two junctions are arranged as shown in Figure A1.2.6 to form a superconducting loop. Simple quantum mechanics shows that, if a magnetic flux Φ threads the loop, the differences in phase across the two junctions must differ by $2\pi\Phi/\phi_0$. The total critical current across the two junctions, I_0 (the maximum value of $I_{01}\sin\Delta S_1 + I_{02}\sin\Delta S_2$) must therefore be reduced by an amount that is periodic in Φ/ϕ_0. Since ϕ_0 is very small, we have the basis for a very sensitive magnetometer.

A1.2.8 The Role of the Normal Electrons

We have seen that at a finite temperature there are condensed electrons in the superconductor and electronic excitations, which form, respectively, the superfluid and normal-fluid components in a two-fluid model. Any steady electric current in the superconductor is carried exclusively by the superconducting electrons; there is no electric field to drive the normal electrons, which exhibit an electrical resistivity due to the scattering of the excitations by lattice defects, as is the case for the electrons in an ordinary metal. At high frequencies, however, an electric field is required to move the superconducting electrons owing to their inertia, and this field will produce a response, and therefore dissipation, in the normal electrons. The superconductor is therefore not free from resistance at high frequencies, especially at microwave frequencies. In calculating this resistance, one must be careful to remember that the normal electrons are really excitations with different properties from electrons in the normal metal. At still higher frequencies (usually in the infrared), the photon energy may be sufficient to break a Cooper pair, leading to greatly increased dissipation.

The excitations play a role also in the thermal properties of the superconductor. Here, one must consider both the electronic excitations and the phonons. Both contribute to the heat capacity. Owing to the presence of the energy gap, the electronic contribution to the heat capacity becomes very small for $T \ll T_c$, but the phonon contribution remains. The condensed electrons carry no entropy, so heat conduction is entirely due to the excitations. In normal pure metals, the heat is carried largely by the electrons; the phonons carry little heat because they travel much more slowly than the electrons and because they are strongly absorbed by the electrons. In a superconductor, the situation is very different, at least for $T \ll T_c$. The excitations disappear at low temperatures, so they themselves carry little heat. At the same time the phonons can interact only with the electronic excitations (they do not have enough energy to break the Cooper pairs), so they are no longer strongly absorbed. Heat is therefore carried largely by the phonons, which can have a long mean free path as in a dielectric; the low-temperature phonon conductivity can therefore be high.

A1.2.9 Alloys

So far we have confined our attention to pure single-element metals, which are described by the BCS theory. The BCS theory is valid only if the electron–phonon interaction that is responsible for formation of the Cooper pairs is weak. In some metals, for example lead, this is not true, and a development of the BCS theory is required to treat such strong-coupling superconductors quantitatively. However, the basic physics remains unchanged.

For the rest of this chapter, we shall consider other types of superconducting material. As far as we know, superconductivity always involves the formation of electron pairs, but the pairs may be formed by a different (nonphonon) mechanism, and they may be formed in more complicated states than in a BCS superconductor, where the electrons in the pair have opposite spins and are in a state with no relative angular momentum (s-state).

We shall make a start, in this section, by considering a very important class of superconducting material formed from an alloy of two or more metals in which there is s-state pairing. In a pure metal, at the low temperatures required for conventional superconductivity, the electrons in the normal state have long mean free paths, much larger than the superconducting coherence length ξ_0. In an alloy, this is generally no longer the case, owing to the scattering of electrons by the disorder in the alloy. At first sight, one might expect that this would have a profound effect, since it would surely affect the way in which the pair wavefunction can be formed. Surprisingly, it has very little effect in the case of nonmagnetic components. The addition of a nonmagnetic alloying element to a pure metal has very little effect on the critical temperature. It turns out that this is because the pairing wavefunction can be set up on the basis of the real normal-state electronic wave functions, whatever they are, but still with the pairing of opposite spins; in the case of an alloy, the wavefunctions include the effect of the scattering. The gap parameter as a function of temperature is essentially the same, as is the critical temperature. In many pure metals, the electronic wavefunctions in the normal state exhibit isotropy associated with the details of the band structure, and this is reflected in anisotropy in the gap parameter, this parameter varying round the Fermi surface in a way that is consistent with the crystal symmetry. Formation of the Cooper pairs from wavefunctions that describe an electron that is being continually scattered removes this anisotropy, but otherwise there is little effect on the gap parameter.

In more detail, however, the properties of an alloy do differ in important ways from those of a pure metal, especially if the electron mean free path in the normal state, ℓ, is less than ξ_0 in the pure metal. The effective density of superconducting electrons is reduced by the scattering, by a factor of ℓ/ξ_0 in the limit $\ell \ll \xi_0$, so that the penetration depth is correspondingly increased. The coherence lengths are modified. The Pippard coherence length (governing the range of the nonlocal relation between $\boldsymbol{J}$ and $\boldsymbol{A}$) is reduced to ℓ in the same limit, while the Ginsburg–Landau coherence length at low temperatures becomes $(\ell\xi_0)^{1/2}$. We see, therefore, that the value of κ is increased. Alloys tend, therefore, to be type II superconductors. Suitably chosen alloys can have large values of κ, with a resulting large value of the upper critical field B_{c2}. Such alloys might therefore be used in applications in which the superconductor is exposed to a high magnetic field.

An important application is the production of high magnetic fields with a superconducting coil. However, a high κ is not sufficient for this purpose. In a high field, the superconducting material will be in the mixed state. When it carries an electric current, the flux lines will be subject to a force (roughly speaking a Lorenz force equal to $\phi_0 \times \boldsymbol{J}$ per unit length of line), which may cause the lines to move in the transverse direction, giving rise to phase slip and dissipation. To prevent such movement, the flux lines must be pinned by suitable microstructure in the alloy. Much of the technology of magnet materials is concerned with the enhancement of this pinning.

A1.2.10 Towards Higher Values of T_c and H_{c2}

Many practical applications would benefit from materials with higher values of both T_c and H_{c2}. Among conventional superconductors, the best that can be done appears to be among intermetallic compounds with the cubic β-W (A15) structure and among the Chevrel phase compounds with composition $M_xMo_6X_8$, where M is a metal and X is either sulphur or selenium. Of these compounds, the best known is probably the A15 compound Nb_3Sn, which has a critical temperature of about 18 K and an upper critical field at low temperatures exceeding 30 T, and which has been used extensively in superconducting magnets. These materials are of course strong-coupling superconductors, and the question arose whether stronger and stronger electron–phonon coupling can lead to higher and higher critical temperatures. Analysis that takes into account both the electron–phonon interaction and the Coulomb repulsion between electrons shows almost certainly that conventional mechanisms of superconductivity cannot lead to critical temperatures exceeding about 30 K, a view that seems to be confirmed by practical experience.

A1.2.11 Unconventional Superconductivity

So far, this introduction has been concerned with 'conventional' superconductors; i.e., those in which superconductivity arises from the formation of a condensate of electron pairs, the attractive electron–electron interaction required to produce the pairs being due to a distortion of the lattice (phonon exchange). In the simplest form of the BCS theory, the electron pairs form in states with zero internal angular momentum (s-states) and therefore with antiparallel spins, and conventional superconductors are correctly described by this form of the theory.

During the past 25 years or so, many superconducting materials have been discovered that appear not to be conventional. It appears that all involve electron pairing and, presumably, the formation of a condensate from the pairs, but they are unconventional in the sense that they involve pairing into states with nonzero angular momentum and/or a pairing mechanism that is not due to electron–phonon interaction.

A search for superconducting materials that do not depend on the electron–phonon interaction was based in part on the wish to find materials with higher critical temperatures. It was suggested by Little in 1964 that superconductivity might be found in solids composed of certain long-chain organic molecules, and that the electron–electron attraction might then arise through the electronic polarization of side-chain molecules. At that time even organic metals were hardly known, but over the years the search for, and study of, such materials has been intense, and it led eventually to the discovery of organic superconductors (although not in polymers). The normal metallic properties of these materials involve much interesting and novel physics (and chemistry!), connected, for example, with their low dimensionality, and the study and understanding of these normal properties have been important.

Much interest has been shown in the possibility that non-s-state pairing might be found in some superconductors. We recall that atoms of the light isotope of helium, ^{3}He, are fermions and have a nuclear spin of one-half. The superfluid phase of liquid ^{3}He was discovered in 1972 at temperatures less than a few mK, and it was shown very quickly that it could be described by a development of BCS theory that incorporated p-state pairing and parallel nuclear spins ($L = 1$, $S = 1$), and which had been developed, at least in part, in earlier years. Pairing in s-states is not possible because of the strong short-range repulsion between two helium atoms; p-state pairing makes use of the long-range attraction. Owing to the p-state pairing, which gives 'structure' to a Cooper pair, superfluid ^{3}He is much more complicated than superfluid ^{4}He, and it exhibits a rich variety of phenomena that have no counterpart in ^{4}He. One aspect of this complication is the existence of (at least) two different superfluid phases of ^{3}He; there is p-state pairing in both phases (A and B), but the orbital and spin angular momenta are differently arranged around the Fermi surface in the two cases. The nontrivial structure of a Cooper pair means that the order parameter has more components than in the case of s-state pairing: for p-state pairing, for example, the order parameter has in principle 18 components (three possible spin orientations; three possible orientations of the orbital angular momentum, and each of the possible nine combinations has its own real and imaginary components of the order parameter); Ginsburg–Landau theory becomes correspondingly more complicated. It should also be mentioned that the study of normal liquid ^{3}He played an important role in the development of our understanding of normal Fermi liquids, a development that carried over into a much better understanding of the electrons in a normal metal than is possible in terms of a model in which the conduction electrons are regarded as a noninteracting Fermi gas. In the case of normal ^{3}He, there are strong exchange interactions that cause the liquid to have a greatly enhanced Pauli susceptibility (due to the nuclear spins) and to be, in fact, almost a nuclear ferromagnet. As a result, there are, within the liquid, strong local fluctuations in the spin magnetic moment (paramagnons). Exchange of paramagnons contributes in an important way to the interatomic interaction leading to pairing and superfluidity, and it also serves to allow types of superfluid phase to exist that would not otherwise be stable (e.g., the A phase).

A superconductor with non-s-state pairing can exhibit properties differing from those of a conventional superconductor, which can allow us to identify it. In some phases of a non-s-state superconductor, the energy gap will vanish at certain points, or on certain lines, on the Fermi surface. This affects the temperature dependence of both the heat capacity and the normal fluid fraction (and hence the penetration depth); it gives rise to a power law dependence at the lowest temperatures instead of the exponential dependence that is characteristic of an energy gap which does not vanish in this way (it should be noted, however, that conventional superconductors containing magnetic impurities can also be gapless). However, this vanishing of the energy gap is not present in all phases of a non-s-state superfluid (e.g., the B-phase of superfluid ^{3}He), and a true vanishing of the energy gap may be difficult to distinguish experimentally from a very strong anisotropy in an s-state system. As we have already noted, a non-s-state superconductor may exist in different superconducting phases, depending on the temperature and the applied magnetic field, as is the case in superfluid ^{3}He; different phases are seen in some superconductors that are based on heavy fermion metals, suggesting non-s-state pairing in these cases (again the heavy fermion metals have interesting normal states, and similarities with liquid ^{3}He are likely, spin fluctuations being important in both the normal and superconducting phases). A fairly clear-cut indication of unconventional pairing comes from the effect of impurity scattering. We noted earlier that in a conventional superconductor, such scattering has the effect of only smoothing out any anisotropy in the energy gap; in the case of unconventional pairing, the scattering mixes states with different arrangements of the spin and/or orbital angular momenta, which destroys the superconductivity, such destruction occurring when the mean free path associated with the scattering is less than the superconducting coherence length. Another fairly clear-cut indication of non-s-state pairing can come from tunneling studies. For example, d-state pairing ($L = 2$, $S = 0$) can lead to a situation where the order parameter has effectively a different sign on different parts of the Fermi surface; this can be observed in the behaviour of a quantum interferometer if the two junctions are formed with a second conventional superconductor on different sides of a single crystal of the unconventional superconductor. Finally, some types of unconventional pairing can involve a violation of time-reversal symmetry, so that the ground state can carry a nonzero (surface) current, even in the absence of an applied magnetic field.

A1.2.12 The High T_c Cuprates

The best known examples of what are almost certainly unconventional superconductors are the high T_c cuprates: $La_{2-x}(Ca,Sr)_x CuO_4$, with T_c in the range 30–40 K, discovered by Bednorz and Müller in 1978; $YBa_2Cu_3O_{7-x}$ (YBCO) with $T_c = 93$ K, discovered by Wu and co-workers in 1978; and all the others, C1, discovered over the subsequent years. Their potential applications have attracted enormous interest. But they are fascinating and important also in their basic physics. They differ from the classic conventional superconductor in at least four respects: their properties in the normal state are anomalous and not described by conventional Fermi liquid theory; they can have very high critical temperatures; they have layer structures, which lead to extreme anisotropy as between properties involving current flow within a layer and those involving current flow between layers; and they have very large (and anisotropic) values of κ, with very small (and anisotropic) coherence lengths, leading to inaccessibly high values of B_{c2}. It seems unlikely that such high values of T_c can result from a phonon-mediated attraction between electrons, but there is as yet no agreement about the mechanism. Indeed, there is no agreement about a theory of the normal state. The extreme anisotropy leads to interesting flux line structures, and this fact combined with the high T_c allows observation of new effects in the mixed state, such as the melting of the flux line lattice and the decoupling of the parts of a vortex in different layers (formation of pancake vortices). Finally, some, at least of the cuprate superconductors, are almost certainly not s-state superconductors: tunneling experiments of the type described in the preceding section show that YBCO probably involves d-state pairing. The physics of these materials continues to pose major challenges, as does the development of the practical applications of them.

Many conventional and unconventional materials are discussed in this handbook, and the reader is encouraged to explore those here.

Further Reading

In a survey of this kind, it is neither possible nor desirable to refer to original papers. Even the number of books on superconductivity is too large. In any case, detailed references can be found in subsequent chapters. It seems wise, therefore, to confine the references here to a few (somewhat arbitrarily) selected books and reviews, which can be used by a newcomer to the field to pursue further study.

Annett J F 2004 *Superconductivity, Superfluids, and Condensates* (Oxford: Oxford University Press).
London F 1950 *Superfluids* Vol 1 *Macroscopic Theory of Superconductivity* (New York: Wiley).
de Gennes P G 1966 *Superconductivity of Metals and Alloys* (New York: Benjamin).
Parks R D (ed) 1969 *Superconductivity* Vols 1 and 2 (New York: Dekker).
Tilley D R and Tilley J 1990 *Superfluidity and Superconductivity* 3rd edn (Bristol: Institute of Physics Publishing).
Tinkham M 1996 *Introduction to Superconductivity* 2nd edn (New York: McGraw-Hill).
Waldram J R 1996 *Superconductivity of Metals and Cuprates* (Bristol: Institute of Physics Publishing).
Ishiguro T, Yamaji K and Saito G 1997 *Organic Superconductors* (Berlin: Springer).
Ketterson J B and Song S N 1999 *Superconductivity* (Cambridge: Cambridge University Press).
Schofield A J 1999 Non-fermi liquids *Contemporary Physics* **40** 95.
Heffner R H and Norman M R 1996, Heavy fermion Superconductivity, *Comments in Condensed Matter Physics* **17** 361–408.
Kleiner R and Buckel W 2016 *Superconductivity: An Introduction*, 3rd edn (New York: Wiley).

A1.3

The Polaronic Basis for High-Temperature Superconductivity*

K. Alex Müller

In 2014, a review of 56 panels was published by the present author, with the aim to describe, using the concept of polarons, the properties of the superconductivity in the hole-doped cuprates [1]. From that review, here I extract and rearrange those of them suitable for possibly giving an interested reader a shorter access to the understanding achieved. A decade earlier, ten relevant articles were published in a book [2] indicating the status, which, in the present effort, has become more concrete, thanks to new experimental and theoretical progress.

A1.3.1 Bipolarons

The discovery of superconductivity in hole-doped $LaCuO_4$ [3] resulted from the concept of Jahn–Teller (JT) polarons, as theoretically postulated in the group of Thomas at the University of Basilea† [4]. In such a polaron, a hole carrier is trapped near a Cu^{2+} ion, which in octahedral oxygen symmetry degenerates regarding its *d*-orbitals. Locally, the oxygen ions are displaced according to the JT active conformations. Very early after the discovery of this superconductivity, it was recognized that such a polaron with effective spin $S = ½$ would be nearly immobile in the antiferromagnetic (AFM) insulating lattice of the Cu^{2+} ions of a cuprate at low temperatures (Mott AFM). This is because if it moved, the spins of the Cu^{2+} would be forced to be turned around, which requires a large amount of energy. Maintaining the polaronic concept, but avoiding the $S = ½$ difficulty, the bipolaron with $S = 0$, i.e., two polarons with antiparallel spin S, was proposed quite early [5]. In case this would be an amenable way to understand the much-investigated copper oxide superconductor, the structure of such a possible quasiparticle nevertheless remained open. This situation lasted until Kabanov and Mihailovic [6] proposed the intersite JT bipolaron in 2000, see Figure A1.3.1.

In brief, this quasiparticle consists of two JT polarons next to each other in the lattice, having an O^{2-} ion in common and the spins of the two JT polarons antiparallel, such that $S = 0$ results. Furthermore, the spins of the two JT polarons would also be oriented antiparallel to those on the Cu^{2+} ions of the bipolaron. The spins of the latter remain oriented antiferromagnetically to the Cu^{2+} spins in the AFM lattice, see Figure A1.3.2.

The two authors arrived at their proposal from three experiments: (1) The local structure determination with x-ray scattering in Bianconi's group in Rome [7], Figure A1.3.3, right panel; (2) the structure as determined by inelastic neutron scattering in the group of Egami in Oak Ridge [8] was also relevant, left panel, and (3) especially, the analysis of the electron paramagnetic resonance results of Sichelschmitt in Elschner's group in Darmstadt by Kochelaev in terms of a three-spin JT polaron, center panel [9].

The three-spin JT polaron consists of one-hole carrier trapped near two Cu^{2+} ions, with only an oxygen ion between them. Trapping a second hole carrier results in the intersite JT bipolaron, which has a *lower* energy than the three-spin JT polaron, despite the fact that the two carriers with the same charge repel each other. The bipolaron is basically an effective negative-U center as proposed earlier by Anderson to occur in disordered semiconductors [10]. The reason for this negative U is the motion of the ions of the center, which overcomes the Coulomb repulsion between the two charges. The proof of the existence of such centers was a substantial success. With it, the absence of magnetic resonance in doped disordered semiconductors (spin $S = 0$) was understood at once. It could also explain the absence of conductivity in an interval with the presence of a mobility edge, etc. Anderson assumed in his paper that the mobility of the negative-U center could be neglected because of the presence of the random electrical potential present in disordered materials. This is, however, not the case in the cuprates, which are basically regular crystals. Because of their mobility, bipolarons, as they are called in the high-temperature superconductors (HTS), can interact and as such clusters are charged, they repel each other by the Coulomb force. However, there is also the elastic force present,

* Figure captions and minor textual modifications have been added by the editors

† Original Roman name of Basel.

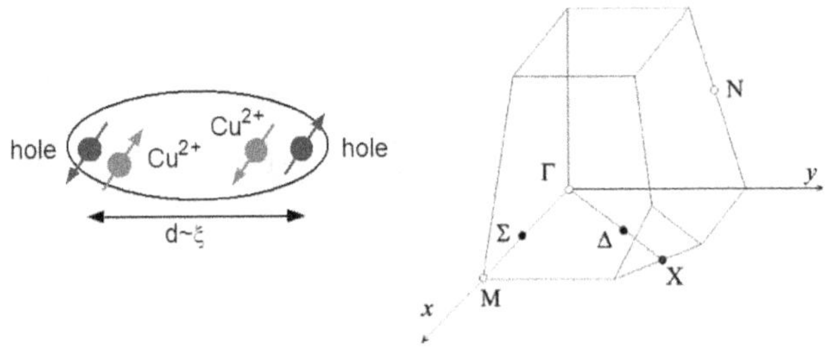

FIGURE A1.3.1 Structure of intersite bipolaron. (From ref [6].)

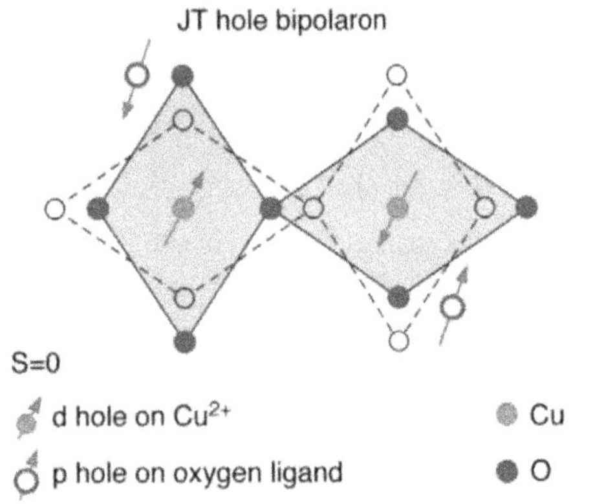

FIGURE A1.3.2 Model of JT Bipolaron. (Adapted from ref [6].)

which overcomes the Coulombic one and leads to clustering and stripe formation as impressive simulations at the Josef Stefan Institute in Ljubljana have shown [11]. Actually, the interaction is such that the bipolarons align along their {10} axis rather than perpendicularly to each other [12]. The well-known stripes result, see Figure A1.3.4.

Here are snapshots of simulations published in Ref. 11, and obtained for $t = 0.04$, $n = 0.2$, and $v_l(1,0)$, as a function of $v_l(1,1)$, where t is the reduced temperature (T/T_c), n is the density, and $v_l(1,0)$, $v_l(1,1)$ stand for the short-range elastic nearest and next-nearest neighbor interactions. Clearly visible is the formation of elongated clusters/stripes, which, depending on the interaction of $v_l(1,1)$, are aligned along {1, 0} or {1, 1} in the plane.

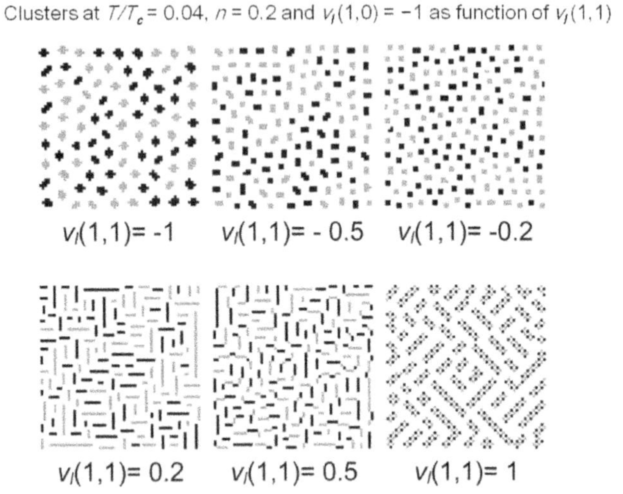

FIGURE A1.3.4 Stripe structures from models of interacting bipolarons. (Reprinted from ref [11b].)

Interestingly, it was found that the hole-rich clusters with an even number of particles are more stable than those with an odd number.

A1.3.2 The Phase Diagram

With the bipolaron model, it is possible to reproduce many features of the hole-doped cuprate phase diagram. It enables not only the *quantitative* prediction of the critical concentration at 6% hole doping for the occurrence of superconductivity on the underdoped side, but also an estimate of the T_c from experimentally measured values of the pairing energy [13]. Furthermore, there exists a quasimetallic region below a temperature $T^*(n) > T_c(n)$ first detected via specific-heat experiments and showing a gap in tunneling data. $T^*(n)$ decreases nearly linearly with n, as was first reported by Deutscher [14] from Andreev reflections. A more recent diagram including data from many experiments is shown in Figure A1.3.5 [15].

This figure reproduces data from many structurally different cuprate systems showing the progression of T^* and T_c, where the former curve approaches the latter *tangentially*

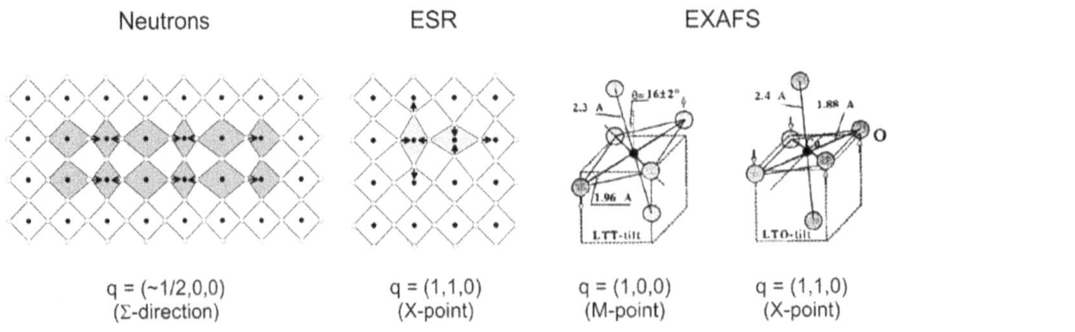

FIGURE A1.3.3 Models of the JT bipolaron as extracted from three experiments. (Left and center panel from ref [6]. Right (EXAFS) results from ref [7].)

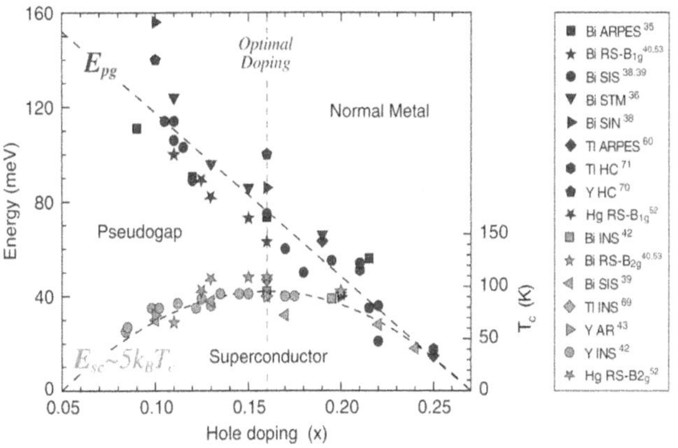

FIGURE A1.3.5 Phase diagram of the cuprates. (Reprinted from ref [15].)

above optimum doping. This excludes all magnetism-based theories, which yield a quantum critical point at $T = 0$ near optimum doping, because the T^* line would have to intersect that of T_c near this doping level.

With the bipolaron model, a number of properties could be clarified. For instance, the nature of the pseudogap shown: At a given doping level n as a function of decreasing T above $T^*(n)$, single JT bipolarons exist, which below this temperature start to aggregate and form clusters or better stripes, see Figure A1.3.6.

In these cluster stripes, type 0D superconductivity is present as a number of tunneling experiments have shown gap-type signatures [16]. Upon further lowering the temperature, phase coherence and 3D superconductivity set in at $T_c(n)$. Important in this process is that the metallic-type stripes are by no means static entities. They emerge in the AFM lattice at a time, move and disappear. This *dynamic* behavior has been documented in the group of Mihailovic

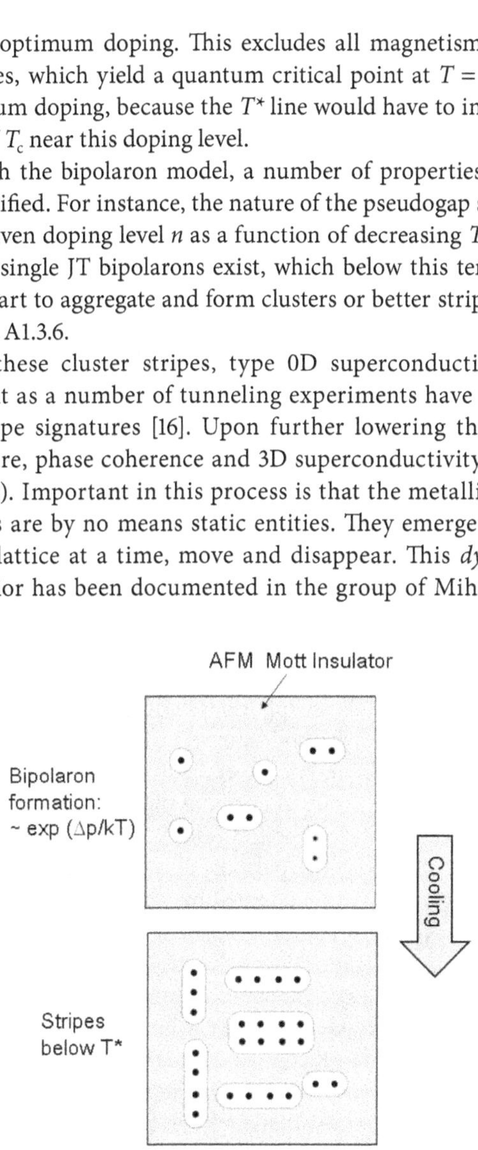

FIGURE A1.3.6 Schematic of the emergence of bipolaronic stripes in the pseudogap phase. (From [1].)

in Ljubljana [17], i.e., these cuprates are intrinsically heterogeneous in a dynamic way. The pulse probe work at the Joseph Stefan Institute, more recent investigations by Oyanagi using EXAFS to record Cu-O in-plane bond lengths as a function of temperature, extensive EPR of Mn probes, and the oxygen isotope effect on the penetration depth have all been shown in a recent review by Shengelaya and Müller [18] to yield quite a clear picture of the dynamics of the heterogeneous state present in the layered hole-doped copper oxides.

A1.3.3 Oxygen Isotope Effects

In the classical superconductors, the isotope effect was substantial in pointing to and supporting the Bardeen–Cooper–Schrieffer (BCS) theory. The key point is summarized in Figure A1.3.7: The shift in T_c is proportional to the mass of an elemental superconductor to an exponent α, which is ½, as predicted by BCS and observed in most of the cases.

In cuprates, relevant oxygen isotope effects, i.e., substitution of the naturally abundant ^{16}O in the oxide by the ^{18}O isotope, have been reported on $T_c(n)$ and $T^*(n)$. We start with the former, in which two groups carried out nearly simultaneously experiments [19, 20], both done at optimal doping, i.e., with n yielding a maximal T_c. A nearly vanishing effect was found. This indicated a purely electronic origin of the superconductivity observed in the cuprates, and supported the RVB theory of Anderson [21] and the t-J model of Zhang and Rice [22]. These theories were since then followed also with similar models by a large part of the community. In contrast, using slightly underdoped samples, the group of Frank reported a clear isotope effect [23]. Since then, the group of Keller at the University of Zurich, with a substantial effort, not only confirmed the Canadian results, but even measured the oxygen isotope effect as a function of the doping for four different compounds [24], see Figure A1.3.8.

This figure was obtained by Steven Weyeneth, whose picture appears on the right side, and the author [25]. It shows the vanishing effect at optimum doping plus its growing to a value of 1, twice as large as the one obtained by BCS theory! The curve shown follows the data very closely and was obtained by Kresin and Wolf in 1994. Its formula is quite simple, see Figure A1.3.9 [26].

Therein, γ(n) is a slowly varying function of n and has been taken as constant in obtaining the curve in Figure A1.3.8. The formula was obtained under the assumption that the polarons are aligned along the c-axis of the crystal, as shown in

$$T_c \propto M^{-\alpha}$$

$$\text{Isotope effect exponent: } \alpha = -\frac{\Delta T_c / T_c}{\Delta M / M} = \frac{d\ln(T_c)}{d\ln(M)}$$

Within the BCS theory: $\alpha = 0.5$

FIGURE A1.3.7 The isotope effect in superconductors.

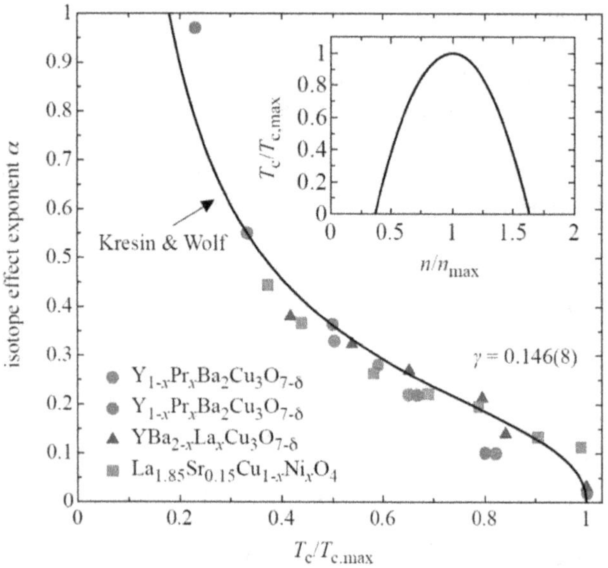

FIGURE A1.3.8 Isotope effect in cuprates. (Reprinted from [25].)

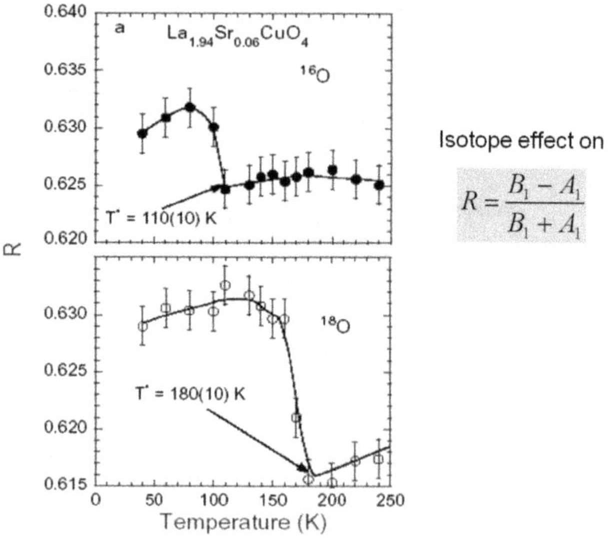

FIGURE A1.3.10 Isotope effect on T^* from X-ray near Absorption Spectroscopy. (Reprinted from [29].)

Figure A1.3.9. This theory has been overlooked because it considered polarons with their axis along c and because when it was published, experiments were scarce. Moreover, the derivation was mathematically not complete. Later, the group of Keller, by selectively doping the ^{18}O, especially in and out of the CuO_2 planes — a real tour de force — showed that the main contribution to the oxygen isotope effect resulted from the oxygens *in the* CuO_2 *planes* [27]. Therefore, the relevance of the formula was not obvious. The present author then recognized that the formula should remain correct for polarons lying in-plane, and the excellent agreement with the data in Figure A1.3.8 supports this view. Furthermore, a mathematically amenable re-derivation of the formula was accomplished recently by Kochelaev, Müller and Shengelaya [28]. This re-derivation in a reproducible manner is based on the original paper of Kresin and Wolf [26]. Therefore, Figure A1.3.8, in the opinion of the author, unambiguously supports the polaronic origin of the superconductivity in the hole-doped copper oxides.

Regarding the oxygen isotope effect present on $T^*(n)$, quite early two entirely different types of experiments were

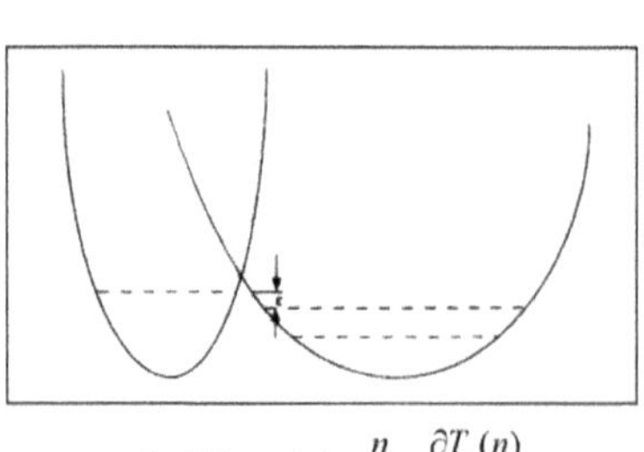

$$\alpha(n,T_c) = \gamma(n)\frac{n}{T_c(n)}\frac{\partial T_c(n)}{\partial n}$$

FIGURE A1.3.9 Polaronic model of the isotope effect. (Reprinted from [26].)

carried out: First, via x-ray near-edge absorption spectroscopy (XANES) and then with inelastic neutron scattering. We start with the results of XANES done by Lanzara *et al.* in Rome [29], see Figure A1.3.10.

Shown in Figure A1.3.10 is the ratio R of the fluorescence counts A_1, and B_1 near the absorption edge as a function of temperature. R depends on the nearest neighbors either out of plane by A_1 or in-plane by B_1. One can therefore monitor the change in the Cu^{2+} to local O neighbor distance either in- or out of plane on an extremely short time, thanks to the x-rays. At T^*, a substantial change is detected for the in-plane distance, with the *largest oxygen isotope effect of 70 K ever published to the day.* This effect was discussed away by the followers of the entirely electronic theories by proposing that, in fact, at this temperature, a structural phase transition (SPT) was present in the $La_{1.94}Sr_{0.06}CuO_4$. However, no such SPT was found there.

To confirm the above important finding, experiments with an entirely different method, namely, inelastic neutron scattering and the stoichiometric superconductor YBCO 1248 were carried out in the group of Furrer at the PSI in Würenlingen (Switzerland). In the compound YBCO, the rare-earth Y^{3+} was replaced by the Ho^{3+} with partially filled $4f$ shell, whose internal transitions could be detected by neutron scattering [30], see Figure A1.3.11.

Shown is the transition of the ground state to the first excited Γ4 of Ho^{3+} as a function of temperature. It is proportional to the A_2^2 crystal field. At T^*, a jump occurs with a ^{16}O to ^{18}O isotope shift of 50 K comparable to that observed in LSCO with XANES (Figure A1.3.10). Also shown is a picture of D. Rubio Temprano, whose thesis work was performed under the direction of Prof. Albert Furrer at the ETH Zurich.

To substantiate the enormous oxygen isotope shift at T^* further, experiments in LSCO were carried out in which

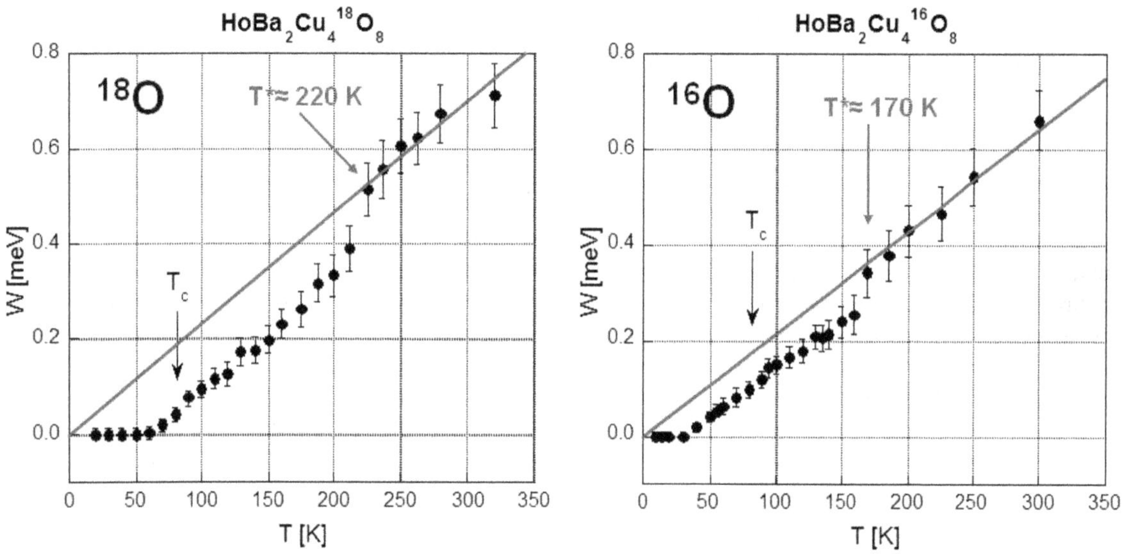

FIGURE A1.3.11 Isotope effect at T^* from inelastic neutron scattering. (From Ref [30].)

the La^{3+} was substituted in part by Ho^{3+}, whose Γ1 to Γ4 transition was recorded for both ^{16}O to ^{18}O and ^{63}Cu to ^{65}Cu substitution [31]. Whereas for the former an isotope shift of 10 K occurs, there is none for the latter. The reason is that in LSCO the Cu^{2+} nearest neighbors are oxygen ions on octahedral lattice sites. There exists an inversion symmetry in their motion that is not crystal-field-active. In contrast, in YBCO, an isotope shift is observed for ^{63}Cu to ^{65}Cu substitution, as expected, because here the Cu^{2+} has five nearest neighbors, located on a pyramid, and the copper lacks inversion symmetry (the so-called Röhler mode is active). Note also that the isotope effect on T^* is negative, i.e., opposite to the one observed on $T_c(n)$. This is what one expects from the Kresin–Wolf formula, in which the derivative of the transition temperature in question appears. See Figure A1.3.12.

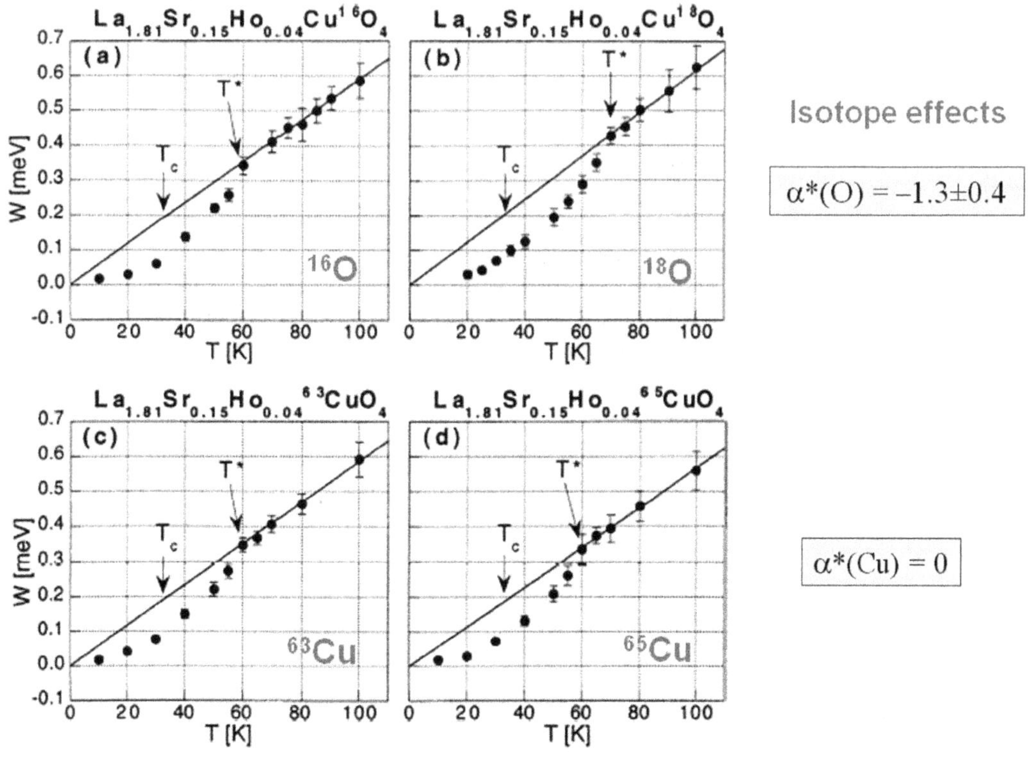

FIGURE A1.3.12 Comparison of oxygen and copper isotope effects. (Reprinted from [1].)

Ignoring consistently and over many years the oxygen isotope effects presented above, which clearly support the polaronic basis of the superconductivity in hole-doped cuprates, is scientifically not acceptable.

A1.3.4 The Vibronic Theory

In quantum physics, there are two basic ways to deal with these phenomena. The particle description of Heisenberg and the Schrödinger wave equation, which are, as Pauli has shown [32], equivalent. This appears to be also the case in the superconductivity of the copper oxides. The first, the quasiparticle one, has been outlined in Section A1.3.1, the second is the vibronic one by Bussmann-Holder and Keller [24]. In it, a lower d-band is coupled via phonon interaction to a higher-lying p-band. The former is at the basis of the single-band electronic models of RVB [21] and t-J [22]. However, it is just this coupling which yields the high transition temperature. The essence of this theory is given in Figure A1.3.13.

The Hamiltonian consists of four terms: The first encompasses the electronic d-functions of the Cu^{2+} ions present, found in the t-J model [22]. The second is due to the oxygen p-band. The third is the vibronic interaction between the two bands: In it, the first is the linear interaction between the electronic densities and the respective ionic displacements, and the second is the exchange between the two bands. The latter is responsible for only one transition temperature and not for two, as predicted by the author almost two decades ago [33]. The last term describes the usual lattice dynamics. This model yields polarons with the binding energy amount shown. They are proportional to the square of the interaction constant γ and inversely proportional to the ionic mass and the phonon frequency squared, see [24].

Actually, from this theory, with sufficiently large vibronic coupling, the effective interaction energy U_{eff} becomes negative, i.e., bipolarons are formed [24]. From the vibronic theory, the oxygen isotope effects on $T_c(n)$ could be well described with the local JT conformation t_2. Also, the oxygen isotope effect on $T^*(n)$ followed [34]. The results obtained from the vibronic theory use the mean field approach. Thus, in the regions where the carrier doping is low and the charge distribution becomes more grainy, the experiments deviate from this theory. This was the case for the oxygen isotope effect on $T_c(n)$, but the earlier theory by Kresin and Wolf of 1994 yielded a spectacular agreement, as outlined in Section A1.3.3.

$$
\begin{aligned}
H &= H_1 + H_2 + H_{e-l} + H_l, \qquad H_l = \hbar\omega \sum_i b_i^+ b_i \\
H_1 &= H_{tJ}(d) \\
H_2 &= H(p) \\
H_{e-l} &= -\gamma \sum_{i,\sigma} [x_i n_{ip} + x_j n_{jd}] - \tilde{\gamma} \sum_{i,j} x_i (d_i^+ p_j + H.c.)
\end{aligned}
$$

Polaron binding energy $E_b = \gamma^2 / 2M\omega^2$

FIGURE A1.3.13 Hamiltonian for the vibronic theory.

A1.3.5 The Symmetry of the Superconducting Wavefunction

The above overview constitutes the previous four headings of this review. Under this heading, attention is turned to the symmetry of the condensate and then in the next one to the superconducting behavior of the electron-doped copper oxides. The electronic single-band t-J theory and those related to it predicted a d-wave symmetry for the superfluid. Photoemission, tunneling, and especially the much hailed tricrystal experiments of Tsuei and Kirtley [35] apparently proved the reality of this prediction and contributed to the acceptance of these theories. What was overlooked and pointed out by the present author was that all these experiments probed *the surface only!* [36] because of the extremely short coherence length. Experiments sensitive to the bulk, such as susceptibility, very early and recent NMR, photo-reflectivity, and especially muon spin rotation, yielded a different answer, see [1]. Actually, because the muons stop near the surface or in the bulk, depending on their incoming kinetic energy, the crossover from pure d-wave at the surface to $s + d$ symmetry in the bulk obtained from group theory by Iachello [37] could be quantitatively demonstrated. Therefore, and in contrast to the long followed electronic single-band theories, these experiments prove unequivocally the vibronic character in which a lower-lying d-band is coupled to a higher-lying oxygen-based p-band as described in Section A1.3.4.

We start by summarizing, in Figure A1.3.14, the essential observations obtained so far regarding the symmetry of the superconducting wavefunction in the hole-doped copper oxides [38]. They will then be shown in more detail in the subsequent panels.

In the CuO_2 plane, the symmetry is 100% d at the surface, and inside it is so far up to 75% d and 25% s, in agreement with the group-theoretical analysis of Iachello [37], and, surprisingly, 100% s along the tetragonal c-axis. The latter property could be derived from the vibronic theory discussed in Figure A1.3.13 [34].

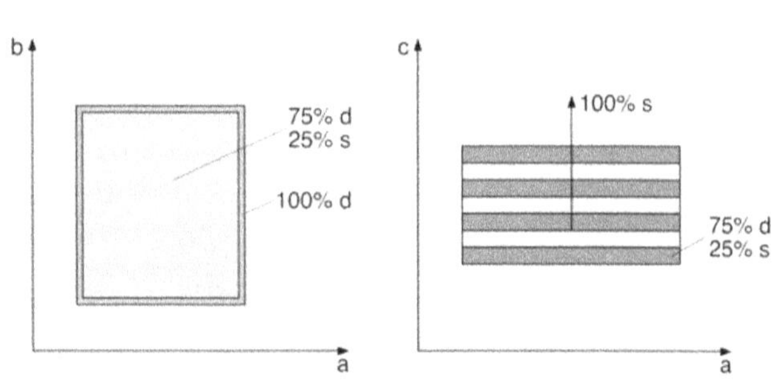

FIGURE A1.3.14 Anisotropic nature of superconductivity in, and perpendicular, to CuO_2 planes. (Reprinted from [1].)

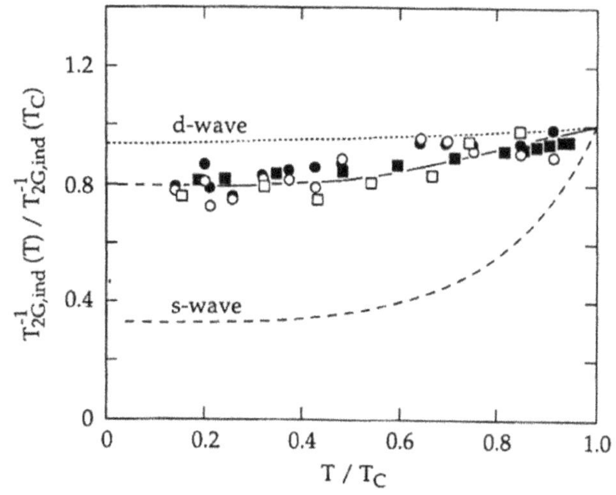

FIGURE A1.3.15 NMR relaxation times T_2. (Theoretical curves from Ref [39] and data from ref [40].)

The time span regarding the symmetry of the superconducting wavefunction started quite early in 1991, when Bulut and Scalapino [39] published their calculation on the inverse NMR T_{2G} relaxation times as a function of temperature down from the phase transition, as shown in Figure A1.3.15, i.e., for *d*- and *s*-wave symmetry.

Included in the figure are also the data on the T_{2G} relaxation times of LSCO of Brinkmann's group at the University of Zürich from 1995 [40]. As seen, they lie in between the theoretically calculated two curves. Interpolating linearly between the two, one arrives at 20 to 25% *s* for the bulk. However, as noted later, this constitutes a lower limit for the *s*-wave-component since, owing to the magnetic field, which is necessary for NMR, vortices are present, which yield an enhanced *d*-wave-component at their border. It was in 1995 that the present author proposed the presence of *s*- and *d*-symmetry for the bulk superconducting wavefunction [33].

Since the experiments which yielded the above-described property of the bulk already in the mid-1990s, other experiments supported this and differences from the surface have been emphasized by the author in Phil. Mag. [36] as well in a conference review [41]. Both are in full agreement with a group-theoretical effort by Iachello [37], in which he deduced a 100% *d*-wave symmetry on the surface towards *a* and *s* + *d* in the bulk *a*-*b* plane of the cuprate superconductor. To stay with the intention of a compact presentation, the reader interested historically or more deeply may look into the references cited, and we present here the results obtained later at the Swiss Muon Source at the PSI in Villigen by R. Khasanov and collaborators, initiated by Prof. H. Keller using muon spin rotation. Muon rotation is a local probe so that the muons, depending on their energy, stop either near the surface or in the bulk, and thus are ideally suited for investigating the wavefunction as a function of the distance from the surface. The relaxation rate $\sigma_{sc}(T)$ is inversely proportional to the square of the London penetration depth $\lambda(T)$. At low magnetic field, the two *s*- and *d*-components are clearly visible in Figure A1.3.16, but less so at higher magnetic fields. This is due to the presence of more vortices at the surface and therefore has *d*-character [41], see also Figure A1.3.14. On the left, the decomposition of $\sigma_{sc}(T)$ into the two components is shown.

In the bulk, the symmetry was investigated at the PSI by the group of Khasanov: In Figure A1.3.17, we show the square of the inverse of the London penetration depth as detected by muons stopped in the bulk of LaSCO along the crystallographic *a*-, *b*-, and *c*-direction. For *a* and *b*, because of the nearly tetragonal crystal, they are the same, as also seen in the left figure. *But along the c-axis, it is pure s!!* This is borne out by the vibronic theory, see Figure A1.3.13.

In NMR as well as in muon spin rotation experiments, a magnetic field has to be present which automatically manifests that vortices are also present. At their surface, for normal conductivity, the character is *d*-wave. Therefore, the average deduced symmetry is an enhanced *d*-wave one as compared to the bulk without magnetic field [41]. A more reliable *s*-to-*d* ratio can be deduced from the x-ray data

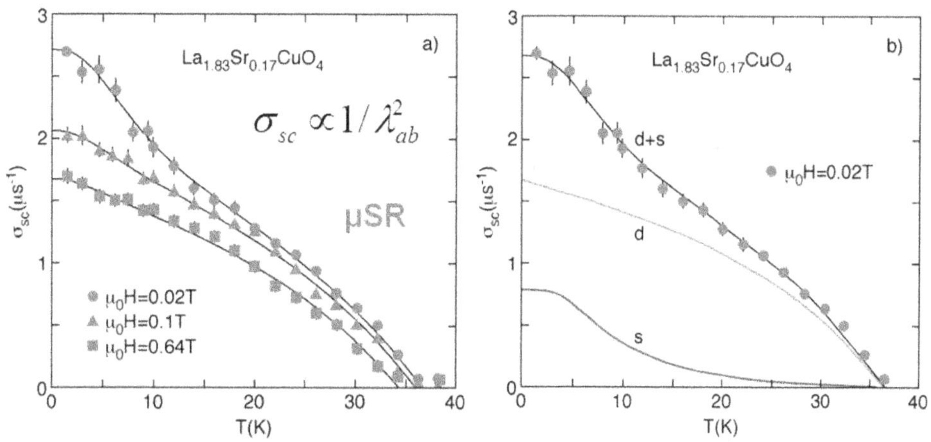

FIGURE A1.3.16 μSR measurements of the penetration depth. (Reprinted from [42b].)

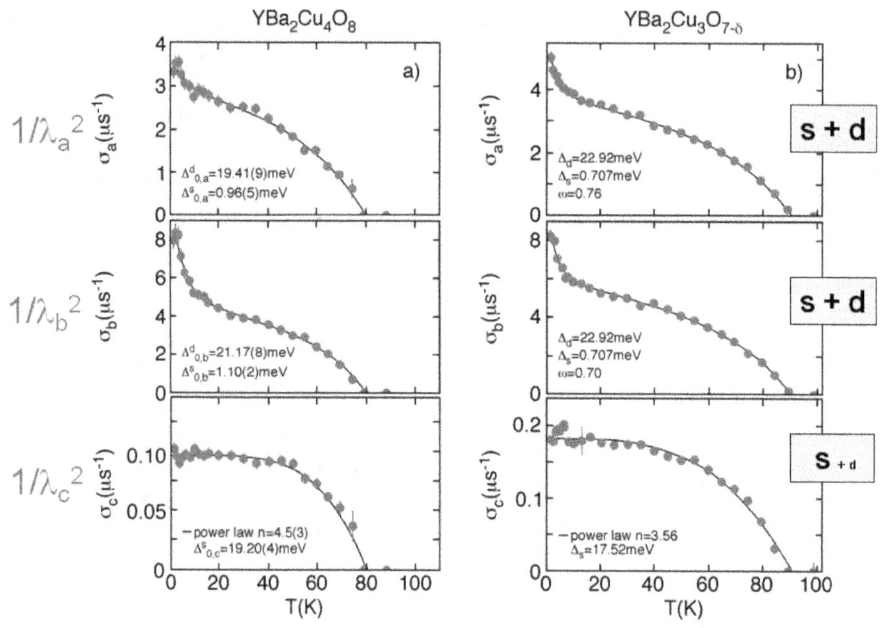

FIGURE A1.3.17 μSR measurements of the penetration depth. (Left from [42a]; right adapted from [43].)

of Oyanagi [44], obtained, of course, in the absence of a magnetic field. He recorded the local distance fluctuations between the in-plane Cu^{2+} and ligand O^{2-} in YBCO as a function of temperature and found substantial enhancement, i.e., peaks, towards the temperatures T^* and T_c. The analysis of Bussmann-Holder with her vibronic theory gave a 40% *s*- and 60% *d*-wave character of the superconducting wavefunction in the bulk [45] as illustrated in Figure A1.3.18. These amounts may be regarded as the intrinsic ones.

To obtain superconductivity in electron-doped copper oxides, a special material treatment was required. In their paper, Dagan and Greene [46] summarized some of these efforts in which the importance of the presence of additional holes was mentioned. In this paper, they reported superconductivity in electron-doped $Pr_{2-x}Ce_xCuO_4$, which was tested with resistivity and Hall angle measurements. Basically, they found that the mobility of the electrons is larger than that of the holes present and mask the latter in the normal conducting phase, but the holes present induce the superconductivity. They arrived at the conclusion that "the electrons have no (or a very small) contribution to superconductivity in the electron-doped cuprate superconductors" [46]. This important conclusion justifies the restriction of the present review to the hole-doped copper oxide superconductors.

A1.3.6 Summary and Conclusions

In Sections A1.3.1 and A1.3.3, the theories for the polaronic properties of hole-doped copper oxide superconductors are described with their particle and Schrödinger wavefunction properties, due to Kabanov and Mihailovic [6] and Bussmann-Holder [24], respectively. These two theories have been shown to yield the following properties of the superconductors in question. a) In the latter theory, the coupling of the lowest (Cu) *d*-type band (occurring also in the t-J model) to the next higher (O) *p*-band yields the high transition temperatures observed. b) The phase diagram of all hole-doped cuprates results from the particle picture, see Section A1.3.2. c) In Section A1.3.3, the observed, in part substantial oxygen isotope effects at both, the cross-over temperature $T^*(n)$ to the pseudogap region and from the latter at $T_c(n)$ to superconductivity is obtained. d) In Section A1.3.4, an overview regarding the symmetry of the superconducting wavefunction is given. Here, the essential

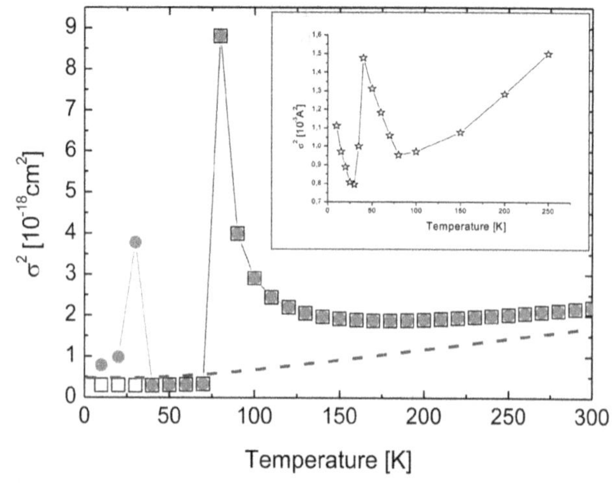

FIGURE A1.3.18 Local distance fluctuations in cuprates. (Main panel is theory from ref [45]; Inset is X-ray data adapted from ref [44].)

property is that the latter is *different* in the *a-b* plane at the surface from that in the bulk and nearly 100% *s* along the *c*-axis. d) In final note, the situation in the electron-doped cuprates is sketched, in that also there the presence of hole pockets is the key.

In conclusion, the two quite equivalent polaronic theories described deliver the main properties of the superconductivity in the copper oxides known. In these theories, the Born–Oppenheimer principle, in which the electronic degrees of freedom can be decoupled from the motion of atomic nuclei, mostly valid in condensed matter physics, is not valid, as evident from the essential presence of bipolarons. However, the single-band electronic theories, such as the RVB or the t-J model, assume at their outset the validity of this approximation, therefore they do not apply, and are unable to explain most of the observations presented in Sections A1.3.3 and A1.3.4. However, the progression from the t-J single-band model to the vibronic theory by coupling the ground-state single *d*-band to the one above it, which is the oxygen *p*-band, yields the valid vibronic theory. This *essential* step is described in Ref. 2 in a didactic manner in the last chapter by Bussmann-Holder et al.

References

1) K.A. Müller, *J. Supercond. Nov. Magn.* **27**, 2163 (2014).
2) *Superconductivity in Complex Systems*, K.A. Müller and A. Bussmann-Holder, Eds., in *Structure and Bonding*, Volume 114 (Springer Verlag, 2005).
3) J.G. Bednorz and K.A. Müller, *Z. Phys. B.* **64**, 189 (1986); *idem, Adv. Chem.* **100**, 757 (1988), Nobel Lecture.
4) K.H. Höck, H. Nickisch, and H. Thomas, *Helv. Phys. Acta.* **56**, 237 (1983).
5) J.E. *Hirsch, Phys. Rev. Lett.* **59**, 228 (1987).
6) V.V. Kabanov and D. Mihailovic, *J. Supercond.* **13**, 959 (2000); D. Mihailovic and V.V. Kabanov, *Phys. Rev. B.* **63**, 054505 (2001).
7) A. Bianconi, N.L. Saini, A. Lanzara, M. Missori, T. Rosetti, H. Oyanagi, H. Yamaguchi, K. Oka, and T. Itoh, *Phys. Rev. Lett.* **76**, 3412 (1996).
8) R.J. McQueeny, Y. Petrov, T. Egami, M. Yethiraj, M. Shirane, and Y. Endoh, *Phys. Rev. Lett.* **82**, 628 (1999); Y. Petrov, T. Egami, R.J. McQueeny, M. Yethiraj, H.A. Mook, and F. Dogan, *J. Phys.: Condens. Mat.* 400034, (2000).
9) B.J. Kochelaev, J. Sichelschmidt, B. Elschner, W. Lemor, and A. Loidl, *Phys. Rev. Lett.* **79**, 4274 (1997).
10) P.W. Anderson, *Phys. Rev. Lett.* **34**, 953 (1975).
11) a) T. Mertelj, V.V. Kabanov, and D. Mihailovic, *Phys. Rev. Lett.* **94**, 147003 (2005); b) V.V. Kabanov, T. Mertelj, and D. Mihailovic, *J. Supercond. Nov. Magn.* **19**, 67 (2006).
12) B.J. Kochelaev, private commun.
13) D. Mihailovic, V.V. Kabanov, and K.A. Müller, *Europhs. Lett.* **57**, 254 (2002).
14) G. Deutscher, *Nature* **397**, 410 (1999).
15) S. Hüfner, M.A. Hossain, A. Damascelli, and G.A. Sawatzky, *Rep. Progr. Phys.* **71**, 062501 (2008).
16) C. Renner *et al.*, *Phys. Rev. Lett.* **80**, 149 (1998).
17) D. Mihailovic, *Phys. Rev. Lett.* **94**, 207001 (2005).
18) A. Shengelaya and K.A. Müller, *EPL (Europhysics Letters)* **109**, 27001 (2015).
19) B. Batlogg *et al.*, *Phys. Rev. Lett.* **58**, 2333 (1987).
20) L.C. Bourne *et al.*, *Phys. Rev. Lett.* **58**, 2337 (1987).
21) P.W. Anderson, *Science* **235**, 1196 (1987).
22) F.C. Zhang and T.M. Rice *Phys. Rev. B.* **37**, 3759 (1988).
23) J.P. Frank, J. Jung, and A.K. Mohamed, *Phys. Rev. B.* **44**, 5318 (1991).
24) A. Bussmann-Holder and H. Keller, *Europ. Phys. J. B.* **44**, 487 (2005); A. Bussmann-Holder and H. Keller, in *Polarons in Advanced Matter*, A.S. Alexandrov, Ed. (Springer Verlag, 2007). The first author's contribution was the culmination of a series of efforts with several collaborations, especially Polish ones. See [2].
25) S. Weyeneth and K.A. Müller, *J. Supercond. Nov. Magn.* **24**, 1235 (2011).
26) V.Z. Kresin and S.A. Wolf, *Phys. Rev. B.* **49**, 3652 (1994).
27) D. Zech, H. Keller, K. Conder, E. Kaldis, E. Liarokapis, and N. Poulakis, *Nature (London)* **371**, 681 (1994).
28) B.I. Kochelaev, K.A. Müller, and A. Shengelaya, *J. Mod. Phys.* **5**, 473 (2014).
29) A. Lanzara, Guo-meng Zhao, N. L. Saini, A. Bianconi, K. Conder, H. Keller, and K.A. Müller, *J. Phys.: Condens. Mat.* **11**, L541 (1999).
30) D. Rubio Temprano, J. Mesot, S. Janssen, K. Conder, A. Furrer, H. Mutka, and K.A. Müller, *Phys. Rev. Lett.* **84**, 1990 (2000).
31) D. Rubio Temprano, K. Conder, A. Furrer, V. Trouvon, and K.A. Müller, *Phys. Rev. B.* **66**, 184506 (2002).
32) W. Pauli, "Die allgemeinen Prinzipien der Wellenmechanik", *Handbuch der Physik* (Springer, 1933).
33) K.A. Müller, *Nature* **377**, 133 (1995).
34) A. Bussmann-Holder, private commun. (2015).
35) C.C. Tsuei and J. R. Kirtley, *Rev. Mod. Phys.* **72**, 969 (2000).
36) K.A. Müller, *Phil. Mag. Lett.* **82**, 279 (2002).
37) F. Iachello, *Phil. Mag. Lett.* **82**, 289 (2002).
38) From Panel 50 of Ref. 1.
39) N. Bulut and D.J. Scalapino, *Phys. Rev. Lett.* **67**, 2898 (1991).
40) R. Stern, M. Mali, J. Roos, and D. Brinkmann, *Phys. Rev. B.* **51**, 15478 (1995).
41) K.A. Müller, in *Applied Superconductivity 2003*: Proc. 6th European Conf. on Applied Superconductivity, Sorrento, Italy, Sept. 2003, pp. 3–9, *Conf. Series No.* 181, A. Andreone, G. Piero Pepe, R. Cristiano, and G. Masullo, Eds. (IOP Publishing, London, 2004).
42) a) R. Khasanov, A. Shengelaya, J. Karpinsiki, A. Bussmann-Holder, H. Keller, and K.A. Müller, *J. Supercond. Nov. Magn.* **21**, 81 (2008); b) R. Khasanov, A.

Shengelaya, A. Maisuradze, F. La Mattina, A. Bussmann-Holder, H. Keller, and K.A. Müller, *Phys. Rev. Lett.* **98**, 057007 (2007).
43) R. Khasanov, S. Strässle, D. Di Castro, T. Masui, S. Miyasaka, S. Tajima, A. Bussmann-Holder, and H. Keller, *Phys. Rev. Lett.* **99**, 237601 (2007).
44) C.J. Zhang and H. Oyanagi, *Phys. Rev. B.* **79**, 064521 (2009).
45) A. Bussmann-Holder, H. Keller, A.R. Bishop, A. Simon, and K.A. Müller, *J. Supercond. Nov. Magn.* **21**, 353 (2008).
46) Y. Dagan and R.L. Greene, *Phys. Rev. B.* **76**, 024506 (2007).

A2

Introduction to Section A2: Fundamental Properties

Alexander V. Gurevich

Superconducting state of materials cooled below a critical transition temperature T_c has distinct physical properties, such as the loss of electrical resistivity to the flow of dc current, the Meissner–Ochsenfeld (M–O) effect, and anomalies in magnetization, thermal conductivity, microwave response, and specific heat. These features result from the fundamental macroscopic phase coherence of the superconducting state, which also manifests itself in the Josephson effect, magnetic flux quantization, and the vortex state in type II superconductors. While the BCS theory, in which superconductivity results from the Bose condensation of Cooper pairs glued together by the electron–phonon interaction, captures the essential physics of conventional low-T_c materials, quantitative calculations of superconducting properties require taking into account effects of strong electron–phonon coupling in the Eliashberg theory. These theories are generally not applicable to the high-T_c cuprates or iron-based superconductors in which superconductivity primarily occurs due to strong electron and magnetic correlations. Section A2 provides an introduction to superconducting phenomena and their relationship to the normal state materials properties. Nine sections describe the main features of the superconducting state, including the phenomenology of superconductivity, the M–O effect, the loss of superconductivity due to the applied magnetic field, the microscopic BCS theory, thermodynamic, thermal and high-frequency electromagnetic properties of superconductors, flux quantization, magnetic vortices, and the Josephson effects.

The main differences between the superconducting and the normal state and the historical evolution of the key concepts are introduced in Chapter A2.1. Chapter A2.2 outlines the fundamentals of the BCS microscopic theory of superconductivity and introduces such key concepts as the condensate of Cooper pairs coupled together by the electron–phonon interaction, and the energy gap in the spectrum of quasiparticles. The energy gap plays a crucial role in the behavior of superconductors under high-frequency electromagnetic fields, and the operation of superconducting tunneling structures for electronic device applications. Thermodynamic and thermal properties of BCS superconductors, particularly the temperature-dependent energy gap, quasiparticle energy spectrum and density of states, specific heat and thermodynamic critical field, thermal conductivity, thermal expansion, and fluctuations of the superconducting order parameter are addressed in Section G3.

Chapter A2.3 presents the essential phenomenology of superconductivity in the framework of the London and Ginzburg–Landau theories. This approach explains such key properties and characteristics of superconductors as the M–O effect, the magnetic penetration depth, the coherence length, the thermodynamic critical magnetic field, the critical depairing current density, the magnetic flux quantization, the type I and type II superconductivity, the upper and lower critical magnetic fields, and quantized superconducting vortices. In Chapters A2.4, the magnetic penetration depth and its connection to the M–O effect are discussed. Chapter A2.5 addresses the effect of magnetic field on the superconducting state, which represents a critical area for many applications of both low- and high-T_c superconductors. In particular, the key concepts of type I and type II superconducting materials, the upper, lower, and thermodynamic critical fields, pinning and flux flow of vortices are introduced, and some of the critical material parameters related to practical application of superconductors are discussed. Chapter A2.6 extends the discussion of magnetic properties in Chapters A2.2 to A2.4 to the high-frequency electromagnetic response of superconductors between 100 MHz and 500 GHz in the framework of the BCS theory. The chapter introduces a frequency-dependent Mattis–Bardeen complex conductivity and its relation to the surface impedance, kinetic inductance, and surface resistance, which are the main parameters of merit for superconducting resonant cavities and devices operating in the radiofrequency and microwave regions. Deviations of high-T_c cuprates from the BCS model as well as the effects of magnetic field on the surface impedance are discussed.

Chapter A2.7 introduces the magnetic flux quantization in superconductors as an example of a macroscopic quantum phenomenon and discusses the origin of flux quantization, the key experiments, which demonstrate flux quantization in thin film cylinders, and magnetic imaging of quantized vortices, including fractional vortices on grain boundaries in the d-wave high-T_c cuprates. Many of the concepts from the previous chapters are incorporated into a discussion of Josephson effects in Chapter A2.8. The fundamental phase coherence of the superconducting state is employed in many analogue and digital devices, such as SQUIDs and digital logic elements. The Chapter addresses the Josephson effect and single-particle tunneling and their applications in a variety of Josephson junction structures. The effects of magnetic field and junction geometry are also discussed. Finally, other Josephson-related effects, including the Andreev reflection in SNS junctions and peculiarities of superconducting weak links with ferromagnetic layers resulting in unusual dependencies of the Josephson critical current on magnetic field are discussed in Chapter A2.9.

Overall, Section A2 presents an overview of the underlying physics of superconductivity and the effects of material parameters, temperature, applied magnetic field, and current on superconducting properties.

A2.1

Phenomenological Theories

Archie M. Campbell

A2.1.1 Introduction

Phenomenological theories take some experimental results, make a simple assumption and calculate the consequences. If they agree with experiment, the phenomenological theory is considered a success. With this definition all theories in physics, including BCS, are phenomenological theories, but conventionally the term is used in superconductivity to describe the London theory [1] and Ginzburg–Landau theory (G–L) [2] in which the assumptions do not involve the interactions between electrons. An early phenomenological theory is the 'two-fluid model' [3], which splits the electrons into two separate populations: the normal electrons and the superconducting electrons. Like many other concepts in superconductivity, this model originated in work on liquid helium and, in spite of being contrary to the spirit of the quantum mechanical picture of electrons as identical particles, remains an important part of the language of superconductivity.

A2.1.2 The London Theory

Both London and Ginzburg–Landau based their ideas on a macroscopic wave function (see Chapter A1), but it is instructive to see how the London equations arise from a simple picture of free electrons, (although strictly such a system should be treated in the extreme non-local limit).

For free electrons $\boldsymbol{eE} = m\dot{\boldsymbol{v}}$ and $\boldsymbol{J} = n_s e\boldsymbol{v}$, where n_s is the number of superconducting electrons. Here $\boldsymbol{J}$ refers to the supercurrent. There may also be a normal current which behaves conventionally and independently of the supercurrent. (This is important not only at high frequencies and in temperature gradients where they carry entropy and determine the thermomagnetic coefficients [4]). In this section, n_s is assumed constant, in contrast to the G–L theory where it can vary in space and time for a number of reasons. The effect of variation of n_s due to composition variations is treated in Reference [5]. Another important situation in which n_s varies is when a normal current is injected into a superconductor and the normal electrons gradually convert to superelectrons (or vice versa). This is the proximity effect.

Ignoring these complications it follows that:

$$\boldsymbol{E} = (m/n_s e^2)\dot{\boldsymbol{J}} \tag{A2.1.1}$$

and

$$\nabla \times ((m/n_s e^2)\dot{\boldsymbol{J}}) = -\dot{\boldsymbol{B}}. \tag{A2.1.2}$$

We lump the material parameters into a single parameter $\Lambda = (m/n_s e^2)$ and integrate Equation (A2.1.2) to get:

$$\boldsymbol{E} = \Lambda\dot{\boldsymbol{J}} \tag{A2.1.3}$$

and

$$\nabla \times (\Lambda\boldsymbol{J}) = -\boldsymbol{B} + f(r). \tag{A2.1.4}$$

where Λ is a material parameter.

Equation (A2.1.4) allows an arbitrary time-independent magnetic field $f(r)$ to be present, the trapped field we would expect in a perfect conductor. However, the discovery of the Meissner effect suggested that in equilibrium there was no flux in the superconductor, and London made the hypothesis that in thermal equilibrium $f(r) = 0$, so that the local field was always uniquely related to the local current density. This removed the multiple equilibrium states possible if the superconductor were merely a perfect conductor (and which of course occur in practice due to pinning effects in both Type I and Type II superconductors). It was therefore much more consistent with classical thermodynamics. The London equations are therefore:

$$\boldsymbol{E} = \Lambda\dot{\boldsymbol{J}} \tag{A2.1.5}$$

and

$$\nabla \times (\Lambda\boldsymbol{J}) = -\boldsymbol{B} \tag{A2.1.6}$$

It is worth pointing out that there is a close relationship between the London equations and conventional eddy currents. For harmonic changes of frequency ω in copper $E = i\omega\sigma J$, where ω is the frequency and σ is the conductivity. By replacing the conductivity in an eddy current problem by $\Lambda/i\omega\sigma$ we get the solution to the same problem for a London superconductor. This can be helpful in seeing how the inductance of loops can determine how a supercurrent is distributed between them.

Taking the curl of Equation (A2.1.6) leads immediately to the exponential decay of current and field with distance from the surface of the superconductor.

$$\nabla^2 J = J/\lambda^2 \quad \text{(A2.1.7)}$$

where

$$\lambda = (\Lambda/\mu_o,)H^{1/2} = (m/\mu_o n_s e^2)^{1/2} \quad \text{(A2.1.8)}$$

λ is the London penetration depth, the equivalent of the skin depth in eddy currents.

A2.1.2.1 Multiply Connected Systems

Multiply connected samples are less straightforward. The term means samples with holes containing trapped flux, or solid samples containing vortices. Suppose we have a hole in a superconductor and apply Faraday's law and Equation (A2.1.5) to a loop round it within the superconductor:

$$\oint \dot{B}.dS = -\oint E.dl = -\oint \mu_o\lambda^2 \dot{J}.dl \quad \text{(A2.1.9)}$$

Hence the integral $\oint (B\, dS + \mu_o\lambda^2 J\, dl)$ is constant with time. This integral is the 'fluxoid', which reduces to the flux in macroscopic samples when the contour can be taken in a region deep in the superconductor where $J = 0$. In small samples, or around vortices in type II superconductors, the flux is less but this is compensated for by the fact that there will be a current density J. For a loop which does not contain the hole the fluxoid is zero.

An extension to the London Equation (A2.1.6) is the equation if there is a vortex present at a point r_o. A vortex has a singularity at its centre so the equation is:

$$\nabla \times (\mu_o\lambda^2 J) + B = \phi_o\delta(r - r_o) \quad \text{(A2.1.10)}$$

It can be seen that this is Equation (A2.1.6) at all points except the vortex core. However, by integrating over a surface spanning r_o we see that it also fulfills the condition that the fluxoid of the vortex is equal to a constant ϕ_o, shown below to be quantised in units of h/2e (Section A2.1.3.3).

The single vortex solution is:

$$B(r) = \phi_o K_o(r/\lambda)/(2\pi\lambda^2) \quad \text{(A2.1.11)}$$

Here, K is the Hankel function, equivalent to the exponential decay in planar geometry.

A vortex lattice is made up by the superposition of such vortices at various values of r_o and since in high-T_c superconductors the core is extremely small, the London limit is an excellent approximation for calculating the properties of the vortex lattice (see Chapter A3.1).

A2.1.2.1.1 Flux in a Thin-Walled Cylinder

To illustrate the London equations in a multiply connected system we consider a hollow cylinder of radius a with a wall thickness much less than the penetration depth. We start with zero field and current and increase the external field H_o parallel to the axis. We assume that the wall is sufficiently thin for the current density to be treated as uniform and the interior field to be assumed equal to the external field as a first approximation.

From Faraday's law, at the wall, $2\pi aE = \pi a^2\mu_o\dot{H}_o$ so $E = \frac{1}{2}a\mu_o\dot{H}_o$. Substituting this in Equation (A2.1.5) and integrating over time gives:

$$J = aH_o/(2\lambda^2). \quad \text{(A2.1.12)}$$

The degree of screening, i.e. the drop in field across the wall, is J times the wall thickness, d,

$$\delta H/H_o = ad/(2\lambda)^2. \quad \text{(A2.1.13)}$$

Significant flux is excluded when λ^2 becomes comparable to ad. The penetration depth is then intermediate in size between the wall thickness and radius, a thickness comparable to λ is not needed for nearly complete screening. (This is a rather general result for thin sheets, another example is a film perpendicular to an applied field. It also applies to the effective skin depth in copper sheets normal to a field and to the spreading of vortices in thin films).

This situation cannot continue indefinitely, there comes a point at which the current density is too great, and we now find this value. We can argue that the material will go normal when its change in free energy is equal to the work done by the solenoid as the flux enters. The free energy per unit volume is $\frac{1}{2}\mu_o H_c^2$, and the work done is $\pi a^2 H_o\Delta B = \pi a^2 H_o\mu_o Jd$.

Substituting for H_o from Equation (A2.1.12) gives:

$$J = H_c/(\lambda\sqrt{2}). \quad \text{(A2.1.14)}$$

This is the maximum current density the material can carry and is equal to the surface current density when the surface field in a bulk samples reaches H_c. It is an overestimate as the high current density reduces the superelectron density, and hence free energy, from its value at zero field and current. (Also the thermodynamics is slightly suspect as the transition is irreversible, the cylinder will immediately become superconducting again once the current has died away, so it is only valid in the limit of thin walls).

A better figure, although only differing by a numerical constant, comes from the G–L theory below (Section A2.1.3.2).

The analysis there brings out the point that although this geometry is topologically multiply connected, provided we have not exceeded the critical current, there is no trapped flux, and it is simply connected as far as the London equations are concerned. Trapped flux is considered in Section A2.1.3.4.2.

High-*Tc* superconductors have an extremely small coherence length compared with the penetration depth. This means that for many purposes the London equations are a good approximation for calculating fields and currents.

A2.1.2.2 The Gauge of A

We can express the London equations in terms of the vector potential defined by $\nabla \times \boldsymbol{A} = \boldsymbol{B}$. With this definition, we can add the gradient of a scalar to $\boldsymbol{A}$ without changing $\boldsymbol{B}$, and this is called a gauge transformation. It must not change any observable quantity. To define $\boldsymbol{A}$ more precisely, we often define $\nabla.\boldsymbol{A} = 0$, although Equation (A2.1.27) shows that this gauge may not always be consistent with the G–L theory. $\boldsymbol{A}$ is still not defined uniquely until we define the boundary conditions, and this is not trivial. There are many different boundary conditions for the definition of a unique electrostatic potential, (there are many situations in which there are electrostatic charges on the surface and within superconductors), and since the vector potential has three components, there are three times as many possible boundary conditions to define a unique value for $\boldsymbol{A}$.

However, if we define $\nabla.\boldsymbol{A} = 0$ it follows from Equation (A2.1.6) that $\boldsymbol{J} + \boldsymbol{A}/(\mu_o\lambda^2) = \nabla\phi$, where ϕ is any scalar function which satisfies $\nabla^2\phi = 0$. To define $\boldsymbol{A}$ uniquely in a simply connected body, London imposed the boundary condition that $\mu_o\lambda^2\boldsymbol{J}_n + \boldsymbol{A}_n = 0$, where n denotes the normal component, i.e the normal component of $\nabla\phi$ is zero. This is the boundary condition for the London gauge. The electrostatic analogy is a body with no charge density (since $\nabla^2\phi = 0$), and no normal component of the electric field at any point on the surface. In this case, the only solution for the electrostatic potential is a constant. Hence in the London gauge we can assume $\phi = 0$ since the addition of a constant to $\boldsymbol{A}$ cannot affect the magnetic field. In multiply connected samples this boundary condition does not define $\boldsymbol{A}$ uniquely, unless there is neither trapped flux nor vortices, see Section A2.1.3.4.

Now Equations (A2.1.5) and (A2.1.6) become:

$$\boldsymbol{E} = -\dot{\boldsymbol{A}} \quad \text{(A2.1.15)}$$

and

$$\text{and } \mu_o\lambda^2\boldsymbol{J} = -\boldsymbol{A} \quad \text{(A2.1.16)}$$

Equation (A2.1.16) can be interpreted as saying that the current density at any point is proportional to the flux that has crossed that point, in contrast to a normal material where it is proportional to the rate at which flux crosses the point, i.e. the electric field. Note that we have now defined $\boldsymbol{A}$ uniquely, we can no longer add even a constant without affecting the current density.

Two-dimensional problems with fields in the xy plane are often solved using the A–V formulation. In this case, $\boldsymbol{A}$ has only a *z* component so is effectively a scalar. Then contouring A_z gives a picture of the flux lines. However, if we have several wires carrying currents perpendicular to the plane, we will need a *V* for each wire if the wires are magnetically decoupled. This is essentially the electrostatic voltage at the ends of the wires due to the flux change between them and gives the self-inductances between the wires.

In the mixed state, *A* is directly proportional to the distance the flux lines move, which can be measured directly from the electric field.

The importance of the vector potential brings out the involvement of wave mechanics in superconductivity, something that London realised from the outset since he postulated a Bose–Einstein condensation as the cause of superconductivity. Equation (A2.1.16) can be put in the form $\boldsymbol{p} = 0$, where $\boldsymbol{p}$, the momentum vector in wave mechanics of a particle of charge *e* and mass *m*, is given by:

$$\boldsymbol{P} = m\boldsymbol{v} + e\boldsymbol{A} \quad \text{(A2.1.17)}$$

For superconductivity to occur this momentum vector must remain zero when a magnetic field is applied, and this is what is meant by the 'rigidity' of the superconducting state.

Since the phase of a wave function θ is related to the momentum by:

$$\boldsymbol{p} = \hbar\boldsymbol{k} = \hbar\nabla\theta \quad \text{(A2.1.18)}$$

It follows that:

$$m\boldsymbol{v} + e\boldsymbol{A} = \hbar\nabla\theta \quad \text{(A2.1.19)}$$

Integrating round a closed loop:

$$\oint (m\boldsymbol{v} + e\boldsymbol{A})/\hbar\, \boldsymbol{di} = \nabla\theta. \quad \text{(A2.1.20)}$$

The quantity in the integral is proportional to the fluxoid defined above, and London pointed out that this produced a natural unit for the fluxoid, $\phi_o = h/e$. Since we would expect the change in phase round a closed loop $\Delta\theta$ to be an integer multiple of 2π, the equation also implies the quantisation of flux, but this does not seem to have been stated explicitly until it was derived from the Ginzburg-Landau theory.

A2.1.3 Ginzburg–Landau Theory

The Ginzburg–Landau theory of superconductors has been an astonishingly rich source of new physics. It has been much more widely productive than even the original authors expected since one sentence in their paper stated that the solutions of the Ginzburg–Landau equations for the limit κ tends to infinity offer no intrinsic interest and would not be discussed. [However, it should be added that they did derive the maximum field for a solution to exist, which we now know as B_{c2}, and recognised that an instability occurs if $\kappa>1/\sqrt{2}$. κ is defined later in Equation (A2.1.35)]. Although proposed for temperatures near T_c, the theory can be derived from the microscopic theory near B_{c2} where the order parameter can be expanded in a power series, but it can be used at all temperatures and lower fields, at least semi-quantitatively. Different authors use different solutions of these equations to make their points and therefore use different simplifications. Only an outline can be given here, with some examples, which are chosen because they do not appear in the standard texts, but more details can be found in the books by Tinkham [6] and de Gennes [7]. The book by St. James, Thomas and Sarma [8] probably has the widest range of solutions, particularly those involving vortices, and the book by van Duzer and Turner [9] is the most relevant to thin films and devices. A recent book by Waldram [4] explains in detail the connection between Ginzburg–Landau and the microscopic theory and includes the applications to cuprate superconductors.

The G–L theory is an extension of Landau's general theory of second-order phase transitions which is based on a free energy expanded in a power series of some kind of order parameter. For example, in a ferromagnetic transition it is the number of aligned electron spins.

In a superconductor they made the plausible assumption that since the energy is lowered when the electrons become superconducting, we can write the free energy near the critical temperature as a power series in the number of superconducting electrons, n_s. (The fact that fluctuation effects mean that near T_c the expansion as a power series is invalid has in no way diminished the predictive value of the G–L theory).

However, this assumption is not sufficient to describe a superconductor in a magnetic field, since in this situation the lowest energy would be obtained by splitting the sample into a series of laminae small compared with the penetration depth, so that we gain the energy of the superconducting electrons without the disadvantage of having to exclude the field. It was therefore postulated that there was an energy term dependent on the gradient of the order parameter, (squared since energies must be positive), as well as its magnitude. This will produce a surface energy, which prevents the subdivision lowering the energy.

The London theory had been based on the idea of the superconducting state as a macroscopic quantum state, so it seems natural (with hindsight) to postulate a macroscopic wave function ψ such that $n_s = |\psi|^2$ is the order parameter. Now a wave function has both magnitude and phase and can be written as $\psi = f.\exp(i\theta)$, where f and θ are real functions of position. f must be single valued. θ can be multivalued, but only by a multiple of 2π. This has important consequences. The phase θ is related to the momentum $\boldsymbol{p}$ by $\boldsymbol{p} = \hbar\boldsymbol{k} = \hbar\nabla\theta$. We suppose the current carriers have mass m^* and charge e^*. (Note that some authors use e as the electronic charge while others use $-e$, which causes differing signs in the G–L equations if expressed in terms of the electronic charge). In a magnetic field $\boldsymbol{p} = m^*\boldsymbol{v} + e^*\boldsymbol{A}$ and the observable quantity, the velocity $\boldsymbol{v}$, cannot depend on the gauge of $\boldsymbol{A}$. (The phase is not directly observable, only the gauge-independent phase difference, see Section A2.1.3.4.3). Hence if we change $\boldsymbol{A}$ by adding an arbitrary $\nabla\phi$, then we must add $e^*\phi/\hbar$ to the phase of ψ. This means that in the expression for the free energy, which must also be gauge invariant, we cannot have a simple $(\nabla f)^2$ or ∇f^2 but the gauge invariant expression $(1/2m^*)|(i\,\hbar\nabla + e^*A)\psi|^2$. (Although ∇f^2 might seem the simplest answer, it is not compatible with the idea of a macroscopic wave function).

The physical significance of this term in the free energy is brought out by substituting $f\,\exp(i\theta)$ for ψ, and $\boldsymbol{v}$ from Equation (A2.1.19). The energy is then $\frac{1}{2}m^*n_s v^2 + (1/2m^*)(\hbar.\nabla f + e^* fA)^2$. This shows that as well as the gradient and field term, we have automatically included the kinetic energy of the electrons, which is an important term in a superconductor.

To these terms we add the magnetic energy of the field in the superconductor, $\boldsymbol{b}$, giving the following expression for the free energy per unit volume in the superconducting state F_s:

$$F_s(b) = F_n(O) + \alpha|\psi|^2 + \tfrac{1}{2}\beta|\psi|^4 + (1/2m^*)|(i\,\hbar\nabla + e^*\boldsymbol{A})\psi|^2 + \boldsymbol{b}^2 2\mu_o. \quad \text{(A2.1.21)}$$

α and β are the only material parameters, which will be temperature dependent, so only two parameters are needed to define all the properties at a given temperature. In order to get a transition as we go through T_c, α must be positive above T_c and negative below it. To ensure an energy minimum as opposed to a maximum, β must be positive at all temperatures. From the case where $\psi = 0$ it follows that $F_n(b) = F_n(0) + b^2/2\mu_o$.

Equation (A2.1.24) is the most fundamental and general expression of the G–L free energy. Later authors have sometimes added terms in the magnetisation $\boldsymbol{M}$, a field $\boldsymbol{H}$ or the external field $\boldsymbol{H}_o$. These terms are restrictive, ill defined and should be avoided. If it is necessary to determine the effect of an external field or current source, then this must be included as a δw, the external work for a small change. The free energy then becomes the 'availablity'.

G–L then used the calculus of variations to turn the minimum energy integrated over the sample volume into two differential equations. References [4] and [8] give the details. Although it is normally assumed that the material is uniform, we get the same equations if α and β are functions of position so we can include pinning centres and the proximity effect. This allows us to calculate the critical current of a region of reduced gap, such as a grain boundary [10]. Anisotropy can be added by making m^* an effective mass tensor instead of a scalar, (although strictly tensors can only be used in linear systems of equations, and the equations are not linear). These equations are the famous Ginzburg–Landau equations which we will just quote here. The first is:

$$-\boldsymbol{j} = (ie^*\hbar/2m^*)\,(\psi^*\nabla\psi - \psi\nabla\psi^*) + (e^{*2}/m^*)|\psi|^2\boldsymbol{A} \quad \text{(A2.1.22)}$$

where $\boldsymbol{j}$ is the local current density. This is identical to the general equation for the current density in wave mechanics. The second contains the material parameters and is:

$$\alpha\psi + \beta|\psi|^2\psi + (1/2m^*)(i\hbar\nabla + e^*\boldsymbol{A})^2\psi = 0 \qquad \text{(A2.1.23)}$$

The surface contribution to the energy gives the G–L boundary condition on ψ [4]:

$$(i\hbar\nabla_n + e^*\boldsymbol{A}_n)\psi = 0 \qquad \text{(A2.1.24)}$$

where n means the normal component.

If the boundary is a vacuum, there can be no current across it which implies that $d\psi/dx = 0$ at the boundary [Equation (A2.1.22)]. If the boundary is with a normal metal, the boundary condition must be derived from the microscopic theory, see References [7] and [4] which discuss the boundary conditions in some detail.

Since the units of ψ are arbitrary we can chose m^* as an arbitrary parameter before fixing the material parameters α and β. m^* is usually chosen as twice the electron mass, which means that ψ^2 is a number per unit volume. The value of e^* is not arbitrary since it determines the size of the flux quantum, and from experiment (and the microscopic theory) is twice the electronic charge (negative for electrons and positive for holes). This means that the properties of a superconductor at any given temperature are determined by only two material parameters: α and β. We usually replace these with two directly measurable parameters, for example, the critical field and penetration depth, which are derived below in terms of α and β. Further simplifications are to normalise the variables with these parameters to give a dimensionless equation with only one parameter, κ, and to use a gauge transformation to remove the phase of the order parameter where possible. This is done below.

Although Maxwell's equations for the magnetic field might appear to be an independent set of equations, the equations for $\boldsymbol{b}$ can be derived, like the G–L equations, by minimising the magnetic energy $b^2/2\mu_0$ which is included in the G–L free energy so the G–L equations are automatically consistent with the field equations. The free energy is a local expression for the energy density on the scale of several atoms. By convention, lower case letters are usually used for local fields.

An exception is $\boldsymbol{A}$, the vector potential, which is also a local field. Upper case letters are usually used for averages, or integrals over the sample.

If we are interested in the internal structure of the superconductor, it is easiest to minimise the Helmholtz free energy, $F = U - TS$, with no external work done, i.e. constant total flux. This can be done both for wires or films carrying a current, or an isolated sample in a magnetic field. However, if we want to know the external field, $\boldsymbol{H}_0$, in equilibrium with the resulting structure in the latter case we must vary $\boldsymbol{b}$ and put the change in F equal to the work done by an external solenoid.

The free energy F includes the free space magnetic energy, so for long cylinders parallel to the external field we must use a work term $\delta w = H_0.\delta\phi$, where ϕ is the flux entering the sample, in order to be consistent with the change in F in the normal state. If we have a sample of unit cross-sectional area, this can be written $\boldsymbol{H}_0\boldsymbol{\delta B}$, where $\boldsymbol{B}$ is the average of $\boldsymbol{b}$ over the sample cross-section. We therefore need to minimise F–$\boldsymbol{H}_0\boldsymbol{B}$ at constant T and $\boldsymbol{H}_0$. If the sample is in thermodynamic equilibrium, we can then, if we wish, define a local $H = H_0$ and minimise a local Gibbs function F–$\boldsymbol{H.B}$. Note, however, that in this expression $\boldsymbol{A}$ and $\boldsymbol{b}$ are fields on a scale between the atomic and the coherence length, while $\boldsymbol{H}$ is uniform across the sample, and $\boldsymbol{B}$ is an average over the sample cross-section.

If the sample has a finite demagnetising factor, the calculation of the magnetic energy requires the integration of b^2 over all space. This is very intractable in most cases, so normally minimising a Gibbs function can only be done for a sample with no demagnetising factor. However, also soluble is the other extreme for thin small samples perpendicular to the field. Extensive examples of numerical solutions can be found in References [11] and [12] and references therein.

Although in principle we can find the equilibrium state of a superconductor by minimising the free energy numerically with respect to all variables at all points, this requires too much computing power for macroscopic samples and to obtain analytic solutions we usually do the minimisation in several steps with increasing length scales. We illustrate this with the steps used to derive the Abrikosov vortex lattice [13]. First we minimise F with respect to $\boldsymbol{A}$ and ψ. This leads to the standard G–L equations above, Equations (A2.1.22) and (A2.1.23), and ensures that $\boldsymbol{b}$ and ψ are in equilibrium on the scale of many atoms. There are many solutions to these equations, of which one is an array of vortices in arbitrary positions. We now try a periodic solution with vortices on a regular lattice, and minimise F by moving a fixed number of vortices around. This is equivalent to surrounding the sample containing the vortices with a Type I material, to keep the total flux fixed, and allowing the vortices to settle into their equilibrium structure. It tells us the lattice structure for a given interior flux density, and does not involve anything outside the superconductor, so avoiding complications with $\boldsymbol{H}$ or applied fields. It is the equivalent to a constant volume experiment in gases.

Having determined the structure of the vortex state for a given vortex density, we then change the vortex density $\boldsymbol{B}$, keeping the same structural symmetry (for example, a hexagonal array), and allow the external field to do work. Then putting $\delta F = \boldsymbol{H}_0\boldsymbol{\delta B}$ gives the equilibrium flux density with an external field $\boldsymbol{H}_0$, which is then the local relation between $\boldsymbol{B}$ and $\boldsymbol{H}$ in equilibrium on the scale of many flux lines. In gases this gives the relation between the volume and the external pressure.

Finally, if we have high-T_c crystals with a large demagnetising factor, (and no pinning), we use the relation between $\boldsymbol{B}$ and $\boldsymbol{H}$ we have derived, combined with classical electromagnetism, to find the magnetic properties of the crystal, just as if it were a ferromagnetic crystal.

A2.1.3.1 Material Properties

It is necessary to relate the empirical material parameters α and β to measurable quantities which is done by considering a long sample parallel to the external field (no demagnetising effects). To do this, we separate the amplitude and phase of the order parameter by a gauge transformation. We put:

$$\psi = f.\exp(i\theta) \quad \text{(A1.2.25)}$$

Here f and θ are real. Then put:

$$\boldsymbol{A} = \boldsymbol{A}' + (\hbar/e^*)\nabla\theta. \quad \text{(A1.2.26)}$$

Since in what follows we will use this new vector potential, we now drop the ‘after $\boldsymbol{A}$. Equation (A2.1.22) becomes:

$$-\boldsymbol{j} = (e^{*2} f^2/m^*)\,\boldsymbol{A} \quad \text{(A1.2.27)}$$

With a good deal of algebra, [including the use of Equation (A1.2.27) to show that $\nabla(f^2\boldsymbol{A}) = 0$] Equation (A1.2.23) becomes:

$$(\hbar^2/2m^*)\,\nabla^2 f = (\alpha + \beta f^2 + (e^{*2}/2m^*)\boldsymbol{A}^2)f \quad \text{(A2.1.28)}$$

The free energy is Equation (A2.1.21) with f written for ψ.

$$F_s(b) = F_n(0) + \alpha f^2 + \tfrac{1}{2}\beta f^4 + (1/2m^*)\left|(i\,\hbar\nabla + e^*\boldsymbol{A})f\right|^2 + \boldsymbol{b}^2/2\mu_o. \quad \text{(A2.1.29)}$$

We now derive a number of measurable parameters from these equations.

Inside a uniform bulk material in zero field $\boldsymbol{b} = 0$, $\boldsymbol{A} = 0$, and there are no gradients so the order parameter is found by putting $dF/d\psi = 0$ and is given by $f^2 = \psi_o^2 = -\alpha/\beta$. It is common to normalise ψ by this factor.

In low magnetic fields, the order parameter will not be very different from this value, so putting $\psi = \psi_o$ in Equation (A2.1.22) and putting:-

$$\mu_o\boldsymbol{j} = \nabla\times\boldsymbol{B} = \nabla\times\nabla\times\boldsymbol{A} = -\nabla^2\boldsymbol{A} \quad \text{(A2.1.30)}$$

Then,

$$-\nabla^2\boldsymbol{A} = (\mu_o e^{*2}\,\alpha/\beta m^*)\boldsymbol{A} \quad \text{(A2.1.31)}$$

This is the same as the London equation and the penetration depth is given by:

$$\lambda^2 = -m^*\beta/(\mu_o e^{*2}\alpha), \quad \text{(A2.1.32)}$$

(note that α is negative below T_c)

In zero field, if we ignore the second order in $|\psi|^2$, Equation (A2.1.28) becomes:

$$(\hbar^2/2m^*)\nabla^2 f = \alpha\, f \quad \text{(A2.1.33)}$$

This shows that above T_c (where α is positive and ψ small) a fluctuation in the order parameter varies over a characteristic length ξ, the coherence length, where:

$$\xi^2 = \left|\hbar^2/2\alpha m^*\right|. \quad \text{(A2.1.34)}$$

Small changes in ψ decay exponentially over this distance. This is the temperature dependent, or G–L, coherence length which describes the length scale over which fluctuations in the order parameter occur. At absolute zero it corresponds to the size of a Cooper pair, but tends to infinity at T_c while the Cooper pair size remains constant. (Below T_c α changes sign, and the solutions are more complex, but ξ is still the characteristic length for changes in order parameter, see References. [4, 6]).

We now define the G–L dimensionless parameter κ.

$$\kappa = \lambda/\xi = (2m^{*2}\beta/(\mu_o e^{*2}\hbar^2))^{1/2} \quad \text{(A2.1.35)}$$

To find the critical field we put the sample in a solenoid with a field B_c, and allow the material to change reversibly from the superconducting state to the normal state. In the superconducting state $B = 0$ in most of the sample so:

$$F_s(B) = F_n(0) + \alpha\psi_0^2 + \tfrac{1}{2}\beta\psi_0^4 \quad \text{(A2.1.36)}$$

In the normal state:

$$F_n(B) = F_n(0) + B_c^2/2\mu_0. \quad \text{(A2.1.37)}$$

The work done by the solenoid as the flux enters the sample is (B_c^2/μ_0)

Putting the change in F equal to the work done gives:

$$B_c^2 = \mu_0\alpha^2/\beta. \quad \text{(A2.1.38)}$$

Since there are only two independent material parameters (initially α and β), the parameters we have derived can be expressed in terms of any two of them. For solution the equations are usually put in dimensionless form by dividing distances by λ, fields by $\sqrt{2}B_c$ and the order parameter by ψ_o. In this form, the equations depend only on κ.

The secondary dimensionless variables are then:

$$\text{Vector Potential } \boldsymbol{a} = \boldsymbol{A}/\sqrt{2}B_c\lambda \quad \text{(A2.1.39)}$$

$$\text{flux } \Phi = (2\pi/\kappa)\Phi/\phi_0 \quad \text{(A2.1.40)}$$

$$\text{Current density } \boldsymbol{j} = (\lambda\pi_0/\sqrt{2}B_c)\boldsymbol{J} \quad \text{(A2.1.41)}$$

$$\text{Free energy difference } \Delta E = \Delta F/2\Delta F_0 \quad \text{(A2.1.42)}$$

ΔF_0 is the free energy difference in zero field.

Equations (A2.1.22) and (A2.1.23) become:

$$\boldsymbol{j} = i/(2^*\kappa)(\psi^*\nabla\psi - \psi\nabla\psi^*) - \boldsymbol{a}\psi^*\psi \quad \text{(A2.1.43)}$$

$$(\kappa^{-1}\nabla + \boldsymbol{a})^2\psi + (\psi^*\psi - 1)\psi = 0 \quad \text{(A2.1.44)}$$

and Equations (A2.1.27) and (A2.1.28) become:

$$j = -af^2 \quad \text{(A2.1.45)}$$

$$-\kappa^{-2}\nabla^2 f + (|a|^2 + f^2 - 1)f = 0 \quad \text{(A2.1.46)}$$

The dimensionless free energy is:

$$\Delta E = -f^2 + \tfrac{1}{2}f^4 + \kappa^{-2}|\nabla f^2| + |a^2| f^2 + b^2. \quad \text{(A2.1.47)}$$

The fact that the G–L equations can be written in dimensionless form with only one parameter, κ, means that if we vary temperatures and materials, we can plot the curves on top of each other by suitable scaling. Thus all materials with the same value of κ will give the same magnetisation curve at all temperatures if scaled. If we now add pinning centres, the energies involved are those of the G–L theory so most theories for simple structures lead to another dimensionless parameter, and the hysteretic magnetisation curves will also scale, but now with two adjustable parameters. This has been found to be true for many systems, (e.g. in Reference [14]) but is not universal. Scaling of experimental results is an important technique, but as the number of free parameters increases, the physical significance decreases. Most experimental curves can be fitted to any reasonable power law with two or three adjustable parameters.

A2.1.3.2 The Maximum Current Density

Equation (A2.1.22) shows that if we have a current density, there is a gradient in ψ and a corresponding energy term attributable to the kinetic energy of the electrons. Therefore, the kinetic energy of the electrons at high current densities has a similar effect on the order parameter to that of a magnetic field. High current densities reduce ψ, and if high enough, cause a transition to the normal state. We assume a wire with a thickness much less than λ so that the current density j is uniform and gradient terms are zero. We substitute the expression for A in Equation (A2.1.22) into Equation (A2.1.23). From Equations (A2.1.32) and (A2.1.38), $(H_c/\lambda)^2 = -(e^{*2}\alpha^3)/(m^*\beta^2)$. If we also put $s = (\psi/\psi_o)^2 = -\psi^2\beta/\alpha$, then Equation (A2.1.23) becomes:

$$s^2 - s^3 = -j^2\lambda^2/2H_c^2 \quad \text{(A2.1.48)}$$

We need a solution for s between 0 and 1. For zero j the low energy solution is $s = 1$, and the solution for s decreases as j increases. The largest value of j for which a solution exists is when the solution is at the maximum of $s^2 – s^4$ which is when $s^2 = 2/3$.

Then:

$$j = \sqrt{(8/27)}H_c/\lambda. \quad \text{(A2.1.49)}$$

This is the maximum possible current density and is similar to the previous result, Equation (A2.1.14).

It is worth pointing out that this procedure is rather different from the way we derive critical fields. These are derived from the G–L equations assuming reversibility and putting the change in F equal to the work done by a solenoid. This is a conventional phase transition. A transport current is not a macroscopic equilibrium state and so cannot be part of a conventional phase transition. It is not possible to equate a change in free energy to the work done by the battery since the battery works continuously in the normal state. What this derivation does is minimise energy on a local scale, i.e. start with the G–L equations, but then show that these have no solutions if the current density is larger than a limiting value.

This maximum current is close to the depairing current of the microscopic theory and was thought to be an absolute maximum critical current density. It can be interpreted as the current at which the kinetic energy of the superelectrons reaches the condensation energy, so that it becomes favourable to revert to the normal state.

However, recently Matsushita has shown that it can be considerably higher [15].

An ideal pinning array consisting of a hole down the core of each vortex can cause currents of this size. If we take an energy of $\tfrac{1}{2}\mu_0 H_c^2$ over a volume $\pi\xi^2$ per unit length, this will produce a force of $\tfrac{1}{2}\pi\mu_0 H_c^2\xi$ per unit length. Putting this equal to the Lorentz force on a vortex, $J_c\phi_0$, and using the relations in Section A2.1.4.1, gives a critical current density of $1.1H_c/\lambda$, which is in fact above the depairing current. Hence, although at first sight the maximum critical current due to pinning centres appears to be quite different from the depairing current, an ideal pinning array should be able to produce critical currents comparable with the depairing current. This can be of the order of 10^7–10^8 A/cm^2 in both low- and high-T_c superconductors and values within a factor of 10 of the depairing current have been achieved. Many high-T_c superconductors can carry extremely high current densities; it is only the grain boundaries which limit the current for practical applications.

A2.1.3.3 Flux Quantisation

In the London theory, the current density $\boldsymbol{j}$ is proportional to $\boldsymbol{A}$, but since $\boldsymbol{A}$ well away from the hole is proportional to the total flux in a hole, while $\boldsymbol{j}$ tends to zero, this expression cannot hold in a multiply connected system with trapped flux, nor if there are vortices present.

If we put $\psi = f\exp(i\theta)$ (but do not change the gauge) Equation (A2.1.22) becomes:

$$j = (e^*\hbar/m^*)f^2\nabla\theta - (e^{*2}/m^*)f^2\boldsymbol{A} \quad \text{(A2.1.50)}$$

Putting $f^2 = -\alpha/\beta$ and $\lambda^2 = -m^*\beta/(\mu_0 e^{*2}\alpha)$ this can be written:

$$\boldsymbol{A} + \mu_0\lambda^2\boldsymbol{J} = \hbar\nabla\theta/e^* \quad \text{(A2.1.51)}$$

If observables are to be single valued, then integrating round a closed loop the integral of $\nabla\theta$ is $2n\pi$, where n is an integer, so:

$$\int \dot{\boldsymbol{B}}.d\boldsymbol{S} + \oint \mu_o \lambda^2 \boldsymbol{j}.d\boldsymbol{l} = n\phi_o \quad \text{(A2.1.52)}$$

where $\phi_0 = h/e^*$.

Hence the G–L theory leads to the conclusion that the fluxoid defined by London is quantised in units of h/e^*. The flux in any macroscopic hole must be an integer number of the flux quantum $\phi_0 = h/e^*$. In small systems, the flux is less but the quantum is made up by the term in $\boldsymbol{J}$. Since this result is also derivable from the London theory, it only requires that the properties depend on the electrons being described by a macroscopic wave function. It is not dependent on the power law expansion of the G–L theory but is a more general result. The BCS theory and experiment are consistent in showing that e^* is 2e.

A2.1.3.4 The Phase of the Order Parameter and the Gauge of the Vector Potential

The question of gauge invariance in the Ginzburg–Landau equations is a difficult one and needs the study of a number of treatments for a full understanding. In Reference [8], the problem is largely avoided by taking the curl of the equations and working in terms of the magnetic field. Clem [16] recommends using the superfluid current density, which is gauge invariant, rather than the phase and vector potential. This is essentially the gauge invariant phase difference of Equation (A2.1.56) below. However, most texts use the phase and vector potential, so some remarks on the subject are appropriate.

Let us return to the transformation of the G–L equations by putting $\psi = f\exp(i\theta)$ and $\boldsymbol{A} = \boldsymbol{A}' + (\hbar/e^*)\nabla\theta$, [Equations (A2.1.25) and (A2.1.26). One consequence is that since $\nabla.\boldsymbol{j} = 0$, Equation (A2.1.27) shows that $\nabla(f^2\boldsymbol{A}) = 0$. Therefore, if the order parameter is not uniform, we cannot use a gauge in which $\nabla.\boldsymbol{A} = 0$. However, in most cases f is a constant, so we can put $\nabla.\boldsymbol{A} = 0$.

More important is the fact that we have removed the phase from the G–L equations which suggests that we can always use a gauge with a real order parameter. Unfortunately, this is not true, because in making the substitution we have implicitly assumed that the phase is single valued.

If the phase is single valued, the substitution in Equations (A2.1.23) and (A2.1.24) to obtain Equations (A2.1.27)–(A21.29) is straightforward. From Section A2.1.3.3 there is no trapped flux, and there are no vortices. Furthermore since the phase no longer appears in the equations, we can ignore it and assume it to be zero. (This is probably a better definition of the London gauge, i.e. a real order parameter but see below Sections A2.1.3.4.1 and A2.1.3.4.2).

The other possibility is a multivalued phase of the order parameter, which is a physically different situation. In general, the potential θ used in the gauge transformation of Equations (A2.1.21)–(A2.1.23) must be single valued. If this is not the case, different values lead to a different integral of A round a circuit, and hence a change in the flux enclosed, which is observable and must not be changed by a gauge transformation. Hence we cannot use this substitution to get rid of the phase.

In spite of this, theoreticians have been able to simplify derivations by using a real order parameter in multiply connected systems. This can be done for some purposes if the phases at a point differ by an integer times 2π. Taking the curl removes the constant and we get single-valued currents and fields. Thus, using Equations (A2.1.27) and (A2.1.28) when there are vortices present gives correct values for most of the variables of interest.

However, we can expect problems with anything that depends on the flux enclosed, such as the electric field of moving vortices, since this will depend on the integral of $\boldsymbol{A}$ round a contour loop, rather than its differential. Waldram [17] has described the technique in physical terms as the insertion down the centre of a vortex of a very small solenoid containing one flux quantum in the opposite sense to that of the vortex. Since this solenoid has no external field, it has no influence on the currents and fields outside the vortex. Therefore to find local currents and fields it may be permissible to use a real order parameter if vortices are present. However, this is clearly a different physical situation from the original problem since if the vortex moves, there is no net flux transfer and no voltage induced, so any results which depend on the flux of the vortex will be wrong. These problems should be made clearer by the analysis of the two geometries which follow.

A2.1.3.4.1 A Single Vortex

A single vortex at the centre of a cylinder has been analysed in two gauges by Tinkham [3]. A single vortex has cylindrical symmetry, so it is physically reasonable to assume an order parameter in which the phase is equal to the azimuthal angle. This is multivalued in differences of 2π. Since the currents are all in the theta direction, this must also be true of the vector potential. A_θ is a function of r, and other components are zero. Near the centre of the vortex:

$$A_\theta(r) = \tfrac{1}{2}b(0)r \quad \text{(A2.1.53)}$$

This describes a uniform field b(0).

At large distances:

$$A_\theta(r) = \phi_0/(2\pi r) \quad \text{(A2.1.54)}$$

This gives the correct flux enclosed, ϕ_0.

The solution satisfies the magnetic conditions for the London gauge since $\nabla.\boldsymbol{A} = 0$ and the normal component of the superfluid velocity at the surface of the cylinder is zero. However, the order parameter is not real, and a real order parameter is the best definition of the London gauge. As pointed out above, the normal boundary condition is not sufficient to define the gauge uniquely.

If on the other hand we impose the condition that order parameter is real, which is the London gauge, then $\boldsymbol{j}$ is proportional to $\boldsymbol{A}$ so that $\boldsymbol{A}$ vanishes exponentially at large radii. Now the new vector potential $\boldsymbol{A}'$ is given by:

$$A'_\theta(r) = A_\theta(r) - \phi_0/(2\pi r) \qquad \text{(A2.1.55)}$$

In this case, we still get the same currents and magnetic fields. However, if we integrate $\boldsymbol{A}$ round a circle enclosing the vortex core, we conclude there is no flux enclosed. This is because the physical situation requires a multivalued phase of the order parameter so the gauge transformation involves a potential, which is not single valued. The fields and currents in the vortex are correct, but the flux is not, so we do not have the same physical situation as before the transformation in all respects. The G–L equations have many solutions, and it is always necessary to check their consistency with all the boundary conditions, which means more than just matching fields at material boundaries.

A2.1.3.4.2 A Thin-Walled Cylinder

As a second illustration, we consider a field B_0 applied parallel to the axis of a hollow circular cylinder of radius a, initially uniform, but later with a weak link to make the connection to the chapter on SQUIDS. We assume a thin-walled cylinder so that the current density is uniform, and the field inside is approximately equal to that outside. The London treatment of this is in Section A2.1.2.1.1 and we now consider the G–L treatment, which is very similar, and in particular the phase. To do this, we rewrite Equation (A2.1.51) as:

$$\mu_0\lambda^2 e^* \boldsymbol{J}/\hbar = \nabla\theta - e^* \boldsymbol{A}/\hbar \qquad \text{(A2.1.56)}$$

$\int(\nabla\theta - e^*\boldsymbol{A}/\hbar)\, \boldsymbol{dl}$ is defined as the gauge invariant phase difference between two points, $\Delta\theta_g$. Locally this phase is related to the superfluid current density, $\boldsymbol{J}$, (see Section A2.9.5).

If we first look at a solution with θ zero, i.e. a real order parameter, then integrating Equation (A2.1.56) round the cylinder:

$$\mu_0\lambda^2 2\pi a J = -\pi a^2 B_0 \qquad \text{(A2.1.57)}$$

This is the situation if we start with no trapped flux in the cylinder, the fluxoid is zero.

In this case, the Coulomb and London gauges coincide, and although there is flux within the cylinder, it is not quantised and will disappear if the external field is removed. The system is a simply connected system since the fluxoid is zero. The current density is:

$$J = -aB_0/(2\pi_0\lambda^2) \qquad \text{(A2.1.58)}$$

This will apply as the external field is increased until the critical current is reached, when flux will enter the cylinder and be trapped.

If the order parameter is not single valued, the fluxoid is not zero, and there is trapped flux in the cylinder. We try a phase, which differs by $2n\pi$ on going round the cylinder. Now integrating Equation (A2.1.56) gives:

$$2\mu_0\pi a\lambda^2 J = n\phi_0 - \pi a^2 B_0 \qquad \text{(A2.1.59)}$$

This means that if in the normal state the applied field puts an integer number of flux quanta, n, into the circle defined by the cylinder, there will be no screening current on cooling to the superconducting state. If this is not the case, the current will be positive or negative depending on the sign of the difference between the applied number of fluxoids, and the fluxoid in the sample. If B_0 is increased, the screening current density increases until the critical value is reached. There will then be a sudden change in n, and the trapped flux changes by a whole number of quanta.

For a circular cylinder the London and Coulomb gauges coincide, since there are no electrostatic charges. If the cylinder is a different shape, the derivation above is similar in the London gauge. For the initial application of the field the current density is $-B_0A/(\mu_0\lambda^2L)$, where A/L is the ratio of the area to circumference. However, there will now be electrostatic charges on the surface of the cylinder to direct the current round any corners so that in the conventional Coulomb gauge we must include an electrostatic potential term.

A2.1.3.4.3 Weak Links

The language used to describe Josephson junctions is very similar to that of the G–L equations but was originally derived from the tunnelling of Cooper pairs. However, it soon became clear that the Josephson equations did not only apply to tunnel junctions, but to any small weak section of a superconductor. There is therefore a continuum starting with Josephson tunnel junctions, extended to other weak links, which are small pinning centres for Abrikosov vortices, and therefore also to the bulk pinning of Abrikosov vortices in type II superconductors. For example, Reference [18] links the RSJ model of a Josephson junction with an isolated pinning centre. In this section, we make the connection between the phase of the order parameter in the G–L theory and the phase difference across a weak link.

We introduce a weak link into the cylinder of the previous section as a thin line parallel to the axis, to make a SQUID, and revert to a real order parameter with $\theta = 0$, i.e. the initial application of a field. (Real SQUIDS have a very different geometry, but this does not alter the principles of the argument). Equation (A2.1.56) becomes:

$$\mu_0\lambda^2\boldsymbol{J} = -\boldsymbol{A} \qquad \text{(A2.1.60)}$$

If the weakness is in the material, then λ will be increased at this point since λ is proportional to $n_s^{-1/2}$. Alternatively, if there is a constriction, $\boldsymbol{J}$ is increased. In either case the effect is the same, that is, from Equation (A2.1.56) the gradient of the gauge invariant phase difference, $\nabla\theta_g = \mu_0\lambda^2 J e^*/\hbar$ will be greatly increased at the weak link compared with the rest of the circuit. This is the phase that appears in the Josephson equations, not the phase of the order parameter θ which is zero everywhere until a quantum enters the ring. (A close analogy is if we were to introduce a thin but very resistive line down the cylinder. If a current is induced, the voltage drop will take place almost entirely across the resistive strip).

To take a specific example, suppose the subscript *s* refers to the superconductor and *w* to the weak link, and that these are of length l_s and l_w, respectively. We make a weak link consisting of a thin strip of reduced order parameter, and so increased λ. Then integrating Equation (A2.1.60) round the ring:

$$\mu_0 J(\lambda_s^2 l_s + \lambda_w^2 l_w) = \pi a^2 B_0 \quad \text{(A2.1.61)}$$

The change in gauge independent phase in the two regions is:

$$\Delta\theta_{gs} = \lambda_\sigma^2 l_s e^* / \hbar((\lambda_s^2 l_s + \lambda_w^2 l_w) \text{and} \, \Delta\theta_{gw} = \lambda_\omega^2 l_w e^* / \hbar((\lambda_s^2 l_s + \lambda_w^2 l_w) \quad \text{(A2.1.62)}$$

It can be seen that if $\lambda_w^2 l_w$ is much greater than $\lambda_s^2 l_s$, as will be the case for a weak link, nearly all the phase difference occurs across the weak link. Similarly, if we impose an alternating current, the voltage drop will appear almost entirely at the junction since either λ or ***J*** is increased at this point. Across the link:

$$\Delta V = E l_w = -\dot{A} l_w = \mu_0 \lambda^2 \dot{J} l_w = (\hbar/e^*) d(\Delta\theta_{gw})/dt \quad \text{(A2.1.63)}$$

It can be seen that the electric field is proportional to the rate of change of θ_g, as deduced by Josephson.

Equation (A2.1.63) leads to the 'kinetic inductance'. The origin of the term can be seen by considering a junction of area *S* and thickness *d*. In terms of the voltage *V* and current *I*, Equation (A2.1.63) becomes:

$$V/d = \mu_0 \lambda^2 \dot{I}/S \quad \text{(A2.1.64)}$$

This implies an inductance $\mu_0\lambda^2 d/S$ associated with the junction (or indeed any section of superconducting material).

The energy in the inductance is $\frac{1}{2}LI^2$ which can then be written $\frac{1}{2}n_s mv^2$ per unit volume. This is the kinetic energy of the electrons, and hence the name kinetic inductance.

Weak links made from narrow sections (microbridges) and thin sections of reduced order parameter (SS's or SNS junctions) are the easiest type of Josephson junction to understand within the framework of the G–L theory since they are essentially pinning centres which operate on single vortices. From the point of view of this section, the kinetic inductance is due to the reversible motion of a vortex in the potential well of the junction.

It is perhaps unfortunate for the understanding of the subject that the brilliant intuition of Josephson led to his equations being derived for the most sophisticated type of junction, a tunnel junction. Many physicists working on the much easier route of small type II superconductors would have reached the same conclusions later and less elegantly, but the subject might have been easier to understand. For example, the famous relation between voltage and frequency follows at once from Faraday's law of induction. If *f* vortices enter per second, then $V = f\phi_o = fh/e^*$. Josephson derived this relation from tunnelling theory as $Ve^* = hf$, [19], which can be interpreted as the Planck relation between frequency and energy.

Differences in V–I characteristics between junctions and bulk type II superconductors arise from several sources. One is that a junction is a single pinning centre acting on one vortex at a time, while in a bulk sample, the large number of vortices and pinning centres leads to collective effects. This causes the difference in curvature of the V–I curve at the critical current. There are two kinds of long Josephson junctions. One is an ideal smooth junction when the length is greater than the Josephson penetration depth, but a single quantum is pinned. This only requires a more complex solution of the Josephson equations. However, in grain boundaries in high-T_c superconductors, there are likely to be several Josephson vortices in the boundary, and the critical current depends on the pinning of this linear vortex array by inhomogeneities in the boundary. In this case, the V–I characteristics approach the characteristics of type II superconductors due to the collective pinning of Josephson vortices along the junction.

More fundamental is the hysteresis in the V–I characteristic of S-I-S tunnel junctions. This arises because the Josephson vortex has no normal core in which dissipation takes place, so there is almost no damping, in contrast to an Abrikosov vortex, which is very heavily damped. Therefore, the mass term in the equation of motion dominates the viscous term, and inertial effects cause hysteresis. This is the only measurement which can show that the coupling between layers is due to Josephson tunnelling. (Type II superconductors can also show hysteretic V–I curves, but for quite different reasons). For the case of microbridges, most Josephson effects appear if the bridge is smaller than a penetration depth since this forces vortices to enter one at a time. However, to show the hysteretic behaviour in the V–I characteristic found in tunnel junctions, the microbridge must be smaller than the coherence length. But having pointed out that the parallel between junctions and pinning centres is useful in understanding the effects qualitatively, it must be added that most quantitative results can only be obtained using Josephson's theory.

A2.1.4 Type II Superconductors

The G–L theory introduced the idea of a surface energy to explain the Meissner effect, which requires the surface energy to be positive. G–L recognised that if $\kappa > 1/\sqrt{2}$, an instability of some kind would occur. It was Pippard who introduced the idea of a coherence length, a distance over which the order parameter could vary. He pointed out that if this was less than the penetration depth, a boundary could gain the magnetic energy by allowing flux penetration while costing little in condensation energy because the coherence length was short [20]. The surface energy would then be negative and lead to a fine subdivision of the normal and superconducting phases. (An analogy is the surface energy between sugar and water which is also negative and causes a similar subdivision called solution). A laminar model was developed from this idea, but clearly if a laminar structure lowers the energy, it can be further subdivided by perpendicular boundaries to produce more surface, and the result is essentially the vortex lattice arrived at earlier by Abrikosov [13].

A2.1.4.1 The Abrikosov Theory

The most celebrated solution of the Ginzburg–Landau equations is the vortex lattice derived by Abrikosov [13]. This is covered in more detail elsewhere. For completeness we quote some of his results here. These can be derived approximately by physical arguments which are given here, while the equations quoted are the more accurate versions from the Abrikosov theory.

Firstly, for a hexagonal lattice (Figure A2.1.1) each unit cell contains flux ϕ_0. Hence if the spacing of the vortices is a, then:

$$Ba^2 = \phi_0(2/\sqrt{3}) \qquad \text{(A2.1.65)}$$

For example, if we cool a film in the Earth's field, there will be trapped vortices every six microns. This can have a major impact on phenomena such as microwave losses.

Since the vortex needs energy to enter the superconductor, an external field B_{c1} is needed to push them in. Once it is strong enough and a vortex enters, it will migrate to the centre of a macroscopic specimen so a large number can enter until the internal field is comparable with the external field to equalise the magnetic pressure. This leads to a sharp drop in the magnetisation. Since they stop entering when their circulating currents overlap and the sample is full, a flux density of B_{c1} corresponds to a vortex spacing of about λ.

$$B_{c1}\lambda^2 = \phi_0((\ln\kappa + .08)/4\pi) \qquad \text{(A2.1.66)}$$

As the external field increases, the vortices are pushed closer together until at the upper critical field B_{c2} the cores overlap and the spacing is about ξ. Since this is a second-order transition, there is no sudden change between the superconducting state and the normal state at this point.

$$B_{c2}\xi^2 = \phi_0(1/2\pi). \qquad \text{(A2.1.67)}$$

From the vortex spacing at B_{c1} and B_{c2}, it follows that B_{c1}/B_{c2} is approximately $(\xi/\lambda)^2 = 1/\kappa^2$.

$$B_{c1}/B_{c2} = \ln(\kappa + .08)/(2\kappa^2) \qquad \text{(A2.1.68)}$$

The resulting magnetisation curve is roughly a straight line, (ignoring the initial sharp drop), between B_{c1} and B_{c2} and the area under it is $B_c^2/2\mu_0$ (Figure A2.1.2)

Hence:

$$B_{c1}/B_{c2} = B_c^2\ln(\kappa + .08) \qquad \text{(A2.1.69)}$$

The gradient of the magnetisation curve at high fields is approximately $-B_{c1}/(B_{c2} - B_{c1})$ so:

$$\mu_0 M = (B_{c2} - B)/(1 + 1.16(2\kappa^2 - 1)). \qquad \text{(A2.1.70)}$$

In the London approximation (i.e. small coherence length) the magnetisation at intermediate fields is given by [7]:

$$\mu_0 M = -(\phi_0/8\pi\lambda^2)\ln(0.368 B_{c2}/B) \qquad \text{(A2.1.71)}$$

The area under the magnetisation curve in Figure A2.1.2 must equal that of the triangle based on B_c. This means that many of the results of Abrikosov can be found approximately from Euclidean geometry and the vortex spacing at B_{c1} (about λ) and B_{c2}, (about ξ)

From these results:

$$B_{c1} - B_c\ln(\kappa + .08)/\sqrt{2}\kappa \qquad \text{(A2.1.72)}$$

$$\text{and } B_{c2} = \sqrt{2}\kappa B_c \qquad \text{(A2.1.73)}$$

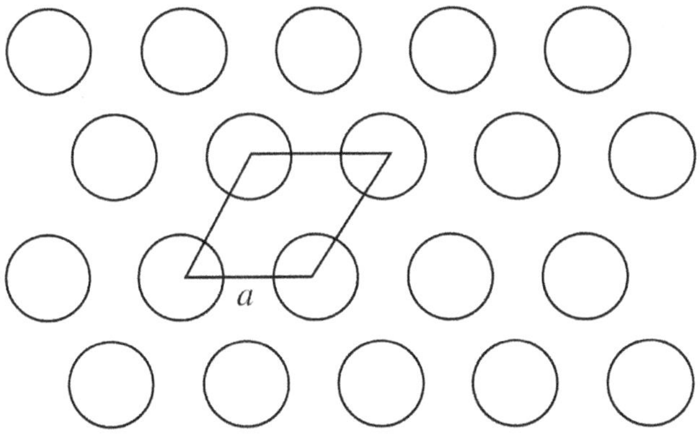

FIGURE A2.1.1 A hexagonal array with vortex spacing a. The unit cell area is $a^2\sqrt{3}/2$.

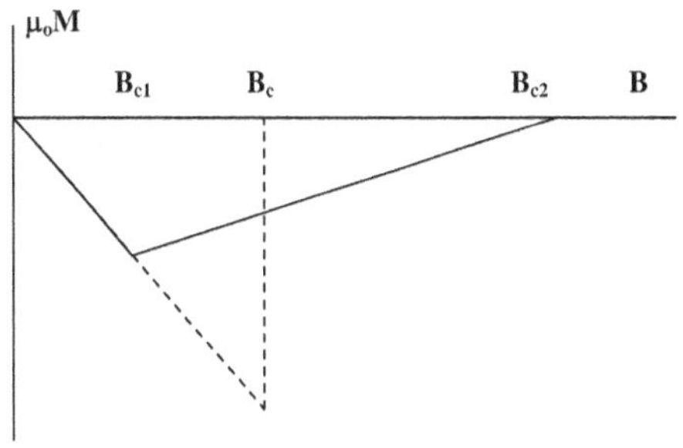

FIGURE A2.1.2 The magnetisation curve based on the qualitative vortex picture.

There are many combinations of these parameters, but there are only two independent ones, ultimately related to α and β in the G–L free energy. It can be seen that if any one parameter behaves strangely, it has implications for all the others. For example, some underdoped oxide superconductors appear to have a B_{c2} which increases exponentially at very low temperatures [21]. If this is true, and if B_c stays fairly constant, it follows that B_{c1} must decrease exponentially, and the penetration depth must increase similarly. Alternatively, if the penetration depth behaves normally, then B_c must increase exponentially. Neither scenario seems likely which forces us to look for explanations outside the conventional Abrikosov theory, such as fluctuation effects.

Acknowledgements

This article has concentrated on aspects of the Ginzburg-Landau theory more relevant to applications, and I am grateful to a number of people who have helped me to link these results with the more fundamental theory, which they understand much better than I do. In particular, I would like to thank Dr. J.R.Waldram and the late Dr. E. H. Brandt who have put me right on a number of occasions.

References

[1] F. London, 'Superfluids', Dover, New York, (1961).
[2] V.L. Ginzburg and L.D. Landau, 'On the Theory of Superconductivity', Zh. Eksperim. i. Teor. Fiz. **20**, 1064 (1950), English Translation in 'Collected papers of L.D. Landau', D. Ter Haar editor, Pergamon, Oxford, (1965).
[3] Gorter and Casimir, 'Superconductivity', D. Shoenberg, Cambridge University Press, (1965).
[4] J.R. Waldram, 'Superconductivity', Cambridge University Press, (1997).
[5] J.R. Cave and J.E. Evetts, 'Critical Temperature Profile Determination Using a Modified London Equation for Inhomogeneous Superconductors', J. Low Temp. Phys., **63**, 35–55, (1986).
[6] M. Tinkham, 'Introduction to Superconductivity', McGraw-Hill, New York, (1996).
[7] P.G. de Gennes, 'Superconductivity of Metals and Alloys', W.A. Benjamin, New York, (1966).
[8] D. Saint James, E.J. Thomas and G. Sarma, 'Type II Superconductors', Pergamon, Oxford, (1969).
[9] T. van Duzer and C.W. Turner, 'Principles of Superconductive Devices and Circuits', Elsevier, New York, (1981).
[10] A.M. Campbell, 'Ginzburg-Landau Calculations of the Critical Current Densities of Grain Boundaries', Physica C, **162-164**, 1609–1610, (1989).
[11] V.A. Schweigert and F.M. Peeters, 'Transitions between Different Superconducting States in Mesoscopic Disks', Physica C, **144**, 266–271, (2000).
[12] L.F. Chibotaru, A. Ceulemans, V. Bruyndoncx and V.V. Moshchalkov, 'Vortex Entry and Nucleation of Anti-vortices in a Mesoscopic Superconducting Triangle', Phys. Rev. Lett., **86**, 1323–1326, (2001).
[13] A. Abrikosov, 'On the Magnetic Properties of Superconductors of the Second Group', Sov. Phys. JETP, **5**, 1174–1182, (1957).
[14] R.I. Coote, A.M. Campbell and J.E. Evetts, 'Vortex Pinning by Large Precipitates', Canad. J. Phys., **50**, 421–427, (1972).
[15] T. Matsushita and M. Kiuchi, 'Depairing Current Density in Superconductors', Appl. Phys. Express, **12**, 063003, (2019).
[16] J.R. Clem Lecture notes, (Unpublished).
[17] J.R. Waldram Private Communication.
[18] A.M. Campbell, 'Pinning and Critical Currents in Type II Superconductors', Proc. 18th Int. Conf. on Low Temperature Physics, Kyoto. Jpn. J. Appl. Phys., **26**, Supplement 26-33, 2053–2058, (1987).
[19] B.D. Josephson, 'Possible New Effects in Superconducting Tunelling', Phys. Lett., **1**, 251–253, (1962).
[20] A.B. Pippard, 'Trapped Flux in Superconductors', Phil. Trans. Royal Soc., **248**, 97–129, (1955).
[21] A.P. Mackenzie, S.R. Julian, G.G. Lonzarich, A. Carrington, S.D. Hughes, R.S. Liu and D. Sinclair, 'Resistive Upper Critical Field of $Tl_2Ba_2CuO_6$', J. Supercond., **7**, 271–277, (1994).

A2.2

Microscopic Theory

Anthony J. Leggett

The essentials of the microscopic theory of superconductivity presented by BCS [1] in their classic 1957 paper, and generally believed to describe at least a large subclass (referred to hereafter as 'classic') of the currently known superconductors will be outlined in this chapter. Exactly which aspects of it, if any, are relevant to the cuprates and ferropnictides is, at present, an open question.

A2.2.1 Normal State: Cooper Instability

By the time a metal has been cooled down to the critical temperature T_c for the onset of superconductivity, the electrons are already highly degenerate, that is, their distribution is profoundly affected by the Pauli principle, which states that no more than one electron can occupy a given single-particle state. Under these conditions, the normal state of the 'classic' superconductors is usually fairly well described by the standard 'textbook' model of a metal. In the very simplest version of this model, usually associated with the name of Sommerfield, the conduction electrons move freely in a constant potential; thus, they occupy plane-wave states with wave vector **k**, momentum $\hbar\boldsymbol{k}$ and energy $\varepsilon(\boldsymbol{k}) = \hbar^2\boldsymbol{k}^2/2m$. At the next (Bloch) level, one takes into account the effect of the periodic potential of the static ionic lattice, and, as a result, the relevant electron states are no longer plane waves but 'Bloch waves'; the energy of the state in general depends on the direction as well as the magnitude of the wave vector, giving the usual band structure. [Also, the electrons can be scattered by small vibrations of the ionic lattice (phonons) as well as by impurities]. Finally, at the most sophisticated (Landau–Silin) level, it is recognized that the interactions between conduction electrons, up to now neglected, are actually very important and mean that the true energy eigenstates correspond to the occupation of a Bloch-wave state not by a single electron but by a 'quasiparticle', that is, an electron surrounded by a 'screening cloud' of other electrons. The resulting picture, while not quite in one–one correspondence with the Bloch scheme, is close enough to it that in the present context we may treat a quasiparticle as effectively equivalent to an actual electron.

There is one fundamental property which persists throughout this increasingly sophisticated description: imagine that we could cool the normal phase to zero temperature without it becoming superconducting. Then the basic 'fermionic' entities, be they real electrons, Bloch waves or quasiparticles, would fill up the available states one by one up to the Fermi surface, that is the locus in **k**-space of states with the maximum (Fermi) energy. At finite temperatures of the order of T_c, most of this 'Fermi sea' is inert and, for any phenomenon which involves only relatively weak excitation of the system, all the action comes from electrons in a narrow shell of states, of width $\sim k_B T$ in energy, close to the Fermi surface. Although superconductivity is not quite of this type, it is still true that the states mainly involved are indeed close to the Fermi surface, and the bulk of the Fermi sea can be ignored.

The historical jumping-off point for BCS theory was the following observation by Cooper [2]: consider a gas of $N - 2$ free electrons at $T = 0$, so that all states below the Fermi energy ε_F are filled and all those above empty as illustrated in Figure A2.2.1. Then, imagine we introduce two more electrons, with opposite spin and total momentum zero, which, to maintain the Pauli principle, are allowed to occupy only the vacant states and which are subject to a weak mutual interaction whose matrix elements for scattering from $(\boldsymbol{k}\uparrow : -\boldsymbol{k}\downarrow)$ to $(\boldsymbol{k}'\uparrow : -\boldsymbol{k}'\downarrow)$ are constant whenever the states $\boldsymbol{k}$ and $\boldsymbol{k}'$ both have energy (relative to the Fermi energy) less than ε_c and zero otherwise. If we denote this constant by $-V_0$, then for $V_0 < 0$ (repulsive interaction), nothing interesting happens. If, however, $V_0 > 0$ (attraction), we find that the Schrödinger equation for the two added electrons always has a solution which lies *below* the minimum energy $2\varepsilon_F$, which they would have if non-interacting, by an amount E given for small V_0 by

$$E = 2\varepsilon_c \exp-(2/N(0)V_0) \qquad \text{(A2.2.1)}$$

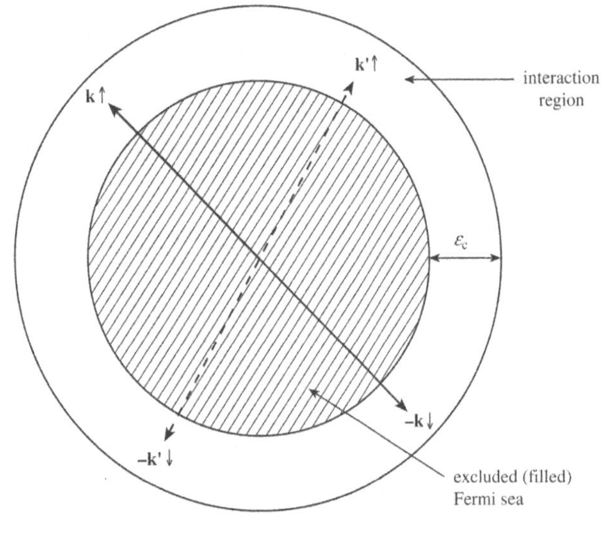

FIGURE A2.2.1 The Cooper problem.

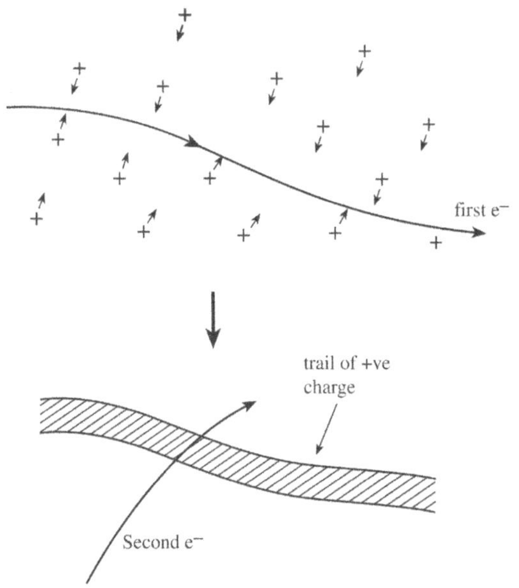

FIGURE A2.2.2 The effective electron–electron interaction induced by polarization of the ionic background.

where $N(0)$ is the density of single-particle states of one spin per unit energy near the Fermi surface (for a free gas this is $3N/4\varepsilon_F$). Moreover, the wave function corresponding to this negative-energy state indeed corresponds to a 'bound' state; the two-electron wave function is constant as a function of the center-of-mass coordinate, but falls off fast as a function of their relative coordinate $\boldsymbol{r}$, so that the result is a sort of 'di-electronic molecule' with a radius $\xi' \sim \hbar v_F/E$ (where $v_F = (2m\varepsilon_F)^{1/2}$ is the Fermi velocity). This bound state is called a *Cooper pair*; note that it exists for any attraction, however weak.

A2.2.2 Effective Interaction

If Cooper's observation is to have any relevance to real metals, it is necessary that the effective interaction between electrons in states close to the Fermi surface be attractive. At first sight, this seems unlikely, since the 'bare' Coulomb interaction $e^2/(4\pi\varepsilon_0 r)$ between any two electrons is certainly repulsive. However, it turns out that in a metal this bare interaction is strongly screened by the collective effects of the other electrons, and the resulting 'effective' interaction falls off exponentially with a characteristic length which can be of the order of the mean distance between electrons.

Equally important, the interaction with phonons (lattice vibrations) can actually generate an effective *attraction* between electrons. One way of seeing this is that the first electron attracts the positive ions towards its path and thereby leaves behind a trail of (slowly relaxing) positive charge, which subsequently attracts a second electron; thus, we generate an effective electron–electron attraction, as illustrated in Figure A2.2.2. An alternative point of view is that the exchange of virtual phonons leads to an attraction between electrons, in the same way as in the Yukawa theory of nuclear forces exchange of virtual mesons leads to one between nucleons. In any event, detailed calculation shows that this phonon-generated attraction, when added to the screened Coulomb repulsion, can generate a net interaction, the relevant matrix elements of which are attractive. This is not invariably so—if it were, all metals should be superconductors at low enough temperature—and it is actually not at all trivial to determine the sign of the net interaction from first principles.

The actual form of the interaction as a function of the wave vectors $\boldsymbol{k}$ and $\boldsymbol{k}'$ involved in the scattering process is quite complicated, and this is taken into account in more sophisticated calculations; however, it turns out to be a surprisingly good approximation for many purposes to replace it, as BCS did in their original work, by something close to the simple form used above for the Cooper problem, with the 'cutoff' energy ε_c taken to be of the order of the characteristic (Debye) phonon energy.

A2.2.3 Nature of the Superconducting Groundstate

While Cooper's simple calculation gives much insight into the basic mechanism of superconductivity (at least in the classic superconductors), it is obvious that it is internally inconsistent in treating the last two 'special' electrons which form the Cooper pair on a different footing from the remaining $N - 2$, whose only function is to fill up the Fermi sea and thereby exclude the paired electrons from it. A complete calculation which treats all N electrons on the same footing was given by BCS in [1]. A somewhat intuitive interpretation of the principal BCS results will now be presented.

For most purposes, and in particular, when one wishes to calculate the *changes* that various physical properties undergo on passing from the normal to the superconducting state, it is adequate to visualize the superconducting groundstate as having all the N electrons bound into di-electronic 'molecules' (Cooper pairs) which are all described by *the same* two-particle wave function. This fundamental property, that all Cooper pairs have to occupy the same pair state, is characteristic not only of the groundstate but of all low-energy states in which a finite fraction of electrons are paired; it is tempting to think of it as a kind of 'Bose condensation' (the pairs have total spin zero and thus are indeed bosons!), but, irrespective of the terminology, it is the essential key to understanding the abnormal properties of superconductors such as the Meissner effect (see Chapter A2.4).

What is the basis of the assertion that all Cooper pairs must occupy the same two-particle state? Why, for example, could we not put half of them into a pair state with center-of-mass momentum $\boldsymbol{K}$ equal to zero and the other half into one with finite $\boldsymbol{K}$? The answer is that we could, but this would lose a large fraction of the gain in potential energy obtained by the 100% 'condensation' into a single state described above, a loss which is not adequately compensated by entropic or other factors. (This is because to enjoy the maximum benefit of the attractive interaction, all pairs of electrons must be able to scatter into the same set of states, something that is only possible if they have a common value of $\boldsymbol{K}$. For details see [3].)

Given, then, that all Cooper pairs must be described by the same two-particle wave function F, what does this function look like? For the classic superconductors, at least, it is believed to be a product of space and spin functions, with the latter corresponding to the singlet state:

$$F = F(\mathbf{r}_1, \mathbf{r}_2)\psi_{\text{spin}}(S = 0). \qquad \text{(A2.2.2)}$$

If, furthermore, we restrict our attention for the moment to the groundstate (in zero magnetic field), then the spatial wave function $F(\boldsymbol{r}_1, \boldsymbol{r}_2)$ should correspond to the center of mass (COM) of the pair being at rest, i.e. it should be independent of the COM coordinate $\boldsymbol{R} \equiv (1/2)(\mathbf{r}_1 + \mathbf{r}_2)$ and a function only of the *relative* coordinate $\boldsymbol{r}_1 - \boldsymbol{r}_2 = \boldsymbol{\rho}$. Moreover, for the classic superconductors, it is believed that the net orbital angular momentum of the pairs is zero, which means that $F(\boldsymbol{\rho})$ is independent of the direction of $\boldsymbol{\rho}$ and a function only of the relative distance $|\boldsymbol{\rho}| \equiv \rho$. An approximate representation of $F(\rho)$, valid for most distances of interest, is

$$F(\rho) \sim \text{const}[\sin(2k_F\rho)/2k_F\rho]\exp[-(\rho/\xi_p)] \qquad \text{(A2.2.3)}$$

where $\xi_p \gg 1/k_F$ (see below). Thus, at 'short' distances ($\rho \ll \xi_p$), the relative wave function is indistinguishable from that of two free electrons each with magnitude of momentum k_F, with COM at rest and in a relative s-state (the term in square brackets), but if one goes out to longer distances, one sees that the pair is indeed bound, forming a 'molecule' with radius $\sim \xi_p$ as illustrated in Figure A2.2.3. This 'pair radius' is of the same order as the more familiar 'Pippard coherence length'

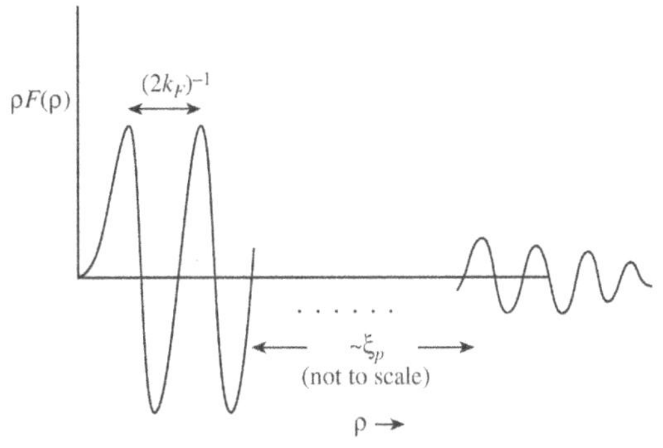

FIGURE A2.2.3 Qualitative behavior of the Cooper pair wave function. Note that the factor $(2k_F\rho)^{-1}$ has been extracted for clarity.

ξ_0, which is traditionally used to describe the electrodynamics (see Chapters A4.2 and A2.3), i.e. of order $\hbar v_F/\Delta(0)$, where the quantity $\Delta(0)$, whose physical significance will be explained in the next section, is given in the simple model used by BCS by the formula

$$\Delta(0) = 2\epsilon_c \exp-\{1/[N(0)V_0]\}. \qquad \text{(A2.2.4)}$$

Thus, for weak attraction V_0, the quantity ξ_p can be very large, in practice, as much as $\sim 10^{-6}$ m; the different 'molecules' (Cooper pairs) therefore overlap one another very substantially, to the extent that within the volume of a single pair one may find 10^9–10^{12} electrons belonging to 'other' pairs (though the fundamental indistinguishability of electrons makes this form of words of dubious meaning!). That ξ_p and ξ_0 are of the same order is to be expected, since the non-local effects described by the Pippard coherence length arise in some sense from the fact that the two electrons of a pair cannot be treated as independent.

As yet, the constant in Equation (A2.2.4) has not been specified, and this raises a rather delicate point about the 'number' of Cooper pairs. It turns out that the *changes* in 'two-particle' properties such as the interaction energy induced by pair formation may be correctly calculated by treating the 'pair-wave function' [Equation (A2.2.3)] (or a slightly more accurate version) exactly like an ordinary molecular wave function, that is, by multiplying the two-particle quantity in question [e.g. $V(\rho)$] by the square of $F(\rho)$ and integrating over ρ. However, it is then necessary to know the overall constant; we cannot assume *a priori* that F is normalized to one! In fact, it turns out that the correct choice is such that the integral of F^2 is of order $N(0)\boldsymbol{\Delta}(0)$; since $N(0)$ is of order N/ε_F, this means that in some sense, although the pair wave function [Equation (A2.2.3)] characterizes all N particles, the 'number of Cooper pairs' Nc is only a fraction of order $\boldsymbol{\Delta}(0)/\varepsilon_F \sim 10^{-4}$ of N. Further, since, as we shall see, the quantity $\boldsymbol{\Delta}(0)$ is a measure of the binding energy of a pair, the total energy gained by condensation into the superconducting state is of order $N(0)[\boldsymbol{\Delta}(0)]^2$—typically, a fraction of order 10^{-8} of the total energy of the normal groundstate.

A2.2.4 BCS Theory at Finite Temperature

Having established the nature of the superconducting groundstate in BCS theory, we will now investigate the excited states. As already mentioned, the states obtained by relaxing the condition of 'uniqueness' of the condensate (i.e. of the pair wave function) have energies far too large to be thermodynamically relevant. However, there is another way of obtaining excited states, as follows: formation of a 'complete' pair state requires partial occupation of each pair of plane-wave states (***k***, −***k***) by a *pair* of electrons with opposite spins. If a given state ***k*** is occupied by an electron which does not have an opposite-spin 'partner' in the state −***k***, it turns out that this costs, relative to the groundstate, an energy $\left\{\epsilon_k^2+[\Delta(0)]^2\right\}^{1/2}$, where ε_k is the normal-state energy relative to the Fermi energy. Thus, the minimum energy for such excitation is $\mathbf{\Delta}(0)$, as a result of which, this quantity is called the (zero-temperature) 'energy gap'. The electron states which are pushed out from the energy region below $\mathbf{\Delta}(0)$ accumulate at energies slightly greater than $\mathbf{\Delta}(0)$, giving a larger density of states there than in the normal state.

At finite temperature, entropy considerations make it thermodynamically favorable for a finite number of such 'broken pairs' to exist in the system. As a result, while the condensate wave function is still unique, the effective number N_c of Cooper pairs is reduced, and this in turn decreases the energy gap $\mathbf{\Delta}$, which is thus temperature-dependent. When the temperature reaches a 'critical' value T_c given by $\mathbf{\Delta}(0)/1.76$, the gap tends to zero, as $(T_c - T)^{1/2}$, illustrated in Figure A2.2.4. It should be emphasized that the 'pair radius' which characterizes the fall-off of the Cooper-pair wave function F is, like the Pippard coherence length which is of the same order, only rather weakly temperature-dependent and does *not* diverge as T approaches Tc from below. [Neither of these lengths should be confused with the 'Ginzburg–Landau (GL) correlation length' (healing length) ξ (T), which does diverge as $(T_c - T)^{-1/2}$; see Chapter A3.1.]

At finite T, the condensate is still 'inert' just as at $T = 0$ (see next section), but the broken pairs form a 'normal component' which behaves qualitatively like the electrons in a normal metal (although their dynamics and scattering rates are somewhat different). Thus, for example, quantities like the specific heat or the Pauli spin susceptibility receive contributions only from the normal component (the latter because the Cooper pairs, having total spin zero, cannot be polarized by an external magnetic field).

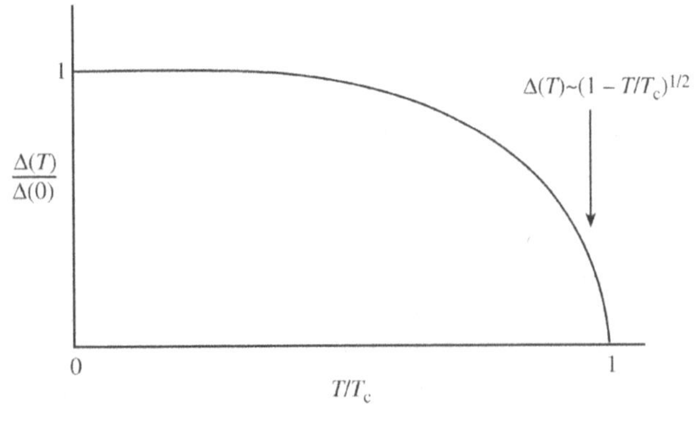

FIGURE A2.2.4 Temperature-dependence of the energy gap $\mathbf{\Delta}(T)$.

A2.2.5 Meissner Effect: Relation of BCS and GL Descriptions

The most fundamental property of superconductors is the Meissner effect, that is, the property of excluding a weak magnetic field: see Chapter A2.4. For simplicity, let us consider an 'extreme type II' superconductor (see Chapter A2.3), for which the electrodynamics is local and well described by the original London theory (see Chapter A2.1). In this case, the fundamental equation, which when coupled with the standard Maxwell equations, leads to exclusion of a weak field, can be written, in the simplest case, as a relation between the local electric current density $\boldsymbol{j}(\boldsymbol{r})$ and the electromagnetic vector potential $\boldsymbol{A}(\boldsymbol{r})$:

$$\boldsymbol{j}(\boldsymbol{r}) = -\Lambda(T)\boldsymbol{A}(\boldsymbol{r}) \qquad \text{(A2.2.5)}$$

Equations (A3.2.1 and A3.2.2) follow from Equation (A2.2.5) on taking the time derivative and the curl, respectively. London's expression for the coefficient $\Lambda(T)$ was $n_s e^2/m$, where the 'superfluid density' $n_s(T)$ tends to the total electron density as T tends to zero and to zero as T tends to T_c. As we shall see, Equation (A2.2.5) is a direct consequence of the uniqueness and single-valuedness of the pair wave function F.

The argument proceeds by analogy with one which may be familiar in the context of atomic diamagnetism. Consider a single particle of charge e in the presence of a weak magnetic vector potential $\boldsymbol{A}(\boldsymbol{r})$ at $T = 0$. The standard gauge-invariant expression for the electric current density $\boldsymbol{j}(\boldsymbol{r})$ is

$$\boldsymbol{j}(\boldsymbol{r}) = \mathrm{Im}[(e/m)(-i\hbar\psi\nabla\psi)] - (e^2/m)\backslash\psi(\boldsymbol{r})|^2\,\boldsymbol{A}(\boldsymbol{r}) \qquad \text{(A2.2.6)}$$

where $\psi(\boldsymbol{r})$ is the Schrödinger wave function. Suppose that, as is normally the case, the wave function is real for $\boldsymbol{A} = 0$ and thus $\boldsymbol{j}(\boldsymbol{r}) = 0$. If, on the application of a weak potential $\boldsymbol{A}(\boldsymbol{r})$, the wave function does not change its form, then the gradient terms still do not contribute and we find

$$\boldsymbol{j}(\boldsymbol{r}) = -(e^2/m)\rho(\boldsymbol{r})\boldsymbol{A}(\boldsymbol{r}) \qquad \text{(A2.2.7)}$$

where $\rho(\boldsymbol{r}) \equiv |\psi(\boldsymbol{r})|^2$ is the probability density. An exactly similar result follows for a many-electron system, interacting or not, provided $\rho(\boldsymbol{r})$ is now interpreted as the total electron density at point $\boldsymbol{r}$ (this is the result used in deriving the standard formula for the diamagnetism of rare gas atoms). At finite temperature T, a similar result will follow, provided, (a) the thermal average of the current density is zero for $\boldsymbol{A} = 0$ and,

(b) a finite but weak value of $\boldsymbol{A}$ produces no change in either the wave functions or in the occupation factors for the different states; however, $\rho(\boldsymbol{r})$ is now the density only of the condensed electrons (see below).

We thus see that at $T = 0$, we recover Equation (A2.2.5), with London's value of the constant $\Lambda(T)$, *provided* that the groundstate many-body wave function is 'rigid' (inert) against application of a weak vector potential. But why should the groundstate possess this kind of rigidity? For definiteness, consider a very thin ring of radius R, so that application of a flux ϕ through the ring leads to a circumferential vector potential $\mathbf{A}_\theta = \phi/2\pi R$. For $\mathbf{A} = 0$, it is obvious from the symmetry that the pair wave function $F(\mathbf{r}_1, \mathbf{r}_2)$ is constant as a function of the COM coordinate $\boldsymbol{R} = (1/2)(\boldsymbol{r}_1 + \boldsymbol{r}_2)$. Now, a weak but finite vector potential acts symmetrically on the two electrons of the Cooper pair and thus cannot change the dependence of F on the relative coordinate; the only possibility is that it changes its dependence on the COM coordinate and, from the symmetry of the problem, the only possibility is to multiply the latter by a phase factor $\exp(i\alpha\theta)$, where θ is the angular component of the vector $\boldsymbol{R}$ in cylindrical polar coordinates. But the crucial point, now, is that since the wave function must return to its original value when the two electrons of a pair are carried together once around the ring (the 'single-valuedness' condition), the only possible values of α are integers n (including zero). Since a non-zero value of n costs a large extra kinetic energy (actually of order $N\hbar^2/2mR^2$, although to discuss why the factor is N and not N_c would take more space than I can afford here), this state cannot be energetically favorable for small $\boldsymbol{A}$, and the wave function must remain unchanged from its $\boldsymbol{A} = 0$ value, precisely the 'rigidity' necessary to justify Equation (A2.2.5). At finite temperatures, the same argument goes through for those electrons which are still condensed, but not all are, and the quantity $\Lambda(T)$ is therefore proportional, crudely speaking, to the density of condensed electrons ('superfluid density').

Why does a similar argument not go through for a normal metal? It turns out that the simplest case to analyze is that of low but not ultra-low temperature, so that $k_B T \gg \hbar^2/mR^2$. The point is that although the single-particle wave functions, like those of the Cooper-pair COM, must of course be single-valued and thus of the form $\exp(in\theta)$ with n integral, even for $\boldsymbol{A} = 0$, there is a considerable population of finite-n states, and it is only the thermal average of $\boldsymbol{j}(\boldsymbol{r})$ which is zero. A finite value of $\boldsymbol{A}$ now can (and does) induce shifts in the relative populations of these states in just such a way as to preserve zero average current. It is essential to appreciate that this argument would work equally for an uncondensed system of bosons, and that the only reason it does not work for the Cooper pairs is that to distribute the latter between many different pair states (corresponding to different n-values) would be prohibitively costly in energy, i.e. that even at finite temperature the pairs must still be 'Bose-condensed'. (Actually, the phenomenon of superfluidity in liquid ^{4}He, which is the analogue of superconductivity for a neutral system, is usually attributed to the onset of Bose condensation in that system, see e.g. [4]).

It is appropriate, finally, to indicate the connection between the microscopic BCS theory and the phenomenological GL one, which historically preceded it. Actually, the complex scalar order parameter $\Psi(\mathbf{r})$ of GL theory turns out to be nothing but the pair wave function $F(\boldsymbol{r}_1, \boldsymbol{r}_2) = F(\boldsymbol{R}, \boldsymbol{\rho})$ of BCS theory evaluated, at $\boldsymbol{\rho} = 0$, as a function of $\boldsymbol{R} \equiv \boldsymbol{r}$, i.e. it is nothing but *the COM wave function of the Cooper pairs.* This statement is true up to a normalization factor which in GL theory is arbitrary; moreover, while the correspondence can be made formally for arbitrary conditions, it tends not to be very useful unless the variation of $\boldsymbol{\Psi}$ with $\boldsymbol{r}$ is slow on the scale of the pair radius—a condition which is usually assumed in applications of GL theory. Once this is realized, one can obtain the free energy functional $F\{\boldsymbol{\Psi}(\boldsymbol{r})\}$ of GL theory from BCS theory by allowing $\boldsymbol{\Psi}(\boldsymbol{r})$ to have arbitrary (slow) variations in amplitude and/or phase but requiring that, subject to this constraint, the system be in thermal equilibrium. The resulting expression is, in general, extremely messy, but for temperatures close to critical takes the simple form of Equation (A2.1.1), where the coefficients α, β and γ can be calculated from BCS theory.

References

[1] Bardeen J, Cooper L N and Schrieffer J R 1957 *Phys. Rev.* **108** 1175

[2] Cooper L N 1956 *Phys. Rev.* **104** 1189

[3] Leggett A J 1997 *Electron*, ed M Springford (Cambridge, UK: Cambridge University Press) pp 148–181 see especially pp 159–162

[4] Leggett A J 1995 *Twentieth Century Physics* vol. 2, ed L M Brown, A Pais and B Pippard (Bristol, UK: IOP Publishing and AIP Press) pp 913–966

Further Reading

de Gennes P G 1966 *Superconductivity of Metals and Alloys* trans. Pincus P A (New York: Benjamin)

Tinkham M 1996 *Introduction to Superconductivity* (New York: McGraw-Hill)

A2.3

Normal-State Metallic Behavior in Contrast to Superconductivity: An Introduction

David Welch

A2.3.1 Introduction

In recent years, there has been a tremendous expansion in our knowledge of the classes of materials which exhibit what is generally called 'metallic behavior' and in the subclasses of these which exhibit superconductivity. In fact, it is not a trivial matter to define what constitutes a 'metal,' given that a number of organic compounds with structures which contain various low-dimension features, including doped C_{60} fullerenes, a wide variety of intermetallic compounds, many ceramic oxides and even suitably doped liquid ammonia are now considered to be metallic in character and even normally gaseous elements such as iodine and hydrogen have been made to become metallic at high pressures. A working definition of 'metallic behavior' is that the electrical resistivity (or at least one element of the resistivity tensor in anisotropic materials) remains finite, or falls to zero in the case of superconductors, as the temperature approaches absolute zero, as shown in Figure A2.3.1.

At the turn of the 20th century, the behavior of 'good metals' such as Cu, Ag, Al, etc., began to be explained in terms of the behavior of independent, non-interacting 'free' electrons by Drude and Lorentz, although these pioneering efforts were hampered by the necessity of using classical Boltzmann statistics. In the 1930s, the then new quantum mechanics, and their attendant Fermi–Dirac statistics, were used with great success by Sommerfeld, Bloch, Wilson and Brillouin to shed great illumination on what constituted semiconductors, metals and insulators. This state of understanding was well-described in the classic work of Mott and Jones, *The Theory of the Properties of Metals and Alloys,* originally published in 1936, and still in print [1], and is still well worth consulting even today for a cogent introduction to the normal-state properties of simple metals, transition metals and their alloys.

In these early quantum theories, the electrons were considered to act in an independent manner, although this approximation was usually justified only by its success. The notion of independent electron behavior received theoretical justification in Landau's elegant theory of the Fermi liquid [2, 3], which showed that under certain circumstances, not 'bare' electron, but 'quasiparticles' (i.e. electrons together with their accompanying regions of perturbed electron density, lattice distortions, etc.) acted in an independent manner, and in many circumstances, the effects of electron–electron interactions on electron dynamics and thermodynamics could be accounted for by means of an effective mass. Large effective masses arising from electron–electron interaction seem to be vital in understanding the properties of so-called 'heavy fermion' superconductors [4].

Despite such advances in the understanding of metallic properties such as the low-temperature heat capacity and the temperature-dependent electrical resistivity in normal metals, certain experimental facts pointed to the vital importance of correlations between electronic states and dynamics, namely the existence in the metallic state of ferromagnetism and antiferromagnetism (spin correlations) and superconductivity (momentum correlations). Furthermore, the importance of correlated electron behavior was recognized by Wigner and Seitz in 1933 to be crucial in describing cohesion, even in the simplest monovalent metals such as Na, Li, etc. [1, 2]. This important point was further developed by Mott in 1949 when he showed that below a certain volume density monovalent atoms such as Na, which should always be metallic, according to independent electron models, cannot sustain delocalized, metallic behavior and that electron–electron correlations and screening are required to understand the onset of metallic behavior at a certain critical density; this approach was developed much more thoroughly by Hubbard in the 1960s (see discussion in [2, 3]). The basic ideas of Mott and Hubbard are of great importance as a basis for understanding the metallic, insulating and superconducting states in cuprate high-temperature superconductors [5].

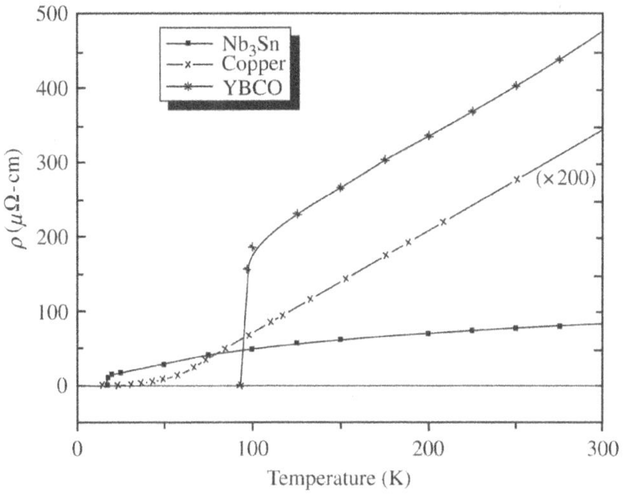

FIGURE A2.3.1 The temperature dependence of the resistivity of a normal metal (Cu), a conventional superconductor (Nb_3Sn), and a high-temperature superconductor ($YBa_2Cu_3O_7$). [Note: The resistivity of Cu is multiplied by 200 for the sake of visibility.] (Data are from: (1) Cu; [13] (2) Nb_3Sn; [14] (3) $YBa_2Cu_3O_7$; [15].)

As the brief discussion above indicates, the characterization of the electronic states and properties of metals, including superconductors, utilizes concepts ranging from the behavior of 'nearly free,' nearly independent electrons (quasiparticles) to the correlations in their dynamics and spatial arrangements which result from the interaction between quasiparticles and which lead to ferromagnetism, antiferromagnetism and superconductivity. In the latter case, the crucial correlation effect that underlies the formation of the superconducting state is the formation of paired electronic states, either in momentum space (Cooper pairs) or in 'real' space (e.g. bipolarons). This pairing results in composite particles which obey Bose–Einstein statistics, which then permits Bose condensation of the paired electrons into a quantum superfluid, i.e. superconductivity occurs.

In the following two sections, we will give a brief sketch of some characteristics of the electronic states and behavior in the normal metallic state and in the superconducting state.

A2.3.2 Normal Metallic State Characteristics

In the normal state of metals, it is a matter of practical necessity, as well as being a rather good approximation, to describe the electronic states as one-electron Bloch functions, which for a given state labelled by **k**, are of the form:

$$\psi(r) = u_k(r)\exp(k.r) \qquad \text{(A2.3.1)}$$

where $\boldsymbol{k}$ is the wave vector (which describes the momentum associated with the state), $\boldsymbol{r}$ is the position coordinate of the electron and u is a function which reflects the effect of the periodic potential field experienced by the electron and includes the effect of the ion cores of the atoms and other electrons (in an averaged way). Each such state can accommodate two electrons of opposite spin. For a constant potential field (free electrons), the function u is a constant, and the various wave functions are plane waves. In this case, the energy of an electron in state k is given simply by a parabola:

$$E(k) = \hbar^2 k^2 / 2m \qquad \text{(A2.3.2)}$$

where $\hbar$ is Planck's constant divided by 2π, and m is the electronic mass.

In the more general case of periodic potentials due to interaction with other electrons and ions, the energy versus k relation is not a simple parabola. Such non-constant potentials give rise to bands of electron energies periodic in the wave vector $\boldsymbol{k}$, and these bands of electrons may be separated by insulating energy gaps between them. For partly filled bands, the effects of the periodic lattice potential and electron–electron interactions can be approximated by the use of an effective mass tensor given by [2]:

$$m^*_{ij} = \frac{\hbar^2}{\left(d^2E/dk_i dk_j\right)}. \qquad \text{(A2.3.3)}$$

The one-electron states are filled with the available electrons according to the Pauli principle for fermions until all available electrons are exhausted; the degree of filling of the bands determines whether the system under consideration is a metal, semiconductor or insulator [1–3]. In the present case, we assume the system to be metallic, which means that at least one of the available bands is only partially filled and that many nearby empty states exist with energies very close to the maximum energy of filled states, E_F, the so-called 'Fermi level.' The boundary in the space of wave vectors $\boldsymbol{k}$ (k-space) between filled and empty states is called the Fermi surface. For a free electron system with a parabolic energy relation (Equation [A2.3.2]), the Fermi surface is a sphere with radius k_F, which is a simple function of the electron density, as discussed in numerous solid-state physics texts [6]. In less-idealized materials, including A15 compounds, heavy fermion superconductors and cuprate superconductors, the Fermi surface does not have a simple spherical geometry in k-space [7]. This is shown in Figure A2.3.2 for a hypothetical metal with a simple cubic structure. However, in many metals and as is assumed in the Bardeen–Cooper–Schrieffer (BCS) theory of superconductivity [5, 8], the Fermi surface can be assumed to be approximately spherical and to be characterized by a Fermi energy E_F or, equivalently a Fermi wave vector $\boldsymbol{k}_F$ or an equivalent Fermi velocity v_F given by:

$$v_F = \frac{\hbar k_F}{m^*}. \qquad \text{(A2.3.4)}$$

[This Fermi velocity enters into expressions for the BCS superconducting coherence length, see Equations (A2.3.8) and (A2.5.2) of Chapter A2.5.]

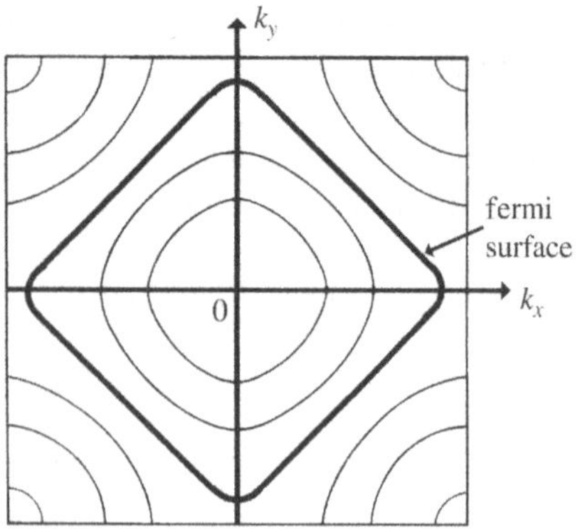

FIGURE A2.3.2 A cross-section through surfaces of constant energy, at equal energy intervals, in k-space for a hypothetical metal in a simple cubic structure. Note that the Fermi surface is not spherical.

Another characteristic parameter which describes the electronic structure and which appears in the BCS theory (and other theories) of superconductivity is the density of states in energy, evaluated at the Fermi surface, $N(E_F)$. The density of states $N(E)dE$ is defined as the number of states per atom which have energies between E and $E + dE$. [Sometimes the density of states at the Fermi level appearing in superconductivity literature is written $N(0)$ when the zero of energy is taken as the Fermi level and usually the density of states per atom and per spin is used.]

The parameters above [$N(E_F)$, m^*, v_F, etc.] can be evaluated from electronic structure theory (either *ab initio* or semi-empirical) [7] or by the interpretation of experimental data, such as paramagnetic susceptibilities and low-temperature heat capacities, as discussed in Chapter A2.4. Such parameters are used as input in descriptions of the superconducting state, such as the BCS theory, etc.

Another important characteristic of the normal state is that deviations from crystalline perfection, such as impurity atoms or structural defects, as well as atomic displacements due to thermal vibration, cause electrons to be scattered from filled to empty states in the vicinity of the Fermi surface. Such scattering gives rise to electrical resistance, and measurements of the electrical resistivity in the normal state, ρ_N, can be used to deduce valuable information about electron dynamics [9]. One important characteristic parameter of this scattering is the mean free path, ℓ, between scattering events caused by impurities and defects for electrons near the Fermi surface. This is an important parameter in determining certain superconducting properties such as the coherence length.

A2.3.3 The Superconducting State

It is implicit in the description of the normal metallic state discussed above that the only interaction between the electrons is the long-range Coulomb repulsion. Although Landau showed that the quasiparticle concept allows for correlations in the electron positions, which reduce the range of the repulsions by means of screening, this does not qualitatively change the picture. However, in the early 1950s, Leon Cooper made a major advance by demonstrating that if there is any mechanism to cause an attractive interaction, no matter how small, between electrons, then the independent quasiparticle becomes unstable with respect to the formation of bound pairs of quasiparticles, now called Cooper pairs, in momentum space (k-space); a succinct description of this can be found in [2] and [8]. Bardeen, Cooper and Schrieffer showed how the electron–phonon interaction can give rise to the necessary attraction and developed the first successful, and now standard, theory of superconductivity. In this theory, even though the creation of a phonon by an electron scattering event costs an energy of $\hbar\omega$, the subsequent interaction of the phonon with a second electron can result in a reduction in energy, the magnitude of which we denote by V, and that for a free electron system with a spherical Fermi surface, the energy of the resulting bound pair of electrons at the Fermi surface is reduced by a binding energy $\mathbf{\Delta}_0$ given by

$$\Delta_0 = \frac{2\hbar\omega}{\exp(1/N(0)V - 1)}. \quad \text{(A2.3.5)}$$

Note that this binding energy depends not only on the attractive energy V but also on the magnitude of the density of electronic states at the Fermi surface, $N(0)$.

These bound Cooper pairs of electrons formed from one-electron states at the Fermi surface are bosons and thus can undergo Bose condensation into a single coherent quantum superfluid state for which the wave function

$$\psi = \psi_0 \exp(i\theta) \quad \text{(A2.3.6)}$$

is characterized by an amplitude ψ_0 where $|\psi_0|^2$ is the density of Cooper pairs, and a phase θ. This quantum state is macroscopic in nature, with a size dictated by the size of the superconducting body. At zero temperature, the ground state of the electronic system in this superconducting state can be described as a coherent mixture of occupied states within a somewhat diffuse Fermi surface (in k-space), separated from excited states which include unpaired quasiparticles by an energy gap given by 2Δ, where, at zero temperature, Δ is given by Equation (A2.3.5). The difference, produced by the gap, between the energy of excitations in a normal metal and in a superconductor is shown in Figure A2.3.3. The coherent character of the condensed state of paired electrons within a spread of k-values at the Fermi surface and the exclusion of nearby available empty states because of the energy gap means that scattering of electrons by impurities, defects, phonons, etc. does not occur. The gap is attached to the Fermi surface, and still exists even when the net momentum is not zero, i.e. even when the center of the Fermi surface is shifted so that a net current of electrons is flowing: this state has no resistance;

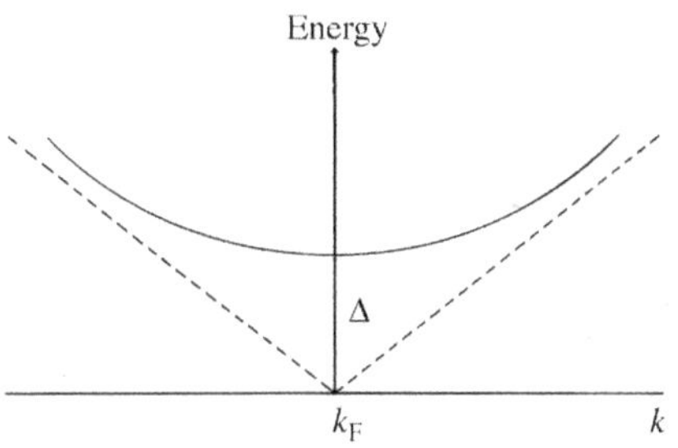

FIGURE A2.3.3 The excitation spectrum for quasiparticles in a BCS superconductor (solid line) compared with that for the normal state (dashed lines) [5].

it exhibits superconductivity (see Figure A2.3.4). A consideration of the thermodynamics of this state (including the effects of temperature), as discussed in Chapter A2.4, reveals an even more remarkable property: the superconducting state is perfectly diamagnetic (the so-called 'Meissner effect').

Including the effects of a non-zero temperature in the formation and properties of the condensed superconducting state shows that the magnitude of the energy gap 2Δ, as well as the density of paired electrons, diminishes continuously with increasing temperature and vanishes in a second-order transition (at zero magnetic field) at a critical value of the temperature T_c. (See [8] for details.) The value of T_c is determined by the size of the electron pair binding energy Δ_0:

$$k_B T_c = \alpha \Delta_0 \tag{A2.3.7}$$

where k_B is the Boltzmann constant, and α is a number of order unity; the BCS theory yields a value for α of 0.568. The value of the energy gap Δ_0 can be measured by a variety of

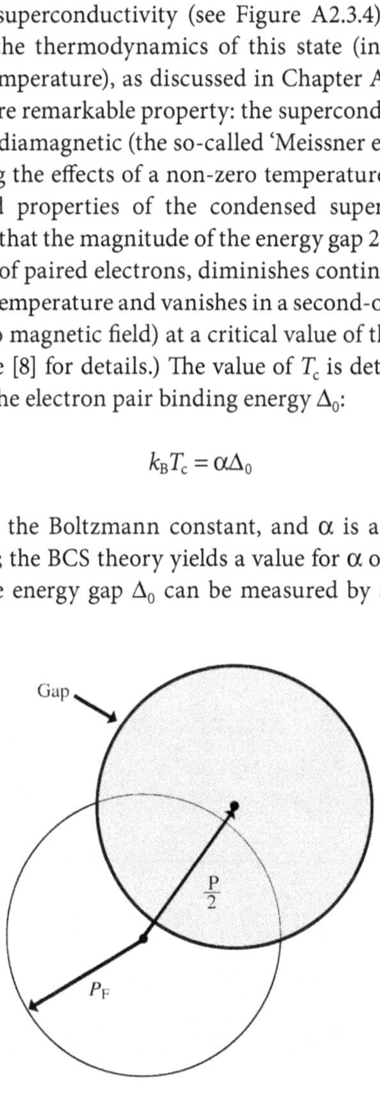

FIGURE A2.3.4 The momentum distribution in a current-carrying superconductor (shaded circle) compared to that in the normal state (open circle). *P* is the total momentum of a Cooper pair. In both cases, the momentum vectors are uniformly distributed in spheres of radius p_F, the surfaces of which are the Fermi surface. Note that the energy gap in the superconducting state is carried by the Fermi surface.

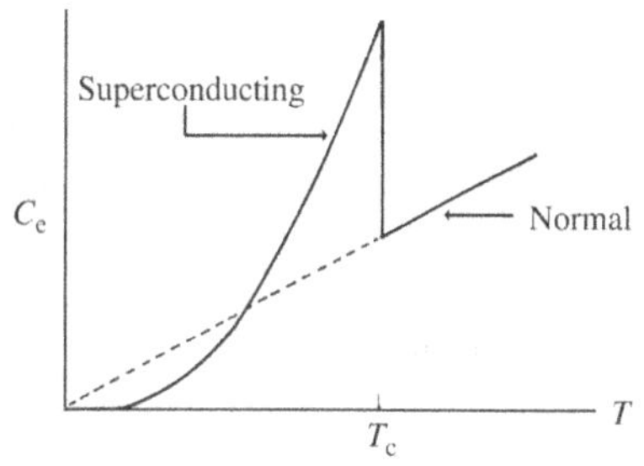

FIGURE A2.3.5 The temperature dependence of the electronic contribution to the specific heat of a conventional superconductor.

experimental methods [10], including heat capacity measurements, shown in Figure A2.3.5, as discussed in Chapter G3.1. This permits evaluation of the constant α and is thus one test of the validity of particular theories, e.g. BCS theory, for particular classes of superconducting materials, e.g. cuprates.

The critical temperature T_c is one important thermodynamic characteristic of the superconducting state. Another is the critical magnetic field H_c, and its variants H_{c1} and H_{c2}, as discussed in Chapter A2.5. There are also two important length scales which characterize the superconducting state. The first is the coherence length, ξ, first introduced by Pippard in 1953, which characterizes the distance which is required for the density of superconducting electrons n_s to change appreciably in an inhomogeneous superconductor [5, 8, 10]. In pure and defect-free superconductors, this length, ξ_0, is a characteristic of the material and depends on T_c and the Fermi velocity v_F (Equation [A2.3.4]); the BCS theory result is:

$$\xi_0 = 0.18 \frac{\hbar v_F}{k_B T_c} \tag{A2.3.8}$$

Electron scattering by impurities, defects, etc. can reduce the coherence length, and in 'dirty' superconductors, ξ is given approximately by $(\xi_0 \ell)^{1/2}$, where ℓ is the mean free path for electron scattering in the normal state; thus, alloying or plastic deformation can be used to reduce the coherence length when desired.

The second characteristic length associated with the superconducting state is the so-called 'penetration depth,' λ. This length describes the distance required for magnetic fields to decay in going from a region of normal material into the perfectly diamagnetic superconductor [8]. This length arose from the pioneering work by F. London and H. London on the electrodynamics of superconductors and is sometimes called the London penetration depth. For essentially pure superconductors, the London penetration depth is controlled by the density of superconducting electrons n_s and is given by:

$$\lambda_L = \left(\frac{mc^2}{4\pi n_s e^2} \right)^{1/2} \tag{A2.3.9}$$

where *c* is the speed of light in vacuum and, *e* and *m* are the electronic charge and mass. Where the electron mean free path ℓ is limited by scattering of impurities or defects, the penetration depth is given approximately by $\lambda_L(\xi_0/\lambda)^{1/2}$. The nature of the behavior (Type I or Type II) of superconductors in magnetic fields is determined in a very important way by the value of the ratio of the penetration depth to the coherence length. This is discussed in detail in Chapter A2.5.

The present state of understanding of the nature of the superconducting state is in an active state of development. It is not yet clear whether or not a 'BCS-like' theory of superconductivity will suffice to describe high-T_c superconductors. A review [11] of the present state of the theoretical development has recently appeared, which discusses some of the issues. Other aspects of the situation are described in [5, 10]. Not only are the superconducting state properties of high-T_c cuprate superconductors different in many respects from those of conventional superconductors, but also there are important differences between the normal states of these two classes of superconductors. For example, there is now considerable evidence that the symmetry of the pair state in cuprate superconductors is not s-wave, as in conventional BCS superconductors, but rather is d-wave in nature [12]. This symmetry is reflected in the character of the superconducting gap, for example, resulting in lines of nodes (zeros of the gap) on the Fermi surface, and this, in turn, is reflected in many properties which depend on the excitation spectrum [5]. Regarding differences between the normal states, one of the most important is the existence of a gap or at least a substantial reduction in the density of states, a 'pseudogap,' near the Fermi surface in the normal state of underdoped cuprates. The pseudogap is manifested in anomalies in the heat capacity and the paramagnetic susceptibility of underdoped cuprates [5].

References

[1] Mott N F and Jones H 1936 *The Theory of the Properties of Metals and Alloys* (New York: Dover)

[2] Cottrell A H 1988 *Introduction to the Modern Theory of Metals* (London: Institute of Materials)

[3] Mott N F 1990 *Metal-Insulator Transitions* 2nd edn (London: Taylor and Francis)

[4] Fisk Z, Hess D W, Pethick C J, Pines D, Smith J L, Thompson J D and Willis J O 1988 Heavy-electron metals: new highly correlated states of matter *Science* **239** 33

[5] Waldram J R 1996 *Superconductivity of Metals and Cuprates* (Bristol: Institute of Physics Publishing)

[6] Quéré Y 1998 *Physics of Materials* (Amsterdam: Gordon and Breach)

[7] Harrison W A 1999 *Elementary Electronic Structure* (Singapore: World Scientific)

[8] de Gennes P G 1966 *Superconductivity of Metals and Alloys* (Redwood City, CA: Addison-Wesley)

[9] Rossiter P G 1987 *The Electrical Resistivity of Metals and Alloys* (Cambridge: Cambridge University Press)

[10] Kresin V G, Morantz H and Wolf S A 1993 *Mechanisms of Conventional and High T_c Superconductivity* (Oxford: Oxford University Press)

[11] Ruvalds J 1996 Theoretical prospects for high-temperature superconductors—topical review *Supercond. Sci. Technol.* **9** 905

[12] Tsuei C C and Kirtley J R 2000 Pairing symmetry in cuprate superconductors *Rev. Mod. Phys.* **72** 969

[13] Bass, J and Fischer K H eds 1982 *Landolt-Börnstein Numerical Data and Functional Relationships in Science and Technology, vol 15a Metals: Electronic Transport Phenomena* (Berlin: Springer)

[14] Woodward D W and Cody G D 1964 *Phys. Rev.* **136** 166

[15] Iye Y, Tamegai T, Sakakibara T, Goto T, Miura N, Takeya H and Takei H 1988 *Physica C* **153–155** 26

A2.4

The Meissner–Ochsenfeld Effect

Rudolf P. Huebener

Following Kamerlingh Onnes' discovery of superconductivity, in 1933 Walther Meissner and his collaborator Robert Ochsenfeld at the Physikalisch Technische Reichsanstalt in Berlin discovered the most fundamental property of a superconductor, namely its ability to expel magnetic flux from its interior [1]. The perfect diamagnetism associated with this Meissner–Ochsenfeld effect results from electric shielding currents flowing without resistance near the surface of the superconductor. In Figure A2.4.1, we show schematically how the superposition of the applied magnetic field and the field generated by the shielding current result in zero magnetic flux density B inside the superconductor. We define $B(r)$ in terms of the local field $H(r)$ produced by the superposition of the external field H_a, produced by external currents (e.g., the field by a long solenoid), and the field generated by currents flowing within the superconductor. From Ampere's theorem, we can see that for $B(r) = \mu_0 H(r) = 0$, the total current I per unit length flowing around the outer surface of a superconducting cylinder in a parallel field is $I = -H$. Since the supercurrent-flow without resistance is a necessary consequence of the existence of the Meissner–Ochsenfeld effect, whereas the inverse conclusion does not hold, the Meissner–Ochsenfeld effect is clearly more fundamental than just the disappearance of the electric resistance (although only the latter phenomenon is suggested by the name 'superconductivity').

It is interesting to note that for about two decades after the discovery of superconductivity one had concentrated exclusively on the electric conductivity and had ignored completely the magnetic properties of superconductors. In the early 1930s, Max von Laue had turned his attention to the demagnetizing effects in the case of a perfect electrical conductor in a weak magnetic field. Since 1925, von Laue had an appointment at the Reichsanstalt in Berlin as a theoretical consultant, in addition to his position at the Friedrich-Wilhelms University. Consequently, von Laue had to be in close contact with Walther Meissner and likely played some role in his plans for experiments dealing with the magnetic behavior of superconductors. According to von Laue, the discovery of the Meissner–Ochsenfeld effect represented a turning point in the field of superconductivity.

Flow of current without resistance is a necessary consequence of the Meissner–Ochsenfeld effect, whereas a transition to zero resistance does not imply flux exclusion. Indeed, a transition to zero resistance would simply trap within the superconductor any field previously present; the final state would then depend on the magnetic and thermal history of the sample. This is shown in Figure A2.4.2, where the critical magnetic field $H_c(T)$, separating the superconducting state from the normal state, is plotted versus temperature. We consider two different ways to pass from point (1) in the normal state to point (4) in the superconducting state. First, we assume only infinite electric conductivity and the absence of the Meissner–Ochsenfeld effect in the superconductor. Then, along the path 1–2–4 at point (4), we have $B = 0$ in the superconductor. In contrast, on path 1–3–4 at the transition to zero resistance, the applied field would be trapped inside the superconductor, so that the magnetic induction would remain constant with $B = \mu_0 H$. The exclusion of magnetic flux associated with the Meissner effect ensures that the final state is in practice independent of the thermodynamic path. This means that superconductivity is a thermodynamic state of the system, so that equilibrium thermodynamics can be applied to the superconducting phase transition.

The magnetic energy per unit volume required for achieving magnetic flux expulsion is given by

$$-\mu_0 \int_0^H M(H)\mathrm{d}H = \frac{\mu_0}{2} H^2 \qquad \text{(A2.4.1)}$$

since for perfect diamagnetism we have $M = -H$. Here, μ_0 is the vacuum permeability, and $M(H)$ is the magnetization. The magnetic energy of Equation (A2.4.1) must be overcompensated by the gain in free energy density for the superconducting state to be energetically favorable. At the thermodynamic critical magnetic field $H_c(T)$, the energy gain of the superconducting state vanishes. Denoting the free energy densities in

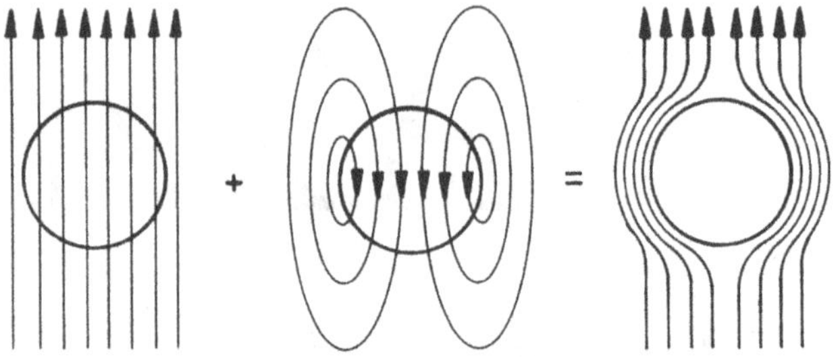

FIGURE A2.4.1 The superposition of the applied magnetic field and the magnetic field generated by the shielding supercurrent results in zero magnetic flux density inside the superconductor.

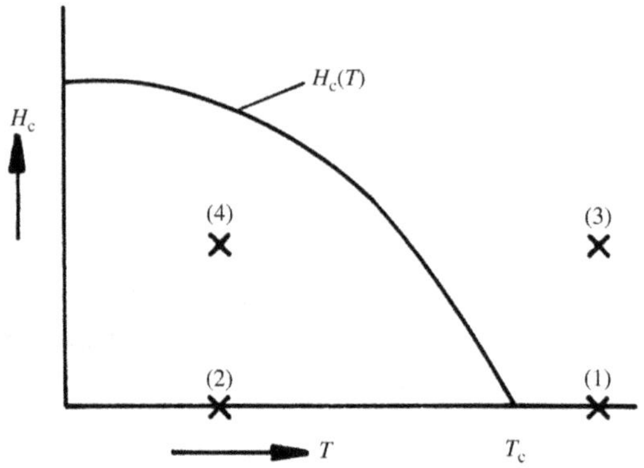

FIGURE A2.4.2 Critical magnetic field H_c versus temperature. From point (1) one can reach point (4) via point (2) or point (3).

zero magnetic field in the normal and in the superconducting state by $f_n(T)$ and $f_s(T)$, respectively, we obtain for their difference [2]

$$f_n(T) - f_s(T) = \frac{\mu_0 H_c^2(T)}{2}. \quad \text{(A2.4.2)}$$

For a typical value of $H_c = 10^4$ Am^{-1}, we have $f_n - f_s = 40$ J m^{-3}.

Since the density of the Meissner shielding current cannot become infinite, the shielding currents are spread over a distinct distance from the surface. As a consequence, the magnetic field drops from its value $\mu_0 H_a$ to zero only within a characteristic length scale given by the London penetration depth λ_L. The first theoretical discussion of this finite magnetic field penetration was given in the London theory, yielding the exponential decay of the local magnetic flux density $B(x)$ within the superconductor with increasing distance x from the surface

$$B(x) = B(0)\exp(-x/\lambda_L) \quad \text{(A2.4.3)}$$

The London penetration depth λ_L for electrons flowing without dissipation is given by [3]

$$\lambda_L = \left(\frac{m}{\mu_0 e^2 n_s}\right)^{1/2} \quad \text{(A2.4.4)}$$

where m, e, and n_s are the electron mass, charge, and number density, respectively. For superconductors, both the effective mass and the charge are doubled; the mass is the effective mass involved in electrical transport, and $n_s(T)$ is the number density of the superconducting electrons, which increases from zero at T_c to the normal-state density at low temperatures. Because of the appearance of the density n_s in the denominator of Equation (A2.4.4), λ_L is temperature dependent and diverges for $T \to T_c$. The experimental values of the temperature-dependent penetration depth λ can be well fitted by the empirical relation

$$\lambda(T) = \lambda(0)\left[1 - \left(\frac{T}{T_c}\right)^4\right]^{-1/2}. \quad \text{(A2.4.5)}$$

A similar relation has been predicted in an early two-fluid model by Gorter. For pure metallic superconductors, such as Pb, Sn, In, Al, and Hg, the value of $\lambda(0)$ is typically around 50 nm. From this we see that the perfect diamagnetism from the Meissner–Ochsenfeld effect is well established only if both sample dimensions perpendicular to the magnetic field are much larger than this value of $\lambda(0)$. From Maxwell's equation, $j = (1/\mu_0)$ curl B, and from the fact that B drops almost to zero within the distance λ from the surface, we find the approximate expression for the maximum density of the shielding supercurrent $j_c = H_c/\lambda$.

From Equation (A2.4.4), we note that $\lambda^{-2}(T)$ is proportional to the density of superconducting electrons $n_s = n_n - n_{qp}$, where n_{qp} is the density of quasiparticles thermally excited across the gap in a superconductor. Measurements of $\lambda(T)$ are therefore very important in testing microscopic models for superconductivity. For conventional BCS superconductors, the existence of a near isotropic, s-wave energy gap gives rise to a vanishingly small exponential variation in $\lambda(T)$ at low temperatures. In contrast, for the recently discovered cuprate superconductors, one finds approximately

$$\lambda^{-2}(T) = \lambda^{-2}(0)[1 - aT] \quad \text{(A2.4.6)}$$

over a very wide range of temperatures, for screening currents flowing in the cuprate planes. This is consistent with the now firmly established d-wave symmetry of the cuprate superconductors, which has nodes in the energy gap along certain directions. At low temperatures, the presence of such nodes leads to power law dependences of many of the thermal and transport properties instead of the exponential dependences expected for conventional BCS superconductors with an isotropic energy gap in all directions (s-wave). However, as shown recently by Schopohl and Dolgov, in the low-temperature limit, the temperature dependence of λ must vanish in order to remain consistent with the third law of thermodynamics.

The experimental values $\lambda(0)$ of the penetration depth at zero temperature are up to five times larger than the quantity $\lambda_L(0)$ in Equation (A2.4.4). This deviation has been explained by Pippard in terms of the superconducting coherence length ξ_0 and by extending the London theory accordingly. According to Pippard, in dirty superconductors, in which the electron mean free path $l \ll \xi_0$, the magnetic penetration depth is increased by a factor $\sim(\xi_0/l)^{1/2}$. For discussion of the length ξ_0, see Chapters A2.5 and A3.1.

The Meissner–Ochsenfeld effect and the associated screening currents flowing within the penetration depth of the surface are the intrinsic properties of all the superconductors at sufficiently small fields. In Chapter A2.5, we will introduce the concept of two types of superconductors: type I in which, on application of a sufficiently strong field, a direct transition is made from the Meissner state with $B = 0$ to the normal state with $B = \mu_0 H_a$, and type II, in which partial flux penetration is nucleated at a smaller field H_{c1} in the form of quantized flux lines, so that the Meissner state only exists in the range $0 < H_a < H_{c1}$.

In the previous discussion, we have essentially been assuming superconductors in the form of long rods or cylinders aligned parallel to the applied field, where the maximum field at the surface is simply the applied field H_a. However, for an arbitrary-shaped sample, flux exclusion implies a concentration of flux around the surface and a higher local surface field $H_i = H_a - \Sigma_i n_{ij} M_j$ (with $i, j = x, y, z$). Here, n_{ij} is the demagnetization tensor with unit trace, such as this can be defined for uniformly magnetized samples (such as superconductors in the Meissner state). If the magnetic field is applied parallel to the principal axis x_i of a sample shaped as a rotational ellipsoid, a description of demagnetization in terms of a single quantity n_i (i.e., one of the principal values of n_{ij}) is possible. One can then write $H_i = H_a + H_{dem} = H_a - n_i M$; for a superconductor in the Meissner state, $H_i = H_a + n_i H_i$, i.e., $H_i = H_a/(1 - n_i)$. For a cylinder with its axis parallel to $x_1 \equiv x$, in a field transverse to its length, $n_y = n_z = 1/2 \equiv D$; for a sphere, $n_x = n_y = n_z \equiv D = 1/3$. Here, D is known as the demagnetization factor. When the surface field exceeds $H_c(1-D)$ or $H_{c1}(1-D)$ for a type I or a type II superconductor, flux will begin to penetrate and we no longer have a complete Meissner effect (see further discussion in Chapter A2.5).

Because of the fundamental importance of the Meissner–Ochsenfeld effect and since magnetization measurements can be performed relatively easily, more recently, the detection of this effect has played an important role in the discovery and confirmation of new superconductors and, in particular, in the discovery of high-temperature superconductivity by Bednorz and Müller [4].

References

[1] Meissner W and Ochsenfeld R 1933 *Naturwiss.* **21** 787

[2] Gorter C J and Casimir H B G 1934 *Physica* **1** 306

[3] London F and London H 1935 *Proc. R. Soc. A* **149** 71

[4] Bednorz J G, Takashige M and Müller K A 1987 *Europhys. Lett.* **3** 379

Further Reading

Huebener R P 2001 *Magnetic Flux Structures in Superconductors 2nd edition* (Berlin: Springer)

Shoenberg D 1965 *Superconductivity* (Cambridge: Cambridge University Press)

Tilley D R and Tilley J *Superfluidity and Superconductivity* (New York: Van Nostrand Reinhold)

Tinkham M 1996 *Introduction to Superconductivity* (New York: McGraw Hill)

Waldram J R 1996 *Superconductivity of Metals and Cuprates* (Bristol: Institute of Physics Publishing)

A2.5

Loss of Superconductivity in Magnetic Fields

Rudolf P. Huebener

A2.5.1 The Superconducting Coherence Length

In addition to the magnetic penetration depth λ discussed in Chapter A2.4, there is another important length scale: the superconducting coherence length ξ_0 first introduced by Pippard [1]. The length ξ_0 is a measure of the spatial extent of the wave function describing Cooper pairs and is therefore the minimum distance over which the density n_s of superconducting electrons can change significantly. It indicates the spatial rigidity of the superconducting wave function. An estimate based on the Heisenberg uncertainty principle yields

$$\xi_0 \approx \frac{\hbar v_F}{\Delta} \approx \frac{\hbar v_F}{k_B T_c} \quad \text{(A2.5.1)}$$

where $\hbar$ is Planck's constant divided by 2π, v_F is the Fermi velocity, Δ is the superconducting energy gap, and k_B is Boltzmann's constant. A more accurate result can be obtained from the BCS theory [2]

$$\xi_0 = \frac{\hbar v_F}{\pi\Delta} = 0.18\frac{\hbar v_F}{k_B T_c} \quad \text{(A2.5.2)}$$

where we have substituted the BCS relation $\Delta = 1.76 k_B T_c$. For pure metallic superconductors, such as Pb, Sn, In, and Hg, the experimental values of ξ_0 range around 100–300 nm. For Al, a value as large as $\xi_0 = 1.4$ μm has been reported.

These experimental values given are for the clean limit with the electron mean free path $l \gg \xi_0$. In the presence of strong impurity scattering, and in most alloys, $l \ll \xi_0$. Pippard [3] introduced an effective coherence length to account for electron scattering of the form

$$\frac{1}{\xi} = \frac{1}{\xi_0} + \frac{1}{l}. \quad \text{(A2.5.3)}$$

Many of the newly discovered cuprate superconductors are believed to be in the clean limit with mean free paths within the CuO planes much greater than the coherence length of typically 2–3 nm. The out-of-plane coherence length is believed to be less than ~0.1 nm, reflecting the highly localized nature of the electronic wave functions largely confined to the CuO planes, which are responsible for the metallic and superconducting properties.

Although the coherence length describes the extent of the superconducting wave function of a Cooper pair, the co-operative nature of the superconducting transition involves interactions between the pairs which lead to strongly temperature-dependent coherence and penetration lengths. This can most conveniently be considered using the phenomenological Ginzburg–Landau (GL) theory [4]. While Gorkov has shown that near T_c it is equivalent to the BCS theory [5], GL is believed to be applicable over a wide range of temperatures and fields. In the clean limit ($l \gg \xi_0$), it can be shown that

$$\xi(T) = 0.74\xi_0\left(\frac{1}{1-t}\right)^{1/2} \quad \text{(A2.5.4)}$$

$$\lambda(T) = \frac{\lambda_L(0)}{\sqrt{2}}\left(\frac{1}{1-t}\right)^{1/2}. \quad \text{(A2.5.5)}$$

where $t = T/T_c$.

In the dirty limit ($l \ll \xi_0$), scattering modifies the prefactors with $\xi_{eff}(0) \sim l$ and $\lambda_{eff}(0) \sim \lambda_L(0) \times (\xi_0/l)^{1/2}$ leaving the temperature factors unchanged.

The dimensionless GL parameter,

$$\kappa = \frac{\lambda(T)}{\xi(T)} \quad \text{(A2.5.6)}$$

is therefore independent of temperature but is strongly dependent on scattering. In Sections A2.5.2 and A2.5.3, we will show how the GL parameter differentiates between type I (when $\kappa < 1\sqrt{2}$) and type II (when $\kappa > 1\sqrt{2}$) superconductors (see Chapter A3.1).

A2.5.2 Type I Superconductors

With the exception of Nb, V, and *Tc*, all elemental superconductors are type I, exhibiting a full Meissner flux expulsion before making a transition to the normal state at $H_c(T)$.

However, as indicated in Chapter A2.4, this is only true for samples with a negligible demagnetizing factor.

For a type I superconductor with a finite demagnetizing factor *D*, the critical field at the surface will reach the thermodynamic transition field when the external field is $H_c(1 - D)$. Normal regions will be nucleated at the surface and flux will enter generating a domain structure of normal regions in which the field is equal to H_c and superconducting regions excluding flux. A schematic example of this 'intermediate state' for a thin superconducting slab in a field perpendicular to its surface is shown in Figure A2.5.1. The intermediate state is established in the magnetic field regime $H_c(1 - D) < H < H_c$.

An important feature of the domain configuration of Figure A2.5.1 is the wall energy associated with the interface between the normal and superconducting phases. This wall energy is somewhat analogous to the Bloch wall energy separating regions of opposite magnetization in a ferromagnet. It can be obtained in the following way. At the interface, the transition from the superconducting to the normal state can take place only over the distance $\xi(T)$ because of the rigidity of the superconducting wave function. $\xi(T)$ is the temperature-dependent coherence length, discussed in Section A2.5.1. Over this distance $\xi(T)$, the superconducting condensation energy is lost, yielding the contribution $[\mu_0 H_c^2(T)/2]\xi(T)$ per unit area (see Equation [A2.5.2]). However, because of the magnetic flux penetration into the superconducting domain over the distance $\lambda(T)$, no gain and also no loss of condensation energy occur. Therefore, by subtracting this part, we find the wall energy α per unit area

$$\alpha = \frac{\mu_0 H_c^2(T)}{2}[\xi(T) - \lambda(T)]. \qquad \text{(A2.5.7)}$$

The difference $\delta \equiv \xi(T) - \lambda(T)$ is referred to as the wall energy parameter. In type I superconductors, we have $\xi(T) > \lambda(T)$ and, hence, the quantities δ and α are positive. More careful calculation shows that the crossover between positive and negative wall energies occurs when $\lambda(T)/\xi(T) = 1/\sqrt{2}$.

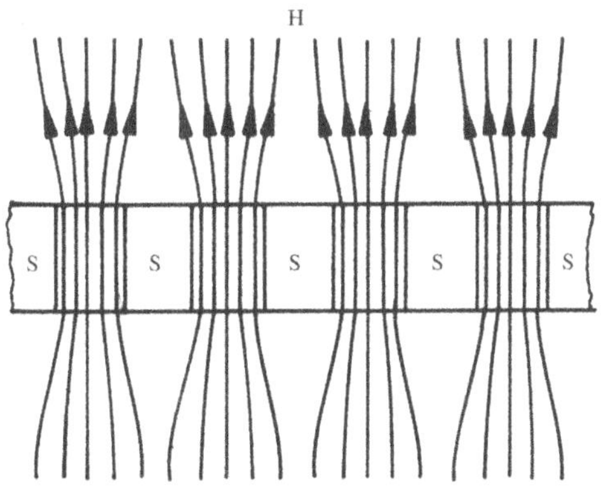

FIGURE A2.5.1 Intermediate state of a type I superconductor. The normal domains carry the flux density $\mu_0 H_c$ and are separated from each other by the superconducting phase with zero flux density.

The first theoretical treatment of the domain structure in the intermediate state of a type I superconductor was given by Landau [5]. He minimized the total free energy taking into account the interface wall energy, the magnetic energy within and outside the superconductor, and the energy associated with changes in domain structure as the field leaves the sample. The Landau domain theory yields the periodicity length *a* of the domain configuration. The length *a* is the sum of the lengths α_n and α_s of the normal and superconducting domains, respectively: $\alpha = \alpha_n + \alpha_s$. For the length α one finds

$$a = \left[\frac{\delta d}{f(H/H_c)}\right]^{1/2} \qquad \text{(A2.5.8)}$$

where δ is the wall energy parameter, $f(H/H_c)$ is a numerical function of the normalized field H/H_c, and *d* is the thickness of the plate measured along the magnetic field direction (see Figure A2.5.1). Similar concepts are also used for calculating the domain size in other cases, such as a ferromagnet.

In the discussion leading to Equation (A2.5.8), a structure of long laminar domains is assumed, such that the spatial variations can be restricted to two dimensions, as is often observed at intermediate fields. This is shown in Figure A2.5.2

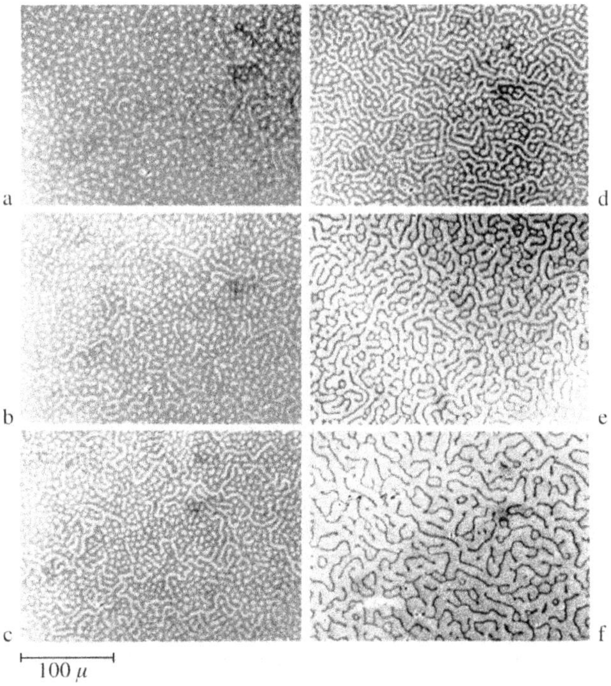

FIGURE A2.5.2 Intermediate state structure of a lead film in a perpendicular magnetic field observed magneto-optically for the following field values: (a) 7.6 kAm^{-1}; (b) 10.5 kAm^{-1}; (c) 14.2 kAm^{-1}; (d) 17.4 kAm^{-1}; (e) 27.8 kAm^{-1}; (f) 32.7 kAm^{-1}. The superconducting phase is dark, $T = 4.2$ K, film thickness = 9.3 μm.

for a type I superconducting lead film in a perpendicular field. However, in addition, normal domains with nearly circular cross-section ('flux tubes') can also be found, in particular, at low magnetic fields. In the intermediate state, the volume fraction filled by the normal domains is equal to H/H_c, since the normal domains carry the field value H_c. For the laminar domain pattern, this yields the relation $a_n/a = H/H_c$. From Equation (A2.5.8), we see that the length scale a of the laminar domain structure varies proportional to $d^{1/2}$, decreasing with decreasing thickness of the superconductor. Similarly, with decreasing d the diameter of the flux tubes also becomes smaller.

If the magnetic field is oriented exactly perpendicular to the surface of a large flat plate or film, the long, laminar domains are statistically oriented in the plane of the superconductor, as shown in Figure A2.5.2. However, if the magnetic field is inclined at an angle, the domains become oriented parallel to this field component in the plane. This effect was first demonstrated by Sharvin. The Sharvin geometry is advantageous for accurately measuring the periodicity length a of the magnetic domain structure.

Superconductivity can also be destroyed, even in zero applied field, if the current flowing through a superconductor creates a field at its surface in excess of H_c. For a cylindrical wire of radius r, this leads to a critical current $I_c = H_c r/2$, known as Silsbee's rule.

A2.5.3 Type II Superconductors

Type I superconductivity, discussed so far, is established in metals of high purity, where the electron scattering in the normal state is relatively weak. However, an increasing amount of evidence has accumulated, indicating the existence of another type of superconductors with various 'unusual' properties. It was Shubnikov in the early 1930s who provided this evidence. The 'unusual' behavior such as the appearance of electric resistance at magnetic fields below the critical field $H_c(T)$ or the observation of an incomplete Meissner effect was found in metals with a large impurity concentration and in alloys, where the electron scattering is relatively strong. It was not until Abrikosov's 1955 derivation of the magnetic properties of superconductors with κ values $> 1/\sqrt{2}$ that such properties were recognized as intrinsic and characteristic of what are now known as type II superconductors [6].

Type II superconductors are distinguished by the fact that the penetration depth $\lambda(T)$ is larger than the coherence length $\xi(T)$. In this case, the wall energy α from Equation (A2.5.7) becomes negative, and the magnetic domains within the superconductor are reduced to the smallest possible unit of magnetic flux, namely individual magnetic flux quanta Φ_0 (see Chapter A2.8).

In type II superconductors, the Meissner–Ochsenfeld effect is established only up to what is known as the lower critical magnetic field $H_{c1} < H_c$ (see Chapter A2.4). It then becomes energetically favorable for the sample to undergo a transition in which flux tubes (flux lines) carrying a single quantum of flux Φ_0 are nucleated at the surface, forming a flux line lattice in the bulk. This is known as the 'mixed state', representing a mixture between the superconducting and normal phases. In this state, the spatially averaged magnetic flux density B is

$$B = n \cdot \Phi_0 \quad \text{(A2.5.9)}$$

where n is the areal density of the flux lines. The magnetic flux quantum is

$$\Phi_0 = \frac{h}{2e} = 2.07 \times 10^{-15}\,\text{Tm}^2. \quad \text{(A2.5.10)}$$

The mixed state extends to the upper critical magnetic field H_{c2} above which the normal state is formed. The thermodynamic critical field H_c defined from the difference between the free energy densities of the normal and superconducting phases according to Equation (A2.5.2) lies between H_{c1} and H_{c2}: $H_{c1} < H_c < H_{c2}$. As the field is increased above H_{c1}, the density of flux lines increases until their cores with radius $\sim\xi(T)$ start overlap leading to suppression of superconductivity and a second-order transition to the normal state at a field $H_{c2}(T) = \Phi_0/2\pi\xi(T)^2$. Figure A2.5.3(a) shows magnetic flux density versus magnetic field, and Figure A2.5.3(b) shows magnetization versus magnetic field for a type II superconductor. If the sample geometry causes demagnetization effects, the mixed state of a type II superconductor is established in the magnetic field range $H_{c1}(1 - D) \le H \le H_{c2}$.

A phenomenological understanding of type II superconductivity has been provided by the GL theory [4]. Here, the spatial and temporal properties of the superconductor are described by a macroscopic complex wave function,

$$\psi = |\psi| \exp(i\varphi) \quad \text{(A2.5.11)}$$

acting as an order parameter. The squared amplitude $|\psi|^2$ is identified as the Cooper pair density n_s or the superconducting energy gap. Both the absolute value $|\psi|$ and the phase φ can be space and time dependent. Strictly speaking, the GL

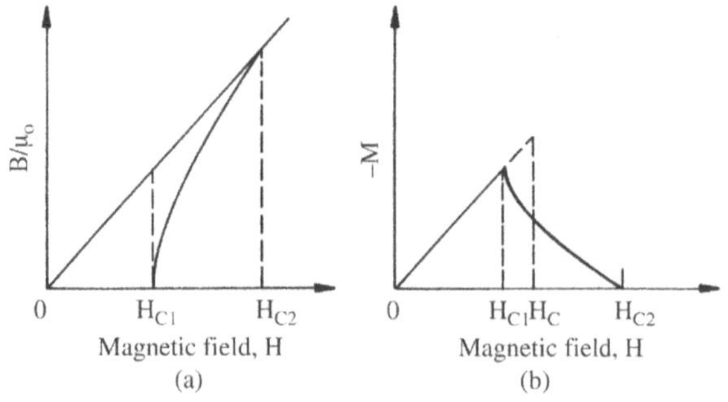

FIGURE A2.5.3 (a) Magnetic flux density B and (b) magnetization M versus magnetic field for a type II superconductor with a demagnetization factor D = 0.

theory is only applicable close to the second-order phase transition at $H_{c2}(T)$, though in practice, it appears to provide an excellent description of the superconducting state over a large part of the magnetic phase diagram. Within this theory, magnetic flux quantization results from the requirement that the wave function ψ must be single-valued at any point in the superconductor and, hence, that the phase φ can change only by multiples of 2π following a complete closed path within the superconductor. Therefore, flux quantization is the macroscopic analog of the Bohr–Sommerfeld quantum condition in atomic physics and of the quantization of hydrodynamic circulation or vorticity in superfluid helium, first observed by Vinen in 1958, $\oint v_s \, dl = nh/m$. The formation of a regular lattice of magnetic flux quanta was first predicted theoretically by Abrikosov from an ingenious solution of the GL equations. Therefore, this lattice is often referred to as the Abrikosov vortex lattice.

In a type II superconductor, flux penetrates the bulk in the form of flux lines with line cores along which ψ drops to zero. This allows solutions of the form $|\psi(r)|e^{in\theta}$, as in a multiply connected superconducting ring. These vortex-like solutions involve currents circulating around the flux core with associated magnetic flux $n\Phi_0$. Abrikosov showed that singly quantized flux lines would be nucleated in a type II superconductor at a field H_{c1}. The structure of an isolated magnetic flux line carrying a single flux quantum and representing the building block of the mixed state is shown schematically in Figure A2.5.4. At the center of the flux line, $|\psi(r)|$ rises to its equilibrium bulk value over a distance $\sim\xi(T)$. The local flux density $h(r)$ is a maximum at the center of the core and falls to zero as $\sim e^{-r/\lambda(t)}$ at large distances. The circulating current peaks at a distance $\sim\xi(T)$ and falls off with distance $\sim e^{-r/\lambda(t)}/r$. Crudely speaking, the core of a flux line represents a tube of normal phase with radius ξ imbedded in the superconducting phase. At least for the classical superconductors, this is a reasonable picture, since these materials reside in the dirty limit where the electron mean free path l is small compared to the radius ξ of the vortex core. However, such a picture does not hold any more for the cuprate superconductors, since they are in the clean limit with $l \gg \xi$. In the latter materials, the electronic vortex structure is strongly affected by the Andreev bound states in the core. For clarification of this area, further experimental and theoretical work is needed at present. The role of the d-wave symmetry of the order parameter [7] in the electronic vortex structure of many cuprate superconductors represents another intriguing question which has not yet been answered completely.

For fields in excess of H_{c1}, Abrikosov showed that flux lines penetrate to form a lattice, held apart by their mutual repulsion. On increasing the field, the magnetic fields of the flux lines overlap, leading to weaker variations in the relative values of the internal fields, though the absolute magnitude of the variations in internal field remains approximately constant. In materials with κ close to unity, the interactions between flux lines can become positive, giving rise to an instability in which flux lines clump together in regions at a constant flux density within a matrix of superconducting material free of flux lines. This is known as the 'intermediate mixed' state.

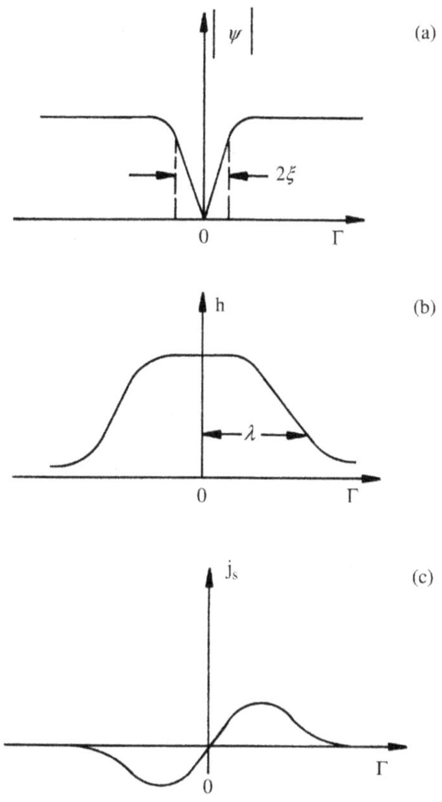

FIGURE A2.5.4 Structure of an isolated magnetic flux line carrying a single flux quantum. (a) Amplitude $|\psi|$ of the superconducting wave function, (b) local magnetic flux density h, and (c) supercurrent density j_s versus the distance from the vortex axis.

The following are useful relations for the lower and upper critical fields in a type II superconductor:

$$H_{c1} \approx \frac{\Phi_0}{4\pi\mu_0\lambda^2}\ln\kappa \approx \frac{H_c}{\sqrt{2}\kappa}\ln\kappa \quad \text{(A2.5.12)}$$

$$H_{c2} = \frac{\Phi_0}{2\pi\mu_0\xi^2} = \sqrt{2}\kappa H_c. \quad \text{(A2.5.13)}$$

An electric transport current passing through a superconductor residing in the mixed or intermediate state always generates a Lorentz force acting on the magnetic flux structure. An induced motion of the flux structure results in an emf and dissipation. The superconducting properties of a type II superconductor can only be maintained if the flux structure is purposely pinned. Investigating the pinning of flux lines and the development of metallurgical microstructures to produce efficient pinning is essential for the successful application of both conventional and cuprate superconductors for magnets and other power engineering applications.

A2.5.4 Electronic Vortex Core Structure in the Cuprates: Recent Developments

Since the cuprate superconductors are in the clean limit, the electronic structure of the core of an isolated vortex is characterized by discrete energy levels ε_n of the quasiparticles given by

$$\varepsilon_n = (n + \tfrac{1}{2})\frac{\Delta^2}{E_F} = (n + \tfrac{1}{2})\frac{2\hbar^2}{m\xi^2} \qquad \text{(A2.5.14)}$$

Here n is an integer, m the quasiparticle mass, Δ the superconducting energy gap, and E_F the Fermi energy. This has been shown by Caroli et al. [8], and later by Bardeen et al. [9]. In the clean limit, the energy smearing $\delta\varepsilon = \hbar/\tau < \Delta$ is small compared to the superconducting energy gap Δ (τ = mean electron scattering time). We see from Equation (A2.5.14) that there exists a minigap $\varepsilon_0 = \Delta^2/2E_F$ between the Fermi energy and the lowest bound state in the vortex core. Both the minigap and the level separation $\varepsilon_n - \varepsilon_{n-1}$ are proportional to ξ^{-2}. Because of the relatively large coherence length in the classical superconductors (typically $\xi \approx 100$ nm), this electronic quantum structure of the vortex core is negligible, and the core can be treated as an energetic continuum of quasiparticles, i.e., as a cylinder of normal material with radius ξ. However, in the cuprate superconductors, the coherence length can be as small as 1–2 nm. Hence, the minigap and the level separation can be up to 10^4 times larger than in the classical superconductors. A detailed theory of the physics of the quasiparticle excitations in the core of isolated 'pancake vortices' has been presented by Rainer et al. [10]. The pancake structure of the vortices in the cuprates results from the strong anisotropy of these materials, where the superconductivity resides within the CuO_2 planes.

At sufficiently high magnetic fields, where the intervortex distance becomes small, and the interaction between vortices becomes appreciable, the discrete energy levels in the vortex cores develop into minibands (similar to the concepts of the approximation of bound electrons in a crystal, introduced by Felix Bloch in 1928).

Experimentally, the electronic quantum structure of the vortex core in the cuprates has been investigated by scanning tunneling microscopy. The first results have been reported by Maggio-Aprile et al. for single crystals of $YBa_2Cu_3O_{7-\delta}$ with B oriented along the c-axis [11]. However, subsequently, these results were shown to be unrelated to vortices [12]. Further details and references can be found in [13–15].

The implications of the electronic vortex core structure in the cuprates in the case of nonlinear effects during flux motion at high vortex velocities represent an interesting subject, and an early summary can be found in [15].

References

[1] Pippard A B 1950 *Proc. R. Soc. A* **203** 210
[2] Bardeen J, Cooper L N and Schrieffer J R 1957 *Phys. Rev.* **108** 1175
[3] Pippard A B 1953 *Proc. R. Soc. A* **216** 547
[4] Ginzburg V L and Landau L D 1950 *Zh. Eksp. Teor. Fiz.* **20** 1064
[5] Landau L D 1937 *Zh. Eksp. Teor. Fiz.* **7** 371
[6] Abrikosov A A 1957 *Zh. Eksp. Teor. Fiz.* **32** 1442
[7] Tsuei C C and Kirtley J R 2000 *Rev. Mod. Phys.* **73** 969
[8] Caroli C, de Gennes P G and Matricon J 1964 *Phys. Lett.* 9 307
[9] Bardeen J, Kümmel R, Jacobs A E and Tewordt L 1969 *Phys. Rev.* **187** 556
[10] Rainer D, Sauls J A and Waxman D 1996 *Phys. Rev.* **B54** 10094
[11] Maggio-Aprile I, Renner Ch, Erb A, Walker E and Fischer O (1995) *Phys. Rev. Lett.* **75** 2754
[12] Bruér J, Maggio-Aprile I, Jenkins N, Ristic Z, Erb A, Berthod C, Fischer O and Renner C 2016 *Nat. Commun.* 7 11139
[13] Fischer O, Kugler M, Maggio-Aprile I and Berthod C 2007 *Rev. Mod. Phys.* **79** 353
[14] Berthod C, Maggio-Aprile I, Bruér J, Erb A and Renner C 2017 *Phys. Rev. Lett.* **119** 237001
[15] Huebener R P 2001 *Magnetic Flux Structures in Superconductors*, 2nd edition, Springer

Further Reading

Brandt E H 1995 *The Flux-Line Lattice in Superconductors* Rep. Prog. Phys. 58, 1465–1594
Huebener R P 2001 *Magnetic Flux Structures in Superconductors 2nd edition* (Berlin: Springer)
Saint-James D, Sarma G, Thomas E J 1969 *Type II Superconductivity* (New York: Pergamon)
Shoenberg D 1965 *Superconductivity* (Cambridge: Cambridge University Press)
Tinkham M 1996 *Introduction to Superconductivity* (New York: McGraw Hill)
Waldram J R 1996 *Superconductivity of Metals and Cuprates* (Bristol: Institute of Physics Publishing)

A2.6

High-Frequency Electromagnetic Properties

Adrian Porch, Enrico Silva, and Ruggero Vaglio

A2.6.1 Introduction

Measurements of the electromagnetic response of superconductors in the microwave spectrum (for the purposes of this chapter defined to be in the range from 100 MHz to 500 GHz) and in the infrared spectrum (> 500 GHz) allow the simultaneous investigation of the dynamics and energy states of both paired carriers and unpaired carriers (quasiparticles) in a superconductor. This allows information to be deduced regarding the nature of the superconducting pairing state, the energy gap, the dynamics of quasiparticles and the effects of defects and impurities [1–3]. High-frequency measurements of conventional (i.e. low T_c) superconductors [4] provided the springboard for the BCS theory, and later confirmation of many of the theory's predictions [1–3]. Measurements in static applied magnetic fields [5, 6] probe the dynamics of flux lines, yielding flux line viscosities, pinning strengths and information regarding quasiparticle states within the flux line cores.

While the electrodynamics of conventional superconductors are well understood, there are many open issues regarding HTS materials (including Fe-based superconductors) that warrant further investigation using high-frequency techniques. Examples of these are the origins of the nonlinear effects at high microwave field levels (discussed in detail in Chapter G2.9), of direct relevance to potential HTS microwave device applications (see Section H2), and fundamental issues regarding the symmetry and mechanism of the pairing in HTS materials.

A2.6.2 Surface Impedance, Complex Conductivity and the Two-Fluid Model

The response of a superconducting sample at microwave frequencies is characterized by its *surface impedance*, which is a function of the electrical conductivity, and can be measured by a number of standard microwave techniques (see Chapter G2.9). At infrared frequencies, the conductivity is extracted directly from measurements of the power transmitted through very thin film samples or reflected from thicker samples [7, 8].

When a high-frequency electromagnetic wave falls on the surface of a good conductor, the electromagnetic fields are confined within a shallow *skin depth*, which for a superconductor is the magnetic penetration depth λ. If the conductor is flat on the scale of the skin depth, the *surface impedance* Z_s of the conductor is defined to be

$$Z_s = \frac{E}{H} = R_s + iX_s \qquad \text{(A2.6.1)}$$

where E and H are the magnitudes of the tangential electric and magnetic fields, respectively, at the surface. Physically, the *surface resistance* of the conductor R_s quantifies the rate of energy dissipation within the skin depth, while the *surface reactance* X_s quantifies the peak electromagnetic energy stored within the skin depth. Consequently, the measurement of these parameters is relevant to high-frequency applications (see Section H2).

The general theory of the surface impedance of superconductors in low-amplitude ac fields was developed by Mattis and Bardeen [9, 10] using the BCS theory. The full theory is complicated and beyond the scope of this handbook. However, in certain limits, simple results emerge depending on the relative sizes of the superconducting coherence length ξ, the quasiparticle mean free path l and λ; typical values of ξ and λ are listed in Table A2.6.1.

Of most relevance are the *local limits* (i.e. $\lambda > l$) of the Mattis–Bardeen theory, of which there are two. The *clean local limit* (often called the *London limit*) is attained when $\lambda > l > \xi$. In this limit, the supercurrent density J_s (i.e. that part of the current associated with paired carriers) at some point in the superconductor is related to the electric field **E** *at the same point* by the first London equation

$$\frac{\partial \mathbf{J}_s}{\partial t} = \frac{1}{\mu_0 \lambda_L^2}\mathbf{E} \qquad \text{(A2.6.2)}$$

TABLE A2.6.1 Typical Values of the Material Parameters of a Selection of Superconducting Materials

Material	$\lambda(0)$ (nm)	$\xi(0)$ (nm)	$2\Delta(0)/kT_c$	T_c(K)
Al	16	1500	3.4	1.2
In	25	400	3.5	3.3
Sn	28	300	3.6	3.7
Pb	28	110	4.1	7.2
Nb	32	39	3.7	9.2
Nb_3Sn	50	5	4.4	18
$YBa_2Cu_3O_7$ (*ab*-plane)	140	2	Unclear	93

where the subscript '*L*' is used to denote the London limit[1]. From Table A2.6.1 it can be seen that metals like Pb and Nb fall outside this limit and require a rigorous solution of the Mattis–Bardeen theory. However, HTS materials with *ab*-plane currents fall within the London limit and exhibit local electrodynamics; the same is also true for A-15 superconductors like Nb_3Sn. Consequently, the high-frequency electrodynamics of both the main classes of superconducting materials for microwave applications (i.e. HTS and A-15) can be modelled quite simply using the first London equation. The other local limit is the *dirty local limit*, attained when $\lambda > l \approx \xi$; Equation (A2.6.2) still applies, but with λ_L^2 replaced by $\lambda_{\text{dirty}}^2 = \lambda_L^2(1+\xi/l)$. This limit is appropriate for alloy superconductors.

The effect of the electric field within the penetration depth is to accelerate both quasiparticles and pairs, so that in either of the local limits the current density is the sum of the quasiparticle current and supercurrent densities (denoted $\mathbf{J}_n$ and $\mathbf{J}_s$, respectively). Applying Equation (A2.6.2) for an incident electromagnetic wave of angular frequency ω gives

$$\mathbf{J} = \mathbf{J}_n + \mathbf{J}_s = (\sigma_1 - i\sigma_2)\mathbf{E} \qquad \text{(A2.6.3)}$$

where the conductivity of the superconductor is a complex, frequency- and temperature-dependent quantity

$$\sigma(\omega,T) = \sigma_1(\omega,T) - i\sigma_2(\omega,T). \qquad \text{(A2.6.4)}$$

The real part of Equation (A2.6.4), σ_1 is associated with the response of the quasiparticles, and the imaginary part $\sigma_2 = 1/\omega\mu_0\lambda^2$ with the response of the supercurrent. This description is called the *two-fluid model* and is appropriate for local electrodynamics ($\lambda > l$).

The surface impedance of any good conductor with a local conductivity σ can be derived using the *classical skin effect* theory

$$Z_s = R_s + iX_s = \sqrt{\frac{i\omega\mu_0}{\sigma}}. \qquad \text{(A2.6.5)}$$

Unsurprisingly, for a superconductor at microwave frequencies and temperature not too close to T_c, it is found that $\sigma_2 >> \sigma_1$, in which case, expanding Equation (A2.6.5) to the first order in σ_1/σ_2 yields the following results for the surface resistance and surface reactance, respectively,

$$R_s \approx \sqrt{\frac{\mu_0\omega}{\sigma_2}}\frac{\sigma_1}{2\sigma_2} \equiv \frac{1}{2}\omega^2\mu_0^2\lambda^3\sigma_1,\; X_s \approx \sqrt{\frac{\mu_0\omega}{\sigma_2}} \equiv \omega\mu_0\lambda. \qquad \text{(A2.6.6)}$$

Unlike a normal metal where $R_s \propto \sqrt{\omega}$, for a superconductor $R_s \propto \omega^2$, this result has considerable importance for microwave applications, as discussed in Section H2. Equation (A2.6.6) also states that a superconductor is only truly lossless (i.e. $R_s = 0$) at non-zero frequencies when the temperature is reduced to absolute zero, since then $\sigma_1 = 0$.

A2.6.3 The Conventional (Mattis–Bardeen) Picture of the High-Frequency Conductivity

The conductivity $\sigma(\omega,T)$ is most easily calculated in the dirty local limit [9, 10], and some numerical results for σ/σ_n are shown in Figure A2.6.1, where σ_n is the corresponding (real) conductivity of the normal metal at the same frequency and temperature.

The dirty limit conductivity ratios of Figure A2.6.1 can be qualitatively understood in terms of the energy gap Δ in the quasiparticle density of states. The gap frequency is defined as $\omega_g = 2\Delta/\hbar$, of the order of 1 THz for conventional superconductors at low temperatures. Referring to Figure A2.6.1(a), $\sigma_1 = 0$ when $T = 0$ for frequencies $\omega < \omega_g$ since there are no quasiparticles present under these conditions. However, when $\omega > \omega_g$, σ_1 increases rapidly even at $T = 0$, corresponding to an energy loss mechanism as a result of the incident infrared photons creating pairs of quasiparticles from the ground state. At very high frequencies, σ_1 approaches σ_n.

At low temperatures when $\omega < \omega_g$, the presence of the energy gap ensures that $\sigma_1 \propto \exp(-\Delta/kT)$ (Figure A2.6.1[b]), and σ_1 exhibits a broad peak just below T_c as a result of two characteristics of the BCS theory: the enhanced density of states at the gap edge and the *coherence factors* associated with electromagnetic absorption, the latter feature often called the *coherence peak*.

For frequencies well below ω_g, it is found that $\sigma_2 \propto 1/\omega$ for all limits of the Mattis–Bardeen theory, so that λ is approximately independent of frequency. In the normal state $\sigma_2 = 0$, and there is a rapid rise in σ_2 on entering the superconducting state which ensures that $\sigma_2 >> \sigma_1$, apart from at temperatures very close to T_c (Figure A2.6.1[c]), and which causes a precipitous drop in surface resistance. The quantity $\sigma_2(0) - \sigma_2(T)$ is called the quasiparticle *backflow*, where $\sigma_2(0) - \sigma_2(T) \propto \exp(-\Delta/kT)$ at low temperatures, thus giving rise to $\lambda_L(T) - \lambda_L(0) \propto \exp(-\Delta/kT)$; this is closely linked to the low T form of $\sigma_1(T)$, when also $\sigma_1(T) \propto \exp(-\Delta/kT)$.

The conductivity in the London limit can be expressed in terms of the Mattis–Bardeen dirty limit conductivity ratios

$$\left(\frac{\sigma}{\sigma_n}\right)_{\text{London}} = \left(\frac{\sigma_1}{\sigma_n}\right)_{\text{dirty}} - i\left(\frac{\sigma_2}{\sigma_n}\right)_{\text{dirty}}. \qquad \text{(A2.6.7)}$$

[1] Mean free paths vary significantly with sample quality and are not quoted, but a large value of λ/ξ is usually sufficient to push the material into the local limit.

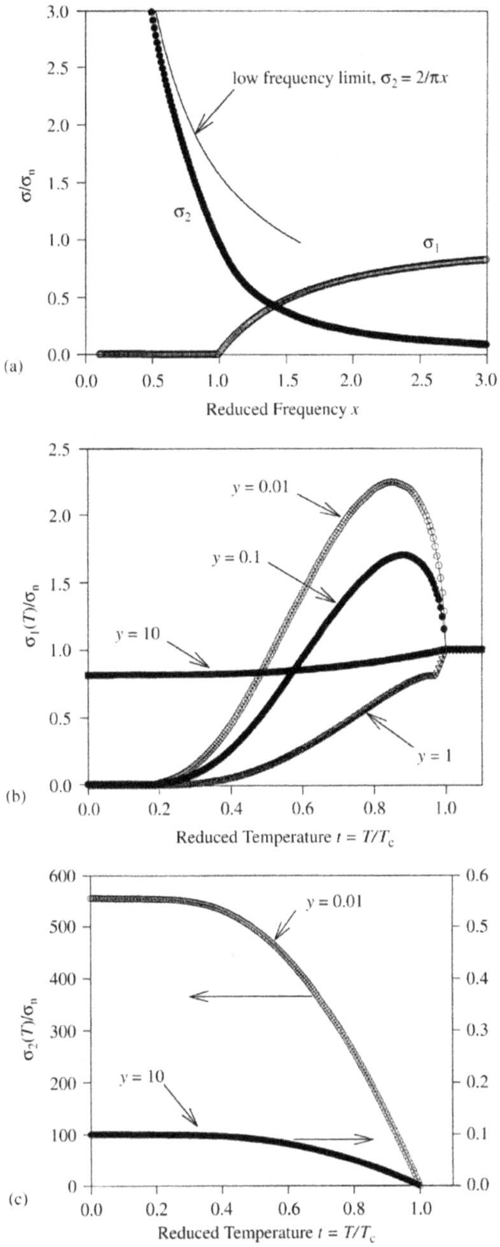

FIGURE A2.6.1 (a) The dirty limit conductivity ratios at $T = 0$ as a function of reduced frequency $x = \hbar\omega/2\Delta$. Not shown is the δ-function in σ_1 at zero frequency, which arises as a result of the presence of the energy gap Δ in the quasiparticle density of states, and which ensures that $\sigma_2 \propto 1/\omega$ at low frequencies. The solid line is the low-frequency BCS prediction for σ_1. (b) The temperature dependence of σ_1 in the dirty limit for various frequencies $y = \hbar\omega/kT_c$ above and below the gap frequency (which has a BCS value of $y = 3.52$ at low T). (c) The temperature dependence of σ_2 in the dirty limit for various frequencies $y = \hbar\omega/kT_c$.

Since $\sigma_n \propto l$ in the London limit, σ_1 is independent of l, as we expect from Equation (A2.6.2). The effect of increasing l (thus pushing the material from the dirty to the clean limit) is to suppress the coherence peak in σ_1, of relevance to HTS materials.

However, the characteristic low-temperature features of both σ_1 and σ_2 in the dirty limit are retained in the London limit, as are the characteristic frequency dependences.

In the *Pippard* limit ($l,\xi > \lambda$), the electrodynamics are nonlocal (see Chapter A2.5). The limiting result of the Mattis–Bardeen theory for the surface impedance is that calculated using the *anomalous skin effect* theory

$$\frac{Z_s}{Z_n} = \left(\frac{\sigma}{\sigma_n}\right)_{\text{dirty}}^{-1/3} \tag{A2.6.8}$$

where $Z_n \propto (i\omega\mu_0)^{2/3}(\sigma_n/l)^{-1/3}$ is the surface impedance of the normal metal in the extreme anomalous limit (i.e. when l is much greater than the skin depth). A number of pure superconductors with very low T_c (e.g. Al) are in this limit.

The Mattis–Bardeen predictions for $\sigma(\omega,T)$ have been successfully verified by experiments at microwave and infrared frequencies for conventional superconductors under a wide range of conditions [1–3]. The temperature dependence of the surface resistance of a Nb sample is shown in Figure A2.6.2, where at low temperatures $R_s \propto \sigma_1\omega^2 \propto \omega^2 \exp(-\Delta/kT)$. In practice, R_s reaches a limiting value $R_{s,res}$ at very low temperatures, the "residual surface resistance", which is highly dependent on sample purity and microstructure, in addition to frequency and applied magnetic field. In high-quality Nb samples, it is possible to reduce $R_{s,res}$ to a few nΩ at 10 GHz. A peculiar behavior, also at high frequencies, is exhibited by magnesium diboride (MgB_2), due to its two-gap nature [11]. In MgB_2, the low-temperature experimental results appear in agreement with the simple BCS behavior, but with a very small gap value. This is explained by the fact that the high-frequency properties at low temperature depend on the smallest-energy excitation, so they are sensitive only to the smallest gap [12].

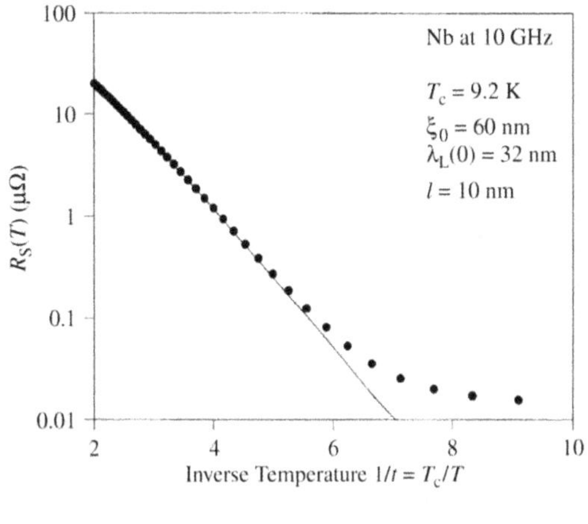

FIGURE A2.6.2 The surface resistance at 10 GHz for a pure Nb sample as a function of T_c/T at low temperatures. The residual resistance of this sample is 12 nΩ. The solid line is the fit to $\exp(-\Delta/kT)$, with $2\Delta(0) = 3.7\ kT_c$.

A2.6.4 The High-Frequency Conductivity of HTS Materials

Without an accepted pairing mechanism to describe the superconductivity in HTS materials, it is impossible to predict the precise form of the high-frequency conductivity $\sigma(\omega,T)$. However, since these materials are in the London limit, it is straightforward to determine $\sigma(\omega,T)$ experimentally from measurements of the surface impedance, followed by inversion of Equation (A2.6.6).

There are a number of reasons why the conductivity of HTS materials should differ from that predicted by Mattis–Bardeen theory, even if the pairing mechanism is BCS-like (i.e. phonon mediated with s-wave pairing). Any anisotropy of the energy gap would cause rounding of the quasiparticle density of states at the gap edge, thus suppressing the coherence peak in σ_1 just below T_c; similar effects occur as a result of strong phonon coupling, and also due to the fact that l is strongly temperature dependent, increasing very rapidly below T_c and pushing HTS materials well into the London limit.

There is conclusive experimental evidence to suggest that hole-doped HTS materials exhibit an order parameter with $d_{x^2-y^2}$ symmetry ('d-wave pairing') [13, 14]. This opens up an interesting new set of phenomena at low temperatures owing to the presence of line nodes in the energy gap on the Fermi surface. Such a symmetry predicts a limiting value of σ_1 at absolute zero [15], independent of the quasiparticle scattering rate

$$\sigma_{00} \approx \frac{ne^2\hbar}{mn\Delta(0)} \approx \frac{\hbar}{2\pi\mu_0\lambda_L^2(0)kT_c}. \qquad \text{(A2.6.9)}$$

This is of the order of $10^5\Omega^{-1}m^{-1}$ for $YBa_2Cu_3O_7$. Correspondingly, there is a fundamental limit to the low-temperature value of R_s of around 1 $\mu\Omega$ at 10 GHz; however, this is difficult to distinguish from the sample dependent residual resistance $R_{s,res}$, which is present even in conventional materials. Within the London limit at low temperatures, for *ab*-plane currents the quasiparticle backflow for d-wave symmetry is proportional to T, resulting in $\sigma_1(T) \approx \sigma_{00} + \alpha T$ and $\sigma_2(T) \approx \sigma_2(0) - \beta T$, where α and β are constants. Since $\sigma_2 \propto 1/\lambda^2$, the latter of these results in $\lambda_L(T) - \lambda_L(0) \propto T$ [16, 17]. The behavior in Fe-based superconductors is still uncertain, since the T-linear backflow does not appear to be a general property, even within the same family of compounds [18].

The most appropriate phenomenological description of the *ab*-plane conductivity of HTS materials is a two-fluid model (consistent with the London limit) within the context of d-wave pairing symmetry [19–21]. The quasiparticle dynamics can be described using the simple Drude model, including a temperature-dependent quasiparticle scattering rate $\tau(T) = l(T)/v_F$, where v_F is the Fermi velocity. The temperature dependences of the pair and quasiparticle densities can be introduced in a nonrigorous manner using the dimensionless parameters $x_s(T)$ and $x_n(T)$, respectively, which can be calculated from $\lambda_L(T)$, since $x_s(T) = \sigma_2(T)/\sigma_2(0) = [\lambda_L(0)/\lambda_L(T)]$ and $x_n(T) = 1 - x_s(T)$; the two-fluid conductivity is then [19, 21]

$$\sigma(\omega,T) = \frac{1}{\mu_0\lambda_L^2}\left[\frac{x_n(T)\tau(T)}{1+i\omega\tau(T)} - i\frac{x_s(T)}{\omega}\right] + \sigma_{00}. \qquad \text{(A2.6.10)}$$

Simulated results for the conductivity of $YBa_2Cu_3O_7$ single crystals and thin films for *ab*-plane microwave currents are plotted in Figure A2.6.3, assuming that $\sigma_1(0)$ is very small. The broad peak in $\sigma_1(T)$ is *not* a coherence peak, but is due to the competition between the rapidly increasing $l(T)$ and the rapidly decreasing $x_n(T)$ as the temperature is reduced, since from Equation (A2.6.10) it is found that approximately $\sigma_1(T) \propto x_n(T)l(T)$. For thin films, $l(T)$ is reduced owing to the increased quasiparticle scattering rate due to the greater density of defects compared to high-quality crystals, although it is not possible to associate this behavior to any particular type of defect. This is sufficient to suppress the peak in $\sigma_1(T)$ and shift it to higher temperatures, in addition to changing

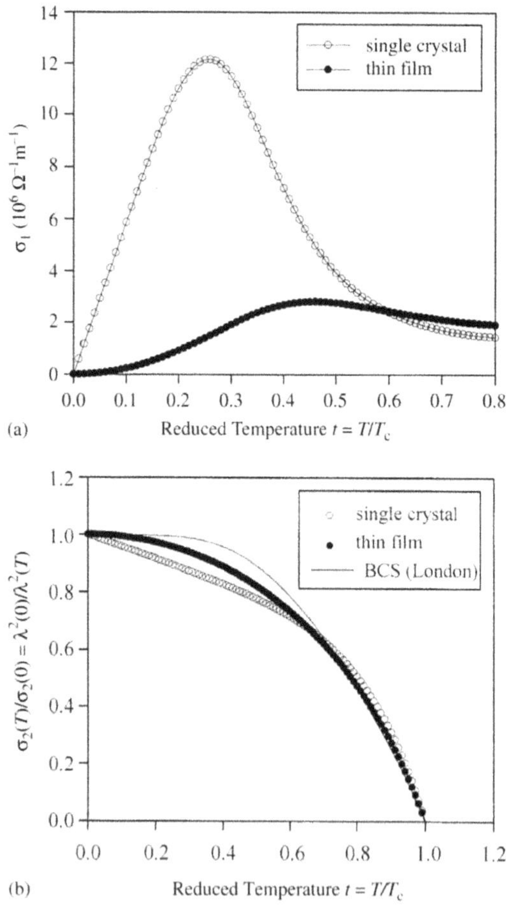

FIGURE A2.6.3 (a) $\sigma_1(T)$ of $YBa_2Cu_3O_7$ single crystals and thin films modelled using Equation (A2.6.10), ignoring the effects of any residual conductivity at $T = 0$. (b) The corresponding $\sigma_2(T)$. The quasiparticle backflow at low T is $\propto T$ for crystals, but $\propto T^2$ for thin films. Also shown are the results expected from the Mattis–Bardeen theory in the London limit.

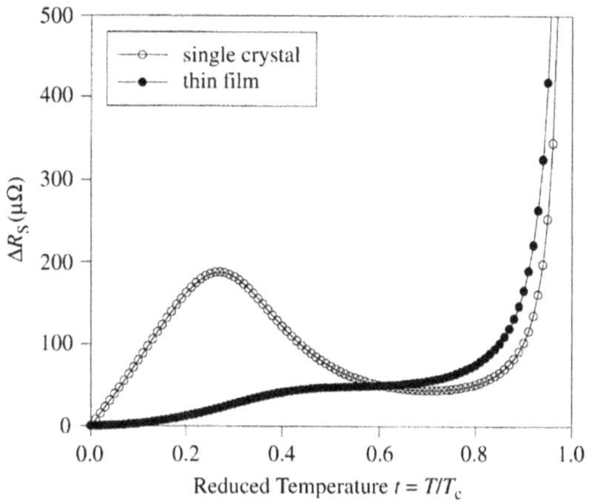

FIGURE A2.6.4 $\Delta R_s(T)$ at 10 GHz for $YBa_2Cu_3O_7$ single crystals and thin films using the conductivities of Figure A2.6.3 after subtracting the residual surface resistance. The peak in $\sigma_1(T)$ is reflected in the surface resistance of the single crystal. For $T < T_c/2$, high-quality thin films have lower surface resistance than high-quality single crystals owing to the reduced quasiparticle mean free path l. Note that the surface resistance values are much higher than those for Nb (Figure A2.6.2) at the same reduced temperature.

the quasiparticle backflow term from $\propto T$ to $\propto T^2$ within the d-wave framework [19, 21].

The corresponding *ab*-plane surface impedance is obtained by substituting the conductivity of Equation (A2.6.10) in Equation (A2.6.6), which is plotted in Figure A2.6.4, having first subtracted the residual resistance at low temperatures. The form of $R_s(T)$ reflects that of $\sigma_1(T)$, where the increased amount of scattering present in thin films can lead to lower values of $R_s(T)$ than those of the best quality crystals at low temperatures, a significant result for thin film microwave applications.

A2.6.5 The Surface Impedance in Large dc Magnetic Fields

Static flux lines are nucleated within a type II superconductor by applying a dc magnetic field whose magnitude exceeds the lower critical field B_{c1}. A low-amplitude, high-frequency applied field will then cause these flux lines to oscillate reversibly about their pinning centers, leading to a redistribution of magnetic energy and additional energy dissipation associated with their motion; both effects can be studied by measuring the surface impedance.

Modelling of the dc flux line response to high-frequency fields was first performed by Gittleman and Rosenblum [22]. Despite its simplicity, this model yields a clear physical picture of the processes occurring and is based on the ac Lorentz force on an individual flux line. For the simple case of parallel dc and ac magnetic fields (ac current density perpendicular to the dc magnetic field), they proposed an equation of motion for individual flux lines of the form $\mathbf{F} = \eta\mathbf{v} + \kappa_p\mathbf{x}$, where $\boldsymbol{F} = \boldsymbol{F}_L$ is the Lorentz force on the flux line due to the microwave current, $\mathbf{v}$ and $\mathbf{x}$ are the flux line velocity and displacement, respectively, η is the flux line viscosity and κ_p is the pinning force constant. The vortex viscosity (sometimes called drag coefficient) represents the power dissipation arising from freely moving vortices subject to the drag force $\eta\mathbf{v}$, due to the nonequilibrium relaxation processes of quasiparticles or Joule dissipation inside the cores [5]. It should be noted that the term $\kappa_p\mathbf{x}$ comes from the simple quadratic (elastic) approximation of the potential well, which under certain circumstances can be an oversimplification [23]. From the force equation above, and the expression of the electric field due to moving fluxons [5, 22], one derives the following expression for the effective high-frequency complex resistivity in a static magnetic field B due to the motion of flux lines:

$$\rho_f = \frac{B\Phi_0}{\eta}\left(1 - i\frac{\omega_p}{\omega}\right)^{-1} \tag{A2.6.11}$$

where Φ_0 is the magnetic flux quantum (Chapter A2.5) and $\omega_p \equiv \kappa_p/\eta$ is called the *depinning frequency*. At intermediate fields, $\mu_0 H_{c1} \ll B \ll \mu_0 H_{c2}$, when this contribution exceeds the quasiparticle and superfluid term, Equation (A2.6.11) can be inserted into the local expression for surface impedance $Z_s = R_s + iX_s = \sqrt{i\omega\mu_0\rho_f}$ to quantify the high-frequency response of the flux lines.

There are several important limiting regimes, in particular, if one includes at least approximately the two-fluid conductivity in the analysis [24]. When the two-fluid contribution is small, at very high frequencies ($\omega > \omega_p$) one enters the *flux flow regime*, which is highly dissipative. The effective resistivity is $\rho_f \approx B\Phi_0/\eta$ (i.e. it is almost real), and the surface resistance and reactance are approximately equal

$$R_s \approx X_s \approx \sqrt{\frac{\omega\mu_0 B\Phi_0}{2\eta}} \tag{A2.6.12}$$

with both R_s and X_s large, and both proportional to $\sqrt{\omega B}$. Surface impedance measurements performed in this flux flow regime are a useful means of determining η.

At low frequencies ($\omega < \omega_p$), one enters the *pinned regime*, in which case the flux line impedance is almost purely reactive and $\rho_f = i\omega B\Phi_0/\kappa_p \equiv i\omega\mu_0\lambda_c^2$, where $\lambda_c = \sqrt{B\Phi_0/\mu_0\kappa_p}$ is the *Campbell penetration depth.* If B is large $X_s \approx \omega\mu_0\lambda_c$, but for smaller values of B (close to B_{c1}), one has to resort to the Coffey–Clem model [25] to take into account the contributions of two-fluid and vortex motion together. At low frequencies, the quasiparticle contribution can be neglected, see Equation (A2.6.10). The surface reactance is then $X_s \approx \omega\mu_0\sqrt{(\lambda^2 + \lambda_c^2)}$, and the surface resistance differs little from its zero field value $R_s(0) \approx \sigma_1\omega^2\mu_0^2\lambda^3/2$. The small surface resistance change $\Delta R_s(B) = R_s(B) - R_s(0)$ on applying a small field B is approximately proportional to $\omega^2 B$; furthermore, in the small field limit, it can be shown that the ratio $r = \Delta X_s(B)/\Delta R_s(B) = \omega_p/\omega$. Therefore, single-frequency measurements of the surface impedance in the pinned regime can be used to determine the depinning frequency $\omega_p = \kappa_p/\eta$. From this η can be found, since κ_p can be determined directly from $X_s(B)$. In practice, it is found that κ_p is itself a function of B at large fields owing to the effects of collective flux pinning [5].

When measurements in extended ranges of temperature, magnetic field and frequency are involved, it is necessary to resort to more complete models that take into account thermally induced phenomena and, in some cases, the comparable weight of the fluxon motion and the quasiparticle/superfluid contributions. The models by Coffey and Clem [25] and by Brandt [26] included the thermal effects, by explicitly writing in the thermal stochastic force or by allowing a phenomenologically relaxing pinning constant, respectively. Both models, and a set of other models with relaxational dynamics, can be cast under the same framework for the vortex motion resistivity [27] as follows

$$\rho_f = \frac{B\Phi_0}{\eta_{eff}}\left(\chi + i\frac{\omega}{\omega_0}\right)\left(1 + i\frac{\omega}{\omega_0}\right)^{-1} \qquad (A2.6.13)$$

where η_{eff} is an effective viscosity, χ is a creep factor and $\omega_0 \approx \omega_p$ when thermal effects are negligible. The frequency dependence of Equation (A2.6.13) is shown in Figure A2.6.5, and the role of the various parameters is depicted. Some simulated results for the frequency dependence of the surface impedance of $YBa_2Cu_3O_7$ in an applied magnetic field of 5 T are shown in Figure A2.6.6. The simulated results are computed in absence of thermal activation effects (Gittleman and Rosenblum model), and with a non-zero creep factor to show the effect of the thermal activation on the measured surface impedance. The Coffey–Clem has been used for the specific

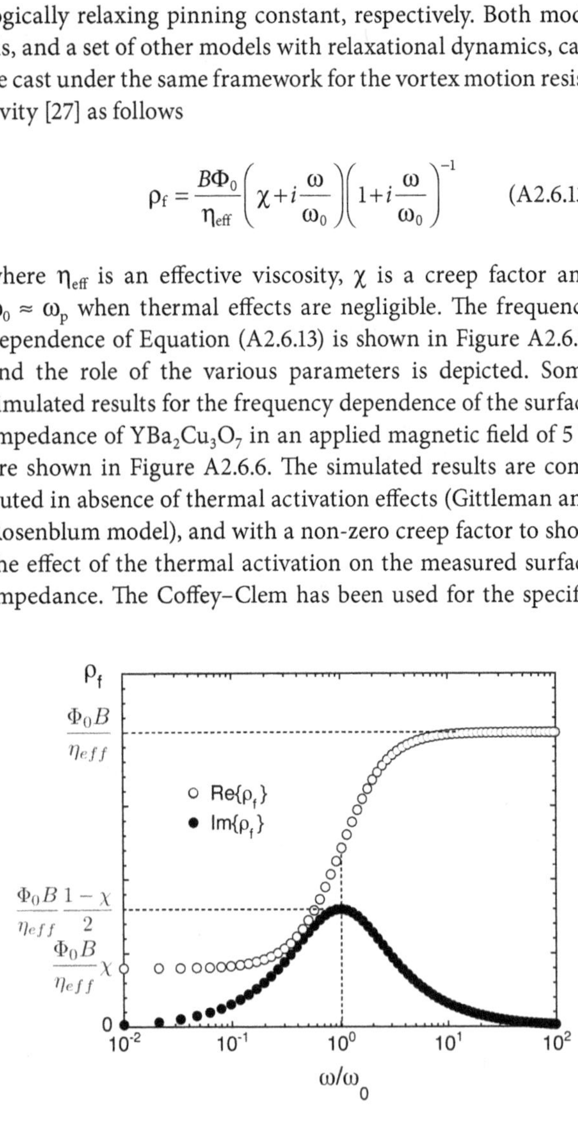

FIGURE A2.6.5 Frequency dependence of the complex resistivity in the general vortex motion model [25–27], Equation (A2.6.13), in terms of normalized units. The width of the bell-shape curve in the imaginary part of ρ_f shows that the crossover at ω/ω_0 from the Campbell or thermally activated regime to flux flow is relatively slow. The main vortex parameters are labelled.

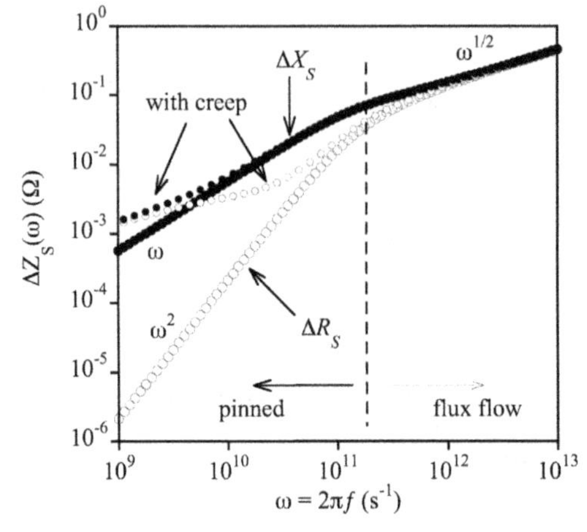

FIGURE A2.6.6 The change in surface resistance ΔR_s and surface reactance ΔX_s as a function of frequency on applying a static magnetic field of magnitude 5 T calculated without flux creep (Gittleman and Rosenblum model [22], large symbols), and with flux creep (Coffey and Clem model [25], small symbols); here $\kappa_p = 4 \times 10^4$ Nm^{-2} and $\eta = 3 \times 10^{-7}$ Nsm^{-2}, appropriate for $YBa_2Cu_3O_7$ thin films around 60 K [5], yielding a depinning frequency $\omega_p = 1.3 \times 10^{11}$ $s^{-1} \equiv 21$ GHz. Small symbols: simulation with the same parameters, with a creep factor $\chi = 0.1$. Qualitatively similar results are obtained for conventional superconductors, but with lower depinning frequencies [22]. The characteristic frequency dependences above and below the depinning frequency are labelled. The low-frequency results are affected significantly by even a small flux creep contribution.

relations between ω_p, ω_0 and χ, the Brandt model would give only slightly different results (see [27] for the explicit formulas). The pinned regime (low frequencies) is most influenced [26], thus affecting the characteristic frequency dependences. It should be emphasized that Equation (A2.6.13) represents many models [27]. To assess the validity of a specific model one should analyze from the experimental data the specific field or temperature dependences of the parameters involved (η, ω_p, χ). Finally, it must be mentioned that at low frequencies, a glassy dynamics can set in, at least in HTS [28], so that the very same validity of Equation (A2.6.13) can be questioned. Depinning frequencies for conventional superconductors are typically < 100 MHz [22], while those for HTS materials are typically in the range 10–100 GHz [5, 6, 27] due to large values of κ_p and small values of η. Again, a general behavior is not yet recognized in Fe-based HTS, with values of $\omega_p/2\pi \approx 3$ GHz in LiFeAs [29] but larger in other compounds. HTS exhibits then a wide range of vortex dynamics depending on the measuring frequency: standard 10 GHz measurements can be in the flux flow limit in some materials, and then the presence of flux creep is irrelevant (see Figure A2.6.6), or they can be in the pinned regime, where the dynamics is severely affected by the flux creep.

Measurements of the flux line viscosity are of fundamental importance since they probe the quasiparticle dynamics within the flux line cores. The simplest model for flux line dissipation is that of Bardeen and Stephen [5], which predicts that the normal-state core conductivity is $\sigma_n = \eta\Phi_0 / B_{c2}$. This is valid for a continuum of core quasiparticle energy states, and measurements of η and σ_n for conventional superconductors agree well with this model. However, measurements of η in HTS materials with fields applied parallel to the *c*-axis imply that at low temperatures, η exceeds 10^{-6} Nsm^{-2} [5] putting them in the so-called *superclean limit*; here the quasiparticle scattering time τ within the cores is long enough for the core energy states to become discrete, and the Bardeen–Stephen model becomes invalid (since this assumes a continuum of core states, as in the bulk metal). In some Fe-based materials, $\eta \sim 10^{-7}$ Nsm^{-2} at low temperature, giving rise to the intermediate *moderately clean* limit. The quasiparticle dynamics are further complicated by the possibility of *d*-wave pairing and the necessity to include a Hall effect term in the flux line equation of motion. In MgB_2, there is the additional need to include the sensitivity of the small band to the magnetic field, which implies that the field-induced quasiparticle contribution cannot be neglected [30]. Needless to say, despite the continued effort to derive reliable phenomenological models for the high-frequency vortex motion response [22, 25–27], a complete microscopic description of the high-frequency dynamics of flux lines in HTS materials at low temperatures has yet to have been established.

References

[1] Waldram J R 1964 Surface impedance of superconductors *Adv. Phys.* **13** 1

[2] Halbritter J 1974 On surface resistance of superconductors *Z. Phys.* **238** 466

[3] Klein O, Nicol E J, Holczer K and Grüner G 1994 Conductivity coherence factors in the conventional superconductors Nb and Pb *Phys. Rev. B* **50** 6307

[4] Pippard A B 1953 An experimental and theoretical study of the relation between magnetic field and current in a superconductor *Proc. R. Soc. A* **216** 547

[5] Golosovsky M, Tsindlekht M and Davidov D 1996 High frequency vortex dynamics in $YBa_2Cu_3O_7$ *Supercond. Sci. Technol.* **9** 1

[6] Maeda A, Kitano H and Inoue R 2005 Microwave conductivities of high-T_c oxide superconductors and related materials *J. Phys. Condens. Matter* **17** R143

[7] Palmer L H and Tinkham M 1968 Far infrared absorption in thin superconducting lead films *Phys. Rev.* **165** 588

[8] Holmes C C, Kamal S, Bonn D A, Liang R, Hardy W N and Clayman B P 1998 Determination of the condensate from optical techniques in unconventional superconductors *Physica C* **230** 230–240

[9] Mattis D C and Bardeen J 1958 The theory of the anomalous skin effect in normal and superconducting metals *Phys. Rev.* **111** 412

[10] Turneaure J P, Halbritter J and Schwettman H A 1991 The surface impedance of superconductors and normal metals: the Mattis–Bardeen theory *J. Supercond.* **4** 341

[11] Giubileo F, Roditchev D, Sacks W, Lamy R, Thanh D X, Klein J, Miraglia S, Fruchart D, Marcus J and Monod P 2001 Two-gap state density in MgB_2: a true bulk property or a proximity effect? *Phys. Rev. Lett.* **87** 177008; Gonnelli R S, Daghero D, Ummarino G A, Stepanov V A, Jun J, Kazakov S M and Karpinski J 2002 Direct evidence for two-band superconductivity in MgB2 single crystals from directional point-contact spectroscopy in magnetic fields *Phys. Rev. Lett.* **89** 247004

[12] Jin B B, Klein N, Kang W N, Hyeong-Jin Kim, Eun-Mi Choi, Sung-Ik Lee, Dahm T and Maki K 2002 Energy gap, penetration depth, and surface resistance of MgB_2 thin films determined by microwave resonator measurements *Phys. Rev. B* **66** 104521

[13] Kirtley J R, Tsuei C C, Sun J Z, Chi C C, Yujahnes L S, Gupta A, Rupp M and Ketchen M B 1995 Symmetry of the order parameter in the high T_c superconductor $YBa_2Cu_3O_7$ *Nature* **373** 6511

[14] Tsuei C C, Kirtley J R, Ren Z F, Wang J H, Raffy H and Li Z Z 1997 Pure d_{x2-y2} order parameter symmetry in the tetragonal superconductor $Tl_2Ba_2CuO_6$ *Nature* **387** 6632

[15] Lee P A 1993 Localised states in a d-wave superconductor *Phys. Rev. Lett.* **71** 1887

[16] Kamal S, Ruixing Liang, Hosseini A, Bonn D A and Hardy W N 1998 Magnetic penetration depth and surface resistance in ultrahigh purity crystals *Phys. Rev. B* **58** R8933

[17] Panagopoulos C, Cooper J R, Xiang T, Peacock G B, Gameson I and Edwards P P 1997 Probing the order parameter and the *c*-axis coupling of high-T_c cuprates by penetration depth measurements *Phys. Rev. Lett.* **79** 2320

[18] Hashimoto K, Yamashita M, Kasahara S, Senshu Y, Nakata N, Tonegawa S, Ikada K, Serafin A, Carrington A, Terashima T, Ikeda H, Shibauchi T and Matsuda Y 2010 Line nodes in the energy gap of superconducting $BaFe_2(As_{1-x}P_x)$ single crystals as seen via penetration depth and thermal conductivity *Phys. Rev. B* **81** 220501(R

[19] Bonn D A, Kamal S, Kuan Zhang, Ruixing Liang, Baar D J, Klein E and Hardy W N 1994 Comparison of the influence of Ni and Zn impurities on the electromagnetic properties of $YBa_2Cu_3O_7$ *Phys. Rev. B* **50** 4051

[20] Waldram J R, Theopistou P, Porch A and Cheah H M 1997 Two-fluid interpretation of the microwave conductivity of $YBa_2Cu_3O_{7-\delta}$ *Phys. Rev. B* **55** 3222

[21] Fink H J 1998 Residual and intrinsic surface resistance of $YBa_2Cu_3O_7$ *Phys. Rev. B* **58** 9415

[22] Gittleman J I and Rosenblum B 1966 Radio frequency resistance in the mixed state for subcritical currents *Phys. Rev. Lett.* **16** 734

[23] Embon L, Anahory Y, Suhov A, Halbertal D, Cuppens J, Yakovenko A, Uri A, Myasoedov Y, Rappaport M L, Huber M E, Gurevich A and Zeldov E 2015 Probing dynamics and pinning of single vortices in superconductors at nanometer scales *Sci. Rep.* **5** 7598

[24] Calatroni S and Vaglio R 2017 Surface resistance of superconductors in the presence of a dc magnetic field: frequency and field intensity limits *IEEE Trans Appl. Supercond.* **27** 3500506; Corrections 2017 *ibid.* **27** 9700501

[25] Coffey M W and Clem J R 1991 Unified theory of effects of vortex pinning and flux creep upon the rf surface impedance of type II superconductors *Phys. Rev. Lett.* **67** 386

[26] Brandt E H 1991 Penetration of ac fields into type II superconductors *Phys. Rev. Lett.* **67** 2219

[27] Pompeo N and Silva E 2008 Reliable determination of vortex parameters from measurements of the microwave complex resistivity *Phys. Rev. B* **78** 094503

[28] Wu Dong-Ho, Booth J C and Anlage S M 1995 Frequency and field variation of vortex dynamics in $YBa_2Cu_3O_{7-\delta}$ *Phys. Rev. Lett.* **75** 525

[29] Okada T, Takahashi H, Imai Y, Kitagawa K, Matsubayashi K, Uwatoko Y and Maeda A 2012 Microwave surface-impedance measurements of the electronic state and dissipation of magnetic vortices in superconducting LiFeAs single crystals *Phys. Rev. B* **86** 064516

[30] Sarti S, Amabile C, Silva E, Giura M, Fastampa R, Ferdeghini C, Ferrando V and Tarantini C 2005 Dynamic regimes in MgB_2 probed by swept-frequency microwave measurements *Phys. Rev. B* **72** 024542

Further Reading

Bonn DA *et al.* 1996 Surface Impedance Studies of YBCO *Proc. 21st Conf. on Low Temperature Physics, Czech. J. Phys.* **46** (S6) 3195–3202

A summary of the pioneering surface impedance studies of HTS crystals undertaken at the University of British Columbia, including the effects of impurities

Portis A M 1993 *Lecture Notes in Physics—Vol. 48: Electrodynamics of High-Temperature Superconductors* (Singapore: World Scientific)

Discusses all aspects of microwave studies of HTS. Comprehensive discussions of nonlinear effects at high field amplitudes.

Tinkham M 1996 *Introduction to Superconductivity* 2nd edn, (New York: McGraw-Hill)

The classic educational text on superconductivity, recently updated to include HTS and more recent developments in conventional superconductivity. Thorough discussions of BCS theory and dirty limit conductivity ratios.

Waldram J R 1996 *Superconductivity of Metals and Cuprates* (Bristol and Philadelphia: Institute of Physics Publishing)

An excellent, topical account of all aspects of superconductivity, with much discussion of the high frequency properties. Contains particularly good accounts of both the theoretical and experimental results of conventional materials, together with comparisons with HTS.

A2.7

Flux Quantization

Colin Gough

A2.7.1 Introduction

The quantization of magnetic flux is a direct manifestation of the macroscopic quantum description of the superconducting state. Flux quantization is a defining property of any superconductor, as is the closely related Meissner effect (see Chapter A2.4).

Almost as an afterthought, while discussing the quantum-mechanical, wave-like properties of superconductors, London (1950), in his monograph on Superfluids [1], added a footnote predicting that flux in a superconducting ring would be quantized in units of h/q, where q was the charge of the superconducting electrons. London assumed that q was equal to the single electron charge. The subsequent BCS theory [2] showed that superconductivity involved paired electron states (Cooper pairs). Flux is therefore quantized in units of

$$\phi_0 = h/2e\left(2.07\times10^{-15}\,\mathrm{Tm}^2 \text{ or in equivalent units of Vs}\right) \tag{A2.7.1}$$

The flux Φ within a thick superconducting ring or cylinder is therefore given by

$$\Phi = \int_{\text{area}} B.ds = n\phi_0 \tag{A2.7.2}$$

where n is an integer.

Although London assumed that the quantization of flux was of purely academic interest, worthy only of a footnote, this property now underpins many of the most important device applications of superconductors. These include:

a. superconducting quantum interference devices (SQUIDs), where the response to a magnetic field is periodic in the flux quantum (see Chapter H3.2);
b. various versions of single flux quantum logic (SFQ), where the quantum of flux in a superconducting ring is used as the elementary bit of information (see Chapters H3.5 and H3.6); and
c. the primary voltage standard, where the volt is now defined in terms of a frequency, $f = V/\phi_0$, equivalent to the number of flux quanta per second passing through a Josephson junction, where V is the voltage across the junction. This is known as the ac Josephson relation [3], where the frequency $f = 484$ MHz μV^{-1}. Using an array of superconducting junctions in series, the frequency and, hence, the voltage can be measured to very high precision (see Chapters A2.8 and H5.1).
d. flux quanta as the Q-bits used in current embodiments of quantum computing (see Chapter H3.7).

The observation of quantized flux in units of $h/2e$ provided the first convincing evidence for the pairing of electrons in conventional superconductors [4, 5]. A similar experiment using an early ceramic $YBa_2Cu_3O_8$ sample [6] confirmed electron pairing in the cuprate high-temperature superconductors. Because such superconductors are believed to have d-wave symmetry, superconducting circuits that result in quantized flux trapping with $\Phi = (n + 1/2)\phi_0$ can be devised. The observation of trapped states involving half a flux quantum provides the most convincing demonstration to date for d-wave superconductivity in the cuprate high-temperature superconductors [7–9].

In 1955, Abrikosov showed that the magnetic state of type II superconductors involved singly quantized flux lines penetrating the bulk of the material [10]. At the centre of the flux line, the superconducting order parameter is reduced to zero, so that states involving a circulating vortex of current can exist within the bulk, the flux line core essentially acting as line singularity around which the quantized current can flow. The flux associated with the circulating current is ϕ_0, so that their number density is $B/\phi_0 m^{-2}$. In many cuprate high-temperature superconductors, the coupling between the superconducting CuO_2 planes is so weak that they are essentially two-dimensional. Flux penetrates such layers in the form of two-dimensional vortices, or flux pancakes, again involving a single flux quantum ϕ_0. As the interlayer coupling is increased, the flux pancakes align along the field direction to form the equivalent of a conventional flux line in a three-dimensional, but strongly anisotropic type II superconductor (see Chapter A3.1).

Single electron flux quantum states were later found in normal metals via the Aaronov–Bohm effect [11].

A2.7.2 Theory

A2.7.2.1 Quantization of Flux in a Ring or Cylinder

Flux quantization is a direct consequence of a macroscopic wave-mechanical description of the superconducting state. This was first suggested by London and was further developed by Abrikosov [10] within the context of the Ginsburg–Landau (GL) theory [12], based on Landau's earlier model of second-order phase transitions. In such a model, the superconducting state is described by a macroscopic wave function $\psi(r)e^{i\theta(r)}$ where $|\psi|^2 = n_s$ is the number density of superconducting electrons, and $\theta(r)$ is a position-dependent phase. By analogy with the conventional wave-mechanical treatment of particles in a magnetic field, the supercurrent density J_s can be written as

$$J_s = \frac{e^*}{2m^*}\left[\Psi^*\left(-i\hbar\nabla - e^* A\right)\Psi + \text{complex conjugate}\right] \quad \text{(A2.7.3)}$$

where the term involving the vector potential A arises from the generalized momentum, $p \rightarrow p - e^* A$, of a particle with effective mass m^* and charge e^* in a magnetic field $B = \text{curl}\, A$. The above expression can be written more simply as

$$J_s = \frac{e^* n_d}{m^*}\left[\hbar\nabla\theta(r) - e^* A\right]. \quad \text{(A2.7.4)}$$

We now consider a superconducting ring or cylinder and evaluate the line integral of the above expression well inside the superconductor, where the screening currents associated with the Meissner effect are effectively zero, as shown in Figure A2.7.1(Left) (see Chapter A2.4). Along this path $J_s = 0$, so that

$$A = \frac{\hbar}{e^*}\nabla\theta(r). \quad \text{(A2.7.5)}$$

On integrating around the path, we have

$$\oint_{\text{path}} A.ds = \frac{\hbar}{e^*}\oint_{\text{path}} \nabla\theta(r).ds. \quad \text{(A2.7.6)}$$

Now,

$$\oint_{\text{path}} \nabla\theta(r).\, ds = \oint_{\text{path}} \frac{\partial\theta(r)}{\partial s} ds = [\theta(r)]_A^B = n2\pi \quad \text{(A2.7.7)}$$

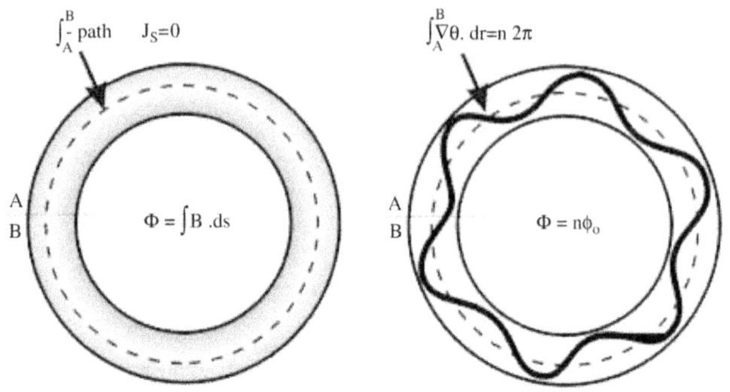

FIGURE A2.7.1 (Left) A superconducting ring or cylinder in an external field B, illustrating the path of integration well inside the superconductor along which $J_s = 0$; (Right) an allowed wave solution with $n = 6$ waves around the loop, so that the enclosed flux is $6\phi_0$.

since $\psi(r)$ must be single-valued, as shown by a typical wave solution with $n = 6$ in Figure A2.7.1(Right). Furthermore,

$$\oint_{\text{path}} A.ds = \oint_{\text{area}} \text{curl}\, A\, .\, ds = \int_{\text{area}} B.\, ds \quad \text{(A2.7.8)}$$

is simply the magnetic flux Φ enclosed within the integration path. Combining these two results, we have the principal result that

$$\Phi = \frac{\hbar n2\pi}{e^*} = n\frac{h}{2e} \quad \text{(A2.7.9)}$$

where we have assumed Cooper pairing with $e^* = 2e$. The experimental confirmation of flux quantization in units of $h/2e$ therefore provides a direct confirmation of electron pairing in the superconducting state.

The Meissner state with $\Phi = 0$ is a special case of the above relationship with $n = 0$. This is the only allowed solution for a 'singly connected' superconductor—a superconductor with no hole or line singularity, such as a flux line core, passing through it.

If the thickness of the superconducting ring or cylinder is less than the penetration depth, or if we take an integration path close to the inner surface, the screening current will not be zero. It is then not the total flux that is quantized, but what is termed the fluxoid. In this more general case, we can write

$$\int_{\text{area}} B.ds + \mu_0\lambda^2 \oint_{\text{path}} j.ds = n\phi_0 \quad \text{(A2.7.10)}$$

where we have substituted the relationship $m^*/n_s e^{*2} = \mu_0\lambda_L^2$, where λ_L is the London penetration depth over which the surface screening currents decay.

Fundamental to the operation of any SQUID is the periodicity of its properties with the flux in units of the flux quantum ϕ_0. This may easily be shown by considering the energy of a superconducting ring intersected by a single Josephson junction, as shown in Figure A2.7.2.

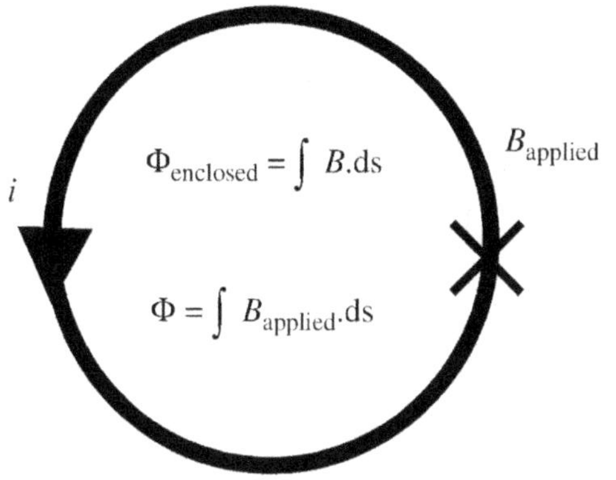

FIGURE A2.7.2 A superconducting ring intersected by a Josephson junction.

The stored energy E in a ring with inductance L can be written as

$$E = \left(\Phi - \Phi_{\text{enclosed}}\right)^2 / 2L - E_J \cos(\theta_A - \theta_B). \qquad \text{(A2.7.11)}$$

The first term is the stored magnetic energy $(1/2)Li^2 = (1/2)(\Phi - \Phi_{\text{enclosed}})^2/L$, where $(\Phi - \Phi_{\text{enclosed}})$ is the difference between the applied flux Φ (B_{applied} × area of ring) and the flux enclosed within the superconducting ring, while the second term is the Josephson energy $-E_J\cos(\theta_A - \theta_B)$, where $\theta_A - \theta_B$ is the difference in phase of the superconducting order parameter across the Josephson junction and $E_J = I_c\phi_0 2\pi$ (see Chapter A2.8). If we again consider a line integral well within the superconducting ring, where $J_s = 0$, we obtain $\theta_A - \theta_B = 2\pi\Phi_{\text{enclosed}}/\phi_0$.

The total energy E can therefore be written as

$$E = \left(\Phi - \Phi_{\text{enclosed}}\right)^2 / 2L - E_J \cos\left(2\pi\Phi_{\text{enclosed}} / \phi_0\right). \qquad \text{(A2.7.12)}$$

Minimizing the energy with respect to Φ_{enclosed}, we obtain

$$\Phi - \Phi_{\text{enclosed}} = LI_c \sin\left(2\pi\Phi_{\text{enclosed}} / \phi_0\right). \qquad \text{(A2.7.13)}$$

The supercurrent flowing around the ring, $(\Phi - \Phi_{\text{enclosed}})/L$, is therefore periodic in Φ_{enclosed} in units of the flux quantum ϕ_0. This is the basis of operation of both the rf and dc SQUID. For a dc SQUID, the critical current is periodic in the applied flux Φ such that

$$I_c = 2I_0 \cos(\pi\Phi / \phi_0) \qquad \text{(A2.7.14)}$$

as discussed in more detail in the section on Josephson junction devices (Chapter H3.2).

A2.7.3 Experimental

A2.7.3.1 Flux Quantization in Conventional Superconductors

The confirmation of flux quantization in conventional superconductors was published independently by two groups in 1961 [4, 5]. The same journal issue also included a number of related theoretical papers. Both experiments involved the use of very small-diameter superconducting cylinders coated on the surface of a nonmagnetic core.

In the measurements of Deaver and Fairbank [4], a thin cylinder of tin was deposited on a 1 cm length of 13-μm-diameter copper wire. This was cooled in a small axial field, and the magnetic moment trapped was deduced from the voltage induced when the sample was vibrated inside a pick-up coil. Many successive measurements were made on cooling in different fields. The trapped flux was shown to be a stepped function of applied field consistent with flux quantization in units of $h/2e$, as shown in Figure A2.7.3. Doll and Näbauer

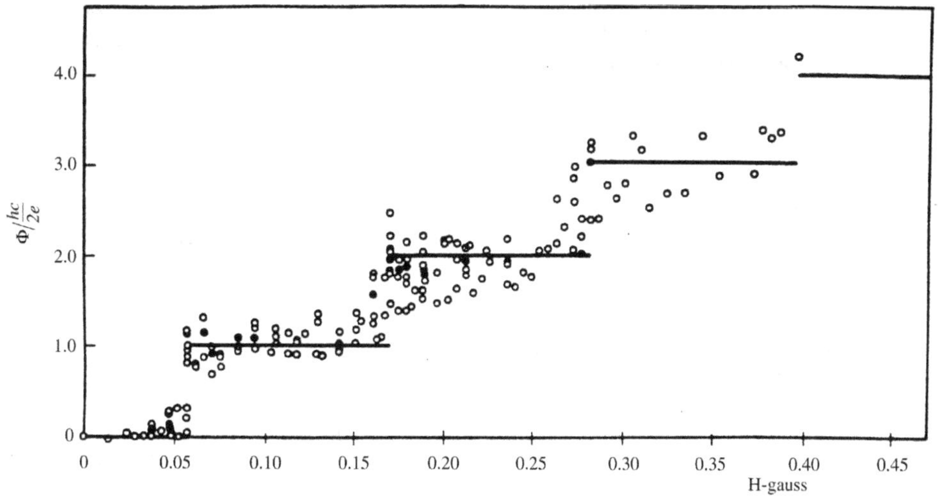

FIGURE A2.7.3 The first experimental confirmation of flux quantization in units of $h/2e$, demonstrated by the discrete values of flux trapped in a small tin cylinder when cooled in a small axial magnetic field [4].

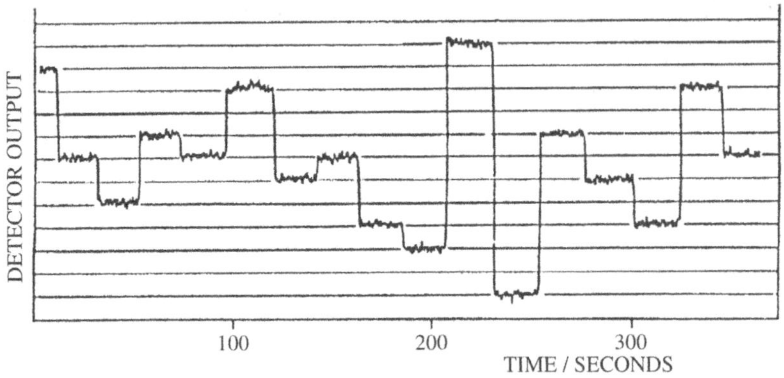

FIGURE A2.7.4 Demonstration of flux quantization in units of $h/2e$ in a ceramic YBCO ring [6]. The lines are spaced exactly $h/2e$ apart and the regular transitions between the quantum states were induced by short bursts of electromagnetic radiation.

[5] performed a slightly different experiment, trapping field in a lead cylinder coated on the surface of a 10-mm quartz fibre. This was supported on a very light torsion fibre and the trapped magnetic moment deduced from the torsional oscillations of the sample in the applied field. To their surprise, they observed the flux trapped in units of about half the value predicted by London, but consistent with later ideas of electron pairing and the quantization of flux in units of $h/2e$.

A2.7.3.2 Flux Quantization in the Cuprate High-Temperature Superconductors

Many exotic theories were initially proposed to account for the very high transition temperatures of the cuprate superconductors, including superconducting states involving 2, 4, 8 or even 16 electrons. At the same time, several experienced researchers remained highly sceptical of the 'superconducting' transition, believing it could simply be a transition to a very much lower-resistance state. Early measurements [6] on a multiphase ceramic ring of Y-Ba-Cu-O demonstrated the existence of very long-lived, trapped flux states differing in flux by $h/2e$, as shown in Figure A2.7.4. These measurements not only confirmed the existence of a truly superconducting, zero-resistance, quantum mechanical state, but also confirmed electron pairing, just as in conventional superconductors. However, such measurements cannot reveal anything about the nature of the microscopic mechanisms leading to such pairing.

A2.7.3.3 Flux Trapping in the Cuprate HTc Superconductors

It is now firmly established that the cuprate HTc superconductors have a predominantly d-wave symmetry, with nodes in the superconducting wave function at 45° to the in-plane CuO bonds. The sign of the wave function therefore reverses on changing between the a- and b-directions, as illustrated schematically in Figure A2.7.5(a). Several experiments have confirmed this sign change, consistent with d-wave symmetry of the HTS wave function [7–9].

In one such experiment, Wellstood and co-workers used a conventional s-wave superconductor to make a superconducting loop connecting adjacent edges of a square HTS crystal. Because

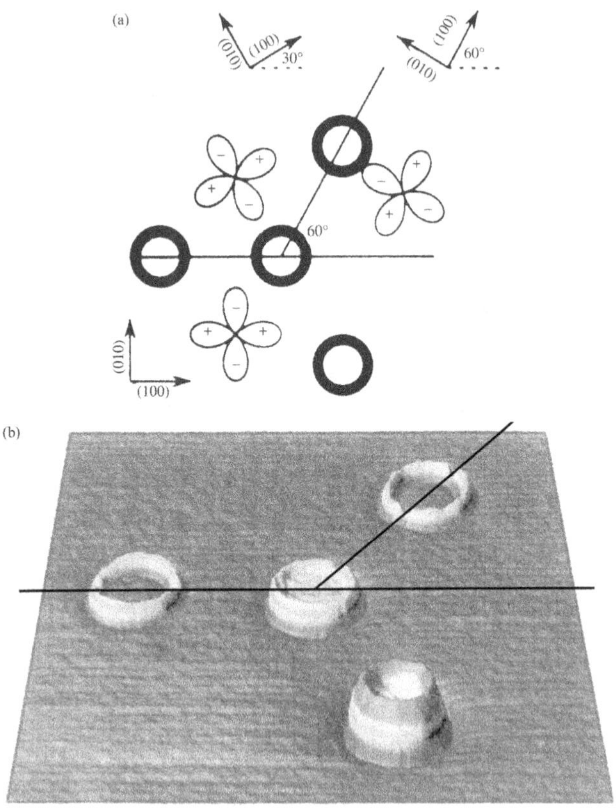

FIGURE A2.7.5 Scanning flux microscope measurements illustrating the d-wave symmetry of cuprate superconductors by Tsuei and co-workers. (a) the orientations of the tricrystal and epitaxially grown high T_c superconductors and the position of the lithographically patterned rings, which are intersected by weak-link, Josephson junctions, on crossing the substrate grain boundaries. (b) The trapping of half a flux quantum in the central ring crossing three grain boundaries, with no flux quanta trapped in the two rings crossing single grain boundaries, and one flux quantum (though it could have been any integer) in the isolated ring. (Courtesy of Tsuei.)

of the symmetry of the *d*-wave function, the weak-link Josephson junctions formed at the boundaries between the *d*-wave and *s*-wave superconductors have to accommodate the change in sign with direction of the superconducting *d*-wave function. This implies that the flux trapped in such a loop is given by

$$\Phi_{\text{enclosed}} = (n + 1/2)\phi_0. \quad \text{(A2.7.15)}$$

The smallest flux that can be trapped is, therefore, half a flux quantum, and all trapped flux states are offset from the origin by half a flux quantum, as Mathai *et al.* [8] confirmed in a series of careful measurements.

Half quantum flux trapping was also observed by Kirtley *et al.* [9] in a series of measurements in which thin film HTc superconductors were deposited on a tricrystal substrate (three crystals cut in specific crystalline orientations and fused together along their carefully cut edges). As shown in Figures A2.7.5(a) and (b), a series of thin film rings were lithographically patterned, one enclosing the intersection of all three interfaces, two rings crossing a single interface and another ring well away from the interface areas. The tricrystal was carefully designed with crystalline orientations chosen so that, over the three junctions crossing the central ring, symmetry dictates that there has to be a π phase change in the superconducting phase. The lowest flux that such a ring can trap is therefore $\pm 1/2\phi_0$, as shown in Figure A2.7.5(b), taken from the cover of Science (Vol **271**, 1995).

Experiments demonstrating 1/2 quantum flux trapping provide the strongest evidence for the *d*-wave symmetry of the cuprate superconductors. Such trapping has been confirmed for all cuprate superconductors investigated to date, but has never been observed for conventional superconductors.

A2.7.3.4 The SQUID and Flux Quantization

As indicated in the previous section, the flux enclosed by a superconducting ring containing one or more Josephson junctions is periodic in the flux quantum ϕ_0. SQUIDs (described in Chapter H3.2) can measure changes in applied flux ($B \times$ area of SQUID) to an accuracy approaching $10^{-7}\ \phi_0\ H/z^{-1/2}$, making a SQUID the most sensitive of any broadband magnetic sensor. A single Josephson junction has a Fraunhofer pattern with a strong flux quantum dependence, which can also be used to measure magnetic fields, see Schilling *et al.* [13].

Figure A2.7.6 illustrates the voltage response of a HTS *rf* SQUID as a function of the externally applied flux ϕ_e, illustrating the response periodic in ϕ_0 [14]. The curves are obtained for different values of *rf* bias.

A2.7.4 Flux Quanta and the Mixed State of Type II Superconductors

Unlike type I superconductors, which exhibit the Meissner effect excluding the penetrations of flux into the bulk, flux enters type II superconductors in a semi-regular array of

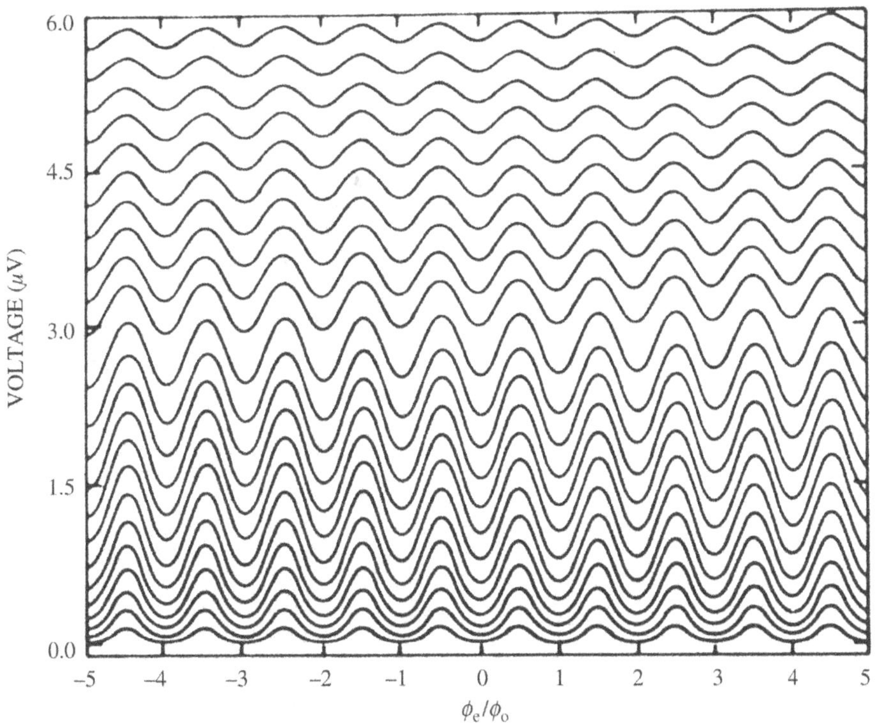

FIGURE A2.7.6 The periodic response of an *rf* HTS SQUID as a function of applied field for various levels of *rf* bias. The response can be shown to be periodic in ϕ_0 over many thousands of flux quanta. (Courtesy of Gross.)

quantized Abrikosov vortices each carrying a single quantum of flux.

Although quantized flux lines in the bulk of a superconductor cannot be visualized directly, their existence can be inferred from small-angle neutron diffraction. In such measurements, the magnetic moment of the neutron probes the spatial variations of the internal magnetic field associated with the circulating currents around each flux line. Such measurements by Cribier *et al.* [15] provided the first direct evidence for the flux line lattice in the mixed state of type II superconductors.

Figure A2.7.7 shows a more recent example of a very high-resolution neutron diffraction pattern obtained from a triangular flux lattice in the mixed state of a large single crystal of niobium. Note the similarity with an X-ray diffraction pattern. Neutron scattering provides detailed information on the spatial variations of the local flux density $B(x)$, whereas X-rays probe the variation in electron density $\rho(x)$. Neutron diffraction has proved to be a very powerful tool for looking at the phase transitions and rich flux lattice states of type II and high-T_c superconductors as a function of temperature and applied magnetic field.

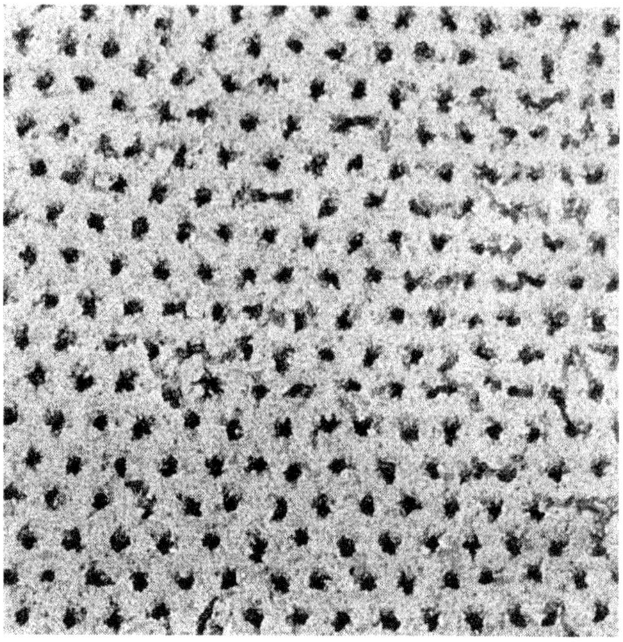

FIGURE A2.7.8 Magnetic flux decoration measurements illustrating a near-perfect triangular lattice [14].

A2.7.5 Flux Decoration Measurements

It is possible to decorate individual flux lines emerging from the surface of a superconductor with very fine magnetic powder, which can then be visualized using a scanning electron microscopy (SEM). The magnetic grains tend to segregate in the regions where flux lines leave the surface of a superconductor. This technique was pioneered by Essmann and Trauble [16] in the 1970s. For a more recent review of the Bitter decoration technique, see Fasano *et al.* [17]. Figure A2.7.8 shows an example of such a measurement, which illustrates a near-perfect triangular lattice of singly quantized flux lines emerging from the surface of a superconductor. Occasional defects, such as dislocations in the flux line lattice, can be identified.

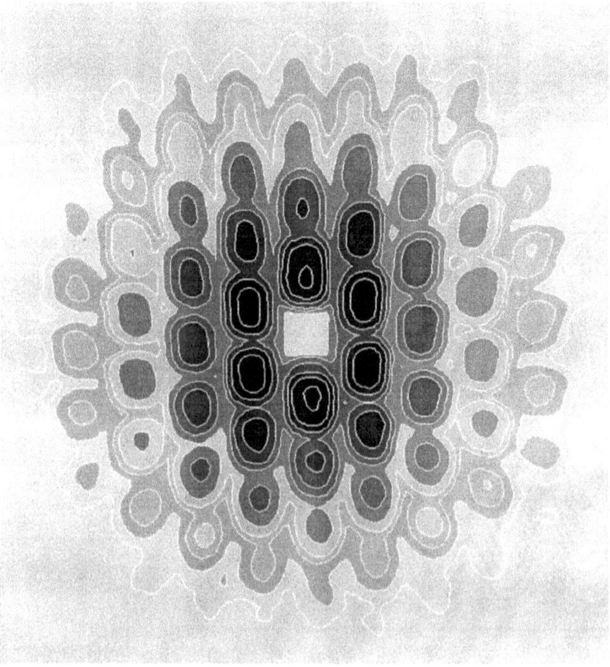

FIGURE A2.7.7 Small-angle neutron diffraction from the lattice of quantized flux lines in a niobium single crystal. (ILL Grenoble, by courtesy of Forgan.)

A2.7.6 Scanning SQUID and Hall Probe Microscopes

More recently, microscopes using dc SQUIDs with small pick-up coils (< 10 μm) as magnetic pick-up coils have been used to map out the fields produced by quantized flux lines emerging from the surface of thin films and single crystals. Figure A2.7.5(b) shows the use of such a microscope to measure the half flux quantum of flux in a HTS thin film ring on a tricrystal substrates (see Figure A2.7.5[a]). A small Hall probe can also be used to probe flux line structures with almost the same sensitivity [18].

A2.7.7 Direct Measurement of Quantized Flux by Electron Holography

Another exciting modern advance has been the development by Tonomura's group of transmission electron microscopy using interfering coherent electron beams to image quantum vortices in thin cross-sections of superconducting materials

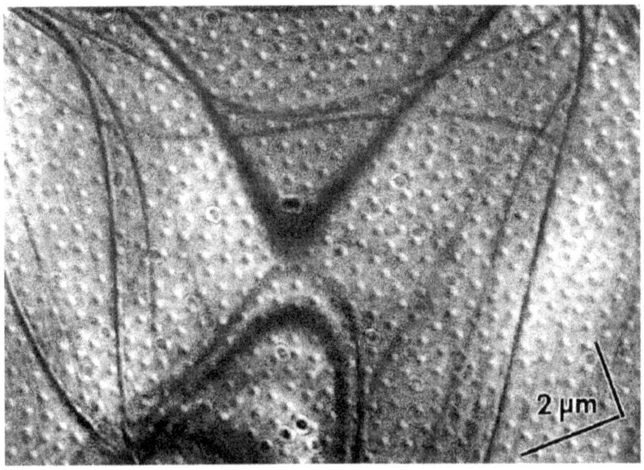

FIGURE A2.7.9 Image of individual flux lines in a thinned single crystal of niobium. Individual vortices are images as 'dimples' with one side darker than the other. The larger streaks crossing the image are interference effects associated with crystal bending and differences in thickness. (Courtesy of Tonomura.)

[19]. An example of such an image in a thin section of a Nb superconductor is shown in figure A2.7.9, where every 'dimple' represents an individual flux line.

A2.7.8 Scanning Tunneling Microscopy

Hess *et al.* [20] used Lorentz scanning tunneling microscopy (STM) to investigate the electronic core states of Abrikosov flux lines. See also the review of such measurement by Suderow *et al.* [21].

STM/STS measurements can also be used to investigate flux lines. See, for example, Cottet *et al.* [22] investigation of flux vortices in the type II superconductor MgB_2.

A2.7.9 Magneto-Optics

Researchers from the University of Oslo and Bell Laboratories were the first to observe flux lines using an optical microscope [23]. Flux lines in $NbSe_2$ were imaged by Hess *et al.* with ~ 1 µm resolution making use of the magneto-optic Faraday effect, involving rotation of the plane of polarisation of light passing through a ferrite garnet film placed in close proximity to the surface (Figure A2.7.10).

Magneto-optic methods are reviewed by Jooss *et al.* [24]. Even real-time magneto-optical imaging of the vortex lattice and individual vortices is now possible [25]. Such techniques enable the dynamics of both correlated groups of vortices and individual flux quanta to be investigated as a function of time, temperature and applied field. It is important to be able to visualize such events, because such processes are almost certainly present though often ignored in theoretical models for thermally activated and quantum creep. For a more detailed review of magneto-optic methods, see Chapter G3.5.

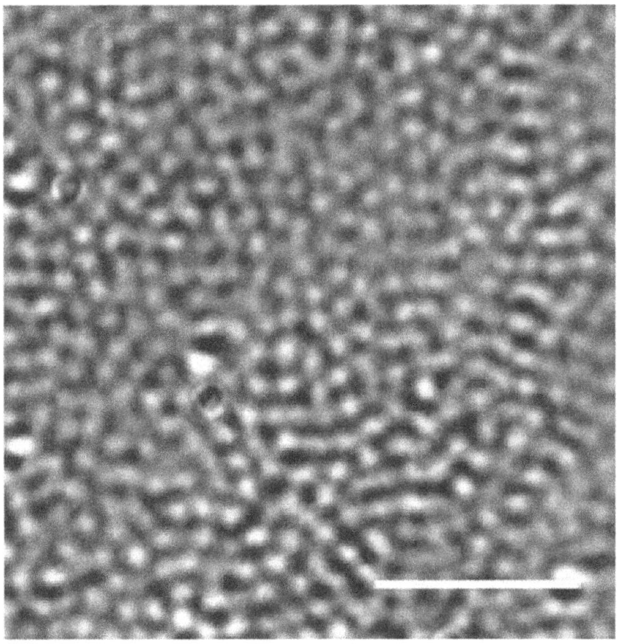

FIGURE A2.7.10 Magneto-optic images of vortices in $NbSe_2$ at 4.2 K in a field of 8 gauss. The white marker represents 10 µm.

A2.7.10 Superconducting Logic Circuits and Quantum Computing

Electronic circuits use Josephson junctions with single flux quanta as the quanta of information form the basis of various ultra-fast, low-power, computer circuits; see Chapters H3.5 and H3.6. Currently, there is a great interest in the use of superconducting circuits using the switching of flux quantum states to act as Q-bits to carry information; see Chapter H3.7.

References

[1] London F 1950 *Superfluids* **Vol 1**, (New York: Wiley) p 152

[2] Bardeen J, Cooper L N and Schrieffer J R 1957 *Phys. Rev.* **108** 1175

[3] Josephson B D 1962 *Phys. Lett.* **1** 251

[4] Deaver B S and Fairbank W M 1961 *Phys. Rev. Lett.* **7** 43

[5] Doll R and Nabauer M 1961 *Phys. Rev. Lett.* **7** 51

[6] Gough C E, Colclough M S, Forgan E M, Jordan R G, Keene M, Muirhead C M, Rae AIM, Thomas N, Abell J S and Sutton S 1987 *Nature* **326** 855

[7] Wollman D A, Van Harlingen D J, Giapintzakis J and Ginsberg D M 1995 *Phys. Rev. Lett.* **74** 797

[8] Mathai A, Gim Y, Black R C, Amar A and Wellstood F C 1995 *Phys. Rev. Lett.* **74** 4523

[9] Kirtley J R, Tsuei C C, Rupp M, Sun J Z, Yu-Jahnes L -S, Gupta A, Ketchen M B, Moler K A and Bhushan M 1996 *Phys. Rev. Lett.* **76** 1336

[10] Abrikosov A A 1957 *Sov. Phys. JETP* **5** 1174
[11] Aaronov Y and Bohm D 1959 *Phys. Rev.* **115** 485
[12] Ginsburg V L and Landau L D 1950 *Zh. Eksperim. I. Teor. Fiz.* **20** 1064 (in Russian)
[13] Gross R, Chaudhari P, Kawasaki M, Ketchen M B and Gupta A 1990 *Appl. Phys. Lett.* **57** 727
[14] Schilling M, Barthelmess H J, Krey S and Ludwig F 2007 *Advances in Solid State Physics* **40** 769
[15] Cribier D, Jacort B, Rao L M and Farnoux B 1964 *Phys. Lett.* **9** 106
[16] Essmann U and Trauble H 1967 *Phys. Lett.* **A24** 526
[17] Fasano F and Menghini M 2008 *Supercond. Sci. Technol.* **2** (2) 023001
[18] Oral A, Bending S J, Humphreys R G and Heneni M 1996 *J. Low Temp. Phys.* **105** 1135
[19] Harada K, Matsuda T, Bonevich J, Igarisho M, Kopndo S, Pozzi G, Kawabe U and Tomura A 1992 *Nature* **360** 51
[20] Hess H F, Robinson R B, Dynes R C, Valles J M and Waszczak J V 1989 *Phys. Rev. Lett.* **62** 214
[21] Suderow H, Guillamon I, Rodrigo J G and Viera S 2014 *Supercond. Sci. Technol.* **27**(6) 063001
[22] Cottet M J G, Cantoni M, Mansart B, Alexander D T L, Hebert C, Zhigadlo N D, Karpinski J and Carbone F 2013 *Phys. Rev. B* **88** 014505
[23] Goa P E, Haughlin H, Baziljevich M, Il'yashenko E, Gammel P and Johansen T H 2001 *Supercond. Sci. Technol.* **14** 729
[24] Jooss Ch, Albrecht J, Kuhn H, Leonhardt s and Kronmüller H 2002 *Rep. Prof. Phys.* **I65** 651
[25] Gao P, Hauglin H, Olson Å A, Baziljevich M and Johansen T H 2003 *Rev. Sc. Instrum.* **74** 141

Further Reading

Fujita S et al *Quantum Theory of Conducting Matter – Superconductivity* (2009 New York: Springer)
Kleiner R and Godoy K *Supercondutivity - An Introduction* (3rd edition) (2015 Berlin: Wiley-VCH)
Rose-Innes A C and Rhoderick E H *Introduction to Superconductivity* (1978 Oxford: Pergamon)
Tinkham M *Introduction to Superconductivity* (2nd edition) (1996 Singapore: McGraw Hill)
Waldram JR *Superconductivity of Metals and Cuprates* (1990 Bristol: IOP Publishing)

A2.8

Josephson Effects

Edward J. Tarte

A2.8.1 Introduction

The Josephson effects [1] are very important to many areas of modern superconductivity research. Their most obvious and direct relevance is to the study and development of active superconducting electronic devices. Here, Josephson junctions form the basic elements of superconducting quantum interference devices (SQUIDs) [2], single flux quantum (SFQ) logic [3], junction arrays for the voltage standard [4] and for high-frequency radiation sources [5]. However, for most high-temperature superconductors (HTS), they govern the transport of supercurrent in a significant number of situations. For example, a number of HTS materials appear to be Josephson coupled perpendicular to the copper oxide planes [6]. In addition, the Josephson effect often limits the flow of current in poorly aligned polycrystalline material [7].

A wide variety of structures exhibit Josephson effects, but in general, they can be classified into three groups: systems in which superconducting electrodes are separated by insulating layers; conducting (but not superconducting) layers; or a region where the superconductivity is weakened because it has an extremely narrow cross-section. In this chapter, tunnel junctions are concentrated upon, because these form the basis of most of the established applications of Josephson effects. To this end, the properties of superconducting tunnel junctions will also be described first. Following this, some of the most basic features of Josephson effects will be described.

A2.8.2 Single-Particle Tunnelling

Tunnel junctions usually consist of electrode layers of superconductor and/or normal metal separated by an insulating barrier, although it is important to note that other structures such as grain boundaries may exhibit tunnelling phenomenon [8]. The insulator may be another material sandwiched between the electrodes or may be a vacuum. The transmission of electrons through the barrier is determined by a transmission coefficient D, which according to elementary quantum mechanics should be of the form $\exp[-l\sqrt{(2m\Gamma/\hbar^2)}]$, where Γ is the barrier height, l the barrier width, m the electron mass and $\hbar$ Planck's constant. If the insulator is simply a vacuum layer, the barrier height Γ is simply the work function of the metal, whereas, if it consists of a dielectric layer, Γ is fixed by the bandgap in the barrier's electronic structure. A number of factors may complicate this simple picture, but the overall features are generally preserved and as a rule of thumb we can take a transmission probability D^2 to be approximately given by $\exp(-l)$ if l is measured in Å [9].

A very important feature of superconducting tunnel junctions is that their current–voltage characteristics can be used to determine the energy gap Δ and to explore the density of single-particle excitations of one spin in the superconductor $N(E)$. Somewhat surprisingly, it turns out that the structure can be treated using a relatively simple 'semiconductor model', and that the tunnelling current L as a function of voltage V is given by Equation (A2.8.1) where $f(E)$ is the Fermi function, E the excitation energy referred to the Fermi level and e the charge on the electron. The R and L subscripts refer to the right or left side of the barrier.

$$I = \frac{8\pi D^2 e}{\hbar}\int_{-\infty}^{\infty}[f_R(E-eV)-f_L(E)]N_R(E-eV)N_L(E)\mathrm{d}E. \quad \text{(A2.8.1)}$$

The simplicity of this equation is surprising because it totally ignores the presence, in the superconductors, of the Cooper pairs to which the single-particle excitations are intimately connected and which might have been expected to affect the tunnelling rates. The explanation for this is rather subtle and can be found in the review article by Waldram [9].

The three simplest examples of tunnelling systems are shown in Figure A2.8.1: normal–insulator–normal (NIN),

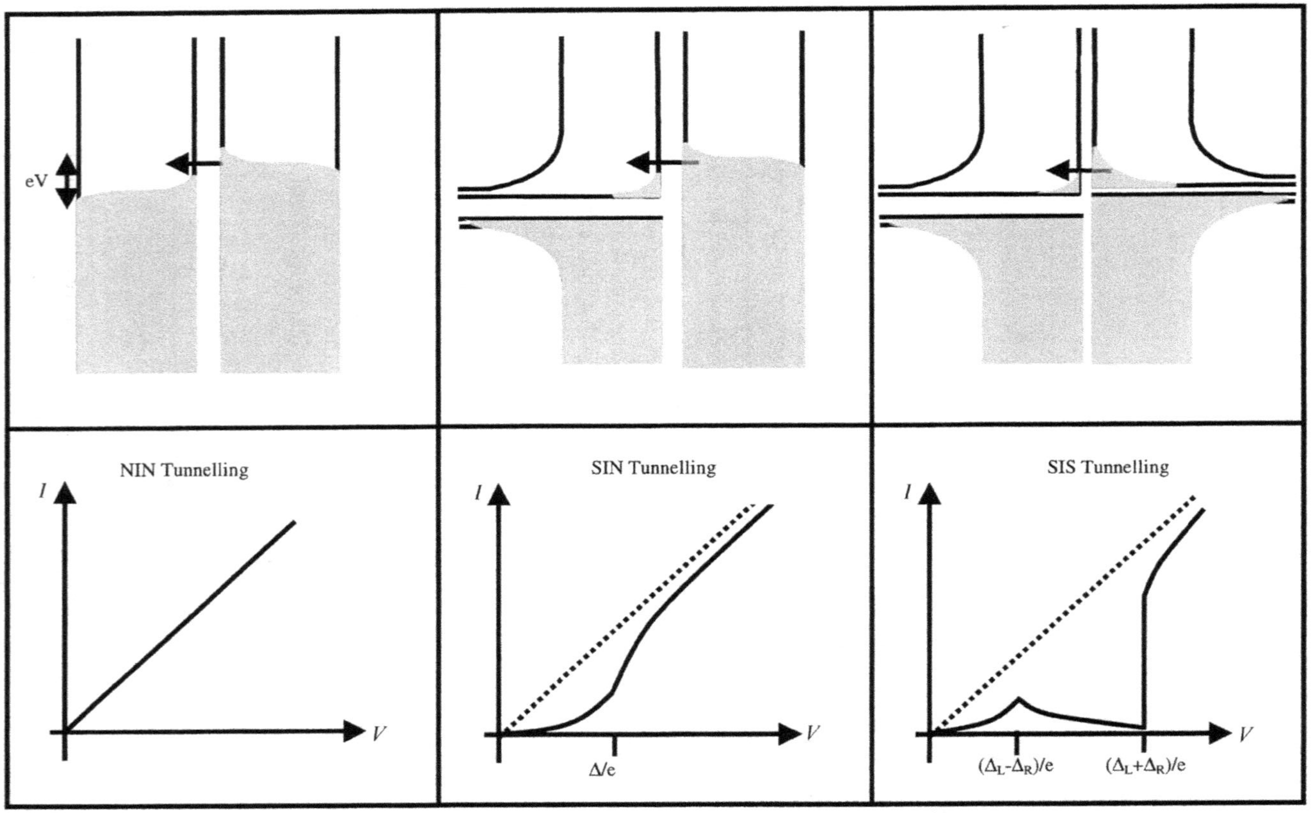

FIGURE A2.8.1 Single-particle tunnelling showing the densities of states and occupation at finite temperature with the corresponding current–voltage characteristic.

superconductor–insulator–normal (SIN) and superconductor–insulator–superconductor (SIS) tunnelling. Using the semiconductor model, the current-voltage characteristic is constructed by shifting the equilibrium density of states by an energy equal to eV and comparing the occupation of states on either side of the barrier. Experimental data for an SIN tunnel junction and an SIS tunnel junction are shown in Figure A2.8.2.

For NIN tunnelling, the density of states can be taken to be constant (at least for small voltages) in the vicinity of the Fermi level, where $f_R(E - eV) - f_L(E)$ is finite, and Equation (A2.8.1) reduces to Ohm's law $I = V/R$ where R is the tunnel junction resistance.

For SIN tunnelling, the current is very small until the voltage reaches the energy gap voltage Δ/e in the superconductor and then rises very rapidly. The amount of current at voltages $< \Delta/e$ is very sensitive to temperature since, in equilibrium, the occupation of the levels above the gap is determined by the thermal tail of the Fermi function. For an SIN junction, it is easy to show that

$$\frac{dI}{dV} = \frac{8\pi D^2 e N_N(0)}{\hbar} \int_{-\infty}^{\infty} \frac{df_R(E-eV)}{dE} N_s(E)dE \qquad \text{(A2.8.2)}$$

where $N_N(0)$ is the density of states in the normal metal at the Fermi level. At very low temperatures, the derivative of the Fermi function becomes a delta function, and then it is clear that the voltage dependence of the differential conductance gives the form of the superconducting density of states. As the density of states at the edge of the tunnel barrier on the superconductor side is the quantity measured, tunnelling can be used to study the anisotropy of the gap in single crystals [10] or to compare tunnelling from the superconducting and normal parts of an SN bilayer. SIN tunnelling is important in verifying the prediction of the BCS theory for the form of the superconducting density of states.

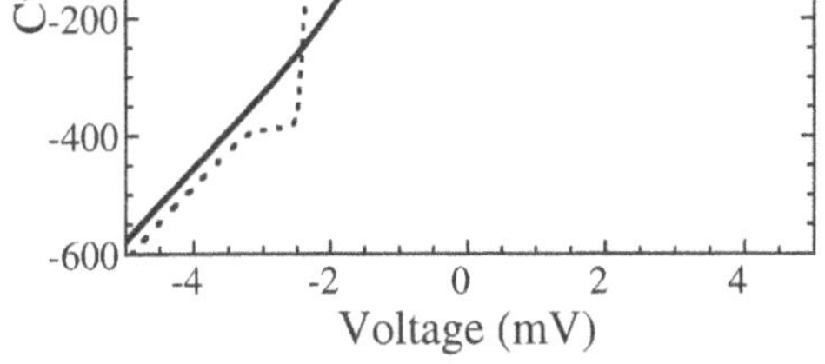

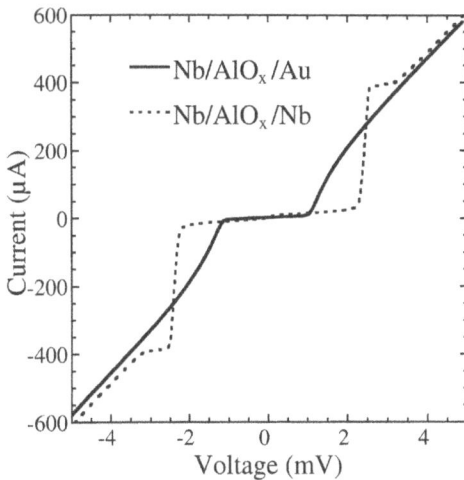

FIGURE A2.8.2 Experimental I–V characteristics for a Nb/AlOx/Nb SIS tunnel junction and a Nb/AlOx/Au SIN tunnel junction (courtesy of Dr Gavin Burnell and Professor Mark Blamire).

In Figure A2.8.1, tunnelling between dissimilar superconductors, known as SIS tunnelling, is shown. The current-voltage characteristic contains two important features. As the voltage across the junction is increased, the current rises because the states become available in the high gap superconductor S for excitations in the low gap superconductor S' to tunnel into. At a voltage where the peaks in the densities of states are aligned in energy $(\Delta_S - \Delta_{S'})/e$, the current reaches a maximum and then, as the voltage is increased, declines again. When the voltage reaches $(\Delta_S + \Delta_{S'})/e$, the heavily occupied peak in the density of states of S' is aligned with the upper sparsely occupied peak in, S and the current rises very rapidly. For a symmetric SIS tunnel junction, the difference gap peak at $(\Delta_S - \Delta_{S'})/e$ is obviously absent. Sufficiently below the superconducting transition temperature, the current flowing at voltages below $2\Delta_S/e$ is very small, and the differential conductance at this voltage is extremely small. The latter feature is very useful in creating quasiparticle mixer devices for mm-wave signals.

A2.8.3 The dc Josephson Effect and Quantum Interference

The Josephson effects are consequences of the macroscopic quantum nature of the superconducting state and provide some of the most striking demonstrations of this. We can define a single wave function $\Psi(r)$ for the superfluid of pairs, of the form $|\Psi(r)| \exp i\theta(r)$, where $\theta(r)$ is the phase of the wave function. When two superconductors are weakly coupled in one of the ways described before, a supercurrent whose density depends on the difference in the phases on the two sides of the barrier ϕ flows between them. In the absence of an external magnetic field, ϕ is $(\theta_2 - \theta_1)$.

Often, we can assume that the coupling is mediated by the weak penetration of the wave function from one superconducting electrode to the other side of the barrier and vice versa. In this case, we can consider the total wave function Ψ_{tot} to be a linear superposition of the two contributions, then we can use Ψ_{tot} to calculate the current as follows. The quantum mechanical equation for the current is

$$J = \frac{ie\hbar}{2m}\left(\psi^* \nabla\psi - \psi\nabla\psi^*\right) - \frac{2e^2}{m}\psi\psi^* A \qquad \text{(A2.8.3)}$$

but, in zero applied field, we can take $A = 0$ and the two contributions to Ψ_{tot}, Ψ_1 and Ψ_2 have phases which do not vary individually with position. Associating Ψ_2 with the left (negative x) side of the barrier and Ψ_1 with the right (positive x) side, if we write $\Psi_{tot} = \Psi_1 + \Psi_2 \exp(i\phi)$ and calculate the current at a point in one of the electrodes, then we obtain the simple Josephson current-phase relationship:

$$J_s = J_0 \sin\phi. \qquad \text{(A2.8.4)}$$

Note that ϕ is defined to be *minus* the phase difference measured across the junction in the direction of current flow, so that J_0 is positive. The critical supercurrent density is J_0, and the junction critical current I_0 is its integral over the junction area. Thus, a Josephson junction can support a dc supercurrent I_s with $I_0 < I_s < I_0$, dependent on the phase difference across the junction.

Equation (A2.8.4) holds for a wide variety of weak links. However, the current-phase relationship can deviate from this in the microbridge case [11]. It is this relationship between the supercurrent density and the quantum mechanical phase difference which gives rise to quantum interference effects. These can be observed when a magnetic field is applied to a single isolated junction or to a circuit containing Josephson junctions and superconducting loops. The latter forms the basis for the SQUID. The response of a single junction to an external magnetic field encompasses a wide range of behaviour determined by its geometry as described below.

A2.8.3.1 Short Junctions in a Magnetic Field

When an external magnetic field is applied to the superconductor, the critical current of the junction can be modified because the field makes the phase difference across the junction vary with position. This can be understood by considering Figure A2.8.3. When the magnetic field is applied in the y direction, a screening current is set up in the surface of the superconductor, flowing within the magnetic penetration depths on either side of the barrier. We can relate this to the phase of the wave function and the magnetic vector potential using the quantum mechanical equation for the current (A2.8.3) which, on substituting $\Psi = |\Psi(r)|\exp[i\theta(r)]$, gives

$$\nabla\theta = -\frac{2\pi}{\Phi_0}\left(\frac{m}{|\psi|^2 e^2}J + A\right) \tag{A2.8.5}$$

where Φ_0 is the magnetic flux quantum. However, we must be careful of what we mean by the phase difference ϕ, because we cannot simply write it as the difference of the phases of the two wave functions ($\theta_2 - \theta_1$), since this is not gauge invariant. In fact, according to Equation (A2.8.5), the only way to ensure $\boldsymbol{J}$ is invariant under the gauge transformation $A \rightarrow A + \nabla\chi$ is if the corresponding transformation for θ is $\theta \rightarrow \theta - (2\pi/\Phi_0)\chi$. This suggests that we should use the following definition for the gauge invariant phase difference:

$$\phi = \theta_2 - \theta_1 - \frac{2\pi}{\Phi_0}\int_2^1 A \cdot dl \tag{A2.8.6}$$

where 1 and 2 refer to points on opposite sides of the junction barrier, e.g. A_1 and D_2 or D_1 and A_2. Remember that ϕ is defined to be positive in the opposite direction to the Josephson current flow. Hence, in Figure A2.8.3, between points at x and $x + dx$, the change in the gauge invariant phase difference is

$$\phi(x+dx) - \phi(x) = (\theta_{A2} - \theta_{D1}) - (\theta_{D2} - \theta_{A1})$$
$$-\frac{2\pi}{\Phi_0}\left(\int_{A2}^{D1} A \cdot dl - \int_{D2}^{A1} A \cdot dl\right). \tag{A2.8.7}$$

Using Equation (A2.8.5), we can write the change in phase between the points A_1 and D_1 caused by the magnetic field as

$$\theta_{D_1} - \theta_{A_1} = -\frac{2\pi}{\Phi_0}\int_{ABCD_1}\left(A + \frac{m}{|\psi|^2 e^2}J\right)\cdot dl \tag{A2.8.8}$$

and a similar expression for the contour $(ABCD)_2$. Here $\boldsymbol{J}$ is the screening current flowing in the electrodes as a response to the applied magnetic field. In carrying out this integration, we make sure that the contours $(ABCD)_1$ and $(ABCD)_2$ extend a distance into the electrodes beyond the penetration depth where $\boldsymbol{J}$ is equal to zero. Then $\boldsymbol{J}$ makes no contribution at all to the integral, because inside the penetration regions, it is perpendicular to the contour. Thus, we can rewrite Equation (A2.8.7) as

$$\phi(x+\mathrm{d}x) - \phi(x) = \frac{2\pi}{\Phi_0}\oint A \cdot dl. \tag{A2.8.9}$$

$$\phi(x+dx) - \phi(x) = \frac{2\pi}{\Phi_0}\oint \boldsymbol{A} \cdot \boldsymbol{dl}$$

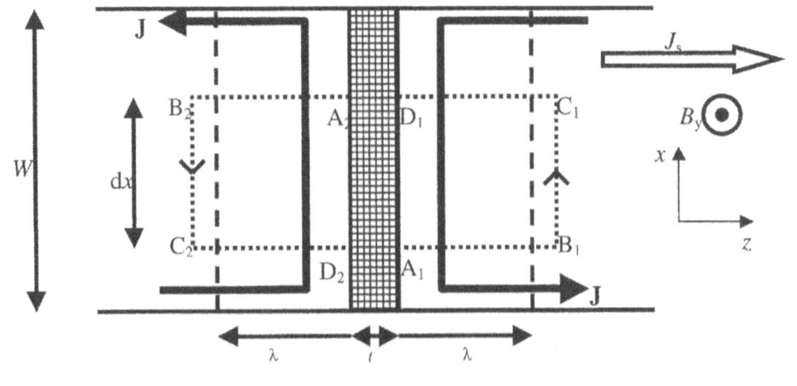

FIGURE A2.8.3 Contours of integration used to derive the magnetic field dependence of the gauge invariant phase difference. The screening current J is also indicated.

The integral represents the flux enclosed by the contours and the paths across the junction which is equal to $B(2\lambda + t)dx$, so the phase difference across the junction can be obtained from

$$\frac{d\phi(x)}{dx} = 2\pi\frac{B(2\lambda+t)}{\Phi_0}. \quad \text{(A2.8.10)}$$

If Equation (A2.8.10) is integrated and $\phi(x)$ is substituted in Equation (A2.8.4), then we can integrate J_s over the junction area to obtain the current I and maximize I with respect to the constant of integration ϕ_0 to determine the critical current. For a uniform critical current density J_0, the junction critical current is

$$I_c(\Phi) = I_0\left|\frac{\sin\left(\frac{\pi\Phi}{\Phi_0}\right)}{\frac{\pi\Phi}{\Phi_0}}\right| \quad \text{(A2.8.11)}$$

where $\Phi = BLd$ and $d = (2\lambda + t)$ is the magnetic width of the barrier. This function is plotted in Figure A2.8.4(a) for a range of applied flux values. An important feature to note is that the central maximum of this curve is twice as wide as the secondary maxima. This curve is often called the 'Fraunhofer' pattern due to the analogy between the quantum interference effects, which define its shape, and Fraunhofer diffraction from a single slit. We can also define a magnetic area for the junction equal to Ld. In general, for an arbitrary critical current density distribution $J_0(x, y)$, the critical current as a function of magnetic field is related to the Fourier transform of a function $j(x)$ as follows:

$$I_c(k) = \left|\int\int_{\text{area}} dx\,dy J_0(x,y)\exp ikx\right| = \left|\int_{-L/2}^{L/2} dx j(x)\exp(ikx)\right| \quad \text{(A2.8.12)}$$

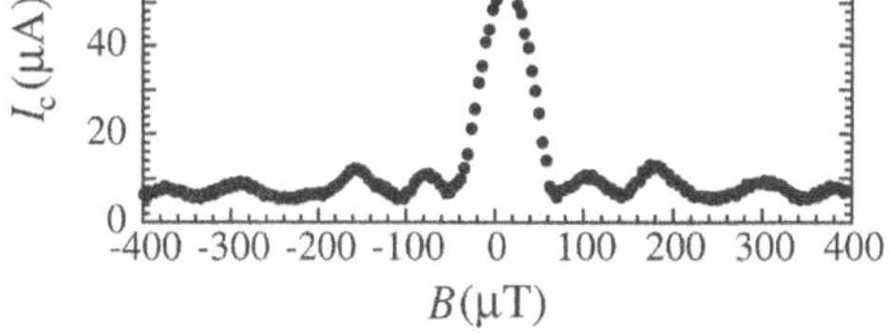

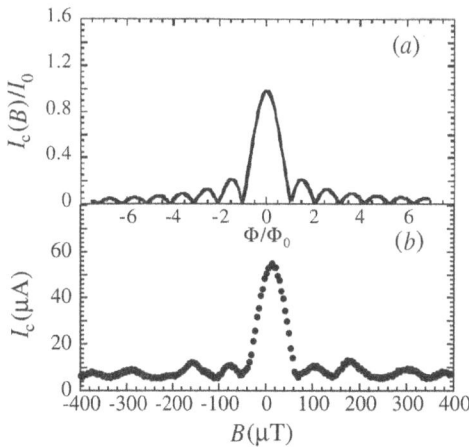

FIGURE A2.8.4 (*a*) The ideal critical current versus applied flux curve (Fraunhofer's pattern) for a uniform junction. (*b*) Critical current versus field, for a 6-μm-wide 24° misoriented grain boundary junction in a $YBa_2Cu_3O_{7-\delta}$ film.

where $k = 2\pi Bd/\Phi_0$. Thus, the variation of the junction critical current with applied magnetic field is a very important measure of their quality since it gives information about the uniformity of $J_0(x, y)$ over the junction area. Deviations from a uniform critical current density distribution are easily detected by comparing the observed Fraunhofer pattern with the ideal one.

An example of the variation of critical current with applied field is shown in Figure A2.8.4(b), for a 6-μm-wide 24° misoriented grain boundary junction in a 100-nm-thick $YBa_2Cu_3O_{7-\delta}$ film. At first sight, the agreement between these data and the theoretical curve above it is reasonable, but the experiment and theory disagree in two important ways. Firstly, the critical current never goes to zero at the minima, and this is a common feature of grain boundary junctions in high-temperature superconductors. In contrast, complete suppression may be observed with low-T_c tunnel junctions [9] or high-T_c junctions produced by focussed electron beam irradiation [12]. The offset seen here is of the type predicted by Yanson [13] for a junction whose barrier is extremely non-uniform on a very fine scale. It has recently been shown by Hilgenkamp *et al.* [14] that the behaviour of these devices can be understood in this way, due to the dx^2–y^2 pairing state of the high-temperature superconductors and the facetting of the grain boundary.

Secondly, using the theory described before, the magnetic field required to reach the first minimum should be equal to $B_{min} = \Phi_0/L(t + 2\lambda) = 1$ mT, assuming a penetration depth of 150 nm, but the actual value is much smaller, 50 μT. This discrepancy arises because this junction has its dimension parallel to the field direction less than the penetration depth. In this case, the screening currents set up in response to the applied field penetrate much further into the electrodes, and it is not possible to draw the contour $(ABCD)_1(ABCD)_2$ so that J makes no contribution to the phase difference in Equation (A2.8.8). Hence, the magnetic thickness of the barrier is no longer $d = 2\lambda + t$, and the position of the minima is then more difficult to estimate. This problem has been solved for a very thin film where the first minimum occurs at $B_{min} = 1.8\Phi_0/L^2$ [15], which in this case would give a value close to 100 μT.

A2.8.3.2 Junctions in Superconducting Loops

The above argument can easily be extended to treat the behaviour of one or more Josephson junctions in superconducting loops. This is important because such structures form the basis of SQUIDs and SFQ electronics which are dealt with in Chapters E4.2 and E4.5. However, at this point we will only give the results for the simplest cases.

For a single junction, in a loop the gauge invariant phase difference is given by

$$\phi = -2\pi\frac{\Phi}{\Phi_0} \quad \text{(A2.8.13)}$$

where Φ is the flux inside the loop. If the loop has inductance L_s, then $\Phi = \Phi_{ext} + L_s I_s$, where Φ_{ext} is the externally applied flux, and I_s is the screening current set up in the loop equal to $I_0 \sin \Phi$. The relationship between Φ and Φ_{ext} is

$$\phi + \frac{2\pi L_s I_0}{\Phi_0}\sin\phi + \frac{2\pi\Phi_{ext}}{\Phi_0} = 0 \quad \text{(A2.8.14)}$$

and this equation governs the behaviour of rf SQUIDs.

The result equivalent to Equation (A2.8.13) for a loop containing two junctions is

$$\phi_1 - \phi_2 = 2\pi\frac{\Phi}{\Phi_0} \quad \text{(A2.8.15)}$$

and this can be used to determine the magnetic field dependence of the critical current if the junctions are connected electrically in parallel. However, if the loop has a finite inductance, this cannot be done analytically. For two junctions in parallel with a very low inductance loop, the critical current is given by

$$I_c(\Phi) = 2I_0\left|\cos\frac{\pi\Phi}{\Phi_0}\right| \quad \text{(A2.8.16)}$$

as may be shown using Equation (A2.8.12).

A2.8.3.3 Long Junctions

When one or both of the lateral dimensions (width and length) of a Josephson junction become large, then the simple quantum interference effects described above are modified by the screening of magnetic fields out of the junction region. This applies to both externally applied fields and the self-fields associated with the current flowing in the leads for the junction. The detailed behaviour depends on the geometry of its electrical connections, which determine the magnitude and direction of self-fields. The five important classes are shown in Figure A2.8.5. Junctions (a)–(d) can be fabricated using thin film tunnel junction technologies, whereas the slab geometry corresponds to grain boundaries in bulk material. For the in-line, slab and cross geometries, the leads generate appreciable magnetic fields in the plane of the junction barrier. Grain boundaries in linear strips of HTS film have been shown to behave like overlap junctions [16] where the self-field is negligible.

Since the magnetic field is screened out of the junction region, we cannot assume that it is uniform within the barrier, as we did earlier. Hence, we use the Maxwell equation $dB_y/dx = \mu_0 J_z$ and Equation (A2.8.10) to obtain an equation for the spatial variation of the phase along the junction:

$$\frac{d^2\phi}{dx^2} = \frac{2\pi\mu_0 d}{\Phi_0}J_z = \frac{2\pi\mu_0 d}{\Phi_0}J_0\sin\phi = \frac{\sin\phi}{\lambda_J^2} \quad \text{(A2.8.17)}$$

where $\lambda_J = [\Phi_0/(2\pi\mu_0 d J_0)]^{1/2}$ is the Josephson penetration depth. When the junction's largest dimension is significantly longer than the Josephson penetration depth, the device is said to be in the long limit. It should be noted that this expression is not strictly valid for grain boundaries in HTS films because of the problem of defining d. However, the behaviour of long HTS grain boundary junctions can be understood at least qualitatively using the solutions of Equation (A2.8.17).

The critical current as a function of applied field for two types of long junction is shown in Figure A2.8.6. Both are

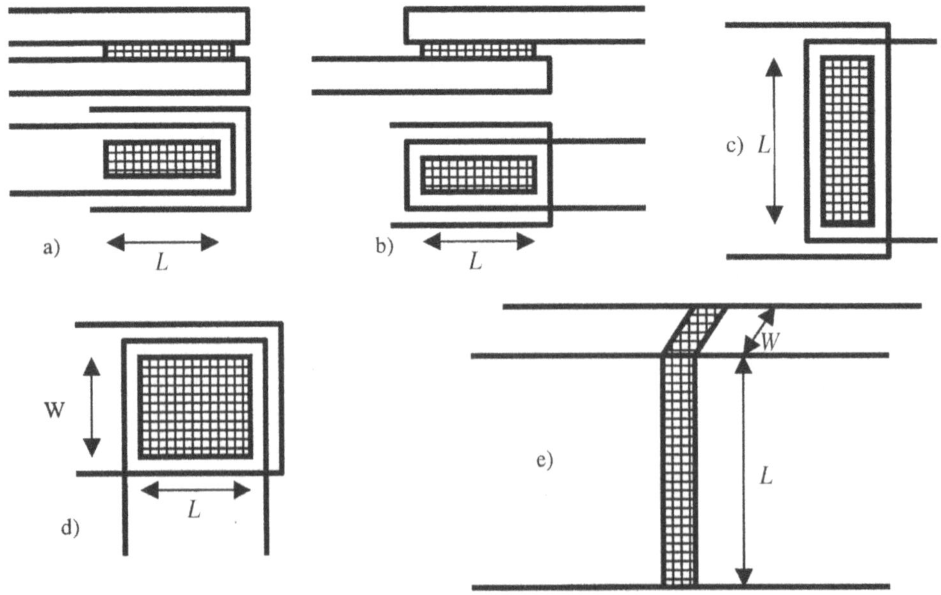

FIGURE A2.8.5 Josephson junction geometries: (a) symmetric in-line, (b) asymmetric in-line, (c) overlap, (d) cross and (e) slab. The upper parts of (a) and (b) are side views, whilst the lower parts are top views. The shaded area represents the active area of the junction barrier.

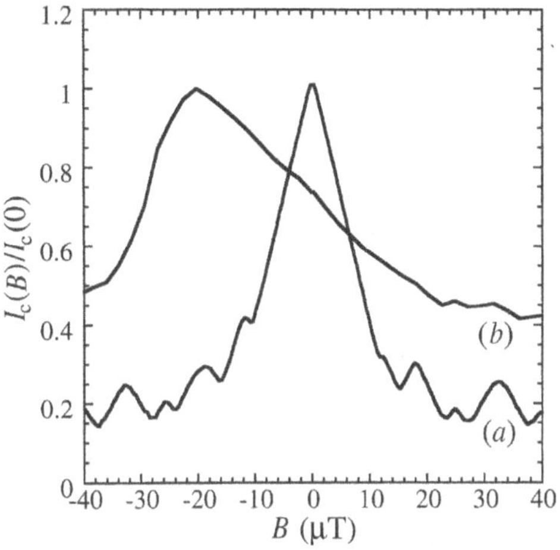

FIGURE A2.8.6 Critical current versus field curves for YBCO thin film grain boundary along Josephson junctions. Curve (*a*) is for a junction formed in a strip of film whilst curve (*b*) is for an asymmetric in-line junction (courtesy of Dr Stephen Isaac and Professor Mark Blamire).

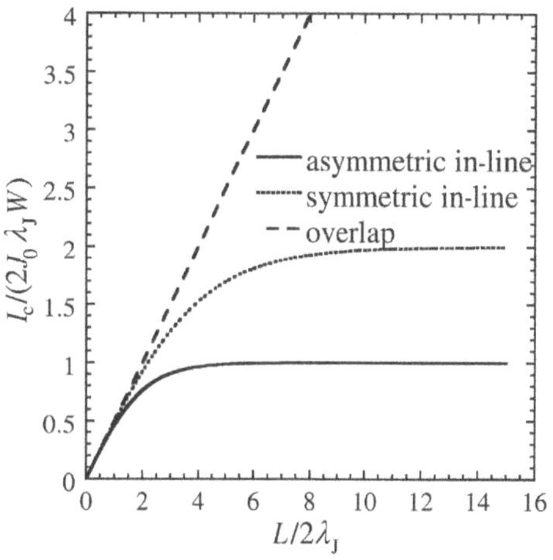

FIGURE A2.8.7 Variation of maximum critical current in zero external field with junction length for overlap geometry, symmetric in-line and slab geometry and asymmetric in-line geometry.

grain boundary junctions in YBCO films, but in one case, the film is patterned to form an asymmetric in-line device. In the other, the film is patterned into a linear strip crossing the grain boundary. Both curves have a triangular form for low fields with a number of small triangular features appearing for larger fields. The shape for small fields is associated with the screening of the applied field from the junction, and the curve extrapolates to a critical field value of $4\mu_0 J_0\lambda_J$, where one Josephson vortex enters the junction. For the asymmetric in-line junction, the self-field is reflected in its critical current versus field curve, and the maximum critical current no longer occurs at zero field. The smaller triangular features represent states with one or more Josephson vortices in the junction and several of these states can occur at the same field value with the $I_c(B)$ curve exhibiting hysteresis.

The form of the $I_c(B)$ curve for all types of junction listed is similar to those shown in Figure A2.8.6. However, if we plot the critical current in zero field $I_c(0)$ versus length L for the devices, then major differences appear, as shown in Figure A2.8.7. Initially, all of the junctions are in the small limit and the critical current increases in proportion to length. When the length of the asymmetric in-line junction becomes comparable to $2\lambda_J$, the self-field begins to be excluded from the interior of the junction and the rate of increase of $I_c(0)$ with L decreases eventually saturating at $2WJ_0\lambda_J$. This also happens for the symmetric in-line and the slab geometry, but not until L becomes comparable to $4\lambda_J$, and $I_c(0)$ saturates at $4WJ_0\lambda_J$. In contrast, although the overlap junction shows long junction behaviour for the same range of L values as the other devices, its $I_c(0)$ value never saturates because its self-field is so small.

HTS grain boundaries provide examples of both overlap and slab geometry long junctions. For grain boundaries in linear strips of thin HTS film [16], W is typically < 300 nm whilst L is ≥ 2 μm. These dimensions result in a negligible self-field, and it has been shown experimentally that $I_c(0)/2WJ_0\lambda_J$ does not saturate with increasing L/λ_J. However, single grain boundaries in bulk material having dimensions of order 100 μm × 100 μm are clearly slab-like and their critical currents are of order $4WJ_0\lambda_J$, assuming the same critical current densities as equivalent thin films' grain boundaries [17].

A2.8.4 ac Josephson Effects

In order to understand what happens when a Josephson junction is in the finite voltage state, we must look at the energy of the superfluid of pairs. Just as we can apply the quantum mechanical equation for the current to the pair wave function, we can use the quantum mechanical energy operator $\left(\hat{E} = i\hbar\partial/\partial t\right)$ to calculate its energy eigenvalue. If we assume that the superconductor is in equilibrium with a reservoir of particles with an electrochemical potential of μ, then when a pair of electrons enters the superfluid, they must do so with an energy of 2 μ. Thus, in a steady state of the form $|\Psi(r)|\exp[i\theta(r)]$, we have $\hbar\partial\theta/\partial t = -2\mu$. If a junction is sufficiently weakly coupled, we can apply this result to both its electrodes independently and since the voltage across the junction $V = (\mu 1 - \mu 2)/e$, we can write:

$$\hbar\frac{\partial\phi}{\partial t} = 2eV. \qquad \text{(A2.8.18)}$$

A2.8.4.1 Voltage-Biased Josephson Junctions and the ac Josephson Effect

When a constant finite voltage is applied across a junction, the contribution of the supercurrent to the total current can be found by integrating Equation (A2.8.18) and substituting for the phase in Equation (A2.8.4) to obtain

$$I_s = I_0 \sin\left(\frac{2eV}{\hbar}t + \phi_0\right) = I_0 \sin(\omega_J t + \phi_0). \quad \text{(A2.8.19)}$$

Thus, the supercurrent flowing through the junction at a finite voltage oscillates at a frequency $f_J = \omega_J/2\pi = V/\Phi_0$, which is 0.48 GHz/μV. This is known as the ac Josephson effect.

The normal current I_n that flows in addition to the supercurrent depends on the nature of the barrier. As we have seen in Section A2.8.1, for a tunnel junction, the normal current is a very strong function of the voltage. However, the essential physics can be understood by taking the junction resistance to be ohmic, so that $I_n = V/R$. This gives the resistively shunted junction model which can also be applied to other types of junction such as microbridges and normal barrier devices.

Figure A2.8.8(a) shows a schematic diagram of the current–voltage characteristic of a voltage-biased junction based on this model. The time averaged dc current at each voltage value is simply the normal current flowing in the resistance, but the presence of the ac component suggests that the device could be used as a tunable oscillator. However, each junction can only provide power of order RI0, which is typically less than 1 μW, so usually an array of such devices is used [5].

If we apply a combination of dc and ac voltage to the device of the form $V = V_0 + V_{rf}\cos(\omega_{rf}t)$, using a microwave source for example, then the phase itself oscillates:

$$\phi = \frac{2eV}{\hbar}t + \frac{2eV_{rf}}{\hbar\omega_{rf}}\sin\omega_{rf}t + \phi_0. \quad \text{(A2.8.20)}$$

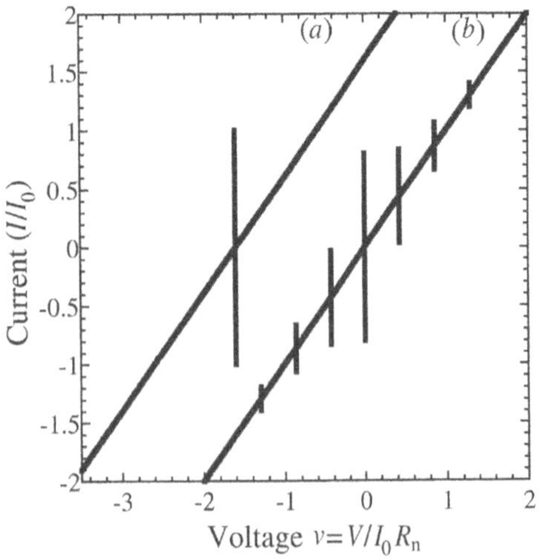

FIGURE A2.8.8 Schematic diagram of the *I*–*V* curve for (*a*) a voltage-biased Josephson junction (shifted) and (*b*) a voltage-biased Josephson junction under microwave irradiation.

This generates ac supercurrents at the Josephson frequency ω_J and the side frequencies $\omega_J \pm n\omega_{rf}$, so that:

$$I_s = I_0 \sum_n J_n\left(\frac{2eVrf}{\hbar\omega_{rf}}\right)\sin(\omega_J t \pm n\omega_{rf}t + \phi_0) \quad \text{(A2.8.21)}$$

where the amplitude functions J_n are Bessel functions. The ac supercurrents have no effect at most voltages but, when $2eV = n\hbar\omega_{rf}$, one of the side frequencies will be zero, and there is a corresponding vertical spike in the *I*–*V* characteristic. This is shown schematically in Figure A2.8.8(b). The appearance of these Shapiro spikes in the *I*–*V* characteristic is known as the inverse ac Josephson effect. Arrays of Josephson junctions biased on the same spike are currently used in standard laboratories to maintain the International Standard Value of the volt [4].

A2.8.4.2 Current-Biased Josephson Junctions

Although in the previous section, we have assumed that we could apply a constant dc voltage to the junction, in practice, this is rather difficult. This is because Josephson junctions typically have resistances of a few ohms, whereas the transmission lines used to connect the device to a source have an impedance of 50 Ω. Therefore, it is much more realistic to treat the junction as biased by a constant dc current. We are then interested in the voltage across the junction; this is determined by the Josephson element and the components of its impedance in parallel (capacitance and resistance). Both the current flowing in each of these components and the voltage are then time dependent, oscillating at the Josephson frequency corresponding to the average dc voltage, as will be shown later.

Equation (A2.8.18) enables us to generate a model for the junction in the finite voltage state with a constant current bias: the resistively and capacitively shunted junction model (RCSJ). Here, we treat the Josephson current, quasiparticle current and displacement current as parallel contributions to the total current through the device. This gives

$$I = I_0 \sin\phi + \frac{V}{R} + C\frac{dV}{dt}. \quad \text{(A2.8.22)}$$

We can write this in normalized form in terms of the phase as:

$$i = \sin\phi + \frac{d\phi}{d\tau} + \beta_c\frac{d^2\phi}{d\tau^2} \quad \text{(A2.8.23)}$$

where $i = I/I_0$, $\tau = (2\pi I_0 R/\Phi_0)t = \omega_c t$ and $\beta_c = \omega_c RC$. The normalized voltage is $v(\tau) = V(t)/I_0R = d\phi/d\tau$.

We can now use Equation (A2.8.18) to calculate the voltage for a particular bias current from the phase difference $\phi(t)$. This gives a voltage which is a periodic function of time, but the dc voltage across the junction is proportional to the time average of $d\phi/dt$ so that

$$V_{dc} = \frac{\Phi_0}{2\pi}\left\langle\frac{d\phi}{dt}\right\rangle = \frac{\Phi_0}{2\pi T}\int_0^{T_J}\frac{d\phi}{dt}dt = \frac{\Phi_0}{T_J} = \Phi_0 f_J \quad \text{(A2.8.24)}$$

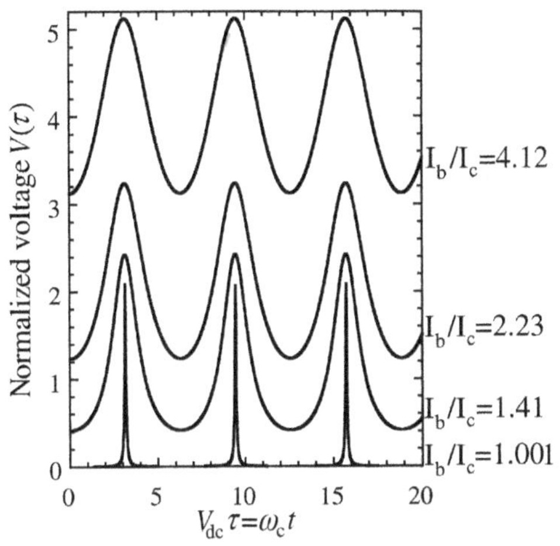

FIGURE A2.8.9 The time dependence of the voltage across a Josephson junction for three different bias currents.

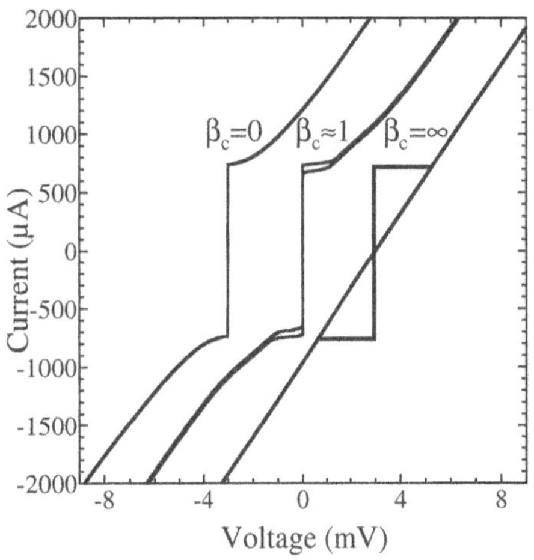

FIGURE A2.8.10 *I–V* characteristics of current-biased Josephson junctions with different values of β_c. The curve for $\beta_c \approx 1$ is experimental data measured using a bicrystal grain boundary Josephson junction. The curves for βc = 0 and βc = ∞ have been calculated using the resistance and critical current of the real junction.

where T_J is the period of the Josephson oscillation and f_J is its frequency. Figure A2.8.9 shows the time dependence of the voltage across the junction for different values of bias current. For large biases, the voltage reduces to a sinusoidal waveform with a constant voltage equal to V_{dc} superimposed upon it. As the bias current is decreased, the voltage waveform changes to a series of narrow pulses which contain a large number of harmonics, but whose fundamental frequency decreases in proportion to V_{dc}. However, it is apparent from Equation (A2.8.24) that the area under each pulse is equal to Φ_0. This is also true for other Josephson circuits where the waveform is more complicated, e.g. a SQUID, and the pulses are therefore often known as SFQ pulses (see Chapter E4.5).

A2.8.4.3 Current–Voltage Characteristics of Current-Biased Junctions

Using Equations (A2.8.14) and (A2.8.15), we can obtain the current–voltage (*I–V*) characteristics of the junction which depend on the value of the junction capacitance and hence β_c. For a junction with no capacitance we have an analytic solution:

$$I = \sqrt{I_0^2 + \frac{V^2}{R^2}}. \quad \text{(A2.8.25)}$$

However, for a finite β_c value, the detailed form of the *I–V* characteristic must be calculated numerically, which is complicated by the appearance of hysteresis for $\beta_c \geq 1$. Figure A2.8.10 shows examples of *I–V* characteristics for junctions with β_c values of 0, 1 and ∞. The curves for $\beta_c = 0$ and ∞ are theoretical, whereas the curve for $\beta_c = 1$ is experimental data. For the latter, hysteresis can be readily observed with a well-defined discontinuity in the voltage at $I = I_0$ and an increase in voltage as the bias current increases which is slower than the $\beta_c = 0$ case. As the bias current decreases through I_0, the voltage does not return to zero, but persists until a smaller value called the return or retrapping current *Ir*, which is a non-linear function of β_c. In the limit of $\beta_c = \infty$, $I_r = 0$ which means that the *I–V* characteristic reduces to a supercurrent branch and an ohmic branch, as shown.

Here the resistance is ohmic, which is clearly not in the case of a Josephson tunnel junction and to a first approximation we can usually replace the ohmic branch by the quasiparticle tunnelling characteristics described in Section A2.8.1. In general, LTS Josephson tunnel junctions always show hysteresis, whereas HTS Josephson junctions are only hysteretic at low temperatures (with the exception of intrinsic Josephson junctions between copper oxide planes [6]). This is mainly due to the difference in their resistance-area products, which are typically smaller than 1 Ω μm² for HTS junctions as compared to 100 Ω μm² above the gap voltage for LTS tunnel junctions.

Just as is the case with finite capacitance, many other properties of current-biased Josephson junctions are difficult to derive analytically. Important cases which have been treated numerically and using other techniques are the *I–V* characteristics of SQUIDs [18], the inverse ac Josephson effect and the effect of thermal noise. Under microwave irradiation, the Shapiro spikes which appear in the voltage-biased case are replaced by Shapiro steps at the same voltages, but the step heights are only equal to that of the spikes when $\hbar\omega_{rf} > 2eI_0R$. An example is shown in Figure A2.8.11. Following Ambegaokar and Halperin [19], the effect of thermal noise is quantified by $\gamma = \hbar I_0/ekT$. This causes the rounding of the current–voltage characteristic in the vicinity of the critical

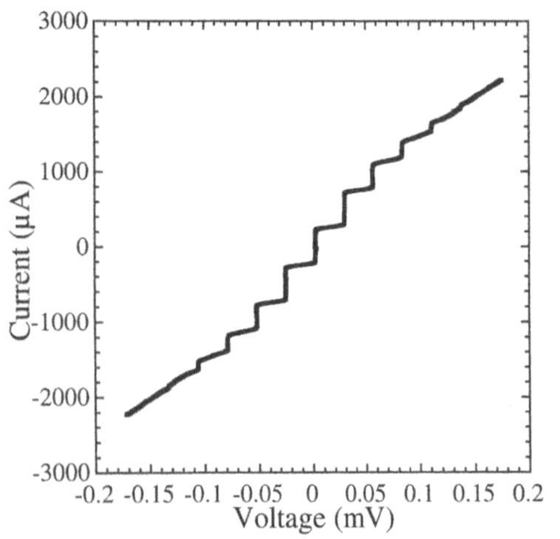

FIGURE A2.8.11 The *I–V* characteristic of a Nb-Au-Nb junction under microwave irradiation showing Shapiro steps. (Courtesy of Dr Richard Moseley and Professor Mark Blamire.)

current. This is illustrated by Figure A2.8.12, where the *I–V* characteristic of a grain boundary Josephson junction at 77 K can be compared to the $\beta_c = 0$ case in Figure A2.8.10. For high-temperature superconductors, thermal noise is a serious problem because of the much higher operation temperature.

A2.8.5 The Magnitude of the Critical Current

This subject has been left until last because, unlike the other topics covered earlier, it is very model dependent and can only be determined using microscopic theory. We will give the results for two important cases that may be said to represent the extremes: classic tunnel junctions and superconductor–normal metal–superconductor (SNS) junctions. There is a range of behaviour in between these examples which is very sensitive to the device structure.

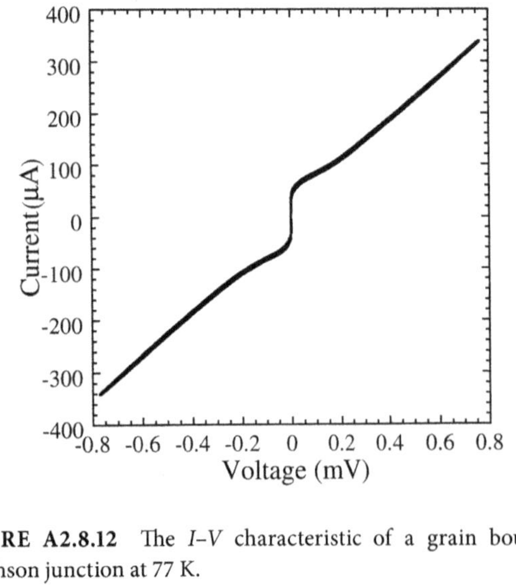

FIGURE A2.8.12 The *I–V* characteristic of a grain boundary Josephson junction at 77 K.

Tunnel junctions were the case originally treated by Josephson, but the form of $I_0(T)$ was derived by Ambegoakar and Baratoff [20]:

$$I_0 R = \frac{\pi\Delta}{2e}\tanh\left(\frac{\Delta}{2kT}\right) \qquad \text{(A2.8.26)}$$

which near T_c can be approximated by

$$I_0 R = \frac{2.34\pi k}{e}(T_c - T) \qquad \text{(A2.8.27)}$$

The temperature dependence of I_0 is largely determined by that of the gap $\Delta(T)$ and the behaviour near T_c is determined by the fact that in this region $I_0 \propto \Delta^2(T)$. If the junction had electrodes with different transition temperatures, we would have $I_0 \propto \sqrt{(T_c - T)}$. At $T = 0$, $I_0R = \pi\Delta(0)/2e$, which sets an upper limit on ω_c and, hence, on the range of frequencies over which the ac Josephson effects can be observed. Close to T_c, for microbridges, I_0R is given by Equation (A2.8.27), but at lower temperatures rises above the value for a tunnel junction and has a form which depends on the mean free path in the superconductor [11]. At $T = 0, I_0R = \pi\Delta(0)/e$ in the clean limit and $I_0R = 1.3\pi(0)/2e$ in the dirty limit.

A detailed summary of theory of SNS junctions is given in review [21]. A simple expression can only be written down for a number of special cases. Close to T_c, the temperature dependence of I_0R is given by [22]

$$I_0 R \approx \frac{\pi\Delta(0)}{e}\frac{\rho_n^2 \xi_n l}{\rho_s^2 \xi_s^2}\left(1 - \frac{T}{T_c}\right)^2 \exp\left(\frac{-l}{\xi_n}\right) \qquad \text{(A2.8.28)}$$

for junctions whose barrier length l is longer than the normal metal coherence length ξ_n, where ξ_s is the superconductor's coherence length and the barrier and superconductor have normal state resistivities ρ_n and ρ_s, respectively. The exponential dependence of I_0R upon the barrier length clearly makes its value typically much smaller than that of a tunnel junction or a microbridge near T_c.

However, Equation (A2.8.28) becomes invalid when l is shorter than ξ_n and, for very small barrier lengths, an SNS junction approaches the microbridge limit [11]. In this regime, the form of $I_0(T)$ was derived by Kulik and Omelyanchuk who provided a simple expression for $I_0(T)$ for the clean limit [23]

$$I_0 R = \frac{\pi\Delta}{e}\sin\left(\frac{\varphi}{2}\right)\tanh\frac{\Delta\cos\left(\frac{\varphi}{2}\right)}{2T} \qquad \text{(A2.8.29)}$$

where R is the junction resistance (called in literature the Sharvin resistance). More complex models of SNS junctions are overviewed in Ref. [21] including double-barrier SINIS structures as well as SFS junctions where F is a metallic ferromagnet.

The detailed behaviour of a particular device is determined by its structure and the materials used in its fabrication, which give a wide variety of behaviour even with conventional superconducting systems, but all of the results given in this section have been verified for systems which are close to the models based on which they have been derived. An excellent survey of the experimental situation for high-temperature superconductors is given in Ref. [24].

References

[1] Josephson B D 1962 Possible new effects in superconductive tunnelling *Phys. Lett.* **1** 251

[2] Ryhänen T, Seppä H, Ilmoniemi R and Knuutila J 1989 SQUID magnetometers for low-frequency applications *J. Low Temp. Phys.* **76** 287

[3] Likharev K K 1996 Ultrafast superconductor digital electronics: RSFQ technology roadmap *Czech. J. Phys.* **46** 3331

[4] Kohlmann J, Muller F, Gutmann P, Popel R, Grimm L, Dunschede F W, Meier W and Niemeyer J 1997 Improved 1V and 10 V Josephson voltage standard arrays *IEEE Trans. Appl. Supercond.* **7** 3411

[5] Wan K, Jain A K and Lukens J E 1989 Submillimeter wave generation using Josephson junction arrays *Appl. Phys. Lett.* **54** 1805

[6] Kleiner R and Muller P 1994 Intrinsic Josephson effects in high T_c superconductors *Phys. Rev. B* **49** 1327

[7] Chaudhari P, Mannhart J, Dimos D, Tsuei C C, Chi J, Oprysko M M and Scheuermann M 1988 Direct measurement of the superconducting properties of single grain-boundaries in $YBa_2Cu_3O_{7-\delta}$ *Phys. Rev. Lett.* **60** 1653

[8] Gross R and Mayer B 1991 Transport processes and noise in $YBa_2Cu_3O_{7-\delta}$ grain-boundary junctions *Physica C* **180** 235

[9] Waldram J R 1976 The Josephson effects in weakly coupled superconductors *Rep. Prog. Phys.* **39** 751

[10] Wolf EL 1996 Scanning tunneling spectrometry of a superlattice superconductor: $Bi_2Sr_2CaCu_2O_8$ *Superlattice Microstruct.* **19** 305

[11] Likharev K K 1979 Superconducting weak links *Rev. Mod. Phys.* **51** 101

[12] Pauza A J, Campbell A M, Moore D F, Somekh R E and Broers A N 1994 Josephson-junctions in $YBa_2Cu_3O_{7-\delta}$ by electron-beam irradiation *Physica B* **194** 119

[13] Yanson I K 1970 Effect of fluctuations on the dependence of the Josephson current on the magnetic field *Sov. Phys. -JETP* **31** 800

[14] Hilgenkamp H, Mannhart J and Mayer B 1996 Implications of $d_{x^2-y^2}$ symmetry and faceting for the transport properties of grain boundaries in high-T_c superconductors *Phys. Rev. B* **53** 14586

[15] Rosenthal P A, Beaseley M R, Char K, Colclough M S and Zaharchuk G 1991 Flux focusing effects in planar thin-film grain-boundary Josephson-junctions *Appl. Phys. Lett.* **59** 3482

[16] Mayer B, Schuster S, Beck A, Alff L and Gross R 1993 Magnetic-field dependence of the critical current in $YBa_2Cu_3O_{7-\delta}$ bicrystal grain-boundary junctions *Appl. Phys. Lett.* **62** 783

[17] Gray K E, Field M B and Miller D J 1998 Explanation of low critical currents in flat, bulk versus meandering, thin-film [001] tilt bicrystal grain boundaries in $YBa_2Cu_3O_{7-\delta}$ *Phys. Rev. B* **58** 9543

[18] Tesche C D and Clarke J 1977 dc SQUID: Noise and Optimisation *J. Low Temp. Phys.* **29** 301

[19] Ambegaokar V and Halperin B I 1969 Voltage due to thermal noise in the dc Josephson effect *Phys. Rev. Lett.* **22** 1364

[20] Ambegaokar V and Baratoff A 1963 Tunneling between superconductors *Phys. Rev. Lett.* **10** 486

[21] Golubov A A, Kupriyanov M Yu and Il'ichev E 2004 The current-phase relation in Josephson junctions, *Rev. Mod. Phys.* **76** 411

[22] de Gennes P G 1964 Boundary effects in superconductors *Rev. Mod. Phys.* **36** 225

[23] Kulik I O and Omelyanchuk A N 1977, *Fiz. Nizk. Temp.* **3** 945 [**Sov. J. Low Temp. Phys.** **3** 459 (1977)]

[24] Delin K A and Kleinsasser A W 1996 Stationary properties of high-critical-temperature proximity effect Josephson junctions *Supercond. Sci. Technol.* **9** 227

A2.9

Other Josephson-Related Phenomena

Alexander A. Golubov and Francesco Tafuri

A2.9.1 Josephson Effect from Quasi-Particle Andreev Bound States

Andreev reflection (*AR*) is the scattering mechanism describing how an electron excitation slightly above the Fermi level in the normal metal is reflected at the interface as a hole excitation slightly below the Fermi level [1]. The missing charge of 2e is removed as a Cooper pair. This is a branch-crossing process which converts electrons into holes and vice versa, and therefore changes the net charge in the excitation distribution. The reflected hole (or electron) has a shift in phase compared to the incoming electron (or hole) wave function: $\phi_{hole} = \phi_{elect}+\phi_{superc}+\arccos(E/\Delta)$ ($\phi_{elect} = \phi_{hole}-\phi_{superc}+\arccos(E/\Delta)$), where Δ and ϕ_{superc} are the gap value and the superconducting phase of the *S*. The macroscopic phase of the *S* and the microscopic phase of the quasiparticles are therefore mixed through *AR*. To provide an intuitive idea of effects related to *AR*, the Andreev-reflected holes act as a parallel conduction channel to the initial electron current, thus doubling the normal-state conductance of the *S/N* interface for applied voltages less than the superconducting gap $eV < \Delta$ [2]. Blonder, Tinkham and Klapwijk [2] (*BTK*) introduced the dimensionless parameter *Z*, proportional to the potential barrier at the interface, to describe the barrier transparency. This allows the continuous passage from the tunnel limit to a transmissive barrier since the barrier transparency is defined as $\bar{D} = 1/(1+Z^2)$. More elements on *AR* can be found in [3].

The Landauer conductance expression has been extended to the case of an S–N interface through scattering matrix theory [4]:

$$G_{NS} = \frac{2e^2}{\pi\hbar}\sum_{n=1}^{N}\frac{D_n^2}{(2-D_n)^2} \qquad \text{(A2.9.1)}$$

Here the D_n's are the transmission eigenvalues of the disordered normal part. The difference in the behavior of the transmission eigenvalues D_n will lead to different mesoscopic behaviors of tunnel junctions and metallic weak links. While in the former case many small D_n's are relevant, in the latter most D_n's are close to zero or unity. This expression is valid at zero voltage and zero magnetic field. Application of either a voltage or a magnetic field reduces the contact resistance of the *NS* junction by a factor of 2.

The physics of the *dc* Josephson effect can be understood in terms of Andreev reflection. An electron with momentum *k* impinging on one of the interfaces is converted into a hole moving in the opposite direction, thus generating a Cooper pair in a superconductor. This hole is consequently Andreev-reflected at the second interface and is converted back to an electron, leading to a destruction of a Cooper pair (see Figure A2.9.1). As a result of this cycle, a pair of correlated electrons is transferred from one superconductor to another creating a supercurrent flow across a junction. Since Andreev reflection amplitudes depend on corresponding phases $\phi_{1,2}$, the resulting current depends on the phase difference ϕ thus leading to the *dc* Josephson effect [5-8].

Due to electron–hole interference in the quantum well formed by the pairing potentials of the superconducting electrodes, standing waves with quantized energy, E_{AB}, appear in weak link region The corresponding quantum states are referred to as Andreev bound states (ABS). The physics of ABS in Josephson junctions has been studied extensively starting with the pioneering work of Kulik (1969) [8].

It follows from the microscopic theory of superconductivity (see reviews [5–7]) that in stationary situations the supercurrent across Josephson junctions

$$I_S(\phi) \propto \int_{-\infty}^{\infty} dE[1-2f(E)]Im\{I_E(\phi)\} \qquad \text{(A2.9.2)}$$

depends on the electron energy distribution function $f(E)$ and the spectral current $Im\{I_E\}$. Spectral current incorporates the information on the energy distribution of the ABS in a junction. $Im\ I_E$ depends on the distance between the superconductors *d*, transport parameters of the junction's materials (resistivities $\rho_{1,2}$, Fermi velocities $v_{F\,1,2}$ and parameters of interfaces).

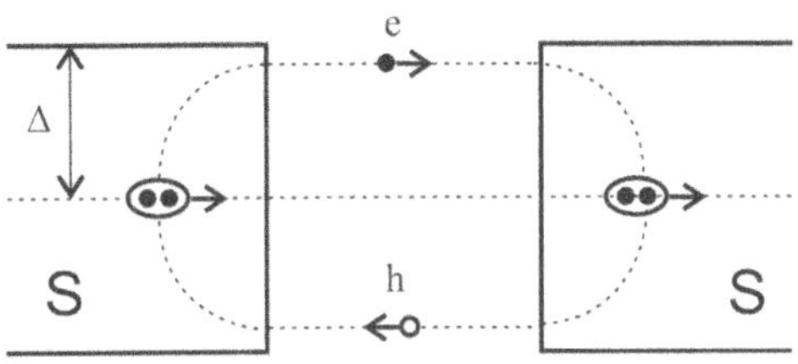

FIGURE A2.9.1 The diagram illustrating the physical mechanism of formation of a Josephson current across a junction. An electron *e* and the Andreev-reflected hole *h* are shown. A Cooper pair is transferred from the left superconductor to the right one.

A very interesting property of the Andreev reflection in a S_1–N–S_2 structure is that the electron obtains an extra phase of $\phi_1 - \phi_2 + \pi$ in each period. The Josephson effect can be reformulated in terms of this property and of quasi-particle bound states. The spectrum of the elementary excitations of a *N* layer in contact with *S* on both sides is quantized for $E < \Delta$. The Josephson current in each channel will result from two Andreev bound states with specific phase-dependent energies, lying inside the gap region $-\Delta, +\Delta$. The energy of the Andreev ground state will determine the Josephson coupling energy. The imbalance in the populations of the two Andreev bound levels finally determines the contribution to the net supercurrent flowing in each channel. The expression of the bound state energy in a S–N–S one-dimensional system, in the short junction limit $L \ll \xi_N$ [8]: $E = \pm\Delta\sqrt{1 - D\cdot\sin^2(\phi/2)}$. There is a general relation between the current through the Andreev state and the phase dispersion of the energy of the Andreev state, $I_s = (2e/\hbar)dE/d\phi$. This equation can be derived directly from the Bogoliubov–deGennes equation or deduced from the thermodynamical equation by using a microscopic expression for the junction free energy [4, 9, 10]. The total supercurrent is given by a summation over the contributions of the current-carrying states which all depend on the phase difference between the two superconductors. The *dc* Josephson current is a resonant effect, where the Josephson current flows through resonant Andreev levels.

As follows from the above discussion, in the structure where the momentum of an electron in the weak link region is a good quantum number (so-called "clean" Josephson junctions), ABS form a regular structure in energy, and $Im\ I_E(\phi)$ is peaked at the corresponding energies [11–18]. An increase of the degree of disorder in the weak link leads to broadening and decreasing of the amplitudes of the peaks. The disorder generates a distribution of the lengths of the electron trajectories in the weak link. Therefore $Im\ I_E(\phi)$ for a disordered junction is a weighted average of the ballistic result over the corresponding distribution. This makes the spectral current $Im\{I_E(\phi)\}$ a continuous function of energy.

The electron energy distribution function in Equation (A2.9.2) defines the population of ABS at a given temperature. Thus the Equation (A2.9.2) shows that the whole supercurrent $I_S(\phi)$ is the sum of the partial currents transported via ABS. Therefore one can modify the shape of $I_S(\phi)$ in two ways: (a) by modifying the spectral current $Im\ I_E(\phi)$, changing the material parameters or geometry of a junction; (b) by manipulating the occupation numbers of ABS, i.e. creating a nonequilibrium distribution function $f(E)$ in a weak link.

Something special with Andreev reflection happens for graphene/superconductor (*G/S*) interfaces, because of the unusual electronic properties of the charge carriers in graphene (no Fermi surface at zero doping and conical band structure) [10]. Differently from the usual case, where the electron and hole both lie in the conduction band, at a *G/S* interface specular *AR* happens if an electron in the conduction band is converted into a hole in the valence band. In undoped graphene, when $E_F = 0$, Andreev reflection is interband at all excitation energies. This has obvious consequences on the Josephson coupling. I_c is $\approx e\Delta/max(W/L, W/\lambda_F)$ where *W* and *L* are the width and length of the junction barrier and λ_F is the Fermi wavelength. At Dirac point $E_F = 0$ ($\lambda_F \to \infty$), I_c reaches its minimal value $\approx e\Delta/(W/L)$. Instead of being independent on *L*, as expected for a short ballistic Josephson junction, I_c at the Dirac point has the diffusion-like scaling $\propto 1/L$. Near the Dirac point in the ballistic regime there are few bound states, the I_s-ϕ relation is identical to the expression obtained in a disordered metal barrier. When finite gate voltages are applied to the graphene, the number of bound states increases and they dominate the supercurrent [10].

Junctions with graphene barrier fall in the emerging category of coplanar hybrid devices, where the barrier is not a thin film but an exfoliated flake. Pre-built components of the junctions, produced through different techniques and mechanically assembled in the last stage of fabrication (see Section H3.1.2). Nanowires (NWs) can be also used as barriers. *S-NW-S* junctions have been proposed as host and sensor of phenomena associated to the presence of Majorana fermions [19–22]. Majorana fermions enable the tunneling of single electrons (with a larger probability $D^{1/2}$). The switch from 2e to e as the unit of transferred charge between the superconductors amounts to a doubling of the fundamental periodicity of the Josephson energy, from $E \propto \cos\phi$ to $E\cos(\phi/2)$ [21, 22]. In contrast to ordinary Josephson currents, this contribution reflects tunneling of half of a Cooper pair across the junction. Such a fractional Josephson effect has been later established in other systems supporting Majorana modes and in direct junctions between *p*-wave superconductors. If the superconductors form a ring, enclosing a flux Φ, the period of the flux dependence of the supercurrent I_s doubles from 2π to 4π. Evidence for an anomalous current phase relation in topological insulator Josephson junctions has been given in Refs. [23, 24].

A2.9.2 Junctions with Magnetic Barriers

Ferromagnetic (*F*) barriers give new functionalities to *JJs*. An established example is the phase battery, a device which provides a constant phase shift between the two superconductors, taking advantage of the tunable equilibrium phase difference offered by S–F–S *JJs*. These junctions are the ambitious playground of two macroscopic forms of order, superconductivity and magnetism, which originally believed to be antagonist, seem more and more to cooperate for new types of order [25, 26]. Both *Nb* and *NbN* technologies have been employed to produce junctions with ferromagnetic barriers.

The physics of the co-existence of superconductivity and magnetism in a S–F bilayer can be naively explained in terms of the Andreev-correlated electrons and holes, having opposite spin directions, under the exchange field H of a ferromagnet. This results in an energy shift between these quasiparticles and the creation of a nonzero momentum Q of Cooper pairs and thus in an energy shift between these quasiparticles and creation of a nonzero momentum of Cooper pairs [27]. The amplitude of superconducting correlations oscillates spatially in the F-metal as cos Qx. Sign change of this amplitude is equivalent to periodic 0-π phase jumps at certain points in the F-metal. Such oscillating Cooper pair amplitude is an analog of the so-called Larkin–Ovchinnikov–Fulde–Ferrel (LOFF) state in bulk magnetic superconductors [28, 29].

The oscillations in F decay with the distance from the S–F interface. There is quantitative difference between clean and diffusive ferromagnets regarding the decay length. At $T = 0$, the decay length in the dirty limit exactly coincides with the oscillation period [30, 31]. In the clean limit the decay length is infinite $T = 0$ [32] and is limited only by elastic impurity scattering [33] or spin-orbit scattering [27] and typically exceeds the oscillation period. Therefore, the spatial oscillations are easier to observe in cleaner systems. The crossover between these two limits has been recognized as an important problem for the theoretical description of S–F–S structures [34–38]. In a Josephson junction this turns into damped oscillations of I_c as a function of the thickness of the ferromagnet layer d_F, where the negative values of I_c correspond to a π-junction [25, 26, 39].

The possibility of π-state was first predicted by Bulaevskii *et al.* [40] in a Josephson tunnel junction with magnetic impurities localized in the barrier. Bulaevskii *et al.* [40] also predicted that a superconducting ring containing π-junction can generate a spontaneous current and magnetic flux.

During long time, the observation of Josephson coupling in S–F–S junctions was a serious challenge. The solution of the problem of measuring supercurrents in S–F–S Josephson junctions was found by employing a dilute ferromagnetic alloy. The first experimental observation of supercurrents in S–F–S junctions and the crossover from 0- to π-state in Nb/Cu_xNi_{1-x}/Nb Josephson junctions was reported by Ryazanov *et al.* [41–44], and further phase-sensitive measurements were reported by Ryazanov *et al.* [45]. The π-state is characterized by the phase shift of π in the ground state of junction and is formally described by the negative critical current I_C in the CPR $I_S(\phi) = I_C \sin(\phi)$.

A 0-π crossover as a function of temperature has been first observed in Nb/Cu_xNi_{1-x}/Nb *JJs* [42]. For specific values of d_F, with decreasing temperature a maximum of I_c is found followed by a strong decrease down to zero, after which I_c rises again. This is a neat signature of a 0-π transition, which occurs because of the temperature dependence of ξ_F. Another set of experiments used a barrier thickness interval ranging from 8 to 28 nm with an enhancement of J_c of six orders of magnitude [46]. These discoveries stimulated further experimental activity in this field and led to observation of new phenomena in Josephson junctions with interlayers made from various ferromagnetic alloys [47–96].

Since then a variety of systems exhibiting π-states have been found including planar S–F–S proximity-effect structures, tunnel junctions with a magnetic insulator or magnetically active interfaces, and structures with barriers containing more than one magnetic layer. Most of the junctions use F-metallic barrier and are overdamped. The CuNi alloy has been used in S–I–F–S *JJs* for phase shifters [97] due to its stable magnetic domain structure (high coercive field) and out-of-plane magnetic anisotropy. A magnetically soft PdFe alloy with low Fe-content because of an in-plane magnetic anisotropy and small coercive field is more convenient for the implementation of switching devices. These S–I–F–S *JJs* have allowed an increase of the I_cR_n quality factor from a few tens of nV to 700 μV (at $T = 4.2$ K), with J_c values up to 3.3 kA/cm^2 [99]. These values are promising for being compatible with high-speed SFQ digital and mixed-signal integrated circuits which in the Hypres standard use 4.5 kA/cm^2 S–I–S process [99].

Stimulated by progress in experimental realization of S–F–S π-junctions, extensive theoretical studies of various types of such structures have been performed in recent years. The physics of 0- to π crossover in S–F–S junctions was studied theoretically using different approaches by [30–32, 100–112]. In addition to π-transitions, new intriguing predictions have been made of complex CPR.

Recently, the generation of the long-range triplet order parameter was predicted in structures with inhomogeneous magnetization or with non-collinear orientations of magnetization in different F-layers [113–116].

Variety of systems exhibiting π-states includes planar S–F–S proximity effect structures, tunnel junctions with magnetic insulator or magnetically active interfaces and structures with the barriers containing more than one magnetic layer.

In junctions with F barriers, the $I_c(H)$ keeps memory of how the magnetic field has been applied [117, 118]. This behavior cannot be framed in any of the canonical deviations of the Fraunhofer pattern [119] or expressed in terms of π-states.

A2.9.3 Magnetic Patterns

The anomalies in the magnetic pattern are a hint to review most of the unconventional magnetic patterns. In Figure A2.9.2 an overview of a few significant patterns is given, referring to [119] for all different types of more canonical patterns of LTS *JJs* and to [120–123] for the more unconventional ones measured mostly in HTS and coplanar hybrid *JJs*. For long junctions, patterns significantly depend on the ratio between the width W and the Josephson penetration depth λ_J, a rough measure of how the current is not uniformly distributed along

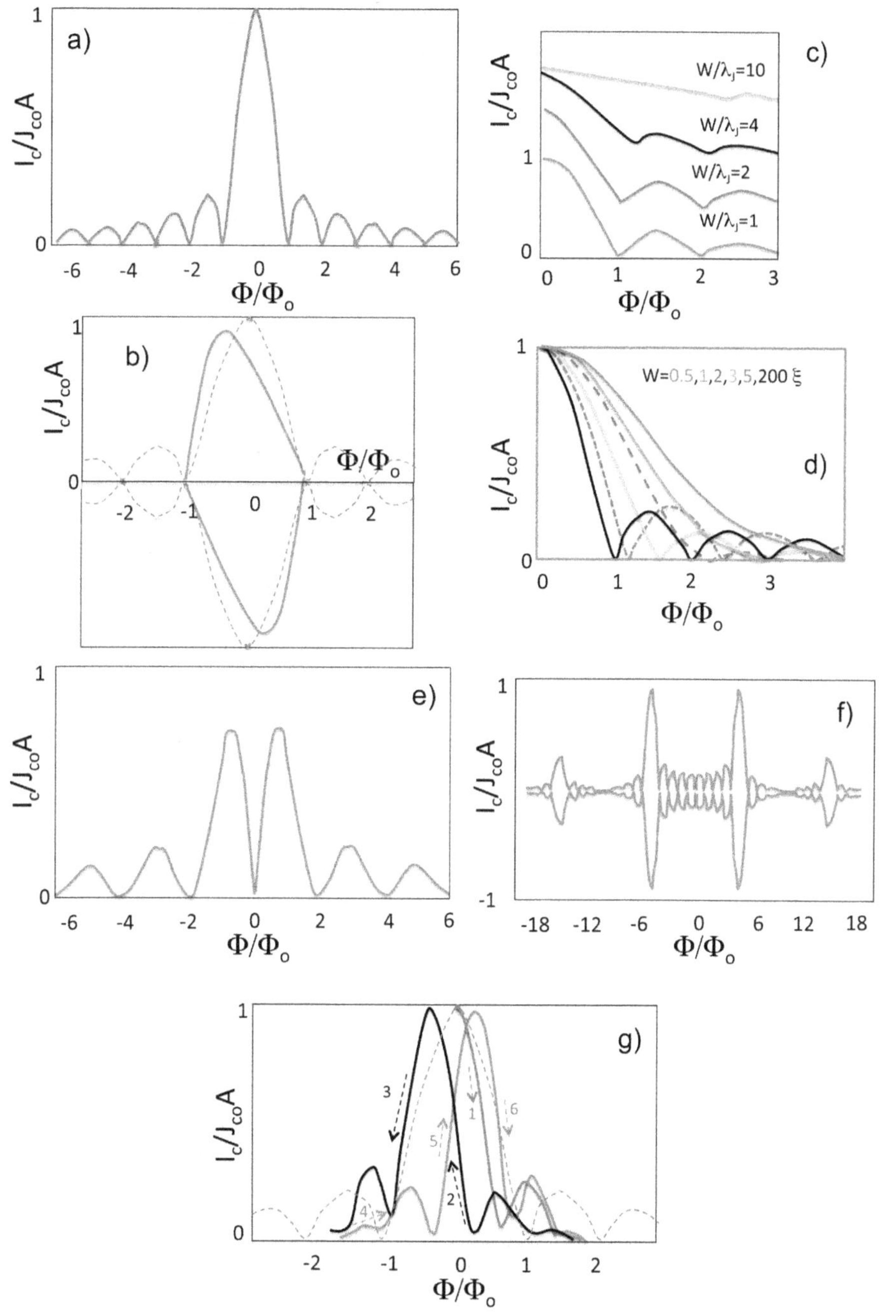

FIGURE A2.9.2 Schematic dependence of the I_c on the magnetic field for an ideal rectangular tunnel junction (Fraunhofer pattern) (a) and for long junctions as a function of W/λ_J, (b) long junctions with current leads from the same side of the device and (c) long junctions with transverse current leads from opposite sides or with longitudinal leads), respectively. In S–N–S junctions the $I_c(H)$ strongly depends on the barrier dimensions (d). The characteristic pattern of a 0-π corner junction (e) and of an array of 0-π junctions (f). Characteristic features of the pattern in (f) are the occurrence of sharp maxima in I_c for an applied magnetic flux $\Phi_{max} = N\Phi_o/2$ and the vanishing I_c at $\Phi = 0$ for an even number of facets N. The number of minima in I_c in the flux range $-\Phi_{max} < \Phi < \Phi_{max}$ is predicted to be N-1. In this limit, the facet length $L_f < \lambda_J$ and self-generated magnetic flux is thus not taken into account. The corner junction (e) is the limit for $N = 1$. In (g) the I_c keeps memory of the history of how the magnetic field has been applied, as for instance in spin filter S–F–S *JJs*.

the width of the junction (see Chapters A.2.12 and H.3.1). In long junctions with current leads from the same side of the device the central lobe of the Fraunhofer pattern is even distorted and shifted [see Figure A2.9.2(b)]. For junctions with transverse current leads from opposite sides or with longitudinal leads, there is no shift and distortion of the maximum, but the amplitude of the lobes significantly decreases when W/λ_J increases [see Figure A2.9.2(c)]. These are the classical notions of tunnel junctions [119]. In diffusive S–N–S junctions the Fraunhofer pattern turns into a monotonic decay when the width of the normal wire W is smaller than the magnetic length $\xi_H = \sqrt{\Phi_0/H}$ as shown in Figure A2.9.2(d) [124]. This behavior is intimately related to the appearance of a linear array of vortices in the middle of the normal wire, the properties of which are very similar to those in the mixed state of a type II superconductor [124]. In Figure A2.9.2(e) the characteristic pattern of a 0-π corner junction with two symmetric maxima at finite H is shown, as commonly measured in HTS *JJs* [120–122, 125, 126]. When increasing the number of 0-π facets [122] symmetric maxima move to higher H and a number of small I_c oscillations proportional to the number of facets appear [see Figure A2.9.2(f)] [127]. If the *OP* were to comprise an imaginary *s*-wave admixture, the pattern for the arrays would display distinct asymmetries, especially for low fields. This is a clear example of how intrinsic phase variations are induced, even in the absence of externally applied fields, by momentum-dependent pairing wave functions. Magnetic interactions in the barrier region, as shown in the next example, or locally applied currents produce phase variation as well. In Figure A2.9.2(g) the pattern for a spin filter junction is shown. The absolute maximum of I_c shifts to finite fields depending on how H is swept [117, 118]. When H is first increased from $H = 0$ a standard Fraunhofer pattern is measured (red continuous curve). When decreasing the field the I_c maximum shifts to negative finite H values, and the pattern shifts in the negative direction of H (black continuous curve). When increasing H from the maximum negative value, the pattern is shifted again. This time the maximum I_c value occurs at positive finite values of H (blue continuous curve). A number of other effects and anomalous behaviors in $I_c(H)$ is listed in [119, 121, 122]. Flux focusing effects can also play a relevant role and change the periodicity between two minima of the magnetic pattern [123, 128, 129]. In the thin limit approximation, for instance, the effective area of the Josephson junction scales as the square of ($W \propto 1/W^2$) rather than as the usual $1/(W\,(2\lambda + L)$ dependence, where λ is the London penetration depth [129]. A prevailing second harmonic can also induce a dramatic change in the flux periodicity as occurring in spin filter junctions [117, 118].

Ferromagnetic insulators (F_i) give tunnel-like barriers, and junctions are underdamped [117]. J_c values in NbN-GdN-NbN range from 60 A/cm^2 to 1.7×10^3 A/cm^2, while R_nA values range from 5×10^{-7} Ω cm^2 to 5×10^{-6} Ω cm^2, respectively [117, 118]. These junctions have also shown spin filter properties. Below the Curie temperature the spin-dependent splitting of the band structure causes a spin polarization in the supercurrent, because the spin-up and spin-down electron of the Cooper pair experience different effective barrier heights, and one channel is partially suppressed. Spin filter efficiency can be estimated through conductance measurements [117]. I_cR_n are about 1 mV for non-spin filter devices and about one order of magnitude lower for spin filter devices with spin efficiency larger than 90% [118]. For thicker barriers ($d > 1.5$ nm), these junctions present spin filter behavior, and a pure second harmonic $I_s(\phi)$ relation [130]. For these lower J_c devices, dissipation is well below to damping threshold to observe macroscopic quantum tunneling (MQT) [118].

A2.9.4. Hints on Triplet Superconductivity and Multi-Layered Magnetic Junctions

The prediction of generation of a long-range triplet order parameter in structures with inhomogeneous magnetization or with noncollinear orientations of magnetization in different *F*-layers stimulated intensive experimental work. In Nb–Co (a strong ferromagnetic material)–Nb junctions, when thin layers of either PdNi or CuNi weakly ferromagnetic alloys are inserted between the Co and the two superconducting Nb electrodes, I_c hardly decays for Co thicknesses in the range of 12–28 nm, whereas it decays very steeply in similar junctions without the alloy layers supporting the notion of induced spin-triplet pair correlations [131]. In another set of experiments, in the stack composed by Nb(250 nm)/Cu(5 nm)/Py(1.5 nm)/Cu(5 nm)/Co(5.5 nm)/Cu(5 nm)/Py(1.5 nm)/Cu(5 nm)/Nb (250 nm), 4.5-nm thick Ho layers at Nb/Co interface replacing Py layers, generate spin-polarized supercurrents [132]. These are two examples on a field where major activities are expected [25, 26, 133]. Recently, an incomplete 0-π transition has been measured in highly spin-polarized tunnel ferromagnetic junctions [134]. This observation is consistent with an unconventional magnetic activity of the barrier and represents a key tool to disclose the presence of spin-triplet correlations in JJs with ferromagnetic–insulator (IF) barriers [134].

Another type of magnetic junctions, which is very attractive for applications, are multilayered SFsFS and SIsFS structures since they permit to combine high critical current with the possibility to control its magnitude. The SFsFS devices [135–137] operate as pseudo spin-valve structures, where the mutual magnetization of the F-layers determines the effective exchange and the critical current of the device. On the other hand, SIsFS junctions [138–144] provide high performance of tunnel S–I–S junction with possibility of switching between 0- and π-states. These systems are also characterized by specific hysteretic current–phase relations with possible multiple branches and ambiguity [137, 145, 146]. Especially, these peculiarieties are important in the vicinity of the 0-π transition, where the second harmonic in CPR of single junction plays a dominant role.

Acknowledgements

We would like to thank M. Yu Kupriyanov, S. V. Bakurskiy, V. V. Ryazanov and D. Massarotti for valuable discussions.

Refernces

[1] A. F. Andreev. The thermal conductivity of the intermediate state in superconductors. *Zh. Eksp. Teor. Fiz. (Sov. Phys. JETP)*, 46 (19):1823–1828 (1228–1231), 1964.

[2] G. E. Blonder, M. Tinkham, and T. M. Klapwijk. Transition from metallic to tunneling regimes in superconducting microconstrictions: Excess current, charge imbalance, and supercurrent conversion. *Phys. Rev. B*, 25:4515–4532, Apr 1982.

[3] A. A. Golubov, S. V. Bakurskiy, and M. Yu Kupriyanov. Basic properties of the Josephson effect. In Francesco Tafuri, editor, *Fundamentals and Frontiers of the Josephson Effect*, pages 81–116, Chapter 3. Springer Nature Swizterland, 2019.

[4] C. W. J. Beenakker. Random-matrix theory of quantum transport. *Rev. Mod. Phys.*, 69:731–808, Jul 1997.

[5] C. J. Lambert and R. Raimondi. Phase-coherent transport in hybrid superconducting nanostructures. *J. Phys. Condens. Matter*, 10(5):901–941, Feb 9 1998.

[6] W. Belzig, F. K. Wilhelm, G. Schon, C. Bruder, and A. D. Zaikin. Quasiclassical green's function approach to mesoscopic superconductivity (vol 25, pg 1251, 1999). *Superlattices Microstruct.*, 35(1–2):157, Jan-Feb 2004.

[7] N. B. Kopnin. *Theory of Nonequilibrium Superconductivity.* Number Vol. 110 in International Series of Monographs on Physics. Clarendon Press, 2001.

[8] I. O. Kulik. Macroscopic quantization and the proximity effect in s-n-s junctions. *Zh. Eksp. Teor. Fiz. (Sov. Phys. JETP)*, 57 (30):1745–1759 (944–950), 1969.

[9] A. A. Golubov, M. Yu Kupriyanov, and E. Il'ichev. The current-phase relation in Josephson junctions. *Rev. Mod. Phys.*, 76:411–469, Apr 2004.

[10] C. W. J. Beenakker. *Colloquium:* Andreev reflection and klein tunneling in graphene. *Rev. Mod. Phys.*, 80:1337–1354, Oct 2008.

[11] I. O. Kulik. Macroscopic quantization and proximity effect in s-n-s junctions. *Sov. Phys. JETP-USSR*, 30(5):944, 1970.

[12] C. Ishii. Josephson currents through junctions with normal metal barriers. *Prog. Theor. Phys.*, 44(6):1525–1547, 1970.

[13] J. Bardeen and J. L. Johnson. Josephson current flow in pure superconducting-normal-superconducting junctions. *Phys. Rev. B (Solid State)*, 5(1):72–78, 1972.

[14] P. F. Bagwell. Suppression of the Josephson current through a narrow, mesoscopic, semiconductor channel by a single impurity. *Phys. Rev. B*, 46(19):12573–12586, Nov 15 1992.

[15] U. Schussler and R. Kummel. Andreev scattering, Josephson currents, and coupling energy in clean superconductor-semiconductor-superconductor junctions. *Phys. Rev. B (Condens. Matter)*, 47(5):2754–2759, 1993.

[16] H. X. Tang, Z. D. Wang, and Y. Zhang. Josephson current in a clean superconductor normal-metal superconductor junction. In *Czechoslovak Journal of Physics*, volume 46 of Part *S2; Superconductivity I*, pages 565–566, 1996.

[17] H. X. Tang, Z. D. Wang, and Y. H. Zhang. Normal reflection effect on the critical current in a clean-limit superconductor-normal-metal-superconductor junction. *Zeitschrift Fur Physik B (Condens. Matter)*, 101(3):359–366, Nov 1996.

[18] H. X. Tang, Z. D. Wang, and J. X. Zhu. Supercurrent and quasiparticle interference between two d-wave superconductors coupled by a normal metal or insulator. *Phys. Rev. B*, 54(17):12509–12516, Nov 1 1996.

[19] Liang Fu and C. L. Kane. Superconducting proximity effect and majorana fermions at the surface of a topological insulator. *Phys. Rev. Lett.*, 100:096407, Mar 2008.

[20] V. Mourik, K. Zuo, S. M. Frolov, S. R. Plissard, E. P. A. M. Bakkers, and L. P. Kouwenhoven. Signatures of majorana fermions in hybrid superconductor-semiconductor nanowire devices. *Science*, 336(6084):1003–1007, 2012.

[21] C. W. J. Beenakker. Random-matrix theory of majorana fermions and topological superconductors. *Rev. Mod. Phys.*, 87:1037–1066, Sep 2015.

[22] Jason Alicea. New directions in the pursuit of majorana fermions in solid state systems. *Rep. Prog. Phys.*, 75(7):076501, 2012.

[23] C. Kurter, A. D. K. Finck, Y. S. Hor, and D. J. Van Harlingen. Evidence for an anomalous current-phase relation in topological insulator Josephson junctions. *Nat. Commun.*, 6:06, 2015.

[24] I. Sochnikov, L. Maier, C. A. Watson, J. R. Kirtley, C. Gould, G. Tkachov, E. M. Hankiewicz, C. Brüne, H. Buhmann, L. W. Molenkamp, and K. A. Moler. Nonsinusoidal current-phase relationship in Josephson junctions from the 3d topological insulator hgte. *Phys. Rev. Lett.*, 114:066801, Feb 2015.

[25] A. I. Buzdin. Proximity effects in superconductor-ferromagnet heterostructures. *Rev. Mod. Phys.*, 77:935–976, Sep 2005.

[26] J. Linder and J. W. A. Robinson. Superconducting spintronics. *Nat. Phys.*, 11(4):307–315, 2015.

[27] E. A. Demler, G. B. Arnold, and M. R. Beasley. Superconducting proximity effects in magnetic metals. *Phys. Rev. B (Condens. Matter)*, 55(22):15174–15182, 1997.

[28] A. I. Larkin and Y. N. Ovchinnikov. Ai larkin and yn ovchinnikov, zh. eksp. teor. fiz. 47, 1136. *Zh. Eksp. Teor. Fiz.*,47:1136, 1964.

[29] P. Fulde and R. A. Ferrell. Superconductivity in a strong spinexchange field. *Phys. Rev.*, 135(3A):A550, 1964.

[30] Z. Radovic, M. Ledvij, L. Dobrosavljevic-Grujic, A. I Buzdin, and J. R. Clem. Transition temperatures of superconductor-ferromagnet superlattices. *Phys. Rev. B*, 44(2):759, 1991.

[31] A. I. Buzdin and M. Y. Kupriyanov. Josephson junction with a ferromagnetic layer. *JETP Lett.*, 53(6):321–326, Mar 25 1991.

[32] A. I. Buzdin, L. N. Bulaevskii, and S. V. Panyukov. Critical-current oscillations as a function of the exchange field and thickness of the ferromagnetic metal (f) in an s-f-s Josephson junction. *JETP Lett.*, 35(4):178–180, 1982.

[33] F. S. Bergeret, A. F. Volkov, and K. B. Efetov. Local density of states in superconductor-strong ferromagnet structures. *Phys. Rev. B*, 65(13), Apr 1 2002.

[34] F. S. Bergeret, A. F. Volkov, and K. B. Efetov. Josephson current in superconductor-ferromagnet structures with a nonhomogeneous magnetization. *Phys. Rev. B (Condens. Matter Mater. Phys.)*, 64(13):1345061, 2001.

[35] F. Born, M. Siegel, E. K. Hollmann, H. Braak, A. A. Golubov, D. Yu Gusakova, and M. Yu Kupriyanov. Multiple 0-pi transitions in superconductor/insulator/ferromagnet/superconductor Josephson tunnel junctions. *Phys. Rev. B (Condens. Matter Mater. Phys.)*, 74(14):140501-1-4, 2006.

[36] D. Yu Gusakova, A. A. Golubov, and M. Yu Kupriyanov. Superconducting decay length in a ferromagnetic metal. *JEPT Lett.*, 83(9):418–422, 2006.

[37] J. Linder, M. Zareyan, and A. Sudbo. Proximity effect in ferromagnet/superconductor hybrids: from diffusive to ballistic motion. *Phys. Rev. B (Condens. Matter Mater. Phys.)*, 79(6):064514, 2009.

[38] N. G. Pugach, M. Yu Kupriyanov, E. Goldobin, R. Kleiner, and D. Koelle. Superconductor-insulator-ferromagnet-superconductor Josephson junction: From the dirty to the clean limit. *Phys. Rev. B (Condens. Matter Mater. Phys.)*, 84(14):144513, 2011.

[39] L. N. Bulaevskii, V. V. Kuzii, and A. A. Sobyanin. Superconducting system with weak coupling to the current in the ground state. *Pis'ma Zh. Eksp. Teor. Fiz. (JETP Lett.)*, 25 (25):314–318 (290–294), 1977.

[40] L. N. Bulaevskii, V. V. Kuzii, and A. A. Sobyanin. A superconductive system with weak coupling and with a current in the ground state. *Zh. Eksp. Teor. Fiz., Pis'ma V Redaktsiyu*, 25(7):314–318, 1977.

[41] V. V. Ryazanov, A. V. Veretennikov, V. A. Oboznov, A. Yu Rusanov, V. A. Larkin, A. A. Golubov, and J. Aarts. Reentrant superconducting behavior of the Josephson sfs junction. Evidence for the π-phase state. *Physica C*, 341(Part 3):1613–1614, 2000.

[42] V. V. Ryazanov, V. A. Oboznov, A. Y. Rusanov, A. V. Veretennikov, A. A. Golubov, and J. Aarts. Coupling of two superconductors through a ferromagnet: evidence for a pi junction. *Phys. Rev. Lett.*, 86(11):2427–2430, 2001.

[43] V. V. Ryazanov, V. A. Oboznov, A. Y. Rusanov, A. V. Veretennikov, A. A. Golubov, and J. Aarts. Coupling of two superconductors through a ferromagnet. sfs pi-junctions and inner frustrated networks. In T. Martin, G. Montambaux, and J. T. Thanh Van, editors, *Electronic Correlations: From Meso- to Nano-physics*, pages 143–146. CNRS; Commiss Energie Atom; Union Europeenne; Reg Rhone Alpes, 2001. 36th Rencontres de Moriond on Electronic Correlations, Savoie, FRANCE, Ja 20–27 2001.

[44] A. V. Veretennikov, V. V. Ryazanov, V. A. Oboznov, A. Yu Rusanov, V. A. Larkin, and J. Aarts. Super-currents through the superconductor-ferromagnet-superconductor (sfs) junctions. *Physica B Condens. Matter*, 284(Part 1):495–496, 2000.

[45] V. V. Ryazanov, V. A. Oboznov, A. V. Veretennikov, and A. Y. Ru-sanov. Intrinsically frustrated superconducting array of superconductor-ferromagnet-superconductor pi junctions. *Phys. Rev. B*, 65(2), Jan 1 2002.

[46] V. A. Oboznov, V. V. Bol'ginov, A. K. Feofanov, V. V. Ryazanov, and A. I. Buzdin. Thickness dependence of the Josephson ground states of superconductor-ferromagnet-superconductor junctions. *Phys. Rev. Lett.*, 96:197003, May 2006.

[47] S. M. Frolov, D. J. Van Harlingen, V. A. Oboznov, V. V. Bolginov, and V. V. Ryazanov. Measurement of the current-phase relation of supercon-ductor/ferromagnet/superconductor π Josephson junctions. *Phys. Rev. B*, 70:144505, Oct 2004.

[48] V. A. Oboznov, V. V. Bol'ginov, A. K. Feofanov, V. V. Ryazanov, and A. I. Buzdin. Thickness dependence of the Josephson ground states of superconductor-ferromagnet-superconductor junctions. *Phys. Rev. Lett.*, 96(19):1970031–1970034, 2006.

[49] S. M. Frolov, D. J. Van Harlingen, V. V. Bolginov, V. A. Oboznov, and V. V. Ryazanov. Josephson interferometry and shapiro step measurements of superconductor-ferromagnet-superconductor 0-π junctions. *Phys. Rev. B*, 74(2):020503, 2006.

[50] T. Kontos, M. Aprili, J. Lesueur, F. Genet, B. Stephanidis, and R Boursier. Josephson junction through a thin ferromagnetic layer: negative coupling. *Phys. Rev. Lett.*, 89(13):137007, 2002.

[51] H. Sellier, C. Baraduc, F. Lefloch, and R. Calemczuk. Temperature-induced crossover between 0 and π states in s/f/s junctions. *Phys. Rev. B*, 68(5):054531, 2003.

[52] W. Guichard, M. Aprili, O. Bourgeois, T. Kontos, J. Lesueur, and P. Gandit. Phase sensitive experiments in ferromagnetic-based Josephson junctions. *Phys. Rev. Lett.*, 90(16):167001, 2003.

[53] Y. Blum, A. Tsukernik, M. Karpovski, and A. Palevski. Oscillations of the superconducting critical current in nb-cu-ni-cu-nb junctions. *Phys. Rev. Lett.*, 89(18), Oct 28 2002.

[54] C. Suürgers, T. Hoss, C. Schoünenberger, and C. Strunk. Fabrication and superconducting properties of nano-structured sfs contacts. *J. Magn. Magn. Mater.*, 240(1-3):598–600, 2002.

[55] C. Bell, R. Loloee, G. Burnell, and M. G. Blamire. Characteristics of strong ferromagnetic Josephson Junctions with epitaxial barriers. *Phys. Rev. B*, 71(18):180501, 2005.

[56] V. Shelukhin, A. Tsukernik, M. Karpovski, Y. Blum, K. B. Efetov, A. F. Volkov, T. Champel, M. Eschrig, T. Lüofwander, G. Schoen, et al. Observation of periodic π-phase shifts in ferromagnet-superconductor multilayers. *Phys. Rev. B*, 73(17):174506, 2006.

[57] M. Weides, K. Tillmann, and H. Kohlstedt. Fabrication of high quality ferromagnetic Josephson junctions. *Physica C*, 437:349–352, 2006.

[58] M. Weides, M. Kemmler, E. Goldobin, D. Koelle, R. Kleiner, H. Kohlstedt, and A. Buzdin. High quality ferromagnetic 0 and π Josephson tunnel junctions. *Appl. Phys. Lett.*, 89(12):122511, 2006.

[59] M. Weides, M. Kemmler, H. Kohlstedt, R. Waser, D. Koelle, R Kleiner, and E. Goldobin. 0-π Josephson tunnel junctions with ferromagnetic barrier. *Phys. Rev. Lett.*, 97(24):247001, 2006.

[60] J. Pfeiffer, M. Kemmler, D. Koelle, R. Kleiner, E. Goldobin, M. Weides, A. K. Feofanov, J. Lisenfeld, and A. V. Ustinov. Static and dynamic properties of 0, π, and 0-π ferromagnetic Josephson tunnel junctions. *Phys. Rev. B*, 77(21):214506, 2008.

[61] H. Sellier, C. Baraduc, F. Lefloch, and R. Calemczuk. Half-integer shapiro steps at the 0-π crossover of a ferromagnetic Josephson junction. *Phys. Rev. Lett.*, 92(25):257005, 2004.

[62] F. Born, M. Siegel, E. K. Hollmann, H. Braak, A. A. Golubov, D. Yu Gusakova, and M. Yu Kupriyanov. Multiple 0-π transitions in superconductor/insulator/ferromagnet/superconductor Josephson tunnel junctions. *Phys. Rev. B*, 74(14):140501, 2006.

[63] J. W. A. Robinson, S. Piano, G. Burnell, C. Bell, and M. G. Blamire. Critical current oscillations in strong ferromagnetic π junctions. *Phys. Rev. Lett.*, 97(17):177003, 2006.

[64] J. W. A. Robinson, S. Piano, G. Burnell, C. Bell, and M. G. Blamire. Zero to n transition in superconductor-ferromagnet-superconductor junctions. *Phys. Rev. B*, 76(9):094522, 2007.

[65] G. B. Halasz, J. W. A. Robinson, J. F. Annett, and M. G. Blamire. Critical current of a Josephson junction containing a conical magnet. *Phys. Rev. B*, 79(22):224505, 2009.

[66] T. S. Khaire, W. P. Pratt Jr, and N. O. Birge. Critical current behavior in Josephson junctions with the weak ferromagnet pdni. *Phys. Rev. B*, 79(9):094523, 2009.

[67] J. W. A. Robinson, Z. H. Barber, and M. G. Blamire. Strong ferromagnetic Josephson devices with optimized magnetism. *Appl. Phys. Lett.*, 95(19):192509, 2009.

[68] M. A. Khasawneh, W. P. Pratt Jr, and N. O. Birge. Josephson junctions with a synthetic antiferromagnetic interlayer. *Phys. Rev. B*, 80(2):020506, 2009.

[69] A. A. Bannykh, J. Pfeiffer, V. S. Stolyarov, I. E. Batov, V. V. Ryazanov, and M. Weides. Josephson tunnel junctions with a strong ferromagnetic interlayer. *Phys. Rev. B*, 79(5):054501, 2009.

[70] J. W. A. Robinson, G. B. Halasz, A I. Buzdin, and M. G. Blamire. Enhanced supercurrents in Josephson junctions containing nonparallel ferromagnetic domains. *Phys. Rev. Lett.*, 104(20):207001, 2010.

[71] T. S. Khaire, M. A. Khasawneh, W. P. Pratt Jr, and N. O. Birge. Observation of spin-triplet superconductivity in co-based Josephson junctions. *Phys. Rev. Lett.*, 104(13):137002, 2010.

[72] M. Kemmler, M. Weides, M. Weiler, M. Opel, S. T. B. Goennenwein, A. S. Vasenko, A. A. Golubov, H. Kohlstedt, D. Koelle, R. Kleiner, and E. Goldobin. Magnetic interference patterns in 0-π superconductor/insulator/ferromagnet/superconductor Josephson junctions: Effects of asymmetry between 0 and π regions. *Phys. Rev. B*, 81:054522, Feb 2010.

[73] G. B. Halász, M. G. Blamire, and J. W. A. Robinson. Magnetic-coupling-dependent spin-triplet supercurrents in helimagnet/ferromagnet Josephson junctions. *Phys. Rev. B*, 84(2):024517, 2011.

[74] M. A. Khasawneh, T. S. Khaire, C. Klose, W. P. Pratt Jr, and N. O. Birge. Spin-triplet supercurrent in co-based Josephson junctions. *Supercond. Sci. Technol.*,24(2):024005, 2011.

[75] K. Senapati, M. G. Blamire, and Z. H. Barber. Spin-filter Josephson junctions. *Nat. Mater.*, 10(11):849, 2011.

[76] V. V. Bol'ginov, V. S. Stolyarov, D. S. Sobanin, A. L. Karpovich, and V. V. Ryazanov. Magnetic switches based on nb-pdfe-nb_Josephson junctions with a magnetically soft ferromagnetic interlayer. *JETP Lett.*, 95(7):366–371, 2012.

[77] J. D. S. Witt, J. W. A. Robinson, and M. G. Blamire. Josephson junctions incorporating a conical magnetic holmium interlayer. *Phys. Rev. B*, 85(18):184526, 2012.

[78] C. Klose, T. S. Khaire, Y Wang, W. P. Pratt Jr, N. O. Birge, B. J. McMorran, T. P. Ginley, J. A. Borchers, B. J. Kirby, B. B. Maranville, et al. Optimization of spin-triplet supercurrent in ferromagnetic Josephson junctions. *Phys. Rev. Lett.*, 108(12):127002, 2012.

[79] J. W. A. Robinson, F. Chiodi, M. Egilmez, G. B. Haláasz, and M. G. Blamire. Supercurrent enhancement in bloch domain walls. *Sci. Rep.*,2:699, 2012.

[80] E. C. Gingrich, P. Quarterman, Y. Wang, R. Loloee, W. P. Pratt Jr, and N. O. Birge. Spin-triplet supercurrent in co/ni multilayer Josephson junctions with perpendicular anisotropy. *Phys. Rev. B*, 86(22):224506, 2012.

[81] M. G. Blamire, C. B. Smiet, N. Banerjee, and J. W. A. Robinson. Field modulation of the critical current in magnetic Josephson junctions. *Supercond. Sci. Technol.*, 26(5):055017, 2013.

[82] M. G. Blamire and J. W. A. Robinson. The interface between superconductivity and magnetism: understanding and device prospects. *J. Phys. Condens. Matter*, 26(45):453201, 2014.

[83] B. M. Niedzielski, S. G. Diesch, E. C. Gingrich, Y. Wang, R. Loloee, W. P. Pratt, and N. O. Birge. Use of pd-fe and ni-fe-nb as soft magnetic layers in ferromagnetic Josephson junctions for nonvolatile cryogenic memory. *IEEE Trans. Appl. Supercond.*, 24(4):1–7, 2014.

[84] M. Abd El Qader, R. K. Singh, S. N. Galvin, L. Yu, J. M. Rowell, and N. Newman. Switching at small magnetic fields in Josephson junctions fabricated with ferromagnetic barrier layers. *Appl. Phys. Lett.*, 104(2):022602, 2014.

[85] J. W. A. Robinson, Niladri Banerjee, and Mark Giffard Blamire. Triplet pair correlations and nonmonotonic supercurrent decay with cr thickness in nb/cr/fe/nb Josephson devices. *Phys. Rev. B*, 89(10):104505, 2014.

[86] N. Banerjee, J. W. A. Robinson, and M. Giffard Blamire. Reversible control of spin-polarized supercurrents in ferromagnetic Josephson junctions. *Nat. Commun.*, 5:4771, 2014.

[87] W. M. Martinez, W. P. Pratt Jr, and N. O. Birge. Amplitude control of the spin-triplet supercurrent in s/f/s Josephson junctions. *Phys. Rev. Lett.*, 116(7):077001, 2016.

[88] E. C. Gingrich, B. M. Niedzielski, J. A. Glick, U. Wang, D. L. Miller, R. Loloee, W. P. Pratt Jr, and N. O. Birge. Controllable 0- π Josephson junctions containing a ferromagnetic spin valve. *Nat. Phys.*, 12(6):564, 2016.

[89] D. Massarotti, R. Caruso, A. Pal, G. Rotoli, L. Longobardi, G. P. Pepe, M. G. Blamire, and F. Tafuri. Low temperature properties of spin filter nbn/gdn/nbn Josephson junctions. *Physica C*, 533:53–58, 2017.

[90] J. A. Glick, R. Loloee, W. P. Pratt, and N. O. Birge. Critical current oscillations of Josephson junctions containing pdfe nanomagnets. *IEEE Trans. Appl. Supercond.*, 27(4):1–5, 2017.

[91] J. A. Glick, M. A. Khasawneh, B. M. Niedzielski, R. Loloee, W. P. Pratt Jr, N. O. Birge, E. C. Gingrich, P. G. Kotula, and N. Missert. Critical current oscillations of elliptical Josephson junctions with single-domain ferromagnetic layers. *J. Appl. Phys.*, 122(13):133906, 2017.

[92] J. A. Glick, S. Edwards, D. Korucu, V. Aguilar, B. M. Niedzielski, R. Loloee, W. P. Pratt Jr, N. O. Birge, P. G. Kotula, and N. Missert. Spin-triplet supercurrent in Josephson junctions containing a synthetic antiferromagnet with perpendicular magnetic anisotropy. *Phys. Rev. B*, 96(22):224515, 2017.

[93] S. Mesoraca, S. Knudde, D. C. Leitao, S. Cardoso, and M. G. Blamire. All-spinel oxide Josephson junctions for high-efficiency spin filtering. *J. Phys.: Condens. Matter*, 30(1):015804, 2017.

[94] F. Li, H. Zhang, L. Zhang, W. Peng, and Z. Wang. Ferromagnetic Josephson junctions based on epitaxial nbn/ni60cu40/nbn trilayer. *AIP Advances*, 8(5):055007, 2018.

[95] M. J. A. Stoutimore, A. N. Rossolenko, V. V. Bolginov, V. A. Oboznov, A. Y. Rusanov, N. Pugach, S. M. Frolov, V. V. Ryazanov, and D. J. Van Harlingen. Second-harmonic current-phase relation in Josephson junctions with ferromagnetic barriers. *arXiv preprint arXiv:1805.*12546, 2018.

[96] B. Baek, M. L. Schneider, M. R. Pufall, and W. H. Rippard. Anomalous supercurrent modulation in Josephson junctions with ni-based barriers. *IEEE Trans. Appl. Supercond.*, 28(7):1–5, Oct 2018.

[97] A. K. Feofanov, V. A. Oboznov, V. V. Bol'ginov, J. Lisenfeld, S. Poletto, V. V. Ryazanov, A. N. Rossolenko, M. Khabipov, D. Balashov, A. B. Zorin, P. N. Dmitriev, V. P. Koshelets, and A. V. Ustinov. Implementation of superconductor/ferromagnet/superconductor [pi]-shifters in superconducting digital and quantum circuits. *Nat. Phys.*, 6(8):593–597, 2010.

[98] T. I. Larkin, V. V. Bolginov, V. S. Stolyarov, V. V. Ryazanov, I. V. Vernik, S. K. Tolpygo, and O. A. Mukhanov. Ferromagnetic Josephson switching device with high characteristic voltage. *Appl. Phys. Lett.*, 100(22), 2012.

[99] S. K. Tolpygo, D. Yohannes, R. T. Hunt, J. A. Vivalda, D. Donnelly, D. Amparo, and A. F. Kirichenko. 20 ka/cm^2 process development for superconducting integrated circuits with 80 ghz clock frequency. *IEEE Trans. Appl. Supercond.*, 17(2):946–951, June 2007.

[100] A. I. Buzdin, B. Vujicic, and M. Y. Kupriyanov. Superconductor ferromagnetic structures. *Zh. Eksp. Teor. Fiz.*, 101(1):231–240, Jan 1992.

[101] A. Buzdin and I. Baladie. Theoretical description of ferromagnetic pi junctions near the critical temperature. *Phys. Rev. B*, 67(18), May 1 2003.

[102] A. Buzdin. Peculiar properties of the Josephson junction at the transition from 0 to π state. *Phys. Rev. B*, 72:100501, Sep 2005.

[103] Y. Tanaka and S. Kashiwaya. Theory of Josephson effects in anisotropic superconductors. *Phys. Rev. B*, 56(2):892, 1997.

[104] M. Fogelstrom. Josephson currents through spin-active interfaces. *Phys. Rev. B*, 62(17):11812, 2000.

[105] Y. S. Barash and I. V. Bobkova. Interplay of spin-discriminated andreev bound states forming the 0-pi transition in superconductor-ferromagnet-superconductor junctions. *Phys. Rev. B*, 65(14), Apr 1 2002.

[106] M. L. Kulic and I. M. Kulic. Possibility of a pi Josephson junction and switch in superconductors with spiral magnetic order. *Phys. Rev. B*, 63(10), 2001.

[107] E. Koshina and V. Krivoruchko. Spin polarization and π-phase state of the Josephson contact: critical current of mesoscopic sfifs and sfis junctions. *Phys. Rev. B*, 63(22):224515, 2001.

[108] N. M. Chtchelkatchev, W. Belzig, Yu V. Nazarov, and C. Bruder. π-0 transition in superconductor-ferromagnet-superconductor junctions. *J. Exp. Theor. Phys. Lett.*, 74(6):323–327, 2001.

[109] Z. Radovic, N. Lazarides, and N. Flytzanis. Josephson effect in double-barrier superconductor-ferromagnet junctions. *Phys. Rev. B*, 68(1):014501, 2003.

[110] Ya V. Fominov, N. M. Chtchelkatchev, and A. A. Golubov. Nonmonotonic critical temperature in superconductor/ferromagnet bilayers. *Phys. Rev. B*, 66(1):014507, 2002.

[111] K. Halterman and O. T. Valls. Layered ferromagnet-superconductor structures: The π state and proximity effects. *Phys. Rev. B*, 69(1):014517, 2004.

[112] K. Halterman and O. T. Valls. Stability of π-junction conigurations in ferromagnet-superconductor heterostructures. *Phys. Rev. B*, 70(10):104516, 2004.

[113] F. S. Bergeret, A. F. Volkov, and K. B. Efetov. Long-range proximity effects in superconductor-ferromagnet structures. *Phys. Rev. Lett.*, 86(18):4096–4099, Apr 30 2001.

[114] A. Kadigrobov, R. I. Shekhter, and M. Jonson. Quantum spin fluctuations as a source of long-range proximity effects in diffusive ferromagnet-super conductor structures. *Europhys. Lett.*, 54(3):394, 2001.

[115] A. F. Volkov, F. S. Bergeret, and K. B. Efetov. Odd triplet superconductivity in superconductor-ferromagnet multilayered structures. *Phys. Rev. Lett.*, 90(11):117006, 2003.

[116] Ya V. Fominov, A. F. Volkov, and K. B. Efetov. Josephson effect due to the long-range odd-frequency triplet superconductivity in s f s junctions with náeel domain walls. *Phys. Rev. B*, 75(10):104509, 2007.

[117] K. Senapati, M. G. Blamire, and Z. H. Barber. Spin-filter Josephson junctions. *Nat. Mater*, 10(11):849–852, 2011.

[118] D. Massarotti, A. Pal, G. Rotoli, L. Longobardi, M. G. Blamire, and F. Tafuri. Macroscopic quantum tunnelling in spin filter ferromagnetic Josephson junctions. *Nat. Commun.*, 6, 2015.

[119] A. Barone and G. Paternoá. *Physics and Applications of the Josephson Effect.* John Wiley & Sons, Inc. New York, NY, USA, 1982.

[120] C. C. Tsuei and J. R. Kirtley. Pairing symmetry in cuprate superconductors. *Rev. Mod. Phys.*, 72:969–1016, Oct 2000.

[121] H. Hilgenkamp and J. Mannhart. Grain boundaries in high-T_c superconductors. *Rev. Mod. Phys.*, 74:485–549, May 2002.

[122] F. Tafuri and J. R. Kirtley. Weak links in high critical temperature superconductors. *Rep. Prog. Phys.*, 68(11):2573–2663, 2005.

[123] J. R. Kirtley. Magnetic field effects in Josephson junctions. In Francesco Tafuri, editor, *Fundamentals and Frontiers of the Josephson Effect*, pages 209–233, Chapter 6. Springer Nature Swizterland, 2019.

[124] F. S. Bergeret and J. C. Cuevas. The vortex state and Josephson critical current of diffusive sns junction. *J Low Temp. Phys.*, 153(5-6):304–324, 2008.

[125] D. J. Van Harlingen. Phase-sensitive tests of the symmetry of the pairing state in the high-temperature superconductors~evidence for d_{x2-y2} symmetry. *Rev. Mod. Phys.*, 67:515–535, Apr 1995.

[126] J. R. Kirtley, K. A. Moler, and D. J. Scalapino. Spontaneous flux and magnetic-interference patterns in 0- π Josephson junctions. *Phys. Rev. B*, 56:886–891, Jul 1997.

[127] H. J. H. Smilde, A. A. Golubov, Ariando, G. Rijnders, J. M. Dekkers, S. Harkema, D. H. A. Blank, H. Rogalla, and H. Hilgenkamp. Admixtures to d-wave gap symmetry in untwinned $yba_2cu_3o_7$ superconducting films measured by angle-resolved electron tunneling. *Phys. Rev. Lett.*, 95:257001, Dec 2005.

[128] E H. Brandt and J. R. Clem. Superconducting thin rings with finite penetration depth. *Phys. Rev. B*, 69:184509, May 2004.

[129] P. A. Rosenthal, M. R. Beasley, K. Char, M. S. Colclough, and G. Za-harchuk. Flux focusing effects in planar thin film grain boundary Josephson junctions. *Appl. Phys. Lett.*, 59(26):3482–3484, 1991.

[130] A. Pal, Z. H. Barber, J. W. A. Robinson, and M. G. Blamire. Pure second harmonic current-phase relation in spin-filter Josephson junctions. *Nat. Commun.*, 5, Feb 2014.

[131] T. S. Khaire, M. A. Khasawneh, W. P. Pratt, and N. O. Birge. Observation of spin-triplet superconductivity in co-based Josephson junctions. *Phys. Rev. Lett.*, 104:137002, Mar 2010.

[132] N. Banerjee, J. W. A. Robinson, and M. G. Blamire. Reversible control of spin-polarized supercurrents in ferromagnetic Josephson junctions. *Nat. Commun.*, 5, Aug 2014.

[133] F. S. Bergeret, A. F. Volkov, and K. B. Efetov. Odd triplet superconductivity and related phenomena in superconductor-ferromagnet structures. *Rev. Mod. Phys.*, 77:1321–1373, Nov 2005.

[134] R. Caruso, D. Massarotti, G. Campagnano, A. Pal, H. G. Ahmad, P. Lucignano, M. Eschrig, M. G. Blamire, and F. Tafuri. Tuning of magnetic activity in spin-filter Josephson junctions towards spin-triplet transport. *Phys. Rev. Lett.*, 122:047002, 2019.

[135] M. Alidoust and K. Halterman. Spin-controlled coexistence of 0 and π states in *sfsfs* Josephson junctions. *Phys. Rev. B*, 89:195111, May 2014.

[136] K. Halterman and M. Alidoust. Josephson currents and spin-transfer torques in ballistic sfsfs nanojunctions. *Supercond. Sci. Technol.*, 29(5):055007, 2016.

[137] J. Ali Ouassou and J. Linder. Spin-switch Josephson junctions with magnetically tunable sin (δ ϕ/n) current-phase relation. *Phys. Rev. B*, 96(6):064516, 2017.

[138] T. I. Larkin, V. V. Bolginov, V. S. Stolyarov, V. V. Ryazanov, I. V. Vernik, S. K. Tolpygo, and O. A. Mukhanov. Ferromagnetic Josephson switching device with high characteristic voltage. *Appl. Phys. Lett.*, 100(22):222601, 2012.

[139] I. V. Vernik, V. V. Bol ginov, S. V. Bakurskiy, A. A. Golubov, M. Y. Kupriyanov, V. V. Ryazanov, and O. A. Mukhanov. Magnetic Josephson junctions with superconducting interlayer for cryogenic memory. *IEEE Trans. Appl. Supercond.*, 23(3):1701208, 2013.

[140] S. V. Bakurskiy, N. V. Klenov, T. Yu Karminskaya, M. Yu Kupriyanov, and A. A. Golubov. Josephson y-junctions based on structures with complex normal/ferromagnet bilayer. *Supercond. Sci. Technol.*, 26(1):015005, 2013.

[141] S. V. Bakurskiy, N. V. Klenov, I. I. Soloviev, V. V. Bol ginov, V. V. Ryazanov, I. V. Vernik, O. A. Mukhanov, M. Yu Kupriyanov, and A. A. Golubov. Theoretical model

of superconducting spintronic sisfs devices. *Appl. Phys. Lett.*, 102(19):192603, 2013.

[142] N. Ruppelt, H. Sickinger, R. Menditto, E. Goldobin, D. Koelle, R. Kleiner, O. Vavra, and H. Kohlstedt. Observation of 0-π transition in sisfs Josephson junctions. *Appl. Phys. Lett.*, 106(2):022602, 2015.

[143] R. Caruso, D. Massarotti, A. Miano, V. V. Bolginov, A. B. Hamida, L. N. Karelina, G. Campagnano, I. V. Vernik, F. Tafuri, V. V. Ryazanov, et al. Properties of ferromagnetic Josephson junctions for memory applications. *IEEE Trans. Appl. Supercond.*, 28(7):1–6, 2018.

[144] R. Caruso, D. Massarotti, V. V. Bolginov, A. B. Hamida, L. N. Karelina, A. Miano, I. V. Vernik, F. Tafuri, V. V. Ryazanov, O. A. Mukhanov, et al. RF-assisted switching in magnetic Josephson junctions. *J. Appl. Phys.*, 123(13):133901, 2018.

[145] S. V. Bakurskiy, V. I. Filippov, V. I. Ruzhickiy, N. V. Klenov, I. I. Soloviev, M. Yu Kupriyanov, and A. A. Golubov. Current-phase relations in sisfs junctions in the vicinity of 0-π transition. *Phys. Rev. B*, 95(9):094522–1–094522–11, 2017.

[146] S. V. Bakurskiy, N. V. Klenov, I. I. Soloviev, N. G. Pugach, M. Yu Kupriyanov, and A. A. Golubov. Protected 0-π states in sisfs junctions for Josephson memory and logic. *Appl. Phys. Lett.*, 113(8):082602, 2018.

A3

Introduction to Section A3: Critical Currents of Type II Superconductors

David A. Cardwell

The current-carrying properties of type II superconductors are of considerable importance for both fundamental and applied studies of these technologically important materials. The critical current density, J_c, in particular, is regarded by engineers as the key parameter to the practical application of low- and high-temperature superconductors in a diversity of engineering applications, such as MRI, Maglev, energy storage devices and motors and generators. As a result, the optimization of J_c for the different superconducting materials fabricated in the form of wires and tapes, thin films and bulks has been the focus of extensive research over many decades. J_c is essentially a characteristic of the so-called mixed state in type II materials, in which normal and superconducting phases co-exist thermodynamically. Magnetic flux enters type II superconductors in the form of lines of magnetic flux quanta, equivalent to Faraday's lines of force. These thread current vortices that circle predominantly normal regions of the material and give rise to some extraordinary magnetic behaviour, driven entirely by the ability of the material to resist the motion of these flux lines.

The critical current density is a direct measure of the resistance to motion of flux quanta in type II materials. In general, a wide diversity of microscopic features such as dislocations, lattice strains and second phases, form effective barriers, or pins, to flux motion under a variety of conditions. The basic physics of the effects of flux pinning are straightforward. Individual flux lines will remain pinned as long as the Lorentz force they experience, given by $\boldsymbol{B} \times \boldsymbol{J}$, remains below the temperature and field-dependent potential energy of the pinning site. The current density at which this potential is exceeded at a given field, therefore, corresponds directly to J_c. The subsequent motion of these flux lines on a microscopic scale will generate an associated electric field and cause dissipation in the material. The nature and properties of the various flux pinning centres and how individual vortices interact with one another, on the other hand, are generally complex and influence the resultant J_c in a variety of ways. This is the subject of this section.

The terminology used to describe the behaviour of flux lines in type II superconductors can be the source of some confusion. The magnetization, or M–H, curve of defect-free single crystals, for example, is often described as ideal. Clearly, such materials are unable to pin magnetic flux and consequently exhibit zero J_c, which makes them entirely unsuitable and far from ideal for most engineering applications. A recent and more useful trend, therefore, has been to describe the magnetization of type II materials in terms of their magnetic reversibility, or hysteresis. J_c is directly proportional to the latter and may be extracted using a simple model from the M–H loop or, even more simply, estimated from the gradient of the trapped field profile, particularly in bulk samples.

This section contains two very different but entirely complementary articles that address the mixed state, vortices and their interaction and flux pinning in general. There is necessarily some overlap between the two articles that has been retained, given the slightly different emphasis by the three authors on the various theoretical and experimental elements of flux pinning in each case. Each article illustrates the complexity of critical currents in type II high- and low-temperature superconductors and provides a window into how this fundamental property determines critically the applied properties of these materials.

A3.1

Vortices and Their Interaction

E. Helmut Brandt

A3.1.1 Introduction

The existence of vortices in type II superconductors was predicted first by Alexei A. Abrikosov when he discovered a two-dimensional (2D) periodic solution of the Ginzburg–Landau (GL) equations. Abrikosov correctly interpreted this solution as a periodic arrangement of flux lines, the so-called flux-line lattice (FLL). Each flux line (or, alternatively, fluxon or vortex line) carries one quantum of magnetic flux $\Phi_0 = h/2e = 2.07 \times 10^{-15}$ Tm^2, generated by a vortex of circulating supercurrents (see Chapter A2.6). Consequently, the magnetic field peaks at the vortex positions. The vortex core is a tube in which the superconductivity is weakened with its centre defined by the line at which the superconducting order parameter vanishes. For well-separated or isolated vortices, the radius of the tube of magnetic flux is the magnetic penetration depth, λ, and the core radius approximates to the superconducting coherence length, [1, 2] (see Chapter A2.3). The spacing a of the vortices decreases with increasing applied magnetic field and the average flux density, $\bar{B} = 2\Phi_0/(\sqrt{3}a^2)$, increases for the triangular FLL (see Figure A3.1.1 and Chapter A2.6). The flux tubes begin to overlap with further increase in $\bar{B}$ such that the periodic induction $B(x,y)$ is nearly constant, with only a small relative modulation around its average value. Eventually, the vortex cores begin to overlap such that the amplitude of the order parameter decreases until it vanishes when $\bar{B}$ reaches the upper critical field, $B_{c2} = \Phi_0/(2\pi\xi^2)$, where the superconductivity disappears. Figure A3.1.2 shows profiles of the induction $B(x,y)$ and of the superconducting order parameter $|\psi(x,y)|^2$ for two values of $\bar{B}$ corresponding to flux-line spacings $a = 4\lambda$ and $a = 2\lambda$.

The periodic solution which describes the FLL exists when the GL parameter $\kappa = \lambda/\xi$ exceeds the value $1/\sqrt{2}$ (this condition defines type II superconductors), and when the applied magnetic induction $B_a = \mu_0 H_a$ ranges between the lower and upper critical fields, $B_{c1} \approx \Phi_0 \ln\kappa/(4\pi\lambda^2)$ (at which, $\bar{B} = 0$) and B_{c2} (where $\bar{B} = B_a = B_{c2}$). The superconductor is in the Meissner state for $|B_a| < B_{c1}$, in which all magnetic flux is expelled, so that $B \equiv 0$ inside the superconductor (see Chapter A2.2). With increasing $B_a > B_{c1}$, the induction $\bar{B}$ inside the superconductor increases monotonically. The volume magnetization for long cylinders with field applied parallel to their axes (i.e., for which demagnetizing or shape effects are negligible) is defined as $M = (\bar{B} - B_a)/\mu_0$. In this case, the negative magnetization $-M$ initially increases linearly, $M = -H_a$ for $B_a \le B_{c1}$; $-M$ decreases sharply at $B_a = B_{c1}$ as flux lines start to penetrate; it then decreases approximately linearly at higher B_a until it vanishes at B_{c2}, see also Section A3.1.5 and Chapter A3.2 by C.J. van der Beek and P.H. Kes. The area under the magnetization curve $-M(H_a)$ is $B_c^2/2\mu_0$ where $B_c = \Phi_0/\left(\sqrt{8}\pi\xi\lambda\right) = B_{c2}/\kappa$ is the thermodynamic critical field. The three critical fields coincide (i.e., $B_{c1} = B_c = B_{c2}$) in superconductors with $\kappa = 1/\sqrt{2}$ (this is almost realized exactly in pure niobium).

The properties of the vortex lattice may be calculated from London theory at low inductions $\bar{B} \ll B_{c2}$ and large $\kappa \gg 1$, to which the GL theory reduces when the magnitude of the order parameter is nearly constant. In the London limit, $B(x,y)$ is the linear superposition of the fields of isolated vortices in which case the London expressions for $B(x,y)$ and for the energy also apply to a non-periodic arrangements of vortices. The London theory was further extended to describe curved vortices and to anisotropic superconductors. The GL theory, however, has to be used at larger $\bar{B} > 0.25B_{c2}$ or smaller $\kappa < 2$, although analytical solutions are available only for the periodic FLL near B_{c2}. This theory cannot be applied to a distorted FLL or at lower $\bar{B}$.

The elastic moduli of the vortex lattice may be obtained by expanding the energy of the superconductor with respect to small displacements of the vortices from their ideal lattice positions. The elasticity of the vortex lattice is *non-local* in contrast to the local elasticity of atomic lattices, i.e., the energy of compressional and tilt deformations of the FLL is strongly reduced when the wave vector k of the strain field is large (i.e., $k > \lambda^{-1}$) [3].

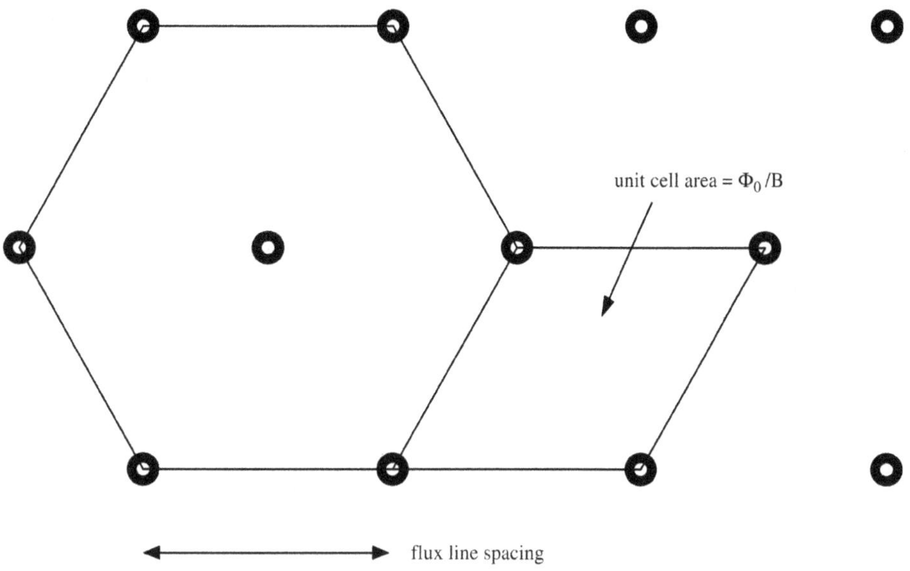

FIGURE A3.1.1 The triangular flux-line lattice.

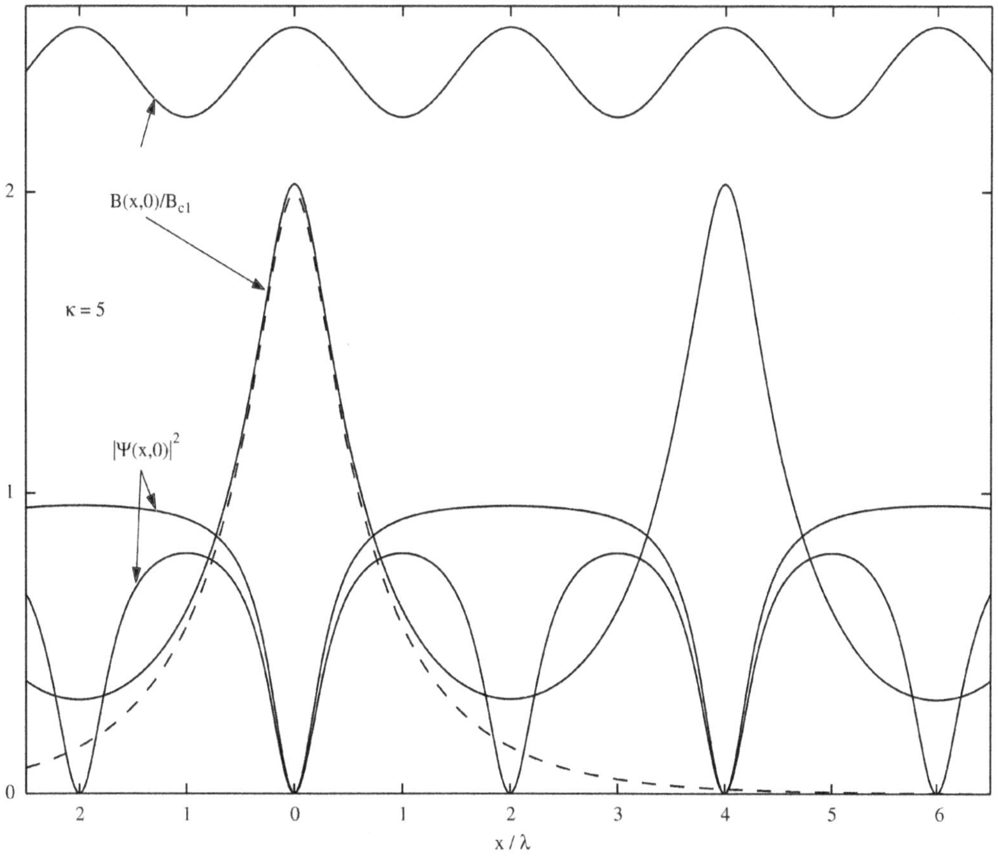

FIGURE A3.1.2 Profiles of the magnetic field $B(x,y)$ and order parameter $|\Psi(x,y)|^2$ along the x-axis (a nearest neighbour direction) for two flux-line lattices with lattice spacing $a = 4\lambda$ (solid lines) and $a = 2\lambda$ (thin lines). The dashed line shows the magnetic field of an isolated flux line. The data are derived from the Ginzburg–Landau theory for $\kappa = 5$.

A3.1.2 Results from London Theory

The London theory may be formulated by minimizing the sum F of the potential energy of the magnetic field $\boldsymbol{B}(\boldsymbol{r})$ [$E_p = \mathbf{B}^2(\boldsymbol{r})/2\mu_0$ per unit volume] and the kinetic energy of the supercurrent density $\boldsymbol{J}(\boldsymbol{r}) = \mu_0\, \nabla \times \boldsymbol{B}(\boldsymbol{r})$ [$E_k = \lambda^2(\nabla \times \boldsymbol{B})^2/2\mu_0$ per unit volume]

$$F = \frac{1}{2\mu_0}\int_V [\boldsymbol{B}^2 + \lambda^2(\nabla\times\boldsymbol{B})^2]d^3r \qquad \text{(A3.1.1)}$$

with respect to $\boldsymbol{B} = \nabla \times \boldsymbol{A}$, where $\mathbf{A}$ is the vector potential. This yields the homogeneous London equation $B - \lambda^2\nabla^2\boldsymbol{B} = 0$ or $\boldsymbol{J} = -\mu_0^{-1}\lambda^{-2}\boldsymbol{A}$, assuming the Maxwell equations $\nabla\cdot\boldsymbol{B} = 0$ and $\nabla \times \boldsymbol{B} = \mu_0\mathbf{J}$.

A3.1.2.1 Parallel Vortices

In the presence of vortices, one has to add singularities to the formulation which describe the vortex core. The result is the modified London equation for straight, parallel vortex lines aligned along $\hat{z}$

$$\boldsymbol{B}(\boldsymbol{r}) - \lambda^2\nabla^2\boldsymbol{B}(\boldsymbol{r}) = \hat{z}\Phi_0\sum_\nu \delta_2(\boldsymbol{r}-\boldsymbol{r}_\nu). \qquad \text{(A3.1.2)}$$

Here $\boldsymbol{r}_\nu = (x_\nu, y_\nu)$ are the two-dimensional (2D) vortex positions and $\delta_2(\boldsymbol{r}) = \delta(x)\,\delta(y)$ is the 2D delta function. This linear equation may be solved by the Fourier transform using $\int \exp(i\boldsymbol{kr})d^2k = 4\pi^2\delta_2(\boldsymbol{r})$ and $\int \exp(i\boldsymbol{kr})(k^2+\lambda^{-2})^{-1}d^2k = 2\pi K_0(|\mathbf{r}|/\lambda)$. Here $K_0(x)$ is a modified Bessel function with the limits $K_0(x) \approx -\ln(x)$ for $x \ll 1$ and $K_0(x) \approx (\pi/2x)^{1/2}\exp(-x)$ for $x \gg 1$. The resulting magnetic field of any arrangement of parallel vortices is then the sum of individual vortex fields centred at the positions $\mathbf{r}_\nu$,

$$\boldsymbol{B}(\boldsymbol{r}) = \hat{z}\frac{\Phi_0}{2\pi\lambda^2}\sum_\nu K_0\left(\frac{|\boldsymbol{r}-\boldsymbol{r}_\nu|}{\lambda}\right). \qquad \text{(A3.1.3)}$$

The energy F_{2D} of this 2D arrangement of vortex lines with length L is obtained by inserting Equation (A3.1.2) into Equation (A3.1.1). The London energy is determined by the magnetic field values at the vortex positions and is obtained by integrating over the delta function

$$\begin{aligned} F_{2D} &= L\frac{\Phi_0}{2\mu_0}\sum_\mu B(\boldsymbol{r}_\mu) \\ &= L\frac{\Phi_0^2}{4\pi\mu_0\lambda^2}\sum_\mu\sum_\nu K_0\left(\frac{|\boldsymbol{r}_\mu-\boldsymbol{r}_\nu|}{\lambda}\right). \end{aligned} \qquad \text{(A3.1.4)}$$

This expression shows that the energy is composed of the self-energy of the vortices (i.e., terms $\mu = \nu$) and a pairwise interaction energy (i.e., terms $\mu \neq \nu$). To avoid the divergence of the self-energy, the logarithmic infinity of B has to be terminated at the vortex centres $\boldsymbol{r}_\nu$ by introducing a finite radius of the vortex core of order ξ, the coherence length of the GL theory. This cut-off may be achieved by replacing the distance $r_{\mu\nu} = |\boldsymbol{r}_\mu - \boldsymbol{r}_\nu|$ in Equation (A3.1.4) by $\tilde{r}_{\mu\nu} = (r_{\mu\nu}^2 + 2\xi^2)^{1/2}$ and multiplying by a normalization factor ≈ 1 to conserve the flux Φ_0 of the vortex. This analytical expression suggested by Clem [4] for a single vortex, and later generalized to the vortex lattice [5], is an excellent approximation as was shown numerically [6] by solving the GL equation for the periodic FLL in the entire range of $\bar{B}$ and κ, $0 \le \bar{B} \le B_{c2}$, $\kappa \ge 1/\sqrt{2}$.

A3.1.2.2 Curved Vortices

Arbitrary three-dimensional (3D) arrangements of curved vortices at positions $\boldsymbol{r}_\nu(z) = [x_\nu(z), y_\nu(z), z]$ satisfy the following 3D London equation [3]

$$\boldsymbol{B}(\boldsymbol{r}) - \lambda^2\nabla^2\boldsymbol{B}(\boldsymbol{r}) = \Phi_0\sum_\nu\int d\boldsymbol{r}_\nu\delta_3(\boldsymbol{r}-\boldsymbol{r}_\nu). \qquad \text{(A3.1.5)}$$

Here the integral is along the vortex lines and $\delta_3(\boldsymbol{r}) = \delta(x)\,\delta(y)\,\delta(z)$. The resulting magnetic field and energy are, with $\tilde{r}_{\mu\nu} = [|\boldsymbol{r}_\mu(z) - \boldsymbol{r}_\nu(z)|^2 + 2\xi^2]^{1/2}$,

$$\begin{aligned} \boldsymbol{B}(\boldsymbol{r}) &= \frac{\Phi_0}{4\pi\lambda^2}\sum_\nu\int d\boldsymbol{r}_\nu\frac{\exp(-\tilde{r}_{\mu\nu}/\lambda)}{\tilde{r}_{\mu\nu}}, \\ F_{3D} &= \frac{\Phi_0}{2\mu_0}\sum_\mu\int d\boldsymbol{r}_\mu B(\boldsymbol{r}_\mu) \\ &= \frac{\Phi_0^2}{8\pi\mu_0\lambda^2}\sum_\mu\sum_\nu\int d\boldsymbol{r}_\mu\int d\boldsymbol{r}_\nu\frac{\exp(-\tilde{r}_{\mu\nu}/\lambda)}{\tilde{r}_{\mu\nu}}. \end{aligned} \qquad \text{(A3.1.6)}$$

This indicates that all vortex segments interact with each other in a similar manner to the interaction between magnetic dipoles or tiny current loops, but with their long-range magnetic interaction ($\propto 1/r$) screened by a factor $\exp(-\tilde{r}_{\mu\nu}/\lambda)$. The 3D interaction between curved vortices is illustrated schematically in Figure A3.1.3.

A3.1.2.3 Vortices Near a Surface

The solutions of Equations (A3.1.6) apply to vortices in the bulk and have to be modified near the surface of the superconductor. In simple geometries, such as for superconductors with one or two planar surfaces surrounded by vacuum, the magnetic field and energy of a given vortex arrangement are obtained by adding the field of appropriate images (in order to satisfy the boundary condition that no current leaves the surface) and a magnetic stray field generated by a fictitious surface layer of magnetic monopoles to ensure continuity of the total magnetic field across the surface [7]. The magnetic field and interaction of straight vortices oriented perpendicular to a superconducting film of arbitrary thickness are calculated in [8].

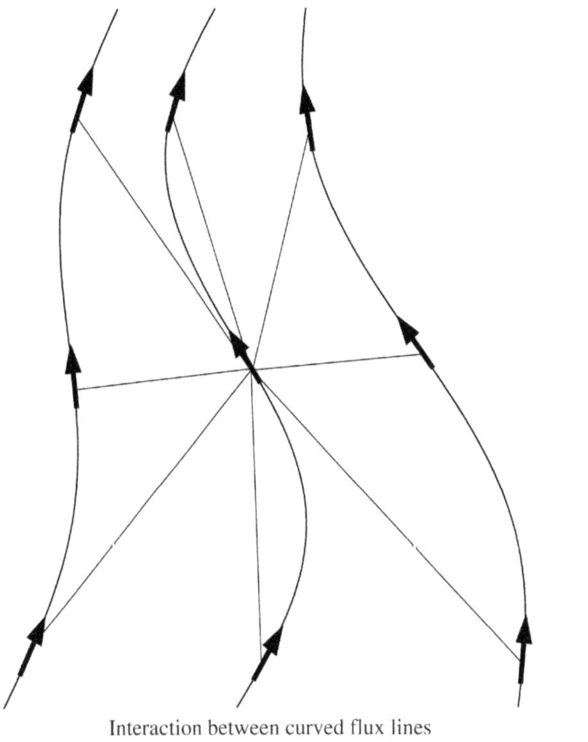

FIGURE A3.1.3 Visualization of the pairwise interaction between the line elements (arrows) of curved flux lines within London theory.

A3.1.2.4 Thin Films and Layered Superconductors

The self- and interaction energies are modified near the surface of a superconductor. In films of thickness $d \ll \lambda$, the short 2D vortices interact mainly via their magnetic stray field outside the superconductor over an effective penetration depth $\Lambda = 2\lambda^2/d$. At short distances $r \ll \Lambda$, this interaction is logarithmic as in the bulk; it decreases as $\exp(-r/\Lambda)$ at large $r \gg \Lambda$. The Fourier transform of the 2D vortex interaction $V(r) = \int (d^2k/4\pi^2)\tilde{V}(k)$ $\exp(i\mathbf{kr})$ changes with decreasing thickness d, from $\tilde{V}(k) = E_0(k^2 + \lambda^{-2})^{-1}(d \gg \lambda)$ to $\tilde{V}(k) = E_0(k^2 + k\Lambda^{-1})^{-1}(d \ll \lambda)$, where $E_0 = d\Phi_0^2/(\mu_0\lambda^2)$. A similar (but 3D) magnetic interaction exists between the 2D pancake vortices in the superconducting CuO layers of high-temperature superconductors (HTS), $\tilde{V}(\mathbf{k}) = E_0 d k_3^2 k_2^{-2}(\lambda^{-2} + k_3^2)^{-1}$, where d is now the distance between the layers, $k_2 = k_x^2 + k_y^2, k_3^2 = k_2^2 + k_z^2$ and $\lambda = \lambda_{ab}$ is the penetration depth for currents flowing within them [3].

A3.1.2.5 Anisotropic Superconductors

For many purposes, HTS may be considered as uniaxially anisotropic materials. Thus, within the London theory, they may be characterized by two penetration depths $\lambda_a \approx \lambda_b \approx \lambda_{ab}$ (for currents in the *ab* plane) and λ_c (for currents along the *c*-axis, i.e., along the anisotropy axis). The anisotropy ratio $\varepsilon = \lambda_{ab}/\lambda_c = \xi_c/\xi_{ab} \leq 1$ describes also the anisotropy of the GL coherence lengths, ξ_{ab} and ξ_c, which define the inner cut-off lengths in the anisotropic London theory. The general solution for arbitrarily arranged straight or curved vortex lines is [3]

$$B_\alpha(\boldsymbol{r}) = \Phi_0 \sum_\mu \int dr_\mu^\beta f_{\alpha\beta}(\boldsymbol{r} - \boldsymbol{r}_\mu)$$
$$F_{3D} = \frac{\Phi_0^2}{2\mu_0} \sum_\mu \sum_\nu \int dr_\mu^\alpha \int dr_\nu^\beta f_{\alpha\beta}(\boldsymbol{r}_\mu - \boldsymbol{r}_\nu) \qquad \text{(A3.1.7)}$$

with the tensorial interaction ($\alpha, \beta = x, y, z$)

$$f_{\alpha\beta}(\boldsymbol{r}) = \int \frac{d^3k}{8\pi^3} \exp(i\boldsymbol{kr})\, f_{\alpha\beta}(\boldsymbol{k}) \qquad \text{(A3.1.8)}$$
$$f_{\alpha\beta}(\boldsymbol{k}) = \frac{\exp[-2g(k,q)]}{1 + \Lambda_1 k^2}\left(\delta_{\alpha\beta} - \frac{q_\alpha q_\beta \Lambda_2}{1 + \Lambda_1 k^2 + \Lambda_{2q^2}}\right).$$

Here $g(k, q) = \xi_{ab}^2 q^2 + \xi_c^2(k^2 - q^2) = (\Lambda_1 k^2 + \Lambda_2 q^2)\xi_c^2/\lambda_{ab}^2$ enters the cut-off factor $\exp(-2g)$; $\mathbf{q} = \mathbf{k} \times \hat{\mathbf{c}}$, $\hat{\mathbf{c}}$ is the unit vector along the *c*-axis, $\Lambda_1 = \lambda_{ab}^2, \Lambda_2 = \lambda_c^2 - \lambda_{ab}^2 \geq 0$, and the sums and integrals are over the μth and νth vortex line. The contribution to $\boldsymbol{B}(\boldsymbol{r})$ of the segment dr_μ is now in general not parallel to dr_μ due to the tensorial character of $f_{\alpha\beta}(\boldsymbol{r})$.

A3.1.3 Ginzburg–Landau, Pippard and BCS Theories

The London theory was extended in two ways, both of which introduce a second length ξ. These are the descriptions of Ginzburg and Landau (GL), which is non-linear, and of Pippard, which is non-local. All three descriptions were shown later to follow from the microscopic BCS theory in limiting cases.

A3.1.3.1 Ginzburg–Landau Theory

The GL theory of 1950 introduces a complex order parameter $\psi(\mathbf{r})$ in addition to the magnetic field $\boldsymbol{B}(\boldsymbol{r}) = \nabla \times \boldsymbol{A}(\boldsymbol{r})$. The GL function $\psi(\boldsymbol{r})$ is proportional to the BCS energy-gap function $\Delta(\mathbf{r})$, and its square $|\psi(\mathbf{r})|^2$ is proportional to the density of Cooper pairs. The superconducting coherence length ξ gives the scale over which $\psi(\mathbf{r})$ can vary, while the penetration depth λ governs the variation of the magnetic field as in London theory. Both λ and ξ diverge at the superconducting transition temperature T_c according to $\lambda \propto \xi \propto (T_c - T)^{-1/2}$, but their ratio, the GL parameter $\kappa = \lambda/\xi$, is nearly independent of the temperature T. The GL theory reduces to the London theory (which is valid down to $T = 0$) in the limit $\xi \ll \lambda$, which means that the magnitude of $|\psi(\mathbf{r})|$ is constant, except in the vortex cores, where it vanishes. The GL equations are obtained by minimizing a free energy functional

$F\{\psi, A\}$ with respect to the GL function $\psi(r)$ and the vector potential $A(r)$. With the length unit λ and magnetic field unit $\sqrt{2}B_c$ the GL functional becomes

$$F\{\psi, A\} = \frac{B_c^2}{\mu_0} \int \left[-|\psi|^2 + \frac{1}{2}|\psi|^4 + \left| \left(-\frac{i\nabla}{\kappa} - A \right) \psi \right|^2 + (\nabla \times A)^2 \right] d^3 r. \quad \text{(A3.1.9)}$$

$F\{\psi, A\}$ and the resulting GL equations may be expressed in terms of the real function $|\psi|$ and the gauge-invariant super-velocity $\nabla\varphi/\kappa - A$, where $\varphi(r)$ is the phase of $\psi = |\psi|\exp(i\varphi)$. The supercurrent density is $J = \mu_0^{-1}\lambda^2(\nabla\varphi/\kappa - A)|\psi|^2$. In an external field H_a, it is necessary to minimize $G = F - BH_a$, rather than F, which yields the equilibrium field $H_a = \partial F/\partial \mathbf{B}$. The reversible magnetization curves $B(H_a)$ of pin-free superconductors are calculated in this way from GL theory [6], as were the field profiles shown in Figure A3.1.2.

The GL theory modifies the London interaction between vortices in two ways: firstly the range of the magnetic repulsion at large inductions $\bar{B}$ becomes larger, $\lambda' = \lambda/(1 - \bar{B}/B_{c2})^{-1/2}$ and, secondly, a weak attraction of range $\xi' = \xi/(2 - 2\bar{B}/B_{c2})^{-1/2}$ is added, caused by the condensation energy gained by the overlap of the vortex cores. Parallel vortex lines then interact by an effective potential $V(r) \propto K_0(r/\lambda') - K_0(r/\xi')$ which no longer diverges at zero distance r [9].

A3.1.3.2 Pippard's Theory

Inspired by Chamber's non-local generalization of Ohm's law, Pippard introduced a superconductor coherence length ξ in 1953 by generalizing the London equation $\mu_0 J = -\lambda_L^2 A$ to a non-local relationship [2] (see also Chapter A2.3)

$$\mu_0 J(r) = \lambda_P^{-2} \frac{3}{4\pi\xi^2} \int \frac{r'(r'A(r - r'))}{r'3} e^{-r'/\xi} d^3 r'. \quad \text{(A3.1.10)}$$

In the presence of electron scattering with mean free path l, the Pippard penetration depth $\lambda_P = (\lambda_L^2 \xi_0/\xi)^{1/2}$ exceeds the London penetration depth λ_L of a pure material with coherence length ξ_0, since the effective coherence length ξ is reduced by scattering, $\xi^{-1} \approx \xi_0^{-1} + l^{-1}$ [2]. In the limit of small $\xi \ll \lambda_P$, Equation (A3.1.10) reduces to the local relation $\mu_0 J(r) = -\lambda_P^{-2} A(r)$. Pippard's Equation (A3.1.10) in the Fourier space becomes $\mu_0 J(k) = -Q_P(k) A(k)$ with

$$Q_P(k) = \lambda_P^2 h(k\xi), \quad h(x) = \frac{3}{2x^3}[(1 + x^2)\text{atan}\, x - x], \quad h(0) = 1. \quad \text{(A3.1.11)}$$

A3.1.3.3 BCS Theory

The microscopic BCS theory (in the Green function formulation of Gor'kov) for weak magnetic fields yields a similar non-local relation $\mu_0 J(k) = -Q(k)A(k)$ as suggested by Pippard, in which the Pippard kernel $Q_P(k)$ is replaced by the BCS kernel [10]

$$Q_{BCS}(k) = \lambda^{-2}(T) \sum_{n=1}^{\infty} \frac{h[k\xi_K/(2n+1)]}{1.0518\,(2n+1)^3}. \quad \text{(A3.1.12)}$$

Here $h(x)$ is defined in Equation (A3.1.11); $\lambda(T) = Q_{BCS}(0)^{-1/2} \approx \lambda(0)(1 - T^4/T_c^4)^{-1/2}$ is the temperature-dependent magnetic penetration depth and $\xi_K = \hbar v_F/(2\pi k_B T) \approx 0.844\lambda(T)T_c/(\kappa T)$ (v_F = Fermi velocity, κ = GL parameter). The range of the BCS Gor'kov kernel is of the order of the BCS coherence length $\xi_0 = \hbar v_F/(\pi\Delta_0)$, where $\mathbf{\Delta}_0$ is the BCS energy gap at $T = 0$.

With the non-local relation $\mu_0\,(k) = -Q(k)A(k)$, Equation (A3.1.2) for a vortex line at $\mathbf{r}_v = 0$ now becomes $[1 + Q(k)^{-1}k^2]\tilde{B}(k) = \Phi_0$ with the solution

$$\tilde{B}(k) = \frac{\Phi_0 Q(k)}{Q(k) + k^2}, \quad B(r) = \frac{\Phi_0}{2\pi} \int_0^{\infty} \frac{Q(k)}{Q(k) + k^2} J_0(kr) k\, dk \quad \text{(A3.1.13)}$$

where $J_0(x)$ is a Bessel function. The Pippard or BCS field $B(r)$ Equation (A3.1.13) of an isolated vortex line is no longer monotonic as compared with the London field, [cf Equation (A3.1.3)], but exhibits a field reversal with a negative minimum at large distances $r \gg \lambda_P$ from the vortex core. This effect should be observable if $\xi \approx \lambda$, i.e., for clean superconductors with small GL parameter κ at low temperatures.

The field reversal of the vortex field is partly responsible for the attractive interaction between flux lines at large distances, which has been observed in clean niobium at temperatures significantly below T_c and which follows from BCS theory at $T < T_c$ for pure superconductors with GL parameter κ close to $1/\sqrt{2}$. This attraction leads to abrupt jumps in the magnetization curve and to an agglomeration of flux lines that can be observed in superconductors with demagnetization factor $N \neq 0$ as FLL islands surrounded by Meissner state, or Meissner islands surrounded by FLL [11]. For the definition of N see Chapter A3.1 by A. M. Campbell and Section A3.1.5.

Another BCS effect which differs from the GL result is that in clean superconductors at low temperatures, the periodic magnetic field of the FLL near B_{c2} is not a smooth spatial function as in GL theory, but has sharp conical maxima and minima such that the profile $B(x, 0)$ along a nearest neighbour direction has zig-zag shape [3].

A3.1.4 Elasticity of the Vortex Lattice

The flux-line displacements caused by pinning forces and by thermal fluctuations may be calculated using the elasticity theory of the FLL. Figure A3.1.4 illustrates the three basic distortions of the triangular FLL, namely shear, uniaxial compression and tilt. The linear elastic energy F_{elast} of the FLL is obtained by expanding its free energy F with respect to small displacements $u_v(z) = r_v(z) - R_v = (u_{vx}, u_{vy})$ of the flux lines

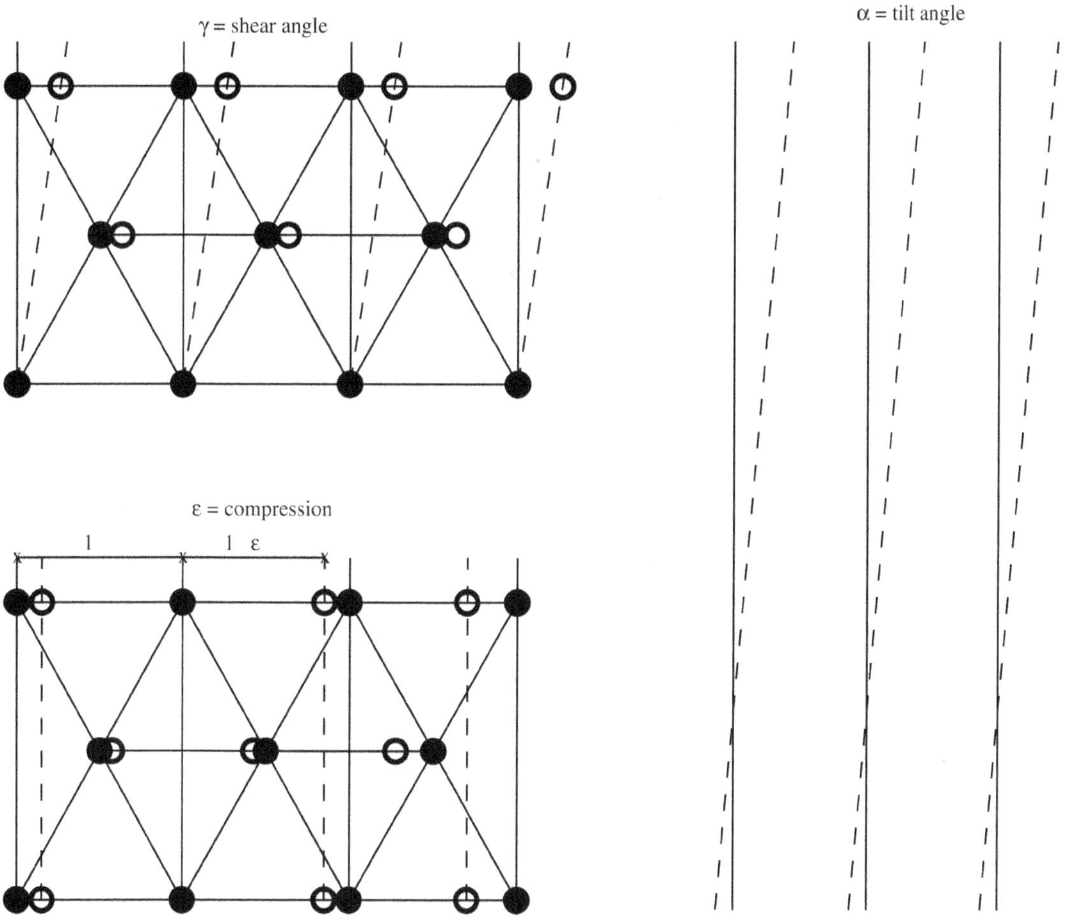

FIGURE A3.1.4 The three basic homogeneous elastic distortions of the triangular flux-line lattice. The full dots and solid lines mark the ideal lattice and the hollow dots and dashed lines the distorted lattice.

from their ideal parallel lattice positions $\boldsymbol{R}_v$ and keeping only the quadratic terms. This yields [3]

$$F_{\text{elast}} = \frac{1}{2}\int_{\text{BZ}} \frac{d^3k}{8\pi^3} u_\alpha(\boldsymbol{k})\Phi_{\alpha\beta}(\boldsymbol{k})u_\beta^*(\boldsymbol{k}) \qquad \text{(A3.1.14)}$$

where $\boldsymbol{u}(\boldsymbol{k})$ is the Fourier transform of the displacement field $\boldsymbol{u}_v(z)$, now $(\alpha, \beta) = (x, y)$ and $\boldsymbol{k} = (k_x, k_y, k_z)$. The k-integral in Equation (A3.1.14) is over the first Brillouin zone (BZ) of the FLL, since the 'elastic matrix' $\Phi_{\alpha\beta}(\boldsymbol{k})$ is periodic in the k_x, k_y plane. The finite vortex core radius restricts the k_z integration to $|k_z| \le \xi^{-1}$. For an elastic medium with uniaxial symmetry, the elastic matrix becomes

$$\Phi_{\alpha\beta}(\boldsymbol{k}) = (c_{11} - c_{66})k_\alpha k_\beta + \delta_{\alpha\beta}[(k_x^2 + k_y^2)c_{66} + k_z^2 c_{44}]. \qquad \text{(A3.1.15)}$$

The coefficients c_{11}, c_{66} and c_{44} are the elastic moduli of uniaxial compression, shear and tilt, respectively. $\Phi_{\alpha\beta}(\boldsymbol{k})$ has been calculated from GL and London theories [3] for the FLL. The result, a sum over reciprocal lattice vectors, should coincide with Equation (A3.1.15) in the continuum limit, i.e., for small $|\mathbf{k}| \ll k_{BZ}$, where $k_{BZ} = (4\pi B/\Phi_0)^{1/2}$ is the radius of the circularized (actually hexagonal) Brillouin zone of the triangular vortex lattice with area πk_{BZ}^2. In the London limit, the elastic moduli for isotropic superconductors becomes

$$c_{11}(k) \approx \frac{B^2/\mu_0}{1+k^2\lambda^2}, \; c_{66} \approx \frac{B\Phi_0/\mu_0}{16\pi\lambda^2}, \; c_{44}(\boldsymbol{k}) \approx c_{11}(k) + 2c_{66}\ln\frac{\kappa^2}{1+k_z^2\lambda^2}. \qquad \text{(A3.1.16)}$$

The GL theory yields an additional factor $(1 - B/B_{c2})^2$ in c_{66} [i.e., $c_{66} \propto B(B - B_{c2})^2$]. The $\mathbf{k}$ dependence (dispersion) of the compression and tilt moduli $c_{11}(k)$ and $c_{44}(\boldsymbol{k})$ means that the elasticity of the vortex lattice is *non-local*, i.e., strains with short wavelengths $2\pi/k \ll 2\pi\lambda$ have a much lower elastic energy than a homogeneous compression or tilt (corresponding to $\boldsymbol{k} \to 0$). This elastic non-locality comes from the fact that the magnetic interaction between the flux lines typically has a range λ much longer than the flux-line spacing a. Each flux line, therefore, interacts with many other flux lines.

Both the compressional modulus c_{11} and the typically much smaller shear modulus $c_{66} \ll c_{11} \approx c_{44}$ originate from the flux-line interaction. The last term in the tilt modulus c_{44}

in Equation (A3.1.16), however, originates from the line tension of isolated flux lines, defined by $P = \lim_{B\to 0}(c_{44}\Phi_0/B)$. In isotropic (or cubic) superconductors like Nb and its alloys, the line tension coincides with the self-energy of a flux line, $P = F_s, F_s = \Phi_0 B_{c1}/\mu_0 \approx (\Phi_0^2/4\pi\mu_0\lambda^2)(\ln\kappa + 0.5)$ for $\kappa \gg 1$. In anisotropic materials, the line tension and line energy of flux lines in general are different and depend on the angle θ of the vortex line with respect to the c-axis, with $P(\theta) = F_s(\theta) + \partial^2 F_s/\partial\theta^2$. Using $F_s(\theta) = F_s(0)(\cos^2\theta + \varepsilon^2\sin^2\theta)^{1/2}$ with $\varepsilon = \lambda_{ab}/\lambda_c \le 1$, one obtains $P(0) = \varepsilon^2 F_s(0)$ and $P(\pi/2) = (\pi/2)\varepsilon^{-2}F_s(\pi/2) = \varepsilon^3/2P(0)$ [3]. In isotropic superconductors, the uniaxial symmetry of the ideal vortex lattice (i.e., the appearance of a preferred axis) is induced by the applied magnetic field. This induced anisotropy leads to a small difference between the compressional and tilt moduli c_{11} and c_{44}, but not to a difference between line energy and line tension.

As a consequence of non-local elasticity, the flux-line displacements $\boldsymbol{u}_v(z)$ caused by local pinning forces, and also the space and time averaged thermal fluctuations $\langle \boldsymbol{u}_v(z)^2\rangle$, are much larger than they would be if $c_{44}(\boldsymbol{k})$ had no dispersion, i.e., if it were replaced by $c_{44}(0) \approx B^2/2\,\mu_0$. The maximum displacement $u(0) \propto f$ caused at $\boldsymbol{r} = 0$ by a point force of density $f\delta_3(\mathbf{r})$, and the thermal fluctuations $\langle u^2\rangle \propto k_B T$, is given by similar expressions [3],

$$\frac{2u(0)}{f} \approx \frac{\langle u^2\rangle}{k_B T} \approx \int_{BZ} \frac{d^3k}{8\pi^3}\frac{1}{(k_x^2+k_y^2)c_{66} + k_z^2 c_{44}(\boldsymbol{k})} \approx \frac{k_{BZ}^2\lambda}{8\pi[c_{66}c_{44}(0)]^{1/2}}. \quad \text{(A3.1.17)}$$

In this result, a large factor $[c_{44}(0)/c_{44}(k_{BZ})]^{1/2} \approx k_{BZ}\lambda \approx \pi\lambda/a \gg 1$ originates from the elastic non-locality. In anisotropic superconductors with $B\|c$, the length λ in the numerator of Equation (A3.1.17) is replaced by the larger length λ_c, which effectively enhances the thermal fluctuations by an additional factor $\varepsilon^{-2} = \lambda_c/\lambda_{ab}$.

A3.1.5 Continuum Description of the Vortex State

A continuum description of the vortex state may be used to calculate the distributions of magnetic field and current in superconductors of arbitrary shape for length scales large compared with the vortex spacing a and the penetration depth λ. Two different algorithms have been proposed [12,13] which, in principle, allow computation of the electromagnetic behaviour of superconductors of arbitrary shape with and without vortex pinning.

Apart from the Maxwell equations, a continuum description requires the constitutive laws of the superconductor. These may be obtained, for example, from the London equation by taking the limits $a \to 0$, $\lambda \to 0$, and from appropriate models of vortex dynamics. One constitutive law is the reversible magnetization curve of a pin-free, or ideal, superconductor, $M(B_a)$ or $B(B_a)$. In the simplest case, this may read $\boldsymbol{M} = 0$ or $\boldsymbol{B} = \mu_0\boldsymbol{H}$, which is valid if everywhere B is larger than several times the lower critical field B_{c1}. In general, however, the reversible $M(B_a)$ computed from London or GL theories should be used.

The correct $M(B_a)$ for finite B_{c1} and general geometry will lead to an irreversible magnetization loop, even in complete absence of vortex pinning. This irreversibility is caused by a geometric barrier for the penetration of magnetic flux, which is absent only if the superconductor has the shape of an ellipsoid or is a cone with a sharp cusp or edge where flux lines can penetrate easily. In superconductor cylinders or strips with rectangular cross-section $2a \times 2b$ ($2a$ = diameter or width, $2b$ = height) in increasing Ba, for example, the magnetic flux penetrates first reversibly at the four corners in the form of nearly straight flux lines, as illustrated in Figure A3.1.5.

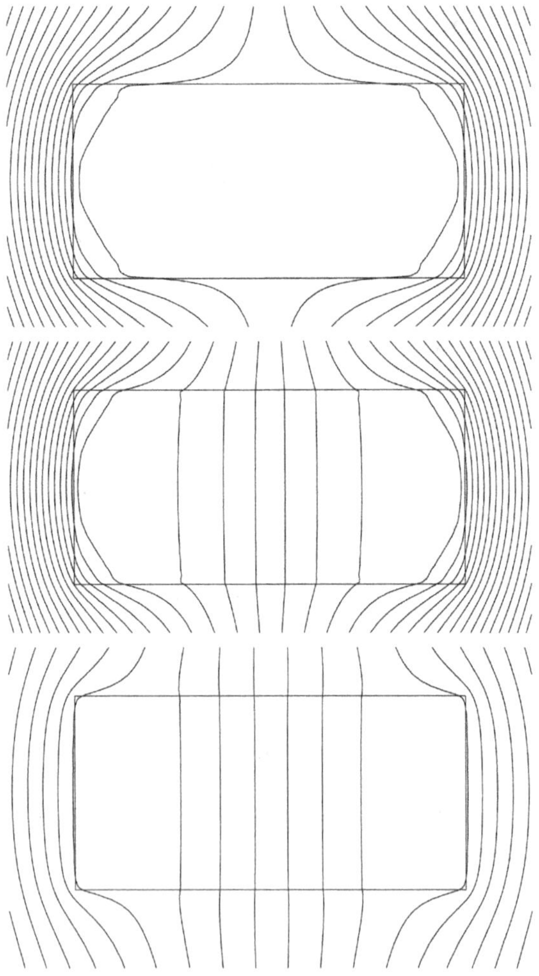

FIGURE A3.1.5 The magnetic field lines in a pin-free strip of aspect ratio $b/a = 0.5$ in an increasing applied perpendicular field at two field values $B_a/B_{c1} = 0.38$ (top) and 0.44 (middle) just below and above the entry field $B_{en} = 0.40B_{c1}$, and in decreasing field at $B_a/B_{c1} = 0.20$ (bottom). Calculation by the method [13].

When the field of first flux entry B_{en} [13] is reached, these flux lines join at the equator and jump to the specimen centre, from where they gradually fill the entire superconductor. On decreasing B_a, some flux initially exits reversibly, until a reversibility field $B_{rev} > B_{en}$ is reached at which the magnetization loop opens since the barrier for flux exit is weaker than the barrier for flux entry. In the limit of thin films, there is no barrier for flux exit. For arbitrary aspect ratio b/a, the entry field is given by

$$B_{en} \approx B_{c1} \tanh \sqrt{cb/a} \qquad \text{(A3.1.18)}$$

where $c = 0.36$ for strips and $c = 0.67$ for discs or cylinders. This geometric barrier should not be confused with the Bean–Livingston barrier for the penetration of a straight vortex line into the planar surface of a superconductor, which would lead to a similar asymmetric magnetization loop (see Chapter A4.3). The geometric barrier is caused by the line tension of the vortices penetrating at sharp [12, 13] or rounded [14] corners. This line tension (equal to $\Phi_0 B_{c1}/\mu_0$ in isotropic superconductors [3]) is balanced by the Lorentz force exerted on the vortex ends by the surface screening currents and directed towards the specimen centre.

The calculated irreversible magnetization loops of pin-free cylindrical superconductors with various aspect ratios b/a = 0.08 (thin disc) to $b/a = \infty$ (long cylinder) in an axial field B_a are illustrated in Figure A3.1.6, together with the corresponding reversible magnetization curves of ellipsoids which have the same initial (Meissner state) slope as the cylinders. The magnetization loops of these cylinders (like those of other non-ellipsoids) have a maximum at $B_a \approx B_{en}$, and are reversible at $B_a > B_{rev}$ at which field they coincide with the magnetization of the corresponding ellipsoid. All pin-free magnetization loops are symmetric, $M(-B_a) = -M(B_a)$ with $M(0) = 0$, i.e., no remanent flux can remain at $B_a = 0$ since bulk pinning is absent, and there is no barrier for flux exit when $B_a = 0$. The small finite $M(0)$ in Figure A3.1.6 is caused by the finite ramp rate used in these calculations, leading to a viscous drag force on the vortices.

The reversible magnetization curves of pin-free ellipsoids $M(B_a, N)$ follow from the magnetization curve $M(B_a, N = 0)$ of long cylinders in parallel field B_a by the concept of a demagnetization factor N. N takes the values $0 \le N \le 1$, $N = 0$ for parallel geometry, $N = 1/2$ for infinite cylinders in perpendicular field, $N = 1/3$ for spheres and $N = 1$ for thin films in perpendicular field. The implicit equation for an effective internal field B_i has to be solved for ellipsoids with $N \neq 0$

$$B_i = B_a - N_{\mu 0} M(B_i, N = 0) \qquad \text{(A3.1.19)}$$

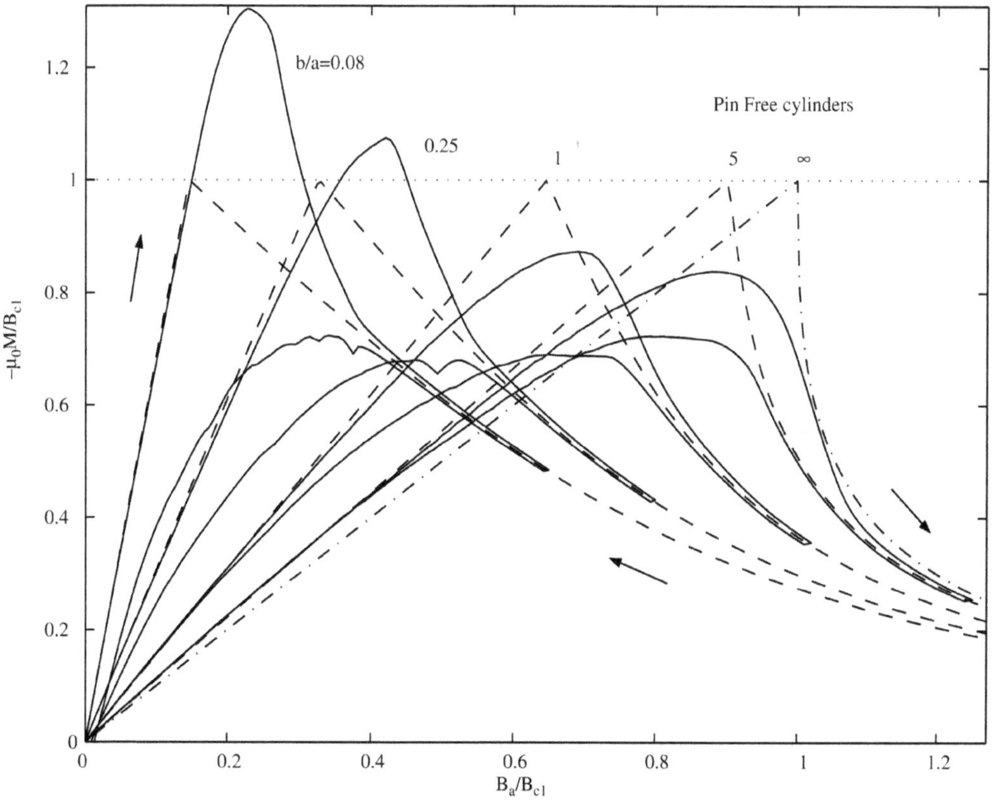

FIGURE A3.1.6 The irreversible magnetization of pin-free cylinders with various aspect ratios b/a in a cycled axial magnetic field B_a (solid lines). The dashed lines show the reversible magnetization of ellipsoids with the same initial slope as the cylinder.

to obtain the reversible magnetization $M(B_a, N) = M(B_i; N = 0)$ (dashed lines in Figure A3.1.6).

A second constitutive law is based on the local electric field **E(J,B)** which is generated by moving vortices and which in a compact way can describe flux flow and the Hall effect, vortex pinning and thermally activated depinning. A simple but still quite general isotropic model is $\boldsymbol{E} = \rho(J, B)\mathbf{J}$ with $\rho = \text{const}{\cdot}B{\cdot}|J/J_c|^{n-1}$, where J_c is the critical current density, and n is the creep exponent [3, 15]. Within this realistic model, flux flow is described by $n = 1$, flux creep (relaxation with approximately logarithmic time law) by $n \gg 1$, and the critical state model of vortex pinning by the limit $n \to \infty$ (see also Chapters A4.3 and A3.1). In general, both $J_c(B,T)$ and $n(B,T)$ may depend on the local induction B and on the temperature T. Figure A3.1.7 shows some magnetization loops of a short cylinder calculated in [15] for two different $J_c(B)$ models and two creep exponents. The finite London penetration depth A can be accounted for by adding to $E(J, B)$ a term $E = \mu_0\lambda^2\dot{J}$, where $\dot{J}$ is the time derivative of the current density. This generalization is described in [16].

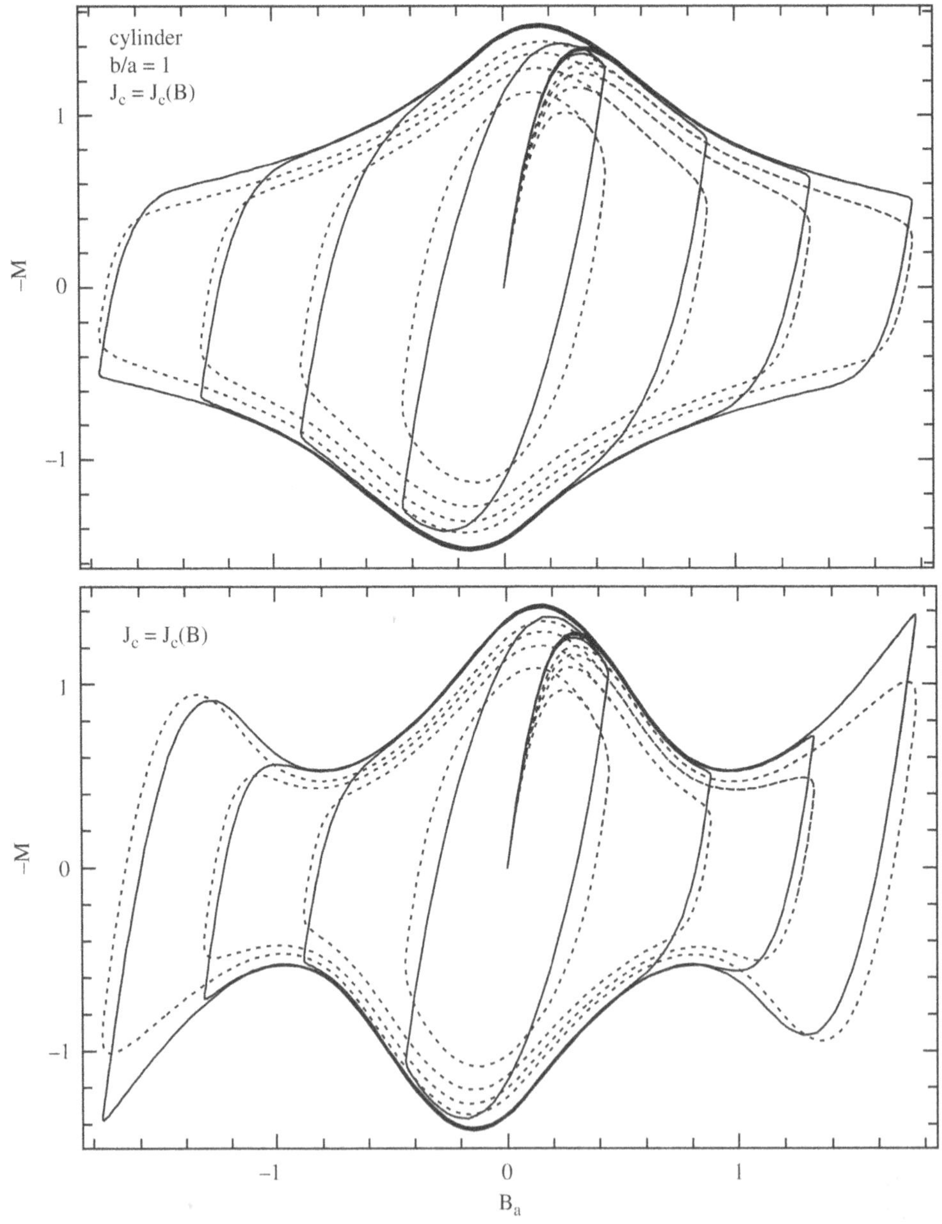

FIGURE A3.1.7 Magnetization loops of cylinders with aspect ratio $b/a = 1$, creep exponents $n = 51$ (bold lines) and $n = 5$ (dashed lines), with $B_{c1} = 0$, in cycled applied field B_a for two induction dependent critical current densities $J_c(B) = J_{c0}/(1 + 3\beta)$ (top, Kim model) and $J_c(B) = J_{c0}(1 - 3\beta + 3\beta^2)$ (bottom, a 'fish-tail' model), where $\beta = B_a/(\mu_0 J_{c0} a)$. The magnetization M is in units $J_{c0}a/(2\pi b)$ and B_a in units $\mu_0 J_c a$, where a, b are the radius and half height of the cylinder, respectively.

References

[1] DeGennes P G 1966 *Superconductivity of Metals and Alloys* (New York: Benjamin)

[2] Tinkham M 1975 *Introduction to Superconductivity* (New York: McGraw-Hill)

[3] Brandt E H 1995 The flux-line lattice in superconductors *Rep. Prog. Phys.* **58** 1465–1594

[4] Clem J R 1975 Simple model for the vortex core in type II superconductors *J. Low Temp. Phys.* **18** 427–434

[5] Hao Z, Clem J R, Mc Elfresh M W, Civale L, Malozemov A P and Holtzberg F 1991 Model for the reversible magnetization of high-κ type II superconductors: application to high-T_c superconductors *Phys. Rev. B* **43** 2844–2852

[6] Brandt E H 1997 Precision Ginzburg-Landau solution of the flux-line lattice with arbitrary induction and symmetry *Phys. Rev. Lett.* **78** 2208–2211

[7] Brandt E H 1981 Properties of the distorted flux-line lattice near a planar surface *J. Low Temp. Phys.* **42** 557–584

[8] Jung-Chun W and Tzong-Jer Y 1996 Current distribution and vortex-vortex interaction in a superconducting film of finite thickness *Jpn. J. Appl. Phys.* **35** 5696–5700

[9] Brandt E H 1986 Elastic and plastic properties of the flux-line lattice in type II superconductors *Phys. Rev. B* **34** 6514–6517

[10] Abrikosov A A, Gorkov L P and Dzyaloshinski I E 1963 *Methods of Quantum Field Theory in Statistical Physics* (Englewood Cliffs: Prentice Hall)

[11] Brandt E H and Essmann U 1987 The flux-line lattice in type II superconductors *Phys. Stat. Solidi b* **144** 13–38

[12] Labusch R and Doyle T B 1997 Macroscopic equations for the description of the quasi-static magnetic behaviour of a type II superconductor of arbitrary shape *Physica C* **290** 143–160

[13] Brandt E H 1999 Geometric barrier and current string in type II superconductors obtained from continuum electrodynamics *Phys. Rev. B* **59** 3369–3372

[14] Benkraouda M and Clem J R 1998 Critical current from surface barriers in type II superconducting strips *Phys. Rev. B* **58** 15103–15107

[15] Brandt E H 1998 Superconductor disks and cylinders in an axial magnetic field. I. Flux penetration and magnetization curves *Phys. Rev. B* **58** 6506–6522

[16] Brandt E H 2001 Theory of type II superconductors with finite London penetration depth *Phys. Rev. B* **64** 024505

A3.2

Flux Pinning

Kees van der Beek and Peter H. Kes

A3.2.1 Introduction

Type II superconductors with high upper critical fields H_{c2} have considerable potential for practical applications, including high magnetic field solenoids, permanent magnets and energy storage devices, as well as magnetic field detectors (SQUIDs) and superconducting components for electronics and communications. However, the interplay between electrical currents and lines of quantised magnetic flux (the "flux lines", "vortex lines", or simply "vortices" of Section A3.1) in superconducting materials results in a driving force that puts the latter in motion. Vortex motion leads to dissipation of energy, manifested as an electric potential within the material, which therefore can no longer be considered to be superconducting, i.e. its electrical resistance becomes non-zero (Bardeen and Stephen 1965). Preventing vortex motion up to a high critical current density j_c is therefore essential. This is achieved through *pinning* of the vortex lines by imperfections of the material. Fortunately, such flux pinning is a general phenomenon in commonly produced materials. In the following sections, an overview is given of this interesting phenomenon and some related issues.

A3.2.2 Origin of Flux Pinning: Material Defects and Flux Pinning Interactions

Material imperfections providing flux pinning occur in many different varieties, and include inhomogeneities, defects, and engineered structures of sizes ranging from the atomic scale to the macroscopic scale. They are either naturally present as a result of the composition of the material, artificially introduced as the byproduct of the growth or preparation method, or tailored by micro- or nano-engineering. It is safe to assume that any imperfection of the material will lead to some extent of flux pinning and therefore affect the macroscopic physical properties of the superconductor, even if the effects may be very small. However, for the vast majority of superconducting materials, including technological superconductors, flux pinning will completely determine the electrical transport and magnetic properties in the superconducting state. One may classify flux pinning defects through the origin of the interaction with the vortex lines, through their "strength", and through their shape: a cylindrical defect extending along the length of a vortex line will generally be more effective in arresting it than a point-like imperfection.

Imperfections locally alter material properties and, consequently, superconductivity in their environment. Such changes couple to the periodic variations of both the order parameter and the local electromagnetic field, which are characteristic of the mixed state (see Chapter A3.1). In principle, the interaction can be derived by solving the Ginzburg–Landau equations with the appropriate boundary conditions imposed by the defects. Most commonly, the vortex–defect interaction force $f(r)$ is attractive, so that the presence of imperfections results in a random "landscape" $U_{\text{pin}}(\mathbf{r})$ of attractive pinning potential wells for the vortices. However, repulsive pins also exist. An example of such are engineered magnetic centres (see *e.g.* Marchiori et al. 2017) of polarity equal to that of the vortices.

A3.2.2.1 The Core Interaction

The local modification of the order parameter (related to the density of Cooper pairs) by the presence of defects has an incidence on the electronic structure and the diameter of the vortex core. One therefore speaks of "core pinning". Defects deviate from the surrounding material by differences in density, elasticity, or electron–phonon coupling. These give rise to a local change in T_c – and hence of the condensation energy $\frac{1}{2}\mu_0 H_c^2$ – which may range from a minor, secondary effect (due to local strain, for example) to the complete suppression of superconductivity in or around the defect. The defects may also manifest themselves through their scattering effect on quasiparticles in the vortex core, and the subsequent changes in quasiparticle mean free path ℓ, quasiparticle conductivity, and vortex core level structure. One thus distinguishes between δT_c– and $\delta\ell$– pinning.

The coupling to the variation of the order parameter is the main origin of flux pinning by many types of defects, including dislocations, point defects of various kinds, voids, grain boundaries, precipitates, and ion-irradiation-induced damage such as amorphous columnar defects. The typical length scale r_f of the core interaction depends on the spatial variations of the order parameter. Therefore, $r_f \approx \xi$ for low $B < 0.2B_{c2}$ and $r_f \approx a_0/2$ for high flux densities, respectively, where a_0 is the vortex lattice parameter ($\frac{1}{2}a_0^2\sqrt{3} = \Phi_0/B \equiv n_v^{-1}$ where n_v is the vortex line areal density. The elementary pinning force $f(u)$ generally assumes its maximum value f_p for a vortex displacement $u \approx r_f$.

A3.2.2.2 The Electromagnetic Interaction

Imperfections of size comparable to or larger than the London penetration depth λ_L (see Equation [A2.2.4]) will necessarily alter the supercurrent flow in the material, including that of vortex currents. The associated spatially dependent reduction in the kinetic energy provides the attractive pinning interaction. Examples of the magnetic interaction are the attraction of vortex lines to surfaces parallel to the applied magnetic field H_a (this might be the external surface, see Section A3.2.3.6, as well as some large precipitate interface within the sample bulk) and thickness variations of thin films for fields normal to the film. In the latter case, the vortices are trapped at the sites of least thickness where the line energy of the vortex is minimum.

Evidently, the typical length scale related to the magnetic interaction is λ. In materials with a large $\kappa = \lambda/\xi$, this kind of interaction is therefore relatively small; also, it rapidly vanishes as the magnetic field increases beyond several H_{c1}. The electromagnetic interaction with large defects and the surface can be readily modelled using the image vortex method (Clem 1974). A secondary effect of small defects is the depletion of the Cooper pair (superfluid) density, which may entail a spatially modulated increase of λ. Inhomogeneities of the defect distribution will therefore contribute to flux pinning, with vortices being attracted to regions of larger average λ (Demirdis et al. 2013).

A3.2.3 Effect on Electromagnetic and Transport Properties in the Mixed State

A3.2.3.1 Electric Conductivity and the $E(j)$ Relation

Experimentally, flux pinning manifests itself in many ways which provide one with different opportunities to measure the critical current density j_c. Given that the local current density $\boldsymbol{j}$ is proportional to the force $\boldsymbol{F}_L = \boldsymbol{j} \times \boldsymbol{B}$ on the vortices, and the electric field $\boldsymbol{E} = \boldsymbol{v} \times \boldsymbol{B}$ (averaged over a distance larger than several times a_0) is given by the cross product of average vortex velocity $\boldsymbol{v}$ and the flux density $\boldsymbol{B}$, the local relation between electric field and current density $E(j)$ reflects the average force–velocity curve characterising the motion of the vortices through the pinning "landscape". Ideally, $|\mathbf{E}| = |\mathbf{v}| = 0$ up to $|\mathbf{j}| = j_c$ and then rises almost linearly, with a slope slightly greater than the flux flow resistivity ρ_f for $j > j_c$ (see below and Figure A3.2.1). Thus $j_c \equiv F_p/B$ is defined as the critical current density that will overcome the net "volume" pinning force F_p (unit [N/m^3]) exerted by the material imperfections on the vortex ensemble, thereby setting the latter in motion. A reasonable approximation for ρ_f is $\rho_f \approx \rho_n(B/B_{c2})$ (Bardeen and Stephen 1965), where ρ_n is the electrical resistivity in the normal state, suggesting that the dissipation takes place mainly in

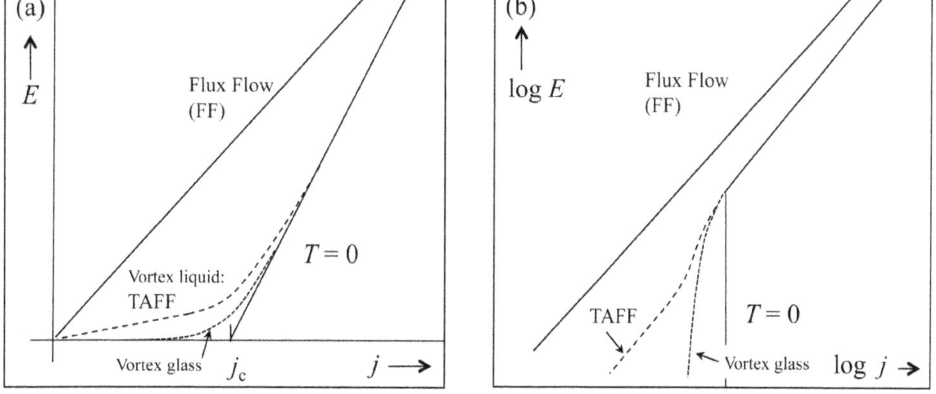

FIGURE A3.2.1 Typical $E(j)$ characteristics in the mixed state in the absence of pinning (FF), with bulk pinning at zero temperature ($T = 0$) and at finite temperature for finite energy barriers (TAFF, dashed, appropriate for the vortex liquid phase) and infinite barriers (glass, dotted) for $j \to 0$, rendered as a linear (a) and double logarithmic plot (b). Note that the differential resistivity above the critical current density j_c is larger than or equal to the flux flow (FF) resistivity in the absence of pinning. This effect is due to velocity fluctuations and is more than often ignored.

the core of the moving flux lines. In strong pinning materials, j_c can be as large as a few tenths of the depairing current density, $j_0 \approx H_c / \lambda$ (see Section A2.2), which is typically 10^8 A/cm^2. To achieve such large current densities in electrical transport measurements is difficult due to heating effects and contact problems, and therefore inductive probes (magnetic measurements) are frequently used.

A3.2.3.2 Inductive Measurements and Magnetic Hysteresis

In the absence of pinning centres, surface or geometrical effects, the flux distribution inside a superconductor in a magnetic field $H > H_{c1}$ is uniform with density B. The Meissner screening currents in a surface layer of thickness λ produce a (diamagnetic) moment m and an associated volume magnetisation $M = m/V$ which is in equilibrium with H, as described by the reversible magnetisation curve of Abrikosov (see Chapter A3.1, section A3.1.3.1), i.e. $M = M_{rev}(H)$ and $B_{rev}(H) = \mu_0(H + M_{rev})$ (Indenbom et al. 1994).

Since their electromagnetic response is dictated by their nonlinear $E(j)$ curve rather than by the London equation, type II superconductors with flux pinning in the material bulk behave as perfect conductors rather than superconductors. Their response to low-frequency time-varying electromagnetic fields is described by a set of macroscopic nonlinear diffusion equations for the (electro-)magnetic field and the induced bulk electrical shielding current $\mathbf{j}(\mathbf{r})$, obtained by combining Maxwell's equations with $E(j)$:

$$-\frac{\partial \boldsymbol{B}}{\partial t} = \nabla \times \boldsymbol{E}(j) \equiv \nabla \times \hat{\rho}(j) \nabla \times \boldsymbol{B} \quad \text{(A3.2.1)}$$

$$-\frac{\partial \boldsymbol{j}}{\partial t} = \nabla \times \nabla \times \boldsymbol{E}(j) \quad \text{(A3.2.2)}$$

(see also Section A3.1.5) In direct analogy with Ohmic conductors, magnetic flux entry and exit into the superconductor bulk is delayed with respect to the variation of the applied electromagnetic field, leading to flux density gradients and hysteresis of the local flux- and current destiny $\boldsymbol{j}(\boldsymbol{r})$ (here $\boldsymbol{r}$ is the position measured from the sample centre). When the field is increased, vortex lines enter at the sample edge so that the vortex density is highest there, and lowest in the sample centre. When the magnetic field is decreased, vortices exit first through the sample boundary, while those in the centre remain trapped, leading to magnetic remanence. Thus, the magnetic moment

$$m = \int_V \left| \boldsymbol{j}(\boldsymbol{r}) \times \boldsymbol{r} \right| d^3 r \quad \text{(unit [Am}^2\text{])} \quad \text{(A3.2.3)}$$

is irreversible: the value of m depends on the path taken to reach the measurement field, as illustrated in Figure A3.2.2 (V is the sample volume). The width ΔM_{sat} of the main hysteresis loop is a direct measure of j_c, or, more generally, of the sustainable lossless current density, as will be discussed below.

A3.2.3.3 Self-Organisation of the Flux Distribution and the Flux Front

The hysteresis is the result of vortex lines being (drastically) slowed down by repeated trapping by successive pinning centres as the screening current induced by the time-varying field pushes them into or out of the material. On first flux penetration (zero-field cooled condition), the nonlinear nature of the vortex–pin interaction leads to the presence of a well-defined flux front to which the first vortices have

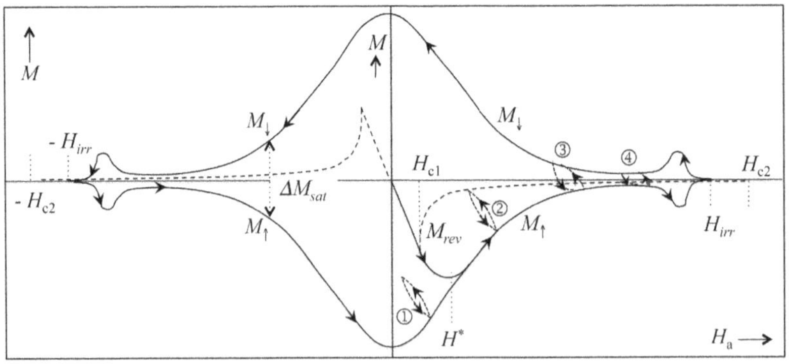

FIGURE A3.2.2 Model magnetisation curves for a type II superconductor in the absence of any pinning mechanism (M_{rev}, dashed) and in presence of bulk pinning only (drawn line) for increasing field ($M_\uparrow$) and decreasing field ($M_\downarrow$). H^* is the characteristic field above which vortices completely permeate the sample when the field is applied after zero-field cooling. Also shown are minor magnetisation loops (dotted) traversed when a field modulation or ac-field of moderate amplitude h_0 and frequency is superimposed on the dc-field. When the ac-field amplitude increases from $h_0 < H^*$ to $h_0 \gg H^*$ the minor hysteresis loops change shape from lenticular (minor loops [1] and [2]) to parallelogramatic (minor loop [4]). H_{irr} is the irreversibility field above which the effects of pinning can no longer be discerned. The little "humps" below H_{irr} illustrate the peak effect. In this example, the ratio between the upper and lower critical fields H_{c2} and H_{c1} corresponds to $\kappa \approx 2.5$.

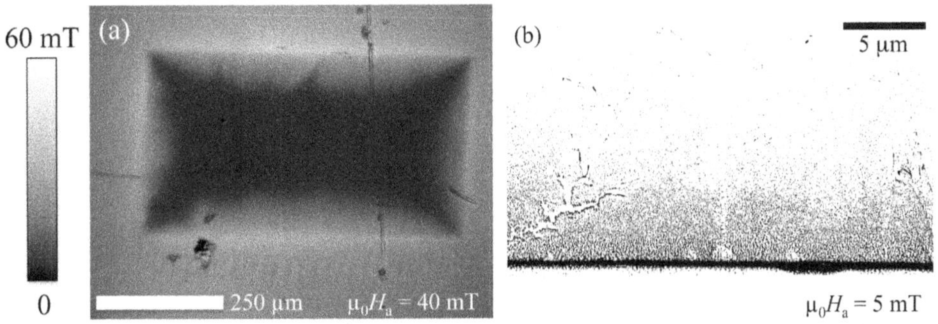

FIGURE A3.2.3 Imaging of initial vortex penetration into a $Ba(Fe_{0.93}Co_{0.07})_2As_2$ single crystal with strong pinning, at 10 K, after zero-field cooling. (a) A low-resolution magneto-optical image of the magnetic flux density shows how the latter penetrates from the four edges of the rectangular sample, up to a well-defined flux front (bright areas correspond to high flux density B, black areas to $B=0$). The inner region is in the Meissner state and therefore appears as black (spots are artefacts, and wedges are magnetic domains of the ferrimagnetic film used for imaging). (b) A Bitter decoration image reveals the individual vortices. After Grisolia et al. 2013.

advanced. Behind the flux front, the vortices self-organise so as to maintain a flux density gradient $\nabla \times \boldsymbol{B}$ equal to the critical (or sustainable) current density. The local advances of the vortices into the superconductor are affected by the spatial variations of the "pinning landscape" and proceed in an avalanche-like manner. The flux front therefore has a rough profile, described by a nonlinear diffusion equation (Edwards and Wilkinson 1982). At the characteristic field H^*, flux fronts entering from opposite sides of the sample merge, and the description in terms of self-organisation is no longer appropriate. The dynamics of vortices entering a type II superconductor with pinning has been compared to that of a sandpile (de Gennes 1989).

A3.2.3.4 Modelling Using a Power Law $E(j)$-Curve

The macroscopic hysteretic behaviour of (the local values of) $\boldsymbol{B}$ and $\boldsymbol{j}$ can be conveniently modelled by considering a model power law $E \propto (j/j_c)^n$ (Brandt 1996 and Section A3.1.5). This corresponds to an empirical law sometimes used to characterise the quality of a wire, where a large value of n signifies a high material uniformity, and is also well-suited to quantitatively understand the effect of flux creep on magnetic hysteresis (see below). Note that $E \propto (j/j_c)^n$ smoothly interpolates between Ohm's law and the case of ideal flux pinning. In the first case, $n=1$, and the material shows a linear electromagnetic response described by the skin effect; the effective depth to which the electromagnetic field penetrates the material is the skin depth $\delta = \sqrt{\rho/\mu_0\omega}$ where $\rho = \rho_n, \rho_f$ for a normal metal and a superconductor without pinning, respectively, and ω is the angular frequency of the time-varying field. In the second case, $n \to \infty$, the response is very strongly nonlinear and adequately described by the Bean model.

A3.2.3.5 Bulk Pinning and the Bean Model

The Bean model (Bean 1962) consists of approximating the electromagnetic response of the superconductor by taking the working point on the materials' $E(j)$ (or IV) curve as fixed. In the case of near-perfect pinning without any measurable dissipation, i.e. $|\mathbf{v}|=|\mathbf{E}|=0$ for $|\mathbf{j}|<j_c$, the working point is fixed at $|\mathbf{j}|=j_c$. As a consequence, the modulus $|\mathbf{j}|=\pm j_c$ throughout all regions of the material in which vortices are present. That is, the magnetic flux density gradients obey

$$\left(\frac{\partial B}{\partial H_a}\right)_{rev}^{-1}|\nabla\times\mathbf{B}| = \left(\frac{\partial B}{\partial H_a}\right)_{rev}^{-1}\left|\left(\frac{\partial B_x}{\partial z}\right)-\left(\frac{\partial B_z}{\partial x}\right)\right| = j_c(\mathbf{B}), \quad \text{(A3.2.4)}$$

the "critical state equation". In all other regions of the material $\boldsymbol{B}=0$, and $\boldsymbol{j}$ takes on the value required by the Biot–Savart law and the material geometry. In Equation (A3.2.4), $\nabla \times \boldsymbol{B}$ has been written out to describe the flux gradients in an (xz)–planar section of a sample subjected to a magnetic field applied along the z-direction. One sees that for samples that are long in the direction of the applied magnetic field, the current density corresponds to a constant transverse flux gradient $-\partial B_z/\partial x$ (the flux density decreases as a straight line from the sample edge), while for a thin film in perpendicular field, it corresponds to the gradient $\partial B_x/\partial z$ of the parallel flux across the sample thickness. For fields well above H_{c1} and for a Ginzburg–Landau parameter $\kappa = \lambda/\xi \gg 1$, the prefactor $(\partial B/\partial H_a)_{rev}$ (which follows from the slope of the reversible magnetisation curve) can be replaced by μ_0. One can then extract the modulus j_c of the current density from the integral in Equation (A3.2.3), so that the difference between the (saturated) magnetic moments measured in increasing and decreasing magnetic field $\Delta m_{sat} = |m_\uparrow - m_\downarrow| = Cj_c$, with C a constant that depends only on sample geometry. Hence, the Bean model can be applied to irreversible magnetisation

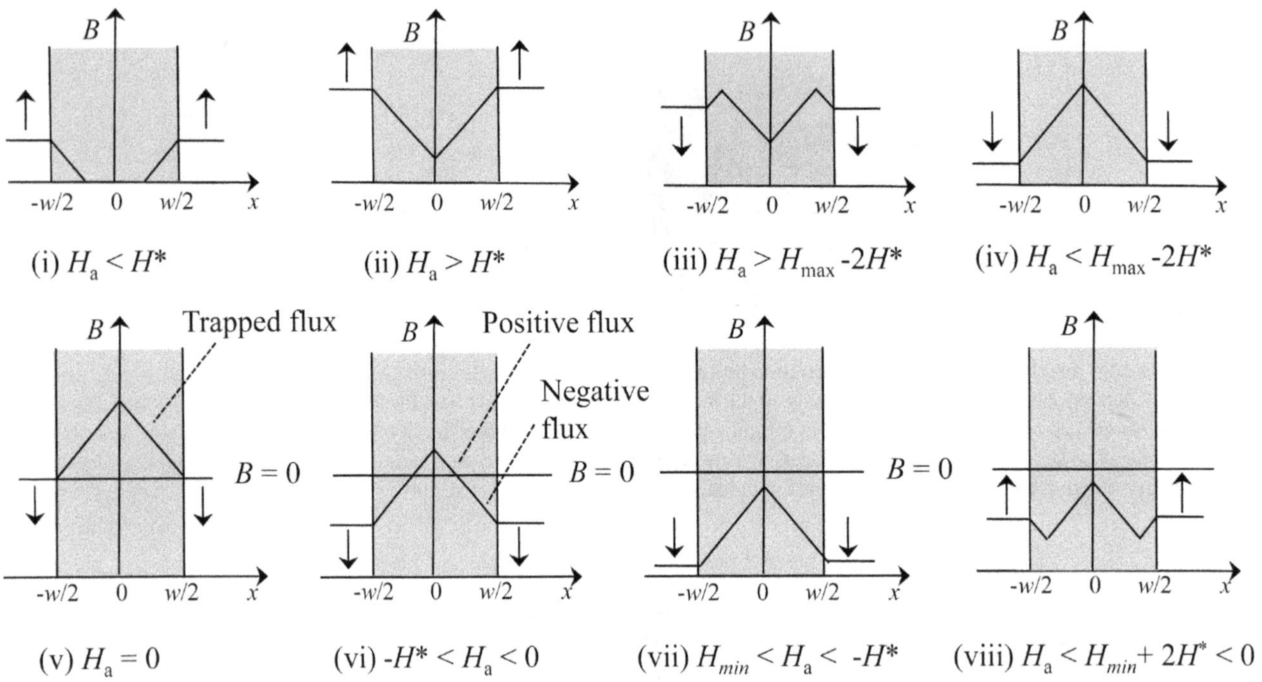

FIGURE A3.2.4 Distribution of the flux density B (or vortex density B/Φ_0) in a slab-shaped type II superconducting sample occupying $|x| < w/2$, with the external magnetic field H_a applied along z, according to the Bean model, at various stages of the hysteresis loop (the y coordinate points into the plane). (i) $H_a < H^*$, the flux fronts penetrating from opposite edges have not yet met in the sample centre; (ii) once $H_a > H^*$, $j = \pm j_c$ throughout the sample and the flux distribution does not change its shape upon further increase of the applied magnetic field; (iii) upon reduction of H_a from the maximum applied field H_{max}, the screening current density at the sample boundaries reverses sign and flux starts leaving the sample; (iv) $H_a < H_{max} - 2H^*$: the direction of the flow of the screening (critical) current and therefore the shape of the flux density profile is now reversed throughout the sample; the flux profile does not change shape as H_a is reduced further; (v) when H_a is reduced to 0, trapped or "remnant" flux (vortices) remains in the sample; (vi) when H_a is reduced below 0, positive trapped flux co-exists with magnetic flux (vortices) of the opposite sign penetrating from the boundaries. If the Bean model does not imply any discontinuity of the screening current density at the interface between positive and negative flux, in real superconductors a vortex-free region appears because no vortices can exist when $H_a < H_{c1}$. In superconductors of thickness $2\lambda \ll d \ll w$, vortex lines wrap around this flux-free zone to form a so-called "current string" or "Meissner hole" (Indenbom et al. 1995). (vii) and (viii) The further decrease of H_a to a minimum (negative) value H_{min} and its subsequent increase entail flux distributions that mirror those in stages (ii) and (iii).

curves to determine j_c from Δm_{sat}. It also follows that the reversible magnetisation lies midway between the increasing and decreasing field branches of the hysteresis loop (indicated in Figure A3.2.2). Several important geometries relevant for experiments are summarised in Table A3.2.1.

A3.2.3.6 Bean–Livingston Barrier

The vortex ensemble in a type II superconductor is stabilised by the inward force provided by diamagnetic screening currents circulating in the surface layer of thickness λ, which effectively forms a magnetic container. This force, which, as discussed in Section A.3.1.2.2, decays as $e^{-x/\lambda}$, with x the distance between a vortex line and an outer surface of the sample, counteracts the (outward) attractive force $\propto e^{-2x/\lambda}$ to the very same surface (see Section A3.2.2.2). The superposition of the two forces results in an energy barrier acting against vortex entry into the sample as the applied magnetic field is increased.

As a result of this "Bean–Livingston barrier", vortices are only admitted into the superconductor when the field reaches the first penetration field $H_p \simeq H_c > H_{c1}$. In the absence of bulk pinning, the surface screening current then pushes the vortices to the sample centre. Contrary to the case of bulk pinning and the Bean model, the flux density on increasing the applied field is thus higher in the sample centre that is near its boundary. As more and more vortices accumulate in the sample, the screening current is increasingly compensated by the vortex currents, yielding a magnetisation that takes the general form $M_{\uparrow} \simeq -H_p^2/2H_a$ (Clem 1974). Upon decreasing the magnetic field, vortex exit is counteracted by the surface screening current until the magnetic flux densities inside and outside the sample are equal, and the barrier disappears. Vortices can then exit freely, and the magnetisation $M_{\downarrow} \simeq H_{c1}/(\pi \ln \kappa) \approx 0$ (Clem 1974). The resulting magnetic hysteresis loop, with near-zero magnetic moment for decreasing field as a hallmark, is shown in Figure A3.2.5. The Bean–Livingston barrier is significantly

TABLE A3.2.1 Results of the Bean Model Applied to Various Geometries[1]

Geometry	$\Delta M_{sat} \equiv \Delta m_{sat} / V$ [A/m]	H^* [A/m]	I_c [A]	Hysteretic losses	R_s
Long thin sheet of thickness d and height $h \gg d$, H || to the surface	$j_c d / 2$	$j_c d$	$j_c dh$	$\frac{4\mu_0\omega}{3\pi}\frac{h_0^3}{j_c d}$	$\frac{\sqrt{4\pi\mu_0}}{j_c}P_0^{1/2}$
Thin strip of thickness d and width w, in perpendicular applied field	$j_c d / 2$	$j_c d / \pi$	$2 j_c dw$	$\frac{\pi^2\mu_0\omega}{12}\frac{h_0^4}{j_c^2 d^2}$	$\frac{4\pi}{j_c^2 d}P_0$
Long thin cylinder or wire of radius R, H || to the surface	$2 j_c R / 3$	$j_c R$	$\pi j_c R^2$	$\frac{2\mu_0\omega}{3\pi}\frac{h_0^3}{j_c R}$	$\frac{\sqrt{4\pi\mu_0}}{j_c}P_0^{1/2}$
Thin disk of thickness d and radius $R \gg d$ in perpendicular applied field	$2 j_c R / 3$	$j_c d / 2$		$\frac{16\mu_0\omega}{3\pi^3}\frac{R}{d}\frac{h_0^4}{j_c^2 d^2}$	

[1] Here $\Delta M_{sat} \equiv \Delta m_{sat}/V$ is the width of the magnetisation loop at saturation ($H > H^*$), V is the sample volume, H* is the field of full flux penetration, I_c is the critical current (for current applied along the longest direction of the sample), the hysteretic losses (for the situation when the sample is subjected to an ac field) are expressed per ac field cycle (in the limit $h_0 \ll H^*$) [Clem and Sanchez 1994], ω and h_0 are the angular frequency and the amplitude of the ac magnetic field, R_s is the surface resistance, and P_0 is the incident microwave power density (when the electric field and the current are along the longest direction of the sample).

diminished in the presence of sharp corners on the sample, and can be quelled by appropriate surface treatment such as abrasion or irradiation, yielding so-called "gates" for flux entry, or by treatment that smoothly diminishes T_c to zero at the surface, such as an oxygen diffusion layer at the surface of superconducting Nb or V.

A3.2.3.7 Geometrical Barrier

A second barrier effect appears in type II superconducting samples of non-ellipsoidal cross-section; it concerns the vast majority of low-field transport and magnetisation experiments carried out on low-pinning materials, for these are typically carried out on rectangular strip- or platelet-shaped samples. Whereas the external field H_i at the surface (and therefore at the equator) of an ellipsoidal sample is enhanced according to the expression $H_i = H_a - n_x M_{rev}(H_i)$ (with the field H_a applied parallel to one of the principal axes (x) and n_x the corresponding demagnetisation coefficient), the magnetic field at the equator of non-ellipsoidal samples is effectively shielded by the screening currents circulating at the vertices (see Figure A3.2.5). Vortex line segments first (partially)

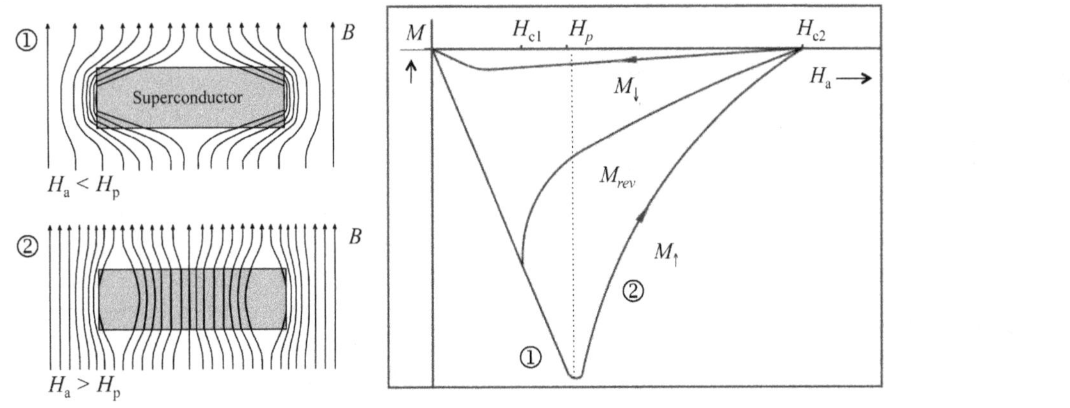

FIGURE A3.2.5 The effect of a Bean–Livingston surface barrier on magnetisation in the absence of bulk pinning. Note the characteristics: the first penetration of flux occurs at $H_p \approx H_c$, the magnetisation is zero at $H_a = 0$, and the magnetisation for decreasing field $M_\downarrow$ is close to zero. The left-hand diagram (1) shows the situation for a platelet-like sample with a rectangular cross-section at $H_a < H_p$, in which vortex line segments partially penetrate through the sharp corners (vertices) of the sample. Diagram (2) shows the situation for $H_a > H_p$: vortex lines accumulate in the centre of the sample.

penetrate through the sharp corners (vertices) of the sample. Full penetration only occurs when segments penetrating from the top and the bottom meet at the sample equator, whence they can enter freely. For strip- or bar-shaped samples of thickness d smaller than the width w, the field of first penetration $H_p \simeq H_{c1}(2d/w)^{1/2}$ is larger than the penetration field $H_p^{ell} = H_{c1}(1-n_x) \approx H_{c1}(2d/w)$ expected for an ellipsoidal sample of the same aspect ratio. The sample geometry, or shape, thus creates a geometrical barrier for increasing field (Zeldov et al. 1994). For decreasing field, this barrier does not exist, leading to strong magnetic hysteresis at low fields $H_a \lesssim H_{c1}$. The shape of the hysteresis loop can be distinguished from that controlled by the Bean–Livingston barrier through the behaviour of the magnetic moment in decreasing field. This approaches the reversible moment (Section A3.2.3.2) and is therefore not zero.

Thus, the geometrical barrier dominates the low-field magnetic irreversibility and electrical transport properties of high aspect ratio, low-j_c strip, and platelet-shaped superconducting materials in perpendicular field. The geometrical barrier is suppressed by careful abrading of the sample to an ellipsoidal or pyramidal cross-section, or by the presence of burrs or sharp wedges on the sample surface (which act as vortex entry gates). Such treatment will reveal the bulk pinning properties. When the sample height in the field direction is larger than the width, flux penetration happens at H_{c1}, or is determined by the Bean–Livingstone barrier.

A3.2.3.8 Simultaneous Presence of Bulk Pinning and a Barrier

In type II superconductors with intermediate j_c (so that the characteristic field $H^* \cong H_{c1}$), both bulk pinning and the barrier will be manifest. Upon first flux penetration, the competition between surface screening currents and bulk pinning will lead to "flux puddles": there is neither a well-defined front of penetrating vortices from the edges nor an accumulation of vortices in the sample centre. Another effect of this intermediate situation is "edge contamination", whereby the vortex ensemble arrangement and the local critical current density are determined by disorder introduced at vortex entry because of local variations of the geometrical barrier or edge roughness (Paltiel et al. 2000).

A3.2.3.9 ac and High-Frequency Electromagnetic Fields

The analysis of the ac response is a useful means to determine the critical current density in conditions where dc magnetic measurements are insufficiently sensitive, but j_c is still too high to allow direct electrical transport measurements. The exposure of a type II superconductor to a time-varying (ac) magnetic field of amplitude h_0 superposed on the background dc field H_a results in a minor magnetisation loop being traversed, as indicated in Figure A3.2.2 by M_{ac}. The sample response can be measured by various methods probing, *e.g.* the global complex ac susceptibility, $\chi \equiv (2\pi h_0)^{-1} \int M(H_a + h_0(t))e^{-i\omega t} d\omega t$, the local complex "transmittivity", $T \equiv (2\pi\mu_0 h_0)^{-1} \int B(H_a + h_0(t))e^{-i\omega t} d\omega t$, the response of a mechanical oscillator to which the sample is attached, or, at higher frequencies, through the frequency shift and phase shift of a resonant electromagnetic circuit or cavity containing the sample. In all cases, the time-varying field results in a time-varying force on the vortex ensemble, which is made to oscillate in the pinning potential "landscape" in which it is embedded. The nature of the response depends on the magnitude of both h_0 and ω, and can be described by combining the (overdamped) vortex equation of motion

$$\gamma\dot{u} + f(u) = \boldsymbol{j} \times \boldsymbol{B} + \eta(t,T) \quad \text{(A3.2.5)}$$

with $\boldsymbol{E} = \boldsymbol{v} \times \boldsymbol{B}$ and Maxwell's equations. Here, $\gamma = B^2/\rho_f$ is the friction coefficient due to flux flow losses, $\eta(t,T)$ is a stochastic noise term, and $\dot{u} \equiv v$.

The linear response regime of small h_0 is defined by the regime of vortex excursion u for which $f(u) \approx ku$, where the pinning restoring force constant k (unit [N/m^4]) is known as the Labusch constant. At frequencies $\omega > \omega_0 \equiv k/\gamma$, response is dominated by the friction force $\gamma\dot{u}$. Vortices perform periodic oscillations of amplitude much smaller than r_f. Therefore, the details of pinning are irrelevant to the macroscopic ac response, which is entirely described by the flux flow skin effect (with a skin-depth $\delta = \sqrt{\rho_f/\mu_0\omega}$). For $\omega < \omega_0$, the details of pinning come into play. For small h_0, the response is purely elastic; vortices perform reversible oscillations near their equilibrium pinned positions. The (lossless) macroscopic ac response in this so-called "Campbell regime" is in phase with the driving field and therefore "Meissner-like", but with a penetration depth $\lambda_C = \sqrt{B^2/\mu_0 k}$ determined by flux pinning (Campbell and Evetts 1972). At very low frequencies $\omega \lesssim \omega_0 e^{-U(j)/k_BT}$, the ac response is determined by thermally activated flux jumps (see below).

For larger $h_0 > F_p/\sqrt{\mu_0 k}$, the vortices can be depinned by the ac field. Depending on the strength of pinning, the low-temperature electromagnetic response is determined by the Bean model, by the Bean–Livingstone barrier, or by the geometrical barrier. The response is hysteretic and strongly nonlinear; therefore, strong higher harmonic response appears. The nonlinearity can be detected most easily by measuring the third harmonic of the ac susceptibility. The conditions of temperature and magnetic field at which flux pinning becomes unobservable due to thermal activation define the so-called irreversibility line $T_{irr}(B,\omega)$ in the (T,B) (temperature-field) phase diagram. This can be readily determined from the (ω-dependent) demise of the third harmonic ac response. Note that the harmonic content of the ac susceptibility is very similar, irrespective of whether bulk pinning or a barrier is dominant. Only imaging measurements of the local flux density (Paltiel et al. 2000), or an in-depth analysis of the harmonic susceptibilities (van der Beek et al. 1996) can distinguish which situation is relevant.

Hysteretic losses and associated heating effects are very important because they determine the dissipation in superconducting transformers and motors, as well as the quality factor of radiofrequency (RF) and high-frequency (HF) superconducting devices such as detectors, resonators, and reception chains. The power loss mechanism can be characterised through the measurement (at low frequency) of the ac losses or (at high frequency) of the surface resistance R_s as function of RF or HF power (see Section A2.7.2). Some useful relations are summarised in Table A3.2.1.

A3.2.3.10 Measurement of the I–V Curve

In thin films, strips, or wires, or in low-j_c materials, the $E(j)$ curve can be directly assessed by the measurement of the current–voltage (IV) characteristic of the superconductor. It should be always borne in mind that the application of a (non-equilibrium) electrical transport current I to the superconductor in the mixed state gives rise to the same kind of irreversible properties as the application of a magnetic field. In the presence of bulk pinning, the Bean model applies (Zeldov et al. 1994). As the magnitude of the electrical current is increased, this first flows along the sample edges, provoking vortex entry. As vortices move inwards, the current density is limited to $|\mathbf{j}| = j_c$ in the flux-penetrated regions; thus the total current grows through the increase of the width of the flux-penetrated regions rather than that of the local current density. The critical current I_c of the sample is reached when the flux fronts from opposite edges meet in the sample centre. For $I > I_c$ and $H \gg H^*$, vortices transit freely through the sample. If $H \ll H^*$ and $I > I_c$, vortices carrying positive and negative flux enter from opposite edges and annihilate in the sample interior.

In the case of slab-shaped samples (of thickness greater than 2λ) with weak bulk pinning, the Bean–Livingstone and/or the geometrical barrier can sustain the flow of lossless current at the sample boundaries as long as the magnetic "self-field" produced by the current flow does not exceed the penetration field H_p. For currents larger than $I_c \sim H_p / d$, vortices of positive and negative flux enter from opposite sample boundaries and annihilate in the interior, leading to resistive losses.

It is important to note that while j_c is a property of the superconducting material, I_c is a property of the sample. Since non-uniformities give rise to rounding of the IV curve, it has been common practice to define j_c (rather arbitrarily) by a voltage criterion of 1 μV dropped over 1 cm.

A3.2.3.11 Thermal Fluctuations and Flux Creep

Thermal fluctuations cause the vortex lattice to explore different metastable configurations in the pinning potential. In the absence of driving currents, this results in effectively larger pinning wells and an exponential reduction with increasing temperature of the pinning force, of the energy gain U_p due to pinning, and of the critical current density (Feigel'man and Vinokur 1990), well beyond the "trivial" temperature dependence entering via $H_c(T)$, $\xi(T)$ and $\lambda(T)$ (see Equations [A2.3.4] and [A2.3.5]).

In the presence of a driving force, vortices will be driven between metastable configurations of almost equal energy, a process that leads to a unidirectional creep of the vortex ensemble, accompanied by the reappearance of resistance for $|\mathbf{j}| < j_c$. The average velocity of the vortices is determined by the effective jump rate of the system between such states, which is a thermally activated process. The calculation of the average vortex velocity in the random pinning potential is a complicated problem of statistical physics; usually, the velocity is modelled using an activated law, such that

$$v = v_0 e^{-U(j)/k_B T}, \qquad (A3.2.6)$$

where $U(j)$ is the magnitude of the thermal activation barriers relevant for a driving current of density j, and v_0 is a prefactor related to the so-called "attempt rate". Formulated as such, the critical current density j_c corresponds to the driving force $\mathbf{j}_c \times \mathbf{B}$ at which all creep barriers vanish, i.e. $U(j_c) = 0$. For current densities smaller than but comparable to j_c, the average activation barrier is small, and the vortex system can easily find new most favourable metastable states near to its actual one. The smaller the current density, the lower the probability of a jump to an equivalent pinned configuration becomes, and the higher the average activation barrier $U(j)$. At low temperatures, the elastic properties of the vortex ensemble lead to nucleation-type creep and $U(j) \propto j^{-\mu}$ (Feigel'man et al. 1989), which diverges at low current densities (see Subsection A3.2.6.1 below). The current density-dependence of the average activation barrier gives rise to the highly nonlinear $E(j)$-curve of Subsection A3.2.3.1,

$$E = \rho_0 e^{-U(j)/k_B T}, \qquad (A3.2.7)$$

which should be combined with Equation A3.2.2 in order to describe the low-frequency electrodynamics of the superconductor. As far as the magnetic moment is concerned, the Bean model is usually still a good approximation, but j_c should be replaced by a time- or field sweep rate-dependent "sustainable current density" j_s as the new work point on the $E(j)$-curve. As a result, the width Δm_{sat} of the magnetic hysteresis loop is no longer indicative of the "true" critical current density but, rather, of the flux creep rate at different magnetic fields. An elementary rearrangement of terms shows that j_s satisfies

$$U(j_s) = k_B T \ln\left(\frac{t_i + t}{\tau}\right), \qquad (A3.2.8)$$

where t is the time scale of the experiment, τ is a normalisation time, and t_i is the time on which transient effects are important. Information about the energy barrier $U(j)$ can be obtained by measuring the temporal relaxation, as a result of flux creep, of the saturated magnetic moment $m_{sat}(t)$ or the local induction $B(t)$ over the sample. From such measurements, the $U(j)$-curve can be directly obtained by plotting $k_BT\ln\left(c-\left|\partial m_{sat}/\partial t\right|\right)$ versus $\Delta m_{sat} \propto j_s$. Repeating the experiment at slightly different temperatures yields a series of curves over finite j intervals which can be overlapped so as to determine the constant c (van der Beek et al. 1992). In this way the $U(j)$ dependence can be determined over several decades in j. In principle, this can also be achieved by waiting long enough (~10^{40} years) for the current to decay! The true j_c is obtained by extrapolating to $U = 0$. It is interesting to note that the $U(j)/k_BT$ ratio always falls within the range 10−30. This remarkable observation can be understood in terms of the self-organising mechanism of creep and the experimental time window of the experiments.

The $U(j)$-curve can also be obtained from frequency-dependent measurements of the low-frequency ac response of the superconductor; then, the relevant barriers satisfy $U(j) = k_BT\ln(1/\omega\tau)$. Note that activation-type creep over the Bean–Livingston barrier is also possible and leads to a time- or field sweep rate-dependent first penetration field $H_p(t)$. By contrast, the geometrical barrier is proportional to the product of the vortex line energy $\varepsilon_0 = 4\pi\xi^2\left(\frac{1}{2}\mu_0H_c^2\right)$ (with unit [$\mathrm{Jm^{-1}}$]) and the sample thickness. It is therefore of macroscopic nature, and cannot be overcome by thermally activated vortex motion.

A3.2.3.12 Pinning Energies and Activation Barrier

In very many reports in literature, the flux creep activation barrier $U(j)$, which is related to non-equilibrium dynamics and transport properties, is confused with the pinning energy U_p, which corresponds to the free energy gain of the pinned vortex system with respect to the hypothetical unpinned vortex lattice in the same material. It is important to stress that these quantities are not trivially related. Notably, U_p cannot be extracted from electrical or inductive transport measurements, while $U(j)$ can.

A3.2.4 Statistics of Pinning: From the Elementary Force to the Critical Current Density

Vortex lines are made of only magnetic flux and electrical current. They are therefore flexible objects – in fact, the softest "matter" in nature – that can be characterised by a set of elastic constants. These are the (single) vortex line tension $\varepsilon_l \approx \varepsilon^2\varepsilon_0$, and the vortex lattice compression, tilt, and shear moduli $c_{11}(\boldsymbol{k})$, $c_{44}(\boldsymbol{k})$, and c_{66} (see Section A3.1.4). The compression- and tilt moduli are dispersive and depend on the wave vector $\boldsymbol{k} = (k_{xy}, k_{\parallel})$ of the elastic distortion. For many estimates in the intermediate field regime $B_{c1} \ll B \lesssim 0.2B_{c2}$, the approximations $c_{44}(\boldsymbol{k}) \sim \varepsilon^2\varepsilon_0a_0^{-2}(1-b)$ and $c_{66} \approx (\varepsilon_0/4a_0^2)(1-b)^2(1-0.29b)$ are useful. $\varepsilon \equiv \xi_c/\xi_{ab}$ is the anisotropy parameter for uniaxially anisotropic superconductors, and $b \equiv B/B_{c2}$. In anisotropic and layered superconductors, ε, and, by consequence, the vortex tilt modulus is small. Vortices can therefore easily bend to better adapt to the pinning landscape.

In the absence of bulk flux pinning, vortex lines form the regular Abrikosov lattice. In the presence of pinning, the vortex ensemble distorts as a result of the forces f_p exerted by each inhomogeneity or defect. The vortex displacement field $\boldsymbol{u}(\boldsymbol{r})$ mirrors the local equilibrium between the pinning forces and the elastic restoring force of the vortex ensemble, and corresponds to a given metastable configuration, and therefore to a local energy minimum in "vortex configuration space". The free energy gain (per unit volume) with respect to the unpinned Abrikosov lattice is the pinning energy U_p. The volume pinning force F_p is the force needed to drive the vortex ensemble to another energy minimum.

An essential ingredient of any calculation of F_p is the determination of the volume of vortex ensemble involved in a particular change of configuration, i.e. the volume of the vortex ensemble that can be considered as "pinned independently". When the pinning forces f_p are very weak, the vortex lattice is only very slightly perturbed. Any change of configuration will entail slight readjustments of vortex lines over a large "correlation volume" $V_c = R_c^2L_c$. Here R_c and L_c are correlation lengths transverse and parallel to the field direction. The larger the pinning forces are, the smaller the V_c, since a local adjustment will be insufficient to entail rearrangements of well-pinned far-away vortices.

A3.2.4.1 Point Defects: Weak Collective Pinning

"Weak" or "collective" pinning corresponds to the situation of a large density $n_d \gg r_f^{-2}L_c^{-1}$ of randomly distributed weak and small pinning centres (i.e. of dimensions $\ll r_f$) depicted in Figure A3.2.6(a). (Blatter et al. 1994, Blatter, Geshkenbein, and Koopman 2004). Such defects typically correspond to vacancies, interstitials, dopant atoms, small voids, small dislocation loops, or other atomic-sized impurities. The seminal

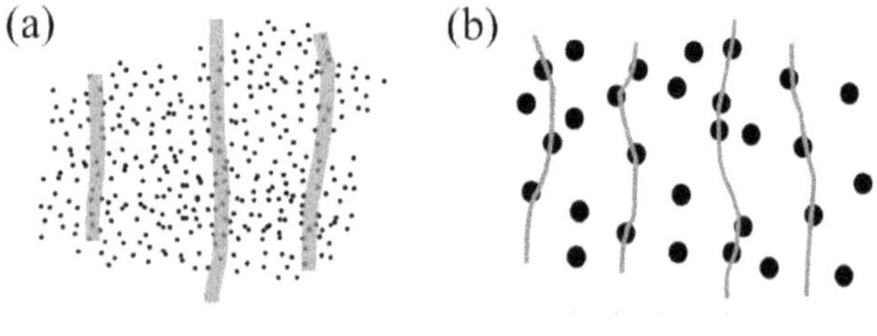

FIGURE A3.2.6 Cartoon of weakly (a) versus strongly pinned vortex lines (b). Black dots depict pinning centres, and grey lines represent the vortex cores.

collective pinning theory of Larkin and Ovchinnikov (Larkin and Ovchinnikov 1979) introduces the correlated volume as that over which the average relative vortex displacement $\langle|\boldsymbol{u}(\boldsymbol{r})-\boldsymbol{u}(0)|^2\rangle^{1/2}$ induced by pinning remains smaller than r_f. The forces exerted by all pinning centres in the volume V_c compete, so that the average (i.e. the first moment of the) total force F_c vanishes. Hence, the vortex ensemble is only pinned by the fluctuations of F_c. The largest non-zero moment $\delta F_c = (n_d V_c \langle f^2\rangle)^{1/2}$, where $\langle f^2\rangle \approx \frac{1}{2} f_p^2$ is the mean square of $f(r)$ averaged over a primitive cell of the vortex lattice with area Φ_0 / B. A pinning strength, or mean-squared pinning force density, can be defined by $W \equiv n_d\langle f^2\rangle \approx \frac{1}{2} n_p f_p^2$ in order to express the volume pinning force as

$$F_p = \delta F_c / V_c = (W / V_c)^{1/2}. \quad \text{(A3.2.9)}$$

Note that W is related to the strength of the short-range correlations $\langle U_{pin}(\mathbf{r}) U_{pin}(\mathbf{r}')\rangle = \gamma_U \delta(\mathbf{r}-\mathbf{r}')$ of the local energy gain $U_{pin}(\mathbf{r})$ introduced by Blatter et al. (1994) as $W = \gamma_U / a_0^2$. The notation of Blatter et al. is meaningful for fields $B < 0.2 B_{c2}$ and expresses the fact that only the vortex core area is involved in pinning.

To proceed further, one has to compute V_c from the balance between the energy loss due to tilt- and shear deformations (see Section A3.1.4), on the one hand, and the work done by the pinning centres on the other hand:

$$\frac{1}{2} c_{44}\left(k_{xy} \approx \frac{\pi}{R_c}\right)\left(\frac{r_f}{L_c^2}\right) = \frac{1}{2} c_{66}\left(\frac{r_f}{R_c^2}\right) = \left(\frac{W}{R_c^2 L_c}\right)^{\frac{1}{2}} r_f. \quad \text{(A3.2.10)}$$

It should be noted that the dispersion of the tilt modulus c_{44} plays an important role. This leads to

$$L_c = \left[c_{44}(0) / c_{66}\right]^{\frac{1}{2}} R_c; \qquad \lambda_h \ll R_c \ll L_c \quad \text{(A3.2.11)}$$

$$L_c = \left[c_{44}(0) / c_{66}\right]^{\frac{1}{2}} R_c^2 / \pi\lambda_h; \qquad R_c \ll L_c \ll \lambda_h \quad \text{(A3.2.12)}$$

for local and non-local elasticity, respectively. Here $\lambda_h = \lambda / (1-b)^{1/2}$. The limit of so-called "single vortex pinning" is reached when the pin energy gain (the right-hand member of Equation [A3.2.10]) becomes too large to be accommodated by shear distortions. This can be expressed by putting $R_c = a_0$. The independently pinned objects are now single flux line segments of length

$$L_c \simeq \left(\frac{c_{44}(0) a_0^3 r_f}{\lambda_h^2 W^{\frac{1}{2}}}\right)^{\frac{2}{3}} \approx \varepsilon r_f \left(\frac{j_0}{j_c}\right)^{1/2}. \quad \text{(A3.2.13)}$$

The second equality allows one to rapidly estimate the pinning regime that one is dealing with from the measured value of j_c. In practice, single vortex pinning occurs quite frequently, notably in cuprate- and iron-based high-temperature superconductors, in the intermediate field regime above 1–2 T, but below $0.2B_{c2}$ (van der Beek et al. 2010); in this field regime, it gives rise to a field-independent

$$j_c \approx j_0 \left(\frac{27 n_d \langle f_p^2\rangle r_f^3}{16\varepsilon\varepsilon_0}\right)^{2/3}. \quad \text{(A3.2.14)}$$

At higher fields, relevant for disordered superconducting films and alloys, single vortex pinning results in dome-shaped F_p versus B curves, well-described by a scaling relation $F_p \propto B_{c2}^n(T) b^p (1-b)$ with $n \approx 2.5$ and $p = \frac{7}{6}$ for δT_c pinning and $p = \frac{15}{6}$ for δl pinning (Kes 1992).

In cuprate- and iron-based high-temperature superconductors, the magnitude of the single vortex critical current density is typically some 10^4–10^5 Acm^{-2} at 4 K. In the $Bi_2Sr_2Ca_yCu_{1+y}O_{6+2y}$ family of layered cuprate superconductors, some organic superconductors, or in artificial superconductor multilayers, $\varepsilon \ll 1$ and the longitudinal correlation length L_c can be smaller than the spacing s between superconducting layers. So-called "pancake vortices" corresponding to the intersection of the vortex lines with individual superconducting layers are then pinned (and depinned) independently. The critical current density in this "single pancake" regime may exceed 10^6 Acm^{-2}. For example, pancake vortices are individually pinned by oxygen vacancies in $Bi_2Sr_2CaCu_2O_{8+\delta}$ (Li et al. 1996), a mechanism that accounts for j_c values at low temperatures as large as $0.1 j_0 \approx 4\times10^{10}$ Am^{-2}). However, these current densities are not accessible because the correspondingly small pinning creep barriers result in very strong flux creep effects as well as a low-lying irreversibility line.

In weakly pinning superconductors, such as MgB_2, dichalcogenides, or amorphous metallic thin films, L_c can, on the contrary, be very large and the three-dimensional collective pinning (3DCP) scenario described by Equation (A3.2.10) is realised. Critical current densities are then, typically, of the order $j_c \lesssim 10^2 - 10^3$ Acm^{-2}.

In thin films, the situation where $L_c \gg d$ can arise when the magnetic field is perpendicular to the film. Vortex line cores then remain undistorted across the film thickness d, and disorder of the vortex ensemble only develops because of shear deformations. As regards flux pinning, the materials then behave two-dimensionally. In this limit of two-dimensional collective pinning (2DCP), R_c follows by substituting $L_c = d$ in Equation (A3.2.10) giving

$$R_c \approx 2 r_f c_{66}\left[2\pi d / W \ln(w / R_c)\right]^{1/2}. \quad \text{(A3.2.15)}$$

Here w is, again, the width of the film. Good agreement is obtained (Kes 1992, Wordenweber and Kes 1986) between this expression and experiment. The "single pancake" regime in layered superconductors can be considered a peculiar realisation of 2DCP.

A3.2.4.2 Point Defects: Strong Pinning

The limit of so-called strong pinning (Ovchinnikov and Ivlev 1991) corresponds to sparse ($n_d \ll r_f^{-2} L_c^{-1}$) randomly distributed large, strong pinning centres. It is schematically illustrated in Figure A3.2.6(b) and concerns the vast majority of type II superconductors, and nearly all technological superconductors, in which strong pinning centres are responsible for critical current density values above 10^6 or even 10^7 Acm^{-2}. Strong pinning centres are typically nm-sized second phase inclusions, voids, or nm-scale modulations of the superconducting properties; their (maximum) pinning force f_p usually results from the combined effect of core- and electromagnetic pinning. In such situations, the net pinning force F_c over a correlated volume does not average out. Therefore, the volume pinning force $F_p = (a_0^2 d)^{-1} \Sigma_i^N f_p^i = f_p / a_0^2 \bar{\mathcal{L}}$ is the direct sum of the pinning force exerted by each defect the vortex line encounters over the sample thickness d. Any vortex line will be pinned on average by $N = d / \bar{\mathcal{L}}$ defects, where $\bar{\mathcal{L}}$ is the average distance between "effective" pinning centres that the line can reach. N is again limited by the elastic properties of the vortex ensemble. For low fields, vortex line excursions are limited by the line tension ε_1, and (Demirdis et al. 2013, Blatter, Geshkenbein, and Koopman 2004, van der Beek et al. 2002)

$$j_c(0) = \frac{f_p}{\Phi_0 \bar{\mathcal{L}}} = \pi^{1/2} \frac{f_p}{\Phi_0 \varepsilon} \left(\frac{f_p n_d r_f}{\varepsilon_0} \right)^{1/2} \quad (B \ll \tilde{B}). \qquad \text{(A3.2.16)}$$

Above the crossover field $\tilde{B} = \Phi_0 \left(\varepsilon \varepsilon_0 r_f / f_p \right)^2$, the repulsion between vortices is limiting, and the critical current density follows the behaviour expected for three-dimensional strong pinning (Demirdis et al. 2013, Blatter, Geshkenbein, and Koopman 2004, van der Beek et al. 2002),

$$j_c(B) = \frac{f_p}{\Phi_0 \bar{\mathcal{L}}^2} \frac{\varepsilon a_0}{\pi} = \frac{f_p}{\Phi_0 \varepsilon} \left(\frac{f_p n_d r_f}{\varepsilon_0} \right) \left(\frac{\Phi_0}{B} \right)^{1/2} \quad (B \gg \tilde{B}), \qquad \text{(A3.2.17)}$$

i.e. $j_c \propto B^{-a}$ with $\alpha = \frac{1}{2}$. Such behaviour is ubiquitous in high-temperature superconducting thin films and composite conductors (van der Beek et al. 2002, Mele et al. 2019). The set of Equations (A3.2.16) and (A3.2.17) can be combined to obtain an experimental estimate of $f_p = (\Phi_0 \varepsilon / \pi) j_c^2(0) / \left[\partial j_c(B) / \partial B^{-1/2} \right]$ from the low-temperature, low-field current density $j_c(0)$ and the slope $\partial j_c(B) / \partial B^{-1/2}$ at intermediate fields.

A crossover from weak collective to strong pinning can occur through various scenarios. In the presence of both strong and weak pins, the former will determine the critical current density at low fields. When all strong pins are occupied by vortices, the weak background pinning will take over and determine j_c. If pinning is determined by a single type of point defect, long-range fluctuations (on length scales $\gg r_f$) of the average defect density can give rise to strong pinning, while short-range randomness is responsible for weak collective pinning (Demirdis et al. 2013). Finally, the presence of weak pins can slow down thermally activated vortex motion between strong pins (see *e.g.* Sadovsky et al. 2015).

A3.2.4.3 Pinning by Correlated Disorder

A particular form of strong pinning concerns extended correlated defects. These comprise "one-dimensional" pins in the form of edge- and screw dislocation cores such as are commonly found in high-temperature superconducting films showing Vollmer–Weber (island) or Stransky–Krastanov (island plus layer) type growth, and purposely included linear defects such as columnar second phase inclusions ("nanorods") (Mele et al. 2019) and heavy-ion-irradiation–induced amorphous latent tracks (Bourgault et al. 1989, Civale et al. 1991, Nelson and Vinokur 1992). Two-dimensional pinning defects include platelet-like inclusions, grain boundaries, anti-phase boundaries, twin boundaries in superconductors with an orthorhombic crystal structure, as well as the intrinsic structure of layered superconductors in a magnetic field applied parallel to the layers (Kwok et al. 1991). Because of their ability to trap vortex lines along their entire length, such defects give rise to a profound anisotropy of pinning-related quantities, distinct from any anisotropy related to the crystal structure or electronic properties of the host material (see *e.g.* Bartolomé et al. 2019). The defect-induced anisotropy manifests itself through a cusp-like behaviour of the critical current density and the flux creep activation barriers for field aligned with the defect direction.

A3.2.5 Thermodynamics of the Mixed State

In addition to the work associated with the Meissner expulsion $\frac{1}{2}\mu_0 H_a^2$ and the energy loss $n_v \varepsilon_v$ associated with admitting vortex lines (Brandt 2003), the Gibbs free energy (per unit volume) of the mixed state (with respect to the normal state)

$$G = f_s(T) - f_n(T) + \frac{1}{2}\mu_0 H_a^2 - B H_a + n_v \varepsilon_v - U_p - k_B T S \qquad \text{(A3.2.18)}$$

also includes the contribution U_p from flux pinning and the entropy contribution. The latter is the sum of the configurational entropy due to positional disorder of the vortices and the contribution from vortex thermal fluctuations. Both pinning and entropy lower the energy per vortex and, therefore, the free energy of the mixed state. As a consequence, flux pinning lowers H_{c1}, and increases H_{c2}. It also modifies the specific heat $c_p = -T(\partial S / \partial T)_p$ (van der Beek et al. 2005) and the reversible magnetisation

$$M_{rev} \equiv -\frac{1}{\mu_0} \left. \frac{\partial G}{\partial H_a} \right)_T = -\frac{1}{\Phi_0} \frac{\partial G}{\partial n_v} \left(\frac{\partial B}{\partial H_a} \right) \equiv -\frac{\mu_v}{\Phi_0} \left(\frac{\partial B}{\partial H_a} \right) \qquad \text{(A3.2.19)}$$

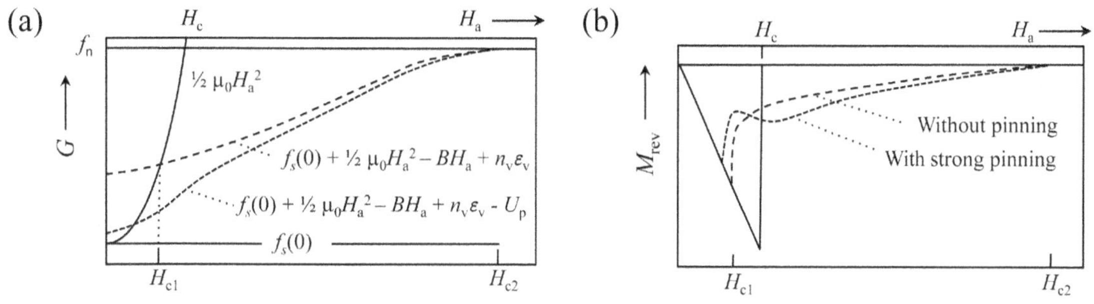

FIGURE A3.2.7 Schematic evolution of (a) the free energy G and (b) the reversible magnetisation $M_{rev} = -\partial G / \partial \mu_0 H_a$ for a type I superconductor (drawn line), a type II superconductor without pinning (dashed lines) and a type II superconductor with strong pinning centres (dotted lines). In the latter case, the free energy paid per vortex is less than in the absence of pinning. The s-shape of $M_{rev}(H_a)$ is due to all pinning centres becoming occupied at intermediate field strengths.

which plays the role of the vortex chemical potential μ_v. In the presence of pinning, the equilibrium density of vortices in the superconductor is higher, and $|M_{rev}|$ lower than when pinning is absent. Therefore, a careful measurement of the reversible magnetisation yields direct information on the pinning energy U_p.

While this task is usually difficult due the hysteresis described in Subsection A3.2.3.2, it can be used in the magnetically reversible vortex liquid state (see below), as well as through vortex imaging in materials with large local modulations of the pinning energy (Demirdis et al. 2011). The effect of pinning on thermodynamic properties is large in superconductors with very strong pinning, e.g. that induced by amorphous columnar defects introduced by heavy-ion irradiation. An example is the reversible magnetisation curve of heavy-ion-irradiated single crystalline $Bi_2Sr_2CaCu_2O_{8+\delta}$, which has an s-shape due to the fact that at small fields $B \ll B_\phi \equiv \Phi_0 n_d$ vortices can profit from the pinning energy, whereas this is prohibited for $B \gg B_\phi$ because all columnar defects are occupied (van der Beek et al. 1996) (see Figure A3.2.7). From such measurements, the pinning energy and the configurational entropy of the vortex ensemble can be obtained. Similar results where obtained on heavy-ion-irradiated single crystalline $YBa_2Cu_3O_{7-d}$ and $(Ba,K)BiO_3$ (van der Beek et al. 2005).

A3.2.6 Vortex States and Phase Transitions

In a "clean" type II superconductor without any flux pinning, inter-vortex repulsion leads to an arrangement into the triangular Abrikosov lattice. While details of the superconductors' electronic structure may modify the structure and even the symmetry of the Abrikosov lattice, the change of the free energy of the mixed state due to pinning as well as to the entropy related to thermal fluctuations induce a more profound change into disordered states of "vortex matter".

As all elastic objects in spatial dimension $d < 4$, the vortex ensemble in type II superconductors is inherently unstable to the presence of a random disorder potential (Larkin and Ovchinnikov 1979, Imry and Ma 1975). Therefore, any amount of pinning, however small, will disrupt the long-range positional and orientational order of the Abrikosov lattice. For very weak pinning, the length scale on which longe-range order is undone can be excessively large and entail thousands of lattice parameters a_0 (Kim et al. 1999). In the opposite limit of very strong pinning, the vortex ensemble is entirely amorphous; it is even possible to observe local density fluctuations of the vortex ensemble due to pinning (Demirdis et al. 2011). However, it is not the breaking of long-range positional and orientational order, but the breaking of gauge symmetry that properly classifies vortex states.

A3.2.6.1 The Vortex Glass

In the (low-temperature) vortex glass state (Fisher, Fisher, and Huse 1991), all vortices are localised in a pinning-induced metastable state. Even if local thermal fluctuations are present (Feigel'man and Vinokur 1990), no thermally activated jumps are possible in the absence of a driving force. Thus, the global configuration of the phase of the superconducting order parameter is fixed; there is long-range phase order, and gauge symmetry is broken with respect to the normal metallic state. The localisation of vortices implies the divergence of creep barriers for vanishing driving force, i.e. $U(j) \to \infty$ for $j \to 0$ and a truly zero resistivity $\rho \sim \rho_f e^{-U(j)/k_B T}$. In practice, this is achieved because of the elastic nature of the vortex ensemble. As the driving force decreases, only vanishingly rare and large activation nucleii – large vortex "bundles" in the language of the collective creep theory (Feigel'man et al. 1989, Blatter et al. 1994) or very large scale vortex re-arrangements in the language of the vortex glass theory (Fisher, Fisher, and Huse 1991à) – can be made to expand and bring the vortex ensemble to another metastable state. For vanishing driving force, no activation nuclei can expand sufficiently, whatever their size, and the vortex ensemble always falls back into its original configuration. As a result, the energy barriers for creep grow according to $U(j) \sim U_c (j_c / j)^\mu$. The exponent μ depends on the dimensionality of the elastic medium and of the environment, and ranges from $\frac{1}{7}$ for single vortex lines to $\frac{16}{9}$ for large "bundles". When the vortex glass is subjected to an ac driving force, periodic jumps between pairs of metastable states are allowed since these do not break long-range phase order (Koshelev and Vinokur 1991).

A3.2.6.2 The Bragg Glass

In the case of weak collective pinning, the average relative vortex displacements characterised by the correlation function $B(r) \equiv \langle |\boldsymbol{u}(r) - \boldsymbol{u}(0)|^2 \rangle$ are small, and grow with vortex separation r as (Giamarchi and Le Doussal 1994, Giamarchi and Le Doussal 1995)

$$B(r) \propto r \quad (r < R_c) \quad \text{- the so-called collective pinning or "Larkin" regime} \tag{A3.2.20}$$

$$B(r) \propto r^{2\zeta} \quad (r < R_a) \quad \text{- the random manifold regime, with the wandering exponent } \zeta \sim 0.44 \tag{A3.2.21}$$

$$B(r) \propto \ln r \quad (r > R_a) \quad \text{- the charge density wave regime} \tag{A3.2.22}$$

R_a is the distance at which relative displacements are larger than the vortex spacing and dislocations appear. For very weak pinning, R_a may be larger than the sample size (Kim et al. 1999). One then speaks of a "Bragg glass", since the vortex glass will show the usual Bragg diffraction pattern of the Abrikosov lattice, albeit with power law rather than exponential tails of the diffraction peaks (Giamarchi and Le Doussal 1994).

A3.2.6.3 The Bose Glass

In the case of correlated disorder, both the pinning energy and the dynamics of the vortex ensemble are strongly anisotropic. In particular, the anisotropy of the pinning energy leads to an anisotropic phase boundary of the vortex glass, with a higher transition temperature to the vortex liquid (see below) when the field is aligned with the defect direction. Flux creep is also highly anisotropic. For fields aligned with the defects, creep is of nucleation type (Nelson and Vinokur 1992). For misaligned fields, vortex motion proceeds by sliding of vortex "kinks" between defects (Schuster et al. 1994), which is only hindered by background pinning by point-like defects. Thus, the values of the creep exponent μ are strongly dependent on the field orientation angle. This type of anisotropic vortex glass has been termed "Bose glass", due to the analogy between the system of vortex lines interacting with parallel correlated defects and the world lines of interacting bosons in two dimensions in a random potential.

A3.2.6.4 The Vortex Liquid

The vortex liquid state (Brézin, Nelson, and Thiaville 1985) occupies the high-temperature high-field region of the mixed state, and therefore separates the vortex glass from the normal state. The vortex liquid state is characterised by diffusive vortex motion: the vortex ensemble can access (many different) metastable configurations even in the absence of a driving force. As a result, the phase and amplitude of the superconducting order parameter at every position fluctuate in time. The linear resistivity is non-zero even in the limit of small currents, $j \to 0$. Therefore, the vortex liquid state does not break gauge symmetry and has electrodynamic properties that are formally equivalent to the normal state, even if the Cooper pair density is clearly non-zero on average, supercurrents and vortices can still be defined, and vortices can still be pinned (van der Beek et al. 1996) – although only on a finite time scale.

If defects are absent or the effect of thermal fluctuations (Feigel'man and Vinokur 1990) is very pronounced, pinning is very weak and vortex motion in the liquid state gives rise to the usual flux flow resistivity ρ_f. However, the presence of pinning in the liquid state may considerably slow down vortex motion. The resistivity

$$\rho_{TAFF} \approx \rho_f e^{-U_c/k_B T} \tag{A3.2.23}$$

is then determined by thermally assisted flux flow, or TAFF (Kes et al. 1989). The energy barrier U_c is related to the plastic deformations of the vortex ensemble; it provides a current-independent upper bound (cut-off) for the diverging glassy barriers at low driving force, thus enabling flux flow. Whereas in very thin ($d \ll L_c$) films U_c is the activation barrier for the motion of unbound dislocations or disclinations in the vortex lattice (Nelson and Halperin 1979), it is generally admitted that in disordered three-dimensional superconductors vortex diffusion is made possible through the mechanism of vortex cutting and reconnection. This process is very much facilitated in layered superconductors, comprising certain organic materials and the high-temperature cuprates and iron-based superconductors. Namely, the layeredness of these materials leads to a strong reduction of the vortex line tension and the vortex lattice tilt modulus (see Section A3.1.4), which promotes vortex line wandering and bending, processes that in turn lower the activation barrier for flux cutting. In extremely anisotropic materials such as $Bi_2Sr_2CaCu_2O_{8+\delta}$, the cutting and reconnection process can take place on the scale of a single layer, involving the exchange of single "pancake vortices" between vortex lines.

A3.2.6.5 Depinning Transitions

When the thermal vortex displacements (Feigel'man and Vinokur 1990) exceed the range of the elementary pinning force r_f, the pinning energy and the critical current density will decrease exponentially by thermal smearing. This rapid "softening" of pinning, marked by a sudden decrease of the

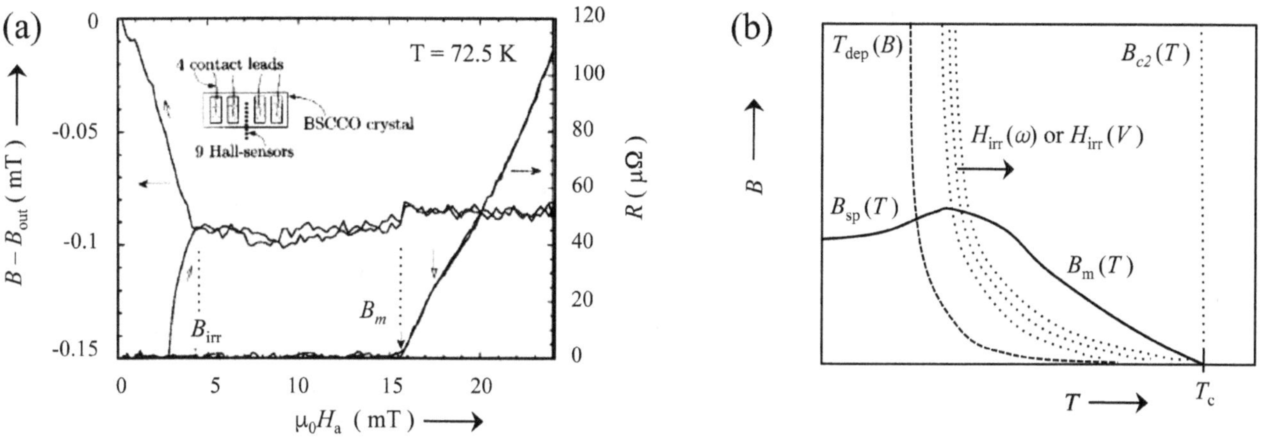

FIGURE A3.2.8 (a) Typical curves of the local magnetisation (defined as the difference between the local flux density B and the applied field $B_{out} = \mu_0 H_a$) and the resistivity of $Bi_2Sr_2CaCu_2O_{8+\delta}$ single crystal. The curves show the first order from the Bragg glass to the vortex liquid phase at the ("melting") field B_m as a simultaneous discontinuity in both physical quantities. Flux pinning in the sample bulk can no longer be discerned above the irreversibility field B_{irr}. After Fuchs et al. 1998. (b) Schematic vortex matter (B, T) phase diagram in a generic clean type II superconductor. The "melting" field B_m constitutes a first-order phase transition from the vortex solid to the vortex liquid state. It is prolonged into the low-temperature regime in which flux pinning manifests itself as the so-called "second peak transition" at B_{sp}, which also overlies a first-order phase transition. The depinning temperature T_{dep} marks the onset of an exponential decrease of j_c with temperature, and the demise of pinning. Also marked are various irreversibility lines $B_{irr}(T, \omega)$ above which pinning in the bulk becomes indiscernible, depending on the measuring frequency ω or the voltage threshold criterion/working point V.

irreversibility (van der Beek et al. 1996, Thompson et al. 1998, Fuchs et al. 1998), can be described by a "depinning line" or "depinning temperature" T_{dep} in the (T, B) phase diagram (Feigel'man and Vinokur 1990). By comparing length scales, it is clear that at low fields (i.e. $B < 0.2B_{c2}$) the depinning line, which as such does not constitute a phase transition, should lie below the melting line (see below), whereas the opposite is true at high fields. The "depinning line" is to be distinguished from the "irreversibility line" of Section A3.2.3.9 – the first is associated with the intrinsic temperature dependence of pinning properties, while the second denotes the temperature at which pinning becomes unobservable using a given experimental technique.

A3.2.6.6 The Bragg Glass to Vortex Liquid Transition

In superconductors with very weak bulk pinning, the vortex Bragg glass phase transforms to the vortex liquid through a first-order transition, often denoted "vortex lattice melting" (Brezin, Nelson, and Thiaville 1985, Brandt 1989, Zeldov et al. 1995). The transition to the liquid state is driven by the excess entropy of the latter; as a result, the liquid has a higher vortex density than the Bragg glass. On the contrary, upon cooling the Bragg glass is stabilised through the gain in elastic deformation energy associated with the arrangement of vortices into a regular array. The first-order melting transition is hysteretic, with the vortex liquid showing supercooling. The transition is enabled by thermal fluctuations of vortex segments (or kinks) around their equilibrium positions, until at the transition, plastic barriers of height U_c can be overcome (Fendrich et al. 1995, López et al. 1997). At that point, thermal cutting and reconnection leads to the demise of the "identity" of vortices, and to entanglement in the vortex liquid state(López et al 1996). The first-order transition has often been described by a Lindemann criterion: the vortex lattice "melts" when the thermal displacements become larger than a fraction $c_L a_0$ of the inter-vortex spacing (Brandt 1989). Here c_L is the Lindemann constant, $c_L \approx 0.2$. This criterion, which is tantamount to a modified Ginzburg criterion for superconducting fluctuations, gives rise to a transition line in the (T, B) phase diagram which, for conventional superconductors, is located near the upper critical field $H_{c2}(T)$. For layered superconductors such as $Bi_2Sr_2CaCu_2O_{8+\delta}$, the transition lies at fields below 1 T.

The "vortex lattice melting" transition is characterised by latent heat (Schilling et al. 1996), by a discontinuous jump in the local induction (vortex density) (Zeldov et al. 1995), and by a sharp jump in the resistivity (Safar et al. 1992, Kwok et al. 1992). Since the transition occurs at a temperature higher than the depinning transition of the Bragg glass (Fuchs et al. 1998), the jump of the resistivity at the transition is to be associated with the demise of the Bean–Livingstone and/or the geometrical barrier due to the collapse of the vortex line tension. The term "vortex lattice melting" is very often misused, through the association with other experimental observations. It should be reserved for the case described here, i.e. a thermodynamic temperature-driven first-order transition from the vortex lattice – or the Bragg glass to the vortex liquid state.

A3.2.6.7 The Vortex Glass to Vortex Liquid Transition

In superconductors with strong pinning, as well as at high magnetic fields where the depinning temperature lies above the first-order melting line, the disordered vortex glass phase transits continuously to the vortex liquid. Since the vortex glass cannot be distinguished from the vortex liquid through the breaking of any translational or orientational symmetries, but only through broken gauge symmetry, scaling procedures have been proposed to identify whether one is actually dealing with a phase transition (Fisher, Fisher, and Huse 1991). Scaling of thermodynamic and transport properties around the vortex glass phase transition line $T_g(B)$ would expose the existence of a diverging length scale $\xi_g \propto |T-T_g|^{-\nu}$ ("the glass correlation length") describing the establishment of long-range order of the superconducting phase. The vortex glass theory predicts a specific behaviour for the resistivity in the critical regime around $T_g(B)$ where all data for different temperatures and fields collapse on two curves, F_+ for $T>T_g$ and F_- for $T<T_g$ (Koch et al. 1989) when plotted as $(E/j)\cdot|T-T_g|^{-\nu(z-1)}$ versus $(j/T)\cdot|T-T_g|^{-2\nu}$. Here, z is a universal dynamic exponent describing critical slowing down. A typical experimental scaling plot is shown in Figure A3.2.9. The exact functional dependence of $F_\pm$ is unknown. $F_+ \rightarrow 1$ for small j denoting linear resistivity, and $F_- \rightarrow \exp(-1/x^\mu)$ for small x, describing true superconductivity. At T_g, E should depend on j according to the power law $E \propto j^{(z+1)/2}$. Although there are many reports of the observation of vortex glass scaling, in particular, in thin films of the high-T_c cuprate $YBa_2Cu_3O_{7-\delta}$, no clear consensus as to the universality of scaling has emerged. This may partially be due to the presence of extended defects such as screw dislocations cores, for which the Bose glass approach is more suited (Nakielski et al. 1996).

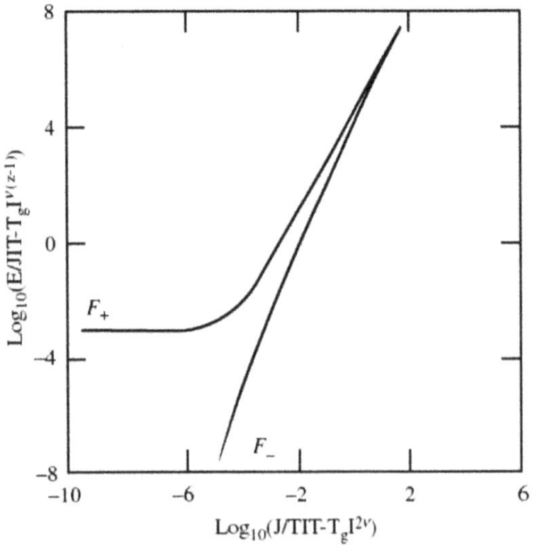

FIGURE A3.2.9 Collapse of over 100 I-V curves in a vortex glass scaling plot for $YBa_2Cu_3O_7$ in a field of 4 T for the parameter values $T_g = 74.5$ K, $z = 4.8$ and $\nu = 1.7$ Koch et al. 1989.

Another issue is the homogeneity of disorder. Heterogeneity of the material disorder (on various length scales) may lead to local variations of parameters such as the superconducting transition temperature and/or the superfluid density and thereby to a spatially inhomogeneous vortex glass to liquid transition (Demirdis et al. 2011, Demirdis et al. 2013). This will appear as concomitantly broadened, when averaged physical quantities such as the resistivity or magnetic moment of the full sample are measured, compromising the verification of any hypothetical scaling laws. At the same time, the spatial inhomogeneity of disorder on the nm scale was itself shown to be a source of pinning (Demirdis et al. 2011, Demirdis et al. 2013).

A3.2.6.8 The Bose Glass to Vortex Liquid Transition

The Bose glass to liquid transition has a phenomenology similar to the vortex glass transition, but is distinguished by its dependence on the orientation of the sample defect structure with respect to the field direction, and by the anisotropy of the vortex dynamics. To account for this, the Bose glass theory introduces the correlations lengths $\xi_\perp$ and $\xi_\parallel = \xi_\perp^\varsigma$ perpendicular and parallel to the linear defects, respectively (Nelson and Vinokur 1992, Lidmar and Wallin 1999). Here ς is the anisotropy exponent. At the Bose glass transition, the linear resistivity for field perpendicular and parallel to defects scales as

$$\rho_\perp = \frac{E_\perp}{j_\perp} \sim \xi_\perp^{D+\varsigma-z-3} \sim \left|\frac{T-T_{BG}}{T_{BG}}\right|^{\nu(2-z)} \quad (A3.2.24)$$

$$\rho_\parallel = \frac{E_\parallel}{j_\parallel} \sim \xi_\parallel^{D-\varsigma-z-1} \sim \left|\frac{T-T_{BG}}{T_{BG}}\right|^{-\nu z} \quad (A3.2.25)$$

where the last step was taken by invoking $\varsigma = 2$ and a spatial dimension $D = 3$. At the Bose glass transition temperature, the $E(j)$-curve again shows a power law,

$$E_\perp = j^{(1+z)/(2-D-\varsigma)} \quad (A3.2.26)$$

$$E_\parallel = j^{(\varsigma+z)/(1-D)}. \quad (A3.2.27)$$

Some experimental evidence for such scaling behaviour has been found in $YBa_2Cu_2O_{7-\delta}$, both as a result of twin-boundary pinning in single crystalline material (Grigera et al. 1998), and as a result of pinning by heavy-ion-irradiation–induced latent tracks in both single crystals (Jiang et al. 1994, Espinosa-Arronte et al. 2007) and films (Nakielski et al. 1996).

A3.2.6.9 Two-Dimensional Vortex Melting

The only available complete theory (Berezinskii 1972, Kosterlitz and Thouless 1978, Nelson and Halperin 1979) for the transition from the low-temperature vortex state to the vortex liquid state describes the situation in thin superconducting films of thickness $d < L_c$, where vortex cores are straight across the thickness (Berghuis, van der Slot, and Kes 1990). The vortex ensemble melts in two stages, accurately described by the theory of Kosterlitz–Thouless for 2D melting (Berezinskii 1972, Kosterlitz and Thouless 1978, Nelson and Halperin 1979). In the first stage, the unbinding of vortex lattice edge dislocation pairs in the lattice at

$$T_m^{2D} = \frac{A\varepsilon_0(T_m^{2D})d}{16k_B} \quad (b \ll 1) \qquad \text{(A3.2.28)}$$

$$= \frac{A\varepsilon_0(T_m^{2D})d}{16k_B}(1-b) \quad (b \lesssim 1) \qquad \text{(A3.2.29)}$$

drives the transition to a hexatic state. In the second stage, the hexatic transits to the liquid through the unbinding of disclinations (pairs of 5-fold and 7-fold coordinated vortices). At the melting line, the shear modulus should fall to zero, although in practice, disorder smears out the transition.

A3.2.6.10 Plasticity of the Vortex Lattice and the Peak Effect

A sudden increase of the sustainable current density is often apparent in measurements of the irreversible magnetic moment of type II superconductors, or in plots of j_c versus H or T. This phenomenon causes a dip in $R(T)$ or $R(H)$, while in ac-susceptibility experiments, a peak in χ'' occurs each time the condition $h_0 \sim H^*$ is fulfilled. Depending on the material, this "peak effect" can occur at rather low magnetic fields, or, more often, just before the vortex glass to liquid transition. At the basis of the "peak effect" is the transition between different regimes of flux pinning, driven by the appearance of plastic deformations of the vortex ensemble (Mikitik and Brandt 2001). This translates to an abrupt transition between (often crossing) $E(j)$-curves characterised by different curvatures and different current density scales j_c. This results, in experiment, to an abrupt change of the working point. It also means that in inductive measurements the peak can appear or disappear as function of field sweep rate, temperature, or waiting time after a field change, as the working points on the $E(j)$-curves shift. The peak effect is only observed in situations where pinning is initially weak, or of intermediate strength.

In many bulk superconductors ($d \gg L_c$), a nearly temperature-independent first-order phase transition underlies the peak effect, which therefore appears primarily on sweeping the applied magnetic field (Kokkaliaris et al. 1999, van der Beek et al. 2000, Avraham et al. 2001, Klein et al. 2010). Neutron scattering measurements (Aragón et al. 2019) as well as vortex decoration experiments (Aragón et al. 2019) show that at the transition, the vortex ensemble changes from a rather ordered state, which may be the Bragg glass, to a vortex polycrystal. This suggests that the peak effect is triggered by the appearance of edge- and screw dislocations that allows the vortex ensemble to better adapt to the pinning potential. The transition to the (denser) high-field disordered vortex state is driven by the gain in pinning energy, while the low-field ordered state is stabilised by the gain in elastic deformation energy and the elimination of vortex lattice dislocations. Above the depinning temperature T_{dep}, the energy gain that can be obtained from pinning decreases sharply, and the transition to the disordered state moves to higher magnetic fields, eventually joining the vortex melting transition in a tricritical point (Safar et al. 1993). This fact poses the as-yet unanswered question whether the high-field disordered state is thermodynamically distinct from the vortex liquid or not.

In materials showing weak collective pinning such as α-Nb_3Ge (Wordenweber and Kes 1986), $NbSe_2$ (Koorevaar et al. 1990, Bhattacharya and Higgins 1993, Banerjee et al. 1998), $CeRu_2$ (Banerjee et al. 1998), MgB_2 (Klein et al. 2010), or $BaFe_2(As_{1-x}P_x)_2$ (Putzke et al. 2014), the peak effect may show only in the vicinity of the upper critical field. In such cases, it is the softening of the elastic moduli (notably c_{44}) of the vortex lattice on approaching B_{c2} that is responsible for the appearance of plastic deformations of the vortex lattice and stronger pinning. A controlled increase of the pinning strength, for example, though electron irradiation, moves the peak effect transition field down (Klein et al. 2010).

In two-dimensional superconducting films, the peak effect is closely associated with the two-dimensional vortex melting (Berghuis, van der Slot, and Kes 1990). Here, it is the unbinding of vortex lattice edge dislocation pairs that allows the vortex ensemble to better adjust to the pinning potential (Wordenweber and Kes 1986).

A3.2.7 Vortex Dynamics at High Driving Force

Even for current densities above j_c, the driven vortex ensemble is still influenced by the presence of the "quenched" pinning potential. Notably, the pinning potential induces fluctuations δv of the velocity of the moving vortex ensemble, and, more rarely, of the vortex density $\delta n_v = \delta B / \Phi_0$. These fluctuations result in a voltage noise contribution $\delta V = \bar{n}_v \delta v + \bar{v} \delta n_v$ characteristic of vortex motion (van Ooijen and van Gurp 1965, Clem 1981). Most often, the voltage noise is related to the (irregularity of the) entrance and exit of vortex lines (Paltiel et al. 2000); its frequency spectrum is therefore related to the (spectrum of) vortex transit time(s) through the sample. However, it has been shown that both the magnitude of the noise and the spectrum are also clearly affected by flux pinning in the superconductor, and depend on the strength of pinning in the superconductor and the ensuing positional order of the vortex ensemble. Most prominently, edge-induced vortex lattice disorder (see Subsection A3.2.3.8) and subsequent annealing at high driving forces have been shown to be directly linked to the noise spectrum (Paltiel et al. 2004).

Strong pinning is prone to lead to more pronounced flux density gradients associated with the drive current, and, thereby, to the associated avalanche-like motion of the vortex ensemble. These occur on time scales smaller than the transit time and have been suggested to result in $1/f$ noise (Bak, Tang, and Wiesenfeld 1986, Altshuler and Johansen 2004). It has been shown that the $1/f$ noise ubiquitous in strongly pinning high-temperature superconducting films and devices can be effectively suppressed by the appropriate incorporation of artificial pinning structures (Wordenweber, Castellanos, and Selders 2000).

When pinning is weak, vortex motion is more regular. In this case, the quasiperiodic nature of the moving Bragg glass entails the motion of vortices in well-ordered rows (Giamarchi and Le Doussal 1996). In this scenario, each vortex line encounters the same disorder configuration and undergoes the same depinning events as its predecessors, which leads to the appearance of a so-called "washboard frequency" $f_w = v / a_0$ and narrow-band noise (Pardo et al. 1998, Troyanovsky, Aarts, and Kes 1999, Togawa et al. 2000). The presence of these spectral features may appear as a distinguishing feature between the different vortex phases outlined above (Togawa et al. 2000).

References

Altshuler E and Johansen T H (2004) Experiments in vortex avalanches, Rev. Mod. Phys. 76: 471

Aragón Sánchez J *et al.* (2019) Unveiling the vortex glass phase in the surface and volume of a type-II superconductor, Nat. Commun. 2: 143

Avraham N *et al.* (2001) 'Inverse' melting of a vortex lattice, Nature (London) 411: 451

Bak P, Tang C, and Wiesenfeld K (1986) Self-organised criticality: and explanation of $1/f$ noise, Phys. Rev. Lett. 59: 381

Banerjee S S *et al.* (1998) Anomalous peak effect in $CeRu_2$ and 2H–$NbSe_2$: fracturing of a flux line lattice, Phys. Rev. B 58(2): 995–999

Bardeen J and Stephen M J (1965) Theory of the motion of vortices in superconductors, Phys. Rev. 140: A1197

Bartolomé E, Vallés F, Palau A, Rouco V, Pompeo N, Balakirev F F, Maiorov B, Civale L, Puig T, Obradors X, and Silva E (2019) Intrinsic anisotropy versus effective pinning anisotropy in $YBa_2Cu_3O_7$ thin films and nanocomposites, Phys. Rev. B 100: 054502

Bean C P (1962) Magnetization of hard superconductors, Phys. Rev. Lett. 8: 250

Berezinskii V L (1972) Destruction of long-range order in one-dimensional and two-dimensional systems possessing a continuous symmetry group, Sov. Phys. JETP **34**: 610

Berghuis P, van der Slot A L F, and Kes P H (1990) Dislocation-mediated vortex-lattice melting in thin films of $_{\alpha}$-Nb_3Ge Phys. Rev. Lett. 65: 2583

Bhattacharya S and Higgins M J (1993) Dynamics of a disordered flux line lattice, Phys. Rev. Lett. 70: 2617

Blatter G, Feigel'man M V, Geshkenbein V B, Larkin A I, and Vinokur V M (1994) Vortices in high-temperature superconductors, Rev. Mod. Phys. 66: 1125

Blatter J, Geshkenbein V B, and Koopman J A G (2004) Weak to strong pinning crossover, Phys. Rev. Lett. 92: 067009

Bourgault D, Groult D, Bouffard S, Provost J, Studer F, Nguyen N, Raveau B, and Toulemonde M (1989) Modifications of the physical properties of the high-T_c superconductors $YBa_2Cu_3O_{7-\delta}$ ($0.1 \leq \delta < 0.7$) by 3.5-GeV xenon ion bombardment, Phys. Rev. B 39: 6549

Brandt E H (1989) Thermal fluctuation and melting of the vortex lattice in oxide superconductors, Phys. Rev. Lett. 63: 1106

Brandt E H (1996) Superconductors of finite thickness in a perpendicular magnetic field: strips and slabs, Phys. Rev. B 54: 4246

Brandt E H (2003) Properties of the ideal Ginzburg-Landau vortex lattice, Phys. Rev. B 68: 054506

Brezin E, Nelsonn D R, and Thiaville A (1985) Fluctuation effects near H_{c2} in type-II superconductors, Phys. Rev. B 31: 7124

Campbell A M and Evetts J E (1972) Flux vortices and transport currents in type II superconductors, Adv. Phys. 21: 199

Civale L, Marwick A D, Worthington T K, Kirk M A, Thompson J R, Krusin-Elbaum L, Sun Y, Clem J R, and Holtzberg FF (1991) Vortex confinement by columnar defects in $YBa_2Cu_3O_7$ crystals: Enhanced pinning at high fields and temperatures, Phys. Rev. Lett. 67: 648

Clem J (1974) A model for Flux Pinning in Superconductors Low Temperature Physics LT-13 3, eds K D Timmerhaus, W J O'sullivan and E F Hammel (Plenum: New York) p 102

Clem J R (1981) Flux-flow noise in superconductors, Phys. Rep. 75(1): 1–55

Clem J R and Sanchez A (1994) Hysteretic ac losses of thin superconducting disks, Phys. Rev. B 50: 9355

de Gennes P G (1989) Superconductivity of Metals and Alloys, (Addison Wesley, New York) p 83

Demirdis S *et al.* (2011) Strong pinning and vortex energy distributions in single-crystalline $Ba(Fe_{1-x}Co_x)_2As_2$, Phys. Rev. B 84: 094517

Demirdis S *et al.* (2013) Disorder, critical currents, and vortex pinning energies in isovalently substituted $BaFe_2(As_{1-x}P_x)_2$, Phys Rev. B 87: 094506

Edwards S F and Wilkinson D R (1982) The surface statistics of a granular aggregate, Proc. R. Soc. Lond. Ser. A 381: 17

Espinosa-Arronte B, Andersson M, van der Beek C J, Nikolaou M, Lidmar J, and Wallin M 2007 Fully anisotropic superconducting transition in ion-irradiated $YBa_2Cu_3O_{7-\delta}$ with a tilted magnetic field, Phys. Rev. B 75: 100504(R)

Feigel'man M V, Geshkenbein V B, Larkin A I, and Vinokur V M (1989) Theory of collective flux creep, Phys. Rev. Lett. 63: 2303

Feigel'man M V and Vinokur V M (1990) Thermal fluctuations of vortex lines, pinning and creep in high-T_c superconductors, Phys. Rev. B 41: 8986

Fendrich J A *et al.* (1995) Vortex liquid state in an electron irradiated untwinned $YBa_2Cu_3O_{7-\delta}$ Crystal, Phys. Rev. Lett. 74: 1210

Fisher D S, Fisher M P A, and Huse D (1991) Thermal fluctuations, quenched disorder, phase transitions and transport in type II superconductors, Phys. Rev. B 43: 130

Fuchs D T, Zeldov E, Tamegai T, Ooi S, Rappaport M, and Shtrikman H (1998) Possible new vortex matter in $Bi_2Sr_2CaCu_2O_8$, Phys. Rev. Lett. 80: 4971

Giamarchi T and Le Doussal P (1994) Elastic theory of pinned flux lattices, Phys. Rev. Lett. 72: 1530

Giamarchi T and Le Doussal P (1995) Elastic theory of flux lattices in the presence of weak disorder, Phys. Rev. B 52:1242

Giamarchi T and Le Doussal P (1996) Moving glass phase of driven lattices, Phys. Rev. Lett. 76: 3408

Grigera S, Morré E, Osquiguil E, Balseiro C, Nieva G, and de la Cruze F (1998) Bose-glass phase in twinned $YBa_2Cu_3O_{7-\delta}$, Phys. Rev. Lett. 81: 2348

Grisolia M N, van der Beek C J, Fasano Y, Forget A, and Colson D (2013) Multifractal scaling of flux penetration in the iron-based superconductor $Ba(Fe_{0.93}Co_{0.07})_2As_2$, Phys. Rev. B 87: 104517

Imry Y and Ma S K (1975) Random-field instability of the ordered state of continuous symmetry, Phys. Rev. Lett. 35: 1399

Indenbom M V, Kronmüller H, Li T W, Kes P H, and Menovsky A A (1994) Equilibrium magnetic properties and Meissner expulsion of magnetic flux in $Bi_2Sr_2CaCu_2O_x$ single crystals, Physica(Amsterdam) C 222: 203

Indenbom M V, Schuster Th, Kuhn H, KronmŸller H, Li T W, and Menovsky A A (1995) Observation of current strings in $Bi_2Sr_2CaCu_2O_8$ single crystals, Phys. Rev. B 51: 15484

Jiang W *et al.* (1994) Evidence of a Bose-glass transition in superconducting $YBa_2Cu_3O_{7-\delta}$ single crystals with columnar defects, Phys. Rev. Lett. 72: 550

Kes P H (1992) Flux pinning and the summation of pinning forces. In: Concise Encyclopedia of Magnetic & Superconducting Materials (J E Evetts ed.) (Pergamon, Oxford) p 163

Kes P H, Aarts J, Van den Berg J, Van der Beek C J, and Mydosh J A (1989) Thermally assisted flux flow at small driving forces, Supercond. Sci. Technol. 1: 242

Kim P, Yao Z, Bolle C A, and Lieber C M (1999) Structure of flux line lattices with weak disorder at large length scales, Phys. Rev. B 60: 12589

Klein T *et al.* (2010) First-order transition in the magnetic vortex matter in superconducting MgB_2 tuned by disorder, Phys. Rev. Lett. 105: 047001

Koch R H, Foglietti V, Gallagher W J, Koren G, Gupta A, and Fisher M P A (1989) Experimental evidence for vortex-glass superconductivity in Y–Ba–Cu–O, Phys. Rev. Lett. 63: 1115; Koch R H, Foglietti V, and Fisher M P A (1990) Reply to comment on experimental evidence for vortex-glass superconductivity in Y–Ba–Cu–O eds Coppersmith S N, Inui M and Littlewood P B, Phys. Lett. 64: 2586

Kokkaliaris S, de Groot P A J, Gordeev S N, Zhukov A A, Gagnon R, and Taillefer L (1999) Onset of plasticity and hardening of the hysteretic response in the vortex system of $YBa_2Cu_3O_{7-\delta}$, Phys. Rev. Lett. 82: 5116

Koorevaar P, Aarts J, Berghuis P, and Kes P H (1990) Tilt-modulus enhancement of the vortex lattice in the layered superconductor 2H-$NbSe_2$, Phys. Rev. B 42: 1004

Koshelev A E and Vinokur V M (1991) Physica C 175: 465

Kosterlitz J M and Thouless D J (1978) Two-dimensional physics. In: Progress in Low Temperature Physics (D F Brewer ed.), vol. VII-B (North-Holland: Amsterdam) pp 371–433

Kwok W K, Fleshler S, Welp U, Vinokur V M, Downney J, Crabtree G W, and Miller M M (1992) Vortex lattice melting in untwinned and twinned single crystals of $YBa_2Cu_3O_{7-\delta}$, Phys. Rev. Lett. 69: 3370

Kwok W K, Welp U, Vinokur V M, Fleshler S, Downey J, and Crabtree G W (1991) Direct observation of intrinsic pinning by layered structure in single-crystal $YBa_2Cu_3O_{7-\delta}$, Phys. Rev. Lett. 67: 390

Larkin A I and Ovchinnikov Yu N 1979 Pinning in type II superconductors, J. Low Temp. Phys. 34: 409

Li T W, Menovsky A A, Franse J J M, and Kes P H (1996) Flux pinning in Bi-2212 single crystals with various oxygen contents, Physica C 257: 179

Lidmar J and Wallin M 1999 Critical properties of Bose-glass superconductors, Europhys. Lett. 47: 494

López D, Righi E F, Nieva G, and de la Cruz F (1996) Coincidence of vortex-lattice melting and loss of vortex correlation along the c direction in untwinned $YBa_2Cu_3O_{7-\delta}$ single crystals, Phys. Rev. Lett. 76: 4034

López D *et al.* (1997) Pinned vortex liquid above the critical point of the first-order melting transition: a consequence of pointlike disorder, Phys. Rev. Lett. 80: 1070

Marchiori E, Curran P J, Kim J, Satchell N, Burnell G, and Bending S J (2017) Reconfigurable superconducting vortex pinning potential for magnetic disks in hybrid structures, Sci. Rep. 7: 45182

Mele P, Prassides K, Tarantini C, Palau A, Badica P, Jha A K, and Endo T (2019) Superconductivity: from materials science to practical applications, Springer Nature ISBN-13 978-3030233020

Mikitik G P and Brandt E H (2001) Peak effect, vortex-lattice melting line, and order-disorder transition in conventional and high-T_c superconductors, Phys. Rev. B 64: 184514

Nakielski G *et al.* (1996) Enhancement of Bose-glass superconductivity in $YBa_2Cu_3O_{7-\delta}$ thin films, Phys. Rev. Lett. 76: 2567

Nelson D R and Halperin B I (1979) Dislocation-mediated melting in two dimensions, Phys. Rev. B. 19: 2457.

Nelson D R and Vinokur V M (1992) Boson localization and pinning by correlated disorder in high-temperature superconductors, Phys Rev Lett. 68: 2398; Phys. Rev. B 48: 13060

Ovchinnikov Yu N and Ivlev B I (1991) Pinning in layered inhomogeneous superconductors, Phys. Rev. B 43: 8024

Paltiel Y *et al.* (2000) Dynamic instabilities and memory effects in vortex matter Nature 403: 398–401

Paltiel Y, Jung G, Myasoedov Y, Rappaport M L, Zeldov E, Ocio M, Higgins M J, and Bhattacharya S (2004) Velocity-fluctuations–dominated flux-flow noise in the peak effect, Europhys. Lett. 66 (3): 412

Pardo F, de la Cruz F, Gammel P L, Bucher E, and Bishop D J (1998) Observation of smectic and moving-Bragg-glass phases in flowing vortex lattices, Nature 396: 348–350

Putzke C *et al.* (2014) Anomalous critical fields in quantum critical superconductors, Nature Communications 5: 5679

Sadovskyy I A, Koshelev A E, Glatz A, Ortalan V, Rupich M W, and Leroux M (2015) Simulation of the vortex dynamics in a real pinning landscape of $YBa_2Cu_3O_{7-\delta}$ coated conductors, Phys. Rev. Applied 5: 014011

Safar H, Gammel P L, Huse D A, Bishop D J, Rice J P, and Ginsberg D M (1992) Experimental evidence for a first-order vortex-lattice-melting transition in untwinned, single crystal $YBa_2Cu_3O_{7-\delta}$, Phys. Rev. Lett. 69: 824

Safar H *et al.* (1993) Experimental evidence for a multicritical point in the magnetic phase diagram for the mixed state of clean untwinned $YBa_2Cu_3O_{7-\delta}$, Phys. Rev. Lett. 70: 3800

Schilling A *et al.* (1996) Calorimetric measurement of the latent heat of vortex-lattice melting in untwinned $YBa_2Cu_3O_{7-\delta}$, Nature 382 (6594): 791

Schuster Th, Indenbom M V, Kuhn H, Kronmüller H, Leghissa M, and Kreiselmeyer G (1994) Observation of in-plane anisotropy of vortex pinning by inclined columnar defects, Phys. Rev. B 50: 9499

Thompson J R, Krusin-Elbaum L, Civale L, Blatter G, and Feild C (1998) Superfast vortex creep in $YBa_2Cu_3O_{7-\delta}$ crystals with columnar defects: evidence for variable-range vortex hopping, Phys. Rev. Lett. 78: 3181

Togawa Y, Abiru R, Iwaya K, Kitano H, and Maeda A (2000) Direct observation of the washboard noise of a driven vortex lattice in a high-temperature superconductor, $Bi_2Sr_2CaCu_2O_y$, Phys Rev Lett. 85(17): 3716–3719.

Troyanovsky A M, Aarts J, and Kes P H (1999) Collective and plastic vortex motion in superconductors at high flux densities, Nature 399: 665

van der Beek C J, Colson S, Indenbom M V, and Konczykowski M (2000) Supercooling of the disordered vortex lattice in $Bi_2Sr_2CaCu_2O_8$, Phys. Rev. Lett. 84: 4196

van der Beek C J, Indenbom M V, Berseth V, Li T W, and Benoit W (1996) Onset of bulk pinning in BSCCO single crystals, J. Low. Temp. Phys. 105: 1047

van der Beek C J, Indenbom M V, D'Anna G, and Benoit W (1996) Physica C 258: 105

van der Beek C J, Kes P H, Maley M P, Menken M J V, and Menovsky A A (1992) Flux pinning and creep in the vortex-glass phase in $Bi_2Sr_2CaCu_2O_{8+\delta}$ single crystals, Physica C 195: 307

van der Beek C J, Konczykowski M, Li T W, Kes P H, and Benoit W (1996) Large effect of columnar defects on the thermodynamic properties of $Bi_2Sr_2CaCu_2O_8$ single crystals, Phys. Rev. B 54: R792

van der Beek C J *et al.* (2002) Strong pinning in high temperature superconducting films, Phys. Rev. B 66: 024523

van der Beek C J *et al.* (2005) Thermodynamics of the vortex liquid in heavy-ion-irradiated superconductors, Phys. Rev. B 72: 214504

van der Beek C J *et al.* (2010) Quasiparticle scattering induced by charge doping of iron-pnictide superconductors probed by collective vortex pinning, Phys. Rev. Lett. 105: 267002

van Ooijen D J and van Gurp G J (1965) Motion and pinning of flux in superconducting vanadium foils, studied by means of noise, Phys. Letters, 17: 230

Wördenweber R, Castellanos A M, and Selders P (2000) Vortex lattice matching effects and $1/f$ noise reduction in HTS films and devices equipped with regular arrays of artificial defects, Physica C: Superconductivity 332 (1–4): 27–34

Wördenweber R and Kes P H (1986) Dimensional crossover in collective flux pinning, Phys. Rev. B 34: 494(R)

Wördenweber R, Kes P H, and Tsuei C C (1986) Peak and history effects in two-dimensional collective flux pinning, Phys. Rev. B 33: 3172

Zeldov E *et al.* (1994) Geometrical barriers in high temperature superconductors, Phys. Rev. Lett. 73: 1428

Zeldov E, Majer D, Konczykowski M, Geshkenbein V B, Vinokur V M, and Shtrikman H (1995) Thermodynamic observation of first-order vortex-lattice melting in $Bi_2Sr_2CaCu_2O_8$, Nature 375: 373

Part B

Low-Temperature Superconductors

B

Introduction to Section B: Low-Temperature Superconductors

Peter J. Lee

In this section, we introduce the fundamental properties of the primary low-temperature superconductors (those typically requiring liquid He cooling to operate), which include the Nb-based superconductors that dominate the commercial application of superconductors as well as MgB_2 and the Chevrel phase compounds (ternary Mo chalcogenides) that offer the potential of lower material costs in the future.

The first successful superconducting solenoid, made by George Yntema at the University of Illinois in 1954, used cold worked Nb wire, and Nb-based superconductors continue to be the workhorse superconductors. Pure Nb continues to be important because of its application to superconducting radio-frequency cavities (SRF), which have become a vital part of particle accelerators like the Large Hadron Collider (LHC) and free electron lasers like the European X-ray free electron laser. Furthermore, for superconducting electronic application, commercial fabrication of Nb-based Josephson junctions has reached the level of maturity that thousands of the junctions can be fabricated into multilayer structures for superconductor integrated digital circuits, and NbN is used extensively for high-frequency superconducting electronics circuits.

The success of multifilamentary wires based on the Nb-47Ti alloy for magnetic resonance imaging (MRI) has brought superconducting technology into the lives of a broad segment of the world's population, and Nb-Ti MRI technology has been extended to 11.7 T with the Iseult whole-body magnet. The technology of Nb-47Ti was first industrialized on a large scale for the Tevatron II Energy Saver project at Fermilab, which was dedicated in 1984 and then pushed toward its limits for the ill-fated Superconducting Super Collider (SSC). The performance developed for the SSC then made the LHC possible and ultimately the discovery of the Higgs boson.

Nb_3Sn can be argued to be the first high magnetic field superconductor with the achievement of high-current wires at 8.8 T by John Kunzler and co-workers at Bell Labs in 1961. Despite its sensitivity to strain and brittleness, its ability to be fabricated into micrometer dimension filaments in high conductivity matrices at relatively low cost compared to HTS technology continues to make it the superconductor of choice for applications between 10 and 20 T. Without Nb_3Sn strands, projects such as the vast ITER magnetic confinement fusion reactor and the series-connected hybrid magnets at the National High Magnetic Field Laboratory and Helmholtz–Zentrum Berlin would not be possible. Remarkably, almost 60 years on from that initial discovery, Nb_3Sn technology continues to be developed to new performance limits with applications of great importance like the high-luminosity (HiLumi) upgrade to the LHC.

Advances in cryocooling and the cost of Nb have continued to drive interest in alternatives to Nb-47Ti and Nb_3Sn with higher critical temperatures or fields and lower materials costs. The Chevrel phases were only discovered in 1971 and superconductivity was discovered in them only a year later. The critical temperatures sit between Nb-Ti and Nb_3Sn but their upper critical fields (40-60 T) far surpass Nb-47Ti and Nb_3Sn, and the raw material costs, being based on Pb, Mo, S and Sn, are relatively low. Development of these conductors suffered from the advent of high-temperature superconductors (HTS), but with the continued difficulty in reducing the cost of HTS, there continues to be interest in their application. Superconductivity was only discovered in MgB_2 in 2001, but its surprisingly high critical temperature of 39 K gained it immediate interest and offered the prospect of a low-cost superconductor that could be operated using a cryocooler. This offers the prospect of MRI systems being introduced into markets that do not have ready access to cryogens. Low-temperature superconductors continue to be of great interest for superconducting applications, and the continued need to improve performance and reduce costs continues to drive new advances.

B1

Nb-Based Superconductors

Gianluca De Marzi and Luigi Muzzi

Due to their high melting temperatures, it was not until the 1930s that group 4 and 5 refractory metals were synthesized and measured at low temperature. Superconductivity in Nb was first observed in 1930 at the cryogenics laboratory of W. Meissner in Berlin, a few years before his studies on the effect of magnetic flux expulsion from the interior of a superconductor (Meissner and Ochsenfeld, 1933). Among the superconducting metallic elements, niobium exhibits the record critical temperature, 9.25 K; it is also one of the few elemental superconductors exhibiting type II magnetic properties and some non-negligible transport critical current capability. Owing to these remarkable properties, the earliest attempts to make magnets out of superconducting materials, as envisaged by Heike Kamerlingh-Onnes, were made of pure Nb.

G. Yntema in 1955 demonstrated a field generation of 0.71 T at 4.2 K in a 3 mm gap (Lee and Strauss, 2012; Yntema, 1987) using a cold-worked Nb wire wound over an iron core. A few years later, S. Autler reached 1.4 T (Lee and Strauss, 2012; Autler, 1960) (corresponding to 0.43 T on the conductor), using a silk-insulated Nb wire.

However, from an application perspective, especially at higher fields, the superconducting domain needed to be extended (i.e. increased T_c and/or H_{c2}), which could be achieved with metallic binary or ternary compounds. B. Matthias demonstrated a relation between T_c in a superconductor and the number of valence electrons per atom (*e/a*) (Matthias, 1955), a number that could be varied by alloying Nb. In addition, it was empirically observed that a cubic crystal structure was highly favorable for superconductivity. Stable body-centered cubic (*bcc*) solid solution alloys are formed between the superconducting group 5 elements, V, Nb, Ta, and up to about 80 at.% of their group 4 neighbors (Ti, Zr, Hf) or over the complete range with their group 6 neighbors (Cr, Mo, W). J. Hulm and R. Blaugher (Hulm and Blaugher, 1961) at Westinghouse Research Laboratories in Pittsburgh have studied the solid solutions of transition elements systematically: they traced the T_c vs. *composition* for binary alloys of neighboring rows, or columns, or for diagonal neighbors in groups 4, 5, 6 transition elements. It was observed that the critical temperature of Nb is increased by additions of Ti, Zr, Hf, whereas it decreases with additions of elements on its right in the periodic table.

Although metals such as Ti, Zr, V, Nb are influenced by small amounts of impurities from dissolved gases and in spite of their extremely high melting temperatures, suggesting possible metallurgical difficulties in the preparation, they have been widely studied because the very wide range of solid solutions available offered a mechanism for varying the electronic configuration within a given crystal structure. This provided insight into the role of the various mass, volume, and electronic structure effects in a superconductor (Hulm and Blaugher, 1961). Of particular importance for such studies was the availability of extremely pure Nb, with O_2 and N_2 gas impurities at the level of a few hundred ppm, group 8 metal impurities well below 1000 ppm, and $Ta < 1500$ ppm.

B. Matthias expanded the path originally undertaken by W. Meissner in Berlin, at first systematically investigating with a chemical approach the nitrides and carbides of groups 4, 5, 6 transition metals (Matthias and Hulm, 1952; Matthias and Geballe, 1963), and measuring T_c magnetically, thus providing less ambiguous results in case of sample non-uniformities.

The real breakthrough for applications is considered to be the demonstration of high-field superconductivity in intermetallic compounds with the A15 crystal structure: V_3Si was the first of this class to be discovered in 1953, with $T_c = 17.1$ K, by J. Hulm and his student G. Hardy at the University of Chicago (Hardy and Hulm, 1954), followed one year later by the discovery of a T_c of 18 K in Nb_3Sn, by B. Matthias and T. Geballe at Bell Labs (Matthias et al., 1954).

It was then J. Kunzler who demonstrated (Kunzler et al., 1961) in late 1960 that Nb_3Sn could effectively be used up to 8.8 T, the actual maximum field available in his experimental facility at Bell Labs. Finally, the highest T_c of all A15 compounds was observed in Nb_3Ge, with $T_c = 23$ K, by J. Gavaler in Pittsburgh in 1972 (Gavaler, 1973).

The first superconducting wires were manufactured in 1962, and used Nb–Zr. Nb–Ti followed shortly after, and became the predominant product, because it was easier to manufacture (it was, for example, simpler to bond stabilizing Cu to this material, whereas Nb–Zr becomes brittle below about 100°C) and because it had a 2 T higher upper critical field. As determined experimentally by T. Berlincourt and R. Hake (Berlincourt and Hake, 1962), Nb–Ti exhibits the highest H_{c2} (14.5 T at T = 0 K).

As of today, the binary alloy Nb–Ti and the intermetallic compound Nb_3Sn are the most technologically advanced materials, employed in most of superconductivity applications. Nb_3Sn is not competitive commercially to displace Nb–Ti in low-field applications. It is instead often the material of choice in high-field applications (roughly > 8 T for accelerator magnets, > 10 T for solenoids) where its superior high-field properties outweigh the processing difficulties associated to its being a brittle intermetallic compound with a low strain tolerance. The reason for, and the limitations to, Nb_3Sn's success is much related to the processing of the compound into wires and tapes. The issue was summed up very well by Bruce Strauss from the U.S. Department of Energy when he challenged the community: *"Anyone can make a sample with exciting properties; can you consistently make multiple km with the same performance in every mm and in tonnage quantity cost effectively?"*.

The annual production of Nb–Ti by different industrial suppliers around the world is of a few thousand tons, mainly for magnetic resonance imaging (MRI) magnets. The Large Hadron Collider (LHC) accelerator project alone required about 1.3 tons of multifilamentary Nb–Ti wires (Rossi, 2010), whereas the International Thermonuclear Experimental Reactor (ITER) project required about 600 tons of Nb_3Sn low-loss wires (Vostner et al., 2017), the largest production ever, to be compared with a pre-ITER worldwide annual supply of about 15 tons. Purely in terms of resource availability, superconducting applications employ less than 0.5% of the total niobium quantity consumed worldwide, which amounts to about 60,000 tons/year (Carneiro, 2014), mainly for the steel industry. In this sense, as far as superconductivity applications are concerned, the price of Nb-based materials is dictated by the specific required purity level and homogeneity and by the transformation processes to manufacture the required multifilamentary and stabilized wires.

B1.1 Superconductivity in Elemental Niobium

B1.1.1 Physical Properties and Phase Diagram

Niobium is a ductile refractory transition metal element with atomic number Z = 41, belonging to the group 5 of the periodic table. This element is characterized by 5 valence electrons, with electronic structure [Kr]$4d^45s^1$: four electrons occupy highly localized *d*-orbitals, whereas the fifth electron occupies a less localized *s*-orbital. Niobium crystallizes in the *bcc* over the whole range of temperatures and at standard pressure conditions; its lattice constant is a = 3.3063 Å at T = 295 K (Roberge, 1975), and the variation of lattice parameter from room temperature to liquid helium temperature is very small (0.143%). Table B1.1 summarizes the main physical properties of elemental niobium.

TABLE B1.1 Physical Properties of Nb

Physical Property	Value
Conduction electron density	$5.56{\cdot}10^{22}$ cm^{-3}
Atom configuration	[Kr]$4d^45s^1$
Atomic weight	92.91
Crystal structure	*Bcc*
Number of valence electrons	5
Density (298 K)	8.57 g cm^{-3}
Melting point	2468 °C
Boiling point	4927 °C
Specific heat at const. press. (298 K)	0.265 kJ kg^{-1} K^{-1}
Specific heat at const. press. (4 K)	0.0004 kJ kg^{-1} K^{-1})
Latent heat of fusion	290 kJ kg^{-1}
Latent heat of vaporization	7490 kJ kg^{-1}
Coefficient of linear thermal expansion (298 K)	7.3 10^{-6} K^{-1}
Thermal contraction, ΔL/L (4 K)	0.143%
Magnetic susceptibility	$241{\cdot}10^{-6}$ SI
Electrical resistivity (77 K)	3.0 μΩ cm
Electrical resistivity (273 K)	15.2 μΩ cm
Drude relaxation time (77 K)	21 fs
Drude relaxation time (273 K)	4.2 fs
Thermal conductivity (300 K)	53.7 W m^{-1} K^{-1}
Electronic specific heat parameter	7.8 mJ mole^{-1} K^2
Effective mass	12 m_e
Debye temperature	265 K
First ionization potential	6.87 eV
Work function	4.3 eV

Source: [data from Miller (1959); "NIOBIUM", Proceedings of the International Symposium, Niobium '81; Ekin (2006); Poole et al., 2007); http://www.periodictable.com/Elements/041/data.html].

B1.1.2 Electronic Properties: Band Structure and Fermi Surface

The physical properties of niobium, and in particular, the electronic, vibrational, and structural properties can be explored by *ab-initio* density functional theory (DFT) methods. As previously stated, four *d*-electrons are highly localized, whereas the remaining *s*-electron is much more delocalized with respect to the others. Consequently, the band structure of niobium is characterized by the overlap and hybridization of a rather large *d*-band with a larger, hybridized *sp*-band (Jani et al., 1988).

The electronic properties of niobium, such as the density of states (DOS) and energy bands were calculated for the equilibrium lattice parameter, obtained for each case by minimizing the electronic total energy with respect to the cell volume.

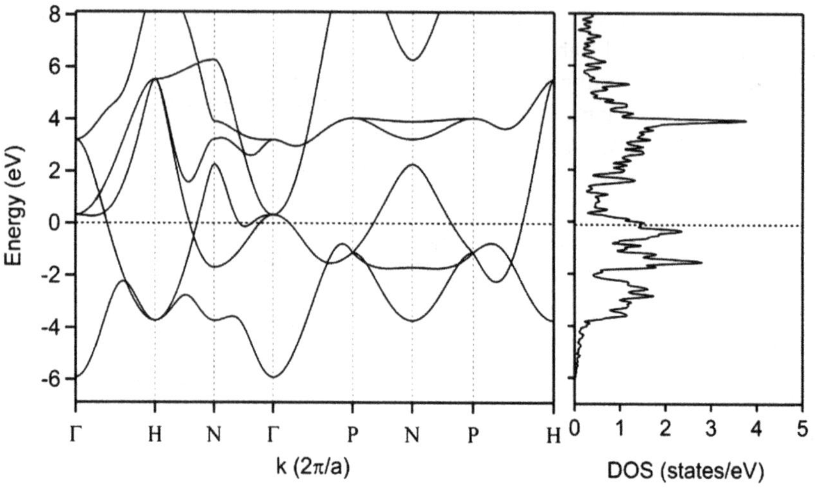

FIGURE B1.1 Band structure (left) and density of states (DOS) of Nb (right). The dotted line represents the Fermi level, which is set to zero.

The equilibrium lattice parameter, together with the bulk modulus, can be obtained by fitting the total energy to the Murnaghan's equation of state (Murnaghan, 1944):

$$E(V) = V\frac{B_0}{B_0'}\left[\frac{(V_0/V)^{B_0'}}{B_0'-1}+1\right]+const, \quad \text{(B1.1)}$$

where V is the volume, B_0 the bulk modulus, and B_0' its pressure derivative.

In Figure B1.1, plots of the E–$\boldsymbol{k}$ relations calculated along several main symmetry directions (Δ, G, Σ, Λ, D, F) of the Brillouin zone are shown. On the right of Figure B1.1, the corresponding electronic DOS is plotted: the Fermi level is conventionally set to zero. As it can be clearly seen, there are several peaks, which can be correlated to measured data from photoemission experiments (a band with a width of ~3 eV comprised three peaks at ~0.4, ~1.1, and~2.3 eV below the Fermi energy, E_F) (Eastman, 1969). The DOS at E_F is found to be equal to $N(E_F)$ = 1.46 states/eV^{-1}. By using this value for $N(E_F)$, the Sommerfeld constant, γ, for the normal-state electron specific heat, C_e, can be calculated as follows:

$$\gamma = C_e/T = \pi^2/3 \cdot k_B^2 \cdot N(E_F)(1+\lambda) \quad \text{(B1.2)}$$

In Equation (B1.2), T is the temperature, k_B the Boltzmann constant, and λ the electron–phonon interaction constant. By considering an estimated value of ≈ 1 for λ (the values determined from de Haas–van Alphen experiments are comprised between 0.75 (Halloran et al., 1970) and 1.14 (Scott and Springford, 1970), the Sommerfeld constant can be easily obtained: γ = 6.89 mJ mol^{-1} K^{-2}, in good agreement with the experimental findings (7.8 mJ mol^{-1} K^{-2} (Poole et al., 2007; van der Hoeven and Keesom, 1964).

In Figure B1.2, the DOS projected onto the atomic orbitals of Nb is plotted as a function of the energy; the total DOS is also reported for comparison. In the considered energy range, the main contribution to the total DOS comes from *4d*-orbitals, for about 82%, with small contributions from the remaining *s*- (5%) and *p*-orbitals (13%).

With five electrons per atom, the Fermi surface extends outside the first Brillouin zone (BZ), the $\boldsymbol{k}$-states being occupied up to the third BZ. Niobium is characterized by a rather topologically complex Fermi surface, which was first calculated in 1970 by Mattheiss by means of the augmented plane-wave method (Mattheiss, 1970). Qualitatively, the topology of the Fermi surface is very similar to that of other transition elements such as, for example, vanadium (Halloran et al., 1970). It is convenient to map the Fermi surface in the reduced zone scheme, in which any wave-vector $\boldsymbol{k}$ belonging to the Fermi surface can be always represented within the first BZ,

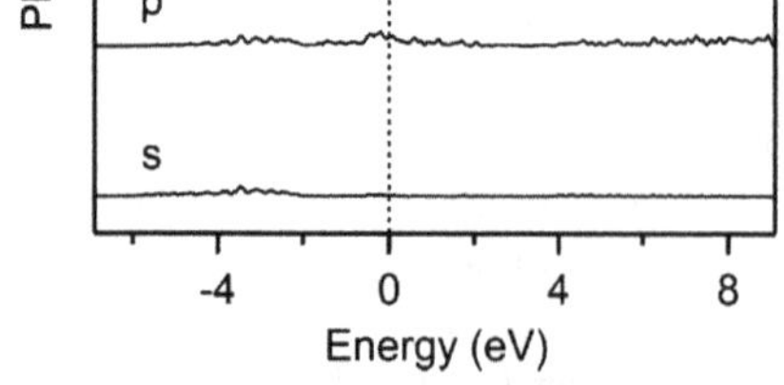

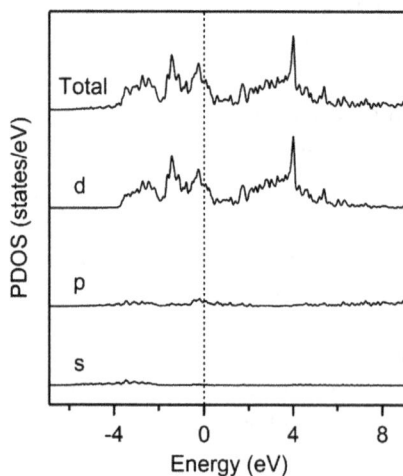

FIGURE B1.2 Density of states projected on atomic orbitals (PDOS). The main contribution comes from *4d*-orbitals (82%). The dotted line represents the Fermi level (conventionally set to zero).

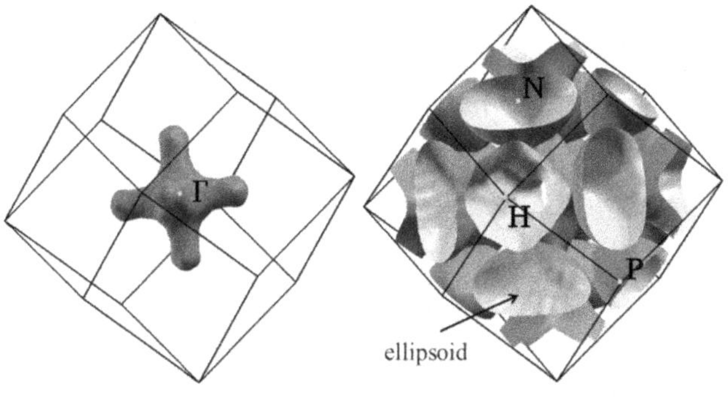

FIGURE B1.3 Holes Fermi surface of niobium: (left) closed structure with octahedron symmetry, centered at the Γ point (second BZ); (right) ellipsoids centered at the *N* points, together with the '*jungle-gym*' geometry (third BZ).

by properly adding a reciprocal vector ***G*** to ***k***. Figure B1.3 depicts, in the reduced zone scheme, the Fermi surface sheets that enclose the unoccupied regions in the second and third BZ (the first BZ is completely filled by electrons). In the second zone, the Fermi surface is characterized by a hole surface with the shape of a distorted octahedron centered at Γ, with rather large effective masses (1.4–2.0 m_0) (Eastman, 1969; Halloran et al., 1970). In the third zone, the hole Fermi surface is comprised of six distorted ellipsoidal structures centered at the points *N*, plus a multiply connected sheet which extends along <100> directions, from Γ to *H*. This is often referred to as '*jungle-gym*' structure.

The knowledge of the Fermi topology allows in many cases to predict the electrical, magnetic, thermal, and optical properties of metals, semimetals, and doped semiconductors. For an extensive description, the reader may refer to Chapters 12 and 13 of Ashcroft and Mermin (1976).

B1.1.3 Vibrational Properties: Phonon Dispersion Curves and Elastic Constants

Transition metals belonging to group 5 are characterized by an anomalous softening in the phonon dispersion curves, which has been correlated with their observed high T_c. In addition, since those materials are BCS superconductors, the study of the vibrational properties can be of great help for the comprehension of those mechanisms leading to superconductivity in Nb and its alloys.

In a DFT self-consistent calculation, the total energy is minimized with respect to the occupation of the Kohn–Sham orbitals. At absolute zero, a band is either occupied or empty and, when dealing with metallic systems, such discrete occupation results in discontinuous changes in energy with changes in occupation. As a result, the self-consistent calculation can run very slow, and a large sampling of the Brillouin zone is required (i.e. large number of ***k*** points) in order to facilitate the convergence of the self-consistent cycle. One way to resolve this issue is to artificially broaden the band occupancies, as if the system were at a higher temperature; consequently, the resulting continuous dependence of energy on the partial occupancy drastically increases the rate of convergence. At absolute zero temperature, the occupation function is a step function; in order to broaden the band occupancies, the step function is replaced with a smoother (i.e. *smeared*) function. There exists a large variety of smearing schemes: in the calculations shown in Figure B1.4 we used the Fermi–Dirac function. The phonon dispersion curves have been calculated for three different *smearing* values (0.005, 0.01, and 0.02 Ry). The dependence of the phonon frequencies from the *smearing* parameter has been observed and discussed in previous works (de Gironcoli, 1995). As it can be clearly seen in Figure B1.4, the

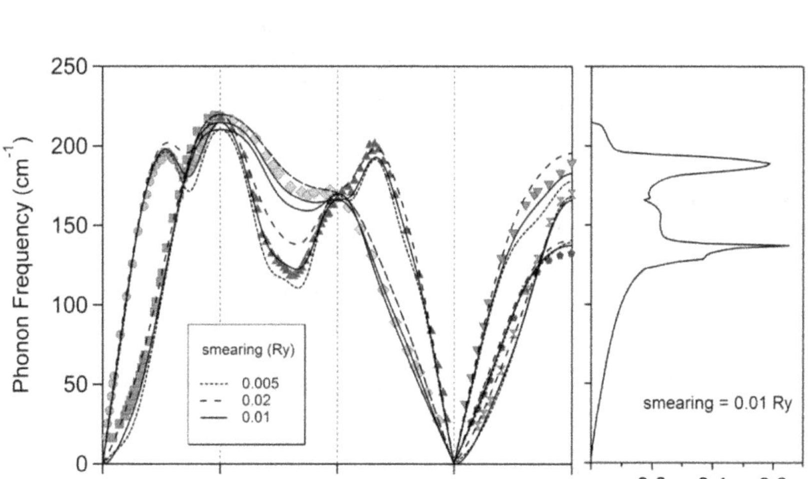

FIGURE B1.4 Phonon dispersion curve of niobium and corresponding phonon DOS. Dotted and continuous lines represent the DFT calculation, in excellent agreement with experimental data (symbols). [From Powell et al. (1977); originally from Nakagawa and Woods (1963).]

phonon dispersion curves calculated with *smearing* = 0.01 Ry are in excellent agreement with the experimental data (Powell et al., 1977; Nakagawa and Woods, 1963), with the exception of the double degenerate transverse acoustic (TA) phonons in the long wavelength limit in the Γ–*H* direction (Salvetti, 2010), and the high-frequency branches along the *H*–*P* direction. All major experimental findings are reproduced by calculations: along the Γ–*H* direction a cross-over of the longitudinal and transverse branches near $\boldsymbol{q}$=(0.0, 0.0, 0.7) is found, with the concomitant presence of a maximum and minimum. Moreover, the transverse branches cross near $\boldsymbol{q}$=(0.3, 0.3, 0.0) along the Γ–*N* direction.

From the phonon dispersion curves, one can infer information about the elastic constants. In fact, in a perfect crystal the elastic constants can be directly obtained from the slopes of the acoustic branches in the long wavelength limit($|\boldsymbol{q}|\to 0$). The relation between the sound velocity v_s along a given direction and the corresponding elastic constant C_{ij} is $C_{ij} = \rho \cdot v_s^2$, where ρ is the density of the crystal. For a cubic lattice, the expressions for C_{ij} associated to the phonon branches are reported in Table B1.2; in particular, for C_{11} and C_{44} the following relations hold:

$$\left.\frac{\partial \omega_{[\xi 00]L}(\mathbf{q})}{\partial \boldsymbol{q}}\right|_{\mathbf{q}\to 0} = \sqrt{\frac{C_{11}}{\rho}}, \qquad \text{(B1.3)}$$

$$\left.\frac{\partial \omega_{[\xi 00]T}(\mathbf{q})}{\partial \boldsymbol{q}}\right|_{\mathbf{q}\to 0} = \sqrt{\frac{C_{44}}{\rho}}, \qquad \text{(B1.4)}$$

where

$$\frac{\partial \omega}{\partial \boldsymbol{q}} = \frac{\partial 2\pi \nu}{\partial \boldsymbol{q}} = \frac{2\pi \partial \nu}{\partial \xi}\frac{a_0}{2\pi} = \frac{\partial \nu}{\partial \xi}a_0 \qquad \text{(B1.5)}$$

In Equation (B1.5), $\omega = 2\pi\nu$ and $\boldsymbol{q}$ represent the frequency and the wave vector of a given phonon mode, respectively, whereas ξ the phonon wave vector renormalized by a factor $a_0/2\pi$, a_0 being the lattice parameter of the cubic cell. It easily follows that:

$$C_{11} = \rho\left(\lim_{|\xi|\to 0}\frac{\partial \upsilon_{[\xi 00]L}}{\partial|\xi|}a_0\right)^2 = \rho v^2_{s,[\xi 00]L}, \qquad \text{(B1.6)}$$

$$C_{44} = \rho\left(\lim_{|\xi|\to 0}\frac{\partial \upsilon_{[\xi 00]T}}{\partial|\xi|}a_0\right)^2 = \rho v^2_{s,[\xi 00]T}. \qquad \text{(B1.7)}$$

By considering a density of 8.57 g/cm³, from the slope of the TA branch in the Γ–*H* direction, a value C_{11} = 2.27 Mbar is calculated, in good agreement with experiments [$C_{11,exp}$ = 2.46 Mbar (Bolef, 1961)]. On the other hand, the elastic constant C_{44} is largely underestimated: C_{44} = 0.179 Mbar for a *smearing* value of 0.02 Ry, whereas the experimental value is $C_{44,exp}$ = 0.287 Mbar (Bolef, 1961). Finally, from the relation which connects the bulk modulus B_0 with the elastic constants C_{11} and C_{12}:

$$B_0 = (C_{11} + 2\cdot C_{12})/3 \qquad \text{(B1.8)}$$

one can obtain a value for C_{12}: 1.45 Mbar, which is overestimated owing to the uncertainty in C_{44}. Table B1.3 reports the experimental values for the elastic constant of niobium, together with DFT calculations obtained with different exchange functionals (local density approximation: LDA; and gradient generalized approximation: GGA).

As mentioned earlier, the TA frequencies in the long wavelength limit along the Γ–*H* direction are largely underestimated by DFT calculations, and this reflects on the value estimated for the elastic constant C_{44}. On the other hand, the longitudinal acoustic (LA) branch is in excellent agreement with the experimental results for any wavelength. The elastic constant associated to those LA phonons, C_{11}, is therefore more accurate. The reason for such inaccuracy in calculating C_{44} is not known. Some authors (Nagasako et al., 2010) suggest that the softening of C_{44} could be due to the concomitant presence of a van Hove singularity near the Fermi surface.

TABLE B1.2 Correspondence between the Acoustic Branches and Elastic Constants for a Cubic Crystal

Phononic Branch	C_{ij} =
[ξ ξ ξ]*L*	⅓(C_{11}+2C_{12} + 4C_{44})
[ξ ξ ξ]*T1*	⅓(C_{11} - C_{12} + C_{44})
[ξ ξ ξ]*T2*	⅓(C_{11} - C_{12} + C_{44})
[ξ ξ 0]*L*	½(C_{11} + C_{12} + 2C_{44})
[ξ ξ 0]T_1	C_{44}
[ξ ξ 0]T_2	½(C_{11} - C_{12})
[ξ 0 0]*L*	C_{11}
[ξ 0 0]*T*	C_{44}

Note: T = degenerate transverse modes; T_i= non-degenerate transverse modes; and *L* = longitudinal modes. [ξ ξ ξ] denotes the directions in the reduced BZ.

TABLE B1.3 Bulk Modulus and Elastic Constants (C_{11} and C_{12}) of Niobium, Calculated with Different Exchange Functionals (LDA; GGA) and Measured (Bolef, 1961; Simmons and Wang, 1971)[a]

	B_0	C_{11}	C_{12}	C_{44}
This work	1.720	2.27	1.45	0.179
GGA (PBE)	1.712	2.5242	1.3006	0.2112
GGA (PW91)	1.7226	2.4499	1.3589	0.1519
LDA	1.9182	2.7619	1.4963	0.1445
Exp. (Bolef)	1.718	2.46	1.347	0.287
Exp. (Simmons, Wang)	1.7303	2.5270	1.3319	0.3097

[a] In the GGA calculations, the correlation energy is introduced by means of the gradient-corrected PW91 [as proposed by Perdew and Wang (Perdew, 1991)] or PBE [Burke-Ernzerhof (Perdew et al., 1996)] functionals.

TABLE B1.4 Superconducting Properties of Nb

Physical Property	Value
Critical temperature	$T_c = 9.25$ K
Thermodynamic critical field	$B_c = 206$ mT
Lower critical field	$B_{c1}(T=0) = 174$ mT
Upper critical field	$B_{c2}(T=0) = 404$ mT
Coefficient of T_c variation with pressure	$dT_c/dP = -2.0$ K GPa^{-1}
Energy gap	$E_g = 2\Delta_0 = 3.0$ meV
Energy gap ratio	$2\Delta_0/k_BT_c = 3.8$
Specific heat jump	$C_s\text{-}C_n/\gamma T_c = 1.93$
Coherence length	ξ (T=0) = 39 nm
Magnetic penetration depth	λ (T=0) = 52 nm
Ginzburg–Landau constant	k (λ/ξ) = 1.28
Electron–phonon coupling constant	$\lambda_{el.\text{-}ph.} = 0.85$
Coulomb pseudo-potential	$\mu_c^* = 0.15$
Electronic specific heat parameter	$\gamma = C_e/T = 7.80$ mJ mole^{-1} K^2
Debye temperature	Θ_D: between 250 K and 276 K
Density of states at Fermi level $D(E_F)$	1.46 states eV^{-1}

Source: [Data from Ekin (2006), Poole et al. (2007) and Finnemore et al. (1966)].

B1.1.4 Superconductivity in Niobium

The main characteristic properties of niobium in its superconducting state are summarized in Table B1.4.

Once the electronic and vibrational properties are known, it is possible to estimate the critical temperature with the formula proposed by McMillan (McMillan, 1968):

$$T_c = \frac{\Theta_D}{1.45} e^{-\frac{1.04[1+\lambda]}{\lambda - \mu^*[1+0.62\lambda]}} \quad \text{(B1.9)}$$

where Θ_D is the Debye temperature, μ^* is the effective Coulomb repulsion parameter, and λ the electron–phonon coupling constant. Later, Allen and Dynes (Allen and Dynes, 1975) refined the McMillan formula by substituting the phenomenological quantity $\Theta_D/1.45$ with the quantity $\omega_{ln}/1.2$, where the logarithmically averaged characteristic frequency ω_{ln} can be calculated by means of *ab-initio* methods:

$$\omega_{ln} = e^{\left\{\frac{2}{\lambda}\int_0^\infty \frac{d\omega}{\omega}\alpha^2(\omega)F(\omega)\ln(\omega)\right\}} \quad \text{(B1.10)}$$

$$\lambda = 2\int \frac{d\omega}{\omega}\alpha^2(\omega)F(\omega) \quad \text{(B1.11)}$$

The function $\alpha^2(\omega)F(\omega)$ is called Eliashberg's spectral function. In the Allen–Dynes formulation, the only free parameter is μ^*, which in general is obtained by a fitting of a given spectral function to the experimental value for T_c. In the vast majority of cases, μ^* varies in a range comprised between 0.1 and 0.25. An estimate for μ^* can be obtained with the empirical relation proposed by Bennemann and Garland (Bennemann and Garland, 1971):

$$\mu^* = \frac{0.26N(E_F)}{1+N(E_F)}. \quad \text{(B1.12)}$$

By considering $N(E_F) = 1.46$ states/eV^{-1} for niobium, the effective Coulomb repulsion is found to be $\mu^* \approx 0.15$.

Unfortunately, the estimation of T_c by means of Equation (B1.9) lacks of accuracy, because of the uncertainty in the Coulomb pseudopotential μ^*, the electron–phonon coupling parameter λ, and the Debye temperature Θ_D. In fact, for μ^*, λ, and Θ_D the experimental and theoretical values are widely spread in a range comprised between ≈0.1-0.15 (μ^*), ≈0.7-1.2 (λ), and 250-276 K (Θ_D). If we consider λ ~0.9 and $\mu^* \approx 0.15$, then T_c varies in the range 8.9 K to 9.8 K for Θ_D between 250 K and 276 K ($T_{c,exp} = 9.25$ K).

B1.2 Nb-Based Interstitial Compounds

A system extensively studied by many of the pioneers in the research of superconducting materials (Matthias et al., 1963) is that of the refractory nitrides and carbides of the transition metals. These are interstitial compounds, where the metal (Nb) exhibits a *bcc* or *hcp* structure, with the small metalloid element (nitrogen or carbon) occupying the interstices of the lattice. Nitrides, i.e. compounds of nitrogen with oxidation state-3, of groups 4, 5, 6 transition metals are all chemically stable, whereas the stoichiometric NbC compound is metastable at room temperature. The compounds NbN and NbC exhibit the NaCl B1-type crystal structure, and are characterized by the superconducting transition temperatures T_c(NbN) = 17.2 K; T_c(NbC) = 12 K. In 1953, Matthias (Matthias, 1953) was able to obtain the record T_c of 17.8 K in the ternary alloy $(NbN)_{0.75}(NbC)_{0.25}$, surpassed one year later by that of the intermetallic Nb_3Sn (Matthias et al., 1954).

The compound NbN, or the more complex similar compound Nb–Ti, has maintained some interest for radiofrequency applications, as in accelerating cavities, where the requirement is that of having a sufficiently high T_c, and of being a good metal in the normal state, thus exhibiting a low normal-state resistivity, ρ_N, in order to reduce the residual losses.

B1.3 Nb-Based Metallic Alloys

At the early stage of Nb-based superconductor developments, ductile solid solution alloys of the transition metals, as for example Nb–Zr, were identified as promising candidates. All the nearest neighbor, body-centered cubic (bcc), binary solid-solution alloys of niobium were investigated (Hulm and Blaugher, 1961). Curiously, the last of the series to be probed, Nb–Ti, represents now the most commercialized superconducting material. Although Nb–Ti market is essentially driven by magnets for magnetic resonance imaging (MRI), its main technological developments and the understanding of the underlying materials science, were mostly pushed by the accelerator community, striving to obtain the highest possible critical current capability in fine filament, stabilized wires (Lee and Strauss, 2012).

This chapter is organized as follows: after an overview of the physical, superconducting, and microstructural characteristic properties of the Nb–Ti alloy system, a discussion about flux pinning and transport current capabilities of the material is reported.

Extensive reviews of Nb–Ti alloy superconductors have been published in the past: Lee (1999), Kreilick (1990) and Lee et al. (1994). These works update the earlier reviews: Larbalestier (1981), Collings (1983) and McInturff (1980), whereas more recent issues are discussed in Wilson (2008). Additionally, the interested reader should also refer to Section E3.2.2, which contains an extremely comprehensive review of basic properties and processing techniques for Nb–Ti by some of the researchers who contributed most to its development and understanding. Finally, access to many pertinent documents can be gained over the internet at: https://nationalmaglab.org/magnet-development/applied-superconductivity-center.

B1.3.1 Physical and Superconducting Properties of Nb–Ti

Nb–Ti is one of the ductile transition metal alloys formed by group 4 elements and elements of either group 5 or group 6, exhibiting a body-centered cubic (*bcc*) structure. The formation of this β alloy takes place at high temperatures (> 882°C), where Ti undergoes an allotropic transformation from the α-phase, characterized by a hexagonal close-packed (*hcp*) structure, to a *bcc* one, thus favoring the solid solution processing with Nb.

The equilibrium phase diagram of the system, originally determined by M. Hansen (Hansen et al., 1951) by micrographic analysis, is reported and widely discussed in Section E3.2.2. Niobium lowers the α to β transformation temperature of Ti, so that the α+β region broadens with decreasing temperature. For composition values around the standard Nb-47wt%Ti, the β/α+β phase boundary lies at temperatures of about 550°C to 600°C.

The superconducting critical temperature, T_c, of the alloy shows a very weak dependence on the Ti concentration value, up to about 50wt%Ti, with a more remarkable decrease about this value: the T_c reaches a maximum of 9.6 K at about 30wt%Ti. The upper critical field, H_{c2}, exhibits a more pronounced dependence on the concentration: it increases with increasing Ti concentration, up to a maximum at 4.2 K of about 11.4 T, for composition values between Nb-44wt%Ti and Nb-50wt%Ti, above which H_{c2} decreases rapidly (Lee, 1999).

Figure B1.5 shows the critical field map for the ternary alloy of Nb, Ti, and Ta at 4.2 K, as reported by Suenaga and

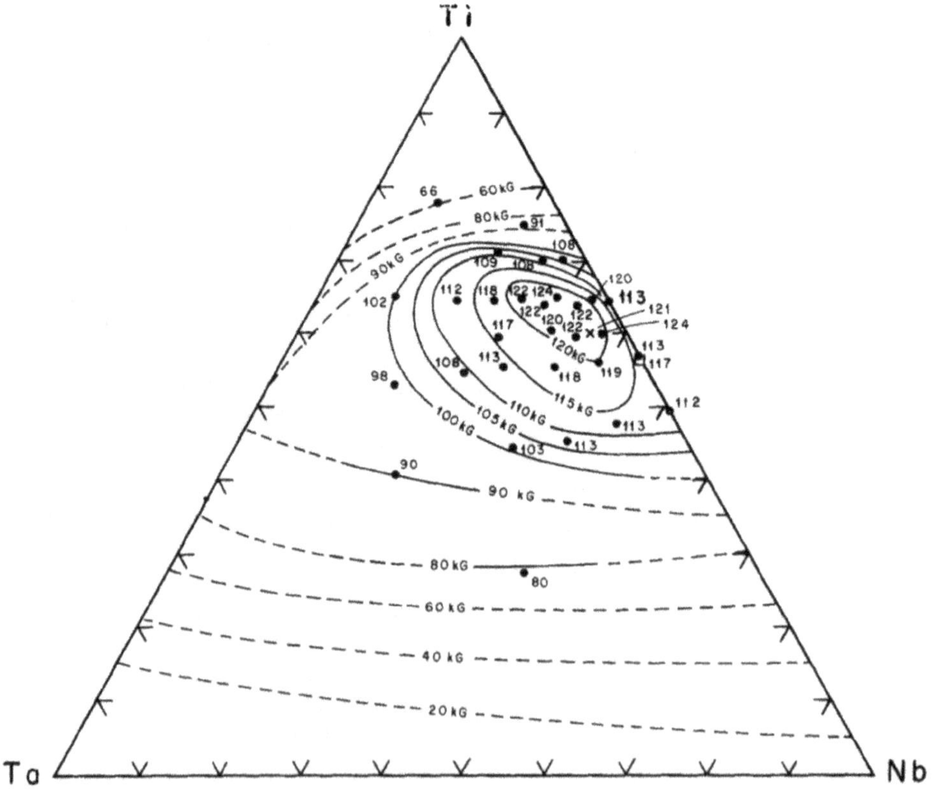

FIGURE B1.5 Critical field map at 4.2 K for the Nb–Ti–Ta alloy system. (From Suenaga and Ralls (1969), *with kind permission of The American Institute of Physics.*) Magnetic field values are in kG. The peak value reported here is 124 kG. The advantage of Ta alloying is primarily the suppression of the paramagnetic limitation, thus more effective for lower temperature (e.g. 2 K) operation (Hawksworth and Larbalestier, 1980).

Ralls (Suenaga and Ralls, 1969). A slight enhancement can be observed with the addition of Ta, which primarily suppresses the strong paramagnetism of Nb (Hawksworth and Larbalestier, 1980) (see Section E3.2.2.2.3). This represents a real advantage mostly for lower temperature (e.g. 2 K) operation.

Further details on the physical and superconducting properties of the Nb–Ti alloy can be found in Section E3.2.2.

B1.3.2 Nb–Ti Microstructural Features

The great success of Nb–Ti as a technological material lies in its highly optimized microstructure, which favors flux pinning, and thus in-field critical current capability. Nb–Ti microstructure is made of a complex, ribbon-like, nanometer-scale structure of α-Ti precipitates, which are created by a series of drawing and heat treatment steps (Lee, 1999). Figure B1.6 reports a transmission electron microscopy (TEM) cross-sectional view of an Nb–Ti optimized multifilamentary wire. The Ti domains exhibit the ribbon-like morphology, and the average thickness of the precipitate domains is of a few nanometers. A filament longitudinal cross-section reveals the morphology of the ribbons, extending three-dimensionally.

A very wide set of images and details are reported in Section E3.2.2.

B1.3.3 Critical Current and Flux Pinning in Nb–Ti

Critical current is not an intrinsic property of a superconducting material, but is a structure-sensitive property, strongly influenced by the metallurgical processing.

When an external magnetic field stronger than H_{C1}, the lower critical field up to which any superconductor is a perfect diamagnet ($\mu_r = -1$) is applied, magnetic flux can penetrate locally in the form of non-superconducting cylindrical

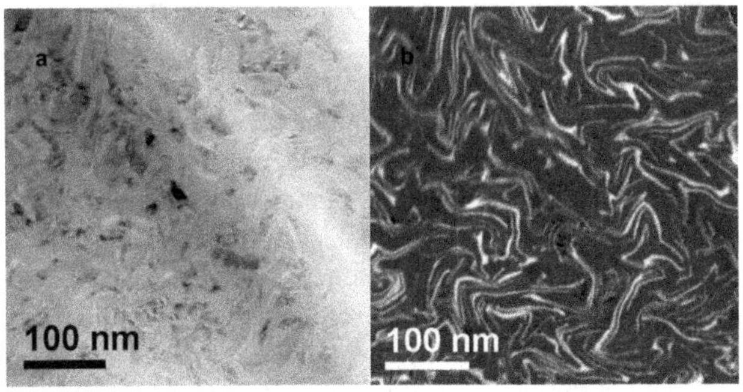

FIGURE B1.6 (left) Bright-field transmission electron microscopy (BF TEM) image of a conventionally processed Nb–Ti wire cross-sectional view. (right) Ti L2,3 ratio map from the same region in (left). Images and analyses by A. Falqui, G. Bertoni, R. Brescia (Istituto Italiano di Tecnologia – iit), with the support of H. Gnaegi (DiATOME) for ultramicrotomy.

'vortices' or fluxoids. A vortex is encircled by a superconducting paramagnetic current loop, and its normal-state core contains a quantum of magnetic flux $\Phi_0 = h/2e$, approximately 2×10^{15} Wb. With a current in a magnetic field, these vortices can move under the Lorentz force and have to be anchored, or 'pinned', to avoid a voltage and hence energy loss. Pinning centers can be crystal lattice defects, such as dislocations found in cold-worked materials, grain boundaries, and precipitates of a second phase; pinning centers can also be introduced artificially in order to target a specific pinning landscape. When the Lorentz force exceeds the pinning force, flux flow can occur and dissipate energy, and that effectively defines $J_c(B)$. Interaction between the vortex currents surrounding each core can immobilize the whole flux line population ('flux-line lattice'), provided the pinning force, F_p, is equal and opposite to the Lorentz force (plus thermal agitation) per unit volume, which is the vector product $\boldsymbol{J} \times \boldsymbol{B}$.

In Nb–Ti the main source of pinning are the nanometer-scale α-Ti precipitates, created during the processing through drawing and heat treatment steps (Lee, 1999). As a matter of fact, a linear increase in critical current density was in fact clearly observed with volume fraction of precipitates (Lee, 1999). In particular for this material, pinning is highly optimized (Lee and Strauss, 2012), since the structure of the normal precipitates can be tuned with specific drawing and heat treatment steps to match the fluxoid spacing corresponding to the magnetic field of interest. In addition, the highest critical current values are obtained when the area reduction at the cold drawing step applied before precipitation heat treatments, is greater than 90%, thereby indicating that the dislocation networks act as nucleation sites for the α-phase. Somerkoski et al. (2012) studied isothermal annealing treatments of different duration applied on heavily cold drawn samples. The annealing time does not affect grain boundary spacing, whereas grain boundary widths increase as a result of the precipitation process. In addition, the bulk pinning force, and thus the critical current, improves with the final drawing strain applied after the last heat treatment step. This in fact causes a reduction in the precipitate size and spacing, down to less than a coherence length in thickness. A qualitatively similar behavior of pinning force was obtained by applying annealing treatments to an optimized, multifilamentary wire. As Figure B1.7 shows, with increasing annealing temperature, corresponding to an increase in precipitate size and separation, and a corresponding decrease in n_p, the pinning-force peak decreases and shifts to lower field values, though the upper critical field is maintained unvaried.

An enormous theoretical as well as experimental work was performed up to the mid 90's, leading to remarkable achievements in the optimization of the properties of Nb–Ti technological strands. At the time the Superconducting Super Collider (SSC) project was cancelled in 1993, highly optimized Nb–Ti multifilamentary wires had reached full maturity, up to the industrial level, and what is presently known as 'conventional process' was fully developed. Since then, as shown

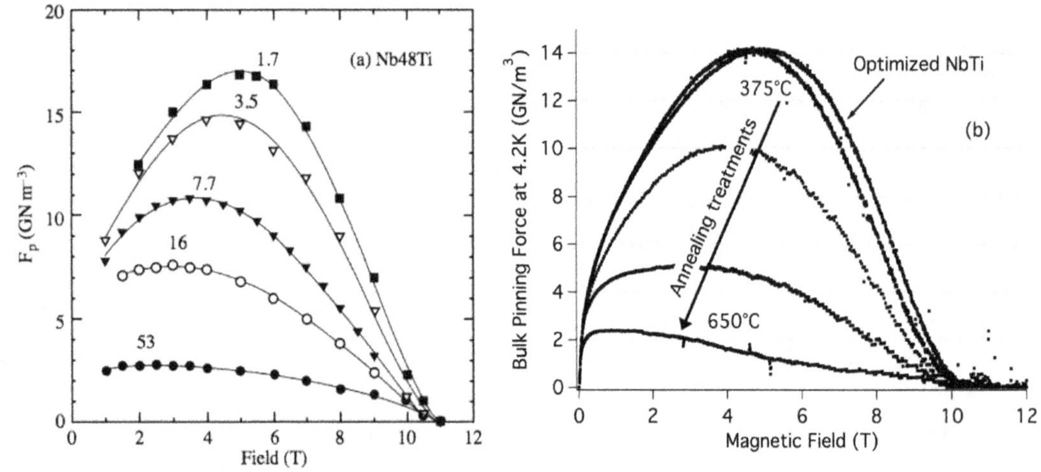

FIGURE B1.7 The change of the bulk pinning-force curve at 4.2 K with decreasing pinning-center thickness is illustrated in these plots. In (a), bulk pinning-force curves are presented for a conventional Nb48Ti composite with α-Ti precipitates (Meingast and Larbalestier, 1989); in (b), similar curves are shown, obtained by applying annealing treatments on an optimized Nb–Ti alloy. In (a), the curves are labelled by the value of the pinning-center thickness (nm), which decreases together with the average pinning-center separation and the overall filament diameter. [Figure B1.7(a) is reproduced from Handbook, 1st edition, p. 674 (Cardwell and Ginley, 2003).]

in Figure B1.8, the critical current capabilities achieved reproducibly and over long lengths in large industrial productions for big accelerator (as for the LHC at CERN) or nuclear fusion (as for ITER) projects have not changed. This reflects the fact that all material characteristics and the pinning landscape controlling its transport properties (Cooley et al., 1996; Lee and Larbalestier, 1987) are well developed and understood.

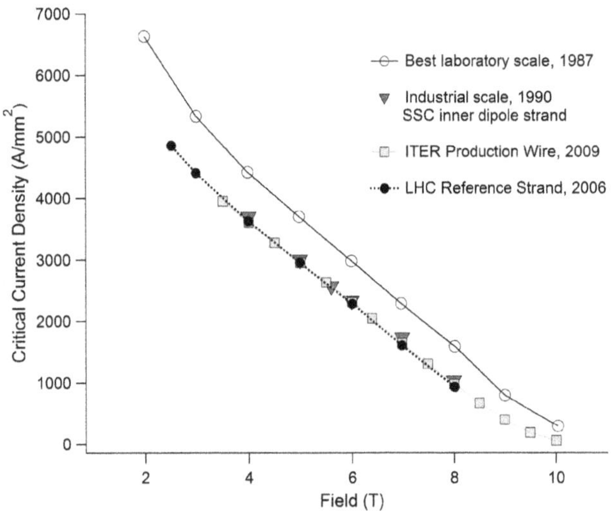

FIGURE B1.8 The critical current density as a function of field for various Nb–Ti wire composites produced with stable properties and over long lengths for large projects. The magnet program for the Superconducting Supercollider (SSC) project (cancelled in 1993), had already brought the wire properties to full maturity: the later Large Hadron Collider (LHC) and International Thermonuclear Experimental Reactor (ITER) production wires show in fact the same behavior. The performance for a laboratory scale wire is also shown for comparison. [Figure readapted from Cardwell and Ginley (2003), p. 604, based on data from Lee and Larbalestier (1983), Meingast and Larbalestier (1989), Muzzi et al. (2011) and Boutboul et al. (2006).]

However, differently from high energy physics requirements, in fusion coils, the nominal operating conditions for Nb–Ti wires are at relatively high temperature and magnetic field: an operating temperature around 6 K or 6.5 K should be considered, as the sum of the operating temperature, typically 4.5 K in forced flow cable-in-conduit conductors, and the required temperature margin for safe magnet operation, with magnetic fields up to about 6 T. In this operating range, not a lot of measured data are available for the critical current of Nb–Ti, and the extrapolation from the standard conditions of, for example, 5 T–4.2 K, is not straightforward. At the same time, it is possible that the optimization of manufacturing processes and thermomechanical production methods of Nb–Ti could be tailored on wire performance in the specific range of interest.

From an engineering perspective, critical current datasets of Nb–Ti optimized wires cannot be accurately fitted over a wide magnetic field and temperature range, using a parameterization that is based on the assumption of a single pinning mechanism (Muzzi et al., 2009). Especially in the low critical current density range of operation for fusion coils, this could lead to appreciable errors in the estimation of magnet performance from strand data (Pong et al., SUST 2012).

Recently, the shape of the pinning-force curves for optimized multifilamentary wires was studied, in relation to the possible contribution of two concurring pinning mechanisms. It was observed that an accurate description of critical current data over a wide temperature and magnetic field range could only be obtained, assuming the concurrence of two elementary pinning mechanisms (Muzzi et al., 2011) (see Figure B1.9), and a full summation hypothesis, so that the total pinning force can be expressed as:

$$F_p = \sum_{i=1}^{2} F_p^{(i)} = \sum_{i=1}^{2} C_i \cdot \left(1-t^n\right)^{\gamma_i} \cdot b^{\alpha_i} \cdot \left(1-b\right)^{\beta_i} \qquad \text{(B1.13)}$$

$$\text{with } b = \frac{B}{B_{irr}(T)} = \frac{B}{B_{irr}(0)\left(1-t^n\right)}; t = \frac{T}{T_{c0}} \qquad \text{(B1.14)}$$

Data on numerous Nb–Ti wires intended for nuclear fusion magnets have been analyzed within this frame (Muzzi et al., 2011), and typical values for the α_i, β_i exponents describing the field dependency of pinning force were: $\alpha_1 = 2.5–3$; $\beta_1 = 2–2.4$, $\alpha_2 = 0.5–0.6$; $\beta_2 = 2$. An example is shown in Figure B1.10: the experimental pinning-force maxima lie at a reduced field b ≅ 0.45 ÷ 0.5, gradually shifting to lower values with increasing temperature. According to the proposed model, two types of pinning centers are mainly operating, or two different pinning mechanisms. Their relative contribution to the total pinning force changes with temperature, so that the relative maxima, at b ≅ 0.25 and b ≅ 0.6, do not shift with temperature, whereas the weight of the low-field pinning component becomes gradually dominant with increasing temperature, with respect to the high-field one. Figure B1.11 shows the

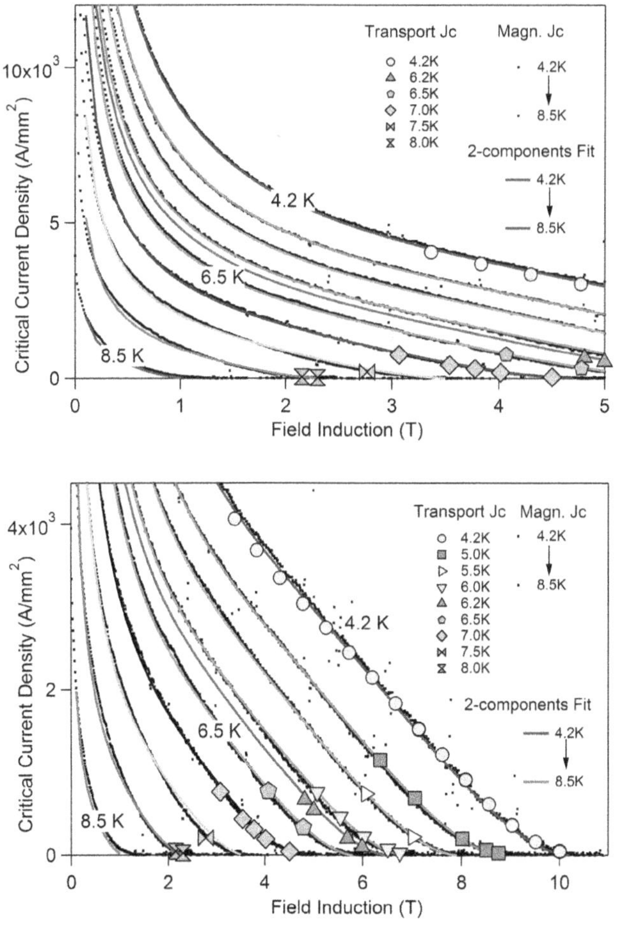

FIGURE B1.9 Superconducting critical current density measured on an optimized multifilamentary Nb–Ti strand for fusion at various temperatures, between 4.2 K and 8.5 K. Symbols report transport J_c data, whereas small dots represent critical current density extracted from magnetization measurements. Thick lines represent fitted data, according to the two-pinning-components model. Upper panel reports data in the magnetic field range 0 T to 5 T and in the whole critical current density measured range, whereas the lower panel reports data up to 11 T, in a lower critical current density range.

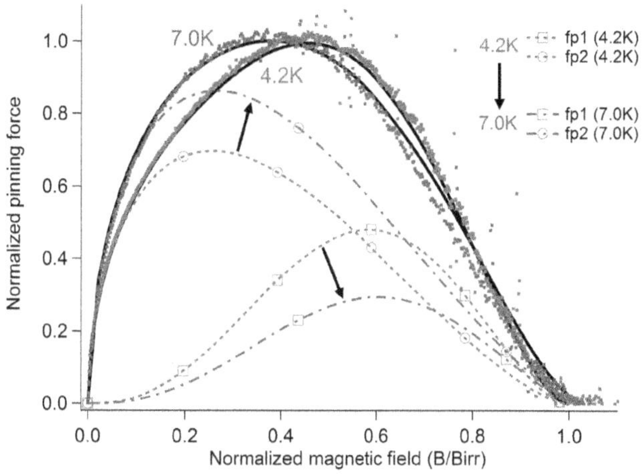

FIGURE B1.10 Reduced pinning-force curves measured on an ITER Nb–Ti wire at 4.2 K and 7 K. Experimental results (small symbols) are compared to fit results (continuous lines); the shape of the two single pinning components, the sum of which provides the overall pinning curve, is evidenced (dotted and dash-dotted lines are for pinning components at 4.2 K and 7 K, respectively); round and square sparse symbols are for the low-field (fp2) and high-field (fp1) pinning components, respectively.

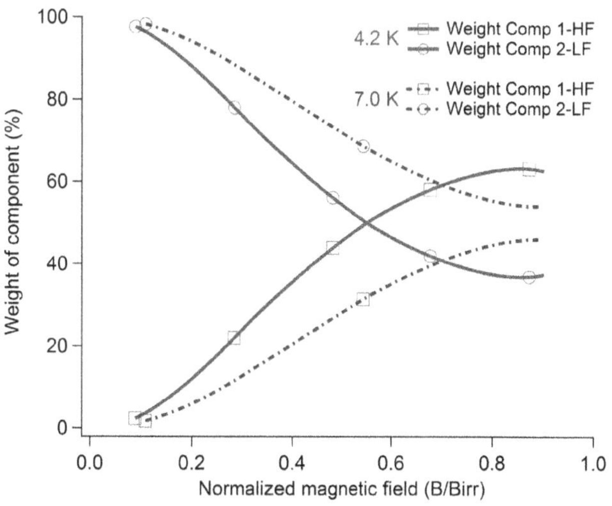

FIGURE B1.11 The relative weight of the two pinning components, acting at low magnetic field (with a pinning-force peak at b = B/B_{c2} ≅ 0.25) and high magnetic field (with a peak at b ≅ 0.60). Continuous and dash-dotted lines refer to the reduced pinning-force curves at 4.2 K and 7 K, respectively; round and square sparse symbols are for the low-field (fp2) and high-field (fp1) pinning components, respectively.

relative contribution of the two reduced pinning-force components, peaking at respectively low- (LF) and high- (HF) reduced magnetic fields, at two different temperatures, 4.2 K and 7 K. The plot evidences the different behavior at the two different temperatures: close to the critical temperature of the material, the low-field pinning component is predominant at all fields, whereas at 4.2 K, the relative weight of the two pinning component changes with field.

The possibility that two mechanisms, of core (δH_c) and $\delta\kappa$ types, respectively, could play a major role was already investigated by Meingast and Larbalestier (1989), who also observed the lack of temperature scaling for the pinning curves. It was noted that, in fact, the α-Ti precipitates cause both fluctuations in Ginzburg–Landau constant (κ) and critical field (H_c). A slightly different perspective can be offered by speculating that, beside core pinning at domain walls caused by the normal (α-Ti) precipitates, particularly effective at relative low reduced magnetic fields, also the Ti concentration gradients around them may be a source of a second pinning mechanism (Somerkoski et al., 2012; Bormio-Nunes et al., 2007), acting in a higher reduced fields range.

Applying annealing treatments at gradually higher temperatures to optimized wires, a shift of the pinning curves to lower fields is observed, as already reported in Figure B1.7. TEM analyses on the same samples revealed that the annealing treatment not only caused the α-Ti precipitates to be coarser and less dense, but also a smoothening in the distribution of the Ti concentration.

Unfortunately, the investigation of such phenomena and the possible explanation of its physical nature has not been developed further. For a long time no application field or project has in fact kept pulling further developments of such ideas into more systematic studies. In fact, in the fields of nuclear fusion and most likely MRI, the currently available performance already fulfills the typical requirements, and future accelerator projects are not pulling this type of research any longer, because they will rely on Nb_3Sn as a material for higher-field magnets.

B1.4 Nb-Based Intermetallic Compounds

Among all the elements, alloys and compounds belonging to the so-called 'conventional' superconductors, the A15 intermetallic compounds have long maintained the record for the highest critical temperatures, T_c (around ~20 K), also showing type II superconductivity with upper critical fields as high as tens of tesla at zero temperature. The term 'A15' came from the Struckturbericht (Ewald, 1977) naming system which is no longer universally used, although 'A15' itself is still often used in the applied superconductivity community. With chemical formula A_3B (where A is a transition metal and B can be any element), the A15 phase is commonly referred as Cr_3Si, but also sometimes as β–tungsten or β-W structure. The latter name refers to the crystal structure of an allotrope of tungsten discovered in 1931 by Hartmann and Fink by electrochemical deposition from the melt of tungstates. Chromium silicide, Cr_3Si, was the first intermetallic compound discovered. Thereafter, several other compounds belonging to the same A15 family were discovered. The first A15 which manifested superconducting properties (T_c about 17 K, see Tables B1.5 and B1.6) was the vanadium silicide, V_3Si; discovered in 1953, V_3Si immediately attracted considerable interest because of its potential superconducting applications, which consequently opened the race for the synthesis of novel A15 compounds. In fact, in the following years, Bernd Matthias teamed up with Ernie Corenzwit, Ted Geballe, and Seymour Geller and successfully synthesized about 30 new A15 compounds, including the today's workhorse triniobium-tin, Nb_3Sn, in 1954 (Matthias et al., 1954). After the discovery of superconducting Nb_3Sn, higher T_c were found in Nb_3Al [T_c = 18.8 K (Willens et al., 1969)], in $Nb_3Al_{0.8}Ge_{0.2}$ [T_c = 20.0 K (Matthias et al., 1967)], in Nb_3Ga [T_c = 20.3 K (Webb et al., 1971)], and in Nb_3Ge sputtered films [T_c = 23.2 K (Gavaler, 1973), and optimized to 23.2 K 1 year later]. The high-performances triniobium–aluminum, Nb_3Al, is currently being considered as a promising candidate for high-field magnets, although 10-km-grade wires are still not available for large-scale applications. Triniobium–germanium (Nb_3Ge) sputtered films, although rarely used for superconducting applications, held the record for the highest temperature of 23.2 K, until the discovery of the cuprate superconductors in 1986 by Bednorz and Müller.

B1.4.1 The A15 Family Compounds

B1.4.1.1 The A15 Crystal Structure

The prototypical simple cubic A15 structure (space group no. 223: $Pm\bar{3}n$, O_h^3), with the ideal formula unit A_3B, is depicted in Figure B1.12. The six A sites (Wyckhoff position 6c: ¼, 0, ½) are occupied by a transition element, with the exclusion of hafnium, whereas the two B sites (Wyckhoff position 2a: 0, 0, 0) are occupied either by a transition element or by an element belonging to row III, IV, or V of the periodic table. The critical temperatures of several A15 compounds are reported in Tables B1.5 and B1.6, for non-transition elements (Table B1.5) and transition elements (Table B1.6) at the B site. The A15 structure is characterized by three mutually orthogonal chains made of transition metal ions A, each chain bisecting a face of the cubic lattice. Compared to the pure A crystal, the interatomic spacing between the atoms of the chain is reduced; for example, in the Nb_3Sn structure, the distance between the Nb atoms is reduced to 265 pm compared to the distance of the elemental structure (286 pm); this is understood to cause a narrow peak in the d-band density of states near the Fermi level, raising the critical temperature from ~9.2 K in Nb to 18.3 K in Nb_3Sn.

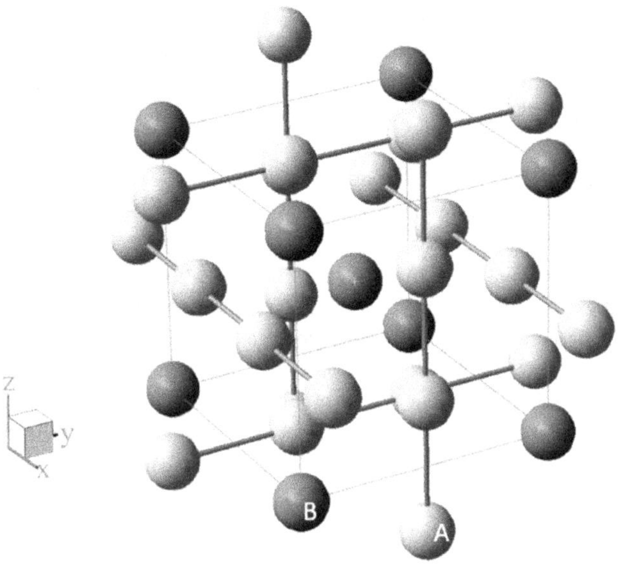

FIGURE B1.12 The ideal A15 crystal structure: the *B*-atoms form a body-centered cube, whereas the *A*-atoms form one-dimensional chains in the three orthogonal directions, with an interatomic spacing along the chains of ½ of the lattice parameter.

As a general occurrence, the *A*-chains generate a narrow *d*-band near the Fermi surface (Weger, 1964), thus resulting in a high density of states at the Fermi level, $N(E_F)$. The critical temperature is related to $N(E_F)$ through the well-known BCS expression:

$$k_B T_c \approx \hbar\omega \exp\left(-\frac{1}{N(E_F)V_0}\right), \quad \text{(B1.15)}$$

where $\hbar\omega$ is a characteristic phonon energy, and V_0 is the attractive potential between electrons (mediated by the cloud of virtual phonons); hence, an increase in $N(E_F)$ can result in an enhanced value of T_c. For most superconductors, there is a correlation between $N(E_F) = 3/2\gamma/(\pi k_B^2)$ and T_c, as shown in Section E3.2.1, Figure E3.2.1.4.

B1.4.1.2 The Dependence of T_c upon Stoichiometric Ratio in A15 Compounds

There is wide experimental evidence that in the A15 compounds the critical temperature is sensitive to stoichiometry (with few exceptions); the T_c values reported in Tables B1.5 and B1.6 are not always referred to the canonical ratio 3:1, but to the highest obtained values. Earlier experiments by Matthias et al. indicated that the '*maximum transition temperature of compounds with the β-W type structure (A15) generally occurs at the stoichiometric composition*' (Matthias et al., 1967). The A15 superconducting phase typically extends in a narrow range centered at 25% composition % *B* where T_c reaches its maximum,

TABLE B1.5 Superconducting Critical Temperatures, T_c, of A_3B Compounds, with B = Non-Transition Elements

A_3B		Ti	Zr	V	Nb	Ta	Cr	Mo
Row III	Al			11.8	18.8			0.58
	Ga			16.8	20.3			0.76
	In			13.9	9.2[b]			
	Tl				9			
Row IV	Si			17.1	19			1.7
	Ge			11.2[c]	23.2[c]	8.0[c]	1.2	1.8
	Sn	5.8	0.92[a]	7.0	18.0	8.4		
	Pb		0.76		8.0	17		
Row V	P							
	As			0.2				
	Sb	5.8		0.8	2.2	0.7		
	Bi		3.4[b]		4.5			

Source: Data from Poole et al. (1995) and Palmieri (2001) and references therein.

[a] Rapid quenching.
[b] High-pressure synthesis.
[c] Film deposition technique.

TABLE B1.6 Superconducting Critical Temperatures, T_c, of A_3B Compounds, with B = Transition Elements

A_3B	Ti	Zr	V	Nb	Ta	Cr	Mo	W
Tc							13.4	
Re			8.4				~15	11[c]
Ru						3.4	10.6	
Os			5.7	1.1		4.7	12.7	
Rh			~1	2.6	10.0	0.3		
Ir	5.4		1.7	3.2	6.6	0.75	9.6	
Pd			0.08					
Pt	0.5		3.7	10.9	0.4		8.8	
Au		0.9	3.2	11.5	16.0			

Source: Data from Poole et al. (1995) and Palmieri (2001) and references therein.

with a sharp fall in T_c away from the optimum 3:1 stoichiometry. Consequently, the earlier efforts were geared to the achievement of the canonical stoichiometric ratio, which was often hampered by difficulties in obtaining stable A15 structure in bulk specimens (especially in Nb_3Ga, Nb_3Ge, and Nb_3Si).

The A15 phase field of two compounds, shown in Section E3.2.1, Figure E3.2.1.5, illustrates that the composition 3:1 is only stable at high temperatures. In other compounds, e.g. Nb_3Ge, Nb_3Si, or V_3Al, the stoichiometric composition can only be obtained by thin-film deposition, i.e. by non-equilibrium methods. Since the beginning, a number of workers tried to correlate the critical temperature of different A15 compounds with the lattice parameter a_0, finding a monotonic rise of T_c with decreasing a_0. However, the increase in T_c is correlated to the variation in composition rather than to the smaller unit cell. In fact, the experimental evidence that T_c is depressed under pressure (Webb et al., 1971), demonstrate that the highest

T_c values ought to be ascribed to the closer approach to 3:1 stoichiometry rather than to the decrease in the interatomic spacing, the smaller a_0 just reflecting the changes in stoichiometry.

The decrease of T_c with variation of stoichiometry from the ideal 3:1 ratio has been found in Nb_3Ge, Nb_3Ga, V_3Sn, V_3Ge, V_3Ga, and Nb_3Sn. In contrast, there are A15 compounds, such as Cr_3Os, Mo_3Ir, Mo_3Pt, and V_3Ir (Poole et al., 2007), showing their highest T_c far from the ideal stoichiometric ratio, e.g. $A_{1-\beta}B_\beta$, with $\beta \neq 0.25$. For example, the V–Re and Mo–Tc systems reach their highest critical temperatures for $\beta = 0.71$ [$T_c = 8.4$ K (Giorgi et al., 1978)] and $\beta = 0.6$ [($T_c = 13.4$ K (Giorgi and Matthias, 1978)], respectively. In these compounds, where $\beta > 0.25$ and the *B*-atoms occupy the *A*-sites, the *A*-chains are broken, and consequently the critical temperature is decreased. In this sense, all deviations from the canonical A15 stoichiometry, which tend to recover the integrity of the *A*-chains can return the highest T_c values. On the contrary, for $\beta < 0.25$, with the *A*-atoms in the *B*-sites, the *A*-chains tend to remain intact, but disorder is introduced in the *bcc* lattice formed by the *B*-atoms. A noticeable example is Nb_3Nb, stabilized by the addition of a small amount of Ge at the *B*-sites [less than 1%, see Stewart et al. (1980)].

Many theoretical works (Weger, 1964; Dew-Hughes, 1975; Clogston and Jaccarino, 1961; Labbé and van Reuth, 1970) explained the effects of order and composition by considering the electronic structure of the A15 superconductors. As stated before, the *A*-chains lead to a narrow peak in *d*-band near the Fermi surface (Weger, 1964), thus the high T_c observed in these materials result from the very high $N(E_F)$ values. Since the peak at the *d*-band originate from the *A*-chains, any deterioration of the integrity of the *A*-chains, for example a chain breakage caused by the presence of vacancies or substitutional *B*-atoms on *A*-sites, would result in a broadening of the *d*-band, with a dramatic decrease of $N(E_F)$ and hence T_c. More generally, a deterioration of the *A*-chains can be induced by deviations from the canonical stoichiometric ratio in the A_3B structure, 3:1. In A15 compounds, this usually implies an excess of either *A* or *B*-atoms. The excess of *B*-atoms affects the integrity of the *A*-chains through substitution of *B*-ions in the *A*-sites, whereas an excess of *A*-atoms create disorder in the body-centered cubic (*bcc*) lattice formed by the *B*-atoms. In either case, a rounding-off of the *d*-band peak is expected and the consequent reduction of T_c, although the disorder in the *bcc* lattice has generally a smaller effect compared to that caused by the adulteration of the *A*-chains. Such interpretation also explains why compounds in which the *B* element is a transition metal generally show reduced T_c values: the *d*-band originating from the *B* elements in the *bcc* structure competes for electrons with the *d*-band from the A-chains, thus causing a rounding-off of the total *d*-band and a reduction of $N(E_F)$.

B1.4.1.3 The Batterman–Barrett Instability in A15

A reversible and spontaneous low-temperature structural transformation was firstly observed in V_3Si by Batterman and Barrett in 1964 (Batterman and Barrett, 1964). Below $T_m \sim 21$ K, a structural and volume-conserving transformation occurs from a cubic to a slightly tetragonal phase, in which $|c - a|/a \approx 0.2–0.3\%$. Three years later, Mailfert (Mailfert et al., 1967; Mailfert et al., 1969) found a low-temperature tetragonal phase below $T_m \sim 43$ K in Nb_3Sn, with $|c - a|/a \approx -0.6\%$. The temperature dependence of the lattice parameter, for both V_3Si and Nb_3Sn (Testardi, 1975), is reported in Figure B1.13. In the tetragonal phase, small displacements of the *A*-atoms cause the dimerization of the two *A*-chains orthogonal to the *c*-axis, as depicted in Figure B1.14. Since the transformation is diffusionless, it is generally referred to as martensitic, although this terminology can be confusing because the transformation does not resemble the conventional martensitic transformation found, for example, in stainless steels. The temperature T_m represents the martensitic or Batterman–Barrett transformation temperature and it has been always found to be

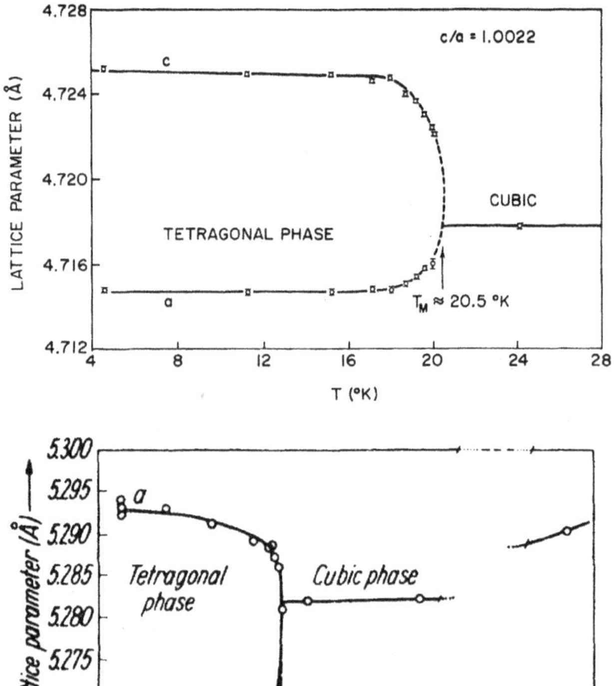

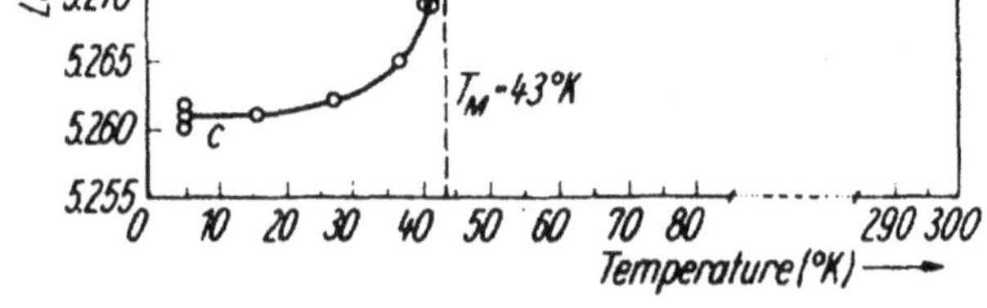

FIGURE B1.13 Temperature dependence of the lattice parameter of V_3Si ($c/a > 1$) and Nb_3Sn ($c/a < 1$) (Testardi, 1975), showing the cubic to tetragonal structural transition in V_3Si ($T_m \sim 21$ K) (Batterman and Barrett, 1964) and in Nb_3Sn ($T_m \sim 43$ K) (Mailfert et al., 1967). The volume of the cell remains constant. In contrast with V_3Si, in Nb_3Sn c/a becomes less than 1 below the martensitic transformation. (Figure reproduced from Testardi (1975), *with kind permission of The American Physical Society.*)

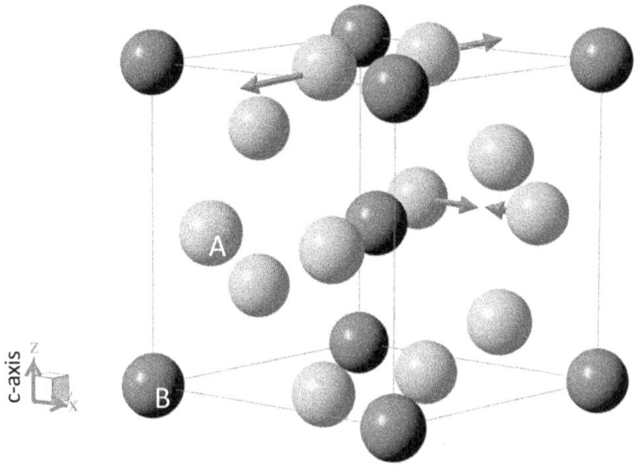

FIGURE B1.14 In the A15 structure, the B-atoms occupy the *bcc* sites, whereas the two A-atoms are located onto each face of the cube. The ideal cubic structure undergoes a diffusionless transformation to a tetragonal phase, in which the volume is conserved and $c \neq a$, with the concomitant sublattice distortions of the A-chains along the direction orthogonal to the *c*-axis; the arrows denote the displacement direction.

higher than T_c. Excellent reviews on the structural cubic-to-tetragonal transition in A15 compounds can be found in literature (Testardi, 1975; Weger and Goldberg, 1973; Allen, 1980). This structural instability has been found in highly ordered stoichiometric samples (*transforming* samples). In disordered samples, with excess of either A-atoms and/or ternary additions, the structural instability is suppressed and the unit cell remains cubic (*non-transforming* samples). The Batterman-Barrett instability has been claimed in a variety of A15 compounds, but there is unambiguous experimental evidence (anomalies in the specific heat, x-ray, and neutron scattering data) only for two binary compounds: Nb_3Sn and V_3Si. The anisotropy ratio and T_m values are reported in Table B1.7 for selected A15 superconductors (Poole et al., 2007; Vonsovsky et al., 1982). In particular, it can be noticed that the sign of the anisotropy ratio is different in these two A15 materials: such a difference is also manifested in the pressure dependences of their T_m and T_c: for Nb_3Sn, $dT_m/dP > 0$ and $dT_c/dP < 0$ (Chu, 1974), whereas for V_3Si, the behavior is just the opposite (Chu and Testardi, 1974).

TABLE B1.7 The Batterman–Barrett Transformation Temperature T_m and Anisotropy $(c - a)/a$ in the Low-Temperature Tetragonal Phase of Several A15-Type Superconductors (Poole et al., 2007; Vonsovsky et al., 1982)

Compound	T_m [K]	T_c [K]	Anisotropy$(c - a)/a$
V_3Si	21	17	0.24%
Nb_3Sn	43	18	−0.61%
V_3Ga	>50	14.5	-
Nb_3Al	80	17.9	-
$Nb_3(Al_{0.75}Ge_{0.25})$	105	18.5	−0.3%
$Nb_3(Al_{0.7}Ge_{0.3})$	130	17.4	-

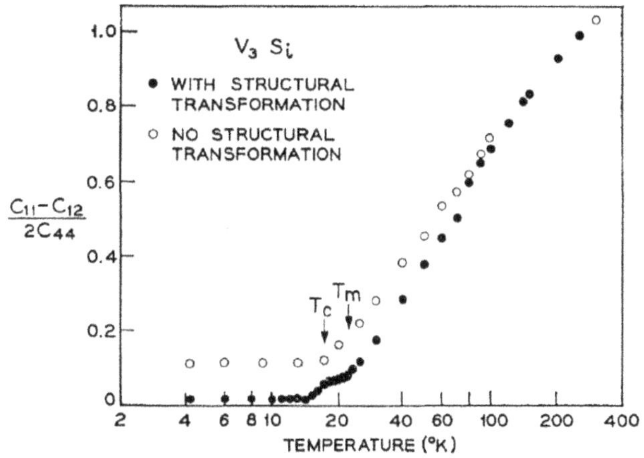

FIGURE B1.15 Temperature dependence of $(C_{11} - C_{12})/2C_{44}$ in V_3Si, from Testardi and Bateman (1967). (Figure reproduced from Testardi (1975), *with kind permission of The American Physical Society.*)

The structural instability has been correlated to the softening of the elastic modulus $(C_{11}-C_{12})/2$ for [110] phonons with $[1\bar{1}0]$ polarization which, in the superconducting state, is reduced to near-zero magnitude (Testardi and Bateman, 1967). A continuous decrease in C_{11} ([001] phonons with [001] polarization) and C_{44} ([110] phonons with [001] polarization) was also observed, although to a lesser extent with respect to $(C_{11} - C_{12})/2$. In Figure B1.15, the temperature dependences of C_{11} and C_{44} in V_3Si show reductions of 36% and 4%, respectively; a dramatic reduction of the elastic modulus $(C_{11} - C_{12})/2$ as a function of temperature is depicted in Figure B1.16, where

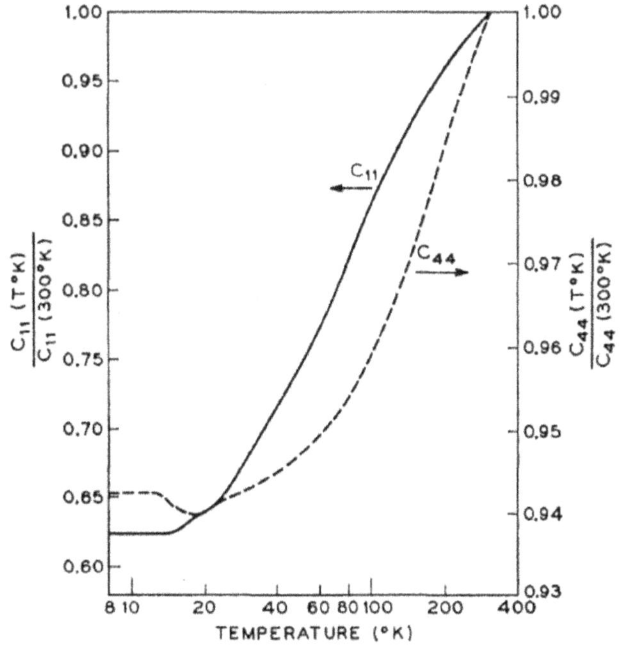

FIGURE B1.16 Temperature dependence of the elastic moduli C_{11} and C_{44} in V_3Si. (Figure reproduced from Testardi and Bateman (1967), *with kind permission of The American Physical Society.*)

TABLE B1.8 Physical Parameters for a Series of A15-Type Superconductors

	μ_0H_{c2} [T]	γ [mJ mol K^{-2}]	$\Theta_D(T\to0/T>T_c)$ (K)	λ	$\Delta C/\gamma T_c$	$2\Delta/k_BT_c$	$N(E_F)$ exp./theory (states/eV atom)
Nb_3Ge	38	30±1 34±1	302±3	1.7±0.2	1.9	4.2	1.2±0.1 / 1.9 1.5±0.1 / 1.8
Nb_3Ga	38	46±8	280/262	1.7±0.2			1.8±0.4/1.8
Nb_3Al	34	36±2	283±5	1.7±0.2	2.1		1.4±0.2/1.8
Nb_3Sn	32	35±3	208/270	1.7±0.2 1.8 1.6±0.1		4.2–4.4	1.4±0.2/1.5
Nb_3Si		24±6	310±40	1.7±0.2			0.95±0.3/0.6
V_3Si	24	53	291-324	1.29±0.2 0.96	2.0	3.5±0.2	2.4±~0.3/1.8
$Nb_3Al_{0.8}Ge_{0.2}$	43	35±2	278±5	1.7±0.2			1.4±0.2

Source: The upper critical fields are taken from Flükiger (2003), the remaining parameters from Stewart (2015) and references therein.

the shear instability leads to the Batterman–Barrett transition around 21 K. Thus, the cubic-to-tetragonal transition can be viewed as a *soft-mode* phase transition driven by an acoustic instability in the long wavelength limit. For details about the link between the Batterman–Barrett structural instability and superconductivity in A15 compounds, the reader can refer to the review by Testardi (Testardi, 1975).

From the theoretical point of view, several works were devoted to the comprehension of such instability. The structural transition may be ascribed to a Jahn–Teller distortion (Jahn and Teller, 1937) with suppression of degeneracy of the *d*-bands (Labbé and Friedel, 1966). As the degeneracy is removed, the crystal rearranges itself in a lower symmetry structure, which is energetically more favorable than the ideal cubic structure. In a similar way, Madar *et al.* (Madar et al., 1979) propose that in transforming A_3B samples, with a composition richer in *B*-atoms than stoichiometric samples, there is a large number of antisite-type point defects, which eventually lead to the formation of local strain resulting from the repulsive interaction between the *B*-atoms. At low temperature, a new equilibrium is reached as soon as the local strain becomes strong enough to trigger the lattice distortion.

To conclude this paragraph, in Table B1.8 we report a list of the physical parameters for a series of A15-type superconductors using the data of Flükiger (2003) and Stewart (2015): the upper critical fields, μ_0H_{c2}, the Sommerfeld constants, γ, the Debye temperature, Θ_D, the electron–phonon (*el–ph*) coupling constants, λ, the BCS superconducting gap, Δ, and the density of states at the Fermi level, $N(E_F)$.

In most conventional superconductors there is a correlation between the *el–ph* interaction and the critical temperature, being:

$$k_BT_c \approx \hbar\omega_D \exp\left(-\frac{1}{\lambda-\mu^*}\right). \quad \text{(B1.16)}$$

Here, μ^* is the effective Coulomb repulsion, ω_D the Debye frequency, and λ is the *el–ph* coupling constant which, in terms of the BCS theory, corresponds to $\lambda - \mu^* = N(E_F)|V_0|$. At zero temperature, the ground state is separated from the excited states by the energy gap $2\Delta(T{=}0) = 2\Delta_0$, and the value of T_c can be determined by the BCS expression $k_BT_c = \alpha\Delta_0$. The A15 compounds are characterized by weak-coupled, BCS-like superconductivity, such as V_3Si ($2\Delta_0/k_BT_c$ =3.5), as well as strong-coupled compounds, such as Nb_3Sn ($2\Delta_0/k_BT_c$ = 4.2-4.4) and Nb_3Ge ($2\Delta_0/k_BT_c$ = 4.2).

B1.4.2 The Binary $Nb_{1-\beta}Sn_\beta$ Phase

Intermetallic triniobium-tin, Nb_3Sn, constitutes the present workhorse for almost all superconducting high-field applications, thus it can be regarded as one of the key materials in superconducting science and technology. In this regard, it is worth mentioning the International Thermonuclear Experimental Reactor (ITER), whose toroidal field and central solenoid coils comprise Nb_3Sn multifilamentary wires (Vostner et al., 2017; Devred et al., 2014), as well as the high-luminosity (HiLumi) upgrade of the LHC machine, which will require a new generation of Nb_3Sn high-field superconducting quadrupoles and dipoles (Bottura et al., 2012). At present, Nb_3Sn is the only superconducting alternative to Nb–Ti for high-field applications (above 10 T) (De Marzi et al., 2016).

B1.4.2.1 The Nb_3Sn Binary Phase Diagram

This compound was discovered in 1954 by Matthias *et al.* (Matthias et al., 1954) and its highest reported critical temperature is T_c = 18.3 K (Hanak et al., 1964). Above T_m = 43 K the structure is cubic, with lattice parameter a = 5.293 Å at room temperature. A list of the main physical parameters of *transforming*, stoichiometric Nb_3Sn is reported in Table B1.9 (Godeke, 2006a). It is important to note that these parameters can be influenced by stoichiometry, mechanical strain, doping,

TABLE B1.9 Characteristic Properties of Nb_3Sn

Superconducting transition temperature	T_c	18.3	[K]
Lattice parameter at room temperature	a	5.293	[Å]
Martensitic transformation temperature	T_m	43	[K]
Tetragonal distortion at 10 K	$(c-a)/a$	−0.26%	
Mean atomic volume at 10 K	V_{Mol}	11.085	[cm^3 mol^{-1}]
Upper critical field at zero temperature	$\mu_0 H_{c2}$	32	[T]
Thermodynamic critical field	$\mu_0 H_c$	0.52	[T]
Lower critical field	$\mu_0 H_{c1}$	38	[mT]
Ginzburg–Landau coherence length	ξ	3.6	[nm]
Ginzburg–Landau penetration depth	λ	124	[nm]
Ginzburg–Landau parameter ξ/λ	κ	34	
Superconducting energy gap	Δ	3.4	[meV]
Electron–phonon interaction constant	λ_{ep}	1.8	

Source: Godeke (2006a) and references therein.

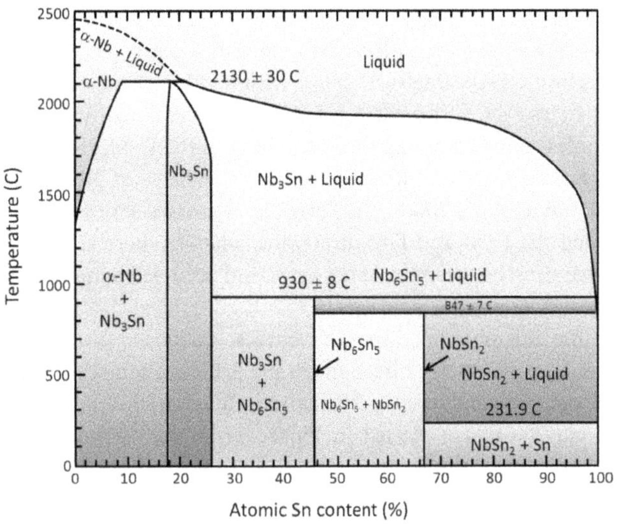

FIGURE B1.17 Equilibrium phase diagram for the $Nb_{1-\beta}Sn_\beta$ binary system in the high-temperature region. (Figure readapted from Cardwell and Ginley (2003), p. 640; Godeke (2006a); Suenaga (1980).)

and other factors, therefore the values in table are merely for reference. The critical current density (J_c) is especially dependent on processing and microstructure, so a reference value is not included in this table. The binary phase diagram of $Nb_{1-\beta}Sn_\beta$ is depicted in Figure B1.17 (Suenaga, 1980; Charlesworth et al., 1970): in thermodynamic equilibrium, three different intermetallic phases can form, depending on the value of β, namely Nb_3Sn, $NbSn_2$, and Nb_6Sn_5. Although superconducting ($NbSn_2$: $T_c \leq 2.68$ K at $\beta \sim 0.67$; Nb_6Sn_5: $T_c \leq 2.8$ K at $\beta \sim 0.45$), these two compounds are not useful for practical applications. The solubility of niobium in liquid tin is small at temperatures below 1000°C and the solid solubility of tin in niobium decreases from about 9 at.% tin at the peritectic temperature of Nb_3Sn to about 1 at.% tin at 1495°C (Suenaga, 1980). While the $NbSn_2$ and Nb_6Sn_5 phases are stoichiometric in a narrow homogeneity range, the brittle Nb_3Sn phase can exist over a wide composition range from about $0.18 \leq \beta \leq 0.25$ at.% tin (although the niobium-rich compositions are not formed on annealing below 1400°C). The A15 phase can be formed from a solid-state reaction between elemental niobium and $NbSn_2$ or Nb_6Sn_5, or above ~930°C from Nb–Sn melt.

B1.4.2.2 Theoretical Hints

The Nb_3Sn intermetallic compound is a strong coupling BCS superconductor: using first principle calculations based on the density functional theory, the electronic, vibrational, and superconducting properties of Nb_3Sn compounds can be investigated on the basis of the Eliashberg approach [see for example, De Marzi et al. (2013) and Tütüncü et al. (2006), and references therein].

The calculated electronic band structure along many high symmetry directions of the simple-cubic Brillouin zone is reported in Figure B1.18. There, the Fermi level is marked by

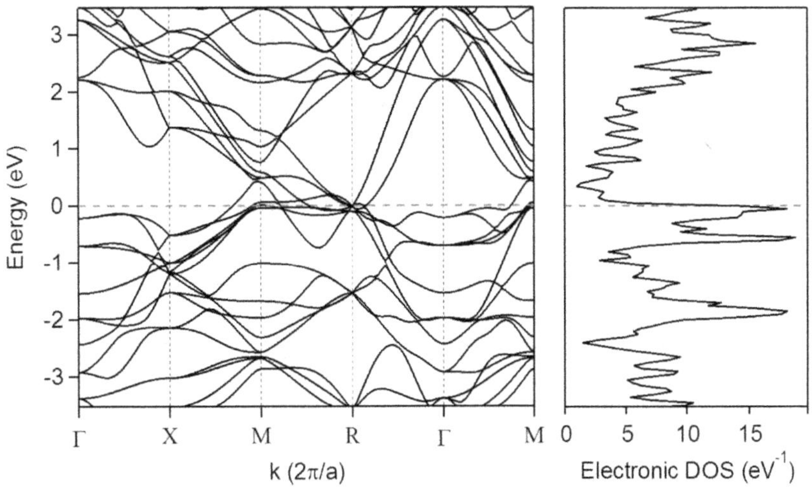

FIGURE B1.18 Electronic band structure and density of states of cubic Nb_3Sn (De Marzi et al., 2013; Tütüncü et al., 2006). The Fermi level is set to 0 eV.

a dashed horizontal line and is set to 0 eV. It is interesting to notice that the Fermi level falls close to a sharp peak in the electronic DOS (Stewart, 2015), with a value of the order of 20 states eV^{-1}. This peak is generated by several nearly dispersionless bands crossing the Fermi level in the Γ–M, Γ–R, and Γ–R directions and deriving from the *d*-states of Nb atoms (Tütüncü et al., 2006). The Fermi level crosses several bands along the Γ–M and Γ–R directions, whereas there is a clear separation between the unoccupied and occupied bands along the Γ–X direction.

The full phonon dispersion curves, calculated along several high-symmetry directions in the Brillouin zone by means of density functional perturbation theory [DFPT, (Baroni et al., 2001)], are plotted in Figure B1.19; the phonon DOS is depicted in the right panel. For the sake of completeness, some inelastic neutron scattering measurements (Pintschovius et al., 1985; Axe and Shirane, 1973; Shirane and Axe, 1978) are also reported in figure, showing good agreement with calculated acoustic branches. Tütüncü *et al.* (Tütüncü et al., 2006) observed that the degenerate TA branch along Γ–X becomes flat at X, whereas the upper transverse acoustic (TA) branch crosses the lowest optical branch along the Γ–M direction. At the M point, the lower TA branch becomes degenerate with the lowest optical branch, and the degenerate TA branch along Γ–*R* is further degenerated with the lowest optical branch near and at *R*.

This kind of *el–ph* coupling can be qualitatively understood from the following considerations. According to the BCS theory, the Fermi sea is unstable against the formation of bound electron pairs (the Cooper pairs) because the exchange of virtual phonons produces an attraction for electrons close to the Fermi level. This is possible because in a crystal the lattice deformations overscreen the electron–electron (*el–el*) interactions. In a very rough figuration of this mechanism, an electron (first electron) moves through the crystal lattice and attracts the nearby positive ions, thus polarizing the medium. The deformation of the crystal lattice in turn attracts another electron (second electron) toward the region with excess of positive charge, giving an effective attractive interaction between the two electrons. As a general rule, the more the deformation is localized in space, the stronger is the *el–ph* interaction: this is the reason why the phonon modes with large ***q***-vectors (i.e. very localized distortions) might give rise to stronger *el–ph* interactions. Looking again at Figure B1.18, one can also see that only the *d*-states crossing the Fermi level along the M–R directions can eventually couple with phonons, because according to the BCS theory only those electrons with energy between E_F *and* $E_F + \hbar\omega_D$ (where $\hbar\omega_D$ is a measure of the amount of energy which can be exchanged by virtual phonons) can participate to the Bose condensation into the BCS ground state.

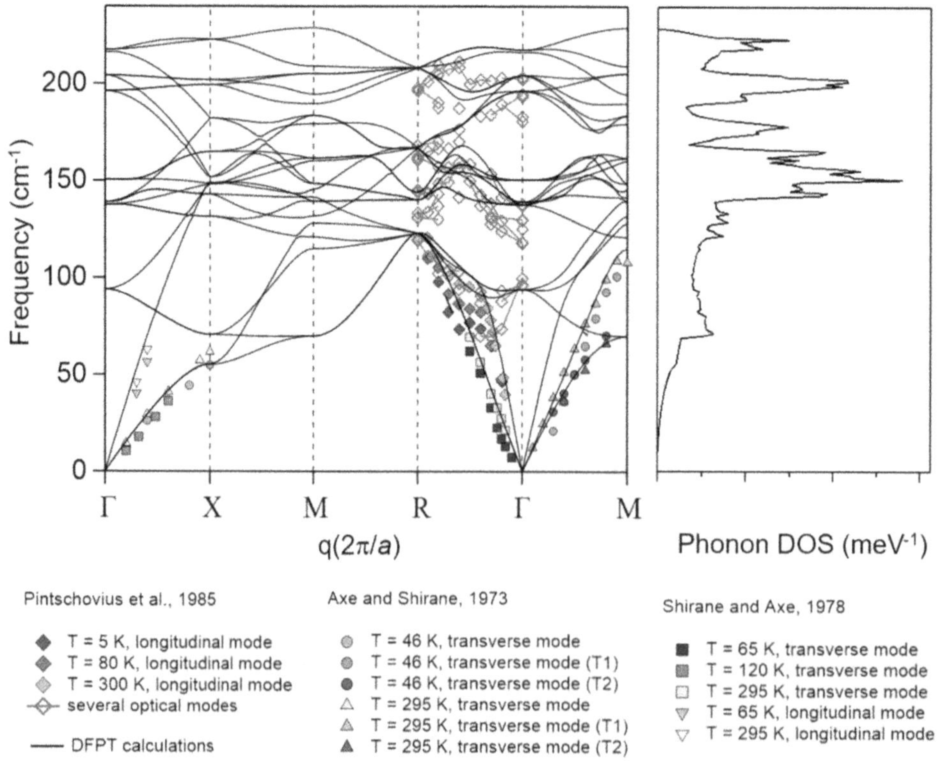

FIGURE B1.19 Phonon dispersion curves and density of states for Nb_3Sn. Continuous lines represent the phonon branches calculated by means of DFPT methods (De Marzi et al., 2013). Markers are experimental data. (Tütüncü et al., 2006; Axe and Shirane, 1973; Shirane and Axe, 1978; Pintschovius et al., 1985.)

Ab-initio methods also allow for the calculation of the *el-ph* coupling constant, which is defined by the matrix elements $g_{ij,v}(\boldsymbol{q},\boldsymbol{k}+\boldsymbol{q})$ corresponding to the interaction between an electron in the i-th band (with energy E_{ik}) and an electron in the j-th band (with energy E_{ik+q}) mediated by a phonon with energy $\hbar\omega_v(\boldsymbol{q})$ and vector $\boldsymbol{q}$, where v is the polarization branch index. The matrix elements $g_{ij,v}(\boldsymbol{q},\boldsymbol{k}+\boldsymbol{q})$ define the Eliashberg spectral function $\alpha^2F(\omega)$ and the *el-ph* coupling constant by the following three equations:

$$\lambda = 2\int \frac{d\omega}{\omega}\alpha^2F(\omega) = \sum_{q\in IBZ}\lambda_q W_q \quad (B1.17)$$

$$\alpha^2F(\omega) = \frac{1}{2\pi N(E_F)}\sum_q\sum_v \delta(\omega-\omega_{qv})\frac{\gamma_{qv}}{\hbar\omega_{qv}} \underline{\underline{\text{def}}} \sum_q\left[\alpha^2F(\omega)\right]_q \quad (B1.18)$$

$$\gamma_{qv} = 2\pi\omega_{qv}\sum_{i,j}\int_{BZ} d^3k\left|g_{ij,v}(\boldsymbol{q},\boldsymbol{k}+\boldsymbol{q})\right|^2\delta(E_{ik}-E_F)\delta(E_{ik+q}-E_F) \quad (B1.19)$$

In Equation (B1.17), the sum is restricted to the irreducible part of the first Brillouin zone (IBZ), and W_q represents the weight of a $\boldsymbol{q}$-point in the IBZ; in Equations (B1.18) and (B1.19), γ_{qn} represents the linewidth of the phonon with polarization mode v and momentum $\boldsymbol{q}$.

Regarding Nb_3Sn compound, from DFPT calculations it appears that the most important contributions to the *el-ph* coupling constant come from three doubly degenerate modes at *R*, four degenerate modes at *X*, six modes at *M*, and three triple degenerate modes at Γ (Gala et al., 2016). Among them, there is clear evidence (De Marzi et al., 2013; Tütüncü et al., 2006) of a strong interaction between the electronic *d*-states near the Fermi level from Nb and several phonon modes (longitudinal acoustic phonons and a group of high-$\boldsymbol{q}$ optical phonon modes with average frequency of 4.5 THz) along the [111] direction. In particular, six modes at the high-symmetry point *R* return the largest contribution to the *el-ph* coupling. More interestingly, these modes at *R* only involve the Nb intrachain vibrations, the Sn atoms being frozen in their equilibrium positions. The other phonon modes, in particular those involving the motion of Sn atoms, contribute much less to λ.

From DFPT calculation on A15 Nb_3Sn, it produces $\lambda = 1.85$ (De Marzi et al., 2013; Gala et al., 2016), a value that is in good agreement with the experiments (Wolf et al., 1980). The aforementioned value can be employed to calculate the critical temperature through Allen–Dynes modification of the McMillan formula (McMillan, 1968; Allen and Dynes, 1975), generalizing the BCS result for intermediate or strong coupling:

$$T_c = \frac{\hbar\omega_{ln}}{1.20}\exp\left(-\frac{1.04(1+\lambda)}{\lambda-\mu^*(1+0.62\lambda)}\right) \quad (B1.20)$$

where the logarithmically averaged phonon frequency, as defined in Equation (B1.10), is of the order of $\hbar\omega_{ln}$ ~ 200 K from *ab-initio* calculations on Nb_3Sn. The effective Coulomb repulsion μ^* depends on the electronic DOS at the Fermi level and it can be estimated as in Equation (B1.12). For Nb_3Sn, by considering a value for $N(E_F)$ ~ 11–12 states/eV, it results μ^* ~ 0.24, thus yielding a transition temperature T_c ~ 18 K.

It is worth mentioning that using first principle calculations based on DFT, the electronic, vibrational, and superconducting properties of compounds have also been studied at different stoichiometry ratios (Gala et al., 2016) or at different hydrostatic (Loria et al., 2017) or uniaxial pressures (De Marzi et al., 2013). These *ab-initio* studies represent a novel approach to understand and optimize the performances of Nb_3Sn materials under the hard operational conditions of high-field superconducting magnets.

B1.4.2.3 The Batterman–Barrett Instability in Highly Ordered Nb_3Sn

The cubic Nb_3Sn phase can undergo a martensitic shear transformation at cryogenic temperatures. This transformation can affect the critical temperature and notably the upper critical field. However, recent research (Zhou et al., 2011) has shown evidence that the upper critical field is independent of it, although the results were questioned by some (Mentink, 2014). The tetragonal phase is only formed within a narrow window of Sn content, and the transformation can be influenced by stress (Cardwell and Ginley, 2003).

As described in Section B1.4.1.3, the cubic phase of the highly ordered stoichiometric Nb_3Sn is unstable toward a tetragonal distortion below approximately 43 K, as evidenced by XRD data (Mailfert et al., 1967; Mailfert et al., 1969). In neutron scattering experiments (Shirane and Axe, 1971), a detailed analysis of the intensity distribution among the Bragg reflections revealed that the structure was determined uniquely as D_{4h}^9—$P4_2/mmc$, with Nb displacements from the special positions of 0.016(3) Angstrom at 4 K. The Batterman–Barrett transition is accompanied by a sublattice distortion, which involves a pairing of the Nb atoms along a chain similar to a certain extent to a Peierls distortion in a one-dimensional system. From accurate X-ray diffraction profile measurements around 45 K, Vieland *et al.* (Vieland et al., 1971) showed that in Nb_3Sn the structural transformation below T_m is a first-order transition, characterized by a discontinuity of the distortion $(c-a)/a$, with a significant spontaneous strain at T_m. It should be noted that such discontinuity is not present in V_3Si, for which a first-order character of the instability depicted in Figure B1.13 has so far escaped any observation. Whether the Batterman–Barrett transformation in V_3Si is a weak first-order or a truly second-order phase transition is still an open question, to which only improved measurements will be able to answer.

Ultrasonic velocity measurements in single-crystal Nb_3Sn have been firstly reported by Keller and Hanak (Keller and

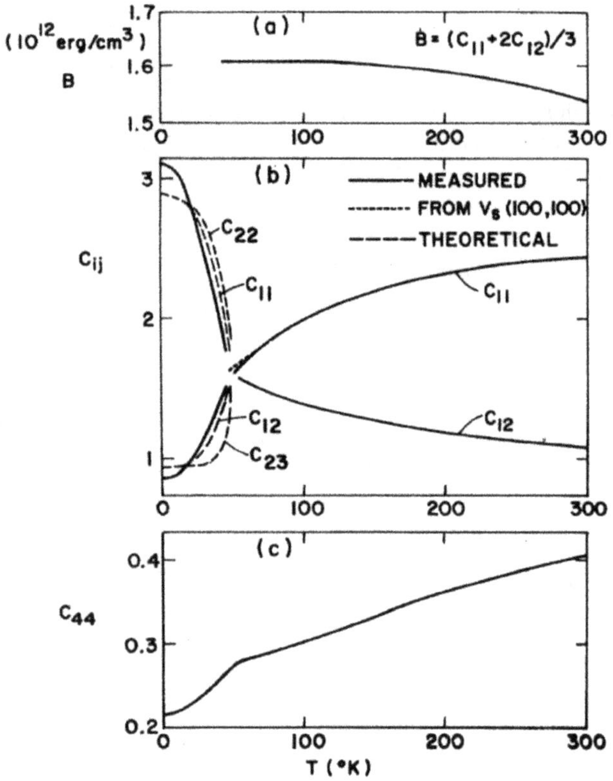

FIGURE B1.20 Nb_3Sn elastic moduli *vs.* temperature. (Figure reproduced from Rehwald et al. (1972), *with kind permission of The American Physical Society.*) The dashed curves have been obtained from theory (Vieland et al., 1971): (a) behavior of the bulk modulus, B; (b) elastic constants C_{11} and C_{12}; (c) elastic constant C_{44}.

Hanak, 1966), who observed large elastic softening with decreasing temperature, similar to what found in V_3Si (Testardi et al., 1965). The elastic constants decrease with temperature, with a reduction in C_{11} and C_{44} from room temperature values of 33.8% and 31.8%, respectively. In particular, the reduction of C_{44} is considerable, when compared to that of V_3Si (Keller and Hanak, 1966). In Figure B1.20, the elastic moduli are plotted as a function of temperature (Rehwald et al., 1972). Like in V_3Si (see Figure B1.15), $C_{11}-C_{12}$ vanishes as $T \to T_m$; however, $(C_{11}-C_{12})/2$ recovers to a higher value below T_m.

It is worth recalling that the elastic softening, which can be regarded as a precursor of the Batterman–Barrett instability, was thought to be a key factor for the enhancement of the *el–ph* coupling λ, thus leading to higher T_c (Testardi, 1975). Again, the theoretical background for relating the instability to the superconducting T_c comes from the BCS theory in the intermediate-strong coupling limit, for which the following expression holds (McMillan, 1968):

$$\lambda = \frac{N(E_F)\langle I^2\rangle}{M\langle\omega^2\rangle} \tag{B1.21}$$

In Equation (B1.21), M is the ionic mass, $<I^2>$ is the average over the Fermi surface of the *el–ph* matrix element squared, and $<\omega^2>$ is the average phonon frequency square given by

$$\langle\omega^2\rangle = \frac{\int \omega\alpha^2 F(\omega)d\omega}{\int \alpha^2 F(\omega)\frac{d\omega}{\omega}} \tag{B1.22}$$

From Equation (B1.21), it is straightforward to see that, as $<\omega^2>$ softens, the *el–ph* coupling constant increases, thus enhancing T_c.

According to calculations by Testardi (Testardi, 1975), the phonon softening could lead to an increase of about 6–8 K in T_c; however, for an accurate calculation it is necessary to get data on the phonon behavior, particularly for the high-frequency phonons, for which $F(\omega)$ is large. As a concluding remark, it is not known to what extent the soft modes involved in the instability could contribute to the enhancement of T_c.

B1.4.2.4 Variation in Structural and Superconducting Properties with Tin Content

The amount of tin in Nb_3Sn strongly influences its structural and superconducting properties, namely: the lattice parameter a, T_c, B_{c2}, the superconducting gap Δ, the normal-state resistivity close to T_c, and the long-range order. The dependence of the long-range order upon Sn content (β) is largely discussed in Section E3.2.1.2 of this Handbook, and also in the review paper (Godeke, 2006a).

The lattice parameter evolution as a function of β was determined by Devantay *et al.* (Devantay et al., 1981; Vieland, 1964). The dependence is linear in the range ~18–25% Sn, with:

$$a(\beta) = 0.136\beta + 5.256 \cdot [\text{Å}] \tag{B1.23}$$

Flükiger (1981) also proposed a linear fit,

$$a(\beta) = 0.176\beta + 5.246 \cdot [\text{Å}] \tag{B1.24}$$

in which the extrapolation to $\beta = 0$ gives the lattice parameter of Nb_3Nb ($a = 5.246$ Å).

The normal-state resistivity as a function of β, $\rho_n(\beta)$, was measured by several groups around the world (Devantay et al., 1981; Orlando et al., 1979; Orlando et al., 1981, Hanak et al., 1964). The behavior of ρ_n vs. β can be described by a fourth-order fit:

$$\rho_n(\beta) = 91[1-(7\beta-0.75)^4] + 3.4[\mu\Omega \text{ cm}] \tag{B1.25}$$

The superconducting gap can be described by a Boltzmann sigmoidal function (Moore et al., 1979):

$$\Delta(\beta) = \frac{\Delta_{\min} - \Delta_{\max}}{1 - \exp\left(\frac{\beta - \beta_0}{d\beta}\right)} + \Delta_{\max} \tag{B1.26}$$

where Δ_{min}, Δ_{max}, β_0, and $d\beta$ are fitting parameters. From the ratio $2\Delta_0/k_BT_c$ it emerges that Nb_3Sn can be considered a strong coupled BCS superconductor only in a limited composition range above $\beta = 0.23$–0.24, i.e. close to stoichiometry.

Considering the behavior of T_c vs. β, in order to fit the experimental data, Devantay (Devantay et al., 1981) originally proposed the following empirical expression:

$$T_c(\beta) = \frac{12}{0.07}(\beta - 0.18) + 6\,[\mathrm{K}] \tag{B1.27}$$

An alternative equation, which is similar to (B1.27), has been proposed by Godeke (Godeke, 2006a):

$$T_c(\beta) = \frac{-12.3}{1+\exp\left(\dfrac{\beta-0.22}{0.9}\right)} + 18.3\,[\mathrm{K}] \tag{B1.28}$$

To conclude this paragraph, we report an analytical expression, which can be used to analyze the dependence of the upper critical field μ_0H_{c2} upon β (Godeke, 2006a):

$$\mu_0H_{c2}(\beta) = -10^{30}\frac{-12.3}{1+\exp\left(\dfrac{\beta-0.22}{0.9}\right)} + 18.3\,[\mathrm{T}] \tag{B1.29}$$

Several authors in the past observed a strong depression of $\mu_0H_{c2}(T = 0\ K)$ from 29 T to 21.4 T in the tetragonal state below T_m (Godeke, 2006a; Foner and McNiff, 1981; Arko et al., 1978). By contrast, recently the National High Magnetic Field Laboratory reported a value of 29.1 ± 0.2 T in *transforming* Nb_3Sn samples (Zhou et al., 2011), thus concluding that the previously reported low values of $\mu_0H_{c2}(0) \sim 21.4$ T for tetragonal Nb_3Sn were not determined by whether they underwent the cubic to tetragonal transformation or not.

B1.4.3 Ternary/Quaternary Nb–Sn Phases

The use of dopants to improve superconducting and transport properties in practical conductors has been widely experimented. The introduction of ternary or quaternary phases, by doping with a single of two additional elements, respectively, has typically an effect on intrinsic superconducting properties of the material (as the critical temperature, T_c and the upper critical field, H_{c2}) or on extrinsic features (as microstructure). In addition, other elements may influence the diffusion kinetics in solid-state reactions, thus leading, for example, to a refinement of the grain structure, or a change in morphology of the grains. All these features have an influence on the in-field current-carrying capability of the material.

For Nb_3Sn, the dopants most effectively used are titanium or tantalum (Suenaga et al., 1986; Godeke, 2006a), and are in fact the standard options now used, among the numerous explored possibilities (Sekine et al., 1981; Suenaga et al., 1986; Takeuchi et al., 1981; Osamura et al., 1986). In composite wires, Ta is supplied by using Nb-7.5wt.%Ta instead of pure Nb in either part of the filaments, or in the antidiffusion barrier. As for Ti, it is introduced by either substituting part of the Nb filaments in the billet assembly with Nb-47wt.%Ti (Nb–Ti) ones, or by using a Ti-doped Sn or bronze core (see Section E3.2.3 for further details). The space available for equilibrium ternary Nb–Sn–Ta and Nb–Sn–Ti phases is shown in Figure B1.21(left) and B1.21(right), respectively.

Ti addition causes different effects on Nb_3Sn. In presence of Ti, and depending on the reaction heat treatment temperature,

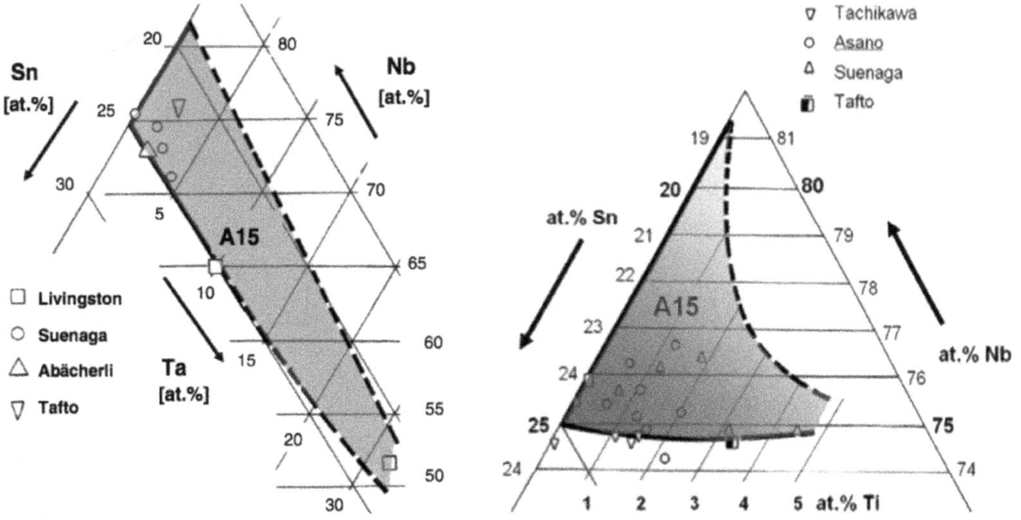

FIGURE B1.21 (left) A15 phase field of the system Nb–Sn–Ta, in the temperature range between 700°C and 750°C. (Figure reproduced from Flükiger et al. (2008), *with kind permission of Elsevier*; data from references therein.) (right) A15 phase field of the system Nb–Sn–Ti, in the temperature range between 700°C and 750°C. (Figure reproduced from Flükiger et al. (2008), *with kind permission of Elsevier*; data from references therein.)

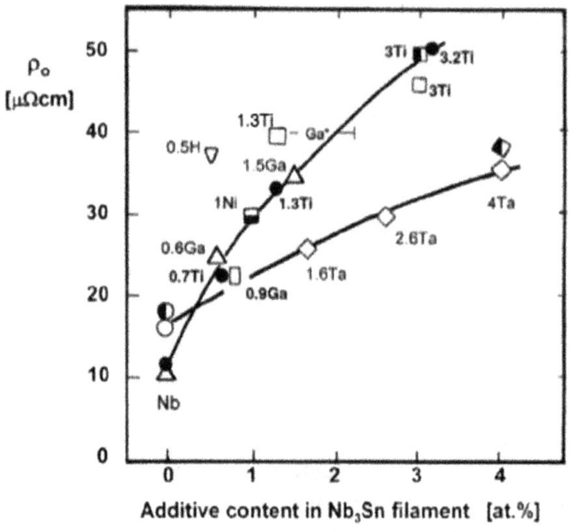

FIGURE B1.22 Normal-state electrical resistivity ρ_0 as function of Ta, Ti, Ga, and Ni additives content. The slopes for Nb_3Sn alloyed with Ti, Ga, and Ni coincide and are considerably steeper than for the Ta additive. (Figure reproduced from Flükiger et al. (2008), *with kind permission of Elsevier*; data from references therein.)

Nb_3Sn forms with a finer grain structure, and thus exhibits a higher grain boundary density. This leads on one side to higher transport properties, being the pinning mechanism in Nb_3Sn dominated by grain boundaries (Godeke, 2006a). On the other side, this has consequences on the reaction kinetics (Hayase and Kajihara, 2006): higher grain boundary density favors Sn diffusion during the heat treatment carried out for the formation of the superconducting phase, and allows to maintain high Sn concentration levels and to improve the material stoichiometry, which directly leads to a beneficial effect on H_{c2}. As shown in Figure B1.22, the increase of H_{c2} is associated with the experimentally observed increase in normal-state resistivity (Suenaga et al., 1986; Sekine et al., 1988). According to the Ginzburg-Landau-Abrikosov-Gorkov (GLAG) formulation in the dirty limit, the upper critical field is proportional to the critical temperature, T_c, and to the normal-state electrical resistivity, ρ_n, according to the following formula (Godeke, 2006a):

$$\mu_0 H_{c2} = \frac{3e}{\pi^2 k_B} \gamma \rho_n T_c(0) \quad \text{(B1.30)}$$

where k_B is the Boltzmann constant, and γ is the Sommerfeld constant, i.e. the linear term of the electronic specific heat constant, and which is a measure of the electronic density of states at the Fermi level, $N(E_F)$. Doping increases the electron scattering and thus ρ_n, but above a certain threshold, of about 2% for Ti or slightly higher for Ta, T_c starts to decrease (Suenaga et al., 1986; Asano et al., 1986; Takeuchi et al. 1981), owing to the detrimental effect on the electron density of states and on the electron–phonon coupling (Cooley et al., 2006; Godeke, (2006a). It should be observed that Ti and Ta additions cause variations in the Nb_3Sn unit cell due to the substitution of dopant atoms (Tafto et al., 1984), occupying niobium 6c chain sites [or, as more recently evidenced for Ti-doping (Flükiger et al., 2008; Mentink et al., 2012a; Tarantini et al., 2016), Tin cubic 2a sites], with a direct effect on superconducting T_c.

As a result of the competition among the different effects, the maximum of T_c and H_{c2} do not occur at the same concentration value, as shown in Figure B1.23.

A direct comparison of superconducting characteristics between Ti- and Ta-doped Nb_3Sn has been carried out not only on bulk samples (Mentink et al., 2012a), but also on composite multifilamentary internal tin wires (Tarantini et al., 2016), where a higher $H_{c2}(0)$ has been observed with Ti-doping, mainly ascribed to the observation that Ti favors Sn diffusion and produces more homogeneous A15 layers in filaments.

Simultaneous addition of Ti and Ta to create a quaternary phase has also been tempted on wires produced following the bronze-route approach, but the effect on the upper critical field H_{c2} has not been completely clarified. Other quaternary phase combinations using different IVa element additions have been probed (Takeuchi et al., 1981), most of which leading, however, to a layer grain coarsening or change of morphology. In a particular case, the addition of 5at.%Hf to the Nb core and 4at.%Ga to the bronze matrix led to a remarkable J_c improvement in magnetic fields above 15 T.

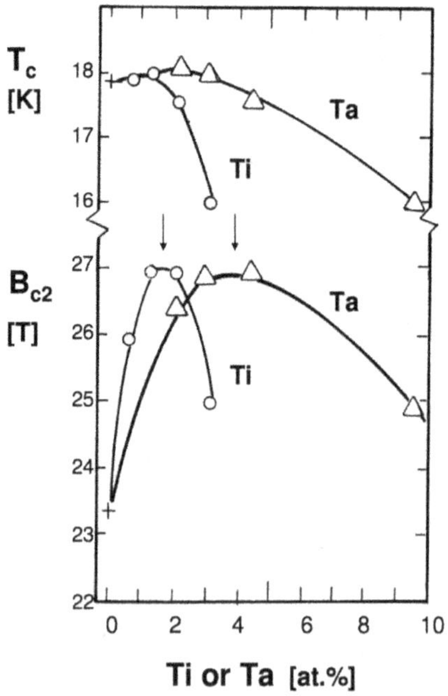

FIGURE B1.23 Variation of T_c and B_{c2} in Nb_3Sn monofilamentary wires as a function of Ta and Ti additive content. The arrows mark the maximum of B_{c2} for Ta and Ti additives (normal-state electrical resistivity ρ_0 as function of Ta, Ti, Ga, and Ni additives content. The slopes for Nb_3Sn alloyed with Ti, Ga, and Ni coincide and are considerably steeper than for the Ta additive. [Figure reproduced from Flükiger et al. (2008), *with kind permission of Elsevier*; based on data from Suenaga et al. (1986).]

B1.4.4 Pinning Properties in Nb_3Sn and Effect of Additions

As anticipated, the dominating pinning mechanism in Nb_3Sn is due to grain boundaries and most of the attempts to improve Nb_3Sn transport properties address the possibility of grain refinement to increase the effectiveness of pinning, especially at high fields. Assuming a hexagonal (regular) flux-line lattice, the pinning center separation, d, – and hence the ideal grain size for maximum pinning – would be a function of the applied magnetic flux density, B:

$$d = \sqrt{\frac{\Phi_0}{B \cdot \sin\left(\frac{\pi}{3}\right)}} = \sqrt{\frac{2\Phi_0}{\sqrt{3} \cdot B}} \cong \sqrt{\frac{1.15 \cdot \Phi_0}{B}} \qquad \text{(B1.31)}$$

At 12 T, the theoretical ideal grain size would be about 14 nm, compared to typical values found in wires of over 100 nm. To compare the grain size with the A15 lattice parameter across 14 nm, there are only about 27 unit cells, and grain boundaries take up approximately 1 nm.

Grain size and morphology are influenced by multiple factors, but can be controlled through the Nb_3Sn reaction heat treatment temperature and duration (Ochiai et al., 1986). However, reaction kinetics influences the homogeneity and stoichiometry of the superconducting phase, so that the recipe providing optimal microstructure might not necessarily coincide with the one providing optimal microchemistry. Very extensive studies have been performed on multifilamentary composite wires, which are intrinsically non-homogeneous systems, characterized by regions with different Sn compositions, or with different grain size and morphology (Lee and Larbalestier, 2008; Flükiger et al., 2008), and the de-convolution of the various effects is very complicated. However, in such systems, extensively discussed in Section E3.2.3, the achievement of an optimal stoichiometry and composition homogeneity seems to play the key role for the achievement of optimal transport properties (Tarantini et al., 2016).

A different approach to the optimization of current-carrying capability was investigated in thin Nb_3Sn films, where different elements were added (Ti, Dy, Y, Sc, La, and Al_2O_3) (Dietderich and Scanlan, 1997; Dietderich et al., 1998), aimed to be incorporated as artificial pinning centers (APC), rather than as substitutional elements within the binary Nb_3Sn crystal lattice. In some cases, a decrease of the average grain size was also observed, and a shift of the pinning-force peak from the typical value $b=B/B_{c2}=0.2$ to a value of about 0.5 (Dietderich and Godeke, 2008) was achieved, as shown in Figure B1.24. A similar effect was more recently observed upon introduction of Cu(Sn) APC (Da Silva et al., 2009), or with the addition of Yttrium as a quaternary phase in Ta-doped wires (Motowidlo et al., 2011), where a refined Nb_3Sn grain structure was observed, as well as a precipitation of Yttrium particles located inside the grains and at grain boundaries, acting as additional pinning centers. Very promising results have been obtained by adding a SnO_2 layer inside a Nb-1Zr alloy tube (Xu et al., 2014), containing the Cu-Sn core in tube-type wire subelements. During reaction heat treatment, nano-sized ZrO_2 precipitates in intragranular regions, causing a refinement of grains, with final size of about 43 nm, i.e. less than half of that normally achieved. For comparison, a flux-line spacing of about 15 nm characterizes Nb_3Sn at 12 T and 4.2 K. The peak in the pinning-force curve consistently shifted to a reduced field of about 0.3 B_{irr} (B_{irr} being the irreversibility field).

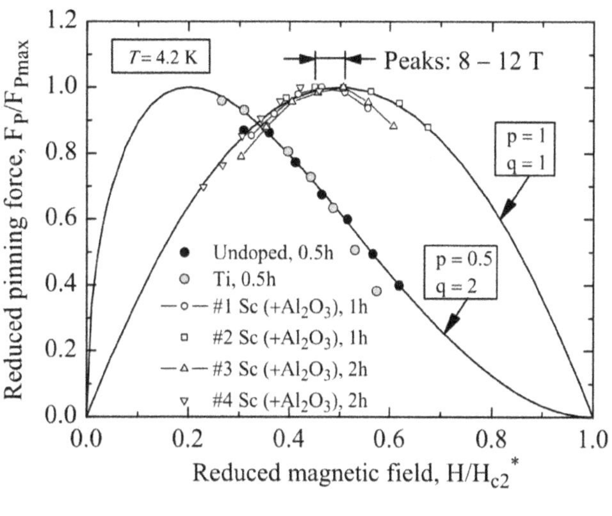

FIGURE B1.24 Reduced pinning force as a function of reduced magnetic field for undoped and doped Nb_3Sn thin films. A shift in the peak of the pinning-force curve to higher reduced fields is achieved with the addition of second phase precipitates. The shape of the curve, as characterized by the low- and high-field exponents (p and q) changes, indicating a change in the dominating pinning mechanism. [Figure reproduced from Dietderich and Godeke (2008), *with kind permission of Elsevier.*]

Following a different route, irradiation of Nb_3Sn with either protons or neutrons although degrading T_c, greatly improves transport properties (Spina et al., 2016), because artificial pinning structures are created in the form of dislocations. An additional pinning-force component has been observed to raise in this case, characteristic of a point pinning mechanism (Baumgartner et al., 2014), i.e. acting at higher reduced fields with respect to the pinning due to grain boundaries.

A different principle, which has been demonstrated to achieve an enhancement of the upper critical field with very limited detrimental effect on T_c, is to introduce disorder instead of alloying elements, by high-energy ball milling Nb and Sn powders before reaction (Cooley et al., 2006). This approach had provided extraordinary results when applied to the $PbMo_6S_8$ superconducting system (Niu and Hampshire, 2003). In this approach, material properties are determined by pinning at grain boundaries as well as by electron scattering within the grains, so that the distinction between optimization of intrinsic and extrinsic effects becomes evanescent (Taylor et al., 2008).

B1.4.5 The Nb_3Al Compound

Nb_3Al is an intermetallic compound belonging to the A15 family, and its general features have been therefore discussed within Section B1.4.1. As for its superconducting properties, it exhibits a T_c very similar to Nb_3Sn (see Table B1.5), but a higher H_{c2} (see Table B1.8), which makes this material very attractive for high-field applications. The phase diagram for the binary Nb–Al phase is reported in Figure B1.25. As one can see, the stoichiometric Nb_3Al is only stable at very high temperatures (above about 1800°C), therefore the growth of the optimal superconducting phase is rather complex. As a matter of fact, various methods have been proposed (Glowacki, 1999): reactive diffusion processes, employing either solid-state or solid–liquid solutions, proved successful at relatively low temperature (below 1000°C) but for small dimensions of the diffusional distance (less than about 1 μm). In terms of current-carrying capability, such techniques do not lead to the best grain structure, and therefore to the optimal critical current, J_c. Further raising the processing temperature above 1500°C, the stoichiometric phase can be obtained by very fast reactions, which can be stabilized by rapid quenching. In samples manufactured applying this technique, T_c is an increasing function of the reaction temperature, up to about 1950°C, above which samples start to be characterized by multiple phases (Glowacki, 1999). Measurements of T_c on samples produced using this technique have clarified the importance of ordering and compositional effects in Nb_3Al: not only is the T_c an increasing function of the reaction temperature, but also a highly non-linear function of the Al atomic concentration: above about 22 at.% Al, in fact, the material exhibits a marked tendency toward saturation (Jorda et al., 1981), with a maximum of about 17.7 K at 25 at.% Al. This behavior is a peculiarity of this A15 structure. In addition, on the same sample it was shown that improving the long-range order by low-temperature heat treatments caused an increase in T_c up to about 19.1 K at stoichiometry. Differently from Nb_3Sn and other A15 compounds, in fact, Nb_3Al exhibits a certain amount of atomic disorder, as characterized by the Bragg–Williams long-range order parameter S, even at perfect stoichiometry, so that normal resistivity and critical temperature values change with annealing or quenching treatments.

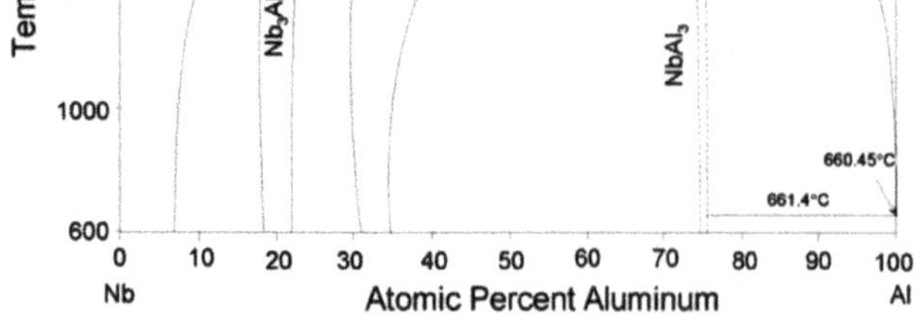

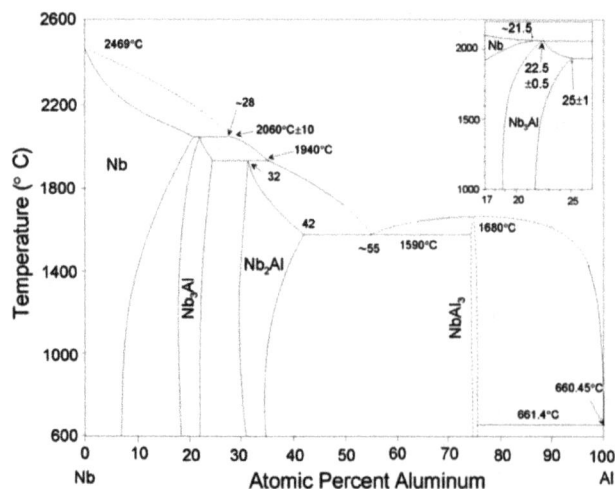

FIGURE B1.25 Nb–Al binary phase diagram. Figure reproduced from 1st edition, p. 674 Cardwell and Ginley (2003).

Another approach to obtain stoichiometric Nb_3Al samples is mechanical alloying by high-energy ball milling of precursor powders followed by annealing treatments: in this case, Nb and Al first form to a large extent the Al-rich Nb_2Al phase, and only partly the required Nb_3Al phase. By subsequent heat treatments, the Nb_2Al phase reacts with the remaining Nb to form the desired A15 phase. Depending on the milling treatment duration, Nb_3Al may also form directly from a $Nb(Al)_{SS}$ supersaturated solid solution during the heat treatment. Bulk samples produced by this technique exhibit a rather low critical temperature, with a maximum onset $T_{c\text{-onset}} = 15.8$ K observed within a wide range of explored values for the various parameters (milling duration, heat treatment temperature, and initial Al content) (Qi et al., 2014).

The application of these formation methods to technical multifilamentary composite wires is not straightforward, and the most promising technique is based on rapid heating to about 2000°C and quenching of fine Nb and Al composites to form a Nb–Al solid solution, which transforms into the A15 phase by subsequent annealing at relatively low temperatures (about 800°C) (Takeuchi, 2000). A further improvement is obtained applying a second rapid heating/quenching treatment to the $Nb(Al)_{SS}$, instead of a low-temperature treatment. All details of these techniques are discussed within Section E3.2.4, but a compound synthesized according to this method exhibited remarkable superconducting properties: $T_c(4.2\text{ K}) = 18.4$ K; $H_{c2}(4.2\text{ K}) = 29.7$ T (Kikuchi et al., 2004), demonstrating an improvement mostly in the final stoichiometry of the A15 phase and a reduction in the presence of stacking faults and other crystallographic defects (Banno et al., 2002), otherwise observed (Kikuchi et al., 2001).

As in Nb_3Sn, intrinsic properties are optimized by achieving stoichiometry, whereas transport properties depend on the microscopic grain structure (Kikuchi et al., 2001). Depending on the heat treatment schedule, the characteristic peak in the pinning force seems to lie at higher values of the reduced magnetic field with respect to what observed in Nb_3Sn (Buta et al., 2005; Glowacki, 2016), although grains are usually larger in Nb_3Al. This indicates that sub-grain structures, such as stacking faults, crystallographic defects, or out-of-stoichiometry regions (Kikuchi et al., 2001; Takeuchi, 2000), might play a leading role in the pinning properties of Nb_3Al at high fields.

Alloying has been also tempted for the Nb–Al system, with various elements such as Ge, Si, Ga, Be, B, Mg, Cu, Ag, Zn, and Cu, aimed to the optimization of its properties. The most remarkable results were achieved using Ge as a doping element to optimize intrinsic material properties, leading to T_c

values as high as about 20 K (Müller, 1970) and upper critical field H_{c2}(4.2 K) above 40 T (Takeuchi, 2000).

B1.4.6 Electromechanical Properties of Nb-Based A15 Superconductors

The investigation of the strain sensitivity of superconducting properties is of paramount importance, since superconductors operate in cryogenic environment and often in high magnetic field, where stresses rise due partly to differential thermal contractions, partly to operational, electromagnetic loads. Effect of deformations and stresses on intrinsic superconducting properties, as critical temperature or upper critical field (T_c or H_{c2}), depends on the material, and reflects on transport capability of technical wires.

The ductile Nb-47wt%Ti alloy is not sensitive to strain (Ekin, 1981), whereas the brittle Nb_3Sn is characterized by an asymmetric bell-shaped critical current density (J_c) curve (Ekin, 1980), typical of all A15 phase superconductors, with a strain-free maximum and a J_c relative reduction, which depends on B/B_{c2}, but which can be as high as about of 50% in Nb_3Sn at typical operation strain values of −0.5% (at 4.2 K, 12 T). Other A15-structured superconductors, as Nb_3Al, are characterized by a much less pronounced sensitivity to strain (Ekin, 1984; Flükiger et al., 1984; Banno et al., 2005).

In multifilamentary wires, deformations cause a reversible degradation of superconducting properties through the characteristic critical current *vs.* strain (I_c *vs.* ε) curve and, beyond a certain critical strain value (ε_{irr}), the fracture of the active filaments, leading to irreversible performance degradation.

Such observations on transport properties have been investigated, both experimentally and theoretically, since the early development stage of materials for applications, in the effort to correlate macroscopic behavior and fundamental material properties.

In situ X-ray diffraction studies carried out on polycrystalline Nb_3Sn tapes subject to uniaxial deformations (Flükiger et al., 2008; ten Haken et al., 1997) have shown a very clear correlation between macroscopic strain and Nb_3Sn lattice cell deformation. This evidences that the strain dependence of the Nb_3Sn critical properties is determined by the lattice distortions at the microscopic level. Similar studies are carried out since a few years on composite multifilamentary wires produced by different techniques, using quantum beams (Awaji, 2013), as neutrons (Oguro et al., 2007) or high-energy synchrotron radiation X-rays (Scheuerlein et al., 2009). With the aid of such techniques, microscopic lattice cell variations along different crystallographic orientations can be studied *in situ*, during the application of loads at room temperature or in cryogenic environment. Phase evolution and texture of the different constituents and of intermediate phases can also be monitored during the reaction heat treatment (Scheuerlein et al., 2010; Scheuerlein et al., 2014a).

Extending experiments originally set up and carried out by the research groups at CERN and at the University of Geneva (Scheuerlein et al., 2014b; Scheuerlein et al., 2009), in (Muzzi et al., 2012) critical current measurements performed at the University of Geneva on a Walters spring system were correlated with Nb_3Sn lattice deformations at 4.2 K in axial and transversal direction with respect to the wire axis, during the application of a tensile load. Diffraction measurements at 4.2 K were performed with high-energy X-rays at the ID15B beamline of the European Synchrotron Radiation Facility (ESRF). Investigated samples were internal tin, Ti-doped, multifilamentary wires, either in a bare configuration or with an external steel reinforcement applied before reaction heat treatment, thus providing an additional quasi-hydrostatic stress component at cryogenic temperatures, and allowing to explore a different range of Nb_3Sn deformations. The linear relation between the lattice strain at the microscopic level and the macroscopic wire axial deformation, with a slope equal to 1, demonstrated that the stress is completely transferred in both cases to the individual Nb_3Sn grains within the structure. Upon cooling, the lattice cell of Nb_3Sn in the bare wire is slightly distorted, the lattice parameter in the transversal direction with respect to the wire axis being 0.3% larger than the *c-axis* cell dimensions. This tetragonal distortion is induced by the stresses due to the differential thermal contraction of the various elements present in the composite structure, and is in fact much larger (of the order of 0.7%) for Nb_3Sn in the steel reinforced wire. Upon application of the tensile loading, this distortion relaxes, until a cubic cell is recovered, as shown in the lower panel of Figure B1.26.

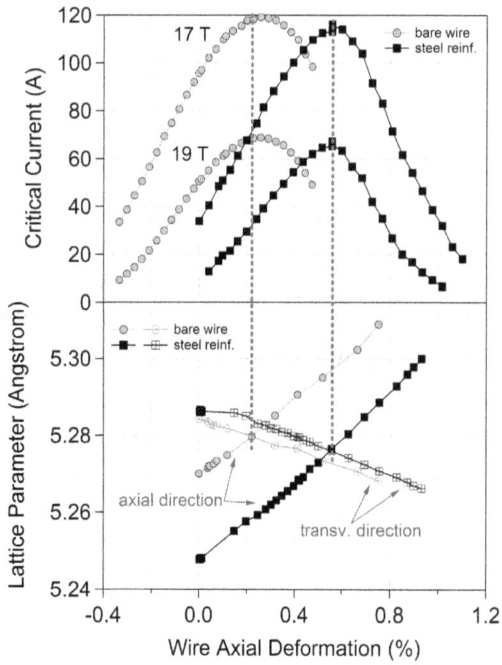

FIGURE B1.26 Comparison between Nb_3Sn lattice cell variation (lower plot) and critical current at 17 T and 19 T (upper plot), with an axial strain applied to a multifilamentary internal tin wire for fusion. Lattice cell data were measured using diffraction from high-energy X-rays at the European Synchrotron Radiation Facility, during *in situ* strain application at 4.2 K; critical currents were measured at the University of Geneva using a Walters Spring system.

In this plot it is also evident that the characteristic peak in the curve of the critical current as function of strain is very close to that of the Nb_3Sn lattice cell being cubic, thus characterized by a minimum in the distortional, or deviatoric, strain component, in agreement with what postulated in (Godeke et al., 2006b). Elaborating further the data in Figure B1.26, it can be observed that the tetragonal distortion of the cell is not symmetric with respect to the central point, where the unit cell is cubic (at the so-called intrinsic strain value, ε_m), which is at the origin of the asymmetry in the shape of the critical current curve with respect to its maximum (Flükiger et al., 2008).

From this, and other experimental observations, as for example the effect of strain on the critical temperature, T_c, or on the upper critical field, B_{c2}, it is now assessed that any stress-induced lattice deformation modifies the density of states at the Fermi level, $N(E_F)$, and also lattice vibration modes, that influence the electron–phonon coupling. However, a full understanding should be able to explain the different strain sensitivity observed, for example, on Nb_3Sn or on Nb_3Al, though these materials are characterized by very similar T_c values, electronic properties and electron–phonon coupling constant (see Tables B1.5, B1.7, and B1.8). On the other hand, the upper critical field, H_{c2}, is different between the two Nb-based A15 materials (see Table B1.8). This, and the observation that stoichiometry heavily influences the strain sensitivity (Flükiger et al., 1984), might represent the key aspects here.

Based on such considerations, some attempts have been carried out, to investigate systematically, using first principle calculations, the evolution of the Nb_3Sn band structure, phonon dispersion curves, and superconducting parameters (electron–phonon mass enhancement parameter, λ, and critical temperature, T_c) as a function of applied strain (De Marzi et al., 2013; Valentinis et al., 2014). In (De Marzi et al., 2013) the measured lattice cell, deformations (Muzzi et al., 2012) were used as inputs to vary the unit cell of a purely binary and homogeneous Nb_3Sn within an *ab-initio* formulation, described with some detail in Section B1.4.2.2. Lattice deformations induce modifications of the frequency-dependent electron–phonon interaction and of the electronic density of states (DOS) at the Fermi level, $N(E_F)$, as well as of the phonon DOS. The results show that both electronic band structure and phonon dispersion curves vary with strain, and contribute to a characteristic T_c vs. ε bell-shaped curve with a zero-strain maximum. It should, however, be mentioned that there is not a complete consensus on the contribution of electronic or lattice degrees of freedom to the strain sensitivity (Valentinis et al., 2014).

For a full description of the strain influences, however, the strain dependence of $H_{c2}(T)$ should also be correlated to the microscopic theory so that the exact change in $N(E_F)$ with strain should be known. To this aim, normal-state resistivity has been studied as function of strain, and used to estimate, through the Werthamer, Helfand, and Hohenberg formulation (Werthamer et al., 1966), the behavior of the electron density of states at the Fermi level, $N(E_F)$ (Qiao et al., 2015).

All models should anyway be compared to sound experimental datasets: ideal systems to be studied should be thin films or single crystals, where possible extrinsic effects are minimal, but there is not a large literature of such measurements. On the other hand, many data are available on multifilamentary composite wires, characterized by a large degree of non-homogeneity in A15 composition and in strain at the microscopic level. As a matter of fact, recent studies on the most sophisticated and highly engineered technological composite wires (see Section E3.2.3) have evidenced that a complete correlation and understanding between strain sensitivity or irreversibility limit (Godeke, 2006a), and microscopic features, with a clear distinction between intrinsic and extrinsic effects, has not been established yet (Cheggour et al., 2014). Motivated by these considerations, more ideal samples, with clear and homogeneous properties have been probed recently, for example by Mentink et al. (2011 and 2012b), but some issues are still open and deserve further attention. For further information, readers may refer to Godeke's topical review (Godeke, 2006a).

B1.4.6.1 Nb_3Sn Critical Current Scaling Functions

Final aim of all studies on the effects of strain on intrinsic material properties is that of deducing a general scaling to describe the material strain sensitivity starting from microscopic parameters, once the exact correlation between the electron–phonon coupling constant and the lattice strain is developed. The three critical parameters (T_c, $\mu_0 H_{c2}$, and J_c) are functions of each other, and together they define the *critical surface*. These critical parameters are strain (ε) sensitive, and the description of the critical surface as a parametric function is commonly called parameterization.

Many semi-empirical formulations have been proposed for a scaling function describing the critical current density of composite wires, and formulations starting purely from first principles have also been attempted (in a more or less refined form). Usually, the axial strain component is the only one considered, mostly because the experimental database is largely based on uniaxial strain tests on multifilamentary wires. However, 3D components are fundamental as well for a thorough description, and to be able to model apparent anomalies of the strain function close to the (T, B, ε) critical surface (De Marzi et al., 2012).

In a general form, the pinning force, F_P, and thus the critical current of a practical high-field superconductor can be expressed using a unified strain-and-temperature scaling law (USL), which, under the assumption of a separable contribution of strain and of temperature, can be written as (Ekin, 2010):

$$F_P = |\boldsymbol{J}_c \times \boldsymbol{B}| = C \cdot g(\varepsilon) \cdot h(t) \cdot f(b) \qquad \text{(B1.32)}$$

where C is a scaling constant, and the three functions $g(\varepsilon)$, $h(t)$, and $f(b)$ describe the dependency on intrinsic strain, reduced temperature, and reduced field, respectively. The explicit form of these functions differs in different models; a summary of all proposals is reported in Ekin (2010) and in Bottura and Bordini (2009), to which more recent formulations (Bordini et al., 2013; Arbelaez et al., 2009) should be added; an extensive discussion of the underlying physical principles of scaling behavior and of comparison between theoretical formulation and experimental data can be found in Godeke, 2006a. In all cases, the following definitions hold: $g(\varepsilon)$ depends or, in some models coincides with the reduced strain function, $s(\varepsilon)$; the strain ε is usually taken as the difference between externally applied strain, ε_a, and intrinsic strain, ε_m, at which the maximum of critical properties is reached.

The reduced temperature is:

$$t = \frac{T}{T_c^*(\varepsilon)} \tag{B1.33}$$

where

$$T_c^*(\varepsilon) = T_c^*(0)\cdot\left[s(\varepsilon)\right]^{1/w} \tag{B1.34}$$

represents the *effective* strain-dependent zero-field critical temperature.

The reduced magnetic field is:

$$b = \frac{B}{B_{c2}^*(T,\varepsilon)} \tag{B1.35}$$

where

$$B_{c2}^*(T,\varepsilon) = B_{c2}^*(0,0)\cdot s(\varepsilon)\cdot\left(1-t^{\kappa}\right) \tag{B1.36}$$

is the strain- and temperature-dependent *effective* upper critical field. The exponent κ is usually fixed to $\kappa = 1.5$. Here, the asterisk (*) superscripts are used to indicate that these are *effective values* obtained by extrapolating critical current data, but do not necessarily coincide with the corresponding physical quantity. Finally, for the effective pinning force, the following relation holds:

$$f(b) = b^{p}\cdot(1-b)^{q} \tag{B1.37}$$

The parameterization Equation (B1.32) at constant strain and temperature reveals Kramer's original scaling law (Kramer, 1973):

$$F_p = \left|\boldsymbol{J}_c \times \boldsymbol{B}\right| = K\cdot b^{p}\cdot(1-b)^{q} \tag{B1.38}$$

where K is some constant.

It can be shown that when $b = p/(p+q)$, F_p attains its maximum value. For Nb_3Sn, typical values for the pinning-force exponents are $p = 0.5$ and $q = 2$, thereby giving the maximum pinning force at $b = 0.2$ (see Figure B1.27). Since the upper

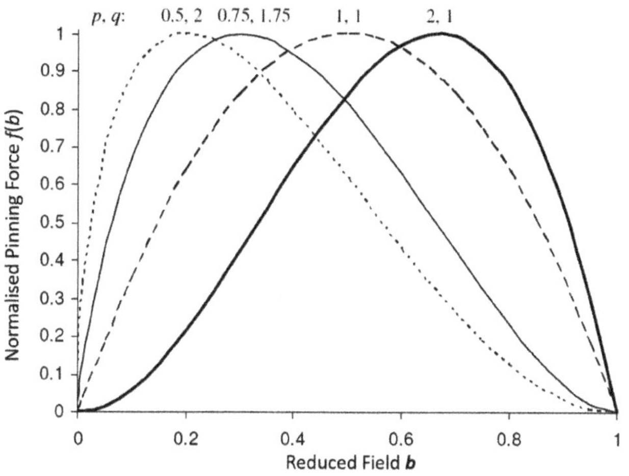

FIGURE B1.27 Example pinning curves with different pinning-force exponents. [Figure reproduced with minor modifications from Hopkins (2007), with kind permission from Simon Hopkins.]

critical field of Nb_3Sn is in the range of 25 T to 30 T, the peak pinning force would occur at approximately 5 T to 6 T, which is much lower than the magnetic field Nb_3Sn is usually used in. Therefore, there is considerable potential for improving the high-field performance of Nb_3Sn (see Figure B1.28).

Nowadays, many efforts are devoted in optimizing and increasing transport properties, which are leading to pinning-force curves that are better described by different values for the low- and high-field pinning exponents, or by the sum of different components, as previously discussed in Section B1.3.3. For example, thin film experiments with Sc and Al_2O_3 doping have shown grains in the region of 15–30 nm using 7 nm Al_2O_3 inclusions, with J_C data suggesting pinning-force exponents of $p = q = 1$ (Dietderich

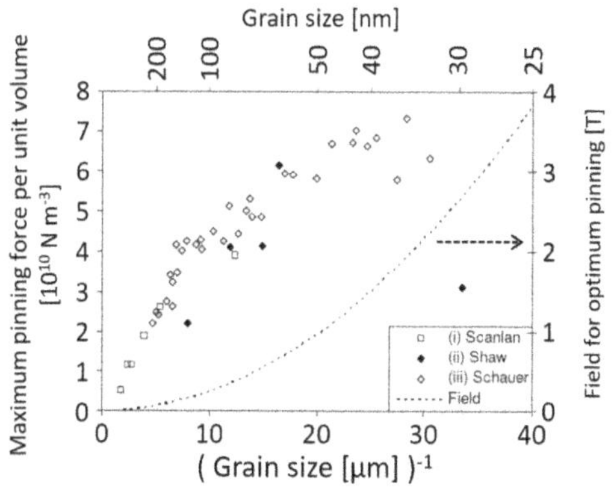

FIGURE B1.28 Variation of the maximum bulk pinning force with the reciprocal of the grain size, using data from (i) Scanlan et al. (1975), (ii) Shaw (1976) and (iii) Schauer and Schelb (1981). For comparison, the magnetic field resulting in a flux-line spacing equal to the grain size is also shown. [Figure and caption reproduced from Hopkins (2007), with kind permission from Simon Hopkins.]

TABLE B1.10 The ESE Core Scaling Parameters, Pinning-Force Shape Parameters, and Individual Wire Scaling Parameters for Extrapolating Transport Data

Core Scaling Parameters	Symbol [unit]	Suggested Values
Maximum effective critical temperature	$T_c^*(0)$ [K]	16.7
Prefactor temperature parameter	η [dimensionless]	2.25 (RRP® Ta-doped) 2.0 (ITER wires)
Prefactor strain parameter	σ [dimensionless]	1.1 (RRP® Ta-doped) 1.4 (ITER wires)
Strain sensitivity parameter	C_1 [dimensionless]	0.7–0.8
Pinning parameters		
Low-field pinning-force shape parameter	p [dimensionless]	0.50
High-field pinning-force shape parameter	q [dimensionless]	2.00
Individual wire scaling parameters		
Prefactor constant	C [A T]	Recommend using linear regression to determine
Maximum *effective* upper critical field	$B_{c2}(0,0)$ [T]	

and Godeke, 2008). There are promising ongoing attempts using oxide precipitates (for a comprehensive discussion, see Section E3.2.3 by Ian Pong) to refine Nb_3Sn grain size in wires (Xu et al., 2014).

If the pinning-force exponents are known or assumed, the flux pinning Equation B1.38 can be rearranged as:

$$J_c^{\frac{1}{q}} \cdot B^{\frac{1-p}{q}} = -\frac{K^{\frac{1}{q}}}{B_{c2}^{\frac{p}{q}+1}} B + \frac{K^{\frac{1}{q}}}{B_{c2}^{\frac{p}{q}}} \quad \text{(B1.39)}$$

which is in the form of $y = mx + c$, and B_{c2} can be determined from the ratio of y-intercept to slope (i.e. c/m) by plotting, for example, $J_c^{0.5} \times B^{0.25}$ (assuming $p = 0.5$ and $q = 2$) against B. The B_{c2} thus determined is known as the Kramer field (conventionally often H_K instead of B_K), after Edward Kramer (Kramer, 1973).

We conclude this section by discussing one of the most popular parameterizations, the extrapolative scaling expression (a.k.a. ESE) (Ekin et al., 2016; Ekin et al., 2017a; Ekin et al., 2017b). The ESE depends on four 'core scaling parameters', two 'pinning-force shape parameters', and two 'individual wire scaling parameters', see Table B1.10; a free, downloadable Microsoft Excel spreadsheet for plotting, data fitting, and magnet load line versus short sample limit calculation is made available (Pong and Ekin, 2017). The separable parts $g(\varepsilon)$, $h(t)$, and $f(b)$ are further defined below:

$$g(\varepsilon) = \left[s(\varepsilon)\right]^{\sigma} \quad \text{(B1.40)}$$

$$h(t) = \left[\left(1 - t^{1.5}\right) \cdot \left(1 - t^2\right)\right]^{\frac{\eta}{2}} \quad \text{(B1.41)}$$

For the strain function, the ESE model uses the exponential scaling law (Bordini et al., 2013):

$$s(\varepsilon) = \frac{e^{-C_1 \frac{J_2+3}{J_2+1} J_2} + e^{-C_1 \frac{I_1^2+3}{I_1^2+1} I_1^2}}{2} \quad \text{(B1.42)}$$

where I_1 and J_2 are the first invariant of the strain tensor and the second invariant of its deviatoric part, respectively. In terms of the principal strains ε_1, ε_2, and ε_3, these tensor invariants can be expressed as follows:

$$I_1 = \varepsilon_1 + \varepsilon_2 + \varepsilon_3 \quad \text{(B1.43)}$$

$$J_2 = \frac{1}{6}\left[(\varepsilon_1 - \varepsilon_2)^2 + (\varepsilon_2 - \varepsilon_3)^2 + (\varepsilon_3 - \varepsilon_1)^2\right] \quad \text{(B1.44)}$$

To specialize the scaling function to the case of uniaxial strain applied to a single wire, a further simplification is introduced, in which the principal strain component ε_1 is along the strand axis (longitudinal component), whereas the other two principal components, ε_2 and ε_3, are equal and perpendicular to the strand axis. Using these assumptions one can then write:

$$\varepsilon_1 = \varepsilon_{l0} + \varepsilon_a \quad \text{(B1.45)}$$

$$\varepsilon_2 = \varepsilon_3 = \varepsilon_{t0} - \nu\varepsilon_a \quad \text{(B1.46)}$$

where ε_a is the axial strain applied to the strand during the measurements; ε_{l0} and ε_{t0} are the longitudinal and transverse strains of the Nb_3Sn that are due to differential thermal contraction between the Nb_3Sn and the other materials of the composite wire; and ν is the 'effective' Poisson ratio of Nb_3Sn, recommended to assume a value of 0.36. The transverse residual strain, expressed in percent, can be computed by means of the following empirical relationship:

$$\varepsilon_{t0} = -\nu\varepsilon_{l0} + 0.1 \quad \text{(B1.47)}$$

Hence, by substituting Equations (B1.45) and (B1.46) into (B1.43) and (B1.44), one can obtain:

$$I_1 = (1 - 2\nu)(\varepsilon_a + \varepsilon_{l0}) + 0.2 \quad \text{(B1.48)}$$

$$J_2 = \frac{1}{3}\left[(1 + \nu)(\varepsilon_a + \varepsilon_{l0}) - 0.1\right]^2 \quad \text{(B1.49)}$$

At the end, all dependence on strain in the pinning force, and thus in the critical current, reduces to the definition of the strain function, $s(\varepsilon)$, which assumes different functional forms within the different models. For further information on parameterization for flux pinning, readers are referred to Ekin's topical reviews (Ekin, 2010; Ekin et al., 2016; Ekin et al., 2017a; Ekin et al., 2017b) and Bordini's work (Bordini et al., 2013).

References

Allen P. B., *Dynamical Properties of Solids*, edited by G. K. Horton and A. A. Maradudin (North-Holland, Amsterdam, 1980), Vol. 3

Allen P. B. and Dynes R. C., "Transition temperature of strong-coupled superconductors reanalyzed", *Physical Review B* 12 (1975): 905

Arbelaez D., Godeke A., and Prestemon S. O., "An improved model for the strain dependence of the superconducting properties of Nb_3Sn", *Superconductor Science and Technology* 22 (2009): 025005

Arko A., Lowndes D., Muller F. A., Roeland L. W., Wolfrat J., van Kessel A. T., Myron H. W., and Muller F. M., "de Haas-van Alphen effect in the high-T_c A15 superconductors in Nb_3Sn and V_3Si", *Physical Review Letters* 40 (1978): 1590

Asano T., Iijima Y., Itoh K., and Tachikawa K., "Effects of titanium addition to the niobium core on the composite-processed Nb_3Sn", *Transactions of the Japan Institute of Metals* 27 (1986): 204–214

Ashcroft N. and Mermin N. D., *Solid State Physics* (Saunders College Publishing, 1976)

Autler S. H., "Superconducting electromagnets", *Review of Scientific Instruments* 31 (1960): 369

Awaji A., "Quantitative strain measurements in Nb_3Sn wire and cable conductors using high-energy x-ray and neutron beams", *Superconductor Science and Technology* 26 (2013): 073001 (Topical Review)

Axe J. D. and Shirane G., "Inelastic-neutron-scattering study of acoustic phonons in Nb_3Sn", *Physical Review B* 8 (1973): 1965

Banno N., Takeuchi T., Fukuzaki T., and Wada H., "Optimization of the TRUQ (transformation-heat-based up-quenching) method for Nb_3Al superconductors", *Superconductor Science and Technology* 15 (2002): 519–525

Banno N., Uglietti D., Seeber B., Takeuchi T., and Flükiger R., "Strain dependence of superconducting characteristics in technical Nb_3Al superconductors", *Superconductor Science and Technology* 18 (2005): 284

Baroni S., de Gironcoli S., Corso A. D., and Giannozzi P., "Phonons and related crystal properties from density-functional perturbation theory", *Reviews of Modern Physics* 73 (2001): 515

Batterman B. W. and Barrett C. S., "Crystal structure of superconducting V_3Si", *Physical Review Letters* 13 (1964): 390

Baumgartner T., Eisterer M., Weber H. W., Flükiger R., Scheuerlein C., and Bottura L., "Effects of neutron irradiation on pinning force scaling in state-of-the-art Nb_3Sn wires", *Superconductor Science and Technology* 27 (2014): 015005

Bennemann K. H. and Garland J. W., *Superconductivity in d- and f-Bands Metals*, edited by Douglass D. H. (American Institute of Physics, New York, 1971)

Berlincourt T. G. and Hake R. R., "Upper critical field of transition metal alloy superconductors", *Physical Review Letters* 9 (1962): 293

Bolef D. I., "Elastic constants of single crystals of the bcc transition elements V, Nb, and Ta", *Journal of Applied Physics* 32 (1961): 100

Bordini B., Alknes P., Bottura L., Rossi L., and Valentinis D., "An exponential scaling law for the strain dependence of the Nb_3Sn critical current density", *Superconductor Science and Technology* 26 (2013): 075014

Bormio-Nunes C., Sandim M. J. R., and Ghivelder L., "Composition gradient as a source of pinning in Nb-Ti and NbTa-Ti superconductors", *Journal of Physics: Condensed Matter* 19 (2007): 446204

Bottura L. and Bordini B., "$J_C(B, T, \varepsilon)$ parameterization for the ITER Nb_3Sn production", *IEEE Transactions on Applied Superconductivity* 19 (2009): 1521

Bottura L., de Rijk G., Rossi L., and Todesco E., "Advanced accelerator magnets for upgrading the LHC", *IEEE Transactions on Applied Superconductivity* 22 (2012): Article #: 4002008

Boutboul T., Le Naour S., Leroy D., Oberli L., and Previtali V., "Critical current density in superconducting Nb–Ti Strands in the 100 mT to 11 T applied field range", *IEEE Transactions on Applied Superconductivity* 16 (2006): 1184–1187

Buta F., Sumption M. D., and Collings E. W., "Flux pinning in RHQT-processed Nb_3Al after various transformation heat treatments", *IEEE Transactions on Applied Superconductivity* 15 (2005): 3380–3384

Cardwell D. A. and Ginley D. S. (ed.), *Handbook of Superconducting Materials*, Institute of Physics (2003)

Carneiro T., 2014 Applied Superconductivity Conference, Charlotte (NC), August 2014. Retrieved July 19, 2019, from: http://ieeetv.ieee.org/mobile/video/niobium-manufacturing-for-superconductivity-asc-2014-plenary-series-4-of-13-tuesday-2014-8-12

Charlesworth J. P., MacPhail I., and Madsen P.E., "Experimental work on the niobium-tin constitution diagram and related studies", *Journal of Materials Science* 5 (1970): 580

Cheggour N., Lee P. J., Goodrich L. F., Sung J. H., Stauffer T. C., Splett J. D., and Jewell M. C., "Influence of the heat-treatment conditions, microchemistry, and microstructure on the irreversible strain limit of a selection of Ti-doped internal-tin Nb_3Sn ITER wires", *Superconductor Science and Technology* 27 (2014): 105004

Chu C. W., "Pressure-enhanced lattice transformation in Nb_3Sn single crystal", *Physical Review Letters* 33 (1974): 1283

Chu C. W. and Testardi L. R., "Direct observation of enhanced lattice stability in V_3Si under hydrostatic pressure", *Physical Review Letters* 32 (1974): 766; Erratum *Physical Review Letters* 32 (1974): 1149

Clogston A. M. and Jaccarino V., "Susceptibilities and negative Knight shifts of intermetallic compounds", *Physical Review* 121 (1961): 1357

Collings E. W., *A Sourcebook of Titanium Alloy Superconductivity* (Plenum, New York, 1983)

Cooley L. D., Hu Y. F., and Moodenbaugh A. R., "Enhancement of the upper critical field of Nb_3Sn utilizing disorder induced by ball milling the elements", *Applied Physics Letters* 88 (2006): 142506

Cooley L. D., Lee P. J., and Larbalestier D. C. "Flux-pinning mechanism of proximity-coupled planar defects in conventional superconductors: evidence that magnetic pinning is the dominant pinning mechanism in niobium–titanium alloy", *Physical Review B* 53 (1996): 6638–6652

Da Silva L. B. S., Rodrigues C. A., Bormio-Nunes C., Oliveira N. F. Jr, and Rodrigues D. Jr, "Influence of the introduction and formation of artificial pinning centers on the transport properties of nanostructured Nb_3Sn superconducting wires", *Journal of Physics: Conference Series* 167 (2009): 012012

de Gironcoli S., "Lattice dynamics of metals from density functional perturbation theory", *Physical Review B* 51 (1995): 6773

De Marzi G., Corato V., Muzzi L., della Corte A., Mondonico G., Seeber B., and Senatore C., "Reversible stress-induced anomalies in the strain function of Nb_3Sn wires", *Superconductor Science and Technology* 25 (2012): 025015

De Marzi G., Muzzi L., and Lee P. J., *Superconducting Wires and Cables: Materials and Processing* in Reference Module in Materials Science and Materials Engineering (Elsevier, 2016) SN 978-0-12-803581-8

De Marzi G., Morici L., Muzzi L. della Corte A., and Buongiorno Nardelli M., "Strain sensitivity and superconducting properties of Nb_3Sn from first principles calculations", *Journal of Physics: Condensed Matter* 25 (2013): 135702

Devantay H., Jorda J., Decroux M., Muller J., and Flükiger R., "The physical and structural properties of superconducting A15-type Nb-Sn alloys", *Journal of Materials Science* 16 (1981): 2145

Devred A., Backbier I., Bessette D., Bevillard G., Gardner M., Jong C., Lillaz F., Mitchell N., Romano G., and Vostner A., "Challenges and status of ITER conductor production", *Superconductor Science and Technology* 27 (2014): 044001

Dew-Hughes D., "Superconducting A-15 compounds: A review", *Cryogenics* 15 (1975): 435

Dietderich D. R. and Godeke A., "Nb_3Sn research and development in the USA - Wires and cables", *Cryogenics* 48 (2008): 331–340

Dietderich D. R., Kelman M., Litty J. R., and Scanlan R. M., "High critical current density in Nb_3Sn films with engineered microstructures - artificial pinning microstructures", *Advances in Cryogenic Engineering* 44 (1998): 951–958

Dietderich D. R. and Scanlan R. M., "Nb_3Sn artificial pinning microstructures", *IEEE Transactions on Applied Superconductivity* 17 (1997): 1201

Eastman D. E., "Photoemission studies of d-band structure in Sc, Y, Gd, Ti, Zr, Hf, V, Nb, Cr and Mo", *Solid State Communications* 7 (1969): 1697

Ekin J. W., "Strain scaling law for flux pinning in practical superconductors. Part 1: Basic relationship and application to Nb_3Sn conductors", *Cryogenics* 20 (1980): 611

Ekin J. W., "Strain scaling law for flux pinning in Nb–Ti, Nb_3Sn, Nb-Hf/Cu-Sn-Ga, V_3Ga and Nb_3Ge", *IEEE Transactions on Magnetics* 17 (1981): 658

Ekin J. W., "Strain effects in superconducting compounds", *Advances in Cryogenic Engineering* 30 (1984): 823–836

Ekin J. W., *Experimental Techniques for Low-Temperature Measurements* (Oxford University Press, 2006) ISBN: 0-19-857054-6 978-0-19-857054-7

Ekin J. W., "Unified scaling law for flux pinning in practical superconductors: I. Separability postulate, raw scaling data and parameterization at moderate strains", *Superconductor Science and Technology* 23 (2010): 083001

Ekin J. W., Cheggour N., Goodrich L., and Splett J., "Unified Scaling Law for flux pinning in practical superconductors: III. Minimum datasets, core parameters, and application of the Extrapolative Scaling Expression," *Superconductor Science and Technology* 30, (2017): 033005

Ekin J. W., Cheggour N., Goodrich L., Splett J., Bordini B., and Richter D., "Unified Scaling Law for flux pinning in practical superconductors: II. Parameter testing, scaling constants, and the Extrapolative Scaling Expression", *Superconductor Science and Technology* 29 (2016): 123002

Ekin J. W., Cheggour N., Goodrich L., Splett J., Bordini B., Richter D., and Bottura L., "Extrapolative scaling expression: a fitting equation for extrapolating full $I_c(B, T, \in)$ data matrixes from limited data," *IEEE Transactions on Applied Superconductivity* 27 (2017): 1–7

Ewald P., "The early history of the International Union of Crystallography", *Acta Crystallographica Section A* 33 (1977): 1–3

Finnemore D. K., Stromberg T. F., and Swenson C. A., "Superconducting properties of high-purity niobium", *Physical Review* 149 (1966): 231

Flükiger R., "Phase diagrams of superconducting materials", in *Superconductor Materials Science Metallurgy*, edited by Foner S. and Schwartz B. B. (Plenum, New York, 1981), pp. 511–603

Flükiger R., Growth of A15 type single crystals and polycrystals and their physical properties, in *Handbook of Superconducting Materials Volume I: Superconductivity, Materials and Processes*, edited by Cardwell D. A. and Ginley D. S. (IOP Publishing, Bristol and Philadelphia, 2003), p. 392

Flükiger R., Isernhagen R., Goldacker W., and Specking W., "Long-range atomic order, crystallographical changes and strain sensitivity of J_c in wires based on Nb_3Sn and other A15 type compounds", *Advances in Cryogenic Engineering* 30 (1984): 851–858

Flükiger R., Uglietti D., Senatore C., Buta F., "Microstructure, composition and critical current density of superconducting Nb_3Sn wires", *Cryogenics* 48 (2008): 293–307

Foner S. and McNiff E., "Upper critical fields of cubic and tetragonal single crystal and polycrystalline Nb_3Sn in DC fields to 30 tesla", *Solid State Communications* 39 (1981): 959

Gala F., De Marzi G., Muzzi L., and Zollo G., "The role of stoichiometry in superconducting $Nb_{1-\beta}Sn_{\beta}$: electronic and vibrational properties from ab-initio calculations", *Physical Chemistry Chemical Physics* 18 (2016): 32840–32846

Gavaler J. R., "Superconductivity in Nb-Ge films above 22K", *Applied Physics Letters* 23 (1973): 480

Giorgi A. L. and Matthias B.T., "Unusual superconducting behaviour of the molybdenum-technetium system", *Physical Review B* 17 (1978): 2160–2162

Giorgi A. L., Matthias B. T., and Stewart G. R., "Discovery of a superconducting A-15 phase in the V-Re system", *Solid State Communications* 27 (1978): 291

Glowacki B. A., "Niobium aluminide as a source of high-current superconductors", *Intermetallics* 7 (1999): 117–140 (Review)

Glowacki B. A., "Pinning Improvement of A15 Applied Superconducting Materials", *Acta Physica Polonica A* 130 (2016): 531–536

Godeke A., ten Haken B., ten Kate H. H. J., and Larbalestier D. C., "A general scaling relation for the critical current density in Nb_3Sn", *Superconductor Science and Technology* 19 (2006): R100–R116 (Topical Review)

Godeke A., "A review of the properties of Nb_3Sn and their variation with A15 composition, morphology and strain state", *Superconductor Science and Technology* 19 (2006): R68–R80 (Topical Review)

Halloran M. H., Condon J. H., Graebner J. E., Kunzler J. E., and Hsu F. S. L., "Experimental study of the Fermi surfaces of niobium and tantalum", *Physical Review B* 1 (1970): 366

Hanak J., Strater K., and Cullen R., "Preparation and properties of vapour-deposited niobium stannide", *RCA Review* 25 (1964): 342

Hansen M., Kamen E. L., Kessler H. D., and Mc Person D. J., "Systems titanium – molybdenum and the titanium – columbium", *Journal of Metals* 3 (1951): 881–888

Hardy G. F. and Hulm J. K., "The superconductivity of some transition metal compounds", *Physical Review* 93 (1954): 1004

Hawksworth D. G. and Larbalestier D. C., "Enhanced values of H_{c2} in Nb–Ti ternary and quaternary alloys", in *Advances in Cryogenic Engineering Materials,* edited by Clark A. F., Reed R. P. (Springer, Boston, MA, 1980), pp. 479–486

Hayase T. and Kajihara M., "Kinetics of reactive diffusion between Cu-8.1Sn-0.3Ti alloy and Nb", *Materials Science and Engineering: A* 433 (2006): 83–89

Hopkins S. C., "Optimisation, characterisation and synthesis of low temperature superconductors by current-voltage techniques," PhD thesis, Department of Materials Science and Metallurgy, University of Cambridge (2007)

Hulm J. K. and Blaugher R. D., "Superconducting solid solution alloys of the transition elements", *Physical Review* 123 (1961): 1569

Jahn H. and Teller E., "Stability of polyatomic molecules in degenerate electronic states I – orbital degeneracy", *Proceedings of the Royal Society* 161 (1937): 220

Jani A. R., Brener N. E., and Callaway J., "Band structure and related properties of bcc niobium", *Physical Review B* 38 (1988): 9425

Jorda J. L., Flükiger R., Junod A., and Muller J., "Metallurgy and superconductivity in Nb-Al", *IEEE Transactions on Magnetics* MAG-17 (1981): 557–560

Keller K. R., Hanak J. J., "Lattice softening in single crystal Nb_3Sn", *Physics Letters* 21 (1966): 263–264

Kikuchi A., Iijima T., Banno N., Takeuchi T., Inoue K., Nimori S., Kosuge M., and Yuyama M., "Microstructure and J_c – B performance of DRHQ processed Nb_3Al Tape with Ag stabilizer", *IEEE Transactions on Applied Superconductivity* 14 (2004): 1008–1011

Kikuchi A., Iijima T., and Inoue K., "Microstructures of rapidly-heated/quenched and transformed Nb_3Al multi-filamentary superconducting wires", *IEEE Transactions on Applied Superconductivity* 11 (2001): 3615–3618

Kramer E. J., "Scaling laws for flux pinning in hard superconductors", *Journal of Applied Physics* 44 (1973): 1360–1370

Kreilick T. S., "Niobium-titanium superconductors", in *ASM Handbook Volume 2: Properties and Selection: Nonferrous Alloys and Special-Purpose Materials*, 10th ed., (ASM International, Materials Park, OH, 1990), p. 1043

Kunzler J. E., Buehler E., Hsu F. S. L., and Wernick J. H., "Superconductivity in Nb_3Sn at high current density in a magnetic field of 88 kgauss", *Physical Review Letters* 7 (1961): 215

Labbé J. and Friedel J., "Effet de la température sur l'instabilité électronique et le changement de phase cristalline des composés du type V_3Si à basse température", *Journal de Physique* 27 (1966): 303–308

Labbé J. and van Reuth E. C., "Model to explain large changes in the electronic density of states with atomic ordering in V_3Au", *Physical Review Letters* 24 (1970): 1232

Larbalestier D. C., "Niobium-titanium superconducting materials", in *Superconductor Materials Science Metallurgy Fabrication and Applications*, edited by Foner S. and Schwartz B. B. (Plenum, New York, 1981), pp. 133–199

Lee P. J. and Larbalestier D. C., "An examination of the properties of SSC Phase II R&D strands", *IEEE Transactions on Applied Superconductivity* 3 (1983): 833–841

Lee P. J. and Larbalestier D. C., "Development of nanometer scale structures in composites of Nb–Ti and their effect on the superconducting critical current density", *Acta Metallurgica* 35 (1987): 2526–2536

Lee P. J., Larbalestier D. C., "Microstructural factors important for the development of high critical current density Nb_3Sn strand", *Cryogenics* 48 (2008): 283–292

Lee P. J., "Abridged metallurgy of ductile alloy superconductors", in *Wiley Encyclopedia of Electrical and Electronics Engineering* (Wiley, New York, 1999)

Lee P. J., Larbalestier D. C., Togano K., Tachikawa K., Suzuki M., Hamasaki K., Noto K., and Watanabe K., "Fabrication methods — 1. BCC alloys", in *Composite Superconductors*, edited by Osamura K. (Marcel Dekker, New York, 1994), pp. 237–258

Lee P. J. and Strauss B., "Nb–Ti - from beginnings to perfection", in *100 Years of Superconductivity*, edited by Rogalla H. and Kes P. H. (CRC Press, Taylor and Francis Group, Boca Raton, FL, 2012)

Loria R., De Marzi G., Anzellini S., Muzzi L., Pompeo N., Gala F., Silva E., and Meneghini C., "The effect of hydrostatic pressure on the superconducting and structural properties of Nb_3Sn: Ab-initio modeling and SR-XRD investigation", *IEEE Transaction on Applied Superconductivity* 27 (2017): 8400305

Madar R., Senateur J. P., and Fruchart R., "On the "martensitic" transformation in the A-15 superconductors compounds", *Journal of Solid State Chemistry* 28 (1979): 59

Mailfert R., Batterman B. W., and Hanak J. J., "Low temperature structural transformation in Nb_3Sn", *Physics Letters* 24A (1967): 315

Mailfert R., Batterman B. W., and Hanak J. J., "Observations related to the order of the low temperature structural transformation in V_3Si and Nb_3Sn", *Physica Status Solidi* 32 (1969): K67

Mattheiss L. F., "Electronic structure of niobium and tantalum", *Physical Review B* 1 (1970): 373

Matthias B. T. and Hulm J. K., "A Search for new superconducting compounds", *Physical Review* 87 (1952): 799

Matthias B. T., Geballe T., Geller S., and Corenzwit E., "Superconductivity of Nb_3Sn", *Physical Review* 95 (1954) 1435

Matthias B. T., Geballe T. H., and Compton V. B., "Superconductivity", *Reviews of Modern Physics* 35 (1963): 1

Matthias B. T., Geballe T. H., Longinotti L. D., Corenzwit E., Hull G. W., Willens R. H., and Maita J. P., "Superconductivity at 20 degrees Kelvin", *Science* 156 (1967): 645

Matthias B. T., "Transition temperatures of superconductors", *Physical Review* 92 (1953): 874

Matthias B. T., "Empirical relation between superconductivity and the number of valence electrons per atom", *Physical Review* 97 (1955): 74

McInturff A. D., *The Metallurgy of Superconducting Materials*, edited by Luhman T. and Dew-Hughes D., (Plenum, New York, 1980), Chapter 3

McMillan W. L., "Transition temperature of strong-coupled superconductors", *Physical Review* 167 (1968): 331

Meingast C. and Larbalestier D. C., "Quantitative description of a very-high critical current density Nb–Ti superconductor during its final optimization strain: II. Flux pinning mechanisms", *Journal of Applied Physics* 66 (1989): 5971–5983

Meissner W., Ochsenfeld R., "Ein neuer Effekt bei Eintritt der Supraleitfähigkeit", *Naturwissenschaften* 21 (1933): 787

Mentink M. G. T., "An Experimental and Computational Study of Strain Sensitivity in Superconducting Nb_3Sn", PhD Thesis, University of Twente (2014)

Mentink M. G. T., Anders A., Dhalle M. M. J., Dietderich D. R., Godeke A., Goldacker W., Hellman F., ten Kate H. H. J., Putnam D., Slack J. L., Sumption M. D., and Susner M. A., "Analysis of bulk and thin film model samples intended for investigating the strain sensitivity of Niobium-Tin", *IEEE Transactions on Applied Superconductivity* 21 (2011): 2550

Mentink M. G. T., Dhalle M. M. J., Dietderich D. R., Godeke A., Goldacker W., Hellman F., Sumption M. D., Susner M. A., and ten Kate H. H. J., "The effect of Ta and Ti additions on the strain sensitivity of bulk Niobium-Tin", *Physics Procedia* 36 (2012): 491

Mentink M. G. T., Dhalle M. M. J., Dietderich D. R., Godeke A., Goldacker W., Hellman F., and ten Kate H. H. J., "Towards analysis of the electron density of states of Nb_3Sn as a function of strain", *Advances in Cryogenic Engineering, AIP Conference Proceedings* 1435 (2012): 225–232

Miller G. L., "Tantalum and Niobium", in *Metallurgy of the Rarer Metals - 6*, (Butterworth Scientific Publications, London, 1959)

Moore D. F., Zubeck R. B., Rowell J. M., and Beasley M. R., "Energy gaps of the A-15 superconductors Nb_3Sn, V_3Si, and Nb_3Ge measured by tunneling", *Physical Review B* 20 (1979): 2721

Motowidlo L. R., Distin J., Lee P. J., Larbalestier D. C., and Ghosh A. K., "New developments in Nb_3Sn PIT strand: the effects of titanium and second phase additions on the superconducting properties", *IEEE Transactions on Applied Superconductivity* 21 (2011): 2546

Müller A. "Supraleitung und Existenzbereich der A-15 Phase im System Nb-Al-Ge", *Zeitschrift für Naturforschung*, 25 (1970): 1659–1669

Murnaghan F. D., "The compressibility of media under extreme pressures", *Proceedings of the National Academy of Sciences* 30 (1944): 244

Muzzi L., Affinito L., Corato V., De Marzi G., Di Zenobio A., Fiamozzi Zignani C., Napolitano M., Turtù S., Viola R., and della Corte A., "Magnetic and transport characterization of NbTi strands as a basis for the design of fusion magnets", *IEEE Transactions on Applied Superconductivity* 19 (2009): 2544

Muzzi L., Corato V., della Corte A., De Marzi G., Spina T., Daniels J., Di Michiel M., Buta F., Mondonico G., Seeber B., Flükiger R., and Senatore C., "Direct observation of Nb_3Sn lattice deformation by high-energy x-ray diffraction in internal-tin wires subject to mechanical loads at 4.2 K", *Superconductor Science and Technology* 25 (2012): 054006

Muzzi L., De Marzi G., Fiamozzi Zignani C., Besi Vetrella U., Corato V., Rufoloni A., and della Corte A., "Test results of a Nb–Ti wire for the ITER poloidal field magnets: a validation of the 2-pinning components model", *IEEE Transactions on Applied Superconductivity* 21 (2011): 3132

Nagasako N., Jahnatek M., and Hafner J., "Anomalies in the response of V, Nb and Ta to tensile and shear loading: *ab-initio* density functional theory calculations", *Physical Review B* 81 (2010): 094108

Nakagawa Y. and Woods A. D. B., "Lattice dynamics of niobium", *Physical Review Letters* 11 (1963): 271

"*NIOBIUM*", Proceedings of the International Symposium, Niobium '81, San Francisco, November 1981, ed. by H. Stuart, The Metallurgical Society of AIME, ISBN: 0-89520-468-1 (1984)

Niu H. J. and Hampshire D. P., "Disordered nanocrystalline superconducting $PbMo_6S_8$ with a very large upper critical field", *Physical Review* 91 (2003): 027002

Ochiai S., Uehara T., and Osamura K., "Tensile strength and flux pinning force of superconducting Nb_3Sn compound as a function of grain size", *Journal of Materials Science* 21 (1986): 1020–1026

Oguro H., Awaji S., Nishijima G., Badica P., Watanabe K., Shikanai F., Kamiyama T., and Katagiri K., "Room and low temperature direct three-dimensional-strain measurements by neutron diffraction on as-reacted and prebent CuNb/Nb_3Sn wire", *Journal of Applied Physics* 101 (2007): 103913

Orlando T. P., Alexander J., Bending S., Kwo J., Poon S., Hammond R., Beasley M., McNiff E., and Foner S., "The role of disorder in maximizing the upper critical field in the Nb-Sn system", *IEEE Transactions on Magnetics* 17 (1981): 368

Orlando T. P., McNiff E. J., Jr., Foner S., and Beasley M. R., "Critical fields, Pauli paramagnetic limiting, and material parameters of Nb_3Sn and V_3Si", *Physical Review B* 19 (1979): 4545

Osamura K., Ochiai S., Kondo S., Namatame M., and Nosaki M., "Influence of 3rd elements on growth of Nb_3Sn compounds and on global pinning force", *Journal on Material Science* 21 (1986): 1509

Palmieri V., "New materials for superconducting radiofrequency cavities", Proceedings of the 10th Workshop on RF Superconductivity, 2001, Tsukuba, Japan

Perdew J. P., *Electronic Structure of Solids*, edited by Ziesche P., and Eschrig H. (Academic Press, Berlin, 1991)

Perdew J. P., Burke K., and Ernzerhof M., "Generalized gradient approximation made simple", *Physical Review Letters*, 77 (1996): 3865–3868

Pintschovius L., Takei H., and Toyota N., "Phonon anomalies in Nb_3Sn", *Physical Review Letters* 54 (1985): 1260

Pong I. and Ekin J. W., *ESE Scaling Spreadsheet* (2017). Available: http://researchmeasurements.schralpit.com/ese-scaling-spreadsheet/

Pong I., Vostner A., Bordini B., Jewell M., Long F., Wu Y., Bottura L., Devred A., Bessette D., and Mitchell N., "Current sharing temperature of NbTi SULTAN samples compared to prediction using a single pinning mechanism parametrization for NbTi strand", *Superconductor Science and Technology* 25 (2012): 054011

Poole C., Farach H. A., and Creswick R. J., *Superconductivity*, 1st edition (Elsevier Academic Press, 2007), p. 71

Poole C., Farach H. A., Creswick R. J., and Prozorov R., *Superconductivity*, 2nd edition, (Academic Press, San Diego, 1995)

Powell B. M., Martel P., and Woods A. D. B., "Phonon properties of niobium, molybdenum, and their alloys", *Canadian Journal of Physics* 55 (1977): 1601

Qi M., Pan X. F., Zhang P. X., Cui L. J., Li C. S., Yan G., Chen Y. l., and Zhao Y., "Fabrication of Nb_3Al superconducting bulks by mechanical alloying method", *Physica C* 501 (2014): 39–43

Qiao L., Yang L., Song J., "Estimate of density-of-states changes with strain in A15 Nb_3Sn superconductors", *Cryogenics* 69 (2015): 58–64

Rehwald W., Rayl M., Cohen R. W., and Cody G. D., "Elastic moduli and magnetic susceptibility of monocrystalline Nb_3Sn", *Physical Review B* 6 (1972): 363

Roberge R., "Lattice parameter of niobium between 4.2 and 300 K", *Journal of Less Common Metals* 40 (1975): 161

Rossi L., "Superconductivity: its role, its success and its setbacks in the Large Hadron Collider of CERN", *Superconductor Science and Technology* 23 (2010): 034001

Salvetti M., "Hyperelastic continuum modeling of cubic crystals based on first-principles calculations", PhD Thesis, Massachussets Institute of Technology (2010)

Scanlan R. M., Fietz W. A., and Koch E. F., "Flux pinning centers in superconducting Nb_3Sn", *Journal of Applied Physics* 46 (1975): 2244–2249

Scheuerlein C., Arnau G., Alknes P., Jimenez N., Bordini B., Ballarino A., Di Michiel M., Thilly L., Besara T., and Siegrist T., "Texture in state-of-the-art Nb_3Sn multifilamentary superconducting wires", *Superconductor Science and Technology* 27 (2014): 025013

Scheuerlein C., Di Michiel M., and Buta F., "Synchrotron radiation techniques for the characterization of Nb_3Sn superconductors", *IEEE Transactions on Applied Superconductivity* 19 (2009): 2653

Scheuerlein C., Di Michiel M., Buta F., Seeber B., Senatore C., Flukiger R., Siegrist T., Besara T., Kadar J., Bordini B., Ballarino A., and Bottura L., "Stress distribution and lattice distortions in Nb_3Sn multifilament wires under uniaxial tensile loading at 4.2K", *Superconductor Science and Technology* 27 (2014): 044021

Scheuerlein C., Di Michiel M., Thilly L., Buta F., Peng X., Gregory E., Parrell J. A., Pong I., Bordini B., and Cantoni M., "Phase formations during the reaction heat of Nb_3Sn superconductors", *Journal of Physics: Conference Series* 234 (2010): 022032

Scott G. B. and Springford M., "The Fermi surface in niobium", *Proceedings of the Royal Society of London A.* 320 (1970): 115–130

Sekine H., Itoh K., and Tachikawa K., "A study of the H_{c2} enhancement due to the addition of Ti to the matrix of bronze-processed Nb_3Sn superconductors", *Journal of Applied Physics* 63 (1988): 2167

Sekine H., Takeuchi T., and Tachikawa K., "Studies on the composite processed Nb-Hf/Cu-Sn-Ga high-field superconductors", *IEEE Transactions on Magnetics* MAG-17 (1981): 383

Schauer W. and Schelb W., "Improvement of Nb_3Sn high-field critical current by a 2-stage reaction", *IEEE Transactions on Magnetics* 17 (1981): 374–377

Shaw B. J, "Grain size and film thickness of Nb_3Sn formed by solid-state diffusion in the range 650–800 °C", *Journal of Applied Physics* 47 (1976): 2143–2145

Shirane G. and Axe J. D., "Neutron scattering study of the lattice-dynamical phase transition in Nb_3Sn", *Physical Review B* 4 (1971): 2957

Shirane G. and Axe J. D., "Phonon softening of Nb_3Sn in $[\zeta\zeta\zeta]T$ modes", *Physical Review B* 18 (1978): 3742

Simmons G. and Wang H., *Single Crystal Elastic Constants and Calculated Aggregated Properties: A Handbook*, 2nd edition (MIT Press, Cambridge, Massachusetts and London, England, 1971)

Somerkoski J., Fiamozzi Zignani C., De Marzi G., and Muzzi L., "Metallurgical processes in Nb–Ti filaments as a function of isothermal annealing time", *Physics Procedia* 36 (2012): 1516

Spina T., Scheuerlein C., Richter D., Ballarino A., Cerutti F., Esposito L. S., Lechner A., Bottura L., and Flükiger R., "Correlation between the number of displacements per atom and T_c after high-energy irradiations of Nb_3Sn wires for the HL-LHC", *IEEE Transactions on Applied Superconductivity* 26 (2016): Article #: 6001405

Stewart G. R., "Superconductivity in the A15 structure", *Physica C* 514 (2015): 28

Stewart G. R., Newkirk L. R., and Valencia F. A., "Impurity stabilized A15 Nb_3Nb – a new superconductor", *Physical Review B* 21 (1980): 5055–5064

Suenaga M., *Metallurgy of Continuous Filamentary A15 Superconductors Superconductor Materials and Science: Metallurgy, Fabrication and Applications*, edited by Foner S. and Schwartz B. B. (Plenum Press, New York, 1980), p. 213

Suenaga M., Welch D. O., Sabatini R. L., Kammerer O. F., and Okuda S., "Superconducting critical temperatures, critical magnetic fields, lattice parameters, and chemical compositions of "bulk" pure and alloyed Nb_3Sn produced by the bronze process", *Journal of Applied Physics* 59 (1986): 840

Suenaga M. and Ralls K. M., "Some superconducting properties of Ti-Nb-Ta ternary alloys", *Journal of Applied Physics* 40 (1969): 4457

Tafto J., Suenaga M., and Welch D. O., "Crystal site determination of dilute alloying elements in polycrystalline Nb_3Sn superconductors using a transmission electron microscope", *Journal of Applied Physics* 55 (1984): 4330

Takeuchi T., "Nb_3Al conductors for high-field applications", *Superconductor Science and Technology* 13 (2000): R101–R119 (Topical Review)

Takeuchi T., Asano T., Iijima Y., and Tachikawa K., "Effects of the IVa element addition on the composite-processed superconducting Nb_3Sn", *Cryogenics* 21 (1981): 585–590

Tarantini C., Sung Z. H., Lee P. J., Ghosh A., and Larbalestier D. C., "Significant enhancement of compositional and superconducting homogeneity in Ti rather than Ta-doped Nb_3Sn", *Applied Physics Letters* 108 (2016): 042603

Taylor D. M. J., Al-Jawad M., and Hampshire D. P., "A new paradigm for fabricating bulk high-field superconductors", *Superconductor Science and Technology* 21 (2008): 125006

ten Haken B., Godeke A., and ten Kate H. H. J., "Investigation of microscopic strain by X-ray diffraction in Nb_3Sn tape conductors subjected to compressive and tensile strains", *Advances in Cryogenic Engineering* 42 (1997): 1463–1470

Testardi L. R., "Structural instability and superconductivity in A-15 compounds", *Reviews of Modern Physics* 47 (1975): 637

Testardi L. R. and Bateman T. B., "Lattice instability of high-transition-temperature superconductors. II. single-crystal V_3Si results", *Physical Review* 154 (1967): 402

Testardi L. R., Bateman T. B., Reed W. A., and Chirba V. G., "Lattice instability of V_3Si at low temperatures", *Physical Review Letters* 15 (1965): 250

Tütüncü H. M., Srivastava G. P., Bağcı S., and Duman S., "Theoretical examination of whether phonon dispersion in Nb_3Sn is anomalous", *Physical Review B* 74 (2006): 212506

Valentinis D. F., Berthod C., Bordini B., and Rossi L., "A theory of the strain-dependent critical field in Nb_3Sn, based on anharmonic phonon generation", *Superconductor Science and Technology* 27 (2014): 025008

van der Hoeven B. J. C., Jr. and Keesom P. H., "Specific heat of niobium between 0.4 and 4.2 K", *Physical Review* 134 (1964): A1320

Vieland L., "High-temperature phase equilibrium and superconductivity in the system niobium–tin", *RCA Review* 25 (1964): 366

Vieland L. J., Cohen R. W., and Rehwald W., "Evidence for a first-order structural transformation in Nb_3Sn", *Physical Review Letters* 26 (1971): 373

Vonsovsky S. V., Izyumov Yu. A., and Kurmaev E. Z., *Superconductivity in Transition Metals* (Springer, New York, 1982).

Vostner A., Jewell M., Pong I., Sullivan N., Devred A., Bessette D., Bevillard G., Mitchell N., Romano G., and Zhou C., "Statistical analysis of the Nb_3Sn strand production for the ITER toroidal field coils", *Superconductor Science and Technology* 30 (2017): 045004

Webb G. W., Vieland L. J., Miller R. E., and Wicklund A., "Superconductivity above 20°K in stoichiometric Nb_3Ga", *Solid State Communications* 9 (1971): 1769–1773

Weger M., "The electronic band structure of V_3Si and V_3Ga", *Reviews of Modern Physics* 36 (1964): 175

Weger M. and Goldberg I. B., *Solid State Physics: Advances in Research and Applications*, edited by H. Ehrenrich, F. Seitz, and D. Turnbull (Academic, New York, 1973), Vol. 28

Werthamer N. R., Helfand E., and Hohenberg P. C., "Temperature and purity dependence of the superconducting critical field, Hc2. III. Electron spin and spin-orbit effects", *Physical Review* 147 (1966): 295

Willens R. H., Geballe T. H., Gossard A. C., Maita J. P., Menth A., Hull G. W., Jr., and Soden R. R., "Superconductivity of Nb_3Al", *Solid State Communication* 7 (1969): 837–841

Wilson M. N., "Nb–Ti superconductors with low ac loss: a review", *Cryogenics* 48 (2008): 381

Wolf E. L., Zasadzinski J., Arnold G. B., Moore D. F., Rowell J. M., and Beasley M. R., "Tunneling and the electron-phonon-coupled superconductivity of Nb_3Sn", *Physical Review B* 22 (1980): 1214

Xu X., Sumption M., Peng X., and Collings E. W., "Refinement of Nb_3Sn grain size by the generation of ZrO_2 precipitates in Nb_3Sn wires", *Applied Physics Letters* 104 (2014): 082602

Yntema G. B., "Niobium superconducting magnets", *IEEE Transactions on Magnetics* 23 (1987): 390

Zhou J., Jo Y., Sung Z. H., Zhou H., Lee P. J., and Larbalestier D. C., "Evidence that the upper critical field of Nb_3Sn is independent of whether it is cubic or tetragonal", *Applied Physics Letters* 99 (2011): 122507

B2

Magnesium Diboride

Chiara Tarantini

Despite having been synthesized as far back as the 1950s, the superconducting properties of magnesium diboride (MgB_2) were only discovered at the beginning of 2001. Akimitsu and his group were attempting to make a chemical analogue of CaB_6 by replacing Ca with Mg (Cava 2001). They intended to use MgB_2, a compound commercially available, as starting material and, during a routine characterization of MgB_2 properties before using it in the synthesis, they discovered that it had a superconducting transition T_c at 39 K (Nagamatsu *et al.*, 2001). Although high-T_c superconductivity in cuprates had already been studied for 15 years, this discovery aroused huge surprise because the highest T_c in a simple intermetallic compound was 23 K (Nb_3Ge) and because the theoretical limit for conventional BCS superconductors, whose properties are mediated by the electron–phonon interaction, was thought to be 30 K. Moreover, from a practical point of view, MgB_2's high critical temperature in principle allowed working with liquid hydrogen or cryocoolers at an operational temperature above 20 K, thus obviating the expense and availability issues of liquid helium.

As discussed in the following, the high T_c is a consequence of the peculiar multiband electronic structure of MgB_2 and of the presence of two distinct gaps clearly identified in tunnelling and specific heat experiments (Giubileo *et al.*, 2001, Bouquet *et al.*, 2001). The theoretical aspects of a multiband and multigap superconductivity were discussed by Suhl *et al.*, since 1959, only a few years after the development of the Bardeen, Copper and Schrieffer theory of superconductivity (Bardeen *et al.*, 1957). The authors considered a disparity of pairing interaction in different bands, predicting the existence of different order parameters and a possible enhancement of critical temperature. Nevertheless, before the discovery of MgB_2, two distinct order parameters were only observed in Nb-doped $SrTiO_3$ at 100 mK (Binnig *et al.*, 1980) and, though few superconductors (like A15 and borocarbides) presented a multiband structure, they seemed to exhibit only one gap.

B2.1 Crystallographic and Electronic Structure

Magnesium diboride (MgB_2) is a binary intermetallic compound with a simple hexagonal structure (a = 3.086 Å, c = 3.524 Å) consisting of alternating layers of boron and magnesium (Nagamatsu *et al.*, 2001). Boron atoms form honeycomb layers stacked without displacement which generate hexagonal prisms. At their centres, large nearly spherical sites are created and occupied by magnesium atoms. As a consequence, magnesium atoms form a triangular lattice halfway between the B layers (see Figure B2.1).

The MgB_2 electronic band structure (Figure B2.2, left panel) consists of three bonding σ bands, corresponding to in-plane strongly covalent sp_xp_y (sp^2) hybridization in the boron layer, and two π bands (bonding and antibonding) formed by hybridized boron p_z orbitals (An and Pickett 2001, Kortus *et al.*, 2001, Mazin and Antropov 2003, Choi *et al.*, 2003). As a consequence of neighbouring boron atoms, a large overlap between all p orbitals occurs producing strong in-plane dispersion in both σ and π bands. On the contrary, the interlayer overlaps are smaller, especially for the p_{xy} orbital, and the k_z dispersion of σ bands remains weak. Only two σ bands cross the Fermi level and, because of the weak k_z dispersion, they form two nearly cylindrical sheets of Fermi surface around the Γ-A line (see Figure B2.2, right panel). For this reason, the holes at the top of these σ bands manifest notably two-dimensional properties and they are localized on the B sheets. Both π bands cross the Fermi level forming two planar honeycomb tubular networks: an antibonding electron-type sheet is centred on the ALH plane and a more compact bonding hole-type sheet is centred on the ΓMK plane. Differently from the σ bands, both π bands have three-dimensional character and are delocalized over the whole crystal, showing metallic behaviour (Kortus *et al.*, 2001, Mazin and Antropov 2003). The MgB_2 electronic band structure was confirmed by various techniques such as angle-resolved photoemission spectroscopy (ARPES) (Uchiyama *et al.*, 2002, Souma *et al.*, 2003) and de Haas–van Alphen effect (Yelland *et al.*, 2002, Carrington *et al.*, 2007).

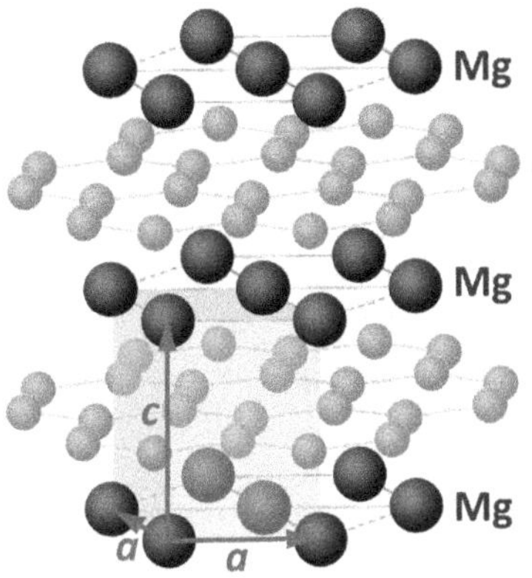

FIGURE B2.1 Crystal structure of magnesium diboride. Boron atoms form a stack of honeycomb layers whereas magnesium atoms sit between at the centre of the B-prisms. The shadowed area represents the unit cell.

An important feature of MgB_2 is that the σ and π bands are formed by different local orbitals and that they are orthogonal. Moreover the p_z orbital has odd parity whereas the B-B bond orbital has even parity with respect to the boron plane, giving a small σ-π interband scattering in the pure material (Mazin *et al.*, 2002). Although the Fermi surface is built up from four distinct sheets, the similarity in the Fermi velocity and the electron–phonon coupling (EPC) of the two σ and π bands and the small interband scattering frequently leads to a simplification of the model into just two effective bands: one 2D σ band and one 3D π band. The σ and π bands can thus be considered as two distinct channels conducting in parallel. This point underpins our discussion of the normal state resistivity.

B2.2 Critical Temperature, Isotope Effect and Electron–Phonon Coupling

In a conventional BCS superconductor driven by electron–phonon coupling, the critical temperature for a strong-coupled material can be expressed by the Allen–Dynes modification (Allen and Dynes 1975) of the McMillan equation (McMillan 1968):

$$k_B T_c = \frac{\hbar\omega_{\ln}}{1.2} e^{-\frac{1.04(1+\lambda)}{\lambda - \mu^*(1+0.62\lambda)}} \tag{B2.1}$$

where $\omega_{\ln}$ is the average phonon energy, λ is the coupling constant and μ^* is the Coulomb repulsion pseudo-potential. As a consequence of the electron–phonon interaction, a BCS superconductor shows an isotope effect α that in a monoatomic material with atomic mass M is given by: $\alpha = -\frac{\partial \ln T_c}{\partial \ln M} \approx -\frac{\Delta T_c}{T_c}\frac{M}{\Delta M}$, with a predicted value of 0.5. In a multiatomic system, the isotope effect is instead given by the contributions of all the elements $\alpha_t = \sum \alpha_i = \sum -\partial \ln T_c / \partial \ln M_i$.

In the case of magnesium diboride, the mechanisms that induce superconductivity were not initially clear because of its high transition temperature and because the first estimation of the parameters in Equation (B2.1) ($\omega_{\ln}$ ~60 meV, μ^*~0.1, λ~ 0.6 [Liu *et al.*, 2001, Choi *et al.*, 2002a, Wang *et al.*, 2001, Bouquet *et al.*, 2001]) led to a predicted T_c less than 20 K. However, measurements of the boron isotope effect performed by Bud'ko *et al.*, 2001 revealed a T_c variation of 1 K in samples prepared with ^{10}B and ^{11}B. Afterward Hinks *et al.*, 2001 also investigated the influence of two Mg isotopes on the critical temperature observing a

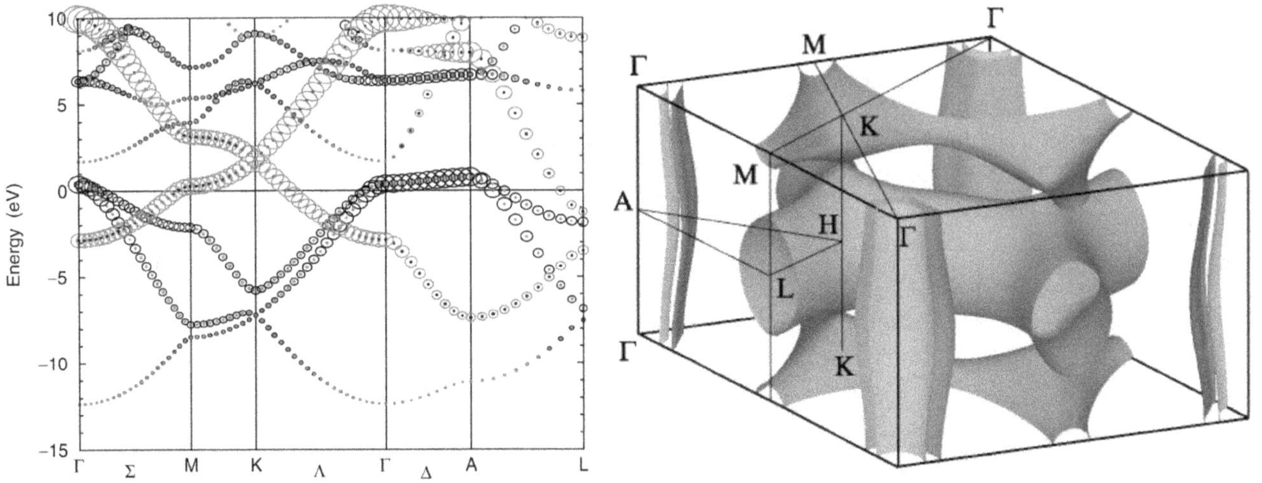

FIGURE B2.2 (Left) Electronic band structure of MgB_2 along the symmetry lines of the Brillouin zone with the B p-character. The radii circles are proportional to the π and σ characters. (Right) Fermi surface of MgB_2: the cylinders are the σ sheets, whereas the tubular networks are the π sheets. [Reprinted figures with permission from J. Kortus, I.I. Mazin, K.D. Belashchenko, V.P. Antropov, L.L. Boyer, Phys. Rev. Lett. 86, 4656 (2001). Copyright 2001 by the American Physical Society (http://dx.doi.org/10.1103/PhysRevLett.86.4656).]

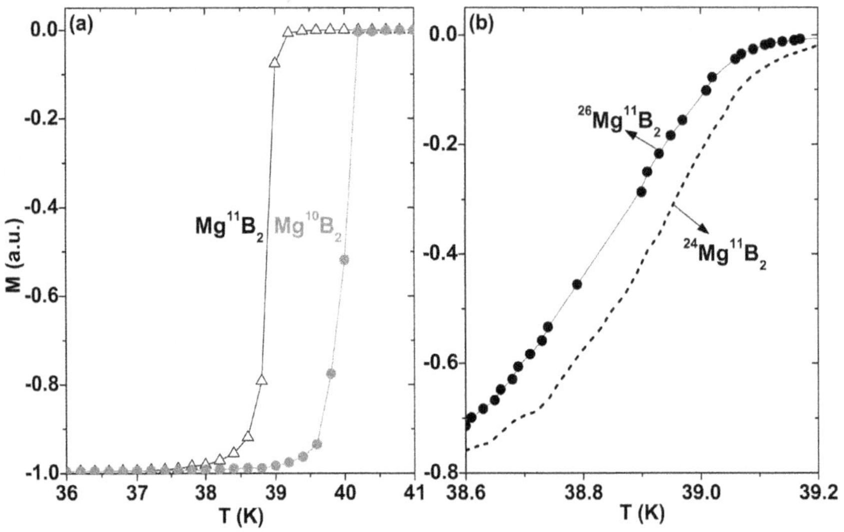

FIGURE B2.3 The superconducting transitions for isotopically substituted MgB_2 samples. (a) Samples prepared with ^{11}B and ^{10}B (data from Bud'ko *et al.*, 2001) show a significant B isotope effect with a T_c variation of 1 K. (b) Sample prepared with ^{24}Mg and ^{26}Mg (data from Hinks *et al.*, 2001) shows a small Mg isotope effect with a T_c variation of 0.1 K.

weak effect of 0.1 K (see Figure B2.3). The presence of the isotope effect clarified that MgB_2 is indeed a phonon-mediated superconductor and the larger boron contribution to the total α_t ($\alpha_B = 0.30$, $\alpha_{Mg} = 0.02$, leading to a total effect of $\alpha_t = 0.32$) indicates that boron atoms are greatly involved in the electron–phonon coupling. The failure of Equation (B2.1) in predicting the correct T_c is due to the fact that this equation is for isotropic single-band superconductors, hence too simple to explain MgB_2. In the case of an anisotropic multiband superconductor, Equation (B2.1) has to be substituted by (Mazin and Antropov 2003):

$$k_B T_c = \frac{\hbar\omega_{\ln}}{1.2} e^{-\frac{1}{(\lambda-\mu^*)_{\mathrm{eff}}}} \tag{B2.2}$$

where $(\lambda\text{-}\mu^*)_{\mathrm{eff}}$ is defined as the maximum eigenvalue of the matrix $\Lambda_{ij}^{\mathrm{eff}} = \frac{\Lambda_{ij} - \mu_{ij}^*\left(1+0.62\sum_n \Lambda_{in}\right)}{1+\sum_n \Lambda_{in}}$. In these expressions, the coupling constant λ and the Coulomb repulsion pseudopotential μ^* have been substantially split in the 2×2 matrices, Λ_{ij} and μ_{ij}^*, respectively, where ij are the band indices. The Λ_{ij} and μ_{ij}^* matrices were calculated by several groups (Liu *et al.*, 2001, Golubov *et al.*, 2002, Mazin and Antropov 2003, Choi *et al.*, 2003, Floris *et al.*, 2007), obtaining slightly different values (for instance Golubov *et al.*, 2002 calculated $\lambda_{\sigma\sigma}$~1.017, $\lambda_{\pi\pi}$~0.448, $\lambda_{\sigma\pi}$~0.213,$\lambda_{\pi\sigma}$~0.155, $\mu_{\sigma\sigma}^*$~0.210, $\mu_{\pi\pi}^*$~0.172, $\mu_{\sigma\pi}^*$ ~0.095, $\mu_{\pi\sigma}^*$~0.069). However, all papers reported a very strong electron–phonon coupling in the σ bands, a weaker coupling in the π bands and even smaller coupling for the interband coupling, and they correctly reproduced the experimental T_c.

The effects of the electron–phonon coupling in MgB_2 were intensively studied by different groups (Yildrim *et al.*, 2001, Liu *et al.*, 2001, An and Pickett 2001, Choi *et al.*, 2002a,b, Choi *et al.*, 2003) and four distinct modes were identified at the zone centre Γ: B_{1g}, A_{2u}, E_{1u} and E_{2g} (see Figure B2.4[a]). Only one of them, the E_{2g} phonon mode, significantly contributes to the coupling. This mode is doubly degenerate on the Γ-A line and involves only in-plane boron motions in opposite directions

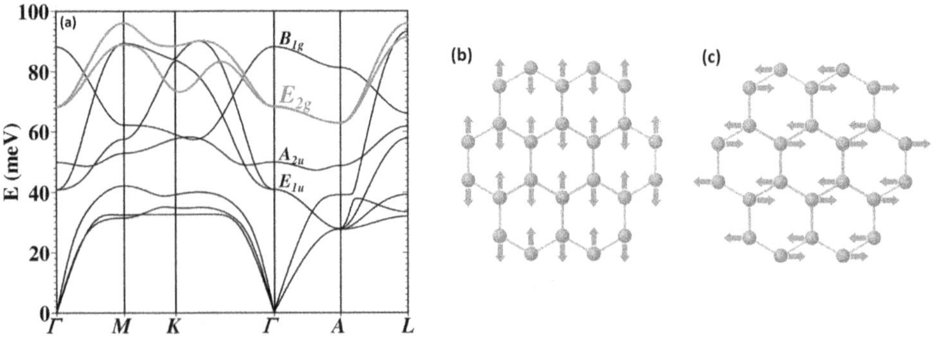

FIGURE B2.4 (a) Phonon dispersion in MgB_2: the E_{2g} modes dominate the electron–phonon coupling (data from Yildrim *et al.*, 2001). (b) and (c) show the E_{2g} vibration modes of the boron atoms in the B-layers.

along x or y axes (Figure B2.4[b] and [c]). Since the in-plane motions of the boron change the boron orbital overlap, a significant electron–phonon coupling was expected in the σ bands at the Fermi level. This result was later confirmed by inelastic X-ray scattering measurements (Shukla *et al.*, 2003). Although the first calculations hypothesized a significant contribution of the anharmonicity of the E_{2g} mode (large fourth-power term in the potential well) with a strong hardening to explain the reduced isotopic effect in MgB_2 (experimental α_t = 0.32 instead of the theoretical BCS 0.5) (Choi *et al.*, 2002a,b, Liu *et al.*, 2001, Yildrim *et al.*, 2001, An and Pickett 2001), more recent calculations that take into account scattering between different phonon modes at different points of the Brillouin zone (Lazzeri *et al.*, 2003, Calandra *et al.*, 2007) indicate that the anharmonicity is fairly small. These calculations are in good agreement with the X-ray scattering measurements (Shukla *et al.*, 2003, Baron *et al.*, 2004) leaving still open the question of the origin of the reduced isotope effect.

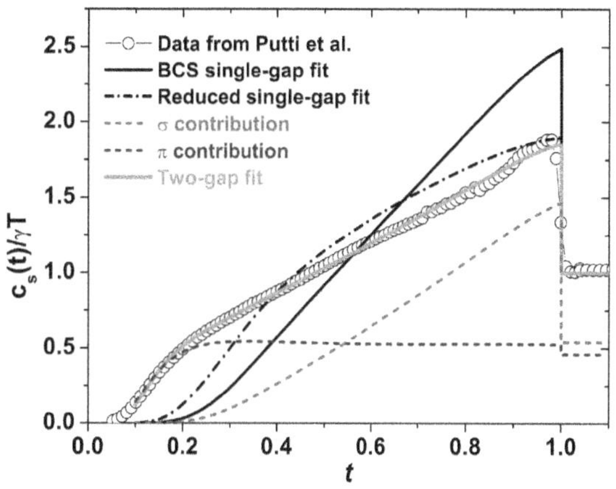

FIGURE B2.6 Experimental data on the normalized electronic contribution to the specific heat $c_s/\gamma T$ as a function of reduced temperature for a pure MgB_2 sample (circles, data from Putti *et al.*, 2006), compared with a BCS curve and a curve with reduced gap amplitude (solid and dash-dot lines). The thick solid line represents the two-gap fit obtained by the σ and π contributions (dashed lines).

B2.3 Multigap Superconductivity

Multigap behaviour is a rare feature of a superconductor. In fact, having different bands crossing the Fermi level is a necessary but not a sufficient condition for the existence of distinct gaps. A further condition is that the bands must not become mixed by interband impurity scattering. This is fulfilled in MgB_2 as a consequence of the different parity of the σ and π bands.

On the base of first-principles calculations, multigap superconductivity in MgB_2 was suggested in several works (Liu *et al.*, 2001, Golubov *et al.*, 2002, Choi *et al.*, 2002b). In particular, starting from the anisotropic Eliashberg formulation, Choi *et al.*, calculated that the energy gap changes significantly on different portions of the Fermi surface. As shown in Figure B2.5, the magnitude of the energy gap at 4 K varies from 6.4 to 7.2 meV on the σ sheets and from 1.2 to 3.7 meV on the π sheets, with average values of 6.8 and 1.8 meV, respectively. The ratio $2\Delta(0)/k_BT_c$ is ~4.0 on the σ sheets and ~1.07 on the π ones, respectively larger and smaller than BCS value (3.53). The spread in the gap values inside each band, due to the anisotropy, is rarely experimentally observable, and typically only two gaps are measured. The first evidence of a multigap behaviour in MgB_2 came from specific heat measurements (Wang *et al.*, 2001, Bouquet *et al.*, 2001). As highlighted by Bouquet *et al.*, (2001,2003), there is a large discrepancy between MgB_2 experimental data and the theoretical BCS curve (see Figure B2.6). The data cannot be described by any single gap curve because of a large excess of specific heat at low temperature (t~0.2). This feature can be explained only considering a second small gap that influences the curve shape in this temperature range. The existence of the two gaps was then verified by scanning tunnelling microscopy and spectroscopy (STM and STS) (Giubileo *et al.*, 2001, Iavarone *et al.*, 2002), SIS tunnel junction (Schmidt *et al.*, 2002) and point contact spectroscopy (Szabó *et al.*, 2001, Gonnelli *et al.*, 2003). In Figure B2.7, the first conductance spectrum shows a clear two-gap structure in MgB_2 as obtained by STM-STS; the temperature dependence of the gaps reveals that both gaps close at the same T_c (Iavarone *et al.*, 2002), similarly to the theoretical prediction (Figure B2.5). Such behaviour was actually predicted already in 1959 by Suhl *et al.*, the first to deal with the multiband/multigap superconductivity: they explained that if off-diagonal elements of coupling matrix λ_{12} and λ_{21} were zero, two gaps with distinct transition temperatures would exist, whereas gap closing at the same T_c is due to non-zero interband coupling ($\lambda_{12}\neq0$ and $\lambda_{21}\neq0$).

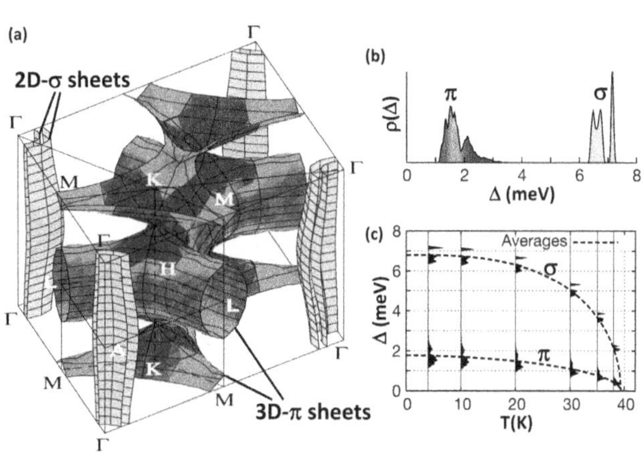

FIGURE B2.5 (a) The superconducting energy gap on the Fermi surface at 4 K in grey scale and (b) the distribution of gap values at 4 K. (c) Calculated temperature dependence of the superconducting gaps. (Calculations from Choi *et al.*, 2002b.)

Mazin *et al.*, 2004 explained that observation of the fine structure within the σ or π bands described by Choi *et al.*, 2002b is extremely hard because intraband impurity scattering suppresses any intraband non-uniformity of the order

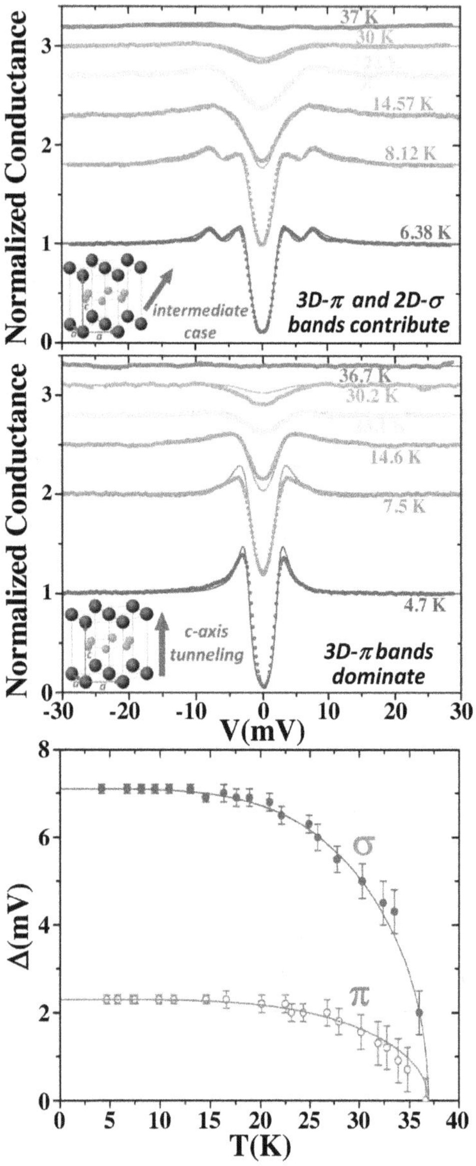

FIGURE B2.7 (Top and middle) Temperature evolution of the two tunneling spectra in two different grains together with theoretical fits: the first spectrum shows a clear two-gap structure, while the second spectrum reveals only the smaller gap because of the grain orientation. (Bottom) Temperature dependence of the two gaps as obtained from the above spectra. (Data from Iavarone *et al.*, 2002.)

parameter: the authors also pointed out that the gap structure could be observable only in extremely clean MgB_2 with a mean free path greater than 150 nm. The only MgB_2 samples that can achieve such a clean limit are thin films deposited by hybrid physical–chemical vapour deposition (HPCVD) (Zeng *et al.*, 2002, Xi *et al.*, 2007), and gap distributions similar to the theoretical prediction (Figure B2.5) were only recently demonstrated in these films (Chen *et al.*, 2012).

Considering the general relationship between the gap amplitude at 0 K and the coherence length, $\xi_0 = \dfrac{\hbar v_F}{\pi\Delta(0)}$, the presence of two gaps in MgB_2 implies the existence of two distinct coherence lengths, ξ_0, as well; moreover, because of the anisotropy of the Fermi velocities, the coherence lengths are also anisotropic. Considering the gap amplitudes measured by Iavarone *et al.*, 2002 and the Fermi velocities calculated by Brinkman *et al.*, 2002, the coherence lengths in the clean limit can be estimated to be ~13 and ~2 nm for the σ band in the ab plane and along c and ~49 and ~57 nm for the π band. The σ band coherence lengths can be also estimated from the upper critical fields of clean single crystals using the formulae $\mu_0 H_{c2}^{//c} = \dfrac{\phi_0}{2\pi\xi_{ab}^2}$ and $\mu_0 H_{c2}^{//ab} = \dfrac{\phi_0}{2\pi\xi_{ab}\xi_c}$. Since the typical clean limit values are $\mu_0 H_{c2}^{//c} \sim$ 3-5 T and $\mu_0 H_{c2}^{//ab} \sim$ 12-15 T, the σ band ξ_0 values are so confirmed.

B2.4 Effects of Disorder

Introducing disorder into a superconductor is frequently employed both to understand the pairing mechanisms and to change the properties, especially with a view towards applications. The unique multiband/multigap nature of MgB_2 implies that disorder affects T_c, the gap amplitudes and the normal state properties differently from any other material. For this reason, disorder introduced by doping or irradiation has been used as a tool to investigate the physical properties of MgB_2. Of particular interest is the influence of the three different scattering channels (the two intrabands and the interband) on the upper critical fields, H_{c2}. In the most favourable field orientation, H_{c2} can be pushed up to ~70 T (Braccini *et al.*, 2005).

B2.4.1 T_c and Gap Amplitudes: Impurity Scattering

The study of the effects of disorder in magnesium diboride is of particular interest because it clearly differs from that observed in single-gap materials, where only magnetic impurities are able to reduce the critical temperature (unless there is smearing of the density of states, as in the A15 phase (Testardi and Mattheiss 1978). The effects of both magnetic and non-magnetic impurity scattering in a multiband superconductor were already treated by Golubov and Mazin in 1997, before the discovery of superconductivity in MgB_2, and they were later extended to MgB_2 by Mazin and Antropov 2003. They predicted that non-magnetic impurities would suppress superconductivity in a multigap material in the same way that magnetic impurities do in a single-gap superconductor. However, only interband scattering is pair-breaking, whereas intraband scattering does not affect T_c. In the weak scattering limit, the T_c suppression by non-magnetic impurities is given by:

$$\frac{\delta T_c}{T_c} = -\frac{\pi\gamma_{12}}{8k_B T_c}\frac{(\Delta_1-\Delta_2)(\Delta_1 N_1-\Delta_2 N_2)}{(\Delta_1^2+\Delta_2^2)N_2} \tag{B2.3}$$

where $\gamma_{12} \equiv \gamma_{21} N_2 / N_1$ is the interband scattering rate, Δ_i are the gap amplitudes and N_i are the densities of states of the two bands at the Fermi level. This relation implies that the T_c suppression is linear in γ_{12} in the weak scattering limit. In MgB_2, a T_c suppression of 1 K would require an interband scattering of order 1 meV. It was also estimated that in the weak scattering limit, the variation of the gap amplitudes can be linearly approximated by: $\delta\Delta_\pi \approx 0.23\gamma_{\sigma\pi}$, $\delta\Delta_\sigma \approx -0.38\gamma_{\sigma\pi}$ (where $\delta\Delta_i$ is expressed in meV and $\gamma_{\sigma\pi}$ in cm^{-1}) (Iavarone *et al.*, 2005). This means that a small interband scattering causes a decrease in the σ gap and an increase in the π gap.

In the case of strong interband scattering, the critical temperature gradually decreases with increasing $\gamma_{\sigma\pi}$ and saturates at the value expected for isotropic BCS coupling when a complete isotropization of the whole Fermi surface occurs. This T_c limit value has been estimated by several groups to lie between 19 and 25 K (Choi *et al.*, 2002b, Liu *et al.*, 2001, Dolgov *et al.*, 2005). The limit is expected to be reached when the interband scattering rate becomes larger than the relevant phonon frequency (600 cm^{-1} ~ 75 meV) (Golubov and Mazin 1997, Mazin and Antropov 2003). If interband scattering were the only pair-breaking mechanism, this value would be the lowest T_c limit. The calculated effect of interband scattering alone on the gap amplitudes is to decrease the σ gap and to increase the π gap until they merge into the isotropic weak coupling BCS value of $\Delta(0)/k_BT_c \sim 1.76$ at the T_c limit (Kortus *et al.*, 2005).

B2.4.1.1 T_c and Gap Amplitudes: Doping and Irradiation

In order to verify the gap merging and to confirm the two-gap model, disorder was introduced in MgB_2 in many experiments. The usual way to change the superconducting properties is by chemical substitution, and this approach was attempted also in MgB_2 in order to increase the interband scattering. The most effective chemical substitutions in MgB_2 are Al on the Mg sites (Putti *et al.*, 2003, Putti *et al.*, 2005, Daghero *et al.*, 2008, Szabó *et al.*, 2007, Klein *et al.*, 2006) and C on the B sites (Hol'anová *et al.*, 2004, Gonnelli *et al.*, 2005, Iavarone *et al.*, 2005, Tsuda *et al.*, 2005). However, both Al and C substitutions dope with electrons producing not only interband scattering but also differently shifting the Fermi level and changing the density of states (DOS), with C substitution mainly affecting the σ bands while Al substitution changes the π bands. Both substitutions induce a T_c suppression that roughly scales with the charge doping. In fact Kortus *et al.*, 2005 pointed out that T_c versus Al and C content shows very similar behaviour if a factor of two scaling is considered in order to take into account the different quantity of introduced electrons per cell (Figure B2.8). Both substitutions suppressed T_c below the limit predicted for solely interband scattering (horizontal lines in Figure B2.8) and Kortus *et al.* were able to reproduce the T_c behaviour (solid line in Figure B2.8) considering the change in the DOS due to the band filling and the hardening of the E_{2g} phonon modes that, in turn, reduces the electron–phonon coupling ($\lambda \sim 1/\omega^2$).

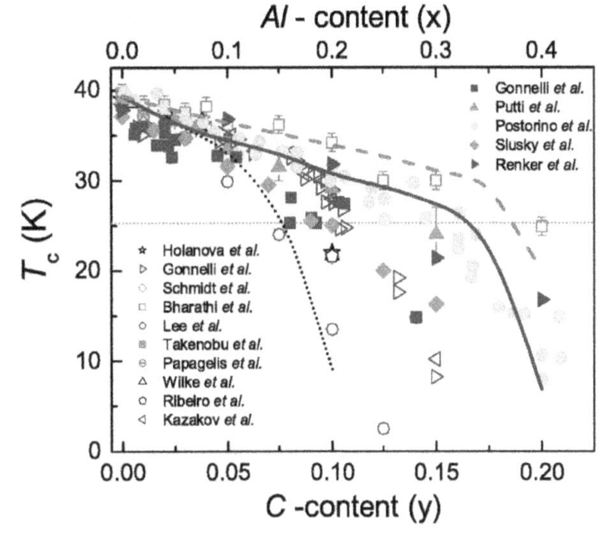

FIGURE B2.8 Critical temperature T_c as a function of Al (solid symbols) and C (open symbols) doping concentration in $Mg_{1-x}Al_x(B_{1-y}C_y)_2$. The lines represent calculations based on Eliashberg's theory: dotted line considers changes in DOS from rigid band model, dashed line DOS obtained by recalculating the electronic band structure and solid line recalculated DOS and phonon renormalization. The horizontal dotted line is the lower limit for interband scattering alone. [Reprinted figure with permission from J. Kortus, O.V. Dolgov, R.K. Kremer, A.A. Golubov, *Phys. Rev. Lett.* 94, 027002 (2005). Copyright 2005 by the American Physical Society (http://dx.doi.org/10.1103/PhysRevLett.94.027002).]

The effect of substitutions on the gap amplitudes differs from that theoretically predicted considering only the interband scattering (dashed line in Figure B2.9). In the Al-doped samples (Figure B2.9[a]), a wide range of values were experimentally obtained, but in every case both π and σ gaps are reduced when T_c is significantly suppressed and merging of the gaps is not observed. This occurs because the dominant effect in the Al case is the reduction of the DOS, although some interband scattering might explain the initial flat trend of the π gap in some samples. A similar behaviour is observed in most of the C-doped samples (Figure B2.9[b]) with both gaps roughly following the theoretical curves obtained considering the reduced DOS only (dash-dot lines). However, there is one sample series (Gonnelli *et al.*, in Figure B2.9[b]) that shows a nearly constant π gap behaviour and the gap merging seems to occur at about 19 K (Gonnelli *et al.*, 2005), indicating that the interband scattering is more effective in the C-doped case than in Al case.

A more clear observation of the gap merging was finally found in neutron irradiated samples (Figure B2.10) (Putti *et al.*, 2006, Wang *et al.*, 2003, Daghero *et al.*, 2006). Since it does not produce charge doping, irradiation is the most effective way to investigate an increase in the interband scattering as the dominant effect. Specific heat data clearly show that for T_c above 20 K, a two-gap model is still necessary to fit the experimental curve, whereas below 11 K, the data are well reproduced by a single-gap fit (Putti *et al.*, 2006). The gap merging was later confirmed by point-contact Andreev reflection spectroscopy on the same sample series (Daghero *et al.*, 2006).

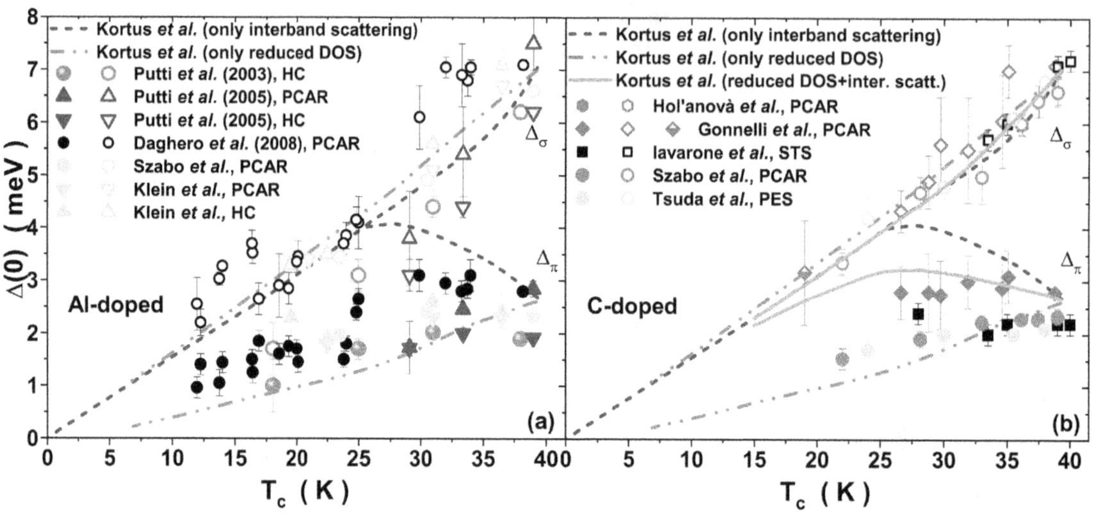

FIGURE B2.9 Gap amplitudes as a function of T_c as determined by heat capacity (HC) point-contact Andreev reflection spectroscopy (PCAR), photoemission spectroscopy (PES) and scanning tunneling spectroscopies (STS) for MgB_2 samples doped with Al (a) and C (b). (The data are taken from Putti *et al.*, 2003, Putti *et al.*, 2005, Daghero *et al.*, 2008, Szabó *et al.*, 2007, Klein *et al.*, 2006, Hol'anová *et al.*, 2004, Gonnelli *et al.*, 2005, Iavarone *et al.*, 2005 and Tsuda *et al.*, 2005.) The theoretical curves correspond to calculations from Kortus *et al.*, 2005 and show the changes in the gap amplitudes in two distinct cases, when gaps and T_c are affected by only a reduced density of states (DOS) (dash-dot curve) or by only interband scattering (dashed curve). The dashed curve is extended at low temperature below the merging point by a BCS curve. In panel (b), a theoretical curve taking into account both reduced DOS and interband scattering is shown (solid line).

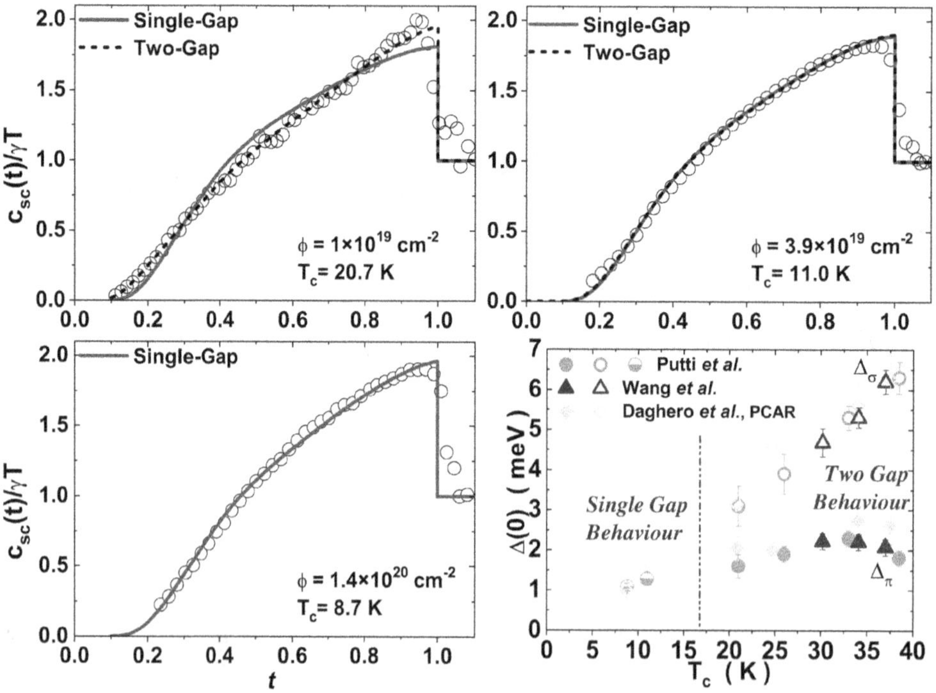

FIGURE B2.10 Superconducting contribution of the specific heat as a function of reduced temperature for three irradiated samples. The solid lines represent the single-gap fit and the dashed lines are obtained by the two-gap fit (data from Putti *et al.*, 2006). In the bottom right panel, the gap amplitudes as a function of T_c is reported for this sample series and compared with other specific heat data (Wang *et al.*, 2003) and point-contact Andreev reflection spectroscopy (Daghero *et al.*, 2006).

B2.4.2 Normal State Resistivity: Intrinsic and Extrinsic Factors

In single-band superconductors, increasing the disorder typically produces an enhancement in the residual resistivity, ρ_0, and a decrease in T_c, but the temperature dependence, $\Delta\rho(T)=\rho(T)-\rho_0$, is usually weakly affected by the impurity scattering and $\Delta\rho = \rho(300\text{ K})-\rho_0$ remains constant, as long as the saturation is reached. The saturation occurs when the mean free path decreases down to the size of the lattice parameters. However, MgB_2 does not usually follow this simple behaviour. No direct $T_c-\rho_0$ correlation is usually found and samples with the same T_c can have ρ_0 values that vary by orders of magnitude. $\Delta\rho$ seems to be significantly enhanced with increasing disorder as well. This variety of normal-state behaviour may be intrinsic, i.e. due to the multiband nature of MgB_2, or extrinsic, i.e. related to the phase purity.

In a two-band superconductor, the resistivity is theoretically determined by a parallel-conductor formula (Mazin and Antropov 2003)

$$1/\rho(T) \propto \left(\frac{\omega_{pl,\pi}^2}{\Gamma_\pi(T)} + \frac{\omega_{pl,\sigma}^2}{\Gamma_\sigma(T)}\right) \quad \text{(B2.4)}$$

where $\omega_{pl,i}$ ($i = \sigma,\pi$) are the plasma frequencies and Γ_i are determined by γ_{ii} and γ_{ij}, the intraband and interband impurity scattering rates, respectively, and by the electron–phonon coupling (EPC) interaction (Mazin and Antropov 2003). As explained earlier, the interband scattering is also small in very disordered samples because of the different band parity, and the resistivity is mainly determined by the intraband impurity scattering. Since T_c only depends on the interband scattering (see Section B2.4.1), this could explain the absence of a direct correlation between T_c and ρ_0.

The previous relation (Equation [B2.4]) could also be responsible for the variation in the temperature dependence of the normal-state resistivity $\Delta\rho(T)$. In the clean limit (all the $\gamma_{ij} \sim 0$), the conductivity of σ band close to room temperature is lower than in the π band because of stronger σ-band EPC. As a consequence, the resistivity behaviour is mainly determined by the π bands. In the dirty limit, two distinct cases have to be considered: when $\gamma_{\pi\pi} \gg \gamma_{\sigma\sigma} \gg \gamma_{\sigma\pi}$, the π bands contribute very little to the conductivity, and the temperature dependence is mainly dominated by the EPC in the σ bands (Mazin and Antropov 2003). In the opposite dirty case, $\gamma_{\sigma\sigma} \gg \gamma_{\pi\pi} \gg \gamma_{\sigma\pi}$, the resistivity should be mainly determined by π bands, and the T-dependence be similar to the clean limit (Mazin and Antropov 2003). However, these considerations can hardly explain the high resistivity values of hundreds of $\mu\Omega$·cm sometimes measured in MgB_2 and other factors have to be taken into account.

Looking for more extrinsic factors, Rowell focused his attention on the connectivity issue (Rowell 2003). He compared the resistivity behaviour of different kinds of samples and observed that in some cases, although T_c remains near 39 K and the transition is sharp, the resistivity at 300 K can exceed 1 mΩ·cm, well beyond the resistivity value normally associated with the metal–insulator transition. Moreover, samples whose resistivity is extremely high do not show a tendency towards a ρ-saturation at high temperature. This behaviour clearly indicates that, even if the intrinsic resistivity does not significantly increase, connectivity issues at grain boundary exist. The presence of insulating precipitates at the grain boundary, as MgO or BO_x, has the effect of disconnecting the MgB_2 grains from each other. Also, porosity, i.e. low sample density, contributes in a similar way. In fact, they both reduce the effective cross-sectional area that is able to carry current, leading to an overestimation of the resistivity and, consequently, to a high value of $\Delta\rho$ that would be otherwise inexplicable. Rowell pointed out that the intrinsic intragrain $\Delta\rho$ should not significantly change with disorder as determined by Mazin *et al.* For this reason, he proposed the rescaling of the resistivity curves with respect to a constant $\Delta\rho$ value, neglecting the weak multiband effect, so as to overcome the connectivity issue and properly estimate the intrinsic resistivity. The resistivity T-dependence of the samples should be rescaled following the expressions:

$$\rho(T) = F\left[\Delta\rho_{sc}(T) + \rho(0)\right], \quad \Delta\rho = F\left(\Delta\rho_{sc}\right)$$

where $1/F$ is the fractional area of the sample that carries current, $\Delta\rho_{sc}(T)$ is the single crystal (i.e. intragrain) temperature-dependent part of the resistivity, $\Delta\rho$ is the measured change of resistivity between 300 and 40 K, $\Delta\rho_{sc}$, which is also the rescaling value, is the single crystal change of resistivity between 300 and 40 K and $\rho(0)$ is the residual resistivity. An opportune choice of $\Delta\rho_{sc}$ can be found between the lowest experimental values: Rowell suggested the pure single crystal value of 4.3 $\mu\Omega$·cm (Eltsev *et al.*, 2002); values between ~6 and 9 $\mu\Omega$·cm were used for pure bulk samples (Tarantini *et al.*, 2006, Jiang et al 2006, Yamamoto *et al.*, 2007, Senkowicz *et al.*, 2008), which are also affected by the anisotropy. Figure B2.11 shows that irradiated samples (Tarantini *et al.*, 2006, Ferrando *et al.*, 2007, Gandikota *et al.*, 2005), whose residual resistivity was corrected following this approach, actually have a roughly linear correlation with T_c and the difference in the $\rho_{Corr}(T_c \rightarrow 0)$ is only related to the choice of the rescaling value $\Delta\rho_{intra}$. The inset reveals that a similar linear correlation exists also in C-doped samples, if the change in $\Delta\rho_{sc}$ due to the C-content is taken into account (Senkowicz *et al.*, 2008).

B2.4.3 Upper Critical Field

The upper critical field H_{c2} is a very important property in a superconducting material in view of potential high-field applications. In a conventional BCS superconductor, H_{c2} is determined by the relation $\mu_0 H_{c2} = \phi_0 / 2\pi\xi^2$, where ϕ_0 is the quantum flux, and ξ is the coherence length, and the H_{c2} anisotropy, γ, is substantially temperature independent. In the clean limit, the coherence length depends on the Fermi velocity and the gap amplitude by $\xi_0 = \hbar v_F / \pi\Delta(0)$. In the dirty limit,

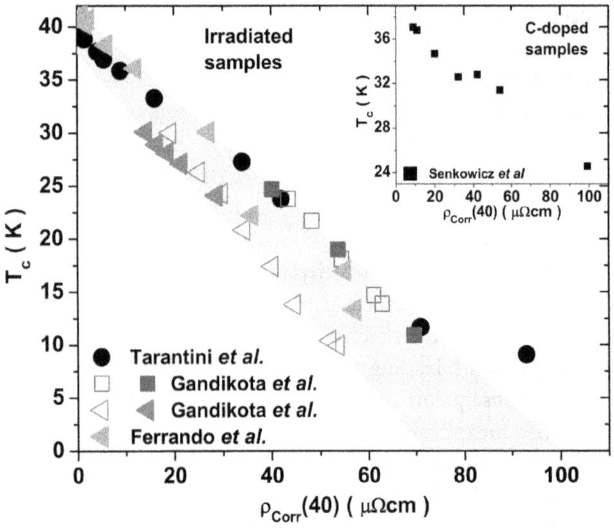

FIGURE B2.11 Critical temperature as a function of residual resistivity corrected by Rowell rescaling from various irradiated samples. (Data from Tarantini *et al.*, 2006, Ferrando *et al.*, 2007) and before and after annealing from Gandikota *et al.*, 2005.) In the inset a similar plot for C-doped samples (Senkowicz *et al.*, 2008).

i.e. when the mean free path is $l < \xi_0$, the upper critical field at zero temperature is determined by the Werthamer–Helfand–Hohenberg (WHH) relation (Werthamer *et al.*, 1966):

$$H_{c2}(0) = 0.69 T_c \left.\frac{dH_{c2}}{dT}\right|_{Tc} \tag{B2.5}$$

where $\left.\frac{dH_{c2}}{dT}\right|_{Tc}$, the slope of $H_{c2}(T)$ at T_c, is proportional to the residual resistivity ρ_0. Because of this direct relation between H_{c2} and ρ_0, the upper critical field of low-temperature superconductors like Nb-Ti and Nb_3Sn can be enhanced by disorder (Fietz and Webb 1967, Orlando *et al.*, 1979). The same approach was also initially attempted on magnesium diboride, introducing disorder by chemical substitutions (Putti *et al.*, 2005, Karpinski *et al.*, 2005, Wilke *et al.*, 2004, Serquis *et al.*, 2007), irradiation (Eisterer *et al.*, 2002, Wang *et al.*, 2003, Gandikota *et al.*, 2005, Tarantini *et al.*, 2006, Ferrando *et al.*, 2007) and in doped or naturally disordered thin films (Braccini *et al.*, 2005, Ferrando *et al.*, 2003). However, the H_{c2} results for MgB_2 were not straightforward. First of all, differently from conventional BCS materials, the H_{c2} anisotropy of pure MgB_2 is strongly temperature dependent, varying from 6–7 at low temperature to 2–3 near T_c (Angst *et al.*, 2002). This is due to the multiband/multigap nature of MgB_2 that also strongly affects the H_{c2} behaviour (Dahm and Schopohl 2003, Golubov and Koshelev 2003, Gurevich *et al.*, 2003). In Al-doped samples (Putti *et al.*, 2005, Karpinski *et al.*, 2005), despite an increase in ρ_0, the $H_{c2}^{//c}(0)$ was almost unchanged, whereas $H_{c2}^{//ab}(0)$ was strongly suppressed, also inducing a decrease of the H_{c2} anisotropy. It was explained that the disorder introduced by Al-doping increases the resistance in the π band, whereas it is unable to establish the dirty limit in the σ band that in this case determined H_{c2}. The H_{c2} suppression in Al-doped samples was ascribed to the change in the Fermi surface topology and the decrease of the gap amplitude upon doping (Putti *et al.*, 2005). Clearly different is the case of C-doped or irradiated MgB_2 where an increase of H_{c2} was observed. However, the H_{c2} curves can reveal very different temperature dependencies. As shown in Figure B2.12(a), irradiated bulk MgB_2 samples have a quite marked upward curvature near T_c and reach an estimated $\mu_0 H_{c2}(0)$ of over 30 T (Tarantini *et al.*, 2006). In contrast, bulk MgB_2 samples doped with carbon nanotubes exhibit an almost linear behaviour down to ~$0.2T_c$ and an upturn at low temperature reaching $\mu_0 H_{c2}(0)$ ~ 44 T (Serquis *et al.*, 2007). Even more striking is the extremely high H_{c2} measured in C-doped thin films (Figure B2.12[b]) with an estimated $\mu_0 H_{c2}(0)$ of over 70 T for H//ab (Braccini *et al.*, 2005). It is also important to notice that in this

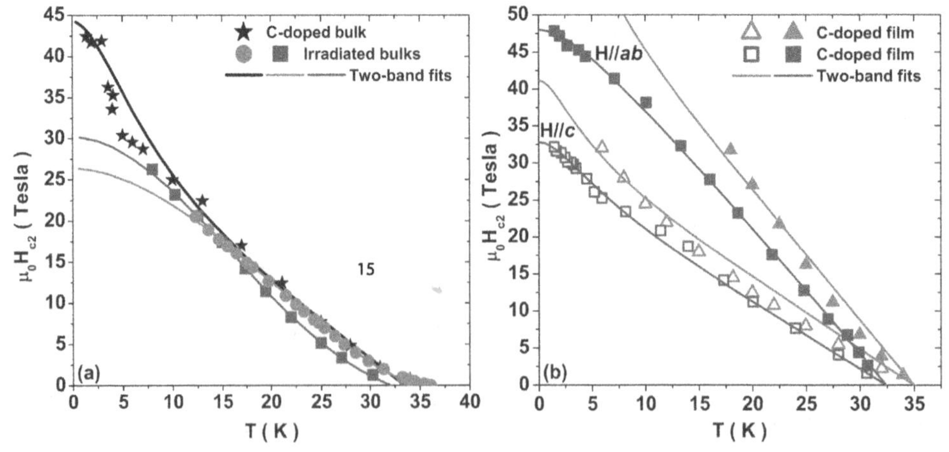

FIGURE B2.12 H_{c2} as a function of temperature in (a) irradiated MgB_2 bulks and MgB_2 bulk doped with carbon nanotubes. (Data from Tarantini *et al.*, 2006, Serquis *et al.*, 2007.) and in (b) C-doped MgB_2 films. (Data from Braccini *et al.*, 2005.) The continuous lines represent fits with the two-band model (Braccini *et al.*, 2005).

case the H_{c2} anisotropy is lower at low temperature, and it increases approaching T_c, the opposite of what is seen in pure MgB_2 crystals (Angst *et al.*, 2002) and in naturally disordered films (Ferrando *et al.*, 2003).

Those H_{c2}(T) curves clearly differ from the conventional WHH curves and the high H_{c2} values also exceed the BCS paramagnetic limit ($\mu_0 H_p$=1.84T_c). These unusual and distinct trends observed in different samples are again related to the multiband nature of MgB_2. A two-band theory of H_{c2} in the dirty limit can describe these H_{c2}(T) shapes by changing the ratio of the electronic diffusivities in the π and σ bands, $\eta=D_\pi/D_\sigma$, and the interband scattering rate $\Gamma_{\pi\sigma}$, which also affects T_c (Gurevich, 2003, Braccini *et al.*, 2005). In fact, the model developed by Gurevich explains that when the π band is dirtier than the σ band (η<1), an upturn develops at low temperature, whereas when the σ band is dirtier than the π band (η>1), H_{c2}(T) has an upward curvature near T_c (Gurevich, 2003). In both cases, the zero-temperature value $H_{c2}(0)$ can be significantly larger than the one for the single-band WHH extrapolation (Equation [B2.5]). The diffusivity ratio is also responsible for the opposite temperature dependence of the H_{c2} anisotropy: in fact γ increases with temperature for η<1 (Braccini *et al.*, 2005) and it decreases for η>1 (Ferrando *et al.*, 2003). Besides affecting T_c, the interband scattering also changes the temperature dependence of H_{c2}: in the η<1 case, an increasing $\Gamma_{\pi\sigma}$ suppresses the low-temperature upturn, whereas when η>1, a large $\Gamma_{\pi\sigma}$ accentuates the high-temperature curvature (Braccini *et al.*, 2005). The fits with the two-band model in Figure B2.12 show that, whereas the neutron irradiation introduced more disorder in the σ band, the C-doping in both bulks and thin films produces the low-temperature upturn indicative of a dirtier π band.

B2.5 Summary

Magnesium diboride is a binary intermetallic compound with a simple hexagonal structure. It has four bands crossing the Fermi level, two quasi-2D σ bands and two 3D π bands, strongly affecting the superconducting properties. Despite the unusual multiband behaviour, the superconductivity is mediated by electron–phonon interaction. The E_{2g} phonon mode, involving the in-plane boron motions in opposite directions along *x* or *y* axes, significantly contributes to the coupling. Magnesium diboride has also a rare multigap nature: on the σ sheets the magnitude of the energy gap ranges from 6.4 to 7.2 meV while on the π sheets it varies from 1.2 to 3.7 meV with both gaps closing at the same T_c.

Because of the multiband/multigap nature, disorder can be introduced in MgB_2 through three different scattering channels, two intrabands and one interband, and only the interband scattering is pair-breaking and able to suppress T_c. It was theoretically predicted that, in case of dominant interband scattering, the amplitude of the σ gap should decrease whereas the π gap should increase, inducing a gap merging at about 20 K. Disorder introduced by doping (Al on Mg site and C on B site) did not clearly show the gap merging because of a significant change in the DOS, but the gap merging was clearly observed in irradiated samples.

Because of the weak interaction between the bands, disorder can be selectively introduced in the σ and π bands producing different temperature dependences of H_{c2}, markedly different from the single-band WHH. This is a unique feature of MgB_2, and it allows $H_{c2}(0)$ to increase to as high as 45 T in bulk samples and over 70 T in thin films.

References

Allen PB, Dynes RC (1975) Transition temperature of strong-coupled superconductors reanalyzed. *Phys. Rev. B* 12: 905.

An JM and Pickett WE (2001) Superconductivity of MgB_2: covalent bonds driven metallic. *Phys. Rev. Lett.* 86: 4366

Angst M et al. (2002) Temperature and field dependence of the anisotropy of MgB_2. *Phys. Rev. Lett.* 88: 167004.

Bardeen J, Copper LN, Schrieffer JR (1957) Theory of superconductivity. *Phys. Rev.* 108: 1175.

Baron AQR et al. (2004) Kohn anomaly in MgB_2 by inelastic X-Ray scattering. *Phys. Rev. Lett.* 92: 197004

Binnig G, Baratoff A, Hoenig HE, Bednorz JG (1980) Two-band superconductivity in Nb-doped $SrTiO_3$. *Phys. Rev. Lett.* 45: 1352.

Bouquet F, Fisher RA, Phillips NE, Hinks DG, Jorgensen JD (2001) Specific heat of $Mg^{11}B_2$: evidence for a second energy gap. *Phys. Rev. Lett.* 87: 047001.

Bouquet F et al. (2003) Unusual effects of anisotropy on the specific heat of ceramic and single crystal MgB_2. *Physica C* 385: 192

Braccini V et al. (2005) High-field superconductivity in alloyed MgB_2 thin films. *Phys. Rev. B* 71: 012504.

Brinkman A et al. (2002) Multiband model for tunneling in MgB_2 junctions. *Phys. Rev. B* 65: 180517

Bud'ko SL, Laperton G, Petrovic C, Cunningham CE, Anderson N, Canfield PC (2001) Boron isotope effect in superconducting MgB_2. *Phys. Rev. Lett.* 86: 1877.

Calandra M, Lazzeri M, Mauri F (2007) Anharmonic and non-adiabatic effects in MgB_2: implications for the isotope effect and interpretation of Raman spectra. *Physica C* 456: 38–44.

Carrington A, Yelland EA, Fletcher JD, Cooper JR (2007) de Haas-van Alphen effect investigations of the electronic structure of pure and aluminum-doped MgB_2. *Physica C* 456: 92–101.

Cava R J (2001) Genie in a bottle. *Nature* 410: 23.

Chen K et al. (2012) Momentum-dependent multiple gaps in magnesium diboride probed by electron tunnelling spectroscopy. *Nat. Commun.* 3: 619

Choi HJ, Cohen ML, Louie SG (2003) Anisotropic Eliashberg theory of MgB_2: T_c, isotope effects, superconducting energy gaps, quasiparticles, and specific heat. *Physica C* 385: 66.

Choi HJ, Roundy D, Sun H, Cohen ML, Louie SG (2002a) First-principles calculation of the superconducting transition in MgB_2 within the anisotropic Eliashberg formalism. *Phys. Rev. B* 66: 020513(R).

Choi HJ, Roundy D, Sun H, Cohen ML, Louie SG (2002b) The origin of the anomalous superconducting properties of MgB_2. *Nature* 418: 758.

Daghero D et al. (2006) Point-contact spectroscopy in neutron-irradiated $Mg^{11}B_2$. *Phys. Rev. B* 74: 174519

Daghero D et al. (2008) Point-contact Andreev-reflection spectroscopy in segregation-free $Mg_{1-x}Al_xB_2$ single crystals up to x = 0.32. *J. Phys.: Condens. Matter.* 20: 085225

Dahm T, Schopohl N (2003) Fermi surface topology and the upper critical field in two-band superconductors: application to MgB_2. *Phys. Rev. Lett.* 91: 017001.

Dolgov OV, Kremer RK, Kortus J, Golubov AA, Shulga SV (2005) Thermodynamics of two-band superconductors: the case of MgB_2. *Phys. Rev. B* 72: 024504.

Eisterer M et al. (2002) Neutron irradiation of MgB_2 bulk superconductors. *Supercond. Sci. Technol.* 15: L9

Eltsev et al. (2002) Anisotropic superconducting properties of MgB_2 single crystals probed by in-plane electrical transport measurements. *Phys. Rev. B* **65**: 140501(R).

Ferrando V et al. (2003) Effect of two bands on critical fields in MgB_2 thin films with various resistivity values. *Phys. Rev. B* 68: 094517

Ferrando V et al. (2007) Systematic study of disorder induced by neutron irradiation in MgB_2 thin films. *J. Appl. Phys.* 101: 043903

Fietz WA, Webb WW (1967) *Phys. Rev.* 161: 4231.

Floris A et al. (2007) Superconducting properties of MgB_2 from first principles. *Physica C* 456: 45–53

Gandikota R et al. (2005) Effect of damage by 2 MeV He ions and annealing on H_{c2} in MgB_2 thin films. *Appl. Phys. Lett.* 87: 072507

Giubileo F et al. (2001) Two-gap state density in MgB_2: a true bulk property or a proximity effect? *Phys. Rev. Lett.* 87: 177008

Golubov AA, Koshelev AE (2003) Upper critical field in dirty two-band superconductors: breakdown of the anisotropic Ginzburg-Landau theory. *Phys. Rev. B* 68: 104503

Golubov AA, Mazin II (1997) Effect of magnetic and nonmagnetic impurities on highly anisotropic superconductivity. *Phys. Rev. B* 55: 15146

Golubov AA et al. (2002) Specific heat of MgB_2 in a one- and a two-band model from first-principles calculations. *J. Phys.: Condens. Matter.* 14: 1353

Gonnelli RS et al. (2003) Independent determination of the two gaps by directional point-contact spectroscopy in MgB_2 single crystals. *Supercon. Sci. Technol.* 16: 171

Gonnelli RS et al. (2005) Evidence for single-gap superconductivity in $Mg(B_{1-x}C_x)_2$ single crystals with x=0.132 from point-contact spectroscopy. *Phys. Rev. B* 71: 060503(R)

Gurevich A (2003) Enhancement of the upper critical field by nonmagnetic impurities in dirty two-gap superconductors *Phys. Rev. B* 67: 184515

Hinks DG, Claus H, Jorgensen JD (2001) The complex nature of superconductivity in MgB_2 as revealed by the reduced total isotope effect. *Nature* 411: 457.

Hol'anová Z, Szabó P, Samuely P, Wilke RHT, Bud'ko SL, Canfield PC (2004) Systematic study of two-band/two-gap superconductivity in carbon-substituted MgB_2 by point-contact spectroscopy. *Phys. Rev. B* 70: 064520.

Iavarone M et al. (2002) Two-band superconductivity in MgB_2. *Phys. Rev. Lett.* 89: 187002

Iavarone M et al. (2005) Effect of disorder in MgB_2 thin films. *Phys. Rev. B* 71: 214502

Jiang J, Senkowicz BJ, Larbalestier DC, Hellstrom EE (2006) Influence of boron powder purification on the connectivity of bulk MgB_2. *Supercond. Sci. Technol.* 19: L33–L36

Karpinski J et al. (2005) Al substitution in MgB_2 crystals: Influence on superconducting and structural properties. *Phys. Rev. B* 71: 174506

Klein T et al. (2006) Influence of Al doping on the critical fields and gap values in magnesium diboride single crystals. *Phys. Rev. B* 73: 224528

Kortus J, Dolgov OV, Kremer RK, Golubov AA (2005) Band filling and interband scattering effects in MgB_2: carbon versus aluminum doping. *Phys. Rev. Lett.* 94: 027002

Kortus J, Mazin II, Belashchenko KD, Antropov VP, Boyer LL (2001) Superconductivity of metallic boron in MgB_2. *Phys. Rev. Lett.* 86: 4656

Lazzeri M, Calandra M, Mauri F (2003) Anharmonic phonon frequency shift in MgB_2. *Phys. Rev. B* 68: 220509(R)

Liu AY, Mazin II, Kortus J (2001) Beyond Eliashberg superconductivity in MgB_2: anharmonicity, two-phonon scattering, and multiple gaps. *Phys. Rev. Lett.* 87: 087005

Mazin II, Andersen OK, Jepsen O, Golubov AA, Dolgov OV, Kortus J (2004) Comment on "First-principles calculation of the superconducting transition in MgB_2 within the anisotropic Eliashberg formalism". *Phys. Rev. B* 69: 056501

Mazin II, Antropov VP (2003) Electronic structure, electron–phonon coupling, and multiband effects in MgB_2. *Physica C* 385: 49

Mazin II et al. (2002) Superconductivity in MgB_2: clean or dirty? *Phys. Rev. Lett.* 89: 107002

McMillan WL (1968) Transition temperature of strong-coupled superconductors. *Phys. Rev.* 167: 331

Nagamatsu J, Nakagawa N, Muranaka T, Zenitani Y, Akimitsu J (2001) Superconductivity at 39 K in magnesium diboride. *Nature* 410: 63.

Orlando TP, McNiff EJ, Foner S, Beasley MR (1979) Critical fields, Pauli paramagnetic limiting, and material parameters of Nb_3Sn and V_3Si. *Phys. Rev. B* 19: 4545.

Putti M, Affronte M, Ferdeghini C, Manfrinetti P, Tarantini C, Lehmann E (2006) Observation of the crossover from two-gap to single-gap superconductivity through specific heat measurements in neutron-irradiated MgB_2. *Phys. Rev. Lett.* 96: 077003

Putti M, Affronte M, Manfrinetti P, Palenzona A (2003) Effects of Al doping on the normal and superconducting properties of MgB_2: a specific heat study. *Phys. Rev. B* 68: 094514.

Putti M et al. (2005) Critical field of Al-doped MgB_2 samples: correlation with the suppression of the σ-band gap. *Phys. Rev. B* 71: 144505

Rowell JM (2003) The widely variable resistivity of MgB_2 samples. *Supercond. Sci. Technol.* 16: R17–R27

Schmidt H, Zasadzinski JF, Gray KE, Hinks DG (2002) Evidence for two-band superconductivity from break-junction tunneling on MgB_2. *Phys. Rev. Lett.* 88: 127002

Senkowicz BJ et al. (2008) Nanoscale grains, high irreversibility field and large critical current density as a function of high-energy ball milling time in C-doped magnesium diboride. *Supercond. Sci. Technol.* 21: 035009

Serquis A et al. (2007) Correlated enhancement of H_{c2} and J_c in carbon nanotube doped MgB_2. *Supercond. Sci. Technol.* 20: L12

Shukla A et al. (2003) Phonon dispersion and lifetimes in MgB_2. *Phys. Rev. Lett.* 90: 095506

Souma S et al. (2003) The origin of multiple superconducting gaps in MgB_2. *Nature* 423: 65

Suhl H, Matthias BT, Walker LR (1959) Bardeen-Cooper-Schrieffer theory of superconductivity in the case of overlapping bands. *Phys. Rev. Lett.* 3: 552

Szabó P et al. (2001) Evidence for two superconducting energy gaps in MgB_2 by point-contact spectroscopy. *Phys. Rev. Lett.* 87: 137005

Szabó P et al. (2007) Point-contact spectroscopy of Al- and C-doped MgB_2: superconducting energy gaps and scattering studies. *Phys. Rev. B* 75: 144507

Tarantini C et al. (2006) Effects of neutron irradiation on polycrystalline $Mg^{11}B_2$. *Phys. Rev. B* 73: 134518

Testardi LR, Mattheiss LF (1978) Electron lifetime effects on properties of A15 and bcc materials. *Phys. Rev. Lett.* 41: 1612

Tsuda S et al. (2005) Carbon-substitution dependent multiple superconducting gap of MgB_2: a sub-meV resolution photoemission study. *Phys. Rev. B* 72: 064527

Uchiyama H et al. (2002) Electronic structure of MgB_2 from angle-resolved photoemission spectroscopy. *Phys. Rev. Lett.* 88: 157002

Yamamoto A, Shimoyama J, Kishio K, Matsushita T (2007) Limiting factors of normal-state conductivity in superconducting MgB_2: an application of mean-field theory for a site percolation problem. *Supercond. Sci. Technol.* 20: 658

Yelland EA et al. (2002) de Haas–van Alphen effect in single crystal MgB_2. *Phys. Rev. Lett.* 88: 217002

Yildrim T et al. (2001) Giant anharmonicity and nonlinear electron-phonon coupling in MgB_2: a combined first-principles calculation and neutron scattering study. *Phys. Rev. Lett.* 87: 037001

Wang Y et al. (2003) Specific heat of MgB_2 after irradiation. *J. Phys.: Condens. Matter* 15: 883

Wang YX, Plackowski T, Junod A (2001) Specific heat in the superconducting and normal state (2–300 K, 0–16 T), and magnetic susceptibility of the 38K superconductor MgB_2: evidence for a multicomponent gap. *Physica C* 355: 179–193.

Werthamer NR, Helfand E, Hohenberg PC (1966) Temperature and purity dependence of the superconducting critical field, H_{c2}. III. Electron spin and spin-orbit effects. *Phys. Rev.* 147: 295

Wilke RHT, Bud'ko SL, Canfield PC, Finnemore DK, Suplinskas RJ, Hannahs ST (2004) Systematic effects of carbon doping on the superconducting properties of $Mg(B_{1-x}C_x)_2$. *Phys. Rev. Lett.* 92: 217003

Xi XX et al. (2007) MgB_2 thin films by hybrid physical-chemical vapor deposition. *Physica C* 456: 22–37

Zeng XH et al. (2002) In situ epitaxial MgB_2 thin films for superconducting electronics. *Nat. Mater.* 1: 35–38

B3

Chevrel Phases

Damian P. Hampshire

B3.1 Introduction

Chevrel phase superconductors are a class of materials that have generated enormous interest in the superconductivity community. There are more than 100 different Chevrel phase compounds that exhibit a wide range of properties of both fundamental and technological interest. They have the chemical formula $M_xMo_6X_8$, where M can be one of more than 25 different elements that are mono-, di- or trivalent, x can range from 1 to 4, and X is usually one of the chalcogenides (S, Se or Te). The structure can incorporate elements of different size, concentration and oxidation state. Some of the most intensively studied Chevrel phase materials with superconducting critical temperatures above 2 K are listed in Table B3.1 (Fischer et al., 1975b; Shelton et al., 1976; Chevrel and Sergent, 1982; Fischer and Maple, 1982a, b). Figure B3.1 shows the critical temperature, magnetic ordering temperature and some structural properties of Chevrel phase materials with rare-earth ions (Ishikawa et al., 1982; Peña and Sergent, 1989; Pena et al., 1999; Perrin and Perrin, 2012; Pena, 2015).

The Chevrel phase compounds were discovered in 1971 (Chevrel et al., 1971). Interest has focused on the $PbMo_6S_8$ and $SnMo_6S_8$ materials with high upper critical fields (B_{C2} ~ 40 - 60 T) and T_C values of ~ 12–15 K. The B_{C2} values of these Chevrel phase compounds lie between those of the high-temperature copper oxide superconductors (e.g. $Bi_2Sr_2Ca_2Cu_3O_x$) and the intermetallic low-temperature superconductors (e.g. Nb_3Sn), which means the superconducting coherence length is sufficiently long to reduce the effects of granularity found in high-temperature superconductors but sufficiently short for very high-field applications. Powder-in-tube wires have been fabricated (Cheggour et al., 1997; Cheggour et al., 1998; Eastell, 1998) with a reasonably high critical current density (J_C) in high magnetic fields (Yamasaki et al., 1992; Seeber et al., 1995; Cheggour et al., 1997) in single lengths up to 1 km long (c.f. Chapter B3.3.5). This has opened the possibility that these materials could be used in high-field applications operating at magnetic fields significantly above 25 T (Cheggour et al., 1998). Nevertheless, a further increase in J_C is probably required for these materials to compete with Nb_3Sn, which is already well-established, or for them to attract industrial R&D interest away from developing the HTS materials, although as discussed below, there are magnet design and stability issues (i.e. rapid quench detection) that may eventually favour use of Chevrel phase materials.

The fundamental interest in these materials has long included studies of the many Chevrel phase compounds that include rare-earth elements that are found to exhibit the coexistence of superconductivity and long-range magnetic order (Matthias et al., 1972; Foner et al., 1974; Fischer, 1978; Niu et al., 2001). Also their presence on the Uemura plot (Uemura et al., 1991) has opened the question of whether Chevrel phase compounds are really Bardeen–Cooper–Schrieffer (BCS) superconductors or are better considered as belonging to the class of [non-BCS (Bardeen et al., 1957; Abrikosov, 2000)] superconductors that includes the cuprates and is characterised by relatively high critical temperature for such small n_s/m^* (carrier density/effective mass) (Uemura et al., 1991).

We note that there is a huge family of $M_6X_{8-x}X'_x$ cluster compounds (Gougeon et al., 1984; Perrin and Perrin, 2012; Pena, 2015) that includes Chevrel phase materials with the general formula $M_{2n-2}M'_{6n}X_{6n-x+2}X'_x$, where the $M'_6X_{8-x}X'_x$ clusters are the basic structural building block. One can consider Chevrel phase materials as $Mo_{6n}X_{6n-x+2}X'_x$ cluster compounds connected in a three-dimensional framework and the other family compounds as connected in 2-, 1- and 0-dimensional frameworks. Some of these compounds include some relatively high-temperature superconductors such as $Mo_6S_6I_2$ (T_C = 14 K) (Perrin and Perrin, 2012; Pena, 2015) although the in-field properties are not well studied. Many of these materials have also been intensively investigated at room temperature, because of their potential to be electrode materials in catalysts (Benson et al., 1995; Kamiguchi et al., 2013; Kamiguchi et al., 2015), thermoelectric devices (Caillat et al., 1999; Nunes et al., 1999;

TABLE B3.1 The Critical Temperature of Chevrel Phase Materials and Related Compounds: The Critical Temperature (T_C) (Fischer et al., 1975a; Shelton et al., 1976); the Rhombohedral Angle and the Rhombohedral Lattice Parameter (Fischer and Maple, 1982a) (Chevrel and Sergent, 1982) (Perrin et al., 1979)

Compound	*Tc*	Rhombohedral Lattice Parameter (Å)	Rhombohedral Angle	Compound	*Tc*	Rhombohedral Lattice Parameter (Å)	Rhombohedral Angle (degrees)
		Sulphides				Selenides	
$PbMo_6S_8$	15	6.55	89.4	$PbMo_6Se_8$	6.7	6.81	89.23
$SnMo_6S_8$	13	6.52	89.7	$SnMo_6Se_8$	6.8	6.78	89.6
$AgMo_6S_8$	9	6.48	92.0	$AgMo_6Se_8$	6	6.73	91.4
$ScMo_6S_8$	3.6	-	-	$Cu_{2.8}Mo_6Se$	6	6.79	94.9
YMo_6S_8	3.0	6.45	89.5			Binaries and doped binaries	
VMo_6S_8	8.2	-	-	Mo_6Se_8	6.4	6.66	91.58
$NbMo_6S_8$	3.5	-	-	$Mo_6Se_{4.8}Te_{3.2}$	2.7	-	-
$LaMo_6S_8$	7.1	6.51	88.9	Mo_6Se_7Br	7.1	-	-
Mo_6S_8	1.6	6.43	91.6	Mo_6Se_7Cl	7.0	-	-
$Cu_{1.8}Mo_6S_8$	11	6.48	94.9	$Mo_6S_6Br_2$	13.8	6.50	94.43
$C_{3.2}Mo_6S_8$	6.4	-	-	$Mo_6S_6I_2$	14.0	-	-
$Cu_4Mo_6S_8$	< 1	6.59	95.6	Mo_6Se_7I	7.6	-	-
$Cd_{1.1}Mo_6S_8$	3.5	6.52	92.8	$Mo_6Te_6I_2$	2.6	-	-
$Li_4Mo_6S_8$	4.4	6.62	94.5	$Mo_4Re_2Te_8$	3.5	-	-
$Mg_{1.14}Mo_6S_8$	3.5	6.51	93.6	$Mo_6S_{4.8}Te_{3.2}$	2.7	-	-
$Cu_{1.2}Mo_6S_8$	5.6	-	-				
$Zn_{1.1}Mo_6S_8$	3.6	6.49	94.7				
$Cu_2Mo_6S_6O_2$	9	6.54	95.51				
$PbMo_6S_8O_2$	11.7	6.52	88.98				

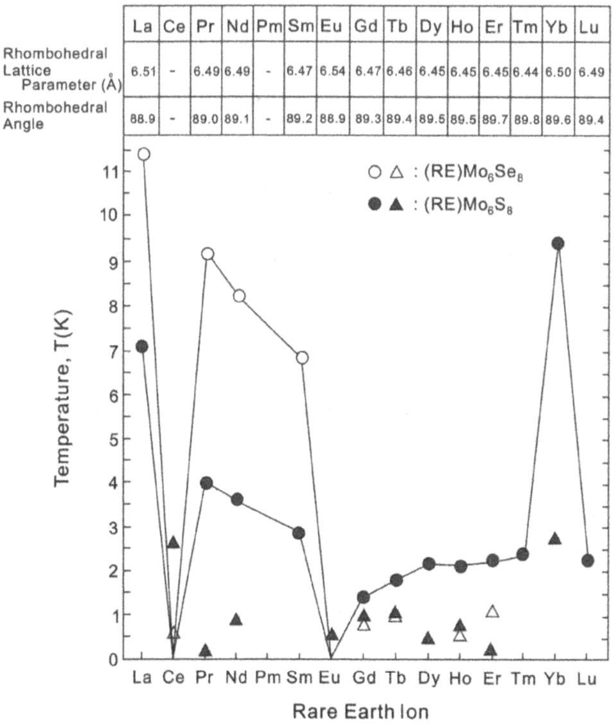

FIGURE B3.1 The rare-earth Chevrel phase superconductors: the critical temperature (T_C: o, •), the magnetic ordering temperatures (Δ, ▲) (Fischer et al., 1975a; Shelton et al., 1976; Ishikawa et al., 1982; Peña and Sergent, 1989); the rhombohedral angle and the rhombohedral lattice parameter (Chevrel and Sergent, 1982) (Pena, 2015). No large high-quality single crystals of heavy rare-earth selenides (i.e. Gd–Yb) have yet been obtained (Pena et al., 1999); two phase samples with Mo_6Se_8 (T_C ~ 6.4 K) are produced.

Fleurial et al., 2002; Ohta et al., 2009b; Ohta et al., 2009a; Goncalves and Godart, 2014) and batteries (Aruchamy et al., 1994; Mitelman et al., 2007; Levi et al., 2010; Kumta et al., 2015; Cheng et al., 2016), most obviously using magnesium in batteries (Aurbach et al., 2000) but also using calcium (Smeu et al., 2016) and aluminium (Geng et al., 2015). In this article, although we point to developments made during that research, it remains broadly focussed on superconducting Chevrel phase materials that may be of use in potential high-field superconducting applications.

B3.2 Structural Properties

The majority of superconducting Chevrel phase superconductors are trigonal with space group $R\bar{3}$ and can be described using either the hexagonal or rhombohedral structure (Nespolo et al., 2018). Figure B3.2 provides a schematic of the structure for $M_xMo_6X_8$, where M is Pb and X is S. The Pb occupies the origin of the unit cell and the tightly bound Mo_6S_8 cluster lies at the centre of the rhombohedron. The clusters are rotated through ~ 25° about the [111] and thus form the channels in which the M atoms are located. The Mo forms an octahedron such that each Mo atom is slightly outside the middle of the faces of the S-cube. The Mo_6S_8 is bound together as a cluster with weak intercluster Mo–Mo bonds. The 4d orbitals of the Mo ions are well extended so they favour the metallic bond. The edges of the sides of the rhombohedron are equal and inclined at the same angle (~90°) which means the structure can also be considered as a slightly distorted pseudo-cubic structure as shown in the bottom of Figure B3.2 (Uchida and Wakihara, 1991). In this cubic description, the Pb

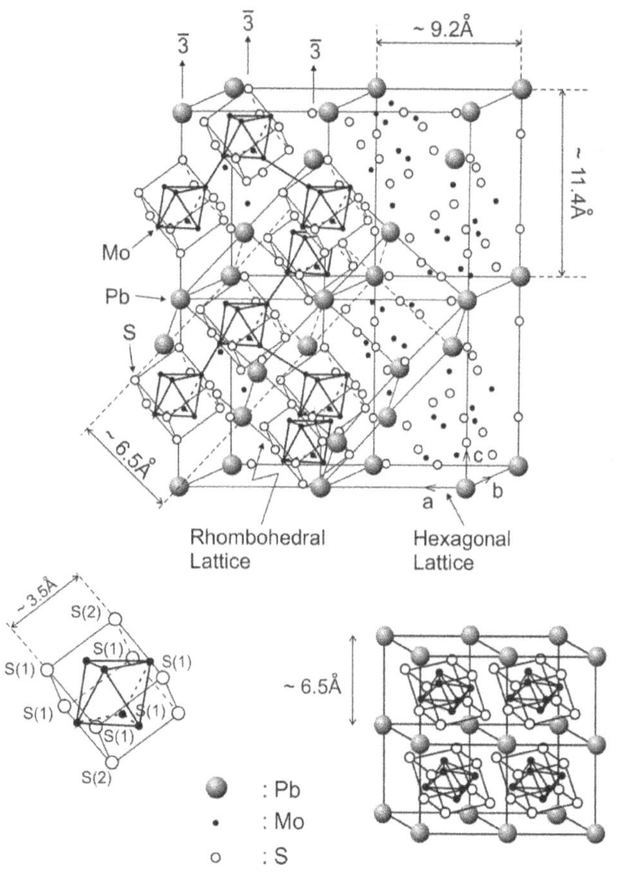

FIGURE B3.2 A schematic of the rhombohedral and pseudo-cubic structure of $PbMo_6S_8$. (Uchida and Wakihara, 1991.)

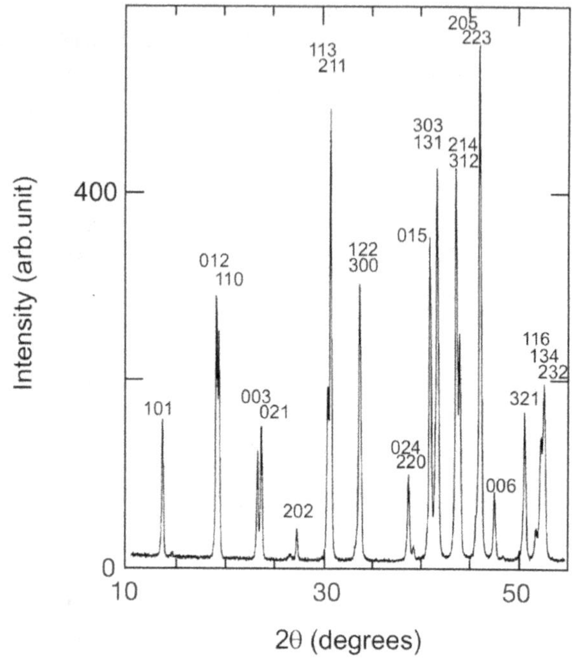

FIGURE B3.3 CuKα powder x-ray diffraction powder pattern for a typical single phase $PbMo_6S_8$ sample. (Niu et al., 2002.)

is almost at the centre of the central cube. At the corners of the central cube are the tightly bonded Mo_6S_8 clusters. The sides of the unit cell are about 6.5 Å and those of the Mo_6S_8 cluster about 3.5 Å. The bond lengths of the Mo_6X_8 cluster in ternary Chevrel phase materials are generally similar to those of the binary parent compounds. All the compounds have metal–metal bonds within the cluster. Important structural features that affect superconducting properties are the Mo–Mo intra-cluster distance, which can vary from 2.7 Å to 2.9 Å and is correlated with the number of valence electrons on the Mo_6S_8 cluster (Yvon and Paoli, 1977), and the intercluster distance, which can vary from 3.1 Å to 3.6 Å. The iono-covalent inter-cluster bond can be doped from an insulating state to produce a metallic bond that is superconducting.

Chevrel phase compounds can be divided into two types: stoichiometric and non-stoichiometric:

i. Stoichiometric compounds of $M_xMo_6X_8$ contain large cations such as Pb, Sn, Ag and the rare-earth elements. The structure basically accommodates 1 M ion in one of six positions which are sufficiently close together to appear as a single site. There is a narrow range of M solubility. All stoichiometric compounds crystallise in the trigonal structure at high temperatures with the rhombohedral angle between 88° and 90°. A typical x-ray diffraction pattern for $PbMo_6S_8$ is shown in Figure B3.3 (Niu et al., 2002). The positions of the main diffraction lines and h, k, l indices for the Chevrel phase compound $PbMo_6X_8$ and the most important second phases are given in Table B3.2.

ii. Non-stoichiometric compounds occur with small ions such as Li, Cu and Zn. The rhombohedral angle is between 92° and 95°. The structure can accommodate more than one ion. Figure B3.4 shows the possible sites for Cu in the inner cube of the Chevrel phase structure (Chevrel and Sergent, 1982). There are six inner sites and six outer sites. The sites for the small cations depend on the particular cation (Mancour-Billah and Chevrel, 2003) (Pena et al., 2009). With In for example, only the six inner sites are available. At sufficiently low temperatures, the M ion will freeze in one position which breaks the trigonal symmetry and can favour a phase transition to a triclinic structure.

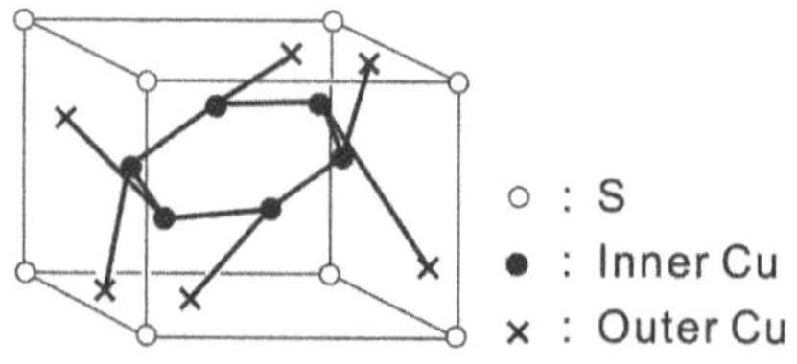

FIGURE B3.4 The possible sites for Cu in the inner cube of $Cu_XMo_6S_8$. [Adapted from (Chevrel and Sergent, 1982).]

TABLE B3.2 The X-Ray Diffraction Peaks for $PbMo_6S_8$ and the Strong Intensity Lines (> 10%) for the Important Impurity Phases (Mo, MoS_2, Mo_2S_3, Pb, PbS and S). Radiation: CuKα1, λ = 1.54056 Å

2-theta	Int.	h k l	2-theta	Int.	h k l	2-theta	Int.	h k l
	$PbMo_6S_8$			MoS_2			Pb	
13.612	50	1 0 1	14.38	100	0 0 2	29.55	100	1 0 0
19.153	100	0 1 2	32.67	22	1 0 0	33.66	80	1 0 1
19.364	80	1 1 0	33.51	12	1 0 1	42.82	75	-
23.303	50	0 0 3	35.87	10	1 0 2	52.23	75	1 1 0
23.700	40	0 2 1	39.53	58	1 0 3	57.95	75	1 0 3
-	-	2 0 2	44.15	11	0 0 6		PbS	
-	-	1 1 3	49.78	29	1 0 5	25.96	84	1 1 1
30.753	100	2 1 1		Mo_2S_3		30.07	100	2 0 0
-	-	1 2 2	10.48	10	0 0 1	43.06	57	2 2 0
33.666	80	3 0 0	16.28	95	1 0 1	50.97	35	3 1 1
38.730	20	0 2 4	21.06	36	0 0 2	53.41	16	2 2 2
-	-	2 2 0	29.61	35	$\bar{2}$ 0 1		S	
40.971	40	0 1 5	29.73	26	0 1 1	16.59	10	1 2 1
-	-	3 0 3	31.66	46	1 1 0	19.62	17	2 1 2
-	-	1 3 1	31.66	46	0 0 3	22.08	17	2 2 0
-	-	2 1 4	32.33	16	$\bar{1}$ 1 1	23.28	100	2 2 2
43.938	40	3 1 2	32.89	12	$\bar{2}$ 0 2	23.58	23	1 3 2
-	-	2 0 5	34.46	19	1 1 1	24.30	25	1 2 5
-	-	2 2 3	35.09	37	0 1 2	25.37	23	1 3 3
-	-	0 0 6	36.39	11	$\bar{1}$ 1 2	26.05	49	0 2 6
-	-	3 2 1	39.04	28	$\bar{2}$ 0 3	26.84	34	3 1 1
-	-	1 1 6	40.16	58	1 1 2	27.87	36	2 0 6
-	-	1 3 4	40.72	39	2 0 2	28.86	36	1 3 5
-	-	2 3 2	41.03	24	$\bar{2}$ 1 1	31.60	24	0 4 4
	Mo		42.27	21	$\bar{1}$ 0 4	34.27	13	4 0 0
40.51	100	1 1 0	42.93	100	0 1 3	34.37	13	1 3 7
58.60	16	2 0 0	42.93	100	0 0 4	35.06	11	3 3 3
			43.58	51	$\bar{2}$ 1 2	37.17	11	4 0 4
			46.18	23	$\bar{3}$ 0 2	42.84	12	3 1 9

Source: [Taken from the International centre for Diffraction data (Fischer and Maple, 1982a).]

The structural transformation from high-temperature rhombohedral to triclinic ($P\bar{1}$) between 100 K and 140 K (Jorgensen et al., 1987) has been observed in $PbMo_6S_8$ (Jorgensen and Hinks, 1985) using neutron scattering. Synchrotron data show a much smaller triclinic distortion, which suggests that sample preparation [possibly oxygen contamination (Wolf et al., 1996)] affects the low-temperature transformation (François et al., 1994). The triclinic phase is the stable low-temperature phase for the non-superconducting divalent (Eu^{2+}, Ba^{2+}, Sr^{2+} and Ca^{2+}) molybdenum sulphide Chevrel phases (Jorgensen and Hinks, 1986). Steric and electronic effects (principally charge transfer to the Mo_6S_8 cluster) are important in determining the equilibrium structure (Pena et al., 1999). Materials with the highest T_C are those metallic compounds adjacent to a structural instability. They may be in a mixed-phase region consisting of both a superconducting rhombohedral phase and an insulating triclinic phase (Jorgensen and Hinks, 1986). The structural instability can result either from changing the cation (chemical pressure) or applying pressure directly (Jorgensen et al., 1987). As discussed below in the context of the Uemura plot (Uemura et al., 1991), a similar structural instability is found in some of the A15 superconductors which undergo a shear martensitic transformation at some temperature above T_C (Batterman and Barrett, 1964).

The structure of $Cu_xMo_6S_{8-y}$ has been studied in detail below 300 K. Four different low-temperature modifications of the rhombohedral phase have been observed (Flükiger and Baillif, 1982). For $x = 1.2$, $T_C = 5.6$ K; for $x = 1.8$, $T_C = 11$ K; for $x = 3.2$, $T_C = 6.4$ K and for $x = 4$, the material is not superconducting (Niu et al., 2001). Less detailed studies have also been completed on Ni molybdenum sulphides and selenides.

Solid solutions of Chevrel phase materials can generally be fabricated if the end compounds exist. The solution can either occur with the chalcogenides (e.g. $M_xMo_6Se_{8-x}S_x$) or with the M-elements (Fischer, 1978). Neutron measurements show that in the ternary Pb- and Sn Chevrel phase sulphides, oxygen can substitute for sulphur which strongly affects the superconducting properties (Hinks et al., 1983). Of note is the $EuMo_6Se_8$ compound in which either a vacancy or oxygen substitution for Se causes the Eu to move 0.9 Å away from the usual central site (Le Berre et al., 1998a). Low-level doping of 1–1.5 at% Pb or rare-earth ions into the Chevrel phase binary compound Mo_6Se_8 has also been reported (Le Berre et al., 1997; Corrignan et al., 1999).

There is some limited work on structures that occur at very high pressures and temperatures (30–80 kbar and 1200°C). Preliminary work suggests that a metastable structure can be formed in the $PbMo_6S_8$ system (Khlybov et al., 1986) at high pressure which has a T_C enhanced by about 1 K but a severely reduced upper critical field (~7 T).

B3.3 Phase Diagrams and Fabrication

The high vapour pressure of the Pb and S makes accurate phase studies difficult. Very careful exclusion of oxygen and water during material fabrication is required to avoid oxygen substituting for the sulphur in the Mo_6S_8 cluster (Hinks et al., 1984) and is necessary for reliable comparisons between structure/composition and superconducting properties. Argon should be used rather than argon–nitrogen since trapped nitrogen can form MoN. Chevrel phase materials start decomposing under vacuum or low helium pressure above 650°C (Miraglia et al., 1987). Most of the important compounds melt in the temperature range from 1500°C to 2000°C, and the vapour pressure of the S or Se (Pb, Sn) is high (Peña and Sergent, 1989; Horyn et al., 1994). The Pb, Sn, Ag and rare-earth sulphide and selenide Chevrel phase materials melt peritectically (Fischer and Maple, 1982a). To make single crystals of the metallic superconductors, a typical off-stoichiometric composition of $Pb_{1.2}Mo_7S_8$ (Flükiger and Baillif, 1982) or for rare-earth sulphides compounds, $RE_{11}Mo_{38}S_{51}$, is used (Holtzberg et al., 1984; Horyn et al., 1989; Peña and Sergent, 1989). These compositions are chosen to produce an excess of the binary chalcogenide which minimises formation of the Mo_2S_3 phase that competes with the Chevrel phase. Figure B3.5 (upper) shows a schematic of the phase diagram for $PbMo_6S_8$ found at ~ 1000°C (Krabbes and Oppermann, 1981; Yamamoto et al., 1985). The Mo_2S_3 phase does not form at lower temperatures (Yamasaki and Kimura, 1986). The high-temperature phase diagram characteristic of the rare-earth compounds is shown in Figure B3.5 (lower) (Peña and Sergent, 1989). Material can be simply cooled from above its melting point to produce single crystals. The final product is generally single crystals in a binary chalcogenide crust which can be separated using HCl diluted with ethyl alcohol (≈ 20 vol.% HCl). The natural cleavage planes are the

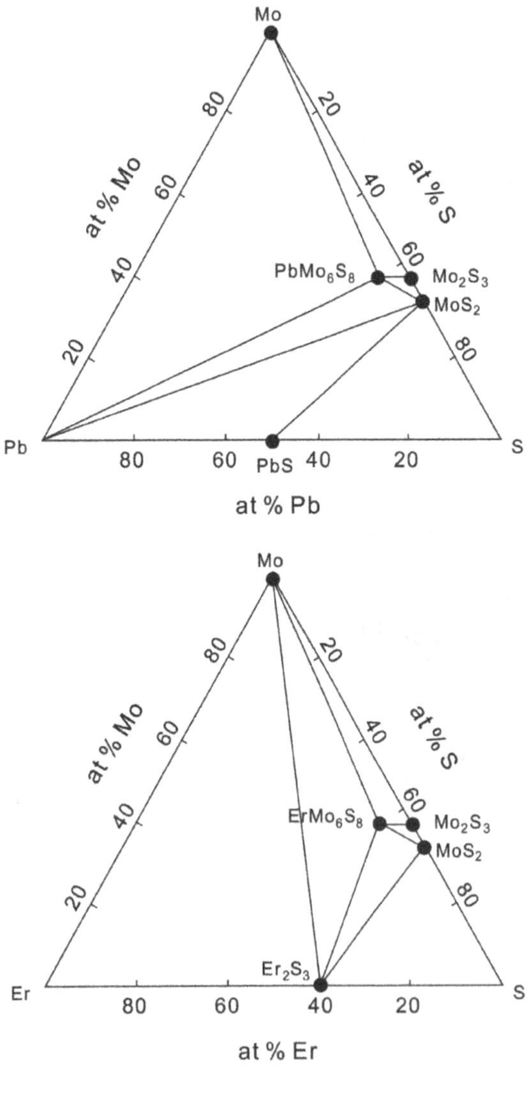

FIGURE B3.5 A schematic of the phase diagram for $PbMo_6S_8$ (upper) (Krabbes and Oppermann, 1981; Yamamoto et al., 1985) and $ErMo_6S_8$ (lower) at temperatures above 1000°C (Peña and Sergent, 1989).

(100) and (110) crystallographic planes (Holmgren et al., 1987). The Cu and Ni sulphide Chevrel phase materials form congruently so that relatively large single crystals can be formed from a stoichiometric melt. In selenium-based materials, only single crystals of the light rare earths have been produced (Horyn et al., 1994; Horyn et al., 1996; Pena et al., 1999; Le Berre et al., 2000). In heavy rare-earth selenides, RE deficiency produces a two-phase sandwich structure of doped binary Mo_6Se_8 and $REMo_6Se_8$ (Le Berre et al., 1995; Horyn et al., 1996; Le Berre et al., 1996; Pena et al., 1999; Hamard et al., 2002).

A number of authors have studied the phase diagrams of the $Pb_xMo_6S_{8-y}$ at 900°C (Yamasaki and Kimura, 1986) and ~ 1000°C (Krabbes and Oppermann, 1981; Yamamoto et al., 1985). There is no general agreement about the composition of the Chevrel phase material in the single-phase region although many studies show the presence of a small amount of sulphur defects, consistent with high-resolution electron microscopy (HREM) studies (Kang et al., 1994). Although as long as there is no oxygen contamination, the onset of T_C is in the range of 14–15 K, the sulphur stoichiometry is very close to 8, and Pb deficiency is present (Decroux et al., 1993). Chevrel phase materials have been fabricated at high pressure (typically up to 2000 atmospheres) to increase density and connectivity. It has been suggested that this enhances the effect of oxygen contamination since the material is pushed into a two-phase Chevrel phase + MoS_2 region (Ingle et al., 1998). Limited work has been completed on decomposition. Since $PbMo_6S_8$ can be formed between 450°C and 1650°C, it is reasonable to assume that the structure is very stable. Electron beam–induced decomposition has been observed in $Ni_2Mo_6S_8$ first showing increased disorder and then, after time, reduction of the Mo_6S_8 clusters giving regions of Mo, Mo–Ni alloy and sulphides (Kang et al., 1994). A detailed investigation of the phase diagram at 1200°C in La–Mo–Se has also been completed and the correlation between structural and superconducting properties investigated (Horyn et al., 1996; Le Berre et al., 1998b; Peña et al., 1998).

Amongst the family of cluster compounds related to Chevrel phase materials (Perrin and Perrin, 2012; Pena, 2015), substitutions of Br, I and O for the X-element and mixed sulphur-selenium compounds have been fabricated. The Br, I and O substitutions can increase T_C. Replacing S by Se or vice versa immediately reduces T_C producing a minimum when there are equal quantities of each element (e.g. $PbMo_6Se_4S_4$). Most materials in which the Mo-element has been substituted (e.g. Nb, Ta, Re, Ru and Rh) are not superconducting (Chevrel and Sergent, 1982).

Fine grain, well-connected bulk samples are required for high J_C applications. Soft chemistry methods have been used to produce ultrafine and amorphous precursors (PbS, MoS_2 and Mo) (Chevrel et al., 1974; Even-Boudjada et al., 1998b, a). Alternatively very fine grains of the binary compound Mo_6S_8, which are almost oxygen free, can be formed by leaching the Ni or Li ions out of the parent Chevrel phase compound using HCl (Selvam et al., 1992; Even-Boudjada et al., 1999), where the Ni or Li ions can migrate over distances of ~100 μm. The fine

Mo_6S_8 grains can then be used in powder route fabrication of fine grain bulk Chevrel phase materials by reacting with PbS or SnS (Cheggour et al., 1993; Selvam et al., 1993). Fast cheap microwave synthesis can also be effective (Murgia et al., 2016).

Thin films have been fabricated using a range of different deposition techniques including sputtering (Koo et al., 1995) and laser ablation (Decroux et al., 1999) as well as spin coating (Boursicot et al., 2012). Multifilamentary wires have also been produced (Sergent et al., 1984; Willis et al., 1995; Flükiger, 2010), as have monocore wires with Ag (Luhman and Dew-Hughes, 1978), Ta (Goldacker et al., 1989a) and Mo sheathing (Yamasaki et al., 1991; Seeber et al., 1995; Seeber, 2015) (c.f. Figure B3.6 and Chapter B3.3.5). The importance of superconducting grain boundaries for achieving high critical current densities in polycrystalline materials is discussed below. Detailed TEM has been performed on a range of bulk samples of (Pb,Gd)Mo_6S_8 and used to show that material can be produced with coherent tilt grain boundaries. Figure B3.7 shows some HREM demonstrating that in good material the grain boundaries are free of second phase. In some cases, dislocations are observed that occur at regular intervals along the boundary (Eastell, 1998). Detailed HREM on $Ni_2Mo_6S_8$ has shown that edge dislocations, which include an extra plane of Mo_6S_8 clusters, can form. Both coherent and incoherent interfaces have been found in this Ni-based compound (Kang et al., 1994). A small number of sulphur defects in the clusters were also observed (i.e. $Ni_2Mo_6S_{7.6}$). It has been several decades since small three-layer coils were fabricated using $PbMo_6S_8$ to demonstrate their potential use in magnet applications (Kubo et al., 1993), but the potential benefits of hot-drawing routes have garnered new interest in using Chevrel phase wires more recently (Seeber, 2015).

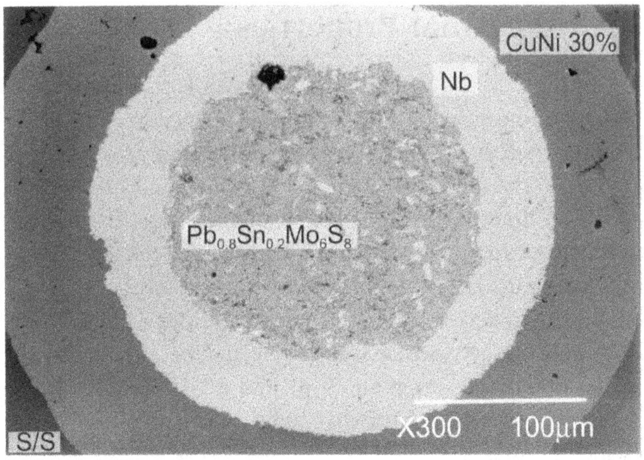

FIGURE B3.6 A $Pb_{0.6}Sn_{0.4}Mo_6S_8$ powder-route wire [measured in (Cheggour et al., 1997) and (Eastell, 1998)] which has a critical current density value of 6.7×10^8 A m^{-2} at 14 T and 4.2 K (Cheggour et al., 1998). The Nb is a diffusion barrier. The CuNi30% is a hard material that reduces damage to the Nb during drawing. The stainless steel (S/S) provides additional mechanical strength and precompresses the Chevrel phase core after cool down.

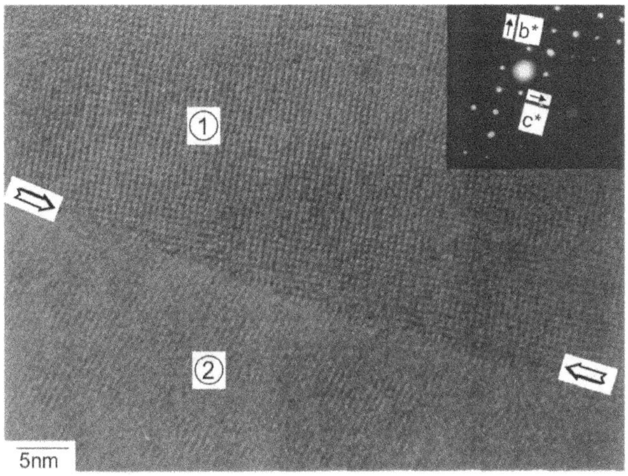

FIGURE B3.7 A high-resolution electron micrograph (HREM) of a grain boundary of $Pb_{0.7}Gd_{0.3}Mo_6S_8$ bulk sample. The selected area diffraction pattern (SADP) is for grain (1) and is close to the [100] zone axis. The grain boundary is very narrow. (Eastell, 1998.)

B3.4 Mechanical Properties

Among Chevrel phase materials, the coefficient of thermal expansion (α) has been studied most extensively for $PbMo_6S_8$. Using x-ray diffraction in the range 10 K–1200 K on bulk material, it has been concluded that α is almost temperature independent up to 900 K (650°C), where $\alpha = 1/L\,(dL/dT) = 9.4 \times 10^{-6}$ K^{-1} (Miraglia et al., 1987). For comparison, at room temperature α for $PbMo_6S_8$ is about half that of Cu or steel, about 25% higher than Nb and about twice that of Mo (Miraglia et al., 1987). Single-crystal measurements show a relatively strong anisotropic variation for α of about a factor of 3 (Alekseevskii et al., 1988). Such considerations are particularly important for optimising Chevrel phase wires (Miraglia et al., 1987). High-resolution thermal expansion measurements have been made using capacitive techniques. These show a very strong change in α at T_C in $PbMo_6S_8$ but not in $SnMo_6S_8$ (Ingle et al., 1998) which was attributed to different coupling between the superconductivity and the trigonal–triclinic structural transition (c.f. Chapter B3.3.5).

Compressibilities have been measured for 11 sulphur and selenide Chevrel phase compounds (Webb and Shelton, 1978) and typical data shown in Figure B3.8. The Young's modulus calculated from these data is around 40 GPa (Miraglia et al., 1987). This makes it similar to indium, about a factor of 5 smaller than steel and eight times smaller than Mo. Elastic constants have also been measured using ultrasonic techniques. The elastic constant for transverse distortions is 90 GPa and for longitudinal distortions is 21 GPa (Wolf et al., 1996).

The strain tolerance of Chevrel phase material has been investigated most comprehensively in the context of wires (Goldacker et al., 1989b). $PbMo_6S_8$ and $SnMo_6S_8$ fracture at about 0.65%, which is typical for a ceramic material.

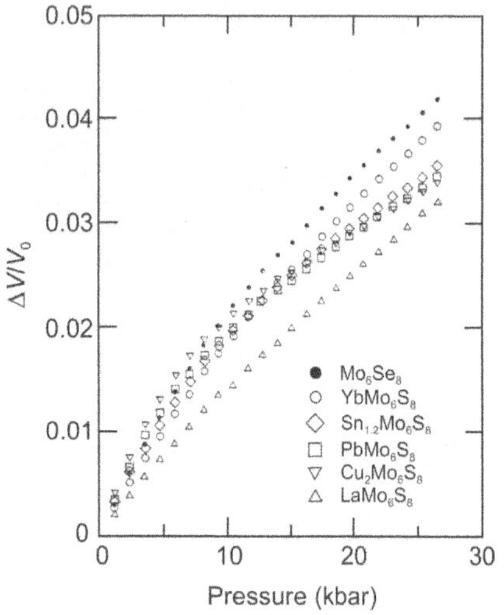

FIGURE B3.8 The compressibility of five ternary molybdenum sulphides $M_XMo_6S_8$ and the binary compound Mo_6Se_8. (Webb and Shelton, 1978.)

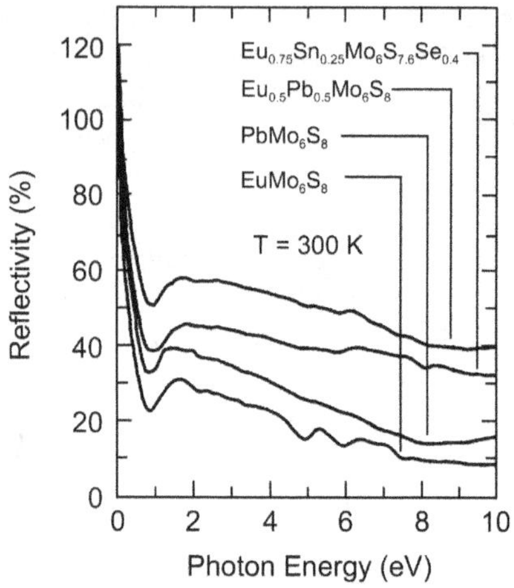

FIGURE B3.10 The near-normal reflectivity at room temperature of $EuMo_6S_8$, $PbMo_6S_8$, $Pb_{0.5}Eu_{0.5}Mo_6S_8$ and $Sn_{0.25}Eu_{0.75}Mo_6S_{7.6}Se_{0.4}$. (Fumagalli and Schoenes, 1991.)

Mechanical properties such as fracture toughness, crack propagation and fatigue properties are strongly dependent on the porosity of the material. Chevrel phase materials, therefore, can be considered as soft ceramics.

B3.5 Optical Properties

Vibrational Raman spectra have been observed for Cu-, Pb-, Ba- and Sn sulphide Chevrel phase compounds in the range from 10 meV to 50 meV (Holmgren et al., 1987) as shown in Figure B3.9. Such measurements give energies and symmetries of the Raman active optical phonons near the Brillouin zero centre. Several bands or peaks are independent of the metal (M) atom even in the non-stoichiometric Cu-compounds, in broad agreement with tunnelling data (Ohtaki et al., 1984). Reflectivity measurements and complex magneto-optical Kerr-effect measurements have been completed in magnetic fields up to 12 T and temperatures down to 0.5 K (Fumagalli and Schoenes, 1991).

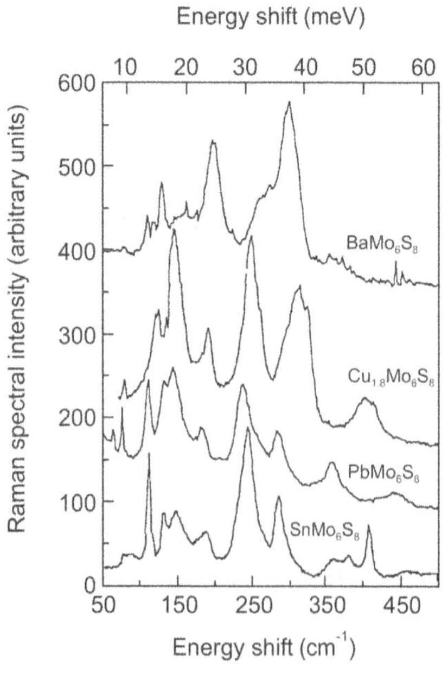

FIGURE B3.9 The Raman spectra for $BaMo_6S_8$, $Cu_{1.8}Mo_6S_8$, $PbMo_6S_8$ and $SnMo_6S_8$. (Holmgren et al., 1987.)

Figure B3.10 shows that the reflectivity (and optical conductivity spectra) are similar for $Eu_{1-x}Pb_xMo_6S_8$ and $Eu_{1-x}Sn_xMo_6S_{8-y}Se_y$ at 300 K. The carrier density was low (~ 6×10^{27} m^{-3}), the mobility ~ 1 cm^2 V^{-1} s^{-1} and the effective masses for the carriers ~10 m_e (Fumagalli and Schoenes, 1991).

B3.6 Thermal Properties

The specific heat capacity (C_P) of $PbMo_6S_8$ has been measured by a number of authors in high magnetic fields up to 24 T (Cattani et al., 1988; Cors et al., 1990; van der Meulen et al., 1995). The electronic contribution in the normal state is ~7 mJ K^{-2} (g-at)$^{-1}$. In the superconducting state, in addition to the BCS exponential gap term, there is also the term linear in temperature that accounts for the normal cores of the fluxons in the mixed state. At the superconducting jump C/T is ~ 70 mJ K^{-2} (g-at)$^{-1}$, which is relatively high because of the soft modes present (Bader et al., 1976), and $\Delta C_e/T_C$ is about 12 mJ K^{-2} (g-at)$^{-1}$. The effective Debye temperature measured using C_P measurements changes by a factor of 2 from ~200 K at 4.2 K up to ~400 K at room temperature (Fradin et al., 1976).

Measurements and analysis of a large number of Chevrel-type superconductors have been completed (Lachal et al., 1984). Figure B3.11 shows a typical Debye plot for a series of compounds of the form $Pb_{1-x}Cu_{1.8x}Mo_6S_8$ (Niu et al., 2002). Ultrasonic measurements give an average Debye temperature of ~245 K (Wolf et al., 1996). Many magnetic Chevrel phase

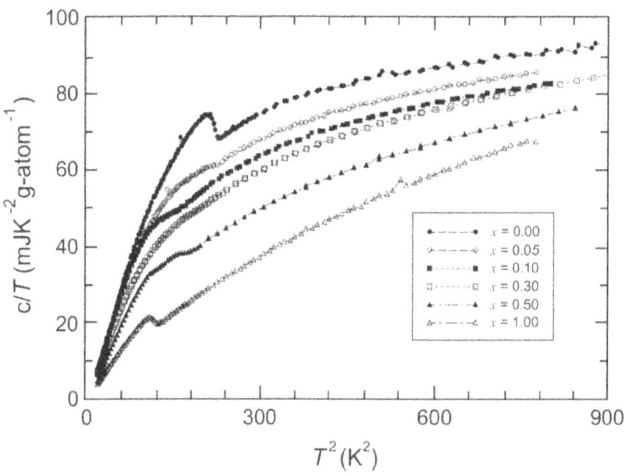

FIGURE B3.11 Debye plot (heat capacity)/(temperature) versus (temperature)^2 for $(Pb_{1-x}Cu_{1.8x})Mo_6S_8$ for different values of x. (Niu et al., 2002.)

superconductors have also been measured (Fischer and Maple, 1982a). Such work includes investigating pressure-induced re-entrant superconductivity (Chen et al., 1993) and the reduction in $\Delta C_e/T_C$ with increased magnetic doping (Leigh, 2001). Nevertheless, there is typically a factor of 2 variation in all the parameters derived from specific heat data in the literature (N.B. for $PbMo_6S_8$, 1 mole = 1037 g = 15 gat.). The differences are attributed to the sensitivity of the materials to the fabrication process.

The consensus on the phonon density of states in Chevrel phase materials is good. The weighted phonon density of states for $PbMo_6S_8$ and $SnMo_6S_8$ measured using neutron scattering measurements and specific heat measurements is consistent with calculations that assume the Mo_6S_8 clusters are tightly bound but only weakly interact with other clusters or the M ion. There is a relatively flat dispersion curve and a strong peak at about 5 meV, associated with the Einstein mode from the M ion. Modes in the energy range up to 18 meV are associated with the soft external modes of the Mo_6S_8 clusters. The hard internal modes are responsible for the energy range from 18 meV to 50 meV (Bader et al., 1976).

B3.7 Normal-State Properties

The normal-state resistive properties of Chevrel phase compounds are rather well described in terms of their chemistry and are consistent with band structure calculations (Mattheis and Fong, 1977). The dominant carriers are holes associated with the $4d_{x2-y2}$ states. The three important factors that contribute to the electronic properties are the charge transfer from the M cation to the Mo_6S_8 cluster and the volume and structure of the unit cell.

The position of the Fermi level in the sulphide system is strongly related to the charge transfer between the M cations and the sulphur anions (Yvon and Paoli, 1977). A characteristic of the Mo_6S_8 cluster is that it is only slightly distorted when filled with 24 electrons and tends to be insulating. The electronic configuration of $Mo4d^55s^1$ contributes six electrons to the Mo_6S_8 cluster. Hence with the S in the −2 valence state, there are 20 electrons in the Mo_6S_8 cluster and there is hole conduction. Adding Pb or Sn to the structure (both have valence +2) increases the number of electrons to 22, equivalent to two holes per cluster which produces the highest value of T_C. For comparison, the valence of the S, Se and Te in Mo_6S_8, Mo_6Se_8 and Mo_6Te_8 are −2, −1.75 and −1.33, respectively. Band structure calculations show that in the rhombohedral $PbMo_6S_8$, the Fermi energy lies below an energy gap about 1 eV wide (Mattheis and Fong, 1977). The states near to the Fermi energy are strongly confined within the Mo_6S_8 cluster (Bullett, 1977). Indeed there are some broad similarities between Chevrel phase materials and the intercalates of TaS_2 (Prober et al., 1980) and MoS_2 (Woollam and Somoano, 1976), where there is a hybridisation of the metal d-bands and an associated energy gap (Mattheis, 1973). The band structures for the Chevrel phase selenides and the telluride have similar properties (Bullett, 1977) (Roche et al., 1999), but one of the marked differences between them and the sulphides is that the density of states at the Fermi energy is higher for trivalent ions than for divalent ions. This is supported by the T_C values which are higher for the rare-earth selenides than for the sulphides as shown in Figure B3.1.

In $PbMo_6S_8$ samples, the resistivity (ρ) at room temperature is about 100 μΩ cm–1 mΩ cm, and typical values for room temperature–resistivity ratio (RRR) are 4 to 6 (Miraglia et al., 1987; Niu et al., 2002). In the $Cu_xMo_6S_8$ single crystals, ρ is similar with a RRR value of about 7 which leads to a scattering length (l) of about 20–30 Å (Fischer and Maple, 1982a). In thin films, RRR values were found in the range 2–6 and l estimated to be ~40 Å (Alterovitz and Woolam, 1978). The temperature dependence of ρ(T) is approximately linear up to 50 K but shows negative curvature at higher temperatures similar to the A15 superconductors. Theories that address the non-linearity utilise a strong peak in the density of states (Cohen et al., 1967) or a scattering length that is comparable to the lattice spacing (Fisk and Webb, 1976; Sunandana, 1979). Experimental data at low temperatures can be misleading, particularly if the material includes pure Mo or Mo_2S_3. Hall effect measurements in the $(Eu_{1-x})Sn_xMo_6S_8$ at room temperature gives Hall coefficients of +0.7 $\times 10^{-3}$ cm^3 G^{-1} confirming a hole carrier concentration of ~9×10^{27} m^{-3} (Meul, 1986) [in agreement with the optical measurements and muon measurements (Birrer et al., 1993)] and implying ~2.5 holes per formula unit. In $REMo_6S_8$ compounds, ρ is typically ~300 μΩ cm and RRR range from 8 to 34 (Beille et al., 1991), although $LuMo_6S_8$ has a resistivity at room temperature of only 50 μΩ cm (Geantet et al., 1990). Limited thermopower measurements on $Cu_{1.8}Mo_6S_{8-x}Te_x$ have also been completed (Kaiser, 1997).

The density of states derived from susceptibility measurements, specific heat measurements and band structure calculations give consistent values. For the Pb- and Sn Chevrel phase sulphide compounds, χ is about 3.5×10^{-5} emu (g-at)$^{-1}$. The (phonon-enhanced) density of states calculated from C_P measurements is about 1 state per (eV-atom-spin). This is about twice that found from χ measurements or band structure calculations which is expected with strong electron–phonon coupling (Fischer, 1978). In materials with relatively high critical temperature $PbMo_6S_8$, $LaMo_6S_8$ and $LaMo_6Se_8$, the susceptibility varies by about a factor of 1.5–2 between room temperature and 20 K (Peña and Sergent, 1989; Peña et al., 1998) and shows an anisotropy (for single crystals) of ~40%. A strongly temperature-dependent susceptibility has also been observed in the high-temperature superconductors. This is taken to be evidence that the Fermi level is situated near a peak in the density of states which standard BCS theory (Bardeen et al., 1957; Abrikosov, 2000) associates with relatively high-T_C values.

B3.8 Superconducting Properties

B3.8.1 Transition Temperature

The BCS theory (Bardeen et al., 1957; Abrikosov, 2000) currently provides the only generally accepted microscopic explanation for superconductivity. Within this framework, the critical temperature is determined by the density of states at the Fermi level, the phonon spectrum and the electron–phonon coupling. Tunnelling measurements on $Cu_{1.8}Mo_6S_8$ and $PbMo_6S_8$ give the ratio of the gap (Δ) to T_C of $2\Delta / k_B T_C = 4-5$ (BCS theory predicts 3.5) showing strong coupling. In the pseudobinary $Mo_6Se_{8-x}S_x$ system, strong coupling in Mo_6Se_8 (T_C = 6.2 K, coupling constant: λ = 1.25, $2\Delta / k_B T_C = 4.2$, $\Delta C_E / C_E = 2.25$) gives way to weak coupling showing BCS behaviour in $Mo_6Se_4S_4$ ($T_C = 1.8\ K$, λ = 0.6, $\Delta C_E / C_E = 1.4$) (Pobell et al., 1982; Furuyama et al., 1989). Although most of the phonon modes associated with the Mo_6S_8 cluster modes contribute to the electron coupling (Poppe and Wuhl, 1981), the materials with higher T_C values approaching 16 K have large values of λ, which are probably most strongly affected by the soft modes (Furuyama et al., 1989). Recent tunnelling spectroscopy provides evidence for two-gap superconductivity in both $PbMo_6S_8$ and $SnMo_6S_8$ (Dubois et al., 2007; Petrovic et al., 2011). The pseudobinary $Mo_6S_6I_2$ compound has the relatively high T_C of 14 K, which suggests that the superconductivity in Chevrel phases of highest T_C is fundamentally associated with the clusters. In this context, the isotope effect observed in Mo_6Se_8 is consistent with BCS theory and suggests that an electron–phonon mechanism operates in Chevrel phase materials (Pobell et al., 1982). For a given structure, T_C also depends on the volume of the unit cell (Hinks et al., 1984) and the valence electron concentration in the Mo_6S_8 cluster (Sergent et al., 1978). The trivalent sulphur-based Chevrel phase materials have uniformly low T_C whereas the divalent trigonal materials have high T_C (e.g. Sn and Pb). Among the trivalent rare-earth ions, there is a correlation between T_C and the volume of the unit cell (Fischer, 1978). When the volume decreases, the intercluster Mo–Mo decreases, so the valence bands are expected to broaden, and the density of states and T_C to fall (Fischer et al., 1975a). Hydrostatic pressure has been used to change the volume of the unit cell, and hence T_C, in divalent sulphides (Shelton, 1976; Capone II et al., 1984; Hinks et al., 1984). The effect of pressure on T_C is about an order of magnitude higher than found in elemental superconductors (Shelton, 1976) $dT_C / dP \sim 10^{-4}$ (kbar)$^{-1}$. The difference in critical temperature (T_C) between $PbMo_6S_8$ and $SnMo_6S_8$ can be explained by the difference in the volume of the unit cell (Hinks et al., 1984).

The Chevrel phase materials that transform fully from a rhombohedral structure at high temperatures to a triclinic structure at low temperatures have low electronic density of states (Lachal et al., 1983) and are non-superconducting. However, the superconductivity can be restored if pressure is applied to prevent the triclinic transition occurring. In $BaMo_6S_8$ for example, applying a pressure of 4 GPa changes the material from a triclinic semiconductor to a mixed triclinic–rhombohedral phase that is metallic with a T_C of 12 K (Yao et al., 1988). The structural instabilities in the Chevrel phase superconductors may enhance the electron–phonon coupling. In $Eu_{1.2}Mo_6S_8$ (Chu et al., 1981), (shown in Figure B3.12) superconductivity is also close to the metal–insulator transition and can be tuned by pressure. In $(Sn_{1-x}Eu_xM)_{1.2}Mo_6S_8$ (Harrison et al., 1981), the transition is tuned by Sn content (or carrier concentration). These properties are found in other Chevrel phase materials and is reminiscent of the HTS materials where high–T_C values also occur in materials with relatively low carrier concentration that are in proximity to the metal–insulator transition. There has been intense research into alternative

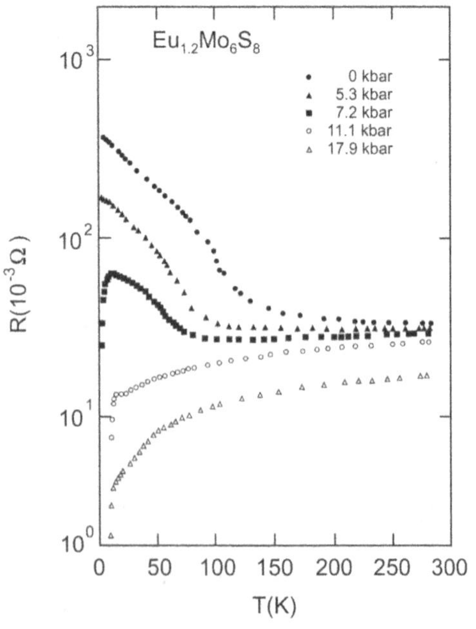

FIGURE B3.12 The resistance versus temperature for $Eu_{1.2}Mo_6S_8$ at various pressures. (Chu et al., 1981.)

microscopic mechanisms for superconductivity following the discovery of the high-temperature superconductors in the late 1980's. The Uemura plot provides empirical evidence that the cuprate and bismuthate high-temperature superconductors, the organic, Chevrel phase and heavy Fermion systems all belong to a single class of superconductors where T_C is proportional to the (small) n_s/m^* (carrier density/effective mass) (Uemura et al., 1991; Harshman and Mills Jr, 1992; Uemura, 2004) as shown in Figure B3.13. In the cuprates, the high values of T_C and the lack of a clear isotope effect suggest a non-phononic mechanism (Batlogg et al., 1987). For Chevrel phase materials, the carrier density ($n_s \sim 2$ holes/unit cell) and the effective mass ($m^* \sim m_e$) are rather robust numbers from both experiment and theory. Their presence on the Uemura plot suggests that Chevrel phase materials, which have a well-known chemistry and electronic structure, may be model systems in which to investigate non-standard mechanisms for superconductivity because of the simplifications which follow from their (almost) cubic (isotropic) structure.

The simple proportionality and universal behaviour in Figure B3.13 provides evidence that a single mechanism may produce superconductivity in many different types of superconductors including the Chevrel phase materials (Hillier and Cywinski, 1997; Sonier et al., 2000; Kiefl et al., 2010). Given that over the last 20 years, most classes of new superconductors appear close to the Universal line in the Uemura plot, it no longer makes much sense to describe them as exotic

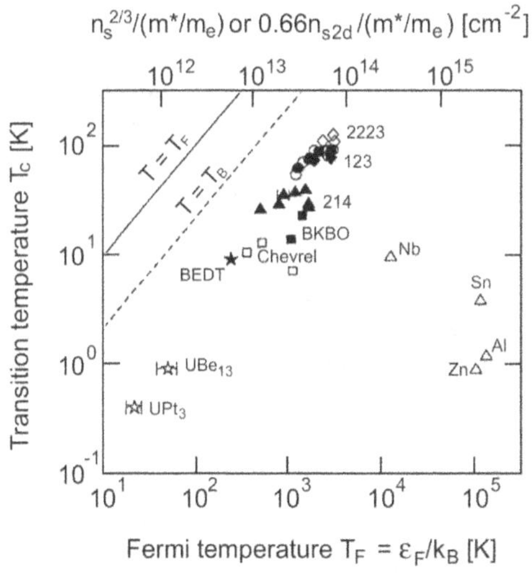

FIGURE B3.13 An Uemura plot – a log–log plot of critical temperature (T_C) versus the Fermi temperature (T_F) estimated from muon spin resonance measurements (combined with the interplanar distance for 2D and the Sommerfeld constant for 3D systems) for Cuprates, BKBO, Chevrel phase, BEDT, heavy fermion and some elemental superconductors. The solid line and the dashed line show when the critical temperature equals the Fermi temperature, T_F, and the dashed line represents the Bose–Einstein condensation temperature, T_B, of the ideal boson gas. (Uemura et al., 1991; Uemura, 2004.)

superconductors or high-temperature superconductors. Nevertheless, the plot does provide a guide as to how to synthesize materials with even higher T_C (Rybicki et al., 2016) and insight into the underlying canvas or phase diagram for the microscopic details for these materials (Uemura, 2009). Also, it naturally opens the questions of how the electrons pair to form bosons and whether the bosons are first preformed and then condense at T_C (in a similar way to superfluid helium) (Dzhumanov et al., 2016). Without a microscopic theory, we don't know whether the temperature at which bosons may preform bears any relation at all to the temperature at which superconductivity appears (c.f. Figure B3.14). Alternatively, akin to the BCS model, the electrons may become pairs and condense at the same temperature. One of the central issues remains the signature linear temperature dependence found for the normal-state resistivity of 'non-BCS' superconductive materials up to temperatures well above the Mott metal–insulator transition temperature (Gurvitch and Fiory, 1987) which has led to the 'strange metal' notation. Some authors associate the strange metal and the superconductivity (Loram et al., 2001; Tallon and Loram, 2001) with the quantum critical point. Others suggest that the pseudogap competes with the superconductivity (Kondo et al., 2009) or that the strange metal behaviour can be explained without any 'exotica' such as the quantum critical point (Anderson, 2006). Were a single mechanism able to explain the Universal superconductivity behaviour in the Uemura plot, it would clearly simplify things. Unfortunately identifying such commonality is not straightforward since even in a single material there are contradictory results reported. For example in Bi-2212, some authors report pseudogap behaviour only in underdoped samples (Ding et al., 1996; Williams et al., 1997; DeWilde et al., 1998), some that T^*

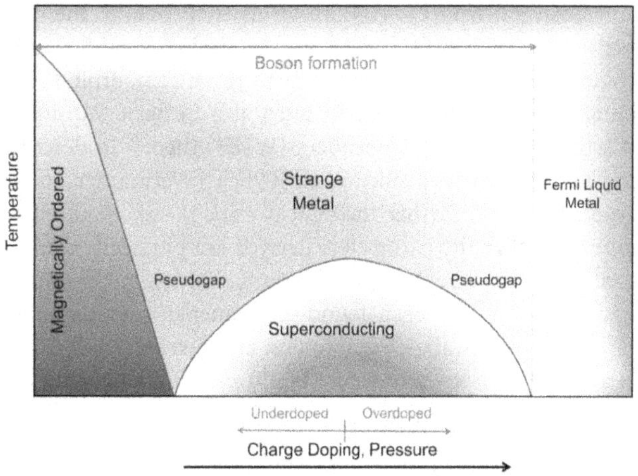

FIGURE B3.14 A schematic phase diagram for a Bose–Einstein type of condensation mechanism for superconductivity that occurs near the metal–insulator transition in metals with low carrier density. In the conducting phase, over a range of carrier concentration, the bosons are preformed at higher temperatures and eventually condense into the superconducting phase.

just gives the temperature scale of the pseudogap (Williams et al., 1997), and others that the same generic tunnelling density of states, or pseudogap behaviour, occurs in both overdoped and underdoped materials (Renner et al., 1998). Although the microscopic mechanism for superconductivity in the Chevrel phase materials has long been considered to be classic BCS because of the isotope effect found in Mo_6Se_8, and the reasonable agreement between microscopic normal-state properties and the predictions of BCS theory for T_C, Figure 13 shows that there is some important microscopic Physics that we just have not discovered yet - an understanding that will explain why superconductors appear together on the Uemura plot. Our current understanding is not complete. Indeed some authors have even argued that the simple empirical proportionality shown in Figure B3.13 does not hold in Chevrel phase materials (Birrer et al., 1993) but is best described using a rather more complex percolation model dependence (Dallacasa and Feduzi, 1992). We conclude that we just do not know yet whether we should classify Chevrel phase materials as classic BCS or non-BCS superconductors (Hirsch et al., 2015a; Hampshire, 2020).

B3.8.2 Upper Critical Field

Given that the only microscopic theory available to the community is BCS theory (Bardeen et al., 1957; Abrikosov, 2000), and so a large body of research has been described using it, we state here that the upper critical field (B_{C2}) can be given by:

$$B_{C2}(0) = \left(8.3\times10^{34}\left[\frac{\gamma T_C}{S}\right]^2 + 3.1\times10^3\gamma\rho_N T_C \right) \quad \text{(B3.1)}$$

where the two terms are the clean and dirty contribution, respectively (Decroux et al., 1993; Morley et al., 2001) and are consistent with $\xi_0 \sim l$. Both terms contribute to the very high upper critical field values in the $PbMo_6S_8$ and $SnMo_6S_8$ compounds. The complexity of the intrinsic spin, orbital coupling and spin-orbit coupling must also be included, using Werthamer–Helfand–Hohenberg (WHH) theory, to describe the temperature dependence of $B_{C2}(T)$ (Werthamer et al., 1966). However, further theoretical work is still required to assess whether the parameters derived are physically significant. The anisotropy of B_{C2} in $PbMo_6S_8$, $PbMo_6Se_8$ $Cu_{1.8}Mo_6S_8$ and $SnMo_6Se_8$ has been found experimentally to be about 15% (Decroux et al., 1978; Decroux and Fischer, 1982; Pazol et al., 1989) (as shown in Figure B3.15) and correlated with the rhombohedral angle (c.f. Chapter B3.3.5). There is currently no adequate explanation for this, since the band structure calculations show nearly cubic symmetry and predict low anisotropy.

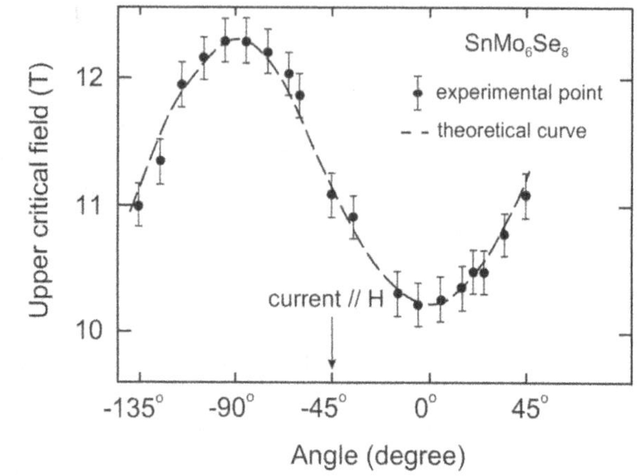

FIGURE B3.15 The anisotropy of the upper critical field of $SnMo_6S_8$ at 4.2 K (Decroux et al., 1978). The angle is measured between the ternary axis and the magnetic field.

B3.8.3 Ginzburg–Landau Description

Ginzburg-Landau (G–L) theory provides a self-consistent explanation for the properties of metallic superconductors in-field (i.e. superconductors that are non-magnetic in the normal state). There are only two free parameters, which can be taken to be the G–L constant (κ), which is broadly temperature independent, and $B_{C2}(T)$ (Cave, 1998). The fundamental properties of the superconducting state can be determined by measuring the reversible magnetisation close to $B_{C2}$and using the G–L relation:

$$M = -\left(\frac{H_{C2} - H}{(2\kappa^2 - 1)\beta_A} \right) \quad \text{(B3.2)}$$

where H_{C2} is the critical field strength, H is the applied field strength and β_A ($\approx$ 1.16) is the Abrikosov constant (Kleiner et al., 1964). Reversible magnetisation measurements similar to those shown in Figure B3.16 can be used to obtain values for κ and $dB_{C2}(T)/dT$, from which the slopes $dB_C(T)/dT$ and $dB_{C1}(T)/dT$ can be calculated. For high κ materials such as

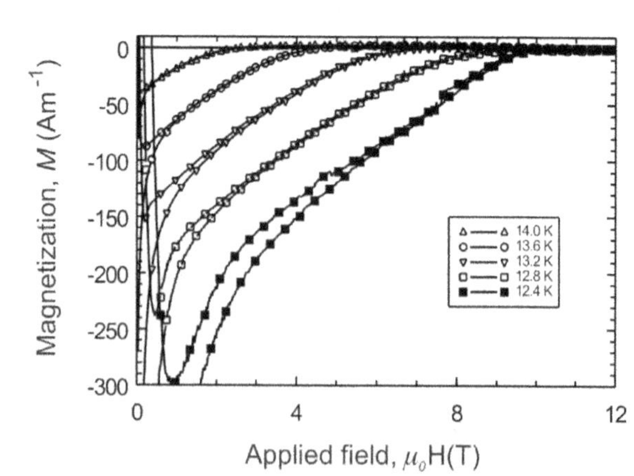

FIGURE B3.16 The magnetisation of bulk $PbMo_6S_8$ as a function of field at different temperatures. From the reversible data, one can use Ginzburg–Landau theory to calculate the Ginzburg–Landau parameter and the upper critical field. (Zheng et al., 1995; Niu et al., 2002.)

TABLE B3.3 The Values of the Fundamental Superconducting Parameters of $PbMo_6S_8$ Derived from Reversible Magnetisation Data (Zheng et al., 1995)[a]

T_C	κ	$B_{C2}(0)$	$B_{C1}(0)$	$B_C(0)$	$\lambda_{GL}(0)$	$\xi_{GL}(0)$
13.7 K	130	56 T	6.4 mT	250 mT	230 nm	2.0 nm

[a] The *h*, *k* and *l* values are provided for the hexagonal structure.

the Chevrel phase superconductors, it is best not to calculate the critical fields at low temperatures using the G–L relations directly, since G–L theory is strictly only valid close to T_C. In order to calculate $B_{C2}(0)$, the WHH relation can be used where:

$$B_{C2}(0) = -0.7T_C \frac{dB_{C2}}{dT}\bigg|_{T_C} \quad \text{(B3.3)}$$

$B_C(0)$ can be calculated using the BCS expression

$$B_C(T) = 1.74B_C(0)\left(1 - \frac{T}{T_C}\right) \quad \text{(B3.4)}$$

and $B_{C1}(0)$ can be calculated using the Gorter–Casimir (Gorter and Casimir, 1934) two-fluid empirical relation:

$$B_{C1}(T) = B_{C1}(0)\left(1 - \left(\frac{T}{T_C}\right)^2\right) \quad \text{(B3.5)}$$

Note that this approach does mean that the G–L relations do not hold at low temperatures (Zheng et al., 1995) – for example $B_{C2}(0) \neq \varphi_0 / 2\pi\xi^2_{G-L}(0)$. However, more reliable values for the critical fields are found at low temperatures using this procedure. The critical parameters for $PbMo_6S_8$ are shown in Table B3.3. The penetration depth has been measured in $SnMo_6S_{8-x}Se_x$ and $PbMo_6S_{8-x}Se_x$ (Birrer et al., 1993) using both magnetic and muon measurements and reasonably good agreement found.

B3.8.4 Irreversibility Field

The concept of the irreversibility field (B_{IRR}) is well documented in the literature both in the high-temperature and low-temperature superconductors (Rossel et al., 1991; Youwen and Suenaga, 1991). B_{IRR} is the magnetic field (below B_{C2}) at which the critical current density falls to zero. An important experimental problem is that measurements can only determine the field at which J_C drops below a minimum detection level. For practical purposes, a number of techniques are used, although the best procedure to measure B_{IRR} has not been generally agreed.

Vibrating sample magnetometry (VSM) can determine the field which delineates the hysteretic and reversible magnetic properties of a material and hence B_{IRR} directly (Cave, 1998). Such measurements have been performed on $PbMo_6S_8$. In bulk $PbMo_6S_8$, it was found that $B_{IRR} = 63(1 - T/T_C)^{1.46}$ (Zheng et al., 1995). A similar power law has been observed in single crystals (Rossel et al., 1991) and derived theoretically using a thermally activated flux-creep model (Yeshurun and Malozemoff, 1988). Measurements have also been completed on $(Pb_{1-x}Gd_x)Mo_6S_8$ that demonstrate B_{IRR} decreased with increased Gd content (Zheng and Hampshire, 1997). However, one should be very careful interpreting such data, since the hysteresis in magnetisation found in VSM measurements can only be simply related to J_C using Bean's model (Bean, 1962) if the field applied to the sample is uniform (c.f. Figures B3.17 and B3.18). Consider a typical material in which J_C reduces as the applied magnetic field increases and approaches B_{c2}. In general, the hysteresis collapses to zero when the variation in applied field

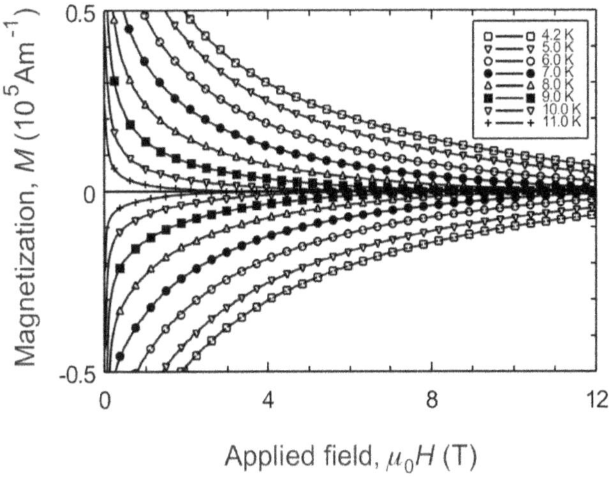

FIGURE B3.17 The magnetisation of bulk $PbMo_6S_8$ as a function of field at different temperatures (Niu et al., 2002). From the magnitude of the hysteresis, one can calculate the critical current density using Bean's model. (Bean, 1962.)

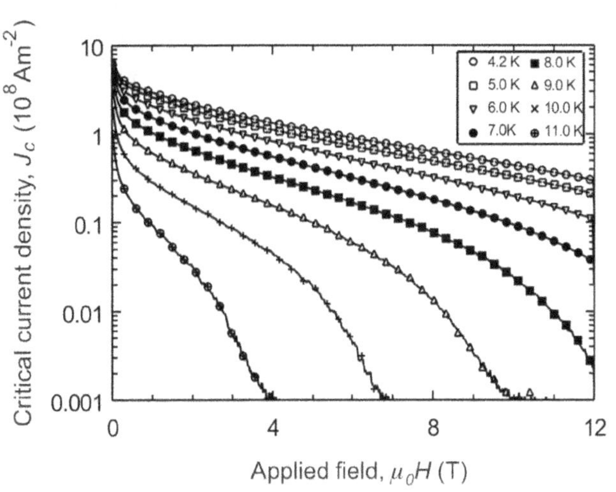

FIGURE B3.18 The critical current density of bulk $PbMo_6S_8$ as a function of field at different temperatures calculated using the data from the magnetisation data in Figure B3.17 and Bean's critical state model. (Niu et al., 2002.)

the sample experiences while oscillating is equal to the self-field of the sample and not when J_C is zero. For example in Figure B3.16, although the hysteresis falls dramatically at 12.8 K when the applied field reaches 0.5 T, it is associated with the inhomogeneity of the applied field and not related to B_{IRR} at all (Daniel and Hampshire, 2000). In such cases, any comparison of such field values with theoretical calculations of B_{IRR} is compromised, although one can say that the comparisons are useful to assess practical limits for high-field applications.

Flux penetration measurements and transport measurements offer alternative means to measure the critical current density (Ramsbottom and Hampshire, 1999). A type of irreversibility field can be determined by extrapolating the functional form of J_C to zero using a Kramer extrapolation (Kramer, 1973). In $PbMo_6S_8$, the irreversibility field has been improved from ~22 T up to nearly 40 T at 4.2 K by fabricating the material using hot isostatic pressing (Ramsbottom and Hampshire, 1997). B_{IRR} values of 35.4 T at 4.2 K have been achieved in high J_C $(Pb,Sn)Mo_6S_8$ wires (Cheggour et al., 1998). Standard resistance or susceptibility measurements can also be used to determine B_{IRR} (Nakamura et al., 1997). However, values obtained using different techniques can differ markedly. For example, increasing the Gd content in $(Pb,Gd)Mo_6S_8$ increases the irreversibility field in high fields measured using the onset of the resistive transition but shows a decrease using the onset of the susceptibility transition (Ramsbottom and Hampshire, 1999).

B3.8.5 Pinning Energy

The apparent pinning energy (U*) of $PbMo_6S_8$ has been measured using the decay of the magnetisation in time (Zheng et al., 1995). U* is calculated to vary from about 40 meV at 2 T to 15 meV at 12 T at 4.2 K. These values are about double the equivalent values found in $YBa_2Cu_3O_7$ and four times that of $Tl_2Ba_2Ca_2Cu_30_{10}$ at low fields. Activation energies derived from Arrhenius plots of resistivity give 130 meV at 9 T for $PbMo_6S_8$ and 186 meV at 9 T and 4.2 K for $SnMo_6S_8$ (Gupta et al., 1994). Comparisons between equivalent measurement techniques suggest that the effect of 'flux creep' is more pronounced in $PbMo_6S_8$ than in NbTi but less than that in high-temperature superconductors.

B3.8.6 Microwave Surface Resistance

Very few microwave measurements have been completed on Chevrel phase superconductors. For a superconducting $Cu_2Mo_6S_8$ thin film, a surface resistance of 4.5 mΩ at 10 GHz and 4.2 K has been obtained (Lemee et al., 1998).

B3.8.7 Critical Current versus Field and Temperature

The mechanism that determines the critical current density has long been a topic of theoretical and experimental research. J_C is determined both by the intrinsic fundamental superconducting properties and by the extrinsic metallurgical and microstructural factors such as the grain size of the material. Fietz and Webb (Fietz and Webb, 1967) suggested parameterising J_C through a scaling law for the volume pinning force ($F_P = J_C \times B$). The Chevrel phase materials can be described using

$$F_P = J_C \times B = \propto \left(B^*_{C2}(T)\right)^n b^{\frac{1}{2}}(1-b)^2 \qquad \text{(B3.6)}$$

where $B^*_{C2}(T)$ is the effective upper critical field, α and n are constants and b is the reduced field $(B/B^*_{C2}(T))$ (Cheggour et al., 1998). The index n is typically between 2 and 3 as found in many A15 materials (Keys et al., 1999; Keys and Hampshire, 2003; Taylor and Hampshire, 2005).

In low fields, the parameter α increases as the grain size decreases as is also found in low-temperature superconductors such as Nb_3Sn (Schauer and Schelb, 1981). There are many different approaches to modelling the pinning including that of Kramer after whom the reduced field dependence is named but which is probably not correct in detail (Kramer, 1973; Hampshire et al., 1985). Other pinning models have also been suggested that give the Kramer dependence and emphasise the importance of the grain boundaries. As yet, however, there is no consensus on the nature of the pinning that causes the ubiquitous Kramer dependence (Dew-Hughes, 1974; Hampshire and Jones, 1987; Gupta et al., 1994).

In materials optimised for high J_C in high fields, there is a much weaker correlation between J_C and grain size. For example, at 4.2 K in $SnMo_6S_8$, J_C is almost independent of grain size at fields above 15 T (Bonney et al., 1995). In $PbMo_6S_8$, it has been reported that J_C at 6 T saturates for grain sizes below 0.3 μm, although it must be noted that these samples are not fully dense (Karasik et al., 1985). Kramer found a similar saturation (or peak effect) close to B_{C2} in many low-temperature superconductors (Kramer, 1975; Daniel et al., 1997). The value of field at which J_C extrapolates to zero ($B^*_{C2}(T)$) is strongly correlated with the properties of the grain boundaries rather than either intragranular properties or the thermodynamic upper critical field (Cattani et al., 1991). Hence for high J_C materials, the standard grain boundary description may not be appropriate in the high-field (or saturation) regime. Whether this is because the efficiency of the grain boundaries falls or because a different [pinning (Le Lay et al., 1991; Gupta et al., 1994) or non-pinning (Hampshire, 1998)] mechanism limits J_C remains unresolved.

There has long been evidence that during dissipation (above J_C) flux flow in Chevrel phase materials can be localised along narrow channels (Herrmann et al., 1992). More recent experimental and (time-dependent Ginzburg–Landau) computational work has shown that the degradation of some critical properties at grain boundaries leads to a Kramer dependence for the volume pinning force. In such materials, flux penetrates the interior of a polycrystalline superconductor by first flowing along grain boundaries (Carty and Hampshire,

2008). This approach has opened the possibility that a general description of dissipation in many high-field polycrystalline superconductors is flux flow along grain boundaries (Carty and Hampshire, 2008, 2013; Sunwong et al., 2013).

The influence of neutron irradiation on J_C of Chevrel phase compounds has been investigated and some limited improvements were found (Rossel and Fischer, 1984). The highest J_C in wires is found in the quaternary (Pb,Sn)Mo_6S_8 for which at 4.2 K and 14 T, J_C is 7×10^8 A m^{-2} (Cheggour et al., 1997; Cheggour et al., 1998) and at 20 T about 2×10^8 A m^{-2} (Rimikis et al., 1991). Although the Pb-based Chevrel phase material has the highest critical field, some Sn is often included in bulk materials. This addition improves the homogeneity of the bulk (Selvam et al., 1995) and the interconnectivity between the grains probably by suppressing formation of MoS_2 (Even-Boudjada et al., 1999). Further improvements in the grain boundaries are still required to increase J_C.

A maximum value of J_C for $PbMo_6S_8$ has been estimated at 10^{10} A m^{-2} at 4.2 K and 20 T using a model which assumes ideal arrangement of the pinning sites (Rossel et al., 1991). Flux penetration measurements with small AC fields have found that $J_C > 10^{10}$ A m^{-2} at 4.2 K and 5 T at the surface of bulk $PbMo_6S_8$ which demonstrates the potential of this material (Kajiyana et al., 1985). Very significant improvements in J_C have been achieved in the upper critical field in Chevrel phase materials (Hampshire and Niu, 2005) by making it nanocrystalline which suggest we are far from optimum properties for polycrystalline materials. Along with most high-field polycrystalline superconductors, J_C in Chevrel phase materials is typically three orders of magnitude lower than its depairing current density (Wang et al., 2017) which leaves open the possibility of improvements – for example, following those of high-temperature superconductors, by fabricating highly textured (or single-crystalline) Chevrel phase material (Grant, 1995) and adding localised artificial pinning sites.

To produce magnetic fields above 25 T using superconductors alone requires rapid quench detection to protect the magnet from burn-out. This usually means that good magnet protection requires that the superconductor becomes as resistive as possible once it is outside its operating temperature range and a quench has occurred. Rapid quench detection is currently a problem for some HTS materials because the superconductor does not become resistive until it has been heated for a comparatively long time. This may be solved using better detection systems (Scurty et al., 2016), but if not, Chevrel phase materials may become attractive for quench control in magnet systems because of their relatively low-T_C values.

B3.9 The Magnetic Chevrel Phase Superconductors

Very small amounts of magnetic impurities at the parts-per-million level or paramagnetic ions at the 1 at% level are known to destroy the superconducting properties of most superconductors. Ginzburg observed that among the elements of the periodic table, superconductivity and magnetism seem to be mutually exclusive (Ginzburg, 1957) although clearly with the development of the Fe pnictide superconductors this observation is not generally true for complex compounds (Kamihara et al., 2008). Early experimental work investigating superconductors with magnetic impurities was compromised by uncertainty over whether or not the superconductivity and the ferromagnetism coexisted in the same region of the sample. Along with Chevrel phase superconductors, there are several other classes of superconductor that have magnetic ions within the superconducting unit cell. It is important in surveying this literature to distinguish magnetic order within the conducting layer and magnetic order produced by say magnetic rare-earth ions that hardly overlap with the conduction electrons/holes. The structure of magnetic Chevrel phase superconductors locates the rare-earth element a relatively large distance from the Mo atoms. This leads to a weak overlap between the 4d-electrons of the Mo and the 4f-electrons of the rare-earth element. This is similar to the cuprates where the rare-earth ions also hardly overlap with the conducting carriers. In contrast, the 3d-elements in Chevrel phase materials are located close to the Mo_6S_8 cluster, and the superconductivity is destroyed.

Susceptibility measurements show that the ternary magnetic superconductors with rare-earth elements have an effective Bohr magneton number that is close to the theoretical values for isolated ions at temperature above about 50 K (Johnston and Shelton, 1977; Pellizone et al., 1977). At lower temperatures, deviations from the Curie–Weiss law occur because of crystal field and magnetic correlation effects. There are many similarities in the superconducting and magnetic properties of the rare-earth Chevrel phase materials and the strongly magnetic nickel–boron–carbide materials (Eisaki et al., 1994), the rare earth–rhodium–borides and the cuprates because of the potential for the spatial separation of the rare-earth ions and the superelectrons. However, consider the properties of $GdMo_6S_8$ shown in Figure B3.19 (Fischer and Maple, 1982a) (Ishikawa et al., 1982). The re-entrant resistance is correlated with the antiferromagnetic ordering which occurs at 0.82 K as shown by the heat capacity measurements and neutron scattering measurements (Majkrzak et al., 1979; Thomlinson et al., 1981). The nature of the magnetic ordering is dependent on the particular rare-earth element in the compound and can be antiferromagnetic, ferromagnetic or oscillatory. For example in $HoMo_6S_8$, tunnelling spectroscopy suggests that superconductivity coexists with ferromagnetism (Morales et al., 1996). In high fields, so called re-entrant superconductivity can occur in some systems (Hampshire, 2001). It has been suggested that in high fields, the applied external field compensates for the negative exchange interaction between the rare-earth ion and the conduction electrons so that the material becomes superconducting (Jaccarino and Peter, 1962) in a limited part of B–T phase space. The degree to which ordering of the ions is exchange driven (Blount and

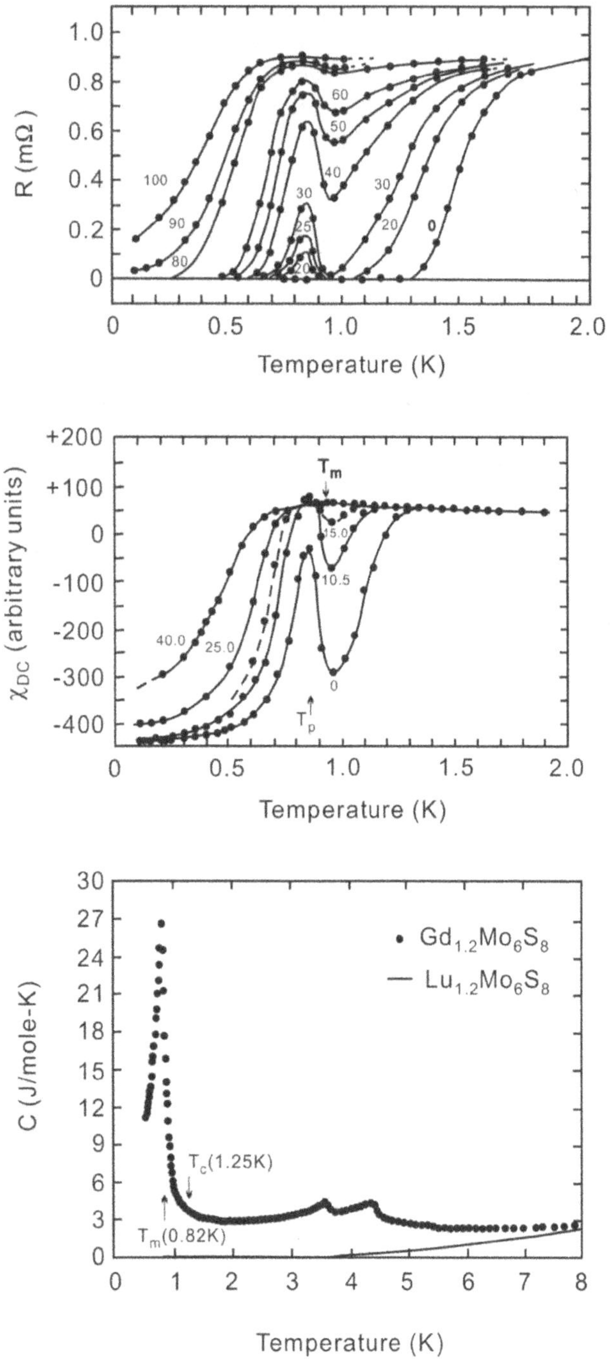

FIGURE B3.19 Selected properties of the magnetic rare-earth compound $GdMo_6S_8$. (Ishikawa et al., 1982.)

Varma, 1979) and driven by dipolar interactions (Hampshire, 2001) is still open to discussion.

The range of phenomena of the rare-earth Chevrel phase superconductors continues to fascinate the scientific community. The complexity arises because in order to understand the properties of these materials we must understand how magnetism and superconductivity operate at the atomic level.

B3.10 Concluding Comments

Since their discovery in 1971, Chevrel phase materials have been of interest to the whole superconductivity community from engineers, who want to make high-field magnets, to physicists, who want to understand the microscopic mechanism that causes superconductivity, magnetism and coexistence of the two. Interest in these materials inevitably waned with the discovery of the high-temperature cuprate superconductors. Massive interest in HTS materials was driven by the possibility of discovering a new mechanism producing superconductivity and the potential for new applications operating at liquid nitrogen temperatures. Related driving forces are now increasing the research activity into Chevrel phase superconductors. Fundamental interest arises because these materials may offer a model (almost) cubic system in which to address non-BCS superconductivity without the strong anisotropy or layering present in the HTS materials. Furthermore Chevrel phase superconductors have interesting fundamental properties that are intermediate between the HTS and LTS materials. Technological interest arises because of the high values of B_{C2} in these materials and improvements in cryocooler technology, which facilitates operating very high-field magnets at ~ 4 K (Watanabe et al., 1998). Indeed if J_C in the wires of these materials can be improved by say a factor of ~ 8, it will open the possibility of using them in the next generation of high-field magnet systems operating in fields significantly above 25 T.

Acknowledgements

The author acknowledges helpful comments and literature from Dr. Raine as well as the research groups in Durham, Geneva and Rennes. He would also like to thank the referee who did a very welcome thorough job. Thanks also to Amanda, Emily, Peter, Alex and Michael for their support. The data are available at: http://dx.doi.org/10.15128/r1h702q636g

Further Reading

There is a very good volume of review articles dedicated to Ted Geballe for his 95th birthday - Physica C:514 1 - 444 July 2015 edited by J E Hirsch, M B Maple and F Marsiglio (Hirsch et al., 2015b). Relevant articles include:

Hirsch J E, Maple M B, and Marsiglio F, 2015. Superconducting materials classes: Introduction and overview, *Physica C: Superconductivity* vol. 514, pp. 1–8 (Hirsch et al., 2015a). An interesting approach to classifying many of the materials that appear on the Uemura plot.

Pena O, 2015. Chevrel phases: Past, present and future, *Physica C - Superconductivity and Its Applications*, 514, 95–112. DOI: 10.1016/j.physc.2015.02.019. An excellent review of the superconductivity properties of Chevrel phase materials (Pena, 2015).

Wolowiec C T, White B D, and Maple M B, 2015. Conventional magnetic superconductors, *Physica C: Superconductivity*, vol. 514, pp. 113–129 (Wolowiec et al., 2015). An interesting review of conventional magnetic superconductors that is focussed on rhodium borides, Chevrel phases and nickel-borocarbides.

Shimizu K, 2015. Superconductivity from insulating elements under high pressure, *Physica C: Superconductivity*, vol. 514, pp. 46–49 (Shimizu, 2015). Hamlin, J J 2015. Superconductivity in the metallic elements under pressure, 514, 59–76. (Hamlin, 2015) A pair of interesting reviews that focus on elements that show a substantial increase in critical temeprature with pressure.

Chu W, Canfield P C, Dynes R C, Fisk Z, Batlogg B, Deutscher G, et al. 2015. Epilogue: Superconducting materials past, present and future, *Physica C*, vol. 514, pp. 437–443 (Chu et al., 2015). Some interesting discussions about future materials.

King R B, 1999. Chemical structure and superconductivity, *Journal of Chemical Information and Computer Science*, vol. 39, pp. 180–191 (King, 1999). An excellent review of structure and superconductivity in Chevrel phase materials and lanthanide rhodium borides.

Also:

Perrin A and Perrin C, 2012. The molybdenum and rhenium octahedral cluster chalcohalides in solid state chemistry: from condensed to discrete cluster units, *Comptes Rendus Chimie*, 15, 8125–836. An excellent and detailed review of the solid state chemistry of these materials (Perrin and Perrin, 2012).

King R B, 1999. Chemical structure and superconductivity. *Journal of Chemical Information and Computer Science*, 39, 180–191. An excellent review of chemical structure in superconducting materials (King, 1999).

Evetts J, 1992. Concise Encyclopaedia of Magnetic and Superconducting Materials (Pergamon Press UK). A series of compact, introductory articles predominantly on the science and technology of magnetic and superconducting materials (Evetts, 1992).

Pena O and Sergent M, 1989. Rare earth based Chevrel phases REMo6X8: Crystal growth, physical and superconducting properties, *Progress in Solid State Chemistry*, 19, 165–281. A very interesting comprehensive review of single crystal growth and properties. Great detail is provided about the chemistry and materials science of rare earth materials (Peña and Sergent, 1989).

Fischer Ø and Maple M B, 1982, Superconductivity in Ternary Compounds Vol. I - Structural, Electronic and Lattice Properties and Vol. II - Superconductivity and Magnetism (Springer-Verlag Berlin Heidelberg). These two-volume texts were written by many of the individual researchers involved in the intensive research of the seventies. The texts include an excellent compendium of many of the material properties for Chevrel phase superconductors - rather than extend our reference list considerably we have quoted these texts as a source rather cite the several hundred papers quoted therein (Fischer and Maple, 1982a, b).

Fischer Ø, 1978. Chevrel phases: Superconducting and normal state properties, *Applied Physics*, 16, 1–28. Excellent older review of the properties of Chevrel phase superconductors (Fischer, 1978).

References

Abrikosov AA (2000) Theory of high-T_C superconducting cuprates based on experimental evidence. Physica C 341-348:97–102.

Alekseevskii NE, Nizhankowski VL, Beille J, Lacheisserie E (1988) Investigation of the thermal expansion, electrical resistivity, and magnetic susceptibility anisotropy in $PbMo_6S_8$ single crystals. Journal of Low Temperature Physics 72:241–246.

Alterovitz SA, Woolam JA (1978) Upper critical field of copper molybdenum sulphide. Solid State Communications 25:141–144.

Anderson PW (2006) The 'strange metal' is a projected Fermi liquid with edge singularities. Nature 2:626–630.

Aruchamy A, Tamaoki H, Fujisjhima A, Berger H, Speziali NL, Levy F (1994) Photoelectrochemical characterization of a rhenium octahedral cluster compound $[Re_6Se_7Br_4]$. Materials Research Bulletin 29:359–368.

Aurbach D, Lu Z, Schechter A, Gofer Y, Turgeman R, Cohen Y, Moshkovich M, Levi E (2000) Protoype systems for rechargeable magnesium batteries. Nature 407:724–727.

Bader SD, Knapp GS, Sinha SK, Schweiss P, Renker B (1976) Phonon spectra of Chevrel-phase lead and tin molybdenum sulfides: A molecular-crystal model and its implications for superconductivity. Physical Review Letters 37:344–348.

Bardeen J, Cooper LN, Schrieffer JR (1957) Theory of superconductivity. Physical Review 108:1175–1204.

Batlogg B, Cava RJ, Jayaraman A, van Dover RB, Kourouklis GA, Sunshine S, Murphy DW, Rupp LW, Chen HS, White A, Short KT, Mujsce AM, Rietman EA (1987) Isotope effect in the High-T_C superconductors $Ba_2YCu_3O_7$ and $Ba_2EuCu_3O_7$. Physical Review Letters 58:2333–2336.

Batterman BW, Barrett CS (1964) Crystal structure of superconducting V_3Si. Physical Review Letters 13:390–392.

Bean CP (1962) Magnetization of hard superconductors. Physical Review Letters 8:250–253.

Beille J, Schmitt H, Padiou J, Sergent M (1991) The superconducting state of $Sr_xMo_6S_8$ (x<1) by simultaneous application of chemical and external pressure. Journal of Physics C - Solid State 3:2471–2477.

Benson JW, Schrader GL, Angelici RJ (1995) Studies of the mechanism of thiophene hydrodesulfurization:2H NMR and mass spectral analysis of 1,3-butadiene produced in the deuterodesulfurization (DDS) of thiophene over $PbMo_6S_8$ catalyst. Journal of Molecular Catalysis A: Chemical 96:283–299.

Birrer P, Gygax FN, Hitti B, Lippelt E, Schenck A, Weber M, Cattani D, Cors J, Decroux M, Fischer Ø (1993) Magnetic penetration depth in the Chevrel-phase superconductors $SnMo_6S_{8-x}$ and $PbMo_6S_{8-x}Se_x$. Physical Review B 48:16589–16599.

Blount EI, Varma CM (1979) Electromagnetic effects near the superconductor-to-ferromagnet transition. Physical Review Letters 42:1079–1082.

Bonney LA, Willis TC, Larbalestier DC (1995) Dependence of critical current density on microstructure in the $SnMo_6S_8$ Chevrel-phase superconductor. Journal of Applied Physics 77:6377–6387.

Boursicot S, Bouquet V, Péron I, Guizouarn T, Potel M, Guilloux-Viry M (2012) Synthesis of $Cu_2Mo_6S_8$ powders and thin films from intermediate oxides prepared by polymeric precursor method. Solid State Sciences 14:719–724.

Bullett DW (1977) Relation between electronic structure and T_C in binary and ternary molybdenum chalcogenides. Physical Review Letters 39:664–666.

Caillat T, Fleurial JP, Snyder GJ (1999) Potential of Chevrel phases for thermoelectric applications. Solid State Science 1:535–544.

Capone II DW, Guertin RP, Foner S, Hinks DG, Li HC (1984) Effect of pressure and oxygen defects in divalent Chevrel-phase superconductors. Physical Review B 29:6375–6377.

Carty GJ, Hampshire DP (2008) Visualising the mechanism that determines the critical current density in polycrystalline superconductors using time-dependent Ginzburg-Landau theory. Physical Review B 77:172501.

Carty GJ, Hampshire DP (2013) The critical current density of an SNS junction in high magnetic fields. Superconductor Science and Technology 26:065007.

Cattani D, Cors J, Decroux M, Fischer Ø (1991) Intra- and intergrain critical current in $PbMo_6S_8$ sintered samples. IEEE Transactions on Magnetics 27:950–953.

Cattani D, Cors J, Decroux M, Seeber B, Fischer Ø (1988) Calorimetric determination of H_{C2} of $PbMo_6S_8$. Physica C 153:461–462.

Cave J (1998) Handbook of Applied Superconductivity, 1st Edition. Bristol: Institute of Physics.

Cheggour N, Decroux M, Fischer Ø, Hampshire DP (1998) Irreversibility line and granularity in Chevrel phase superconducting wires. Journal of Applied Physics 84:2181–2183.

Cheggour N, Deroux M, Gupta A, Fischer Ø, Pederboom JAAJ, Bouquet V, Sergent M, Chevrel R (1997) Enhancement of the critical current density in Chevrel phase superconducting wires. Journal of Applied Physics 81:6277–6284.

Cheggour N, Decroux M, Gupta A, Ritter S, Schrœter V, Seeber B, Flükiger R, Fischer Ø, Boudjada S, Burel L, Bouquest V, Chevrel R, Sergent M, Massat H, Langlois P, Genevey P (1993) Superconducting properties of $PbMo_6S_8$ wires with micron grain size obtained from the decomposition of Mo_6S_8. Proceedings of ICMAS 93:403–408.

Chen X, Perel AS, Brooks JS, Guertin RP, Hinks DG (1993) Specific heat measurements of pressure-induced reentrant superconductivity in $Eu_{0.9}Ho_{0.1}Mo_6S_8$. Journal of Applied Physics 73:1886–1891.

Cheng Y, Luo L, Zhong L, Chen JT, Li B, Wang W, Mao SX, Wang CY, Sprenkle VL, Li G, Liu J (2016) Highly reversible Zinc-ion intercalation into Chevrel phase Mo_6S_8 nanocubes and applications for advanced Zinc-ion batteries. Applied Materials and Interfaces 9:13673–13677.

Chevrel R, Sergent M (1982) Chemistry and structure of ternary molybdenum chalcogenides. In: Superconductivity in ternary compounds I: Structural, electronic and lattice properties (Fischer Ø, Maple MB, eds), pp. 25–86.

Chevrel R, Sergent M, Prigent J (1971) Sur de Nouvelles Phases Sulfurées Ternaires du Molybdène. Journal of Solid State Chemistry 3:515–519.

Chevrel R, Sergent M, Prigent J (1974) Un nouveau sulfure de molybdène: Mo_3S_4 preparation, propriétés et structure cristalline. Materials Research Bulletin 9:1487–1498.

Chu CW, Canfield PC, Dynes RC, Fisk Z, Batlogg B, Duetscher G, Geballe TH, Zhao ZX, Greene RL, Hosono H, Maple MB (2015) Epilogue: Superconducting materials past, present and future. Physica C 514:437–443.

Chu CW, Huang SZ, Lin CH, Meng RL, Wu MK, Schmidt PH (1981) High-pressure study of the anomalous rare-earth ternaries $Eu_{1.2}Mo_6S_8$ and $Eu_{1.2}Mo_6Se_8$. Physical Review Letters 46:276–279.

Cohen RW, Cody GD, Halloran JJ (1967) Effect of Fermi-level motion on normal state properties of Beta-tungsten superconductors. Physical Review Letters 19:840–844.

Corrignan A, Hamard C, Pena O (1999) Superconducting properties of Solid Solutions $(Mo_6Se_8)Pb_x$ and $Pb_xMo_6Se_8$ in the ternary system Pb-Mo-Se. Journal of Alloys and Compounds 289:260–264.

Cors J, Cattani D, Decroux M, Stettler A, Fischer Ø (1990) The critical field of $PbMo_6S_8$ measured by specific heat up to 14 T. Physica B 165:1521–1522.

Dallacasa V, Feduzi R (1992) Quantum percolation of high Tc materials: Tc dependence on carrier density. Physics Letters A 170:153–158.

Daniel IJ, Hampshire DP (2000) Harmonic calculations and measurements of the irreversibility field using a vibrating sample magnetometer. Physical Review B 61:6982–6993.

Daniel IJ, Zheng DN, Hampshire DP (1997) An investigation of the peak effect in the Chevrel phase superconductor tin molybdenum sulphide. IOP Appl Super Conf 2:1169–1172.

Decroux M, Antognazza L, Kugler M, Koller E, Fischer Ø, Lemee N, Guilloux-Viry M, Perrin A (1999) Investigation of vortex dynamics close to B_{C2} in $Cu_2Mo_6S_8$ quasi epitaxial thin films. Solid State Sciences 1:585–589.

Decroux M, Fischer Ø (1982) Critical fields of ternary molybdenum chalcogenides. In: Superconductivity in Ternary Compounds II (Maple MB, Fischer Ø, eds), pp. 57–98: Springer-Verlag.

Decroux M, Fischer Ø, Flükiger R, Seeber S, Delesclefs R, Sergent M (1978) Anisotropy of H_{C2} in the Chevrel phases. Solid State Communications 25:393–396.

Decroux M, Selvam P, Cors J, Seeber B, Fischer Ø, Chevrel R, Rabiller P, Sergent M (1993) Overview on the recent progress of Chevrel phases and their impact on the development of $PbMo_6S_8$ wires. IEEE Transactions on Applied Superconductivity 3:1502–1509.

Dew-Hughes D (1974) Flux pinning mechanisms in type II superconductors. Philosophical Magazine 30:293–305.

DeWilde Y, Miyakawa N, Guptasarma P, Iavarone M, Ozyuzer L, Zasadzinski JF, Romano P, Hinks DG, Kendziora C, Crabtree GW, Gray KE (1998) Unusual strong-coupling effects in the tunneling spectroscopy of optimally doped and overdoped $Bi_2Sr_2CaCu_2O_8$. Physical Review Letters 80:153–156.

Ding H, Yokoya T, Campuzano JC, Takahashi T, Randeria M, Norman MR, Mochiku T, Kadowaki K, Giapintzakis J (1996) Spectroscopic evidence for a pseudogap in the normal state of underdoped high-Tc superconductors. Nature 382:51–54.

Dubois C, Petrovic AP, Santi G, Berthod C, Manuel AA, Decroux M, Fischer Ø, Potel M, Chevrel R (2007) Node-like excitations in superconducting $PbMo_6S_8$ probed by scanning tunneling spectroscopy. Physical Review B 75:104501.

Dzhumanov S, Karimboev EX, Djumanov SS (2016) Underlying mechanisms of pseudogap phenomena and Bose-liquid superconductivity in high-Tc cuprates. Physics Letters A 380:2173–2180.

Eastell C (1998) Microstructure and properties of high temperature superconducting wires. In: Materials Department: Thesis: Oxford University.

Eisaki H, Takagi H, Cava RJ, Batlogg B, Krajewski JJ, Peck WF, Mizuhashi K, Lee JO, Uchida S (1994) Competition between magnetism and superconductivity in rare-earth nickel boron carbides. Physical Review B 50:647–650.

Even-Boudjada S, Burel L, Chevrel R, Sergent M (1998a) New synthesis route of $PbMo_6S_8$ superconducting Chevrel phase from ultrafine precursor mixtures: II PbS, Mo_6S_8 and Mo powders. Materials Research Bulletin 33:419–431.

Even-Boudjada S, Burel L, Chevrel R, Sergent M (1998b) New synthesis route of $PbMo_6S_8$ superconducting Chevrel phase from ultrafine precursor mixtures: I. PbS, MoS_2 and Mo powders. Materials Research Bulletin 33:237–252.

Even-Boudjada S, Tranchant V, Chevrel R, Sergent M, Crosnier-Lopez M, Laligant Y, Retoux R, Decroux M (1999) One of the possible explanations of the major J_c limiting factor at the $PbMo_6S_8$ granular superconductor grains surface. Materials Letters 38:90–97.

Evetts J (1992) Concise Encyclopedia of Magnetic and Superconducting Materials: Pergamon.

Fietz WA, Webb WW (1967) Magnetic properties of some Type-II alloy superconductors near the upper critical field. Physical Review 161:423–433.

Fischer Ø (1978) Chevrel phases - superconducting and normal state properties. Applied Physics 16:1–28.

Fischer Ø, Decroux M, Roth S, Chevrel R, Sergent M (1975b) Compensation of the paramagnetic effect on H_{C2} by magnetic moments: 700 kG superconductors. Journal of Physics C - Solid State 8:L474–L477.

Fischer Ø, Maple MB (1982a) Superconductivity in Ternary Compounds Vol I-Structural, Electronic and Lattice Properties, 1st Edition. Berlin: Springer-Verlag.

Fischer Ø, Maple MB (1982b) Superconductivity in Ternary Compounds Vol II-Superconductivity and Magnetism. Berlin: Springer Verlag.

Fischer Ø, Treyvaud A, Chevrel R, Sergent M (1975a) Superconductivity in the $RE_xMo_6S_8$. Solid State Communications 17:721–724.

Fisk Z, Webb GW (1976) Saturation of the high temperature normal state electrical resistivity of superconductors. Physical Review Letters 36:1084–1086.

Fleurial JP, Snyder G, Borshchevsky A, Caillat T (2002) Thermoelectric materials formed based on Chevrel phases. In: Google Patents.

Flükiger R (2010) Procedure of densifying filaments for a superconductive wire. In: Google Patents.

Flükiger R, Baillif R (1982) Metallurgy and structural transformations in ternary molybdenum chalcogenides. In: Superconductivity in Ternary Compounds Vol I-Structural, Electronic and Lattice Properties (Fischer Ø, Maple MB, eds), pp. 113–141.

Foner S, McNiff EJ, Jr, Alexander EJ (1974) 600kG superconductors. Physics Letters 49A:269–270.

Fradin FY, Knapp GS, Bader SD, Cinader G, Kimball CW (1976) Electron and Phonon Properties of A-15 Compounds and Chevrel Phases. In: Superconductivity in d- and f-Band Metals: Second Rochester Conference (Douglass DH, ed.), pp. 297–312. Boston, MA: Springer US.

François M, Yvon K, Cattani D, Decroux M, Chevrel R, Sergent M, Boudjada S, Wroblewski T (1994) Synchrotron powder diffraction study of the low-temperature lattice distortion of $PbMo_6S_8$. Journal of Applied Physics 75:423–430.

Fumagalli P, Schoenes J (1991) Magneto-optical Kerr-effect study of the high-field superconductors $Eu_{1-x}Pb_xMo_6S_8$ and $Eu_{1-x}Sn_xMo_6S_{8-y}Se_y$. Physical Review B 44:2246–2262.

Furuyama M, Kobayashi N, Muto Y (1989) Electron-phonon interactions in the superconducting Chevrel phase compounds $Mo_6Se_{8-x}S_x$. Physical Review B 40:4344–4354.

Geantet C, Horyn R, Padiou J, Pena O, Sergent M (1990) Single crystal studies of $REMo_6S_8$ (RE = Er, Lu). Physica B 163:431–434.

Geng L, Lv G, Xing X, Guo J (2015) Reversible electrochemical intercalation of aluminum in Mo6S8. Chemistry of Materials 27:4926–4929.

Ginzburg VL (1957) Ferromagnetic superconductors. Soviet Physics JETP 4:153–160.

Goldacker W, Rimikis G, Specking W, Weiss F, Flukiger R (1989a) Jc versus strain investigations of $PbMo_6S_8$ and $SnMo_6S_8$ wires. Advances in Cryogenic Engineering 36A:353-360.

Goldacker W, Specking W, Weiss F, Rimikis G, Flükiger R (1989b) Influence of transverse, compressive and axial tensile stress on the superconductivity of $PbMo_6S_8$ and $SnMo_6S_8$ wires. Cryogenics 29:955–960.

Goldacker W, Specking W, Weiss F, Rimikis G, Flükiger R (1990) Effect of axial tensile and transverse compressive stress on J_c and B_{c2} of $PbMo_6S_8$ and $SnMo_6S_8$ wires. In: Advances in Cryogenic Engineering Materials. An International Cryogenic Materials Conference Publication, vol 36 (Reed RP, Fickett FR, eds). Springer, Boston, MA. https://doi.org/10.1007/978-1-4613-9880-6_46

Goncalves AP, Godart C (2014) New promising bulk thermoelectrics: Intermetallics, pnictides and chalcogenides. European Physics Journal B 87:42.

Gorter CJ, Casimir HBG (1934) 2 fluid model. Physikalische Zeitschrift 35:963–966.

Gougeon P, Pena O, Potel M, Sergent M, Brusetti R (1984) New ternary $M_{(2n-2)}Mo_{6n}X_{(6n+2)}$ chalcogenides (M= Rb, Cs; X= S, Se) with condensed Mo_{6n} clusters (n= 2, 3, 4, 5). Annales de chimie et de physique(Paris) 9:1079–1082.

Grant PM (1995) Superconducting superwires. Nature 375:107–108.

Gupta A, Decroux M, Selvam P, Cattani D, Willis TC, Fischer Ø (1994) Critical currents and pinning in powder metallurgically processed Chevrel phase bulk superconducting samples. Physica C 234:219–228.

Gurvitch M, Fiory AT (1987) Resistivity of $La_{1.825}Sr_{0.175}CuO_4$ and $YBa_2Cu_3O_7$ to 1100K: absence of saturation and its implications. Physics Review Letters 59:1337–1340.

Hamard C, Lancin M, Marhic C, Pena O (2002) Intergrowth between binary and ternary phases in Chevrel-phase compounds $REMo_6Se_8$ containing heavy rare-earth elements. Materials Science and Engineering: A A333:250–261.

Hamlin JJ (2015) Superconductivity in the metallic elements under pressure. Physica C: Superconductivity 514:59–76.

Hampshire D, Niu H (2005) High-field superconductors. In: Google Patents.

Hampshire DP (1998) A barrier to increasing the critical current density of bulk untextured polycrystalline superconductors in high magnetic fields. Physica C: Superconductivity and Its Applications 296:153–166.

Hampshire DP (2001) The non-hexagonal flux-line lattice in superconductors. Journal of Physics: Condensed Matter 13:6095–6113.

Hampshire DP (2020) Parameterisation of critical temperature versus carrier concentration in superconductors. In progress.

Hampshire DP, Jones H (1987) A detailed investigation of the E-J characteristic and the role of defect motion within the flux-line lattice for high-current-density, high field superconducting compounds with particular reference to data on Nb_3Sn throughout its entire field-temperature phase space. Journal of Physics C – Solid State 20:3533–3552.

Hampshire DP, Jones H, Mitchell EWJ (1985) An in-depth characterisation of $(NbTa)_3Sn$ filamentary superconductor. IEEE Transactions on Magnetics 21:289–292.

Harrison DW, Lim KC, Thompson JD, Huang CY, Hambouger PD, Luo HL (1981) Observation of the transition from semiconductor to high-T_C superconductor in $(Sn_xEu_{1-x})_y Mo_6S_8$ under high pressure. Physical Review Letters 46:280–283.

Harshman DR, Mills Jr AP (1992) Concerning the nature of high-Tc superconductivity: Survey of experimental properties and implications for interlayer coupling. Physical Review B 45:10684–10712.

Herrmann PF, Schellenberg L, Zuccone J, Seeber B, Fischer Ø, Grill R, Perenboom JAAJ (1992) Investigation of voltage steps of U(I) curves in $PbMo_6S_8$ wires. Physica C 202:61–68.

Hillier AD, Cywinski R (1997) The classification of superconductors using muon spin rotation. Applied Magnetic Resonance 13:95–109.

Hinks DG, Jorgensen JD, Li HC (1983) Structure of the oxygen point defects in $SnMo_6S_8$ and $PbMo_6S_8$. Physical Review Letters 51:1911–1914.

Hinks DG, Jorgensen JD, Li HC (1984) Oxygen impurity in the Chevrel-phase $SnMo_6S_8$. Solid State Communications 49:51–54.

Hirsch JE, Maple MB, Marsiglio F (2015a) Superconducting materials classes: Introduction and overview. Physica C: Superconductivity 514:1–8.

Hirsch JE, Maple MB, Marsiglio F (2015b) Superconducting materials: Conventional, unconventional and undetermined. Physica C: Superconductivity 514:1–444.

Holmgren DJ, Demers RT, Klein MV, Ginsberg DM (1987) Raman study of phonons in Chevrel-phase crystals. Physical Review B 36:1952–1955.

Holtzberg F, LaPlaca SJ, McGuire TR, Webb RA (1984) Magnetic susceptibility and superconductivity of single-crystal Ho-Mo-S Chevrel phase. Journal of Applied Physics 55:2013–2015.

Horyn R, le Berre F, Wojakowski A, Pena O (1996) Phase equilibria in the La-Mo-Se system at 1200°C in the vicinity of $LaMo_6Se_6$ and Mo_3Se_4. Superconductor Science and Technology 9:1081–1086.

Horyn R, Pena O, Geantet C, Sergent M (1989) Kinetics of destruction of Mo-S binary phases and crystal growth of rare-earth molybdenum chalcogenides. Superconductor Science and Technology 2:71–90.

Horyn R, Pena O, Wojakowski A, Sergent M (1994) The growth of single crystals of some $REMo_6Se_8$ superconductors. Superconductor Science and Technology 7:146–153.

Ingle NJC, Willis TC, Larbalestier DC, Meingast C (1998) Effects of hot isostatic pressing on the lattice parameters and the transition temperature of $Pb_{0.8}Sn_{0.2}Mo_6S_8$. Physica C 308:191–197.

Ishikawa M, Fischer Ø, Muller J (1982) Superconductivity and magnetism in $(RE)Mo_6S_8$ type compounds. In: Superconductivity in Ternary Compounds II (Maple MB, Fischer Ø, eds), pp. 143–165: Springer-Verlag.

Jaccarino V, Peter M (1962) Ultra-high-field superconductivity. Physical Review Letters 9:290–292.

Johnston DC, Shelton RN (1977) Magnetic properties of $RE_xMo_6Se_8$ compounds between 0.7 and 295 K. Journal of Low Temperature Physics 26:561–572.

Jorgensen J, Hinks DG (1985) Low temperature structural distortions in the high Tc Chevrel-phase superconductors $PbMo_6S_8$ and $SnMo_6S_8$. Solid State Communications 53:289–292.

Jorgensen JD, Hinks DG (1986) Rhombohedral-to-Triclinic phase transition in $BaMo_6S_8$. Physica B 136:485–488.

Jorgensen JD, Hinks DG, Felcher GP (1987) Lattice instability and superconductivity in the Pb, Sn and Ba Chevrel phases. Physical Review B 35:5365–5368.

Kaiser AB (1997) Comparison of thermopower behaviour in different superconductors. Physica C 282-287:1251–1252.

Kajiyana K, Matsushita T, Yamafuji K, Hamasaki K, Komata T (1985) International Symposium on Flux Pinning and Electromagnetic Properties in Superconductors. Fukuoka: Matsukuma Press Co.

Kamiguchi S, Seki Y, Satake A, Okumura K, Nagashima S, Chihara T (2015) Catalytic cracking of Methyl tert-Butyl Ether to Isobutene over Bronsted and Lewis acid sites on solid-state Molydenum and Sulfide clusters with an octahedral metal framework. Journal of Cluster Science 26:653–660.

Kamiguchi S, Takeda K, Kajio R, Okumura K, Nagashima S, Chihara T (2013) Application of solid-state molybdenum sulfide clusters with an octahedral metal framework to catalysis: Ring-opening of Tetrahydrofuran to Butyraldehyde. Journal of Cluster Science 24:559–574.

Kamihara Y, Watanabe T, Hirano M, Hosono H (2008) Iron-based layered superconductor $LaO_{1-x}F_x$ FeAs (x=0.05-0.12) with T_c=26 K. Journal of the American Chemical Society 130:3296.

Kang ZC, Eyring L, Hinode H, Uchida T, Wakihara M (1994) Structures, structural defects and reactions in a Nickel Chevrel-phase sulphide: A high resolution electron microscopy study. Journal of Solid State Chemistry 111:58–74.

Karasik V, Rikel MO, Togonidze TG, Tsebro VI (1985) Investigation of current-carrying capacity of bulk single-phase $PbMo_6S_8$ samples with grains of 0.1 micron size. Soviet Physics Solid State 27:1889–1890.

Keys SA, Cheggour N, Hampshire DP (1999) The effect of hot isostatic pressing on the strain tolerance of the critical current density found in modified jelly roll Nb_3Sn wires. IEEE Transactions on Applied Superconductivity 9:1447–1450.

Keys SA, Hampshire DP (2003) A scaling law for the critical current density of weakly and strongly-coupled superconductors, used to parameterise data from a technological Nb_3Sn strand. Superconductor Science and Technology 16:1097–1108.

Khlybov EP, Kuzmicheva GM, Evdokimova VV (1986) The structure of Mo_2S_3 and of high-pressure modifications of ternary molybdenum sulphides. Russian Journal of Inorganic Chemistry 31:627–630.

Kiefl RF, Hossain MD, Wojek BM, Dunsiger SR, Morris GD, Prokscha T, Salman Z, Baglo J, Bonn DA, Liang R, Hardy WN, Suter A, Morenzoni E (2010) Direct measurement of the London penetration depth in $YBa_2Cu_3O_{6.92}$ using low-energy mu SR. Physical Review B 81:4.

King RB (1999) Chemical structure and superconductivity. Journal of Chemical Information and Computer Science 39:180–191.

Kleiner WH, Roth LM, Autler SH (1964) Bulk solution of Ginzburg-Landau equations for Type II superconductors: Upper critical field region. Physical Review 133:A1226–A1227.

Kondo T, Khasanov R, Takeuchi T, Schmalian J, Kaminski A (2009) Competition between the pseudogap and superconductivity in the High-T_C copper oxides. Nature Letters 457:296–300.

Koo KF, Schewe-Miller IM, Schrader GL (1995) Reactive sputter deposition of lead Chevrel phase thin films. In: Google Patents.

Krabbes G, Oppermann H (1981) The phase diagram of the Pb-Mo-S system at 1250 K and some properties of the superconducting $PbMo_6S_8$. Crystal Research and Technology 16:777–784.

Kramer EJ (1973) Scaling laws for flux pinning in hard superconductors. Journal of Applied Physics 44:1360–1370.

Kramer EJ (1975) Microstructure-critical current relationships in hard superconductors. Journal of Electronic Materials 4:839–879.

Kubo Y, Uchikawa F, Utsunomiya S, Noto K, Katagiri K, Kobayashi N (1993) Fabrication and evaluation of small coils using $PbMo_6S_8$ wires. Cryogenics 33:883–888.

Kumta PN, Saha P, Datta MK, Manivannan A (2015) Cathodes and electrolytes for rechargeable magnesium batteries and methods of manufacture. In: Google Patents.

Lachal A, Junod A, Muller J (1984) Heat capacity analysis of a large number of Chevrel-type superconductors. Journal of Low Temperature Physics 55:195–232.

Lachal B, Bailif R, Junod A, Muller J (1983) Structural instabilities of Chevrel phases: the alkaline earth molybdenum sulphide series. Solid State Communications 45:849–851.

Le Berre F, Hamard C, Pena O, Wojakowski A (2000) Structural studies on single crystals of Chevrel phase selenides $REMo_6Se_8$ (RE: La, Ce, Pr, Nd, or Sm) at 298 K. Inorganic Chemistry 39:1100–1105.

Le Berre F, Maho F, Pena O, Horyn R, Wojakowski A (1995) Physical and structural properties of Chevrel-phase selenides Mo_3Se_4 and $RE_xMo_6Se_8$: Crystal growth and mutual solubility. Journal of Magnetism and Magnetic Materials 140–144: 1171–1172.

Le Berre F, Pena O, Hamard C, Corrignan A, Horyn R, Wojakowski A (1997) New superconducting materials based upon doping of the Chevrel phase binary compound Mo_6Se_8. Journal of Alloys and Compounds 262:331–334.

Le Berre F, Pena O, Perrin C, Padiou J, Horyn R, Wojakowski A (1998a) Novel crystal structure in the Chevrel-phase compound $EuMo_6Se_8$. Transport and magnetic properties. Journal of Alloys and Compounds 280:85–93.

Le Berre F, Pena O, Perrin C, Sergent M, Horyn R, Wojakowski A (1998b) Single-crystal studies of the Chevrel-phase superconductor $La_xMo_6Se_8$–I: Correlation between T_C and the interatomic distances. Journal of Solid State Chemistry 136:151–159.

Le Berre F, Tshimanga D, Giulloux AL, Leclercq J, Sergent M, Pena O, Hiryn R, Wojakowski A (1996) Rare-earth doping of the Mo_3Se_4 superconductor. Physica B 228:261–271.

Le Lay L, Willis TC, Larbalestier DC (1991) Magnetization properties of a $SnMo_6S_8$ single crystal. IEEE Transactions on Magnetics 27:954–957.

Leigh NR (2001) Specific heat measurements on Chevrel phase materials exhibiting coexistence of superconductivity and magnetism. In: Department of Physics, p. 143. Durham: University of Durham.

Lemee N, Guilloux-Viry M, Perrin A, Sergent M (1998) Superconducting $Cu_2Mo_6S_8$ thin films deposited in-situ by laser ablation on R-plane sapphire. The European Physical Journal Applied Physics 1:197–201.

Levi E, Gofer Y, Aurbach D (2010) On the way to rechargeable Mg batteries: The challenge of new cathode materials. Chemistry of Materials 22:860–868.

Loram JW, Luo J, Cooper JR, Liang WY, Tallon JL (2001) Evidence on the pseudogap and condensate from the electronic specfic heat. Journal of Physics and Chemistry of Solids 62:59–64.

Luhman T, Dew-Hughes D (1978) Superconducting wires of $PbMo_{5.1}S_6$ by a powder technique. Journal of Applied Physics 49:936–938.

Majkrzak CF, Shirane G, Thomlinson W, Ishikawa M, Fischer Ø, Moncton DE (1979) A neutron diffraction study of the coexistence of antiferromagnetism and superconductivity in $GdMo_6S_8$. Solid State Communications 31:773–775.

Mancour-Billah A, Chevrel R (2003) A new increasing delocalization of M=3d-elements (TI, Fe, Co) in the channels network of the ternary $M_yMo_6Se_8$ Chevrel phases. Journal of Solid State Chemistry 170:281–288.

Mattheis LF (1973) Band structures of transition-metal-dichalcogenide layer compounds. Physical Review B 8:3719–3740.

Mattheis LF, Fong CY (1977) Cluster model for the electronic structure of the Chevrel-phase comnpound $PbMo_6S_8$. Physical Review B 15:1760–1768.

Matthias BT, Marezio M, Corenwitt E, Cooper AS, Barz HE (1972) High-temperature superconductors, the first ternary system. Science 175:1465–1466.

Meul HW (1986) On the unusual physical properties of europium-based molybdenum chalcogenides and related Chevrel compounds. Helvetica Physica Acta 59:417–489.

Miraglia S, Goldacker W, Flükiger R, Seeber B, Fischer Ø (1987) Thermal expansion studies in the range 10 K–1200 K in $PbMo_6S_8$ by means of X-ray diffraction. Materials Research Bulletin 22:795–802.

Mitelman A, Levi MD, Lancry E, Levi E, Aurbach D (2007) New cathode materials for rechargeable Mg batteries; fast Mg ion transport asnd reversible extrusion in $Cu_yMo_6S_8$ compounds. Chemical Communications 41:4212–4214.

Morales F, Escudero R, Briggs A, Monceau P, Horyn R, Le Berre F, Pena O (1996) Point contact spectroscopy on the ferromagnetic superconductor $HoMo_6S_8$. Physica B 218:193–196.

Morley NA, Leigh NR, Niu H, Hampshire DP (2001) High upper critical field in the Chevrel phase superconductor lead-molybdenum-sulphide doped with europium. IEEE Transactions on Applied Superconductivity 11:3599–3602.

Murgia F, Antitomaso P, Stievano L, Monconduit L, Berthelot R (2016) Express and low-cost microwave synthesis of the ternary Chevrel phase $Cu_2Mo_6S_8$ for application in rechargeable magnesium batteries. Journal of Solid State Chemistry 242:151–154.

Nakamura T, Hanayama Y, Kiss T, Vysotsky V, Okamoto H, Matsushita T, Takeo M, Irie F, Yamafuji K (1997) Current-voltage characteristics in YBaCuO thin films over more than 13 decades of electric-field. Institute of Physics Conference Series No 158:1017–1020.

Nespolo M, Aroyo MI, Souvignier B (2018) Crystallographic shelves: space-group hierarchy explained. Journal of Applied Crystallography 51:1481–1491.

Niu H, Leigh N, Hampshire D (2002) Superconducting properties of quaternary Chevrel phase $Pb_{1-x}Cu_{1.8x}Mo_6S_8$ - Private Communication.

Niu HJ, Morley NA, Hampshire DP (2001) Chevrel phase $(Pb_{1-x}Cu_{1.8x})Mo_8S_6$ with a mixed structure and high critical parameters. IEEE Transactions on Applied Superconductivity 11:3619–3622.

Nunes RW, Mazin II, Singh DJ (1999) Theoretical search for Chevrel-phase-based thermoelectric materials. Physical Review B 59:7969–7972.

Ohta M, Obara H, Tamamoto A (2009b) Preparation and thermoelectric properties of Chevrel-phase $CuxMo_6S_8$ $(2.0 < x < 4.0)$. Materials Transactions 50:2129–2123.

Ohta M, Yamamoto A, Obara H (2009a) Thermoelectrfic properties of Chevrel-phase sulfides $M_xMo_6S_8$ (M: Cr, Mn, Fe, Ni). Journal of Electronic Materials 39:2117–2121.

Ohtaki R, Zhao BR, Luo HR (1984) Superconducting tunneling on $Cu_xMo_6S_8$ films. Journal of Low Temperature Physics 54:119–127.

Pazol BG, Holmgren DJ, Ginsberg DM (1989) Upper critical field anisotropy in the Chevrel phase compound $PbMo_6S_8$. Journal of Low Temperature Physics 74:133–140.

Pellizone M, Treyvaud A, Spitzli P, Fischer Ø (1977) Magnetic susceptibility of (Rare Earth)$_xMo_6S_8$. Journal of Low Temperature Physics 29:453–465.

Pena O (2015) Chevrel phases: Past, present and future. Physics C 514:95–112.

Pena O, Geantet C, Schmitt H, Le Berre F, Hamard C (1999) Structural properties of rare-earth molybdenum chalcogenides $REMo_6X_8$. Solid State Sciences 1:577–584.

Pena O, Le Berre F, Padiou J, Marchand T, Horyn R, Wojakowski A (1998) Single-crystal studies of the Chevrel-phase superconductor $La_xMo_6Se_8$–II: Physical and superconducting properties. Journal of Solid State Chemistry 136:160–166.

Pena O, Mancour-Billah A, Chevrel R (2009) Single crystal studies of ternary transition-metal molybdenum selenides MMo_6S_8 (M=Ti, Fe, Mn, Cr and Co). Journal of Cluster Science 20:51–62.

Pena O, Sergent M (1989) Rare earth based Chevrel phases $REMo_6X_8$: Crystal growth, physical and superconducting properties. Progress in Solid State Chemistry 19:165–281.

Perrin A, Perrin C (2012) The molybdenum and rhenium octahedral cluster chalcogenides in solid state chemistry: From condensed to discrete cluster units. Comptes Rendus Chimie 15:815–836.

Perrin C, Chevrel R, Sergent M (1979) Etude structurale d'un thiohalogenure supraconducteur derivant du Mo II: $Mo_6S_6Br_2$. Materials Research Bulletin 14:1505–1515.

Petrovic AP, Lortz R, Santi G, Berthod C, Dubois C, Decroux M, Demuer A, Antunes AB, Pare A, Salloum D, Gougeon P, Potel M, Fischer O (2011) Multiband superconductivity in the Chevrel phases $SnMo_6S_8$ and $PbMo_6S_8$. Physics Review Letters 106:017003.

Pobell F, Rainer D, Wuhl H (1982) Electron-phonon interaction in Chevrel phase compounds. In: Topics in Current Physics-Superconductivity in Ternary Compounds Vol I (Fischer Ø, Maple MB, eds), pp. 251–277. Berlin: Springer-Verlag.

Poppe U, Wuhl H (1981) Tunneling spectrocopy on the superconducting Chevrel-phase compounds $Cu_{1.8}Mo_6S_8$ and $PbMo_6S_8$. Journal of Low Temperature Physics 43:371 370–382.

Prober DE, Schwall RE, Beasley MR (1980) Upper critical fields and reduced dimensionality of the superconducting layered compounds. Physical Review B 21:2717–2733.

Ramsbottom HD, Hampshire DP (1997) Improved critical current density and irreversibility line in HIP'ed Chevrel phase superconductor $PbMo_6S_8$. Physica C 274:295–303.

Ramsbottom HD, Hampshire DP (1999) Flux Penetration measurements and the harmonic magnetic response of hot isostatically pressed (Pb,Gd)Mo_6S_8. Journal of Applied Physics 85:3732–3739.

Renner C, Revaz B, Genoud JY, Kadowaki K, Fischer Ø (1998) Pseudogap precursor of the superconducting gap in under- and overdoped $Bi_2Sr_2CaCu_2O_{8+d}$. Physical Review Letters 80:149–152.

Rimikis G, Goldacker W, Specking W, Flükiger R (1991) Critical currents in $Pb_{1.2-x}Sn_xMo_6S_8$ wires. IEEE Transactions on Magnetics 27:1116–1119.

Roche C, Chevrel R, Jenny A, Pecheur P, Scherrer H, Scherrer S (1999) Crystallography and density of states calculation of MxMo6Se8 (M=Ti, Cr, Fe, Ni). Physical Review B 60:16442–16447.

Rossel C, Fischer Ø (1984) Critical current densities in bulk Chevrel-phase samples. Journal of Physics F: Metal Physics 14:455–472.

Rossel C, Pena O, Schmitt H, Sergent M (1991) On the irreversibility line in the Chevrel phase superconductors. Physica C 181:363–368.

Rybicki D, Jurkutat M, Reichardt S, Kapusta C, Haase J (2016) Perspective on the phase diagram of cuprate high-temperature superconductors. Nature Communications:1–6.

Schauer W, Schelb W (1981) Improvement of Nb_3Sn high field critical current by a two-stage reaction. IEEE Transactions on Magnetics 17:374–377.

Scurty F, Ishmael S, Flanagan G, Schwartz J (2016) Quench detection for high temperature superconductor magnets: a novel technique based on Rayleigh-backscattering interrogated optical fibres. Superconductor Science and Technology 29:03LT01.

Seeber B (2015) Ternary molybdenum chalcogenide superconducting wire and manufacturing thereof. In: Google Patents.

Seeber B, Erbuke L, Schroeter V, Perenboom JAAJ, Grill R (1995) Critical current limiting factors of hot isostatically pressed (HIP'ed) $PbMo_6S_8$ wires. IEEE Transactions on Applied Superconductivity 5:1205–1208.

Selvam P, Cattani D, Cors J, Decroux M, Junod A, Niedermann P, Ritter S, Fischer Ø, Rabiller P, Chevrel R (1992) Superconducting, microstructural and grain boundary properties of hot-pressed $PbMo_6S_8$. Journal of Applied Physics 72:4232–4239.

Selvam P, Cattani D, Cors J, Decroux M, Niedermann P, Fischer Ø, Chevrel R, Pech T (1993) The role of Sn addition on the improvement of Jc in $PbMo_6S_8$. IEEE Transactions on Applied Superconductivity 3:1575–1578.

Selvam P, Cors J, Cattani D, Decroux M, Fischer Ø, Seibt EW (1995) Homogeneity and critical current density of Sn-doped $PbMo_6S_8$ superconductors. Applied Physics A 61:615–621.

Sergent M, Chevrel R, Padiou J, Pena O, Barathe R, Hirrien M, Massat H, Pech T, Turck B, Dubots P, Couach M (1984) Preparation and predevelopment of superconductors so called Chevrel phases. Annales de Chimie Science des Materiaux 9:1069–1074.

Sergent M, Chevrel R, Rossel C, Fischer Ø (1978) On the superconductivity of PbMo6S_8 and the series $M_xPbMo_6S_8$ and $M_xPb_{1-x}Mo_6S_8$. Journal of the Less Common Metals 58:179–193.

Shelton RN (1976) The effect of high pressure on superconducting ternary molybdenum chalcogenides. In: 2nd Rochester Conference on Superconductivity in d- and f-Band Metals (Douglas DH, ed), pp. 137–160. New York: Plenum Press.

Shelton RN, McCallum RW, Adrian H (1976) Superconductivity in rare earth molybdenum selenides. Physics Letters 56A:213–214.

Shimizu K (2015) Superconductivity from insulating elements under high pressure. Physica C: Superconductivity 514:46–49.

Smeu M, Hossain MS, Wang ZZ, Timoshevskii V, Bevan KH, Zaghib K (2016) Theoretical investigation of Chevrel phase materials for cathodes accommodating Ca^{2+} ions. Journal of Power Sources 306:431–436.

Sonier JE, Brewer JH, Kiefl RF (2000) Mu SR studies of the vortex state in type-II superconductors. Review of Modern Physics 769–811.

Sunandana CS (1979) On the electrical resistivity of Chevrel phases. Journal of Physics C 12:L165–L168.

Sunwong P, Higgins JS, Tsui Y, Raine MJ, Hampshire DP (2013) The critical current density of grain boundary channels in polycrystalline HTS and LTS superconductors in magnetic fields. Superconductor Science and Technology 26:095006.

Tallon JL, Loram JW (2001) The doping dependence of T* - what is the real high-T-c phase diagram? Physica C – Superconductivity and Its Applications 349:53–68.

Taylor DMJ, Hampshire DP (2005) The scaling law for the strain dependence of the critical current density in Nb_3Sn superconducting wires. Superconductor Science and Technology 18:S241–S252.

Thomlinson W, Shirane G, Moncton DE, Ishikawa M, Fischer Ø (1981) Magnetic order in superconducting $TbMo_6S_8$, $DyMo_6S_8$ and $ErMo_6S_8$. Physical Review B 23:4455–4462.

Uchida T, Wakihara M (1991) Thermal behaviour of the Chevrel phase sulphides. Thermochimica Acta 174:201–221.

Uemura YJ (2004) Condensation, excitation, pairing, and superfluid density in high-T-c superconductors: The magnetic resonance mode as a roton analogue and a possible spin-mediated pairing. Journal of Physics: Condensed Matter 16:S4515–S4540.

Uemura YJ (2009) Commonalities in phase and mode. Nature Materials 8:253–255.

Uemura YJ, Le LP, Luke GM, Sternlieb BJ, Wu WD, Brewer JH, Riseman TM, Seaman CL, Maple MB, Ishikawa M, Hinks DG, Jorgensen JD, Saito G, Yamochi H (1991) Basic similarities among cuprate, bismuthate, organic, Chevrel-phase, and heavy-fermion superconductors shown by penetration-depth measurements. Physical Review Letters 66:2665–2668.

van der Meulen HP, Perenboom JAAJ, Berendschot TTJM, Cors J, Decroux M, Fischer Ø (1995) Specific heat of $PbMo_6S_8$ in high magnetic field. Physica B 211:269–271.

Wang G, Raine MJ, Hampshire DP (2017) How resistive must grain-boundaries be to limit J_C in polycrystalline superconductors? Superconductor Science and Technology 30:104001.

Watanabe K, Awaji S, Motokawa M, Mikami Y, Sakuraba J, Watzawa K (1998) 15 T cryocooled Nb_3Sn superconducting magnet with a 52 mm room temperature bore. Japanese Journal of Applied Physics 37:L1148–L1150.

Webb AW, Shelton RN (1978) Compressibilities and volume dependence of Tc for eleven Chevrel phase superconductors. Journal of Physics F: Metal Physics 8:261–269.

Werthamer NR, Helfand E, Hohenberg PC (1966) Temperature and purity dependence of the superconducting critical field, H_{c2}. III. Electron spin and spin-orbit effects. Physical Review 147:295–302.

Williams GVM, Tallon JL, Haines EM, Michalak R, Dupree R (1997) NMR evidence for a d-wave normal-state pseudogap. Physical Review Letters 78:721–724.

Willis TC, Jablonski PD, Larbalestier DC (1995) Hot isostatic pressing of Chevrel phase bulk and hydrostatically extruded wire samples. IEEE Transactions on Applied Superconductivity 5:1209–1213.

Wolf B, Molter J, Bruls G, Luthi B, Jansen L (1996) Elastic properties of superconducting Chevrel-phase compounds. Physical Review B 54:348–352.

Wolowiec CT, White BD, Maple MB (2015) Conventional magnetic superconductors. Physica C: Superconductivity 514:113–129.

Woollam JA, Somoano RB (1976) Superconducting critical fields of alkali and alkaline-earth intercalates of MoS_2. Physical Review B 13:3843–3853.

Yamamoto S, Wakihara M, Taniguchi M (1985) Phase relations in the Pb-Mo-S ternary system at 1000°C and the superconductivity of $PbMo_6S_{8-y}$. Materials Research Bulletin 20:1493–1500.

Yamasaki H, Kimura Y (1986) The phase field of the Chevrel phase $PbMo_6S_8$ at 900°C and some superconducting and structural properties. Materials Research Bulletin 21:125–135.

Yamasaki H, Umeda M, Kosaka S (1992) High critical current densities reproducibly observed for hot-isostatic-pressed $PbMo_6S_8$ wires with Mo barriers. Journal of Applied Physics 72:1–3.

Yamasaki H, Umeda M, Kosaka S, Kimura Y, Willis TC, Larabalestier DC (1991) Poor intergrain connectivity of $PbMo_6S_8$ in sintered Mo-sheathed wires and the beneficial effect of hot-isostatic-pressing treatments on the transport critical current density. Journal of Applied Physics 70:1606–1613.

Yao YS, Guertin RP, Hinks DG, Jorgensen J, Capone DW (1988) Superconductivity of divalent Chevrel phases at very high pressures. Physical Review B 37:5032–5037.

Yeshurun Y, Malozemoff AP (1988) Giant flux creep and irreversibility in an Y-Ba-Cu-O crystal: An alternative to the superconducting-glass model. Physical Review Letters 60:2202–2205.

Youwen X, Suenaga M (1991) Irreversibility temperature in superconducting oxides: The flux-line-lattice melting, the glass-liquid transition, or the depinning temperatures. Physical Review B 43:5516–5525.

Yvon K, Paoli A (1977) Charge transfer and valence electron concentration in Chevrel phases. Solid State Communications 24:41–45.

Zheng DN, Hampshire DP (1997) The effect of Gd doping on the critical current of the Chevrel phase superconductor $PbMo_6S_8$. IEEE Transactions on Applied Superconductivity 7:1755–1759.

Zheng DN, Ramsbottom HD, Hampshire DP (1995) Reversible and irreversible magnetization of the Chevrel-phase superconductor $PbMo_6S_8$. Physical Review B 52:12931–12938.

Part C

High-Temperature Superconductors

C

Introduction to Section C: High-Temperature Superconductors

Jeffery L. Tallon

The discovery of cuprate high-temperature superconductors (HTS) by Bednorz and Mueller in 1986 [1] brought about a revolution in both materials development and in the techniques used to characterize them. Coming as it did at a time when interest and funding investment in superconductivity was in decline, the discovery of HTS injected a huge resurgence of research activity and the establishment of national programs directed toward a coordinated, multidisciplinary, multi-institutional approach to investigating, understanding and exploiting HTS. The study of HTS has greatly enhanced the sensitivity of techniques like angle-resolved photoelectron spectroscopy (ARPES) and promoted the shift from synchrotron-based ARPES to laser-based ARPES with its higher resolution and greater probing depth. The promise of applications has also accelerated the development of refrigerator systems, both in terms of efficiency and miniaturization.

The term 'high-temperature superconductors' loosely refers to a group of materials with critical temperatures (T_c) exceeding 23 K, the pre-1986 record T_c for the Nb_3Ge intermetallic superconductors. Following the discovery of $La_{2-x}Ba_xCuO_4$ by Bednorz and Mueller (and later its substitutional counterpart $La_{2-x}Sr_xCuO_4$ with $T_c \approx 39$ K), a collaboration between the Universities of Houston and Alabama soon followed with the discovery in 1987 of the Y/Ba/Cu/O system [2] with $T_c \approx 90$ K. At the time the authors did not identify the structure and composition of the active phase, namely $YBa_2Cu_3O_{7-x}$, but this was discovered and reported almost immediately by several other groups [3–5]. Recognition that Y was replaceable by almost any lanthanide rare-earth element opened up the idea of a class of cuprate superconductors [6], and it was quickly shown that the active structure within the cuprates was the square-planar CuO_2 planes in which superconductivity is nucleated. The YBCO system is reviewed in Chapter C1.

Then, in the following year, new cuprate superconductors were discovered by Maeda and coworkers within the Bi/Sr/Ca/Cu/O system [7]. This had a multistep transition and, again, it was only later that the active phases were identified [8] and found to comprise an homologous series of $Bi_2Sr_2CuO_6$ ($T_c \approx 27$ K [9]), $Bi_2Sr_2CaCu_2O_8$ ($T_c \approx 95$ K [10]) and $Bi_2Sr_2Ca_2Cu_3O_{10}$ ($T_c \approx 110$ K [8]) – often referred to generically as BSCCO. The second of these superconductors has subsequently been almost as widely studied as $YBa_2Cu_3O_{7-x}$, due to its cleavability, large available doping range and availability of large high-quality single crystals. The third in the series (with Pb partially substituted for Bi to accelerate synthesis), $Bi_{1.65}Pb_{0.35}Sr_2Ca_2Cu_3O_{10}$, eventually became the first-generation HTS wire material using the powder-in-tube method. This is still being used for commercial magnets, cables and other HTS-based products. The key structural element in this system is the weakly bonded Bi_2O_2 double layer which allows a slip system much like graphite or mica so that randomly oriented grains in the precursor power can readily be aligned and densified by a suitable rolling deformation. The BSCCO system is reviewed in Chapter C2.

The cuprate template established in BSCCO of an homologous series exhibiting progressively increasing numbers of CuO_2 planes per repeat unit led quickly to the discovery of similar structures with Bi replaced by Tl and Sr replaced by Ba [11]: $Tl_2Ba_2CuO_6$ ($T_c \approx 90$ K), $Tl_2Ba_2CaCu_2O_8$ ($T_c \approx 118$ K) and $Tl_2Ba_2Ca_2Cu_3O_{10}$ ($T_c \approx 128$ K) – often referred to generically as TBCCO [12, 13]. Unlike BSCCO, the Tl_2O_2 double layer in TBCCO is strongly bonded, and despite its higher T_c value, it is not amenable to deformation-induced texturing and so has never been used as a commercial wire product. But the structure of TBCCO is more versatile, and the double-layer Tl_2O_2 motif may be replaced by a single-layer TlO motif giving the structural series $TlBa_2CuO_5$, $TlBa_2CaCu_2O_7$, $TlBa_2Ca_2Cu_3O_9$, ... Another important variant occurs when the TlO layer is replaced by a $Tl_{0.5}Pb_{0.5}O$ layer giving compounds such as $Tl_{0.5}Pb_{0.5}Sr_2CaCu_2O_7$. This compound may be doped across the entire superconducting phase diagram by substituting La for Sr or Y for Ca, thus $Tl_{0.5}Pb_{0.5}Sr_{2-x}La_xCaCu_2O_7$ and $Tl_{0.5}Pb_{0.5}Sr_2Ca_{1-x}Y_xCu_2O_7$. Despite their compositional complexity, these compounds are extremely stable, can be synthesized very rapidly, unlike other cuprates, they are oxygen stoichiometric, they can be widely doped and they exhibit high transition temperatures maximizing at $T_c = 107$ K. The TBCCO system is reviewed in Chapter C3.

Four years later, superconductivity was discovered in $HgBa_2CuO_4$ with $T_c = 94$ K [14] and soon after in the general series $HgBa_2Ca_{n-1}Cu_nO_{2n+2}$ [15] with the current record under ambient pressure held by the third member of the series $HgBa_2Ca_2Cu_3O_8$ with $T_c = 134$ K [15]. That record has remained to the present, but since then many new HTS have been discovered, and we name just a few. The HBCCO system is reviewed in Chapter C4.

In 2001, the group of Akimitsu discovered superconductivity in MgB_2 with $T_c = 39$ K [16]. This has since matured as a wire technology, and MgB_2-based magnetic resonance imaging is now available [17]. The MgB_2 system is reviewed in Chapter B2. Then a major new revolution began with the discovery of what eventually proved to be a wide class of iron pnictides in 2008 by the group of Hosono [18]. Their observation of $T_c = 26$ K in $La(O_{1-x}F_x)FeAs$ led in short time to the record T_c value of 56 K in $Gd_{1-x}Th_xFeAsO$ [19] and subsequently to the very important systems $BaFe_2As_2$ (and its various substituted forms) and FeSe. In bulk form, the latter system has $T_c = 8$ K [20], but astonishingly, when disposed as a single atomic layer on $SrTiO_3$, it exhibits superconductivity at substantially elevated temperatures, up to 77 K in the original report [21] and up to 108 K as later reported by *in situ* resistivity measurements [22]. These are remarkable developments which have inspired a broad new endeavor in studying superconductivity in single-unit cell crystalline samples. The Fe-based superconductors are reviewed in Chapter C5.

Finally, the recent discovery of superconductivity at 203 K in highly compressed H_2S and its subsequent dramatic developments has to be mentioned [23]. Long ago, Ashcroft predicted very high-temperature superconductivity in compressed hydrogen with T_c around 300 K [24]. That search remains to be completely fulfilled, but more recently, he predicted that in various hydrides the extreme requisite pressure for superconductivity would be reduced to more manageable scales by virtue of effective chemical pressure [25]. The various hydrides he discussed included SiH_4, SnH_4, GeH_4, ScH_3, YH_3, GaH_3 and SH_2. The last very familiar compound was investigated by Drozdov *et al.* and at 155 GPa was found to superconduct at 203 K. Subsequently, superconductivity above 260 K was predicted [26] and reported [27, 28] in clathrate structures of LaH_{10}, thus demonstrating the accuracy and power of modern structural and electronic computation. An updated article showed the presence of superconductivity in this system at 280 K [29] and, just as this Handbook goes to press, a carbonaceous sulphur hydride compound was shown to superconduct up to 288 K at a pressure of about 270 GPa [30]. While the active composition and structure is yet to be determined, the long-standing quest for room-temperature superconductivity would appear to have finally been achieved. The ongoing challenge now is how to reduce the enormous pressures required. Doubtless the "high hydrides" will remain a fertile field of superconductivity research at high pressure and one even wonders about future prospects for a yet more familiar compound, OH_2. Hydrides are reviewed in Chapter C6.

References

[1] Bednorz G J and Müller K A 1986 Possible high T_c superconductivity in the Ba-La-Cu-O system *Z. Phys. B* **64** 189

[2] Wu M K, Ashburn R, Torng J, Hor P H, Meng L, Huang J, Wang Y Q and Chu C W 1987 Superconductivity at 93 K in a new mixed-phase Y-Ba-Cu-O compound system at ambient pressure *Phys. Rev. Lett.* **58** 908

[3] Cava R J, Batlogg B, van Dover R B, Murphy D W, Sunshine S, Siegrist T, Remeika J P, Rietman E A, Zahurak S and Espinosa G P 1987 Bulk superconductivity at 91 K in single-phase oxygen-deficient perovskite $Ba_2YCu_3O_{9-\delta}$ *Phys. Rev. Lett.* **58** 1676

[4] Grant P M, Beyers R B, Engler E M, Lim G, Parkin S S P, Ramirez M L, Lee V Y, Nazzal A, Vazquez J E and Savoy R J 1987 Superconductivity above 90 K in the compound $YBa_2Cu_3O_x$: structural, transport, and magnetic properties *Phys. Rev. B* **35** 7242(R)

[5] Qadri S B, Toth L E, Osofsky M, Lawrence S, Gubser D U and Wolf S A 1987 X-ray identification of the superconducting high-T_c phase in the Y-Ba-Cu-O system *Phys. Rev. B* **35** 6868

[6] Murphy D W, Sunshine S, van Dover R B, Cava R J, Batlogg B, Zahurak S M and Schneemeyer L F 1987 New superconducting cuprate perovskites *Phys. Rev. Lett.* **58** 1888

[7] Maeda H, Tanaka Y, Fukutomi M and Asano T 1989 A new high-T_c oxide superconductor without a rare earth element *Jpn. J. Appl. Phys.* **27** L209

[8] Tallon J L, Buckley R G, Gilberd P W, Presland M R, Brown I W M, Bowden M E, Christian L A and Goguel R 1988 Superconducting phases in the series $Bi_{2.1}(Ca,Sr)_{n+1}Cu_nO_{2n+4}$ *Nature* **333** 153

[9] Michel C *et al.* 1987 Superconductivity in the Bi - Sr - Cu - O system *Z. Phys. B* **68** 421

[10] Subramanian M A *et al.* 1988 A new high-temperature superconductor: $Bi_2Sr_{3-x}Ca_xCu_2O_{8+y}$ *Science* **239** 1015

[11] Sheng Z Z and Hermann Z M 1988 Bulk superconductivity at 120 K in the Tl-Ca/Ba-Cu-O system *Nature* **332** 138

[12] Parkin S S P, Lee V Y, Engler E M, Nazzal A I, Huang T C, Gorman G, Savoy R and Beyers R 1988 Bulk superconductivity at 125 K in $Tl_2Ca_2Ba_2Cu_3O_x$ *Phys. Rev. Lett.* **60** 2539

[13] Torardi C C, Subramanian M A, Calabrese J C, Gopalakrishnan J, Morrissey K J, Askew T R, Flippen R B, Chowdhry U and Sleight A W 1988 Crystal structure of $Tl_2Ba_2Ca_2Cu_3O_{10}$, a 125 K superconductor *Science* **240** 631

[14] Putilin S N, Antipov E V, Chmaissem O and Marezio M 1993 Superconductivity at 94 K in $HgBa_2Cu0_{4+\delta}$ *Nature* **362** 226

[15] Schilling A, Cantoni M, Guo J D and Ott H R 1993 Superconductivity above 130 K in the Hg-Ba-Ca-Cu-O system *Nature* **363** 56

[16] Nagamatsu J, Nakagawa N, Muranaka T, Zenitani Y and Akimitsu J 2002 Superconductivity at 39 K in magnesium diboride *Nature* **410** 63

[17] Patel D, Al Hossain M S, Qiu W, Jie H, Yamauchi Y, Maeda M, Tomsic M, Choi S and Kim J H 2017 Solid cryogen: a cooling system for future MgB_2 MRI magnet *Sci. Reports* **7** 43444

[18] Kamihara Y, Watanabe T, Hirano M and Hosono H 2008 Iron-based layered superconductor $La(O_{1-x}F_x)FeAs$ ($x = 0.05–0.12$) with $T_c = 26$ K *J. Am. Chem. Soc.* **130** 3296

[19] Wang C, Li L, Chi S, Zhu Z, Ren Z, Li Y, Wang Y, Lin X, Luo Y, Jiang S, Xu X, Cao G and Xu Z 2008 Thorium-doping–induced superconductivity up to 56 K in $Gd_{1-x}Th_xFeAsO$ *Europhys. Lett.* **83** 67006

[20] Hsu F C, Luo J Y, Yeh K W, Chen T K, Huang T W, Wu P M, Lee Y C, Huang Y L, Chu Y Y, Yan D C and Wu M K 2008 Superconductivity in the PbO-type structure α-FeSe *Proc. Natl. Acad. Sci.* **105** 14262

[21] Q. Y. Wang *et al.* 2012 Interface-induced high-temperature superconductivity in single unit-cell FeSe films on $SrTiO_3$ *Chin. Phys. Lett.* **29** 037402

[22] Ge J F, Liu Z L, Liu C, Gao C L, Qian D, Xue Q K, Liu Y and Jia J F 2015 Superconductivity above 100 K in single-layer FeSe films on doped $SrTiO_3$ *Nature Materials* **14** 285

[23] A. P. Drozdov A P, Eremets M I, Troyan I A, Ksenofontov V and Shylin S I 2015 Conventional superconductivity at 203 kelvin at high pressures in the sulfur hydride system *Nature* **525** 73

[24] Ashcroft N W 1968 Metallic hydrogen: a high-temperature superconductor? *Phys. Rev. Lett.* **21** 1748

[25] Ashcroft N W 2004 Hydrogen dominant metallic alloys: high temperature superconductors? *Phys. Rev. Lett.* **92** 187002

[26] Peng F, Sun Y, Pickard C J, Needs R J, Wu Q and Ma Y 2017 Hydrogen clathrate structures in rare earth hydrides at high pressures: possible route to room-temperature superconductivity *Phys. Rev. Lett.* **119** 107001

[27] Somayazulu M, Ahart M, Mishra A K, Geballe Z M, Baldini M, Meng Y, Struzhkin V V and Hemley R J 2019 Evidence for superconductivity above 260 K in lanthanum superhydride at megabar pressures *Phys. Rev. Lett.* **122** 027001

[28] Drozdov A P, Kong P P, Minkov V S, Besedin S P, Kuzovnikov M A, Mozaffari S, Balicas L, Balakirev F, Graf D, Prakapenka V B, Greenberg E, Knyazev D A, Tkacz M and Eremets M I 2019 Superconductivity at 250 K in lanthanum hydride under high pressures *Nature* **569** 528

[29] Somayazulu M, Ahart M, Mishra A K, Geballe Z M, Baldini M, Meng Y, Struzhkin V V and Hemley R J 2019 Evidence for superconductivity above 260 K in lanthanum superhydride at megabar pressures https://arxiv.org/abs/1808.07695

[30] Snider E, Dasenbrock-Gammon N, McBrideR, Debessai M, Vindana H, Vencatasamy K, Lawler K V, Salamat A and Dias R P 2020 Room-temperature superconductivity in a carbonaceous sulfur hydride *Nature* **586** 373

C1

YBCO[1]

Jeffery L. Tallon

C1.1 Introduction

Probably the most extensively studied high-temperature superconductor (HTS) is $YBa_2Cu_3O_{7-x}$, abbreviated YBCO or just Y-123 (T_c = 93 K). This compound was discovered in early 1987 by Wu *et al* [1] and is the first compound found for which T_c exceeded the boiling point of liquid nitrogen (77 K at 1 atm). It seems fitting that this material exhibits one of the most interesting and complex relationships between chemistry, crystal structure and physical properties of any ceramic material ever studied.

C1.1.1 Oxygenation

The oxygenation procedure for $YBa_2Cu_3O_{7-x}$ turned out to be crucial. In a typical preparation procedure, YBCO is formed at reaction temperatures with an oxygen content close to six ($x = 1$) and then has to be loaded with oxygen in order to display superconductivity. This results in a phase transition from a tetragonal to orthorhombic material at $x \approx 0.65$ [2], resulting in the characteristic formation of twin boundaries. With oxygenation T_c rises from zero at $x \approx 0.65$, exhibits a plateau (now known to be associated with short-range charge ordering) in the range $0.3 \leq x \leq 0.5$ and maximizes at $x \approx 0.1$ [2]. This is optimum doping as far as T_c is concerned, but critical fields and superfluid density go on increasing as x is reduced further towards 0 while T_c falls a little. In practice, for single crystals it is extremely difficult to reach the $x = 0$ state within reasonable time frames [3–5] because of the relatively low oxygen diffusion coefficient.

This high sensitivity of the electronic properties of YBCO to oxygen content leads to a rich variety of behaviours, including, on the one hand, those associated with oxygen ordering and, on the other hand, those associated with oxygen inhomogeneity, which contributes to pinning and causes the so-called fishtail effect in the magnetization loops [6].

C1.1.2 Applications of YBCO

YBCO, despite the other superconducting compounds with higher critical temperatures, is now the most promising material for applications aimed to operate either at 77 K or at lower temperatures in high field. It is currently being used in cable manufacture and applications such as power transformers [7]. This enhanced performance of YBCO arises from a structural element unique to this system, linear chains of Cu–O extending in the *b*-direction. These, and their effects, are discussed in more detail below.

Soon after the discovery of YBCO, it was found that the resulting bulk materials are particularly sensitive to the presence of high-angle grain boundaries, which made the fabrication of silver-clad YBCO wires using the 'powder-in-tube' method [8, 9] practically impossible. Instead coated conductors, comprising long tapes coated in YBCO films a few microns thick, have been successfully developed into what is referred to as second-generation (2G) conductors. Typically, these tapes comprise metal alloy substrates with oxide buffer layers, such as Zr_2O_3, CeO_2, Y_2O_3 and/or MgO, coated by YBCO using laser ablation, ion-beam–assisted deposition (IBAD) or liquid-phase metal-organic deposition (MOD). Values of the critical current density, J_c, now typically reach 3–5 MA/cm^2 at 77 K (self-field) in lengths of hundreds of metres [10]. This 2G conductor is currently being used in commercial applications [7, 11]. Further research is directed towards thicker films while still maintaining J_c, increased pinning for enhanced in-field performance, longer length conductors and generally reducing manufacturing costs.

Melt texturing was found to be a promising way to produce bulk YBCO samples, thereby eliminating most of the high-angle grain boundaries. Currently, bulk samples of YBCO with diameters up to 10 cm and J_c of 10^6 A/cm^2 at 77 K (self-field) can be produced [12]. The record trapped flux in such 'bulks' is 17.6 Tesla [13].

C1.1.3 Atomic Substitution

Atomic substitution in YBCO was studied in considerable detail in a quest to increase T_c even further and to better understand the fundamental behaviour of cuprates both in

[1] Revised and extended from the 1st Edition chapter C1 by A. Koblischka-Veneva, N. Sakai, S. Tajima & M. Murakami.

the normal and superconducting states. Another important aspect of such substitution is to obtain a denser ceramic body approaching the theoretical density and a microstructure with an optimal grain size and pinning site density for high mechanical strength and high J_c. It is useful to distinguish between *dopants*, which alter the electron or hole concentration and *substituents*, which do not necessarily alter the carrier concentration but modify the superconductivity or microstructure in some way.

One can classify these effects into four groups [14]. Firstly, elements which substitute in the copper sublattice. These act as scattering centres which, because of the *d*-wave symmetry, rapidly suppress T_c and even more rapidly suppress the superfluid density [15]. Examples include Zn, Mg, Fe and Ni. Elements in the second category substitute in the yttrium, barium or chain copper sublattices and possess a valence different from the atom they substitute. These substituents are electronic dopants which alter the carrier concentration and enable different parts of the doping phase diagram to be accessed. Examples include Ca^{2+} on the Y^{3+} site (or more generally on the R^{3+} rare-earth site) which is a hole dopant; La^{3+} on the Ba^{2+} site which is an electron dopant or Co on the chain copper site which acts as an electron donor by modifying the chain oxygen content and order. The third category is like the second but the substituents are not altervalent (i.e. not of a different valence), and while they do not change the hole or electron concentration, they are of a different ionic size and hence induce internal (or chemical) pressure which can play an important role in its own right. Thus, replacing Y^{3+} by a larger lanthanide element induces a tensile stress in the CuO_2 plane and raises T_c (up to 98.5 K) while progressively replacing Ba^{2+} by Sr^{2+} induces a compressive in-plane stress and lowers T_c (down to 70 K [16]). Elements in the fourth category have a very limited solubility in YBCO; they are virtually non-reactive with YBCO, and so they are often present only as a second phase. Typical examples of these dopants are silver and gold which, despite their apparent non-reactivity, act as reaction-rate enhancers and sintering agents. Thus, introduction of Ag at a few percent level enables synthesis of $Y_2Ba_4Cu_7O_{15-x}$ under oxygen at 1 bar [17]. This compound is a high-pressure variant of YBCO and is discussed further below. In the absence of Ag, oxygen pressures in excess of 50 bar are needed to synthesize this compound as a pure phase.

Despite their harmful effect on T_c the doping of YBCO with small amounts of impurity scatterers like Pr or Zn was found to be beneficial for flux pinning, and there is scope for further improvement [12]. Because of the high reactivity of YBCO with most elements, it was important for the synthesis of single crystals to develop inert crucibles to minimize or eliminate impurities. The use of $BaZrO_3$ crucibles [18, 19] eventually enabled ultra-pure YBCO single crystals to be prepared, which allowed a variety of measurements until then hampered by the presence of impurities. These include thermal conductivity, magneto-resistance and quantum oscillation [20] experiments.

In this chapter, the structural, thermal, mechanical, chemical, optical, normal-state and superconducting properties of YBCO will be described.

C1.2 Structural Properties of Y-123, Y-124 and Y-247

C1.2.1 Crystal Structure

When the crystal structure of $YBa_2Cu_3O_{7-x}$ was first determined, it was a unique and highly unusual atomic arrangement among known complex oxides. Subsequently it was found that YBCO-123 is one member of a family of structurally related YBCO compounds. Other members of the family $YBa_2Cu_4O_8$ (Y-124) and $Y_2Ba_4Cu_7O_{15}$ (Y-247) require more exotic synthesis conditions and therefore were discovered and identified later [21–27].

YBCO has a layered defect-perovskite structure and as a consequence, is highly anisotropic in its electronic properties. Figure C1.1 shows the crystal structure of Y-123, which

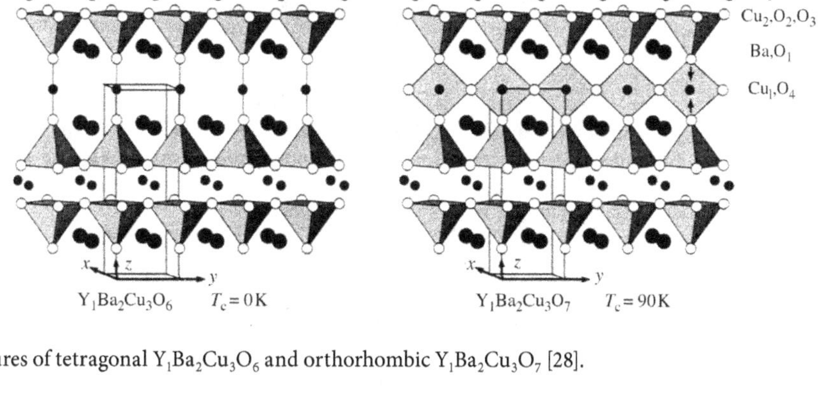

FIGURE C1.1 Structures of tetragonal $Y_1Ba_2Cu_3O_6$ and orthorhombic $Y_1Ba_2Cu_3O_7$ [28].

TABLE C1.1 Lattice Parameters, Space Group and Bond Lengths for YBCO-123 Materials

Abbreviation	Compound	Space Group	Lattice Parameters (Å)			Ref.
Y-123	$YBa_2Cu_3O_{7-x}$	*Pmmm*	$a = 3.8227$	$b = 3.8872$	$c = 11.6802$	[32]
Y-123	$YBa_2Cu_3O_7$	*Pmmm*	$a = 3.8185(4)$	$b = 3.8856(3)$	$c = 11.6804(7)$	[33]
Y-123	$YBa_2Cu_3O_{6.8}$	*Pmmm*	$a = 3.8214(7)$	$b = 3.8877(7)$	$c = 11.693(2)$	[33]
Y-123	$YBa_2Cu_3O_{6.56}$	*Pmmm*	$a = 3.8336(4)$	$b = 3.8807(4)$	$c = 11.7355(10)$	[33]
		Bond Lengths (Å)				
		Ref. [29]	Ref. [30]	Ref. [35]	Ref. [2]	
YBCO-123	Cu1–O1	1.836	1.876	1.848(4)	1.857	
	Cu1–O4	1.944	1.921	1.926	1.944	
	Cu2–O1	2.306	2.293	2.302(5)	2.296	
	Cu2–O2/3	1.946	1.931	1.944(1)	1.946	
	Ba–O1	2.744	2.735	2.744(1)	2.742	
	Ba–O2/3	2.972	3.168	2.951(5)	2.976	
	Ba–O4	2.879	2.835	2.908(3)	2.879	

is an oxygen-deficient tripled perovskite with ordered vacancies. The ideal fully oxygenated $x = 0$ structure consists of three sub-unit cells with four different layers stacked in the sequence BaO–CuO–BaO–CuO_2–Y–CuO_2–BaO–CuO–BaO [2]. In the CuO layer, each Cu atom is coordinated in-plane by two oxygen atoms forming a linear chain in the *b*-direction which is the cause of the structural orthorhombicity. This coordination differs from the active CuO_2 layers in which the copper is five-coordinated forming corner-shared square pyramids which are common to most of the two-layer cuprate superconductors. With deoxygenation, the chain layer in $YBa_2Cu_3O_{7-x}$ becomes depleted so that it is better represented as CuO_{1-x}. At the same time, there is an increasing tendency for the chain oxygens to disorder onto the otherwise vacant O5 sites between the chains and the structure becomes tetragonal when $x \geq 0.65$ with equal occupancy of both the O1 and O5 sites in the chain layer. At full oxygen depletion ($x = 1.0$), the 'chain layer' comprises only bare Cu atoms, as shown in Figure C1.1, which are two-coordinated by the apical oxygens above and below. The relation between crystal structure parameters and oxygen content in Y-123 has been reported [2, 28–31]. Data for the lattice parameters and bond lengths are given in Table C1.1.

Typical X-ray diffraction patterns for Y-123 with various oxygen contents are presented in Figure C1.2 [36], and the position, intensity and (h k l) indices of the main diffraction lines for $YBa_2Cu_3O_7$ are shown in Table C1.2 [33].

The structures of Y-124 and Y-247 compounds are similar to that of Y-123. They are shown schematically in Figure C1.3 [34]. Data for the lattice parameters and bond lengths are given in Table C1.3.

The superconducting phase $YBa_2Cu_4O_8$ ($T_c = 80$–81 K) was originally discovered as an intergrowth defect in $YBa_2Cu_3O_7$ [37]. The structure of Y-124 is closely related to that of Y-123, but with one additional Cu–O chain layer in the unit cell. Because the position of Cu in adjacent Cu–O chains is staggered by *b*/2 along the *b*-axis [37], the c-axis length is doubled (see Table C1.3). Two distinct Cu sites exist in the crystal structure: Cu(1), lying in the Cu–O chains, with four-fold square-planar coordination of oxygen and Cu(2) with a five-fold pyramidal coordination of oxygen in the CuO_2 planes. Unlike Y-123, this compound is stable with respect to oxygen stoichiometry which is fixed at 8 per formula unit. The fixed stoichiometry, combined with absence of twinning, means that Y-124 generally is the most defect-free HTS. This is most important for NMR studies where electric field gradients and chemical shifts due to disorder can seriously broaden out the resonance as in the Bi family or $La_{2-x}Sr_xCuO_4$. Y-124 usually has to be synthesized in high oxygen pressure at a high temperature to stabilize the structure. When the oxygen content is not stoichiometric, the solid decomposes into Y-123 and CuO, or $Y_2Ba_4Cu_7O_{15}$ (Y-247) and CuO [37].

The Y-247 superconductor was first observed by Karpinski et al [38] as an impurity phase in the investigation of the pressure-temperature-composition phase diagram in the system of $YBa_2Cu_3O_{6+x}/O_2$. The structure in the *c*-direction consists of alternating blocks of CuO_{1-x} single chains (123 units) and CuO double chains (124 units) [39]. When the single chains are fully oxygenated ($x = 0$), Y-247 has a rather high T_c of 92 K [24] (later revised up to 97 K). Due to the presence of the Y-124 blocks in its structure, Y-247 remains orthorhombic even when fully deoxygenated, and unlike Y-123, when fully deoxygenated Y-247 remains superconducting with $T_c = 25$ K [24] [see Figure C1.6(b)].

Typical X-ray diffraction patterns of Y-124 and Dy-247 are presented in Figure C1.4 [40–42], and the position, intensity and h k l indices of the main diffraction lines of Y-124 and Dy-247 samples are shown in Tables C1.4 [43] and C1.5 [44].

C1.2.2 Compositional Phase Diagram

These YBCO compounds exhibit different domains of stability in high oxygen pressure and elevated temperature. The

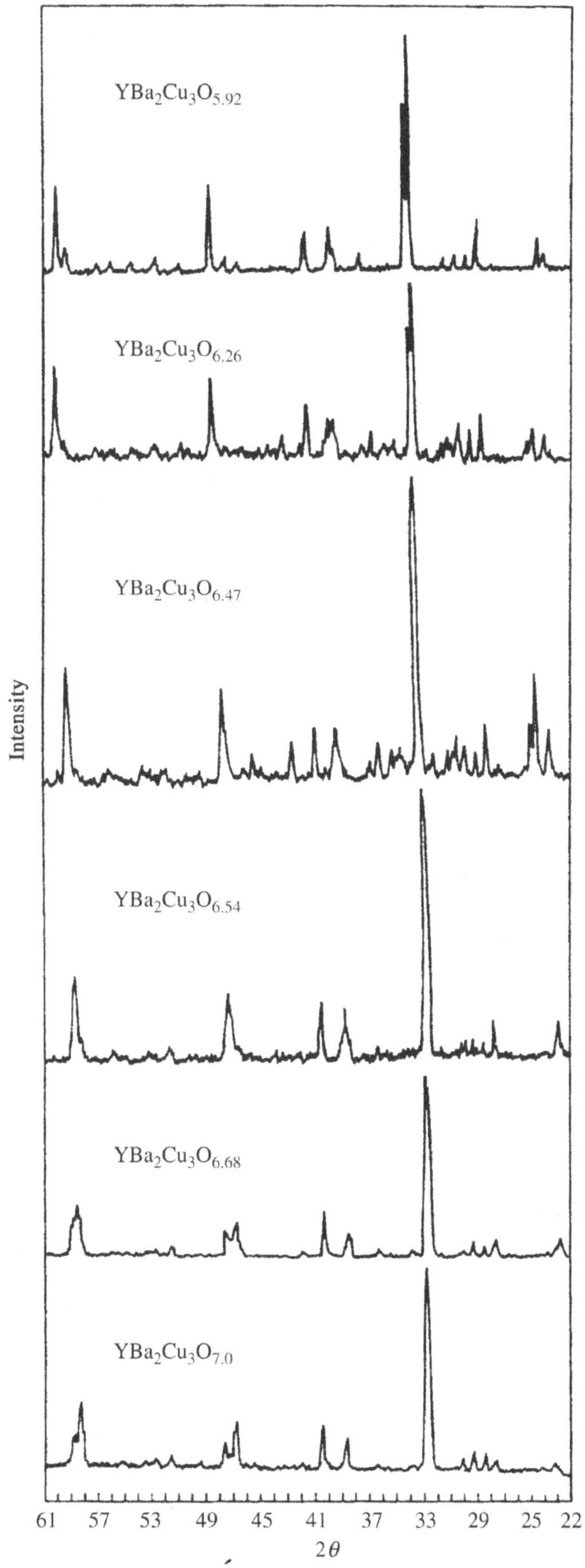

FIGURE C1.2 X-ray diffraction patterns of YBCO samples with different oxygen contents [37].

TABLE C1.2 $YBa_2Cu_3O_7$ X-Ray Diffraction Lines

2-θ	Int.	*h k l*	2-θ	Int.	*h k l*	2-θ	Int.	*h k l*
7.557	< 1	0 1 0	53.400	2	0 3 2	77.480	4	1 9 0
15.169	4	0 2 0	54.997	2	0 7 0	77.654	6	0 9 1
22.835	10	0 3 0	55.313	1	2 2 1	77.827	5	3 0 1
23.274	4	0 0 1	58.207	26	1 6 1	79.088	5	0 3 3
27.554	3	1 2 0	58.826	13	1 3 2	79.749	3	2 7 1
25.893	5	0 2 1	60.309	<1	1 7 0	81.141	1	1 2 3
30.618	<1	0 4 0	60.494	1	0 7 1	81.812	2	2 5 2
32.538	55	1 3 0	62.080	2	2 5 0	82.335	<1	3 3 1
32.842	100	0 3 1	62.262	2	2 4 1	82.498	1	0 10 0
33.757	2	1 1 1	62.809	3	0 5 2	83.650	1	1 3 3
36.370	3	1 2 1	65.571	2	1 7 1	87.028	3	1 10 0
38.512	13	0 5 0	68.134	5	2 6 0	87.287	6	2 8 1
38.799	5	0 4 1	68.618	5	1 8 0	87.748	4	1 8 2
40.384	14	1 3 1	68.797	13	0 8 1	90.351	1	3 5 1
45.524	2	1 4 1	68.889	12	2 0 2	91.089	1	2 9 0
46.633	22	0 6 0	72.820	1	0 9 0	91.772	1	0 9 2
46.725	21	2 0 0	72.995	<1	3 0 0	93.027	1	0 11 0
47.580	12	0 0 2	73.561	2	1 6 2	93.786	1	2 7 2
51.485	4	1 5 1	74.997	<1	2 7 0	95.853	4	3 6 1
52.526	3	1 6 0	75.615	1	0 7 2	96.394	4	3 3 2
52.733	4	0 6 1	77.247	<1	2 4 2	97.145	4	1 6 3

Radiation: Cu K α 1, λ = 1.540598 Å [33].

phase boundaries for Y-123, Y-247 and Y-124 are shown in Figure C1.5 together with the location of the orthorhombic/tetragonal transition in metastable Y-123 (marked O/T), the decomposition line for Y-123 to sub-oxides, and the boundary (A-A′) between Cu_2O and CuO. It is fortunate that Y-247 has a small stability window at 1 bar oxygen pressure between 860 and 870°C. This allows synthesis of this compound at atmospheric pressure [24], and this low-temperature synthesis material exhibits a remarkably high T_c (= 95–97 K). In contrast, high-pressure synthesized Y-247 shows rather low-T_c values (40–60 K) [18], and no amount of annealing seems to be able to correct this. This is due to the formation of 90° *c*-axis rotation twins [45]. The reaction rate at such low temperatures is very slow, and synthesis is only feasible with suitable reaction rate enhancers. These include $NaNO_3$, KNO_3 [17,24] or Ag additions in the form of Ag_2O or $AgNO_3$. Such additions also enable the synthesis of Y-124 under oxygen at ambient pressure at a temperature of 825°C [17].

C1.2.3 Electronic Phase Diagram

As noted, the transition temperature, T_c, is fixed at 81 K for Y-124 but varies with oxygen composition for Y-123 and Y-247. Figure C1.6 shows the variation in T_c with oxygen deficiency, *x*, for (a) $YBa_2Cu_3O_{7-x}$ and (b) $Y_2Ba_4Cu_7O_{15-x}$. The former reveals a clear plateau due to charge ordering which seems absent in the latter. However, the Y-247 data was very early and oxygen content was determined by mass change of polycrystalline pellets after quenching. If carefully repeated using, for example, thermoelectric power measurements to determine the doping state, then it is expected that a plateau would also be observed in Y-247, revealing a possibly universal behaviour associated with charge ordering around a doping state of 0.125 holes/Cu in which T_c is lightly suppressed.

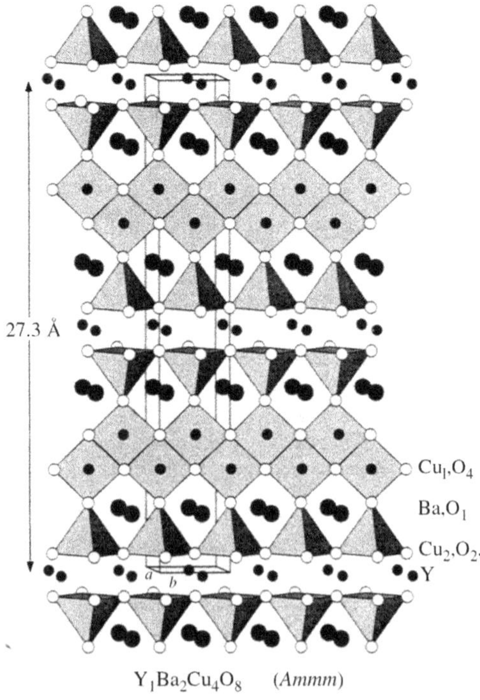

FIGURE C1.3 (Part 1) Structure of $Y_1Ba_2Cu_4O_8$ (Y-124) [28]. *(Continued)*

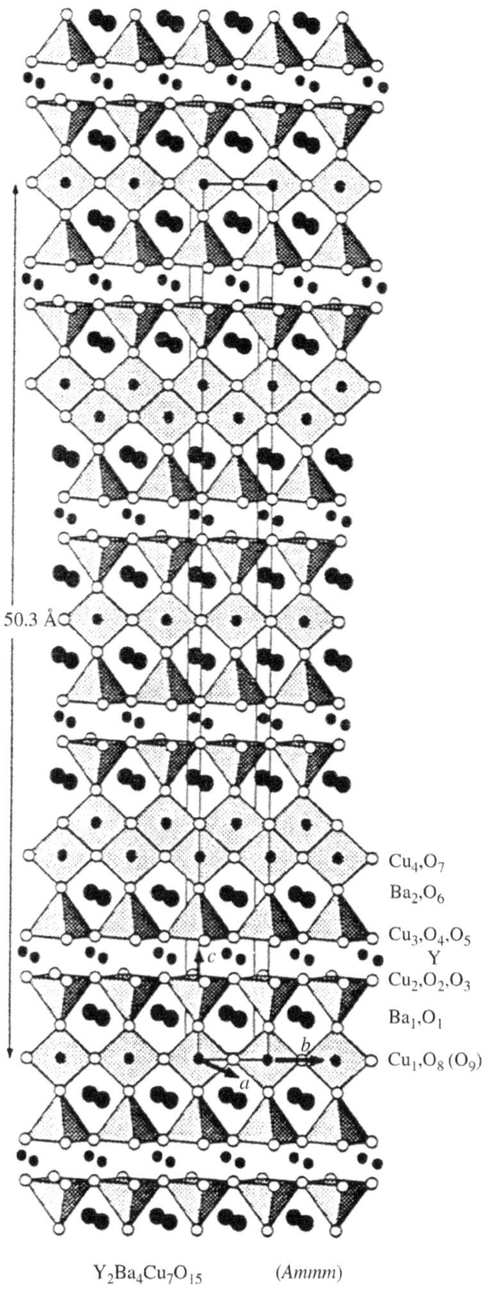

FIGURE C1.3 (Continued). (Part 2) Structure of $Y_2Ba_4Cu_7O_{15}$ (Y-247) [28].

TABLE C1.3 Lattice Parameters, Space Group and Bond Lengths for $YBa_2Cu_4O_8$ and $Y_2Ba_4Cu_7O_{15}$ Materials

Abbreviation	Compound	Space group	Lattice parameters (Å)			Ref.
YBCO-124	$YBa_2Cu_4O_8$	*Ammm*	$a = 3.838$	$b = 3.868$	$c = 27.20$	[43]
YBCO-247	$Y_2Ba_4Cu_7O_{15}$	*Ammm*	$a = 3.833$	$b = 3.878$	$c = 50.59$	[44]
		Bond lengths (Å) for Y-124, [23]				
Y–O2*x*4	Y–O3*x*4	Ba–O1*x*4	Ba–O2*x*2	Ba–O3*x*2	Ba–O4*x*2	
2.400(2)	2.396(2)	2.740(0)	2.962(3)	2.939(3)	2.980(2)	
Cu1–O1	Cu1–O4*x*2	Cu1–O4	Cu2–O1	Cu2–O2*x*2	Cu2–O3*x*2	
1.830(4)	1.943(0)	1.876(4)	2.276(4)	1.935(0)	1.950(0)	

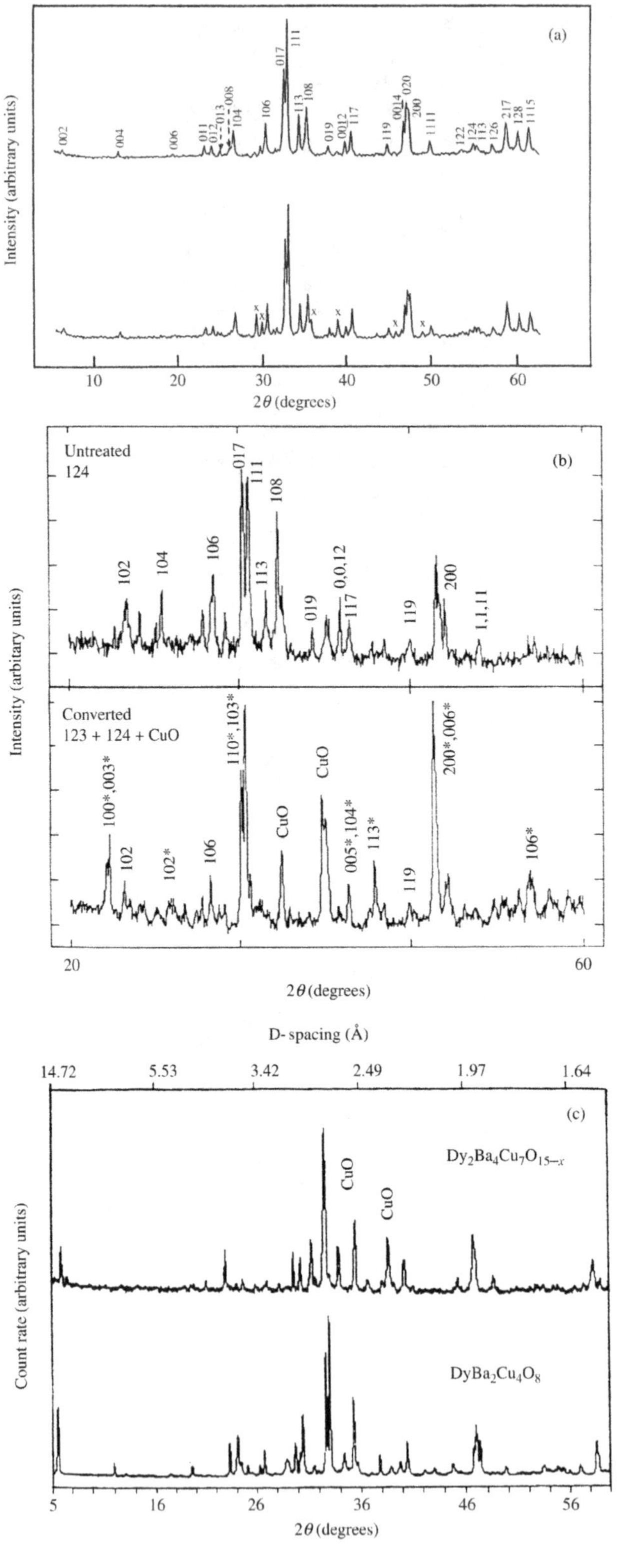

FIGURE C1.4 (Top) X-ray diffraction patterns of ground Y-124 pellets. The Y-124 peaks are indexed and impurity peaks are marked with crosses. The upper plot is a starting composition of Y/Ba/Cu = 1/2/4; the lower plot a starting composition of Y/Ba/Cu = 1/2/3. (Middle) X-ray diffraction pattern for Y-124 starting material. Appropriate peaks are marked with their *h k l* reflection indices. The lower panel shows the X-ray diffraction pattern after partial conversion into Y-123 + CuO. The asterisks (*) identify the Y-123 peaks [23, 41]. (Bottom) X-ray powder diffraction patterns for the Y-124 and Y-247 phases in the Dy–Ba–Cu–O system. Each pattern shows a single phase, except for additional peaks identified as CuO [42].

TABLE C1.4 $YBa_2Cu_4O_8$ X-Ray Diffraction Lines

2-θ	Int.	*h k l*	2-θ	Int.	*h k l*	2-θ	Int.	*h k l*
6.494	275	0 0 2	52.963	4	1 2 0	74.215	3	0 3 3
13.00	88	0 0 4	53.348	25	2 1 1	74.411	3	3 0 2
19.566	20	0 0 6	53.422	27	1 2 2	75.045	4	2 2 8
23.203	111	1 0 0	54.258	11	2 1 3	75.552	10	3 0 4
23.203	111	0 1 1	54.534	7	0 2 8	75.737	6	0 3 5
24.070	139	1 0 2	54.783	65	1 2 4	76.250	1	2 1 15
25.003	46	0 1 3	54.887	37	2 0 8	77.443	10	3 0 6
26.188	21	0 0 8	55.311	5	1 1 13	77.678	6	0 1 21
26.645	305	1 0 4	56.050	5	2 1 5	78.002	29	0 3 7
28.283	21	0 1 5	57.003	58	1 2 6	78.211	68	1 3 1
30.490	217	1 0 6	58.670	147	2 1 7	78.444	32	2 2 10
32.635	490	0 1 7	58.838	79	2 0 10	78.798	57	3 1 1
33.010	999*	1 1 1	59.412	3	1 0 16	78.961	33	1 3 3
34.336	254	1 1 3	60.021	83	1 2 8	79.547	13	3 1 3
35.233	242	1 0 8	61.360	8	1 1 15	80.067	15	3 0 8
36.862	4	1 1 5	62.053	8	2 1 9	80.994	2	0 3 9
37.739	22	0 1 9	62.794	1	0 1 17	80.994	2	315
39.733	12	0 0 12	63.134	4	0 2 12	81.768	1	1 0 22
40.396	69	1 1 7	63.456	4	2 0 12	82.239	1	2 1 17
43.394	1	0 1 11	63.772	1	1 2 10	82.545	3	2 2 12
44.744	29	1 1 9	66.141	1	2 1 11	82.685	6	1 3 7
46.489	6	1 0 12	66.435	1	1 0 18	83.265	6	3 1 7
46.715	24	0 0 14	67.874	1	1 1 17	84.699	1	0 3 11
46.932	339	0 2 0	68.200	2	1 2 12	85.259	1	1 2 18
47.324	325	2 0 0	68.377	8	0 2 14	85.641	3	0 0 24
47.433	183	0 2 2	68.856	159	2 2 0	85.641	3	1 3 9
47.822	7	2 0 2	68.996	83	0 0 20	85.857	2	0 1 23
48.913	6	0 2 4	70.019	1	01 19	86.219	3	3 1 9
49.294	6	2 0 4	70.424	4	2 2 4	87.343	5	2 2 14
49.484	3	01 13	70.883	1	2 1 13	87.633	2	0 2 20
49.751	19	1 1 11	72.364	2	2 2 6	88.879	1	21 19
51.308	3	0 2 6	73.449	4	0 3 1	89.321	2	1 3 11
51.676	3	2 0 6	74.029	1	3 0 0	89.897	2	3 1 11
52.764	1	1 0 14	74.215	3	0 2 16			

Radiation: Cu Kα1, λ = 1.54060 Å [43].

That this plateau is associated with electronic ordering and not oxygen ordering is demonstrated by Ca and La substitution in $YBa_2Cu_3O_{7-x}$. By using 0.2La, 0.1La, 0.0La, 0.1Ca and 0.2Ca, the plateau can be arbitrarily shifted in x from 0.11 to 0.73 (0.11, 0.22, 0.45, 0.62, 0.73, respectively), but when hole doping, p, is measured by thermoelectric power, the plateau is always centred on $p = 0.125$ thus revealing that this is not associated with oxygen ordering around $x = 0.5$ [47]. Figure C1.6(b) shows T_c plotted versus doping, p, determined in this way for $Y_{0.8}Ca_{0.2}Ba_2Cu_3O_{7-x}$ which, unlike the Ca-free material, can be significantly overdoped. Apart from the different magnitude of the maximum T_c the detailed phase curve is very similar to that observed for $La_{2-x}Sr_xCuO_4$ and may be considered generic to the cuprates.

The doping phase diagram for YBCO (combined with Ca substituted for Y to enable overdoping) is actually much more complex than the simple T_c versus p diagram shown in Figure C1.6. The full electronic phase diagram as currently envisaged (e.g. see Grissonnanche *et al* [48] and many other papers cited below) is summarized in Figure C1.7. At low doping, YBCO is an antiferromagnetic (AF) insulator. Short-range AF correlations persist at higher doping and freeze out into a spin glass at low temperature. With increasing T, superconductivity sets in and rises to optimum doping around $p \approx 0.16$ holes/Cu. In the shaded region short-range charge-density wave order competes with superconductivity resulting in the so-called '60-K plateau'. The pseudogap probably arises from short-range AF order and is responsible for the suppression of T_c in the underdoped region. It disappears rather abruptly at $p \approx 0.19$ holes/Cu, where the Fermi surface reconstructs from hole pockets in the underdoped region to a large hole-like Fermi surface on the overdoped

TABLE C1.5 $Y_2Ba_4Cu_7O_{15}$ X-Ray Diffraction Lines

2-θ	Int.	*h k l*	2-θ	Int.	*h k l*	2-θ	Int.	*h k l*
6.986	11	0 0 4	34.061	45	1 1 5	51.862	1	0 2 12
10.645	3	0 0 6	34.061	45	1 0 14	52.422	1	2 0 12
13.997	8	0 0 8	35.164	4	1 1 7	52.705	1	1 0 26
21.055	4	0 0 12	35.451	2	0 0 20	52.855	1	1 2 0
22.980	10	0 1 1	36.836	6	1 0 16	53.314	2	2 1 1
23.186	5	1 0 0	38.083	7	0 1 17	53.411	2	1 2 4
23.447	4	1 0 2	38.370	3	1 1 11	53.590	2	0 2 14
24.238	11	1 0 4	39.133	13	0 0 22	54.055	2	1 2 6
24.599	7	0 1 5	40.358	24	1 1 13	54.346	4	0 1 27
24.599	7	0 0 14	42.611	2	1 1 15	54.346	4	0 0 30
25.479	3	1 0 6	42.801	2	1 0 20	54.981	3	1 2 8
26.064	1	0 1 7	44.232	2	0 1 21	56.595	5	1 1 25
27.181	13	1 0 8	45.067	3	1 1 17	57.541	6	1 2 12
27.920	4	0 1 9	46.640	33	0 0 26	58.234	6	2 0 18
29.198	1	1 0 10	46.793	36	020	58.335	7	0 0 32
31.520	25	1 0 12	47.391	24	200	58.689	22	2 1 13
31.830	7	0 0 18	47.663	4	1 1 19	59.169	11	1 2 14
32.569	100	0 1 13	49.115	1	028	59.853	13	1 1 27
32.864	100	1 1 1	49.684	1	208	59.853	13	1 0 30
33.254	11	1 1 3	50.517	5	1 1 21			

Radiation: Cu Kα1, λ = 1.54056 Å [44].

side. Here superconductivity is much more BCS-like, and T_c falls smoothly with further increase in doping.

C1.2.4 Defects in YBCO

All YBCO compounds can exhibit highly defective structures. Several types of defects are easily incorporated: oxygen vacancies, cation vacancies, twin boundaries, intergrowths, dislocations, stacking faults, interstitials, Schottky defects, anti-site defects and Frenkel defects [49–54].

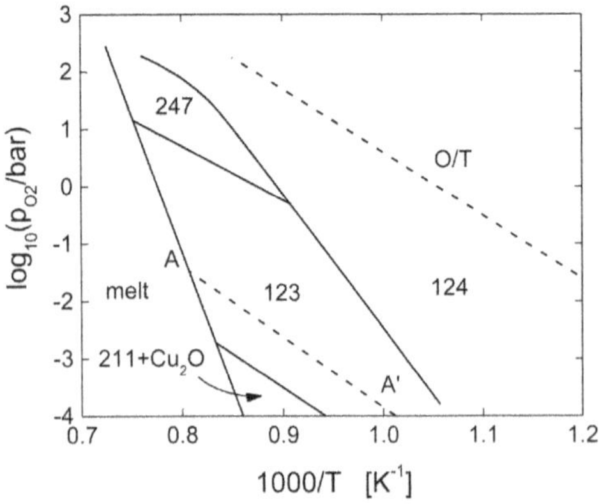

FIGURE C1.5 The oxygen phase diagram for the Y/Ba/Cu/O system with Y-247 composition. The line O/T marks the orthorhombic/tetragonal transition in metastable Y-123, and the line A-A′ marks the phase boundary between Cu_2O and CuO. The line below this shows where Y-123 decomposes to $Cu_2O + Y_2BaCuO_5$ (referred to as 211) [46].

In Y-123 and Y-247, the most common defects are oxygen atoms that occupy the ideally vacant O5 lattice sites in the chain layers. The range of such defect concentrations is

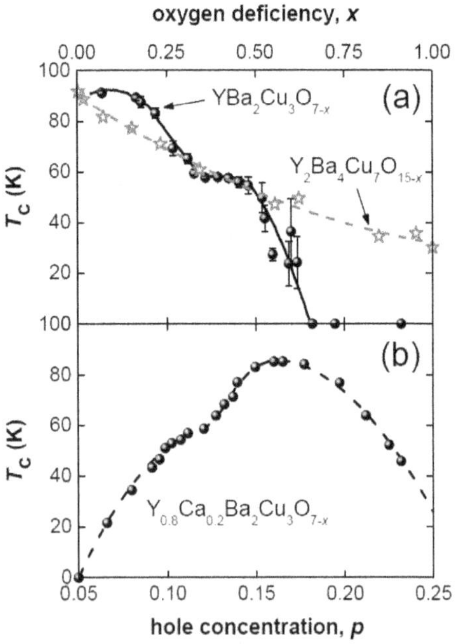

FIGURE C1.6 (a) T_c plotted versus oxygen deficiency, x, for $YBa_2Cu_3O_{7-x}$ (spheres) and $Y_2Ba_4Cu_7O_{15-x}$ (stars). Note that T_c remains finite for Y-247 even when fully deoxygenated ($x = 1.0$) [24, 47]. (b) T_c plotted versus hole concentration, p, for $Y_{0.8}Ca_{0.2}Ba_2Cu_3O_{7-x}$. Note that Ca substitution allows significant overdoping, whereas the phase curve for the Ca-free Y-123 extends only to $p \approx 0.185$.

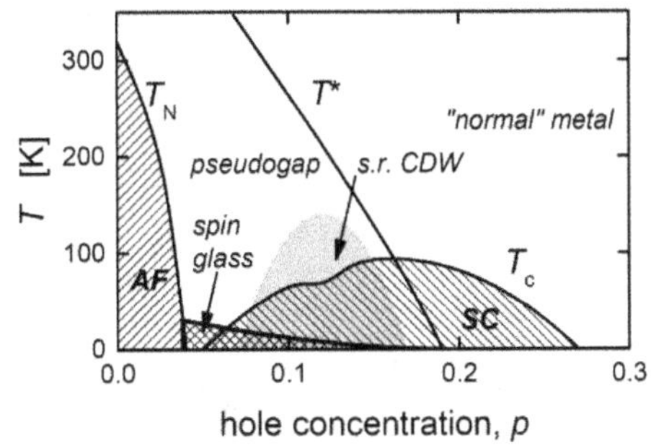

FIGURE C1.7 Electronic phase diagram of $YBa_2Cu_3O_{7-x}$ and $Y_{0.8}Ca_{0.2}Ba_2Cu_3O_{7-x}$ which enables overdoping.

unusually large, allowing the material to be changed from insulator to superconductor [50]. The influence of oxygen stoichiometry on YBCO properties, especially with respect to the tetragonal (T) to orthorhombic (O) phase transformation, has been the subject of much research. The simplest phase diagram for Y-123 as a function of temperature and x has a T field at high temperature and high x, an O field at lower temperature and low x, and a T + O two-phase field that separates the two at low temperature (Figure C1.8). At the ideal composition $YBa_2Cu_3O_7$, the orthorhombic phase has a perfect ordering of oxygen atoms and vacancies forming CuO chains in the b-direction. At the composition $YBa_2Cu_3O_6$, the basal planes are completely depleted of oxygen, and therefore the tetragonal crystal structure is observed (having space group *P4/mmm*). Near the intermediate composition $YBa_2Cu_3O_{6.5}$, careful annealing can result in the ordered formation of alternating full and empty chains referred to as the OII phase (Figure C1.8), and as a consequence, this structure can be relatively free of disorder while being significantly underdoped.

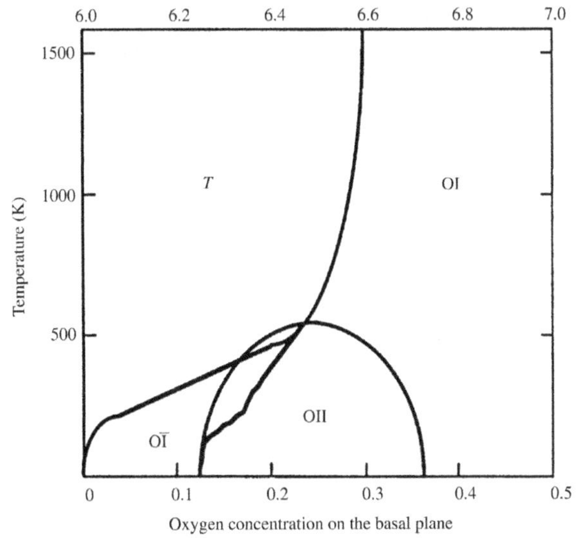

FIGURE C1.8 Phase diagram for Y-123 as a function of oxygen stoichiometry.

FIGURE C1.9 A TEM image showing a Y-124 planar stacking fault within the dominantly Y-123 structure [51]. The stacking fault arises from the insertion of an additional CuO chain layer staggered adjacent to the CuO chain layer in Y-123.

The most frequently observed planar stacking fault in YBCO is a $(CuO)_2$ double-chain layer formed by insertion of an extra CuO chain layer, leading to a local composition of Y-124 in a background matrix of Y-123 (or Y-125 in Y-124) [52, 53]. Figure C1.9 shows a typical example of planar defects [51]. Such intergrowths tend to lower the average doping state in proportion to their density [55] and, despite their potential for pinning, typically rapidly reduce the critical current density due to this reduced doping.

In YBCO, twinning is associated with the tetragonal-to-orthorhombic order-disorder phase transformation. This transformation nucleates in different parts of the crystal at random with equal likelihood for nucleation of the b-axis in either of the available orientations. The orthorhombic nuclei grow and eventually meet other growing nuclei and any orthogonal interfaces result in twin boundaries in the basal plane. Because the difference in the orthorhombic a- and b-parameters is of the order of 1% or less, such twin boundaries can remain metastable. These twins exhibit a 90° (0 0 1) rotation of the lattice. Another twin structure sometimes appears in Y-123 due to the fact that $c \approx 3\times a \approx 3\times b$ resulting in a 90° (1 0 0) rotation of the lattice at the boundary. Across the boundary of these twins the c-axis of one sub-grain continues as an a- or b-axis in the adjacent sub-grain, as can be seen from Figure C1.10 [52]. Such 90° (1 0 0) rotation twins are responsible for the reduced doping state and T_c of high-pressure

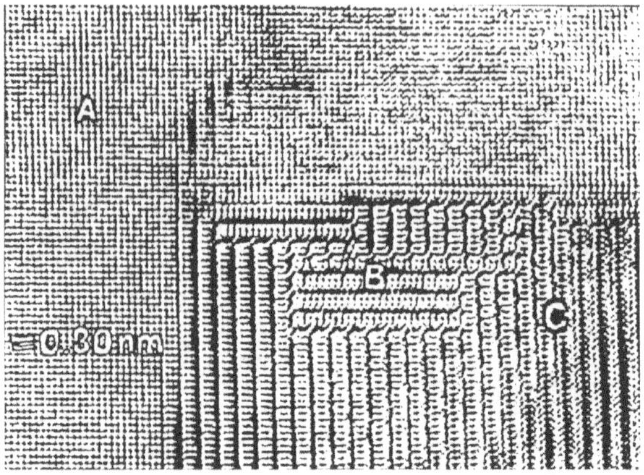

FIGURE C1.10 A TEM micrograph of $YBa_2Cu_3O_{7-\delta}$ showing [1 0 0] 90° rotation twin boundaries [52].

synthesized Y-247 [45]. High-resolution transmission electron microscopy (HRTEM) studies have led to the suggestion that some twin boundaries are oxygen deficient [49] and therefore underdoped. Again, this can lead to reduced intergranular critical current density.

Three types of dislocations have been observed in YBCO. The first type is an edge dislocation running along (1 0 0) with a Burgers vector of a [0 1 0], the second—with a Burgers vector of a [0 0 1] and third—with a Burgers vector of a/2 + b/2 [0 0 1]. The last two types of dislocations are observed more often in materials containing many (1 0 0) 90° rotation twins. The lattice image of Y-124 in the region indicated by arrow C in Figure C1.11 [53] shows the presence of an edge dislocation with a Burgers vector of either $b = \langle 010 \rangle$ or $\langle 100 \rangle$ on [0 0 1].

C1.2.5 Anisotropy

Besides high transition temperatures, the HTS are characterized by their pronounced anisotropy. Such a high anisotropy originates from the fact that the superconductivity occurs mainly in the CuO_2 planes, and there are weak interlayer couplings between them. The anisotropy is usually expressed as a dimensionless parameter γ, which is defined as $\gamma = \lambda_c/\lambda_{ab}$, where λ_{ab} and λ_c are the London penetration depths parallel and perpendicular to the a–b plane, respectively. It is this anisotropy that insists that practical conductors with high J_c must have a strong *c*-axis alignment of grains and the Josephson coupling between CuO_2 layers is evidenced, for example, in infrared plasmons [56] whose energy scales with both the superconducting condensation energy [57] and transition temperature in multilayers cuprates [56]. This Josephson coupling between layers also enables the generation of THz radiation from mesas fabricated on the surface of HTS which comprises just 10 to 20 CuO_2 layers [58]. *c*-axis Josephson tunneling spectra reveal features possibly attributable to the density of states of the active pairing boson [59].

In the case of the Y-123 compounds, a further anisotropy is evident. Because of the orthorhombic crystal structure and the existence of Cu–O chains along the *b*-axis, many properties

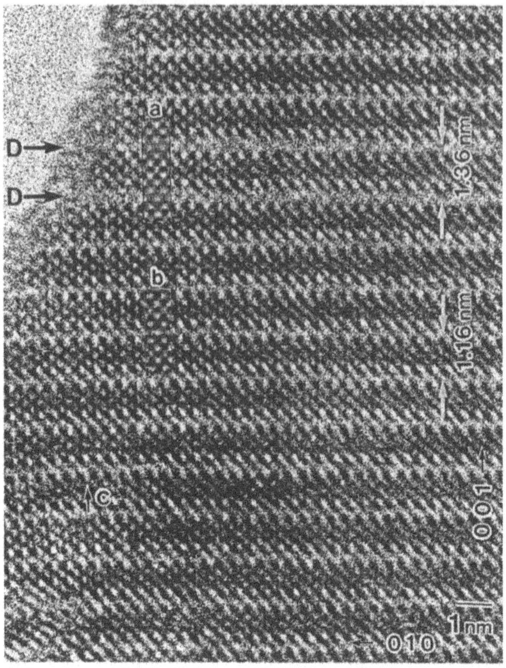

FIGURE C1.11 A high-resolution TEM image of a Y-123 crystal observed along the [1 0 0] direction [53]. Arrow D indicates thick and bright bands parallel to the basal plane. A simulated image of Y-124 in a [1 0 0] projection is inserted at (a) and a simulated image of Y-123 in the same projection is inserted at (b). Arrow C indicates an edge dislocation in Y-123 with a Burgers vector **B** = **a**(1 0 0) or **B** = **b**(0 1 0) on [0 0 1].

reveal a so-called in-plane anisotropy [60–62]. For example, the penetration depth along the *b*-direction is found to be substantially smaller than that along the *a*-direction, as seen in both infrared measurements [63] and muon spin relaxation measurements [64]. Incidentally, one consequence of this is that one would expect a larger critical current density in the *b*-direction than in the *a*-direction, though this remains to be demonstrated. Because of this anisotropy, there is remarkable difference in the effect of uniaxial strain in the two directions on T_c. In fact, as shown in Figure C1.12, they are opposite to each other [65].

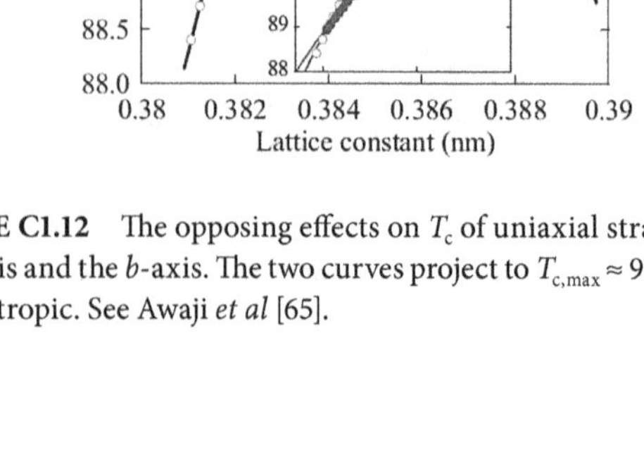

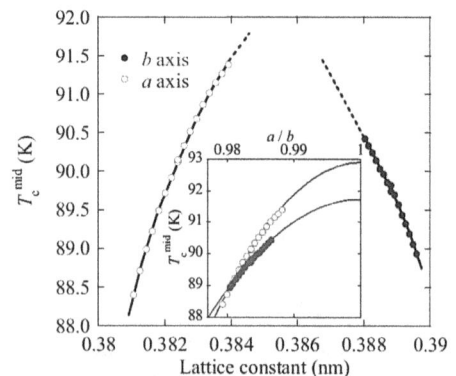

FIGURE C1.12 The opposing effects on T_c of uniaxial strain along the *a*-axis and the *b*-axis. The two curves project to $T_{c,max} \approx 93$ K when fully isotropic. See Awaji *et al* [65].

C1.3 Thermal Properties

C1.3.1 Specific Heat of YBCO

The electronic specific heat represents an energy integral of the electronic density of states and therefore captures the full integrated excitation spectrum, including both spin and charge degrees of freedom. It is therefore an important window on the full complement of electronic interactions and the way in which they evolve with temperature and doping. An early review on the specific heat of HTS cuprates was given by Junod [66]. Unfortunately, at elevated temperatures the electronic specific heat represents only a few percent of the total specific heat and is dominated by the lattice (phonon) contributions. This is not such a difficulty with low-T_c superconductors where T_c is well below the Debye temperature and the lattice term is suppressed. But for HTS where T_c is so high that lattice modes are excited, it is difficult to extract the electronic term with any accuracy. This is illustrated in Figure C1.13 which shows typical examples of the temperature dependence of the specific heat for Y-123 and Y-124 [67]. The electronic anomaly at the superconducting transition for Y-123 (90 K) is not easily seen on the phonon background and it is not discernible for Y-124. An additional challenge is that the cuprates exhibit correlations (such as the pseudogap) which extend high above T_c, indeed even to room temperature so an alternative strategy is required to accurately determine the electronic part of the specific heat. Fortunately, this is available.

High-precision differential specific heat measurements have been developed by Loram and coworkers [68, 69]. The technique involves use of a reference sample similar to the target sample, importantly, having the same number of atoms. A typical reference is Zn-substituted Y-123 in which T_c is shifted far below that of the pure material but at its simplest Cu may be used. This backs off the phonon contribution except for a small residual. The residual can be precisely determined by repeating this measurement for many different doping states as the oxygen content x in $YBa_2Cu_3O_{6+x}$ is reduced from 1 to 0. This is illustrated in Figure C1.14.

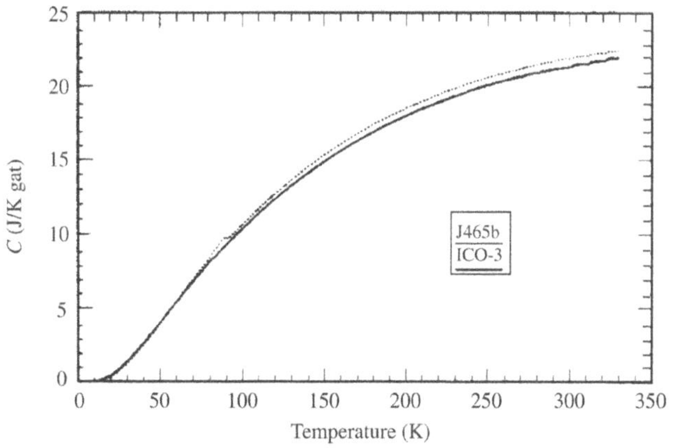

FIGURE C1.13 Specific heat of Y-123 (sample J465b) and Y-124 (sample JCO-3) up to room temperature [67].

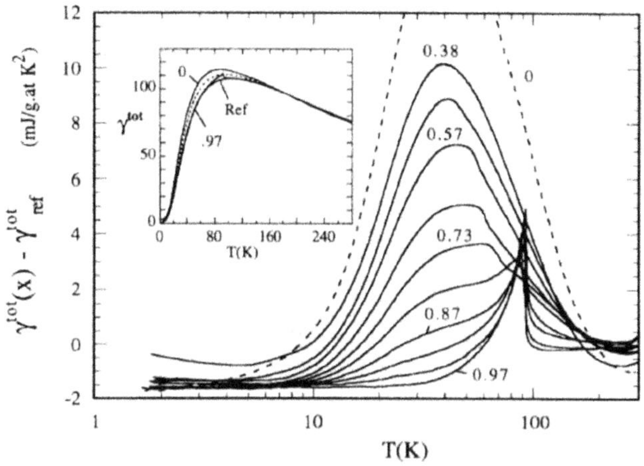

FIGURE C1.14 The difference in specific heat coefficient $\gamma \equiv C_p/T$ between sample and reference for various values of x in $YBa_2Cu_3O_{6+x}$ (note the modification in formula definition to that used in this chapter). The rising hump around 40 K is due to the residual phonon term which can be evaluated as described in the text. The inset shows the absolute value of γ^{tot}, and the small electronic anomaly is just visible at T_c for $x = 0.97$ [68].

There is a residual phonon term, $\Delta\gamma^{ph}$, which grows as x is reduced. However, it is found to scale very closely with the oxygen content $\Delta\gamma^{ph}(x_1, x_2;T) = (x_2 - x_1) f(T)$, so by scaling in this way the residual can be determined and subtracted, leaving the bare electronic $\gamma(T)$. Key to this analysis is to work in units of gram atom (g.at) instead of moles. The electronic entropy, $S(T)$, is obtained by integrating $\gamma(T).dT$, requiring that $S(T{=}0) = 0$. Figure C1.15 shows (a) $S(T)$ and (b) $\gamma(T)$ data measured in this way for $Y_{0.8}Ca_{0.2}Ba_2Cu_3O_{7-x}$ where the Ca substitution allows the Y-123 compound to be substantially overdoped down to $T_c = 50$ K. The dashed curve is for optimal doping at $T_{c,max}$ and the thick curve is at the critical doping where the pseudogap closes, apparently rather abruptly. Arrows indicate increasing hole concentration, p.

Where direct, absolute measurements of the specific heat barely resolve the anomaly at T_c. These differential measurements expose the entire electronic term, not just near T_c but, importantly, both the low- and high-temperature behaviour. Figure C1.15 reveals a wealth of features: (i) the linear-in-T behaviour at low T is consistent with the d-wave gap, where the slope is inversely proportional to the gap amplitude. Evidently, the gap remains fairly constant on the underdoped side but falls with T_c on the overdoped side. (ii) Strong fluctuations are evident in the curvature above and below T_c. Analysis of this data enables the mean-field transition temperature, T_c^{mf}, to be determined [70]. At maximum, it is about 30 K above the observed T_c. The BCS ratio thus formed, $2\Delta_m/(k_B T_c^{mf})$ is found to be close to the weak-coupling d-wave value of 4.28 over the overdoped region [70]. (iii) The weight of the specific heat anomaly remains fairly constant in the overdoped region where the broadening arises from inhomogeneity from Ca substitution. However, the anomaly falls

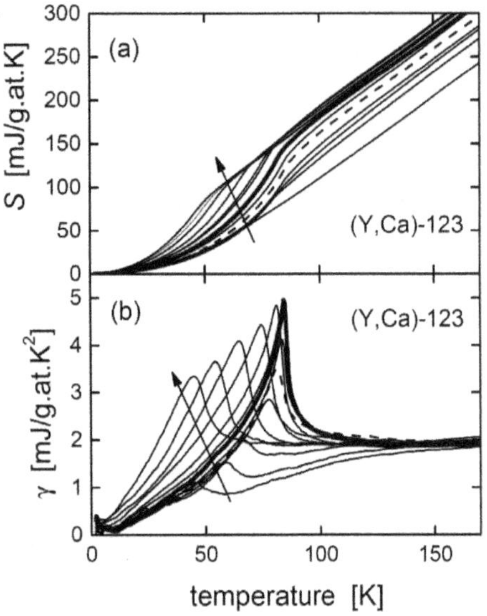

FIGURE C1.15 (a) The electronic entropy, $S(T)$, and (b) the electronic specific heat coefficient $\gamma \equiv C_p/T$ for various values of x in $Y_{0.8}Ca_{0.2}Ba_2Cu_3O_{7-x}$ measured using high-precision differential specific heat measurements. Arrows indicate increasing hole concentration. Dashed curves are at optimum doping, $T_c = T_{c,max}$, and bold curve is at critical doping, $p \approx 0.19$ holes/Cu, where the pseudogap opens. (Adapted from [69].)

away rapidly in the underdoped region due to the opening of the pseudogap beginning at the bold curve when $p \approx 0.19$ holes/Cu. (iv) The presence of the pseudogap above T_c is very evident in the entropy, which, like a 2D metal, is roughly linear to the origin in the overdoped region, but a negative intercept develops starting abruptly at critical doping. For a triangular gap, this intercept is $2\ln 2\ k_B\ N_0\ E_g$, where N_0 is the electronic density of states at the Fermi level, and E_g is the amplitude of the pseudogap. This grows monotonically below $p \approx 0.19$. (v) By integrating the entropy one can calculate the condensation energy U_0. It is found that the BCS ratio $U_0/(\gamma_n\ T_c^{mf\,2})$ is again rather close to the weak-coupling value of 0.17 for a d-wave superconductor across the entire overdoped region [69, 71]. The ratio collapses abruptly below $p \approx 0.19$ signaling the opening of the pseudogap. (vi) The fact that $\gamma \rightarrow 0$ as $T \rightarrow 0$ shows that all quasiparticles are paired at $T = 0$, contrary to earlier reports [67]. The substitution of Zn (or other impurity scatterers) breaks Cooper pairs, rapidly reduces T_c and introduces a finite value of $\gamma(0)$ [69]. It turns out that the critical Zn concentration to just suppress superconductivity is numerically equal to $S(T_c)/k_B$ across the entire superconducting phase diagram. This implies that each Zn atom breaks one Cooper pair [72].

Differential specific heat measurements on pure and Zn-doped $YBa_2Cu_4O_8$ [70] clearly reveal the evolution of the full electronic $\gamma(T)$ including the anomaly at T_c with progressive Zn substitution. Because Y-124 is naturally underdoped, the pseudogap is evident from the suppression of the normal-state $\gamma(T)$ at low temperature.

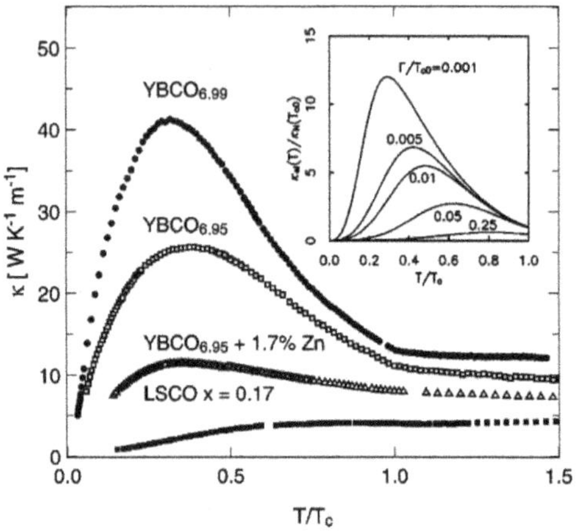

FIGURE C1.16 The low-temperature thermal conductivity, κ, for Y-123 at two doping levels and additionally with Zn substitution and compared with $La_{1.83}Sr_{0.17}CuO_4$. The inset shows the calculated effect of impurity scattering on the thermal conductivity. (From ref. [74].)

C1.3.2 Thermal Conductivity of YBCO

Studies of the thermal conductivity, κ, give valuable information on the interaction between charge carriers and phonons and on the scattering of both defects and impurities. An early review was given by Uher [73]. One basic characteristic of copper oxide superconductors is their poor thermal conductivity and this is utilized in their application as current leads into cryogenic environments for superconducting magnets. Experimental data [73–75] from thermal conductivity studies of Y-123 show that in the range $T > 10$ K values of $\kappa(T)$ differ by almost an order of magnitude depending on sample microstructure, oxygen content, admixture and crystallite borders. The various temperature dependencies of κ obtained for Y-123 are all qualitatively similar, in that κ shows a maximum in the temperature range 40–90 K and a step change of slope at T_c. The temperature of the maximum depends on the oxygen content. All the results show a dominant phonon contribution above T_c. For the YBCO system in the 90–240 K temperature range, the contribution of the electronic part is from 1 to 20% depending on the sample. The electrons contribute 60% of the total thermal conductivity at 300 K in single YBCO crystals [76, 77]. Examples of $\kappa(T)$ for Y-123 up to just above T_c, including the effect of Zn substitution, can be seen in Figure C1.16 [74]. The evolution of the peak is very similar to the calculated effect of impurity scattering shown in the inset.

Because cuprates are d-wave superconductors with nodes in the gap, then even small amounts of disorder or other scattering centres result in pair-breaking and delocalized quasiparticle excitations. These contribute a universal residual linear thermal conductivity term, κ_0, given by [78]:

$$\frac{\kappa_0}{T} = \frac{k_B^{\,2}}{3\hbar}\frac{n}{d}\left(\frac{v_F}{v_2} + \frac{v_2}{v_F}\right) \quad \text{(C1.1)}$$

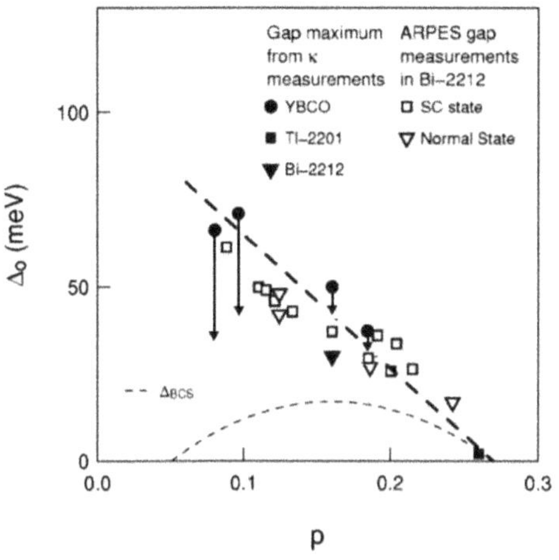

FIGURE C1.17 The doping dependence of the superconducting gap in Y-123 and other cuprates measured from the low-temperature thermal conductivity. (From ref. [74].) Arrows show corrections based on renormalization of v_F close to te Fermi surface as observed in Bi-2212 [81].

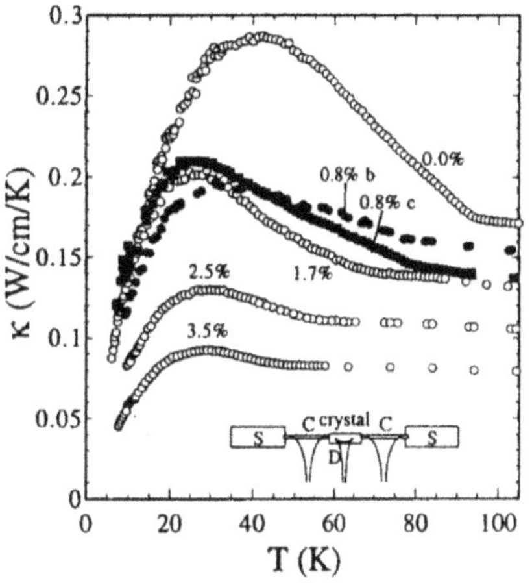

FIGURE C1.18 The thermal conductivity as a function of progressive Zn substitution in single crystals of $YBa_2(Cu_{1-x}Zn_x)_3O_{7-\delta}$. (From [90].) The peak below T_c is progressively reduced.

Here v_F and v_2 are the quasiparticle velocities normal and tangential to the Fermi surface at the gap node, n is the number of CuO_2 planes per unit cell and d is the c-axis lattice spacing. Because v_2 is the slope of the nodal gap at the node, such measurements enable the superconducting gap amplitude to be determined from $\frac{1}{2}\hbar k_F v_2$. The result of such measurements on Y-123 is shown in Figure C1.17 for different doping states and compared with other HTS cuprates and measurements using ARPES [74].

The data in Figure C1.17 remained a puzzle for some time. The gap magnitude appears to rise uniformly with underdoping to large values of the order of 80 meV. This is substantially larger than that observed in specific heat [69] and infrared spectroscopy [79]. The gap magnitude looks more like the pseudogap, which does indeed rise uniformly to such large values. Subsequent ARPES studies succeeded in distinguishing between the superconducting gap and the pseudogap at the antinode resulting in values of Δ_0, which, as found in the specific heat, plateau out in the underdoped region at about 40 meV. Moreover, the above analysis using Equation (C1.1) required input values of the Fermi velocity v_F, and early ARPES studies indicated this was doping independent [80] with $v_F \approx 1.8$ eV.Å. More recently, however, an additional renormalization of the dispersion was discovered very close to the Fermi level [81] yielding values of v_F which fall sharply on underdoping. The gap values in Figure C1.17, as shown by the arrows, do indeed level out much as seen in ARPES, specific heat and infrared spectroscopy. It is satisfying to see these troubling discrepancies eventually being resolved.

The Wiedemann–Franz law (namely that $\kappa/\sigma T = L_0$ as $T \to 0$, where σ is the electrical conductivity and $L_0 = (\pi^2/3)(k_B/e)^2$ is the Lorenz number) was found to be satisfied for many cuprates across the normal-state phase diagram beyond the insulator-to-metal transition [82]. The opening of the pseudogap in the underdoped region can increase the Lorenz number [83]. The data for κ in Y-124 differ to a surprisingly large extent from those for Y-123, despite their structural similarity, and κ is dominated by phonon–phonon interactions [84, 85]. At all temperatures, phonon–phonon interactions give rise to a much larger thermal resistivity in Y-124 than do electron–phonon interactions, while point-defect scattering was found to be negligible, probably because of the stable oxygen stoichiometry of Y-124. In contrast, point-defect scattering dominates κ completely in Y-123, even for single crystals [86, 87]. Single crystal Y-124, as noted, is naturally detwinned and thermal conductivity in these crystals show a huge enhancement over Y-123 by up to an order of magnitude in the peak below T_c [88]. The in-plane anisotropy, κ_b/κ_a, of up to 4.5 for Y-124 is also much larger than for Y-123 (up to 1.4). This reflects the additional conductance of the relatively defect-free double chains extending in the b-direction.

Only a few experimental results have been published concerning thermal conductivity in doped YBCO [89–91]. Figure C1.18 shows the thermal conductivity as a function of temperature for progressive Zn substitution in single crystals of $YBa_2(Cu_{1-x}Zn_x)_3O_{7-\delta}$ [90]. The characteristic enhancement of κ just below T_c in pure Y-123 arising from gapped-out quasiparticles was suppressed by the Zn substitution. This suppression of the κ enhancement is attributed to the depressed phonon–electron scattering caused by the Zn substitution [89].

C1.4 Mechanical Properties of RE123

For practical applications, it is important to improve mechanical properties, since superconductors experience various kinds of forces such as thermal stress due to the heat cycles and electromagnetic forces due to fields arising from current flow. High-temperature superconductors are brittle and sometimes fracture or delaminate, therefore a good understanding of mechanical properties is important for all engineering applications.

In this section, mechanical properties and related physical parameters of $REBa_2Cu_3O_{7-\delta}$ (RE123, where RE is a rare-5earth element or Y) will be reviewed along with the methods to measure mechanical properties.

C1.4.1 Thermal Expansion Coefficient

Cracking in ceramics is often caused by differential thermal expansion between different phases in contact.

The thermal expansion coefficient is defined as the change of length or volume per degree of temperature. The coefficient of linear thermal expansion (α) is defined by:

$$\alpha = \mathrm{d}l / \mathrm{d}T \qquad \text{(C1.2)}$$

and the volume thermal expansion (α_V) is defined by:

$$\alpha_V = \mathrm{d}V / \mathrm{d}T \qquad \text{(C1.3)}$$

where l is the length, V is the volume and T is temperature. It is usually understood that the derivative is at constant pressure. Although these values are temperature dependent, the mean value averaged over a certain temperature range is often used as the coefficient value.

With increasing temperature, the amplitude of atomic vibration at the equilibrium position increases, leading to an increase in bond length and, thus, lattice expansion. The volume change with the lattice vibration will cause an increase in the total free energy of the lattice. Therefore, the temperature dependence of the thermal expansion coefficient is somewhat similar to that of the specific heat. At low temperatures, the thermal expansion coefficient increases rapidly with temperature and becomes almost constant above the Debye temperature Θ_D on the assumption that the lattice simply expands. However, for most materials, it can increase even above Θ_D through the formation of lattice defects such as vacancies and interstitials. Then the concentration of the defects is directly related to the thermal expansion.

Because of the structural and electronic anisotropy of RE123, the thermal expansion coefficients also exhibit both out-of-plane and in-plane anisotropy.

The thermal expansion coefficient can be determined from the temperature dependence of lattice parameters using X-ray or neutron diffraction [92–96]. It is also common to measure the coefficient directly with dilatometers [97–103].

Figure C1.19 shows the thermal expansion coefficients of the principal crystalline axes of a detwinned Y-123 single crystal at low temperatures [97]. Thermal expansion is greatest along the c-axis and least along the b-axis. The temperature dependence of α for Y-123 follows conventional behaviour and rapidly increases with increasing temperature up to Θ_D of

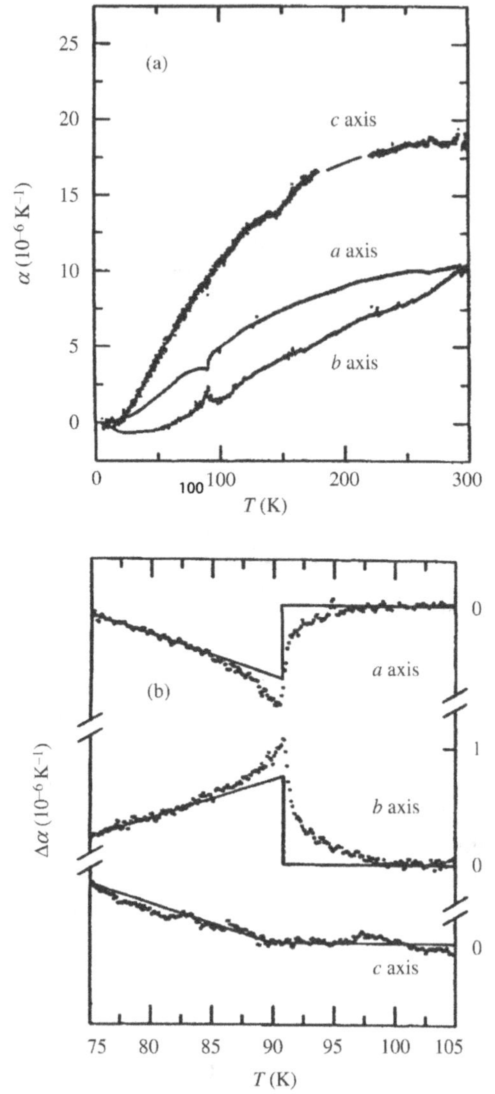

FIGURE C1.19 Linear thermal expansion coefficients α for Y-123: (a) along principal axis and (b) magnified views of Δα near T_c [97].

Y-123, which is reported to be around 370–450 K [95, 98], and thereafter only gradually increases.

Table C1.6 shows typical reported values of the thermal expansion coefficients for RE123 compounds [92–95, 103, 104]. Relatively large scatter in the literature data probably arises from differences in sample quality such as density, porosity, oxygen content and misalignment in crystal orientation.

C1.4.2 Elastic Modulus

For uniaxial compression (or extension) the elastic modulus is defined as the slope of an initial linear portion of the stress–strain curve and is represented by the following equation:

$$\sigma = E\varepsilon \qquad \text{(C1.4)}$$

TABLE C1.6 Thermal Expansion Coefficients, α_a, α_b, α_c, for RE123 Compounds

Sample	α_a ($10^{-5}K^{-1}$)	α_b ($10^{-5}K^{-1}$)	α_c ($10^{-5}K^{-1}$)	Temp. range (K)	Method	Ref.
Y-123	1.1	1.1	2.5	298–873	XRD	[9287]
Y-123	0.7	0.7	1.5	298–1200	XRD	[9388]
Y-123 (O = 6.95)	0.52	0.92	1.57	150–300	XRD	[9489]
Y-123	1.69	–	–	303–1173	Dilatometer	[103]
Y-123	1.67	–	–	–	–	[104]
Y-123/15 vol% Ag	1.72	–	–	–	–	[104]
Nd123	1.3	–	1.9	273–1073	Neutron	[96]
Yb123	1.5	–	–	303–1173	Dilatometer	[102]

where σ is the uniaxial stress, E is Young's modulus and ε is the uniaxial strain. Shear stress τ is proportional to the shear strain (γ):

$$\tau = G\gamma \tag{C1.5}$$

where G is the modulus of rigidity, or shear elastic modulus. Under uniaxial compression, there is generally a transverse expansion, and this is represented by Poisson's ratio (ν) defined by:

$$\nu = (\Delta l / l) / (\Delta t / t) \tag{C1.6}$$

where $\Delta l/l$ is the change in the sample length, and $\Delta t/t$ is the change in its thickness. In plastic deformation and creep, the volume is constant, then v is 0.5. Generally, in elastic deformation, the Poisson ratio varies from 0.2 to 0.3. For isotropic or cubic systems Poisson's ratio can be expressed in terms of the Young's modulus (E) and the shear elastic modulus (G) as follows:

$$\nu = E/(2G-1) \tag{C1.7}$$

While limited to isotropic or cubic systems, this formula can be a good approximation for polycrystalline ceramics. More generally, the elastic stress and strain should each be expressed as a second-rank tensor and the moduli connecting them as a fourth-rank tensor [105].

Elastic modulus values are usually determined by measurement of the velocity of ultrasonic waves using the pulse-echo technique [106–115], or the vibrating reed technique [4, 116–118].

Here, the shear modulus (G) and the bulk modulus (B) are expressed using the longitudinal sound velocity, v_L, transverse sound velocity, v_S, and the density, ρ, as follows:

$$G = \rho v_S^2; \quad B = (3\rho v_L^2 - 4G)/3 \tag{C1.8}$$

When the material is isotropic:

$$E = 9BG/(3B+G) \tag{C1.9}$$

Table C1.7 shows reported data for various elastic moduli E, G and B and Poisson's ratio for RE123 [104–108]. Not unexpectedly, these values are strongly sensitive to the oxygen content, because the orthorhombic phase is significantly stiffer than the tetragonal phase [102, 112]. The temperature dependence of both the Young's modulus and shear modulus reveal soft-mode behaviour in the neighbourhood of the orthorhombic to tetragonal transition due to fluctuations in orthorhombicity [4]. There is also an elastic damping peak arising from chain oxygen atoms jumping resonantly between O1 and O5 sites which can be used to measure the oxygen diffusion coefficient [4,119]. It was found in this way that increasing the size of the rare-earth element greatly accelerated oxygen diffusion due to the reduction of hopping barriers [119].

TABLE C1.7 Elastic Moduli E, B, G and Poisson Ratio v for Various RE123

Sample	Density (g/cm³)	Porosity (%)	x	E [GPa]	B [GPa]	G [GPa]	V	Ref.
YBCO[a]	5.952			95.89	41.79	45.3	0.14	[106]
Y-123	6.07	4.2		104		51.6	0.163	[108]
Y-123/15 vol% Ag	6.58	5.4		119		91	0.282	[108]
Y-123	6.171		6.2	91.3	38.5	48.6	0.187	[107]
Y-123	6.381		6.91	126.2	50.5	84.5	0.25	[107]
Y-123				116			0.26	[104]
Y-123/15 vol% Ag				105			0.285	[104]

[a] Melt-processed Y-123/Y211.

C1.4.3 Fracture Toughness

Fracture toughness describes the resistance of a material against crack propagation. For a brittle material like RE123, this is generally characterized using the Vickers indentation method (ID) [116–125] or the single-edge notch beam techniques (SENB) [104, 126–131].

In the indentation method, the threshold stress intensity factor K_{th} (=K_c^{air}), which is calculated from an impression half-diagonal (a), a radial/median crack length (c) and indentation peak load (P) as follows [125]:

$$K_{th} = x(E/H_v)^{1/2} \times P/c^{3/2} \quad \text{(C1.10)}$$

$$H_v = 1.854P/(2a)^2 \quad \text{(C1.11)}$$

where x is 0.016 and a substantially material-independent constant, E is Young's modulus and H_v is the Vickers hardness. Cooks et al [124] chose the value of $E/H = 40$ for ceramics with small elastic recovery, and $K_{th} = 0.74$ [MPam$^{1/2}$] was estimated for Y-123.

Table C1.8 shows the reported fracture toughness and hardness obtained by the ID and SENB methods. Hardness is unaffected by the presence of twin boundaries and moisture whereas the fracture toughness is enhanced by twin boundaries, and it was found that moisture in the air promotes crack growth, thereby degrading K_c [121]. Compared with single crystals, relatively low H_v values in ceramic materials are attributed to the typically high porosity [123]. It is reported that addition of Ag and excess RE211 phase is effective in increasing the fracture toughness of RE123 [102, 99].

C1.4.4 Fracture Strength

Unlike metals, the absence of multiple active slip systems makes ceramic RE123 materials very brittle and they fracture at the elastic limit, which defines the fracture strength. When stressed in tension, the stress at which the sample breaks is called the tensile strength. When stressed in bending or in compression, the stresses are called the flexural strength and the compression strength, respectively. Generally, the compression strength of conventional ceramics is an order of magnitude larger than the tensile strength, and the flexural strength is slightly larger than the tensile strength, which depends on the amount of defects. A bending test is usually employed to estimate the fracture strength for brittle materials using three-point or four-point bending test techniques [131–136]. The flexural strength (σ_f) is given by [135]:

$$\sigma_f = 14.72LP/Bt^2 \quad \text{(C1.12)}$$

where L, P, B and t are the span, load, width and thickness of a bar-shaped sample, respectively. Table C1.9 shows reported

TABLE C1.8 Hardness H, Fracture Toughness K_c and Young's Modulus E for RE123

Sample	H [GPa]	K_c^{air} [MPa m]	E [GPa]	Method	Ref.
Y-123[a] twinned (1 0 0)/(0 1 0) on {0 0 1}	8.7	0.74	($E/H = 40$)	ID	[124]
Y-123[a] detwinned (1 0 0) on {0 0 1}	9.4	0.59	157	ID	[121]
Y-123[a] detwinned (0 1 0) on {0 0 1}	9.4	0.47	157	ID	[121]
Y-123[a] twinned (1 0 0)/(0 1 0) on {0 0 1}	9.5	0.66	157	ID	[121]
Y-123[a] twinned (0 0 1) on {1 0 0}/{0 1 0}	9.8	0.8	157	ID	[121]
Y-123[a] twinned (1 0 0) on {1 0 0}/{0 1 0}	9.8	0.32	89	ID	[121]
Y-123[b] (1 0 0)/(0 1 0) on {0 0 1}	6.7	0.67	182	ID	[122]
Y-123[b] (1 0 0)/(0 1 0) on {0 0 1}	7.35	0.84	($E/H = 40$)	ID	[119]
YBCO[c] (1 0 0)/(0 1 0) on {0 0 1}		1.48	($E/H = 40$)	ID	[121]
Nd123[a] (1 0 0)/(0 1 0) on {0 0 1}	7.79	0.703	158	ID	[120]
Y-123		1.8	116	SENB	[104]
Y-123/15 wt% Ag		3.6	105	SENB	[104]
Y-123/19 wt% Ag		3.6		SENB	[129]
Y-123 ($n = 18.4\%$)	2.1	0.71	66.8	ID	[123]
Y-123 ($n = 11.8\%$)	4.4	0.87	76.5	ID	[123]
Y-123/0.1 mol% Ag_2O ($n = 9.0\%$)	3.9	1.37	80.6	ID	[123]
Y-123 ($n = 33\%$)		1.05		SENB	[130]
Y-123 ($n = 13\%$)		1.4		SENB	[130]
Y-123/30 wt% Ag ($n = 13\%$)		2.4		SENB	[130]

ID, indentation; SENB, single-edge notch beam; n, porosity.

[a] Single crystal.

[b] Melt-processed sample.

[c] Melt-processed Y-123 with a large amount of Y-211.

TABLE C1.9 Fracture Strength σ_f of RE123

Sample	σ_f (MPa)	E (Gpa)[a]	Method	Ref.
Y-123	40	81	3-Point	[135]
Y-123/15 wt% Ag	60	96	3-Point	[135]
YBCO/15 wt% Ag[b]	70	100	3-Point	[135]
Y-123	87	–	4-Point	[133]
Y-123/10 vol% Ag	116	–	4-Point	[133]
Y-123/30 vol% Ag	136	–	4-point	[133]
Y-123	38	–	3-Point	[136]
Y-123/10 vol% Ag + 0.8 mol% Zr	280	–	3-Point	[136]
YBCO[b]	28	–	3-Point	[134]
YBCO[b] unannealed	110	130	3-Point	[132]
YBCO[b] annealed	77	128	3-Point	[132]

[a] Calculated from stress–strain curves.
[b] Melt-processed sample. (Y-123 + Y-211).

data for the flexural strengths of RE123 compounds. Here, E is calculated from the slope of a linear portion of the load–displacement curve for bent bars.

The fracture strength of Y-123 is reported to range from 40 to 200 MPa, depending on the sample quality. Again, the scatter in the data is attributed to variation in the porosity, oxygen content and the amount of cracking in the virgin sample [132–136].

Additions of second-phase material is often effective in increasing the fracture strength of the RE123 matrix. For instance, the flexural strength is substantially increased with Ag addition [104, 135]. When metal inclusions like Ag or Au are dispersed in the matrix, crack tips can be blunted since these metals are ductile materials [133, 137]. It is also probable that Ag particles induce a compressive stress which resists crack propagation at the crack tip [133]. Addition of the RE211 phase or ZrO_2 has also been reported to be effective in improving mechanical properties [136].

C1.4.5 Plastic Deformation

Oxide superconductors are generally so brittle that plastic deformation such as drawing or rolling is impossible at room temperature. (The noted exception is the BSCCO family, $Bi_2Sr_2Ca_{n-1}Cu_nO_{4+2n}$, which have a micaceous structure due to the weak bonding between adjacent BiO layers. This enables the powder-in-tube technique for manufacturing BSCCO wires). However, RE123 materials are known to deform plastically at temperatures above 800°C where the yield stress markedly decreases with increasing temperature. Y-123 deforms more than 50% in compressive strain at temperatures above 840°C as shown in Figure C1.20 [138]. The flow stress gradually decreases as the deformation proceeds after yielding in a temperature range between 840 and 900°C. This seems to be a kind of yield–drop phenomenon where the density of mobile dislocations increases during deformation.

In general, in order to deform polycrystals, it is necessary for five independent slip systems to be activated, as pointed out by von Mises [139]. Since the major slip system of ⟨0 1 0⟩ (0 0 1) has only two independent slip systems, secondary slip systems must operate to satisfy the von Mises condition. As suggested by Suzuki et al [140], a possible secondary system is the ⟨0 3 1⟩ {1 1 3}, which is active above 817°C. Above this temperature, plastic deformation occurs with the motion of dislocations in these two slip systems.

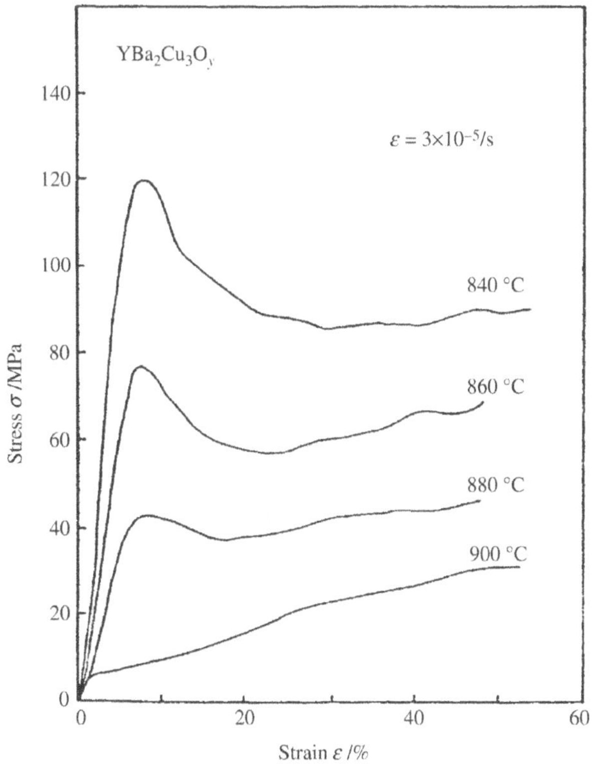

FIGURE C1.20 Stress–strain curves for Y-123 in compression above 840°C [138].

C1.5 Chemical Properties

C1.5.1 Phase Diagrams for YBCO Materials

Knowledge of the phase diagram of the YBCO system is of great importance for preparation procedures for single crystals and films [141–152]. The most familiar diagrams published are top-down projections of the three-component plane. Figure C1.21 is typical and shows the phase equilibrium relations in the ternary system $YO_{1.5}$–BaO–CuO [150]. It is an isotherm at 900°C (in air), and it includes the tie lines connecting the most important phases. The stable compounds in the composition triangle at 900°C are found to be Y-123, Y_2BaCuO_5, $YBa_3Cu_2O_x$, $BaCuO_2$ and $Y_2Ba_2O_5$ [150]. YBCO is a quaternary system, which makes its investigation rather complicated. In order to make phase relationships more clear, the sections of the ternary Y_2BaCuO_5 (2 1 1), $Ba_2Cu_3O_5$ (0 2 3), CuO (0 0 1) system along the 123-CuO tie line have been presented. Figure C1.22 shows a part of the ternary system with vertical *T* (temperature) axes at $P(O_2)$ = 1 bar [149]. All three superconducting YBCO phases, Y-123, Y-124 and Y-247, are stable at $P(O_2)$ = 1 bar but for a limited temperature range only. In reality, the phase diagram changes depending on whether the YBCO is being prepared in air or in oxygen at various partial pressures.

C1.5.2 Temperature/Pressure/Composition Phase Behaviour

In order to synthesize ceramic samples of high quality and to grow single crystals of YBCO, a knowledge of phase relations and phase stability as a function of temperature and oxygen partial pressure is of considerable importance. In particular, the occurrence of melts is either a restriction or a precondition to guarantee optimum manufacturing conditions. The peritectic melting point, T_m, of Y-123 has been investigated as a function of oxygen partial pressure [153–156]. It was concluded that the melting point is a kind of reduction reaction in the form [155]:

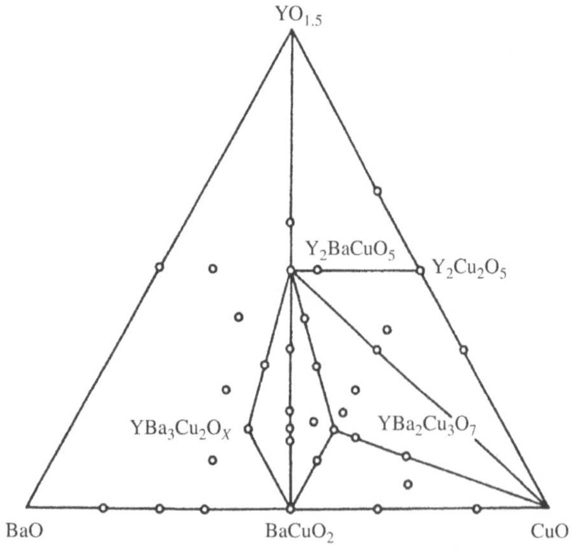

FIGURE C1.21 The phase equilibrium relations at 900°C in the ternary system YO1.5–BaO–CuO. Open circles indicate the compositions at which the phase identification experiments were performed. The tie-like triangles are shown by solid lines in the composition triangle [150].

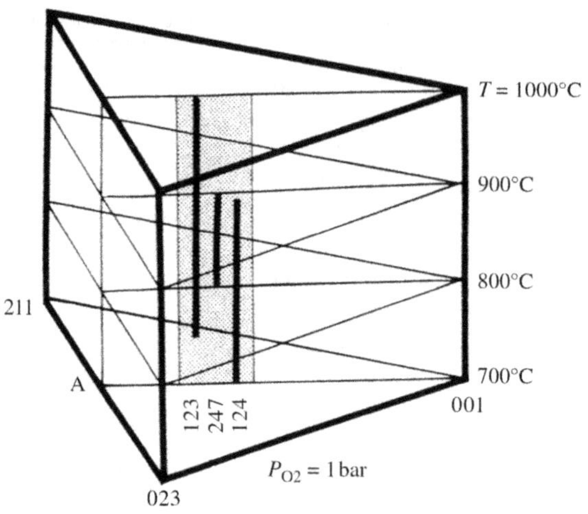

FIGURE C1.22 T–x section of ternary Y_2BaCuO_5–$Ba_2Cu_3O_5$–CuO system at $P(O_2)$ = 1 bar [149].

$$\text{Solid(I)} = \text{Solid(II)} + \text{Liquid} + O_2 \qquad \text{(C1.13)}$$

Since the equilibrium is shifted to the left when the oxygen partial pressure is increased, the melting point increases under these conditions. The melting point of Y-123 for different oxygen partial pressures was found to be: T_m = 1055°C under P_{O2} = 1.0 atm; T_m = 1017°C under P_{O2} = 0.1 atm; and T_m = 984°C at P_{O2} = 0.01 atm [155]. However, different values have been reported [153–156] for the peritectic melting point of Y-123 using the same atmosphere. The difference in the reported values is probably due to experimental factors and powder characteristics including impurity, particle size and CO_2-content. The phase diagram of YBCO with the composition of Y-247 was summarized in Figure C1.5 as a function of temperature and oxygen partial pressure over the range 600–1060°C and $10^{-4} \leq P(O_2) \leq 10^{+2}$ atm [46]. Other studies report somewhat similar results [143, 145, 149, 151, 153] with the following conclusions: (i) from the starting composition of 123, the 247 phase is not formed at any temperature, except for a narrow coexistence range where Y-247 and Y-124 both appear. (ii) From starting compositions of 247, 124 and 125, the 247 phase is only formed at intermediate temperature between the 123 and 124 stability regions. (iii) Under $P(O_2)$ = 1 atm, the phase boundary temperatures of 123/247 and 247/124 are about 870 and 850°C, respectively, so that the stability range for Y-247 is remarkably narrow. Finally, (iv) the tetragonal–orthorhombic phase boundary of 123 lies to the right of the 123/124 phase boundary within the 124 phase stability region. This result is evidence that 123 is stable only in its tetragonal phase (insulating), and the orthorhombic phase (superconducting) is metastable at all temperatures. The 124 phase is more stable than the 123 orthorhombic phase.

The relationship between YBCO phase stability temperature, oxygen partial pressure and *absolute pressure* has been reported by Sawai et al [157].

C1.5.3 Phase Diagram for Y-123 Materials Versus Oxygen: Chemical Compatibility

The influence of oxygen stoichiometry on the properties of YBCO, especially with respect to the tetragonal (T) to orthorhombic (O) phase transformation, has been the subject of much research. The relationship between phase stability of Y-123 type superconductors and oxygen content for various values of $P(O_2)$ in the form of a phase diagram has been reported [21, 149, 158–162]. For example, the temperature–oxygen content diagram for Y-123 is shown in Figure C1.23 [162]. Long-dashed isobars in this figure represent the temperature dependence of equilibrium of the solid solution, and short dashed lines represent metastable equilibrium of the orthorhombic and tetragonal phases as well as a metastable miscibility gap of solid solution. Bold lines show the equilibrium conditions for several decomposition reactions of 123, which outline its thermodynamic stability range. These reactions are [162]:

$$9YBa_2Cu_3O_{6+z} = 4Y_2BaCuO_5 + YBa_4Cu_3O_{8.5+q} + 10BaCu_2O_2 + [(9z+5.5-q)/2]O_2 \quad (C1.14)$$

$$2YBa_2Cu_3O_{6+z} = Y_2BaCuO_5 + BaCuO_2 + 2BaCu_2O_{2+[(2z+1)/2]}O_2 \quad (C1.15)$$

$$6YBa_2Cu_3O_{6+z} = Y_2BaCuO_5 + 3BaCuO_2 + 2Y_2Ba_4Cu_7O_{14+w} + [(6z-2w-3)/2]O_2 \quad (C1.16)$$

$$6YBa_2Cu_3O_{6+z} = 2Y_2BaCuO_5 + 3Ba_2Cu_3O_{5+y} + Y_2Ba_4Cu_7O_{14+w} + [(6z-3-3y-w)/2]O_2 \quad (C1.17)$$

$$YBa_2Cu_3O_{6+z} \rightarrow Y_2BaCuO_5 + \text{Liquid} + O_2 \quad (C1.18)$$

$$4YBa_2Cu_3O_{6+z} = 2Y_2BaCuO_5 + 3Ba_2Cu_3O_{5+y} + CuO + [(4z-2-3y)/2]O_2 \quad (C1.19)$$

$$2YBa_2Cu_3O_{6+z} = Y_2BaCuO_5 + 3BaCuO_2 + 2CuO + [(2z-1)/2]O_2 \quad (C1.20)$$

The theoretical and experimental relationship between temperature, oxygen partial pressure and oxygen content, x, in $YBa_2Cu_3O_{7-x}$ is discussed in [161].

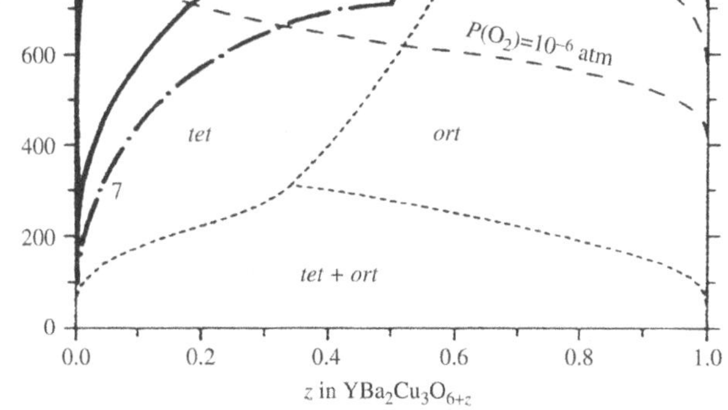

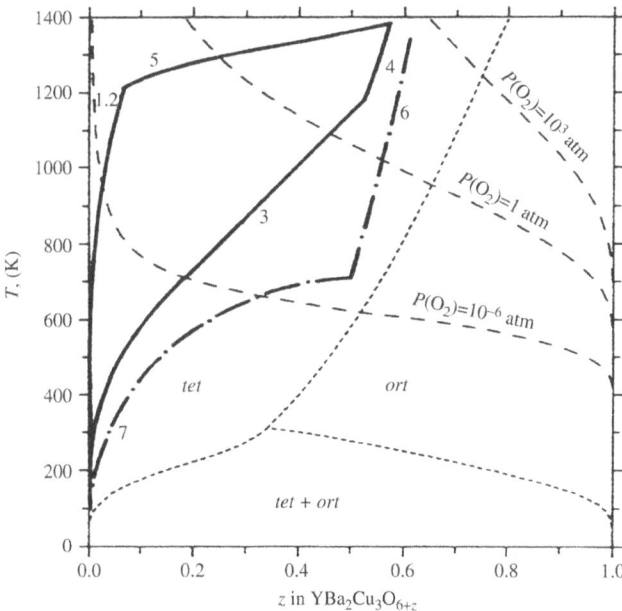

FIGURE C1.23 Stability field of the Y-123 phase. This phase is thermodynamically stable between the bold lines 1–5. The lines 1–7 represent decomposition reactions to Equations (C1.15)–(C1.21) [162].

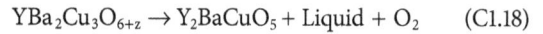

C1.6 Optical Properties

As is commonly observed in all high-temperature superconductors, the reflectivity spectrum $R(\omega)$ and, thus, the conductivity spectrum $\sigma_1(\omega)$ change dramatically with hole doping for all polarization directions. Figure C1.24 shows the doping dependence (namely, oxygen content dependence)

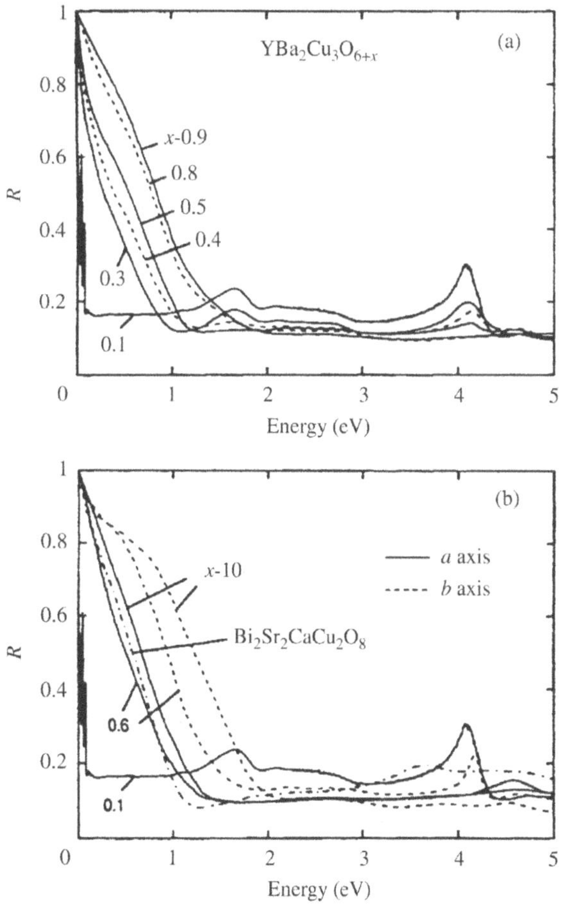

FIGURE C1.24 Room temperature reflectivity spectra of twinned (a) and untwinned $YBa_2Cu_3O_{6+x}$ crystals (b) with $E//a$ and $E//b$ for various x [163].

of the reflectivity spectrum for twinned and untwinned $YBa_2Cu_3O_{7-x}$ for *E//a* and *E//b* [163]. For the low oxygen content (x = 0.1), the reflectivity is very low, and the spectrum is flat below 1 eV except for the phonon absorption peaks below 0.1 eV, which are typical in an insulator. At higher frequencies, the spectrum is characterized by a charge transfer gap excitation at around 1.7 eV and some interband excitations at higher energies. With increasing oxygen content, the reflectivity increases below 1 eV for *E//a* and below 2 eV for *E//b*, forming a Drude-like metallic spectrum, compensating a reduction in the charge transfer absorption. Similar growth of the low-ω spectrum is also seen for *E//c* (see Figure C1.25) [164]. As the oxygen content increases, the insulator-like spectrum at $x \sim 0.1$, dominated by the far-infrared phonon peaks, changes into a Drude-like spectrum at $x \sim 0.9$. All these changes indicate that the material shifts with increasing doping from an insulator to a metal.

When the temperature is lowered, the far-infrared reflectivity and/or conductivity increase in highly oxygenated Y-123 consistent with the metallic resistivity $\rho(T)$. In the superconducting state, the reflectivity is further enhanced and the conductivity suppressed below 2Δ (where Δ is the maximum gap amplitude). The *T*-dependencies of the far-infrared conductivity spectra are shown in Figure C1.26 for *E//a* [165] and in Figure C1.27 for *E//c* [79].

At high doping ($x \sim 0.9$), the conductivity suppression just sets in at T_c both for *E//a* and *E//c*, which indicates that the observed suppression is related to the opening of a superconducting gap. Although there is still an argument that the conductivity suppression for *E//a* is a result of Drude narrowing of quasiparticles below T_c and therefore, the absorption edge around 500 cm^{-1} is not related to a gap, one can roughly estimate a maximum 2Δ gap energy to be 600 cm^{-1} ($2\Delta \sim 74$ meV)

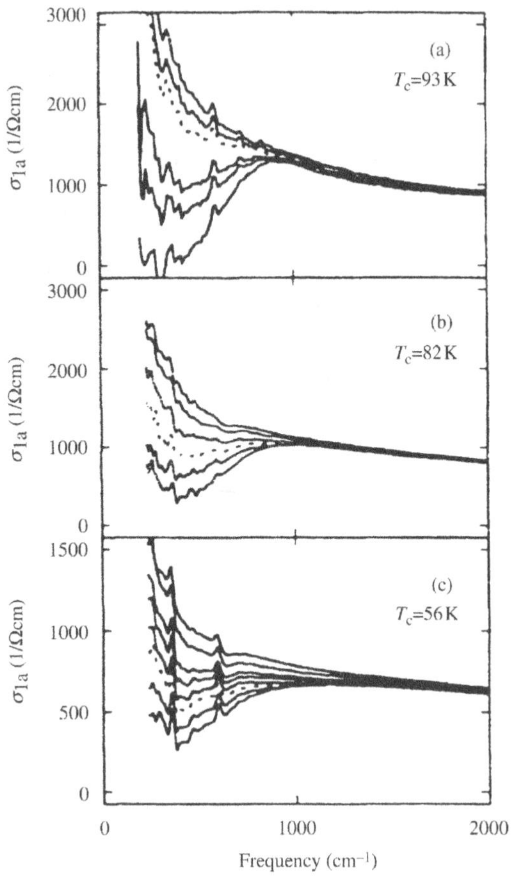

FIGURE C1.26 Temperature dependence of conductivity spectrum of $YBa_2Cu_3O_y$ with *E//a*. (a) Optimally doped crystal with T_c = 93 K for *T* = 120, 100, 90 (dashed), 80, 70 and 30 K (from top to bottom). (*b*) Underdoped crystal with T_c = 82 K for *T* = 150, 120, 90, 80 (dashed), 70 and 20 K (from top to bottom). (*c*) Underdoped crystal with T_c = 56 K for *T* = 200, 150, 120, 100, 80, 60 (dashed), 50 and 20 K (from top to bottom) [165].

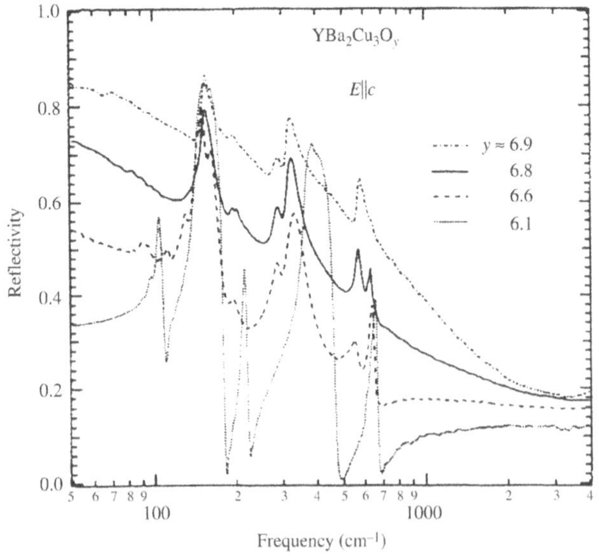

FIGURE C1.25 Room temperature reflectivity spectra of $YBa_2Cu_3O_y$ with *E//c* for various oxygen contents [164].

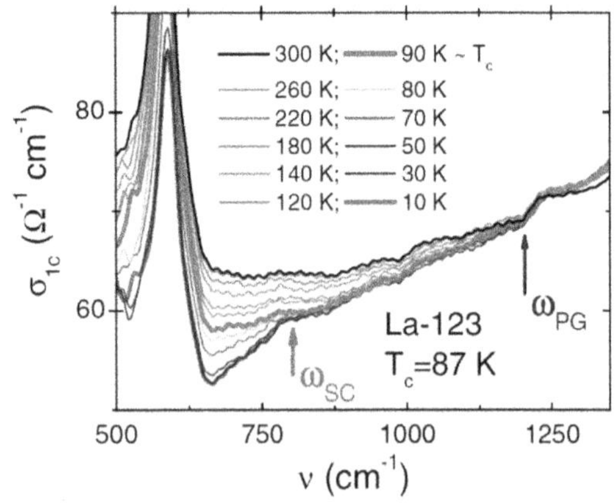

FIGURE C1.27 The frequency dependence of the *c*-axis infrared conductivity of La-123 as measured by ellipsometry [79]. The suppression of spectral weight below $\omega_{SC} \approx 780$ cm^{-1} only below T_c and below $\omega_{PG} \approx 1150$ cm^{-1} from temperatures as high as 300 K clearly identify the superconducting pair-breaking gap, 2Δ, and the distinct pseudogap.

consistent with ARPES [81]. The long tail below 2Δ is consistent with an anisotropic gap such as *d*-wave.

For underdoped crystals with $T_c \sim 50$–70 K, the suppression of conductivity begins well above T_c. As seen in Figure C1.27, this is particularly notable for $E//c$ where the matrix elements for *c*-axis hopping are maximum at the $(0,\pi)$ and $(\pi,0)$ points on the Brillouin zone. This is precisely where the pseudogap is located and the suppression in $\sigma(\omega)$ commencing already at room temperature is caused by the presence of the pseudogap [79] which is observed in specific heat [69] and ARPES [81]. The figure shows the separate and distinct opening of the superconducting gap at ω_{SC} and the pseudogap at ω_{PG}, where the former opens at T_c and the latter above room temperature. The data is for La-123 and the authors have mapped out the doping dependence of the two gaps in this way. It is very consistent with that obtained from the specific heat [69]. This pseudogap feature is closely correlated with the dc *c*-axis resistivity behaviour which exhibits an upturn at a certain temperature that is also associated with the opening of the pseudogap.

C1.7 Normal-State Properties

C1.7.1 Transport Properties

As summarized in the electronic phase diagram shown in Figure C1.7, when the oxygen content *y* is close to six, $YBa_2Cu_3O_y$ is an antiferromagnetic (AF) insulator due to strong electron–electron correlation within the CuO_2 plane. As holes are doped into the CuO_2 planes with increasing oxygen content, the Néel temperature falls and the material changes from an insulator to a metal, but short-range AF correlations persist to much higher doping. This electronic change caused by hole doping cannot be described within a rigid band picture for a doped Mott insulator, but suggests a more radical change in the electronic structure. All anomalies in the normal-state properties originate from this puzzling electronic state in the crossover region, which is predominantly characterized by the so-called 'pseudogap' which exists well above T_c. The pseudogap regime is usually delimited on the electronic phase diagram by a temperature T^*, and though there is some suggestion, this might identify a phase transition line is more useful and possibly more accurate to describe this as an energy scale $E^* = k_B T^*$. E^* descends with increasing doping. A number of recent studies show that the pseudogap closes rather abruptly at a critical doping $p^* \approx 0.19$ holes/Cu which happens in the case of $YBa_2Cu_3O_{7-x}$ to correspond to $x \approx 0$. This is probably a happy coincidence because the *x* value to reach p^* can easily be shifted by as much as 0.6, for example, by Ca or La substitution [47]. But for pure Y-123, one need to fully oxygenate the sample in order to close the pseudogap and maximize critical fields and critical currents [55].

With increasing oxygen content, resistivity decreases along all crystallographic directions as shown in Figures C1.28 and C1.29 [166,167]. For optimal doping around $y \sim 6.88$,

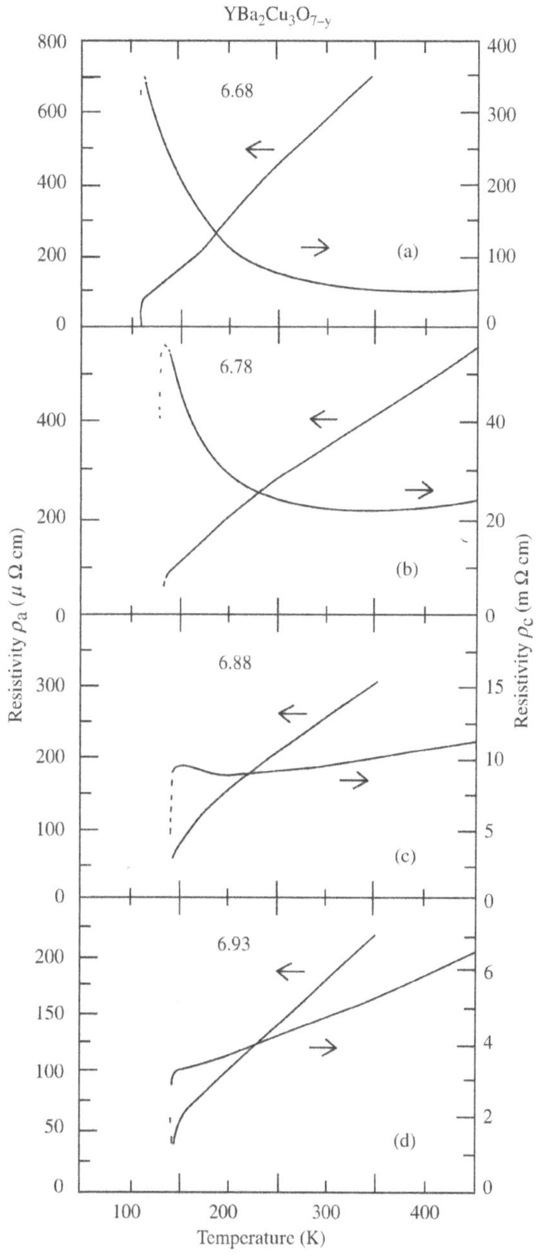

FIGURE C1.28 Temperature dependence of in-plane (ρ_α) and out-of-plane (ρ_β) resistivity of untwinned $YBa_2Cu_3O_{7-y}$ for various oxygen contents [166].

the in-plane resistivity $\rho_a(T)$ is almost linearly temperature dependent with a small downturn just above T_c due to superconducting fluctuations, and $\rho_b(T)$ contains a small T^2-component arising from the conductivity of the chains, while the *c*-axis resistivity $\rho_c(T)$ has a large residual component and shows a slight upturn just above T_c. The effect of the pseudogap manifests itself in the downward bending of $\rho_a(T)$ (see Figure C1.29), which is due to a reduction in carrier scattering caused by a normal-state energy gap [166] (there are less states available to scatter into). At a low doping level, $\rho_c(T)$ shows

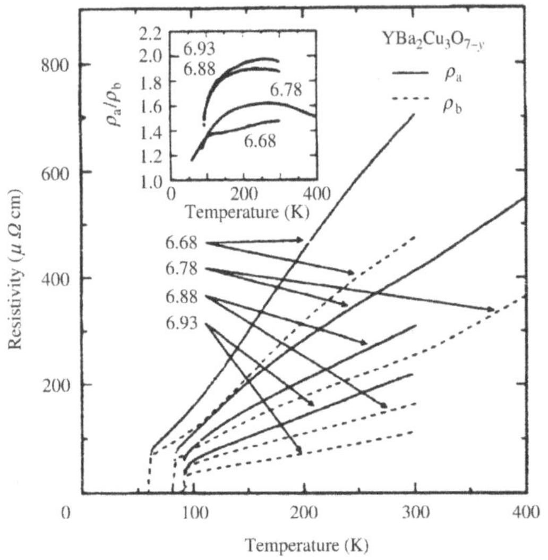

FIGURE C1.29 Temperature dependence of two in-plane resistivity components in untwinned $YBa_2Cu_3O_{7-y}$ for various oxygen contents [166].

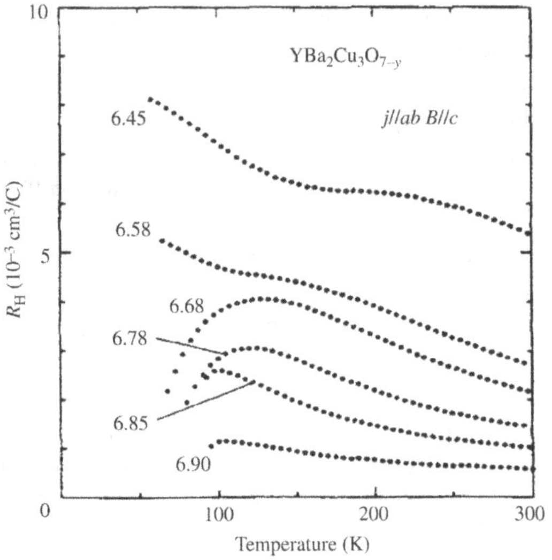

FIGURE C1.30 Temperature dependence of Hall coefficient of two in-plane resistivity components in untwinned $YBa_2Cu_3O_{7-y}$ single crystals for various oxygen contents [169].

a more pronounced semiconductor-like upturn, whereas $\rho_a(T)$ remains still metallic. This anisotropy in the temperature dependence of resistivity is regarded as one of the most anomalous properties of high-temperature superconductors, which cannot be understood in the framework of a conventional metal. The strange anisotropy can be seen in the ratio ρ_c/ρ_a as well. First, the ratio is much larger than the effective mass ratio estimated from the band calculation even in the overdoped crystals. Second, it changes dramatically with temperature and hole doping. This implies that the effective mass model does not hold in this compound. As intimated above, the anisotropy probably arises from the anisotropy of the pseudogap, which is localized at the Brillouin zone boundary near the (0,π) points, combined with the matrix elements for interlayer hopping which maximize at the same points of the Brillouin zone [168]. Thus, the pseudogap has a profound effect on *c*-axis transport and a lesser effect on in-plane transport.

In a conventional metal, the Hall coefficient is almost constant for all temperatures. But, as is shown in Figure C1.30, the Hall coefficient of YBCO increases with decreasing temperature, forming a bump at an intermediate temperature [169]. On the other hand, the Seebeck coefficient, which should change linearly with temperature in a conventional metal, is only weakly temperature dependent (see Figure C1.31) [167]. Both quantities strongly depend on oxygen content. Anomalous T-dependencies are typically observed in the underdoped crystals with low oxygen contents. These non-monotonic behaviours are again attributed to the presence of the pseudogap observable below the temperature T^*, which decreases with increasing oxygen content.

C1.7.2 Pseudogap

What then is the nature and origin of the pseudogap? For a long time it was thought to be some form of precursor superconductivity in the form of real-space pairing above T_c or a phase-incoherent superconducting state. The alternative view was that the pseudogap coexisted and competed with superconductivity [71, 170]. It is probably fair to say that the current consensus is building around the latter view.

Recent very high-field resistivity and Hall effect measurements [171] indicate that the normal-state ground state

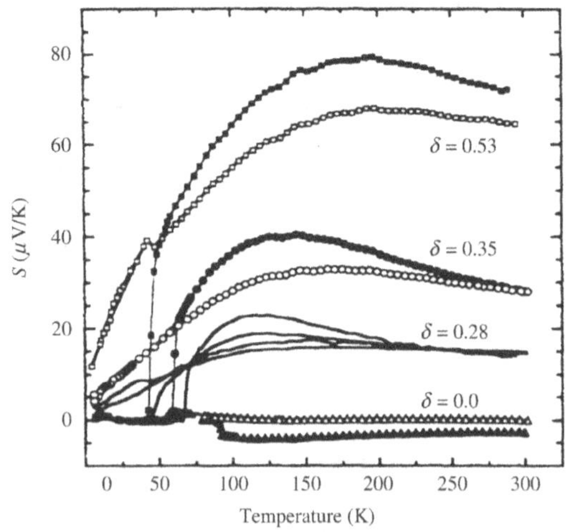

FIGURE C1.31 Temperature dependence of thermopower in $YBa_2Cu_3O_{7-\delta}$ polycrystals for various oxygen contents. Solid symbols denote non-substituted and open symbols denote 3.0% Zn-substituted YBCO. The family of solid curves for $\delta = 0.28$ in the downward order denote 0, 1.0, 3.0 and 5.0% Zn-substitution [167].

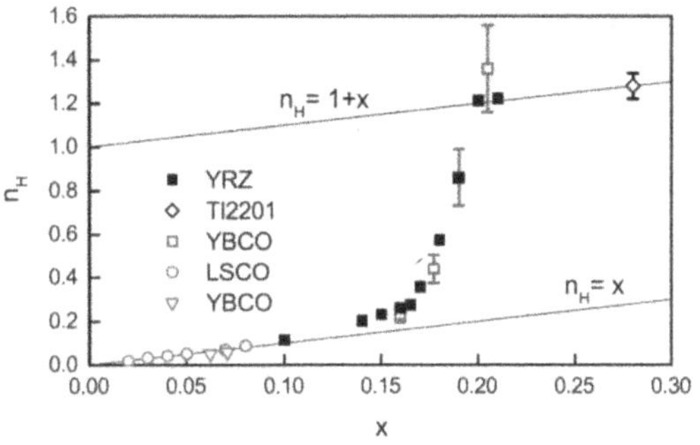

FIGURE C1.32 The doping dependence of the Hall number, n_H, for Y-123 and other cuprates (open symbols) showing the crossover from $n_H = p$ to $n_H = (1{+}p)$ at the pseudogap critical point $p^* \approx 0.19$ holes/Cu. Solid squares are the values calculated [172] using the Yang, Rice and Zhang model [173].

crosses over from an effective carrier concentration of p carriers per Cu atom in the underdoped region to (1+p) carriers per Cu in the overdoped region when the pseudogap closes. As shown in Figure C1.32, the crossover is precisely at the pseudogap critical point $p^* \approx 0.19$ holes/Cu. Theoretical calculations [172] show that the behaviour is quantitatively consistent with a reconstruction of the Fermi surface from a large hole-like Fermi surface (of area 1+p) to one comprising smaller hole pockets (of total area p) - see solid square data points in Figure C1.32. Such a reconstruction is consistent with short-range AF fluctuations or Umklapp scattering at the (0,π) points in the Brillouin zone [173]. The gradualness of the crossover (spread over ≈ 0.04 holes/Cu) is due to the intermediate appearance of additional small electron pockets at (0,π) and (π,0).

C1.8 Superconducting Properties

C1.8.1 Fundamental Superconducting Parameters

The large anisotropy in Y-123 has an important effect on the equilibrium flux-lattice properties, and hence the flux-pinning behaviour and field-dependent critical current. The key defining parameters here are the London penetration depth, λ, and the coherence length, ξ. The penetration depth, which measures the length over which magnetic fields are attenuated near the surface of a superconductor, is related to the effective mass and density of superconducting pairs. The inverse square of the penetration depth, $\rho_s = \lambda^{-2}$, is often loosely referred to as the superfluid density because $\lambda^{-2} = \mu_0 e^2(n_s/m^*)$, where n_s is the true superfluid density. The magnitude of $\rho_s = \lambda^{-2}$ plays a central role in controlling fluctuations [174] - the larger the ρ_s, the smaller is the fluctuation amplitude. The coherence length measures the effective size of Cooper pairs, or perhaps more usefully, the size of flux vortex core. The penetration depth and coherence length are the fundamental characteristic lengths of superconductors [175] that define their superconducting properties. The temperature dependence of the coherence length ξ(T) in the critical region is described by a simple power law [176]:

$$\xi(T) = \xi(0)[1-(T/T_c)]^{-1/2}, \quad \text{(C1.21)}$$

and the penetration depth has the same T-dependence so that κ near T_c is T-independent.

The conventional Ginzburg–Landau (GL) theory is very successful in describing the temperature dependence of several thermodynamic properties of superconductors, such as the thermodynamic critical field, $H_c = \phi_0/(2\sqrt{2}\mu_0\pi\lambda\xi)$, and other characteristic parameters [177]. For an isotropic material, the GL parameter is defined as:

$$\kappa = \lambda/\xi. \quad \text{(C1.22)}$$

For anisotropic superconductors, $\xi = (\xi_a\ \xi_b\ \xi_c)^{1/3}$ and $\lambda = (\lambda_a\lambda_b\lambda_c)^{1/3}$ [175]. Since $\xi_a \approx \xi_b$ and $\lambda_a \approx \lambda_b$ for YBCO, the following definitions are introduced:

$$\lambda_{ab} = \lambda_c/\gamma \quad \text{(C1.23)}$$

$$\xi_{ab} = \gamma\xi_c \quad \text{(C1.24)}$$

$$\kappa_{ab} = \gamma^{-2}\kappa_c \quad \text{(C1.25)}$$

where γ is the anisotropy factor. These definitions ensure that the condensation energy $U_0 = (1/2)\mu_0 H_c^2 \propto (\lambda\xi)^{-2}$ is fully isotropic, as it must. This leads to the relations:

$$\xi_a\lambda_a = \xi_b\lambda_b = \xi_{ab}\lambda_{ab} = \xi_c\lambda_c \quad \text{(C1.26)}$$

The characteristic lengths also define the lower and upper critical fields, $H_{c1} = \phi_0(\ln\kappa+½)/(4\pi\mu_0\lambda^2)$ and $H_{c2} = \phi_0/(2\pi\mu_0\xi^2)$, respectively [174], so that for the anisotropic case

$$B_{c2,ab} = \gamma B_{c2,c} \quad \text{(C1.27)}$$

and

$$B_{c1,ab} = \gamma^{-1}[(\ln\kappa_{ab}+½)/(\ln\kappa_c+½)]B_{c1,c} \quad \text{(C1.28)}$$

Table C1.10 summarizes the full anisotropic data for the most important parameters of YBCO: T_c, ξ, λ, κ, H_{c1} and H_{c2}. These values are for fully oxygenated Y-123, which is overdoped, and the T_c value of 89 K has already reduced from its maximum value 93 K at optimum doping. The doping dependence of λ (or $\rho_s = \lambda^{-2}$) is reported in [64, 174, 178]. Note that, typically, a:b anisotropy in orthorhombic Y-123 crystals is not always easily detectable in measurements of ξ_i and λ_i, but it is in fact quite large (up to a factor of 2), and λ_a:λ_b anisotropy has been successfully extracted from muon spin relaxation experiments [64]. As noted earlier, it agrees well with optical measurements [63].

TABLE C1.10 Transition Temperature (T_c), and Anisotropic Coherence Length (ξ), Penetration Depth (λ), GL Coefficient (κ), and Lower (H_{c1}) and Upper (H_{c2}) Critical Field for Fully Oxygenated YBCO[a]

Parameter	Value	Ref.	Notes
T_c (K)	89	[179]	(Fully oxygenated Y-123 is overdoped)
Penetration depth (nm):			
λ_{ab}	125	[178]	(Muon spin relaxation)
λ_a	155	[64]	(Muon spin relaxation)
λ_b	80	[64]	(Muon spin relaxation)
λ_c	450	[180]	(Magnetization)
	1100	[181]	(Microwave)
Coherence length (nm):			
ξ_{ab}	1.3	[180]	(Magnetization)
	2.0	[178]	(Muon spin relaxation)
ξ_a	1.05		(Calculated from $\xi_a\lambda_a = \xi_b\lambda_b = \xi_{ab}\lambda_{ab}$)
ξ_b	2.0		(Calculated from $\xi_a\lambda_a = \xi_b\lambda_b = \xi_{ab}\lambda_{ab}$)
ξ_c	0.2	[180]	(Magnetization)
	0.15		(Calculated from $\xi_c\lambda_c = \xi_{ab}\lambda_{ab}$)
GL parameter			
κ_{ab}	462	[180]	(From λ_c above)
	723	[181]	(From λ_c above)
κ_c	95	[182]	
Lower critical field (mT)			
B_{c1}(ab)	18	[180]	(Magnetization)
B_{c1}(c)	53	[180]	(Magnetization)
Upper critical field (T)			
B_{c2}(ab)	256	[183]	(Magnetization projected to T=0)
B_{c2}(c)	53	[183]	(Magnetization projected to T=0)
	150	[48]	(Thermal conductivity)
Thermo critical field (T)			
B_c	1.08	[69]	(Specific heat)

[a] Because data reported in the literature will not always be for fully oxygenated samples and there always remains the key issue of the precision of sample alignment in field, there is considerable variation. This table attempts a self-consistent set, but there remain considerable variations in, for example, values of λ_c and the anisotropy ratio $\gamma = \lambda_c/\lambda_{ab}$

C1.8.2 Pinning Energies

The field-dependent critical current density in type II superconductors is governed by vortex pinning induced by small regions in the material, where the superconducting order parameter is suppressed [184]. If a vortex core passes through one of these defects, the system gains an amount of pinning energy equal to the fraction of the condensation energy in the volume V of the vortex core contained in the pinning site:

$$U_0V = \tfrac{1}{2}\mu_0 H_c^2.V = \tfrac{1}{2}\mu_0 H_c^2.\pi\xi^2.\ell \quad \text{(C1.29)}$$

where H_c is the thermodynamic critical field, and ℓ is the length of the pinning defect. In the case of Y-123, the condensation energy is 48.2 J/mole [69] or 0.063 k_BT_c per unit cell. Using Equation (C1.29), this implies a thermodynamic critical field of 1.08 T. Taking the minimum effective length ℓ of pinning defect to be the separation of CuO_2 bilayers (0.84 nm), this gives a pinning energy at zero temperature of $1.4k_BT_c$. Clearly, closer to T_c, thermally activated depinning and associated flux creep will play an important role in the pinning dynamics. In order to determine the flux-pinning energies, different methods were used, such as magnetic relaxation measurements [185], dc-resistance measurements [186], ac-resistance measurements [187], noise measurements [188] and 90° rotating sample magnetic measurements [189]. Zeldov *et al* [190] have studied the resistive transition in Y-123 and reported pinning energies as high as 6 eV in a magnetic field of 0.5 T. On the other hand, Hagen and Griessen [191] deduced a much lower value of the activation energy in YBCO from flux creep measurements. They found a distribution of activation energies with a peak near 0.06 eV. Ferrari et al [188] used the temperature and frequency dependence of the noise to determine the pinning energies of individual flux vortices in thermal equilibrium. They found a distribution of pinning energies with two peaks: a low energy peak below 0.1 eV and a higher energy peak near 0.35 eV. The higher energy peak has been associated with grain boundaries or *a*-axis grains, while the lower energy peak may represent an intrinsic, intragranular pinning energy. Generally, an average pinning energy was obtained, which depends on temperature and field in a complicated way [192]. Since there are various defects in a material, a single value of pinning energy cannot completely describe the interaction between flux lines and the defect landscape. There should be a distribution of pinning energies in the YBCO material [191, 193].

C1.8.3 Hysteresis Curves, J_c Versus Field and Temperature

Plots of measured magnetic moment *versus* magnetic field show 'hysteresis' [194–197]. The Anderson 'flux-pinning' model [198] is usually used to explain hysteresis; in this model, bundles of magnetic flux are pinned to crystalline disorder. When the bundles are depinned, flux creeps from the sample, and the magnetization falls. Typical magnetic hysteresis loops for YBCO at 77 K are shown in Figure C1.33 [199]. The low-field magnetization behaviour of the sintered material can be understood by the presence of a weakly coupled region between grains.

The measurement of magnetic hysteresis plays a vital role in providing the information regarding the temperature and field dependence of critical current density, $J_c(H,T)$, needed to design HTS magnet and coil systems. The relationship between critical current density and isothermal magnetic hysteresis has been established by Bean [200, 201]. The variation of critical current density of YBCO with magnetic field and temperature has been extensively investigated [202]. The results are analyzed on the basis of Bean's critical-state model [200] and its various extensions. According to the critical-state model, the current density $J_c(H)$ is proportional to ΔM. Figure C1.34 (a), (b) show the field and temperature dependence of ΔM for Y-124 [202]. From this figure, it is evident that $J_c(T)$ follows a power law in the field region between the full penetration field and the field where the remnant magnetization saturates. $J_c(T)$ varies exponentially in a limited temperature and field region. It is important to note that magnetization J_c is not the same as transport J_c. For example, the magnetization J_c at zero field still involves non-zero trapped flux with counter-rotating screening currents, whereas transport measurements in zero external field only involve the self-field generated by the transport current itself. Figure C1.35 shows field-dependent transport J_c measurements of a (Nd,Eu,Gd)-123 coated conductor [203]. The plateau at low field is the so-called self-field region where $J_c(sf,T)$ seems to follow a universal dependence on superfluid density and may be used to determine the London penetration depth [204]. Here $J_c(sf,T\to 0)$ reaches values around 30 MA/cm^2. At higher field, $J_c(sf,T,H)$ falls off rapidly with increasing field with a power-law behaviour typically around $J_c(sf,T,H) \propto H^{-2/3}$.

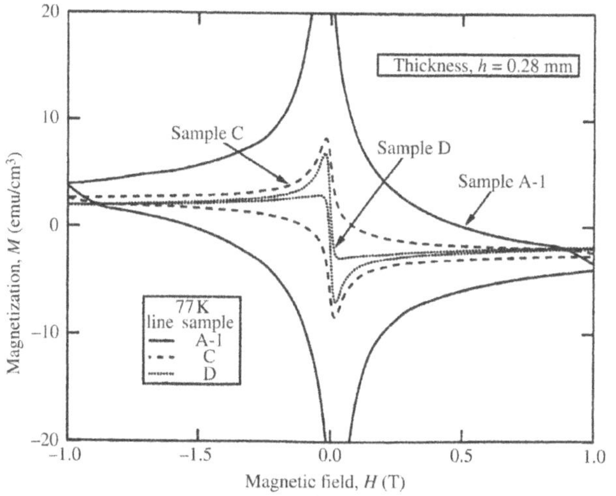

FIGURE C1.33 Magnetic hysteresis loops of YBCO samples measured at 77 K [199].

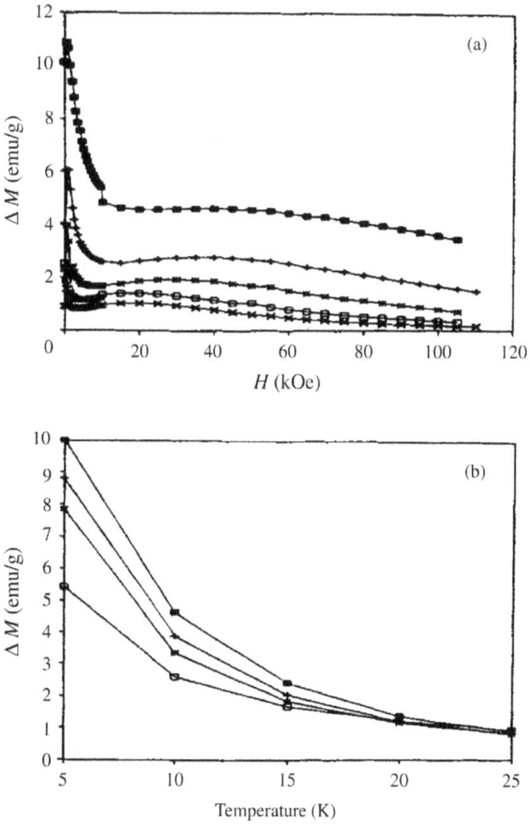

FIGURE C1.34 (a) The plot of ΔM as function of applied field for different temperatures (filled box at 5 K, plus sign at 10 K, star at 15 K, open box at 20 K and cross at 25 K) obtained from hysteresis curves; (*b*) variation of ΔM *versus* temperature for different values of applied fields (filled box at 2 kOe, plus sign at 3 kOe, star at 4 kOe and open box at 10 kOe) for Y-124 specimen [202].

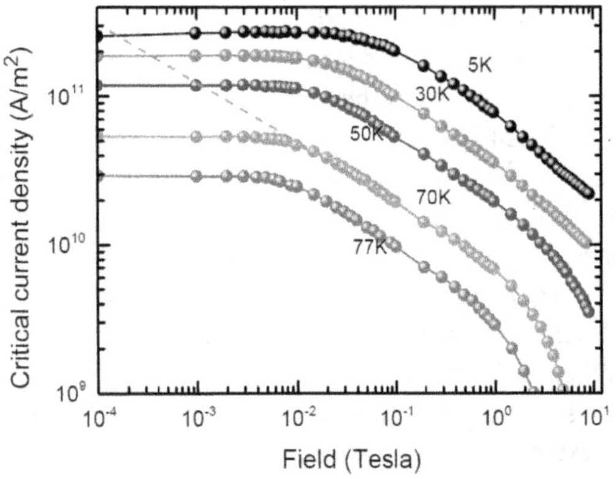

FIGURE C1.35 Transport $J_c(T,H)$ data for thin films of (Nd,Eu,Gd)$Ba_2Cu_3O_7$ reported by Cai *et al* [203].

C1.8.4 Irreversibility Data

The measurement of ac complex susceptibility remains one of the most powerful methods to obtain important information on dissipation mechanisms in polycrystalline YBCO. It is commonly accepted that bulk sintered superconductors generally consist of two components: anisotropic grains with a quite large lower critical field $H_{c1,g}$ and their coupling matrix having a very low $H_{c1,m}$ [205]. Two peaks in the temperature dependence of the imaginary part of the complex ac-susceptibility reflecting the intra- and intergranular losses can be distinguished at high ac-fields. Generally speaking, a maximum in $\chi''(T)$ appears when the supercurrents penetrate just to the centre of their circumferential paths, which may be the centre of the superconducting grain, or the coupling matrix.

An important feature of HTS is the existence of the so-called 'irreversibility line' (IL) where magnetic irreversibility sets in [206–208]. Brandt [209] has listed eight experiments for monitoring the 'irreversibility line'. ac-susceptibility is pointed out as one of the most sensitive methods among them. By this method, the IL is defined either as the relation between the temperature, at which the maximum in the imaginary component appears, and the dc-field H_{DC} ($H_{dc} \gg H_{ac}$), at which the maximum is measured, or as a relation between T_m and the ac-field amplitude ($H_{DC} = 0$) [205, 210–213]. The difference between $T_m(H_{AC})$ for $H_{DC} = 0$ and $T_m(H_{DC})$ for $H_{dc} \gg H_{ac}$, i.e. between the two cases mentioned above, has been studied by Muller [214]. For both the inter- and intragranular $\chi''(T)$ peak temperatures, a linear dependence of T_m and H_{AC} has been found in agreement with the experimental data reported in [205]. Figure C1.36(a), (b) show T_m and T_g as a function of the ac-field, H_m, at which the χ'' peak temperature from the matrix and the grains, respectively, is observed for YBCO samples with different thickness [205]. The onset of irreversible behaviour at the 'irreversibility line' $T_{irr}(H)$ is described, in most experiments, by a power law [215, 216]:

$$1 - T/T_c = AH^n, \tag{C1.30}$$

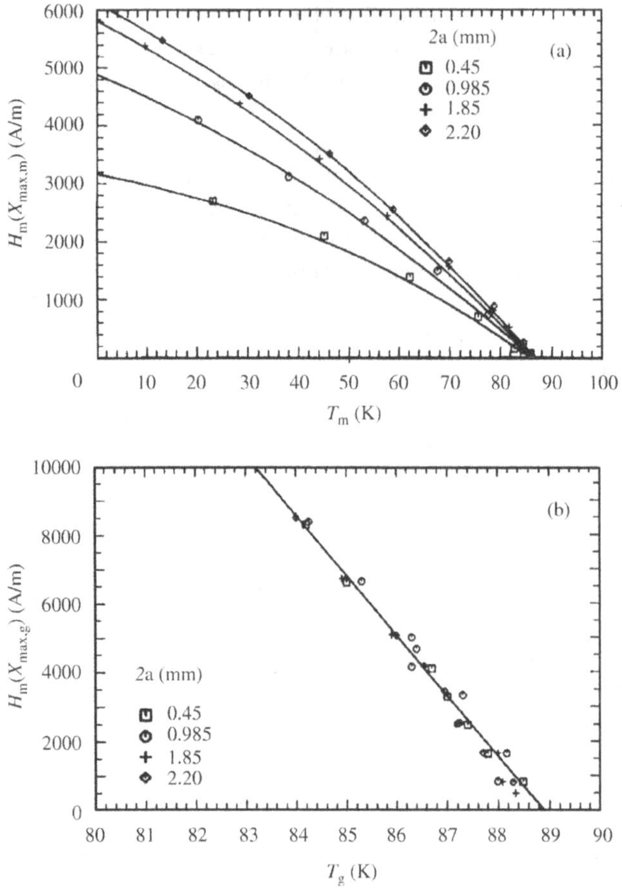

FIGURE C1.36 (a) Intergranular χ'' peak temperature T_m versus intergranular ac-field amplitude $H_m(\chi_{max}, m)$ for samples with different thicknesses; (b) intragranular χ'' peak temperature T_g versus intragranular ac-field amplitude $H_m(\chi_{max}, m)$ for samples with different thicknesses [205].

where A is a frequency-dependent constant, and the exponent n (in dc measurements) is approximately 2/3. Some experimental investigations have shown that, in many cases, the exponent n is different from the theoretical values of 2/3 or 3/4, and a change in the value of n is observed in different regions of applied magnetic field strength [217, 218].

Investigations of Y-124 have shown that the irreversibility line obeys a power-law behaviour similar to that of Y-123 Equation (C1.30), with n ~ 3/2 [213] probably reflecting the strong coupling along the *c*-axis due to the double chains.

C1.8.5 Grain Boundaries in YBCO

With the discovery of YBCO, it seemed that the vision of superconducting power cables operating at liquid nitrogen temperature might be close to realization. The critical current density, J_c, however, is in general suppressed at grain boundaries by phenomena such as interface charging and bending of the electronic band structure [219, 220] as is also known from

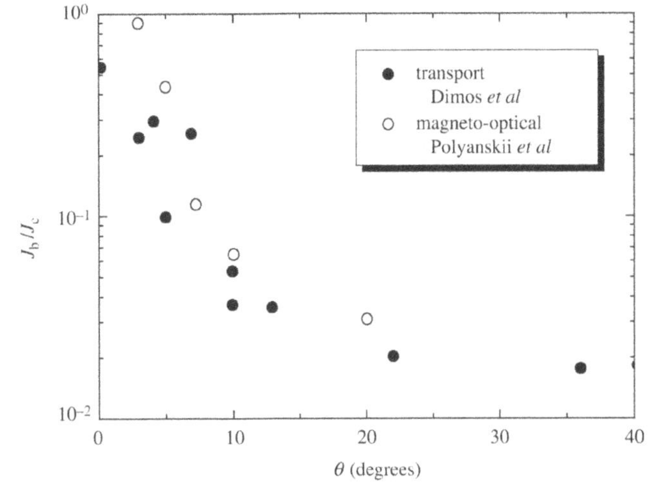

FIGURE C1.37 Plot of the ratio intergrain J_b and intragrain J_c critical currents in YBCO thin-film bicrystals versus the misorientation tilt angle θ. The measurements [223, 226, 237] were taken at $T = 6$ K.

other perovskite ceramics [221, 222]. Grain boundaries also tend to be underdoped so the pseudogap is likely to be present locally in these boundaries even if YBCO is fully oxygenated. Thus, it was many years before an effective coated-conductor technology could produce the transport J_c values reported in Figure C1.35. But that technology is now mature [7, 223]. For YBCO, the ever-present high-angle grain boundaries are the most severe problem for applications. In experiments on bicrystals and bicrystalline thin films [224–227], the critical current density across a grain boundary has been shown to be strongly dependent on the misorientation angle as indicated in Figure C1.37. A rapid decrease is evident in the critical current density of the boundary (J_b), normalized in these cases to the intragranular current density (J_c) of the adjacent grains, with increasing angle of crystallographic misorientation.

In Figure C1.38, a SEM photograph of a typical polycrystalline YBCO sample is shown; Figure C1.39 presents an electron backscatter diffraction (EBSD) mapping of a similar YBCO sample [228], indicating the crystallographic directions of the

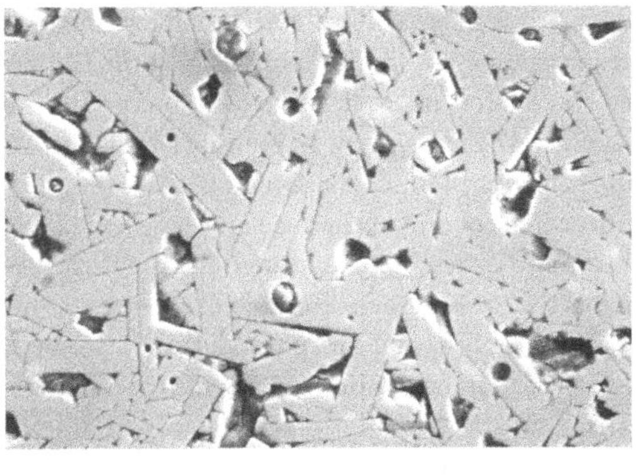

FIGURE C1.38 SEM image of the grain structure of a typical polycrystalline YBCO sample [229].

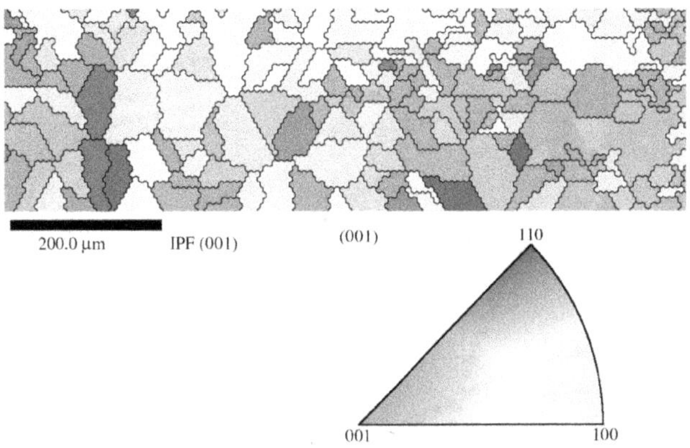

FIGURE C1.39 EBSD mapping of a polycrystalline YBCO sample with Rb-doping [228], presenting the crystal orientations normal [0 0 1] to the sample surface. The individual crystallographic directions are indicated in the colour-coded triangle.

individual grains. As a result, the flux penetration into such a material occurs firstly via the grain boundary regions and, as soon as the lower critical field of the grains is reached, the flux enters in the grains (Figure C1.40). This granular flux penetration behaviour was discussed in detail by Koblischka *et al* [229, 230], based on magneto-optical imaging (see Section D3.4). The high-angle grain boundaries basically act as weak links or Josephson junctions; thus a transport current across these grain boundaries ('intergranular' current density) is severely reduced as compared to the current density inside the grains ('intragranular' current density). By means of a local measurement technique, these two current densities can be measured directly [229]. The effect of the granularity is also seen in the magnetization loops (Figure C1.33) or in ac-susceptibility

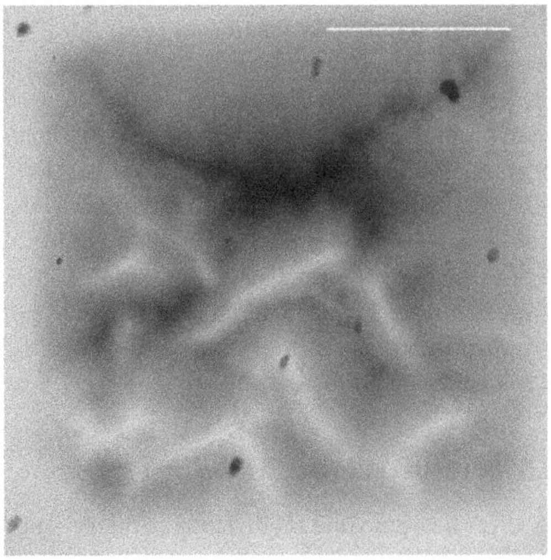

FIGURE C1.40 MO flux patterns of a $KClO_3$-doped YBCO sample at $T = 50$ K and an applied field of 100 mT. The magnetic field is imaged as bright areas; the Meissner phase remains dark. Only in the lower part of the image, flux entering along grain boundaries can be observed. The marker is 500 µm long [229].

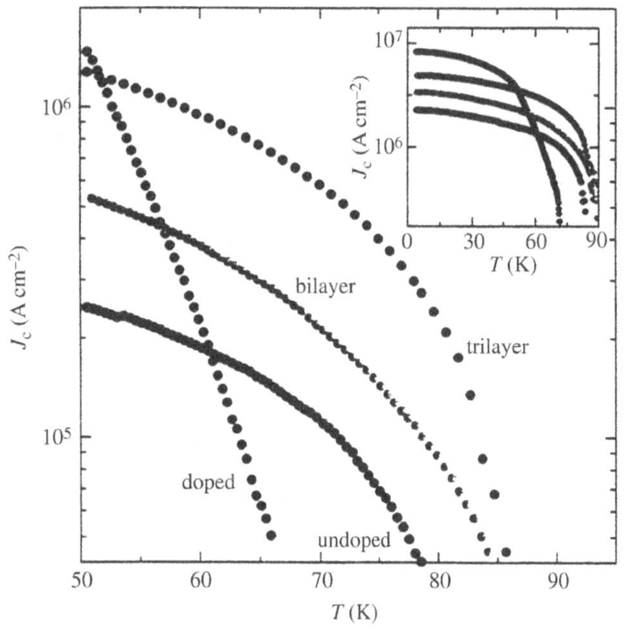

FIGURE C1.41 Temperature dependence of the critical current densities J_c of grain boundaries with various doping configurations, indicating the strong enhancement of J_c in doping heterostructures. Displayed are the $J_c(T)$ dependencies of symmetric 24° [0 0 1] tilt grain boundaries in various bicrystalline samples: a YBCO film (undoped), a YBCO film (Ca-doped), a bilayer (YBCO doped/undoped) and a trilayer (doped/undoped/doped) [236].

measurements. In transport current measurements, it was observed that the current density decreases rapidly with a small applied magnetic field [231]. In some experiments, however, the grain boundaries seem not to exhibit these detrimental effects [232, 233], e.g. high-angle grain boundaries of YBCO with a misorientation relationship near 90° [0 1 0] that are not parallel to the (0 0 1) plane of one crystal do not show weak-link behaviour.

Recent progress in the understanding of the behaviour of the grain boundaries was achieved by Mannhart *et al* [234–237]. The charging of the grain boundaries was shown to be removed by means of overdoping YBCO by adding Ca. The results of these measurements are illustrated in Figure C1.41 [238]. This doping of YBCO material, however, does not seem to carry through to significantly increased performance of YBCO coated conductors [55].

C1.8.6 Microstructure and Pinning

A huge effort has gone into modifying the micro- and nanostructure of YBCO films with a view to increased pinning and enhanced in-field J_c [238]. Much of the recent work has focussed on incorporation of $BaZrO_3$ particles or rods [239–241] because of the passive nature of the inclusions. YBCO is very reactive, and the introduction of $BaZrO_3$ as an inert crucible material profoundly improved crystal growth, and these benefits were successfully transferred to $BaZrO_3$ as a pinning inclusion [229].

Because the angular dependence of J_c in YBCO is quite variable with a strong intrinsic peak with field parallel to the *a–b* plane, much effort has been devoted to enhancing the perpendicular field J_c and achieving a more isotropic response. One of the keys to the success of $BaZrO_3$ is the ability to form rods in the YBCO matrix which align along the *c*-axis [240, 241]. Thus pulsed-laser-deposited films of ErBa2Cu3Oy and YBa2Cu3Oy with BaZrO3 nano-rods aligned along the *c*-axis ('*c*-axis–correlated pinning centres') were found to exhibit strongly enhanced J_c with field normal to the *a–b* plane [241].

Many other pinning methodologies have been explored with varying success [238] and will not be reviewed in further detail here. Interestingly, incorporation of J_c-enhancing gold nanoparticles in YBCO SQUID gradiometer films resulted in significant reduction of noise [242] while, despite early promise [236], Ca-doping of MOD coated conductors seem to provide only small J_c enhancements [243].

There remains a question as to whether any of these pinning strategies for enhanced in-field J_c are effective in the absence of an external field, i.e. under self-field transport conditions. There is quite a literature on the subject [244–247], but only two studies are mentioned in detail here. Lin *et al* [248] used a focussed electron beam to fabricate a regular array of damage nanocolumns on a YBCO microbridge. The columns were 10–20 nm in diameter, and the spacing of 90 nm was tuned for a 0.25-T matching field. The enhanced *c*-axis–correlated pinning resulted in a 60% increase in J_c at 1 T, however, this yielded essentially no change in J_c in self-field (indeed there was a small decrease consistent with the small reduction in area). A more recent example is that of Leroux *et al* [249] who use 3.5 MeV oxygen ions irradiation to rapidly enhance pinning and in-field J_c. While J_c at 10 T is doubled there is, again, no change in self-field J_c.

C1.8.7 Radiation Effects and Pinning

In addition to what has been just mentioned, a large number of results on the effects of particle irradiation on the flux pinning of HTS are now available [249–271]. The general conclusion from those studies is that defects introduced by irradiation can act as effective pinning sites in HTS materials. Electron, proton and light ion irradiations generate mostly point defects, which are effective pinning sites [250, 269–271]. The influence of proton irradiation on the flux pinning and J_c of YBCO coated conductors has been investigated [250]. J_c was shown to more or less double at high field, but the required irradiation times were several orders of magnitude too long to be commercially viable. In contrast, the above-mentioned 3.5 MeV oxygen ion irradiation [249] achieved the same result at a realistic rate of conductor processing of about 1 cm/s. The defects generated by the irradiation included small dislocation loops [251]. Neutrons and some heavier ions produce more distributed defects, like cascades or clusters of a few nanometres in diameter [256–267]. Irradiation with very heavy ions can create columnar defects, which offer more effective pinning at high temperatures and fields and can be used to strongly modify the angular dependence of $J_c(H)$ depending on the angle of irradiation [272]. Information regarding the pinning energy (U) of defects can be obtained from measurements of the thermal relaxation of magnetization. Maley et al [265] use a method, which allows one to build up the function U(J) over an extended range of J values by matching small portions of the curve obtained from relaxation measurements at various temperatures. Using this type of analysis, Lessure et al [252] have obtained U(J) for Y-123 ceramics before and after irradiation with fast neutrons (to a fluence of 2.1×10^{18} neutrons cm^{-2}). The increase in U after neutron irradiation is apparent, as is the deviation from linearity and the diverging behaviour of U at low currents.

C1.9 Summary

In summary, YBCO exhibits a wide range of properties varying according to the oxygen content, so the oxygenation procedure of 123-superconductors is a very crucial one. YBCO is also the material with the smallest anisotropy among the high T_c compounds and the one with the highest critical current densities at elevated temperatures, especially in high magnetic field. Therefore, YBCO and its derivatives of the 123-family (NdBCO, SmBCO, GdBCO and binary and ternary rare-earth compounds) are still the best suited high-T_c superconducting materials for a large variety of applications including all thin-film applications and many bulk applications. Concerning basic research, YBCO has served as the model system to explore a vast array of physical effects in high-T_c compounds concerning structural, chemical, mechanical, electronic, magnetic and optical properties, both in the normal and the superconducting state. As a consequence, a vast literature is available which we have only briefly sampled. It has proved to be a compound rich in complexity and interest, and despite a huge research effort, this material with its various competing correlations is still not fully understood.

References

[1] Wu M K, Ashburn J R, Torng C J, Hor P H, Meng R L, Huang Z J, Wang Y Q and Chu C W 1987 Superconductuvity at 93 K in a new mixed-phase Y–Ba–Cu–O compound system at ambient pressure *Phys. Rev. Lett.* **58** 908

[2] Jorgensen J D, Veal B W, Paulikas A P, Nowicki L J, Crabtree G W, Claus H and Kwok W K 1990 Structural Properties of oxygen-deficient $YBa_2Cu_3O_{7-\delta}$ *Phys. Rev.* B **41** 1863

[3] Pinol S, Gomis V, Gou A, Martinez B, Fontcuberta J and Obradors X 1994 Oxygenation and aging processes in melt textured $YBa_2Cu_3O_{7-\delta}$ *Physica* C **235–240** 3045

[4] Tallon J L, Schuitema A H and Tapp N E 1988 Soft mode behavior in the orthorhombic to tetragonal transition in the high T_c superconductor $YBa_2Cu_3O_7$ *Appl. Phys. Lett.* **52** 507

[5] Chen T G, Li S, Gao W, Xianyu Z, Liu H K and Dou S X 1998 The oxygenation kinetics of $YBa_2Cu_3O_{7-\delta}$–(0–30 percent) Ag superconductors *Supercond. Sci. Technol.* **11** 1193

[6] Erb A, Manuel A A, Dhalle M, Marti F, Genoud J Y, Revaz B, Junod A, Vasumathi D, Ishibashi S, Shukla A, Walker E, Fischer O, Flukiger R, Pozzi R, Mali M and Brinkmann D 1999 Experimental evidence for fast cluster formation of chain oxygen vacancies in $YBa_2Cu_3O_{7-\delta}$ as the origin of the fishtail anomaly *Solid State Commun.* **112** 245

[7] Glasson N, Staines M P, Jiang Z and Allpress N 2013 Commissioning testing of a 1 MVA 3-phase demonstration transformer using 2G HTS Roebel cable *IEEE Trans Appl. Supercon.* **11** 5500206

[8] Okada M, Okayama A, Matsumoto T, Aihara K, Matsuda S, Ozawa K, Morii Y and Funahashi S 1988 Neutron diffraction study on preferred orientation of Ag-sheathed YBCO superconductor tape with $J_c = 1000–3000$ A/cm^2 *Jpn. J. Appl. Phys.* **27** L1715

[9] Okada M, Okayama A, Morimoto T, Matsumoto T, Aihara K and Matsuda S 1988 Fabrication of Ag-sheathed YBCO superconductor tapes *Japan. J. Appl. Phys.* **27** L185

[10] Fleshler S *et al.* 2008 Scale-up of 2G wire manufacturing at American Superconductor Corporation *Physica* C **469** 1316

[11] Rey C, ed. *Superconductors in the Power Grid*, Woodhead Publishing, 2015, ISBN 9781782420293

[12] Krabbes G, Fuchs G, Schätzle P, Gruss S, Park J W, Hardinghaus F, Stover G, Hayn R, Drechsler S L and Fahr T 2000 Zn doping of $YBa_2Cu_3O_7$ in melt textured materials: peak effect and high trapped fields *Physica* C **330** 181

[13] Durrell J H, Dennis A R, Jaroszynski J, Ainslie M D, Palmer K G B, Shi Y-H, Campbell A M, Hull J, Strasik M, Hellstrom E E and Cardwell D A 2014 A trapped field of 17.6 T in melt-processed, bulk Gd-Ba-Cu-O reinforced with shrink-fit steel *Superconductor Sci. Technol.* **27**, 082001

[14] Yan M F, Rhodes W W and Gallagher P K 1988 Dopant effects on the superconductivity of YBCO ceramics *J. Appl. Phys.* **63** 821

[15] Bernhard C, Tallon J L, Bucci C, de Renzi R, Guidi G, Williams G V M and Niedermayer Ch 1996 Suppression of the superconducting condensate in high-T_c cuprates by Zn substitution and overdoping - evidence for an unconventional pairing state, *Phys. Rev. Lett.* **77**, 2300

[16] Mallett B P P, Wolf T, Gilioli E, Licci F, Williams G V M, Kaiser A B, Suresh N, Ashcroft N W and Tallon J L 2013

Dielectric versus magnetic pairing mechanisms in high-temperature cuprate superconductors investigated using Raman scattering *Phys. Rev. Lett.* **111**, 237001

[17] Pooke D M, Buckley R G, Presland M R and Tallon J L 1990 Bulk superconducting $YBa_2Cu_4O_8$ and $Y_2Ba_4Cu_7O_{15-\delta}$ prepared in oxygen at one atmosphere. *Phys. Rev.* B. 41, 6616

[18] Erb A, Walker E and Flükiger R 1995 $BaZrO_3$: the solution for the crucible corrosion problem during the single crystal growth of high- T_c superconductors $REBa_2Cu_3O_{7-\delta}$; RE = Y, *Pr Physica* C **245** 245

[19] Erb A, Walker E, Genoud J-Y and Flükiger R 1997 10 years of crystal growth of the 123- and 124-high T_c superconductors: from Al_2O_3 to $BaZrO_3$. Progress in crystal growth and sample quality and its impact on physics *Physica* C **282–287** 459

[20] Sebastian S E and Proust C 2015 Quantum oscillations in hole-doped cuprates. *Ann. Rev. Cond. Matt. Phys.* **6** 411

[21] Karpinski J, Kaldis E, Jilek E, Rusiecki S and Bucher B 1988 Bulk synthesis of the 81-K superconductor $YBa_2Cu_4O_8$ at high oxygen pressure *Nature* **336** 660

[22] Bordet P, Chaillout C, Chenavas J, Hodeau J L, Marezio M, Karpinski J and Kaldis E 1988 Structure determination of the new high-temperature superconductor $Y_2Ba_4Cu_7O_{14+x}$ *Nature* **334** 596

[23] Fisher P, Karpinski J, Kaldis E, Jilek E and Rusiecki S 1989 High pressure preparation and neutron-diffraction study of the high T_c superconductor $YBa_2Cu_4O_{8+x}$ *Solid State Commun.* **69** 531

[24] Tallon J L, Pooke D M, Buckley R G, Presland M R and Blunt J 1990 $R_2Ba_4Cu_7O_{15-\delta}$ – a 92 K bulk superconductor *Phys. Rev.* B. 41, 7220

[25] van Eenige E N, Griessen R and Wijngaarden 1990 Superconductivity at 108 K in $YBa_2Cu_4O_8$ at pressure up to 12GPa *Physica* C **168** 482

[26] Kourtakis K, Robbins M, Gallagher K P and Tiefel T 1989 Synthesis of $Ba_2YCu_4O_8$ by anionic oxidation-reduction *J. Mater. Res.* **4** 1289

[27] Wada T, Suzuki N, Yamaguchi K, Ichinose A, Yaegashi Y, Yamauchi H, Koshizuka N and Tanaka S 1991 Superconductive $(Y_{1-x}Ca_x)Ba_2Cu_4O_8$ (x = 0.0 and 0.05) ceramics prepared by low and high oxygen partial pressure techniques *J. Mater. Res.* **6** 18

[28] Capponi J J, Chaillout C, Hewat A W, Lejay P, Marezio M, Nguyen N, Raveau B, Soubeyroux J L, Tholence J L and Tournier R 1987 Structure of the 100 K superconductor $Ba_2YCu_3O_7$ between (5 ÷ 300) K by neutron powder diffraction *Europhys. Lett.* **3** 1301

[29] Cava R J, Hewat A W, Hewat E A, Batlogg B, Marezio M, Rabe K M, Krajevski J J, Peck W F Jr and Rupp L W 1990 *Physica* C **165** 419

[30] Brown ID 1991 The influence of internal strain on the charge distribution and superconducting transition temperature in $Ba_2YCu_3O_x$ *J. Solid State Chem.* **90** 155

[31] Kruger Ch, Conder K, Schwer H and Kaldis E 1997 The dependence of the lattice parameters on oxygen content in orthorhombic $YBa_2Cu_3O_{6+x}$: high precision reinvestigation of near equilibrium samples *J. Solid State Chem.* **134** 356

[32] Shaked H, Keane P M, Rodriguez J C, Owen F F, Hitterman R L and Jorgensen J D 1994 *Crystal Structures of the High-T_c Superconducting Copper-oxides, Physica C* (Amsterdam, The Netherlands: Elsevier) p 44

[33] *PCPDFWIN, Version 2.0 1998* JCPDS-ICDD, *PDF-2 Data Base (Sets 1-48 plus 70–85), File no. 38-1433, 39-486, 39-1434* JCPDS—International Centre for Diffraction Data.

[34] Hewat A W 1994 Neutron powder diffraction on the ILL high flux reactor and high T_c superconductors *Materials and Crystallographic Aspects of HT_c-Superconductivity* NATO ASI Series E: Applied Sciences **263**, ed E Kaldis (Dordrecht, The Netherlands: Kluwer) p 17

[35] Shibata H, Kinoshita K and Yamada T 1990 Crystal structure of low-oxygen-defect tetragonal $Ba_2YCu_3O_{6.75}$ *Jpn. J. Appl. Phys.* **29** L423

[36] Dharwadkar S R, Jakkai V S, Yakhmi J V, Gopalakrishnan I K and Iyer R M 1987 X-ray diffraction coupled thermogravimetric investigations of $YBa_2Cu_3O_{7-x}$ *Solid State Commun.* **64** 1429

[37] Zandbergen W H, Gronsky R, Wang K and Thomas G 1988 Structure of $(CuO)_2$ double layers in superconducting $YBa_2Cu_3O_7$ *Nature* **331** 596

[38] Karpinski J, Beeli C, Kaldis E, Wisard A and Jilek E 1988 Crystallization of YBCO from nearly stoichiometric melts, under oxygen pressures up to 2800 bar *Physica* C **153–155** 830

[39] Schwer H, Kaldis E, Karpinski J and Rossel C 1993 Effect of structural changes on the transition temperature in $Y_2Ba_4Cu_7O_{14+x}$ single crystals *Physica* C **211** 165

[40] Murakami H, Yaegashi S, Nishino J, Shiohara Y and Tanaka S 1990 Synthesis of $YBa_2Cu_4O_8$ powders by sol-gel method under ambient pressure *Jpn. J. Appl. Phys.* **29** L445

[41] Morris D E, Markelz A G, Fayn B and Nickel J H 1990 Conversion of 124 into 123 + CuO and 124, 123 and 247 phase regions in the Y–Ba–Cu–O system *Physica* C **168** 153

[42] Morris D E, Asmar N G, Nickel J H, Sid R L, Wie J Y T and Post J E 1989 Stability of 124, 123, and 247 superconductors *Physica* C **159** 287

[43] *PCPDFWIN, Version 2.0 1998 JCPDS-ICDD, PDF-2 Data Base (Sets 1-48* plus 70–85), File no. 84-*2460* JCPDS—International Centre for Diffraction Data.

[44] *PCPDFWIN, Version 2.0 1998 JCPDS-ICDD, PDF-2 Data Base (Sets 1-48 plus 70–85), File no. 43-0410* JCPDS—International Centre for Diffraction Data.

[45] Williams G V M, Staines M P, Tallon J L and Meinhold R 1996 NMR and transport studies of $Y_2Ba_4Cu_7O_{15-\delta}$ having different T_c values from different synthesis conditions *Physica* C **258** 273

[46] Tallon J L, Pooke D M, Buckley R G, Presland M R, Gibson S and Gilberd P W 1991 Enhanced flux pinning by combined atomic substitution and phase decomposition in cuprate superconductors *Appl. Phys. Lett.* **59** 1239
[47] Tallon J L, Williams G V M, Flower N E and Bernhard C 1997 Phase separation, pseudogap and impurity scattering in the HTS cuprates *Physica* C **282**, 236
[48] Grissonnanche G *et al.* 2014 Direct measurement of the upper critical field in cuprate superconductors *Nature Comms.* **5** 3280
[49] McHenry M E and Sutton R A 1994 Flux pinning and dissipation in high temperature oxide superconductors *Progress in Materials Science* Vol 38, eds J W Christian and T B Massalski (Oxford: Pergamon) p 160
[50] Jorgensen J D 1991 Defects and superconductivity in the copper oxides *Phys. Today* **44** 34
[51] Murakami M 1992 *Melt-Processed High-Temperature Superconductors* (Singapore: World Scientific) p 87
[52] Zandbergen H W and Tendeloo G 1991 Microstructures in high temperature superconductors, in *High-temperature Superconductors—Materials Aspects* Vol 2, eds H C Freyhardt, R Flükiger and M Peuckert (Verlag: Informationsge-sellschaft) p 544
[53] Hashimoto K, Akiyoshi M, Wisniewski A, Jenkins M L, Toda Y and Yano T 1996 A high-resolution electron microscopy study of structural defects in $YBa_2Cu_4O_8$ superconductor *Physica* C **269** 139
[54] Domenges B, Hervieu M, Raveau B, Karpinski J, Kaldis E and Rusiecki S 1991 High-resolution electron microscopy study of defects in 'high oxygen pressure' $YBa_2Cu_4O_8$ *J. Solid State Chem.* **93** 316
[55] Talantsev E F, Strickland N M, Wimbush S C, Storey J G, Tallon J L and Long N J 2014 Hole doping dependence of critical current density in $YBa_2Cu_3O_{7-\delta}$ conductors *Appl. Phys. Lett.* **104**, 242601
[56] Hirata Y, Kojima K M, Ishikado M, Uchida S, Iyo A, Eisaki H and Tajima S 2012 Correlation between the interlayer Josephson coupling strength and an enhanced superconducting transition temperature of multilayer cuprate superconductors *Phys. Rev.* B **85**, 054501
[57] Munzar D, Bernhard C, Holden T, Golnik A, Humlı́c̆ek J and Cardona M 2001 Correlation between the Josephson coupling energy and the condensation energy in bilayer cuprate superconductors *Phys. Rev.* B, **64** 024523
[58] Benseman T M, Gray K E, Koshelev A E, Kwok W -K, Welp U, Minami H, Kadowaki K and Yamamoto T 2013 Powerful terahertz emission from $Bi_2Sr_2CaCu_2O_{8+\delta}$ mesa arrays *Appl. Phys. Lett.* **103** 022602
[59] Benseman T M, Cooper J R and Balakrishnan G 2015 Interlayer tunnelling evidence for possible electron-boson interactions in $Bi_2Sr_2CaCu_2O_{8+\delta}$ http://lanl.arxiv.org/abs/1503.00335
[60] Buan J, Zhou B, Huang C C, Liu J Z and Shelton R N 1994 Anisotropy of the thermodynamic response along the *a* and *b* axes of the 1:2:3 compounds *Phys. Rev.* B **49** 12220
[61] Schlesinger Z, Collins R T, Holtzberg F, Feild C, Blanton S H, Welp U, Crabtree G W, Fang Y and Liu J Z 1990 Superconducting energy gap and normal state conductivity of a single-domain YBCO crystal *Phys. Rev. Lett.* **65** 801
[62] Zibold A, Widder K, Geserich H P, Scherer T, Marienhoff P, Neuhaus M, Jutzi W, Erb A and Müller-Vogt G 1992 Optical anisotropy of $YBa_2Cu_3O_{7-\delta}$ films on $NaGaO_3$ (001) substrates: a comparison with single crystals *Appl. Phys. Lett.* **61** 345
[63] Basov D N, Liang R, Bonn D A, Hardy W N, Dabrowski B, Quijada M, Tanner D B, Rice J P, Ginsberg D M and Timusk T 1995 In-plane anisotropy of the penetration depth in $YBa_2Cu_3O_{7-x}$ and $YBa_2Cu_4O_8$ superconductors *Phys. Rev. Lett.* **74** 598
[64] Tallon J L, Bernhard C, Binninger U, Hofer A, Williams G V M, Ansaldo E J, Budnick J I and Niedermayer Ch 1995 In-plane anisotropy of the penetration depth due to superconductivity on the CuO chains in $YBa_2Cu_3O_{7-\delta}$, $Y_2Ba_4Cu_7O_{15-\delta}$ and $YBa_2Cu_4O_8$ *Phys. Rev. Lett.* **74** 1008
[65] Awaji S, Suzuki T, Oguro H, Watanabe K and Matsumoto K 2016 Strain-controlled critical temperature in $REBa_2Cu_3Oy$-coated conductors *Sci. Rep.* **5** 11156
[66] Junod A 1990 Specific heat of high temperature superconductors: a review, in *Physical Properties of High Temperature Superconductors* Vol II, ed D M Ginsburg (Singapore: World Scientific) p 13
[67] Junod A, Eckert D, Graf T, Kaldis E, Karpinski E, Rusiecki S, Sanchez D, Triscone G and Muller J 1990 Specific heat of the superconductor $YBa_2Cu_4O_8$ from 1.5 to 300 K *Physica* C **168** 47
[68] Loram J W, Mirza K A, Cooper J R and Liang Y W 1993 Electronic specific heat of $YBa_2Cu_3O_{7-x}$ from 1.8 to 300 K *Phys. Rev. Lett.* **71** 1740
[69] Loram J W, Luo J, Cooper J R, Liang W Y and Tallon J L 2001 Evidence on the pseudogap and condensate from the electronic specific heat *J. Phys. Chem. Solids* **62** 59
[70] Tallon J L, Storey J G and Loram J W 2011 Fluctuations and critical temperature reduction in cuprate superconductors *Phys. Rev.* B **83** 092502
[71] Tallon J L, Barber F, Storey J G and Loram J W 2013 Coexistence of the superconducting energy gap and pseudogap above and below the transition temperature of cuprate superconductors *Phys. Rev.* B **87** 140508(R)
[72] Tallon J L, Loram J W, Cooper J R, Panagopoulos C and Bernhard C 2003 Superfluid density in cuprate high-Tc superconductors: a new paradigm *Phys. Rev.* B **68** 180501(R)
[73] Uher C 1992 Thermal conductivity of high temperature superconductors, in *Physical Properties of High Temperature Superconductors* Vol III, ed D M Ginsburg (Singapore: World Scientific) p 159
[74] Sutherland M *et al.* 2003 Thermal conductivity across the phase diagram of cuprates: low-energy quasiparticles and doping dependence of the superconducting gap *Phys. Rev.* B **67** 174520

[75] Zhu D M, Anderson A C, Friedmann T A and Ginsberg D M 1990 Thermal conductivity of polycrystalline $YBa_2Cu_3O_{7-\delta}$ in a magnetic field *Phys. Rev.* B **41** 6605

[76] Terzijska B M, Wawryk R, Dimitrov D A, Marucha C, Kovachev V T and Rafalowic Z 1992 Thermal conductivity of YBCO and thermal conductance at YBCO/ruby boundary Part 1: joint experimental set-up for simultaneous measurements and experimental study in the temperature range 10–260 K *Cryogenics* **32** 53

[77] Terzijska B M, Wawryk R, Dimitrov D A, Marucha C, Kovachev V T and Rafalowic Z 1992 Thermal conductivity of YBCO and thermal conductance at YBCO/ruby boundary Part 2: hysteresis behaviour between 40 and 230 K *Cryogenics* **32** 60

[78] Durst A C and Lee P A 2000 Impurity-induced quasiparticle transport and universal-limit Wiedermann-Franz violation in *d*-wave superconductors *Phys. Rev.* B **62** 1270

[79] Yu L, Munzar D, Boris A V, Yordanov P, Chaloupka J, Wolf Th, Lin C T, Keimer B and Bernhard C 2008 Evidence for two separate energy gaps in underdoped high-temperature cuprate superconductors from broadband infrared ellipsometry *Phys. Rev. Lett.* **100** 177004

[80] Zhou X J, *et al.* 2003 Universal nodal Fermi velocity *Nature* **423** 398

[81] Vishik I M, *et al.* 2010 Doping-dependent nodal Fermi velocity of the high-temperature superconductor $Bi_2Sr_2CaCu_2O_{8+\delta}$ revealed using high-resolution angle-resolved photoemission spectroscopy *Phys. Rev. Lett.* **104** 207002

[82] Grissonnanche G *et al.* 2016 Wiedemann-Franz law in the underdoped cuprate superconductor $YBa_2Cu_3O_y$ *Phys. Rev.* B **93** 064513

[83] Minami H, Wittorff V W, Yelland E A, Cooper J R, Changkang C and Hodby J W 2003 Influence of the pseudogap on the thermal conductivity and the Lorenz number of $YBa_2Cu_3O_x$ above T_c *Phys. Rev.* B **68** 220503(R)

[84] Andersson B M and Sundqvist B 1993 Thermal conductivity of $YBa_2Cu_4O_8$ dominated by phonon-phonon interactions *Phys. Rev.* B **48** 3575

[85] Williams R K, Scarbrough J O, Schmitz J M and Thompson J R 1998 Thermal conductivity of polycrystalline $YBa_2Cu_4O_8$ from 10 to 300 K *Phys. Rev.* B **57** 10923

[86] Cohn J L, Wolf S A, Vanderah T A, Selvamanickam V and Salama K 1992 Lattice thermal conductivity of $YBa_2Cu_3O_{7-s}$ *Physica* C **192** 435

[87] Peacor S D, Richardson R A, Nori F and Uher C 1991 Theoretical analysis of the thermal conductivity of $YBa_2Cu_3O_{7-\delta}$ single crystals *Phys. Rev.* B **44** 9508

[88] Cohn J L and Karpinski J 1998 Anisotropic in-plane thermal conductivity of single-crystal $YBa_2Cu_4O_8$ *Phys. Rev.* B **58** 14617

[89] Fujishiro H, Ikebe M, Nakasato K and Noto K 1996 Influence of Cu site impurities on the thermal conductivity of $YBa_2Cu_3O_{7-\delta}$ *Physica* C **263** 305

[90] Wittorff V W, Hussey N E, Cooper J R, Changkang C and Hodby J W 1996 Thermal conductivity of single crystals of $YBa_2(Cu_{1-x}Zn_x)_3O_7$ *Physica* C **282–287** 1287

[91] Bougrine H, Sergeenkov S, Ausloos M and Mehbod M 1993 Thermal conductivity of twinned $YBa_2Cu_3O_{7-x}$ and tweeded $YBa_2(Cu_{0\cdot95}Fe_{0\cdot05})_3O_{7-x}$ in a magnetic field: evidence for intrinsic proximity effect *Solid State Commun.* **86** 513

[92] Yukino K, Sato T, Ooba S, Ohta M, Okamura F P and Ono A 1987 Studies on the thermal behaviour of $Ba_2YCu_3O_{7-x}$ by X-ray powder diffraction method *Jpn. J. Appl. Phys.* **26** L869

[93] Momin A C, Mathews M D, Jakkal V S, Gopalakrishnan L K, Yakhmi J V and Iyer R M 1987 High temperature X-ray powder diffractometric studies of the superconducting compound $YBa_2Cu_3O_{7-x}$ from room temperature to 1300 K in air *Solid State Commun.* **64** 329

[94] Usami K, Kobayashi N and Doi T 1990 Temperature dependence of lattice parameters of $YBa_2Cu_3O_x$ superconductor at low temperature *Jpn. J. Appl. Phys.* **30** L96

[95] You H, Axe J D, Kan X B, Hashimoto S, Moss S C, Liu J Z, Crabtree G W and Lam D J 1988 Phase constitution and thermal expansion of $YBa_2Cu_3O_{7-\delta}$ single crystals *Phys. Rev.* B **38** 9213

[96] Marti W, Altorfer F and Fischer P 1993 Thermal expansion coefficients of $NdBa_2Cu_3O_{7-\delta}$ *Physica* C **206** 158

[97] Meingast C, Kraut O, Wolf T and Wühl 1991 Large a–b anisotropy of the expansivity anormaly at T_c in untwinned $YBa_2Cu_3O_{7-\delta}$ *Phys. Rev. Lett.* **67** 1634

[98] Haetinger C, Castillo I A, Kunzler J V, Ghivelder L, Pureur P and Reich S 1996 Thermal expansion and specific heat of non-random YBCO/Ag composites *Supercond. Sci. Technol.* **9** 639

[99] Kund M and Andres K 1993 Anisotropic stress dependence of T_c in $YBa_2Cu_3O_{7-\delta}$ single crystals deduced from thermal expansion, measured with a capacitive quartz-dilatometer *Physica* C **205** 32

[100] Schnelle W, Braun E, Broicher H, Domel R, Ruppel S, Braunisch W, Harnischmacher J and Wohlleben D 1990 Fluctuation specific heat and thermal expansion of YBaCuO and DyBaCuO *Physica* C **168** 465

[101] Ruan Y Z, Li L P, Hu X L, Peng D K, Hu J B and Zhang Y H 1989 Thermal expansion coefficients of Y with orthorhombic and tetragonal phases *Mod. Phys. Lett.* B **3** 325

[102] Jericho M H, Simpson A M, Tarascon J M, Green L H, McKinnon R and Hall G 1988 Thermal expansion and velocity of ultrasonic waves in single phase $YBa_2Cu_3O_{7-x}$ *Solid State Commun.* **65** 978

[103] Hashimoto T, Fueki K, Kishi A, Azumi T and Koinuma H 1988 Thermal expansion coefficients of high T_c superconductors *Jpn. J. Appl. Phys.* **27** L214

[104] Singh J P, Joo J, Singh D, Warzynski T and Poeppel R B 1993 Effects of silver additions in resistance to thermal shock and delayed failure of $YBa_2Cu_3O_{7-\delta}$ superconductors *J. Mater. Res.* **8** 1226

[105] Nye J F 2000 *Physical Properties of Crystals* (Oxford: Clarendon Press)

[106] Reddy R R, Murakami M, Tanaka S and Reddy P V 1996 Elastic behaviour of a Y–Ba–Cu–O sample prepared by MPMG method *Physica* C **257** 137

[107] Ledbetter H 1992 Elastic constants of polycrystalline $Y_1Ba_2Cu_3O_x$ *J. Mater. Res.* **7** 2905

[108] Cankurtaran M and Saunders G A 1992 Ultrasonic determination of the elastic moduli and their temperature and pressure dependences in $YBa_2Cu_3O_{7-x}$/Ag(15 vol.%) composite *Supercond. Sci. Technol.* **5** 210

[109] Holcomb D J and Mayo M J 1990 Effect of microcracking on the measured moduli of bulk $YBa_2Cu_3O_x$ *J. Mater. Res.* **5** 1827

[110] Ledbetter H M, Austin M W, Kim S A and Lei M 1987 Elastic constants and Debye temperature of polycrystalline $Y_1Ba_2Cu_3O_x$ *J. Mater. Res.* **2** 786

[111] Shindo Y, Ledbetter H and Nozaki H 1995 Elastic constants and microcracks in $YBa_2Cu_3O_7$ *J. Mater. Res.* **10** 7

[112] Suasmoro S, Smith D S, Lejeune M, Huger M and Gault C 1992 High temperature ultrasonic characterization of intrinsic and microstructural changes in ceramic $YBa_2Cu_3O_{7-x}$ *J. Mater. Res.* **7** 1629

[113] Almond D P, Wang Q, Freestone J, Lambson E F, Chapman B and Saunders G A 1989 An ultrasonic study of superconducting and non-superconducting $GdBa_2Cu_3O_{7-x}$ *J. Phys: Condens. Matter.* **1** 6853

[114] Cankurtaran M, Saunders G A, Willis J R, Kheffaji A A and Almond D P 1989 Bulk modulus and its pressure derivative of $YBa_2Cu_3O_{7-x}$ *Phys. Rev.* B **39** 2872

[115] Cankurtaran M, Saunders G A, Goretta K C and Poeppel R B 1992 Ultrasonic determination of the elastic properties and their pressure and temperature dependences in very dense $YBa_2Cu_3O_{7-x}$ *Phys. Rev.* B **46** 1157

[116] Bonetti E, Campari E G, Manfredini T and Mantovani S 1991 Orthorhombic to tetragonal phase transition in $YBa_2Cu_3O_{7-x}$ observed by dynamic Young's modulus measurements *Physica* C **179** 381

[117] Kusz B, Barczynski R, Gazda M, Sadowski W, Murawski L, Ozowski O, Davoli I and Stizza S 1990 Elastic constant and internal friction in $YBa_2Cu_3O_x$ single crystal *Solid State Commun.* **76** 357

[118] Shi X D, Yu R C, Wang Z Z, Ong N P and Chaikin P M 1989 Sound velocity and attenuation in single-crystal $YBa_2Cu_3O_{7-\delta}$ *Phys. Rev.* B **39** 827

[119] Tallon J L and Mellander B E 1992 Large enhancement in oxygen mobility in the superconductors $RBa_2Cu_3O_7$ with increasing rare-earth size *Science* **258** 781

[120] Murayama T, Sakai N, Yoo S I and Murakami M 1996 Mechanical properties of OCMG-processed Nd–Ba–Cu–O bulk superconductors, in *Proceedings of International Symposium on Advances in Superconductivity: New Materials, Critical Currents and Devices (ASMCCD'96)* (Mumbai, India) p 321

[121] Fujimoto H, Murakami M and Koshizuka N 1992 Effect of Y_2BaCuO_5 on fracture toughness of YBCO prepared by a MPMG process *Physica* C **203** 103

[122] Leeders A, Ullrich M and Freyhardt H C 1997 Influence of thermal cycling on the mechanical properties of VGF melt-textured YBCO *Physica* C **279** 173

[123] Ochiai S, Osamura K and Takayama T 1988 Fracture toughness measurements of $Ba_2YCu_3O_{7-x}$ superconducting oxide by means of indentation technique *Jpn. J. Appl. Phys.* **27** L1101

[124] Cook R F, Dinger T R and Clarke D R 1987 Fracture toughness measurements of $YBa_2Cu_3O_x$ single crystals *Appl. Phys. Lett.* **51** 454

[125] Anstis G R, Chantikul P, Lawn B R and Marshall D B 1981 *J. Am. Ceram. Soc.* **64** 532

[126] Joo J, Singh J P, Warzynski T, Grow A and Poeppel R B 1994 Role of silver addition in mechanical and superconducting properties of high T_c superconductors *Appl. Supercond.* **2** 401

[127] Oka T, Itoh Y, Yanagi Y, Tanaka H, Takashima S and Mizutani U 1992 Metallurgical reactions and their relationships to enhanced mechanical strength in Zr-bearing YBCO composite superconductors *Jpn. J. Appl. Phys.* **31** 1760

[128] Xu J A 1994 Elastic properties of high T_c superconductors: elastic systematics in $YBa_2Cu_3O_{7-x}$/Ag composites Supercond *Sci. Technol.* **7** 1

[129] Yeou L S and White K W 1992 The development of high fracture toughness $YBa_2Cu_3O_{7-x}$/Ag composites *J. Mater. Res.* **7** 1

[130] Yeh F and White K W 1991 Fracture toughness behaviour of the $YBa_2Cu_3O_{7-x}$ superconducting ceramic with silver oxide additions *J. Appl. Phys.* **70** 4989

[131] Singh J P, Leu H J, Poeppel R B, Voorhees E V, Goudey G T and Winsley K 1989 Effect of silver and silver oxide additions on the mechanical and superconducting properties of $YBa_2Cu_3O_{7-\delta}$ superconductors *J. Appl. Phys.* **66** 3154

[132] Yu F, White K W and Meng R 1997 Mechanical characterization of top-seeded melt-textured $YBa_2Cu_3O_{7-\delta}$ single crystal *Physica* C **276** 295

[133] Joo J, Kim J G and Nah W 1998 Improvement of mechanical properties of YBCO-Ag composite superconductors made by mixing with metallic Ag powder and $AgNO_3$ solution *Supercond. Sci. Technol.* **11** 645

[134] Oka T, Itoh Y, Yanagi Y, Tanaka H, Takashima S, Yamada Y and Mizutani U 1992 Critical current density and mechanical strength of $YBa_2Cu_3O_{7-\delta}$ superconducting composites containing Zr, Ag and Y_2BaCuO_5 dispersions by melt-processing *Physica* C **200** 55

[135] Lee D and Salama K 1990 Enhancements in current density and mechanical properties of Y–Ba–Cu–O/Ag composites Japan *J. Appl. Phys.* **29** L2017

[136] Oka T, Ogasawara F, Itoh Y, Suganuma M and Mizutani U 1990 Mechanical and superconducting properties of

Ag/YBCO composite superconductors reinforced by the addition of Zr *Jpn. J. Appl. Phys.* **29** 1924

[137] Wu N L, Hsu C C and Lee C H 1998 Reduced cracking in melt-grown $RBa_2Cu_3O_7$ (R = Nd, Sm)/gold composites *Jpn. J. Appl. Phys.* **37** L438

[138] Higashida K and Narita N 1991 High temperature deformation and textures in oxide superconductors, in *Advances in Superconductivity III Proceedings of the 3rd ISS'90, Sendai* (Springer: Tokyo) p 805

[139] von Mises R and Angew Z 1928 Mechanics of plastic shape change of crystals *Math. Mech.* **8** 161

[140] Suzuki T and Takeuchi S 1989 *JJAP Series 2, Lattice Defects in Ceramics* (Tokyo: Publication Office, JJAP) pp 9–15

[141] Guangcan C, Jingkui L, Wei C, Sishen X, Yude Y, Qiansheng Y, Yuengming N, Guirueng L and Genguha C 1987 Study on phase diagram of BaO–Y_2O_3–CuO ternary system *Int. J. Mod. Phys.* B **1** 363

[142] Hinks D G, Soderholm L, Cappone I D W, Jorgensen J D, Schuller I K, Segre C U, Zhang K and Grace J D 1987 Phase diagram and superconductivity in the Y–Ba–Cu–O system *Appl. Phys. Lett.* **50** 1688

[143] Karpinski J and Kaldis E 1988 Equilibrium pressures of oxygen above $YBa_2Cu_3O_{7-x}$ up to 2000 bar *Nature* **331** 242

[144] de Leeuw D M, Mutsares A H A C, Langereis C, Smoorenburg H C A and Rommers P J 1988 Compounds and phase compatibilities in the system Y_2O_3–BaO–CuO at 950°C *Physica* C **152** 39

[145] Karpinski J, Rusiecki S, Kaldis E, Bucher B and Jilek E 1989 Phase diagrams of $YBa_2Cu_4O_8$ and $YBa_2Cu_{3.5}O_{7.5}$ in the pressure range 1 bar ≤ PO_2 ≤ 3000 bar *Physica* C **160** 449

[146] Chandrachood M R, Morris D E and Sinha AP B 1990 Phase diagram and new phases in the Y-Ba-Cu-O system *Physica* C **171** 187

[147] Sestak J 1992 Phase diagrams in CuO_x based superconductors *Pure Appl. Chem.* **64** 125

[148] Krabbes G, Bieger W, Wiesner U, Ritschel M and Teresiak A 1993 Isothermal sections and primary crystallization in the quasiternary $YO_{1.5}$–BaO–CuO_x system at $P(O_2) = 0.21 \times 10^5$ *Pa J. Solid State Chem.* **103** 420

[149] Karpinski J, Conder K, Kruger Ch, Schwer H, Mangeschots I, Jilek E and Kaldis E 1994 Phase diagram, synthesis and crystal growth of YBaCuO phases at high oxygen pressures PO_2 < 3000 bar, *in Materials and Crystallographic Aspects of HT_c-Superconductivity* NATO ASI Series E: Applied Sciences **263**, ed E Kaldis (Dordrecht, The Netherlands: Kluwer) p 555

[150] Maeda M, Kadoi M and Ikeda T 1989 The phase diagram of the $YO_{1.5}$–Bao–CuO ternary system and growth of $YBa_2Cu_3O_7$ single crystals Japan. *J. Appl. Phys.* **28** 1417

[151] Murakami H, Suga T, Noda T, Shiohara Y and Tanaka S 1990 Phase diagram of $YBa_2Cu_3O_{7-x}$, $Y_2Ba_4Cu_7O_{15-x}$ and $YBa_2Cu_4O_8$ superconductors Japan. *J. Appl. Phys.* **29** 2720

[152] Moiseev K G, Vatolin A N, Zaizeva I S, Ilyinych I N, Tsagareishvily S D, Gvelesiani B I and Sestak J 1992 Calculation of thermodynamic properties of the phases in the Y–Ba–Cu–O system *Thermochim. Acta* **198** 267

[153] Barus A M M and Taylor J A T 1994 The effect of particle size distribution on the phase composition in $YBa_2Cu_3O_{7-x}$ as determined by DTA *Physica* C **225** 374

[154] Rodriguez M A, Snyder R L, Chen B J, Matheis D P, Misture S T and Frechette K 1993 The high-temperature reactions of $YBa_2Cu_3O_{7-\delta}$ *Physica C* **206** 43

[155] Idemoto Y and Fueki K 1990 Melting point of superconducting oxides as a function of oxygen partial pressure *Jpn. J. Appl. Phys.* **29** 2729

[156] Ono A and Tanaka T 1987 Preparation of single crystals of the superconductor $Ba_2YCu_3O_{6.5+x}$ Japan. *J. Appl. Phys.* **26** 1825

[157] Sawai Y, Ishizaki K and Takata M 1991 Stability of $YBa_2Cu_3O_7$, $Y_2Ba_4Cu_7O_{15}$ and $YBa_2Cu_4O_8$ superconductors under varying oxygen partial pressure, total gas and pressure and temperature *Physica* C **176** 147

[158] Hohlwein D 1994 Superstructures in 123 compounds. X-ray and neutron diffraction, in *Materials and Crystallographic Aspects of HT_c-Superconductivity* NATO ASI Series E: Applied Sciences **263**, ed E Kaldis (Dordrecht, The Netherlands: Kluwer) p 65

[159] Lindemer T B, Washburn F A, MacDougall C S, Feenstra R and Cavin O B 1991 Decomposition of $YBa_2Cu_3O_{7-x}$ and $YBa_2Cu_4O_8$ for $p_{O2} \le 0.1$ MPa *Physica* C **178** 93

[160] Ceder G, Asta M and Fontaine D 1991 Computation of the OI–OII–OIII phase diagram and local oxygen configuration for $YBa_2Cu_3O_z$ with *z* between 6.5 and 7 *Physica* C **177** 106

[161] Tallon J L 1989 Oxygen stoichiometry, order disorder and the orthorhombic-tetragonal transition in $YBa_2Cu_3O_{7-\delta}$ *Phys. Rev.* B **39** 2784

[162] Voronin G F 1994 Thermodynamic stability of superconductors in the Y–Ba–Cu–O system, in *Materials and Crystallographic Aspects of HT_c-Superconductivity* NATO ASI Series E: Applied Sciences **263**, ed E Kaldis (Dordrecht, The Netherlands: Kluwer) p 585

[163] Cooper S L, Reznik D, Kotz A, Karlow M A, Liu R, Klein M V, Lee W C, Giapintzakis J and Ginsberg D M 1993 Optical studies of the *a*-, *b*-, and *c*-axis charge dynamics in YBCO *Phys. Rev.* B **47** 8233

[164] Tajima S, Schützmann J, Miyamoto S, Teraski I, Sato Y and Hauff R 1997 Optical study of *c*-axis charge dynamics in YBCO: carrier self-confinement in the normal and the superconducting states *Phys. Rev.* B **55** 6051

[165] Rotter L D, Schlesinger Z, Collins R T, Holtzberg F, Feild C, Welp U, Crabtree G W, Liu J Z, Fang Y, Vandervoort K G and Fleshler S 1991 Dependence of the infrared properties of single-domain YBCO on oxygen content *Phys. Rev. Lett.* **67** 2741

[166] Takenaka K, Mizuhashi K, Takagi H and Uchida S 1994 Interplane charge transport in YBCO: spin-gap effect on in-plane and out-of-plane resistivity *Phys. Rev. Lett.* **50** 6534

[167] Tallon J L, Cooper J R, de Silva P S I P N, Williams G V M and Loram J W 1995 Thermoelectric power: a simple instructive probe of high- T_c superconductors *Phys. Rev. Lett.* **75** 4114

[168] Xiang T and Wheatley J M 1995 *c*-Axis superfluid response of copper oxide superconductors *Phys. Rev. Lett.* **77** 4632

[169] Ito T, Takenaka K and Uchida S 1993 Systematic deviation from T-linear behaviour in the in-plane resistivity of $YBa_2Cu_3O_{7-y}$: evidence for dominant spin scattering *Phys. Rev. Lett.* **70** 3995

[170] Tallon J L and Loram J W 2001 The doping dependence of T^* – what is the real high-T_c phase diagram? *Physica* C **349** 53

[171] Badoux S *et al.* 2016 Change of carrier density at the pseudogap critical point of a cuprate superconductor *Nature* **531** 210

[172] Storey J G 2016 Hall effect and Fermi surface reconstruction via electron pockets in the high-T_c cuprates *Europhys. Lett.* **113** 27003

[173] Yang K Y, Rice T M and Zhang F C 2006 Phenomenological theory of the pseudogap state *Phys. Rev.* B **73** 174501

[174] Tallon J L 2015 Thermodynamics and critical current density in high-T_c superconductors *IEEE Trans. Appl. Supercon.* **25** 8000806

[175] Jiang H, Yuan T, How H, Widom A, Vittoria C and Drehman A 1993 Measurements of anisotropic characteristic lengths in YBCO films at microwave frequencies *J. Appl. Phys.* **73** 5865

[176] Rose-Innes A C and Rhoderick E H 1978 *Introduction to Superconductivity* (Oxford: Pergamon) p 67

[177] Ginzburg V L and Landau D 1950 *Zh. Eksp. Teor. Fiz.* **20** 1064 English translation in Landau LD 1965 *Men of Physics* Vol 1, ed D Ter Haar (New York: Pergamon Press)

[178] Sonier J E *et al.* 2007 Hole-doping dependence of the magnetic penetration depth and vortex core size in $YBa_2Cu_3O_y$: evidence for stripe correlations near 1/8 hole doping *Phys. Rev.* B **76** 134518

[179] Tallon J L and Flower N E 1993 Stoichiometric $YBa_2Cu_3O_7$ is overdoped *Physica* C **204** 237

[180] Krusin-Elbaum, Malozemoff A P, Yeshurin Y, Kronemeyer D C and Holtzberg F 1989 *Phys. Rev.* B **39** 2396

[181] Bonn D A, Kamal S, Zhang K, Liang R and Hardy W N 1995 The microwave surface impedance of $YBa_2Cu_3O_{7-\delta}$ *J. Phys. Chem. Solids* **56** 1941

[182] Poole C P, Farach H A, Creswick R J and Prozorov R 2007 *Superconductivity* (2nd edition) (Amsterdam: Academic Press) p 343

[183] Moodera J S, Meservey R, Tkaczyk J E, Hao C X, Gibson G A and Tedrow P M 1988 Critical-magnetic-field anisotropy in single-crystal $YBa_2Cu_3O_7$ *Phys. Rev.* B **37** 619

[184] Campbell A M and Evetts J E 1972 Flux vortices and transport current in type-II superconductors *Adv. Phys.* **21** 199

[185] Yeshurun Y and Malozemoff A P 1988 Direct measurement of the temperature-dependent magnetic penetration depth in Y–Ba–Cu–O crystals *Phys. Rev. Lett.* **60** 2202

[186] Palstra T T M, Batlogg B, van Dover R B, Schneemeyer L F and Waszczak J V 1989 Critical currents and thermally activated flux motion in high-temperature superconductors *Appl. Phys. Lett.* **54** 763

[187] Kes P H, Berguis P, Guo S Q, Dam B and Stallman G M 1989 *J. Less-Common. Met.* **151** 325

[188] Ferrari M J, Johnson M, Wellstood F C, Clarke J, Mitzi D, Rosenthal P A, Eom C B, Geballe T H, Kapitulnik A and Beasley M R 1990 Distribution of flux-pinning energies in $YBa_2Cu_3O_{7-y}$ and $Bi_2Sr_2CaCu_2O_{8+y}$ from flux noise *Phys. Rev. Lett.* **64** 72

[189] Li G, Sun Y, Liu S, Liu G, Yan S, Xiao L and Fu X 1993 Distribution of flux pinning energies in powder-melt-textured-grown $YBa_2Cu_3O_{7-\delta}$ *Solid State Commun.* **88** 451

[190] Zeldov E, Amer M N, Koren G, Gupta A, Gambino R J and McElfresh M W 1989 Optical and electrical enhancement of flux creep in $YBa_2Cu_3O_{7-\delta}$ epitaxial films *Phys. Rev. Lett.* **62** 3093

[191] Hagen C W and Griessen R 1989 Distribution of activation energies for thermally activated flux motion in high T_c superconductors: an inversion scheme *Phys. Rev. Lett.* **62** 2857

[192] Sun Y R, Thompson J R, Christen D K, Holtzberg F, Marwick A D and Ossandon J G 1992 *Physica* C **194** 403

[193] Yan S, Liang S, Ma H, Feng Q, Sun Y, Gao Y and Zhang H 1989 *Solid State Commun.* **70** 553

[194] Parish J L 1994 Note on magnetic hysteresis in HTSC *Physica* C **226** 325

[195] Andrä W, Bruchlos H, Eick T, Hergt R, Michalke W, Schuppel W and Steenbeck K 1991 Critical current density and flux pinning determined by different methods *Physica* C **180** 184

[196] Nojima T and Fujita T 1991 Universality of magnetization curves in superconducting Bi–Sr–Ca–Cu–O and Y–Ba–Cu–O films *Physica* C **178** 140

[197] Schlenker C, Liu C J, Buder R, Schubert J and Stritzker B 1991 Magnetic properties and critical currents in $YBa_2Cu_3O_7$ thin films *Physica* C **180** 148

[198] Anderson P W and Kim Y B 1964 Hard superconductors: theory of the motion of Abrikosov flux lines *Rev. Mod. Phys.* **36** 39

[199] Kamiya H, Kondo A, Yokoyama T, Naito M, Jimbo G, Nagaya S, Miyajima M and Hirabayashi I 1994 Effect of Y_2BaCuO_5 particle size on the properties of $YBa_2Cu_3O_{7-\delta}$ superconductor *Adv. Powder Technol.* **5** 339

[200] Bean C P 1962 Magnetisation of hard superconductors *Phys. Rev. Lett.* **9** 309

[201] Bean C P 1964 Magnetisation of high-field superconductors *Rev. Mod. Phys.* **36** 31

[202] Kumar R, Walia R, Oussena M, de Groot P A J, Lanchester P C, Currie D B and Weller M T 1994 High field magnetization study of $YBa_2Cu_4O_8$ *Solid State Commun.* **9** 783

[203] Cai C, Holzapfel B, Hänisch, Fernández L and Schultz L 2004 High critical current density and its field dependence in mixed rare earth (Nd,Eu,Gd)$Ba_2Cu_3O_{7-\delta}$ thin films *Appl. Phys. Lett.* **84** 377

[204] Talantsev E F and Tallon J L 2015 Universal self-field critical current for thin-film superconductors *Nature Comms.* **6** 7820

[205] Skumryev V, Koblischka M R and Kronmüller H 1991 Sample size dependence of the AC-susceptibility of sintered $YBa_2Cu_3O_{7-\delta}$ superconductors *Physica* C **184** 332

[206] Müller K A, Takashige M and Bednorz J G 1987 Flux trapping and superconductive glass state in La_2CuO_{4-y}: Ba *Phys. Rev. Lett.* **58** 1143

[207] Malozemoff A P, Worthington T K, Yeshurun Y, Holtzberg F and Kes P H 1988 Frequency dependence of the ac susceptibility in an YBaCuO crystal: a reinterpretation of H_{c2} *Phys. Rev.* B **38** 7203

[208] Brandt E H 1989 Thermal fluctuation and melting of the vortex lattice in oxide superconductors *Phys. Rev. Lett.* **63** 1106

[209] Brandt E H 1991 Thermal depinning and melting of the flux-line lattice in high T_c superconductors *Int. J. Mod. Phys.* B **5** 751

[210] Yeshurun Y, Malozemoff A P and Shaulov A 1996 Magnetic relaxation in high-temperature superconductors *Rev. Mod. Phys.* **68** 911

[211] Suryanarayanan R, Leelaprute S and Niarchos D 1993 Irreversibility line of $Y_{1-x}Ca_xSrBaCu_{2.9}Co_{0.1}O_{6+z}$ ($0 < x < 0.25$). A comparative study of DC magnetization and complex AC susceptibility *Physica* C **214** 277

[212] Zheng D N, Campbell A M, Johnson J D, Cooper J R, Blunt F J, Porch A and Freeman P A 1994 Magnetic susceptibilities, critical fields, and critical currents of Co- and Zn-doped $YBa_2Cu_3O_7$ *Phys. Rev.* B **49** 1417

[213] Lee W C and Ginsberg D M 1992 Magnetic measurements of the upper critical field, irreversibility line, anisotropy, and magnetic penetration depth of grain-aligned $YBa_2Cu_4O_8$ *Phys. Rev.* B **45** 7402

[214] Müller K H 1989 AC susceptibility of high temperature superconductors in a critical state model *Physica* C **159** 717

[215] Yacoby E R, Shaulov A, Yeshurun Y, Konczykowski M and Rullier-Albenque F 1992 Irreversibility line in $YBa_2Cu_3O_7$ samples. A comparison between experimental techniques and effect of electron irradiation *Physica* C **199** 15

[216] El-Abbar A A, King P J, Maxwell K J, Owers-Bradley J R and Roys W B 1992 The irreversibility line in polycrystalline $YBa_2Cu_3O_7$ superconductors with Y_2BaCuO_5 inclusions *Physica* C **198** 81

[217] Almasan C C, Seaman C L, Dalichaouch Y and Maple M B 1991 Irreversibility line and magnetic relaxation in a $Sm_{1.85}Ce_{0.15}CuO_{4-y}$ single crystal *Physica* C **174** 93

[218] Sagdahl L T, Laegreid T, Fossheim K, Murakami M, Fujimoto H, Gotoh S, Yamaguchi K, Yamauchi H, Koshizuka N and Tanaka S 1990 Restricted reversible region and strongly enhanced pinning in MPMG $YBa_2Cu_3O_7$ with Y_2BaCuO_5 inclusions *Physica* C **172** 495

[219] Mannhart J and Hilgenkamp H 1998 Possible influence of band bending on the normal state properties of grain boundaries in high- T_c superconductors *Mater. Sci. Eng.* B **56** 77

[220] Gurevich A and Pashitskii E A 1998 Current transport through low-angle grain boundaries in high-temperature superconductors *Phys. Rev.* B **57** 13878

[221] Buessem W R, Cross L E and Goswami A K 1966 Phenomenological theory of high permittivity in fine-grained $BaTiO_3$ *J. Am. Ceram. Soc.* **49** 33

[222] Waser R 1995 Electronic properties of grain boundaries in $SrTiO_3$ and $BaTiO_3$ ceramics *Solid State Ionics* **75** 89

[223] Matthews J N A 2008 Next-generation high-Tc superconducting wires debut in the power grid *Phys. Today* January 30

[224] Dimos J, Chaudhari P and Mannhart J 1990 Superconducting transport properties of grain boundaries in YBCO bicrystals *Phys. Rev.* B **41** 4038

[225] Babcock S E, Cai X Y, Kaiser D L and Larbalestier D C 1990 Weak-link free behaviour of high-angle YBCO grain boundaries in high magnetic fields *Nature* **347** 167

[226] Heinig NF, Redwing RD, Tsu I F, Gurevich A, Nordman J E, Babcock S E and Larbalestier D C 1996 Evidence for channel conduction in low misorientation angle [0 0 1] tilt YBCO bicrystal films *Appl. Phys. Lett.* **69** 577

[227] Polyanskii A A, Gurevich A, Pashitski A E, Heinig N F, Redwing R D, Nordman J E and Larbalestier D C 1996 Magneto-optical study of flux penetration and critical current densities in [0 0 1] tilt YBCO thin-film bicrystals *Phys. Rev.* B **53** 8687

[228] Koblischka-Veneva A and Koblischka M R 2002 *EBSD on High T_c Superconductors, in Studies of High Temperature Superconductors* Vol 41, ed A Narlikar (Commack, NY: Nova Science Publishers) p 1

[229] Koblischka M R, Schuster T and Kronmüller H 1994 Flux penetration in granular YBCO samples: a magneto-optical study *Physica* C **219** 205

[230] Koblischka M R, van Dalen A J J and Ravikumar G 1994 Influence of melt-processing on flux behaviour and critical current densities, in *Proc. 7th IWCC Conference, 24–27 Jan. 1994, Alpbach, Austria*, ed H W Weber (Singapore: World Scientific) p 399

[231] Schuster T, Koblischka M R, Reininger T, Ludescher B, Henes R and Kronmüller H 1992 Influence of low magnetic fields on the transport properties of sintered YBCO with different grain sizes *Supercond. Sci. Technol.* **5** 614

[232] Salama K, Mironova M, Stolbov S and Sathyamurthy S 2000 Grain boundaries in bulk YBCO *Physica* C **341–348** 1401

[233] Eom C B, Marshall A F, Suzuki Y, Boyer B, Pease R F W and Geballe T H 1991 Absence of weak-link behaviour in YBCO grains connected by 90° (0 10) twist boundaries *Nature* **353** 544

[234] Mannhart J, Bielefeldt H, Goetz B, Hilgenkamp H, Schmehl A, Schneider C W and Schulz R R 2000 Grain boundaries in high-T_c superconductors: insights and improvements *Phil. Mag.* B **80** 827

[235] Mannhart J, Bielefeldt H, Goetz B, Hilgenkamp H, Schmehl A, Schneider C W and Schulz R R 2000 Doping induced enhancement of the critical currents of grain boundaries in high T_c superconductors *Physica* C **341–348** 1393

[236] Hammerl G, Schmehl A, Schulz R R, Goetz B, Bielefeldt H, Schneider C W, Hilgenkamp H and Mannhart J 2000 Enhanced supercurrent density in polycrystalline YBCO at 77 K from calcium doping of grain boundaries *Nature* **407** 162

[237] Babcock S E 1999 Roles for electron microscopy in establishing structure-property relationships for high T_c superconductor grain boundaries *Micron* **30** 449

[238] Foltyn S R, Civale L, MacManus-Driscoll J L, Jia Q X, Maiorov B, Wang H and Maley M 2007 Materials science challenges for high-temperature superconducting wire *Nat. Mater.* **6** 631

[239] MacManus-Driscoll J L, Foltyn S R, Jia Q X, Wang H, Serquis A, Civale L, B. Maiorov B, Hawley M E, Maley M P and Peterson D E 2004 Strongly enhanced current densities in superconducting coated conductors of $YBa_2Cu_3O_{7-x}$ + $BaZrO_3$. *Nat. Mater.* **3** 439

[240] Yamada Y, Takahashi K, Kobayashi H, Konishi M, Watanabe T, Ibi A, Muroga T, Miyata S, Kato T, Hirayama T and Shiohara Y 2005 Epitaxial nanostructure and defects effective for pinning in $Y(RE)Ba_2Cu_3O_{7-x}$ coated conductors *Appl. Phys. Lett.* **87** 132502

[241] Fujiyoshi T, Haruta M, Sueyoshi T, Yonekura K, Watanabe M, Mukaida M, Teranishi R, Matsumoto K, Yoshida Y, Ichinose A, Horii S, Awaji S and Watanabe K 2008 Flux pinning properties of $REBa_2Cu_3O_y$ thin films with BaZrO3 nano-rods *Physica* C **468** 1635

[242] Katzer C *et al.* 2011 Increased flux pinning in $YBa_2Cu_3O_7$–δ thin-film devices through embedding of Au nano crystals *Europhys. Lett.* **95**, 68005

[243] Rutter N A, Durrell J H, Mennema S H, Blamire M G and MacManus-Driscoll J L 2005 Transport properties of Ca-doped YBCO coated conductors *IEEE Trans. Appl. Supercon.* **15** 2570

[244] Vostner A A, Sun Y F, Weber H W, Cheng Y S, Kursumovic A and Evetts J E 2003 Neutron irradiation studies on Y-123 thick films deposited by liquid phase epitaxy on single crystalline substrates *Physica* C **399** 120

[245] Withnell T D, Schoppl K R, Durrell J H and Weber H W 2009 Effects of irradiation on vicinal YBCO thin films *IEEE Trans. Appl. Supercon.* **19** 2925

[246] Eisterer M, Zehetmayer M, Weber H W, Jiang J, Weiss J D, Yamamoto A, Hellstrom E E, Larbalestier D C, Zhigadlo N D and Karpinski J 2010 Disorder effects and current percolation in FeAs-based superconductors *Supercond. Sci. Technol.* **23** 014009

[247] Roas B, Hensel B, Saemann-Ischenko G and Schulz L 1989 Irradiation-induced enhancement of the critical current density of epitaxial $YBa_2Cu_3O_{7-x}$ thin films *Appl. Phys. Lett.* **54** 1051

[248] Lin, J Y *et al.* 1996 Flux pinning in $YBa_2Cu_3O_{7-\delta}$ thin films with ordered arrays of columnar defects *Phys. Rev.* B **54**, 12717(R)

[249] Leroux M *et al.* 2015 Rapid doubling of the critical current of $YBa_2Cu_3O_{7-\delta}$ coated conductors for viable high-speed industrial processing *Appl. Phys. Lett.* **107** 192601

[250] Jia J *et al.* 2013 Doubling the critical current density of high temperature superconducting coated conductors through proton irradiation *Appl. Phys. Lett.* **103** 122601

[251] Sheng H *et al.* 2016 Irradiation induced defects in $YBa_2Cu_3O_{7-\delta}$ coated conductors *Microsc. Microanal.* **22** 1492

[252] Lessure H S, Simizu S, Sankar S G, McHenry M E, Cost J R and Maley M P 1991 Critical current density and flux pinning dominated by neutron irradiation induced defects in $YBa_2Cu_3O_{7-x}$ *J. Appl. Phys.* **70** 6513

[253] Sickafus K E, Willis J O, Kung P J, Wilson W B, Parkin D M, Maley M P, Clinard F W Jr, Salgado C J, Dye R P and Hubbard K M 1992 Neutron-radiation-induced flux pinning in Gd-doped $YBa_2Cu_3O_{7-x}$ and $GdBa_2Cu_3O_{7-x}$ *Phys. Rev.* B **46** 11862

[254] Bulaevskii L N, Vinokur V M and Maley M P 1996 Reversible magnetization of irradiated high T_c superconductors *Phys. Rev. Lett.* **77** 936

[255] Ghigo G, Gerbaldo R, Gozzelino L, Mezzetti E, Minetti B and Wisniewski A 1997 Influence of irradiation-induced correlated and random disorder on flux pinning in bulk YBCO melt-textured samples *J. Supercond.* **10** 541

[256] Lee J-W, Lessure H S, Laughlin D E, McHenry M E, Sankar S G, Willis J O, Cost J R and Maley M P 1990 Observation of proposed flux pinning sites in neutron-irradiated $YBa_2Cu_3O_{7-x}$ *Appl. Phys. Lett.* **57** 2150

[257] Cost J R, Willis J O, Thompson J D and Peterson D E 1988 Fast-neutron irradiation of YBa_2Cu_3O *x* *Phys. Rev.* B **37** 1563

[258] Sauerzopf F M, Wiesinger H P, Weber H W and Crabtree G W 1995 Analysis of pinning effects in $YBa_2Cu_3O_{7-\delta}$ single crystals after fast neutron irradiation *Phys. Rev.* B **51** 6002

[259] Sauerzopf F M, Wiesinger H P, Kritscha W, Weber H W, Crabtree G W and Liu J Z 1991 Neutron-irradiation effects on critical current densities in single-crystalline $YBa_2Cu_3O_{7-y}$ *Phys. Rev.* B **43** 3091

[260] Wisniewski A, Czurda C, Weber H W, Baran M, Reissner M, Steiner W, Zhang P X and Zhou L 1996 Influence of oxygen deficiency and of neutron-induced defects on flux pinning in melt textured bulk $YBa_2Cu_3O_{7-x}$ samples *Physica* C **266** 309

[261] Werner M, Sauerzopf F M, Weber H W, Veal B D, Licci F, Winzer K and Koblischka M R 1994 Fishtails in 123-superconductors *Physica* C **235–240** 2833

[262] Frischherz M C, Kirk M A, Farmer J, Greenwood L R and Weber H W 1994 Defect cascades produced by neutron irradiation in $YBa_2Cu_3O_{7-\delta}$ *Physica* C **232** 309

[263] Wisniewski A, Schalk R M, Weber H W, Reissner M and Steiner W 1992 Comparison of neutron irradiation effects in the 90 K and 60 K phases of YBCO ceramics *Physica* C **197** 365

[264] Wisniewski A, Brandstätter G, Czurda C, Weber H W, Morawski A and Lada T 1994 Comparison of fast-neutron irradiation effects in $YBa_2Cu_3O_{7-x}$ (123) and $YBa_2Cu_4O_8$ (124) ceramics *Physica* C **220** 181

[265] Maley M P, Willis J O, Lessure H and McHenry M E 1990 Dependence of flux-creep activation energy upon current density in grain-aligned $YBa_2Cu_3O_{7-x}$ *Phys. Rev.* B **42** 2639

[266] Fleischer R L, Hart H R Jr, Lay K W and Luborsky F E 1989 Increased flux pinning upon thermal-neutron irradiation of uranium-doped $YBa_2Cu_3O_7$ *Phys. Rev.* B **40** 2163

[267] Luborsky F E, Arendt R H, Fleischer R L, Hart H R Jr, Lay K W, Tkaczyk J E and Orsini D 1991 Critical currents after thermal neutron irradiation of uranium doped superconductors *J. Mater. Res.* **6** 28

[268] Wheeler R, Kirk M A, Marwick A D, Civale L and Holtzberg F H 1993 Columnar defects in $YBa_2Cu_3O_{7-\delta}$ induced by irradiation with high energy heavy ions *Appl. Phys. Lett.* **63** 1573

[269] Yacoby E R, Shaulov A, Yeshurun Y, Konczykowski M and Rullier-Albenque F 1992 Irreversibility line in $YBa_2Cu_3O_7$ samples. A comparison between experimental techniques and effect of electron irradiation*Physica* C **199** 15

[270] Civale L, Marwick A D, McElfresh M W, Worthington T K, Malozemoff A P, Holtzberg F H, Thompson J R and Kirk M A 1990 Defect independence of the irreversibility line in proton-irradiated Y–Ba–Cu–O crystals *Phys. Rev. Lett.* **65** 1164

[271] Civale L, McElfresh M W, Marwick A D, Holtzberg F, Feild C, Thompson J R and Christen D K 1991 Scaling of the hysteretic magnetic behaviour in $YBa_2Cu_3O_7$ single crystals *Phys. Rev.* B **43** 13732

[272] Strickland N M, Talantsev E F, Long N J, Xia J A, Searle S D, Kennedy J, Markwitz A, Rupich M W, Li X and Sathyamurthy S 2009 Flux pinning by discontinuous columnar defects in 74 MeV Ag-irradiated $YBa_2Cu_3O_7$ coated conductors *Physica* C **469** 2060

C2

Bismuth-Based Superconductors

Jun-ichi Shimoyama

C2.1 Introduction

The bismuth-based superconductors are one of the representative layered cuprate superconductors, distinguished by their high anisotropy and the ability to manufacture long superconducting tapes for various applications. In addition, this system can be regarded as model compounds characteristic of many other homologous series of cuprate superconductors.

In general, the chemical formula of bismuth-based superconductors is expressed as $Bi_2Sr_2Ca_{n-1}Cu_nO_y$(n = 1,2,3,.., y~2n+4), and each phase is abbreviated as Bi22(n-1)n. Figure C2.1 shows the ideal crystal structures of bismuth-based superconductors with n = 1~3. All bismuth-based superconductors have a layered crystal structure as it is common for cuprate superconductors, and their stacking period along the c-axis is $-(BiO)_2$-SrO-CuO_2-(CaO-CuO_2)$_{n-1}$-SrO-. In the stacking unit, -CuO_2-(CaO-CuO_2)$_{n-1}$- is called as the superconducting layer, while –SrO-$(BiO)_2$-SrO- is the blocking layer. The valence state of bismuth ions is trivalent in all phases, while those of strontium, calcium and copper are divalent. Therefore, the total nominal valence of cations is 4n+8, which is equal to the total valence of anions, divalent oxygen ions, assuming that the oxygen content y is expressed as 2n+4. In the actual bismuth-based superconductors, a certain amount of excess oxygen ions located within the Bi-O double layer provides hole carriers to the CuO_2 plane, resulting in superconductivity. Characteristic features of the crystal structure of this system are a blocking layer with a thickness of approximately 1.2 nm containing a weakly coupled Bi-O double layer, which results in quite large anisotropies in electromagnetic properties based on the large value of m_c^*/m_{ab}^* and in critical current density, where $J_c(//ab) \gg J_c(//c)$. Furthermore, a modulation structure along the b-axis direction is confirmed for all undoped Bi(22(n-1)n, which originates in wavy Bi-O layers. Crystal structures of the bismuth-based superconductors are categorized to be orthorhombic, however, the a- and b-axis lengths are almost identical and they are ~0.538 nm for Bi2201 and ~0.541 nm for Bi2212 and Bi2223. The distance between copper ions in the CuO_2 plane is found to be ~0.380 nm for Bi2201 and ~0.383 nm for Bi2212 and Bi2223. These values are slightly shorter than barium-containing layered cuprates discussed in the next section because of the smaller ion size of strontium relative to that of barium. The c-axis lengths of Bi2201, Bi2212 and Bi2223, which are dependent on the excess oxygen content, are approximately 2.46 nm, 3.08 nm and 3.71 nm, respectively, and they are roughly expressed as 1.64+0.62n nm. This means that the thickness of (CaO-CuO_2) layer is 0.31 nm. Note that the mean spacing between double Bi-O layers is ~0.32 nm almost independent of n. This relatively wide space can accept excess oxygen. The unique physical properties of the bismuth-based superconductors described below are mainly based on these structural characteristics.

C2.2 Brief History of Bismuth-Based Superconductors

The first bismuth-based superconductor is $Bi_2Sr_2CuO_y$ [Bi2201] independently discovered in 1987 in France and Japan [1, 2]. Since T_c of this phase was reported to be ~20 K at most, which was much lower than that of the already discovered superconductor $YBa_2Cu_3O_y$ (Y123: T_c > 90 K) and even lower than $(La_{1-x}Sr_x)_2CuO_4$ (T_c ~38 K), the discovery of the rare-earth free cuprate superconductor had attracted much less attention in the early stage. However, important features of the bismuth-based superconductors were found in Bi2201, namely the modulation structure along the b-axis with a period of ~4.8b and different T_c values between a quenched sample from 900°C (T_c = 22 K) and a sample cooled in the furnace (T_c ~ 8 K). The former feature is common for all $Bi_2Sr_2Ca_{n-1}Cu_nO_y$ and results in in-plane anisotropy, *i.e.*, lower normal-state resistivity in the a-axis direction than in the b-axis direction. The second feature suggested oxygen nonstoichiometry in this system, which affects carrier doping state.

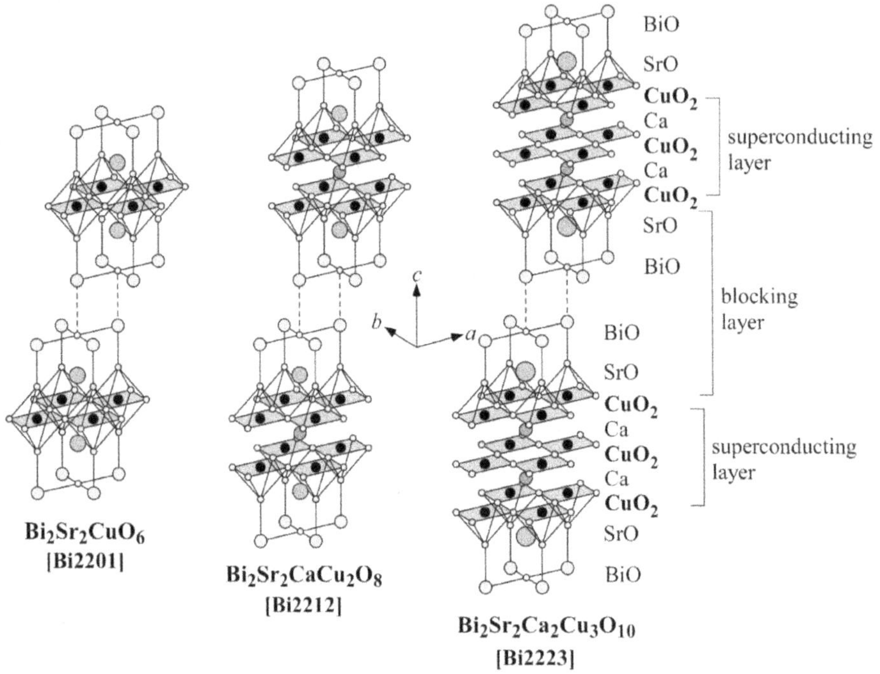

FIGURE C2.1 Ideal crystal structures of Bi2201, Bi2212 and Bi2223.

A breakthrough occurred in December 24, 1987 in resistivity measurements for samples prepared by Dr. H. Maeda at National Research Institute for Metals in Japan. His attempts to precisely control lattice constants by mixing isovalent alkaline earth elements, strontium and calcium, for bismuth and copper containing oxides resulted in the first high T_c superconductivity exceeding 100 K. Although the relevant superconducting phases were not identified in the first paper [3], superconductivity above liquid nitrogen temperature 77 K was clearly confirmed for the Bi-Sr-Ca-Cu-O system and a sample sintered at relatively high temperature for longer time showed a sharp resistivity drop at ~105 K. This paper was submitted almost one month after the first observation of high-T_c superconductivity in the Bi-Sr-Ca-Cu-O system. However, zero resistance above 100 K had not been achieved before submission, suggesting that the synthesis of high-purity 105 K phase is not easy just by the control of starting composition and sintering conditions. Note that the first paper was accepted for publication the next day after submission, and the new rare-earth free high-T_c superconductors attracted much attention.

Vigorous studies promptly carried out at many laboratories revealed that the active superconducting phases of Bi-Sr-Ca-Cu-O reported in the first paper [3] were Bi2212 and Bi2223 [4]. In addition, high-resolution TEM observation revealed the existence of Bi2234 layers as stacking faults [5]. As shown in Figure C2.1, the crystal structures of Bi2212 and Bi2223 reveal that calcium and strontium ions occupy different cation sites; small calcium ions are located between two CuO_2 planes, while larger strontium ions exist between CuO_2 and BiO planes. This new concept, using two alkaline earth elements, for material design of new cuprate superconductors eventually triggered discoveries of various homologous series of superconductors, such as $Tl_2Ba_2Ca_{n-1}Cu_nO_y (n = 1,2,3,..)$ [6, 7] and $HgBa_2Ca_{n-1}Cu_nO_y (n = 1,2,3,..)$ [8, 9]. Thus far, compounds with $n \geq 4$ in $Bi_2Sr_2Ca_{n-1}Cu_nO_y$ have been prepared only as thin films [10, 11], while sintered bulk samples with $n \geq 4$ of thallium- and mercury-based superconductors can be synthesized.

Partial substitution of lead for the bismuth site was found to be effective for the synthesis of highly pure Bi2223 sintered bulks [12–14], leading to zero resistance at 105~110 K. Multiple effects of lead substitution for physical properties of the bismuth-based superconductors will be described later.

The crystal shape of the bismuth-based superconductors is thin and plate-like with a wide *ab*-plane reflecting their layered crystal structure. In addition, crystals are easily cleaved at the weakly coupled Bi-O double layer. These characteristics are preferable for both investigations of fundamental physical properties using single crystals with fresh surfaces and the development of superconducting tapes with strongly *c*-axis–oriented oxide filaments. In the 1990's, understanding of the basic properties in highly anisotropic superconductors had been promoted mainly by studies on Bi2212 single crystals, and fabrication techniques of long length tapes had been established at various manufacturing companies. Especially, progress of multi-filamentary silver-sheathed Bi2223 tapes have been prominent in these past 25 years and they are extensively used as for cables and magnets at present.

C2.3 Flexible Chemical Composition and Phase Stability of Bismuth-Based Superconductors

C2.3.1 Cation Composition

The generic chemical formula of the bismuth-based superconductors is described as $Bi_2Sr_2Ca_{n-1}Cu_nO_y$, however, those with integral cation ratio, Bi:Sr:Ca:Cu = 2:2:*n*-1:*n*, are difficult to prepare. In general, samples with bismuth-rich and strontium-poor composition form, even from a nominal composition of integral cation ratio. Furthermore, calcium ions can partially substitute for the strontium site. Therefore, the actual chemical compositions are expressed as $Bi_{2+\alpha}Sr_{2-\alpha-\beta}Ca_{n-1+\beta}Cu_nO_y$, where α is 0.1~0.2 and β is controlled by the nominal composition. Such nonstoichiometric cation compositions in this system were well recognized soon after its discovery through compositional analysis leading to many efforts to synthesize samples with high phase purity starting from various compositions. Systematic studies on the relationship among nominal composition, sintering temperature in air and constituent phases in the resulting samples by Majewski *et al.*[15] showed a very wide range of cation composition for Bi2212 and Pb-doped Bi2223 and clearly suggested that samples with integral cation ratios, such as Bi:Sr:Ca:Cu = 2:2:1:2 and (Bi,Pb):Sr:Ca:Cu = 2:2:2:3, cannot be prepared by heat treatment in air. Figures C2.2(a)–(c) show the

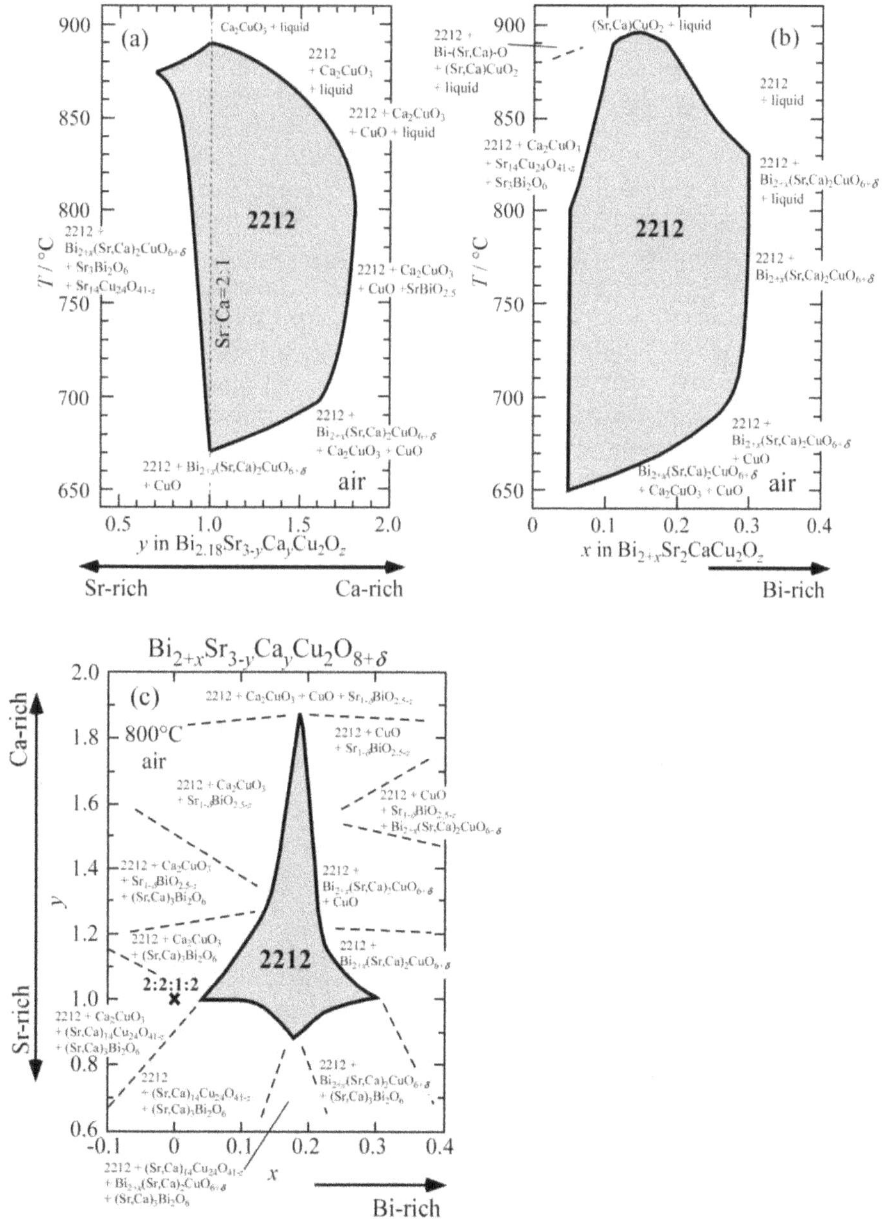

FIGURE C2.2 Constituent phases of Bi-Sr-Ca-Cu-O sintered in air starting from various cation compositions around Bi2212 (a) and (b). Constituent phases of Bi-Sr-Ca-Cu-O sintered in air at 800°C starting from various cation compositions around Bi2212 (c). These figures are replotted from ref. [15].

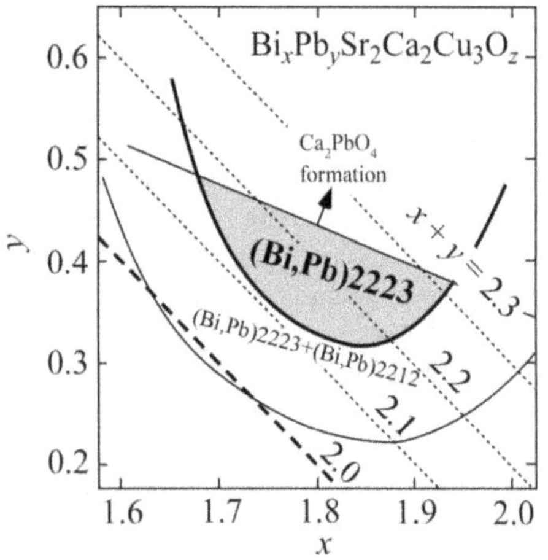

FIGURE C2.3 Phase formation range of (Bi,Pb)2223 sintered at 835°C for 84 h under P_{O2} = 1/13 atm [16]. Almost single-phase (Bi,Pb)2223 is synthesized from the gray-colored region.

constituent phases of samples sintered in air starting from various nominal compositions around Bi2212 [15].

As described before, lead substitution is essential to synthesize high purity (Bi,Pb)2223 polycrystalline bulks. In addition, sintering under a moderately reducing atmosphere of P_{O2} ~1/13 atm was found to be effective for promoting Bi2223 phase formation under a wider temperature range and also from wider cation compositions as shown in Figure C2.3 [16]. This study substantially allowed mass production of (Bi,Pb)2223 superconducting materials. Lead substitutes also for the bismuth site in Bi2201 and Bi2212 accompanying dramatic changes of electromagnetic properties, which are shown later. Substitution levels of lead for the bismuth site in Bi2212 and Bi2223 sintered bulks synthesized in air are shown in Figures C2.4(a) and C2.4(b), respectively [15].

C2.3.2 Oxygen Nonstoichiometry

In the case of $YBa_2Cu_3O_{7-\delta}$, oxygen annealing after sintering is indispensable for achieving 90 K class superconductivity. Since samples of Bi2212 and Bi2223 reproducibly showed bulk superconductivity with T_c's of ~80 K and ~105 K, respectively, without oxygen annealing, oxygen composition was not at first considered to play a significant role for determining their superconducting properties in the early stage. However, as noted excess oxygen located between the double Bi-O layers is the source of hole carriers in the bismuth-based superconductors, and its concentration has been found to be an important factor in determining superconducting properties, especially electromagnetic properties, pinning strength, grain coupling, irreversibility field, H_{irr}, and J_c. Equilibrium changes in oxygen content in a $Bi_{2.1}Sr_{1.9}CaCu_2O_y$ sintered bulk and a (Bi,Pb)2223 tape evaluated as functions of temperature and partial pressure of oxygen are shown in Figures C2.5(a) and C2.5(b), respectively. It should be noted that the absolute value of the oxygen content cannot be precisely determined due to cation nonstoichiometry and coexisting stacking faults. The range of oxygen composition of Bi2212 is less than 0.1 [17] and that of Pb-doped Bi2223 is ~0.2. These values are much smaller than that of Y123, in which oxygen content *y* changes from 6 to 7 [18–20].

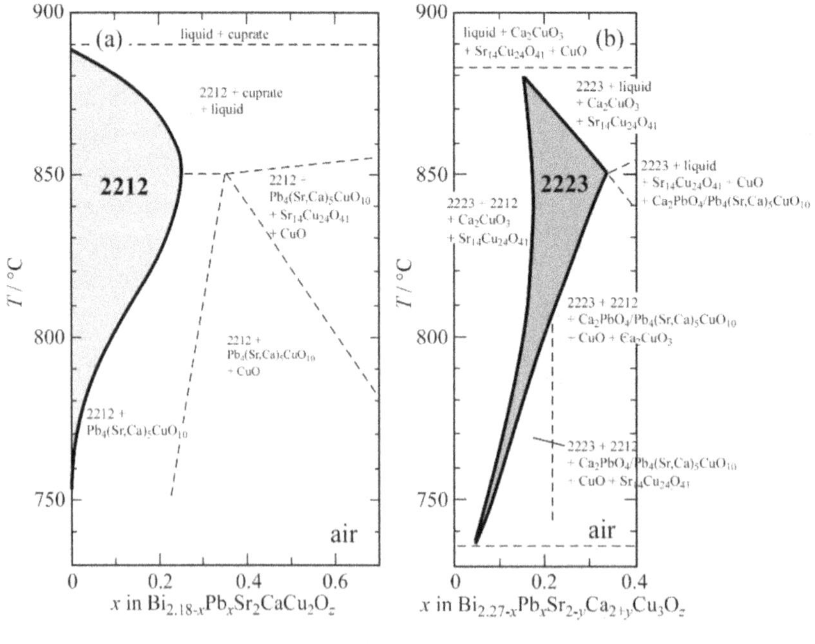

FIGURE C2.4 Ranges of temperature and lead substitution levels for phase formation of Bi2212 (a) and Bi2223 (b) with single phase [15].

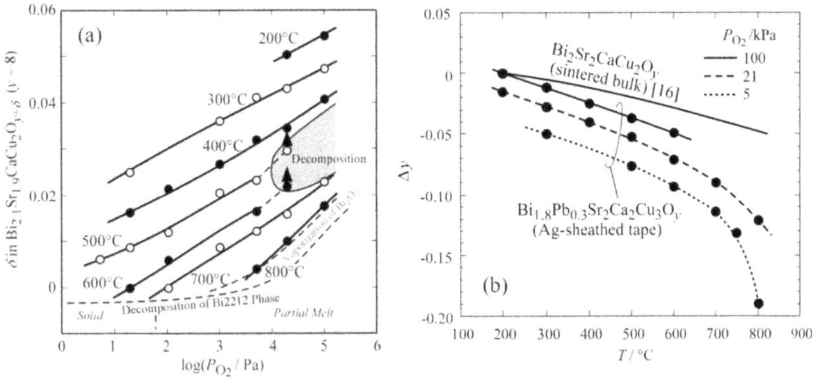

FIGURE C2.5 Oxygen nonstoichiometry of Bi2212 sintered bulk (a) and (Bi,Pb)2223 silver-sheathed tape (b).

C2.3.3 Phase Stability

Once each Bi22(*n*-1)*n* phase is formed during heat treatment, each is stable at lower temperatures than the sintering temperature in general. In addition, the bismuth-based superconductors are chemically stable in humid air even when the polycrystalline materials contain impurity phases, such as (Sr,Ca)-Cu-O. This feature is one of the advantages of this system compared with the barium-containing cuprate superconductors, such as thallium- and mercury-based superconductors or Y123. In these materials, a small amount of barium-containing impurities tends to react with carbon dioxide and moisture, resulting in degradation of superconducting properties and further decomposition of superconducting phases.

However, as indicated in Figure C2.3, Bi2212 phase is unstable under high P_{O2} atmospheres at ~600°C. Under such conditions, a quite sluggish decomposition of this phase to Bi2201 and other impurity phases occur accompanying oxygen absorption [17]. Similar phase decomposition of Bi2212 was also reported for single-crystalline samples [21]. Thermodynamic analysis of the formation enthalpy suggested that the order of phase stability is Bi2201>Bi2212>Bi2223 reflecting higher stability of the Sr-O layer than the Ca-O layer [22]. On the other hand, the (Bi,Pb)2223 phase had been considered to be unstable at ~700°C under relatively high P_{O2} atmospheres, such as in air and in pure oxygen gas, because resistivity largely increases after annealing under such conditions and the transition near T_c becomes much broader [23]. Such degradation in transport properties originates from generation of $(Pb,Bi)_3Sr_2Ca_2CuO_y$ between the plate-like (Bi,Pb)2223 crystals. In this phase, which is the so-called Pb3221, lead ions are tetravalent, while they are divalent in (Bi,Pb)2223. Since in general, a higher valence state is more stable in oxides under high P_{O2} atmospheres, this partial decomposition of (Bi,Pb)2223 is reasonable. Although the substitution level of Pb in (Bi,Pb)2223 is reduced by the formation of Pb3221 precipitates, the (Bi,Pb)2223 phase does not decompose to Bi2212 and/or Bi2201.

The sintered bulks of the bismuth-based superconductors are usually synthesized by sintering at 800~880°C in air. Figure C2.6 shows a schematic phase diagram of the $Bi_2Sr_2CuO_6$-$CaCuO_2$ system at high temperature in air [24]. All the Bi2201, Bi2212 and Bi2223 phases melt incongruently at ~890°C. Therefore, high-quality crystal boules of these phases are grown only under extremely slow growth rate, which is described in Chapter E2.2.4. Since Pb substitution has the effect of decreasing the partial melting temperature, the (Bi,Pb)2223 phase decomposes at ~860°C in air to a (Bi,Pb)2212 phase and other impurities.

The bismuth-based superconductors decompose under strongly reducing atmospheres without generating liquid phase below ~750°C. The decomposition condition as a function of temperature and P_{O2} is close to the phase boundary between CuO and Cu_2O similar to the case of Y123 [25]. As clearly seen in Figure C2.5, the decomposition in reducing atmosphere limits the lowest amount of excess oxygen in the structure.

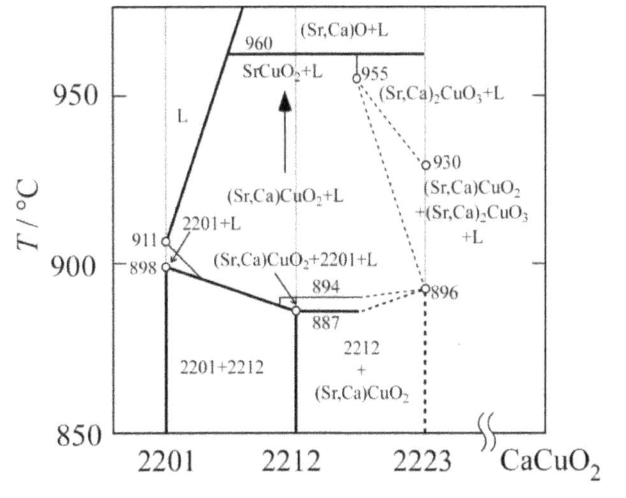

FIGURE C2.6 Phase diagram of $Bi_2Sr_2CuO_y$-$CaCuO_2$ around 900°C in air.

C2.4 Superconducting Properties

C2.4.1 Critical Temperature

Empirically, T_c of layered cuprate superconductors are determined by the following four factors.

1. Crystal structure (flatness of CuO_2 plane [26], apical oxygen height from CuO_2 plane [27], lattice distortion by external force [28])
2. Carrier doping level
3. Deviation of cation composition from integral ratio [29, 30]
4. Effects of impurity elements

In the case of the bismuth-based superconductors, the superconducting layer is sandwiched by Sr-O layers, which is intrinsically smaller than the Ba-O layers, resulting in compressive force along the *ab*-plane and relatively low apical oxygen. As described before, this system has a modulation structure along the *b*-axis. This characteristic structure originates from the distortion of Bi-O double layers with a typical period of ~4.8*b* and causes large distortion in CuO_2 plane with an amplitude of approximately 0.1 nm in the *c*-axis direction. Such structural features substantially suppress T_c in this system. Although the range of excess oxygen concentration is not large in this system, T_c is largely dependent on the oxygen composition especially for Bi2201 and Bi2212, where a change of oxygen content strongly affects the carrier concentration per copper ion. A schematic illustration showing the relationship between T_c and carrier doping level for Bi2201, Bi2212 and Bi2223 is shown in Figure C2.7 [31]. Open circles correspond to furnace-cooled samples in air after sintering, and closed circles indicate T_c for quenched samples from ~800°C in air.

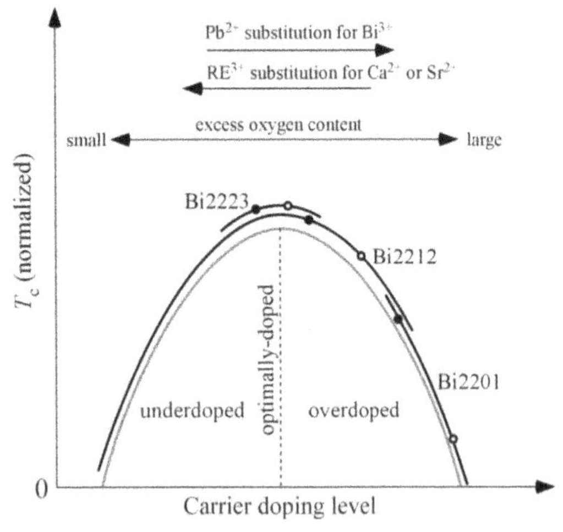

FIGURE C2.7 Schematic illustration showing relationship between T_c and carrier doping level of the bismuth-based superconductors. Open circles correspond to the typical T_c of sintered bulk samples cooled in the furnace after heat treatment in air. Closed circles show T_c of the sample quenched to room temperature from ~800°C in air (see e.g. ref. [31]).

The carrier doping state of Bi2201 is basically strongly overdoped, *i.e.*, excess hole carriers provided at the mono-CuO_2 plane. Sintered bulk samples cooled in the furnace after sintering in air reproducibly show T_c at 5~8 K. Quenching from high temperature or post-annealing in reducing atmosphere to decrease the excess oxygen content is effective to enhance T_c above 20 K, however, those samples are still in the overdoped region. Partial substitution of relatively large trivalent rare-earth element for the strontium site further decreases the carrier concentration. On the other hand, partial substitution of lead for bismuth leads to hole carrier doping, while an increase in the modulation period simultaneously occurs. The latter means that the CuO_2 plane becomes flatter by lead substitution, which is preferable for achieving higher T_c. 40 K-class superconductivity in the Bi2201 phase up to T_c = 40.3 K has been reported for co-doped samples with lead and lanthanum after reductive annealing [32, 33]. In addition, a single crystal with chemical composition of $Bi_{2.0}Sr_{1.61}La_{0.39}CuO_y$ showed T_c = 38 K at optimal doping achieved by quenching from 650°C in a pure oxygen atmosphere [34]. Excess lanthanum substitution above 20% for the strontium site decreases T_c by underdoping, and superconductivity disappears at a lanthanum substitution level above ~35%. A large amount of lead substitution increases T_c of Bi2201 above 20 K [35], despite the increase in carrier doping.

The Bi2212 phase has been called the "80 K phase", because sintered bulk samples of Bi2212 show T_c ~80 K when they are cooled in the furnace after sintering in air. However, these samples are slightly overdoped, and decreasing the oxygen content by quenching from high temperature above 800°C or annealing under moderately reducing atmosphere increases T_c up to more than 90 K. The highest T_c at optimal doping is dependent on the cation composition of Bi2212. As mentioned before, bismuth ions spontaneously substitute for the strontium site and a large amount of calcium can occupy the same site. Numerous studies suggested that the highest T_c for Bi-rich and/or Ca-rich Bi2212 is suppressed by several K compared with Bi2212 with nearly integral cation ratio, Bi:Sr:Ca:Cu = 2:2:1:2, which can be obtained by heat treatment under moderately reducing atmosphere. The highest T_c in undoped Bi2212 is 96 K, achieved by sintering under a reducing atmosphere [36]. Higher T_c values above 96 K were reported for doped Bi2212 in two ways. One is lead doping for the bismuth site to improve the flatness of the CuO_2 plane [37, 38], and the other is a small amount of yttrium doping for the calcium site [39] to suppress substitution of bismuth for the strontium site [40].

On the other hand, Bi2223 has been called the "110 K phase", because sharp superconducting transitions are reproducibly observed at 105~110 K regardless of cooling conditions and Pb-substitution levels. In other words, the carrier doping level of Bi2223 is usually near optimal doping. As described above, post-annealing at ~700°C in high P_{O2} decreases the Pb-substitution level in (Bi,Pb)2223 because of formation

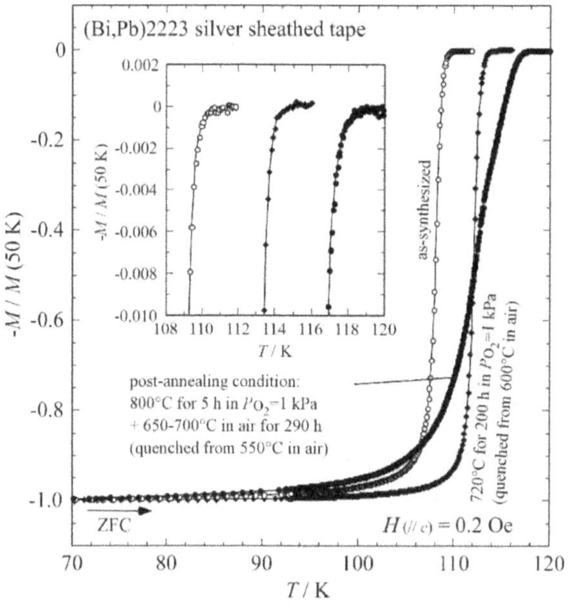

FIGURE C2.8 Zero-field-cooled magnetization curves of (Bi,Pb)2223 silver-sheathed tapes before and after post-annealing.

of the Pb3221 phase. Reduction of the Pb concentration in (Bi,Pb)2223 accompanies an increase in the *c*-axis length and T_c [41, 42]. Control of cation composition by post-annealing under moderately reducing atmosphere is also effective for enhancing T_c up to 115 K [43]. Figure C2.8 shows the zero-field-cooled magnetization curves for (Bi,Pb)2223 silver-sheathed tapes, post-annealed under various conditions. The highest T_c(onset) thus far is ~118 K, which is almost identical to the T_c measured under a high pressure of 8 GPa [28].

It is notable that of the three CuO_2 layers in the unit cell of Bi2223, the inner layer is crystallographically dissimilar to the two outer layers. In particular, it lacks an apical oxygen which may be critical to its higher T_c value [25]. NMR studies show that this inner layer tends to be somewhat underdoped relative to the outer CuO_2 layers [44]. Quite generally, the effect of external pressure in cuprates is to transfer holes from the blocking layer to the active CuO_2 layers. In the case of Bi2223, high external pressure reveals a most interesting effect as holes are transferred onto both the two outer layers and the inner layer. T_c is observed first to rise to a maximum of about 124 K, then fall to about 119 K, then rise again to 135 K [45]. Presumably, first the outer layers pass through optimal doping, then the inner layer approaches optimal doping, resulting in the second peak with its higher T_c value.

T_c values of Bi22(*n*-1)*n* phases with $n \geq 4$ have been reported only for thin films [10, 11]. The variation of T_c with *n* in Bi22(*n*-1)*n* is summarized in Figure C2.9. Even for thin films with $n = 3$ [46], T_c is largely suppressed possibly due to lattice strain caused by the epitaxial growth on the single-crystalline substrates and large deviation of cation composition from integral ratios.

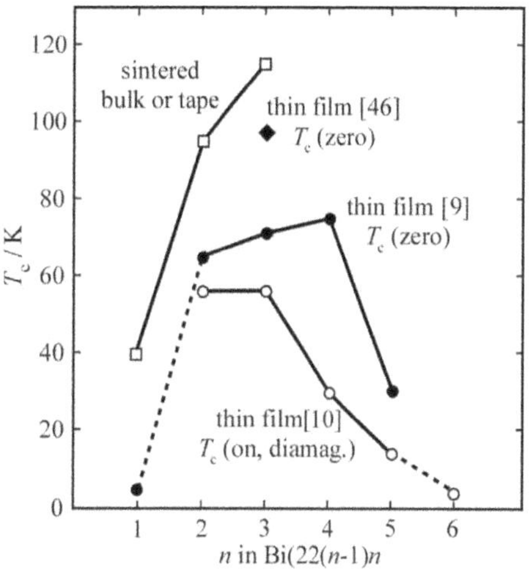

FIGURE C2.9 Relationship between T_c and *n* for Bi22(*n*-1)*n*.

C2.4.2 Electromagnetic Anisotropy and Normal-State Resistivity

Reflecting the layered crystal structure with thick blocking layers and hole carriers confined to the superconducting layer, the electromagnetic properties of bismuth-based superconductors exhibit quite large anisotropy between the *ab*- and *c*-directions. The electromagnetic anisotropy parameter γ is defined by the effective mass ratio, $\gamma \equiv (m_c^*/m_{ab}^*)^{1/2}$, which corresponds to $(\rho_c/\rho_{ab})^{1/2}$ in the normal state and H_{c2} [//*ab*]/ H_{c2} [//*c*] ~ $\xi_{ab}^2/\xi_{ab}\xi_c$. In addition, the peak field H_{pk} of the secondary peak effect in magnetization hysteresis loops indicates γ based on the relationship, $H_{pk} = \Phi_0/(\gamma^2 d^2)$ [47, 48], where Φ_0 and *d* are, respectively, the flux quantum and the interlayer distance between CuO_2 planes, *i.e.*, thickness of the blocking layer. This sharp peak effect originates from a first-order transition of the vortex system [49]. Figure C2.10 shows the dependence of T_c and γ (estimated from H_{pk}) on the

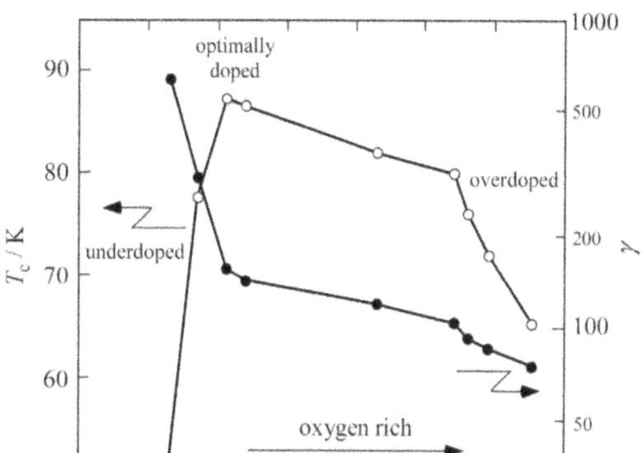

FIGURE C2.10 Variation of T_c and γ with oxygen content of Bi2212 single crystals.

oxygen content in a Bi2212 single crystal. With an increase in the oxygen content, *i.e.*, from underdoped state to overdoped state, T_c increases up to 86 K and decreases down to 66 K, while γ monotonically decreases from ~600 to ~80. Note that the relatively low T_c at optimal doping is due to a non-ideal cation composition as described above. Absolute values of γ in the overdoped state, which is the usual state for Bi2212 cooled in the furnace, and at the optimally doped state are ~100 and ~150, respectively. These are approximately 20 times higher compared with those for Y123, which has the smallest electromagnetic anisotropy among the cuprate superconductors. γ for lead-free Bi2223 is evaluated to be ~80 in the optimally doped state [50]. A slightly smaller value of γ in Bi2223 (c.f. Bi2212) is possibly due to the difference in the thickness of the superconducting layer.

As shown in Figure C2.10, oxygen content strongly affects the electromagnetic anisotropy of the bismuth-based superconductors. The temperature dependence of resistivity along *a*-, *b*- and *c*-axes, ρ_a, ρ_b and ρ_c, for Bi2212 single crystals (nominal composition: $Bi_{2.1}Sr_{1.8}CaCu_2O_y$) with oxygen content ranging from underdoped to overdoped is shown in Figure C2.11 [51]. ρ_b is higher than ρ_a for all samples due to the modulation structure along the *b*-axis direction. The ratio ρ_b/ρ_a is ~2, or less, at optimal doping, and this value increases up to 3 in single crystals having cation compositions close to the integral ratio, 2:2:1:2 [52]. ρ_a and ρ_b show metallic temperature dependence at high temperature, and ρ_b of underdoped samples exhibits slightly semiconducting behavior at low temperature, while absolute values of ρ_a and ρ_b systematically decrease with an increase in oxygen content. The out-of-plane resistivity, ρ_c, basically exhibits semiconductor-like behavior at low temperatures, and its absolute value largely decreases with increasing carrier concentration. Therefore, ρ_c/ρ_{ab} ($=\rho_c/(\rho_a^2+\rho_b^2)^{1/2}$) continuously decreases with oxygen content. This is consistent with the behavior of γ estimated from H_{pk} with oxygen composition. The semiconducting behavior of ρ_c changes to metallic accompanying dramatic decreases in absolute magnitude by the partial substitution of lead for bismuth in Bi2212 [53, 54], and a similar change was also confirmed for Bi2201 [35]. This suggests that the local conductivity at the blocking layer largely improved by lead substitution. Combining oxygen overdoping and lead substitution, γ of Bi2212 decreased down to ~35 [54]. This value was estimated from ρ_c/ρ_{ab}, because the second peak in magnetization hysteresis loop becomes broader with lead substitution [55], and is not caused in this case by the first-order transition of the vortex system. The relationship between T_c and γ for Bi2212, Bi2223 and (Bi,Pb)2212 single crystals is summarized in Figure C2.12.

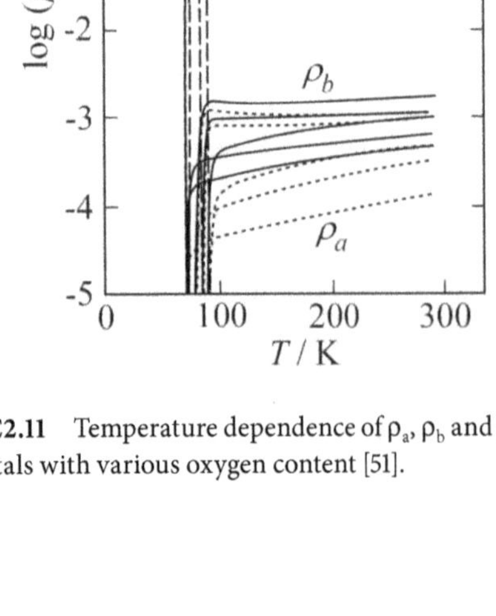

FIGURE C2.11 Temperature dependence of ρ_a, ρ_b and ρ_c for Bi2212 single crystals with various oxygen content [51].

C2.4.3 Coherence Length and Penetration Depth

In general, the GL coherence length ξ of cuprate superconductors is approximately 2 nm in the *ab*-plane near optimal doping, while ξ along the *c*-axis is largely dependent on the system because of the electromagnetic anisotropy. In the case of Bi2212, ξ_c is roughly estimated to be 0.02 nm by assuming $\gamma = 100$, which corresponds to the overdoped state. This is much shorter than the thickness of the blocking layer 1.2 nm. Therefore, superconducting layers are considered to be almost decoupled in the *c*-axis direction, resulting in large anisotropy in J_c in which $J_c\,(ab) \gg J_c\,(c)$. When the magnetic field is applied parallel to the *c*-axis, *H*//*c*, one finds an extraordinary vortex state comprising stacked pancake vortices, where the pancakes are confined to each superconducting layer. The pinning volume at each superconducting layer for one vortex is expressed

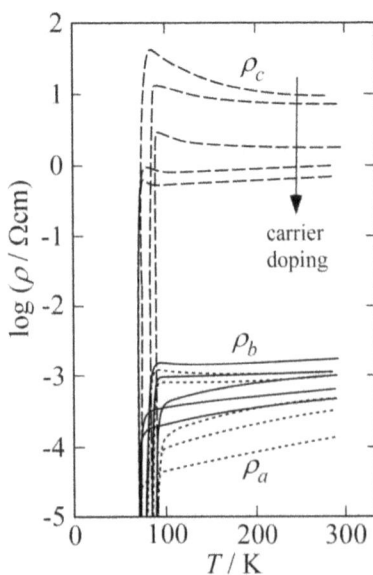

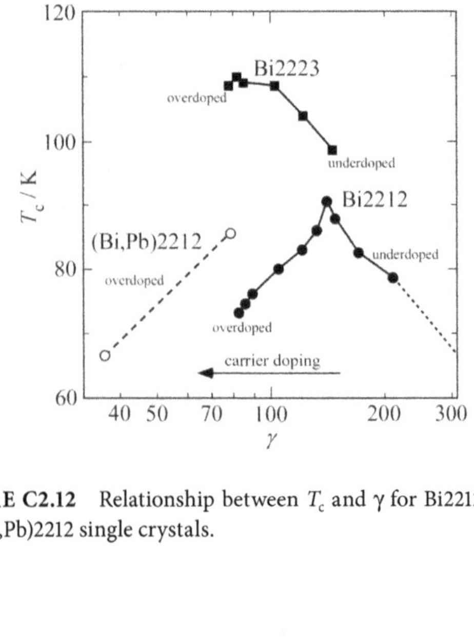

FIGURE C2.12 Relationship between T_c and γ for Bi2212, Bi2223 and (Bi,Pb)2212 single crystals.

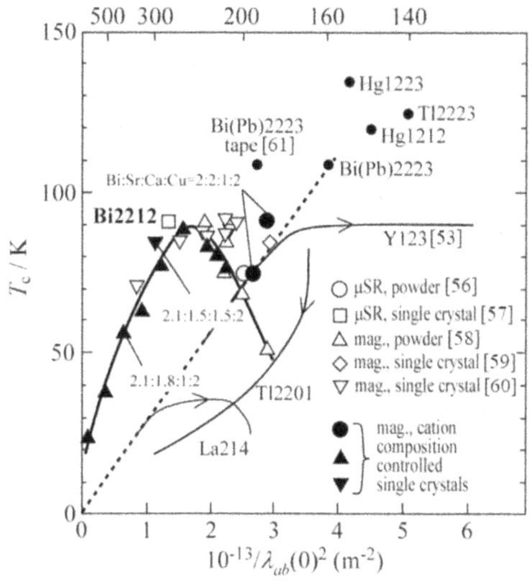

FIGURE C2.13 Relationship between T_c and $\lambda_{ab}(0)$ for cuprate superconductors.

as $\xi_{ab}{}^2\xi_c = \xi_{ab}{}^3/\gamma$. This means that the pinning volume is very small in the bismuth-based superconductors, and the pinning force of this system is intrinsically weak relative to the other cuprate superconductors with smaller γ.

The penetration depth for $H//c$ at 0 K, $\lambda_{ab}(0)$, of the cuprate superconductors is usually 140~180 nm near optimal doping. However, reported values of $\lambda_{ab}(0)$ were exceptionally long ~250 nm for Bi2212, suggesting dirty superconductivity. Figure C2.13 shows the relationship between T_c and $1/\lambda_{ab}(0)^2$ for representative superconductors with various carrier doping state. The quantity $1/\lambda_{ab}(0)^2$ is proportional to the superfluid density. Except for Bi2212, cuprate superconductors typically show universal behaviour from the underdoped to optimally doped state [56]. A so-called Uemura plot of T_c versus $1/\lambda_{ab}(0)^2$ for Bi2212 collected from several papers [57–60] reveals a dome shape. The highest T_c of the dome is located at smaller $1/\lambda_{ab}(0)^2$, while a single crystal with almost integral cation ratio showed apparently larger superfluid density, and it lies on the universal line. Such a single crystal is obtained only by crystal growth under reducing atmosphere pf P_{O2} ~1 kPa starting from Bi:Sr:Ca:Cu = 2:2:1:2 [51]. This fact indicates that most studies on the basic physical properties of Bi2212 are performed for samples with cation compositions far from the integral ratio. However, it should be noted that the intrinsically high electromagnetic anisotropy between the *ab*-plane and *c*-axis directions does not change greatly by controlling cation composition. On the other hand, reported values of $\lambda_{ab}(0)$ for lanthanum-doped Bi2201 are also long >260 nm [61, 62] possibly due to nonstoichiometric cation composition. For (Bi,Pb)2223, relatively short $\lambda_{ab}(0)$ values ~140 nm are reported for sintered bulks [63], while values around ~190 nm are reported for silver-sheathed tapes [64].

C2.4.4 Pinning Characteristics and Vortex Phase Diagram

The bismuth-based superconductors are representative materials showing very weak pinning characteristics at high temperatures especially under $H//c$. There are two major reasons for the poor pinning properties, one is the small pinning volume for very weakly coupled or decoupled vortices as described above, and the other is the frequently nonstoichiometric cation composition. The former had been considered as the predominant reason why Bi2212 and Bi2223 materials show rapid decrease in J_c with increase in magnetic field along with a very low irreversibility field, H_{irr}, above ~30 K. However, recent studies revealed that the latter reason strongly affects H_{irr} and suppresses J_c in the irreversible region under $H//c$. This means that the cation composition as well as carrier doping state strongly affects the pinning force of this system, though the predominant intrinsic pinning site has not been well identified in crystals of the bismuth-based superconductors. Since the superconducting condensation energy is one of the most important determining factors for the elementary pinning force, a large condensation energy is preferable for enhanced pinning force of defects. For Bi2212 single crystals, the condensation energy depends on cation composition, and it is larger in crystals having cation composition close to Bi:Sr:Ca:Cu = 2:2:1:2 [65]. Corresponding to this, the critical current properties of Bi2212 single crystals systematically improve by changing cation composition toward integral ratios, and a similar enhancement of J_c was also confirmed for (Bi,Pb)2212 single crystals [66]. Figure C2.14 shows the magnetic field dependence of J_c for various Bi2212 single crystals in the optimal or slightly overdoped state. A J_c-H curve for Bi2212/Ag composite tape prepared by the melt-solidification method is also shown [67]. As clearly seen, crystals with cation

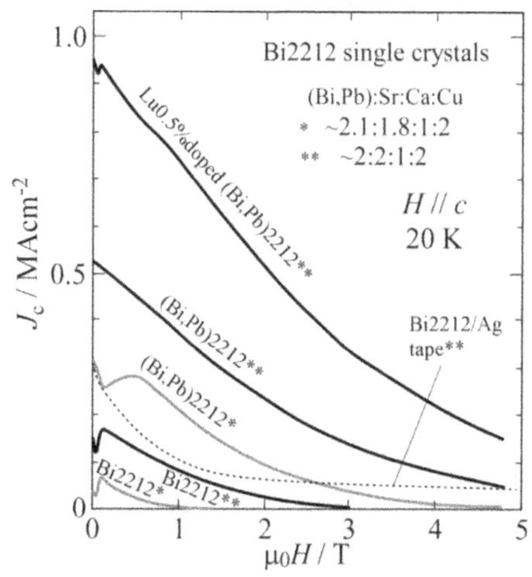

FIGURE C2.14 J_c-H curves for Bi2212 and (Bi,Pb)2212 single crystals at 20 K under $H//c$. The dashed curve for Bi2212/Ag tape is from reference [67].

composition close to integral ratios show higher J_c and H_{irr}. In addition, Pb substitution largely improves J_c in magnetic field due to the decrease in electromagnetic anisotropy and the generation of effective pinning sites originating from the inhomogeneous distribution of Pb ions at the bismuth site [68] as well as micro-phase segregation of Pb-rich and Pb-poor regions [55, 69]. On the other hand, the irreversibility lines for Bi2212 and Bi(Pb)2212 under *H//c* can be scaled by γ and temperature normalized by T_c to give H_{irr} (Oe) = $4 \times 10^7 \gamma^{-2} (1-T/T_c)^{1.5}$ for $T > 0.7\ T_c$. This is a universal relationship for layered cuprates [70].

The magnetic phase diagram of Bi2212 had been studied extensively during the 1990's because it shows unconventional magnetization behavior, such as a large reversible region, very poor pinning characteristics accompanying fast flux creep and first-order vortex melting in the reversible region [71]. Through the numerous studies on the magnetic properties of Bi2212 single crystals, a quite complex magnetic phase diagram for *H//c* was established [72], and it has been considered as a representative example of highly anisotropic cuprate superconductors. However, those studies were carried out for bismuth-rich and strontium-poor Bi2212 single crystals, because large single crystals with such cation nonstoichiometric composition are easily grown in air. As shown in Figure C2.14, Bi2212 single crystals with cation compositions close to the integral ratio show much stronger flux pinning behavior, and the magnetic phase diagram was found to be less complex. Magnetic phase diagrams of conventional and cation-composition-controlled Bi2212 single crystals are compared in Figure C2.15.

C2.4.5 Introduction of Effective Pinning Site

Many attempts have been made to introduce effective pinning sites in crystals of bismuth-based superconductors by various methods. Most of these have not so far been applied to practical polycrystalline materials because the degree of grain alignment, the contact area between grains and the resulting critical current properties across grain boundaries dominate the performance of the materials. We focus then on single-crystal behavior.

Particle irradiation using heavy ions, neutrons, light ions and electrons is one of the popular methods to introduce artificial defects acting as pinning sites in cuprate superconductors. These are effective due to the characteristically short coherence length. For the bismuth-based superconductors, heavy ion irradiation, which produces columnar defects, is well known to enhance J_c in a magnetic field when it is applied parallel to the defect columns. Vortex cores tend to immobilize along the length of the columnar defects because they are effectively pinned at all points along the core. In contrast, cascade-type defects created by neutron irradiation and atomic scale point defects introduced by light ion or high-energy electron irradiation act more as three-dimensional pinning centers. The magnitude of enhancement in J_c and H_{irr} also depends on cation composition as well as the carrier doping state reflecting the magnitude of the condensation energy in the crystal. In the case of Bi2212 single crystals, columnar defects act as stronger pinning sites when the crystal is overdoped with cation composition close to the integral ratio. In addition, Pb substitution enhances the pinning force of artificial defects. It is also found that the pinning force of columnar defects is larger in Bi2223 than in Bi2212 probably due to the additional thickness of the superconducting layer within the unit cell [73].

Columnar-type defects were also successfully introduced as nano-MgO rods [74] and carbon nanotubes [75] in Bi2212 crystals, resulting in improved pinning characteristics. In addition, titanium doping was also reported to enhance J_c through formation of columnar-type defects in Bi2212 crystals [76]. However, J_c enhancement by Pb substitution is comparatively much larger, as described above [55, 77–79]. Dilute impurity doping is a universally effective method to improve flux pinning properties [80]. For example, low-level lutetium doping on the calcium site of (Bi,Pb)2212 single crystals as shown in Figure C2.14. Similarly, cobalt doping on copper

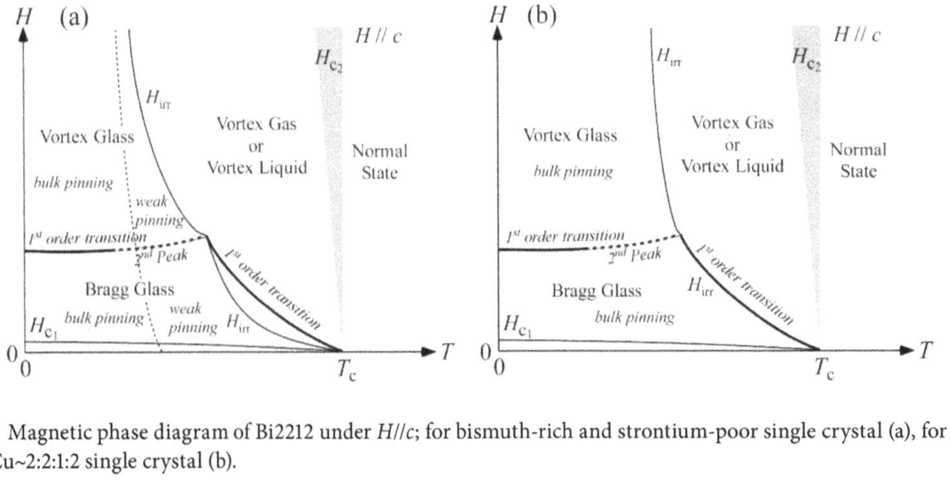

FIGURE C2.15 Magnetic phase diagram of Bi2212 under *H//c*; for bismuth-rich and strontium-poor single crystal (a), for cation composition of Bi:Sr:Ca:Cu~2:2:1:2 single crystal (b).

sites, and light rare earths, such as lanthanum or neodymium, on strontium sites is found to enhance J_c in a magnetic field provided the doping level is less than 1% [81].

C2.5 Summary

The bismuth-based superconductors have provided many important insights regarding the basic features of cuprate superconductors with a thick blocking layer. These include a very large anisotropy in coherence length, normal-state resistivity and J_c, and a weakly coupled, or decoupled, vortex state when *H//c* with relatively weak pinning because of the small pinning volume. Local distortion in the CuO_2 plane suppresses T_c of this system. The highest T_c values of Bi2201, Bi2212 and Bi2223 phases under ambient pressure are ~40 K, ~96 K, ~118 K, respectively, which are much lower than ~95 K [7], ~128 K [8] and ~138 K [82, 83] reported for the $HgBa_2Ca_{n-1}Cu_nO_y$ system due, among other things, to the local lattice distortion of the superconducting CuO_2 plane. However, the chemical stability and thin plate-like crystal shape, with easy cleavage along the *ab*-plane, are favorable characteristics for the development of *c*-axis-oriented long (Bi,Pb)2223 superconducting tapes. In addition, precise control of cation composition to close to integral ratios is a promising strategy for further improvement of superconducting properties of the materials and is also important to clarify the systematics in physical properties of the bismuth-based superconductors.

References

1. C. Michel, M. Hervieu, M. M. Borel, A. Grandin, F. Deslandes, J. Provost and B. Raveau, *Z. Phys. B* **68** (1987) 421–423.
2. J. Akimitsu, A. Yamazaki, H. Sawa and H. Fujiki, *Jpn. J. Appl. Phys.* **26** (1987) L2080.
3. H. Maeda, Y. Tanaka, M. Fukutomi and T. Asano, *Jpn. J. Appl. Phys.* **27** (1988) L209.
4. J. L. Tallon, R. G. Buckley, P. W. Gilberd, M. R. Presland, I. W. M. Brown, M. E. Bowden, L. A. Christian and R. Goguel, *Nature* **333** (1988) 153–156.
5. N. Kijima, H. Endo, J. Tsuchiya, A. Sumiyama, M. Mizuno and Y. Oguri, *Jpn. J. Appl. Phys.* **26** (1988) L821–L823.
6. Z. Z. Sheng and A. M. Hermann, *Nature* **332** (1988) 55–58.
7. Z. Z. Sheng and A. M. Hermann, *Nature* **332** (1988) 138–139.
8. S. N. Putilin, E. V. Antipov, O. Chmaissem and M. Marezio, *Nature* **362** (1993) 226–228.
9. A. Schilling, M. Cantoni, J. D. Guo and H. R. Ott, *Nature* **363** (1993) 56–58.
10. Y. Nakayama, I. Tsukada and K. Uchinokura, *J. Appl. Phys.* **70** (1991) 4371–4377.
11. H. Narita, T. Hatano and K. Nakamurta, *J. Appl. Phys.* **72** (1992) 5778–5785.
12. S. A. Sunshine, T. Siegrist, L. F. Schneemeyer, D. W. Murphy, R. J. Cava, B. Batlogg, R. B. van Dover, R. M. Fleming, S. H. Glarum, S. Nakahara, R. Farrow, J. J. Krajewski, S. M. Zahurak, J. V. Waszczak, J. H. Marshall, P. Marsh, L. W. Rupp, Jr. and W. F. Peck, *Phys. Rev. B* **38** (1988) 893–896.
13. M. Takano, J. Takada, K. Oda, H. Kitaguchi, Y. Miura, Y. Ikeda, Y. Tomii and H. Mazaki, *Jpn. J. Appl. Phys.* **27** (1988) L1041–L1043.
14. S. M. Green, C. Jiang, Yu Mei, H. L. Luo and C. Politis, *Phys. Rev. B* **38** (1988) 5016–5018.
15. P. J. Majewski, *Supercond. Sci. Technol.* **10** (1997) 453–468.
16. U. Endo, S. Koyama and T. Kawai, *Jpn. J. Appl. Phys.* **27** (1988) L1476–L1479.
17. J. Shimoyama, J. Kase, T. Morimoto, J. Mizusaki and H. Tagawa, *Physica C* **185–189** (1991) 931.
18. K. Kishio, J. Shimoyama, T. Hasegawa, K. Kitazawa and K. Fueki. *Jpn. J. Appl. Phys.* **27** (1987) L1228–L1230.
19. T. B. Lindemer, J. F. Hunley, J. E. Gates, A. L. Sutton Jr., J. Brynestad, C. R. Hubbard and O. K. Gallagher, *J. Am. Chem. Soc.* **72** (1989) 1775–1788.
20. J. Shimoyama, Y. Yokota, M. Shiraki, Y. Sugiura, S. Horii and K. Kishio, *MRS Proceedings* **848** (2004), http://dx.doi.org/10.1557/PROC-848-FF10.8
21. W. Wu, L. Wang, X. G. Li, G. Zhou, Y. Qian, Q. Qin and Y. Zhang, *J. Appl. Phys.* **74** (1993) 7388–7392.
22. Y. Idemoto, K. Shizuka, Y. Yasuda and K. Fueki, *Physica C* **211** (1993) 36–44.
23. W. G. Wang, J. Horvat, J. N. Li, H. K. Liu and S. X. Dou, *Physica C* **297** (1998) 1–9.
24. T. Suzuki, K. Yumoto, M. Mamiya, M. Hasegawa and H. Takei, *Physica C* **301** (1998) 173–184.
25. A. Mawdsley, J. L. Tallon and M. R. Presland, *Physica C* **190** (1992) 437–443.
26. O. Chmaissem, J. D. Jorgensen, S. Short, A. Knizhnik, Y. Eckstein and H. Shaked, *Nature* **397** (1999) 45–48.
27. S. Maekawa, J. Inoue and T. Toyama, *The Physics and Chemistry of Oxide Superconductors* **60** (1992), ed. Iye Y and Yasuoka H (Berlin: Springer) pp 105–115.
28. H. Takahashi and T. Tomita, *J. Cryo. Soc. Jpn.* **46** (2011) 203–211.
29. T. Kajitani, K. Kusaba, M. Kikuchi, N. Kobayashi, Y. Syono, T.B. Williams and M. Hirabayashi, *Jpn. J. Appl. Phys.* **27** (1988) L587–L590.
30. U. Endo *et al.*, *Jpn. J. Appl. Phys.* **27** (1988) L1476.
31. M. R. Presland, J. L. Tallon, R. G. Buckley, R. S. Liu and N E Flower, *Physica C* **176** (1991) 95–105.
32. T. Amano, M. Tange, M. Yokoshima, T. Kizuka, S. Nishizaki and R. Yoshizaki, *Physica C* **412–414** (2004) 230–234.

33. Y. Murakoshi, S. Kambe and M. Kawai, *Physica C* **178** (1991) 71–74.
34. S. Ono, Y. Ando, T. Murayama, F. F. Balakirev, J. B. Betts and G. S. Boebinger, *Phys. Rev. Lett.* **85** (2000) 638–641.
35. I. Chong, T. Terashima, Y. Bando, M. Takano, Y. Matsuda, T. Nagaoka and K. Kumagai, *Physica C* **290** (1997) 57–62.
36. M. Kato, K. Yoshimura and K. Kosuge, *J. Solid State Chem.* **106** (1993) 514–516.
37. S. Kambe, T. Matsuoka, M. Takahashi, M. Kawai and T. Kawai, *Phys. Rev. B* **42** (1990) 2669.
38. J. Shimoyama, Y. Nakayama, K. Kitazawa, K. Kishio, Z. Hiroi, I. Chong and M. Takano, *Physica C* **281** (1997) 69–75.
39. J. L. Tallon, R. G. Buckley, M. P. Staines, M. R. Presland and P. W. Gilberd, *Appl. Phys. Lett.* **54** (1989) 1591–1594.
40. H. Eisaki, N. Kaneko, D. L. Feng, A. Damascelli, P. K. Mang, K. M. Shen, Z.-X. Shen and M. Greven, *Phys. Rev. B* **69** (2004) 064512.
41. J. Wang, M. Wakata, T. Kaneko, S. Takano and H. Yamauchi, *Physica C* **208** (1993) 323–327.
42. J. Shimoyama, A. Tanimoto, T. Nakashima, T. Asanuma, S. Horii, K. Kishio, T. Kato, S. Kobayashi, K.Yamazaki, K. Hayashi and K. Sato, *Physica C* **460-462** (2007) 1405–1406.
43. M. Watanabe, J. Shimoyama, K. Obata, K. Kishio, S. Kobayashi and K. Hayashi, *IEEE Trans. Appl. Supercond.* **21** (2011) 2812–2815.
44. A. Trokiner, L. LeNoc, J. Schneck, A. M. Pougnet, R. Mellet, J. Primot, H. Savary, Y. M. Gao and S. Aubry, *Phys. Rev. B* **44** (1991) 2426–2429.
45. X.-J. Chen, V. V. Struzhkin, Y. Yu, A. F. Goncharov, C. T. Lin, H. K. Mao and R. J. Hemley, *Nature* **466** (2010) 950–953.
46. K. Endo, H. Yamasaki, S. Misawa, S. Yoshida and K. Kajimura, *Nature* **355** (1992) 327–328.
47. V.M. Vinokur, P.H. Kes and A.E. Koshelev, *Physica C* **168** (1990) 29–39.
48. T. Tamegai, Y. Iye, I. Oguro and K. Kishio, *Physica C* **213** (1993) 33–42.
49. Y. Matsuda, M. B. Gaifullin, N. Chikumoto, J. Shimoyama and K. Kishio, *Physica C* **357-360** (2001) 432–434.
50. K. Shimizu, T. Okabe, S. Horii, K. Otzschi, J. Shimoyama and K. Kishio, *MRS Proceedings* **689** (2001), http://dx.doi.org/10.1557/PROC-689-E3.5.
51. Y. Kotaka, T. Kimura, H. Ikuta, J. Shimoyama, K. Kitazawa, K. Yamafuji, K. Kishio and D. Pooke, *Physica C* **235-240** (1994) 1529–1530.
52. T. Makise, S. Uchida, S. Horii, J. Shimoyama and K. Kishio, *Physica C* **460-462** (2007) 772–773.
53. F. X. Régi, J. Schneck, H. Savary, R. Mellet and C. Daguet, *Appl. Supercond.* **1** (1993) 627.
54. T. Motohashi, Y. Nakayama, T. Fujita, K. Kitazawa, J. Shimoyama and K. Kishio, *Phys. Rev. B* **59** (1999) 14080–14086.
55. I. Chong, Z. Hiroi, M. Izumi, J. Shimoyama, Y. Nakayama, K. Kishio, T Terashima, Y. Bando and M. Takano, *Science* **276** (1997) 770–773.
56. Y. J. Uemura *et al.*, *Phys. Rev. Lett.* 62 (1989) 2665.
57. C. Bernhard, Ch. Niedermayer, U. Binninger, A. Hofer, Ch. Wenger, J. L. Tallon, G. V. M. Williams, E. J. Ansaldo, J. I. Budnick, C. E. Stronach, D. R. Noakes and M. A. Blankson-Mills, *Phys. Rev. B* **52** (1995) 10488–10498.
58. J. Y. Genoud, G. Triscone, A. Junod, T. Tsukamoto and J. Muller, *Physica C* **242** (1995) 143–154.
59. G. Villard, D. Pelloquin, A. Maignan and A. Wahl, *Physica C* **278** (1997) 11–22.
60. X. Zhao, X. Sun, X. Fan, W. Wu, X. G. Li, S. Guo and Z. Zhao, *Physica C* **307** (1998) 265–270.
61. R. Jin, H. R. Ott and A. Schilling, *Physica C* **228** (1994) 401–407.
62. M. Akamatsu *et al.*, *Proc. of ISS'93: Adv. in Supercond.* **5** (1994) 125–128.
63. J. G. Ossandon, J. R. Thompson, Y. C. Kim, Yang Ren Sun, D. K. Christen and B. C. Chakoumakos, *Phys. Rev. B* **51** (1995) 8551–8559.
64. Qiang Li, M. Suenaga, J. Gohng, D. K. Finnemore, T. Hikata and K. Sato, *Phys. Rev. B* **46** (1992) 3195–3198.
65. T. Matsushita, M. Kiuchi, T. Haraguchi, T. Imada, K. Okamura, S. Okayasu, S. Uchida, J. Shimoyama and K. Kishio, *Supercond. Sci. Technol.* **19** (2006) 200–205.
66. S. Uchida, J. Shimoyama, T. Makise, S. Horii and K. Kishio, *J. Phys. Conf. Series* **43** (2006) 231–234.
67. J. Shimoyama, K. Kadowaki, H. Kitaguchi, H. Kumakura, K. Togano, H. Maeda and K. Nimura, *Appl. Supercond.* **1** (1993) 43–52.
68. S. Nakao, K. Ueno, T. Hanaguri, K. Kitazawa, T. Fujita, Y. Nakayama, T. Motohashi, J. Shimoyama, K. Kishio and T. Hasegawa, *J. Low Temp. Phys.* **117** (1999) 341–345.
69. Z. Hiroi, I. Chong and M. Takano, *J. Solid State Chem.* **138** (1998) 98–110.
70. J. Shimoyama, K. Kitazawa, K. Shimizu, S. Ueda, S. Horii, N. Chikumoto and K. Kishio, *J. Low Temp. Phys.* **131** (2003) 1043–1052.
71. E. Zeldov, D. Majer, M. Konczykowski, V. B. Geshkenbein, V. M. Vinokur and H. Shtrikman, *Nature* **375** (1995) 373.
72. D. T. Fuchs, E. Zeldov, T. Tamegai, S. Ooi, M. Rappaport and H. Shtrikman, *Phys. Rev. Lett.* **80** (1998) 4971–4974.
73. J. Shimoyama, K. Shimizu, S. Ueda, S. Horii, K. Otzschi, K. Kishio and N. Chikumoto, *Physica C* **378-381** (2002) 457–461.
74. P. Yang and C. M. Lieber, *Science* **273** (1996) 1836–1840.
75. K. Fossheim, E. D. Tuset, T. W. Ebbesen, M. M. J. Treacy and J. Schwartz, *Physica C* **248** (1995) 195–202.

76. T. W. Li, R. J. Drost, P. H. Kes, C. Træholt, H. W. Zandbergen, N. T. Hien, A. A. Menovsky and J. J. M. Franse, *Physica C* **274** (1997) 197–203.
77. Y. L. Wang, X. L. Wu, C. C. Chen and C. Lieber, *Proc. Nat. Acd. Sci. USA* **87** (1990) 7058.
78. W. D. Wu, A. Keren, L. P. Le, B. J. Sternlieb, G. M. Luke, Y. J. Uemura, P. Dosanjh and T. M. Riseman, *Phys. Rev. B* **47** (1993) 8172–8186.
79. J. Shimoyama, Y. Nakayama, K. Kitazawa, K. Kishio, Z. Hiroi, I. Chong and M. Takano, *Physica C* **281** (1997) 69–75.
80. J. Shimoyama, T. Maruyama, M. Shigemori, S. Uchida, S. Ueda, A. Yamamoto, Y. Katsura, S. Horii and K. Kishio, *IEEE Trans. Appl. Supercond.* **15** (2005) 3778–3781.
81. M. Shigemori, T. Okabe, S. Uchida, T. Sugioka, J. Shimoyama, S. Horii and K. Kishio, *Physica C* **408–410** (2004) 40–41.
82. G. F. Sun, K. W. Wong, B. R. Xu, Y. Xin and D. F. Lu, *Phys. Lett. A* **192** (1994) 122–124.
83. K. A. Lokshin, D. A. Pavlov, S. N. Putilin, E. V. Antipov, D. V. Sheptyakov and A. M. Balagurov, *Phys. Rev. B* **63** (2001) 064511.

C3

TlBCCO

Emilio Bellingeri and René Flükiger

C3.1 Introduction

Thallium-based superconductors form one of the largest families of high-temperature superconductors; up to the discovery of mercury-based cuprates, Tl2223 held the record for the highest critical temperature, T_c.

These compounds are very interesting, for applications as well as for fundamental studies. In particular, the Tl1223 compound presents a high irreversibility field allowing, in principle, the transport of high currents at high magnetic fields. Indeed, thin films grown on single crystals were found to exhibit even higher J_c values than YBCO thin films (Tönies *et al.*, 2003). In view of these outstanding properties, Tl1223 was recently suggested as a possible superconducting coating of the beam screen, to mitigate the beam impedance in the design of the Future Circular Collider (Calatroni *et al.*, 2017). However, up to the present, the transport current, J_c, of practical Tl-based superconductors is dramatically limited by weak-link phenomena, even if some works show, in some cases, a simple *c*-axis texture allows high J_c values in high magnetic fields (Deinhofer and Gritzner, 2004).

Other compounds, e.g. Tl1212 and Tl2212, are used for the fabrication of microwave devices exploiting their low surface resistance at frequencies up to 10 GHz. The compound Tl1212 is also interesting for fundamental studies: when Ca is partially substituted by Y, the phase evolves continuously, from a superconductor with T_c = 110 K to an antiferromagnetic insulator exhibiting a metal-insulator transition, without any structural modification, thus offering an exceptional opportunity for doping-dependent investigations. Highly influential studies on the pairing symmetry were performed on Tl2201 thin films.

In contrast to Bi-based and YBCO-based superconductors, the Tl-based phases are very 'tolerant' to substitution as well as to deviations from stoichiometry. This fact, on the one hand, makes an organic and complete description of their properties difficult but on the other hand, offers opportunities for improving both the properties and the synthesis.

The toxicity and volatility of Tl is often reported as an obstacle to the preparation of this class of compounds. For this reason, some precautions must be adopted during handling and processing, but subject to these precautions the preparation of Tl-based compounds can now be considered as safe.

C3.2 Structural Properties (Gladyshevskii and Galez, 1999)

The first Tl-based superconducting material was reported in the Tl–Ba–Cu–O system by (Kondoh *et al.*, 1988). A critical temperature of 19 K was measured for a sample of nominal composition $Tl_{1.2}Ba_{0.8}CuO$. Independently, Sheng and Hermann (1988a) reported T_c values up to 90 K for the nominal compositions $Tl_2Ba_2Cu_3O_{8+\delta}$, $TlBaCu_3O_{5.5+\delta}$ and $Tl_{1.5}Ba_2Cu_3O_{7.3+\delta}$: where the active superconducting compound was later identified as Tl2201. A short time later, the same authors (Sheng and Hermann, 1988b) succeeded in preparing superconductors with T_c = 120 K for the nominal compositions $Tl_2BaCa_{1.15}Cu_3O_{8.5+\delta}$ and $Tl_{1.86}BaCaCu3O_{7.8+\delta}$.

At present, Tl-based cuprates constitute one of the largest chemical families of high-temperature superconductors, forming two distinct structural series with the general formulae Tl(Ba or Sr)$_2$Ca$_{n-1}$Cu$_n$O$_{2n+3}$ and $Tl_2Ba_2Ca_{n-1}Cu_nO_{2n+4}$. The two families can both be described by the general formula $Tl_mBa_2Ca_{n-1}Cu_nO_{2n+m+2}$, where *m* is 1 or 2 and *n* = 1, 2, 3 or 4 for bulk samples prepared by solid-state chemistry, and even 5 or 6 for samples synthesized under very high pressure or for thin films.

For simplicity, it is common to abbreviate the formulae for Tl compounds using the suffixes *n* and *m*. Thus, the sequence (*m* 2 *n*–1 *n*) denotes $Tl_mBa_2Ca_{n-1}Cu_nO_{2n+m+2}$.

These series of compounds include one or more two-dimensional CuO_2 layers (*conducting layers*). If the number *n* of these CuO_2 layers is > 1, they are separated by *n*–1 Ca layers (*separating layers*). Together, these layers form the conducting block, on both sides of which two layers are always present: BaO (or SrO) (*bridging layers*), one on each side. The structure is completed by *m* (1 or 2) TlO layers (*additional layers*). All the layers constitute an approximately square mesh:

- in the CuO_2 layers, the Cu atom lies on the center of the square and the O atoms on the center of the square edges
- the Ca atom in the separating layer occupies the corner of the square
- in the bridging layer, the Ba (or Sr) atom is positioned in the corner of the square and the O atom in the center, above or below the underlying Cu atom
- the TlO layer is identical to the bridging layer but is shifted by [½,½,0] in the plane

These structures are obtained by stacking the layers one on top of each other so that the cations of two consecutive layers are always shifted by [½,½,0]. As a consequence of these systematic shifts, the translation period in the stacking direction must contain an even number of layers. It should be noted that in single Tl layer compounds ($m = 1$), the conventional structural cell contains one stacking unit, i.e. one formula unit ($Z = 1$), whereas in double Tl layer compounds ($m = 2$), it contains two stacking units ($Z = 2$). In the first case, the tetragonal cell of the structure is primitive, in the second case, it is body-centered because of the shift by [½,½,0] within the plane when two additional layers are present. The crystal structures of Tl1201 and Tl2201 are shown in Figures C3.1 and C3.7.

The stacking of the layers for the $n = 1$ compounds forms an octahedral CuO_6 block (see Figure C3.1), while the $n = 2$ compounds present two pyramidal CuO_5 blocks (see Figure C3.3) and the $n = 3$ compounds have two pyramidal and one square planar CuO_4 blocks (Figure C3.5).

In the single-layer compound, it is possible to partially or completely substitute the Ba by Sr in the bridging layer. For both structure series, the Ca sites are mainly occupied by Ca atoms with small amounts of Tl. A common feature of the structures of Tl-based superconductors is the displacement of the Tl and O sites in the Tl layer from the ideal position on four-fold rotation axes. The resulting tetrahedral coordination of Tl atoms is typical for Tl^{3+}. Vacancies may occur on the O sites in the Tl layer.

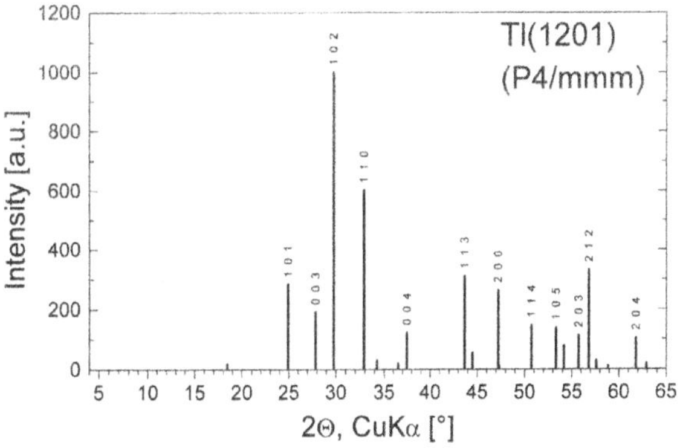

FIGURE C3.2 Calculated X-ray diffraction pattern for powder sample of Tl1201. Indices are given for the stronger reflections.

C3.2.1 The Tl1201 Phase

The structure of stoichiometric $TlBa_2CuO_5$ is tetragonal (space group P4/mmm) (Parkin *et al.*, 1988) whereas that of the analogous $TlSr_2CuO_5$ is reported to be either tetragonal (Kim *et al.*, 1989) or orthorhombic (Pmmm) (Ganguli and Subramanian, 1991), depending on the oxygen content.

Superconductivity may be induced by a partial substitution of Ba^{2+} or Sr^{2+} by La^{3+} or other rare-earth elements (Manako *et al.*, 1989; Subramanian, 1990) or by preparing samples under reducing atmosphere (Gopalakrishnan *et al.*, 1991). When increasing the Pb content in $Tl_{1-x}Pb_xSr_2CuO_5$, a transformation from orthorhombic to tetragonal is observed for $x = 0.12$. A superstructure with doubling of the *b* parameter and an ordered arrangement of oxygen vacancies in the CuO_2 layer is observed in $TlSr_2CuO_{4.515}$ (Chshima *et al.*, 1994). A schematic drawing of the structure is represented in Figure C3.1; calculated peak positions and intensities in a XRD diagram are plotted in Figure C3.2 (Table C3.1).

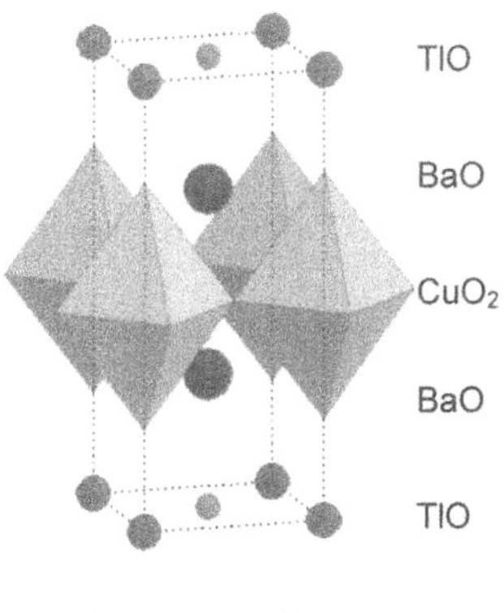

FIGURE C3.1 Crystal structure of Tl1201.

TABLE C3.1 Atomic Coordinates of $Tl_{0.92}Ba_{1.2}La_{0.8}CuO_{4.864}$ P4/mmm $a = 3.8479$, $c = 9.0909$ Å, $Z = 1$

Atom	Site	*x*	*y*	*z*	Occ
Tl	4 (l)	0.0801	0	0	0.230
Ba(La)	2 (h)	1/2	1/2	0.2942	
Cu	1 (b)	0	0	1/2	
O1	4 (n)	0.4281	1/2	0	
O2	2 (g)	0	0	0.2250	
O3	2 (e)	0	1/2	1/2	

Source: Subramanian *et al.* (1990).

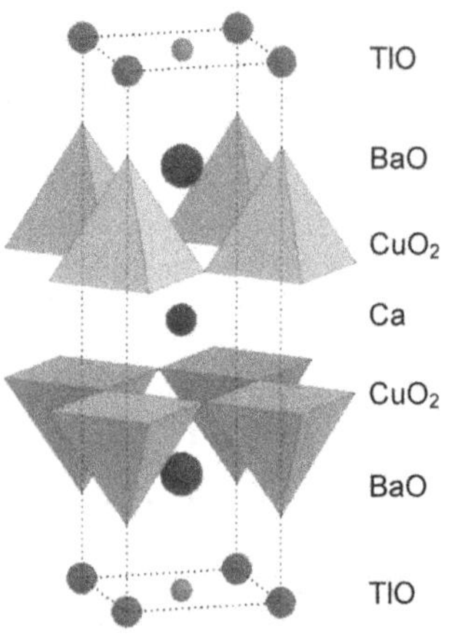

FIGURE C3.3 Crystal structure of Tl1212.

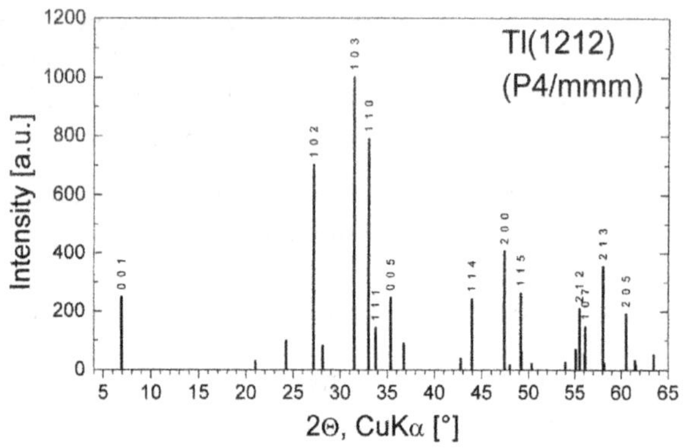

FIGURE C3.4 Calculated X-ray diffraction pattern for powder sample of Tl1212. Indices are given for the stronger reflections.

C3.2.2 The Tl1212 Phase

The structure of $TlBa_2CaCu_2O_7$ is tetragonal (P4/mmm) with $a =$ 3.8472 and $c = 12.721$ Å (Figure C3.3, Table C3.2); the O site in the Tl layer was found to be only ¾ occupied and Tl and Ba atoms were found to be displaced toward the oxygen vacancies (Morosin *et al.*, 1988). In the analogous family $(Tl_{0.5}M_{0.5})Sr_2CaCu_2O$, $M =$ Pb, Bi both the O and Tl site in the Tl layer are displaced from their ideal positions resulting in a tetrahedral coordination of the Tl atoms. The calculated XRD pattern is shown in Figure C3.4.

C3.2.3 The Tl1223 Phase

Superconductivity with $T_c = 110$ K was first identified for the composition $TlBa_2Ca_2Cu_3O_9$ with a structure in the group P4/mmm with $a = 3.8429$ Å, $c = 15.871$ Å (Parkin *et al.*, 1988) (Figures C3.5 and C3.6). The Tl site was found to be displaced from the ideal position along the short translation vectors by about 0.35 Å. Vacancies were observed on Ba sites (occupancy 0.94) and a small amount of Tl was observed on Ca site (5 at%). For $TlBa_2Ca_2Cu_3O_{8.62}$, two partly occupied sites, one in the ideal position (0,0,0) and the other, displaced in (0.1138,0,0) were considered in (Morosin *et al.*, 1991) and associated with the oxidation states Tl^{+} and Tl^{+3}, respectively. The O site in the Tl layer was found to be only partly occupied (occupancy 0.62). In the analogous Sr superconductor,

TABLE C3.2 Atomic Coordinates of $Tl_{1.13}Ba_2Ca_{0.87}Cu_2O_7$ P4/mmm $a = 3.8472$, $c = 12.721$ Å, $Z = 1$

Atom	Site	*x*	*y*	*z*	Occ
Tl	4 (l)	0.0877	0	0	0.250
Ba	2 (h)	1/2	1/2	0.2155	
Cu	2 (g)	0	0	0.3740	
Ca	1 (d)	1/2	1/2	1/2	
O1	1 (c)	1/2	1/2	0	
O2	2 (g)	0	0	0.1582	
O3	4 (i)	0	1/2	0.3797	

Source: Kolesnikov *et al.* (1989).

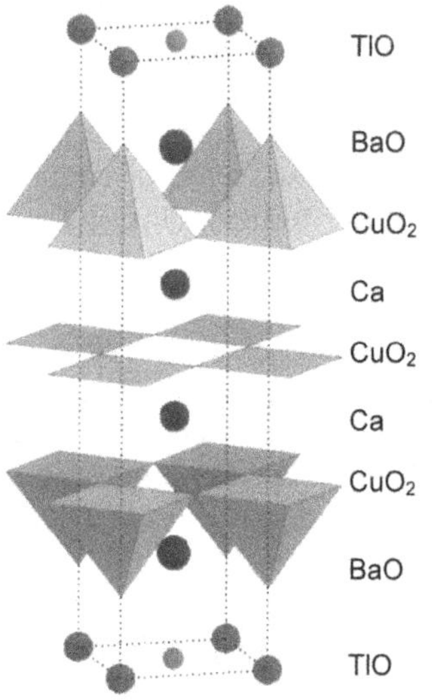

FIGURE C3.5 Crystal structure of Tl1223.

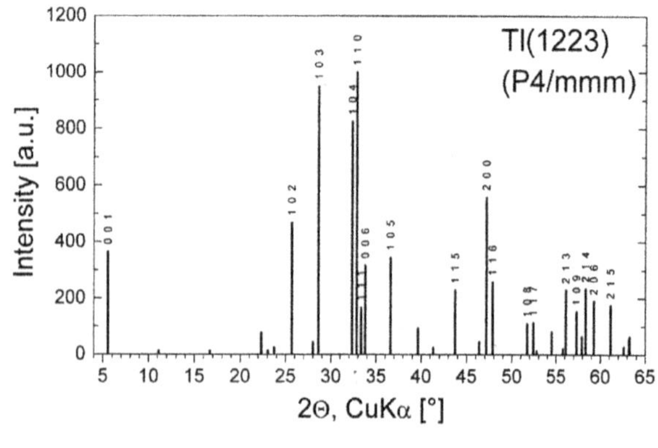

FIGURE C3.6 Calculated X-ray diffraction pattern for powder sample of Tl1223. Indices are given for the stronger reflections.

with half Tl replaced by Pb (Subramanian *et al.*, 1988), the Tl and O sites in the Tl layer and the O site in the Sr layer were found to be off-centered in the [1 1 0] direction by about 0.24, 0.29 and 0.27 Å, respectively. Partial substitution of Ca by Tl was also observed in this material (see Table C3.3).

TABLE C3.3 Atomic Coordinates of $Tl_{0.56}Pb_{0.56}Sr_{2-}Cal_{1.88}Cu_3O_9$, P4/mmm, a = 3.808, c = 15.232 Å, Z = 1

Atom	Site	x	y	z	Occ
Tl	4 (l)	0.067	0	0	0.250
Sr	2 (h)	1/2	1/2	0.1709	
Cul	2 (g)	0	0	0.2868	
Ca	1 (h)	1/2	1/2	0.3928	
Cu2	1 (b)	0	0	1/2	
O1	4 (n)	0.4	1/2	0	
O2	2 (g)	0	0	0.1582	0.250
O3	4 (i)	0	1/2	0.3797	
O4	2 (e)	0	1/2	1/2	

Source: Subramanian *et al.* (1988).

C3.2.4 The Tl2201 Phase

Tl2201 crystallizes with two structural modifications, one orthorhombic, and the other tetragonal, and the latter being superconducting with T_c up to 90 K for an optimal oxygen content. In tetragonal $Tl_2Ba_2CuO_6$ (space group I4/mmm a = 3.866 c = 23.239 Å) (Table C3.4), the Tl site was found to be displaced from its ideal position and a partial substitution of up to 7 at% Tl by Cu was observed (Liu *et al.*, 1992; Opagiste *et al.*, 1993). The presence of an extra oxygen with very low occupation (from 0.0005 to 0.028) located between the two Tl layers was suggested by (Shimakawa *et al.*, 1990). A non-superconducting Tl2201 phase with orthorhombic structure (a = 5.4451, b = 5.4961, c = 23.153 Å) was first reported and the space group Fmmm was proposed by (Huang *et al.*, 1988). A schematic structure is presented in Figure C3.7, while Figure C3.8 shows calculated XRD patterns for tetragonal (a) and orthorhombic (b) Tl2201. As shown in Figure C3.9 in the space group Fmmm, the a and b parameters are rotated by 45° with respect to those in the other space groups considered here. The crystal structure was also refined in subgroups of Fmmm, in order to account for positional disorder in additional TlO layers: A2aa (Hewat *et al.*, 1988) and Abma (Parise et al., 1988; Ström *et al.*, 1994). It is generally agreed that the formation of the tetragonal phase is favored by a Tl-deficient composition, and the formation of the orthorhombic phase by a high Tl content (Shimakawa, 1993). To a minor extent, the tetragonal-orthorhombic transition depends also on the oxygen content,

TABLE C3.4 Atomic Coordinates of $Tl_2Ba_2CuO_6$, T_c = 90K, *I4/mmm*, a = 3.866, c = 23.239 Å, Z = 2

Atom	Site	*x*	*y*	*z*	Occ
Tl	4 (e)	0	0	0.2974	
Ba	4 (e)	1/2	1/2	0.4170	
Cu	2 (b)	0	0	1/2	
O(1)	16 (n)	0.405	1/2	0.2829	0.25
O(2)	4 (e)	0	0	0.3832	
O(3)	4 (c)	0	1/2	1/2	

Source: Torardi et al. (1988).

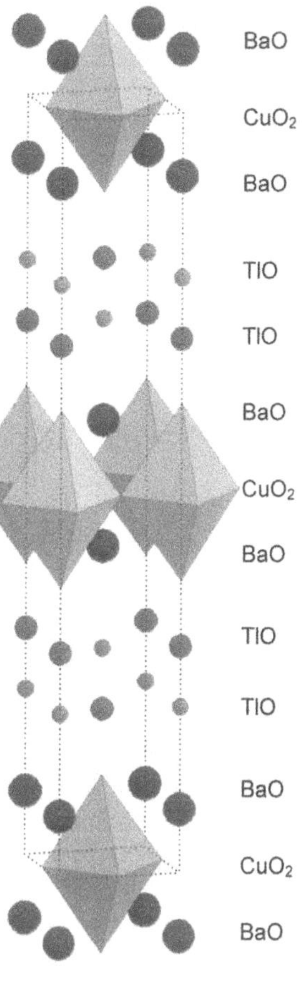

FIGURE C3.7 Crystal structure of Tl2201.

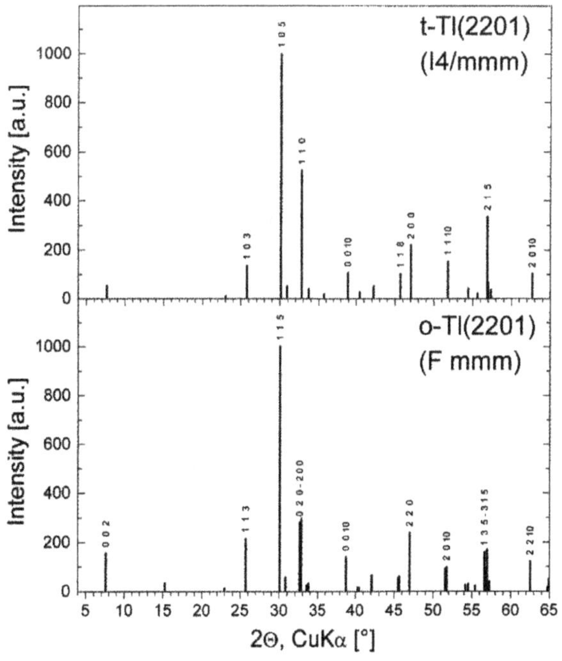

FIGURE C3.8 Calculated X-ray diffraction pattern for powder sample of tetragonal (top) and orthorhombic (bottom) Tl2201. Indices are given for the stronger reflections.

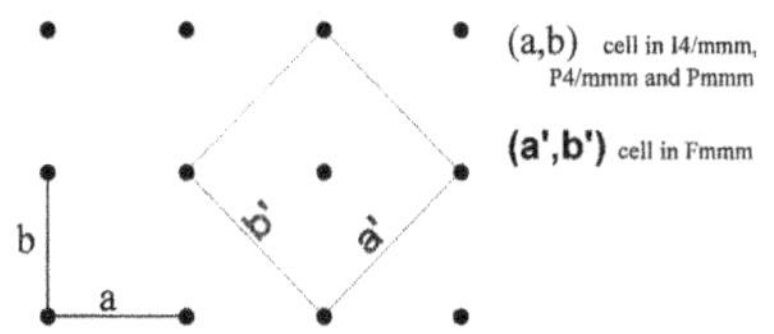

FIGURE C3.9 In-plane crystallographic axis in I4/mmm, P4/mmm, Pmmm (a, b) and in Fmmm (a′, b′).

a higher oxygen content favoring the orthorhombic modification (Ström *et al.*, 1994; Jorda *et al.*, 1993). Thallium based, stoichiometric tetragonal and orthorhombic phases were obtained by high-pressure synthesis (Opagiste *et al.*, 1993; Jorda *et al.*, 1994) (see Table C3.5).

C3.2.5 The Tl2212 Phase

The Tl2212 phase, $Tl_2Ba_2CaCu_2O_{8+\delta}$, was first identified by Hazen *et al.* (1988) and the structural refinement carried out by Subramanian *et al.* (1988). In the majority of structural studies, the Ca site was found to be partly occupied by Tl (12–28 at%), whereas reduced scattering density on the Tl site was attributed to either partial substitution of Tl by Ca (10–11 at%) (Maignan *et al.*, 1988; Kikuchi *et al.*, 1989; Johansson *et al.*, 1994) or by Cu (9 at%) (Onoda et al., 1988), or simply by Tl deficiency (6–13 at%) (Morosin *et al.*, 1991; Ogborne and Weller, 1992; Molchanov *et al.*, 1994). Vacancies on the O site in *additional* layers were detected: (5–6 at% by (Ogborne and Weller, 1992) and (12–16 at%) by (Johansson *et al.*, 1994). Displacement of Tl sites from four-fold rotation axis, split into 32(o) 0.013 0.041 0.28635 was reported by Molchanov *et al.* (1994). The structure and calculated XRD pattern are shown in Figures C3.10 and C3.11 (see Table C3.6).

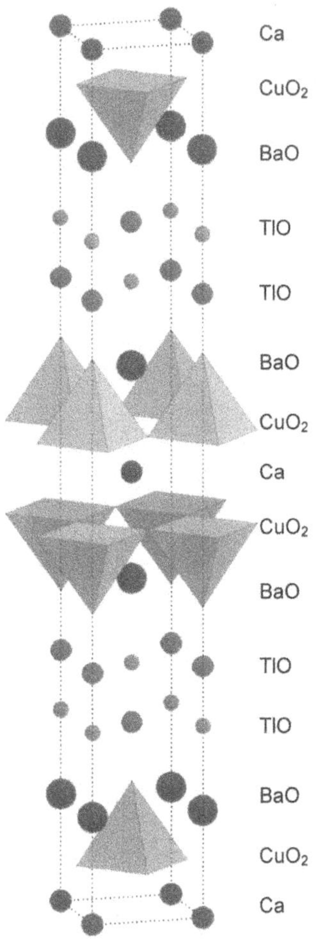

FIGURE C3.10 Crystal structure of Tl2212.

C3.2.6 The Tl2223 Phase

The Tl2223 phase was identified by Politis and Luo (1988), and the first structural refinement was carried out by Torardi *et al.* (1988), which showed partial disorder between the Tl and Ca sites. The Ca site was found to be partly occupied by Tl atoms (3–7 at%) by Hervieu *et al.* (1988) and Sinclair *et al.* (1994). Vacancies on the Tl site (6–12 at%) were detected by Hervieu *et al.*, 1988; Morosin *et al.*, 1991; Ogborne and Weller, 1992), whereas partial substitution of Tl by Cu (14 at%) was reported by Sinclair *et al.* 1994). The O site in the *additional* layers was found to be split from the ideal position into 16(n) 0.6112 1/2 0.2753, with an occupancy of 0.234 (Morosin *et al.*, 1991; Ogborne and Weller, 1992). The Tl and O sites in

TABLE C3.5 Atomic Coordinates of $Tl_2Ba_2CuO_{6.10}$, *Fmmm*, $a = 5.4604$, $b = 5.4848$, $c = 23.2038$ Å, $Z = 4$

Atom	Site	*x*	*y*	*z*	Occ
Tl	16 (m)	0	− 0.025	0.2976	0.5
Ba	8 (i)	1/2	0	0.4172	
Cu	4 (b)	0	0	1/2	
O(1)	8 (f)	1/4	1/4	1/4	0.07
O(2)	16 (m)	1/2	− 0.059	0.2895	0.49
O(3)	8 (i)	0	0	0.3837	
O(4)	8 (e)	1/4	1/4	1/2	

Source: Parise *et al.* (1989).

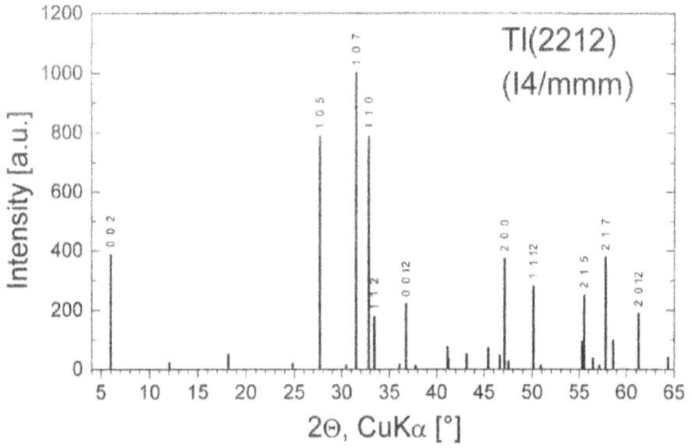

FIGURE C3.11 Calculated X-ray diffraction pattern for powder sample of Tl2212. Indices are given for the stronger reflections.

TABLE C3.6 Atomic Coordinates of $Tl_2Ba_2CaCu_2O_8$, *I4/mmm*, $a = 3.8550$, $c = 29.318$ Å, $Z = 2$

Atom	Site	*x*	*y*	*z*	Occ
Tl	4 (e)	0	0	0.2864	
Ba	4 (e)	1/2	1/2	0.3782	
Cu	4 (e)	0	0	0.4460	
Ca	2 (a)	1/2	1/2	1/2	
O(1)	16 (n)	0.396	1/2	0.2815	0.25
O(2)	4 (e)	0	0	0.3539	
O(3)	8 (g)	0	1/2	0.4469	

Source: Subramanian *et al.* (1988).

additional layers were refined by Sinclair *et al.* (1994), both in 16(m) 0.0276 0.0276 0.27921 (occupancy 0.215) for Tl and 0.5819 0.5819 0.2756 (occupancy 0.25) for O. The structure and the calculated XRD spectra are shown in Figures C3.12 and C3.13 and in Table C3.7.

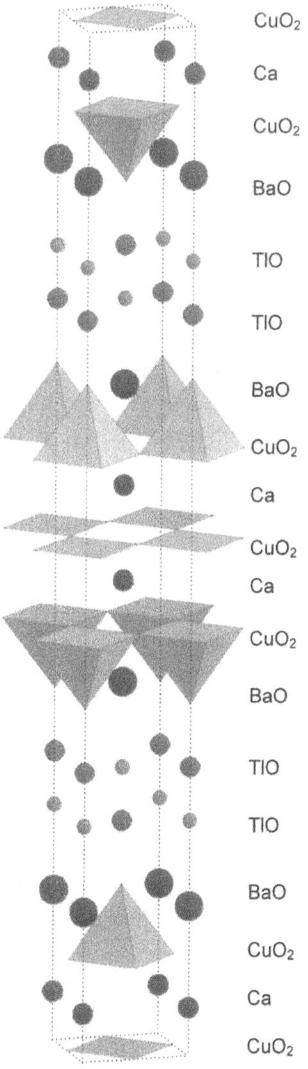

FIGURE C3.12 Crystal structure of Tl2223.

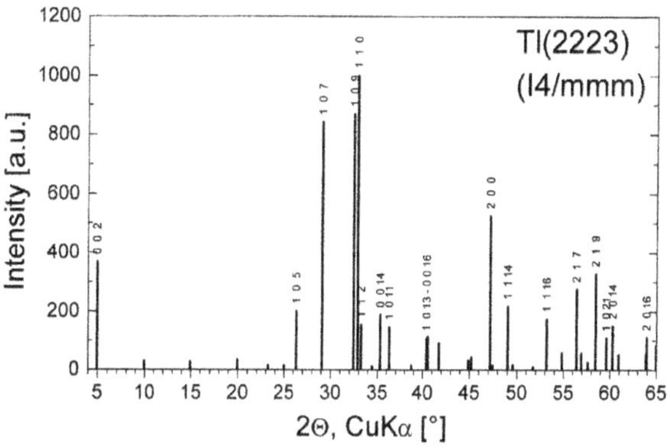

FIGURE C3.13 Calculated X-ray diffraction pattern for powder sample of Tl2223. Indices are given for the stronger reflections.

A change of the superconducting properties of Tl1223 was found by varying the thermal annealing conditions. Shipra et al. (2015) annealed polycrystalline samples in flowing Ar+4%H_2, H_2 or N_2 gases and found an increase of T_c from 104 to 120 K. They found that an excess of Tl_2O_3 is required for obtaining almost single-phase samples. A substitution of Ca by Tl was found in the samples after final annealing, while no such imperfection was observed in the as-prepared samples. Specific heat measurements showed that after heat treatment, the peak at T_c is shifted to higher temperature, but shows some broadening (Ogborne *et al.*, 1992).

C3.2.7 Exotic Structures

The structure of less common superconducting phases of the Tl family are reported in this paragraph: Tl1222 and the two four-copper-layer compounds Tl1234 and Tl2234. Tl1222-based phases were first reported by Martin *et al.* (1989) for $Tl(Ba_{0.84}Tl_{0.16})_2Pr_2Cu_2O_9$ and $Tl_{0.9}(Sr_{0.8}Tl_{0.2})_2Pr_2Cu_2O_9$. Structural refinements were carried out in space group *I4/mmm* with $a = 3.900$ Å, $c = 30.273$ Å and $a = 3.8635$ Å, $c = 29.535$ Å, respectively. In both structures, the Tl site, deficient in the structure of the Sr-containing phase, was found to be located on the four-fold rotation axis in 2(*a*) 0 0 0.

TABLE C3.7 Atomic Coordinates of $Tl_{1.94}Ba_2Ca_{2.06}Cu_3O_{10}$, *I4/mmm*, $a = 3.8503$, $c = 35.88$ Å, $Z = 2$

Atom	Site	*x*	*y*	*z*	Occ
Tl[a]	4 (e)	0	0	0.2799	
Ba	4 (e)	1/2	1/2	0.3552	
Cu(1)	4 (e)	0	0	0.4104	
Ca[b]	4 (e)	1/2	1/2	0.4537	
Cu(2)	2 (b)	0	0	1/2	
O(1)	4 (e)	1/2	1/2	0.2719	
O(2)	4 (e)	0	0	0.3412	
O(3)	8 (g)	0	1/2	0.4125	
O(4)	4 (c)	0	1/2	1/2	

Source: Torardi *et al.* (1988).

TABLE C3.8 Atomic Coordinates of $TlBa_2Eu_{1.5}Ce_{0.5}Cu_2O_9$, *I4/mmm*, $a = 3.8782$, $c = 30.423$ Å, $Z = 2$

Atom	Site	*x*	*y*	*z*	Occ
Tl	8(i)	0.082	0	0	0.25
Ba	4(e)	1/2	1/2	0.0872	
Cu	4(e)	0	0	0.1520	
Eu[a]	4(e)	1/2	1/2	0.2087	
O(1)	2(b)	1/2	1/2	0	
O(2)	4(e)	0	0	0.0831	
O(3)	8(g)	0	1/2	0.1554	
O(4)	4(d)	0	1/2	1/4	

Source: Liu *et al.* (1992).

[a] $Eu = Eu_{0.75}Ce_{0.25}$.

TABLE C3.10 Atomic Coordinates of $Tl_{1.64}Ba2Ca_3Cu_4O_{12}$, *I4/mmm*, $a = 3.84877$, $c = 42.0494$ Å, $Z = 2$

Atom	Site	*x*	*y*	*z*	Occ
Tl	4 (e)	0	0	0.2757	0.82
Ba	4 (e)	1/2	1/2	0.3387	
Cu(l)	4 (e)	0	0	0.3866	
Ca(1)	4 (e)	1/2	1/2	0.4236	
Cu(2)	4 (e)	0	0	0.4636	
Ca(2)	2 (a)	1/2	1/2	1/2	
O(1)	16 (n)	0.394	1/2	0.2664	0.25
O(2)	4 (e)	0	0	0.3234	
O(3)	8 (g)	0	1/2	0.3874	
O(4)	8 (g)	0	1/2	0.4626	

Source: Siegal *et al.* (1997).

Superconductivity ($T_c = 40$ K) was first reported by Iqbal *et al.* (1991) for the composition $(Tl_{0.5}Pb_{0.5})Sr_2(Eu_{0.9}Ce_{0.1})_2Cu_2O_9$ and subsequently by Liu *et al.* (1992) for Pb-free $TlBa_2(Eu_{0.75}Ce_{0.25})_2Cu_2O_9$ (Table C3.8).

The superconducting Tl1234 phase ($T_c = 122$ K) with tetragonal structure ($a = 3.85$ Å and $c = 19.1$ Å) was first reported by Ihara *et al.* (1988) for the nominal composition $TlBa_2Ca_3Cu_4O_{11}$ (Table C3.9). The superconducting Tl2234 phase ($T_c = 114$ K) was first reported by Hervieu *et al.* (1988) for the nominal composition $Tl_2Ba_2Ca_3Cu_4O_{12}$ and a structural model was proposed in the space group *I4/mmm* ($a = 3.852$ Å, $c = 42.00$ Å). Ogborne and Weller (1994) reported disorder between the Tl and Ca site, with 40 at% Ca on the Tl site and 10–15 at% Tl on the Ca sites. In addition, vacancies on the sites of Tl (6–7 at%) and O (0–5 at%) in the TlO layers were detected, the latter depending on the annealing conditions (Table C3.10).

C3.3 Chemical Properties (Ogborne and Weller, 1994)

One of the more serious difficulties in processing Tl-based superconductors is the high volatility of Tl and Tl oxides at the temperatures required for the synthesis and sintering of the superconducting phases. Furthermore, Tl is highly toxic, thus adding significant safety problems for the operator.

There are basically three methods of synthesis of Tl-based superconductors: in a sealed (or well closed) crucible (an example is reported by Ruckenstein and Wu (1994), in a two-zone furnace (see DeLuca *et al.*, 1991) providing a Tl oxide source, or in a (moderately) high-pressure furnace preventing evaporation of Tl oxides (Flükiger *et al.*, 1998). A precise knowledge of the vapor/solid equilibrium of thallium oxide is necessary for the first two methods, the reaction being mainly accomplished between the Tl oxide vapor and a solid precursor. The vapor pressure of thallous oxide over condensed thallium oxides rises rapidly with increasing temperature. At high temperature, the condensed thallium oxide phase may be Tl_2O_3, Tl_4O_3, Tl_2O or a solid or liquid solution of two, depending on temperature and on the oxygen partial pressure, whereas the vapor phase is Tl_2O. The vapor pressure of TlO is plotted as a function of temperature in Figure C3.14 (Ruckenstein and Wu, 1994; Holstein, 1993).

TABLE C3.9 Atomic Coordinates of $Tl_{0.996}Ba_2Ca_{2.96}Cu_4O_{11}$, *P4/mmm*, $a = 3.84809$, $c = 19.0005$ Å, $Z = 1$

Atom	Site	*x*	*y*	*z*	Occ
Tl	4 (l)	0.086	0	0	0.249
Ba	2 (h)	1/2	1/2	0.1437	
Cu(1)	2 (g)	0	0	0.2487	
Ca(1)	2 (h)	1/2	1/2	0.3292	0.98
Ca(2)	2 (g)	0	0	0.4158	
Ca(2)	1 (d)	1/2	1/2	1/2	
O(1)	1 (c)	1/2	1/2	0	
O(2)	2 (g)	0	0	0.1099	
O(3)	4 (i)	0	1/2	0.2515	
O(4)	4 (i)	0	1/2	0.4156	

Source: Ogborne and Weller (1994).

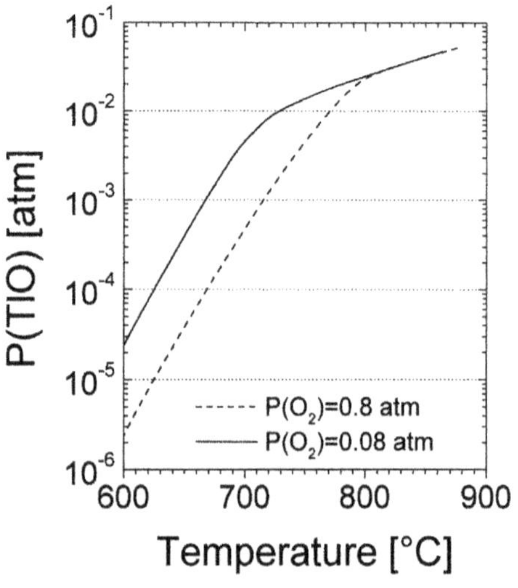

FIGURE C3.14 Vapor pressure of TlO over a solid (or liquid) source of thallium oxides *versus* temperature for different partial pressures of oxygen. (Holstein, 1993.)

The representation of a phase diagram for these superconductors requires at least a three-dimensional space, the minimum number of constituents of the phase (Tl, Ba, Ca, Cu oxides) being 4. Since all the nominal compositions lie in a single plane, the phase diagram can be represented in a pseudo-quaternary tetrahedron with the components $TlO_{1.5}$, Ba_2CuO_3 and $CaCuO_2$ (Figure C3.15). The nominal composition of each quaternary superconductor lies at the intersection of one of the two lines originating at the component $CaCuO_2$ and terminating at the single-copper-layer phases 1201 and 2201, and at one of the three lines originating in the corner $TlO_{1.5}$ and terminating at the Tl-free compositions 0212, 0223, 0234. Moving any of these last lines corresponds to the addition or removal of thallium layers, the amount of Tl moving up or down of any line of this set. Travelling along any of the first two lines corresponds to the addition (or removal) of a CuO_2 layer with its spacing Ca ion.

It has been shown by Aselage *et al.* (1990) that this diagram is too simplified, i.e. it is not possible to find an equilibrium between Tl2212, Tl2223 and Tl2234 and the Tl oxide, since other Tl-rich phases form in an excess of Tl. The diagrams in Figure C3.16 (Aselage *et al.*, 1993) show the stability of different Tl-containing phases at different values of $p(O_2)$ and $p(TlO)$. The compositions Ba:Ca:Cu = 2:1:2 and 2:2:3 are fixed for the diagrams (*a*) and (*b*), respectively; the diagrams are drawn for temperatures lying a few degrees lower than the melting point. The melting temperatures depend on $p(O_2)$, thus the diagrams do not represent isothermal sections where T = constant.

In both diagrams, i.e. for all compositions, Tl2212 results as being the most stable phase at high values of $p(TlO)$ before the decomposition of any layered structure.

Figure C3.16(left) also shows that Tl1212 is only stable for $p(O_2) > 0.1$ atm and for low $p(TlO)$ values; as the latter is increased, Tl2212 immediately starts to form. With a complete occupancy of each cation and oxygen site in the Tl1212

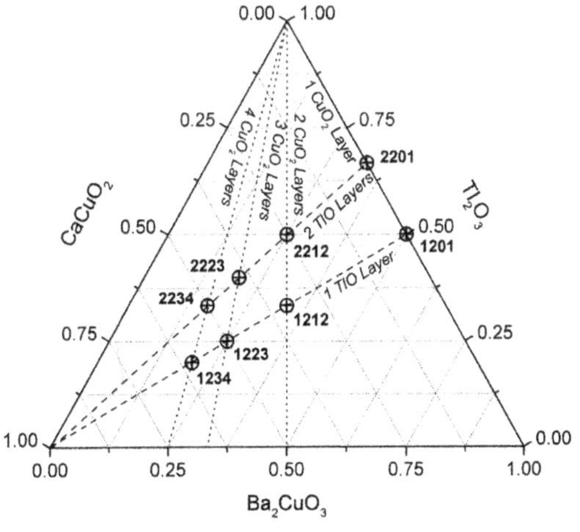

FIGURE C3.15 Pseudo-ternary composition diagram illustrating the position of all the superconducting phases of the Tl system. (Ruckenstein and Wu, 1994.)

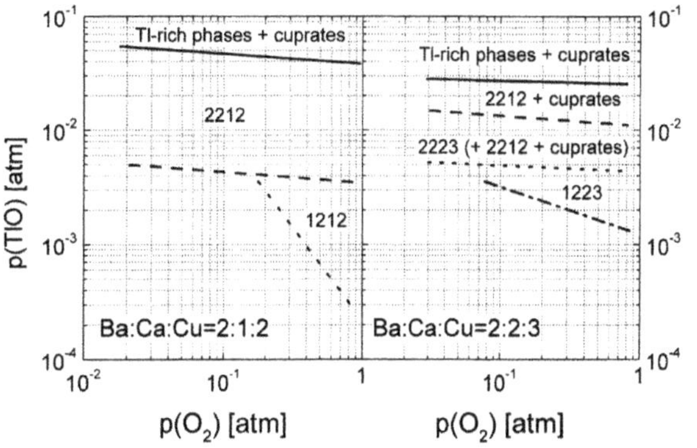

FIGURE C3.16 Phase equilibrium diagrams for superconducting Tl-based compounds at temperature just below melting (diagrams are not sections at constant temperature) in $p(O_2)$, $p(TlO)$ plane. Precursor composition: (left) Ba:Ca:Cu = 2:1:2, (right) Ba:Ca:Cu = 2:2:3. Non-superconducting cuprates are not indicated. (Aselage *et al.*, 1993.)

phase, the formal valence of Cu is calculated to be 2.5. Due to this high formal oxidation state, Tl1212 should become less stable with decreasing $p(O_2)$, confirming the results presented in diagram (a). A similar behavior for the single-TlO-layer phase, Tl1223, is observed in Figure C3.16(right) where Tl1223 transforms into Tl2223 with increasing $p(TlO)$. The region of stability of this latter phase is very narrow, the phase Tl2212 being observed for a further increase in the $p(TlO)$.

Since all these transformations are reversible and thus at thermodynamic equilibrium, one observes a remarkable difference in the kinetics of formation: the phases with 3 CuO_2 layers, e.g. Tl1223 and Tl2223, have a slower growth kinetics than those with double CuO_2 layers, Tl1212 and Tl2212.

In the case of the Sr homologues, the diagrams are different, the double-TlO-layer member of the family being absent. In this case, the phase with the larger stability domain turns out to be Tl1212 through which it is necessary to pass for the formation of the Tl1223. The thermodynamic stability of the Tl1223 phase can be improved by appropriate cationic and anionic substitutions. The compound $Tl_{0.5}Pb_{0.5}Sr_{1.6}Ba_{0.4}Ca_2Cu_3O_x$ has a large stability domain that can be further enlarged by the partial substitution of O by F (Bellingeri et al., 1998).

For synthesis by means of the two-zone furnace method, the $p(TlO)$ can be controlled by setting the temperature of the thallium oxide source, while in a sealed crucible, it is determined (at fixed reaction temperature) by the ratio between the volume of the Tl_2O_3 and the volume of the crucible.

If the synthesis is performed in open or quasi-closed systems, the starting composition usually has an excess of thallium oxide to compensate the Tl losses during the treatment. The reaction is performed moving down, with the time, in the diagrams in Figure C3.16 through the different phases, and the final product can be controlled by accurate setting of the duration of the thermal treatment.

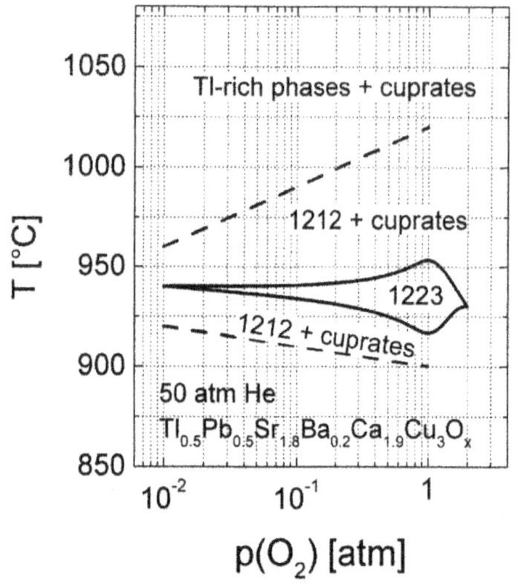

FIGURE C3.17 Phase equilibrium diagrams for the TlSrCCO system where the double Tl layer compounds are absent. The diagram in the ($p(O_2)$, T) plane is drawn for a pressure of 50 bar of He.

The synthesis path is completely different if the reaction is performed under high isostatic pressure. Experiments show that just a moderate high pressure of 50 atm is already sufficient to prevent evaporation of Tl oxides up to 1100°C. In this case, the phase formation is mainly realized through a solid-state diffusion reaction (Gladyshevskii and Galez, 1999). In Figure C3.17, the stability region of the Tl1223 phase is represented in a temperature-oxygen partial pressure diagram. Tl1212 turns out to be the more stable phase at temperatures both lower and higher than the Tl1223 stability region. Since the kinetics of formation of the double CuO layer phase is very high, in normal conditions, Tl1223 is formed through Tl1212.

C3.4 Mechanical Properties

Very few experiments are reported about the mechanical properties of Tl-based superconductors. Our unpublished results on room temperature Vickers microhardness (HV) measured on the core of monofilamentary Ag-sheathed tapes, show that the single-layer compounds Tl1212 and Tl1223 show high hardness values, ~250 HV (in g mm^{-2}). Comparable, but slightly lower values, are observed for the double-layer compounds in Tl2223. The microhardness decreases if the Tl is partially substituted by Bi (about 150 HV for Tl:Bi = 1:1 in Tl1223), approaching the values measured on similar samples of Bi2223 (~ 70 - 120 HV) (see also Chapter D3.3 of this Handbook). This effect can be explained if the stereochemistry of Tl and Bi is considered: indeed, Bi has three short bonds with O mutually perpendicular, two of them being in the additional layer so that only one strong bond with the oxygen of the bridging layer is present. In contrast to Bi, Tl forms strong bonds toward both, the underlying bridging and the upper bridging or additional layer, resulting in a more robust structure.

The different hardness values of the Tl- and Bi-based phases can also partially explain the failure of the powder-in-tube (PIT) technique when applied to prepare high critical current tapes of Tl-based superconductors. In the deformation process of the tape manufacturing, the grains of these phases are broken and not, as in Bi-based superconductors, partially deformed and oriented.

Some deformation studies at high temperature demonstrated that extensive plasticity in Tl1223 is possible (Routbort *et al.*, 1993). Stress *versus* strain curves show that the deformation is controlled by viscous flow for low stresses and high temperature (~ 850°C), whereas at lower temperature (~ 750°C) and higher stress, dislocation gliding or climbing plays a significant role in the deformation process. However, the adoption of a high-temperature deformation process for the preparation of superconducting tapes is severely complicated by the difficulty in manipulating these materials at high temperature, mainly because of evaporation of toxic Tl.

C3.5 Optical Properties: IR Reflectivity (Renk, 1994)

For infrared studies, it is important to have well-defined materials and furthermore high-quality sample surfaces, the penetration depth of the infrared radiation being of the order of 100 nm. Interesting information can be extracted from reflectivity measurements using a Kramers–Kronig analysis that permits the calculation of the dynamic conductivity and, by the use of the lattice dynamical calculations, makes it possible to identify and characterize infrared-active phonon modes (Webb and Sievers, 1986; Thomas *et al.*, 1988; Tinkham, 1970).

Far infrared reflectivity spectra of Tl-based compounds at different temperatures are represented in Figures C3.18 to C3.21. Common to all phases is a high reflectivity at low frequency, a decrease with increasing frequency and pronounced phonon structures. The smooth background is mainly due to electronic excitations in the (*a*,*b*) plane, while the phonon structure is due to the infrared active phonons with displacements in the *c*-direction.

Anomalous behavior of the oscillator strength is also observed in the reflectivity spectra of Tl1212 (Figure C3.18). Comparing the spectra of superconducting $Tl_{0.55}Pb_{0.45}Sr_2CaCu_2O_7$ and non-superconducting $Tl_{0.8}Pb_{0.2}Sr_2CaCu_2O_7$, it is found that the oscillator strength increases strongly with decreasing temperature for all the infrared-active phonons in the superconducting phase, while it remains almost constant for the non-superconducting compound (Zetterer et al., 1990a).

A common feature of all double Tl layer phases is the occurrence of two resonance-like reflection minima near 80 and 140 cm^{-1} that correspond mainly to Ba and Cu vibrations and to "Reststrahlen"-like maxima near 580 cm^{-1} that can be mainly attributed to the vibration of oxygen in the BaO layer against oxygen in the TlO plane.

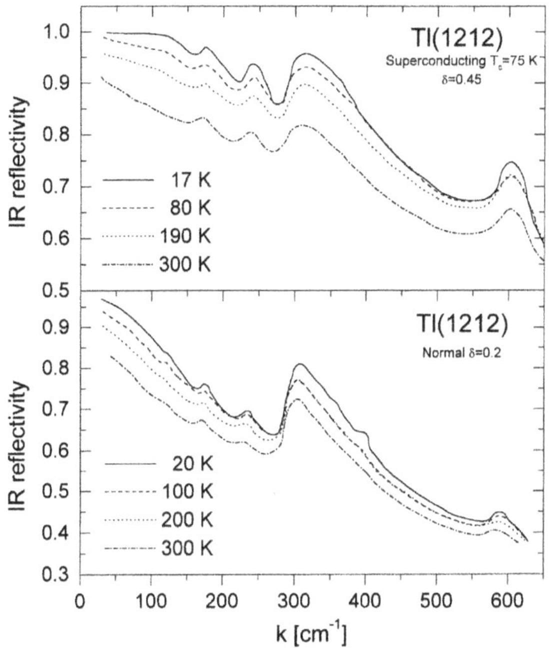

FIGURE C3.18 Infrared reflectivity of superconducting (δ = 0.45) (top) and normal-conducting (δ = 0.2) (bottom) $Tl_{\delta-1}Pb_{\delta}Sr_2Ca_1Cu_2O_z$. (Renk, 1994.)

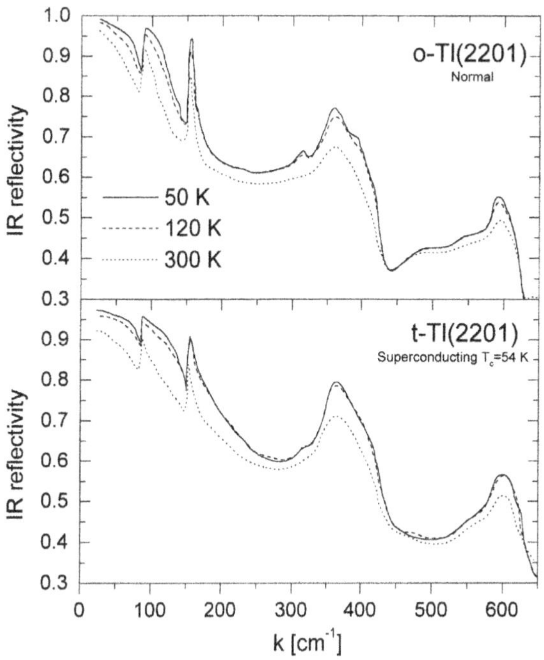

FIGURE C3.19 Infrared reflectivity of superconducting (tetragonal) (top) and normal-conducting (orthorhombic) (bottom) Tl2201. (Renk, 1994.)

The reflectivity of both the superconducting and non-superconducting 2201 does not change much below 100 K (Figure C3.19), while marked changes are seen for the 2212 phase (Figure C3.20). An infrared-active phonon that corresponds mainly to a vibration of the bridging oxygen atoms against the Ca atoms softens in Tl2212 by 3% of the resonance frequency (311 cm^{-1}) and in Tl2223 (Figure C3.21) by 7% of the resonance frequency (305 cm^{-1}) when the temperature is decreased below T_c. The oscillator strengths of most phonons show a strong temperature dependence, with a strong increase below T_c, including the phonons with the anomalous shift in the 2212 phase, while the phonons with the anomalous shift in the 2223 phase shows a strong decrease of strength; also, other phonons at 580 cm^{-1} of the 2223 phase that correspond mainly to vibrations of the bridging oxygen atom against the oxygen atom in the TlO layer show a decrease in strength (Zetterer *et al.*, 1990a).

C3.6 Normal-State Properties

C3.6.1 Electrical and Thermal Transport Properties

A complete and accurate discussion of the transport properties is complicated by the existence of many different stoichiometric compositions for each phase, sometimes with very different properties. Some well-established characteristics can, however, be identified.

Experimentally, the resistivity for optimally doped and overdoped samples can be described by a power law $\rho = \rho_0 + \beta T^n$. For an optimally doped sample, a linear dependence is observed ($n = 1$), whereas in the overdoped case, the resistivity shows a progression toward a quadratic dependence ($n \approx 2$) (Figure C3.22). For underdoped samples, the resistivity is semiconductor-like as clearly shown in Figure C3.23 (Kubo *et al.*, 1991); Datta, 1994). For single-crystal samples, the resistivities ρ_{ab} and ρ_c show strong similarities in the temperature dependence, but ρ_c is two or three orders of magnitude larger than ρ_{ab}. The qualitative similarity of ρ_c and ρ_{ab}, as in some other cuprate superconductors, suggests metal-like

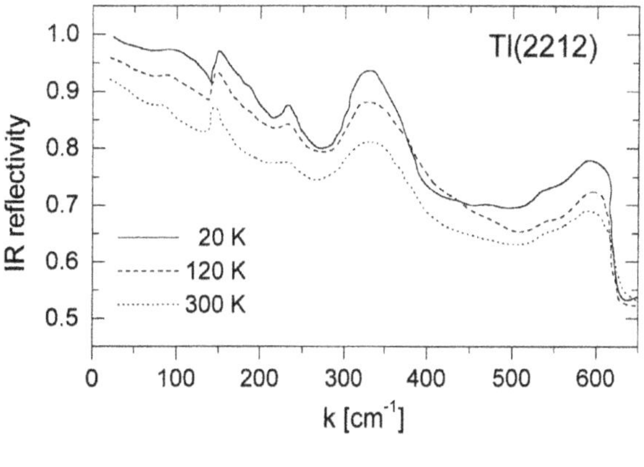

FIGURE C3.20 Infrared reflectivity of superconducting Tl2212. (Renk, 1994.)

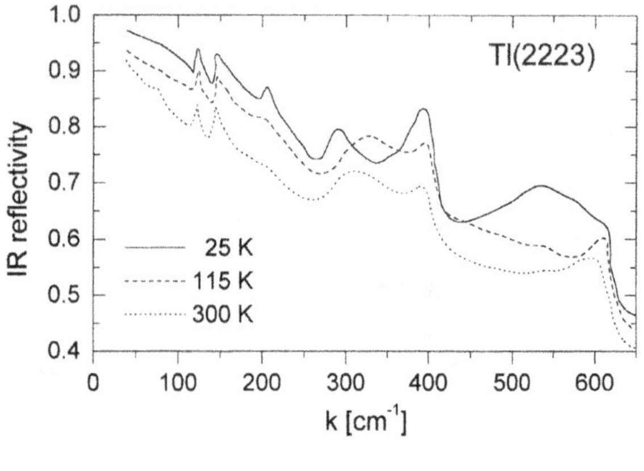

FIGURE C3.21 Infrared reflectivity of superconducting Tl2223. (Renk, 1994.)

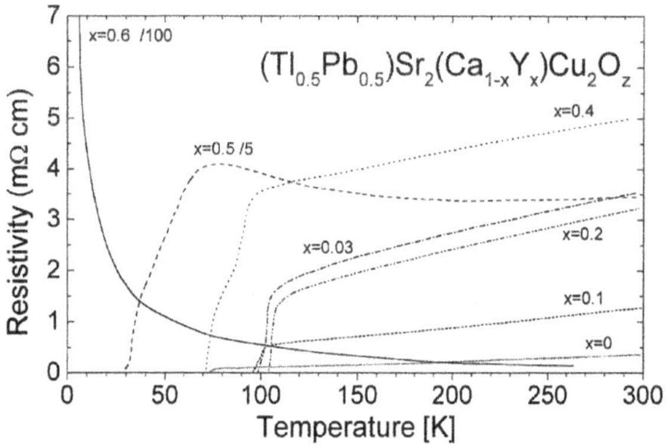

FIGURE C3.22 Electrical resistivity of Tl1212 doped with Y *versus* temperature. Varying the Y content the material changes from a superconductor with T_c up to 105 K to a semiconductor. The resistivity for an Y content of 0.05 and 0.06 per formula unit is divided by 5 and 100, respectively, to fit in the scale. (Liu and Edwards, 1993.)

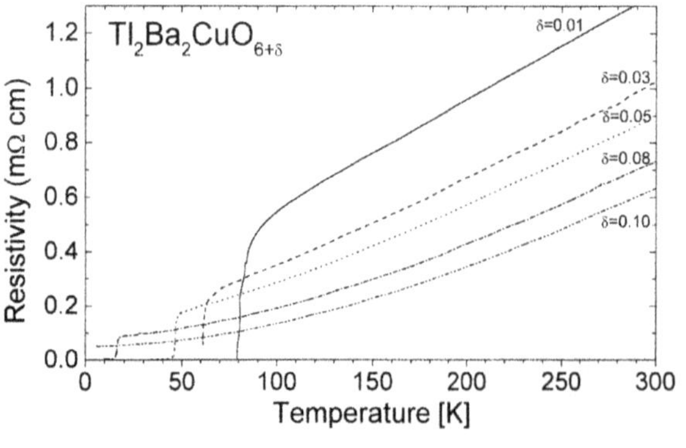

FIGURE C3.23 Electrical resistivity of Tl2201 for different oxygen contents (overdoping). (Kubo *et al.*, 1991.)

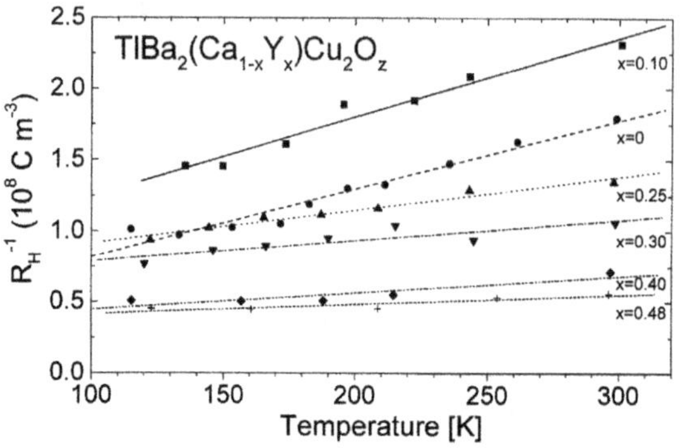

FIGURE C3.24 The Hall coefficient of Tl1212 doped with Y *versus* temperature for different Y contents. (Poddar *et al.*, 1991.)

conductivity, in the out-of-plane direction as well as in CuO_2 planes.

The Hall coefficient in a single-band model is $R_H^{-1} = \pm n_H$ e, where n_H is the number of carriers of charge $\pm e$; the sign of R_H is positive for hole conductors and negative for electron conductors. Great caution should be exercised anyway in equating the Hall number from this simple expression with the actual carrier density n in cases where the carrier density is not homogeneous or more than one conducting band exist, since cancellation effects may occur. In the Tl-based superconductor, R_H is positive for all phases (Figure C3.24): the underdoped and optimally doped samples generally exhibit a linear temperature dependence for R_H^{-1}: $R_H^{1} = A + BT$, with $A > 0$ (Figure C3.25) (Naugle and Kaiser, 1994); in the overdoped case, the temperature dependence of R_H^{-1} is much weaker and shows a minimum (Figure C3.26) (Kubo *et al.*, 1991). A temperature-independent behavior as expected for good metallic samples is not observed. The positive sign of the Hall coefficient indicates that unless some unusual effect reverses its sign, the dominant carriers are hole-like rather than electron-like.

The thermopower S given by $S = \Delta V/\Delta T$, at temperatures above T_c, can be described approximately by a linear equation: $S = A' - B'T$, where A' and B' are always positive. The constant term turns out to be very small for overdoped samples but strongly increases in underdoped samples, whereas

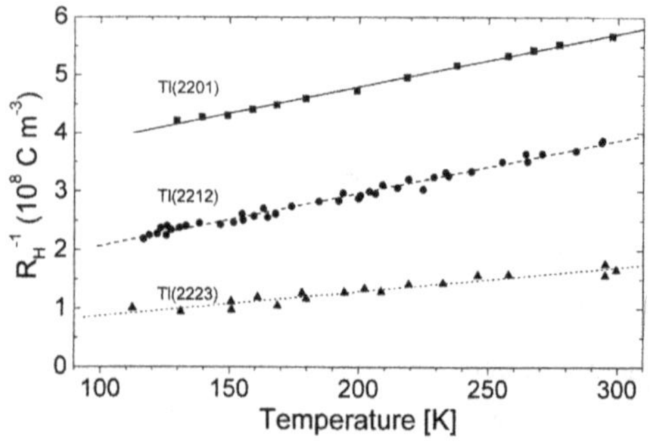

FIGURE C3.25 The Hall coefficient of double Tl layer superconductors *versus* temperature. (Naugle *et al*, 1994.)

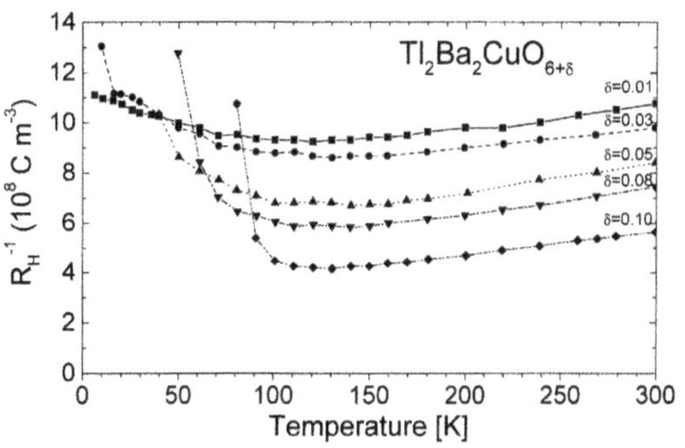

FIGURE C3.26 The Hall coefficient of Tl2201 for different oxygen contents (overdopings). (Kubo *et al.*, 1991.)

the temperature dependence increases with the doping (Figure C3.27). It follows from the linearity of the thermopower that phonon drag does not play a major role in producing the shift of thermopower away from $S \propto T$ that characterizes the superconducting samples. In the underdoped regime where T_c is below its maximum value, the thermopower becomes large and resembles a hopping-type thermopower consistent with the larger resistivity that acquires a semiconductor-like temperature dependence in this regime. In the overdoped regime, the thermopower approaches the traditional metallic behavior described by the Mott formula:

$$S = \frac{\pi^2 k^2 T}{3e}\left[\frac{\partial \ln\sigma(\varepsilon)}{\partial\varepsilon}\right]_{\varepsilon F},$$

where K is Boltzmann's constant, and $\sigma(\varepsilon)$ is a conductivity-like function of electron energy ε (Naugle and Kaiser, 1994; Siri, 1996; Mott and Davis, 1979). The room temperature thermopower for hole-doped cuprate superconductors, including Tl-cuprates, turns out to be a useful universal proxy for the doping level, p, across the underdoped and overdoped regimes (Obertelli *et al.*, 1992).

C3.6.2 Magnetic Properties

A characteristic property of any conductor such as a normal metal or a superconducting solid above T_c, is the Pauli

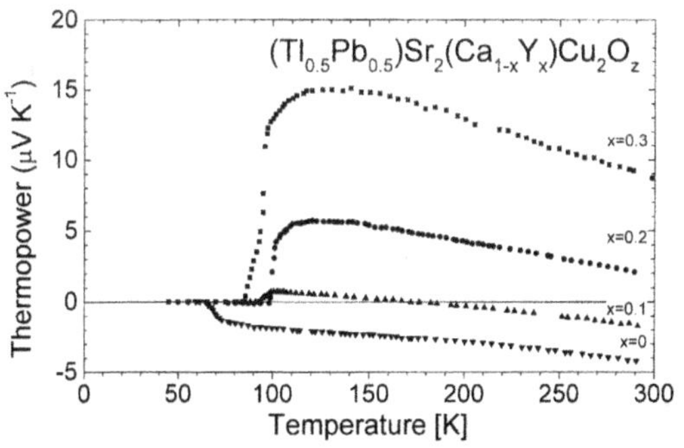

FIGURE C3.27 Thermopower of Tl1212 doped with Y *versus* temperature for different Y contents. (Liu and Edwards, 1993.)

paramagnetism (χ_P). In the paramagnetic state, there is no magnetic order; an external magnetic field can enforce order of the spin of the conduction electrons and produce a net magnetization parallel to the applied field. This contribution to the magnetic susceptibility is small and usually temperature independent since only those electrons that are thermally excitable above the Fermi surface can line up with the field. However, if there is a gap in the density of states at the Fermi level, then the spin susceptibility is reduced on cooling. This is found to be the case in the underdoped cuprates (Alloul *et al.*, 1989) and is an important signature of the so-called *pseudogap* (Naqib *et al.*, 2009), which competes with superconductivity and probably arises from short-range antiferromagnetic correlations (Reymbaut *et al.*, 2019).

A stronger temperature-dependent magnetization originates from localized moments and is well described by a Curie–Weiss model:

$$\chi_{C\text{-}W} = \frac{C}{T - \Theta_C}$$

where Θ_C is the Curie temperature.

In this kind of material, only the two terms (χ_P and $\chi_{C\text{-}W}$) contribute significantly to the total susceptibility, while the other terms like Landau–Peierls, core electron diamagnetism and van Vleck susceptibility account for only a few percent of the already small Pauli contribution (Datta, 1994).

Accurate measurements of χ(T) in the normal state well above T_c allow calculation of the localized moments and their interaction starting from C and Θ_C, respectively.

The density of states at the Fermi surface $D(E_F)$ can be calculated from χ_P by the expression

$$D(E_F) = \frac{3.1\times10^{-4}}{f.u.(\mathrm{Cu})}\chi_P,$$

where $D(E_F)$ is expressed in *states per* eV and χ_p in emu mol^{-1}; *f.u.*(Cu) is the number of Cu in a formula unit [e.g. *f. u.*(Cu) = 3 for Tl1223].

C3.7 Superconducting Properties

C3.7.1 Transition Temperatures

Critical temperatures and their variations for all the Tl-based HTS are plotted *versus* the number of CuO_2 layers in Figure C3.28. Oxygen deficiency seems to be necessary for superconductivity in the single-Tl-layer TBCCO compounds. In the simplest material, $TlBa_2CuO_6$ (1201 phase), all the copper should be nominally Cu^{3+}, resulting in a strong overdoping of the CuO_2 layers. Superconductivity is observed only after annealing in a reducing atmosphere. The progressive reduction of the oxygen content produces at first a semiconductor–metal transition and then a spectacular increase in T_c from 0 to more than 70 K. The T_c value of $TlBa_2CaCu_2O_7$

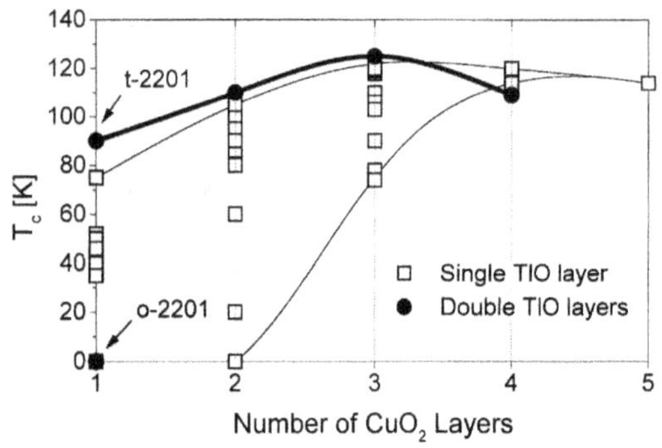

FIGURE C3.28 Critical temperature and variation with chemical substitution in Tl-based superconductors, plotted *versus* the number of CuO_2 layers.

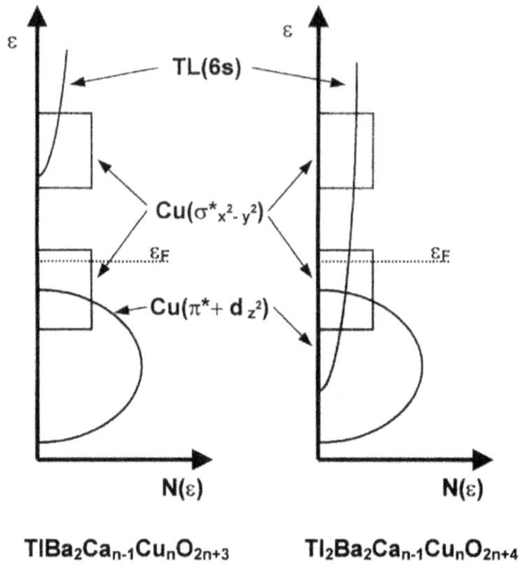

FIGURE C3.29 Schematic diagram of band structures for single (left) and double (right) Tl layer compounds with two copper layers. (Liu and Edwards, 1994.)

(1212 phase) is very variable and depends on the conditions utilized for the synthesis. As for the 1201 phase, 1212 is overdoped and has a formal copper valence of +2.5. Transition temperatures of as-prepared samples normally range from 80 to 90 K, but combinations of oxygen deficiency, and/or thallium substitution on the Ca site, can decrease the copper valence to about +2.2, leading to T_c values exceeding 110 K (see below). The maximum T_c occurs at about 120 K in 1223, where the nominal valence of copper in the fully oxygenated sample is +2.33 near the optimum value. The variation of T_c with the annealing conditions in these phases is consequently more limited. Superconductivity in the $m = 2$ phases of the Tl and Bi system, for which the stoichiometric compositions lead to a nominal copper valence of +2, is still related to oxygen non-stoichiometry and to cationic cross-substitution that occurs spontaneously in these samples. In contrast to the $m = 1$ phases, the oxygen non-stoichiometry consists of an excess rather than a deficiency of oxygen. The insertion of extra oxygen occurs in the Tl compounds on an interstitial site between the double Tl_2O_2 layers, whereas in the Bi phases the extra oxygen is intercalated within the BiO layers along the chains of short Bi–O bonds running along the direction of the modulation wave vector. For a long time, it was believed that the oxygen excess would be simultaneously responsible for both superconductivity in these systems and structural modulation in the Bi system. However, it was demonstrated by (Pham *et al.*, 1992) that stoichiometric Bi2212, prepared in argon, shows an almost commensurate structural modulation and superconductivity at 86 K. For this compound, an increase of the oxygen content led to a shortening of the modulation period and a decrease in the transition temperature. The presence of holes in this compound, as well as in stoichiometric Tl2201 where superconductivity was observed, could be explained by the overlap (hybridization) of the Tl and Bi 6s with Cu $3d_{x^2-y^2}$ bands at the Fermi level, as schematically shown in Figure C3.29 (Liu and Edwards, 1994). This would result in the formation of holes in the CuO_2 layers and in a semi-metallic character of the BiO and TlO layers. This does not happen in the corresponding $m = 1$ Tl compounds, where the 6s bands lie above the Fermi level. The chemical control of superconductivity in these compounds is further complicated by the occurrence of spontaneous cationic cross-substitutions during synthesis, which cannot be easily avoided, even by an accurate control of the reaction atmosphere and of the preparative procedure.

The Tl2201 compound $Tl_2Ba_2CuO_{6+\delta}$, which ranges from a normal metal (as-prepared) to a superconductor with T_c as high as 90 K (after annealing in argon), is very sensitive to small variations of the oxygen stoichiometry. The compound is often Tl deficient and copper rich. In this case, Cu^{2+} was found to substitute Tl^{3+} (about 5%), thus producing an overdoped material for which the maximum T_c can only be achieved by suppressing all the excess oxygen by argon reduction. The corresponding $n = 2$ phase $Tl_2Ba_2CaCu_2O_{8+d}$ also has a critical temperature that is dependent on the oxygen content. Even in this case, an annealing under argon of the as-prepared material, hence a reduction of the hole concentration increases T_c to 112 K. Site disorder was also observed in this phase. In particular, the Tl site contains calcium or copper, whereas the Ca site may be occupied by Tl. The 2223 phase possesses the highest T_c of the family, 128 K, which is rather less dependent on the oxygen content and on the thermal history since only a variation of a few degrees K can be produced. The decreased sensitivity of T_c to varying oxygen content as the number, n, of CuO_2 layers increases can be understood in terms of progressive dilution of the charge transferred from the TlO layers.

C3.7.2 Chemical Control of Superconductivity

Thallium as well as its neighbors in the periodic table, lead and bismuth, present a second stable valency that is lower by two units from the group valency. In addition to the normal group oxidation states, Tl(III), Pb(IV) and Bi(V), stable oxidation states are found in Tl(I), Pb(II) and Bi(III) which present the outer 6 s^2 electronic configuration. These two electrons ionize or participate in covalent bond formation with difficulty because the outer s and p states are separated by a large difference in energy, and are thus called an *inert (lone) electron pair.* On the other hand, the pair often participates in bonding resulting in asymmetrical stereochemistries, its volume being comparable to that of an anion in a solid (Hulliger, 1976).

There is a structural size mismatch between the TlO and the CuO_2 layers, which produces a stretching of the TlO layers resulting in displacements of the atoms of the rocksalt block from their ideal positions. In thallium compounds, the displacements are in most cases statistically distributed, giving rise to an average tetragonal symmetry, but in some cases, a lowering of the symmetry from tetragonal ($a = a_p$) to orthorhombic $\left(a \approx b \approx a_p = \sqrt{2}\right)$ is observed as a result of ordering as previously discussed in Section C3.1 for Tl2201. In bismuth compounds, (see also Chapter C1.2.2 of this Handbook), due to the lone pair character of Bi^{3+}, which produces a remarkable distortion of the coordination (there are three short and three long bonds for the six-fold coordinated Bi), the mismatch is more pronounced than in thallium compounds and gives rise in all cases to an incommensurate unidimensional structural modulation with a modulation period ~ $5a$ of the orthorhombic $\left(a \approx b \approx a_p = \sqrt{2}\right)$ fundamental cell.

The presence of the lone pair on Bi^{3+} results in another important structural difference between Tl-based and Bi-based phases: Tl-based phases are much more three-dimensional in character than the corresponding Bi-based phases as a result of the fact that the three short Bi–O bonds are mutually perpendicular, so that only one strong bond in the direction perpendicular to the BiO layer (the one with the oxygen of the SrO layer) can be formed. Adjacent BiO layers are consequently weakly bound with a spacing of about 3.2 Å, whereas the TlO layers are much more strongly bonded and have an interlayer spacing of about 2.0 Å. This characteristic of the BiO layer is moreover responsible for the impossibility of producing Bi-based phases with $m = 1$. Despite the similarity of the structural sequences and of the formal valence of Bi and Tl in the structures, it is impossible to find a simple unified mechanism to describe the chemical control of superconductivity in these systems. Oxygen non-stoichiometry and cationic substitutions play an important role in defining the superconducting properties, but their effects often overlap and in some cases they seem not to be sufficient for justifying the observed superconductivity. A remarkable example of chemical control of the electronic properties is the compound $Tl_{1-y}Pb_ySr_2Ca_{1-x}Y_xCu_2O_z$ in the family of Tl1212 materials. The compound $TlSr_2CaCu_2O_z$ is itself a metal, but no superconducting transition is observed down to 4 K; the nominal Cu valency of this compound is

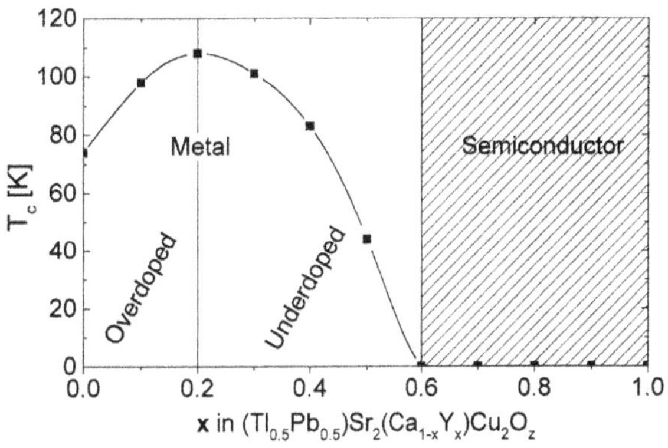

FIGURE C3.30 Critical temperature *versus* Y content in $Tl_{0.5}Pb_{0.5}Sr_2Ca_{1-x}Y_xCu_2O_z$. (Liu and Edwards, 1993.)

+2.5, which indicates an excess of hole carriers in the CuO_2 layers (overdoped state). To induce superconductivity, it is possible to reduce this overdoping by replacing Tl^{3+} by Pb^{4+} and/or Ca^{2+} by Y^{3+}. This substitution is proved to introduce an excess of electrons in the conducting CuO_2 planes, therefore reducing the net hole number (see Figure C3.25) (Liu and Edwards, 1993).

For the phase without Pb ($y = 0$), varying x, the Y content is possible to transform this material from a normal metal $TlSr_2CaCu_2O_z$ ($x = 0$) to a superconductor (with a maximum of $T_c = 80$ K for $x = 0.6$) to an insulator $TlSr_2YCu_2O_z$. The phase with the Tl site half substituted by Pb ($y = 0.5$), as shown in Figure C3.30, is a superconductor with a T_c of ~ 80 K for $x = 0$. It reaches the highest T_c of the family (108 K) for the composition $Tl_{0.5}Pb_{0.5}Sr_2Ca_{0.8}Y_{0.2}Cu_2O_z$ ($x = 0.2$) and finally becomes an antiferromagnetic insulator for $x = 1$ (Liu and Edwards, 1993). A very similar dependence of T_c on La content has been reported for a related compound $Tl_{0.5}Pb_{0.5}Sr_{2-x}La_xCaCu_2O_7$, reaching a similar maximum T_c of 105 K for 0.2 La (Presland and Tallon, 1991).

C3.7.2.1 Substitutions

The conversion of Tl2212 and Tl1212 thin films to Hg1212 was reported in two publications: Wu *et al.* (1999) replaced Tl atoms by Hg reacting the Tl-Ba-Ca-Cu-O films in the presence of a Hg vapor source – see also Chapter C4. The critical current density of the cation-exchanged non-substituted Hg1212 film was impressive ($2.3 - 4 \times 10^6$ A/cm^2 at 77 K), but no marked improvement in its field and temperature dependence was seen. Another study by Bhattacharya *et al.* (2002) reported similar cation exchange experiments using electrodeposited thick films of Tl1212 with Bi, Pb and Sr substitutions, i.e. $(Tl,Bi)(Sr,Ba)_2CaCu_2O_x$. The critical current density at 4.2 K and zero field was the same as reported by (Wu *et al.*, 1999), but its decrease with field was lower: at 0.1T it was still 2×10^6 A/cm^2.

TABLE C3.11 Typical Superconductive Intrinsic Parameters of Tl-Based Compounds Reported in the Literature

Compound	T_c	λ_{ab} (nm)	λ_c (nm)	ζ_{ab} (nm)	ζ_c (nm)
Tl1201	52	–	–	–	–
Tl1212	80	210	–	2.0	–
Tl1223	120	200	–	1.8	–
Tl2201	90	170	–	5.2	0.3
Tl2212	110	215	–	2.2	0.5
Tl2223	125	205	480	1.3	–

C3.7.3 Magnetic Properties

Diamagnetism measurements for single crystals of Tl-based superconductors demonstrate 100% bulk superconductivity with very sharp transitions. In contrast to conventional conductors, the Sommerfeld constant of these compounds is rather small, especially for such a high T_c value. The temperature dependence of magnetic penetration depth indicates a low symmetry non s-wave ground state; a result later confirmed by measurements of 2201 thin films using tricrystal ring magnetometry (Tsuei et al., 1996; Tsuei *et al.*, 1997) and angle-dependent torque magnetometry (Rossel *et al.*, 1997). Possibly there are some regions on the Fermi surface with no superconducting gap or a pseudogap is present. The London penetration depth is large ($\lambda_{ab} \sim 200$ nm), and the coherence length is very small ($\xi_{ab} \sim 2$ nm), so all these compounds are definitively type II superconductors in the clean limit. As for all high T_c superconductors, they present a large effective mass anisotropy ($m_c^* / m_{ab}^* \sim 100$); in addition, a quite pronounced anisotropy in the penetration depth is observed ($\lambda_c/\lambda_{ab} \sim 15$) (Table C3.11).

Consistent with this anisotropy, diamagnetic fluctuations indicate the existence of quasi-two-dimensional superconductivity. The anisotropy turns out to be more pronounced in double Tl layer compounds than in the single Tl layer ones. Softness of the flux line lattice is evidenced in the broadening of the transition and the reduced irreversibility field at the high-field/high-temperature region of the phase diagram. Figure C3.31 shows the temperature dependence of the irreversibility field determined from the 'pinch-off field' in magnetization hysteresis loops for a variety of thallium HTS (Presland *et al.*, 1993). Except for Tl2234, data are reported for samples where T_c has been approximately maximized by atomic substitution or by appropriate oxygen annealing to obtain optimum hole doping. As shown in Figure C3.31, the irreversibility line for Tl1223 exceeds that of any other Tl-based HTS. The Tl1223 irreversibility line is higher than that of Bi2223 and, below approximately 1 T, it is also higher than that of YBCO (see also chapters C1 and C2).

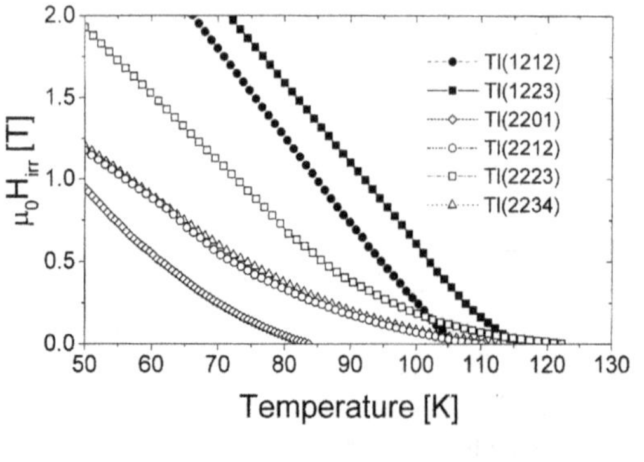

FIGURE C3.31 Irreversibility line from magnetization measurements for Tl-based superconductors. (Rossel *et al.*, 1997.)

Beyond this, it is possible to observe some systematic trends: (a) the irreversibility field for the double-layered TI compounds (2201, 2212, 2223 and 2234) displays a similar temperature dependence, although the number of CuO layers and the values of T_c(max) differ between various compositions. There is a slow steepening of $H_{irr}(T)$ as n decreases; (b) the single-layered Tl-based materials (1212 and 1223) are quite different: both compositions display an irreversibility field that increases more rapidly with decreasing temperature than any other double-layered compositions. Thus $H_{irr}(T)$ is strongly dependent on m and only weakly dependent on n. These data are consistent with a model where pinning of three-dimensional vortex lines is much more effective than that of decoupled two-dimensional pancake vortices in the CuO_2 planes. The anisotropy (two-dimensionality) increases with the number of Tl layers due to the fact that the distance between the CuO_2 layer is increased, and consequently, the Josephson coupling between the superconducting sheets is decreased.

For power applications, such as transmission lines and various magnet systems, two Tl-based compounds may be of special interest, namely the double-layer Tl2223 and particularly the single-layer Tl1223, which has a higher irreversibility line in the $H_{irr}(T)$ representation.

The magnetic properties between 4.2 and 300 K were studied on a series of polycrystalline samples with the formula $TlA_2RCu_2O_{7-x}$, with A = Sr and Ba and R = RE, Y and Ca with the Tl1212 structure (Khlybov *et al.*, 2001). It was found that the magnetic properties depend on the rare-earth metal. The compounds with nonmagnetic Y and Lu showed a sharp peak in the susceptibility *versus* T curve between 15 and 21 K. This behavior was attributed to antiferromagnetic ordering (AFM) of Cu spins. The compounds with magnetic rare-earths ions (R = Nd, Gd, Tb, Dy, Ho and Er) revealed different features in their magnetic characteristics. The superconducting compounds with Tl1212 structure can be regarded as highly degenerate magnetic semiconductors. The phenomenon of phase separation into magnetic and superconducting phases has been observed in a number of compounds.

C3.7.4 Transport Properties

The transport critical current is strongly dependent on the preparation technique utilized for the sample preparation. An overview of transport critical current reached for different kinds of Tl(1223) samples is summarized in Figure C3.32. Two

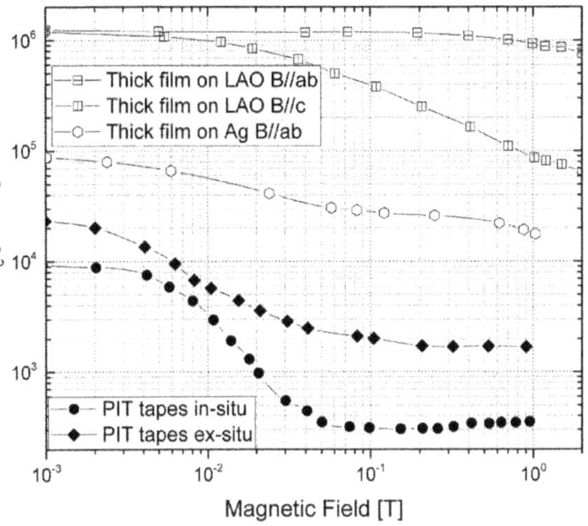

FIGURE C3.32 Transport critical current densities *versus* magnetic field at 77 K for different Tl1223 conductors fabricated by the "open" and "closed" approaches.

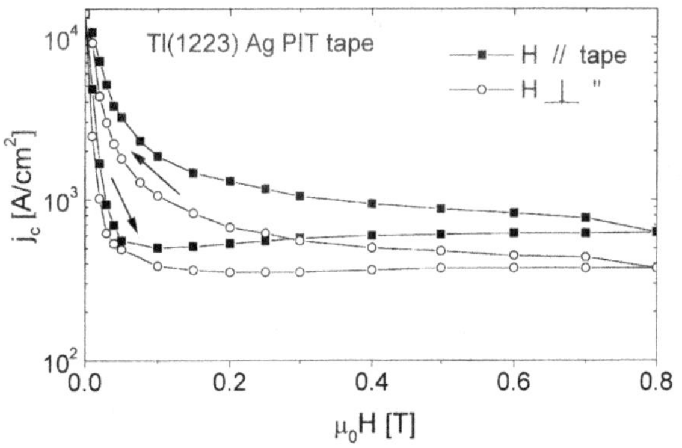

FIGURE C3.33 Transport critical current densities *versus* magnetic field at 77K for a Tl1223/Ag PIT tape with the surface of the tape parallel and perpendicular to the magnetic field. The arrows indicate increasing or decreasing magnetic field.

basic processing procedures for Tl-based conductors are used: the so-called closed approach represented by the PIT method and the more convenient and successful open approach represented by various relatively simple but efficient non-vacuum methods such as aerosol deposition, electrodeposition, sol–gel and screen-printing or painting.

The closed approach PIT process is able to produce conductors with the required core thickness, however, from the point of view of achieving high J_c values, its development has so far plateaued at values around 2×10^4 A cm^{-2} at 77 K/0 T (Ren and Wang, 1993; Bellingeri *et al.*, 1997).

Small increases in J_c may be achieved with improved contacts between the grains obtained by increasing the material density (Jeong *et al.*, 1999) or by the formation of a transient liquid during sintering by means of appropriate substitutions (Bellingeri *et al.*, 1998). However, J_c of polycrystalline samples decreases dramatically in magnetic fields as low as 0.1 T due to weak links arising from lack of grain alignment. Weak links are well evidenced by the fact that the transport current in a magnetic field of 0.2 T decreases by a factor ~ 15 from the zero-field value and, moreover, strong hysteresis is present when the magnetic field is cycled (Figure C3.33). However, it is interesting to note that J_c does not decrease further with increasing magnetic field up to almost 10 T. This suggests that some 'strong-linked' current paths are present and these dominate when the weak-linked paths are quenched (Figure C3.34).

The open system would appear to be a substantially more hopeful approach where more successful methods exist, the results of which have already overtaken those obtained by the PIT procedure. Among these open approach methods, the most successful are aerosol deposition from a solution and the electrodeposition (Bhattacharya *et al.*, 1998) on Ag substrates, with $J_c > 10^5$ A cm^{-2} at 77 K/0 T, and exceeding 10^4 A cm^{-2} in high magnetic fields (5 T) at 77 K. As shown in Figure C3.32, depositions on single-crystalline substrates (Li *et al.*, 1999) present critical current densities higher than 10^6 A cm^{-2} at 77 K and 0 T. The reason for such a tremendous improvement over the PIT method lies in the marked grain alignment of the polycrystalline films, a fundamental condition to obtain high J_c values in magnetic fields. Ideally, the grains should be both *c*- and *a*-axis aligned such that the microstructure approaches that of a single crystal. In this respect, an important discovery has been made demonstrating that a bi-axially textured polycrystalline Ag substrate helps to align the superconducting grains.

For power applications, high I_c values are also required, which requires growth of HTS films with thickness exceeding 10 μm. From this point of view, spray deposition from ink, sol–gel, screening, printing or electrodeposition are of considerable interest. The great challenge associated with thick-film growth involves developing and sustaining the underlying substrate texture up to the very external surface of the thick film.

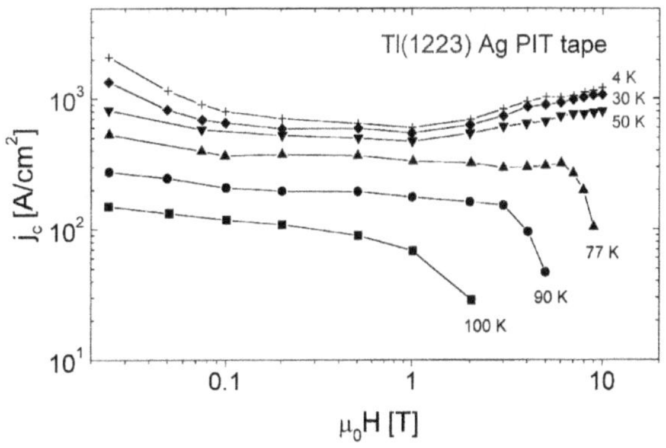

FIGURE C3.34 Transport critical current densities *versus* magnetic field for a Tl1223/Ag PIT tape for different temperature. (Li *et al.*, 1999.)

C3.8 Thallium Safety

Thallium and most of its compounds are potentially toxic. Hundreds of deaths have resulted from accidental, as well as homicidal and suicidal ingestion of Tl compounds. Another problem arises from the fact that soluble Tl compounds (e.g. $TlNO_3$) can penetrate the unbroken skin adding another dimension to their hazard potential. Detailed safety prescriptions are given in "Hazardous substance fact sheets", from the New Jersey Department of Health and Senior Services (1998).

The lowest dose that will cause death is estimated to be about 15 mg of Tl per kg of body weight regardless of the route of administration, thus making Tl as lethal as arsenic. Unlike Pb and Hg, which the human body is unable to remove, Tl is excreted, primarily in the urine but also in the feces, and so considerably reducing the risk of long-term accumulation. The biological half-life (i.e. the time required to excrete half of the remaining body burden of Tl) is estimated to be around 10 days. Exposure to Tl can cause fatigue, weakness, poor appetite, insomnia, confusion and mood changes. Higher exposures can damage the nervous system causing numbness, pain and 'pins and needles' in arms and legs and may affect the liver and kidneys.

The recommended airborne exposure limit is 0.1 mg m^{-3} averaged over an 8-h workshift. The above exposure limits are for air levels only. When skin contact also occurs, overexposure is a risk, even if airborne levels are less than the limit noted above.

Engineering controls are the most effective way of reducing exposure. The best protection is to enclose operations and/or provide local exhaust ventilation at the site of chemical release. Isolating operations can also reduce exposure. Using respirators or protective equipment is less effective than the controls mentioned above, but is sometimes necessary. Good work practices can help to reduce hazardous exposures. The following work practices are recommended:

- On skin contact with Tl, immediately wash or shower to remove the chemical. At the end of the workshift, wash any areas of the body that may have contacted it, whether or not known skin contact has occurred.
- Do not eat, smoke, or drink where Tl is handled, processed, or stored, since the chemical can be swallowed. Wash hands carefully before eating or smoking.
- Use a vacuum or a wet method to reduce dust during clean-up. Do not dry sweep.

Workplace controls are better than personal protective equipment. However, for some jobs or in addition, personal protective equipment may be appropriate.

- Avoid skin contact with Tl or its compounds. Wear protective gloves and clothing. Safety equipment suppliers and manufacturers can provide recommendations on the most protective glove and clothing material for your operation.
- Wear dust-proof goggles and face shield when working with powders or dust, unless full facepiece respiratory protection is worn.
- Use respiratory protection. Be sure to consider all potential exposures in your workplace. You may need a combination of filters, prefilters, cartridges, or canisters to protect against different forms of a chemical (such as vapor and mist) or against a mixture of chemicals.
- Exposure to 15 mg m^{-3} is immediately dangerous to life and health. If the possibility of exposure above 15 mg m^{-3} exists, use a self-contained breathing apparatus with a full facepiece operated in continuous flow or other positive pressure mode.

References

Alloul H, Ohno T and Mendels P (1989) ^{89}Y NMR evidence for a fermi-liquid behavior in $YBa_2Cu_3O_{6+x}$ *Phys. Rev. Lett.* 63: 1700

Aselage T L, Voigt J A and Keefer K D (1990) Instability of thallium-containing superconductor phases in isothermal equilibrium with thallium oxide *J. Am. Ceram. Soc.* 73: 3345

Aselage T L, Venturini E L and Van Deusen S B (1994) Two-zone equilibria of Tl-Ca-Ba-Cu-O superconductors *J. Appl. Phys.* 75: 1023

Bellingeri E, Gladyshevkii R E and Flükiger R (1997) Mono- and multifilamentary Ag-sheathed Tl(1223) tapes *Il Nuovo Cimento D* 19: 1117

Bellingeri E, Gladyshevskii R, Marti F, Dhallé M and Flükiger R (1998) Synthesis and properties of fluorine-doped Tl(1223): bulk materials and Ag-sheathed tapes *Supercond. Sci. Technol.* 11: 810

Bhattacharya R N, Blaugher R D, Ren Z F, Li W, Wang J H, Paranthaman M, Verebelyi D T and Christen D K (1998) Superconducting thallium oxide films by the electrodeposition method *Physica C* 304: 55

Bhattacharya R N, Xing Z, Wu J Z, Chen J, Yang S X, Ren Z F and Blaugher R D (2002) Superconducting thallium oxide and mercury oxide films *Physica C* 377: 327

Calatroni S, Bellingeri E, Ferdeghini C, Putti M, Vaglio R, Baumgartner T and Eisterer M (2017) Thallium-based high-temperature superconductors for beam impedance mitigation in the Future Circular Collider *Supercond. Sci. Technol.* 30: 075002

Datta T (1994) *Thallium-Based High-Temperature Superconductors* ed A M Hermann and J V Yakhmi (New York: Dekker) p 407

Deinhofer C and Gritzner G (2004) Thallium-based high-temperature superconductors for beam impedance mitigation in the *Future Circular Collider Supercond. Sci. Technol.* 17: 1196

DeLuca J A, Garbauskas M F, Bolon R B, McMullin J G, Balz W E and Karas P L (1991). The synthesis of superconducting Tl-Ca-Ba-Cu-oxide films by the reaction os spray deposited Ca-Ba-Cu-oxide precursors with Tl_2O vapour in a two-zone reactor *J. Mater. Res.* 6: 1415

Flükiger R, Gladyshevskii R E and Bellingeri E (1998) Methods to produce Tl(1223) tapes with improved properties *J. Supercond.* 11:23

Ganguli A K and Subramanian M A (1991) Synthesis and characterization of orthorhombic $TlSr_2CuO_5$ *J. Solid State Chem.* 93: 250

Gladyshevskii R E and Galez P (1999) Crystal structures of High-T_c superconducting cuprates, characteristic parameters, thermal properties, electrical properties, magnetic properties, mechanical properties, *Handbook of Superconductivity* ed C P Poole (San Diego: Academic Press)

Gopalakrishnan I K, Yakhmi J V and Iyer R M (1991) Stabilization of superconductivity in $TlBa_2CuO_{5-\delta}$ at 9.5 K and its enhancement to 43 K in $TlBaSrCuO_{5-\delta}$ *Physica C* 175: 183

Hazen R M, Finger L W, Angel R J, Prewitt C T, Ross N L, Hadidiacos C G, Heaney P J, Veblen D R, Sheng Z Z, El Ali A and Hermann A M (1988) 100 K superconducting phases in the Tl-Ca-Ba-Cu-O system *Phys. Rev. Lett.* 60: 1657

Hewat A W, Bordet P, Capponi J J, Chaillout C, Chenavas J, Godinho M, Hewat E A, Hodeau J L and Marezio M (1988) Preparation and neutron diffraction of superconducting "tetragonal" and non-superconducting orthorhombic $Tl_2Ba_2Cu_1O_6$ *Physica C* 156: 369

Hervieu M, Michel C, Maignan A, Martin C and Raveau B (1988) The 125 K superconductor $TI_{2-x}Ba_2Ca_2Cu_3O_{10+\delta}$: A tentative structural model *J. Solid State Chem.* 74: 428

Hervieu M, Maignan A, Martin C, Michel C, Provost J and Raveau B (1988) $Tl_2Ba_2Ca_3Cu_4O_{12}$, a new 104 K superconductor of the family $(AO)_3(A'CuO_{3-y})_m$ *Mod. Phys. Lett. B* 2: 1103

Holstein W L (1993) Thermodynamics of the volatilization of thallium(I) oxide from Tl_2O, Tl_4O_3, and Tl_2O_3 *J. Phys. Chem.* 97: 4224

Huang T C, Lee V Y, Karimi R, Beyers R and Parkin S S P (1988) Preparation and X-ray characterization of superconducting and related $Tl_2Ba_2CuO_6$ phases *Mater. Res. Bull.* 23: 1307

Hulliger F (1976) *Structural Chemistry of Layer-Type Phases ed D Reidel* (Kluwer: Dordrecht)

Ihara H, Sugise R, Hirabayashi M, Terada N, Jo M, Hayashi K, Negishi A, Tokumoto M, Kimura Y and Shimomura T (1988) A new high-T_c $TlBa_2Ca_3Cu_4O_{11}$ superconductor with T_c >120K *Nature* 334: 510

Iqbal Z, Sinha A P B, D Morris E, Barry J C, Auchterlonie G J and Ramakrishna B L (1991) High-pressure oxygen-induced bulk superconductivity in 1222 structure Tl-Pb-Sr- Eu(Ce)-Cu-O *J. Appl. Phys.* 70: 2234

Jeong D Y, Kim H K and Kim Y C (1999) Much enhanced J_c by intermediate rolling in just-rolled Tl-1223/Ag tapes *Physica C* 314: 139

Johansson L G, Ström C, Eriksson S and Bryntse I (1994) A new procedure for the preparation of $Tl_2Ba_2CaCu_2O_{8-\delta}$ characterization by neutron diffraction, electron diffraction and HREM *Physica C* 220: 295

Jorda J L, Jondo T K, Abraham R, Cohen-Adad M Th, Opagiste C, Couach M, Khoder A and Sibieude F (1993) Preparation of pure $Tl_2Ba_2CuO_{6\pm x}$ *Physica C* 205: 177

Jorda J L, Jondo T K, Abraham R, Cohen-Adad M Th, Opagiste C, Couach M, Khoder A F and Triscone G (1994) Thermodynamic and kinetic studies of the phase transitions in $Tl_2Ba_2CuO_{6\pm x}$ *J. Alloys Compd.* 215: 135

Khlybov E P, Kostyleva I E, Nizhankovskii V I, Palewski T, Warschulska J and Nenkov K (2001) Superconductivity and magnetic order in thallium-based cuprates *Physica B* 294-295: 367

Kikuchi M, Kajitani T, Suzuki T, Nakajima S, Hiraga K, Kobayashi N, Iwasaki H, Syono Y and Muto Y (1989) Preparation and chemical composition of superconducting oxide $Tl_2Ba_2Can_{-1}CunO_2n_{+4}$ with *n*=1, 2 and 3 *Japan. J. Appl. Phys.* 28: L382

Kim J S, Swinnea J S and Steinfink H (1989) Cation and oxygen disorder in the structures $TlSr_2CuO_5$ and $(Pb_{0.63}Cu_{0.37})Sr_2CoO_5$ *J. Less Common Met.* 156: 347

Kolesnikov N N, Korotkov V E, Kulakov M P, Lagvenov G A, Molchanov V N, Muradyan L A, Simonov V I, Tanazyan R A, Shibaeva R P and Shchegolev I F (1989) Structure of superconducting single crystals of $TiBa_2(Ca_{0.87}Tl_{0.13})Cu_2O_7$, T_c= *80K* *Physica C* 162-164: 1663

Kondoh S, Ando Y, Onoda M, Sato M and Akimitsu J (1988) Superconductivity in TlBaCuO system, *Solid State Commun.* 65: 132

Kubo Y, Shimakawa Y, Manako T and Igarashi H (1991) Transport and magnetic properties of $Tl_2Ba_2CuO_{6+\delta}$ showing a delta-dependent gradual transition from an 85-K superconductor to a nonsuperconducting metal *Phys. Rev. B* 43: 7875

Li W, Wang D Z, Lao J Y, Ren Z F, Wang J H, Paranthaman M, Verebelyi D T and Christen D K (1999) Epitaxial superconducting $Tl_{0.5}Pb_{0.5}Sr_{1.6}Ba_{0.4}Ca_2Cu_3O_9$ films on $LaAlO_3$ by thermal spray and post-spray annealing *Supercond. Sci. Technol.* 12: L1

Liu R S, Hervieu M, Michel C, Maignan A, Martin C, Raveau B and Edwards P P (1992) $T1Ba_2(Eu, Ce)_2Cu_20_{9+\delta}$, a new member of the double fluorite-type cuprate family structure and possible induced superconductivity by oxygen high-pressure annealing *Physica C* 197: 131

Liu R S, Hughes S D, Angel R J, Hackwell T P, Mackenzie A P and Edwards P P (1992) Crystal structure and cation stoichiometry of superconducting $Tl_2Ba_2CuO_{6+\delta}$ single crystals *Physica C* 198: 203

Liu R S and Edwards P (1993) The chemical control of high-temperature superconductivity; the metal-superconductor-insulator transition in $(Tl_{1-y}Pb_y)Sr_2(Ca_{1-x}Y_x)Cu_2O_7$ *Mater. Sci. Forum* 130-132: 435

Liu R S and Edwards P P (1994) *Thallium-Based High-Temperature Superconductors* ed A M Hermann and J V Yakhmi (New York: Dekker) p 325

Maignan A, Michel C, Hervieu M, Martin C, Groult D and Raveau B (1988) $Tl_2Ba_2CaCu_2O_8$: structure and superconductivity *Mod. Phys. Lett. B* 2: 681

Manako T, Shimakawa Y, Kubo Y, Satoh T and Igarashi H (1989) Superconductivity of $TlBa_{1+x}La_{1-x}CuO_5$ with 1201 structure *Physica C* 158: 143

Martin C, Bourgault D, Hervieu M, Michel C, Provost J and Raveau B (1989) The layered thallium cuprates $Tl_{1+x}A_{2-y}Ln_2Cu_2O_9$: a triple intergrowth of the perovskite, rock salt and fluorite structure *Mod. Phys. Lett. B* 3: 993

Molchanov N, Tamazyan R A, Simonov V I, Blomberg M K, Merisalo M J and Mironov V S (1994) Structure and superconductivity in $Tl_2Ba_2CaCu_2O_8$ *Physica C* 229: 331

Morosin B, Ginley D S, Hlava P F, Carr M J, Baughman R J, Schirber J E, Venturini EL and Kwak J F (1988) Structural and compositional characterization of polycrystals and single crystals in the Bi- and Tl-superconductor systems: Crystal structure of $TlCaBa_2Cu_2O_7$, *Physica C* 152: 413

Morosin B, Venturini E L and Ginley D S (1991) Tl·O charge reservoir changes on annealing Tl-1223 crystals *Physica C* 183: 90

Morosin B, Ginley D S, Venturini E L, Baughman R J and Tigges C P (1991) Structure studies on Tl-2122 and Tl-2223 superconductors *Physica C* 172: 413

Morosin B, Venturini E L and Ginley D S (1991) Annealing study on single crystal Tl-2223 superconductors *Physica C* 175: 241

Mott N F and Davis E A (1979) *Electronic Process in Non-Crystalline Materials* (Oxford: Clarendon) (Singapore: World Scientific) p 445

Naugle D G and Kaiser A B (1994) *Thallium-Based High-Temperature Superconductors* ed A M Hermann and J V Yakhmi (New York: Dekker) p 543

Naqib S H, Cooper J R and Loram J W (2009) Effects of Ca substitution and the pseudogap on the magnetic properties of $Y_{1-x}Ca_xB$ $a_2Cu_3O_{7-\delta}$ *Phys. Rev. B* 79: 104519

New Jersey Department of Health and Senior Services (1998) *Thallium Nitrate: Hazardous Substance Fact Sheet* https://nj.gov/health/eoh/rtkweb/documents/fs/1841.pdf

Obertelli S D, Cooper J R and Tallon J L (1992) Systematics in the Thermoelectric power of high-T_c oxides *Phys. Rev. B* 46: 14928

Ogborne D M, Weller M T and Lanchester P C (1992) Oxygen stoichiometry and the structure of $Tl_2Ba_2Cu_3O_{10-y}$ a high-resolution powder neutron diffraction study *Physica C* 200: 167

Ogborne D M, Weller M T and Lanchester P C (1992) A high-resolution neutron diffraction study on $Tl_2Ba_2CaCu_2O_{8-y}$: y=0, 0.05, 0.06 *Physica C* 200: 207

Ogborne D M and Weller M T (1992) The structure of $Tl_2Ba_2Ca_3Cu_4O_{12}$ *Physica C* 201: 53

Ogborne D M and Weller M T (1994) The structure of $TlBa_2Ca_3Cu_4O_{11}$ *Physica C* 230: 153

Ogborne D M and Weller M T (1994) The structure of $TlBa_2Ca_3Cu_4O_{11}$ *Physica C* 230: 153

Ogborne D M and Weller M T (1994) P Structure and oxygen stoichiometry in $Tl_2Ba_2Ca_3Cu_4O_{12-\delta}$ a high-resolution powder neutron-diffraction study *Physica C* 223: 283

Ohshima E, Kikuchi M, Izumi F, Hiraga K, Oku T, Nakajima S, Ohnishi N, Morii Y, Funahashi S and Syono Y (1994) Structure analysis of oxygen-deficient $TlSr_2CuO_y$ by neutron diffraction and high-resolution electron microscopy *Physica C* 221: 261

Onoda M, Kondoh S, Fukuda K and Sato M (1988) Structural Study of Superconducting Tl-Ba-Ca-Cu-O System *Japan. J. Appl. Phys.* 27: L1234

Opagiste C, Couach M, Khoder A F, Abraham R, Jondo T K, Jorda J-L, Cohen-Adad M Th, Junod A, Triscone G and Muller J (1993) A new elaboration process of the superconducting $Tl_2Ba_2Cu_1O_6$ phase with T_c=90K *J. Alloys Compd.* 195: 47

Opagiste C, Triscone G, Couach M, Jondo T K, Jorda J L, Junod A, Khoder A F and Muller J (1993) Phase diagram of the $Tl_2Ba_2CuO_6$ compounds in the *T*, $p(O_2)$ plane *Physica C* 213: 17

Parise J B, Gopalkrishnan J, Subramanian M A, Sleight A W (1988) Superconducting $Tl_2Ba_2CuO_6$: the orthorhombic form *J. Solid State Chem.* 76: 432

Parise J B, Torardi C C, Subramanian M A, Gopalakrishnan J, Sleight A W and Prince E (1989) Superconducting $Tl_{2.0}Ba_{2.0}CuO_{6+\delta}$: A high resolution neutron powder and single crystal x-ray diffraction investigation *Physica C* 159: 239

Parkin S S P, Lee V Y, Nazzal A I, Savoy R, Huang T C, Gorman G and Beyers R (1988) Model family of high-temperature superconductors: $Tl_mCa_{n-1}Ba_2Cu_nO_{2(n+1)+m}$ (m=1,2; n=1,2,3) *Phys. Rev. B* 38: 6531

Parkin S S P, Lee V Y, Nazzal A I, Savoy R, Beyers R and La Placa S J (1988) $TlCa_{n-1}Ba_2Cu_nO_{2n+3}$ (n=1,2,3): A new class of crystal structures exhibiting volume superconductivity at up to ≃110 K *Phys. Rev. Lett.* 61: 750

Pham A Q, Maignan A, Hervieu M, Michel C, Provost J and Raveau B (1992) Synthesis and characterization of $Bi_2Sr_2CaCu_2O_8$ without excess oxygen *Physica C* 191: 77

Poddar A, Mandal P, Das A N, Ghosh B and Choudhury P (1991) Effect of carrier concentration on the normal transport properties and the superconducting transition temperature in the $Tl_2Ba_2Ca_{1-x}Y_xCu_2O_{8+y}$ system *Phys. Rev B* 44: 2757

Politis C and Luo H L (1988) Superconductivity in Tl-Ca-Ba-Cu-O compounds *Mod. Phys. Lett. B* 2: 793

Presland M R and Tallon J L (1991) *Superconductivity at 105 K in* $Tl_{0.5}Pb_{0.5}CaSr_{2-x}La_xCu_2O_7$ *Physica C* 177: 1

Presland M R, Tallon J L, Flower N E, Buckley R G, Mawdsley A, Staines M P and Fee M G (1993) Flux pinning and critical currents in superconducting thallium cuprates *Cryogenics* 33: 502

Ren Z F and Wang J H (1993) *Enhanced formation of* 1223 phase by partial replacement of Bi for Tl in in-situ synthesized silver-sheathed superconducting $Tl_{1-x}Bi_xSr_{2-y}Ca_2Cu_3O_{9-\delta}$ tape*Physica C* 216: 199

Renk K F (1994) *Thallium-Based High-Temperature Superconductors* ed A M Hermann and J V Yakhmi (New York: Dekker), p 477

Reymbaut A, Bergeron S, Garioud R, Thénault M, Charlebois M, Sémon P and Tremblay A M S (2019) Pseudogap, van Hove singularity, maximum in entropy, and specific heat for hole-doped Mott insulator *Phys. Rev. Research* 1: 023015

Rossel C, Willemin M, Hofer J, Keller H, Ren Z F and Wang J H (1997) Pairing symmetry in single-layer tetragonal $Tl_2Ba_2CuO_{6+\delta}$ from in-plane torque anisotropy *Physica C* 282-287: 136

Routbort J L, Miller D J, Zamirowski E J and Gorretta K C (1993) Plasticity of $TlBa_2Ca_2Cu_3O_x$ *Supercond. Sci. Technol.* 6: 337

Ruckenstein E and Wu N L (1994) *Thallium-Based HighTemperature Superconductors* ed A M Hermann and J V Yakhmi (New York: Dekker), p 119

Shipra R, Idrobo J C, Sefat A S (2015) Structural and superconducting features of Tl-1223 prepared at ambient pressure *Supercond. Sci. Technol.* 28: 115006

Sheng Z Z and Hermann A M (1988a) Superconductivity in the rare-earth-free Tl–Ba–Cu–O system above liquid nitrogen temperature *Nature* 332: 55

Sheng Z Z and Hermann A M (1988b) Bulk superconductivity at 120 K in Tl–Ca/Ba–Cu–O system *Nature* 332: 138

Shimakawa Y, Kubo Y, Manako T, Igarashi H, Izumi F and Asano H (1990) Neutron-diffraction study of $Tl_2Ba_2CuO_{6+\delta}$ with various T_c's from 0 to 73 K *Phys. Rev. B* 42: 10165

Shimakawa Y (1993) Chemical and structural study of tetragonal and orthorhombic $Tl_2Ba_2CuO_6$ *Physica C* 204: 247

Siegal M P, Venturini E L, Morosin B and Aselage T L (1997) Synthesis and properties of Tl-Ba-Ca-O superconductors *J. Mater. Res.* 12: 2825

Siri S (1996) *High Temperature Superconductivity, Models and Measurements* ed M Acquarone

Sinclair D C, Aranda M A G, Attfield P and Rodríguez-Carvajal J (1994) Cation distribution and composition of the Tl-2223 superconductor from combined powder neutron and resonant X-ray diffraction *Physica C* 225 307

Sleight A W (1988) Chemistry of high-temperature superconductors *Science* 242: 249

Ström C, Eriksson S G, Johansson L G, Simon A, Mattausch H J and Kremer R K (1994) The effect of thallium and oxygen stoichiometry on structure and T_c in Tl-2201 and Tl-2212 *J. Sol. State Chem.* 109: 321

Subramanian M A, Torardi C C, Gopalakrishnan J, Gai P L, Calabrese J C, Askew T R, Flippen R B and Subramanian M A, Kwei G H, Parise J B, Goldstone J A and Von Dreele R B (1990) *Physica C* 166: 19

Subramanian M A (1990) Structure property relationship in the single layer superconductor $TlSr_{2-x}R_xCaCu_2O_7$ (R= La, Pr or Nd) *Mater. Res. Bull.* 25: 899

Subramanian M A, Calabrese J C, Torardi C C, Gopalakrishnan J, Askew T R, Flippen R B, Morrissey K J, Chowdhary U and Sleight A W (1988) Crystal structure of the high-temperature superconductor $Tl_2Ba_2CaCu_2O_8$ *Nature* 332: 420

Thomas G A, Orensein J, Rapkine D H, Capizzi M, Millis A J, Bhatt R N, Schneemeyer L F and Waszczak J (1988) $Ba_2YCu_3O_{7-\delta}$: Electrodynamics of crystals with high reflectivity *Phys. Rev. Lett.* 61: 1313

Tinkham M 1970 *Far-Infrared Properties of Solid* ed S S Mitra and S Nudelman (New York: Plenum) p 223

Tönies S, Weber H W, Gritzner G, Heiml O and Eder M H (2003) Tl-1223 thick films - a competitor for Y-123 coated conductors? *IEEE Trans. Applied. Supercond.* 13: 2618

Torardi C C, Subramanian M A, Calabrese J C, Gopalakrishnan J, McCarron E M, Morrissey K J, Askew T R, Flippen R B, Chowdhry U and Sleight A W (1988) Structures of the superconducting oxides $Tl_2Ba_2CuO_6$ and $Bi_2Sr_2CuO_6$ *Phys. Rev. B* 38: 225

Torardi C C, Subramanian M A, Calabrese J C, Gopalakrishnan J, Morrissey K J, Askew T R, Flippen R B, Chowdhry U and Sleight A W (1988) Crystal structure of $Tl_2Ba_2Ca_2Cu_3O_{10}$, a 125 K superconductor *Science* 240: 631

Tsuei C C, Kirtley J R, Rupp M, Sun J Z, Gupta A, Ketchen M B, Wang C A, Ren Z F, Wang J H and Bushan M (1996) Pairing Symmetry in Single-Layer Tetragonal $Tl_2Ba_2CuO_{6+\delta}$ Superconductors *Science* 271: 326

Tsuei C C, Kirtley J R, Ren Z F, Wang J H, Raffy H and Li Z Z (1997) Pure $d_{x^2-y^2}$ order-parameter symmetry in the tetragonal superconductor $Tl_2Ba_2CuO_{6+\delta}$ *Nature* 387: 481

Webb B C, Sievers A J and Mihalisin T, (1986) Observation of an Energy- and Temperature-Dependent Carrier Mass for Mixed-Valence $CePd_3$ *Phys. Rev. Lett.* 57: 1951

Wu J Z, Yan S L, Xie Y Y (1999) Cation exchange: A scheme for synthesis of mercury-based high-temperature superconducting epitaxial thin films *Appl. Phys. Lett.* 74: 1469

Zetterer T, Franz M, Schutzmann J, Ose W, Otto H H and Renk K F (1990a) Strengths of infrared active phonons of a superconducting and a normal conducting $(Tl,Pb)Sr_2CaCu_2O_7$ ceramic *Solid State Commun.* 75: 325

Zetterer T, Franz M, Schutzmann J, Ose W, Otto H H and Renk K F (1990b) Anomalous behavior of phonons in superconducting $Tl_2Ba_2Ca_2Cu_3O_{10}$ detected by far-infrared spectroscopy *Phys. Rev. B* 41: 9499

C4

HgBCCO

Judy Z. Wu

C4.1 Introduction to Hg-Based High-Temperature Superconductors

Hg-based high-temperature superconductors (HgBCCO) encompass a series of materials described by $HgBa_2Ca_{n-1}Cu_nO_{2n+2+\delta}$, where n = 1, 2, 3,..., and many of the chemical elements in the formula could be partially cross-substituted with others. The first report of superconductivity was in $HgBa_2CuO_{4+\delta}$ (Hg-1201) with a superconducting transition temperature (T_c) of 94 K [1]. Subsequent reports on $HgBa_2CaCu_2O_{6+\delta}$ (Hg-1212) [2] and $HgBa_2Ca_2Cu_3O_{8+\delta}$ (Hg-1223) [3] revealed higher T_c values of 125 K and 135 K, respectively. These discoveries placed the HgBCCO at a unique position among the high-temperature superconductors (HTS), since most members of the HgBCCO family have T_c's above 100 K. In particular, the T_c of 138 K discovered in $(Hg_{0.8},Tl_{0.2})$-1223 remains as the highest value so far achieved in superconductors without applying pressure [4], making HgBCCO one of the most interesting systems in the investigation of the fundamental physics underlying the unresolved mechanism of high-temperature superconductivity. For practical applications, a higher T_c implies higher device operating temperatures and lower cost. A particular niche for HgBCCO-related applications is at temperatures exceeding 77 K as illustrated in various prototype devices including superconducting quantum interference devices, photodetectors, microwave passive devices, wires and tapes, fault current limiters, etc.

During the past two decades or so, extensive researches have been carried out in the synthesis, characterization, and application of HgBCCO. Several review articles and book chapters are available to cover different aspects of the research on HgBCCO [5–12]. A few recent topical reviews are particularly worth mentioning. *Mikhailov* reviewed several technologically important HTS materials including $YBa_2Cu_3O_7$ (YBCO), $Bi_2Sr_2CaCu_2O_8$ (Bi-2212), $Bi_2Sr_2Ca_2Cu_3O_{10}$ (Bi-2223), $Tl_2Ba_2Ca_2Cu_3O_{10}$ (Tl-2223), and Hg-1223 regarding their processing phase diagrams, crystal structures, and synthesis techniques, especially for wires/tapes and bulks [8]. The aspects of thermodynamics, crystal structure, and superconducting properties under ambient and high pressure of HgBCCO superconductors are reviewed by *Antipov, Abakumov*, and *Putilin* for exploration of designing new layered materials [6]. This chapter intends to update its predecessor written by *Schwartz and Sastry* in the previous version of the *Handbook of Superconducting Materials* [7], emphasizing material synthesis, physical properties, and applications. This chapter is organized in the following way: after this short introduction in Section C4.1, Section C4.2 will discuss the crystalline structures of HgBCCO. Section C4.3 covers the progress made in synthesis of HgBCCO bulks, films, and conductors, which will be followed with a review of the physical properties of HgBCCO in Section C4.4. Section C4.5 highlights efforts in development of the HgBCCO electric and electronic devices. Section C4.6 includes some concluding remarks on the remaining challenges and future research directions.

C4.2 Crystal Structure of HgBCCO

As noted, HgBCCO can be described by the general chemical formula of $HgBa_2Ca_{n-1}Cu_nO_{2n+2+\delta}$, where n = 1, 2, 3,... While the HgBCCO with lower n numbers of n = 1 (Hg-1201), n = 2 (Hg-1212), and n = 3 (Hg-1223) are the most studied members partly due to the historical reasons of being discovered earlier, the search for newer HgBCCO members has been productive with the highest n = 16 demonstrated stable in experiment [6, 13, 14]. Like other HTS, HgBCCO have layered structures as illustrated in Figure C4.1. An HgBCCO unit cell consists of two basic inter-grown blocks. One is the perovskite block or superconducting block, which contains the Ca-CuO_2 layers (note there is no Ca layer in the n = 1 case as shown in Figure C4.1[left]). The other is the rock-salt block, or charge reservoir block, consisting of three alternating layers of BaO-HgO_δ-BaO. The unit cells of HgBCCO have a tetragonal structure with $a = b$ = 3.85-3.87 Å while the c-axis lattice constant follows the equation: $c = 3.2(n-1)+9.5$ Å. The

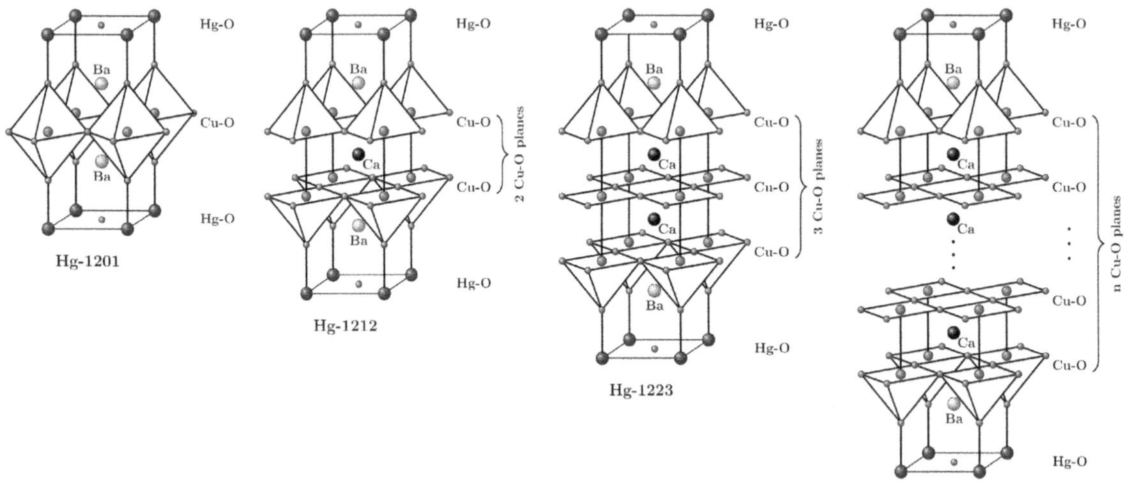

FIGURE C4.1 Crystal structure of $HgBa_2CuO_{4+\delta}$, $HgBa_2CaCu_2O_{6+\delta}$, $HgBa_2Ca_2Cu_3O_{8+\delta}$, and $HgBa_2Ca_{n-1}Cu_nO_{2n+2+\delta}$ at an arbitrary *n* number of CuO_2 layers.

first term is for the superconducting block and the second is for the charge reservoir. This means that the thickness of the superconducting block increases linearly with *n* as depicted schematically in Figures C4.1(a)–(d). The constant thickness ~9.5 Å of the charge reservoir block measures the distance between the superconducting blocks along the *c*-axis and determines the anisotropy of the HgBCCO between parallel (*ab*-plane) and perpendicular (*c*-axis) directions to the layers. Specifically, the value of 9.5 Å is between that for YBCO (8.5 Å) and the double-layered BiSrCaCuO and TlBaCaCuO (11.5 Å). This means the anisotropy of the HgBCCO is between that of YBCO (least anisotropic) and the double-layered BiSrCaCuO and TlBaCaCuO (most anisotropic). The differences in anisotropy of these HTS systems correspond well to the different electrical transport properties relevant to both DC and RF applications as we shall discuss in Section C4.4.

C4.3 Synthesis of HgBCCO

C4.3.1 HgBCCO Bulks

Synthesis of HgBCCO superconductors is certainly more challenging than for most other HTS [7, 8]. The high volatility of mercury necessitates heating the samples in sealed reaction. The resulting lack of independent control of mercury and oxygen partial pressures during the reaction makes it difficult to synthesize phase-pure samples of the individual superconducting phases. The synthesis is also sensitive to trace quantities of moisture and carbon dioxide and demands freshly prepared oxide precursors plus various precautions to prevent pre-exposure of the precursors to air and moisture. Despite these difficulties, much progress has been made in developing various synthesis processes and high-quality bulks and films have been achieved.

Typically, HgBCCO synthesis involves two steps: preparation of Hg-free or Hg-containing precursors in bulk or film form, followed by *ex situ* mercuration in sealed quartz ampules with Hg and O vapors provided from a source pellet of mixed Hg-Ba-Ca-Cu-O powders at temperatures in the range of 750°C-950°C [6]. The sealed quartz ampoules allow over-pressure of the Hg and O vapors to typically 5-10 atmospheric pressures, which is important to stabilize the HgBCCO phases. To avoid formation of intermediate phases, the precursors are often prepared in glove boxes in Ar or other dry gases to minimize their exposure to air and moisture. Since torch sealing of the quartz ampoules become increasingly difficult for large ampoules, the dimension of the HgBCCO samples remains limited to a couple of centimeters [15].

While earlier research on HgBCCO primarily concerned HgBCCO with *n* up to 3, fabrication of $HgBa_2Ca_{n-1}Cu_nO_{2n+2+\delta}$ with $n > 3$ has been a more recent focus. One of the motivations is to explore higher T_c superconductors since T_c increases monotonically with n when $n \leq 3$ [5, 6, 10]. However, lower T_c was observed in Hg-1234 ($n = 4$), which was attributed to modulated structures [16]. In fact, the existence of both modulated and unmodulated Hg-1234 structures was confirmed in these samples, and the structural modulation was found to significantly suppress the T_c. For the Hg-1234 sample without the structural modulation, T_c was found to be close to the highest T_c observed in the Hg-1223 phase [16]. A more general study on $HgBa_2Ca_{n-1}Cu_nO_{2n+2+\delta}$ with $n > 4$ was carried out by Iyo *et al* using high-pressure synthesis [13, 14]. Phases from *n* up to 16 were recognized in the XRD patterns and the *c*-axis lattice constant changes by about 3.17 angstrom with each additional *n*, suggesting that the crystal structure changes by adding a Ca-CuO_2 layer. A large and sharp superconducting transition shown in the susceptibility vs. temperature curve at 105-108 K was observed in the sample of mixed phase of multilayered $HgBa_2Ca_{n-1}Cu_nO_{2n+2+\delta}$ with n up to 16.

While most reported HgBCCO bulks are polycrystalline, single crystals of (Hg, Re)$Ba_2Ca_{n-1}Cu_nO_{2n+2+\delta}$ (n = 2, 3, 4) have been obtained with size up to 1 × 1.1 × 0.1 mm^3 using a flux method in quartz ampoules from flux compositions with excess Ba and Cu [17]. The obtained single crystals were nearly optimally doped as illustrated in the $T_{c,onset}$ of 123 K, 131 K, and 125 K for crystals with n = 2, 3, 4, respectively, although inhomogeneous distribution of rhenium remains an issue. In addition, *c*-axis aligned (Hg-Re)-1223 bulks were obtained with J_c up to 2.4 kA/cm^2 at 77 K using a magnetic field in a slip-casting process [18].

C4.3.2 Chemical Doping of HgBCCO

Chemical doping in HgBCCO has been investigated for improvement of material properties as well as for ease of sample synthesis [7]. A list of dopants has been applied to HgBCCO including rhenium, Re, [19–24], thallium, Tl, [4, 25–27], lead, Pb, [28, 29], bismuth, Bi, [30], and alkali (Li and Na) [31–33]. Adachi *et al* compared Hg-1223 samples with 20% of Hg replaced with other elements including V, Cr, Mn, Mo, Ag, In, Sn, W, Re, Pb, Hf, and Ta [34]. A conventional quartz-tube encapsulation method was employed at ambient pressure and formation of the Hg-1223 phase with most of the dopants mentioned above, except In and W, was observed. On samples with V, Cr, Mn, Mo, and Re dopants, appreciable shortening in the *c*-axis lattice constant was reported. This suggests incorporation of the dopants into the Hg-1223 lattice which is further supported by an enhancement in J_c of the samples with Re and Mo. To understand the role of the doped Re in the (Hg_{1-x}, Re_x)-1223, Re L-III edge X-ray absorption spectroscopy was used in order to determine its valence and the local oxygen coordination in polycrystalline samples prepared with three different oxygen contents [35]. The results indicated that the oxygen local order around Re atoms in the ($Hg_{0.82}$,$Re_{0.18}$)-1223 samples can be described as a distorted ReO_6 octahedron with two different Re-O bound lengths. Synchrotron anomalous X-ray scattering on ($Hg_{0.8}$,$Re_{0.2}$)-1223 suggested that Re distribution on the Hg-O plane does not produce an expected supercell [36, 37]. Even for a high-quality sample (high T_c and single phase), two different superconducting phases were identified and the absence of *2a* × *2b* × *1c* super cell was used to justify the scenario where charge distribution inhomogeneity is present in the outer CuO_2 layers. Y-doping was found to facilitate formation of the Hg-1223 phase in $(Hg_{0.82}Re_{0.18})Ba_2Ca_{1-x}Y_xCu_2O_{6+d}$ with different Y content (0.05 < x < 0. 55) despite T_c decreasing with increasing Y content. A phenomenological model of charge-transfer was proposed to explain the observation, and it was argued that Y doping causes charge carrier overdoping on the inner CuO_2 layer and thus reduced T_c values.

C4.3.3 HgBCCO Thin and Thick Films

A two-step *ex situ* process, similar to that mentioned above, for synthesis of HgBCCO bulks has been applied for fabrication of HgBCCO films. Many groups have reported success in fabrication of *c*-axis–oriented Hg-1212 films [23, 29, 38–43] and Hg-1223 [26, 44, 41, 24, 45]. The deposition of precursor films of nominal composition, with or without Hg, has been explored using physical vapor deposition, chemical vapor deposition, and various non-vacuum-based processes [7, 10]. In most cases, the precursor films are amorphous. Crystalline HgBCCO films are obtained through post-annealing in over-pressured vapors of Hg and oxygen at high temperatures in the range of 600°C–900°C.

The method for deposition of the precursor films is not particularly critical as long as it provides a uniform cation composition. A stringent requirement is, however, that the precursor films should not be exposed to air so that the detrimental effect of moisture and carbon dioxide in air can be minimized. In fact, the sensitivity of the precursors to air has been a more serious problem in fabrication of HgBCCO films than in the case of bulks, resulting in poor sample quality and reproducibility. The problem stems from the much larger surface-to-volume ratio for films. The simple metal oxides in the precursor films, particularly CaO and BaO are chemically reactive even at room temperature and can easily transform to $Ca(OH)_2$ or $CaCO_3$ through reactions with H_2O and CO_2 in air. This can result in a signification proportion of non-superconducting impurity phases in the HgBCCO film since the resultant $Ca(OH)_2$, $Ba(OH)_2$, $CaCO_3$, and $BaCO_3$ are unlikely to decompose at the annealing temperature, typically in the neighborhood of ~800°C.

To eliminate the detrimental effect of air on precursor films, modified precursor films were investigated. Inclusion of Hg into the precursor films using a layer-by-layer mixing of HgO and Ba-Ca-Cu-O, in addition to a thin protective cap layer of either MgO or HgO on the precursor film was found effective in improving film quality in terms of T_c and J_c [40]. For example, the zero-resistance T_c was as high as 124 K and J_c was close to 1 MA/cm^2 at 100 K [46]. Although it was believed that the atomic-scale mixing of HgO and Ba-Ca-Cu-O improves the uniformity of Hg across the film thickness, it was later found that mixing Tl into the Ba-Ca-Cu-O precursor films serves the same purpose and results in even better-quality Hg-1212 films with J_c > 1 MA/cm^2 at 100 K [27]. This suggests that mixing Tl or Hg into the Ba-Ca-Cu-O precursor may chemically stabilize the simple oxides in the precursors from reacting with moisture or carbonates in air.

One of the applications for thin films is passive microwave devices such as resonators and bandpass filters [47]. For microwave applications or high-frequency RF devices in general, substrates with low tangent loss are required. This motivated research on HgBCCO film fabrication on low-loss substrates including $LaAlO_3$, MgO, and sapphire [5, 48]. $LaAlO_3$ seems the most compatible substrate for HgBCCO thin films due to its stability in the HgBCCO fabrication procedure, especially the Hg-vapor treatment process, with the best DC and RF properties demonstrated in Hg-1212 thin film devices as detailed in Sections C4.4 and C4.5 [49, 10, 11].

Thick HgBCCO films are promising for electrical applications and have prompted considerable efforts in synthesis of HgBCCO films of several micrometers in thickness. The effects of Pb and Re doping on microstructure, irreversibility field, and electronic anisotropy of Hg-1223 thick films were investigated on Ag substrates using a simple dip-coating method [50]. Both dopants were found to distribute homogeneously in the Hg-1223 grains and play an important role in promoting grain growth while Pb-doped films have larger colony size compared with Re-doped films. While both have T_c up to 133 K, the irreversible field H_{irr} of (Hg, Re)-1223 is significantly higher than that of (Hg,Pb)-1223 at temperatures below 100 K. The authors argued that Re doping significantly reduced the anisotropy, which enhances flux pinning and consequently improves J_c. A multistep electrolytic process was applied for synthesis of Hg-1223 thick films with T_c ~121.5 K and $J_c = 4.3 \times 10^4$ A cm^{-2} at 77 K [51, 52]. Using a spray process, *c*-axis-orientated $(Hg_{0.8},Re_{0.2})$-1223 films of up to 50 μm in thickness were obtained on MgO (100) substrates [53, 54]. A 1.5-μm-thick interfacial layer between film and substrate was found beneficial to prevent excessive diffusion and dissolution of the Mg^{2+} ions from the substrate and counter-diffusion of Ba, Ca, and Cu ions from the films. In a related work, the effect of thickness on the grain alignment and J_c of $(Hg_{0.8},Re_{0.2})$-1223 films on MgO was studied as the film thickness was varied in the range of 1-80 μm [55]. The best T_c up to 129 K was found for a ~50-μm-thick sample with a J_c ~4.82×10^5 A/cm^2 at 4.5 K.

A cation exchange process was developed to achieve high-quality epitaxial HgBCCO films [10, 49] by using a TlBCCO epitaxial film as the precursor (film or bulk) in a two-step process. Epitaxial HgBCCO films can be obtained through diffusion of Tl cations out of and Hg cations into the (Hg,Tl) BCCO lattice. Compared with HgBCCO, TlBCCO is less volatile, insensitive to air, and easy to be synthesized into epitaxial films on many single-crystal substrates [56]. Epitaxial Hg-1212 (T_c up to 124 K, J_c ~ 2 MA/cm^2 at 100 K) and Hg-1223 (T_c up to 132 K, J_c ~ 1.5 MA/cm^2 at 100 K) films have been obtained using a cation exchange process with high reproducibility [10, 11, 49].

TlBCCO has two series: one has a single Tl-O plane [$TlBa_2Ca_{n-1}Cu_nO_{2(n+1)+1}$, $n = 1,2,3\ldots$] and the other, double Tl-O planes [$Tl_2Ba_2Ca_{n-1}Cu_nO_{2(n+2)}$, $n = 1,2,3\ldots$] in a formula-unit cell. The members of the former have nearly the same structures as their HgBCCO counterparts and the HgBCCO can be obtained by replacing the Tl-cations on Tl-O plane with Hg-cations with little change in lattice structures. The double-layer $Tl_2Ba_2Ca_{n-1}Cu_nO_{2(n+2)}$ may also be employed as the precursor matrices for the HgBCCO by collapsing the two Tl-O planes to a single Hg-O plane. Although it is possible that the two Tl-O planes may be transferred to two Hg-O planes to form, presumably, $Hg_2Ba_2Ca_{n-1}Cu_nO_{2(n+2)+1}$, such a system has not been observed experimentally. The cation exchange process involves diffusion of both cations (Tl^{+3} and Hg^{+2}) and anions (O^{-2}) and has been experimentally tested in thin films of thickness less than 0.5 μm and thicker films of thickness up to 3.0 μm. Epitaxial Tl-2212 and Tl-1212 precursor films were employed as the precursor for Hg-1212 films [57–59] and Tl-2223, for Hg-1223 films [60]. It has been found that the Hg cations in thin films first channel through growth defects across the film thickness and then diffuse into grains along the *ab* planes, while the Tl cations take the opposite path to escape from the lattice. For the conversion from Tl-2212 to Hg-1212, an additional step for Tl-2212 to collapse structurally into Tl-1212 was observed before Tl-Hg cation exchange occurs. In thick films, two different diffusion rates were observed occurring at different depths from the film surface. The faster one of 0.53 μm/h occurred at the top 0.4-0.6 μm thick layer was attributed to the *ab*-plane cation diffusion, and the slower one of ~0.09 μm/h that dominates in the bottom layer, to the *c*-axis diffusion. Oxygen-overdoped Hg-1212 films can be obtained via fluorine-assisted growth with much improved J_c and H_{irr} [58]. In addition to the large processing window as demonstrated in experiment for the cation exchange to be a diffusion process [57–59, 61], the cation exchange has been confirmed to be reversible within the "1212" system (Tl-1212 and Hg-1212) and between Hg-1212 and Tl-2212 [62, 63].

C4.4 Physical Properties of HgBCCO

C4.4.1 Ambient T_c

Figure C4.2 shows a curve of T_c vs. n for HgBCCO with n up to 16. Under ambient pressure, the highest zero-resistance T_c of 135 K was observed in Hg-1223 and 138 K, in $(Hg_{0.8},Tl_{0.2})$-1223. The cupola shape of the T_c vs. n curve is approximately symmetric at low n values below $n = 6$. The value of T_c is almost

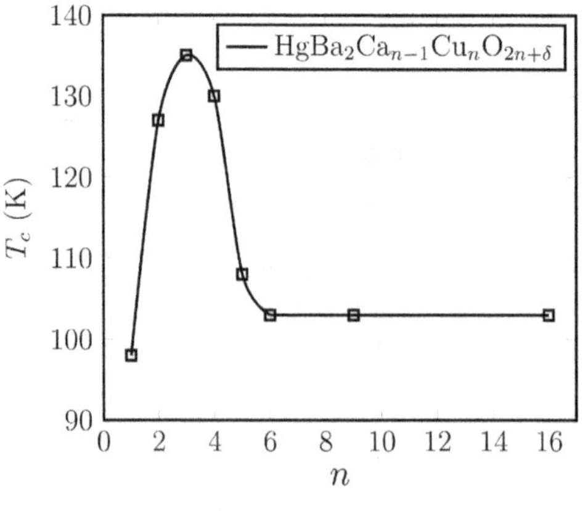

FIGURE C4.2 T_c versus n curve for $HgBa_2Ca_{n-1}Cu_nO_{2n+2+\delta}$ with n ranging from 1 to 16.

constant at ~108 K as n is varied from 6 to 16 (Antipov et al., 2002). Consider the T_c vs. hole concentration curves for HTS materials have a similar cupola shape, an immediate interpretation for the n dependence of the T_c is that the hole concentration varies with n [64]. This argument, however, seems contrary to the experimental observation that T_c values are not improved as the hole doping is systematically varied from the under-doped, to optimally doped, and to over-doped regimes in Hg-1234 and Hg-1245 [65]. Another explanation regards subtle structural changes as the Cu-O apical distance decreases with increasing n, resulting in position shift of the Cu atoms with respect to the charge reservoir block and hence deformation of the CuO_2 planes [6]. This explanation works well in explaining the monotonic increase of T_c as n is increased up to 3. However, how exactly this deformation occurs differently when n is above 3, causing the T_c vs. n curve to deviate from the increasing trend requires further investigation. Some recent studies revealed that the existence of intrinsic inhomogeneities inherent to doping, such as oxygen doping on the Hg-O plane, may affect the superconducting transitions [66, 67].

As the HTS system with the highest T_c, HgBCCO provides an ideal platform for investigation of the pairing mechanism. For example, Cu-NMR (Nuclear Magnetic Resonance) was applied to study a set of five-layered cuprates of $MBa_2Ca_4Cu_5O_y$ (M-1245, M = Hg, Tl, Cu), which have flat CuO_2 planes [68–70]. In Hg-1245 (T_c = 108 K), there are two types of CuO_2 planes in a unit cell; three inner planes (IP's) and two outer planes (OP's). The Cu-NMR study has revealed that the optimally doped OP undergoes a superconducting transition at T_c = 108 K, while the three underdoped IP's experience an antiferromagnetic (AFM) transition below 60 K. The AFM phase was reported as being uniformly coexisting with the superconducting phase in these under-doped planes without any vortex lattice and/or stripe order [69, 71]. Moreover, an AFM metallic phase exists between the AFM insulating phase and the HTS phase for the ideally flat CuO_2 planes in the absence of disorder. The most striking feature of these NMR studies as well as muon-spin rotation [72] studies is that AFM coexists with superconductivity in Hg-1245 below 60 K. In addition, a hysteretic specific heat jump at 41 K was reported as a possible first-order phase transition due to multicomponent superconducting order parameter associated with the five CuO_2 planes in Hg-1245 [73].

Charge carrier concentration relates to T_c directly in the BCS theory (Bardeen–Cooper–Schrieffer), and investigations of the role of hole concentration (n_{hole}) in HgBCCO were reported by quite a few groups. In a comparative study of reversible magnetization as a function of temperature and magnetic field on three-layer cuprate HTS systems with comparable T_c values exceeding 120 K, the effective carrier mass for Hg-1223 was reported to be about three times larger than that for Bi-2223 due to the higher carrier density in Hg-1223 [74]. In a study of thermoelectric power and DC magnetization on $(Hg_{1-x}Re_x)$-1201 (0< x <0.15) samples, the influence of oxygen and Re content on T_c and n_{hole} was extracted [75]. Interestingly, a different T_c *vs* n_{hole} trend was observed from the previously reported common parabola followed by many other HTS materials. The pseudogap phase above T_c of HTS materials presents different energy scales. Gallais *et al* reported a doping-dependent electronic Raman scattering (ERS) study of the dynamics of the antinodal and nodal quasiparticles in Hg-1201 single crystals. When probing the nodal quasiparticles, the energy scale of the pseudogap and that of the superconducting gap was identified [76]. In a related study, Passos *et al,* measured the electrical resistivity, ρ, of Re-doped Hg-1223 samples to extract the values of the pseudogap temperature, the layer coupling temperature between the superconducting layers, the fluctuation temperature, and the critical temperature as a function of the doping level. Based on the experimental results, the authors have derived a phase diagram for Re-doped Hg-1223 [77]. Another study of the pseudogap phase focused on the hole-doping effect in single phase Hg-1223 [78]. The characteristic temperatures describing the role of the pseudogap phenomenon were found to change linearly with n_{hole}. A more recent study of the pseudogap using inelastic neutron diffraction on Hg-1201 revealed a fundamental collective magnetic mode associated with the unusual order [79–81].

C4.4.2 Effect of Pressure on T_c

The effect of pressure on the T_c of HTS including HgBCCO has been studied intensively and a comparison of the T_c vs. pressure curves for four HgBCCO members with n = 1, 2, 3, and 4 is summarized in Figure C4.3. A common monotonic increase of T_c with increasing pressure was reported in the low-pressure range [6]. For Hg-1223, an onset T_c above 160 K was observed at hydrostatic pressures of 25-30 GPa [82, 83]. It should be noted that the mechanism of the pressure effect remains an active research topic [84]. A study

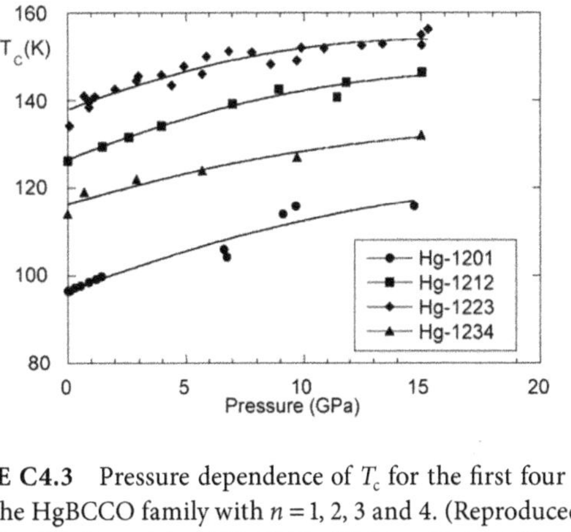

FIGURE C4.3 Pressure dependence of T_c for the first four members of the HgBCCO family with n = 1, 2, 3 and 4. (Reproduced with permission [6]. Copyright 2002, IOP Publishing, United Kingdom.)

of the electronic structure and the hole concentration in the CuO_2 planes of $HgBa_2Ca_{n-1}Cu_nO_{2n+2+\delta}$, n = 1, 2, 3, and 4 under hydrostatic pressures up to 15 GPa suggests possible pressure-induced hole doping and an increasing electronic density of states at the Fermi level [85]. A study of the resistive transition T_c of Hg-1223 at pressures up to 7.8 GPa revealed an inverted dome-shaped pressure dependence T_c, which was attributed to inhomogeneous charge distributions in the inner and outer CuO_2 layer(s) [86]. With improved sample quality, further increased zero-resistance T_c up to 153 K at 15 GPa was recently observed in Hg-1223 [87]. Another study by Monteverde *et al* has revealed that T_c is affected by two main pressure-dependent parameters: the doping level of the CuO_2 planes and structural factor associated with the reduction of the *c*- or *a*-lattice constants [88, 89]. They have measured the pressure sensitivity of T_c in fluorinated Hg-1223 samples with different F contents under applied pressures up to 30 GPa. Fluorine incorporation was found to yield an enhancement of T_c, and the enhancement depends on the F-doping, which may be attributed to a compression of the *a*-axis. A T_c value up to 166 K was obtained at the pressure of 23 GPa in the optimally fluorinated Hg-1223.

C4.4.3 Critical Current Density, J_c

J_c characterization is typically carried out using either electrical transport or a magnetic induction method primarily, in the case of HgBCCO films, at dc or low frequency. The best J_c values have been reported on thin films of thickness in the range of 200-300 nm. Figure C4.4 shows a comparison of the self-field J_c *versus* T curves for epitaxial Hg-1212 and Hg-1223 films on $LaAlO_3$ substrates fabricated using the cation-exchange process [10]. A few other HTS films including YBCO, Tl-2212, Tl-1212, and Tl-1223 on the same substrate are also included in the same figure for comparison. Below 77 K, the YBCO film exhibits the highest J_c while above 77 K, Hg-1212 and Hg-1223 films carry substantially higher J_c values. On epitaxial Hg-1212 thin films, magnetization J_c values up to 4.5 MA/cm^2 and 2 MA/cm^2, respectively, are obtained at 77 K and 100 K [10, 11, 49]. At 110 K, a J_c value close to 1 MA/cm^2 has been observed, which is remarkable and promising for applications above liquid nitrogen temperature. The slightly lower J_c's of the Hg-1223 films suggest that further optimization of the film quality is possible.

The temperature dependence of J_c correlates with the anisotropy of the superconductor since a stronger magnetic vortex pinning is anticipated in less anisotropic superconductors. This trend can be clearly seen in Figure C4.5 for a comparison of normalized J_c as function of reduced temperature for YBCO (less anisotropic), Hg-1212 (medium), and Tl-2212 (more anisotropic). Also included in Figure C4.5 are the microwave critical current density J_{IP3} derived from the third-order intercept for these three types of HTS thin films, set up as band-pass filters at intermodulations between microwave sources at 10.880 GHz and 10.875 GHz. Interestingly, the normalized J_{IP3} and J_c for each of the YBCO, Hg-1212, and Tl-2212 samples follow a similar temperature dependence near T_c [90–93]. This observation illustrates that the magnetic vortex de-pinning in HTS materials is the primary mechanism limiting the J_c at low frequencies and responsible for the microwave nonlinearity. In addition, a comparative study of the J_{IP3}/J_{IP3}(77 K) with J_c/J_c(77 K) in the isomorphic pair of Tl-1212 and Hg-1212 films showed the same behavior when plotted against the reduced temperature ($t = T/T_c$), suggesting the intrinsic pinning strength plays the critical role in determining the dc and microwave power-handling

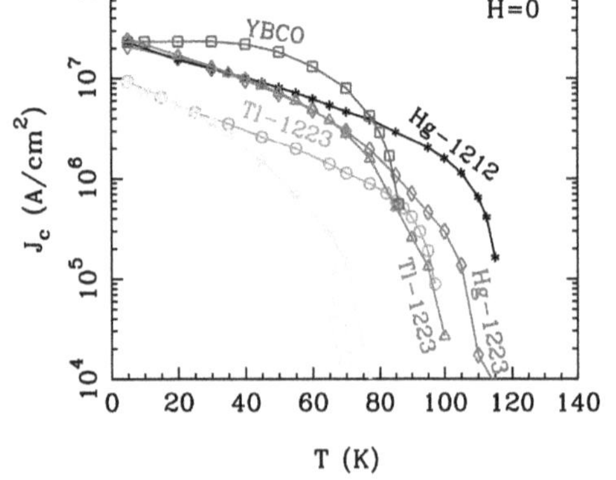

FIGURE C4.4 Comparison of zero-field J_c in thin films of several high-temperature superconductors including YBCO (T_c~90K), Tl-1212 (T_c~87K), Tl-2212 (T_c~105K), Tl-1223 (T_c~105K), Hg-1212 (T_c~125K), and Hg-1223 (T_c~130K). (Reproduced with permission [10]. Copyright 2005, Springer.)

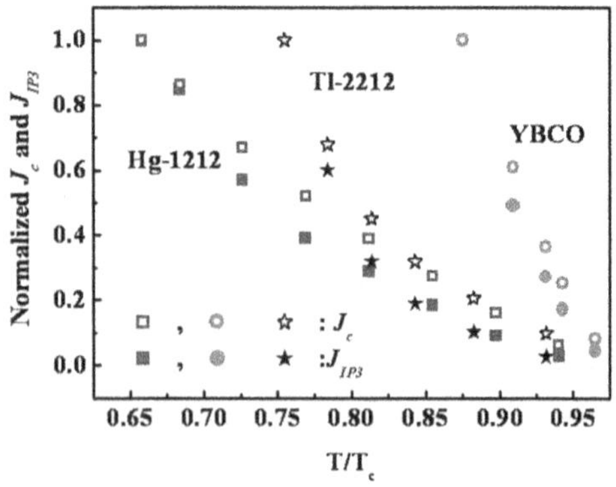

FIGURE C4.5 Normalized J_{IP3}/J_{IP3}(77 K) with J_c/J_c(77 K) *versus* reduced temperature T/T_c for Hg-1212, Tl-2212, and YBCO thin-film microwave band-pass filters. (Reproduced with permission [92]. Copyright 2008, IOP Publishing.)

capability of HTS [91]. This argument may not apply to the cases when artificial pinning centers (APCs) are added to HTS if additional losses are introduced by the non-superconducting APCs [92].

C4.4.4 Irreversible Field H_{irr} and Magnetic Pinning

The irreversible field H_{irr} defines the upper limit of J_c in the presence of a magnetic field and, naturally, a high H_{irr} is desired for in-field applications. For HTS, H_{irr} is found to associate directly with structural anisotropy defined by the spacing "*d*" between the superconducting perovskite blocks in a unit cell. The HgBCCO have a moderately high H_{irr}, somewhere between that of less anisotropic YBCO and more anisotropic Bi-HTS's and double-layer Tl-HTS [94, 95]. On the reduced temperature ($t = T/T_c$) scale, H_{irr} follows a simple power law: $H_{irr} \sim (1-T/T_c)^n$, with the exponent "*n*" being proportional to the spacing *d*. Values of 3/2, 5/2, and 11/2, respectively, were reported for YBCO, HgBCCO, and BiSCCO or double-layered TlBCCO. The $H_{irr}(t)$ curves for HTS materials with the same anisotropy coincide well, as illustrated in a study of the isomorphic pair of Tl-1212 and Hg-1212 films despite a ~40 K difference in their T_c values [96, 97]. Interestingly, a similar trend was reported in $(Hg,Re)Ba_2Ca_{n-1}Cu_nO_{2n+2+\delta}$ (n = 2, 3, 4) single crystals with the electromagnetic anisotropy parameter $\gamma = m_c^*/m_{ab}^*$ were estimated to be 500-700 [98]. The *H-T* phase diagram of Hg-1201 was studied in ac and dc susceptibility measurements and a remarkably high vortex mobility in the mixed state, especially vortex solid-liquid state, was observed [99]. Flux creep is argued to be dominant in the low-field, low-temperature region of the mixed state by the motion of individually pinned flux lines [100]. Through analysis of the relaxation behavior in Hg-1201, a nonlinear relationship between the vortex activation energy and J_c has been reported and the data fit well with the collective pinning model. The mixed-state properties of Hg-1201 single crystals were evaluated in the overdoped regime [101]. A pronounced magnetization "fishtail" is observed in these samples, and the features were found to qualitatively agree with predictions of the order-disorder theory of vortex matter. A similar study of the magnetic relaxation in an epitaxial Hg-1223 thin film revealed a large bundle of flux may act like a single vortex while the size of which was temperature independent, which the authors argued to be a different behavior in Hg-1223 from other HTS [102].

Improving pinning by adding APCs in HgBCCO has been explored via either growth or high-energy particle irradiation. Hg-1212 films on 4 degrees miscut $SrTiO_3$ single crystal substrates was reported to have improved J_c and H_{irr} due to growth defects, such as step-edge dislocations, via a step-flow growth mode on the miscut substrates [103]. H_{irr} up to 2.7 T was observed at 77 K, in contrast to 2.1 T on the nonvicinal counterpart. Correlated defects in the form of randomly oriented columnar tracks were introduced via fission of Hg nuclei in Hg-1212 and Hg-1223 films [104, 105]. Track densities up to a "matching field" of 3.4 T were achieved. On Hg-1201 single crystals, irradiation was applied to vary the anisotropy of the sample. It was found that neutron irradiation reduces the anisotropy and considerably affects the irreversible properties, whereas electron irradiation leads only to small effects [101].

C4.5 Applications of HgBCCO

C4.5.1 Electrical Applications

The high T_c values of most HgBCCO family members suggest their potential for high J_c above liquid nitrogen temperature, making them promising candidates for superconducting wire and tape applications at 77 K or higher. This has prompted efforts in coating HgBCCO thin and thick films on metal substrates. Unlike other HTS-coated conductors, the selection of metal substrates for HgBCCO has been restricted to few, such as Ag, Au, and Ni due to amalgamation of most metals with Hg. For example, platinum and palladium were found to react with Hg-based compounds [106]. Silver (Ag) (or gold, Au) absorbs a significant amount of Hg [107, 108], but the Ag/Hg amalgam was shown to decompose at high temperatures to release Hg, and therefore can be used for HgBCCO tapes [109]. The Hg-1201 phase was obtained with Ag as the sheath material [108, 110] using the powder-in-tube (PIT) method that has been successfully applied for fabrication of Bi-HTS wires [111]. A technical problem associated with these Hg-1201 wires is the porosity that results in poor inter-grain connectivity and therefore low J_c. Reducing the processing temperature was attempted to suppress the reaction and improve the inter-grain connectivity. In Pb-doped Hg-1223 on Ag processed at a lower temperature ~780°C, large regions of aligned grains of $(Hg_{1-x}Pb_x)1223$ were observed with T_c ~133 K [112]. Au seems to be an even better choice due to a reduced Au/Hg reaction. Predominantly single-phase Hg-1223 was obtained with zero-resistance T_c ~127 K and J_c ~1.8 kA/cm^2 at 77 K and zero field [107]. Ni also has low solubility in Hg and a minimal mechanical deformation after the HgBCCO film fabrication. Coating of Ag/Cr buffered Ni substrates with $(Hg_{1-x}Re_x)$-223 resulted in good quality films with zero-resistance T_c up to 117 K and transport J_c ~2.5×10^4 A/cm^2 at 77 K and zero field [113].

Most of these HgBCCO tapes are, however, polycrystalline, and the grain boundaries cause considerable reduction of J_c. To eliminate grain boundaries, biaxially textured metal substrates, such as RABiTs (rolling-assisted bi-axially textured substrates) based Ni or Ni-alloys [114] have been explored. On CeO_2/YSZ/CeO_2 buffered RABiTs Ni tapes, high J_c values close to that achieved on epitaxial Hg-1212 films on dielectric substrates have been obtained on these HgBCCO-coated conductors [115]. For example, the J_c for a 600-nm-thick Hg-1212 on CeO_2/YSZ/CeO_2/Ni is up to 0.8 MA/cm^2 at 100 K and 0.24 MA/cm^2 at 110 K. At 77 K, J_c increases to 2.2×10^6 A/cm^2 in zero field, comparable to the best J_c on YBCO-coated conductors. Irreversibility fields of ~2.4 T and ~0.8 T were observed at 77 K and 100 K, respectively, on Hg-1212-coated conductors, suggesting they may be promising for applications above 77 K [10, 116].

Prototype superconducting fault current limiters based on ($Hg_{0.8}$,$Re_{0.2}$)-1223 ceramic were reported [117–119]. A fault current test at 60 Hz confirmed a reduction in the density of current value at peak from 1.55×10^2 to 0.82×10^2 A/cm^2. The observed prospective/limited current ratio of 1.9 was also obtained on a 0.24-cm-thick sample. A recovery test indicated that the polycrystalline sample kept its superconducting properties and had not altered its stoichiometry. After the fault current event is finished, the device immediately recovered its initial performance without any damage.

C4.5.2 Electronic Applications

One of the practical applications of HTS is passive microwave devices, such as micro-strip resonators, band-pass filters, etc., that have been adopted in wireless communications [47]. Considering that the cost in cryogenics becomes a major portion of the system cost, a higher T_c implies a lower overall cost in both capital investment and maintenance. Therefore, HgBCCO films are particularly suitable for such applications. Hg-1212 micro-strip resonators were obtained by fabricating Tl-2212 thin film micro-strip resonators first using standard photolithography, followed with conversion to Hg-1212 devices using the cation-exchange process [120, 121]. This minimizes the exposure of Hg-1212 films to various chemicals and water-based solutions in the photolithography process that may lead to degradation of Hg-1212 films due to presence of Ba- and Cu-based impurity phases [122, 123]. Measurements of power-handling capability in the Hg-1212 micro-strip transmission lines showed that a stable output power up to 19 dBm at 1 GHz can be attained at 110 K [124]. This means Hg-1212 films are very appealing for microwave applications at 77 K and higher temperatures. A comparative study of Hg-1212 two-pole *X*-band filters revealed an insertion loss of about 0.70 dB at 110 K, which was much lower than that of YBCO (2.3 dB) and copper (3.9 dB) counterparts at 77 K [90, 125]. This means the Hg-1212 filters can provide better filtering performance over both YBCO and copper filters at an operating temperature of 33 K higher.

The nonlinear effect in HTS passive microwave devices is known to be a major limitation to power-handling capability. In order to understand the nonlinear effect in HgBCCO, third-order intermodulation was studied in two-pole *X*-band Hg-1212 micro-strip filters. Interestingly, the third-order intercept (IP3) of the Hg-1212 filters is consistently higher than that of the YBCO counterparts in the temperature range of 77-110 K [90]. At 77 K, the IP3 was 58 dBm for Hg-1212, which was higher than that of the YBCO filter by ~1 dBm. The difference between the IP3 values for Hg-1212 and YBCO devices increases monotonically with increasing temperatures. At 85 K, the IP3 value for Hg-1212 was about 54 dBm, which was ~18 dBm higher than that of YBCO. At 110 K, a substantial IP3 of 38 dBm remained in Hg-1212 filter, demonstrating that Hg-1212 could be a promising alternative material for microwave passive device applications at temperatures above 77 K. In addition, Hg-1212 three-pole hairpin filters of 5% 3-dB bandwidth have also shown promising transmission properties and third-order intermodulation (IM3) [126].

Josephson junctions (JJs) are one of the important applications for superconductors as they serve as the building blocks for a large variety of superconductor microelectronics devices ranging from superconducting quantum interference devices (SQUIDs) to magnetic sensors widely used for magnetic resonance imaging to qubits for quantum computing. HgBCCO JJs are particularly interesting because of their high device operating temperature above liquid nitrogen temperature. Due to the incompatibility of microfabrication with the two-step thin film fabrication process employed for HgBCCO, however, the success in HgBCCO JJs has been limited to grain-boundary Hg-1212 JJs fabricated on commercial bi-crystal $SrTiO_3$ (STO) substrates [127–129]. These Hg-1212 JJs behave typically as resistive shunted junctions. The first Hg-1212 JJ and SQUIDs were obtained on 36.8° STO bi-crystal substrates. The I_cR_n products are in the range of 60-120 μV at 77 K, comparable to that of other HTS JJs on bi-crystal substrates. At 107 K, the flux responsivity, dV/dΦ, is ~8μV/Φ_0. However, the flux noise S_Φ ~10^{-6} Φ_0^2/Hz at 77 K, is about an order of magnitude higher than that of their YBCO and Tl-2212 counterparts. Higher I_cR_n products above 200 μV at 77 K were later reported on Hg-1212 JJs grown on 24° STO bi-crystal substrates by Hitachi/ISTEC and Kansas groups, and the highest I_cR_n is around 450-460 μV, which is in fact the highest reported so far on HTS JJs. Although the Hg-1212 SQUIDs on 24° STO bi-crystal substrates show similar flux responsivity at 110K, their flux responsivity at 77 K is much higher than that on the 36.8° STO bi-crystal substrates. Further, [100]-tilt grain-boundary ($Hg_{0.9}Re_{0.1}$)-1212 JJs were reported on STO bi-crystal substrates (tilt angles of 30° and 36.8°) with a thin buffer layer of YBCO [130]. These JJs exhibited resistively shunted-junction-type I-V characteristics with very low excess current in a wide temperature range from 4.2 to 110 K. The I_cR_n product is 0.2-0.4 mV at 77-100K. A further study in the *ab*-plane of the (Hg,Re)-1212 films across the grain boundary, so-called mountain-type and the valley-type junctions, was carried out on these [100]-tilt grain-boundary JJs as a function of the grain-boundary angle [131]. Although both types of junctions showed resistively shunted junction I-V characteristics over a wide temperature range from 4.2 to 110 K, the valley-type junctions were found to out-perform their mountain-type counterparts in terms of low excess current and homogeneous current distribution. Recently, the intrinsic Josephson effect has also been studied with plasmon frequency reported in the THz regime [132, 133].

Superconductors have been employed for photodetection in the wavelength range of infrared to terahertz, primarily in bolometers that operate below but near T_c, with extraordinarily high sensitivity due to the large resistivity change at the superconducting-normal transition upon photon absorption. Using reduced dimensions for the superconductor detection element, such as microbridges and nanowires, single-photon sensitivity has been demonstrated. The high T_c of HgBCCO allows such detectors to operate at considerably higher temperatures than other superconductor photodetectors. In Hg-1212 microbridges (with T_c around 110 K) incorporated into a 0.1-mm-wide signal line of a coplanar strip transmission line, the photoresponse has been observed in a femtosecond, time-resolved, optical pump-probe spectroscopy measurement [134, 135]. At temperatures much below T_c, these authors observed a faster photoresponse of 90-ps-wide pulse that was related to the kinetic-inductive response. At higher temperatures approaching T_c, a slower bolometric photoresponse was detected. In a related study on (Hg,Re)-1212 films (with a minor volume portion of Hg-1223 phase) with T_c ~122 K, an ultrafast photoresponse was confirmed [136, 137], suggesting a promising application of Hg-based HTS films for photodetectors and optical mixers.

C4.6 Conclusion and Future Prospect

After two decades of extensive research, remarkable progress has been made on HgBCCO in terms of material synthesis, physical property characterization, and applications exploration. In the foreseeable future, HgBCCO will continue to be an important system in superconductor research. In particular, the very high T_c and its large enhancement under applied pressure make the HgBCCO unique for investigation of the mechanism of HTS superconductivity. Moreover, HgBCCO is the only superconductor system that has up to 16 experimentally demonstrated members sharing an otherwise identical atomic structure except the number of CuO_2 planes varying from 1, 2,... to 16. With more advanced experimental approaches emerging for probing and controlling the atomic arrangement, the effect of such an arrangement on superconductivity of this fascinating system will surely shed light on the still mysterious mechanism of HTS superconductivity and possibly lead to discovery of new superconductors with even higher T_c. On the other hand, small-scale applications based on HgBCCO bulks and films may become important for niche applications which require high operating temperatures above 77 K. A remaining technical obstacle for large-scale sample fabrication is to obviate the need for high Hg-vapor pressure – an important focus for future innovation.

Acknowledgements

The author acknowledges support by NSF contracts Nos. NSF-DMR-1105986, NSF-DMR-1337737, NSF-DMR-1508494. She also acknowledges technical assistance from Jack Shi.

References

[1] Putilin S N, Antipov E V, Chmaissem O, Marezio M 1993 Superconductivity at 94 K in $HgBa_2CuO_{4+\delta}$ *Nature* **362** 226

[2] Putilin S N, Antipov E V, Marezio M 1993 Superconductivity above 120 K in $HgBa_2CaCu_2O_{6+\delta}$ *Physica C* **212** 266

[3] Schilling A, Cantoni M, Guo JD, Ott HR (1993) Superconductivity above 130 K in the Hb-BaCaCuO system *Nature* 363:56–58

[4] Sun G F, Wong K W, Xu B R, Xin Y, Lu D F 1994 T_c enhancement of $HgBa_2Ca_2Cu_3O_{8+\delta}$ by Tl substitution. *Physics Letters A* **192** 122

[5] Wu J Z, Tidrow S C 1999 Recent progress in high-T_c superconducting heterostructures, in *Thin Films: Heteroepitaxial Systems* (Liu, W. K. and Santos, M. B., eds), p 267 Singapore: World Scientific.

[6] Antipov E V, Abakumov A M, Putilin S N 2002 Chemistry and structure of Hg-based superconducting Cu mixed oxides *Supercond. Sci. Technol.* **15** R31

[7] Schwartz J, Sastry P V O S S 2003 C4 HgBCCO, in *Handbook of Superconducting Materials*, **1**: Taylor & Francis Books, Inc.

[8] Mikhailov B P 2004 High-temperature superconductors (HTSCs): Investigations, designs, and applications *Russ. J. Inorg. Chem.* **49** S57

[9] Pawar S H, Shirage P M, Shivagan D D, Jadhav A B 2004 Novel room temperature electrochemical deposition of high temperature superconducting films *Mod. Phys. Lett. B* **18** 505

[10] Wu J Z 2005 Epitaxy of Hg-based high temperature superconducting thin films, in *Next generation High Temperature Superconducting Wires* (Goyal, A., ed) Plenum Publishing.

[11] Wu J Z, Zhao H 2010 Recent progress in fabrication, characterization, and application of Hg-based oxide superconductors, in *High Temperature Superconductors* (Bhattacharya, R. and Paranthaman, M. P., eds) Berlin: Wiley-VCH.

[12] Wu J Z 2013 What have we learnt from the highest-T_c superconducting Hg-based cuprates? *Physica C-Superconductivity and Its Applications* **493** 96

[13] Iyo A, Tanaka Y, Kito H, Kodama Y, Shirage P M, Shivagan D D, Matsuhata H, Tokiwa K, Watanabe T 2007 T_c vs *n* relationship for multilayered High-T_c superconductors *J. Phys. Soc. Japan* **76** 094711

[14] Iyo A, Tanaka Y, Kito H, Yasuharu K, Matsuhata H, Tokiwa K, Watanabe T 2007 Variation of T_c in multilayered cuprates of $HgBa_2Ca_{n-1}Cu_nO_y$ *Physica C* **460** 436

[15] Xie Y Y, Wu J Z, Aytug T, Gapud A A, Christen D K, Verebelyi D T, Song K J 2000 Fabrication and physical properties of large-area $HgBa_2CaCu_2O_6$ superconducting films *Supercond. Sci. Technol.* **13** 225

[16] Luo Z P, Li Y, Hashimoto H, Ihara H, Iyo A, Tokiwa K, Cao G H, Ross J H, Larrea J A, Baggio-Saltovitch E 2004 Defective structure in the high-T_c superconductor Hg-1234 *Physica C* **408** 50

[17] Ueda S, Shimoyama J, Horii S, Kishio K 2007 Synthesis of $(Hg,Re)Ba_2Ca_{n-1}Cu_nO_{2+2n+\delta}$ (*n*=2, 3, 4) single crystals and their magnetization properties *Physica C* **452** 35

[18] Ozaki T, Shimoyama J, Ogino H, Yamamoto A, Kishio K 2013 Critical current properties of *c*-axis oriented Hg(Re) 1223 bulks *IEEE Trans. Appl. Supercon.* **23** 6800404
[19] Shimoyama J, Hahakura S, Kobayashi R, Kitazawa K, Yamafuji K, Kishio K 1994 Interlayer distance and magnetic properties of Hg-based superconductors *Physica C: Supercon.* 235–240, (4)2795
[20] Kishio K, Shimoyama J, Yoshikawa A, Kitazawa K, Chmaissem O, Jorgensen J D 1996 Chemical doping and improved flux pinning in Hg-based superconductors *J. Low Temp. Phys.* **105** 1359
[21] Chmaissem O, Guptasarma P, Welp U, Hinks D G, Jorgensen J D 1997 Effect of Re substitution on the defect structure, and superconducting properties of $(Hg_{1-x}Re_x)Ba_2Ca_{n-1}Cu_nO_{2n+2+\delta}$ (n = 2, 3, 4) *Physica C: Supercon.* **292** 305
[22] Yamasaki H, Nakagawa Y, Mawatari Y, Cao B S 1997 Preparation and magnetization properties of rhenium-doped Hg-based oxide superconductors *J. Japan Inst. Metals* **61** 998
[23] Gasser C, Moriwaki Y, Sugano T, Nakanishi K, Wu X J, Adachi S, Tanabe K 1998 Orientation control of ex site $(Hg_{1-x}Re_x)Ba_2CaCu_2O_y$ (x approximate to 0.1) thin films on $LaAlO_3$ *Appl. Phys. Lett.* **72** 972
[24] Moriwaki Y, Sugano T, Tsukamoto A, Gasser C, Nakanishi K, Adachi S, Tanabe K 1998 Fabrication and properties of c-axis Hg-1223 superconducting thin films *Physica C* **303** 65
[25] Brazdeikis A, Flodstrom A S, Bryntse I 1996 Effect of thallium oxide, Tl_2O_3 on the formation of superconducting HgBaCaCuO films *Physica C* **265** 1
[26] Foong F, Bedard B, Xu Q L, Liou S H 1996 *c*-axis oriented (Hg,Ti)-based superconducting films with T_c>125 K. *Appl. Phys. Lett.* **68** 1153
[27] Xie Y Y, Wu J Z, Yan S L, Yu Y, Aytug T, Fang L, Tidrow S C 1999 Elimination of air detrimental effect using Tl-stabilized precursor for epitaxial $HgBa_2CaCu_2O_6$ thin films *Physica C* **328** 241
[28] Higuma H, Miyashita S, Uchikawa F 1994 Synthesis of superconducting Pb-doped $HgBa_2CaCu_2O_y$ films by laser-ablation and postannealing *Appl. Phys. Lett.* **65** 743
[29] Yu Y, Shao H M, Zheng Z Y, Sun A M, Qin M J, Xu X N, Ding S Y, Jin X, Yao X X, Zhou J, Li Z M, Yang S Z, Zhang W L 1997 $HgBa_2CaCu_2O_y$ superconducting thin films prepared by laser ablation. *Physica C* **289** 199
[30] Guo J D, Xiong G C, Yu D P, Feng Q R, Xu X L, Lian G J, Hu Z H 1997 Preparation of superconducting $HgBa_2CaCu_2O_x$ films with a zero-resistance transition temperature of 121 K *Physica C* **276** 277
[31] Gapud A A, Aytug T, Yoo S H, Xie Y Y, Kang B W, Gapud S D, Wu J Z, Wu S W, Liang W Y, Cui X T, Liu J R, Chu W K 1998 Lithium-doping-assisted growth of $HgBa_2Ca_2Cu_3O_{8+\delta}$ superconducting phase in bulks and thin films *Physica C* **308** 264
[32] Kang B W, Gapud A A, Fei X, Aytug T, Wu J Z 1998 Minimization of detrimental effect of air in $HgBa_2CaCu_2O_{6+\delta}$ thin film processing *Appl. Phys. Lett.* **72** 1766
[33] Wu J Z, Yoo S H, Aytug T, Gapud A, Kang B W, Wu S, Zhou W 1998 Superconductivity in lithium- and sodium-doped mercury-based cuprates *J. Supercon.* **11** 169
[34] Adachi S, Sugano T, Moriwaki Y, Tanabe K 2005 Synthesis of high-T_c superconductor $Hg_{0.8}M_{0.2}Ba_2Ca_2Cu_3O_y$ (M=V, Cr, Mn, Mo, Ag, In, Sn, W, Re, Pb, Hf, Ta) *by quartz-tube encapsulation technique J. Ceram. Soc. Jpn.* **113** 678
[35] Orlando M T D, Passos C A C, Passamai J L, Medeiros E F, Orlando C G P, Sampaio R V, Correa H S P, de Melo F C L, Martinez L G, Rossi J L 2006 Distortion of ReO_6 octahedron in the $Hg_{0.82}Re_{0.18}Ba_2Ca_2Cu_3O_{8+d}$ superconductor *Physica C-Supercon. Appl.* **434** 53
[36] Passos C A C, Passamai J L, Orlando M T D 2007 Effects of Y doping on the (Hg,Re)-1212 superconductor properties *Physica C-Supercon. Appl.* **460** 728
[37] Passos C A C, Passamai J L, Orlando M T D, Correa H P S, de Medeiros E F, Martinez L G, Rossi J L, Garcia F, Tamura E, Ferreira F F, de Melo F C L 2007 Phase segregation of (Hg,Re)-1223 superconductor *Physica C-Supercon. Appl.* **460** 1182
[38] Wang Y Q, Meng R L, Sun Y Y, Ross K, Huang Z J, Chu C W 1993 Synthesis of preferred-oriented $HgBa_2CaCu_2O_{6+\delta}$ thin-films *Appl. Phys. Lett.* **63** 3084.
[39] Miyashita S, Higuma H, Uchikawa F 1994 Structure and superconducting properties of $HgBa_2CaCu_2O_y$ films prepared by laser-ablation *Japanese J. Appl. Phys.* **33** L931
[40] Tsuei C C, Gupta A, Trafas G, Mitzi D 1994 Superconducting mercury-based cuprate films with a zero-resistance transition-temperature of 124 kelvin *Science* **263** 1259
[41] Yun S H, Wu J Z, Tidrow S C, Eckart D W 1996 Growth of $HgBa_2Ca_2Cu_3O_{8+\delta}$ thin films on $LaAlO_3$ substrates using fast temperature ramping Hg-vapor annealing *Appl. Phys. Lett.* **68** 2565
[42] Wu J Z, Yun S H, Gapud A, Kang B W, Kang W N, Tidrow S C, Monahan T P, Cui X T, Chu W K 1997 Epitaxial growth of $HgBa_2CaCu_2O_{6+\delta}$ thin films on $SrTiO_3$ substrates. *Physica C* **277** 219
[43] Sun Y, Guo J D, Xu X L, Lian G J, Wang Y Z, Xiong G C 1999 Superconducting $HgBa_2CaCu_2O_y$ thin films growth on $NdGaO_3$, $SrTiO_3$, $LaAlO_3$ and Y-ZrO_2 substrates. *Physica C* 312 197
[44] Yun S H, Wu J Z 1996 Superconductivity above 130 K in high-quality mercury-based cuprate thin films *Appl. Phys. Lett.* **68** 862
[45] Valerianova M, Odier P, Chromik S, Strbik V, Polak M, Kostic I 2007 Influence of the buffer layer on the growth of superconducting films based on mercury *Supercond. Sci. Technol.* **20** 900

[46] Krusin-Elbaum L, Tsuei C C, Gupta A 1995 High current densities above 100K in the high temperature superconductor $HgBa_2CaCu_2O_{6+\delta}$ *Nature* **373** 679

[47] Shen Z-Y 1994 *High-Temperature Superconducting Microwave Circuits* (Artech House Antennas and Propagation Library)

[48] Tidrow S, Tauber A, Wilber W, Lareau R, Brandle C, Berkstresser G, Ven Graitis A, Potrepka D, Budnick J, Wu J 1997 New substrates for HTSC microwave devices *IEEE Trans. Appl. Supercon.* **7** 1766

[49] Wu J Z, Yan S L, Xie Y Y 1999 Cation exchange: a scheme for synthesis of mercury-based high-temperature superconducting epitaxial thin films *Appl. Phys. Lett.* **74** 1469

[50] Su J H, Sastry P V P S S, Schwartz J 2004 Relative effects of Pb and Re doping in Hg-1223 thick films grown on Ag substrates *J. Materials Res.* **19**(9) 2658

[51] Shivagan D D, Shirage P M, Ekal L A, Pawar S H 2004 Synthesis of single-phase $HgBa_2Ca_2Cu_3O_{8+\delta}$ high-T_c superconducting films using the multistep electrolytic process *Supercon. Sci. Tech.* **17** 194

[52] Shivagan D D, Shirage P M, Pawar S H 2004 Studies on the fabrication of Ag/$HgBa_2CaCu_2O_{6+\delta}$/CdSe heterostructures using the pulse electrodeposition technique *Semicond. Sci. Tech.* **19** 323

[53] Yakinci M E, Aksan M A, Balci Y 2005 Fabrication and properties of $(Hg_{0.8}Re_{0.2})Ba_2Ca_2Cu_3O_x$ superconducting thick films *Supercon. Sci. Tech.* **18** 494

[54] Yakinci M E, Altin S 2013 3D-like vortex behavior and the thermally activated flux flow mechanism in $(Hg_{0.8}Re_{0.2})Ba_2Ca_2Cu_3O_x$ superconducting films *J. Mat. Sci.-Matls. Electronics* **24** 3660

[55] Yakinci M E, Aksan M A, Balci Y, Altin S 2007 Effects of thickness on the grain alignment and J_c properties of $(Hg_{0.8}Re_{0.2})Ba_2Ca_2Cu_3O_x$ superconductor thick films Physica C-Supercon. *Appl.* **460** 1386

[56] Siegal M P, Venturini E L, Morosin B, Aselage T L 1997 Synthesis and properties of Tl-Ba-Ca-Cu-O superconductors *J. Mat. Res.* **12** 2825

[57] Xie Y Y, Wu J Z, Aytug T, Christen D K, Cardona A H 2002 Diffusion mechanism of cation-exchange process for fabrication of $HgBa_2CaCu_2O_{6+\delta}$ superconducting films *Appl. Phys. Lett.* **81** 4002

[58] Xie Y Y, Wu J Z 2003 Enhanced critical current density in overdoped $HgBa_2CaCu_2O_{6+\delta}$ superconducting thin films *Appl. Phys. Lett.* **82** 2856

[59] Xing Z W, Xie Y Y, Wu J Z, Cardona A 2004 Fabrication of superconducting $HgBa_2CaCu_2O_x$ thick films using cation exchange process *Physica C* **402** 45

[60] Ji L, Yan S L, Wu J Z 2014 Superconductivity of 132 K in $HgBa_2Ca_2Cu_3O_{8+\delta}$ thin films fabricated using a cation exchange method *Supercond. Sci. Tech.* **27** 015007

[61] Zhao H, Wu J Z 2004 Pinning lattice: effect of rhenium doping on the microstructural evolution from Tl-2212 to Hg-1212 films during cation exchange *J. Appl. Phys.* **96** 2136

[62] Xing Z, Zhao H, Wu J Z 2006 Reversible exchange of Tl and Hg cations on the superconducting "1212" lattice *Adv. Mater.* **18** 2743

[63] Zhao H, Wu J Z 2007 Converting Hg-1212 to Tl-2212 via Tl-Hg cation exchange in combination with Tl cation intercalation *Supercond. Sci. Tech.* **20** 327

[64] Hamdan N M, Hussain Z 2009 Hole doping in high temperature superconductors using the XANES technique *Supercond. Sci. Tech.* **22** 034007

[65] Lokshin K A, Pavlov D A, Kovba M L, Antipov E V, Kuzemskaya I G, Kulikova L F, Davydov V V, Morozov I V, Itskevich E S 1998 Synthesis and characterization of overdoped Hg-1234 and Hg-1245 phases; the universal behavior of T_c variation in the $HgBa_2Ca_{n-1}Cu_nO_{2n+2+\delta}$ series *Physica C* **300** 71

[66] Correia J G, Haas H, Amaral V S, Lopes A M L, Araújo J P, Le Floch S, Bordet P, Rita E, Soares J C, Tröger W 2005 Atomic ordering of the fluorine dopant in the $HgBa_2CuO_{4+\delta}$ high-T_c superconductor *Phys. Rev. B* **72** 144523

[67] Mendonca T M, Correia J G, Haas H, Odier P, Tavares P B, da Silva M R, Lopes A M L, Pereira A M, Goncalves J N, Amaral J S, Darie C, Araujo J P 2011 Oxygen ordering in the high-T_c superconductor $HgBa_2CaCu_2O_{6+\delta}$ as revealed by perturbed angular correlation *Phys. Rev. B* **84** 094524

[68] Kotegawa H, Tokunaga Y, Araki Y, Zheng GQ, Kitaoka Y, Tokiwa K, Ito K, Watanabe T, Iyo A, Tanaka Y, Ihara H (2004) Coexistence of superconductivity and antiferromagnetism in multilayered high T_c superconductor $HgBa_2Ca_4Cu_5Oy$: Cu-NMR study. *Phys. Rev. B* 69:014501

[69] Mukuda H, Abe M, Araki Y, Kitaoka Y, Tokiwa K, Watanabe T, Iyo A, Kito H, Tanaka Y 2006 Uniform mixing of High-T_c superconductivity and antiferromagnetism on a single CuO_2 plane of a Hg-based five-layered cuprate *Phys. Rev. Lett.* **96** 087001

[70] Mukuda H, Shimizu S, Tabata S, Itohara K, Kitaoka Y, Shirage PM, Iyo A 2010 Superexchange interaction and magnetic moment in antiferromagnetic high-T_c cuprate superconductors *Physica C-Supercon. Appl.* **470** S7

[71] Mukuda H, Abe M, Shimizu S, Kitaoka Y, Iyo A, Kodama Y, Tanaka Y, Tokiwa K, Watanabe T 2008 Phase diagram of high-T_c superconductor: Cu-NMR studies on multilayered cuprates *Physica B* **403** 1059

[72] Tokiwa K, Mikusu S, Higemoto W, Nishiyama K, Iyo A, Tanaka Y, Kotegawa H, Mukuda H, Kitaoka Y, Watanabe T 2007 Muon spin rotation study of magnetism in multilayer $HgBa_2Ca_4Cu_5O_y$ superconductor *Physica C-Supercon. Appl.* **460** 892

[73] Tanaka Y, Iyo A, Itoh S, Tokiwa K, Nishio T, Yanagisawa T 2014 Experimental observation of a possible first-order phase transition below the superconducting transition temperature in the multilayer cuprate superconductor $HgBa_2Ca_4Cu_5O_y$ *J. Phys. Soc. Japan* **83** 074705

[74] Kim G C, Kim H, Lee J H, Chae J S, Kim Y C 2007 Comparative analysis of reversible magnetizations for grain-aligned $Bi_{1.84}Pb_{0.34}Sr_{1.91}Ca_{2.03}Cu_{3.06}O_{10+\delta}$ and $HgBa_2Ca_2Cu_3O_{8+\delta}$ with three CuO_2 planes *Sol. State Comms.* **142** 54

[75] Serquist A, Niebieskikwiat D, Sanchez R D, Morales L, Caneiro A 2004 Hole dependence in (Hg, Re)-1201 compound *J. Low Temp. Phys.* **135** 147

[76] Gallais Y, Sacuto A, Devereaux T P, Colson D 2005 Interplay between the pseudogap and superconductivity in underdoped $HgBa_2CuO_{4+\delta}$ single crystals *Phys. Rev. B* **71** 012506

[77] Passos C A C, Orlando M T D, Passamai J L, de Mello E V L, Correa H P S, Martinez L G 2006 Resistivity study of the pseudogap phase for (Hg,Re)-1223 superconductors. *Phys. Rev. B* **74** 094514

[78] Liu S L, Wu G J, Xu X B, Shao H M 2004 Pseudogap phenomenon in single-phase Hg-1223 cuprate superconductors *J. Supercon.* **17** 253

[79] Li Y, Baledent V, Yu G, Barisic N, Hradil K, Mole R A, Sidis Y, Steffens P, Zhao X, Bourges P, Greven M 2010 Hidden magnetic excitation in the pseudogap phase of a high-T_c superconductor *Nature* **468** 283

[80] Das T 2012 Q=0 collective modes originating from the low-lying Hg-O band in superconducting $HgBa_2CuO_{4+\delta}$ *Phys. Rev. B* **86** 054518

[81] Li Y, Yu G, Chan M K, Baledent V, Li Y, Barisic N, Zhao X, Hradil K, Mole R A, Sidis Y, Steffens P, Bourges P, Greven M 2012 Two Ising-like magnetic excitations in a single-layer cuprate superconductor *Nat. Phys.* **8** 404

[82] Gao L, Chen F, Meng R L, Xue Y Y, Chu C W 1993 Superconductivity up to 147 K in $HgBa_2CaCu_2O_{6+\delta}$ under quasi-hydrostatic pressure *Phil. Mag. Lett.* **68** 345

[83] Nu, X F, ez-Regueiro M, Tholence J L, Antipov E V, Capponi J J, Marezio M 1993 Pressure-induced enhancement of T_c above 150 K in Hg-1223 *Science* **262** 97

[84] Orlando M T D, Belich H, Alves L J, Passamai J L, Pires J M, Santos E M, Rodrigues V A, Costa-Soares T 2009 Pressure effect on Hg-12(n-1)n superconductors and Casimir effect in nanometer scale *J. Phys. A Math. Theor.* **42** 025502

[85] Ambrosch-Draxl C, Sherman E Y, Auer H, Thonhauser T 2004 Pressure-induced hole doping of the Hg-based cuprate superconductors *Phys. Rev. Lett.* **92** 18700

[86] Liu S L, Shao H M, Han C Y 2008 Scaling behavior for resistance transition in Hg-1223 superconductor under pressure *Int. J. Mod. Phys. B* **22** 539

[87] Takeshita N, Yamamoto A, Iyo A, Eisaki H 2013 Zero resistivity above 150 K in $HgBa_2Ca_2Cu_3O_{8+\delta}$ at high pressure *J. Phys. Soc. Japan* **82** 023711

[88] Monteverde M, Acha C, Nunez-Regieiro M, Pavlov D A, Lokshin K A, Putilin S N, Antipov E V 2005 High-pressure effects in fluorinated $HgBa_2Ca_2Cu_3O_{8+\delta}$ *Europhys. Lett.* 72 458

[89] Monteverde M, Nunez-Regueiro M, Acha C, Lokshin K A, Pavlov D A, Putilin S N, Antipov E 2004 Fluorinated Hg-1223 under pressure: the ultimate T_c of the cuprates? *Physica C-Supercond. Appl.* **408** 23

[90] Zhao H, Dizon J R, Wu J Z 2007 Third-order intermodulation in two-pole X-band $HgBa_2CaCu_2O_{6+\delta}$ microstrip filters *Appl. Phys. Lett.* **91** 042506

[91] Ji L, Wu J Z 2008 A comparative study of microwave power handling capabilities of Tl-1212 and Hg-1212 superconducting microstrip lines *Supercond. Sci. Tech.* **21** 125027

[92] Zhao H, Wang X, Wu J Z 2008 Correlation of microwave nonlinearity and magnetic pinning in high-temperature superconductor thin film band-pass filters *Supercond. Sci. Tech.* **21** 085012

[93] Ji L, Yan S, Wu J Z 2009 A comparative study of nonlinear microwave properties in $YBa_2Cu_3O_7$, TlBaCaCuO and HgBaCaCuO microstrip resonators *IEEE Trans. Appl. Supercon.* **19** 2913

[94] Welp U, Crabtree G W, Wagner J L, Hinks D G 1993 Flux-pinning and the irreversibility lines in the $HgBa_2CuO_{4+\delta}$, $HgBa_2CaCu_2O_{6+\delta}$ and $HgBa_2Ca_2Cu_3O_{8+\delta}$ compounds *Physica C* **218** 373

[95] Huang Z J, Xue Y Y, Meng R L, Chu C W 1994 Irreversibility line of the $HgBa_2CaCu_2O_{6+\delta}$ high temperature superconductors *Phys. Rev. B* **49** 4218

[96] Gapud A A, Wu J Z, Fang L, Yan S L, Xie Y Y, Siegal M P, Overmyer D L 1999 Supercurrents in $HgBa_2CaCu_2O_{6+\delta}$ and $TlBa_2CaCu_2O_7$ epitaxial thin films *Appl. Phys. Lett.* **74** 3878

[97] Gapud A A, Wu J Z, Kang B W, Yan S L, Xie Y Y, Siegal M P 1999 Giant T_c shift in $HgBa_2CaCu_2O_{6+\delta}$ and $TlBa_2CaCu_2O_{7-\delta}$ superconductors due to Hg-Tl exchange *Phys. Rev. B* **59** 203

[98] Uchoa B, Neto A H C 2007 Superconducting states of pure and doped graphene *Phys. Rev. Lett.* **98** 4

[99] Maurer D, Luders K, Breitzke H, Baenitz M, Pavlov D A, Antipov E V 2006 Weak to strong pinning crossover in Hg-1201 *Physica C-Supercond. Appl.* **445** 219

[100] Maurer D, Luders K, Breitzke H, Baenitz M, Pavlov D A, Antipov E V 2007 Flux creep in Hg-1201. *Physica C-Supercond. Appl.* **460** 382

[101] Zehetmayer M, Eisterer M, Sponar S, Weber H W, Wisniewski A, Puzniak R, Panta P, Kazakov S M, Karpinski J 2005 Magnetic properties of superconducting $HgBa_2CuO_{4+\delta}$ single crystals in the overdoped state before and after particle irradiation *Physica C-Supercond. Appl.* **418** 73

[102] Kim M H, Lee S I, Kim M S, Kang W N 2005 An H-T diagram characterizing the activation barriers obtained from the magnetic relaxation of $HgBa_2Ca_2Cu_3O_{8+\delta}$ thin film *Supercond. Sci. Tech.* **18** 835

[103] Xie Y Y, Wu J Z, Yun S H, Emergo R, Aga R, Christen D K 2004 Magnetic flux pinning enhancement in $HgBa_2CaCu_2O_{6+\delta}$ films on vicinal substrates *Appl. Phys. Lett.* **85** 70

[104] Thompson J R, KrusinElbaum L, Christen D K, Song K J, Paranthaman M, Ullmann J L, Wu J Z, Ren Z F, Wang J H, Tkaczyk J E, DeLuca J A 1997 Current-density enhancements of the highest-T_c superconductors with GeV protons *Appl. Phys. Lett.* **71** 536

[105] Thompson J R, Ossandon J G, Krusin-Elbaum L, Christen D K, Kim H J, Song K J, Sorge K D, Ullmann J L 2004 Pinning action of correlated disorder against equilibrium properties of $HgBa_2Ca_2Cu_3O_x$ *Phys. Rev. B* **69** 104520

[106] Amm K M, Wolters C, Knoll D C, Peterson S C, Schwartz J 1997 Growth of $Hg_{0.9}Re_{0.1}Ba_2Ca_2Cu_3O_{8+x}$ on a metallic substrate *IEEE Trans. Appl. Supercon.* **7** 1973

[107] Lechter W, Toth L, Osofsky M, Skelton E, Soulen R J, Qadri S, Schwartz J, Kessler J, Wolters C 1995 One-step reaction and consolidation of Hg-based high-temperature superconductors by hot isostatic pressing *Physica C* **249** 213

[108] Schwartz J, Amm K M, Sun Y R, Wolters C 1996 HgBaCaCuO superconductors: processing, properties and potential *Physica B* **216** 261

[109] Meng R L, Hickey B, Wang Y Q, Sun Y Y, Gao L, Xue Y Y, Chu C W 1996 Processing of highly oriented $(Hg_{1-x}Re_x)Ba_2Ca_2Cu_3O_{8+\delta}$ tape with x similar to 0.1 *Appl. Phys. Lett.* **68** 3177

[110] Peacock G B, Gameson I, Edwards P P, Khaliq M, Yang G, Shields T C, Abell J S 1997 Fabrication of high-temperature superconducting $HgBa_2CuO_{4+\delta}$ within silver-sheathed tapes *Physica C* **273** 193

[111] Heine K, Tenbrink J, Thoner M 1989 High-field critical current densities in $Bi_2Sr_2CaCu_2O_{8+x}$/Ag wires *Appl. Phys. Lett.* **55** 2441

[112] Sastry P V P S S, Li Y, Su J, Schwartz J 2000 Attempts to fabricate thick HgPb1223 superconducting films on silver *Physica C* **335** 112

[113] Meng R L, Wang Y Q, Lewis K, Garcia C, Cao Y, Chu C W 1998 Hg-1223 thick film on flexible Ni substrates *J. Supercond.* **11** 181

[114] Goyal A 2005 *Second-Generation HTS Conductors* (Springer)

[115] Xie Y Y, Aytug T, Wu J Z, Verebelyi D T, Paranthaman M, Goyal A, Christen D K 2000 Epitaxy of $HgBa_2CaCu_2O_6$ superconducting films on biaxially textured Ni substrates *Appl. Phys. Lett.* **77** 4193

[116] Aytug T, Xie Y Y, Wu J Z, Christen D K 2001 Growth characteristics of $HgBa_2CaCu_2O_6$ superconducting films on CeO_2-buffered YSZ single crystals: an assessment for coated conductors *Physica C* **363** 107

[117] Passos C A C, Orlando M T D, Passamai J L, Medeiros E F, Oliveira F D C, Fardin J F, Simonetti D S L 2006 Superconducting fault current limiter device based on (Hg,Re)-1223 superconductor *Appl. Phys. Lett.* **89** 242503

[118] Passos C A C, Passamai J L, Orlando M T D, Medeiros E F, Sarnpaio R V, Oliveira F D C, Fardin J F, Simonetti D S L 2007 Application of the (Hg,Re)-1223 ceramic on superconducting fault current limiter *Physica C-Supercond. Appl.* **460** 1451

[119] Passos C A C, Rodrigues V A, Pinto J N O, Abilio V T, Silva G M, Machado L C, Machado I P, Marins A A L, Merizio L G, da Cruz P C M, Muri E J B 2014 Development and test of a small resistive fault current limiting device based on Hg,Re-1223 and Sm-123 ceramics *Mat. Res. Ibero-Amer. J. Matls.* **17** 28

[120] Aga R S, Xie Y Y, Wu J Z, Han S 2000 Microwave characterization of $HgBa_2CaCu_2O_{6+\delta}$ thin films *Physica C* **341** 2721

[121] Aga R S, Yan S L, Xie Y Y, Han S Y, Wu J Z, Jia Q X, Kwon C 2000 Microwave surface resistance of $HgBa_2CaCu_2O_{6+\delta}$ thin films *Appl. Phys. Lett.* **76** 1606

[122] Aytug T, Kang B W, Yan S L, Xie Y Y, Wu J Z 1998 Stability of Hg-based high-T_c superconducting thin films *Physica C* **307** 117

[123] Aytug T, Gapud A A, Yoo S H, Kang B W, Gapud S D, Wu J Z 1999 Effect of sodium doping on the oxygen distribution of Hg-1223 superconductors *Physica C* **313** 121

[124] Aga R S, Xie Y Y, Yan S L, Wu J Z, Han S Y S 2001 Microwave-power handling capability of $HgBa_2CaCu_2O_{6+\delta}$ superconducting microstrip lines *Appl. Phys. Lett.* **79** 2417

[125] Dizon J R, Zhao H, Baca J, Mishra S, Emergo R L, Aga R S, Wu J Z 2006 Fabrication and characterization of two-pole X-band $HgBa_2CaCu_2O_{6+\delta}$ microstrip filters *Appl. Phys. Lett.* **88** 092507

[126] Zhao H, Dizon J R, Lu R, Qiu W, Wu J Z 2007 Fabrication of three-pole $HgBa_2CaCu_2O_{6+\delta}$ hairpin filter and characterization of its third-order intermodulation. *IEEE Trans. Appl. Supercond.* **17** 914

[127] Gupta A, Sun J Z, Tsuei C C 1994 Mercury-based cuprate high-transition temperature grain-boundary junctions and SQUIDs operating above 110 kelvin *Science* **265** 1075

[128] Tsukamoto A, Takagi K, Moriwaki Y, Sugano T, Adachi S, Tanabe K 1998 High-performance (Hg,Re) $Ba_2CaCu_2O_y$ grain-boundary Josephson junctions and dc superconducting quantum interference devices *Appl. Phys. Lett.* **73** 990

[129] Yu Y, Yan S L, Fang L, Xie Y Y, Wu J Z, Han S Y, Shimakage H, Wang Z 1999 Fabrication of $HgBa_2CaCu_2O_y$ grain boundary junctions using the cation exchange method *Supercond. Sci. Technol.* **12** 1020

[130] Ogawa A, Sugano T, Wakana H, Kamitani A, Adachi S, Tarutani Y, Tanabe K 2004 (Hg,Re)$Ba_2CaCu_2O_y$ [100]-tilt grain boundary Josephson junctions with high characteristic voltages *Jpn. J. Appl. Phys.* **243** L842

[131] Ogawa A, Sugano T, Wakana H, Kamitani A, Adachi S, Tarutani Y, Tanabe K 2006 Properties of (Hg,Re) $Ba_2CaCu_2O_y$ [100]-tilt grain boundary Josephson junctions *J. Appl. Phys.* **99** 123907

[132] Ueda S, Yamaguchi T, Kubo Y, Tsuda S, Takano Y, Shimoyama J, Kishio K 2009 Switching current distributions and subgap structures of underdoped (Hg, Re) $Ba_2Ca_2Cu_3O_{8+\delta}$ intrinsic Josephson junctions *J. Appl. Phys.* **106** 074516

[133] Hirata Y, Kojima K M, Uchida S, Ishikado M, Iyo A, Eisaki H, Tajima S 2010 Interlayer Josephson couplings in Hg-based multi-layered cuprates *Physica C-Supercond. Appl.* **470** S44

[134] Li X, Xu Y, Chromik S, Strbik V, Odier P, De Barros D, Sobolewski R 2005 Time-resolved carrier dynamics in Hg-based high-temperature superconducting photodetectors *IEEE Trans. Appl. Supercond.* **15** 622

[135] Li X, Khafizov M, Chromik S, Valerianova M, Strbik V, Odier P, Sobolewski R 2007 Ultrafast photoresponse dynamics of current-biased Hg-Ba-Ca-Cu-O superconducting microbridges *IEEE Trans. Appl. Supercond.* **17** 3648

[136] Chromik S, Valerianova M, Strbik V, Gazi S, Odier P, Li X, Xu Y, Sobolewski R, Hanic F, Plesch G, Benacka S 2008 Hg-based cuprate superconducting films patterned into structures for ultrafast photodetectors *Appl. Surf. Sci.* **254** 3638

[137] Cross X L, Zheng X, Cunningham P D, Hayden L M, Chromik S, Sojkova M, Strbik V, Odier P, Sobolewski R 2009 Pulsed-THz characterization of Hg-based, high-temperature superconductors *IEEE Trans. Appl. Supercond.* **19** 3614

C5

Iron-Based Superconductors

Hideo Hosono

C5.1 Introduction: Discovery and Impact

Iron-based superconductors (IBSCs) were discovered by a group led by Hideo Hosono [1, 2] in the course of exploration of magnetic semiconductors as an extension of research on transparent p-type semiconductors. These researchers studied LaTMPnO (TM = 3d transition metal, Pn = pnictogen), which has the same crystal structure as LaCuOCh (Ch = chalcogen) composed of an alternate stack of $(CuCh)^-$ and $(LaO)^+$ layers. LaCuOCh is a wide-gap *p*-type semiconductor that these researchers cultivated. The next objective was to create magnetic semiconductors which can control the magnetic transition temperature by gating. So, Hosono chose LaTMPnO (where TM = 3d transition metal, Pn = P or As) as the candidate materials because this material has a similar crystal structure to LaCuOCh, and TM is expected to work as a magnetic centre. Although the compounds were already reported, almost no information on their electronic and magnetic properties had been examined. His group started to examine the electromagnetic properties of these materials from 2004. It was found that the properties of 3d transition metal (TM) oxypnictides vary drastically with TM. Figure C5.1 summarizes the obtained properties of LaTMPnO.

One may see that the properties of layered TM oxypnictides strongly depend on TM. The synthesis of early TMs and Cu oxypnictides was tried but unsuccessful even using high pressure up to ~9 GPa, and no distinct correlation was found between the stability of the 1111-type compound and the kind of TM. Bulk superconductivity appears in TM = $Fe^{2+}$and Ni^{2+}, both of which have an even number of 3d electrons, but no superconductivity was found in TM = Cr^{2+} [3] with $3d^4$ electronic configuration. Undoped LaFeAsO is an antiferromagnetic metal but does not exhibit superconductivity. For TM = Mn, an exceptionally high electron doping is possible by applying H^- [4] in place of F^- as a substituent for the oxygen site. The transition from antiferromagnetic insulator to ferromagnetic metal was observed but no superconductivity appeared. LaCoAsO is an itinerant ferromagnetic metal [5]. To date no T_c exceeding 10 K has been reported in the 1111 system except for the iron oxyarsenides.

The first IBSC was discovered in 2006 for LaFePO; however, the T_c remained as low as ~4 K [1]. High-T_c materials were subsequently discovered for $LaFeAsO_{1-x}F_x$ with $T_c = 26$ K ($x = 0.08$) in February 2008 [2], surpassing the discovery of LaNiPO with $T_c = 3$ K in 2007 [6]. Then, in April 2008, a T_c of 43 K [7] was reached under a high pressure of 4 GPa, exceeding that of MgB_2 (39 K). Prior to this, it was widely believed that elements with a large magnetic moment were harmful to the emergence of superconductivity because the magnetism arising from the static ordering of magnetic moments competes with superconductivity, which requires the formation of Cooper pairs (dynamic pairing of two conduction electrons with opposite spin). Because iron and nickel are representative magnetic elements, these discoveries were accepted with surprise by the condensed matter physics community, and extensive studies immediately began around the globe. One can see the rapid research progress and enthusiasm at the early stage in the proceedings of the first international conference on IBSCs [8] held in Tokyo in 2008 and in the first special issue on IBSCs [9].

As noted, the dynamic formation of electron pairs is prerequisite for emergence of superconductivity, while (anti)ferromagnetism is associated with long-range static spin ordering. This is the reason why it is widely believed both compete with each other. Iron is a typical magnetic element with a large magnetic spin moment, and had been believed to the most harmful for emergence of superconductivity. However, the situation was totally changed since the discovery of iron oxypnictide high-T_c superconductors in early 2008.

What is the impact of iron-based superconductors? There will be two answers, i.e., the first is the breaking of a widely accepted belief that "iron is antagonistic against superconductivity", which led to the opening of a fruitful frontier in superconducting materials. It has become clear through intense research in the last 10 years that iron can cooperate with high-T_c superconductivity under certain conditions. The second is a rich variety in candidate materials and in pairing interaction. It has turned out that there are many material varieties in

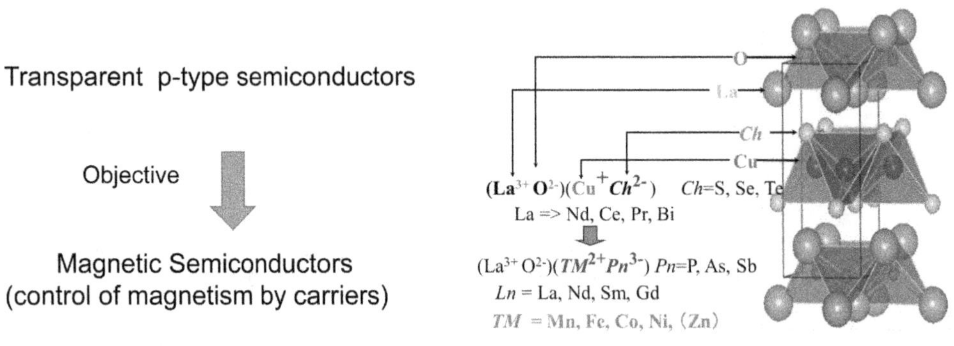

T_M^{2+} (electron configuration)	Cr(3d⁴)		Mn(3d⁵)		Fe(3d⁶)		Co(3d⁷)		Ni(3d⁸)		Cu	Zn(3d¹⁰)	
Pn	P	As	P	As	P	As	P	As	P	As	P, As	P	As
Elect. Prop.	╳	Metal	Mott Insulator		Superconductor		Metal		Superconductor		╳	Semiconductor	
Magnetism	╳	AF(CB)	AF(CB)		Superconductor		FM		Superconductor		╳	Non-magnetic	
E_g	╳	-	~1 eV		-		-		-		╳	~1.5 eV	
T_c(SC)	╳				Undoped: 4K	F-doped: 26K			Undoped: 3K	Undoped: 2.4K	╳		
T_{NC}(Mag)	╳	> 300K	> 400K				43K	66K			╳		
Ref.	S-W. Park et al. IC (2013)		Yanagi et al. JAP (2009)		Kamihara et al. JACS (2006) Kamihara et al. JACS (2008)		Yanagi et al. PRB (2008)		Watanabe et al. IC (2007) Watanabe et al. JSSC (2007)			Kayanuma et al. PRB (2007) Kayanuma et al. TSF (2008)	

╳: impossible to synthesize
CB: checker board type

FIGURE C5.1 From transparent p-type semiconductor LaCuOCh to magnetic LaTMOPn. Both compounds have the same crystal structure and properties of the latter distinctly changes with TM.

iron-based superconductors such as 10 parent materials, 1111, 122, 111, 112, 245, 11 and thick-blocking layer bearing materials (where the number denotes the atomic ratios in the compounds' constituent elements, see Figures C5.2 and C5.3 for crystal structures of each compound). Each type has rather different electrical and magnetic properties including antiferromagnetic semimetal, Pauli paramagnetic metal and antiferromagnetic Mott insulator.

IBSCs have several unique properties such as robustness to impurity, high upper critical field and excellent grain-boundary properties – each advantageous for wire applications. Recent progress in the performance of superconducting wires of IBSC has been remarkable, and the maximal critical current has now reached the level of commercial metal-based superconducting wires.

In this chapter, I describe the material characteristics and recent progress in applications of IBSCs. A vast number of papers on IBSCs have been published to date. Readers are encouraged to refer to the comprehensive reviews and monographs listed in references [10–14].

C5.2 Parent Materials

Since the first paper reporting T_c = 26 K in $LaFeAsO_{1-x}F_x$, several tens of superconducting layered iron pnictides and chalcogenides have been reported. These materials contain a common building block of a square lattice of Fe^{2+} ions with tetrahedral coordination with *Pn* (P and/or As) or Ch ions. To date, ~10 parent compounds are known, and each crystal structure can be derived from the insertion of ions and/or building blocks between the Fe*Pn*(*Ch*) layers [12] as shown in Figure C5.2. Figure C5.3 shows the crystal structures of the parent materials of IBSCs. Because the Fermi level of each parent compound is primarily governed by five Fe 3d orbitals, iron

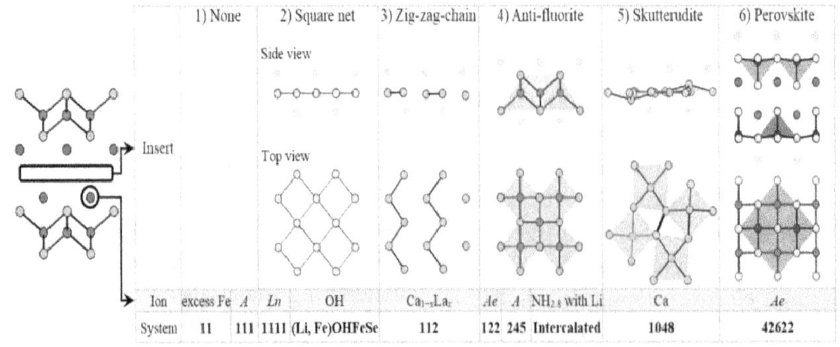

FIGURE C5.2 Structural constitution of IBSCs. These parent compounds are named using an abbreviation of the ratio of the constituent atoms such as "1111" for *Re*FeAsO (*Re* = rare-earth metal).

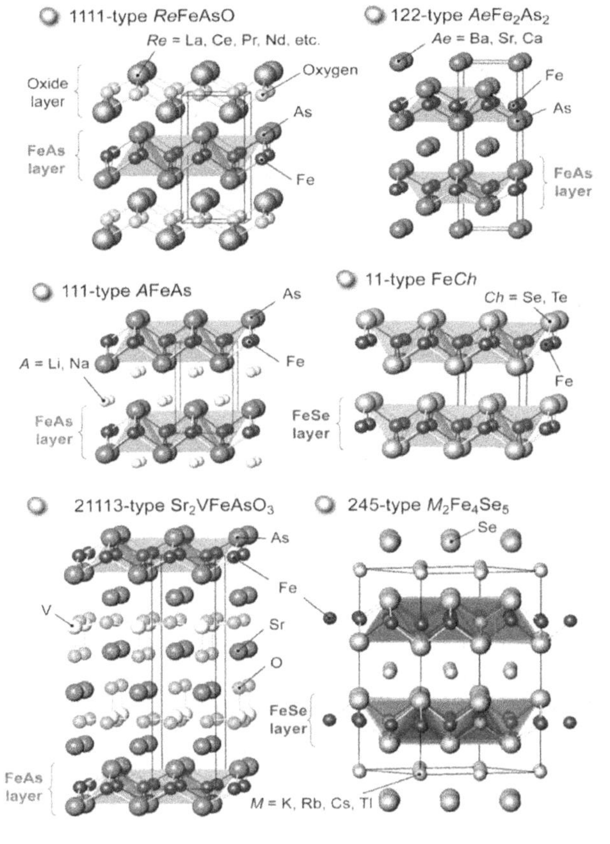

FIGURE C5.3 Crystal structure of parent compounds for IBSCs.

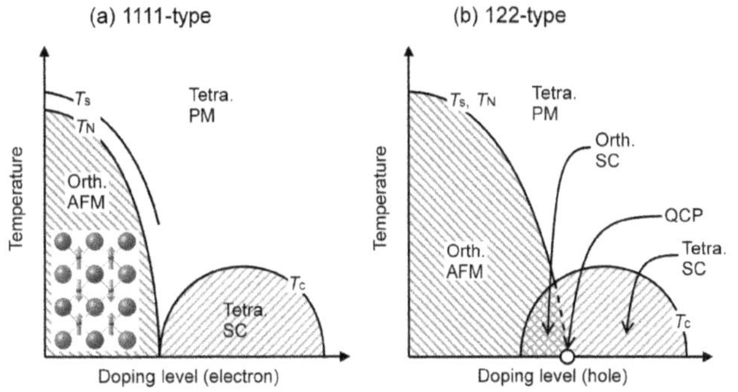

FIGURE C5.4 Schematic phase diagram of (a) 1111-type and (b) 122-type systems. T_s: structural transition temperature, T_N: Néel temperature, SC: superconducting phase, PM: Pauli paramagnetism, QCP: quantum critical point. The distinct differences between the 1111- and 122-type systems are that T_s and T_N are distinctly separated in the low doping region in the 1111 system, whereas both coincide in the parent phase of the 122 system. Further, superconductivity and AFM coexist in the 122 system but not in the 1111 system.

plays a primary role in the superconductivity. This observation is in sharp contrast to that in cuprate superconductors, in which only one 3d orbital is associated with the Fermi level. In addition, these compounds have tetragonal symmetry in the superconducting phase, are Pauli paramagnetic metals in the normal state, and undergo crystallographic/magnetic transitions from the tetragonal to orthorhombic or monoclinic phase with AFM at low temperatures. The exceptions are 11- and 111-type compounds exhibiting Pauli paramagnetism and 245 compounds exhibiting AFM insulating properties.

C5.3 Doping to Induce Superconductivity

Superconductivity emerges when AFM disappears or is diminished by carrier doping, structural modification under external pressure, or chemical pressure via isovalent substitution. In any case, the parent materials are metals with itinerant carriers. Thus, in most cases, removing the magnetism is an experimental step needed for the emergence of superconductivity.

C5.3.1 Aliovalent Doping

The first high-T_c IBSC was discovered through partial replacement of F^- ions at oxygen sites in La-1111 compounds [2]. The 1111-type compounds consist of a metallic conducting FeAs layer sandwiched by insulating LaO layers. When the O^{2-} site is replaced with an F^- ion, the generated electron is transferred to the FeAs layer because of the energy offset between the layers. Figure C5.4 presents a schematic phase diagram of the 1111- and 122-type systems. For the 1111 system, superconductivity appears when the AFM disappears. However, the AFM and superconductivity coexist in the 122 system, and the optimal T_c appears to be achieved at a doping level at which the Néel temperature (T_N) reaches 0 K, suggesting the close relationship between the optimal T_c and quantum criticality (electronic phase transition at 0 K). Electron doping of *RE*-1111 compounds (where *RE* = rare-earth metal) via this substitution was very successful; i.e., the optimal T_c from 26 K to 55 K was achieved with the use of Sm (instead of La) with smaller ionic radius [15, 16].

However, no experimental data on the shape and width of the completely closed T_c-dome in the 1111 system with the highest T_c were obtained until 2011 because the electron-doping level was insufficient to observe the over-doped region. This was attributed to the poor solubility of F^- ions at the oxygen sites (approximately 10–15%), resulting from the preferential precipitation of the stable *RE*OF phase. This obstacle was overcome by using hydride ions (H^-) in place of F^- [17]. Hydrogen is the simplest bipolar element and can take +1 and −1 charge states depending on its local environment [18]. The ionic radius of H^- (110 pm) is also similar to that of F^- (133 pm) or O^{2-} (140 pm). H^--substituted *RE*-1111 compounds, $RE\text{FeAsO}_{1-x}\text{H}_x$, were successfully synthesized with the aid of high pressure and an anvil cell modified for this synthesis. The synthesis was based on the idea that a hydride-substituted state would be more stable than an oxygen vacancy state in the charge blocking layer *RE*O with fluorite structure (an oxygen ion occupies a tetrahedral site) [18, 19]. Based on this idea, the mixture of starting materials was heated with a solid hydrogen source such as CaH_2, which releases H_2 gas at high temperatures under 2 GPa. Figure C5.5 presents electronic phase diagrams of $RE\text{FeAsO}_{1-x}\text{H}_x$ with different *RE*(La, Ce, Sm, and

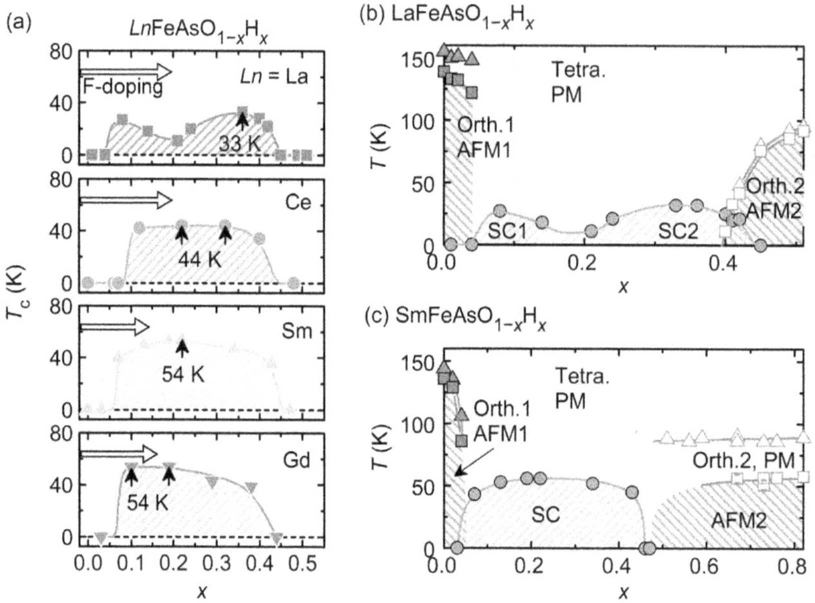

FIGURE C5.5 (a) Phase diagram of $REFeAsO_{1-x}H_x$ and comparison between F- and H-substituted *RE*-1111 system (*RE* = La, Ce, Sm, and Gd) in superconducting region. Two AFM phases are located at the edges of the T_c-domes for the (b) La (with double dome) and (c) Sm (with a single dome) systems. Here, AFM: antiferromagnetism, PM: Pauli paramagnetism.

Gd) [20], which highlight three new findings. First, La-1111 has a two-dome structure. The first dome is the same as that previously reported for $LaFeAsO_{1-x}F_x$; however, the second dome unveiled by H doping has a higher optimal T_c (36 K) and larger width. The temperature dependence of electrical resistivity in the normal state (150 K > *T* > T_c) directly above T_c follows a T^2 behaviour (Fermi liquid like) for the first dome but linear-in-*T* (non-Fermi liquid like) for the second dome. The double-dome structure is not unique to the La-1111 system and has also been observed for chemical compositions with ~30 K > T_c in $SmFeAs_{1-y}P_yO_{1-x}H_x$ [21], as shown in Figure C5.6. Second, although T_c has a single dome for other *RE* systems, its range is much wider than that reported in the F-substituted case for the other *RE* systems. Third, the optimal doping level decreased with deceasing *RE* ion size. The two T_c-dome structures in $LaFeAsO_{1-x}H_x$ became a single dome under high pressure, and the optimal T_c of 52 K was achieved in the valley at ambient pressure [22]. These observations suggest that when the T_c-double-dome structure is transformed into a single dome with a wide width, the optimal T_c > 50 K appears in the *RE*-1111 system regardless of *RE*.

The two-dome structure in the La-1111 system is considered to be derived from the two types of parent compounds with AFM ordering. Both parents are AFM metals; however, their magnetic moments and spin arrangement differ distinctly. Recently, the presence of two parent phases at the extreme edge of the superconducting region was observed for the Sm-1111 system, for which the T_c has a single-dome structure

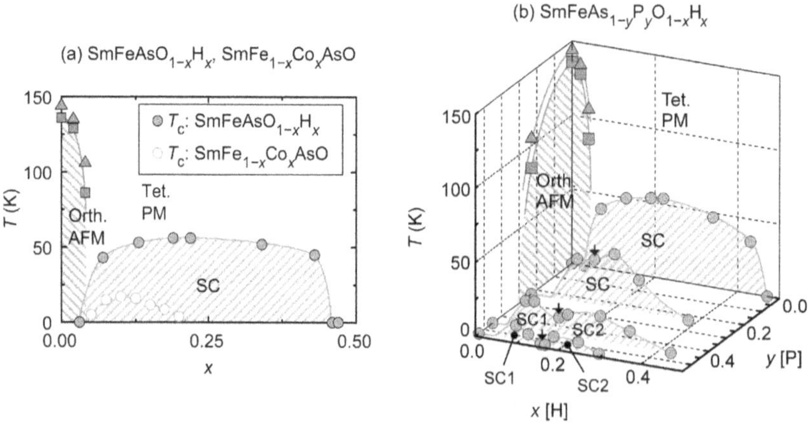

FIGURE C5.6 T_c vs. composition in (a) $SmFe(Co)AsO_{1-x}H_x$ and (b) $SmFeAs_{1-y}P_yO_{1-x}H_x$.

with an optimal T_c > 50 K [23] (see Figure C5.5[c]). These results imply that the high T_c in the 1111 system is realized by the cooperation of two types of fluctuation controlling the two AFM parent phases.

Hole doping of the 122 system is possible by substitution of an alkaline-earth ion site with an appropriate alkali ion (e.g., K substitution of the Ba site) [24], which will be discussed later. However, the hole-doping effect in the *Re*-1111 system remains unclear.

C5.3.2 Electron Doping by Oxygen Vacancies

Another electron-doping method involves the introduction of an oxygen vacancy in the *RE*O layers in *RE*$FeAsO_{1-x}$ [25, 26]. These samples were synthesized by heating a batch of oxygen-deficient compositions under high pressure. If a vacancy substitutes for the oxygen ion site, two carrier electrons per vacancy should be generated; however, the results presented in Figure C5.7(a) differ greatly from this expectation. In addition, the T_c values of nominally oxygen-deficient RE$FeAsO_{1-x}$ samples agree well with those of hydrogen-substituted ones when plotted against their *a*-axis lattice dimension as shown in Figure C5.7(b) [19]. To clarify this contradiction, Muraba *et al.* examined the preferred electron-dopant species at oxygen sites in *RE*FeAsO by changing the atmosphere (H_2, H_2O, or H_2- & H_2O-free), ensuring a high-pressure cell assembly to prevent external contamination [27]. The following observations were made: (1) The samples synthesized under a high-pressure atmosphere of H_2 or H_2O were *RE*$FeAsO_{1-x}H_x$, not *RE*$FeAsO_{1-x}$. (2) The samples with the nominal composition *RE*$FeAsO_{1-x}$ synthesized under H_2- and H_2O-free atmospheres were nearly stoichiometric *RE*FeAsO. These results strongly suggest that the samples of *RE*$FeAsO_{1-x}$ reported thus far are actually *RE*$FeAsO_{1-x}H_x$, which are formed by incorporating hydrogen from the atmosphere and/or starting materials. First-principles calculations substantiated that the hydrogen-substituted samples were more stable than the oxygen-vacancy substituted ones [27]. A similar observation was also recently reported for amorphous oxide semiconductors in which oxygen vacancy is believed to be the dominant defect [28].

C5.3.3 Isovalent Doping

A unique characteristic of doping IBSCs is isovalent doping. Two typical examples are introduced. One is partial substitution of the Fe^{2+} site by Co^{2+} and the other is replacement of the As^{3-} site with P^{3-}. The former example may be understood in terms of electron doping because the Co^{2+} ($3d^7$) has an excess electron compared with Fe^{2+} ($3d^6$) [31]. This finding contrasts sharply with the results of impurity effects in high-T_c cuprates, for which T_c is easily degraded by partial replacement of the Cu^{2+} site. The robustness of T_c to impurities is closely related to the pairing mechanism and symmetry of the order parameter. This type of substitution is often called 'direct doping' because the TM such as Co replaces the iron sites where superconductivity emerges. It is natural to consider that the T_c induced by the direct doping is considerably lower than that induced by indirect doping. Figure C5.8 compares the direct and indirect doping of T_c in the 122 system (also see Figure C5.6[a] for the 1111 system.).

Another effective isovalent substitution is observed in the 122 system such as $BaFe_2(As_{1-x}P_x)_2$ [32]. As the T_N of the parent phase is reduced by *x*, superconductivity appears and reaches a maximum T_c of ~30 K around $x = 0.35$, which appears to correspond to the quantum critical point. The shape of this phase diagram is similar to that obtained by electron doping using Co substitution. The emergence of superconductivity with the similar isovalent substitution of anions directly bonding with iron is observed for $FeSe_{1-x}Te_x$ [33]. Because isovalent anion substitution does not generate carriers (unlike

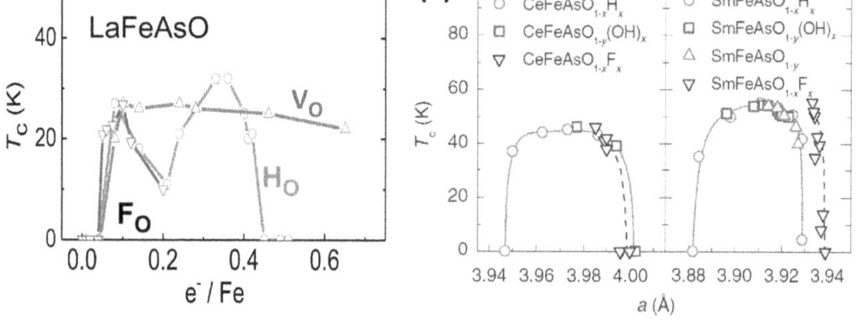

FIGURE C5.7 (a) Relationship between observed T_c and doped carrier concentration in $LaFeAsO_{1-x}H(F)_x$ and nominal $LaFeAsO_{1-x}$. Here, the doped electron concentration was assumed to be 1 per H(F) or 2 per oxygen vacancy. The observed T_c was obtained from ref. [20] for H doping, ref. [29] for F doping, and ref. [30] for oxygen vacancy (Vo) doping. The data on the H-substituted samples (Ho) agree well with the F-substituted ones (Fo); however, the values differed greatly from those on the oxygen-vacancy doped ones (Vo). (b) Lattice constant vs. T_c in three series of electron-doped Ce/Sm1111 samples using different approaches. These data on nominally oxygen-vacancy-doped and OH-doped samples agree well with those on H-substituted ones. Because only the F-doped samples were synthesized in an ambient atmosphere, the lattice constant was shifted from that of the samples synthesized using high pressure [19].

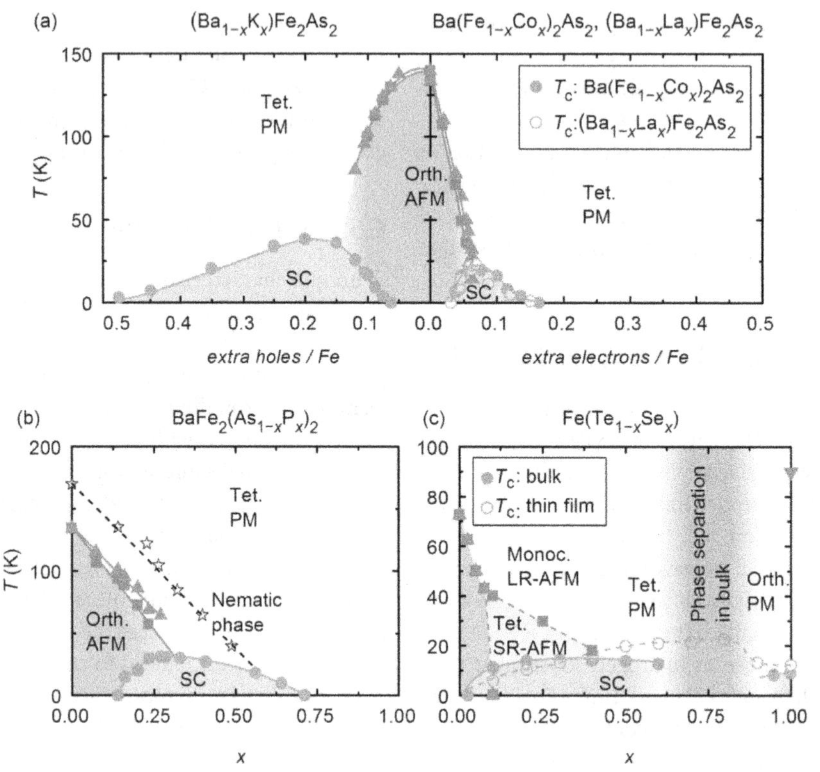

FIGURE C5.8 Effect of doping on superconductivity: (a) Hole doping vs. electron doping; (b) and (c) isovalent doping effect.

Co substitution), it is understood that the anion substitution modifies the local geometry around iron atoms, which in turn leads to weakening of AFM order competing with the emergence of superconductivity. Because the parent materials of IBSCs are metals containing sufficient carriers to induce superconductivity, it is understood that the primary effect of isovalent anion substitution is to weaken the AFM.

C5.3.4 Doping by Intercalation

The parent materials of IBSCs have layered structures. Insertion of ions and/or molecules is possible while maintaining the original Fe*Pn*(*Ch*) layers in some parent materials. Metal–superconductor conversion via this type of doping has been reported for 11 and 122 compounds. The FeSe intercalates obtained from low-temperature alkali metal and NH_3 co-intercalation exhibit higher T_c of 30–46 K than the samples prepared using conventional high-temperature methods [34]. A unique feature of this process is that a small-sized alkali cation such as Li or Na combined with the NH_2^- anion or NH_3 molecules can be intercalated into the FeSe layers [35] because the formation of ion intercalates is restricted to large-sized monovalent cations such as Cs and Tl [36] by conventional high-temperature methods. Recently, the phase diagram of K-intercalated FeSe was reported as shown in Figure C5.9 [37]. The optimal T_c of intercalated FeSe compounds is ~46 K irrespective of intercalating species and lattice dimension.

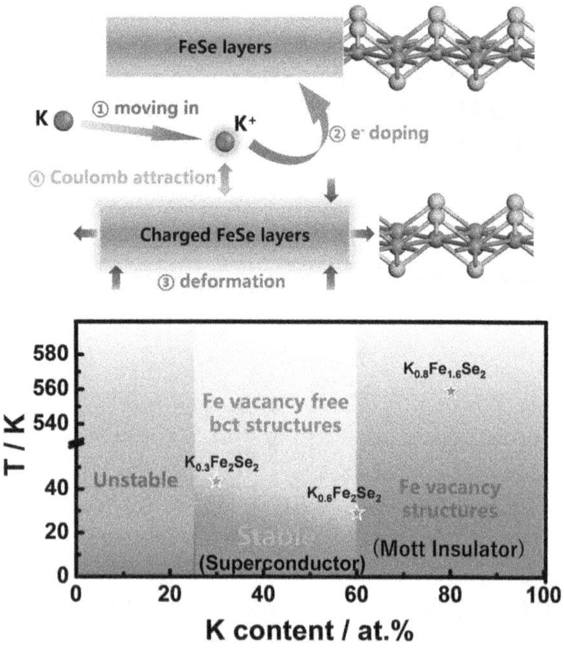

FIGURE C5.9 Phase diagram of K-intercalated FeSe [37]. The phase stability is controlled by energy gain by Coulomb interaction upon K-insertion and energy loss by pushing up the Fermi level upon electron doping.

Referring to the result on gated FeSe thin films using electron-double layer transistor structures [38], T_c appears to increase monotonically with doped electron concentration.

When $SrFe_2As_2$ thin films are placed in an ambient atmosphere, this film is converted into a superconductor accompanying shrinkage of the *c*-axis [39]. Based on an observation that this conversion does not occur in a dry atmosphere, the intercalation of H_2O-relevant species into a vacant site in the Sr layers was suggested. Such a conversion is not observed for $BaFe_2As_2$ [40] where a vacancy allows smaller space than that in $SrFe_2As_2$. This finding led to the shift of thin-film research from $SrFe_2As_2$ to $BaFe_2As_2$, which is less sensitive to ambient atmosphere [41]. Consequently, subsequent research on the 122 system has been performed mainly for $BaFe_2As_2$ to date. A similar conversion was reported after immersing the parent compounds into polar organic solvents including wines [42]. Notably, it has been reported that strain can induce a similar effect in a bulk single crystal [43].

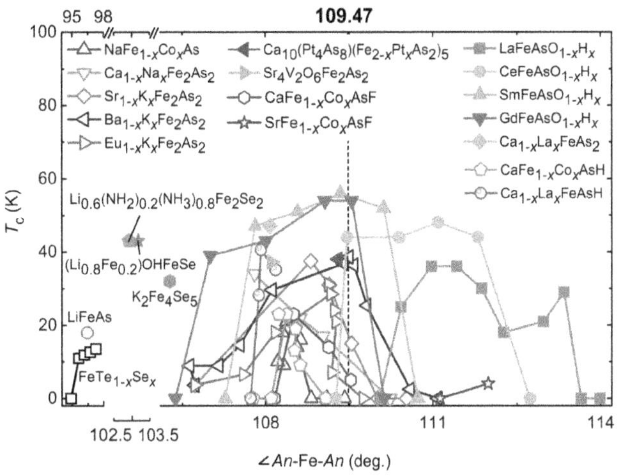

FIGURE C5.10 Correlation between T_c and the bond angle of anion–Fe–anion in various IBSCs. The dotted line denotes the bond angle for a regular tetrahedron.

C5.4 Correlation Between T_c and Local Structure

It is a general trend that the optimal T_c is higher in the order 1111 > 122 > 11. This result implies that the optimal T_c is enhanced by the interlayer spacing of FeAs layers. However, this view is not supported [12]. Instead, there is now a consensus that the T_c of IBSCs is sensitive to the local geometry of the Fe*Pn*(*Ch*)$_4$ tetrahedron. Lee et al. [44] first reported that the optimal T_c is achieved when the bond angle of *Pn*(*Ch*)–Fe–*Pn*(*Ch*) approaches that of a regular tetrahedron (109.5°). Figure C5.10 plots most of the data including the non-optimal T_c values for various types of IBSCs. The phenomenological correlation between T_c and the bond angle becomes worse than that between the optimal T_c and the bond angle; however, the tendency still remains. Exceptions include the 11 system and $LaFeAsO_{1-x}H_x$. This discrepancy stems from the fact that T_c is not determined only by the local structure of Fe*Pn*(*Ch*)$_4$. Kuroki et al. [45] proposed a model in which the pnictogen (chalcogen) height (*h*) from the iron plane is a good structural correlate with the strength of spin fluctuation and that T_c is enhanced by increasing *h*. The correlation between *h* and T_c is comparable to that between T_c and the bond angle around Fe.

C5.5 Perspectives

Table C5.1 summarizes the characteristics of three representative superconductors, IBSCs, MgB_2, and cuprates. Although the T_c of IBSCs is rather low compared with that of cuprates, the IBSCs have distinct advantages in terms of their grain-boundary

TABLE C5.1 Comparison of Three Representative High-T_c Superconductors

	IBSCs	MgB_2	High-T_c Cuprates
Parent materials	AF-semimetal (T_N ~150K)	Pauli paramagnetic metal	AF-Mott insulator (T_N~400K)
Fermi levels	Fe 3d 5-orbitals	B2p 2-orbitals	Cu 3d single orbital
Maximum T_c (K)	56 (1111), 39(122), 47 (11)	40	~93 (YBCO), 110 (Bi2223)
Impurity	Robust	Sensitive	Sensitive
SC gap symmetry	Extended *s*-wave	*s*-wave	*d*-wave
Upper critical field at 0K (T), $H_c^2(0)$	100–200	40	~100
Irreversible Field (T)	> 50 (4 K) > 15 (4 K)	> 25 (4 K) > 10 (20 K)	10 (77 K,YBCO)
Anisotropy	1–2 (for 122 type)	~4.5	5–7 (YBCO), 50–90 (Bi-system)
Crystallographic symmetry in SC state	Tetragonal	Hexagonal	Orthorhombic
Critical GB angle	8–9	No data	~5 (YBCO)
Advantage	High $H_{c2}(0)$, Easy fabrication	Easy fabrication	High T_c and $H_{c2}(0)$
Disadvantage	Toxicity	Low $H_{c2}(0)$	High process cost due to 3D alignment of crystallites

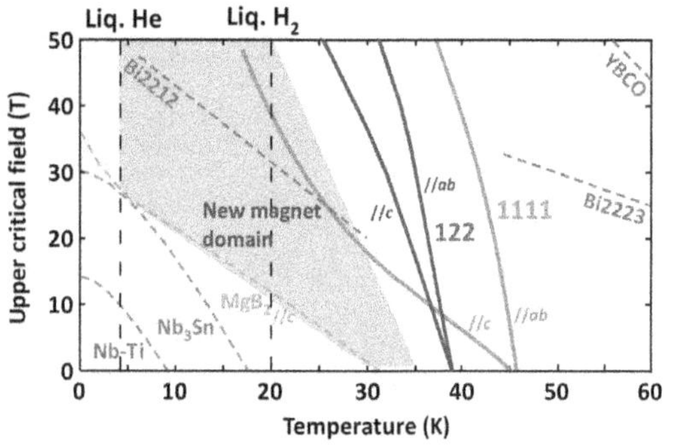

FIGURE C5.11 Comparison of upper critical field H_{c2} for various high-field superconductors, including metallic alloys Nb–Ti, Nb_3Sn, and MgB_2; the IBSCs $NdFeAsO_{1-x}F_x$ (1111) and $Ba_{1-x}K_xFe_2As_2$ (122); and cuprate superconductors YBCO, Bi2212, and Bi2223. For the cuprates, 50% of the in-field resistive transition was regarded as H_{c2}//c. The shaded area highlights the temperature and field range in which the application of IBSCs would be most effective. The vertical broken lines indicate the boiling points of the cryogenic coolants liquid He and H_2.

nature [46], low anisotropy, and high crystallographic symmetry of the superconducting phases. These advantages make it possible to apply the standard processing method for alloy superconductors, the 'powder-in-tube' method, to fabricate wires and tapes [47–49]. Recent success in the fabrication of > 100-m-long wire with a practical J_c by Ma's group in China [50] represents a milestone, casting a bright light for the future. Figure C5.11 shows a summary plot of upper critical field versus temperature for a number of important superconductors. The shaded region shows the range over which IBSCs could be effective for new magnet applications, including high-field devices.

Although research progress in IBSCs has been rapid to date, the time since their discovery has only been 10 years. Before 2008, Fe-based high-T_c superconductors were not contemplated on principle. Extensive research on IBSCs in the past decade has helped to clarify the rich variety of superconducting materials and complex physics arising from the multi-orbital nature of IBSCs [10–14]. The recent discovery of the drastic increase of T_c in extremely thin layer of FeSe [52] suggests that many surprises remain to be unveiled for IBSCs. The discovery of a new superconductor with higher T_c and practical properties appropriate for application is desired. In most cases, high-T_c materials have less desirable properties for application [53]; therefore, metal alloy superconductors are still used for practical application. It is expected that IBSCs with moderate T_c and properties suitable for fabrication would open a new application field for superconductors. The effectiveness of various modes of doping is one of the outstanding characteristics of IBSCs, which results in a rich variety of superconducting materials. We expect that the iron age will become a reality for superconductors with new superconducting materials and improvement of current materials. Iron is still hot!

References

1. Y. Kamihara, H. Hiramatsu, M. Hirano, R. Kawamura, H. Yanagi, T. Kamiya, and H. Hosono, Journal of the American Chemical Society **128**, 10012 (2006).
2. Y. Kamihara, T. Watanabe, M. Hirano, and H. Hosono, Journal of the American Chemical Society **130**, 3296 (2008).
3. S. W. Park, H. Mizoguchi, K. Kodama, S. Shamoto, T. Otomo, S. Matsuishi, T. Kamiya, and H. Hosono, Inorganic Chemistry **52**, 13363 (2013).
4. T. Hanna, S. Matsuishi, K. Kodama, T. Otomo, S. Shamoto, and H. Hosono, Physical Review B **87**, 020401 (2013).
5. H. Yanagi, et al. Physical Review B **77**, 224431 (2008).
6. T. Watanabe, H. Yanagi, T. Kamiya, Y. Kamihara, H. Hiramatsu, M. Hirano, and H. Hosono, Inorganic Chemistry **46**, 7719 (2007).
7. H. Takahashi, K. Igawa, K. Arii, Y. Kamihara, M. Hirano, and H. Hosono, Nature **453**, 376 (2008).
8. Fukuyama H, Takayama-Muromachi E, Terakura K, and Uchida S, eds, "Proceedings of the International Symposium on Fe-Pnictide Superconductors," Journal of the Physical Society of Japan **77** (2008) Suppl. C.
9. P. C. W. Chu, A. Koshelev, W. Kwok, I. Mazin, U. Welp, and H. H. Wen, Physica C **469**, 313 (2009).
10. D. C. Johnston, Advances in Physics **59**, 803 (2010).
11. J. H. Durrell, C. B. Eom, A. Gurevich, E. E. Hellstrom, C. Tarantini, A. Yamamoto, and D. C. Larbalestier, Reports on Progress in Physics **74**, 124511 (2011).
12. H. Hosono and K. Kuroki, Physica C **514**, 399 (2015).
13. N. L. Wang, H. Hosono, and P. Dai, "Iron-Based Superconductors" (Pan Stanford Publishing) (2013).
14. P. D. Johnson, G. Xu, and W.G. Yin, "Iron-Based Superconductivity" (Springer) (2015).
15. X. H. Chen, T. Wu, G. Wu, R. H. Liu, H. Chen, and D. F. Fang, Nature **453**, 761 (2008).
16. Z. A. Ren, et al., Chinese Physics Letters **25**, 2215 (2008).
17. T. Hanna, Y. Muraba, S. Matsuishi, N. Igawa, K. Kodama, S. Shamoto, and H. Hosono, Physical Review B **84**, 024521 (2011).
18. H. Hosono and S. Matsuishi, Current Opinion in Solid State and Materials Science **17**, 49 (2013).
19. S. Matsuishi, T. Hanna, Y. Muraba, S. W. Kim, J. E. Kim, M. Takata, S. I. Shamoto, R. I. Smith, and H. Hosono, Physical Review B **85**, 014514 (2012).
20. S. Iimura, S. Matsuishi, H. Sato, T. Hanna, Y. Muraba, S. W. Kim, J. E. Kim, M. Takata, and H. Hosono, Nature Communications **3**, 943 (2012).
21. S. Matsuishi, T. Maruyama, S. Iimura, and H. Hosono, Physical Review B **89**, 094510 (2014).
22. H. Takahashi, H. Soeda, M. Nukii, C. Kawashima, T. Nakanishi, S. Iimura, Y. Muraba, S. Matsuishi, and H. Hosono, Scientific Reports **5**, 7829 (2015)

23. S. Iimura, H. Okanishi, S. Matsuishi, H. Hiraka, T. Honda, K. Ikeda, T. C. Hansen, T. Otomo, and H. Hosono, Proceedings of the National Academy of Sciences **114**, E4354 (2017).
24. M. Rotter, M. Tegel, and D. Johrendt, Physical Review Letters **101**, 107006 (2008).
25. Z. A. Ren, et al., Europhysics Letters **83**, 17002 (2008).
26. H. Kito, H. Eisaki, and A. Iyo, Journal of the Physical Society of Japan **77**, 063707 (2008).
27. Y. Muraba, S. Iimura, S. Matsuishi, and H. Hosono, Inorganic Chemistry **54**, 11567 (2015).
28. J. Bang, S. Matsuishi, and H. Hosono, Applied Physics Letters **110**, 232105 (2017).
29. C. Hess, A. Kondrat, A. Narduzzo, J. E. Hamann-Borrero, R. Klingeler, J. Werner, G. Behr, and B. Büchner, Europhysics Letters **87**, 17005 (2009).
30. C. H. Lee, K. Kihou, A. Iyo, H. Kito, P. M. Shirage, and H. Eisaki, Solid State Communications **152**, 644 (2012).
31. A. S. Sefat, R. Y. Jin, M. A. McGuire, B. C. Sales, D. J. Singh, and D. Mandrus, Physical Review Letters **101**, 117004 (2008).
32. S. Jiang, H. Xing, G. F. Xuan, C. Wang, Z. Ren, C. M. Feng, J. H. Dai, Z. A. Xu, and G. H. Cao, Journal of Physics: Condensed Matter **21**, 382203 (2009).
33. Y. Imai, Y. Sawada, F. Nabeshima, and A. Maeda, Proceedings of the National Academy of Sciences **112**, 1937 (2015).
34. M. Burrard-Lucas, et al., Nature Materials **12**, 15 (2013).
35. J. G. Guo, H. C. Lei, F. Hayashi, and H. Hosono, Nature communications **5**, 4756 (2014).
36. H. H. Wen, Reports on Progress in Physics **75**, 112501 (2012).
37. Y.Liu, G Wang, T. Ying, X. Lai, S. Jin, N. Liu, J. Hu, and X. Chen, Advanced Science **3**, 1600098 (2016).
38. K. Hanzawa, H. Sato, H. Hiramatsu, T. Kamiya, and H. Hosono, Proceedings of the National Academy of Sciences of the United States of America **113**, 3986 (2016)
39. H. Hiramatsu, T. Katase, T. Kamiya, M. Hirano, and H. Hosono, Physical Review B **80**, 052501 (2009).
40. T. Kamiya, H. Hiramatsu, T. Katase, M. Hirano, and H. Hosono, Materials Science and Engineering B: Advanced Functional Solid-State Materials **173**, 244 (2010).
41. T. Katase, H. Hiramatsu, H. Yanagi, T. Kamiya, M. Hirano, and H. Hosono, Solid State Communications **149**, 2121 (2009).
42. K. Deguchi, Y. Mizuguchi, Y. Kawasaki, T. Ozaki, S. Tsuda, T. Yamaguchi, and Y. Takano, Superconductor Science and Technology **24**, 055008 (2011).
43. S. R. Saha, N. P. Butch, K. Kirshenbaum, J. Paglione, and P. Y. Zavalij, Physical Review Letters **103**, 037005 (2009).
44. C. H. Lee, et al., Journal of the Physical Society of Japan **77**, 083704 (2008).
45. K. Kuroki, H. Usui, S. Onari, R. Arita, and H. Aoki, Physical Review B **79**, 224511 (2009).
46. T. Katase, Y. Ishimaru, A. Tsukamoto, H. Hiramatsu, T. Kamiya, K. Tanabe, and H. Hosono, Nature Communications **2**, 409 (2011).
47. M. Putti, et al., Superconductor Science and Technology **23**, 034003 (2010).
48. J. D. Weiss, C. Tarantini, J. Jiang, F. Kametani, A. A. Polyanskii, D. C. Larbalestier, and E. E. Hellstrom, Nature Materials **11**, 682 (2012).
49. J. D. Weiss, J. Jiang, A. A. Polyanskii, and E. E. Hellstrom, Superconductor Science and Technology **26**, 074003 (2013).
50. X. P. Zhang, H. Oguro, C. Yao, C. H. Dong, Z. T. Xu, D. L. Wang, S. Awaji, K. Watanabe, and Y. W. Ma, IEEE Transactions on Applied Superconductivity **27**, 7300705 (2017).
51. J. D. Weiss, A. Yamamoto, A. A. Polyanskii, R. B. Richardson, D. C. Larbalestier, and E. E. Hellstrom, Superconductor Science and Technology **28**, 112001 (2015).
52. Z. Wang, C. Liu, Y. Liu, and J. Wang, Journal of Physics: Condensed Matter **29**, 153001 (2017).
53. D. Larbalestier, A. Gurevich, D. M. Feldmann, and A. Polyansk, Nature 414, 368 (2001).

C6

Hydrides

Jeffery L. Tallon

C6.1 Introduction

The quest for room temperature superconductivity has obvious practical impetus, but the conceptual rationale is also a powerful motivation in itself. Superconductivity brings quantum mechanics from the atomic scale to the everyday length scale of magnetic resonance imaging (MRI) scanners and power transformers. But to achieve superconductivity under ambient temperatures would accomplish the significant additional triumph of translating quantum mechanics to everyday temperatures. Considerable effort has been expended in that quest especially over the past three decades with the discovery of cuprate, then pnictide, high-temperature superconductors (HTS). But the impetus for this quest for room temperature superconductivity is surely to be found originally in the work of Ashcroft half a century ago [1], when he suggested that metallic hydrogen could be a high-temperature superconductor. Using his numbers, T_c values in excess of 240 K might be anticipated.

The idea that compressed hydrogen might form a metal goes back even earlier. In 1935, Wigner and Huntington [2] proposed that at high pressure, molecular hydrogen would undergo a first-order dissociation into an atomic lattice which is metallic. This was predicted to occur around 25 GPa at a density of about 1.3 g/cm^3 – 18 times that of liquid H_2 at ambient pressure. The density is about correct, but the required pressure assumed a compressibility that remains constant with increasing pressure. Quantum Monte-Carlo calculations suggest the requisite pressure is more like 400–500 GPa. Consistent with this, early in 2017, the discovery of metallic hydrogen was reported at 495 GPa [3] though this has subsequently been questioned [4]. Such high pressures do exist within the cores of the giant planets Jupiter (7000 GPa), Saturn (600 GPa), Uranus (500 GPa), and Neptune (650 GPa), so it is possible that these cores comprise metallic, and even superconducting, hydrogen. Persistent supercurrents might be the origin of the high magnetic field observed around Jupiter [1].

In 2004, Ashcroft additionally suggested that hydrogen-rich compounds can metallize and superconduct at much lower pressures than hydrogen due to "chemical compression" [5]. The systems he considered were SiH_4, SnH_4, GeH_4, ScH_3, YH_3, GaH_3 and, most significantly for subsequent developments, SH_2. Later Li *et al.* [6] carried out *ab initio* density functional theory and electron–phonon coupling calculations suggesting that H_2S, compressed to 160 GPa, would superconduct at 80 K. The culmination of all these predictions was the spectacular discovery of superconductivity in H_2S compressed to a pressure of 155 GPa in a diamond anvil cell. The observed onset of diamagnetism sets in as high as 203 K [7]. This was to be followed by the prediction [8, 9] and independent observation, by two groups [10, 11], of superconductivity setting in at 250–260 K at pressures around 170–200 GPa, the current highest observed superconducting transition temperature.

In this chapter, the structural and superconducting properties of compressed H_2S are described along with proposals for alternative hydrogen-dominant compounds, which calculations suggest will also display high-temperature superconductivity. Several reviews exist [12, 13] for the various high-pressure techniques that are used, especially the diamond anvil technique which is notably pertinent to the present topic and really is the only available apparatus to achieve the necessary pressures. The extreme pressures discussed in the following hardly represent ambient conditions, but nonetheless room temperature superconductivity now seems within reach and, if achieved, the focus will shift to the question as to how to stabilize such high-pressure phases at low or ambient pressures.

C6.2 Structural Properties of Compressed Sulphur Hydride

C6.2.1 Crystal Structure

Under sufficiently high-pressure, H_2S dissociates into multiple phases of composition H_3S [$3(H_2S) \rightarrow 2H_3S + S$] and the two highest pressure phases are the active superconducting ones. Quantum density functional theory calculations [14] suggest

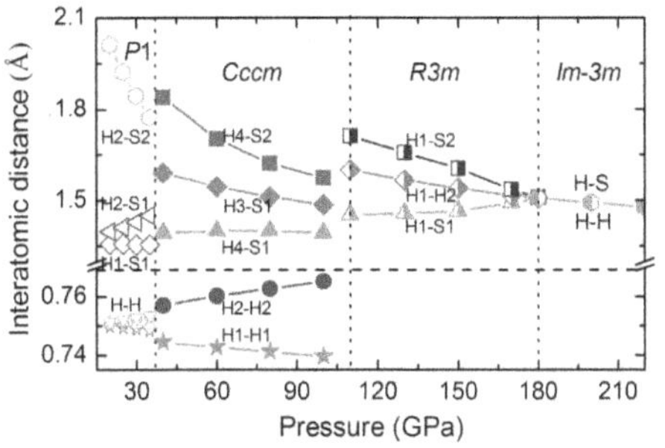

FIGURE C6.1 Calculated interatomic distances and symmetries of the four high-pressure phases of H_3S: triclinic *P*1, orthorhombic *Cccm*, rhombohedral *R*3*m* and bcc *Im*$\bar{3}$*m*. (From ref. [14].)

that there are four relevant high-pressure symmetry phases: triclinic *P*1 (< 37 GPa), orthorhombic *Cccm* (37–110 GPa), rhombohedral *R*3*m* (110–170 GPa) and body-centred cubic *Im*$\bar{3}$*m* (> 170 GPa) (see Figure C6.1).

This picture is confirmed experimentally by synchrotron x-ray scattering. Einaga *et al.* [15] mapped out the phase boundaries and transition temperatures for both H_3S and D_3S, the deuterium analogue. These are reproduced in Figure C6.2 which reveals two important features. Firstly, T_c is reduced for the deuteride consistent with the expected isotope effect arising from the heavier ion, and secondly, the maximum T_c in both cases seems to lie at the boundary between the *R*3*m* and *Im*$\bar{3}$*m* phases. This parallels an apparently common organizing principle, which has emerged in novel superconductivity in the past decade, that of the occurrence and optimizing of superconductivity at a first- or second-order phase boundary between competing phases. Examples include (i) heavy fermion systems, which exhibit a dome of superconductivity around a second-order transition from an antiferromagnetic to paramagnetic state [16], (ii) similarly for iron pnictides [17], (iii) the cuprates exhibit a dome-shaped domain of superconductivity around the closure of the pseudogap [18, 19], (iv) a dome of superconductivity occurring at the ground-state termination of an ordered charge-density-wave state, as in Cu_xTlSe_2 [20] and (v) a dome of superconductivity around a ground-state solid–solid structural phase transition as seen in $(Ca_xSr_{1-x})_3Rh_4Sn_{13}$ [21] and now apparently in the case of compressed H_3S and D_3S. The first examples are likely to be associated with quantum critical fluctuations around a $T = 0$ quantum critical point [22], while the solid–solid transitions involving symmetry change are first-order transitions and may not involve quantum critical fluctuations.

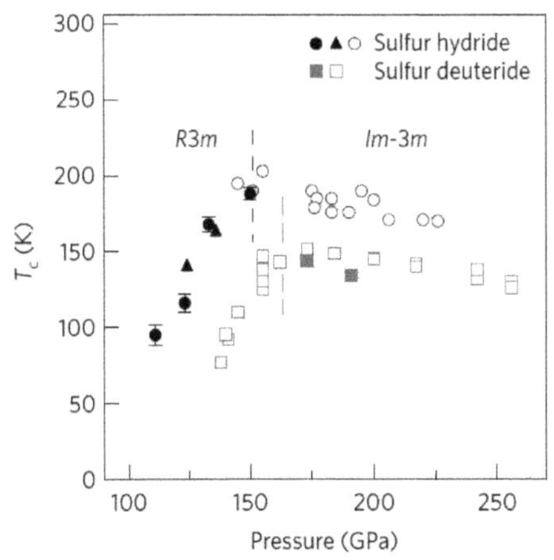

FIGURE C6.2 Pressure dependence of critical temperature, T_c, for H_3S (circles and triangles) and D_3S (squares) from ref. [15]. Circles are for decreasing pressure, squares and triangles for increasing pressure. Broken lines (shorter dashes for H_3S and longer dashes for D_3S) indicate the phase boundary between the *R*3*m* and *Im*$\bar{3}$*m* structural phases.

A further key conclusion to draw from Figure C6.2 is that stabilizing the structure of either the hexagonal *R*3*m* phase or the bcc *Im*$\bar{3}$*m* phase at higher, or lower pressures, respectively, might possibly lead to further increases in transition temperatures. Of particular note is the prospect that stabilizing the bcc phase to lower pressure should enable comparably high T_c values at more accessible pressures. That strategy remains an important challenge to the community.

C6.3 Superconducting Properties of Compressed Sulphur Hydride

C6.3.1 Transition Temperature and Isotope Effect

Figure C6.3(a) shows the resistive transitions as measured in the diamond anvil cell for compressed H_3S and D_3S [7]. As noted, the downward shift in T_c is consistent with the simple nearly weak-coupling BCS picture of phonon-induced pairing, where the pairing boson energy scale is governed by hydrogen-dominated phonon modes. As stated, the pressure dependence of T_c is mapped out in Figure C6.2 for the hexagonal and bcc phases of both isotopes. Note that residual elemental S in these samples arising from the dissociation, $3(H_2S) \rightarrow 2H_3S + S$, should contribute to an additional rise in superfluid density at lower temperatures as this too superconducts under high pressure [23].

C6.3.2 Upper and Lower Critical Fields

Panels (b) and (c) show *M*–*H* magnetization curves at various temperatures and these confirm persistence of the diamagnetic state up to around 200 K. Panel (d) shows the field at the extrema in the magnetization curves plotted versus temperature indicating that diamagnetism vanishes abruptly at $T_c = 203.5$ K. Finally, panel (e) shows the temperature dependence of the upper critical field, H_{c2}, as measured from the in-field resistive transition. The upper data is for H_3S and the lower data for D_3S. The skirting curves indicate

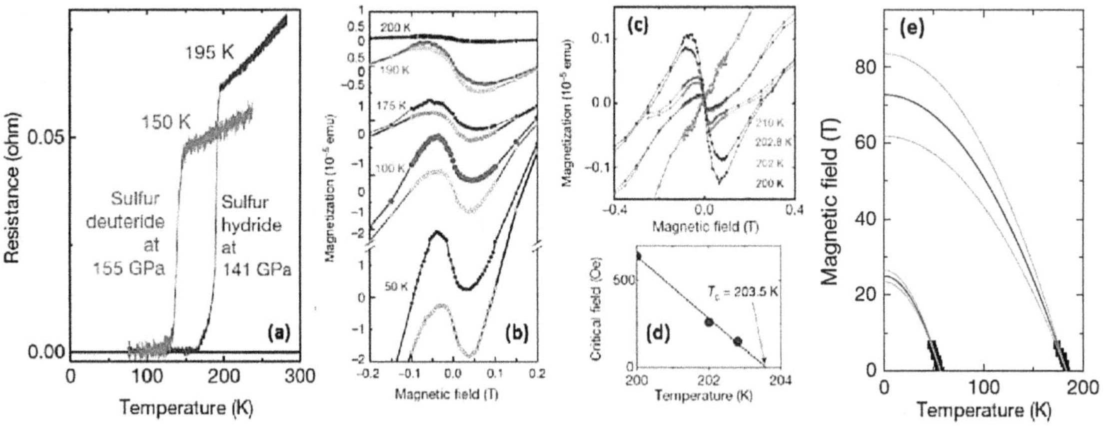

FIGURE C6.3 Superconducting properties of compressed H_3S and D_3S: (a) resistive transitions at 155 Gpa and 141 GPa, respectively, showing an isotope effect consistent with the BCS model of phonon-induced pairing. (b) Static *M–H* magnetization curves for H_3S at various temperatures. (c) Magnetization curves close to T_c. (d) Extrapolation to zero of the extremum magnetization field shown in (c) gives $T_c = 203.5$ K. (e) Shows the upper critical field for H_3O (upper curves) and D_3O (lower curves) obtained from resistive measurements. The two bounding curve-sets show the estimated error bars in extrapolating back to $T = 0$. (From ref. [7].)

the estimated error bars in extrapolating back to the ground state using the usual quadratic expression $H_{c2}(T) = H_{c2}(0)[1 - (T/T_c)^2]$. This gives $B_{c2}(0) \approx 73$ T with a consequent coherence length of $\xi(0) = 2.1$ nm. However, if we extrapolate back these $B_{c2}(T)$ values using the Werthamer–Helfand–Hohenberg (WHH) formula [24], we obtain $B_{c2}(0) \approx 100 \pm 16$ T and $\xi(0) = 1.84 \pm 0.15$ nm [25].

Drozdov *et al.* [7] estimated the lower critical field, B_{c1}, as 30 mT from which they deduce a London penetration depth of $\lambda_L = 125$ nm. However, their formula for B_{c1} inadvertently contained an extra factor of $\sqrt{2}$ so we calculate this afresh. Using $B_{c1} = [\phi_0/(4\pi\mu_0\lambda_L^2)](\ln\kappa + ½)$, where $\kappa = \lambda_L/\xi$ is the Ginzburg–Landau parameter, then, combined with the WHH estimate of $\xi(0) = 1.84 \pm 0.15$ nm, we obtain $\lambda_L = 163$ nm and $\kappa = 89$. It is notable that all these values are not too different from those obtained for the cuprate HTS. As a consequence, the very high T_c value means that H_3S sits very high on the Uemura plot of T_c versus λ_L^{-2}, well above the "universal line" for the cuprates. That is, T_c is very high for its relatively low superfluid density. As superfluid density plays a key role in determining the size of fluctuations, this leads us to the very important question as to the effect of fluctuations on the value of T_c.

Table C6.1 summarizes the deduced superconducting and thermodynamic parameters for compressed H_3S.

C6.3.3 Fluctuations, Superfluid Density and Mean-Field T_c

The obvious question arising from this remarkable discovery is whether yet higher T_c values might be attainable either in this system or in similar hydrogen-rich compounds. A key constraint is the potential role of thermal fluctuations which tend to break Cooper pairs. What are the limits on T_c imposed by such fluctuations and have they already come into play in H_3S? This is a very real question for HTS. In the cuprates, fluctuations near T_c are very strong, impacting primarily on pinning and critical currents and also causing a significant reduction in T_c below the mean-field value, T_c^{mf}, and resulting in an associated residual pairing gap which persists well above T_c [26, 27]. This pairing gap is distinct from the pseudogap [26–28]. These effects are significant, persisting to $1.3T_c$ for $YBa_2Cu_3O_{7-\delta}$ (Y-123) [28] and up to $1.5T_c$ for $Bi_2Sr_2CaCu_2O_{8+\delta}$ [28, 29] (though some suggest a rather smaller effect [30]). If fluctuations were to be comparable in H_3S, then already T_c^{mf} might possibly be as high as 300 K, so the question becomes central to the search for room temperature superconductors.

Writing the order parameter as $\psi = \psi_0.e^{i\varphi}$, we may consider fluctuations in amplitude, ψ_0, or in the phase, φ. The former is considered by Bulaevskii *et al.* [31], who show that the relevant temperature scale for the onset of amplitude fluctuations is given by

$$k_B T_{amp} = \Delta F.\Omega_0 = \Delta F(4\pi/3)\xi_0^3 = \frac{\phi_0^2\xi}{12\pi\mu_0\lambda^2} \quad \text{(C6.1)}$$

where ϕ_0 is the flux quantum, and μ_0 is the permeability of free space. On the other hand, Emery and Kivelsen [32] have discussed the onset of phase fluctuations in terms of a temperature scale

$$k_B T_\varphi = \frac{A\phi_0^2 a}{4\pi^2\mu_0\lambda^2} = \frac{3a}{\sqrt{\pi}} k_B T_{amp} \quad \text{(C6.2)}$$

Here A is a constant of the order of unity which reflects the dimensionality (here 3D). For 3D, we use their value of $A = 2.2$

TABLE C6.1 Parameter Values for Superconducting H_3S at 155 GPa

Parameter	Value	Ref.	Notes
T_c (K)	203	[7]	Extrapolating $H_{c1}(T)$ to zero
T_{amp} (K)	326 ± 26	[this work]	
T_φ (K)	1216 ± 97	[this work]	
Penetration depth (nm):			
$\lambda_L(0)$	163	[7]	Using corrected formula for H_{c1}
$\lambda_L(0)$	191 ± 2	[25]	From zero-field J_c
Coherence length (nm):			
$\xi(0)$	2.1	[7]	Quadratic extrapolation of $H_{c2}(T)$
$\xi(0)$	1.84 ± 0.15	[25]	Using WHH extrapolation
GL parameter:			
κ	78	[7]	Using above parameters, $\kappa = \lambda/\xi$
κ	104	[25]	Using above parameters, $\kappa = \lambda/\xi$
Lower critical field (mT):			
$B_{c1}(0)$	30	[7]	(Magnetization)
Upper critical field (T):			
$B_{c2}(0)$	73	[7]	Quadratic extrapolation of $H_{c2}(T)$
$B_{c2}(0)$	100 ± 16	[25]	Using WHH extrapolation
Thermo. crit. field (T):			
B_c	1.08	[25]	(From B_{c1} and B_{c2})
Molar volume (m^3/mole):			
V_M	8.8×10^{-6}	[15]	
Condensation energy:			
$U(0)$ (J/mole)	2.25 ± 0.72	[25]	
Normal-state γ:			
γ_n (mJ/mole/K^2)	0.3 ± 0.074	[25]	
$\Delta\gamma_n$ (mJ/mole/K^2)	0.427 ± 0.11	[25]	
$N(E_F)$			
(States/eV/fu)	0.063 ±0.015	[25]	

as applicable to the 3D-XY model while for 2D $A \approx 0.9$. For 3D, the constant $a = \sqrt{\pi}\xi_0$, while, for 2D, a is taken to be the mean spacing between superconducting sheets. It is easy to see from Equation (C6.2) that phase fluctuations are less important in 3D than in 2D, where it happens that $T_\varphi \approx T_{amp}$ [28].

Taking the above length scales $\xi_0 \approx 1.84$ nm and $\lambda_L = 163$ nm, we find $T_{amp} \approx 453$ K and $T_\varphi \approx 1687$ K. This illustrates the irrelevance of phase fluctuations for this system, and even amplitude fluctuations for a system with $T_c = 203$ K seem likely to be of minimal consequence. However, the estimate [7] of λ_L from the magnetization critical currents derived from Figure C6.3(b) and (c) gives a large value of $\lambda_L = 191 \pm 2$ nm resulting in $T_{amp} \approx 326 \pm 26$ K and $T_\varphi \approx 1216 \pm 97$ K. Now T_{amp} is getting uncomfortably close to T_c, and it is clear that if these numbers are substantiated, then amplitude fluctuations will already play a role in depressing T_c below its mean-field value, perhaps by about 10 K. That role could be significantly increased in any hydride with higher T_c. As T_{amp} in Equation (C6.1) scales as $\xi \sim T_c^{-1}$, then T_{amp}/T_c scales as T_c^{-2}, and it can be seen that fluctuations could quickly become dominant with any further increase in T_c^{mf}. Whether this ultimately defeats the quest for room temperature superconductivity remains to be seen.

C6.4 Superconductivity in Other Hydrides

While few other hydrides in which superconductivity exists have been experimentally studied, there are an increasing number of theoretical studies on other hydrides which indicate the likely potential for high-T_c superconductivity. These will now be discussed.

C6.4.1 Palladium Hydride

Superconductivity under ambient pressure was discovered long ago in the β-phase of PdH_x for compositions $0.7 < x < 1.0$, with T_c lying in the range 8–10 K when $x \approx 1.0$ [33]. The hydrogen occupies octahedral interstices in the fcc Pd lattice. PdD_x exhibits an inverse isotope effect with T_c rising to 10–12 K when $x \approx 1.0$ [34]. Quite recently, superconductivity was reported in the same systems with $T_c \approx 56$ K in PdH_x and 61 K in PdD_x with, in both cases, $x \approx 1.0$ [35]. The dramatic rise in T_c arises from rapid cooling of the samples from room temperature, and the authors suggest it is due to the freezing in of tetrahedral site occupancy in addition to the usual octahedral occupancy. The superconductivity was resistively measured and exhibits very sharp transitions to a zero-resistance state. Though this work has not yet been duplicated by other groups, the sharpness of the transitions, their reproducibility and the preservation of the inverse isotope effect all seem to validate this discovery.

C6.4.2 Thorium Hydrides

Thorium is a metal with fcc structure and space group $Fm\bar{3}m$. It is a type I superconductor with $T_c = 1.37$ K [36, 37] and critical field $H_c = 15.9$ mT [36, 38]. Thorium forms hydrides ThH_2 and Th_4H_{15}, the latter being the first observed hydride superconductor with metallic conductivity at ambient pressure and a superconducting $T_c = 7.5$–8 K [39, 40]. Thorium hydrides are chemically highly reactive and spontaneously ignite upon exposure to air [41]. Theoretical studies of their electronic structure and crystal stability have been reported [42, 43].

Recently, other high-pressure compositions and phases of thorium hydride have been studied by evolutionary density functional theory [44]. Many new thorium hydrides were predicted to be stable under pressure, including ThH_3, Th_3H_{10}, ThH_4, ThH_6, ThH_7 and ThH_{10}, and their superconducting properties were estimated by first calculating phonon frequencies and electron–phonon coupling coefficients then applying the Allen–Dynes or McMillan formulae for T_c. Of these, the highest calculated T_c was found to be 194 K at 100 GPa for the $Fm\bar{3}m$ structure of ThH_{10}. This phase is stable over a broad range of pressures, but T_c is found to fall with increasing pressure. As for H_3S, there remains the interesting question as to whether this phase can be stabilized to significantly lower pressures. ThH_7 is found also to exhibit high-temperature superconductivity with $T_c \approx 65$ K [44].

C6.4.3 Uranium Hydrides

The same group has also investigated high-temperature superconductivity in uranium hydrides at near-ambient conditions using the same *ab initio* techniques [45]. These authors find nine new hydrides, including U_2H_{13}, UH_7, UH_8, U_2H_{17} and UH_9, all predicted to be high-temperature superconductors. UH_8, for example, is predicted to be stable at pressures above 52 GPa (with predicted $T_c = 156$ K) and remains metastable upon decompression to ambient pressure (where $T_c = 193$ K). This confirms the approach suggested above of seeking metastable structures which enable retention of high-temperature superconductivity at lower or ambient pressure.

C6.4.4 Calcium Hydride Clathrates

The stable hydride of calcium at ambient pressure is CaH_2 with *Pnma* space group and unit-cell parameters $a = 0.596$, $b = 0.360$ and $c = 0.682$ nm [46]. It is a transparent insulator. Two hydrogen sites exist in tetrahedral and square pyramid cavities with the former Ca–H bonds being markedly compressed and the latter significantly stretched. As a consequence, it is not surprising that under pressure a variety of other stable hydride structures appear. Wang *et al.* [47] have investigated the systems Ca + nH and CaH_2 + nH with $2 \le n \le 12$ under pressures 20 to 200 GPa using *ab initio* particle swarm optimization techniques. Stable structures exist for even n but not for odd n. Band structures show that these even-n high-pressure phases are metallic with very strong electron–phonon coupling. Eliashberg calculations indicate that these are expected to be high-temperature superconductors. For CaH_6, in which the hydrogens form a clathrate-like cage around each Ca, T_c values up to 235 K at 150 GPa were deduced with a negative pressure coefficient $dT_c/dP \approx -0.33$ K/GPa. Again, this suggests an even higher T_c at ambient pressure ($\approx$ 285 K) if the structures were to be somehow metastabilized. In all of these estimates, we recall that the effect of fluctuations is ignored, and such extremely high T_c values may well be truncated, possibly well below the mean-field value.

C6.4.5 Group IV Hydrides – Silane

The group IV hydrides have been experimentally investigated for superconductivity [48–58]. Eremets *et al.* [48] reported the occurrence of superconductivity in compressed silane SiH_4 at 96 and 120 GPa with $T_c = 17$ K. They report a metallic/superconducting phase with hexagonal close-packed structure. This result was questioned [49] with the observation that SiH_4 decomposes at pressures above 50 GPa into Si and H_2 and that the observed metallic/superconducting phase may be platinum hydride formed from the reaction of the Pt gasket with the residual H_2 released from this decomposition. A subsequent study by the same group showed that the stable SiH_4 phase at 124 GPa, formed at 300 K, resulted in an insulating $I4_1/a$ structure [50]. An independent, extensive study [51] showed that silane remains in its transparent, insulating phase until at least 150 GPa, and these authors suggest that the previously reported superconductivity [48] may arise from H-doped Si, the latter arising from decomposition of SiH_4.

Despite these conflicting reports, Feng *et al.* [52] have calculated the stability and electronic properties of compressed silane and found six phases which metallize and project to rather high T_c values. All of these metallize when the radius of the free electron sphere, r_s, falls close to the Goldhammer–Herzfeld criterion for cubic systems: $r_s \approx 0.88$ Å, corresponding to $P \ge 91$ GPa. The three most stable phases above 90 GPa are, in order of stability, O3 comprising corner-shared SiH_6 octahedra forming a quasi-2D structure; T3 – an fcc packing of tetrahedrally H-coordinated Si atoms; and O1 comprising 1D chains of edge-sharing SiH_6 octahedra. Estimates of T_c are high because of (i) the high dynamical scale of the protons, (ii) the rapid stiffening of phonon modes with pressure and (iii) the high electronic density of states. The authors find $T_c \approx 166$ K for the energetically most favoured O3 phase. However, we note that fluctuation effects will be more prominent for this quasi-2D system, and it may turn out that the T3 cubic phase might exhibit the highest T_c due to suppressed fluctuations in 3D. Other estimates are somewhat more conservative [53].

The structure and electronic properties of $SiH_4(H_2)_2$ in the range 50–300 GPa were theoretically investigated by Li *et al.* [54]. A layered *Ccca* orthorhombic structure was found to be energetically stable above 248 GPa. This is metallic and comprises hydrogen-shared layers of SiH_8 dodecahedra intercalated by ordered molecular H_2. Using the Allen–Dynes equation, they obtained superconducting transition temperatures of 98–107 K at 250 GPa. Similar T_c values were found by the same group for high-pressure phases of disilane $(SiH_4)_2$ [55].

Other group IV hydrides have been studied. Tse *et al.* [56] found a high-pressure phase in SnH_4, which is stable between 70 and 160 GPa comprising *P6/mmm* layered structure with intercalated H_2 molecules with T_c around 80 K at 120 GPa, while Gao *et al.* [57] studied the high-pressure phases of germane, GeH_4. They found a stable metallic monoclinic structure of *C2/c* symmetry comprising layered motifs containing H_2 units. Perturbative linear-response calculations for this phase at 220 GPa show a large electron–phonon coupling parameter $\lambda \approx 1.12$ and a resultant T_c reaching 64 K. More recently, Esfahani *et al.* [58] find a *C2/m* phase of germane which is stable at pressures above 278 GPa with calculated $T_c \approx$ 67 K at 280 GPa. This then dissociates above 300 GPa to Ge_3H_{11} with space group symmetry $I\bar{4}m2$ and $T_c \approx 43$ K.

C6.4.6 Group III Hydrides

Kim *et al.* [59] have carried out *ab initio* studies on the three trihydrides ScH_3, YH_3 and LaH_3 as a function of pressure. In each case, the fcc phase is stabilized around 20 GPa with T_c values 18 K (ScH_3), 40 K (YH_3) and 20 K (LaH_3) which fall with increasing pressure and the ground state becomes a normal metal by about 35 GPa. Due to the lower proton density, these values are substantially less than found in the tetrahydrides discussed in Section C6.4.5.

Aluminium trihydride, AlH_3, has been investigated experimentally by the Eremets group [60], who discovered in this system a covalent high-pressure metallic phase. The stable ambient pressure α-phase with symmetry $R\bar{3}c$ remains stable to 64 GPa, then transitions to a monoclinic semiconductor, then at 100 GPa becomes cubic bcc with symmetry $Pm\bar{3}n$ in which are found the shortest H–H bonds in any compound except the H_2 molecule itself. This last phase is metallic but with no sign of superconductivity down to 4 K, even at pressures up to 164 GPa. Density functional theory and electron–phonon calculations suggest this phase should be superconducting with $T_c \approx$ 24 K at 110 GPa falling to 6 K at 164 GPa [60]. The absence of observed superconductivity is a salutary reminder that its prediction is not necessarily straightforward. The authors point out that the strong phonon coupling is assisted by nesting of the Fermi surface which can lead to competing correlations.

C6.4.7 Clathrate Rare-Earth Hydrides

In order to exploit the high Debye frequency associated with H-dominated vibrations and the high H-derived electronic density of states at the Fermi level, the obvious strategy is to maximize the H content while avoiding the formation of H_2-like molecular units which tend to appear at high H content. The approach promoted by Peng *et al.* [8] is to explore the formation of clathrate H-cages which are stabilized by electron transfer from an appropriate cation. Focusing on 3+ and 4+ cations derived from the lanthanide rare-earth elements, they found many stable clathrate structures at high pressure comprising H_{24}, H_{29} and H_{32} cages obtained from compositions REH_6, REH_9 and REH_{10}, respectively, as illustrated in Figure C6.4.

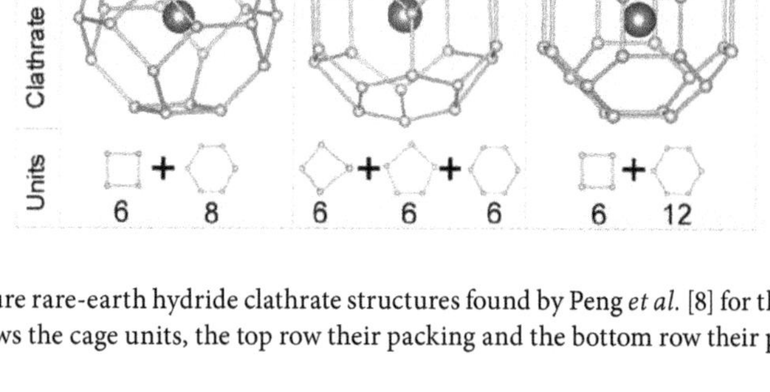

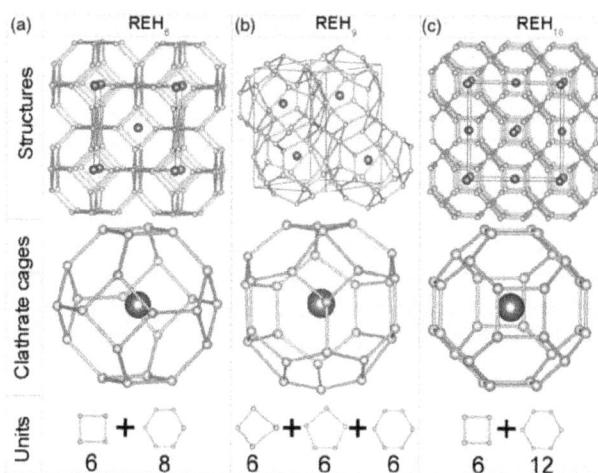

FIGURE C6.4 Stable high-pressure rare-earth hydride clathrate structures found by Peng *et al.* [8] for the compositions (a) REH_6, (b) REH_9 and (c) REH_{10}. The middle row shows the cage units, the top row their packing and the bottom row their polygonal subunits. (From [8].)

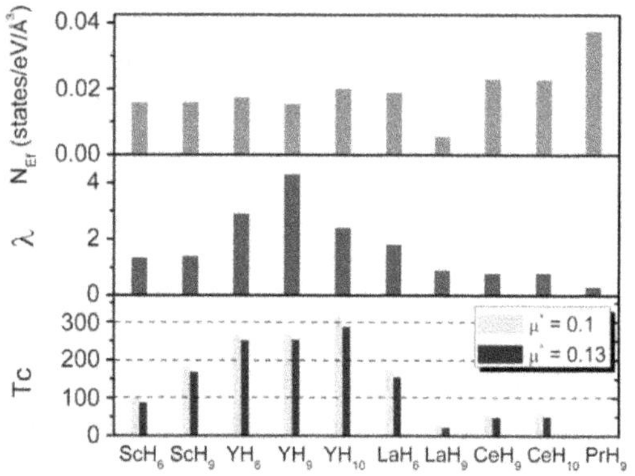

FIGURE C6.5 Calculated electronic densities of states, electron–phonon coupling parameter, λ, and T_c values found by Peng *et al.* [8] for the various rare-earth hydrides REH_n indicated. The light and dark-shaded T_c bars are for two different typical values of the Coulomb pseudopotential parameter, $\mu^* = 0.1$ and 0.13. (From [8].)

Electronic densities of states, electron–phonon coupling parameters, λ, and T_c values obtained by Eliashberg calculations (necessitated by the large λ values) are summarized in Figure C6.5. The light and dark shaded bars are for two different typical values of the Coulomb pseudopotential parameter, $\mu^* = 0.1$ and 0.13. Most notably, T_c for YH_{10} exceeds 300 K at 400 GPa, while for YH_9 at 150 GPa $T_c = 276$ K. LaH_{10} (not shown) yields $T_c = 288$ K at 200 GPa for $\mu^* = 0.1$. Independent calculations by Liu *et al.* give remarkably similar results with $T_c = 286$ K at 210 GPa [9]. These extraordinarily high calculated values certainly appear to validate the strategy of stabilizing H-rich clathrate cages using electron transfer from high valence encaged cations.

These predictions for LaH_{10} have subsequently been spectacularly confirmed in experiments by two independent groups. One, at George Washington University, reported clear onset of superconductivity at $T_c \approx 260$ K at a pressure of 180 – 200 GPa [10], while the Eremets group reported a T_c of 250 K at 170 GPa with a very high upper critical field of $H_{c2} \approx 120$ T [11]. Both revealed a conventional BCS-like isotope effect when hydrogen is replaced by deuterium. While their claims are presented cautiously, the remarkable correspondence with calculation combined with the extremely similar reports from two independent groups would appear to firmly validate these revolutionary results. An update by the former group reports signs of superconductivity up to 280 K [61]. This outstanding agreement between theoretical calculation and experiment must surely give considerable confidence in the modern methods used to compute these high-pressure phases and their superconducting properties. As this Handbook goes to press another group reported very clear evidence of superconductivity in a ternary carbon sulphur hydride system as high as 288 K at pressures around 270 GPa [62]. The active phase is yet to be determined, but it is clear that a number of groups now are reporting what is essentially room-temperature superconductivity, the long-standing quest of superconductivity research. But can this be done at atmospheric pressure? This becomes the new and ultimate mission of our research endeavours.

In the meantime, another even more dramatic theoretical prediction is the appearance of superconductivity in a ternary clathrate Li_2MgH_{16} at ≈ 473 K and a pressure of 250 GPa [63]. At time of writing, this is a rapidly developing field, and we await experimental results on this and other rare-earth clathrate and ternary hydrides with considerable interest.

C6.4.7 Other Hydrides

Papaconstantopoulos [64] has used linearized augmented plane wave calculations to calculate the electronic structure and electron–phonon coupling parameters, λ, to compare the properties of H_3S and H_3F at pressure. The coupling and calculated T_c values are comparable and indeed may well be larger in the case of H_3F, with T_c possibly exceed 220 K around 150 GPa.

Motivated by the high H storage capability of ternary lithium borohydride, Kokail *et al.* [65] have studied the electronic properties of the high-pressure phases of Li_xBH_y. The well-known H-storage compound $LiBH_4$ remains a wide bandgap insulator to high pressures, but between 100 and 200 GPa, a metallic phase of composition Li_2BH_6 and space group $Fm\bar{3}m$ is stable with calculated T_c of the order of 100 K.

C6.5 Summary

To conclude, in the last few years, remarkable progress has been made in realizing Ashcroft's predictions [1, 5] of high-temperature superconductivity in highly compressed hydrogen and hydrogen-rich compounds. These predictions have really kindled the quest for room temperature superconductivity. With the remarkable discovery of superconductivity at 203 K in H_3S at 155 GPa [7], the prediction of similar T_c values in many H-dominant phases, and in particular of T_c exceeding 300 K in YH_{10} at 400 GPa, it is clear that hydrides are sure to present a fertile field of superconductivity research over the next decade.

Some general rules have become clear. Hydrides offer (i) a large H-derived electronic density of states, (ii) a large H-derived phonon Debye energy due to the low proton mass and (iii) a high sensitivity of the electronic structure to H-displacement and hence a large electron–phonon coupling, provided that molecular H_2 units can be avoided. Clathrate-like cages seem to be the ideal response to this limitation. Peng *et al.* [8] also point out the importance of high symmetry. Three-dimensionality is also key to eliminating the role of phase fluctuations in reducing the mean-field T_c. Even so, we have shown that amplitude fluctuations are already impacting on superconducting H_3S to a degree and it remains to be seen what their impact is in systems with still higher T_c. As shown, T_{amp} probably has a quadratic dependence on $1/T_c$, so fluctuations can easily overtake any increase in T_c^{mf} resulting in an actual decrease in observed T_c, despite the increase in pairing order parameter. In order to determine the magnitude of the impact of fluctuations on T_c in systems with very high calculated T_c values, it will be necessary also to calculate the magnitude of the London penetration depth as it plays the central role in determining the impact of fluctuations on the ultimate value of the actual T_c. We advise this approach in any future theoretical calculations of T_c in novel hydride systems. A further essential strategy is to explore ways of stabilizing high-pressure high-T_c phases to lower pressure if such systems are ever to prove amenable to practical application.

References

[1] Ashcroft N W 1968 Metallic hydrogen: a high-temperature superconductor? *Phys. Rev. Lett.* **21** 1748

[2] Wigner E and Huntington H B 1935 On the possibility of a metallic modification of hydrogen. *J. Chem. Phys.* **3** 764

[3] Dias R and Silvera I F 2017 Observation of the Wigner-Huntington transition to solid metallic hydrogen *Science* **355** 715

[4] Eremets M I and Drozdov A P 2017 Comments on the claimed observation of the Wigner-Huntington transition to metallic hydrogen https://arxiv.org/abs/1702.05125

[5] Ashcroft N W 2004 Hydrogen dominant metallic alloys: High temperature superconductors? *Phys. Rev. Lett.* **92** 187002

[6] Li Y, Hao J, Liu H, Li Y and Ma Y 2014 The metallization and superconductivity of dense hydrogen sulfide *J. Chem. Phys.* **140** 174712

[7] Drozdov A P, Eremets M I, Troyan I A, Ksenofontov V and Shylin S I 2015 Conventional superconductivity at 203 kelvin at high pressures in the sulfur hydride system *Nature* **525** 73

[8] Peng F, Sun Y, Pickard C J, Needs R J, Wu Q and Ma Y 2017 Hydrogen clathrate structures in rare earth hydrides at high pressures: possible route to room-temperature superconductivity *Phys. Rev. Lett.* **119** 107001

[9] Liu H, Naumov I I, Hoffmann R, Ashcroft N W and Hemley R J 2017 Potential high-superconducting lanthanum and yttrium hydrides at high pressure *Proc. Natl. Acad. Sci.* **114** 6990

[10] Somayazulu M, Ahart M, Mishra A K, Geballe Z M, Baldini M, Meng Y, Struzhkin V V and Hemley R J 2019 Evidence for superconductivity above 260 K in lanthanum superhydride at megabar pressures *Phys. Rev. Lett.* **122** 027001

[11] Drozdov A P, Kong P P, Minkov V S, Besedin S P, Kuzovnikov M A, Mozaffari S, Balicas L, Balakirev F, Graf D, Prakapenka V B, Greenberg E, Knyazev D A, Tkacz M and Eremets M I 2019 Superconductivity at 250 K in lanthanum hydride under high pressures *Nature* **569** 528

[12] Schilling J S 2006 High pressure effects, in *Treatise on High Temperature Superconductivity*, ed J R Schrieffer (Springer Verlag, Hamburg)

[13] Eremets M I 1996 *High Pressures Experimental Methods* (Oxford, Oxford University Press)

[14] Duan D, Liu Y, Tian F, Li D, Huang X, Zhao Z, Yu H, Liu B, Tian W and Cui T 2014 Pressure-induced metallization of dense $(H_2S)_2H_2$ with high-Tc superconductivity *Sci. Rep.* **4** 6968

[15] Einaga M, Sakata M, Ishikawa T, Shimizu K, Eremets M I, Drozdov A P, Troyan I A, Hirao N and Ohishi Y 2016 Crystal structure of the superconducting phase of sulfur hydride *Nat. Phys.* **12** 835

[16] Gegenwart P, Si Q and Steglich F 2008 Quantum criticality in heavy-fermion metals *Nature Phys.* **4** 186

[17] Shibauchi T, Carrington A and Matsuda Y 2014 A quantum critical point lying beneath the superconducting dome in iron pnictides *Annu. Rev. Condens. Matter Phys.* **5** 113

[18] Tallon J L and Loram J W 2001 The doping dependence of T^* - what is the real phase diagram? *Physica C* **349** 53

[19] Broun D M 2008 What lies beneath the dome? *Nature Phys.* **4** 170

[20] Morosan E, Zandbergen H W, Dennis B S, Bos J W G, Onose Y, Klimczuk C K, Ramirez A P, Ong N P and Cava R J 2006 Superconductivity in Cu_xTiSe_2 *Nature Phys.* **2** 544

[21] Goh S K, Tompsett D A, Saines P J, Chang H C, Matsumoto T, Imai M, Yoshimura K and Grosche F M 2015 Ambient pressure structural quantum critical point in the phase diagram of $(Ca_xSr_{1-x})_3Rh_4Sn_{13}$ *Phys. Rev. Lett.* **114** 097002

[22] Laughlin R B, Lonzarich G G, Monthoux P and Pines D 2011 The quantum criticality conundrum *Adv. Phys.* **50** 361

[23] Struzhkin V V, Hemley R J, Mao H-K and Timofeev Y A 1997 Superconductivity at 10–17 K in compressed sulphur *Nature* **390** 382

[24] Werthamer N R, Helfand E and Hohenberg P C 1966 Temperature and purity dependence of the superconducting critical field, H_{c2}. III. Electron spin and spin-orbit effects *Phys. Rev.* **147** 295

[25] Talantsev E F, Crump W P, Storey J G and Tallon J L 2017 London penetration depth and thermal fluctuations in the sulphur hydride 203 K superconductor *Ann. Phys. (Berlin)* **529** 1600390

[26] Tallon J L, Barber F, Storey J G and Loram J W 2013 Coexistence of the superconducting energy gap and pseudogap above and below the transition temperature of cuprate superconductors *Phys. Rev. B* **87** 140508(R)

[27] Storey J G 2017 Incoherent superconductivity well above T_c in high-T_c cuprates – harmonizing the spectroscopic and thermodynamic data. *New J. Phys.* **19** 073026

[28] Tallon J L, Storey J G and Loram J W 2011 Fluctuations and critical temperature reduction in cuprate superconductors. *Phys. Rev. B* **83** 092502

[29] Jacobs T h, Katterwe S O and Krasnov V M 2016 Superconducting correlations above T_c in the pseudogap state of $Bi_2Sr_2CaCu_2O_{8+\delta}$ cuprates revealed by angular-dependent magnetotunneling *Phys. Rev. B* **94** 220501(R)

[30] Kokanović I, Hills D J, Sutherland M L, Liang R and Cooper J R 2013 Diamagnetism of $YBa_2Cu_3O_{6+x}$ crystals above T_c: evidence for Gaussian fluctuations *Phys. Rev. B* **88** 060505(R)

[31] Bulaevskii L N, Ginzburg V L and Sobyanin A A 1988 Macroscopic theory of superconductors with small coherence length *Physica C* **152** 378

[32] Emery V J and Kivelson S A 1995 Importance of phase fluctuations in superconductors with small superfluid density *Nature* **374** 434

[33] Skoskiewicz T 1972 Superconductivity in the palladium-hydrogen and palladium-nickel-hydrogen systems *Phys. Status Solidi A* **11**, K123

[34] Stritzker B and Buckel W 1972 Superconductivity in the palladium-hydrogen and the palladium-deuterium systems *Z Physik* **257** 1

[35] Syed H M, Gould T J, Webb C J and Gray E MacA 2016 Superconductivity in palladium hydride and deuteride at 52–61 kelvin https://arxiv.org/abs/1608.01774

[36] Griveau J C and Colineau É 2014 Superconductivity in transuranium elements and compounds *Comptes Rendus Phys.* **15** 599

[37] Müller W, Schenkel R, Schmidt H E, Spirlet J E, McElroy D L, Hall R O A and Mortimer M J 1978 The electrical resistivity and specific heat of americium metal *J. Low Temp. Phys.* **30** 561

[38] Decker W R and Finnemore D K 1968 Critical-field curves for gapless superconductors *Phys. Rev.* **172** 430

[39] Dietrich M, Gey W, Rietschel H and Satterthwaite C B 1974 Pressure dependence of the superconducting transition temperature of Th_4H_{15} *Solid State Commun.* **15** 941

[40] Wickleder M S, Fourest B and Dorhout P K 2008 Thorium, in *The Chemistry of the Actinide and Transactinide Elements* ed. L R Morss, N M Edelstein and J Fuger (Springer, Dordrecht) pp 52–160

[41] Greenwood N N and Earnshaw A 1984 *Chemistry of the Elements* (Pergamon Press, Oxford)

[42] Weaver J H, Knapp J A, Eastman D E, Peterson D T and Satterthwaite C B 1977 Electronic structure of the thorium hydrides ThH_2 and Th_4H_{15} *Phys. Rev. Lett.* **39** 639

[43] Shein I R, Shein K I, Medvedeva N I and Ivanovskii A L 2007 Electronic band structure of thorium hydrides: ThH_2 and Th_4H_{15} *Phys. B Condens. Matter* **389** 296

[44] Kvashnin A G, Semenok D V, Kruglov I A, Oganov A R 2017 High-temperature superconductivity in Th-H system at pressure conditions https://arxiv.org/abs/1711.00278

[45] Kruglov I A, Kvashnin A G, Goncharov A F, Oganov A R, Lobanov S, Holtgrewe N and Yanilkin A V 2017 High-temperature superconductivity of uranium hydrides at near-ambient conditions https://arxiv.org/abs/1708.05251

[46] Alonso J A, Retuerto M, Sánchez-Benítez J and Fernández-Díaz M T 2010 Crystal structure and bond valence of CaH_2 from neutron powder diffraction data *Z. Kristallogr.* **225** 225

[47] Wang H, Tse J S, Tanaka K, Iitaka T and Ma Y 2012 Superconductive sodalite-like clathrate calcium hydride at high pressures *Proc. Natl. Acad. Sci.* **109**, 6463

[48] Eremets M I, Trojan I A, Medvedev S A, Tse J S and Yao Y 2008 Superconductivity in hydrogen dominant materials: silane *Science* **319** 1506

[49] Degtyareva O, Proctor J E, Guillaume C L, Gregoryanz E and Hanfland M 2009 Formation of transition metal hydrides at high pressures *Solid State Commun.* **149** 1583

[50] Hanfland M, Proctor J E, Guillaume C L, Degtyareva O and Gregoryanz E 2011 High-pressure synthesis, amorphization, and decomposition of silane *Phys. Rev. Lett.* **106** 095503

[51] Strobel T A, Goncharov A F, Seagle C T, Liu Z, Somayazulu M, Struzhkin V V and Hemley R J 2011 High-pressure study of silane to 150 GPa *Phys. Rev. B* **83** 144102

[52] Feng J, Grochala W, Jaroń T, Hoffmann R, Bergara A and Ashcroft NW 2006 Structures and potential superconductivity in SiH_4 at high pressure: en route to "metallic hydrogen" *Phys. Rev. Lett.* **96** 017006

[53] Martinez-Canales M, *et al.* 2009 Novel structures and superconductivity of silane under pressure *Phys. Rev. Lett.* **102** 087005

[54] Li Y, Gao G, Xie Y, Ma Y, Cui T and Zou G 2010 Superconductivity at ~100 K in dense $SiH_4(H_2)_2$ predicted by first principles *Proc. Natl. Acad. Sci.* **107** 15708

[55] Jin X, Meng X, He Z, Ma Y, Liu B, Cui T, Zou G and Mao H K 2010 Superconducting high-pressure phases of disilane *Proc. Natl. Acad. Sci.* **107** 9969

[56] Tse J S, Yao Y and Tanaka K 2007 Novel superconductivity in metallic SnH_4 under high pressure *Phys. Rev. Lett.* **98** 117004

[57] Gao G, Oganov A R, Bergara A, Martinez-Canales M, Cui T, Iitaka T, Ma Y and Zou G 2008 Superconducting high pressure phase of germane *Phys. Rev. Lett.***101** 107002

[58] Esfahani M M D, Oganov A R, Niu H and Zhang J 2017 Superconductivity and unexpected chemistry of germanium hydrides under pressure *Phys. Rev. B* **95** 134506

[59] Kim D Y, Scheicher R H, Mao H K, Kang T W and Ahuja R 2010 General trend for pressurized superconducting hydrogen-dense materials *Proc. Natl. Acad. Sci.* **107** 2793

[60] Goncharenko I, Eremets M I, Hanfland M, Tse J S, Amboage M, Yao Y and Trojan I A 2008 Pressure-induced hydrogen-dominant metallic state in aluminum hydride *Phys. Rev. Lett.* **100** 045504

[61] Somayazulu M, Ahart M, Mishra A K, Geballe Z M, Baldini M, Meng Y, Struzhkin V V and Hemley R J 2019 Evidence for superconductivity above 260 K in lanthanum superhydride at megabar pressures https://arxiv.org/abs/1808.07695

[62] Snider E, Dasenbrock-Gammon N, McBrideR, Debessai M, Vindana H, Vencatasamy K, Lawler K V, Salamat A and Dias R P 2020 Room-temperature superconductivity in a carbonaceous sulfur hydride *Nature* **586** 373

[63] Sun Y, Lu J, Xie Y, Liu H, Ma Y 2019 Route to a superconducting phase above room temperature in electron-doped hydride compounds under high pressure *Phys. Rev. Lett.* **123**, 097001

[64] Papaconstantopoulos D A 2017 Possible high-temperature superconductivity in hydrogenated fluorine *Nov. Supercond. Mater.* **3** 29

[65] Kokail C, Boeri L and von der Linden W 2017 Prediction of High-T_c conventional superconductivity in the ternary lithium borohydride system https://arxiv.org/abs/1705.06977

Part D
Other Superconductors

D

Introduction to Section D: Other Superconductors

Peter B. Littlewood

A section that is labelled "other" superconductors should explain itself. Presented here is not a grab bag of material that was somehow left out of other sections, but a collage of different families of novel superconductors with some links to each other but many with unique characteristics.

Up until the 1980's, there were only electron–phonon-mediated superconductors, all described microscopically by the BCS theory and macroscopically by the field theory of Ginzburg and Landau. To a great extent, all BCS superconductors are the same – of course, they may have different transition temperatures, different gaps, and different coherence lengths, but they are a unified class. While there was always a continuing search for new superconductors in this class (and of course there have been recent successes – magnesium diboride, hydrogen sulphide, for example), progress was slow and stuttering. Since 1980, this monopoly has been overturned.

Most famously of course are the cuprate superconductors that are so important that this volume allots them a whole part. The cuprates are strongly correlated materials, and their superconductive mechanism is clearly principally electronic, though still not understood in detail. But in parallel and sometimes ahead of the developments in cuprates, other superconducting classes were discovered. They are unified more by their flouting of superconducting convention than by their commonalities, in that pre-1980 all of these systems would have been seen as unlikely candidates for superconductivity. Many of them are "strongly correlated" in that the Coulomb repulsion between electrons is comparable to the electronic kinetic energy. Hence, they are usually poor metals, they coexist and compete with magnetic ground states, with charge- and spin-density waves, with metal–insulator (Mott) transitions, with quantum spin liquids, and they often have non–s-wave symmetry. Our point of view of searching for new superconducting materials or enhancing their phenomena has been turned on its head – now we look for superconductors in what was once seen as the most hostile territory close to magnetic and broken symmetry phases.

This chapter reviews several of these classes. Julian reviews heavy fermion superconductors. These are metals with hybridized f-levels that remarkably become fermi liquids at low temperatures, and since the discovery of superconductivity in UPt_3 in 1984, they have become the poster children for variety in unconventional superconductivity – multiple superconducting phases with finite-angular-momentum pairing. But several years earlier (1980), superconductivity had been discovered in an equally unlikely series of quasi-one-dimensional organic metals, which has now expanded into a large class of several hundred low-dimensional organics, reviewed here by Saito and Yoshida. In 1991, still in the early days of cuprate research, the alkali-intercalated fullerides emerged as a new class, and we now know that their pairing is most strong near to a Mott–Jahn–Teller instability. This story is reviewed by Iwasa and Prassides. Wu, Wang, and Wu bring us up to date on the iron chalcogenide system, discovered in 2008 to be a new class of multiorbital high T_c superconductors living close to magnetic and electronic nematic instabilities. Paul Chu offers us a look forward to future high-temperature superconductors; history suggests there are others surely to be discovered. Separately from studies of bulk materials, two other new waves or research have arrived, topology reviewed by Kotetes, and interfacial superconductors by Schmalian. The discovery of topological materials that have protected states on their surfaces, and the engineering of superconductivity at interfaces, has opened up a new space of quantum engineering that has exciting prospects for science and technology.

D1

Unconventional Superconductivity in Heavy Fermion and Ruthenate Materials

Stephen R. Julian

D1.1 Introduction

The topic of this chapter is non-BCS or 'unconventional' superconductivity in heavy fermion and ruthenate materials. Unconventional superconductivity is important: the high-temperature cuprates (Section C) are unconventional superconductors, and it is likely that any useful room temperature superconductor will also be unconventional. Moreover, it has been proposed that topologically protected qubits could be realized in chiral spin-triplet superconductors, which would be promising for quantum computation.

Heavy fermion and ruthenate superconductors are 'unconventional' in both of the modern senses of the word: their Cooper pairs have non-zero angular momentum, and the pairing mechanism for superconductivity is not the electron–phonon interaction. (For a recent and extensive review comparing the properties of broad classes of unconventional superconductors see Stewart, 2017.) In BCS theory, the Cooper pairs are called '*s*-wave', meaning that they have zero internal orbital angular momentum. In the systems we consider here, in contrast, the Cooper pairs can have $l = 1$ (*p*-wave), $l = 2$ (*d*-wave), etc. Additional degrees of freedom associated with this angular momentum give rise to a rich variety of superconducting ordered states, displaying many and varied exotic properties, including nodes in the superconducting gap, extremely high values of the upper critical field H_{c2}, coexistence with ferromagnetism or antiferromagnetism, half-quantum vortices in the mixed state, spin-triplet superconductivity analogous to superfluid ^{3}He, enhanced superconductivity on the border of magnetism, and more.

The first section of this chapter gives an overview of important themes: the use of symmetry to classify unconventional superconductors, experimental methods that in principle can distinguish different symmetry states (although in practice this is still challenging), and exotic pairing mechanisms that could operate in unconventional superconductors. Quantum criticality, which has been a productive route to discovering new unconventional superconductors, is emphasized.

The remaining sections summarize the properties of particular heavy fermion and ruthenate superconductors.

D1.2 Key Ideas

Here some key ideas in unconventional superconductivity are introduced: how unconventional superconductors are classified, which experimental methods are effective in their study, and what the underlying pairing mechanisms are speculated to be. Readers who are primarily interested in the properties of heavy fermion and ruthenate superconductors can skip to Section D1.3.

D1.2.1 Broken Symmetries in Unconventional Superconductivity

'Unconventional' heavy fermion and ruthenate superconductors break additional symmetries beyond the U(1) gauge symmetry broken by 'conventional', or BCS, superconductors. These additional broken symmetries could, for example, be rotation symmetry, or time-reversal symmetry. A well-known example is the *d*-wave superconducting state of the cuprates, which breaks the four-fold rotation symmetry of the tetragonal crystal structure (see Section C).

In heavy fermion physics UPt_3 is the archetypal example because of its multiple superconducting phases (see Figure D1.1). The upper plot shows that there is a double step in the specific heat, with a splitting of about 50 mK, at the superconducting transition T_c. In an applied field, these two transitions merge at about 0.4 T. Other techniques such as

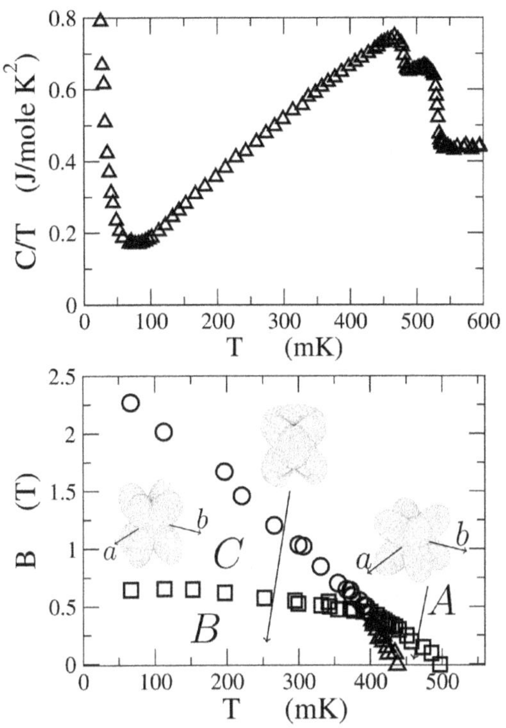

FIGURE D1.1 The multiple superconducting phases of UPt_3. The top figure shows the specific heat *C(T)* divided by temperature *T*, vs. *T* [20]. The superconducting transition is split, with the first transition occurring near 0.55 K, the second transition at approximately 0.49 K. The bottom figure shows the phase diagram as determined by sound velocity measurements [1]. There are three phases, labelled *A*, *B* and *C*, and the possible arrangement of nodes in the superconducting gap is shown for the three phases. The *A* and *C* phases differ by the vertical line nodes being rotated by 45° relative to each other. Note too that the vertical line nodes have four-fold symmetry, not the hexagonal symmetry of the basal plane. [Upper figure reproduced from J.P. Brison et al., J. Low Temp. Phys., (1994) 95:145, with the permission of Springer Publishing. Lower figure reprinted with permission S. Adenwalla et al., Phys. Rev. Lett. (1990) 65:2298, ©(1990) by the American Physical Society.]

sound velocity reveal a further phase boundary, so that three distinct superconducting phases, called *A*, *B*, and *C*, have been mapped out, as shown in the lower plot. Within the different superconducting phases, measurements show that there are nodes in the superconducting gap function, with the different phases characterized by different patterns of nodes (discussed further in Section D1.3.3).

Multiple superconducting phases, and nodes in the gap, indicate that BCS theory does not apply, because BCS predicts a nodeless gap with no internal degree of freedom that would allow more than one superconducting phase. Thus the first challenge of theory is how to go beyond BCS-type superconducting order. This is done by considering the symmetry of the Cooper pair wave-functions.

The Cooper pair wave-function has an orbital part (dependent on the relative position of the two electrons in the Cooper pair) and a spin part. The question of first importance in classifying the symmetry of a Cooper pair wave-function is: how does it behave under inversion of coordinates (i.e. what it its *parity*)? As with atomic wave functions, even-numbered orbital angular momentum states, l = 0,2,4, ... have even parity, meaning that the wave-function does not change sign under inversion of coordinates, while l = 1,3,5,... have odd parity.* Because electrons are fermions, the overall wave-function must be antisymmetric under exchange of the positions of the two electrons in a Cooper pair, thus it must have odd parity. Thus, if l is even, the Cooper pair must be in a spin-singlet state, $\left(|\uparrow\downarrow\rangle-|\downarrow\uparrow\rangle\right)/\sqrt{2}$ in order to be antisymmetric under exchange, while if l is odd, it must be spin-triplet, S = 1. The cuprates, and many heavy fermion superconductors, are l=2, or *d*-wave, superconductors. Spin-triplet superconductvity is found in ferromagnetic heavy fermion superconductors such as UGe_2 (Section D1.3.6). The ruthenate superconductor Sr_2RuO_4 (Section D1.4) may be an l = 1 triplet superconductor, while l = 3 or *f*-wave triplet superconductivity is now believed to apply to superconductivity in UPt_3 (Section D1.3.3).

Going beyond parity, the symmetry classification of $l > 0$ Cooper paired states is well developed, but somewhat technical, so only a brief overview is given here. A review article [63] provides a good introduction with specific application to UPt_3, while the book by Kuramoto and Kitaoka [77] has a more general treatment. A key idea is that the symmetry of the wave-function may be reflected in nodes in the superconducting gap, and a few other properties, which may in turn be observed experimentally.

For a given crystal structure, possible broken symmetries can be described using the mathematical theory of groups. The superconducting order parameter is classified according to how it transforms under the symmetry operations of the crystal. As discussed in reference [63] key points are that (1) the lowest energy Cooper pair wave-function will be a superposition of functions that belong to only one irreducible representation of the symmetry group of the crystal; (2) while each irreducible representation has an infinite number of such functions, these can be broken down into a few (typically one or two) basis functions (whose number defines the 'dimensionality' of the irreducible representation) multiplied by any function that obeys the full point group symmetry of the crystal; and (3) although it is very difficult to calculate or measure the exact form of the Cooper pair wave-function, the basis functions of a given irreducible representation will in general imply that the superconducting gap as a particular pattern of nodes,

* This parity classification breaks down in noncentrosymmetric superconductors (Section D1.3.7).

such as horizontal and/or vertical line nodes and/or point nodes, which can be observed experimentally allowing experiments to determine which irreducible representation the superconducting state belongs to. Although (3) is true in theory, in practice it can be difficult, and there is further work to be done in this area.

In the case of UPt_3, the existence of multiple superconducting phases implies an irreducible representation with more than one basis function (or, less likely, a near-degeneracy of more than one irreducible representation), which immediately takes us beyond $l = 0$ s-wave superconductivity.

D1.2.1.1 Chiral Superconductors

Although for a particular material $l > 0$ may be energetically favourable, nodes in the superconducting gap cost pairing energy[†] because the pairing energy is proportional to the magnitude of the superconducting gap. A nodal superconductor can, however, get rid of its nodes if the pair condensate 'spontaneously' develops macroscopic angular momentum. A relevant example would be a p-wave superconductor (Cooper pairs with spin $S = 1$ and orbital angular momentum $l = 1$), with a cylindrical Fermi surface whose axis is parallel to the p_z axis, so that there can be no nodes at $p_x = p_y = 0$. Such a superconductor could have Cooper pair states that transform like a p_x orbital, having nodes along the y-axis, or like a p_y orbital, having nodes along the x-axis. But a superposition $p_x \pm ip_y$ is also possible, and this would have no nodes. Such superconductors are called 'chiral' superconductors [64] because the phase of the gap function $\Delta(\vec{p})$ advances by $\pm 2\pi$ as $\vec{p}$ rotates once about p_z. This state breaks time-reversal symmetry, which can be detected by μSr or Kerr effect measurements, and chiral superconductors are potentially important because they have topological surface states that have some promise for quantum computing. Chiral p-wave superconductors have particularly exotic properties, including 'half-quantum' vortices containing a Majorana state. The ruthenate superconductor Sr_2RuO_4 may be a chiral p-wave superconductor, although the evidence is not clear (see Section D1.4).

D1.2.2 Experimental Detection of Broken Symmetry

Conventional superconductors have a nodeless gap that has clear experimental signatures, such as a thermally activated specific heat in the $T \to 0$K limit and a coherence peak just below T_c in the spin-lattice relaxation time T_1 as measured by NMR (see Chapter A2.9). Similarly, nodes in the superconducting gap of an unconventional superconductor can be revealed by multiple experimental probes. In the vicinity of a line or point node in the superconducting gap there are low-lying excitations, so probes that are sensitive to quasiparticle excitations, such as the specific heat, thermal conductivity, ultrasound attenuation, or NMR spin-lattice relaxation time, can detect nodes via power law rather than thermally activated temperature dependences. A particularly useful probe is NMR/NQR, for which the Habel–Schlichter coherence peak is suppressed in unconventional superconductors, while the spin-lattice relaxation time T_1 at low temperature shows a power law dependence on temperature (e.g Kuramoto, 1999 [77]).

There are other signatures of unconventional superconductivity: multiple superconducting phases provide 'smoking-gun' evidence; rapid suppression of T_c by non-magnetic impurities (e.g. [88]) indicates a non-BCS superconducting order parameter; unusual vortex lattices can also reflect an unconventional superconducting state (e.g. [3]); and in the case of chiral superconducting states, because they break time-reversal symmetry, μSr spectroscopy and Kerr effect should see a spontaneous magnetization in the superconducting state, as in Sr_2RuO_4 (Section D1.4) and UPt_3 (Section D1.3.3).

Demonstration of *which* of the large number of possible unconventional states actually occurs in a particular material is much more difficult, however. This will be evident in our discussion of Sr_2RuO_4, where even the singlet vs. triplet spin state of the pairs (the most fundamental distinction between species of unconventional superconductor) is in question. As discussed above, the particular pattern of nodes on the Fermi surface and their type (linear, quadratic, etc.) reflects the irreducible representation of the superconducting state. Anisotropy of the thermal conductivity can reveal the pattern of nodes (e.g. [63]). For example, if the only node were a point node along k_z, then the c-axis thermal conductivity would be larger than the in-plane thermal conductivity, and k_c / k_{ab} would diverge in the $T \to 0$ K limit. Similarly, dependence of the specific heat on the direction of an applied external magnetic field can reveal the pattern of nodes on the Fermi surface [168, 169]. Moroever, the power law dependence on temperature of the thermal conductivity, specific heat, etc. can reveal the nature of the nodes, whether they are line or point nodes, and whether the density of states rises linearly or quadratically as a function of distance from the node.

D1.2.3 Pairing Mechanisms

An important question is what microscopic physics favours unconventional superconductivity? $l > 0$ wave-functions have zero amplitude when the separation of the electrons in the pair goes to zero, so they offer a way for the electrons to avoid each other in real space, lowering Coulomb repulsion. In principle then, phonon-mediated superconductors could have $l > 0$ driven by Coulomb repulsion. In practice, however, this is unlikely. The phonon pairing interaction is very local in space, and so the vanishing probability for $l > 0$ pairs as $r \to 0$ would also suppress pairing. Moreover, the phonon pairing mechanism is non-local in time (see Chapter A3.2), which allows the electrons in the Cooper pair to minimize their Coulomb

[†] If the pairing interaction is oscillatory in space, as with antiferromagnetic spin fluctuations, then nodes may actually be helpful, see e.g. [100].

repulsion without resorting to $l > 0$: the electrons avoid each other by occupying the same point in space at different times. Thus, $l > 0$ is strong evidence of a non-phonon pairing interaction that favours finite l because of the microscopic details of the physics. Due to the frequent occurrence of unconventional superconductivity at magnetic quantum critical points (Section D1.2.4), much attention has focussed on pairing by magnetic fluctuations in metals on the border of magnetic order. Nearly magnetic systems have a high magnetic polarizability, and so the spin of one conduction electron can polarize the magnetic background, and the second electron in the Cooper pair can lower its energy if its spin is correctly oriented relative to the induced polarization. For example, pairing by large-amplitude long-wavelength spin fluctuations (so-called 'paramagnons') in a nearly ferromagnetic metal would favour a spin-triplet Cooper pair, which then requires $l = 1,3,5,\ldots$ (i.e. *p*-wave, *f*-wave, ...) for antisymmetrization. Pairing by antiferromagnetic spin fluctuations (e.g. [99, 173]) is more complex, favouring parallel spin pairs along some directions in space, but antiparallel (spin-singlet) states along other directions. Spin-singlet states (with l=2,4,...) win out, but to maximize the pairing energy the superconducting gap must oscillate, and have nodes along the orientations where the pairing interaction is repulsive. Other exotic pairing mechanisms that have been proposed in heavy fermion systems are valence fluctuations [110] and exchange of electric quadrupole fluctuations [13].

D1.2.4 Quantum Criticality

Figure D1.2 shows the temperature–pressure phase diagram of $CePd_2Si_2$ [94], a heavy fermion compound with an antiferromagnetic ground state at ambient pressure. Hydrostatic pressure suppresses antiferromagnetism around 3 GPa, and near this pressure superconductivity appears, with the maximum T_c occurring at the pressure where T_N falls to zero. Several phase diagrams of this kind have been found in heavy fermion systems, and they are also found in the iron-pnictide high-temperature superconductors (see Chapter C5). So common is this phenomenon that locating quantum critical points has become a strategy for discovering new superconductors.

The point at which a magnetic ordering temperature (or other critical temperature) falls to 0 K is called a 'quantum critical point'. This term was introduced by Hertz [44] to denote the fact that, unlike critical points at non-zero temperature, there is no regime in which classical statistical mechanics can be applied to the critical fluctuations of the order parameter, so the theory of second-order phase transitions, famously developed by Kadanoff, Fisher, Wilson, and others, must be recast using quantum statistical mechanics.

The fact that the maximum T_c occurs at the quantum critical point, where the magnetic fluctuations are largest, is strong evidence that the superconductivity is mediated by magnetic fluctuations.

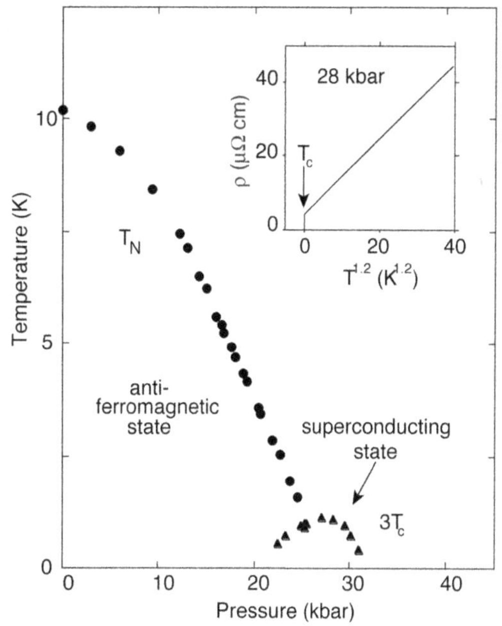

FIGURE D1.2 The phase diagram of $CePd_2Si_2$ [94]. $CePd_2Si_2$ is antiferromagnetic at ambient pressure, but antiferromagnetism is suppressed by hydrostatic pressure between 2.5 and 3 GPa. Around the quantum critical point, a superconducting phase is observed. Note that the superconducting transition temperature has been multiplied by 3, to make it visible on this plot. The inset shows that the resistivity follows a non-Fermi-liquid $T^{1.2}$ power law in the normal state, at the critical pressure. (Reproduced from N.D. Mathur et al., Nature, (1998) 394:39, with the permission of Springer Nature.)

D1.3 Heavy Fermion Superconductors

Heavy fermion systems offer by far the largest number and the largest variety of unconventional superconductors. Moreover, the superconductivity is confined to low temperatures, making it easier for measurements such as specific heat to separate electronic and lattice excitations. Most heavy fermion materials can be purified to a very high degree, removing the complications of disorder, and the normal-state properties are in many cases well characterized and understood, with the notable exceptions of quantum critical superconductors, and UBe_{13}. They are thus a very important proving ground for comparing theory to experiment. A detailed review of heavy fermion superconductivity can be found in reference [119].

D1.3.1 Normal-State Properties of Heavy Fermions

Heavy fermion systems comprise a class of crystalline intermetallic compounds containing a mixture of elements that have partially filled *f*-orbitals (typically Ce, U, or Yb), together with non-magnetic elements. The role of the non-magnetic elements is firstly to separate the *f*-ions thereby suppressing their tendency to order magnetically, and secondly to supply conduction electrons. The essential physics of heavy fermions comes from the coupling of their conduction electrons to partially filled *f*-orbitals.

At high temperature (above 10 to 50 K typically), the partially filled *f*-orbitals in heavy fermion systems produce Curie–Weiss susceptibility, showing that they are behaving as independently fluctuating local moments. Moreover, the electrical resistivity is high (typically $>100\ \mu\Omega$cm) due to strong scattering of the conduction electrons by these incoherently fluctuating local moments. As $T \rightarrow 0$, the $k_B ln2$ entropy of the local moments must fall to zero. In the vast majority of *f*-electron materials this happens through a phase transition to static antiferromagnetic or ferromagnetic long-range order, but this tendency to magnetic order is weakened if the *f*-shell is nearly empty (as in cerium or uranium) or nearly full (as in ytterbium). In heavy fermion systems magnetic order either doesn't set in, or else it is very weak. Instead, the fluctuations of the magnetic moments somehow become coherent, achieving a low entropy state without static order, and the resistivity falls dramatically. Historically, this has been ascribed to the Kondo effect, by which isolated magnetic moments in metals become screened at low temperature by forming a spin-singlet state with an electron from the conduction electron sea. The true picture is probably more complex than this, and recently other models, such as the two-fluid Kondo lattice model of Pines and collaborators [105], have been under development.

In the limit as $T \rightarrow 0$ K, most heavy fermion systems have a 'Fermi liquid' normal state, characterized by a T^2 resistivity $\rho(T) = \rho_o + AT^2$, a Pauli-like susceptibility χ_P ~const., and a linear specific heat $C(T) = \gamma T$ with the values of A, χ and γ being huge: often thousands of times larger than in a conventional metal. The large value of γ shows that the entropy of the local magnetic moments has been converted to itinerant, charged, fermionic degrees of freedom, which remains one of the most beautiful examples of emergence in all of condensed matter physics. It should be noted that there are important exceptions to the Fermi liquid normal state, including systems tuned to a quantum critical point (Sections D1.3.4 and D1.3.5), and UBe_{13} (Section D1.3.8.3).

D1.3.2 Crystal Growth

Heavy fermion metals are very sensitive to impurities and defects. For example, even minute amounts of some impurities can induce large moment antiferromagnetism in paramagnetic heavy fermion systems, while defects wash out the double superconducting transition of UPt_3. Thus crystal growth plays a central role in this field.

Cerium and uranium are highly reactive materials, so during reaction the materials must be protected from oxidizing atmospheres and contamination from crucibles. There are two approaches to this: either the crystals can be grown within a flux which protects the growing crystal from oxygen and other impurities, so that specially pure atmospheres and crucibles are not necessarily needed. Or alternatively, a combination of ultra-high vacuum or ultra-pure argon (which is often required to suppress evaporation if one of the constituents is volatile), with either water-cooled copper crucibles or self-supporting melts as in vertical-floating-zone growth, can be used.

D1.3.2.1 Flux Growth

Growth of single crystals from metal fluxes has been developed mainly by Fisk and co-workers [24]. Very high purity crystals of the 115 compounds (Section D1.3.5) have been grown with this method, and it has been used extensively to grow single crystals of other compounds. This method has the advantage that the crystal growth equipment need not be very specialised, and low growth temperatures are possible, leading to a lower concentration of defects and vacancies. The disadvantage is that sometimes some of the flux grows within the crystal as an inclusion, or it may be a substitutional impurity.

Typically the materials are placed in a crucible which is sealed in a quartz ampoule (for growth below 1200° C). The ampoule is gradually heated, and the flux melts first, and then the higher melting point elements dissolve in the flux. Next, the melt is cooled very slowly, and the crystals grow within the flux, or against the crucible walls. To get rid of the flux at the end, the ampoule can be removed from the furnace and, while the flux is still molten, spun in a centrifuge to throw off the molten flux. Alternatively, crystals may be removed mechanically from the solidified flux, or the flux may be etched away.

D1.3.2.2 Cold Crucible Growth

Very high-quality heavy fermion crystals have been grown in water-cooled copper crucibles, with either ultra-high vacuum or high purity argon atmosphere. The very high thermal conductivity of copper keeps the crucible at or below the boiling point of water, even with a melt at over 2000°C resting on it, thus there is no chemical reaction between the melt and the crucible.

The melt can be heated by radio-frequency induction, with the high purity atmosphere contained within a quartz tube and the rf-coil placed outside. Either vertical crucibles (for Czochralski growth) or horizontal crucibles (for horizontal zone refining or casting polycrystalline ingots) can be used in this way, see e.g. [22]. This is technically difficult, however, and starting in the mid-1990's use of tetra-arc Czochralski pulling (e.g. Haga et al., 1998) has taken over as the preferred method. Some designs are compatible with high purity gases [31, 81]. Some properties, for example the double transition in UPt_3, are very sensitive to annealing conditions [20], and must be explored systematically. Electron-beam zone refining has also been used effectively [79].

D1.3.2.3 Thin Films

Thin films of heavy fermion systems were first produced by magnetron sputtering in $CeCu_2Si_2$ [48] and in UPd_2Al_3 [62]. This is now an active area due to the development of metallic molecular-beam-epitaxy, which has been used, for example, to grow epitaxial layers and superlattices of heavy fermion superconductors [58]. For example, the Matsuda group has studied superlattices of $CeCoIn_5$ and a number of related compounds - $CeRhIn_5$, $CeIn_3$, and $YbCoIn_5$ - which demonstrated the essential role of spin fluctuations in mediating the superconductivity of $CeCoIn_5$ (e.g. [107]).

D1.3.3 UPt_3

UPt_3 is arguably the most important heavy fermion superconductor because it is the first and clearest example of a superconductor with a multi-component order parameter, as revealed by the existence of multiple superconducting phases (Figure D1.1).

Superconductivity in UPt_3 was discovered by Stewart et al. [151] in 1984, who showed using specific heat measurements that bulk superconductivity sets in at $T_c \sim 0.54$ K, within a highly renormalized normal state characterized by a very large linear specific heat coefficient $\gamma > 0.4$ J/mole·K^2. Uniquely among heavy fermion superconductors, UPt_3 also displayed a $T^3 lnT$ term in its electronic specific heat, characteristic of metals with large spin flucuations at low temperature. Soon after the discovery, Louis Taillefer produced high purity single crystals via zone refining in ultra-high vacuum, and carried out the first full de Haas–van Alphen study of a heavy fermion system, furnishing direct proof of the existence of heavy charged electron quasiparticles [157].

The multiple superconducting phases of UPt_3 (Figure D1.1) were uncovered first with ultrasound measurements [102, 121, 133] and changes in the slope of H_{c2} [123, 158] which found the B to C transition line of Figure D1.1, and finally specific heat [30, 38] which found the A to B transition. These measurements are a landmark in the history of superconductivity, offering the first direct proof of non-BCS superconductivity.

UPt_3 has now been investigated in great detail (see Joynt, 2002). A key factor affecting the superconducting phase diagram seems to be a weak form of antiferromagnetism with a Néel temperature $T_N \sim 5$ K and an ordered moment of less than $0.1\mu_B$ [2]. This antiferromagnetism seems to be responsible for the splitting of the superconducting transition: measurements under hydrostatic pressure show that the splitting and the weak antiferromagnetism are suppressed together by hydrostatic presssure of about 0.3 GPa [42, 165].

The current picture of the symmetry of the superconducting states is that the A phase has an equitorial node, and vertical line nodes with four-fold symmetry. Beautiful evidence of the vertical line nodes is given by Strand et al. [155] who placed an array of Josephson Junctions around the edges of a high purity single crystal, and found that the critical current becomes non-zero at $T_c = 0.54$ K except along the 45° direction. A gap of the form $\left(p_x^2 - p_y^2\right)p_z$ would be consistent with this state. Upon entry into the B-phase, this vertical line node becomes gapped, as the superconducting state becomes chiral, for example $\left(p_x^2 - p_y^2 \pm ip_x p_y\right)p_z$ (see e.g. Kallin, 2016). Evidence for the chiral nature of the state comes from observation of time-reversal symmetry breaking in the B-phase by μSr measurements [85] and polar Kerr effect [131]. The observed symmetry of the superconducting gap is consistent with spin-triplet superconductivity, and is now presumed to have a dominant *f*-wave (i.e. $l = 3$) character. The observation of line nodes, notably in field-angle-resolved thermal conductivity measurements [97] appeared to conflict with a symmetry argument that odd-parity superconductors with strong spin–orbit coupling cannot have line nodes [18]. This has recently been resolved with the realization that the symmetry argument does not apply to non-symmorphic crystal structures (i.e. those with a screw axis) (see [109] and [98] and references therein).

D1.3.4 Cerium 122 Compounds

The field of heavy fermion superconductivity began in 1979 with the discovery by F. Steglich of superconductivity with a transition temperature of $T_c \simeq 0.5$ K in $CeCu_2Si_2$ [149]. At low temperature, the normal state is a classic heavy fermion metal, with a very large linear specific heat coefficient of $\gamma \simeq 1$ J/mole·K^2. A crucial finding was that the specific heat jump at T_c is a substantial fraction of the normal-state value, indicating that the heavy fermions form Cooper pairs. Equally surprising was the appearance of superconductivity in a material that displays local moment behaviour at high temperature, since magnetic impurities are very destructive of conventional BCS superconductivity. There was thus strong evidence of a new mechanism of superconductivity. Subsequent studies showed that at ambient pressure $CeCu_2Si_2$ is on the border of antiferromagnetism: slight copper deficiency, or doping with germanium, stabilizes a spin-density-wave ground state [12]. Application of pressure to this antiferromagnetic ground state suppresses magnetic order and reveals superconductivity. Neutron scattering [154] finds that a gap opens in the spin excitation spectrum in the superconducting state, and that the energetics of this gap are consistent with superconductivity mediated by antiferromagnetic spin fluctuations. Indeed, the properties of the superconducting state were long considered to be consistent with a spin-singlet *d*-wave order parameter with line nodes [119]. Very recently, however, it has been found that in fact there are deep minima but no actual nodes in the superconducting gap of $CeCu_2Si_2$ [72], which may indicate chiral rather than nodal *d*-wave, or even so-called $s_\pm$ [55] or s_{++}-wave superconductivity [159], analogous to the superconductivity of iron-pnictide superconductors.

Remarkably, further application of pressure reveals a second superconducting dome in $CeCu_2Si_2$ around 2.5 GPa, with a much higher T_c of over 2 K [177]. Similarly, pressurizing $CeCu_2Ge_2$, which is isoelectronic to $CeCu_2Si_2$ but with a larger lattice parameter, reveals an identical pressure–temperature phase diagram shifted by about 7 GPa [47]. This higher pressure dome coincides with a valence transition seen in x-ray diffraction [111], and it is argued that a new superconducting pairing mechanism, involving 4*f* valence fluctuations, is responsible for this higher temperature superconductivity [47].

$CeCu_2Si_2$ crystallizes in the tetragonal $ThCr_2Si_2$ structure, which seems to be quite favourable for heavy fermion superconductivity (Pfleiderer, 2009, has a table summarizing the properties of Ce122 superconductors). Other notable members of this family include $CePd_2Si_2$ [94] and $CeRh_2Si_2$ [101], in which superconductivity appears at the border of magnetism as $T_N \to 0$ K. Under pressure, antiferromagnetic order in $CePd_2Si_2$ is suppressed from $T_N \simeq 10$ K at ambient pressure to 0 K near a critical pressure of $P_c \simeq 3$ GPa [94]. Centred on P_c is a superconducting dome with a maximum T_c of about 0.4 K. Together with $CeIn_3$ (Section D1.3.5), this was the first observed superconducting dome centred on the critical pressure P_c for suppression of magnetism, and it offers strong evidence that magnetic fluctuations mediate the superconductivity. Such phase diagrams have been found also in pnictide high-temperature superconductors (Chapter C5). At P_c, $CePd_2Si_2$ shows marked non-Fermi-liquid behaviour, having for example a resistivity that varies as $T^{1.2}$over a temperature range of several 10's of K above T_c[94], believed to arise from scattering of conduction electrons from quantum critical magnetic fluctuations.

D1.3.5 115 Compounds

This family of superconductors, which includes $CeIn_3$, $CeCoIn_5$, $CeRhIn_5$ and $PuCoGa_5$, vies with UPt_3 and the cerium-122 family as the most important in heavy fermion physics. These materials provide key examples of superconductivity on the border of antiferromagnetism; they display non-Fermi-liquid properties reminiscent of high-temperature superconducting cuprates and reveal a trend of increasing T_c with increasing quasi-two-dimensionality and increasing bandwidth that have also drawn comparisons with the cuprates.

The 'parent' compound in this family, $CeIn_3$, is similar to $CePd_2Si_2$. T_N falls monotonically from 10 K at $P = 0$ to zero at a quantum critical pressure of $P_c = 2.5$ GPa, with a superconducing dome centred on P_c [94], with a rather low maximum T_c of~ 0.18 K [94].

Superconductivity with a record high T_c for heavy fermions of 2.3 K was discovered in 2001 by Petrovic *et al.* [118], in $CeCoIn_5$. In this crystal structure, layers of $CeIn_3$ are separated by layers of $CoIn_2$, thus it is a lower dimensional version of $CeIn_3$. Increasing T_c with decreasing dimensionality is consistent with theories of antiferromagnetic-spin-fluctuation-mediated superconductivity, for which pairing is frustrated on three-dimensional Fermi surfaces [100].

An interesting property of $CeCoIn_5$ is that its normal state is strongly non-Fermi-liquid at ambient pressure, drawing comparisons with optimally doped cuprate superconductors. For example, the resistivity is linear in temperature over a large range of temperature above T_c [116, 144]. An additional parallel with the cuprates (and with $CeCu_2Si_2$ discussed above) is the observation in neutron scattering of a spin resonance upon entry into the superconducting state [153].

For magnetic field applied parallel to the *c*-axis, H_{c2} is near 5 T [118], the high ratio of H_{c2} / T_c being typical of heavy fermion superconductors and reflecting a very short coherence length. Near H_{c2}, NMR and neutron scattering measurements find that incommensurate spin-density-wave order coexists with finite-momentum-Cooper-paired superconductivity [68, 176], providing further evidence for intimate coupling of superconducivity and magnetism in $CeCoIn_5$. The properties of the superconducting state have been studied with many different probes, as summarized in reference [119], and the consensus is that $CeCoIn_5$ is a spin-singlet superconductor with a line node, indicating a *d*-wave order parameter. Applying a novel scanning microscopy technique of "Bogoliubov quasiparticle interference," the pattern of nodes on individual Fermi surfaces has been mapped [5], and found to be consistent with a $d_{x^2-y^2}$ order parameter. Moreover, comparison with magnetic interactions, also extracted from quasiparticle interference, provides strong evidence that fluctuating *f*-electron magnetism is the pairing mechanism [167].

$CeRhIn_5$ behaves like a 'negative pressure' version of $CeCoIn_5$. At ambient pressure it orders antiferromagnetically, but hydrostatic pressure of $P_c = 2.5$ GPa suppresses T_N and a superconducting dome centered on P_c is observed, with a maximum T_c and normal-state properties very similar to those of $CeCoIn_5$ at zero pressure [43]. de Haas–van Alphen measurements under pressure [143] show that the Fermi surface reconstructs at P_c, from 'small' to 'large', such that the Ce 4*f*-electron is included in the Fermi volume above P_c, but is not included below P_c. A dramatic enhancement of quasiparticle masses is observed as the quantum critical pressure is approached.

Plutonium intermetallics have not been extensively investigated, but $PuCoGa_5$ has the same crystal structure as the Ce115 compounds, with similar non-Fermi-liquid normal-state properties at ambient pressure, and it has a remarkably high superconducting transition temperature, $T_c \sim 18.5$ K [126]. The upper critical field H_{c2} is also very high, and NMR measurements suggest unconventional spin-singlet superconductivity in which antiferromagnetic spin fluctuations are a likely pairing mechanism [28]. Plutonium is a 5*f* element, whose orbitals are more extendended than the 4*f* orbitals of cerium, but less extended than the 3*d* orbitals of copper. The series $CeIn_3$, $CeCoIn_5$, $PuCoGa_5$, which displays monotonically increasing T_c as dimensionality is reduced and bandwidth is increased, has led to suggestions that these materials are points on a theoretical continuum of T_c vs. bandwidth and dimensionality that applies to all antiferromagnetic-fluctuation-mediated superconductors, culminating in the cuprates [126].

D1.3.6 Ferromagnetic Superconductors

A class of superconductors that is unique (as of 2019) to heavy fermion systems is that of ferromagnetic superconductors, in which the itinerant electrons that produce ferromagnetism also become superconducting. There are three systems that have been extensively investigated: UGe_2, URhGe, and UCoGe, all of which become superconducting in the ferromagnetic state, while UCoGe also remains superconducting in its pressure-induced paramagnetic state (for a review of these materials see [10]). Recently, superconductivity has also been found near the border of ferromagnetism in UTe_2 [122]. Remarkably, all of these materials share the structural motif of zig-zag chains of U atoms, an intriguing demonstration that superconductivity tends to run in structural families. This is especially striking when one considers that the magnetic properties of these four materials show considerable variety, as described below.

A material that becomes superconducting within a ferromagnetically ordered state can be expected to have strange properties: even in the absence of an applied field, the internal field due to magnetism will produce a flux lattice, so they will be in the mixed state at $H = 0$. Moreover, it is expected that the Cooper pairs will be in equal-spin triplet states. This is because the magnetism in an itinerant ferromagnet comes from a difference in Fermi volume between spin-up and spin-down electrons. The up- and down-spin Fermi wave-vectors are different, i.e. $k_{F\uparrow} \neq k_{F\downarrow}$ so Cooper pairs involving up and down spin electrons would have non-zero center-of-mass momentum, thus it is energetically favourable for pairs to form between $k_{F\uparrow}$ and $k_{F\uparrow}$ or between $k_{F\downarrow}$ and $k_{F\downarrow}$.

Superconductivity in ferromagnetic UGe_2 was discovered by Saxena et al. [129]. At ambient pressure, UGe_2 is a large-moment ferromagnet below a ferromagnetic ordering temperature $T_{FM} = 52$ K, with strong uniaxial magnetic anisotropy [112]. Within the ferromagnetic state there is another anomaly, at temperature $T_x = 25$ K, at which the resistivity changes slope [113] and below which the ferromagnetism is stronger. The T_x anomaly plays a key role in the superconductivity.

Under hydrostatic pressure, both T_{FM} and T_x are suppressed, falling to zero at approximately $P_c \simeq 1.5$ and $P_x \simeq 1.2$ GPa, respectively [108, 113, 129]. Superconductivity first appears around 0.9 GPa, has a peak near 1.2 GPa, and then falls slowly, before disappearing abruptly at P_c [52, 129], which appears to be a first-order change from a ferromagnetic to a paramagnetic ground state [52]. The maximum T_c is 0.8 K, about 40 times smaller than the magnetic ordering temperature T_{FM} at 1.2 GPa, strongly favouring a spin-triplet pairing scenario as discussed above. Quantum oscillation measurements under pressure [136, 163] confirm that the Fermi surface changes abruptly at P_{FM} from spin-split to paramagnetic. They also show that, at P_x, two Fermi surfaces disappear, and the Fermi surface may become fully spin polarized.

Although superconductivity is observed in resistivity between 0.9 and 1.5 GPa, specific heat measurements [162] reveal that fully developed (i.e. bulk) superconductivity only occurs in a narrow range of pressures around P_x. The implication is that fluctuations associated with the P_x transition mediate the formation of Cooper pairs, but detailed understanding has yet to be achieved.

At ambient field and pressure, URhGe is ferromagnetic below $T_{FM} = 9.6$K with a $T > T_{FM}$ linear coefficient of specific heat of 160 mJ/moleK^2 [37], and superconductivity with $T_c = 0.2$ K is observed at ambient pressure [7]. As in UGe_2, $T_c \ll T_{FM}$. The upper critical field H_{c2} exceeds the Pauli limit for all directions of applied field, consistent with equal-spin spin-triplet pairing [7]. URhGe does not have a T_x transition, and its magnetic anisotropy is different from that of UGe_2, whereas UGe_2 has very strong uniaxial anisotropy, URhGe has a weak uniaxial anisotropy, with an easy axis (the *c*-axis), a slightly harder *b*-axis so that an applied field along *b* can reorient the ferromagnetic moment, and a hard *a*-axis. A remarkable property of this material is that the superconductivity is re-entrant for magnetic fields applied in the *b*–*a*-plane [83, 84]. A magnetic field applied along *b* at low temperature initially suppresses superconductivity so that $T_c \to 0$ K by $H_{c2} \simeq 2$ T. At a higher field of about 11.7 T, the field causes a rapid rotation of the ordered ferromagnetic moment from the *c*- to the *b*-axis [83]. Remarkably, in a narrow range of fields around this moment reorientation field, superconductivity re-appears. Moreover, rotating the field toward the hard *a*-axis increases the field required for the magnetic reorientation transition, and re-entrant superconductivity at fields up to 28 T has been observed [84]! Quantum oscillation measurements have found that a small, heavy pocket of the Fermi surface vanishes near the spin reorientation field, raising the possibility that this Fermi surface reconstruction supports the re-entrant superconductivity [175].

Single crystals of UCoGe at ambient pressure are ferromagnetic below $T_{FM} \sim 3$ K and superconducting below $T_c = 0.8$ K [53, 54]. Like UGe_2, the magnetic state has strong uniaxial anisotropy with an easy *c*-axis and *a* and *b* being magnetically hard. Pressure suppresses ferromagnetism, with $T_{FM} \to 0$ K around 1.0 GPa, but T_c is relatively insensitive to pressure [39]. Thus, unlike UGe_2, superconductivity in this system survives in the paramagnetic state. (Note that the paramagnetic state cannot be reached in URhGe through application of pressure.) One of the key results on this system was an NMR study which correlated a rapid collapse of T_c for fields applied along the easy axis with the suppression of spin fluctuations [41], providing strong evidence that longitudinal spin fluctuations provide the superconducting pairing mechanism.

The T_c of 1.6 K of the recently discovered supereconducting phase of UTe_2 [122] is significantly higher than the ferromagnetic superconductors discussed above. A major difference is that UTe_2 has a paramagnetic normal state, although it is close to being ferromagnetic. Like the other systems, the upper critical field of UTe_2 exceeds the Pauli limit, consistent with equal-spin pairing [11, 122]. Moreover, all samples tested to date have a large residual entropy as $T \to 0$ K, equal to about half of the normal-state entropy, implying that only half of the eletrons become paired. This too is strong evidence for an equal-spin paired state with only one spin direction forming Cooper pairs. But in the paramagnetic normal state there is no spin-splitting at $T > T_c$ (at zero field), so superconductivity would have to lift the spin degeneracy of the Fermi surface, which is a strange idea.

It is fair to say that, of the various materials systems discussed in this chapter, the most interesting at the time of writing is the superconductivity of uranium heavy fermion systems near the border of ferromagnetic order.

D1.3.7 Noncentrosymmetric Heavy Fermion Superconductors

To say that a crystal structure is noncentrosymmetric means that it has no center of inversion symmetry. In this case, parity (symmetry under inversion of coordinates) is not a good symmetry, and the classification of the orbital part of the Cooper pair wave-function by parity, discussed in Section D1.2.1, no longer applies. As a result, noncentrosymmetric superconductors are neither spin-singlet nor -triplet, but some combination of the two.

Conduction electron eigenstates in noncentrosymmetric crystals differ from those in centrosymmetric crystals in that spin-up and spin-down states are not degenerate. Rather, spin–orbit coupling combines with non-inversion-symmetric electric fields to produce an effective magnetic field that is perpendicular to the momentum of the electron. The spin-up and spin-down bands are thereby split as illustrated in Figure D1.3. Note that the spin quantization axis rotates, always being perpendicular to $\vec{k}$ (see e.g. Sigrist, 2009). Such "Rashba-Dresselhaus" splitting of the Fermi surface has been observed in quantum oscillation measurements of a number of noncentrosymmetric heavy fermions (see e.g. Kimura et al., 2007).

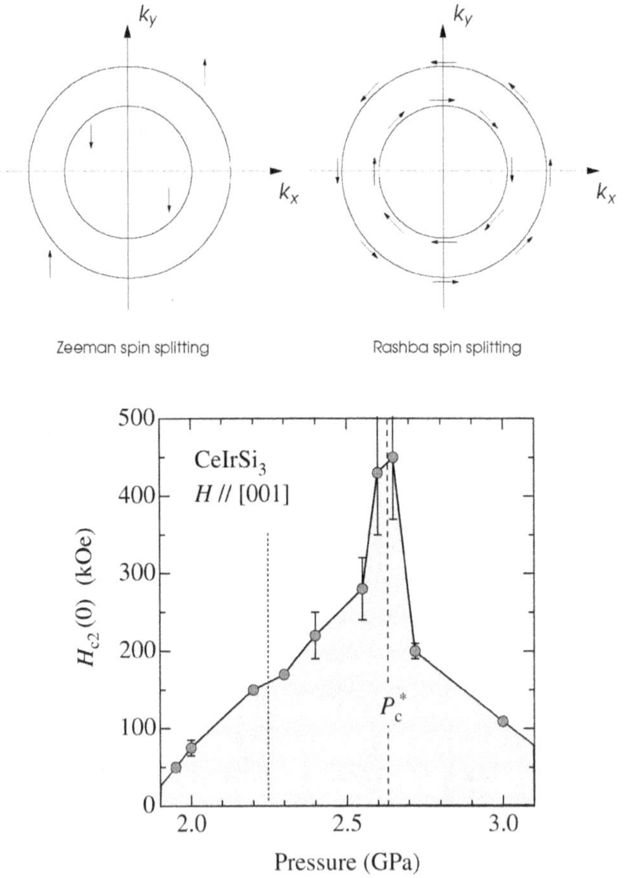

FIGURE D1.3 Top: Comparison, for a two-dimensional Fermi surface, of Zeeman splitting (left) with Rashba spin–orbit splitting (right). Reproduced from M. Sigrist, AIP Conference Proceedings, (2009) 1162:55, with permission of AIP Publishing. The Zeeman splitting lifts the degeneracy of the up- and down-spin Fermi surfaces, suppressing pairing for singlet superconductivity via the Pauli limiting mechanism. For Rashba splitting, if the applied field is smaller than the spin–orbit strength, then the field has little effect on portions of the Fermi surface where the spin is perpendicular to the applied field direction, so superconductivity is comparatively insensitive to the applied field. Lower figure: Upper critical field of CeIrSi$_3$ vs. pressure, showing a very large H_{c2} near 2.6 GPa, which is close to the pressure where T_N would go to zero in the absence of superconductivity (i.e. the antiferromagnetic quantum critical point). The dotted line indicates the pressure where the falling T_N and rising T_c cross. Reproduced from R. Settai et al., J. Phys. Soc. Japan, (2008) 77:073705, © (2008) The Physical Society of Japan.

The physics of noncentrosymmetric superconductors has recently been reviewed [15, 148]. There has been an increase in interest due to the realization that noncentrosymmetric superconductors can host topologically non-trivial surface states that may be applicable to quantum computing [128].

Several noncentrosymmetric heavy fermion superconductors have been discovered, starting with $CePt_3Si$ in 2004 [14], followed by $CeRhSi_3$ [69], $CeIrSi_3$ [156], $CeCoGe_3$ [137], $CeIrGe_3$ [49], and $CePtSi_2$/$CeRhGe_2$ [104].

These systems share key features. They are all antiferromagnetic at ambient pressure, with T_N typically on the order of a few kelvin. Under applied pressure, the antiferromagnetism is suppressed at a critical pressure P_c, and the superconducting dome, approximately centered on P_c, is quite broad. There is strong evidence for nodes in the superconducting gap, which would mean that the superconducting state is unconventional. For example in $CePt_3Si$, magnetic penetration depth [19], thermal conductivity [60], and specific heat [160] measurements in the superconducting state are all consistent with line nodes in the superconducting gap. A striking property is that H_{c2} tends to be quite anisotropic, and sometimes spectacularly high. For example in $CeIrSi_3$, which has a critical pressure for suppression of antiferromagnetism of $P_c \sim 2.26$ GPa, H_{c2} close to P_c is highly anisotropic, and exceeds 40 T for $H \parallel c$, with a T_c of only ~ 1.6 K [138].

The very high ratio of H_{c2}/T_c can be understood if the applied field is weak compared to the Rashba splitting of the bands. For the splitting shown in Figure D1.3, if the magnetic field is applied perpendicular to the plane of the page, then the stronger Rashba coupling will prevent the spins from rotating towards the field, and thus the field has minimal Pauli paramagnetic limiting effect until it becomes comparable to the Rashba splitting, which may require 100's of tesla.

D1.3.8 Miscellaneous Other Systems

It should be clear by now that superconductivity in heavy fermion systems, because it is unconventional, comes in many different species, depending on whether it is spin-singlet or spin-triplet (or neither as in the noncentrosymmetric systems), and that for a particular spin pairing state, the orbital wave-function has further degrees of freedom, giving rise, for example, to different patterns of nodes in the superconducting gap. Moreover, the superconductivity in different materials arises from very different normal states. We have by no means exhausted the variety shown by heavy fermion superconductors, however. There are several more that don't fit into the categories above, and they are briefly described here.

D1.3.8.1 UPd_2Al_3 and UNi_2Al_3

UPd_2Al_3 orders antiferromagnetically at $T_N = 14$ K and is superconducting below $T_c = 2$ K [33], which are both comparatively high ordering temperatures for heavy fermion systems. Moreoever, the antiferromagnetic ordered moment of $0.85\mu_B$ [75] is large compared with most heavy fermion antiferromagnets, but still small compared with the Curie–Weiss moment of ~ 3.2 μ_B measured at high temperature [33], so there is still considerable magnetic entropy below T_N, and the normal state γ is 0.14 J/mol K as $T \to 0$ K. There is evidence that the uranium 5f-electrons have a dual nature, with a more-localized set being responsible for magnetism, and a less-localized set being responsible for superconductivity [25, 127, 178]. From the first discovery there was strong evidence of nodes in the gap [25, 33, 80], and a recent study of specific heat as a function of applied magnetic field direction provides strong evidence for a horizontal line node [142].

The isostructural and isoelectronic system UNi_2Al_3 is also a large-moment antiferromagnetic heavy fermion system, in this case with $T_N = 1.4$ K and $T_c = 1.06$ K [34], but the moment is significantly smaller than UPd_2Al_3, and rather than being commensurate as in UPd_2Al_3 the magnetism of UNi_2Al_3 is incommensurate [134], consistent with itinerant magnetism.

D1.3.8.2 URu_2Si_2

Superconductivity was discovered in URu_2Si_2 by Palstra et al. in 1985 [117], making this one of the earliest heavy fermion superconductors. The superconducting transition temperature is somewhat sample dependent, with T_c as high as 1.53 K in high purity samples, and the upper critical field is large and highly anisotropic: $H_{c2}^{a} = 14$ T vs. $H_{c2}^{c} = 3$ T [21].

Interest in superconductivity in this system has been overshadowed by a 'hidden order' phase transition at 17.5 K at which a large amount of entropy is lost [117]). Despite intense study for many years, the order parameter is still not agreed upon (for a recent update see [103]). Under pressure, the 'hidden order' phase is replaced by a large-moment antiferromagnetic ground state [6], and the superconductivity disappears simultaneously with the appearance of the antiferromagnetic ground state [40].

There are several indications that the superconducting phase is itself exotic, with very interesting properties. For example, thermal conductivity and specific heat measurements are consistent with *d*-wave pairing [65, 174], possibly of the chiral type [66]. The group of Kapitulnik et al. using very sensistive measurements of the polar Kerr effect have recently reported that time-reversal symmetry is broken in the superconducting phase but not in the hidden order phase, providing further strong evidence of chiral superconductivity [132]. Finally, an enormous enhancement of the Nerst effect has been seen in the normal state just above T_c, presumably due to superconducting fluctuations of a chiral nature [172], which also indicates that the superconducting state breaks time-reversal symmetry consistent with a chiral-*d*-wave superconducting state.

D1.3.8.3 UBe_{13}

Superconductivity was first observed in UBe_{13} in 1975 [23], but was only shown to be intrinsic, involving formation of Cooper pairs by heavy quasiparticles, by Ott in 1983 [114]. The T_c is ~ 0.9 K with a very large normal-state specific heat coefficient of $\gamma \sim 1$ J/mole·K^2. The normal state is a non-Fermi-liquid down to a temperature very close to T_c. For example, the resistivity has a peak at about 2 K, and is still very large with a $T^{3/2}$ power law when superconductivity sets in [32]. Such behaviour in heavy fermion systems is normally associated with quantum criticality (Section D1.2.4), but UBe_{13} is not known to be close to ordering magnetically in the $T \to 0$ limit, which has led to suggestions that the famous two-channel Kondo model of Cox *et al.* may apply to the ground state of UBe_{13} [27].

The symmetry of the superconducting wave-function is controversial. Early measurements of specific heat within the superconducting state were interpreted in terms of point nodes [115], but later work suggested line nodes [35, 90]. Recent measurements of specific heat deep in the superconducting state indicate, however, that the gap is nodeless [141]. The balance of evidence is that the superconductivity of UBe_{13} is exotic, so a nodeless gap would indicate that either the symmetry is consistent with point nodes located on open sections of the Fermi surface, or the superconductivity removes the nodes by being chiral.

D1.3.8.4 $PrOs_4Sb_{12}$ and $PrTr_2Al_{20}$

Superconductivity is found in many praseodymium compounds, but heavy fermion behaviour is rare.

$PrOs_4Sb_{12}$ is a 'filled skutterudite' system, in which a cage of 12 Sb ions encloses each Pr ion, leaving the Pr with considerable freedom to oscillate around its equilibrium position. The Pr ion has a $4f^2$ configuration, and the $J = 4$ ground state that would be nine-fold degenerate in the isolated ion is split by the electric field of the surrounding Sb ions such that there is a singlet ground state with a low-lying triplet excited state only 8 K above the ground state. Superconductivity sets in at $T_c = 1.8$K [13], with a very large jump in the specific heat suggesting that heavy fermions become paired at T_c. The properties of the superconducting state are not settled, however, partly due to variations between samples, and partly due to multi-band superconductivity (i.e. the gap has very different magnitude, and possibly different symmetry, on different Fermi surfaces) [139] (see Section B-2). In some single-crystal samples the superconducting transition is split [59], but not in others [140]. Strong evidence for unconventional superconductivity comes from the observation of broken time-reversal symmetry in the superconducting state by both muon spin resonance and Kerr effect measurements [9, 82], and from measurements which show that the penetration depth has power law temperature dependence consistent with point nodes [26]. Specific heat measurements, however, observe no low-lying quasiparticles deep in the superconducting state [74, 140], implying a fully developed gap over the entire Fermi surface.

A most interesting aspect of the superconductivity of $PrOs_4Sb_{12}$ is the possibility that it is mediated by quadrupolar fluctuations: propagating excitations that are coherent superpositions of Pr singlet and triplet states. Neutron scattering has identified quadrupolar excitations as the predominant elementary excitations in this system [78], indeed in an applied field not much stronger than H_{c2} these quadrupolar excitations condense to form an antiferroquadrupolar phase [8, 73], in which alternating Pr sites have different static superpositions of singlet and triplet states.

Two $PrTr_2Al_{20}$ systems, where Tr=Ti or V, are also strong candidates for quadruoplar-fluctuation-mediated superconductivity. Unlike $PrOs_4Sb_{12}$ which has a non-magnetic *singlet* crystal-field ground state, these systems have a non-magnetic *doublet* ground state, involving two orbitals with zero angular momentum [e.g. $(|J_z = 4\rangle + |J_z = -4\rangle)/\sqrt{2}$ or $|J_z = 0\rangle$] that have different spatial shape but are degenerate in energy. PrV_2Al_{20} above 0.7 K has a large linear specific heat coefficient $\gamma \sim 0.9$ J/mole K^2, then at 0.7 K it undergoes a phase transition to an 'antiferroquadrupolar' ordered state which becomes superconducting below $T_c = 50$ mK in high-quality samples. The upper critical field is rather modest: $B_{c2} \sim 0.01$ T [166]. $PrTi_2Al_{20}$ at first glance seems less interesting because the normal-state specific heat is substantially lower, ferroquadrupolar order (in which one of the two degenerate orbitals is predominant on all sites) sets in at a comparatively high temperature $T_c \sim 2$ K, and superconductivity with a T_c of 200 mK has a comparatively low upper critical field of $B_{c2} \sim 0.006$ T [125]. Under pressure, however, it becomes much more interesting: the normal-state specific heat becomes very large, T_Q falls, and the superconductivity strengthens very markedly, so that by 8 GPa T_c is over 1 K, and B_{c2} has risen to over 3 T, which places this material squarely in the category of heavy fermion superconductors [95]. The behaviour suggests that a ferroquadrupolar quantum critical point is approached at high pressure, and the enhancement of T_c and B_{c2} is strong evidence that quadrupolar fluctuations mediate the superconductivity.

D1.3.8.5 β-$YbAlB_4$

Compounds in which the ytterbium (Yb) ion is close to being 3+ should in principle give rise to hole-like analogues of cerium heavy fermion compounds, since they have one hole in the $4f$-shell, as opposed to having one electron. In practice, however, there are very few Yb heavy fermion systems, and of these only two Yb superconductors have been discovered: β−$YbAlB_4$ [106] and, $YbRh_2Si_2$, in which superconductivity was very recently found below 2 mK [135]. β-$YbAlB_4$ has an orthorhombic crystal structure, in which a planar network of boron atoms separates YbAl layers. An important structural motif is zig-zag chains of Yb atoms, reminiscent of the ferromagnetic uranium superconductors (Section D1.3.6). The T_c of 80 mK in this material is very sensitive to impurities, indicative of unconventional superconductivity. A very surprising aspect of β-$YbAlB_4$ is that it has a non-Fermi-liquid normal state at ambient pressure – the resistivity varies as $\rho \propto T^{1.5}$, specific heat as $\frac{C}{T} \propto \mu \ln\left(\frac{T^*}{T}\right)$, and susceptibility as $\chi \propto T^{-\frac{1}{2}}$ [106]. This non-Fermi-liquid state is rapidly suppressed by applied magnetic fields, being replaced by a Fermi liquid ground state under the influence of magnetic fields as small as 200 mT [96, 106], but it is less sensitive to applied pressure [164], surviving up to about 0.4 GPa, above which a Fermi liquid state sets in, to be followed at still higher pressures of about 2.5 GPa by antiferromagnetism. Superconductivity is also suppressed by increasing pressure, disappearing long before the antiferromagnetic quantum critical point is reached. Thus, neither the superconductivity nor the non-Fermi-liquid ground state are associated with the pressure-induced antiferromagnetic quantum critical point. Comparatively little is known about the superconducting state. The upper critical field H_{c2} is approximately 150 mT for $\vec{B} \| ab$ [76], which is large for a T_c of 80 mK, but below the field that suppresses non-Fermi-liquid behaviour.

D1.4 Superconductivity in Sr_2RuO_4

Superconductivity was discovered in Sr_2RuO_4 by Maeno *et al.* in 1994 [91]. This layered perovskite oxide has the same crystal structure as La_2CuO_4, the parent compound of an important family of cuprate superconductors, and to this day Sr_2RuO_4 is one of only a handful of superconducting perovskite oxides that have been found outside of the cuprate family.

Unlike the cuprates, Sr_2RuO_4 is superconducting without doping, indeed T_c is very rapidly suppressed by impurities and defects [88]. Maeno and his co-workers spent several years perfecting crystal growth in this system, ultimately producing crystals of exceptional purity [93] with T_c values up to 1.5 K, using vertical floating zone growth in mirror image furnaces.

Unlike the cuprates, Sr_2RuO_4 has a Fermi liquid normal state, and this state has been well characterized by thermodynamic and transport properties [16]. These reveal the normal state to be a highly anisotropic Fermi liquid [92], having a resistivity anisotropy at low temperature of $\frac{\rho_c}{\rho_{ab}} \sim 1400$ [50]. Moreover, a complete quantum oscillation study has been carried out, which found that the Fermi surface is very simple, consisting of three slightly warped tubes of rather square cross-section, populated by quasiparticles with effective masses as high as $14.4m_e$ [87]. This is a high level of enhancement for a non-*f*-electron system. Early photoemission measurements provided a clear picture of the band structure near the Fermi energy [29] while recent, remarkably detailed, measurements [161] reveal that, beneath it's apparent simplicity, the energy bands display a rich mixture of orbital states, arising from spin–orbit coupling.

Very soon after the discovery of superconductivity in Sr_2RuO_4, Sigrist and Rice suggested that it might be analogous to ^3He [147], and thus it could be a spin-triplet *p*-wave superconductor, and much subsequent interest in this compound arose because it may be a *chiral p*-wave superconductor (for a recent review see Kallin, 2016). As briefly described above in Section D1.2.1.1, in a chiral superconductor the superconducting electrons spontaneously develop a non-zero angular momentum, in order to elminate nodes in the superconducting gap. Such a state breaks time-reversal symmetry, and it is *topological*, with surface states in the superconducting gap for momenta $\vec{p}$ parallel to the surface and perpendicular to the chiral axis.

There is abundant evidence that the superconductivity of Sr_2RuO_4 is unconventional. For example, there is no coherence peak below T_c in NMR measurements [56], and the superconductivity is dramatically suppressed by non-magnetic impurities [88]. Strong evidence that it is chiral comes from μSr [86] and polar Kerr effect measurements [171] which see broken time-reversal symmetry in the superconducting state. There is, moreover, evidence that the unconventional superconducting state is *p*-wave: a thorough exploration of the *B-T* phase diagram finds only square flux lattices, evidence of spin-triplet superconductivity according to Agterberg [3, 124]. On the other hand, Knight shift measurements that saw a slight *enhancement* of the spin susceptibility upon entering the superconducting mixed state [57], which would be strong evidence of *p*-wave superconductivity, have recently been challenged [120]. Moreover, the superconducting transition in applied fields, and below ~ 0.9 K, becomes first-order [71], which is predicted only for singlet, not triplet, superconductors. And recently, measurements of T_c as a function of uniaxial strain [46] found that T_c increases for both compressive and tensile strain applied along the (100) direction, however the T_c vs. strain curve is U-shaped, rather than V-shaped as predicted by *p*-wave theory [146, 170], also suggesting that Sr_2RuO_4 may not be a *p*-wave superconductor. At even larger strains [150], T_c is observed to pass through a maximum close to a Van Hove singularity that, according to theory, would kill off *p*-wave superconductivity, further weakening the *p*-wave case. As noted by Mackenzie et al. [89], this continuing uncertainty regarding spin-singlet vs. spin-triplet superconductivity in what is probably the highest purity material, with the simplest normal state, and the most thoroughly studied superconducting properties, is an indication of how much work remains to be done in developing definitive probes of the symmetry of unconventional superconducting states.

A definite prediction for *chiral p*-wave superconductors is that there should be edge currents at the boundary of the sample that could produce an observable magnetic field. Attempts to observe these states using scanning SQUID microscopy [45, 70] placed the upper limit on these currents of 1000 times smaller than predicted by the simplest theory, however more recent (but still realistic) theories predict much lower edge currents, in agreement with the observations [130]. In-plane tunneling measurements [67] observe distinctive features in the conductance spectra that are expected for a chiral *p*-wave symmetry gap function.

In a chiral $p_x \pm ip_y$ state, the up-spin and down-spin Cooper pairs can be thought of as two interpenetrating superfluids, $|\uparrow\uparrow\rangle$ and $|\downarrow\downarrow\rangle$. In the mixed state an exotic type of 'half-quantum' vortex becomes possible. Cantilever-based torque measurements [61] on samples measuring less than 2 μm × 2 μm in area and with a hole drilled through the centre by focussed ion beam milling, observed half-quantum steps in the magnetization, providing strong evidence for a chiral state.

These half-quantum vortices suggest that vortices in Sr_2RuO_4 support Majorana fermions, which could have relevance in quantum computing (see e.g. [64]).

Finally, if Sr_2RuO_4 is a p-wave superconductor, it is still not clear why it should be so. Neither neutron scattering nor thermodynamic properties seem to place this system especially close to ferromagnetic ordering. However, the discovery of superconductivity in the isoelectronic perovskite Ca_2RuO_4, at high pressure on the boundary of ferromagnetism [4], is consistent with a p-wave scenario in which pairing is mediated by paramagnons.

Sr_2RuO_4 still presents many open questions, despite having probably the best characterized normal state and the most thoroughly studied superconducting state of all of the unconventional superconductors discussed in this chapter, and it is perhaps fitting to close this chapter with this particular demonstration of how far we have come, but how far we have still to go, in understanding the properties of unconventional strongly correlated superconductors.

D1.5 Summary

Unconventional heavy fermion and ruthenate superconductors show a wonderful variety of superconducting states, including spin-singlet and spin-triplet Cooper paired states, many different states with nodes in the superconducting gap, and examples of nodeless chiral superconducting states. These superconducting states emerge from an even greater variety of normal states: non-magnetic Fermi liquids as in Sr_2RuO_4, weakly antiferromagnetic Fermi liquids as in UPt_3, strongly antiferromagnetic Fermi liquids as in UNi_2Al_3, ferromagnetic and antiferromagnetic quantum critical non-Fermi-liquids as in UGe_2 and in $CePd_2Si_2$ respectively, and non-Fermi-liquid normal states of unknown origin, as in UBe_{13} and β-$YbAlB_4$.

While there is still a long way to go in understanding, classifying, and characterizing unconventional superconductors, their rich physics, their promise as an aveue for understanding and discovering new high-temperature superconductors, and their potential applications in quantum computing, make this an important and vital area of research.

References

[1] Adenwalla S et al (1990) Phase-diagram of UPt_3 from ultrasonic velociy-measurements. *Phys. Rev. Lett.* 65:2298–2301.

[2] Aeppli G, Bucher E, Broholm C, Kjems JK, Baumann J, Hufnagl J (1988) Magnetic order and fluctuations in superconducting UPt_3. *Phys. Rev. Lett.* 60:615–618.

[3] Agterberg D F (1998) Vortex lattice structures of Sr_2RuO_4. *Phys. Rev. Lett.* 80:5184–5187.

[4] Alireza PL et al. (2010) Evidence of superconductivity on the border of quasi-2d ferromagnetism in Ca_2RuO_4 at high pressure. *J. Phys. Cond. Mat.* 22:052202.

[5] Allan MP et al. (2013) Imaging Cooper pairing of heavy fermions in $CeCoIn_5$ *Nat. Phys.* 9:468.

[6] Amitsuka H et al. (1999) Effect of pressure on tiny antiferromagnetic moment in the heavy electron compound URu_2Si_2. *Phys. Rev. Lett.* 83:5114–5117.

[7] Aoki D et al. (2001) Coexistence of superconductivity and ferromagnetism in URhGe. *Nature* 413:613–616.

[8] Aoki Y, Namiki T, Ohsaki S, Saha SR, Sugawara H, Sato H (2002) Thermodynamical study on the heavy fermion superconductor $PrOs_4Sb_{12}$: evidence for field-induced phase transition. *J. Phys. Soc. Jpn.* 71:2098–2101.

[9] Aoki Y et al. (2003) Time-reversal symmetry-breaking superconductivity in heavy fermion $PrOs_4Sb_{12}$ detected by muon-spin relaxation. *Phys. Rev. Lett.* 91:067003.

[10] Aoki D, Ishida K, Flouquet J (2019) Review of U-based ferromagnetic superconductors: comparison between UGe_2, URhGe and UCoGe. *J. Phys. Soc. Jpn.* 88:022001.

[11] Aoki, D et al. (2019) Unconventional superconductivity in heavy fermion UTe_2. *J. Phys. Soc. Jpn.* 88:043702.

[12] Assmus W et al. (1984) Superconductivity in $CeCu_2Si_2$ single crystals. *Phys. Rev. Lett.* 52:469–472.

[13] Bauer ED, Frederick NA, Ho P-C, Zapf VS, Maple MB (2002) Superconductivity and heavy fermion behaviour in $PrOs_4Sb_{12}$. *Phys. Rev. B* 65:100506(R).

[14] Bauer E et al. (2004) Heavy fermion superconductivity and magnetic order in noncentrosymmetric $CePt_3Si$. *Phys. Rev. Lett.* 92:027003.

[15] Bauer E, Sigrist M (eds) (2012) Non-centrosymmetric superconductors: introduction and overview. *Lecture Notes in Physics* 847 (Heidelberg: Springer). Chapter 2, by Kimura N, Bonalde I, reviews noncentrosymmetric superconductivity in heavy fermion systems.

[16] Bergemann C, Mackenzie AP, Julian SR, Forsythe D, Ohmichi E (2003) Quasi-two-dimensional Fermi liquid properties of the unconventional superconductor Sr_2RuO_4. *Adv. Phys.* 52:639–725.

[17] Bianchi A, Movshovich R, Vekhter I, Pagliuso PG, Sarrao JL (2003) Avoided antiferromagnetic order and quantum critical point in $CeCoIn_5$. *Phys. Rev. Lett.* 91:257001.

[18] Blount EI (1985) Symmetry properties of triplet superconductors. *Phys. Rev. B* 32:2935–2944.

[19] Bonalde I, Brämer-Escamilla W, Bauer E (2005) Evidence for line nodes in the superconducting energy gap of noncentrosymmetric $CePt_3Si$ from magnetic penetration depth measurements. *Phys. Rev. Lett.* 94:207002.

[20] Brison JP et al. (1994) Magnetism and superconductivity in heavy-fermion systems. *J. Low Temp. Phys.* 95:145–152.

[21] Brison JP et al. (1995) Anisotropy of the upper critical field in URu_2Si_2 and FFLO state in antiferromagnetic superconductors. *Physica C* 250:128–138.

[22] Brown SA, Howard BK, Brown SV, Julian SR (1990) A new design for a UHV compatible Czochralski crystal-growth system. *Rev. Sci. Inst.* 61:2427–2429.

[23] Bucher E, Maita JP, Hull GW, Fulton RC, Cooper AS (1975) Electronic properties of beryllides of rare-earth and some actinides. *Phys. Rev. B* 11:440–449.

[24] Canfield PC, Fisk Z (1992) Growth of single-crystals from metallic fluxes. *Phil. Mag. B* 65:1117–1123.

[25] Caspary R et al. (1993) Unusual ground-state properties of UPd_2Al_3: implications for the coexistence of heavy-fermion superconductivity and local-moment antiferromagnetism. *Phys. Rev. Lett.* 71:2146–2149.

[26] Chia EEM, Salamon MB, Sugawara H, Sato H (2003) Probing the superconducting gap symmetry of $PrOs_4Sb_{12}$: a penetration depth study. *Phys. Rev. Lett.* 91:247003.

[27] Cox DL, Zawadowski A (1998) Exotic Kondo effects in metals: magnetic ions in a crystalline electric field and tunnelling centres. *Adv. Phys.* 47:599–942.

[28] Curro NJ et al. (2005) Unconventional superconductivity in $PuCoGa_5$. *Nature* 434:622–625.

[29] Damascelli A et al. (2000) Fermi surface, surface states, and surface reconstruction in Sr_2RuO_4. *Phys. Rev. Lett.* 85:5194.

[30] Fisher RA et al. (1989) Specific heat of UPt_3 - evidence for unconventional superconductivity. *Phys. Rev. Lett.* 62:1411–1414.

[31] Fort D (1997) A tri-arc system for growing high-purity crystals of metallic materials. *Rev. Sci. Inst.* 68:3504–3511.

[32] Gegenwart P et al. (2004) Non-Fermi liquid normal state of the heavy fermion superconductor UBe_{13}. *Physica C* 408-410C:157–160.

[33] Geibel C et al. (1991) Heavy-fermion superconductivity at $T_c = 2$ K in the antiferromagnet UPd_2Al_3. *Z. Phys. B: Condens. Matter* 84:1–2.

[34] Geibel C et al. (1991) A new heavy-fermion superconductor - UNi_2Al_3. *Z. Phys. B: Condens. Matter* 83:305–306.

[35] Golding B et al. (1985) Observation of a collective mode in UBe_{13}. *Phys. Rev. Lett.* 55:2479–2482.

[36] Haga Y, Yamamoto E, Kimura N, Hedo M, Ohkuni H, Onuki Y (1998) High-quality single crystal growth of uranium-based intermetallics. *J. Magn. Magn. Mater.* 177-181:437–438, and references therein.

[37] Hagmusa IH et al. (2000) Magnetic specific heat of a URhGe single crystal. *Physica B* 281-282:223–225.

[38] Hasselbach K, Taillefer L, Flouquet J (1989) Critical point in the superconducting phase diagram of UPt_3. *Phys. Rev. Lett.* 63:93–96.

[39] Hassinger E, Aoki D, Flouquet J (2008) Pressure-temperature phase diagram of polycrystalline UCoGe studied by resistivity measurement. *J. Phys. Soc. Jpn.* 77:073703.

[40] Hassinger E, Knebel G, Izawa K, Lejay P, Salce B, Flouquet J (2008) Temperature-pressure phase diagram of URu_2Si_2 from resistivity measurements and ac calorimetry: hidden order and Fermi surface nesting. *Phys. Rev. B* 77:115117.

[41] Hattori T et al. (2012) Superconductivity induced by longitudinal ferromagnetic fluctuations in UCoGe. *Phys. Rev. Lett.* 108:066403.

[42] Hayden SM, Taillefer L, Vettier C, Flouquet J (1992) Antiferromagnetic order in UPt_3 under pressure: evidence for a direct coupling to superconductivity. *Phys. Rev. B* 46:8675–8678.

[43] Hegger H et al. (2000) Pressure-induced superconductivity in quasi-2D $CeRhIn_5$. *Phys. Rev. Lett.* 84:4986–4989.

[44] Hertz JA (1976) Quantum critical phenomena. *Phys. Rev. B* 14:1165–1184.

[45] Hicks CW et al. (2010) Limits on superconductivity-related magnetism in Sr_2RuO_4 and $PrOs_4Sb_{12}$ from scanning SQUID microscopy. *Phys. Rev. B* 81:214501.

[46] Hicks CW et al. (2014) Strong increase of T_c of Sr_2RuO_4 under both tensile and compressive strain. *Science* 344:283–285.

[47] Holmes AT, Jaccard D, Miyake K (2004) Signatures of valence fluctuations in $CeCu_2Si_2$ under pressure. *Phys. Rev. B* 69:024508.

[48] Holter G, Adrian H (1986) Thin-film preparation, superconductivity and transport properties of the heavy fermion system $CeCu_2Si_2$. *Solid State Commun.* 58:45–49.

[49] Honda F et al. (2010) Pressure-induced superconductivity and large upper critical field in the noncentrosymmetric antiferromagnet $CeIrGe_3$. *Phys. Rev. B* 81:140507(R).

[50] Hussey NE, Mackenzie AP, Cooper JR, Maeno Y, Nishizaki S, Fujita T (1998) Normal-state magnetoresistance of Sr_2RuO_4. *Phys. Rev. B* 57:5505-5511.

[51] Huxley A, Rodière P, Paul DM, van Dijk N, Cubitt R, Flouquet J (2000) Realignment of the flux-line lattice by a change in the symmetry of superconductivity in UPt_3. *Nature* 406:160–164.

[52] Huxley A et al. (2001) UGe_2: a ferromagnetic spin-triplet superconductor. *Phys. Rev. B* 63:144519.

[53] Huy NT et al. (2007) Superconductivity on the border of weak itinerant ferromagnetism in UCoGe. *Phys. Rev. Lett.* 99:067006.

[54] Huy NT, de Nijs DE, Huang YK, de Visser A (2008) Unusual upper critical field of the ferromagnetic superconductor UCoGe. *Phys. Rev. Lett.* 100:077002.

[55] Ikeda H, Suzuki M-T, Ryotaro A (2015) Emergent loop-nodal $s_\pm$-wave superconductivity in $CeCu_2Si_2$: similarities to the iron-based superconductors. *Phys. Rev. Lett.* 114:147003.

[56] Ishida K et al. (1997) Anisotropic pairing in superconducting Sr_2RuO_4: Ru NMR and NQR studies. *Phys. Rev. B* 56:R505–R508.

[57] Ishida K et al. (2015) Spin polarization enhanced by spin-triplet pairing in Sr_2RuO_4 probed by NMR. *Phys. Rev. B* 92:100502(R).

[58] Izaki M, Shishido H, Kato T, Shibauchi T (2007) Superconducting thin films of heavy-fermion compound $CeCoIn_5$ prepared by molecular beam epitaxy. *Appl. Phys. Lett.* 91:122507.

[59] Izawa K et al. (2003) Multiple superconducting phases in new heavy fermion superconductor $PrOs_4Sb_{12}$. *Phys. Rev. Lett.* 90:117001.

[60] Izawa K et al. (2005) Line nodes in the superconducting gap function of noncentrosymmetric $CePt_3Si$. *Phys. Rev. Lett.* 94:197002.

[61] Jang J et al. (2011) Observation of half-height magnetization steps in Sr_2RuO_4. *Science* 331:186–188.

[62] Jourdan M, Huth M, Adrian H (1999) Superconductivity mediated by spin fluctuations in the heavy-fermion compound UPd_2Al_3. *Nature* 398:47–49.

[63] Joynt R, Taillefer L (2002) The superconducting phases of UPt_3. *Rev. Mod. Phys.* 74:235–294.

[64] Kallin C, Berlinsky J (2016) Chiral superconductors. *Rep. Prog. Phys.* 79:054502.

[65] Kasahara Y et al. (2007) Exotic superconducting properties in the electron-hole-compensated heavy-fermion semimetal URu_2Si_2. *Phys. Rev. Lett.* 99:116402.

[66] Kasahara Y et al. (2009) Superconducting gap structure of heavy-fermion compound URu_2Si_2 determined by angle-resolved thermal conductivity. *New J. Phys.* 11:055061.

[67] Kashiwaya S et al. (2011) Edge states of Sr_2RuO_4 detected by in-plane tunneling spectroscopy. *Phys. Rev. Lett.* 107:077003.

[68] Kenzelmann M et al. (2008) Coupled superconducting and magnetic order in $CeCoIn_5$. *Science* 321:1652.

[69] Kimura N, Ito K, Saitoh K, Umeda Y, Aoki H, Terashima T (2005) Pressure-induced superconductivity in noncentrosymmetric heavy-fermion $CeRhSi_3$. *Phys. Rev. Lett.* 95:247004.

[70] Kirtley JR et al. (2007) Upper limit on spontaneous supercurrents in Sr_2RuO_4. *Phys. Rev. B* 76:014526.

[71] Kittaka S et al. (2014) Sharp magnetization jump at the first-order superconducting transition in Sr_2RuO_4. *Phys. Rev. B* 90:220502(R).

[72] Kittaka S et al. (2014) Multiband superconductivity with unexpected deficiency of nodal quasiparticles in $CeCu_2Si_2$. *Phys. Rev. Lett.* 112:067002.

[73] Kohgi M et al. (2003) Evidence for magnetic field induced quadrupolar ordering in the heavy-fermion superconductor $PrOs_4Sb_{12}$. *J. Phys. Soc. Jpn.* 72:1002–1005.

[74] Kotegawa H et al. (2003) Evidence for unconventional strong-coupling superconductivity in $PrOs_4Sb_{12}$: an Sb nuclear quadrupole resonance study. *Phys. Rev. Lett.* 90:027001.

[75] Krimmel A et al. (1992) Neutron diffraction study of the heavy fermion superconductors UM_2Al_3 (M = Pd, Ni). *Z. Phys. B: Condens. Matter* 86:161–162.

[76] Kuga K, Karaki Y, Matsumoto Y, Machida Y, Nakatsuji S (2008) Superconducting properties of the non-fermi liquid system β-$YbAlB_4$. *Phys. Rev. Lett.* 101:137004.

[77] Kuramoto Y, Kitaoka Y (1999) The Dynamics of Heavy Electrons. Oxford, UK: Oxford University Press.

[78] Kuwahara K et al. (2005) Direct observation of quadrupolar excitons in heavy-fermion superconductor $PrOs_4Sb_{12}$. *Phys. Rev. Lett.* 95:107003.

[79] Kycia JB, Hong JI, Davis BM, Langdo T, Seidman DN, Halperin WP (1996) UPt_3 crystal growth and characterization. *Czech. J. Phys.* 46:775.

[80] Kyogaku M, Kitaoka Y, Asayama K, Geibel C, Schank C, Steglich F (1993) NMR and NQR studies of magnetism and superconductivity in the antiferromagnetic superconductors UM_2Al_3 (M=Ni and Pd). *J. Phys. Soc. Jpn.* 62:4016–4030.

[81] Lejay P, Muller J, Argoud R (1993) Crystal-growth and stoichiometry of the ternary silicides $CeRu_2Si_2$ and $Ce_{1-x}La_xRu_2Si_2$. *J. Cryst. Growth* 130:238–244.

[82] Levenson-Falk EM, Schemm ER, Aoki Y, Maple MB, Kapitulnik A (2018) Polar Kerr effect from time-reversal symmetry breaking in the heavy-fermion superconductor $PrOs_4Sb_{12}$. *Phys. Rev. Lett.* 120:187004.

[83] Lévy F, Sheikin I, Grenier B, Huxley AD (2005) Magnetic field-induced superconductivity in ferromagnet URhGe. *Science* 309:1343-1346.

[84] Lévy F, Sheikin I, Huxley A (2007) Acute enhancement of the upper critical field for superconductivity approaching a quantum critical point in URhGe. *Nat. Phys.* 3:460–463.

[85] Luke GM et al. (1993) Muon spin relaxation in UPt_3. *Phys. Rev. Lett.* 71:1466–1469.

[86] Luke GM et al. (1998) Time-reversal symmetry-breaking superconductivity in Sr_2RuO_4. *Nature* 394:558–561.

[87] Mackenzie AP et al. (1996) Quantum oscillations in the layered perovskite superconductor Sr_2RuO_4. *Phys. Rev. Lett.* 76:3786–3789.

[88] Mackenzie AP et al. (1998) Extremely strong dependence of superconductivity on disorder in Sr_2RuO_4. *Phys. Rev. Lett.* 80:161–164.

[89] Mackenzie AP et al. (2017) Even odder after twenty-three years: the superconducting order parameter puzzle of Sr_2RuO_4. *NJP Quantum Materials* 2:40.

[90] MacLaughlin DE et al. (1984) Nuclear magnetic resonance and heavy-fermion superconductivity in $(U,Th)Be_{13}$. *Phys. Rev. Lett.* 53:1833–1836.

[91] Maeno Y et al. (1994) Superconductivity in a layered perovskite without copper. *Nature* 372:372–534.

[92] Maeno Y et al. (1997) Two-dimensional Fermi liquid behavior of the superconductor Sr_2RuO_4. *J. Phys. Soc. Jpn.* 66:1405–1408.

[93] Mao ZQ, Mori Y, Maeno Y (1999) Suppression of superconductivity in Sr_2RuO_4 caused by defects. *Phys Rev. B* 60:610–614.

[94] Mathur ND et al. (1998) Magnetically mediated superconductivity in heavy fermion compounds. *Nature* 394:39–43; and references therein.

[95] Matsubayashi K, Tanaka T, Sakai A, Nakatsuji S, Kubo Y, Uwatoko Y (2012) Pressure-induced heavy-fermion superconductivity in the nonmagnetic quadrupolar system $PrTi_2Al_{20}$. *Phys. Rev. Lett.* 109:187004.

[96] Matsumoto Y et al. (2011) Quantum criticality without tuning in the mixed valence compound β-$YbAlB_4$. *Science* 331:316–319.
[97] Machida Y et al. (2012) Twofold spontaneous symmetry breaking in the heavy-fermion superconductor UPt_3 *Phys. Rev. Lett.* 108:157002.
[98] Micklitz T, Norman MR (2017) Symmetry-enforced line nodes in unconventional superconductors. *Phys. Rev. Lett.* 118:207001.
[99] Monthoux P, Lonzarich GG (2001) Magnetically mediated superconductivity in quasi-two and three dimensions. *Phys. Rev. B* 63:054529.
[100] Monthoux P, Lonzarich GG (2002) Magnetically mediated superconductivity: crossover from cubic to tetragonal lattice. *Phys. Rev. B* 66:224504.
[101] Movshovich R, Graf T, Mandrus D, Thompson JD, Smith JL, Fisk Z (1996) Superconductivity in heavy-fermion $CeRh_2Si_2$. *Phys. Rev. B* 53:8241–8244.
[102] Müller V et al. (1987) Ultrasonic determination of different phases in superconducting UPt_3. *Phys. Rev. Lett.* 58:1224–1227.
[103] Mydosh JA, Oppeneer PM (2014) Hidden order behaviour in URu_2Si_2 (A critical review of the status of hidden order in 2014) *Philos. Mag.* 94:3642–3662.
[104] Nakano T, Ohashi M, Oomi G, Matsubayashi K, Uwatoko Y (2009) Pressure-induced superconductivity in the orthorhombic Kondo compound $CePtSi_2$. *Phys. Rev. B* 79:172507.
[105] Nakatsuji S, Pines D, Fisk Z (2004) Two fluid description of the Kondo lattice. *Phys. Rev. Lett.* 92:016401.
[106] Nakatsuji S et al. (2008) Superconductivity and quantum criticality in the heavy-fermion system β-$YbAlB_4$. *Nat. Phys.* 4:603–607.
[107] Naritsuka M et al. (2018) Tuning the pairing interaction in a d-wave superconductor by paramagnons injected through interfaces. *Phys. Rev. Lett.* 120:187002.
[108] Nishimura K, Oomi G, Yun SW, Onuki Y (1994) Effect of pressure on the Curie temperature of single crystal UGe_2. *J. Alloys Compd.* 213–214:383–386.
[109] Nomoto Y, Ikeda H (2016) Exotic multigap structure in UPt_3 unveiled by first-principles analysis. *Phys. Rev. Lett.* 117:217002.
[110] Onishi Y, Miyake K (2000) Enhanced valence fluctuations caused by *f-c* Coulomb interaction in Ce-based heavy electrons: possible origin of pressure-induced enhancement of superconducting transition temperature in $CeCu_2Ge_2$ and related compounds. *J. Phys. Soc. Jpn.* 69:3955–3964.
[111] Onodera A et al. (2002) Equation of state of $CeCu_2Ge_2$ at cryogenic temperature. *Solid State Commun.* 123:113–116.
[112] Onuki Y et al. (1992) Magnetic and electrical properties of U-Ge intermetallic compounds. *J. Phys. Soc. Jpn.* 61:293–299.
[113] Oomi G, Kagayama T, Nishimura K, Yun SW, Onuki Y (1995) Electrical resistivity of single crystalline UGe_2 at high pressure and magnetic field. *Physica B* 206-207:515–518.
[114] Ott HR, Rudigier H, Fisk Z, Smith JL (1983) UBe_{13}: an unconventional actinide superconductor. *Phys. Rev. Lett.* 50:1595–1598.
[115] Ott HR, Rudigier H, Rice TM, Ueda K, Fisk Z, Smith JL (1984) *p*-wave superconductivity in UBe_{13}. *Phys. Rev. Lett.* 52:1915–1918.
[116] Paglione J et al. (2003) Field-induced quantum critical point in $CeCoIn_5$. *Phys. Rev. Lett.* 91:246405.
[117] Palstra TTM et al. (1985) Superconducting and magnetic transitions in the heavy-fermion system URu_2Si_2. *Phys. Rev. Lett.* 55:2727–2730.
[118] Petrovic C et al. (2001) Heavy fermion superconductivity in $CeCoIn_5$ at 2.3 K. *J. Phys. Cond. Mat.* 13:L337–L342.
[119] Pfleiderer C (2009) Superconducting phases of *f*-electron compounds. *Rev. Mod. Phys.* 81:1551–1624.
[120] Pustogow A et al. (2019) Constraints on the superconducting order parameter in Sr_2RuO_4 from oxygen-17 nuclear magnetic resonance. *Nature* 574:72–75.
[121] Qian YJ et al. (1987) Longitudinal sound measurements of UPt_3 in a magnetic field. *Solid State Commun.* 63:599–602.
[122] Ran S et al. (2019) Nearly ferromagnetic spin-triplet superconductivity. *Science* 365:684–687.
[123] Rauchschwalbe U, Alheim U, Steglich F, Rainer D, Franse JJM (1985) Upper critical magnetic fields of the heavy fermion superconductors $CeCu_2Si_2$, UPt_3 and UBe_{13}: comparison between experiment and theory. *Z. Phys. B - Condens. Matter* 60:379–386.
[124] Riseman TM et al. (1998) Observation of a square flux-line lattice in the unconventional superconductor Sr_2RuO_4. *Nature* 396:242–245.
[125] Sakai A, Kuga K, Nakatsuji S (2012) Superconductivity in the ferroquadrupolar state in the quadrupolar Kondo lattice $PrTi_2Al_{20}$. *J. Phys. Soc. Jpn.* 81:083702.
[126] Sarrao JL et al. (2002) Plutonium-based superconductivity with a transition temperature above 18 K. *Nature* 420:297–299.
[127] Sato NK et al. (2001) Strong coupling between local moments and superconducting 'heavy' electrons in UPd_2Al_3. *Nature* 410:340–343.
[128] Sato M, Fujimoto S (2009) Topological phases of noncentrosymmetric superconductors: edge states, Majorana fermions and non-Abelian statistics. *Phys. Rev. B* 79:094504.
[129] Saxena SS et al. (2000) Superconductivity on the border of itinerant-electron ferromagnetism in UGe_2. *Nature* 406:587–592.
[130] Scaffidi T, Simon SH (2015) Large Chern number and edge currents in Sr_2RuO_4. *Phys. Rev. Lett.* 115:087003.
[131] Schemm ER, Gannon WJ, Wishne CM, Halperin WP, Kapitulnik A (2014) Observation of broken time-reversal symmetery in the heavy-fermion superconductor UPt_3. *Science* 345:190–193.

[132] Schemm ER, Baumbach RE, Tobash PH, Ronning F, Bauer ED, Kapitulik A (2015) Evidence for broken time-reversal symmetry in the superconducting phase of URu_2Si_2. *Phys. Rev. B* 91:140506(R).

[133] Schenstrom A et al. (1989) Anisotropy of the magnetic-field-induced phase transition in superconducting UPt_3. *Phys. Rev. Lett.* 62:332–335.

[134] Schröder A et al. (1994) Incommensurate magnetic order in the heavy-fermion superconductor UNi_2Al_3. *Phys. Rev. Lett.* 72:136–139.

[135] Schuberth E et al. (2016) Emergence of superconductivity in the canonical heavy-electron metal $YbRh_2Si_2$. *Science* 351:485–488.

[136] Settai R et al. (2001) Quasi-two-dimensional Fermi surfaces and de Haas-van Alphen oscillations in the normal and superconducting mixed states of $CeCoIn_5$. *J. Phys. Cond. Mat.* 13:L627–L634.

[137] Settai R et al. (2007) Pressure-induced superconductivity in $CeCoGe_3$ without inversion symmetry. *J. Magn. Magn. Mater.* 310:844–846.

[138] Settai R, Miyauchi Y, Takeuchi T, Lévy F, Sheikin I, Onuki Y (2008) Huge upper critical field and electronic instability in pressure-induced superconductor $CeIrSi_3$ without inversion symmetry in the crystal structure. *J. Phys. Soc. Jpn.* 77:073705.

[139] Seyfarth G et al. (2005) Multiband superconductivity in the heavy fermion compound $PrOs_4Sb_{12}$. *Phys. Rev. Lett.* 95:107004.

[140] Seyfarth G, Brison JP, Méasson M-A, Braithwaite D, Lapertot G, Flouquet J (2006) Superconducting $PrOs_4Sb_{12}$: a thermal conductivity study. *Phys. Rev. Lett.* 97:236403.

[141] Shimizu Y et al. (2015) Field-orientation dependence of low-energy quasiparticle excitations in heavy-electron superconductor UBe_{13}. *Phys. Rev. Lett.* 114:147002.

[142] Shimizu Y et al. (2016) Omnidirectional measurements of the angle-resolved heat capacity for complete detection of superconducting gap structure in the heavy-fermion antiferromagnet UPd_2Al_3. *Phys. Rev. Lett.* 117:037001.

[143] Shishido H, Settai R, Harima H, Onuki Y (2005) A drastic change in the Fermi surface at a critical pressure in $CeRhIn_5$: dHvA study under pressure. *J. Phys. Soc. Jpn.* 74:1103–1106.

[144] Sidorov VA et al. (2002) Superconductivity and quantum criticality in $CeCoIn_5$. *Phys. Rev. Lett.* 89:157004.

[145] Sigrist M (2009) Introduction to unconventional superconductivity in non-centrosymmetric metals. *AIP Conference Proceedings* 1162:55–97.

[146] Sigrist M, Ueda K (1991) Phenomenological theory of unconventional superconductivity. *Rev. Mod. Phys.* 63:239–311.

[147] Sigrist M, Rice TM (1995) Sr_2RuO_4 – an electronic analog of ^3He? *J. Phys. Cond. Mat.* 7:L643–L648.

[148] Smidman M, Salamon MB, Yuan HQ, Ageterberg DF (2017) Superconductivity and spin-orbit coupling in non-centrosymmetric materials: a review. *Rep. Prog. Phys.* 80:036501.

[149] Steglich F et al. (1979) Superconductivity in the presence of strong Pauli paramagnetism: $CeCu_2Si_2$. *Phys. Rev. Lett.* 43:1892–1896.

[150] Steppke A et al. (2017) Strong peak in T_c of Sr_2RuO_4 under uniaxial pressure. *Science* 355:148

[151] Stewart GR, Fisk Z, Willis JO, Smith JL (1984) Possibility of coexistence of bulk superconductivity and spin fluctuations in UPt_3. *Phys. Rev. Lett.* 52:679–682.

[152] Stewart GR (2017) Unconventional superconductivity. *Adv. Phys.* 66:75–196

[153] Stock C, Broholm C, Hudis J, Kang HJ, Petrovic C (2008) Spin resonance in the *d*-wave superconductor $CeCoIn_5$. *Phys. Rev. Lett.* 100:087001.

[154] Stockert O et al. (2011) Magnetically driven superconductivity in $CeCu_2Si_2$. *Nat. Phys.* 7:119–124.

[155] Strand JD, Bahr DJ, van Harlingen DJ, Davis JP, Gannon WJ, Halperin WP (2010) The transition between real and complex superconducting order parameter phases in UPt_3. *Science* 328:1368–1369.

[156] Sugitani I et al. (2006) Pressure-induced heavy fermion superconductivity in antiferromagnet $CeIrSi_3$ without inversion symmetry. *J. Phys. Soc. Jpn.* 75:043703.

[157] Taillefer L, Lonzarich GG (1988) Heavy-fermion quasiparticles in UPt_3. *Phys. Rev. Lett.* 60:1570–1573.

[158] Taillefer L, Piquemal F, Flouquet J (1988) The anisotropic magnetoresistance of UPt_3 at low temperatures. *Physica C* 153-155:451–452.

[159] Takenaka T et al. (2017) Full-gap superconductivity robust against disorder in heavy-fermion $CeCu_2Si_2$. *Phys. Rev. Lett.* 119:077001.

[160] Takeuchi T et al. (2007) Specific heat and de Haas-van Alphen experiments on the heavy-fermion superconductor $CePt_3Si$. *J. Phys. Soc. Jpn.* 76:014702.

[161] Tamai A et al. (2019) High-resolution photoemission on Sr_2RuO_4 reveals correlation-enhanced effective spin-orbit coupling and dominantly local self-energies. *Phys. Rev. X* 9:021048.

[162] Tateiwa N, Kobayashi TC, Amaya K, Haga Y, Settai R, Onuki Y (2004) Heat-capacity anomalies at T_{sc} and T^* in the ferromagnetic superconductor UGe_2. *Phys. Rev. B* 69:180513(R).

[163] Terashima T et al. (2002) Magnetic phase diagram and the pressure and field dependence of the Fermi surface of UGe_2. *Phys. Rev. B* 65:174501.

[164] Tomita T, Kuga K, Uwatoko Y, Coleman P, Nakatsuji S (2015) Strange metal without magnetic criticality. *Science* 349:506–509.

[165] Trappmann T, von Löhneysen H, Taillefer L (1991) Pressure-dependence of the superconducting phases in UPt_3. *Phys. Rev. B* 43:13714–13716.

[166] Tsujimoto M, Matsumoto Y, Tomita T, Sakai A, Nakatsuji, S (2014) Heavy-fermion superconductivity in the quadrupole ordered state of PrV_2Al_{20}. *Phys. Rev. Lett.* 113:267001.

[167] Van Dyke JS, Massee F, Allan MP, Davis JCS, Petrovic C, Morr DK (2014) Direct evidence for a magnetic *f*-electron-mediated pairing mechanism of heavy-fermion superconductivity in $CeCoIn_5$ *Proc. Natl. Acad. Sci.* 111:11663-11667.

[168] Vekhter I, Hirschfeld PJ, Carbotte JP, Nicol EJ (1999) Anisotropic thermodynamics of *d*-wave superconductors in the vortex state. *Phys. Rev. B* 59:R9023–R9026.

[169] Volovik GE (1993) Superconductivity with lines of gap nodes - density of states in the vortex. *JETP Lett.* 58:469–473.

[170] Walker MB, Contreras P (2002) Theory of elastic properties of Sr_2RuO_4 at the superconducting transition temperature. *Phys. Rev. B* 66:214508.

[171] Xia J, Maeno Y, Beyersdorf PT, Fejer MM, Kapitulnik A (2006) High resolution polar Kerr effect measurements of Sr_2RuO_4: evidence for broken time-reversal symmetry in the superconducting state. *Phys. Rev. Lett.* 97:167002.

[172] Yamashita T et al. (2015) Colossal thermomagnetic response in the exotic superconductor URu_2Si_2. *Nat. Phys.* 11:17–20.

[173] Yang Y-F, Pines D (2014) Emergence of superconductivity in heavy-electron materials. *Proc. Natl. Acad. Sci. U.S.A.* 111:18178–18182.

[174] Yano K et al. (2008) Field-angle-dependent specific heat measurements and gap determination of a heavy fermion superconductor URu_2Si_2. *Phys. Rev. Lett.* 100:017004.

[175] Yelland E, Barraclough JM, Wang W, Kamanev KV, Huxley AD (2011) High-field superconductivity at an electronic topological transition in URhGe. *Nat. Phys.* 7:890–894.

[176] Young B-L et al. (2007) Microscopic evidence for field-induced magnetism in $CeCoIn_5$ *Phys. Rev. Lett.* 98:036402.

[177] Yuan HQ, Grosche FM, Deppe M, Geibel C, Sparn G, Steglich F (2003) Observation of two superconducting phases in $CeCu_2Si_2$. *Science* 302:2104–2107.

[178] Zwicknagl G, Fulde P (2003) The dual nature of 5*f* electrons and the origin of heavy fermions in U compounds. *J. Phys. Cond. Mat.* 15:S1911–S1916.

D2

Organic Superconductors

Gunzi Saito and Yukihiro Yoshida

D2.1 Introduction

High-T_c superconductivity was very recently discovered in a pressurized molecular material, sulfur hydride ($T_c \sim 190$ K at ~150 GPa [1] and 203 K at 155 GPa [2]), the T_c values of which exceed the previous records of cuprate SCs (133 K at AP [3] and 164 K at 30 GPa [4]). The amazing discovery that a simple molecule such as H_2S [SH_n hydrides ($n > 2$) have been predicted to show superconductivity under pressure] can be a high-T_c SC indicates the possibility for discovering RT SCs even in molecular or polymer materials.

This chapter focuses on molecular SCs, especially organic one [5]. Approximately 210 organic SCs, most of which are CT type, have been developed since the first discovery in 1980 [$(TMTSF)_2PF_6$; $T_c = 0.9$ K at 1.2 GPa] [6]. They are summarized in Scheme D2.1 with the molecular structures of the conducting component molecules. The SCs are classified into three groups, namely single-component materials (3 SCs; highest T_c is 5.5 K at 7.5–8.7 GPa), cation radical salts based on 24 types of donor molecules [BEDT-TTF or ET (ca. 80 SCs; highest T_c is 14.2 K at 8.2 GPa), TMTSF (8 SCs; highest T_c is 3 K at 0.5 GPa), BETS (7 SCs; highest T_c is 5.5 K at 3 GPa), and others (48 SCs; highest T_c is 8.6 K at 1.0 GPa)], and anion radical salts based on 8 types of acceptor molecules [C_{60} (ca. 40 SCs; highest T_c is 38 K at 0.7 GPa), aromatic hydrocarbons (14 SCs; highest T_c is 35 K), and $M(dmit)_2$ (11 SCs; highest T_c is 6.2 K at 0.65 GPa)]. Here we describe the development of organic SCs and their characteristics. Most of the superconducting phases including oxide SCs exist next to spin-ordered phases such as SDW and AF. We briefly describe the recent development of SCs having superconducting phases next to spin-disordered QSL state.

D2.2 Common Insulating States near Superconducting States and Dimensionality (Band Structure, Spin Lattice, and Crystal Structure)

The first organic metal TTF·TCNQ has a one-dimensional (1D) band structure, and nesting of its 1D Fermi surface induces the Peierls (CDW) transition [7]. The molecules presented in this section are depicted in Scheme D2.2. For TTF·TCNQ, the lattice distortion corresponding to the nesting vector ($2k_F$) of the 1D Fermi surface produces an energy gap at the Fermi level at 59 K. In order to suppress the Peierls instability, the Se analogue of the TTF molecule, HMTSF, is used, and HMTSF·TCNQ exhibits metallic temperature dependence down to low temperatures under pressure, mainly owing to the increased dimensionality resulting from the short intermolecular Se···N atomic contacts [8]. This is a typical example of the suppression of the Peierls transition by increasing the electronic dimensionality by both chemical modification and increasing the external applied pressure.

Many organic SCs have been obtained through the suppression of several types of phase transitions other than the Peierls, SDW, and AF transitions. The dimensionalities of the electronic band structure, spin lattice, and elasticity of the crystal lattice are important factors that induce or suppress the phase transitions commonly observed in organic conductors. The Peierls and SDW transitions are known as metal–insulator transitions (T_{MI}: metal–insulator transition temperature). The order–disorder (OD) transition of the anion molecule, which is classified within the Peierls transition, and charge ordering (CO; charge localization, charge disproportionation) are sometimes manifested as a metal–insulator transition. These transitions are observed in TMTSF and TMTTF systems. The Mott transition is an insulator–metal transition, and some ET Mott insulators exhibit a superconducting state next to the Mott insulating state. Regarding insulator–insulator transitions, the AF and spin-Peierls transitions derived from a Mott insulator are common in organic conductors. Some ET Mott insulators exhibit the phase sequence such as Mott insulator → AF insulator → SC, Mott insulator → spin-Peierls insulator → AF insulator → SC, or Mott insulator → QSL insulator → SC depending on external conditions.

The metal–insulator transition (Peierls, SDW, or OD) easily occurs in a 1D band structure system, in which a large portion of the Fermi surface diminishes by nesting. Such nesting will be suppressed on increasing the dimensionality of the

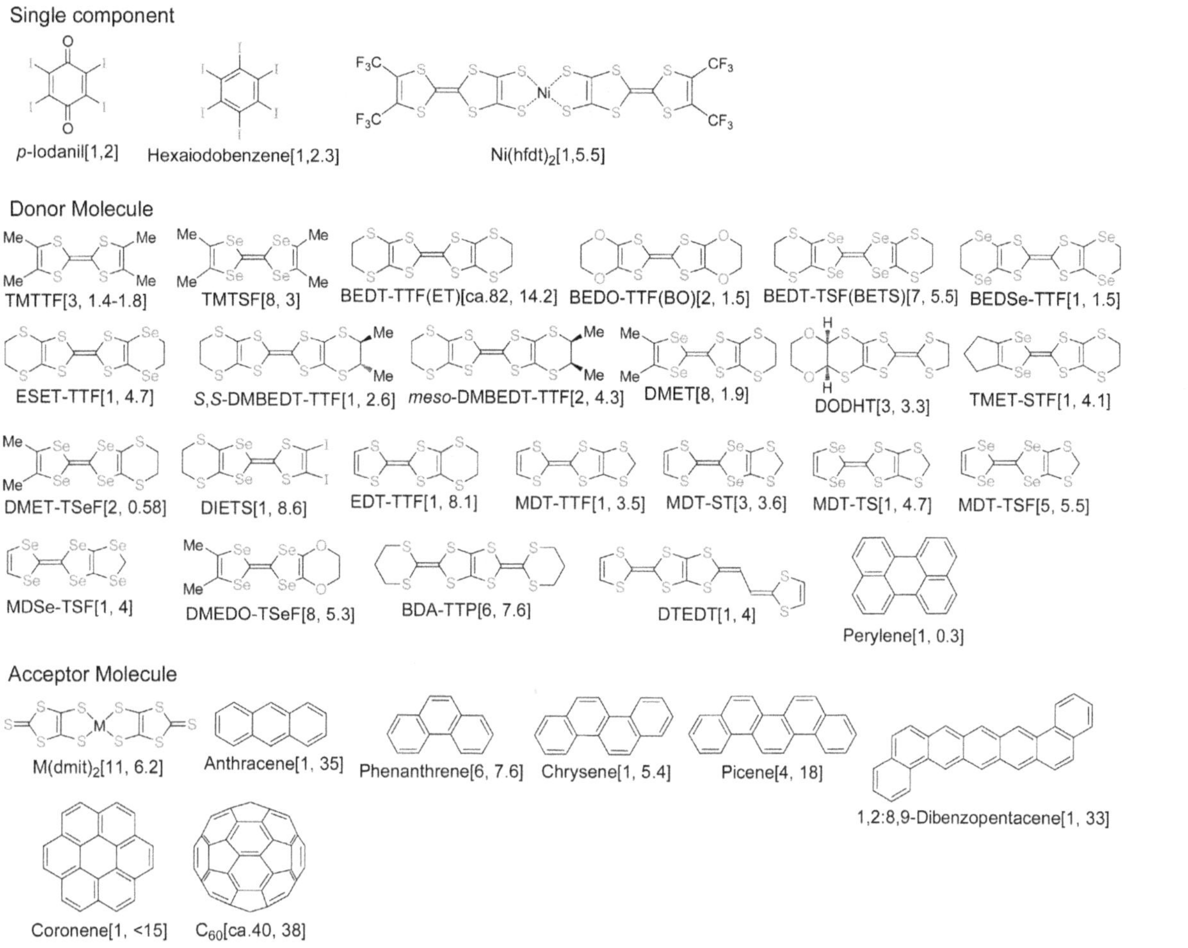

SCHEME D2.1 Component molecules for organic superconductors. Numbers in brackets are the total number of SCs of each molecule and their highest T_c.

band structure. Even though the system keeps the 1D band structure, such metal–insulator transition will be suppressed by forming a superlattice incompatible with the nesting vector (for the Peierls transition and OD of anion molecules) or by enforcing the crystal structure against lattice distortion. The last issue corresponds to the increase in structural dimensionality.

Organic SCs except those composed of C_{60} are commonly composed of segregated stacks of conducting component molecules sandwiched by the insulating counterparts. Therefore, such materials in the spin-localized state form the spin lattice represented in Figure D2.1, where t and t' are the transfer interactions between spins, and the spin geometry is parameterized by t'/t. The linear spin lattice in Figure D2.1(a) can be observed for systems with 1D, quasi-1D, and 2D band structures. Some of them exhibit superconductivity under pressure. Sometimes, the stronger exchange interactions t' alternate as t_1' and t_2' with $t_1' > t_2' \gg t$. The spin lattice in the form of an equilateral triangle [Figure D2.1(b), $t'/t = 1$] has the maximum spin frustration. A spin-disordered insulating state stemming from geometrical spin frustration [Figure D2.1(d)] was observed

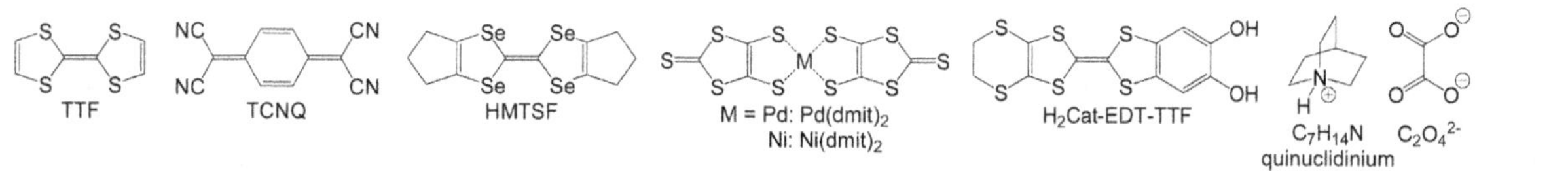

SCHEME D2.2 Other molecules in this chapter.

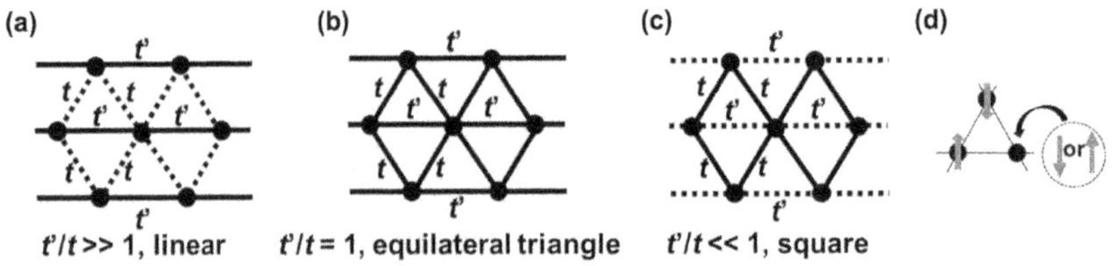

FIGURE D2.1 Changes in spin geometry with the change in t'/t. The 1D linear spin lattice for $t'/t \gg 1$ (a), 2D regular-triangle spin lattice for $t'/t = 1$ (b), and 2D square spin lattice for $t'/t \ll 1$ (c). Stronger and weaker exchange interactions are drawn in solid and dotted lines, respectively. Geometrical spin frustration in a triangular spin lattice will not afford a spin-ordered state in the AF spin-1/2 Heisenberg model (*d*).

in the QSL systems of 2D band structure, κ-$(ET)_2X$ [X = $Cu_2(CN)_3$, $Ag_2(CN)_3$], which exhibit superconductivity under pressure. The square spin lattice in Figure D2.1(c) is observed in systems of 1D [β'-$(ET)_2X$ (X = ICl_2, $AuCl_2$)] and 2D {κ- $(ET)_2Cu[N(CN)_2]Cl$, $EtMe_3P[Pd(dmit)_2]_2$, etc.} band structures. Spin frustration decreases when t'/t departs from unity; therefore, the ground state of β'-$(ET)_2X$ (X = ICl_2, $AuCl_2$), κ-$(ET)_2X$ {X = $Cu[N(CN)_2]Cl$, CF_3SO_3} has spin-ordered (AF) state at AP.

The C_{60} SCs based on the acceptor molecule C_{60} with icosahedral I_h symmetry have characteristic features in the stoichiometry and electronic and structural dimensionality. The LUMO of the C_{60} molecule has triply degenerate orbitals with t_{1u} symmetry. The most common C_{60} SCs are represented as M_3C_{60} (M: alkali metals), and the valence state of −3 is quite different from that of other organic conductors and SCs that are forced to have partial CT valences (non-integer), which are consequences of the Mott criterion. The face-centred-cubic or body-centred-cubic lattices of M_3C_{60} indicate the 3D lattice, band, and spin structures. The superconducting state of M_3C_{60} is close to the AF state. By using bulky organic molecules as counter components, C_{60} conductors have a packing pattern similar to those observed in 2D ET materials, and they have a 2D band structure with a hexagonal spin lattice as in Figure D2.1(b).

D2.3 Quasi-One-Dimensional Superconductors: TMTSF and TMTTF Systems [5, 9]

Figure D2.2(a) shows the crystal structure of $(TMTSF)_2NbF_6$, where TMTSF molecules stack along the a-axis to form a segregated column with small dimerization. Along the b- and c-axis, the transfer interactions are very small, and the Fermi surface is open along the k_y- and k_z-directions with fair warping owing to the lack of adequate side-by-side transfer interactions [Figure D2.2(b)]. The degree of warping depends on the size of X, which affects the intermolecular Se···Se atomic distances along the b-axis. The quasi-1D feature gives rise to a variety of phase transitions, such as a metal–insulator transition caused by the OD of anion molecules and SDW formation and a superconducting transition. The ratio of transfer interactions are t_a: t_b: t_c = 0.25: 0.025: 0.0015 eV (~1: 0.1: 0.01, t_a/t_b = 10) for X = PF_6; hence, no strong spin frustration is expected. $(TMTSF)_2X$ with octahedral anions (X = PF_6, AsF_6, SbF_6, NbF_6, TaF_6) show an SDW transition at 12–17 K with the SDW wave vector $Q_b = 0.24b^*$ and $Q_c = -0.06c^*$ having an amplitude of 0.08 μ_B for X = PF_6. An application of pressure, in general, increases the warping of the Fermi surface and suppresses the metal–insulator transition, resulting in

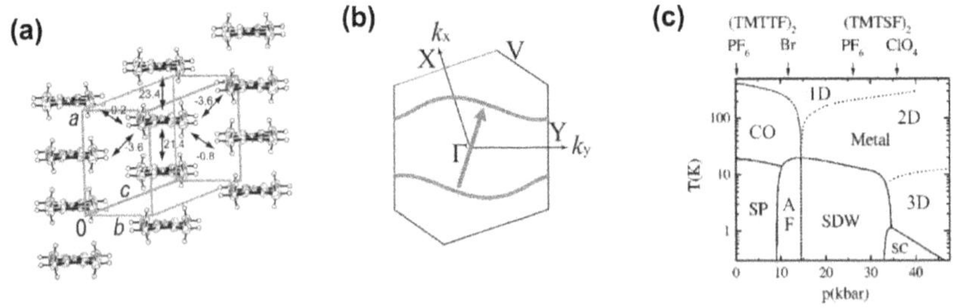

FIGURE D2.2 Crystal structure (a) and calculated Fermi surface of $(TMTSF)_2NbF_6$. (b) The arrow indicates the $2k_F$ nesting vector. The numbers in (a) indicate the overlap integrals in 10^{-3} units. (c) T–P phase diagram of $(TMTTF)_2X$ and $(TMTSF)_2X$ [10]. CO, charge-ordered state; SP, spin-Peierls state; AF, antiferromagnetic state; SDW, spin-density-wave state; SC, superconducting state; 1D~3D, one-dimensional metal~three-dimensional metal. Upper side of the phase diagram shows the positions of $(TMTTF)_2X$ (X = PF_6, Br) and $(TMTSF)_2X$ (X = PF_6, ClO_4) at ambient pressure.

TABLE D2.1 Organic Superconductors of Selected Examples of $(TMTSF)_2X$ and $(TMTTF)_2X$ [5,9]

X	Symmetry	σ_{RT}/ S cm^{-1}	T_{max}[a]/ K	P_c/GPa	T_c[b]/K	Characteristics[c]
TMTSF system						
PF_6	octahedral	540	12–15	0.65	1.1	SDW (12 K), FISDW
AsF_6	octahedral	430	12–15	0.95	1.1	SDW (12 K, $\|J\|k_B = 604$ K)
TaF_6	octahedral	300	15	1.1	1.35	SDW (11 K)
ClO_4	octahedral	700	–	0	1.4	OD (24 K, $a \times 2b \times 2c$), FISDW, ($\gamma = 10.5$, $\beta = 11.4$, $\Theta = 213$), SDW (5 K)
		–	5	–	–	By rapid cool
FSO_3	pseudo-tetrahedral	1000	~88	0.5	3	OD (88 K, $a \times 2b \times 2c$)
TMTTF system						
PF_6	octahedral	20	245	5.2–5.4	1.4-1.8	Spin-Peierls (15 K), $\|J\|/k_B$=210, AF above 0.9 GPa
SbF_6	octahedral	8	150	5.4–9	2.8	CO(150 K), AF (8 K)
BF_4	octahedral	50	190	3.35–3.75	1.38	OD (40 K), SDW, and SC (coexist)
Br	octahedral	260	100	2.6	1.0	C-AF (15 K)

Note: Besides the selected ones, $(TMTSF)_2X$ ($X = ReO_4$, SbF_6, NbF_6) are superconductors with T_c in the range of 0.4–1.3 K.

[a] T_{max}: temperature at maximum conductivity.

[b] Onset value.

[c] OD, order–disorder transition of anion and newly formed superlattice; FISDW, field-induced SDW; CO, charge-order; C-AF, commensurate AF phase. γ and β are important quantities experimentally determined to obtain $D(\varepsilon_F)$ [Equation (D2.1) and Θ Equation D2.2)] which are related with T_c by Equation (D2.3) for the BCS-type superconductors.

the superconducting state. Table D2.1 summarizes selected examples of $(TMTSF)_2X$ SCs.

The anion X resides at the inversion centre, and thus, the OD transition of octahedral anions does not affect the lattice symmetry. On the other hand, the OD transition of tetrahedral ($X = ClO_4$, ReO_4) and pseudo-tetrahedral ($X = FSO_3$) anions modifies the lattice symmetry. When the superlattice generated by the OD transition corresponds to the nesting vector of the Fermi surface ($2a \times 2b$), the metallic salt becomes an insulator ($X = ReO_4$, FSO_3). Since the superlattice for $X = ClO_4$ ($a \times 2b \times 2c$) does not correspond to the nesting vector, the ClO_4 salt remains metallic through the OD transition by slow cooling. Very rapid cooling, however, induces disorder of the anion molecules, and a metallic state remains until an SDW transition at 5 K. An intermediate cooling rate produces an increase in resistivity below 5 K and then an SC transition near 1 K suggesting the coexistence of SDW and superconducting states. $(TMTSF)_2ClO_4$ is the only SC at AP among the TMTSF salts.

The important parameters concerning the superconductivity, *i.e.*, upper critical magnetic field H_{c2}, critical current J_c, Sommerfeld coefficient γ, Debye temperature Θ, etc. [γ and Θ are closely related to the T_c of the BCS-type superconductor, as expressed by Equations (D2.1)–(D2.3) below], have been extensively studied for the ClO_4 salt. In what follows, we present remarkable observations concerning TMTSF SCs.

$$\gamma\text{: Sommerfeld coefficient, mJ mol}^{-1}\text{K}^{-2} \quad \gamma = \pi^2 k_B^2 D(\varepsilon_F)/3 \tag{D2.1}$$

$$\beta\text{: mJ mol}^{-1}\text{K}^{-4} \quad \beta = 48\pi N k_B / 5\Theta^3 \tag{D2.2}$$

$$\Theta\text{: Debye temprature, K} \quad T_c \propto \Theta \exp\left(-1/V_{el\text{-}ph} D(\varepsilon_F)\right) \tag{D2.3}$$

$V_{el\text{-}ph}$: electron–phonon coupling potential.

The H_{c2} of the PF_6 salt [$H_{c2} = 6$ (//b'), 4 (//a) T at 0.1 K] is far beyond the Pauli limit (H_{Pauli}) for the BCS-type SC with weak coupling. The Pauli limit is the critical magnetic field for breaking singlet Cooper pairs with the Zeeman energy of spins. In the free electron–gas model, H_{Pauli} is given by Equation (D2.4):

$$H_{Pauli} = \Delta_0 / \sqrt{2}\mu_B, \tag{D2.4}$$

where Δ_0 is the gap parameter at 0 K. Since $2\Delta_0$ is equal to $3.53 k_B T_c$ for the BCS-type SC with weak coupling, the value of H_{c2} should be limited to H_{Pauli}, which is given by Equation (D2.5).

$$H_{Pauli} = 1.86 T_c (\text{Tesla}). \tag{D2.5}$$

The relaxation rate of ^{1}H NMR absorption T_1^{-1} for the ClO_4 salt does not show the Hebel–Slichter coherence peak, which should be observed immediately below T_c for a normal BCS-type SC having an isotropic gap. The application of magnetic field breaks the superconducting state and induces a sequence of SDW states (field-induced SDW: FISDW) like a cascade above 3 T.

The Fermi surfaces of the sulfur analogues, $(TMTTF)_2X$, are more 1D-like than those of $(TMTSF)_2X$; $(TMTTF)_2X$ exhibit a variety of phase transitions, as summarized in Table D2.1. The generalized phase diagram proposed by Jerome for the quasi-1D system, $(TMTSF)_2X$, and 1D system, $(TMTTF)_2X$, was improved through high-pressure experiments. At AP, $(TMTTF)_2PF_6$ shows the phase sequence 1D metal → CO insulator → spin-Peierls insulator. Under pressure, an AF state appears next to the spin-Peierls state, and the *T–P* phase diagram indicates the sequence spin-Peierls insulator → AF insulator → SDW insulator → SC with increasing pressure [Figure D2.2(c)] [10]. The CO insulating state neighbours the metallic state, but the spin-Peierls state does not directly neighbour the metal and superconducting state. The superconducting phase neighbours the magnetic phase (SDW state), and T_c decreases with increasing pressure owing to the decrease of $D(\varepsilon_F)$.

D2.4 Two-Dimensional Superconductors: ET System [5b–5d]

D2.4.1 General Aspects of ET System

A neutral ET molecule is non-planar and becomes nearly flat with the formation of the partial CT complex except the terminal ethylene groups that are thermally disordered at high temperatures. Consequently, ET molecules tend to pile up one after the other and slide towards each other to minimize the steric hindrance caused by the terminal ethylene groups, leading to a decrease in intermolecular face-to-face interactions. The segregated packing of such molecules leaves cavities along the molecular long axis, which are occupied by counter anions and occasionally solvent molecules. ET molecules also have a strong tendency to form proximate side-by-side intermolecular S···S contacts, leading to an increase in the side-by-side transfer integrals. The 2D conducting layer of ET molecules stems from the increased side-by-side and decreased face- to-face intermolecular interactions.

The ET conductors are composed of alternating 2D conducting layers and insulating anion layers as shown in Figure D2.3(a) for the first 10 K class ET SC, κ-$(ET)_2Cu(NCS)_2$, which has the polymerized anion molecule $[Cu(NCS)_2^-]_\infty$ [Figure D2.3(b)] [11]. Significant donor-anion interactions arise from the short atomic contacts between the ethylene hydrogen atoms of ET and atoms around the anion openings in the anion layer. The hydrogen atom of one ethylene group of the ET molecule fits into the core created by the anion molecules, similar to a key–keyhole relation. The position of such an ethylene hydrogen atom projected onto the anion cores produces unique patterns: called α-type (5 SCs), β-, β'-, and β''-type (about 30 SCs), and κ-type (about 40 SCs). This implies that the ET molecules arrange according to the anion core. Figure D2.3(c) shows the key–keyhole relation observed in κ-$(ET)_2Cu(NCS)_2$, where a spin-1/2 ET dimer depicted by an ellipsoid is located on an anion opening. Consequently, three ET dimers with the transfer interactions t' between two parallel dimers and t between two perpendicularly related dimers form an isosceles-triangular spin lattice, as indicated by the red triangle. The ratio t'/t represents the degree of spin frustration. The frustration of triangular spin is expected to maximize at $t'/t = 1$. The calculated Fermi surfaces of κ-$(ET)_2Cu(NCS)_2$ consist of both 1D warped electron-like and 2D cylindrical hole-like surfaces. For the salt, electrons move along the closed ellipsoid (α-orbit) to exhibit Shubnikov–de Haas (SdH) oscillations [12], and at higher magnetic fields (> 20 T), electrons hop from the ellipsoid to the open Fermi surface with a circular trajectory (β-orbit, magnetic breakdown oscillations), as shown in Figure D2.3(d) [13].

It is known that the ethylene conformation affects the transport properties of ET salts. For example, β-$(ET)_2I_3$ is the first SC at AP (T_c = 1.5 K, the salt is abbreviated as β_L) in the ET family [14], in which ethylene groups have mixed conformations of eclipsed or staggered ones [15]. The T_c increases distinctly to 8.1 K on fixing two ethylene groups in the eclipsed conformation under pressure (the salt is abbreviated as β_H) [16]. Theoretical work has revealed that

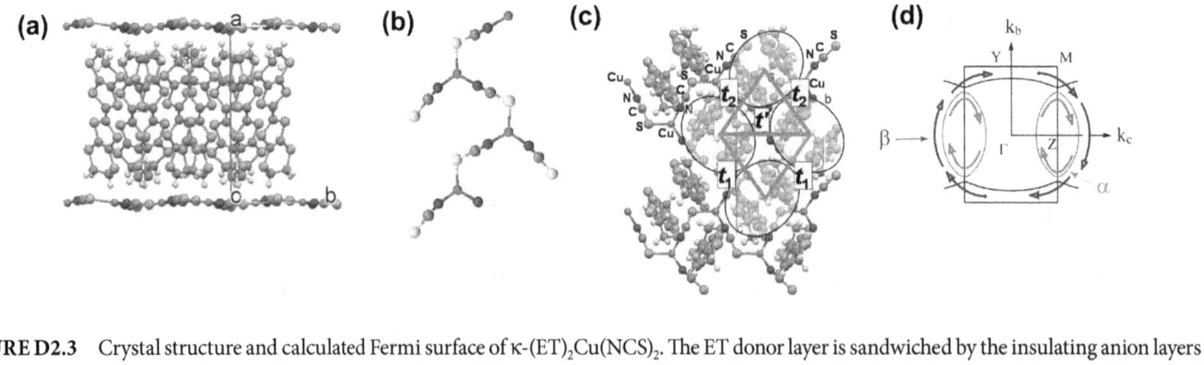

FIGURE D2.3 Crystal structure and calculated Fermi surface of κ-$(ET)_2Cu(NCS)_2$. The ET donor layer is sandwiched by the insulating anion layers (a) composed of a polymerized anion (b). An ET dimer fits on the anion opening in the anion layer with a key–keyhole relation (c). Pairs of dimerized ET molecules are represented by ellipsoids. The triangular spin lattice in red is derived by connecting the centres of ellipsoids and is represented by t'/t, where the averaged $t = (t_1 + t_2)/2$ is employed. (d) Calculated Fermi surface. Arrows in (d) indicate the trajectory of electrons for the Schbunikov–de Haas effect; the α-orbit is around the 2D cylindrical hole-like Fermi surface and the β-orbit is the magnetic breakdown orbit.

the ethylene-end-group conformations influence the electronic bandwidth (W) and switching from an eclipsed to a staggered conformation decreases W and in turn enhances the relative strength of the Hubbard repulsion, bringing the material closer to a Mott insulating state [17]. β-$(ET)_2AuI_2$ and κ-$(ET)_2I_3$ have an eclipsed conformation at RT with T_c (4.9 and 3.6 K, respectively) higher than that of β_L.

The different types of ET···ET (π-π, S···S) and ET···anion (hydrogen bonds) intermolecular interactions, large conformational freedom of ethylene groups, flexible molecular framework, fairly narrow bandwidth (W), and strong electron correlations represented by the ratio U/W, where U is the on-site Coulomb repulsion energy, result in a rich variety of salts with different crystal and electronic structures ranging from insulators to SCs.

D2.4.2 ET Superconductors

Approximately 82 ET SCs have been discovered thus far. Table D2.2 summarizes selected examples of ET SCs and related salts. Their ET layer have a $(ET)_2$ dimer unit. All of them, except **j**, **k**, and **q** in Table D2.2, exhibit superconductivity at AP or under applied pressure. They are classified into four groups based on the transport behaviour at AP. (1) Salts in Group **A** are metallic monotonically down to T_c. (2) Salts in Group **B** are close to a Mott insulating state and exhibit the phase sequence semiconductor → metal → SC. (3) Salts in Group **C** exhibit metal–insulator transitions, and some show superconductivity under pressure. (4) Salts in Group **C** are insulators (Mott, CDW, or CO), and some show superconductivity under pressure. Figure D2.4 compares the temperature dependence of resistivity of the κ-type salts in Groups **A**, **B**, and **D** and β-$(ET)_2AuI_2$ (**e**, Group **A**), which exhibits metallic behaviour down to its T_c of 4.9 K.

The molecular U value of ET (U_0~ 4.5 eV) decreases with the neighbour-site Coulomb repulsion energy V in the solid state as U_{eff}~ U_0–V (= 0.7–1 eV), where U_{eff} is the effective on-site Coulomb energy. According to the dimer model, the U_{eff}value is further reduced, as expressed by Equation (D2.6), and is approximated as ΔE_d because $U_{eff} \gg \Delta E_d$,

$$U_d \sim \Delta E_d + \left\{U_{eff} - (U_{eff}^2 + 4\Delta E_d^2)^{1/2}\right\}/2 \sim \Delta E_d, \qquad \text{(D2.6)}$$

Where U_d and ΔE_d are the on-site Coulomb energy of a dimer and the dimerization energy, respectively [20]. In what follows as well as Table D2.2, U_d is simply denoted by U.

The U values do not vary much among the solids with the same packing (U ~ 0.34–0.51 eV for κ-type packing). The bandwidth W is in the range 0.43–0.61 eV for κ-type salts known thus far. Consequently, ET salts cover a variety of conduction profiles depending on t, t', W, U, J, T, and P.

There are two kinds of anion species, namely discrete and polymerized anions. The polymerized anions are represented as ML_1L_2, where ligand L_1 links transition metal M (Cu^{1+}, Ag^{1+}, Hg^{2+}) to form infinite chains, and ligand L_2 attaches to M as a pendant or connects the infinite chains. Since the extended Hückel method does not include U, virtual Fermi surfaces were obtained for the Mott insulators. However, the calculated Fermi surfaces for Mott insulators **n** and **o** were verified by the quantum oscillations (SdH, de Haas–van Alphen [dHvA]) and/or angular-dependent magnetoresistance oscillations (AMRO) in their metallic state under pressure [19, 20]. The calculated Fermi surfaces of β-type salts (**b**, **d**, **e**) have a 2D cylindrical shape [Figure D2.5(a)], those of κ-type salts (**a**, **c**, **f–q**) have both 1D warped and cylindrical shapes with a gap existing between them for the $P2_1$ space group [Figure D2.3(d)] and no gap existing for others [Figure D2.5(b)], and 1D Fermi surfaces of β'-salts (**r–t**) open along the k_a direction [Figure D2.5(c)].

The T_c values in Table D2.2 are those obtained for the H-salts using h_8-ET, and some of them show a higher T_c for D-salts using d_8-ET (inverse isotope effect for **f** and **h**), but a normal isotope effect was observed in **i**. Currently, β'-$(h_8$-$ET)_2ICl_2$ (**s**; onset T_c = 14.2 K at 8.2 GPa) [21] and the D-salt of **n** (T_c= 13.1 K at 0.03 GPa) [22] show the highest T_c values under pressure. At AP, the D-salt of **f** shows the highest T_c of 12.3 K [23], followed by H-salt of **i** (T_c = 11.6 K) [24]. In what follows, we introduce the characteristic properties of ET conductors with linear trihalide anions and κ-type salts.

D2.4.3 ET Superconductors with Linear Trihalide Anions

One of the most intriguing ET SCs is the salt with the I_3 anion, which afforded α-, α_t-, β_L-, β_H-, δ-, ε-, γ-,θ-, and κ-type salts with different crystal and electronic structures. Among them, α-, α_t-, β_L-, β_H-, γ-, θ-, and κ-type salts are SCs with T_c= 7.2, ~8, 1.5, 8.1, 2.5, 3.6, and 3.6 K, respectively. The β_L-salt was converted to the β_H-salt by applying a pressure (hydrostatic pressure) greater than 0.04 GPa and then depressurizing while keeping the sample below 125 K. The β_H-salt returned to the β_L-salt when kept above 125 K at AP. The β_L-salt is characterized by a superlattice appearing at 175 K with incommensurate modulations of ET and I_3 to each other [15]. The formation of the superlattice was suppressed by an applied pressure greater than 0.04 GPa. Then, the two ethylene groups in an ET molecule were fixed in the eclipsed conformation to give rise to a T_c more than five times higher in the β_H-salt. The T_c of the β_H-salt monotonously decreases with increasing hydrostatic pressure; however, under uniaxial stress, a further increase in T_c with a maximum at a piston pressure of 0.3–0.4 GPa is observed for directions both parallel and perpendicular to the donor stack [25].

The resistivity of α-$(ET)_2I_3$ was nearly independent of temperature down to 135 K, at which CO transition occurs [26]. It has been claimed that α-$(ET)_2I_3$ has a zero-gap state with a Dirac-cone-type energy dispersion similar to that of graphene [27]. Under hydrostatic pressure, it becomes a 2D metal down to low temperatures (2 GPa); however, it becomes a SC under uniaxial pressure along the a-axis (onset T_c = 7.2 K at

TABLE D2.2 Selected ET Conductors and Superconductors. The compound is represented by *Greek alphabet*-$(ET)_2X$ (Greek alphabet: type of donor stacking, L_1, L_2: ligand).

Class, Salts[a]	Space Group	Ligands in Anion L_1, L_2	Transport Behaviour[b] σ_{RT}/Scm^{-1} (ε_a/meV)	T_c / K (Critical Hydrostatic Pressure for SCs/GPa)	Magnetic Behaviour[c]	$\|J\|/k_B$	Band Parameters[d] U/W RT	t'/t	U/W 100 K	U/t	Ground State[e], Characteristics[f]
A (a) κ-$(ET)_2I_3$	$P2_1/c$	Discrete	Good metal, 40–250	3.6	Pauli		0.80	0.54	0.81	6.4_9	SC, γ = 18.9, β = 10.3, Θ = 218, SdH, dHvA
(b) β-$(ET)_2IBr_2$	$P\bar{1}$	Discrete	Good metal, 20	2.7	Pauli		0.83	0.74	0.86	7.0_6	SC, SdH, dHvA, AMRO
(c) κ-$(ET)_2Ag(CN)[N(CN)_2]$	$P2_1$	CN, $N(CN)_2$	Good metal, ~150	6.6			0.84	0.62	0.88	7.0_5	SC
(d) β-$(ET)_2I_3$	$P\bar{1}$	Discrete	Good metal, 60	1.5, 2.0, 8.1	Pauli		0.84	0.72	0.92	7.5_6	SC, SdH, dHvA, AMRO
(e) β-$(ET)_2AuI_2$	$P\bar{1}$	Discrete	Good metal, 20-60	4.9	Pauli		0.87	0.71	0.90	7.3_7	SC, SdH, dHvA
(f) κ-$(ET)_2Cu(CN)[N(CN)_2]$	$P2_1$	CN, $N(CN)_2$	Good metal, 5–50	11.2	Pauli		0.87	0.64	0.91	7.3_0	SC, AMRO
(g) κ-$(ET)_2Ag(CN)_2 \cdot H_2O$	$P2_1$	Discrete	Good metal, 27–37	5.0	Pauli		0.88	0.61	0.90	7.2_0	SC, SdH
B (h) κ-$(ET)_2Cu(NCS)_2$	$P2_1$	SCN, NCS	fuzzy metal, 5–40	10.4	Pauli		0.84	0.80	0.94	7.9_9	SC[*d*], γ = 25, β = 11.2, θ = 215, SdH, dHvA, AMRO
(i) κ-$(ET)_2Cu[N(CN)_2]Br$	*Pnma*	$N(CN)_2$, Br	fuzzy metal, 5–50	11.6	Pauli		0.89	0.67	0.90	7.2_6	SC[*d*], γ = 25, β = 12.8, θ = 210, SdH, AMRO
C (j) κ-$(ET)_2Hg(SCN)_2Cl$	$C2/c$	SCN, Cl	metal, $T_{MI} = 30$ K	No SC	CO or AF		0.69	1.06	0.67	6.1_3	CO or AF
(k) κ-$(ET)_2Hg(SCN)_2Br$	$C2/c$	SCN, Br	metal, 1.5–7, $T_{MI} = 100$ K	No SC	CO		0.71				CO or AF
(l) κ-$(ET)_2Cu[N(CN)_2]I$	*Pnma*	$N(CN)_2$, I	metal, 1, $T_{MI} = 50$~60 K	7.7(0.12)			0.89	0.56	0.89	7.0_9	(SC)
D (m) κ-$(ET)_2CF_3SO_3$	$C2/c \rightarrow$	Discrete	Mott*, 1.8(66)	4.8(1.3)	AF($T_N = 2.5$ K)		0.85	1.50*	0.84*	8.8_9*	AF(SC)
	$P2_1/c$				$J/K_B = 200$ K			1.77*	0.89*	$10._4$*	
(n) κ-$(ET)_2Cu[N(CN)_2]Cl$	*Pnma*	$N(CN)_2$, Cl	Mott*, 2(12–52)	12.8(0.03)	AF($T_N = 27$ K)		0.89	0.73	0.91	7.4_3	AS(SC), SdH, AMRO
(o) κ-$(ET)_2Cu_2(CN)_3$	$P2_1/c$	CN, C/N	Mott*, 3–7(36–43)	3.9(0.06)	$J/K_B = 250$ K		0.93	1.07	1.01	9.1_2	QSL(SC[*d*]), SdH, AMRO
				6.8–7.3							
(p) κ-$(ET)_2Ag_2(CN)_3$	$P2_1/c$	CN, C/N	Mott*, 3(50–88)	5.2(1.05)	175 K		1.04	0.91	1.13	9.6_s	QSL(SC)
(q) κ-$(ET)_2B(CN)_4$	*Pnma*	Discrete	Mott*, 0.5(140)	No SC	$J/K_B = 118$ K		1.11	1.61	1.20	$13._3$	VBS
					$J/K_B = 236$ K						
(r) β′-$(ET)_2BrICl$	$P\bar{1}$	Discrete	Mott*, 1.5 × 10^{-2}(110)	7.2(8.0)	AF($T_N = 19.5$ K)		2.04	0.16**	2.10**	$12._1$**	AF(SC)
(s) β′-$(ET)_2ICl_2$	$P\bar{1}$	Discrete	Mott*, 3 × 10^{-2}(120)	14.2(8.2)	AF($T_N = 22$ K)		2.06	0.17**	2.03**	$11._4$**	AF(SC)
(t) β′-$(ET)_2AuCl_2$	$P\bar{1}$	Discrete	Mott*, 10^{-1}–10^{-2} (120)	No SC	AF($T_N = 28$ K)		2.11	0.18**	2.17**	$10._4$**	AF

[a] TCE: 1,1,2-trichloroethane.

[b] Mott*: under pressure it becomes a fuzzy metal and then a superconductor.

[c] Pauli: Pauli paramagnetic susceptibility, *J*: exchange interaction.

[d] *U*: on-site Coulomb energy. *W*: upper bandwidth. Band parameters t'/t, U/W, and U/t were calculated using a tight-binding model based on the extended Hückel method [18] using crystallographic data at 100 K–RT.

[e] Ground state under pressure is indicated in parenthesis and the superconducting symmetry is shown in square brackets. SC, superconductor; AF, antiferromagnet; QSL, quantum spin liquid; VBS, valence bond solid; CO, charge ordered. γ, β, Θ: see equations (D2.1)–(D2.3).

[f] Observed oscillations. SdH, Schubnikov–de Haas; dHvA, de Haas–van Alphen; AMRO, angular-dependent magnetoresistance oscillation [19].

* Data at 200 K; **Data at 120 K.

Group **A**: Good metal, **B**: Fuzzy metal, **C**: Metal with insulating ground state, **D**: Insulator. They are arranged in the order of U/W at RT in each group.

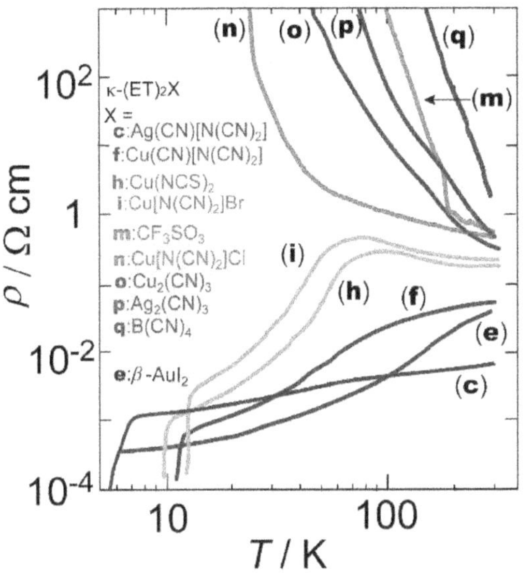

FIGURE D2.4 Temperature dependences of resistivity of typical ET salts at ambient pressure.κ-$(ET)_2Ag(CN)[N(CN)_2]$ (**c**), β-$(ET)_2AuI_2$ (**e**), κ-$(ET)_2Cu(CN)[N(CN)_2]$ (**f**), κ-$(ET)_2Cu(NCS)_2$ (**h**), and κ-$(ET)_2Cu[N(CN)_2]Br$ (**i**) are superconductors at ambient pressure. Mott insulators κ-$(ET)_2X$ [X = CF_3SO_3 (**m**), $Cu[N(CN)_2]Cl$ (**n**), $Cu_2(CN)_3$(**o**), and $Ag_2(CN)_3$ (**p**) show semiconductor–metal–superconductor behaviour under pressure, whereas κ-$(ET)_2B(CN)_4$ (**q**) remains semiconducting behaviour up to 2.5 GPa.

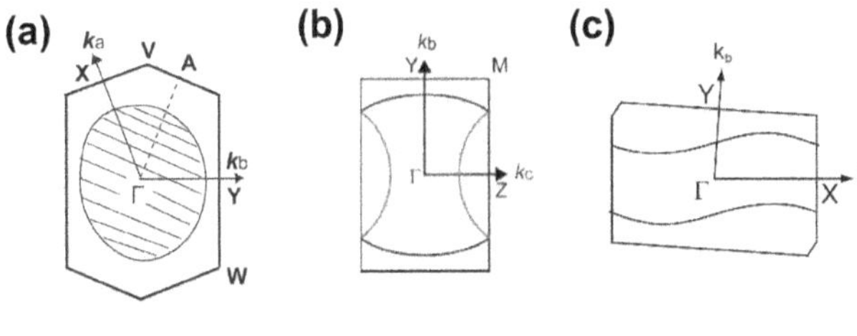

FIGURE D2.5 Calculated Fermi surfaces by using extended Hückel method of (a) β-$(ET)_2I_3$ (**d**), (b) κ-$(ET)_2Ag_2(CN)_3$ (**p**), and (c) β'-$(ET)_2AuCl_2$ (**t**). The shaded region in (a) indicates the hole-like part.

0.2 GPa), though it remains metallic down to low temperatures under uniaxial pressure along the *b*-axis (0.3–0.5 GPa). By tempering at 70–100°C for more than 3 days, α-$(ET)_2I_3$ was successfully converted to a mosaic polycrystal with $T_c \sim 8$ K, giving an α_t-salt exhibiting an NMR pattern similar to that of the β_H-salt. Only one θ-type SC,θ-$(ET)_2I_3$, is known; however, one-third of the obtained crystals are superconducting and others remain metallic. The phase diagram, through a plot of transfer interactions and the dihedral angle between columns, shows that the superconducting θ-$(ET)_2I_3$ is next to the CO state [28]. Tempering of all crystals of θ-$(ET)_2I_3$ at 70°C for 2 hours induces superconductivity with higher T_c (named θ_T-$(ET)_2I_3$: sharp drop of resistivity at 7 K and dull drop at ~5 K) [29].

D2.4.4 κ-Type ET Superconductors

In the κ-type SCs, the ET molecules form a dimer, and the ET dimers are arranged nearly orthogonally to each other to form a 2D conducting ET layer which is sandwiched by the insulating anion layers [Figure D2.3(a)]. For **h**, Cu^+ and SCN form a zigzag infinite chain ···Cu···SCN···Cu···SCN···along the *b*-axis, and the other ligand SCN coordinates to Cu^+ through a N atom to form an open space [indicated by the ellipsoid in Figure D2.3(c)], into which an ET dimer fits. Similar key–keyhole relations are observed in all κ-$(ET)_2X$ salts, as shown for **n**, **o**, and **p** in Figure D2.6. For **n**, one ET dimer is encircled by the ellipsoid surrounded by a unit of two bidentate dicyanamide N–C–N–C–N ligands and one Cl in the upper anion layer and another such unit in the lower anion layer [Figure D2.6(a)]. For **o** and **p**, the anion openings are clearly separated from each other. The neighbouring infinite chains of –M–CN–M–CN– and –M–NC–M–NC– along the *b*-axis are connected by disordered CN^-(indicated by C/N) groups to form a hexagonal anion opening [*M* = Cu; Figure D2.6(b)] and rectangular anion opening [*M* = Ag; Figure D2.6(c)].

Since their *t'*/*t* values (1.07, 0.91 at 100 K) are close to unity, their spin lattices have strong spin frustration. Furthermore,

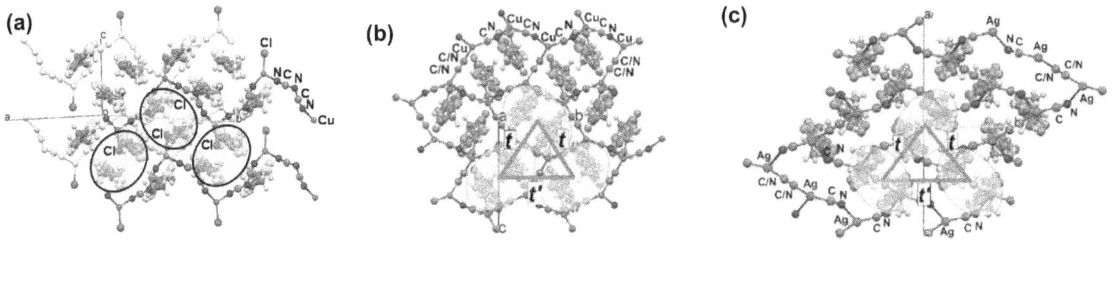

FIGURE D2.6 Two-dimensionally polymerized anions in (a) κ-$(ET)_2Cu[N(CN)_2]Cl$ (**n**), (b) κ-$(ET)_2Cu_2(CN)_3$ (**o**), and (c) κ-$(ET)_2Ag_2(CN)_3$ (**p**). The ET dimer fits into the anion opening in (a) and (b) and on the rim of the anion openings in (c). A good correspondence between an ET dimer (1 spin, ellipsoid) and the anion opening forms the triangle of the spin lattice.

their U/W values (1.01, 1.13 at 100 K) indicate that they have strong electron correlation. Consequently, salts **o** and **p** are Mott insulators at higher temperatures, and magnetic order was not observed down to 20 and 120 mK, respectively, owing to the geometrical spin frustration (QSL state) [30]. Mott insulators having a rather weak spin frustration with t'/t away from unity [$t'/t = 0.17$ for **s** ($T_N = 22$ K), 0.73 for **n** ($T_N = 27$ K), and 1.50 and 1.77 for **m** ($T_N = 2.5$ K)] are stabilized with the spin-ordered AF states showing characteristic divergent peaks in the NMR relaxation rate T_1^{-1} at Néel temperature T_N. These Mott insulators exhibit a temperature-insensitive dependence of T_1^{-1} below approximately 200 K down to T_N. Such behaviour is characteristic of quantum critical behaviour with strong spin frustration [31], in contrast to the metallic behaviour; $T_1T \sim$ constant (Korringa relation). For the Mott insulator **q** ($t'/t = 1.61$, $U/W = 1.20$ at 100 K), after showing the quantum critical behaviour T_1^{-1} decreased sharply below 5 K, indicating a phase transition, and a spin-singlet ground state was confirmed by magnetic susceptibility. The ground state of **q** is the VBS state [32].

The Mott insulator **m** exhibits complicated physical properties associated with the OD of anion molecules and formation of a superlattice. At RT, the anions are disordered by a dynamical rotation around the C–S bond, and at ~230 K, the anions are frozen into an ordered form causing the change of the crystal space group from $C2/c$ to $P2_1/c$. The $P2_1/c$ phase comprises two types of crystallographically independent ET layers (Layer 1 is designated as **m1** and Layer 2 as **m2**). Below approximately 190 K, a further structural transition occurs in connection with the formation of a six-fold superlattice that precludes the structural analysis at lower temperatures. The NMR data confirm the AF state below 2.5 K at AP, whereas the transport data confirm superconductivity with $T_c = 4.8$ K at 1.3 GPa [33].

Since the t'/t values of **j** and **k** are close to unity (1.03–1.05 at RT), they have strong spin frustration. However, their U/W values (0.69–0.71 at RT) indicate that the itinerancy exceeds the localization and that they are metals with strong spin frustration [34].

The H-salt of **n** showed a complicated T–P phase diagram [Figure D2.7(a)] [34]. Thoroughgoing studies under pressure provided firm evidence of the coexistence of superconducting (**I-SC-2** phase: **I-SC** = incomplete superconducting) and AF phases, in which the radical electrons of ET molecules play the roles of both localized and itinerant ones. Under a pressure of ca. 20–30 MPa, another incomplete superconducting phase (**I-SC-1**) appears, and the complete superconducting (**C-SC**) phase neighbours this phase at higher pressures. Below these superconducting phases, a reentrant nonmetallic (**RN**) phase was observed. Similar T–P phase diagrams were obtained for the D-salt of **n** and **i** κ-(d_8-ET)$_2$X (X = Cu[N(CN)$_2$]Cl and Cu[N(CN)$_2$]Br) with a parallel shift of P. They occur at the higher- and lower-pressure sides of **n** for the Br and Cl salts, respectively. Contrary to the H-salt, κ-(d_8-ET)$_2$Cu[N(CN)$_2$]Cl exhibited no coexistence of the superconducting and AF phases. At AP, salt **l** shows a metal–insulator transition at 50–60 K owing to the superlattice formation and becomes superconducting under hydrostatic pressure greater than 0.12 GPa with a T of 7.7 K (onset $T_c = 8.2$ K) [35].

The superconducting characteristics of the κ-type ET SCs, some of which differ from those of the conventional BCS SCs, are as follows.

1. Salt **h** gave higher H_{c2} values for a magnetic field parallel to the 2D plane than H_{Pauli} based on a simple BCS model. At 450 mK, H_{c2} exceeds H_{Pauli} by ca. 50% without any saturation.
2. The superconducting coherent lengths are 29 and 3.1 Å for **h** at 0.5 K along the 2D plane ($\xi_{//}$) and perpendicular to the plane ($\xi_{\perp}$). Whereas $\xi_{//}$ is larger than the lattice constants, $\xi_{\perp}$ is much smaller than the lattice constant, indicating that the conducting layers along this direction is Josephson-coupled.
3. Symmetry of the superconducting state: No Hebel-Slichter coherence peak was observed in **h** and **i** at AP and **o** under pressure in ^{1}H NMR measurements, ruling out the BCS s-wave state. STM spectroscopy shows d-wave symmetry with line nodes along the direction

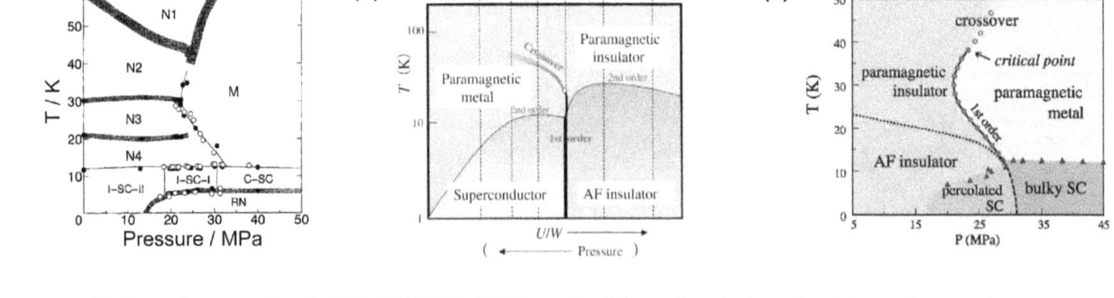

FIGURE D2.7 (a) Phase diagram of κ-(h_8-ET)$_2$Cu[N(CN)$_2$]Cl determined from electrical conductivity and magnetic measurements [34]. **N1–N4**, non-metallic phases; **M**, metallic phase; **RN**, reentrant non-metallic phase; **I-SC-I** and **I- SC-II**, incomplete superconducting phases. **N3** shows the growth of the three-dimensional AF-ordered phase. **N4** is a weak ferromagnetic phase. (b) Proposed simplified phase diagram [36]. **a**, β-(ET)$_2$I$_3$; **b**, κ-(ET)$_2$Cu(NCS)$_2$; **c**, H-salt of κ-(ET)$_2$Cu[N(CN)$_2$]Br; **d**, D-salt of κ-(ET)$_2$Cu[N(CN)$_2$]Br; **e**, κ-(ET)$_2$Cu[N(CN)$_2$]Cl; **f**, β'-(ET)$_2$ICl$_2$. (c) Proposed phase diagram taking into account the critical behaviour near approximately 10 K and 23 GPa in (a) [37].

near $\pi/4$ from the k_a- and k_c-axes ($dx^2–y^2$), which is consistent with thermal conductivity and specific heat measurements [38].

4. Inverse isotope effect: The inverse isotope effect has thus far been observed for **f**, **h**, and **n**, while the normal isotope effect has been observed for **i**. The reason for the observed isotope effects is not yet fully understood in a consistent manner.
5. A simplified *TP* phase diagram was proposed for κ-$(ET)_2X$ [Figure D2.7(b)], in which the parameter U/W is considered as independent of temperature. Figure D2.7(b) includes the salts **d**, **h**, **i** (H- and D-), **n**, and **s** [36]. However, the entire behaviour of **o** and **p** and the low-temperature reentrant behaviour of **i** and **n** [Figure D2.7(a)] cannot be explained on the basis of the diagram. The phase diagram in Figure D2.7(b) was further improved to include the critical behaviour near the triple point in Figure D2.7(a) (around 30 K and 20 MPa) [Figure D2.7(c)] [37]. Still, the diagram needs not only the electron correlation U/W but also both spin frustration and their temperature sensitivity to describe many ET salts.

With increasing distance between the ET dimers in Figures D2.3(c) and D2.6, the transfer interactions between ET dimers decrease; this may correspond to the decrease in W and to the increase in $D(\varepsilon_F)$; consequently, T_c is expected to increase. According to this line of thought, a higher T_c is expected for the salt having a larger anion spacing. Such a κ-type salt may be found near the border between poor metals and Mott insulators. It is true that not only κ-type but also for other types of salts, *e.g.*, β'-$(ET)_2X$ (X = BrICl (**r**) [39] and ICl_2 (**s**) [21]) are Mott insulators at ambient pressure with high T_c greater than 8 GPa.

Among κ-$(ET)_2$ X salts, which are not included in Table D2.2, alternating mixed ET packing motifs were observed in two phases of α'-κ-$(ET)_2Ag(CF_3)_4$ (TCE), the onset T_c of which are 9.5 and 11.0 K for the phase having two-layered (α' + κ) and the four-layered (α' + κ_1 + α' + κ_2) phase, respectively, where TCE is 1,1,2-trichloroethane [40]. Since α'-packing generally imparts a semiconducting state, both systems have a nanoscale heterojunction of a semiconductive/superconductive interface, which is thought to result in a higher T_c in these systems in comparison with κ_L-$(ET)_2Ag(CF_3)_4$(TCE) (T_c = 2.4 K). If this explanation is correct, this is an example of interface superconductivity [41].

D2.4.5 Quantum Spin Liquid State and Superconductivity

The QSL state was proposed for the triangular spin lattice by Anderson *et al.* in 1973–1974 [42]. The triangular and kagome lattices are the main geometries for a QSL system [43]. Since the Curie–Weiss temperature Θ_{CW} [Equation (D2.7)] is a parameter that determines the possibility of realizing QSL state, high $|J|$ and $|\Theta_{CW}|$ values are fundamental requirements to observe the QSL state at an experimentally attainable temperature.Thus far, spin quantum numbers greater than $S = 1/2$ have not afforded the QSL state except one, even when the system has a spin lattice with high $|\Theta_{CW}|$. A system with high spin quantum number preferred spin-ordered Néel states such as $NiGa_2S_4$ (triangle, $S = 1$, $|\Theta_{CW}| = 80$ K, 2D short-range order > 1.5 K) [44], $NaCrO_2$ (triangle, $S = 3/2$, $|\Theta_{CW}| = 290$ K, $T_N = 41$ K) [45], $CoAl_2O_4$ (diamond, $S = 3/2$, $|\Theta_{CW}| = 109$ K, $T_N = 6.5$ K) [46], and $V[C(CN)_3]$ (triangle, $S = 3/2$, $|\Theta_{CW}| = 67$ K, $T_N < 1.7$ K) [47]. These results indicate that the spin system with $S = 1/2$ is the best choice to realize the QSL state.

$$\Theta_{CW} = 2zS(S+1)J/3k_B \quad \text{(D2.7)}$$

Thus far, eleven QSL candidates have been proposed since 2003. Four of them are organic solids with a triangular spin lattice: κ-$(ET)_2Cu_2(CN)_3$(**o**: $|\Theta_{CW}|$ = 375 K, $|J|/k_B$= 250 K) [30a,b], κ-$(ET)_2Ag_2(CN)_3$ (**p**: $|J|/k_B$= 175 K) [30c], $EtMe_3Sb[Pd(dmit)_2]_2$($|\Theta_{CW}|$ = 325–375 K, $|J|/k_B$= 220–250 K) [48], and κ-H_3(Cat-EDT-TTF)$_2$($|J|/k_B$= 80–100 K) [49]. Seven inorganic solids with kagome or hyperkagome spin lattice have been proposed [triangular system: Ba_3CuSbO_9($|\Theta_{CW}|$ = 55 K, $|J|/k_B$= 32 K) [50], $Ba_3NiSb_2O_9$($|\Theta_{CW}|$ = 75.5, $|J|/k_B$= 19 K) [51], kagome system: $ZnCu_3(OH)_6Cl_2$($|\Theta_{CW}|$ = 314 K, $|J|/k_B$ = 180 K) [52], $ZnCu_3(OH)_6SO_4$($|\Theta_{CW}|$ = 79 K, $|J|/k_B$= 65 K) [53], $(NH_4)_2(C_7H_{14}N)(V_7O_6F_{18})$ ($|\Theta_{CW}|$ = 81 K) [54], hexagonal system: $[(C_3H_7)NH]_2\{Cu_2(C_2O_4)_3\}(H_2O)_{2.2}$($|\Theta_{CW}|$ = 120 K) [55], and hyperkagome system: $Na_4Ir_3O_8$($|\Theta_{CW}|$ = 650 K) [56]. Only $Ba_3NiSb_2O_9$ has $S = 1$ while others have $S = 1/2$. The $|\Theta_{CW}|$ and $|J|$ values of some of the systems are not so high, and their lower-temperature states will be examined to verify the QSL state. Since $|J|$ is proportional to $4|t|^2/U$ [Equation (D2.8)], designing the materials to increase $|t|$ and decrease U is essential to have a QSL state next to the superconducting state while keeping the triangular geometry of spin lattice as $t'/t \sim 1$. This means that such a QSL system lies in a Mott insulating regime proximate to the itinerant (metallic and superconducting) regime.

$$|J| \propto 4|t|^2/U. \quad \text{(D2.8)}$$

However, thus far, only ET salts (**o** and **p**) exhibited superconductivity next to the QSL state under pressure (Figure D2.8).

By applying a uniaxial strain to deform the triangular spin lattice of **o**, an anisotropic superconducting state appears at a pressure greater than 0.1 GPa next to the QSL state without passing through a spin-ordered AF state [Figure D2.8(a) and (b)] [57]. The salt **p** needs a higher pressure to convert from the QSL to superconducting state [hydrostatic pressure of 0.9 GPa, Figure D2.8(c)], showing that the QSL state of **p** is more robust than that of **o**. The onset T_c of 5.2 K at 1.05 GPa for **p** is higher than that observed for **o** under hydrostatic pressure; onset T_c = 3.9 K at a hydrostatic pressure of 0.06 GPa [30c].

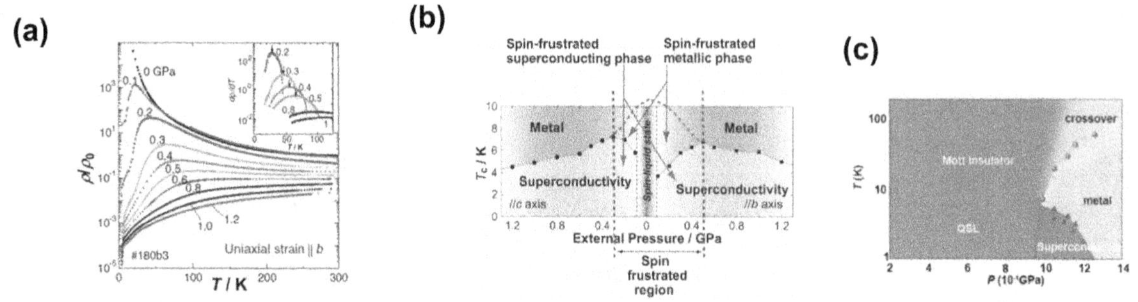

FIGURE D2.8 (a) Temperature dependence of resistivity of κ-$(ET)_2Cu_2(CN)_3$ under uniaxial strain (//*b*). (b) *T*–*P* (uniaxial strain) (//*b*,//*c*) phase diagram in the low-temperature region for κ-$(ET)_2Cu_2(CN)_3$. Two thick dotted lines represent the positions of maximum T_c. Inside the two thin dotted lines is a spin-frustrated region including QSL, spin-frustrated metallic, and spin-frustrated superconducting states. The dotted curve indicates the expected change of T_c in the absence of spin frustration. (c) *T*–*P* (hydrostatic pressure) phase diagram of κ-$(ET)_2Ag_2(CN)_3$ indicating more robust QSL state than that in κ-$(ET)_2Cu_2(CN)_3$.

The Mott transition temperature, defined by the peak of $d\rho/dT$ [inset in Figure D2.8(a)], shows positive pressure dependence. According to the Clausis–Clapeyron relation, the slope of dT_{MI}/dP is represented by the change of the volume (*V*) and spin entropy (*S*) between the insulating and metallic phases, as described by Equation (D2.9).

$$dT_{MT} / dP = \Delta V / \Delta S = (V_{insulator} - V_{metal}) - (S_{insulator} - S_{metal}). \qquad (D2.9)$$

The volumes of the insulating phase ($V_{insulator}$) are larger than the volumes of the metallic phase (V_{metal}) because the metallic phase appears under pressure; hence, $\Delta V > 0$. The observed $dT_{MI}/dP > 0$ indicates that the spin entropy in the insulating phase is higher than in the metallic phase ($\Delta S > 0$), indicating the absence of the AF order in the insulating phase.

The superconducting transition for **o** is anisotropic, reflecting the anisotropic electronic structure even in the 2D conducting *bc* plane [Figure D2.8(b)]. The maximum T_c was observed at 7.2 K under 0.3 GPa along the *c*-axis strain. The experimental results in Figure D2.8(b) indicate that T_c is suppressed when the system approaches the region near $t'/t = 1$ from the higher-pressure region owing to the increase in strong geometrical spin frustration. Therefore, the area between the blue lines in Figure D2.8(b) is classified as QSL, spin-frustrated metallic, and spin-frustrated SC states in this salt.

Anisotropic transport and superconducting behaviours are commonly observed in the 2D ET solids under uniaxial strain [58], but not under hydrostatic pressure. The T_c values under uniaxial strain are much higher than those under hydrostatic pressure, where T_c decreases monotonically with increasing hydrostatic pressure. Thus, the results under hydrostatic pressure for the anisotropic organic SCs are tentative.

D2.4.6 Relation between *U*/*t* and *t'*/*t*

Figure D2.9(a) shows the temperature dependence of the calculated *U*/*W* for κ-type SCs and conductors. The observed temperature dependence of *U*/*W* is diverse in comparison with that presented in Figure D2.9(b). Roughly speaking the Mott boundary between localization and itinerancy exists at $U/W = 0.89$ at RT and $U/W = 0.94$ at low temperatures except in salt **m**. In the quenched state, salt **m** has two kinds of ET layers (**m1, m2**), both of which have enhanced *U*/*W* values

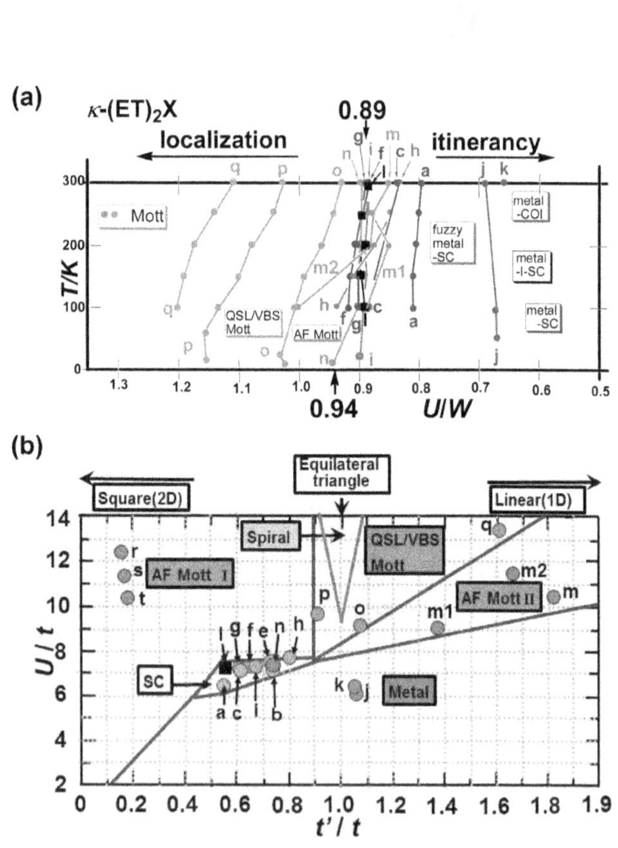

FIGURE D2.9 (a) Temperature dependence of *U*/*W* of κ-$(ET)_2X$. (b) The ground states of dimerized ET salts plotted in the *U*/*t* vs. *t'*/*t* phase diagram. The alphabets are the same as those in Table D2.2. The band parameters for **k** are at RT, those for **r–t** are at 120 K, and those for **m** are at 250 K. Symbol for **d** overlaps with that for **n**.

at low temperatures. Salts **j** and **k** in Figure D2.9(a) do not exhibit superconductivity, indicating that a moderate U/W (≥ 0.8) is crucial to exhibit superconductivity in the ET system. A plot of T_c vs. U/W indicates an enhancement of T_c when the U/W values approach U/W ~ 0.9 for the salt with a dimerized ET unit.

The temperature dependence of t'/t varies from salt to salt. The localized spins on ET dimers of **o** and **p** form nearly equilateral triangular lattices down to low temperatures, resulting in the QSL ground state. The Mott insulator **q** shows a monotonic increase in t'/t down to low temperatures (t'/t = 1.61 at 100 K), exhibiting the QSL phase down to 5 K below which it exhibits the VBS state (indicated as QSL/VBS Mott). Salt **n** has an AF ground state with a temperature-insensitive t'/t value (t'/t = 0.72 at 15 K). Another AF salt **m** shows a drastic decrease in t'/t towards unity below 250 K, resulting in a more spin-frustrated Layer 1 (**m1**: U/W = 0.84, t'/t = 1.50 at 200 K) and a less spin-frustrated Layer 2 (**m2**: U/W = 0.89, t'/t = 1.77 at 200 K).

The competition between itinerancy (metal, SC), localization (Mott insulator), and frustration (AF, QSL) was theoretically discussed using 0 K values of U, t, and t' by several theoreticians [43c,59]. The salts in Table D2.2 are plotted in the U/t vs. t'/t phase diagram in Figure D2.9(b). Instead of the data at 0 K in theories, the band parameters at 100 K were used, except those described in the caption of Figure D2.9. "AF Mott I" is a region containing Mott insulators with an AF ground state and square spin lattice [small t'/t, Figure D2.1(c)]. Salts **r**–**t** are typical examples showing large magnetic susceptibility at RT (EPR spin susceptibility χ = 9.5–11× 10^{-4} emu mol^{-1}), large activation energy for transport (ε_a = 110–120 meV), and poor conductivity (σ_{RT} = 10^{-1}–10^{-2} S cm^{-1}). "AF Mott II" is a region containing Mott insulators with an AF ground state and linear spin lattice [large t'/t, Figure D2.1(a)]; among κ-$(ET)_2X$ salts, only **m** with low χ(5.3×10^{-4} emu mol^{-1}), low ε_a (66 meV), and high σ_{RT}(1.8 S cm^{-1}) belongs to this region. The "QSL/VBS Mott" region includes Mott insulators with QSL (**o**, **p**) or VBS (**q**) ground states. The "Metal" region includes metals with a CO insulating ground state (**j**, **k**). Table D2.2 indicates that the SC state is rather common for dimerized ET salts in a wide range of U/W and t'/t values under pressure. The SC region includes SCs at AP in the range of 6.49 ≤ U/t ≤ 7.79 and in the intermediate range of t'/t (0.54 < t'/t ≤ 0.85), where the upper value of t'/t results from that of κ-$(ET)_2Ag(CF_3)_4(TCE)$ [40c]. According to the t'/t and U/t values for **l** and **n**, they are located in the SC region, though the former is a poor metal (σ_{RT} = 1 S cm^{-1}) with T_{MI} = 50–60 K [35] and the latter is an AF Mott insulator (T_N = 27 K) with moderate conductivity, small activation energy, and small χ(σ_{RT} = 2 S cm^{-1}, ε_a= 12–52 meV, χ = 5.0×10^{-4} emu mol^{-1}) [22b]. These salts become SCs with a small change in external conditions or with a small perturbation [22c]. The salts away from the SC region need stronger perturbations to become SCs.

The blue lines representing phase boundaries are tentative. The SC region is enclosed by the AF, QSL, and metallic regions, which is consistent with theories [59]. Some theories predict the compromise non-frustrated spin configuration such as a spiral order [120°, Figure D2.10(a)] [59b–e] or collinear order [Figure D2.10(b)] [59e, f]. The spiral order is predicted in the region near t'/t = 1 and high U/t as roughly depicted by the lines in Figure D2.9(b). A sequential change from a metal to a QSL above U/t = 7.4 and then to a 120° spiral above U/t = 9.2 at t'/t = 1 was theoretically proposed [59c]. A collinear order is thought to exist between the QSL and metallic phases at high t'/t (> 1.25) [59e]. Thus far, no such 120° spin lattice and collinear systems have been realized in organic materials.

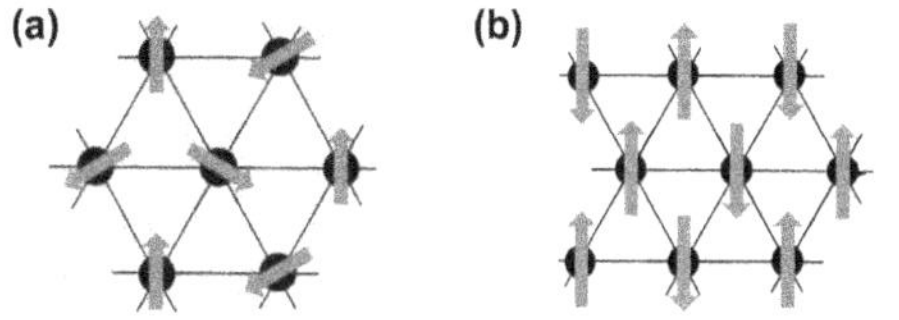

FIGURE D2.10 Non-frustrated spin configurations: (a) 120° spiral order and (b) collinear order of triangular spin lattice.

The QSL Mott salts with magenta circles are allocated in the region 0.91 < t'/t and 9.1 < U/t among them, **q** is in the high-t'/t region (1D linear spin lattice). Thus, Figure D2.9 suggests that the collinear spin order might be realized in **m**.

The superconductivity is associated with the electron correlation and spin fluctuation. It is worth emphasizing that uniaxial strain measurements revealed that salt **m** exhibited a superconducting state when t'/t values approach unity and no superconductivity when t'/t values depart from unity [33]. The appearance of the superconducting state immediately after the release of the strong spin frustration in the QSL state of salts **o** and **p**, as well as when t'/t approaches unity in salt **m** may be an indication of the importance of the spin frustration, though very strong spin frustration suppressed superconductivity.

An anomalous dielectric response such as relaxor-type ferroelectricity in the Mott insulators **n** and **o** is reviewed in ref. 60.

D2.5 Other Organic Superconductors Based on TTF Analogues

The most intriguing phenomenon among the eight BETS SCs (highest T_c = 5.5 K) is the reentrant superconductor–insulator–superconductor transition under a magnetic field for $(BETS)_2FeCl_4$. The λ- or κ-type BETS salts formed with tetrahedral anions FeX_4 (X: Cl and Br) were studied in terms of the competition between magnetic ordering and superconductivity [61]. At 8.3 K, λ-$(BETS)_2FeCl_4$ exhibited coupled AF and metal-insulator transitions. For the $FeCl_4$ salt, a firm nonlinear electrical transport associated with the negative resistance effect in the magnetic-ordered state [62] and a relaxor ferroelectric behaviour in the metallic state below 70 K [63] have been observed. The $FeCl_4$ salt shows a field-induced superconducting transition under a magnetic field of 18–41 T applied exactly parallel to the conducting layers [61]. κ-$(BETS)_2FeX_4$(X = Cl, Br) are AF SCs (T_N = 2.5 K, T_c = 1.1 K for X = Br: T_N = 0.45 K, T_c = 0.17 K for X = Cl) [64]. Similar phenomena, namely AF, ferromagnetic, or field-induced superconductivity have been observed in inorganic solids such as the Chevrel phase [65] and heavy-fermion system [66]. The STM study on λ-$(BETS)_2FeCl_4$ revealed that the minute size of four pairs of $(BETS)_2FeCl_4$ exhibited 2D superconductivity [67].

Besides TMTSF, TMTTF, ET, and BETS SCs, there are other SCs (Scheme D2.1) of CT salts based on symmetric (BO [68], BEDSe-TTF [69] and BDA-TTP [70]) and asymmetric donors (ESET-TTF [71], *S,S*-DMBEDT-TTF [72], *meso*-DMBEDT-TTF [73], DMET [74], DODHT [75], TMET-STF [76], DMET-TSF [77], DIETS [78], EDT-TTF [79], MDT-TTF [80], MDT-ST [81], MDT-TS [82], MDT-TSF [83], MDSe-TSF [84], DTEDT [85], and DMEDO-TSeF [86]).

κ-$(MDT-TTF)_2AuI_2$ (T_c = 3.5 K) exhibited a Hebel–Slichter coherent peak immediately below T_c, indicating a BCS-type gap with *s*-symmetry [80b]. On the other hand, *d*-wave-like superconductivity has been suggested for β-$(BDA-TTP)_2SbF_6$ [70c,d]. β-$(BDA-TTP)_2X$ (X = SbF_6, AsF_6) exhibits a slight increase in T_c at the initial stage of uniaxial strain parallel to the donor stack and interlayer direction while T_c decreases perpendicular to the donor stack [70b]. θ-$(DIETS)_2[Au(CN)_4]$ exhibits superconductivity under uniaxial strain parallel to the *c*-axis (T_c = 8.6 K at 1 GPa), though under hydrostatic pressure, a sharp metal–insulator transition remains even at 1.8 GPa [78]. MDT-ST, MDT-TS, and MDT-TSF SCs [76–78] have non-integer ratios of donor and anion molecules, such as $(MDT-TS)(AuI_2)_{0.441}$, making the Fermi level different from the conventional 3/4 filled band for TMTSF and ET 2:1 salts. The Fermi-surface topology of (MDT-TSF)X (X = $(AuI_2)_{0.436}$, $(I_3)_{0.422}$) and $(MDT-ST)(I_3)_{0.417}$ has been studied by SdH and AMRO [81b,83b-d]. DMEDO-TSeF afforded eight SCs. Six of them are κ-$(DMEDO-TSeF)_2[Au(CN)_2]$(solvent), and their T_c values (1.7-5.3 K) are tuned using cyclic ethers as solvents of crystallization [86b]. The superconducting coherent lengths indicate that β-$(BDA-TTP)_2SbF_6$ shows 2D character ($\xi_{//}$= 105 Å, $\xi_{\perp}$ = 26 Å) while $(DMET-TSeF)_2AuI_2$ shows quasi-1D character (1000, 400, and 20 Å).

D2.6 Two-Dimensional Superconductors: dmit System

Eleven SCs were prepared based on $M(dmit)_2$ (three for *M* = Ni, eight for *M* = Pd), and their T_c values are less than 8.4 K. Only one showed superconductivity at AP (α-EDT-$TTF(Ni[dmit)_2]_2$, T_c = 1.3 K) [87]. This salt exhibits SdH oscillations with coherent lengths $\xi_{//}$= 310 and $\xi_{\perp}$= 24 Å. Superconducting Langmuir–Blodgett (LB) films composed of dimethylbis(tetradecyl)ammonium$[Au(dmit)_2]$ (T_c < 3.9 K) have been reported [88]. In this family of compounds, $(Cation)[Pd(dmit)_2]_2$, a competition between AF, QSL, VBS, metallic and superconducting states has been elucidated by changing the size of cation molecules [89]. The salt with Cation = $EtMe_3Sb$ is a QSL system (*t'/t* = 0.92) without showing superconductivity under pressure. The salt with Cation = $EtMe_3P$ has an electronic structure with 2D nature and nearly equilateral-triangular spin lattice (*t'/t* = 1.02). A spin-Peierls-like transition occurs with a change of dimer to give the tetramer of $[Pd(dmit)_2]_2^{\bullet-}$ at 25 K with a spin-singlet ground state. At higher temperatures, dimers stack parallel to form a π–π column. The dimerized π–π stack deforms into a tetramerized π–π stack with alternating interdimer interplanar distances of 3.762 and 3.854 Å. Furthermore, under pressure, the salt exhibited metallic and SC phases (T_c = 5.5 K at 0.2 GPa) without passing through the CO, AF, and SDW states [89].

D2.7 Three-Dimensional C_{60} Superconductors

The icosahedral C_{60} molecule with I_h symmetry has triply degenerate LUMO and LUMO+1 orbitals with t_{1u} and t_{1g} symmetries, respectively. In 1991, a superconducting phase was observed below 19 K for the K-doped compound [90], immediately after the isolation of macroscopic quantities of C_{60} solid [91]. The powder X-ray diffraction profile revealed that the composition of the superconducting phase is K_3C_{60}, and the diffraction pattern can be indexed as a face-centred-cubic (fcc) structure [92]. The lattice constant (*a* = 14.24 Å) is apparently expanded relative to the undoped cubic C_{60} (*a* = 14.17 Å). The superconductivity has been observed for many A_3C_{60} (A: alkali metals) compounds, e.g., Rb_3C_{60} (T_c = 29 K [93]), Rb_2CsC_{60} (T_c = 31 K [94]), and $RbCs_2C_{60}$ (T_c = 33 K [94]), and their structures were determined to be analogous to that of K_3C_{60} with varying lattice constants. The T_c varies monotonously with the lattice constant, independently of the type of the alkali dopant (Figure D2.11) [94, 95]. The observation of a Hebel–Slichter peak in the relaxation rate immediately below T_c in NMR [96] and μSR [97] indicates the BCS-type isotopic gap. The decrease in T_c due to the isotopic substitution [98] also supports the phonon-mediated pairing in A_3C_{60}, where the value in $T_c \propto (\text{mass})^{-\alpha}$ (ideal value of α predicted by the BCS model is 0.5) is estimated to be 0.30(6) for $K_3{}^{13}C_{60}$ and 0.30(5) for $Rb_3{}^{13}C_{60}$.

Keeping the C_{60} valence invariant (−3), the intercalation of NH_3 molecules (e.g., $(NH_3)K_3C_{60}$) results in the lattice distortion from cubic to orthorhombic accompanied by the appearance of AF ordering instead of superconductivity [99]. Changing the valence in the cubic system also has a pronounced effect on T_c. For example, the T_c in $Rb_{3-x}Cs_xC_{60}$ prepared in liquid ammonia gradually increases as the mixing

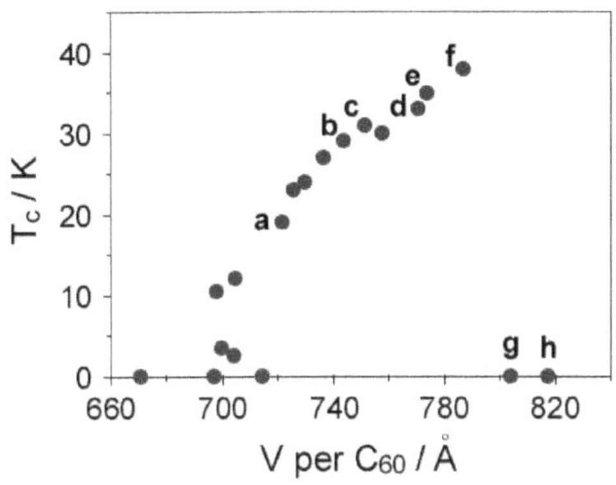

FIGURE D2.11 T_c as a function of volume occupied per C_{60}^{3-} in cubic A_3C_{60} (A: alkali metal). **A,** K_3C_{60}; **b,** Rb_3C_{60}; **c,** Rb_2CsC_{60}; **d,** $RbCs_2C_{60}$; **e,** fcc Cs_3C_{60} at 0.7 GPa; **f,** A15 Cs_3C_{60} at 0.7 GPa; **g,** fcc Cs_3C_{60} at AP; **h,** A15 Cs_3C_{60} at AP.

ratio approaches $x = 2$ [100]. Increasing the nominal ratio of Cs further leads to a sizable decrease in T_c, despite the fact that the lattice keeps the fcc structure for $x < 2.65$. A recent work on the T_c behaviour of $Rb_{3-x}Cs_xC_{60}$ ($1 \le x \le 2.65$) under hydrostatic pressure revealed that the T_c value exhibits a broad maximum at a volume of 755–760 Å^3 occupied per C_{60} at 10 K [101], which seems to be consistent with the boundary observed for the RT-value in Figure D2.11. Band-filling control has also been realized for $Na_2Cs_xC_{60}$ ($0 \le x \le 1$) [102] and Li_xCsC_{60} ($2 \le x \le 6$) [103], which show that the T_c decreases sharply as the valence state on C_{60} deviates from −3.

In 2008, A15 or the body-centred-cubic (bcc) Cs_3C_{60} phase, which shows bulk superconductivity with a T_c of 38 K under hydrostatic pressure, was obtained together with a small amount of T_c phase [104]. Interestingly, the lattice contraction with respect to pressure results in the increase in T_c up to approximately 0.7 GPa, above which T_c gradually decreases. The trend in the initial pressure range is not explicable within the simple BCS theory. At AP, on the other hand, the A15 Cs_3C_{60} shows an AF ordering below 46 K, verified by means of ^{133}Cs NMR and μSR [105]. In 2010, it has been found that the fcc phase also shows an AF ordering at 2.2 K at AP and superconducting transition at 35 K under a hydrostatic pressure of approximately 0.7 GPa [106]. We note that the T_c of the both phases follows the universal relationship for A_3C_{60} superconductors in the vicinity of the Mott boundary, as shown in Figure D2.11.

Thus far, about 40 C_{60}-based SCs have been prepared with the highest T_c of 33 K at AP ($RbCs_2C_{60}$) and 38 K under pressure (A15 Cs_3C_{60}; 0.7 GPa).

D2.8 Polyaromatic Hydrocarbon Superconductors

A new family of organic superconductors began with alkali-metal–doped picene compounds reported in 2010 [107]. Although their shielding fractions are relatively low (< 15%), the bulk superconducting phase was observed below 6.9 K for $K_{3\cdot3}$picene, 18 K for $K_{3.3}$picene, and 6.9 K for $Rb_{3.1}$picene, in which three extra electrons reside in the nearly two-fold degenerate LUMO because of a tiny energy difference between LUMO and LUMO+1 (< 0.1 eV). The T_c of 18 K is significantly higher than that of K-doped graphite (T_c ~ 5.5 K) [108] and comparable to that of K_3C_{60} (T_c = 18 K) [90]. At present, although the crystal structures of the doped compounds are unclear, the refined lattice parameters are indicative of the deformation of the herringbone structure of pristine picene and the intercalation of alkali dopants within the two-dimensional picene layers. Recently, several research groups were successful in reproducing picene-based superconductors by K- [109] and Sm- [110] doping, although showing very small shielding fractions.

Table D2.3 lists the superconductors of charged polyaromatic hydrocarbon molecules. [*n*]Phenacenes (Figure D2.12) with smaller *n* values also become superconducting with electron doping for phenanthrene (n = 3) [110, 111] and for chrysene (n = 4) [110], where the nominal charges of [*n*]phenacenes are approximately −3. The positive pressure dependence of T_c (*ca.* 1 K GPa^{-1}) observed in phenanthrene-based superconductors [111] is not explained by the simple BCS picture of superconductivity. The highest T_c is 33 K for K-doped 1,2:8,9-dibenzopentacene [112] and 35 K for Ba-doped anthracene [109]; the latter seems to be the sole example of [*n*]acene-based doped superconductors (for [*n*]acene, see Figure D2.12). The preparation of high-purity doped compounds (charge state and crystal structure) will provide a clue to develop this material group.

Contrary to the electron-doped system described above, it has been found that a cation radical salt $(perylene)_2Au(mnt)_2$, in which each perylene molecule has an average charge of +0.5 and form segregated columns, shows superconductivity with T_c = 0.3 K when a hydrostatic pressure greater than 0.5 GPa was applied to suppress the CDW phase [114].

Thus far, seven aromatic hydrocarbon molecules form superconductors with the highest T_c of 33 or 35 K at AP.

D2.9 Single-Component Superconductors

Several single-component organic superconductors have been developed. Even though pentacene is known to be the first organic metal (semimetal) showing a decrease of resistivity down to ca. 200 K at 21.3 GPa [115], no superconductivity was reported thus far for the solids composed solely of aromatic hydrocarbons. Electric conductivity increases with the enhancement of intermolecular interactions through the appropriate use of heteroatomic contacts. There are two single-component superconductors under extremely high pressure, *p*-iodanil ($\sigma_{RT} = 1 \times 10^{-12}$ S cm^{-1} at AP, $\sigma_{RT} = 2 \times 10$ S cm^{-1} at 25 GPa, and superconductor at T_c ~ 2 K at 52 GPa) [116] and hexaiodobenzene (T_c = 0.6-0.7 K at around 33 GPa and ca. 2.3 K

TABLE D2.3 Polyaromatic Hydrocarbon Superconductors

Polyaromatic Hydrocarbon	Cation	T_c (K)[a]	Shielding Fraction (%)	Ref.
[*n*]Phenacene				
phenanthrene ($n = 3$)	K_3	4.95	5.3	[111a]
		5.9 (1 GPa)		
	Rb_3	4.75	6.7	[111a]
	$Sr_{1.5}$	5.6	39.5	[111b]
		6.6 (1 GPa)		
	$Ba_{1.5}$	5.4	65.4	[111b]
		6.7 (1 GPa)		
	La_1	6.1	46.1	[111c]
		7.6 (1 GPa)		
	Sm_1	6.0	49.8	[111c]
	Sm_1	5.0-5.2		[110]
chrysene ($n = 4$)	Sm_1	5.4	ca. 1	[110]
picene ($n = 5$)	$K_{3.3}$	6.9	15	[107]
	$K_{3.3}$	18	1.2	[107]
	K_x	22		[109]
	$Rb_{3.1}$	6.9	10	[107]
	$Cm_{1.5}$	7	1.25	[113]
	Sm_1	4		[110]
[*n*]Acene				
anthracene ($n = 3$)	Ba_x	35		[109]
Others				
coronene	K_3	3.5, 7, 11, 15		[113]
1,2:8,9-dibenzopentacene	K_3	7.4	3.6	[112]
	$K_{3.17}$	28.2	5.5	[112]
	$K_{3.45}$	33.1	3.2	[112]
Polyaromatic hydrocarbon	Anion	T_c (K)[a]	Shielding fraction (%)	Ref.
perylene	$Au(mnt)_2$	0.35 (0.53 GPa)		[114]

[a] Onset value.

at 58 GPa) [117]. Both have peripheral chalcogen atoms, iodine, which may cause the increased electronic dimensionality of the solid under pressure owing to intermolecular iodine···iodine contacts. Very recently, a metal complex $Ni(hfdt)_2$ (Scheme D2.1) was added in this class with $T_c = 5.5$ K at 8.1 GPa [118].

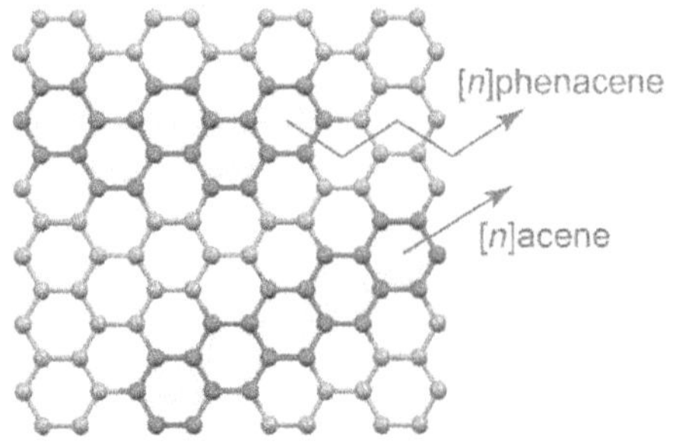

FIGURE D2.12 Molecular structures of [*n*]phenacene and [*n*] acene, where *n* denotes the number of benzene rings.

Acknowledgements

This work was supported by the Japan Society for the Promotion of Science (JSPS) KAKENHI Grant Numbers JP23225005 titled "Development of multi-electronic-functions based on spin triangular lattice".

References

[1] Drozdov AP, Eremets MI, Troyan IA (2014) Conventional superconductivity at 190 K at high pressures. *arXiv*: 1412.0460.

[2] Drozdov AP, Eremets MI, Troyan IA, Ksenofontov V, Shylin SI (2015) Conventional superconductivity at 203 kelvinat high pressures in the sulfur hydride system. *Nature* 525: 73–76.

[3] Schilling A, Cantoni M, Guo JD, Ott HR (1993) Superconductivity above 130 K in the Hg-Ba-Ca-Cu-O system. *Nature* 363: 56–58.

[4] Gao L, Xue YY, Chen F, Xiong Q, Meng RL, Ramirez D, Chu CW, Eggert JH, Mao HK (1994) Superconductivity up to 164 K in $HgBa_2Ca_{m-1}Cu_mO_{2m+2+\delta}$($m$ = 1, 2, and 3) under quasihydrostatic pressures. *Phys. Rev. B* 50: 4260–4263.

[5] a) Jerome D, Schulz HJ (1982) Organic conductors and superconductors. *Adv. Phys.* 31: 299–490. b) Williams JM, Ferraro JR, Thorn RJ, Carlson KD, Geiser U, Wang HH, Kini AM, Whangbo MH (1992) Organic superconductors (including fullerenes). Englewood Cliffs, NJ: Prentice Hall. c) Ishiguro T, Yamaji K, Saito G (1998) Organic superconductors 2nd ed. Berlin: Springer-Verlag. d) Saito G, Yoshida Y (2007) Development of conductive organic molecular assemblies: organic metals, superconductors, and exotic functional materials. *Bull. Chem. Soc. Jpn.* 80: 1–137 and references cited therein.

[6] Jérome D, Mazaud A, Ribault M, Bechgaard K (1980) Superconductivity in a synthetic organic conductor $(TMTSF)_2PF_6$. *J. Phys. Lett.* 41: L95–L98.

[7] Ferraris J, Cowan DO, Walatka V Jr, Perlstein JH (1973) Electron transfer in a new highly conducting donor-acceptor complex. *J. Am. Chem. Soc.* 95: 948–949.

[8] Cooper JR, Weger M, Jérome D, Lefur D, Bechgaard K, Bloch AN, Cowan DO (1976) Semi-metallic behavior of HMTSF-TCNQ at low temperatures under pressure. *Solid State Commun.* 19: 749–754.

[9] a) Jérome D (1991) The physics of organic superconductors. *Science* 252: 1509–1514. b) Jérome D (2004) Organic conductors: from charge density wave TTF-TCNQ to superconducting $(TMTSF)_2PF_6$. *Chem. Rev.* 104: 5565–5591.

[10] a) Dumm M, Loidl A, Fravel BW, Starkey KP, Montgomery LK, Dressel M (2000) Electron spin resonance studies on the organic linear-chain compounds $(TMTCF)_2X$ (C = S, Se; X = PF_6,AsF_6,ClO_4,Br). *Phys. Rev. B* 61: 511–521. b) Itoi M, Kano M, Kurita N, Hedo M, Uwatoko Y, Nakamura T (2007) Pressure-induced superconductivity in the quasi-one-dimensional organic conductor $(TMTTF)_2AsF_2$. *J. Phys. Soc. Jpn.* 76: 053703/1–5.

[11] a) Urayama H, Yamochi H, Saito G, Nozawa K, Sugano T, Kinoshita M, Sato S, Oshima K, Kawamoto A, Tanaka J (1988) A new ambient pressure organic superconductor based on BEDT-TTF with T_c higher than 10 K (T_c= 10.4 K). *Chem. Lett.* 17: 55–58. b) Urayama H, Yamochi H, Saito G, Sato S, Kawamoto A, Tanaka J, Mori T, Maruyama Y, Inokuchi H (1988) Crystal structures of organic superconductor, $(BEDT-TTF)_2Cu(NCS)_2$, at 298 K and 104 K. *Chem. Lett.* 17: 463–466.

[12] Oshima K, Mori T, Inokuchi H, Urayama H, Yamochi H, Saito G (1988) Shubnikov-de Haas effect and the fermi surface in an ambient-pressure organic superconductor $[bis(ethylenedithiolo)tetrathiafulvalene]_2Cu(NCS)_2$. *Phys. Rev. B* 38: 938–941.

[13] Sasaki T, Sato H, Toyota N (1990) Magnetic breakdown effect in organic superconductor κ-$(BEDT-TTF)_2Cu(NCS)_2$. *Solid State Commun.* 76: 507–510.

[14] Yagubskii EB, Shchegolev IF, Laukhin VN, Kononovich PA, Karstovnik MV, Zvarykina AV, Buravov LI (1984) Normal-pressure superconductivity in an organic metal. *JETP Lett.* 39: 12–15.

[15] Emge TJ, Leung PCW, Beno MA, Schultz AJ, Wang HH, Sowa LM, Williams JM (1984) Neutron and X-ray diffraction evidence for a structural phase transition in the sulfur-based ambient-pressure organic superconductor bis (ethylenedithio) tetrathiafulvalene triiodide. *Phys. Rev. B* 30: 6780–6782.

[16] a) Laukhin VN, Kostyuchenko EÉ, Sushko YV, Shchegolev IF, Yagubskii ÉB (1985) Effect of pressure on the superconductivity of ß-$(BEDT-TTF)_2I_3$. *JETP Lett.* 41: 81–84. b) Murata K, Tokumoto M, Anzai H, Bando H, Saito G, Kajimura K, Ishiguro T (1985) Superconductivity with the onset at 8 K in the organic conductor ß-$(BEDT-TTF)_2I_3$ under pressure. *J. Phys. Soc. Jpn.* 54: 1236–1239.

[17] Guterding D, Valentí R, Jeschke HO (2015) Influence of molecular conformations on the electronic structure of organic charge transfer salts. *Phys. Rev. B* 92: 081109(R)/1-6.

[18] Mori T, Kobayashi A, Sasaki Y, Kobayashi H, Saito G, Inokuchi H (1984) The intermolecular interaction of tetrathiafulvalene and bis (ethylenedithio) tetrathiafulvalene in organic metals. Calculation of orbital overlaps and models of energy-band structures.*Bull. Chem. Soc. Jpn.* 57: 627–633.

[19] a) Appendix (pp. 455–458) in Ref. 5c. b) Wosnitza J (1996) Fermi surfaces of low-dimensional organic metals and superconductors. Berlin: Springer-Verlag. c) Singleton J (2000) Studies of quasi-two-dimensional organic conductors based on BEDT-TTF using high magnetic fields. *Rep. Prog. Phys.* 63: 1111–1207.

[20] a) Yamauchi Y, Kartsovnik MV, Ishiguro T, Kubota M, Saito G (1996) Angle-dependent magnetoresistance and Shubnikov-de Haas oscillations in the organic superconductor κ-$(BEDT-TTF)_2CU[N(CN)_2]Cl$ under pressure. *J. Phys. Soc. Jpn.* 65: 354–357. b) Ohmichi E, Ito H, Ishiguro T, Saito G (1998) Shubnikov-de Haas oscillation with unusual angle dependence in the organic superconductor κ-$(BEDT-TTF)_2CU_2(CN)_3$. *Phys. Rev. B* 57: 7481–7484.

[21] Taniguchi H, Miyashita M, Uchiyama K, Satoh K, Môri N, Okamoto H, Miyagawa K, Kanoda K, Hedo M, Uwatoko Y (2003) Superconductivity at 14.2 K in layered organics under extreme pressure. *J. Phys. Soc. Jpn.* 72: 468–471.

[22] a) Schirber JE, Overmyer DL, Carlson KD, Williams JM, Kini AM, Wang HH, Charlier HA, Love BJ, Watkins DM, Yaconi GA (1991) Pressure-temperature phase diagram,

inverse isotope effect, and superconductivity in excess of 13 K in κ-(BEDT-TTF)$_2$CU[N(CN)$_2$]Cl, where BEDT-TTF is bis(ethylenedithio)tetrathiafulvalene. *Phys. Rev. B* 44: 4666–4669. H-salt needs small pressure to observe superconductivity. b) Williams JM, Kini AM, Wang HH, Carlson KD, Geiser U, Montgomery LK, Pyrka GJ, Watkins DM, Kommers JM, Boryschuk SJ, Streiby Crouch AV, Kwok WK, Schirber JE, Overmyer DL, Jung D, Whangbo MH (1990) From semiconductor-semiconductor transition (42 K) to the highest T_c organic superconductor, κ-(ET)$_2$Cu[N(CN)$_2$]Cl (T_c = 12.5 K). *Inorg. Chem.* 29: 3272–3274. However, the following paper reported superconductivity at AP. c) Yagubskii EB, Kushch ND, Kazakova AV, Buravov LI, Zverev VN, Manakov AI, Khasanov SS, Shibaeva RP (2005) Superconductivity at normal pressure in κ-(BEDT-TTF)$_2$CU[N(CN)$_2$]Cl crystals. *JETP Lett.* 82: 93–95.

[23] Saito G, Yamochi H, Nakamura T, Komatsu T, Inoue T, Ito H, Ishiguro T, Kusunoki M, Sakaguchi K, Mori T (1993) Structural and physical properties of two new ambient pressure κ-type BEDT-TTF superconductors and their related salts. *Synth. Met.* 55–57: 2883–2890.

[24] Kini AM, Geiser U, Wang HH, Carlson KD, Williams JM, Kwok WK, Vandervoort KG, Thompson JE, Stupka DL, Jung D, Whangbo MH (1990) A new ambient-pressure organic superconductor, κ-(ET)$_2$Cu[N(CN)$_2$]Br, with the highest transition temperature yet observed (inductive onset T_c= 11.6 K, resistive onset = 12.5 K). *Inorg. Chem.* 29: 2555–2557.

[25] Ito H, Ishihara T, Niwa M, Suzuki T, Onari S, Tanaka Y, Yamada J, Yamochi H, Saito G (2010) Superconductivity of ß-type salts under uniaxial compression. *Physica B* 405: S262–S264.

[26] a) Bender K, Dietz K, Endres H, Helberg HW, Hennig I, Keller HJ, Schäfer HW, Schweitzer D (1984) (BEDT-TTF)$^+{}_2$J-3: a two-dimensional organic metal. *Mol. Cryst. Liq. Cryst.* 107: 45–53. b) Takano Y, Hiraki K, Yamamoto HM, Nakamura T, Takahashi T (2001) Charge disproportionation in the organic conductor, a-(BEDT-TTF)$_2$I$_3$. *J. Phys. Chem. Solids* 62: 393–395.

[27] a) Tajima N, Sugawara S, Tamura M, Nishio Y, Kajita K (2006) Electronic phases in an organic conductor α-(BEDT-TTF)$_2$I$_3$: ultra narrow gap semiconductor, superconductor, metal, and charge-ordered insulator. *J. Phys. Soc. Jpn.* 75: 051010/1–10. b) Katayama S, Kobayashi A, Suzumura Y (2006) Pressure-induced zero-gap semiconducting state in organic conductor α-(BEDT-TTF)$_2$I$_3$ salt. *J. Phys. Soc. Jpn.* 75: 054705/1–6.

[28] Mori H, Tanaka S, Mori T (1998) Systematic study of the electronic state in θ-type BEDT-TTF organic conductors by changing the electronic correlation. *Phys. Rev. B* 57: 12023–12029.

[29] Salameh B, Nothardt A, Balthes E, Schmidt W, Schweitzer D, Strempfer J, Hinrichsen B, Jansen M, Maude DK (2007) Electronic properties of the organic metals Θ-(BEDT-TTF)$_2$I$_3$ and Θ$_T$-(BEDT-TTF)$_2$I$_3$. *Phys. Rev. B* 75: 054509/1–13.

[30] a) Shimizu Y, Miyagawa K, Kanoda K, Maesato M, Saito G (2003) Spin liquid state in an organic Mott insulator with a triangular lattice. *Phys. Rev. Lett.* 91: 107001/1–4. b) Pratt FL, Baker PJ, Blundell SJ, Lancaster T, Ohira-Kawamura S, Baines C, Shimizu Y, Kanoda K, Watanabe I, Saito G (2011) Magnetic and non-magnetic phases of a quantum spin liquid. *Nature* 471: 612–616. c) Shimizu Y, Hiramatsu T, Maesato M, Otsuka A, Yoshida M, Takigawa M, Ono A, Itoh M, Yamochi H, Yoshida Y, Saito G (2016) Pressure-tuned exchange coupling of a quantum spin liquid in the moecuar triangular lattice κ-(ET)$_2$Ag$_2$(CN)$_3$. *Phys. Rev. Lett.* 117: 107203/1–6.

[31] a) Chakravarty S, Halperin BI, Nelson DR (1989) Two-dimensional quantum Heisenberg antiferromagnet at low temperatures. *Phys. Rev. B* 39: 2344–2371. b) Sachdev S (1994) NMR relaxation in half-integer antiferromagnetic spin chains. *Phys. Rev. B* 50: 13006–13008. c) Takigawa M, Motoyama N, Eisaki H, Uchida S (1996) Dynamics in the S = 1/2 one-dimensional antiferromagnet Sr$_2$CuO$_3$via ^{63}Cu NMR. *Phys. Rev. Lett.* 76: 4612–4615.

[32] Yoshida Y, Ito H, Maesato M, Shimizu Y, Hayama H, Hiramatsu T, Nakamura Y, Kishida H, Koretsune T, Hotta C, Saito G (2015) Spin-disordered quantum phases in a quasi-one-dimensional triangular lattice. *Nat. Phys.* 11: 679–683.

[33] Ito H, Asai T, Shimizu Y, Hayama H, Yoshida Y, Saito G (2016) Pressure-induced superconductivity in the antiferromagnet κ-(ET)$_2$CF$_3$SO$_3$ with quasi-one-dimensional triangular spin lattice. *Phys. Rev. B* 94: 020503/1–6.

[34] Ito H, Ishiguro T, Kubota M, Saito G (1996) Metal-nonmetal transition and superconductivity localization in the two-dimensional conductor κ-(BEDT-TTF)$_2$Cu[N(CN)$_2$]Cl under pressure. *J. Phys. Soc. Jpn.* 65: 2987–2993.

[35] a) Kushch ND, Tanatar MA, Yagubskii EB, Ishiguro, T. (2001) Superconductivity of κ-(BEDT-TTF)$_2$Cu[N(CN)$_2$]I under pressure. *JETP Lett.* 73: 429–431. b) Tanatar MA, Ishiguro T, Kagoshima S, Kushch ND, Yagubskii EB (2002) Pressure-temperature phase diagram of the organic superconductor κ-(BEDT-TTF)$_2$Cu[N(CN)$_2$]I. *Phys. Rev. B* 65: 064516/1–5.

[36] Kanoda K (1997) Recent progress in NMR studies on organic conductors. *Hyperfine Interact.* 104: 235–249.

[37] Kagawa F, Miyagawa K, Kanoda K (2009) Magnetic Mott criticality in a κ-type organic salt probed by NMR. *Nat. Phys.* 5: 880–884.

[38] a) Arai T, Ichimura K, Nomura K, Takasaki S, Yamada J, Nakatsuji S, Anzai H (2001) Tunneling

spectroscopy on the organic superconductor κ-$(BEDT\text{-}TTF)_2Cu(NCS)_2$ using STM. *Phys. Rev. B* 63: 104518/1-5. b) Izawa K, Yamaguchi H, Sasaki T, Matsuda Y (2002) Superconducting gap structure of κ-(BEDT-TTF)$_2$Cu(NCS)$_2$ probed by thermal conductivity tensor. *Phys. Rev. Lett.* 88: 027002/1–4. c) Ichimura K, Takami M, Nomura K (2008) Direct observation of d-wave superconducting gap in κ-(BEDT-TTF)$_2$Cu[N(CN)$_2$]Br with scanning tunneling microscopy. *J. Phys. Soc. Jpn.* 77: 114707/1–6.

[39] Uchiyama K, Miyashita M, Taniguchi H, Satoh K, Môri N, Miyagawa K, Kanoda K, Hedo M, Uwatoko Y (2004) Characterization of transport and magnetic properties of a Mott insulator, ß'-(BEDT-TTF)2IBrCl. *J. Phys. IV* 114: 387–389.

[40] a) Schlueter JA, Wiehl L, Park H, de Souza M, Lang M, Koo HJ, Whangbo MH (2010) Enhanced critical temperature in a dual-layered molecular superconductor. *J. Am. Chem. Soc.* 132: 16308–16310. b) Kawamoto T, Mori T, Nakano A, Murakami Y, Schlueter JA (2012) T_c of 11 K identified for the third polymorph of the (BEDT-TTF)$_2$Ag(CF$_3$)$_4$(TCE) organic superconductor. *J. Phys. Soc. Jpn.* 81: 023705/1–4. c) Geiser U, Schlueter JA, Williams JM, Naumann D, Roy T (1995) Anion disorder in the 115–118 K structures of the organic superconductors κL-(BEDT-TTF)$_2$CU(CF$_3$)$_4$(C$_2$H$_3$C$_{13}$) and κL-(BEDT-TTF)$_2$Ag(CF$_3$)$_4$(C$_2$H$_3$C$_{13}$) [BEDT-TTF = 3,4;3',4'-bis(ethylenedithio)-2,2',5,5'-tetrathiafuivalene]. *Acta Cryst. B* 51: 789–797.

[41] Pereiro J, Petrovic A, Panagopoulos C, Bozovic I (2011) Interface superconductivity: history, development and prospects. *Phys. Express* 1: 208–241.

[42] Anderson PW (1973) Resonating valence bonds: a new kind of insulator. *Mater. Res. Bull.* 8: 153–160.

[43] a) Ramirez A (1994) Strongly geometrically frustrated magnets. *Annul. Rev. Mater. Sci.* 24: 453–480. b) Schiffer P, Ramirez AP (1996) Recent experimental progress in the study of geometrical magnetic frustration. *Comments Condens. Matter Phys.* 18: 21–50. c) Greedan JE (2001) Geometrically frustrated magnetic materials. *J. Mater. Chem.* 11: 37–53. d) Balents L (2010) Spin liquids in frustrated magnets. *Nature* 464: 199–208. e) Powell BJ, McKenzie RH (2011) Quantum frustration in organic Mott insulators: from spin liquids to unconventional superconductors. *Rep. Prog. Phys.* 74: 056501/1–60.

[44] Nakatsuji S, Nambu Y, Tonomura H, Sakai O, Jonas S, Broholm C, Tsunetsugu H, Qiu Y, Maeno Y (2005) Spin disorder on a triangular lattice. *Science* 309: 1697–1700.

[45] Hemmida M, Krug von Nidda H-A, Büttgen N, Loidl A, Alexander LK, Nath R, Mahajan AV, Berger RF, Cava RJ, Singh Y, Johnston DC (2009) Vortex dynamics and frustration in two-dimensional triangular chromium lattices. *Phys. Rev. B* 80: 054406/1–5.

[46] MacDougall GJ, Gout D, Zarestky JL, Ehlers G, Podlesnyak A, McGuire MA, Mandrus D, Nagler SE (2011) Kinetically inhibited order in a diamond-lattice antiferromagnet. *Proc. Nat. Acad. Sci.* 108: 15693–15698.

[47] Manson JL, Ressouche E, Miller JS (2000) Spin frustration in $M^{II}[C(CN)_3]_2(M = V, Cr)$. A magnetism and neutron diffraction study. *Inorg. Chem.* 39: 1135–1141.

[48] Itou T, Oyamada A, Maegawa S, Tamura M, Kato R (2007), Spin-liquid state in an organic spin-1/2 system on a triangular lattice, EtMe$_3$Sb[Pd(dmit)$_2$]$_2$. *J. Phys.: Condens. Matter* 19: 145247/1–5.

[49] Isono T, Kamo H, Ueda A, Takahashi K, Kimata M, Tajima H, Tsuchiya S, Terashima T, Uji S, Mori H (2014) Gapless quantum spin liquid in an organic spin-1/2 triangular-lattice κ-H$_3$(Cat-EDT-TTF)$_2$. *Phys. Rev. Lett.* 112: 177201/1–5.

[50] Zhou HD, Choi ES, Li G, Balicas L, Wiebe CR, Qiu Y, Copley JRD, Gardner JS (2011) Spin liquid state in the $S = 1/2$ triangular lattice $Ba_3CuSb_2O_9$. *Phys. Rev. Lett.* 106: 147204/1–4.

[51] Cheng JG, Li G, Balicas L, Zhou JS, Goodenough JB, Xu C, Zhou HD (2011) High-pressure sequence of $Ba_3NiSb_2O_9$ structural phases: new $S = 1$ quantum spin liquids based on Ni^{2+}. *Phys. Rev. Lett.* 107: 197204/1–4.

[52] Mendels P, Bert F, de Vries MA, Olariu A, Harrison A, Duc F, Trombe JC, Lord JS, Amato A, Baines C (2007) Quantum magnetism in the paratacamite family: towards an ideal kagomé lattice. *Phys. Rev. Lett.* 98: 077204/1–4.

[53] Li Y, Pan B, Li S, Tong W, Ling L, Yang Z, Wang J, Chen Z, Wu Z, Zhang Q (2014) Gapless quantum spin liquid in the $S = 1/2$ anisotropic kagome antiferromagnet $ZnCu_3(OH)_6SO_4$. *New J. Phys.* 16: 093011/1–12.

[54] Aidoudi FH, Aldous DW, Goff RJ, Slawin AMZ, Attfield JP, Morris RE, Lightfoot P (2011) An ionothermally prepared $S = 1/2$ vanadium oxyfluoride kagome lattice. *Nat. Chem.* 3: 801–806.

[55] Zhang B, Zhang Y, Wang Z, Wang D, Baker PJ, Pratt FL, Zhu D (2014) Candidate quantum spin liquid due to dimensional reduction of a two-dimensional honeycomb lattice. *Sci. Rep.* 4: 6541/1–6.

[56] Okamoto Y, Nohara M, Aruga-Katori H, Takagi H (2007) Spin-liquid state in the $S = 1/2$ hyperkagome antiferromagnet $Na_4Ir_3O_8$. *Phys. Rev. Lett.* 99: 137207/1–4.

[57] Shimizu Y, Maesato M, Saito G (2011) Uniaxial strain effects on Mott and superconducting transitions in κ-(ET)$_2$Cu$_2$(CN)$_3$. *J. Phys. Soc. Jpn.* 80: 074702/1–7.

[58] For several salts, furthermore, the T_c increases under uniaxial strain along a certain direction, and some exhibited SC only under uniaxial strain but not under

hydrostatic pressure. The followings are some examples describing the increase or appearance of T_c by uniaxial strain but not by hydrostatic pressure. a) Campos CE, Brooks JS, van Bentum PJM, Perenboom JAAJ, Klepper SJ, Sandhu PS, Valfells S, Tanaka Y, Kinoshita T, Kinoshita N, Tokumoto M, Anzai H (1995) Uniaxial-stress-induced superconductivity in organic conductors. *Phys. Rev. B* 52: R7014–R7017. b) Tajima N, Ebina-Tajima A, Tamura M, Nishio Y, Kajita K (2002) Effects of uniaxial strain on transport properties of organic conductor α-$(BEDT\text{-}TTF)_2I_3$ and discovery of superconductivity. *J. Phys. Soc. Jpn.* 71: 1832–1835.

[59] a) Kyung B, Tremblay A-MS (2006) Mott transition, antiferromagnetism, and d-wave superconductivity in two-dimensional organic conductors. *Phys. Rev. Lett.* 97: 046402/1–4. b) Watanabe T, Yokoyama H, Tanaka Y, Inoue J (2008) Predominant magnetic states in the Hubbard model on anisotropic triangular lattices. *Phys. Rev. B* 77: 214505/1–12. c) Yoshioka T, Koga A, Kawakami N (2009) Quantum phase transitions in the Hubbard model on a triangular lattice. *Phys. Rev. Lett.* 103: 036401/1–4. d) Hauke P (2013) Quantum disorder in the spatially completely anisotropic triangular lattice. *Phys. Rev. B* 87: 014415/1–16. e) Tocchio LF, Gros C, Valentí R, Becca F (2014) One-dimensional spin liquid, collinear, and spiral phases from uncoupled chains to the triangular lattice. *Phys. Rev. B* 89: 235107/1–9. f) Yamada A (2014) Magnetic properties and Mott transition in the Hubbard model on the anisotropic triangular lattice. *Phys. Rev. B* 89: 195108/1–9.

[60] Tomic S, Dressel M (2015) Ferroelectricity in molecular solids: a review of electrodynamic properties. *Rep. Prog. Phys.* 78: 096501/1–26.

[61] Uji S, Shinagawa H, Terashima T, Yakabe T, Terai Y, Tokumoto M, Kobayashi A, Tanaka H, Kobayashi H (2001) Magnetic-field-induced superconductivity in a two-dimensional organic conductor. *Nature* 410: 908–910.

[62] Kobayashi H, Kobayashi A, Cassoux P (2000) BETS as a source of molecular magnetic superconductors (BETS = bis(ethylenedithio)tetraselenafulvalene). *Chem. Soc. Rev.* 29: 325–333.

[63] Matsui H, Tsuchiya H, Suzuki T, Negishi E, Toyota N (2003) Relaxor ferroelectric behavior and collective modes in the π-d correlated anomalous metal λ-$(BEDT\text{-}TSF)_2FeCl_4$ *Phys. Rev. B* 68: 155105/1–10.

[64] Fujiwara E, Fujiwara H, Kobayashi H, Otsuka T, Kobayashi A (2002) A series of organic conductors, κ-$(BETS)_2FeBr_xCl_{4-x}$($0 \leq x \leq 4$), exhibiting successive antiferromagnetic and superconducting transitions. *Adv. Mater.* 14: 1376–1379.

[65] Meul HW, Rossel C, Decroux M, Fischer Ø, Remenyi G, Briggs A (1984) Observation of magnetic-field-induced superconductivity. *Phys. Rev. Lett.* 53: 497–500.

[66] Lin CL, Teter J, Crow JE, Mihalisin T, Brooks J, Abou-Aly AI, Stewart GR (1985) Observation of magnetic-field-induced superconductivity in a heavy-fermion antiferromagnet: $CePb_3$. *Phys. Rev. Lett.* 54: 2541–2544.

[67] Clark K, Hassanien A, Khan S, Braun K-F, Tanaka H, Hla S-W (2010) Superconductivity in just four pairs of $(BETS)_2GaCl_4$ molecules. *Nat. Nanotech.* 5: 261–265.

[68] a) Beno MA, Wang HH, Kini AM, Carlson KD, Geiser U, Kwok WK, Thompson JE, Williams JM, Ren J, Whangbo MH (1990) The first ambient pressure organic superconductor containing oxygen in the donor molecule, ßm-$(BEDO\text{-}TTF)_3Cu_2(NCS)_3$, $T_c = 1.06$ K. *Inorg. Chem.* 29: 1599–1601. b) Kahlich S, Schweitzer D, Heinen I, Lan SE, Nuber B, Keller HJ, Winzer K, Helberg HW (1991) $(BEDO\text{-}TTF)_2ReO_4(H_2O)$: a new organic superconductor. *Solid State Commun.* 80: 191–195.

[69] Sakata J, Sato H, Miyazaki A, Enoki T, Okano Y, Kato R (1998) Superconductivity in new organic conductor κ-$(BEDSe\text{-}TTF)_2CuN(CN)_2Br$. *Solid State Commun.* 108: 377–381.

[70] a) Yamada J, Watanabe M, Akutsu H, Nakatsuji S, Nishikawa H, Ikemoto I, Kikuchi K (2001) New organic superconductors ß-$(BDA\text{-}TTP)_2X$ [BDA-TTP = 2,5-bis(1,3-dithian-2-ylidene)-1,3,4,6-tetrathiapentalene; $X^- = SbF_6^-, AsF_6^-$, and PF_{6-}]. *J. Am. Chem. Soc.* 123: 4174–4180. b) Ito H, Ishihara T, Tanaka H, Kuroda S, Suzuki T, Onari S, Tanaka Y, Yamada J, Kikuchi K (2008) Roles of spin fluctuation and frustration in the superconductivity of ß-$(BDA\text{-}TTP)_2X$(X= SbF_6,AsF_6) under uniaxial compression. *Phys. Rev. B* 78: 172506/1–4. c) Shimojo Y, Ishiguro T, Toita T, Yamada J (2002) Superconductivity of layered organic compound ß-$(BDA\text{-}TTP)_2SbF_6$, where BDA-TTP is 2,5-bis(1,3-dithian-2-ylidene)-1,3,4,6-tetrathiapentalene. *J. Phys. Soc. Jpn.* 71: 717–720. d) Nomura K, Muraoka R, Matsunaga N, Ichimura K, Yamada J (2009) Anisotropic superconductivity in ß-$(BDA\text{-}TTP)_2SbF_6$: STM spectroscopy. *Physica B* 404: 562–564.

[71] Okano Y, Iso M, Kashimura Y, Yamaura J, Kato R (1999) A new synthesis of Se-containing TTF derivatives. *Synth. Met.* 102: 1703–1704.

[72] Zambounis JS, Mayer CW, Hauenstein K, Hilti B, Hofherr W, Pfeiffer J, Buerkle M, Rihs G (1992) Crystal structure and electrical properties of κ-((*S*,*S*)-DMBEDT-$TTF)_2ClO_4$. *Adv. Mater.* 4: 33–35.

[73] a) Kimura S, Maejima T, Suzuki H, Chiba R, Mori H, Kawamoto T, Mori T, Moriyama H, Nishio Y, Kajita K (2004) A new organic superconductor ß-(meso-$DMBEDT\text{-}TTF)_2PF_6$. *Chem. Commun.* 2454–2455. b) Kimura S, Suzuki H, Maejima T, Mori H, Yamaura J, Kakiuchi T, Sawa H, Moriyama H (2006) Checkerboard-type charge-ordered state of a pressure-induced superconductor, ß-(*meso*-$DMBEDT\text{-}TTF)_2PF_6$. *J. Am. Chem. Soc.* 128: 1456–1457.

[74] Kikuchi K, Murata K, Honda Y, Namiki T, Saito K, Ishiguro T, Kobayashi K, Ikemoto I (1987) On ambient-pressure superconductivity in organic conductors: Electrical properties of $(DMET)_2I_3$, $(DMET)_2I_2Br$ and $(DMET)_2IBr_2$. *J. Phys. Soc. Jpn.* 56: 3436–3439.
[75] Nishikawa H, Morimoto T, Kodama T, Ikemoto I, Kikuchi K, Yamada J, Yoshino H, Murata K (2002) New organic superconductors consisting of an unprecedented π-electron donor. *J. Am. Chem. Soc.* 124: 730–731.
[76] Kato R, Yamamoto K, Okano Y, Tajima H, Sawa H (1997) A new ambient-pressure organic superconductor $(TMET\text{-}STF)_2BF_4$[TMET-STF = trimethylene (ethylenedithio) diselenadithiafulvalene]. *Chem. Commun.* 947–948.
[77] Kato R, Aonuma S, Okano Y, Sawa H, Tamura M, Kinoshita M, Oshima K, Kobayashi A, Bun K, Kobayashi H (1993) Metallic and superconducting salts based on an unsymmetrical π-donor dimethyl (ethylenedithio) tetraselenafulvalene (DMET-TSeF). *Synth. Met.* 61: 199–206.
[78] Imakubo T, Tajima N, Tamura M, Kato R, Nishio Y, Kajita K (2002) A supramolecular superconductor 6-$(DIETS)_2[Au(CN)_4]$. *J. Mater. Chem.* 12: 159–161.
[79] Lyubovskaya RN, Zhilyaeva EI, Torunova SA, Mousdis GA, Papavassiliou GC, Perenboom JAAJ, Pesotskii SI, Lyubovskii RB (2004) New ambient pressure organic superconductor with T_c= 8.1 K: $(EDT\text{-}TTF)_4Hg_{3\text{-}\delta}I_8$. *J. Phys. IV* 114: 463–466.
[80] a) Papavassiliou GC, Mousdis GA, Zambounis JS, Terzis A, Hountas A, Hilti B, Mayer CW, Pfeiffer J (1988) Low temperature measurements of the electrical conductivities of some charge transfer salts with the asymmetric donors MDT-TTF, EDT-TTF and EDT-DSDTF. $(MDT\text{-}TTF)_2AuI_2$, a new superconductor (T_c= 3.5 K at ambient pressure). *Synth. Met.* 27: 379–383. b) Takahashi T, Kobayashi Y, Nakamura T, Kanoda K, Hilti B, Zambounis JS (1994) Symmetry of the order parameter in organic superconductors: $(MDT\text{-}TTF)_2AuI_2$ vs. $(TMTSF)_2ClO_4$. *Physica C* 235-240: 2461–2462.
[81] a) Takimiya K, Takamori A, Aso Y, Otsubo T, Kawamoto T, Mori T (2003) Organic superconductors based on a new electron donor, methylenedithio-diselenadithiafulvalene (MDT-ST). *Chem. Mater.* 15: 1225–1227. b) Kawamoto T, Mori T, Enomoto K, Konoike T, Terashima T, Uji S, Takamori A, Takimiya K, Otsubo T (2006) Fermi surface of the organic superconductor (MDT-ST) $(I_3)_{0.417}$ reconstructed by incommensurate potential. *Phys. Rev. B* 73: 024503/1–5.
[82] Takimiya K, Kodani M, Niihara N, Aso Y, Otsubo T, Bando Y, Kawamoto T, Mori T (2004) Pressure-induced superconductivity in (MDT-TS) $(AuI_3)_{0.441}$[MDT-TS = 5*H*-2-(1,3-diselenol-2-ylidene)-1,3,4,6-tetrathiapentalene]: a new organic superconductor possessing an incommensurate anion lattice. *Chem. Mater.* 16: 5120–5123.
[83] a) Takimiya K, Kataoka Y, Aso Y, Otsubo T, Fukuoka H, Yamanaka S (2001) Quasi one-dimensional organic superconductor MDT-TSFAuI_2 with T_c= 4.5 K at ambient pressure. *Angew. Chem. Int. Ed.* 40: 1122–1125. b) Kawamoto T, Mori T, Konoike T, Enomoto K, Terashima T, Uji S, Kitagawa H, Takimiya K, Otsubo T (2006) Charge transfer degree and superconductivity of the incommensurate organic superconductor (MDT-TSF) $(I_3)_{0.422}$. *Phys. Rev. B* 73: 094513/1–8. c) Kawamoto T, Mori T, Terakura C, Terashima T, Uji S, Takimiya K, Aso Y, Otsubo T (2003) Incommensurate anion potential effect on the electronic states of the organic superconductor (MDT-TSF)$(AuI_2)_{0.436}$. *Phys. Rev. B* 67: 020508(R)/1–4. d) Kawamoto T, Mori T, Terakura C, Terashima T, Uji S, Tajima H, Takimiya K, Aso Y, Otsubo T (2003) Electronic state anisotropy and the Fermi surface topology of the incommensurate organic superconducting crystal (MDT-TSF)$(AuI_3)_{0.436}$. *Eur. Phys. J. B* 36: 161–167.
[84] Kodani M, Takamori A, Takimiya K, Aso Y, Otsubo T (2002) Novel conductive radical cation salts based on methylenediselenotetraselenafulvalene (MDSe-TSF): a sign of superconductivity in K-(MDSe-TSF)2Br below 4 K. *J. Solid State Chem.* 168: 582–589.
[85] Misaki Y, Higuchi N, Fujiwara H, Yamabe T, Mori T, Mori H, Tanaka S (1995) (DTEDT)$[Au(CN)_2]_{0.4}$: an organic superconductor based on the novel π-electron framework of vinylogous bis-fused tetrathiafulvalene. *Angew. Chem. Int. Ed. Engl.* 34: 1222–1225.
[86] a) Shirahata T, Kibune M, Imakubo T (2006) New ambient pressure organic superconductors κ- and κ_L-$(DMEDO\text{-}TSeF)_2[Au(CN)_4](THF)$. *Chem. Commun.*: 1592–1594. b) Shirahata T, Kibune M, Yoshino H, Imakubo T (2007) Ambient-pressure organic superconductors κ-$(DMEDO\text{-}TSeF)_2[Au(CN)_4]$(solv.): T_c tuning by modification of the solvent of crystallization. *Chem. Eur. J.* 13: 7619–7630.
[87] Tajima H, Inokuchi M, Kobayashi A, Ohta T, Kato R, Kobayashi H, Kuroda H (1993) First ambient-pressure superconductor based on $Ni(dmit)_2$,α-EDT-TTF$[Ni(dmit)_2$. *Chem. Lett.* 22: 1235–1238.
[88] Miura YF, Horikiri M, Saito S-H, Sugi M (2000) Evidence for superconductivity in Langmuir-Blodgett films of ditetradecyldimethylammonium-Au$(dmit)_2$. *Solid State Commun.* 113: 603–605.
[89] a) Tamura M, Nakao A, Kato R (2006) Frustration-induced valence-bond ordering in a new quantum triangular antiferromagnet based on $[Pd(dmit)_2$. *J. Phys. Soc. Jpn.* 75: 093701/1–4. b) Ishii Y, Tamura M, Kato R (2007) Magnetic study of pressure-induced superconductivity in the $[Pd(dmit)_2$. Salt with spin-gapped ground state. *J. Phys. Soc. Jpn.* 76: 033704/1–4.

[90] Hebard AF, Rosseinsky MJ, Haddon RC, Murphy DW, Glarum SH, Palstra TTM, Ramirez AP, Kortan AR (1991) Superconductivity at 18 K in potassium-doped C_{60}. *Nature* 350: 600–601.
[91] Krätschmer W, Lamb LD, Fostiropoulos K, Huffman DR (1990) C_{60}: a new form of carbon. *Nature* 347: 354–358.
[92] Stephens PW, Mihaly L, Lee PL, Whetten RL, Huang S-M, Kaner R, Diederich F, Holczer K (1991) Structure of single-phase superconducting K_3C_{60}. *Nature* 351: 632–634.
[93] Rosseinsky MJ, Ramirez AP, Glarum SH, Murphy DW, Haddon RC, Hebard AF, Palstra TTM, Kortan AR, Zahurak SM, Makhija AV (1991) Superconductivity at 28 K in Rb_xC_{60}. *Phys. Rev. Lett.* 66: 2830–2832.
[94] Tanigaki K, Ebbesen TW, Saito S, Mizuki J, Tsai J-S, Kubo Y, Kuroshima S (1991) Superconductivity at 33 K in $Cs_xRb_yC_{60}$. *Nature* 352: 222–223.
[95] Fleming RM, Ramirez AP, Rosseinsky MJ, Murphy DW, Haddon RC, Zahurak SM, Makhija AV (1991) Relation of structure and superconducting transition temperatures in A_3C_{60}. *Nature* 352: 787–788.
[96] Sasaki S, Matsuda A, Chu CW (1994) Fermi-liquid behavior and BCS *s*-wave pairing of K_3C_{60} observed by ^{13}C-NMR. *J. Phys. Soc. Jpn.* 63: 1670–1673.
[97] Kiefl RF, MacFarlane WA, Chow KH, Dunsiger S, Duty TL, Johnston TMS, Schneider JW, Sonier J, Brard L, Strongin RM, Fischer JE, Smith III AB (1993) Coherence peak and superconducting energy gap in Rb_3C_{60} observed by muon spin relaxation. *Phys. Rev. Lett.* 70: 3987–3990.
[98] Chen CC, Lieber CM (1992) Synthesis of pure $^{13}C_{60}$ and determination of the isotope effect for fullerene superconductors. *J. Am. Chem. Soc.* 114: 3141–3142.
[99] Takenobu T, Muro T, Iwasa Y, Mitani T (2000) Antiferromagnetism and phase diagram in ammoniated alkali fulleride salts. *Phys. Rev. Lett.* 85: 381–384.
[100] Dahlke P, Denning MS, Henry PF, Rosseinsky MJ (2000) Superconductivity in expanded fcc $C_{60}{}^{3-}$ fullerides. *J. Am. Chem. Soc.* 122: 12352–12361.
[101] Zadik RH, Takabayashi Y, Klupp G, Colman RH, Ganin AY, Potočnik A, Jeglič P, Arčon D, Matus P, Kamarás K, Kasahara Y, Iwasa Y, Fitch AN, Ohishi Y, Garbarino G, Kato K, Rosseinsky MJ, Prassides K (2015) Optimized unconventional superconductivity in a molecular Jahn-Teller metal. *Sci. Adv.* 1: e1500059/1–9.
[102] Yildirim T, Barbedette L, Fischer JE, Lin CL, Robert J, Petit P, Palstra TTM (1996) T_c vs carrier concentration in cubic fulleride superconductors. *Phys. Rev. Lett.* 77: 167–170.
[103] Kosaka M, Tanigaki K, Prassides K, Margadonna S, Lappas A, Brown CM, Fitch AN (1999) Superconductivity in $Li_xCs C_{60}$ fullerides. *Phys. Rev. B* 59: R6628–R6630.
[104] Ganin AY, Takabayashi Y, Khimyak YZ, Margadonna S, Tamai A, Rosseinsky MJ, Prassides K (2008) Bulk superconductivity at 38 K in a molecular system. *Nat. Mater.* 7: 367–371.
[105] Takabayashi Y, Ganin AY, Jeglic P, Arcon D, Takano T, Iwasa Y, Ohishi Y, Takata M, Takeshita N, Prassides K, Rosseinsky MJ (2009) The disorder-free non-BCS superconductor Cs_3C_{60} emerges from an antiferromagnetic insulator parent state. *Science* 323: 1585–1590.
[106] Ganin AY, Takabayashi Y, Jeglič P, Arčon D, Potočnik A, Baker PJ, Ohishi Y, McDonald MT, Tzirakis MD, McLennan A, Darling GR, Takata M, Rosseinsky MJ, Prassides K (2010) Polymorphism control of superconductivity and magnetism in Cs_3C_{60} close to the Mott transition. *Nature* 466: 221–225.
[107] Mitsuhashi R, Suzuki Y, Yamanari Y, Mitamura H, Kambe T, Ikeda N, Okamoto H, Fujiwara A, Yamaji M, Kawasaki N, Maniwa Y, Kubozono Y (2010) Superconductivity in alkali-metal-doped picene. *Nature* 464: 76–79.
[108] Avdeev VV, Zharikov OV, Nalimova VA, Pal'nichenko AV, Semenenko KN (1986) Superconductivity of layered compounds C_6K and C_4K. *JETP Lett.* 43: 484–487.
[109] Hillesheim D, Gofryk K, Sefat AS (2014) On the nature of filamentary superconductivity in metal-doped hydrocarbon organic materials. *Nov. Supercond. Mater.* 1: 12–14.
[110] Artioli GA, Hammerath F, Mozzati MC, Carretta P, Corana F, Mannucci B, Margadonna S, Malavasi L (2015) Superconductivity in Sm-doped [*n*]phenacenes (n = 3, 4, 5). *Chem. Commun.* 51: 1092–1095.
[111] a) Wang XF, Liu RH, Gui Z, Xie YL, Yan YJ, Ying JJ, Luo XG, Chen XH (2011) Superconductivity at 5 K in alkali-metal-doped phenanthrene. *Nat. Commun.* 2: 507–513. b) Wang XF, Yan YJ, Gui Z, Liu RH, Ying JJ, Luo XG, Chen XH (2011) Superconductivity in $A_{1.5}$phenanthrene (A = Sr, Ba). *Phys. Rev. B* 84: 214523/1–4. c) Wang XF, Luo XG, Ying JJ, Xiang ZJ, Zhang SL, Zhang RR, Zhang YH, Yan YJ, Wang AF, Cheng P, Ye GJ, Chen XH (2012) Enhanced superconductivity by rare-earth metal doping in phenanthrene. *J. Phys.: Condens. Matter* 24: 345701/1–5.
[112] Xue M, Cao T, Wang D, Wu Y, Yang H, Dong X, He J, Li F, Chen GF (2012) Superconductivity above 30 K in alkali-metal-doped hydrocarbon. *Sci. Rep.* 2: 389–392.
[113] Kubozono Y, Mitamura H, Lee X, He X, Yamanari Y, Takahashi Y, Suzuki Y, Kaji Y, Eguchi R, Akaike K, Kambe T, Okamoto H, Fujiwara A, Kato T, Kosugi T, Aoki H (2011) Metal-intercalated aromatic hydrocarbons: a new class of carbon-based superconductors. *Phys. Chem. Chem. Phys.* 13: 16476–16493.

[114] Graf D, Brooks JS, Almeida M, Dias JC, Uji S, Terashima T, Kimata M (2009) Evolution of superconductivity from a charge-density-wave ground state in pressurized $(Per)_2[Au(mnt)_2]$. *Europhys. Lett.* 85: 27009/1–5.

[115] Aust RB, Bentley WH, Drickamer HG (1964) Behavior of fused-ring aromatic hydrocarbons at very high pressure. *J. Chem. Phys.* 41: 1856–1864.

[116] a) Yokota T, Takeshita N, Shimizu K, Amaya K, Onodera A, Shirotani I, Endo S (1996) Pressure-induced superconductivity of iodanil. *Czech J. Phys.* 46: 817–818. b) Amaya K, Shimizu K, Takeshita N, Eremets MI, Kobayashi TC, Endo S (1998) Observation of pressure-induced superconductivity in the megabar region. *J. Phys.: Condens. Matter* 10: 11179–11190.

[117] Iwasaki E, Shimizu K, Amaya K, Nakayama A, Aoki K, Carlon R P (2001) Metallization and superconductivity in hexaiodobenzene under high pressure. *Synth. Met.* 120: 1003–1004.

[118] Cui HB, Kobayashi H, Ishibashi S, Sasa M, Iwase F, Kato R, Kobayashi A (2014) A single-component molecular superconductor. *J. Am. Chem. Soc.* 136: 7619–7622.

D3

Fullerene Superconductors

Yoshihiro Iwasa and Kosmas Prassides

Nearly 30 years have passed since the discovery of superconductivity in potassium-doped C_{60} at 19 K. Now the critical temperature T_c of fullerene superconductors has reached 38 K. This value is not only the highest among organic/molecule or carbon-based superconductors but also comparable to MgB_2, and is only superseded by cuprate and iron-based superconductors at ambient pressure. This chapter is devoted to fundamental properties of fullerene superconductors, covering from the past understanding to the latest development. We review the crystal structures, electronic states of intercalation compounds, and their relation with the occurrence of superconductivity. We then discuss about the latest discovery of Cs_3C_{60} and its properties with particular focus on the effect of strong electron correlation.

D3.1 Introduction

Fullerenes are a type of cluster molecules composed mainly of carbon. Figure D3.1 shows the most well-known and abundant fullerene, C_{60}, which has a characteristic soccer ball shape. Fullerene clusters were first predicted theoretically by Osawa in 1970 [1] and experimentally confirmed by Kroto, Smalley and coworkers in 1985 [2]. Although the work on fullerenes in the 1980s was limited to gas-phase experiments and theoretical calculations, the discovery in 1990 by Krätchmer *et al* [3] of a simple and inexpensive method to produce large quantities of fullerene clusters provided a great stimulus to the research field of fullerenes. Interestingly, these experimental breakthroughs were made by astrophysicists, not by materials scientists. This interdisciplinary nature is a unique and attractive aspect of fullerene research.

Immediately after the discovery of Krätchmer's method, intercalation of alkali metals into fullerene solids was found to produce a new type of conductor, the first three-dimensional molecular conductor, as reported by Haddon *et al* [4]. Researchers formed thin films of C_{60} and C_{70} by vacuum evaporation, and made *in situ* measurements of resistivity, which was found to decrease dramatically upon doping with alkali metals, such as potassium, rubidium and cesium. Based on Raman scattering data, researchers concluded that the most conductive state appears at the chemical concentration near A_3C_{60}, where A denotes alkali metals [4]. This chemical formula corresponds to the molecular electronic states of $(C_{60})^{-n}$ ($n = 3$). Soon after this, Hebard *et al* [5] cooled down the potassium-doped C_{60} samples and measured resistivity and magnetization, both of which showed unambiguous evidence for bulk superconductivity with the onset temperature at 19 K. Rosseinsky also reported superconductivity of Rb_3C_{60} at 29 K [6]. Through alloying the alkali metal sites, the maximum T_c quickly reached 33 K at ambient pressure in 1991 [7]. In 1994, Cs_3C_{60} was found to superconduct only under high pressure showing a broad onset at 40 K [8]. The volume fraction in this discovery, however, was very tiny, and thus the identification of the superconducting phase remained elusive for a long time. In 2008, Prassides and Rosseinsky found that the high-pressure superconducting phase was the bcc-structured phase Cs_3C_{60}, the T_c of which reached 38 K [9].

Following the pioneering discoveries in 1991, extensive work has been carried out both on the physics and chemistry of fullerene superconductors of 1990's. The most fundamental understanding of the mechanism of superconductivity is the correlation observed between T_c and the interfullerene spacings [10–12], which was established in the early stages, that is, that the T_c of fullerene superconductors increases with the interfullerene spacing. The chemical substitution of alkali metal sites and the application of high pressure yielded the same relation between T_c and interfullerene spacing within the error of 10%. This notable universal behavior strongly suggests that T_c is controlled by the density of states at the Fermi energy $N(\varepsilon_F)$ within the framework of the BCS weak-coupling theory, where the Cooper pairing is mediated by the electron–phonon coupling via intermolecular phonons with rather high frequency (~ 0.1 eV), while the $N(\varepsilon_F)$ is determined simply by the interfullerene spacing. Most of the experiments supported this suggestion at least qualitatively. These developments have been well described in former reviews [13–17]. In this manner, the fullerene intercalation compounds are characterized by the large electron–phonon interaction. The attractive interaction between two electrons induced by this electron–phonon interaction is of the order of 0.1 eV [15]. This considerably

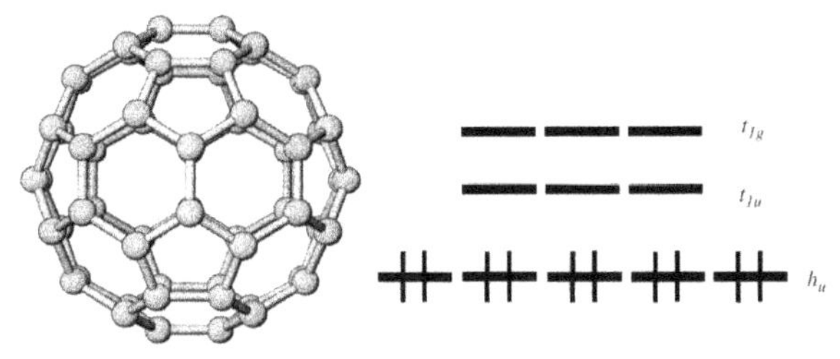

FIGURE D3.1 The molecular structure and electronic energy levels of C_{60}. Doped electrons are introduced into the LUMO (t_{1u}) and, subsequently, to the LUMO + 1 (t_{1g}) states.

large value in comparison to the bandwidth of $W \sim 0.5$ eV causes a rather high T_c.

On the other hand, the experimental estimations of the on-ball electron–electron repulsion energy U are 1.0–1.5 eV, [18, 19], which is much larger than the conduction bandwidth (W ~ 0.5eV) [14, 15, 20]. According to a conventional Mott–Hubbard criterion, fullerene compounds should be Mott insulators rather than superconductors [21, 22], indicating that even the Mott criterion should be reconsidered in the case of fullerene compounds. Because of the strong belief that superconductivity is understood in terms of the conventional BCS scheme, the electron correlation effect in fullerides due to the large U/W value was not well recognized in the beginning. However, a hint of the strong electron correlation effect was presented by the synthesis of ammonia–alkali metal co-intercalated systems exhibiting both superconducting and insulating behavior. Subsequent extensive research on the recently discovered Cs_3C_{60} and the $Rb_xCs_{3-x}C_{60}$ alloy superconductors revealed that superconductivity, particularly high T_c superconductivity, is achieved at the verge of the metal–insulator boundary, displaying a peculiar dome-shaped phase diagram. These features are suggestive of commonalities of fullerides with cuprate, iron-based superconductors, organic superconductors and heavy Fermion superconductors, all of which are regarded as strongly correlated superconductors.

Taking account of the energy scales of the bandwidth, electron–phonon and electron–electron interactions, the superconductivity as well as the other electronic ground states of fullerides should be understood in terms of the competition or cooperation between the electron–phonon interaction and the electron–electron interaction [23, 24]. This is currently an important issue in solid-state physics in general. Also, in light of the importance of light elements for achieving high T_c superconductivity, the chemical and physical criteria for the occurrence of fullerene superconductivity is of great importance for future explorations and design of new high T_c superconductors.

Apart from the competing energy scales, fullerenes have other unique features as building blocks for constructing solids. The first thing to be pointed out is the spherical shape of the molecule. C_{60}, in particular, has an exceptionally high symmetry, which is referred to as icosahedral (I_h) symmetry. The molecular structure is identical to a soccer ball, as shown in Figure D3.1. The I_h symmetry of the C_{60} molecule is directly reflected in the molecular electronic structures. Figure D3.1 also shows the molecular energy levels near the Fermi energy obtained from the Hückel molecular orbital calculation [25]. For instance, the highest occupied molecular orbital (abbreviated as HOMO) is five-fold degenerate, while the lowest unoccupied molecular orbital (LUMO) and the LUMO + 1 state are both triply degenerate. The degeneracy of the latter two levels is quite important, since electrons are introduced into LUMO and LUMO + 1 bands upon doping of alkali metals or alkaline-earth metals. In the forthcoming discussions, these two levels are called t_{1u} and t_{1g} states, respectively, following the symmetry group representation.

The second feature of the fullerene molecule is the capability of intercalation and, at the same time, a large capacity of electrons. As explained in the following sections, alkali metals and alkaline-earth metals can be doped up to six per C_{60} molecule, forming, for example, Ba_6C_{60}. The nominal counting of the valence of the C_{60} molecule in this compound is $n = 12$, assuming that the Ba ion is divalent. Moreover, the valence n can be controlled from 0 to 12 per molecule, by changing the number and species of intercalated elements. Such large capacity and tunability are quite anomalous when one considers the case of TTF or BEDT-TTF-based organic conductors, where the molecular valence can usually be varied between 0 and 1. Here TTF and BEDT-TTF are abbreviations of tetrathiafulvalene and bis(ethylenedithio)tetrathiafulvalene, respectively. However, from a viewpoint of graphite intercalation compounds (GICs), the electron capacity of fullerenes (1/5 electrons/carbon atom) is comparable or still smaller than standard GICs, such as C_8K or C_3Li. Hence, the large electron capacity of fullerenes as molecular clusters is attributed to the enormously large π-electron systems, which behave like a large electron reservoir. In this sense, fullerenes can be viewed as electron sponges.

Combining the two characteristics, the physical interest in the research of C_{60} intercalation compounds is that one is able to tune the electron filling of two triply degenerate electronic levels over a wide range. The most fundamental question related to superconductivity is the correlation between the number of electrons transferred to C_{60} (molecular valence n) and the occurrence of superconductivity. According to the simple band picture, when the formal electron counting of C_{60} molecule is 0, 6 and 12, the materials should be band-insulators, while they are metallic at the other electronic states, where the LUMO (t_{1u}) or LUMO + 1 (t_{1g}) states are partially filled.

In this chapter, we will attempt to provide a background for understanding the electronic properties of fullerene intercalation compounds, and to deduce the chemical, structural and physical criteria for the occurrence of superconductivity. In order to do so, the next section will begin with the characterization of fullerene intercalation compounds in terms of their molecular valence (band filling) and will explain how alkali-doped and alkaline-earth–doped superconductors differ from each other in the conditions that produce superconductivity. The third section will discuss the electronic properties of trivalent fullerides, where superconductivity is destroyed by a Mott–Hubbard transition, possibly induced by the expansion of the interfullerene spacing or reduction in symmetry of the crystal structure. In the fourth section, a comparison between alkali- and alkaline-earth–doped systems focusing on the A_4C_{60} type structure is presented. These arguments demonstrate that, in fullerene intercalation compounds, the competition or cooperation of the electron–electron interaction and the electron–phonon interactions play important roles in determining the electronic ground state including superconductivity. Superconductivity is strongly dependent on the band filling as well as on the symmetry of the crystal structure. The last two sections address the recent developments in the field. These encompass the discovery of a superconductivity dome (the first unambiguous evidence of unconventional non-BCS phenomenology); the revelation that the metallic and superconducting states emerge from an antiferromagnetic insulating precursor (signature of the importance of electronic correlations in determining the pairing interaction); and the visualization of the evolution of the parent Mott insulator into a normal Fermi liquid state via a "bad" Jahn–Teller (JT) metal state in which itinerant metallic electrons co-exist in a dynamic, microscopically heterogeneous fashion with localized electrons, which produce JT on-molecule distortions.

D3.2 Band Filling versus Superconductivity in Intercalated Fullerides

The introduction of itinerant electrons onto C_{60} solids has been successfully made by the intercalation of alkali metals, alkaline-earth metals and rare-earth metals. A vast variety of materials has been synthesized by this method, including a group of superconductors [15, 17, 26]. This section focuses on the correlation between electronic states and molecular valence, particularly on the question of which valency produces superconductors.

Through changing the elements and chemical compositions, one is able to control the number of electrons on C_{60} molecule (molecular valence) from zero to twelve. Since the molecular valence is changed over such a wide range, the crystal structure and molecular valence are usually strongly correlated in the binary materials. The crystal structure of undoped C_{60} will be briefly described first, followed by a structure sequence of K and Ba intercalated compounds, as shown in Figure D3.2. Here, C_{60} and intercalants are shown by large and small spheres, respectively.

The spherical shape of the molecule, which is common to all fullerenes, results in a close packed face-centered-cubic (fcc) structure. Since a C_{60} molecule rotates freely at high temperature, this undoped solid is regarded as a typical example of plastic crystals. When the temperature is reduced below 250 K, the rotation is changed to a ratchet-type motion, associated with the structural transition. In the low-temperature phase, neighboring C_{60}s are not equivalent any more, causing the reduction of crystal symmetry from fcc to simple cubic.

A dominant trend of the structural evolution upon intercalation is that the close packed fcc structure of undoped C_{60} is changed to a less close packed body-centered-cubic (bcc) structure with increasing the chemical concentration of intercalants. In view of the fcc–bcc transformation, only half of the fcc unit cell is displayed in Figure D3.2. Since the ionic radii

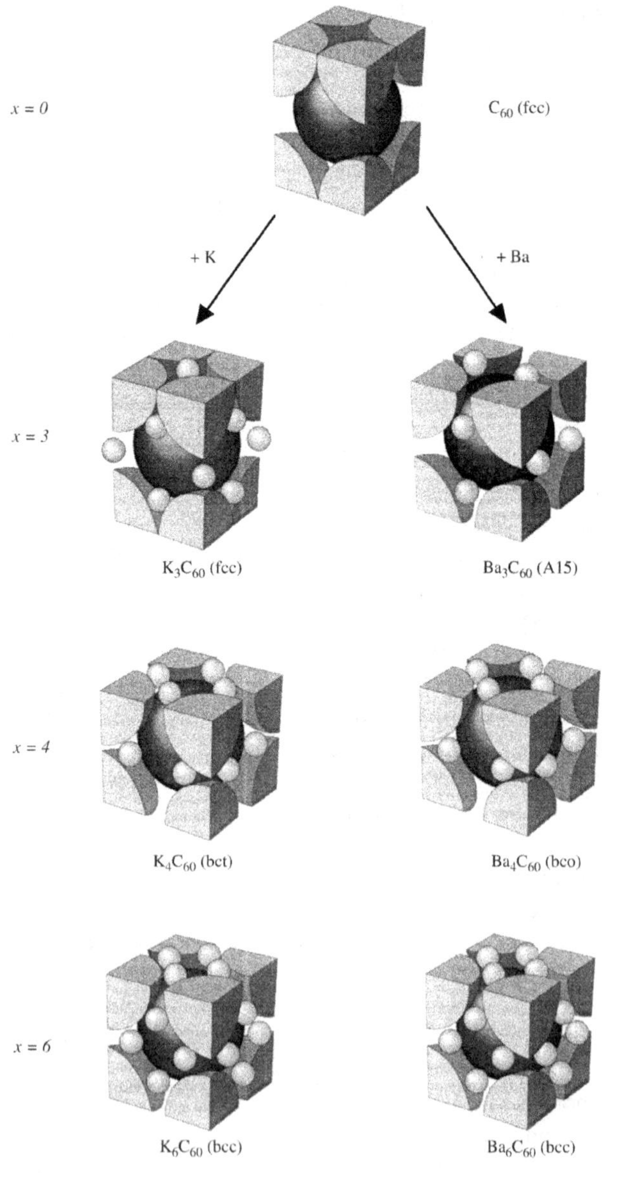

FIGURE D3.2 A schematic illustration of the structural sequence of K_xC_{60} and Ba_xC_{60} with $3 \leq x \leq 6$. The large and small spheres represent the C_{60} molecule and intercalated metal ion, respectively. (Figures are taken from [25].)

of K^+ and Ba^{2+} are almost the same, one may expect that the intercalation compounds of these series are isostructural. In fact, the saturation $x = 6$ phases of both compounds indeed crystallize in the bcc structure ($Im\bar{3}$),, where the molecular rotations are frozen and ordered [27, 28]. For $x = 4$, both phases are again very similar forming a body-centered-tetragonal (bct) structure [29]; however, there exists a slight orthorhombic distortion in Ba_4C_{60} [30, 31].

However, in the case of $x = 3$, a significant difference in the structures is observed. While K_3C_{60} maintains the fcc arrangement of C_{60} molecules, the conversion to the bcc arrangement has already occurred in Ba_3C_{60}. (Due to the analogy to the Nb_3Sn structure,'the structure of Ba_3C_{60} is also known as 'A15'.) Kortan and co-workers [32] attributed the structural difference to the larger Madelung energy in the Ba compounds. Since the divalent Ba ion induces a significant lattice contraction, the tetrahedral hole of hypothetical fcc Ba_3C_{60} becomes too small for the real Ba^{2+} ion. This results in a structural change into the less close packed A15 type structure. The difference in the crystal structure of K_3C_{60} and Ba_3C_{60} shows a simple example of the general rules of how the structure of fullerides is determined. The stable crystal structures seem to be dependent on the relative difference between ionic radii and the size of the interstitial sites. In fact, in the case of Sr_3C_{60}, both fcc and A15 phases are observed, since the ionic radius of Sr^{2+} is smaller than that of Ba^{2+} [33].

The crystal structure of other elements is determined by the same principle. In Ca compounds, the fcc-based structure is maintained at least up to $x = 5$, due to the small ionic radius. A similar trend is also observed in other alkali metal–doped systems. Sodium and lithium intercalates retain the fcc structure up to $x \sim 11$ due to their small ionic radii. Rb compounds display a similar structural sequence to that of K compounds, while the sequence for Cs compounds resembles that of Ba system in terms of fcc–bcc transition possibly due to their large ionic radius.

In the sequence of Figure D3.2, superconductivity appears in K_3C_{60} and in Ba_4C_{60}. While the former compound is well known as the first fullerene superconductor and regarded as a standard material, the latter phase remained unidentified for a long time. Kortan *et al* [28] first reported the synthesis and superconductivity of Ba_6C_{60}. Later, in 1995, Baenitz *et al* [30] claimed that the superconducting phase was Ba_4C_{60}. Since Baenitz's conclusion was based on poorly characterized samples, the superconducting phase of Ba-C_{60} binaries remained an open question. In 1993, single phase material of Ba_4C_{60} was prepared by repeated annealing and intermediate grinding. This work unambiguously proved that the superconducting phase is Ba_4C_{60} [31]. At the same time, superconductivity of isostructural Sr_4C_{60} was established. The crystal structures of most fullerene superconductors are essentially cubic, as seen in fcc A_3C_{60}. In this sense, Ba_4C_{60} is the first non-cubic ambient pressure superconductor in the fulleride family.

The next question about these binary materials is how many electrons are transferred from intercalated metals to C_{60}, in other words, the molecular valence of the electron filling state of the LUMO. The charge state of C_{60} fullerenes can be investigated by various techniques, among which the most sensitive and simple is Raman's spectroscopy. In the case of an alkali-doped C_{60}, such as K_xC_{60}, a relation is established between the Raman shift of the pentagonal pinch mode of the cage and the valence of C_{60}. This totally symmetric $A_g(2)$ mode, which originates from the C=C stretching vibration, is softened by the additional charges on the LUMO state, since this orbital has an antibonding character. The Raman shift displays a downshift approximately linearly to the molecular valence of C_{60} with a slope of ~ 7cm^{-1}/charge [34]. From this experiment, we conclude that the alkali ion is completely ionized in K_xC_{60}. This contrasts with the incomplete charge transfer in the GIC superconductor C_8K, where not only the carbon but also the potassium bands cut the Fermi level. The molecular valence of alkaline-earth metal intercalated systems has not been investigated until recently. Raman's spectroscopy was found to be equally useful for the alkaline earth systems [35]. Figure D3.3(a) shows the Raman spectra of the $A_g(2)$ mode for Ba_xC_{60} ($x = 3$, 4 and 6). This mode shows a large downshift with increasing Ba concentration. The relative Raman shift measured from the position of undoped C_{60} (1468 cm^{-1}) is plotted against the formal valence determined by the simple ionic crystal model. The Raman shift for Ba_3C_{60} is identical to that of K_6C_{60}. Moreover, the shifts observed for Ba_4C_{60} and Ba_6C_{60} approximately follow linear extrapolation of the alkali-doped C_{60} data. These results strongly suggest that charge is fully transferred from Ba to C_{60} even in Ba_6C_{60}. Thus, the C_{60} molecule in Ba_6C_{60} is approximately dodecavalent.

However, it should be pointed out that theories based on the first principle local density approximation (LDA) calculations predict a significant hybridization between Ba and C_{60} orbitals, resulting in an incomplete charge transfer from

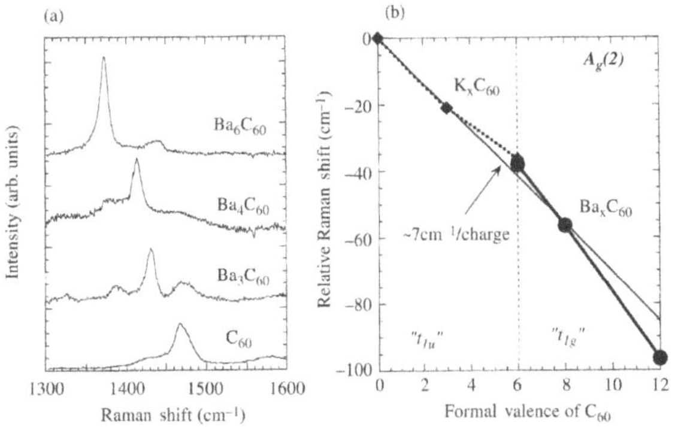

FIGURE D3.3 (a) Raman spectra of the pentagonal pinch [$A_g(2)$] mode of Ba_xC_{60} ($x = 3$, 4 and 6). (b) Raman shift versus formal valence of C_{60} for K_xC_{60} ($x = 0$, 3 and 6) and Ba_xC_{60} ($x = 3$, 4 and 6) determined from the chemical formula assuming that alkali ions are monovalent and alkali earth ions are divalent.

Ba to C_{60} [36, 37]. For example, Erwin *et al* predicts the charge transfer per Ba atom is only 0.7 for Ba_6C_{60}, which is amazingly small compared to that deduced from the Raman shift. Detailed and quantitative analysis of the Raman spectra in the highly doped state has yet to be investigated theoretically taking the electron–phonon interactions into account.

Despite the disagreement with the band calculations, in this discussion, we shall assume the ionic crystal model, where the charge is fully transferred from metals to C_{60} in all intercalation compounds. Therefore, superconductors in Figure D3.2, K_3C_{60} and Ba_4C_{60}, have the molecular valence of $n = 3$ and 8, respectively. The former state corresponds to the half-filling of the t_{1u} band, while the latter is away from the half-filling of the t_{1g} band.

What about other valence states? Are there any other valence states that yield superconductors? Hundreds of materials have been synthesized by intercalation of metals into C_{60}. Particularly, for the alkali intercalated systems, extensive investigations have been made using binary or ternary compounds. In the alkali binary systems A_xC_{60} ($x = 1$, 2, 3, 4 and 6), the molecular valence can be changed discretely, while a quasi-continuous control of band filling is achieved by using ternary compounds. Among A_xC_{60}, superconductivity appears only at $x = 3$.

Control of band filling across the half-full state of the t_{1u} band ($n = 3$) has been achieved by Yildirim *et al* [38] using ternary systems. Researchers succeeded in tuning the molecular valence from 2 to 3 using $Na_2Cs_xC_{60}$ ($0 < x < 1$), and from 3 to 4 by $A_{3-x}Ba_xC_{60}$ ($0 < x < 1$) systems. The T_c band filling correlation is summarized in Figure D3.4. The most important feature of their materials is that only the molecular valence was changed, keeping the crystal structure essentially at fcc (for $x < 3$, the structure was a simple cubic with the ($Pa\bar{3}$), space group). The sharp peak structure in Figure D3.4 clearly shows that the superconductivity is destroyed by only slightly shifting the Fermi energy from the half-filling of the t_{1u} states.

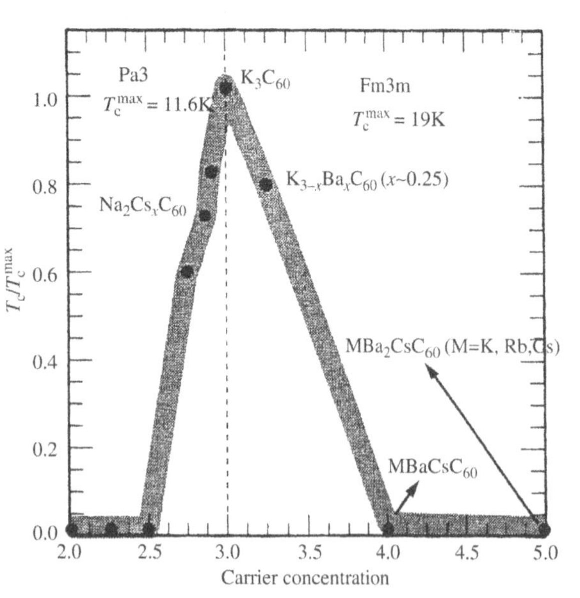

FIGURE D3.4 The relationship between T_c and the band filling across the half-full state of the t_{1u} band.

A current conclusion on alkali intercalated C_{60} compounds is that superconductivity appears only at the half-filled state in the t_{1u}. Superconductivity at different valence states is strictly prohibited for the alkali-doped systems.

In sharp contrast, the criteria for superconductivity are significantly different in the t_{1g} band. As for the alkaline earth binary materials, Ba_4C_{60} and Ca_5C_{60} are superconductors with onset temperatures of 6.5 and 8.4 K, respectively. Though the crystal structure of Ca_5C_{60} is not fully solved yet, it is believed to be a simple cubic structure with the same molecular arrangement as fcc [39]. The formal valence of C_{60} is $n = 8$ and 10, respectively. Another possible way to change the molecular valence in the t_{1g} band is to synthesize alkali–alkaline-earth ternary materials. For example, when we consider the structural sequence shown in Figure D3.2, one may notice that K_6C_{60} and Ba_6C_{60} are isostructural, but the electronic states are totally different. In K_6C_{60}, only the t_{1u} level is fully occupied, while, in Ba_6C_{60}, both t_{1u} and t_{1g} levels are fully occupied. Since the ionic radii of K and Ba are similar to each other, one is able to synthesize a solid solution $K_xBa_{6-x}C_{60}$, in principle. By changing the K concentration, we can theoretically tune the molecular valence quasi-continuously. Based on this idea, $K_3Ba_3C_{60}$ has been synthesized by intercalation of alkali metals into Ba_3C_{60}, which is a very stable A15 phase [40]. The formal valence of C_{60} in $K_3Ba_3C_{60}$ is $n = 9$, which corresponds to the half-filling of the t_{1g} band. The position of K and Ba ions is disordered and, hence, $K_3Ba_3C_{60}$ is regarded as a 1:1 solid solution of K_6C_{60} and Ba_6C_{60}. This nonavalent compound was found to be superconducting with T_c at 5.6 K. Isostructural $Rb_3Ba_3C_{60}$ was also found to be a superconductor at $T_c = 2$ K [41]. Optimization of the synthesis route may produce a rich variety of ternary compounds, in which one is able to tune the t_{1g} band filling continuously, keeping the crystal structure unchanged.

Figure D3.5 summarizes various fullerene superconductors in terms of the relation between the T_c and the molecular

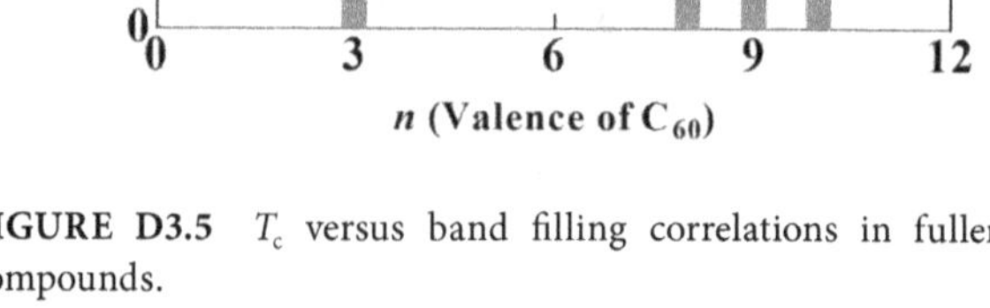

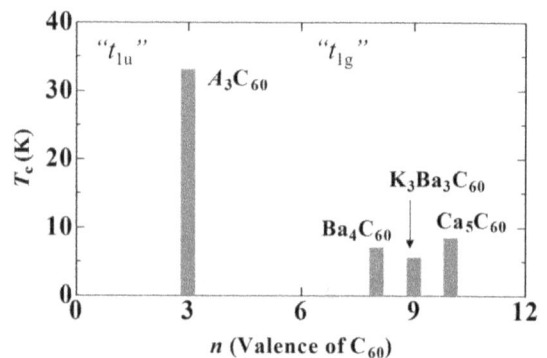

FIGURE D3.5 T_c versus band filling correlations in fulleride compounds.

valence (band filling). Although there exists a vast variety of fullerene intercalation compounds, Figure D3.5 provides a simple insight into the chemical criteria for fullerene superconductivity. (a) When C_{60} is intercalated only with alkali metals, superconductivity occurs only at the trivalent state with the fcc structure. When superconductivity is realized despite this strict constraint, T_c reaches above 30 K. (b) When C_{60} is intercalated with alkaline-earth or rare-earth metals, superconductivity occurs irrespective of molecular valence and crystal structure. Particularly, as far as the t_{1g} band is concerned, superconductivity always appears once the partial band filling is achieved. However, T_c stays lower than 10 K.

Hence, fullerene superconductors are classified into two groups: alkali-doped systems (t_{1u} systems) and alkaline-earth–doped systems (t_{1g} systems). Since understanding of rare-earth–doped systems is still poor, hereafter only the alkali- and alkaline-earth–doped fullerides are discussed. The following sections will concentrate on the t_{1u} systems, where superconductivity is suppressed by the Mott–Hubbard transition. In Section D3.4, comparison with the t_{1u} and t_{1g} fullerides is given with a brief discussion of the origins of the differences.

D3.3 A Hint of Strong Correlation Effect in Trivalent Fullerides A_3C_{60}

The trivalent systems are the best known and studied family of fullerene superconductors due to their early discovery and highest T_c values. An important result established in this class of materials is the scaling of T_c by the interfullerene spacing, where T_c increases with the lattice parameter. This result is strongly indicative of the BCS mechanism and also suggests a chemical method for increasing T_c. Motivated by this empirical relation, extensive efforts have been made to expand the unit cell in order to achieve higher T_cs. Figure D3.6 summarizes the relation between T_c and the volume per C_{60} for various compounds with the nominal C_{60} valence of $n = 3$.

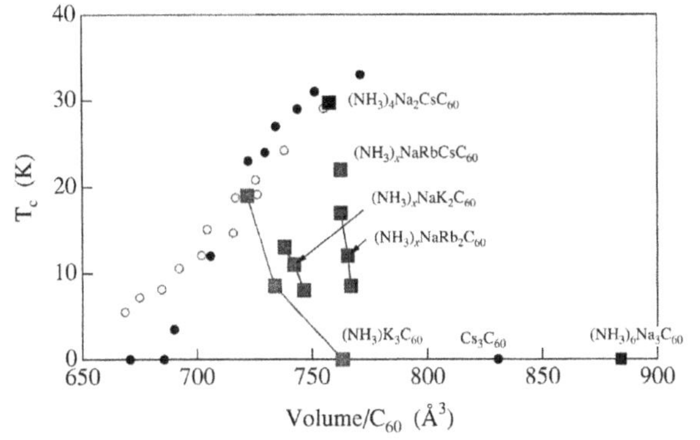

FIGURE D3.6 The relation between T_c and the volume per C_{60} molecule for trivalent compounds. Both alkali-doped (black dots for ambient pressure and white dots for high pressure) and alkali–ammonia co-intercalated compounds (filled squares) are plotted.

The black and white dots show the alkali-doped systems recorded at ambient pressure and at high pressure, respectively. The notable feature is that if you consider the group of alkali-doped fullerides, T_c is simply scaled with the unit volume (related to the interfullerene spacing) for both chemical substitution and application of hydrostatic pressure. This indicates that T_c is controlled by the density of states at the Fermi level $N(E_F)$: a larger lattice parameter decreases the C_{60} overlap integrals reducing the width of the t_{1u} band, hence the $N(E_F)$ value increases, because the total number of states in the band is independent of the bandwidth. This variation of T_c with $N(E_F)$ has been confirmed by various measurements, such as magnetic susceptibility, electron spin resonance and nuclear magnetic resonance experiments.

However, structural expansion using alkali metals alone in the interstitial sites is limited by the size of metal ions available for intercalation. This difficulty was overcome by Zhou, Rosseinsky and coworkers at Bell Labs, who discovered that additional structural expansion can be achieved by using both alkali atoms, for charge transfer, and neutral molecules such as ammonia, as structural spacers [42, 43]. In Figure D3.6, the volume dependence of T_c of the ammonia including intercalates $(NH_3)_xA_3C_{60}$ and $(NH_3)_xA_{3-y}A_yC_{60}$ is also plotted. Ammonia intercalation affords numerous compounds, which can be classified into three types when compared to their parent compounds; superconductors that show an increase in T_c concurrent with lattice expansion, non-superconductors and superconductors with lower T_cs showing a dramatical deviation from the conventional trend. An example of the first type is $(NH_3)_4Na_2CsC_{60}$, which is synthesized by exposing Na_2CsC_{60} ($T_c \sim 10$ K) to ammonia gas. In this case, the fcc structure is maintained and the compound exhibits an almost tripled T_c (29.6 K). Type 2 is exemplified by the non-superconducting $NH_3K_3C_{60}$ which is synthesized by the intercalation of NH_3 into K_3C_{60} [43]. The third type is $(NH_3)_xNaA_2C_{60}$ (A = K or Rb), which shows lower T_cs with a negative correlation between T_c and the interfullerene spacings. Since these types of compounds are unstable without NH_3, Shimoda *et al* [44] synthesized materials directly from the liquid ammonia solution. Including these three typical examples, various compounds co-intercalated with alkali metals and ammonia have been synthesized as shown in Figure D3.6.

Generally, compounds with very large interfullerene spacings are not superconducting, in contradiction to the simple expectation that T_c is increased by expanding the size of the unit cell. This is possibly explained by the metal–insulator transition resulting from too much lattice expansion. However, this interpretation is too primitive, as lattice distortions are not considered. It is to be noted that all the superconducting compounds in Figure D3.6 have fcc or similar cubic structures. Even the lower T_c $(NH_3)_xNaA_2C_{60}$ is fcc. Conversely, all the non-superconducting compounds have non-cubic structures. Superconductivity of the trivalent fullerides occurs exclusively in fcc or related structures, irrespective of the intercalation of ammonia. In other words, superconductivity of the $n = 3$ state seems to be controlled by the symmetry of crystal structure.

Non-superconductors commonly have non-fcc structures, indicating that not only the expansion of the lattice but also the symmetry reduction of the crystal caused the localization of electrons and suppression of superconductivity. Interestingly, the constraint on the crystal structure is not so strict under high pressure. Pressure-induced superconductivity has been observed in two of the ambient pressure non-superconducting compounds. $NH_3K_3C_{60}$ [45] and Cs_3C_{60} [8] become superconducting under a pressure of about 15 kbar at $T_c = 28$ and 40 K, respectively. It is evident that the superconductor–non-superconductor boundary, which was crossed by lattice expansion or symmetry reduction, was re-crossed by application of external pressure.

Another important issue for the non-superconductors is their electronic ground states at ambient pressure. Very recently, a detailed study of the electronic properties has been performed on the typical compound $NH_3K_3C_{60}$. The crystal structure of $NH_3K_3C_{60}$ is schematically described in Figure D3.7 and compared to that of K_3C_{60}. The NH_3 molecule is inserted in the octahedral site, forming a $K(NH_3)$ group. This causes a slight structural distortion from fcc to orthorhombic. An electron paramagnetic resonance (EPR) experiment on $NH_3K_3C_{60}$ revealed that a phase transition takes place at 40 K as shown in the lower panel of Figure D3.7. Comparison between EPR and magnetic susceptibility data suggested that the low-temperature state is antiferromagnetic [46, 47]. Subsequently, muon spin rotation (μSR) and nuclear magnetic resonance (NMR) experiments confirmed long-range antiferromagnetic ordering below 40 K [48, 49]. These results clearly indicate that the suppression of superconductivity in $NH_3K_3C_{60}$ is associated with effects of a magnetic origin, providing an important analogy with the well-established phenomenology in cuprate and organic superconductors. The superconductivity–antiferromagnetic transition seen when ammonia is intercalated into K_3C_{60} to form $NH_3K_3C_{60}$ is a result that has far-reaching consequences for the understanding of fullerene superconductivity, since the electron correlation effect, which plays a crucial role in the insulating phase, has not been correctly considered. Although the superconductivity of fullerides appears to be simply explained by a weak-coupling BCS picture, this issue should be revisited taking the above results into account.

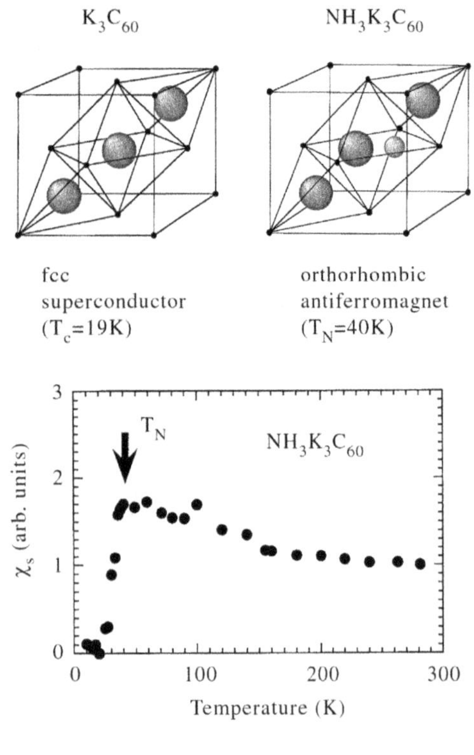

FIGURE D3.7 Upper panel: Schematic crystal structure of K_3C_{60} and $NH_3K_3C_{60}$. C_{60} is depicted by the small dot, while the large and small spheres are intercalated K and NH_3, respectively. Lower panel: Temperature dependence of spin susceptibility determined from the electron paramagnetic resonance measurements for $NH_3K_3C_{60}$. The sharp decrease of susceptibility shows the occurrence of metal–insulator transition. The low-temperature state is a Mott–Hubbard insulating state with an antiferromagnetic ordering.

The effect of symmetry reduction of the crystal structure causes a lowering of symmetry in the local field surrounding each C_{60} molecule. Since the local field on each C_{60} molecule has a cubic symmetry in the fcc structure, the triple degeneracy of the t_{1u} band is maintained. In the non-cubic structure, the degeneracy is lifted. The effect of triple degeneracy of the molecular orbitals has been discussed first by Lu [21] and then by Gunnarsson *et al* [22]. They pointed out that the triple degeneracy is crucial to keep the metallic state in fullerene intercalation compounds, where there is a rather large electron correlation effect. In the simplest criterion of the Mott–Hubbard transition, most of the alkali-doped fullerene solids should be only in the insulating side, since the experimentally estimated on-ball Coulomb interaction $U = 1.0$–1.5 eV is much larger than the conduction bandwidth W ~ 0.5 eV. However, Lu and Gunnarsson conjectured that, in the triply degenerate and half-full system, there are three paths for electron transfer (Figure D3.8), so that the bandwidth is effectively

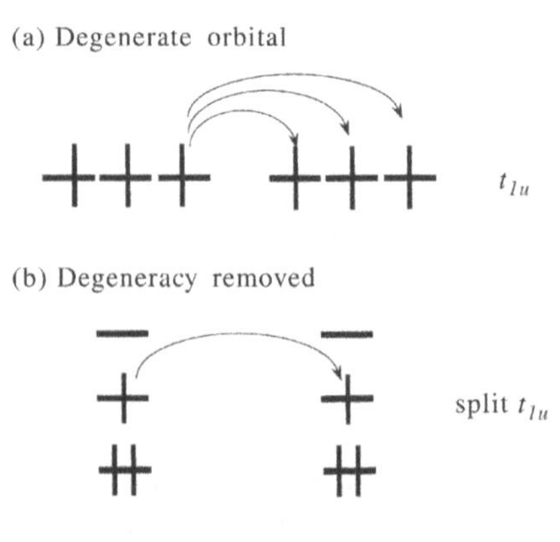

FIGURE D3.8 Electron hopping between neighboring molecules through the t_{1u} level: (a) degenerate case and (b) undegenerate case.

larger compared to the non-degenerate system, where there is only one path for electron hopping. Due to the larger effective transfer, the metallic state is maintained even for the large *U*/*W* values at which the Mott–Hubbard transition takes place in the non-degenerate system. These theories explain the metallic state in the half-filled state of the t_{1u} band, and the Mott–Hubbard state in the non-cubic structures. The antiferromagnetic state found in $NH_3K_3C_{60}$ is evidence for the strong correlation effect, which appears explicitly in non-degenerate systems.

D3.4 Comparison of Alkali- and Alkaline-Earth Metal–Doped Systems: A_4C_{60} and Ae_4C_{60}

This section discusses the comparison between the alkali- and alkaline-earth–doped systems, with particular focus on the A_4C_{60} type structure [29]. Among various issues in the intercalated fullerides, the electronic ground state of the non-superconducting compounds, particularly of the tetravalent compound A_4C_{60}, has been extensively investigated since their discovery. Firstly, the muon spin resonance (μSR) experiments gave evidence of an electronic band gap of about 0.5 eV [50]. A similar gap was observed by the optical spectra, depicted in Figure D3.9 [51], which shows optical conductivity spectra for undoped C_{60}, K_3C_{60}, K_4C_{60} and Rb_6C_{60}. While K_3C_{60} is metallic, C_{60}, K_4C_{60} and Rb_6C_{60} display gap-like structures, indicating insulating behaviors. Only the spectra of K_4C_{60} cannot be explained by band theories. Meanwhile, the electron paramagnetic resonance (EPR) experiment confirmed that K_4C_{60} is a nonmagnetic insulator. However, this experimental finding, established at a relatively early stage, has not been explained consistently.

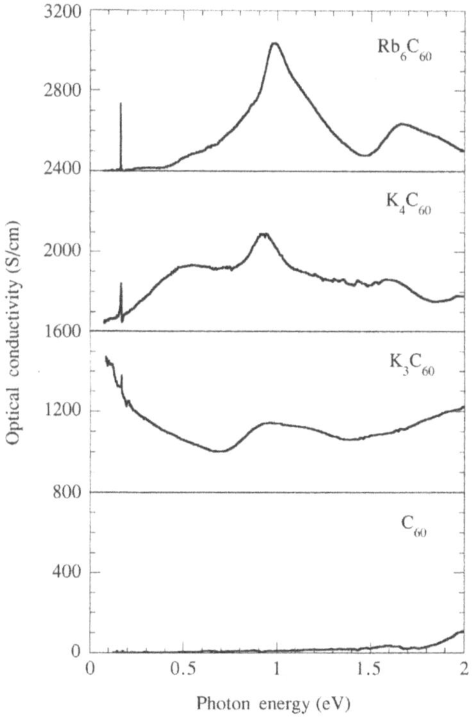

FIGURE D3.9 Optical conductivity for undoped C_{60}, K_3C_{60}, K_4C_{60} and Rb_6C_{60}. K_4C_{60}, which is expected to be metallic from band theories, displays a clear gap structure peaking at about 0.5 eV.

A band calculation using pseudopotential plane wave density functional techniques failed to explain the gap in the bct structure [52]. Thus, the most plausible explanation for the nonmagnetic insulating state is the Jahn–Teller effect, in which the C_{60} molecules are distorted so as to stabilize the electronic energy against the energy loss from molecular distortion. However, theoretical investigations suggest that a simple Jahn–Teller effect is not sufficient to explain the gap opening nor, moreover, the experimentally observed gap. Suzuki and Nakao pointed out that the combined effect of Jahn–Teller and electron correlation could explain the insulating state of K_4C_{60} [23]. A recent systematic experiment revealed that the absolute value of the electronic gap is reasonably explained by electron correlation, producing a Mott insulating state [53]. However, since A_4C_{60} is a nonmagnetic insulator, they suggested the combined effect of electron correlation and the Jahn–Teller effect, where the charge gap observed in the electronic spectra is determined by the Mott–Hubbard gap, while the possibly smaller spin gap is determined by the Jahn–Teller effect. Although no experiments have succeeded in detecting the Jahn–Teller distortion in A_4C_{60} so far, the cooperation of Jahn–Teller and electron correlation effects is the most likely explanation for the insulating state of A_4C_{60}.

The above argument on the electronic ground state of K_4C_{60} implies that cooperation of the electron–phonon interaction and the electron correlation effect destroy the metallic state expected from the band theory. In other words, metallic states of tetravalent fullerides are very unstable. This is also consistent with theories of Lu and Gunnarsson, who predicted that the metallic state is most stabilized at the half-filled state of the t_{1u} band. In tetravalent fullerides, the number of paths for electron transfer is decreased from three to two, facilitating the transition to Mott–Hubbard insulator [15].

In contrast to A_4C_{60} (A = K and Rb), Ae_4C_{60} (Ae = Sr and Ba) are superconductors despite the similar structures [30, 31]. Since all the fullerene superconductors at ambient pressure had been limited to cubic or slightly distorted cubic structures, the superconductivity of Ae_4C_{60} raised a serious question about the structural criteria for superconductivity in fullerenes. It is interesting to compare the properties of Ae_4C_{60} to those of non-superconducting A_4C_{60} and superconducting A_3C_{60}. The density of states at the Fermi level, N(E_F), determined by magnetic susceptibility measurements, is 6.0 and 2.5 states/eV(mol C_{60})$^{-1}$ for Ba_4C_{60} and Sr_4C_{60}, respectively. These values are considerably smaller than those for the t_{1u} superconductors, K_3C_{60} and Rb_3C_{60}. This indicates that the conduction band in the Ae_4C_{60} compounds is rather broad, suggesting that the metallic state is much more stable than the alkali-doped systems. What causes the broad conduction band in Ae_4C_{60}?

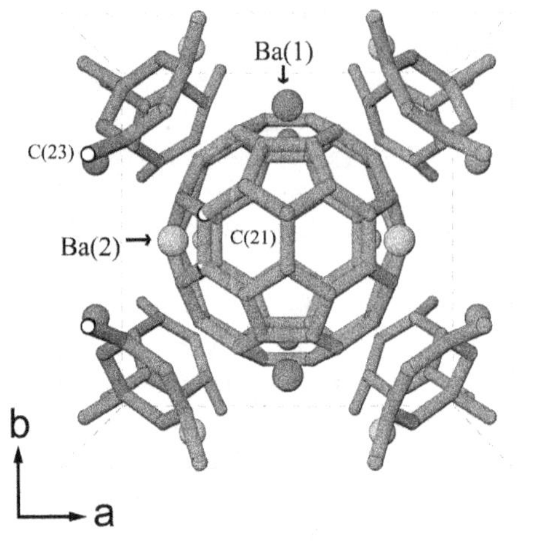

FIGURE D3.10 The detailed crystal structure of Ba_4C_{60}.

Figure D3.10 shows a perspective view of the orthorhombic structure of Ba_4C_{60} (space group *Immm*, $a = 11.610$ Å, $b = 11.235$ Å and $c = 10.883$ Å at room temperature). The orthorhombic unit cell of Ba_4C_{60} is realized by the orientational ordering of the C_{60} units, which contrasts with the merohedral disorder in K_4C_{60} and Rb_4C_{60}. The increased Madelung energy of Ba ions over that in A_4C_{60} produces a lattice contraction in Ba_4C_{60}. This contraction results in anomalously small interatomic distances. The shortest Ba(2)–C contacts are 2.99 and 3.04 Å to hexagon C(21) and pentagon C(23) atoms, respectively. Here, the atomic positions are depicted in Figure D3.10. These values are smaller than the sum of the ionic radius of Ba^{2}+ (1.35 Å) and the van der Waals radius of C (1.70 Å), indicating a strong hybridization between the 5d orbitals of Ba(2) and the 2p orbitals of carbon. The effect of short Ba–C contact is observed in the thermal expansion, which was found to be severely anisotropic ($\delta a/a = 0.080\%$, $\delta b/b = 0.456\%$ and, importantly, $\delta c/c = -0.007\%$ between 5 and 295 K). This unusual property may be attributed to the bonding character between the shortest Ba–C atoms.

The orbital mixing revealed by detailed structural analysis is evidently responsible for the broad t_{1g} conduction band. Broad t_{ig} bands have been suggested experimentally for other Ba-doped fullerides. For instance, Sr_6C_{60} and Ba_6C_{60}, which should be semiconductors according to simple electron counting, show small but finite $N(E_F)$, indicating that these compounds are semimetals due to the band overlap. Another example of broad conduction band was demonstrated by the susceptibility measurement on the nonavalent fulleride superconductors $K_3Ba_3C_{60}$ and $Rb_3Ba_3C_{60}$ [41]. We conjecture that the low $N(E_F)$ commonly observed in Ba-doped C_{60} is a direct result of the hybridization effect.

The low crystal symmetry of A_4C_{60} and Ae_4C_{60} is not favorable for superconductivity. In fact, the electrons in A_4C_{60} are localized, forming a nonmagnetic insulator. In contrast, Ae_4C_{60} maintains the metallic state and superconductivity due to the band broadening, which suppresses the instability to insulators. This scenario might explain why, in the t_{1g} state, superconductivity appears apparently irrespective of band filling and crystal structure.

D3.5 Overexpanded Cubic Fullerides – A New Era in Fullerene Superconductivity

Despite the insight provided by the expanded ammoniated/aminated alkali fullerides through the suppression of the metallic state and the emergence of AFM insulators at large interfullerene separations (Section D3.3), there was no definitive experimental evidence for a non-BCS origin for superconductivity in fullerides, where correlation would play a role. The origin of the metal–insulator transition could not be strictly traced to bandwidth control but could be associated with the symmetry lowering, which lifted the orbital degeneracy. Therefore, what had been still missing since 1992 were ideal materials in which the metal–insulator transition could be traversed in a purely electronic manner without the complications of structural transitions, which masked the role of electronic correlations (Figure D3.6). In such materials, the cubic site symmetry required for orbital degeneracy to survive in all competing electronic states should be maintained across the transition. These were finally discovered after 2008 in the form of the A15- and fcc-structured polymorphs of the most expanded, yet binary, cubic fulleride, Cs_3C_{60} (Figure D3.11) [9, 54].

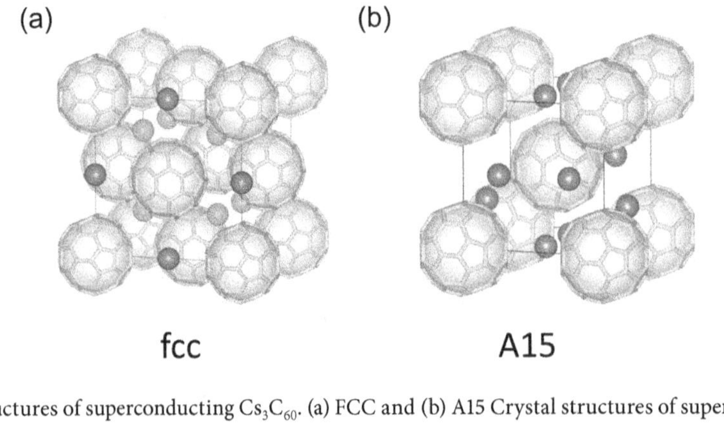

FIGURE D3.11 Crystal structures of superconducting Cs_3C_{60}. (a) FCC and (b) A15 Crystal structures of superconducting Cs3C60.

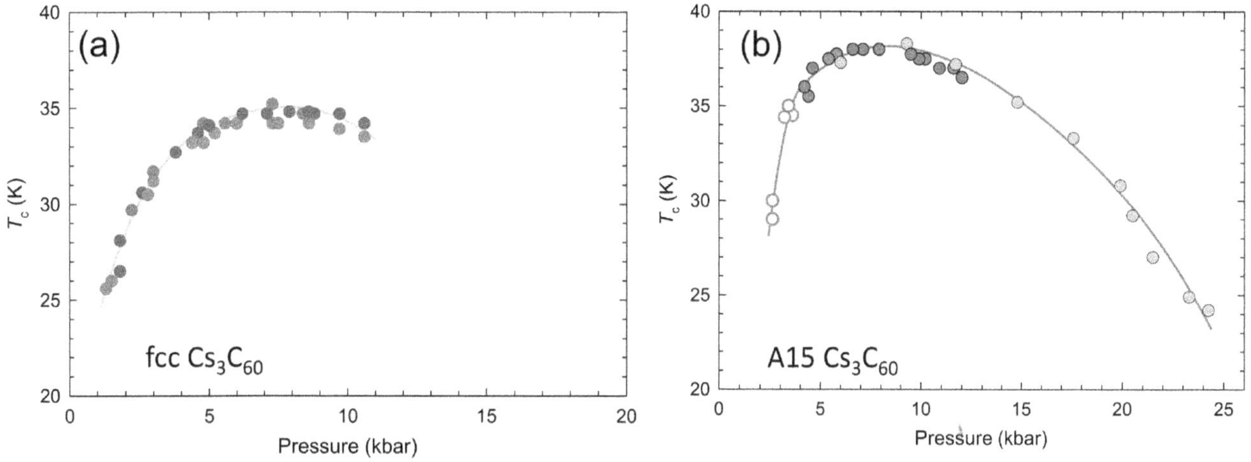

FIGURE D3.12 Superconducting transition temperature, T_c, as a function of pressure for (a) fcc- and (b) A15-structured Cs_3C_{60}.

Both Cs_3C_{60} polymorphs are insulators at ambient pressure but metallicity and superconductivity can be switched on without crystal structure change [9, 54] by the application of moderate external pressures [9, 54, 55]. As superconductivity emerges upon pressurization, T_c rapidly increases and reaches a maximum (35 K in fcc-structured Cs_3C_{60}, 38 K in A15-structured Cs_3C_{60} - this is the highest T_c observed in a bulk molecular material) at ~7 kbar before decreasing upon further increase in pressure (Figure D3.12). These results transformed the entire field of fullerene superconductivity as they established the existence of a key commonality with other unconventional superconductors - the presence of a conduction-bandwidth-controlled superconductivity dome in their electronic phase diagram. DFT calculations in the absence of electron correlations find that the density-of-states at the Fermi level, $N(E_F)$ and the conduction bandwidth, W change monotonically as a function of interfullerene separation, V even when experimentally $\partial T_c/\partial V$ changes sign [56]. This suggests that superconducting pairing in fullerides is more complex than simply originating from BCS-like electron–phonon coupling and necessitates the presence of an additional parameter.

Evidence for the identity of this extra ingredient came from the observation that both parent non-superconducting Cs_3C_{60} polymorphs are antiferromagnetic ($S = ½$) Mott insulators - the hallmark of strong electron correlations. The Néel temperature, T_N of bipartite A15-structured Cs_3C_{60} is 46 K [57], while that of geometrically frustrated fcc is significantly suppressed to 2.2 K [54]. Therefore, the AFM insulator to superconductor transition is purely electronically driven without any crystal structure change by pressure. It is of first order as evidenced by the co-existence of the two competing electronic states over a finite pressure range [57] and antiferromagnetism is transformed into superconductivity solely by changing an electronic parameter: the extent of overlap between the outer wave functions of the constituent anions (Figure D3.13). Thus, unlike in the less expanded systems with smaller alkali metals, the electron correlation effects represented by U dominate despite the retained orbital degeneracy in the expanded cubic fulleride phases, providing firm evidence that correlation directly competes with superconductivity in cubic A_3C_{60} materials. Therefore, the fullerides should be considered as correlated electron systems controlled by Mott–Hubbard physics in the region corresponding to the highest T_c.

Theoretical models of the fullerides beyond conventional BCS theory, which treated electron correlation effects and conventional coupling of electrons to phonons by Jahn–Teller-active intramolecular C_{60} vibrations on an equal footing, predicted correlation enhancement of superconductivity near the metal–insulator transition [58–60]. Indirect evidence of the relevance of the JT effect in Mott-insulating Cs_3C_{60} is provided by the observation of a low-spin $S = ½$ ground state [54, 57, 61] despite the retention of cubic symmetry which should have led to a high-spin $S = 3/2$ state for the C_{60}^{3-} anion (Figure D3.8). In addition, infrared and magic-angle-spinning NMR spectroscopy revealed the existence of subtle changes in the shape of the C_{60}^{3-} ion due to dynamic Jahn–Teller distortions with an interconversion rate in the range, $10^5 < k_{JT} < 10^{11}\ s^{-1}$ in both insulating fcc and A15 Cs_3C_{60} polymorphs [62, 63]. Therefore, it appears that the JT interaction exceeds Hund's rule coupling leading to a physical regime of 'inverted Hund's rule coupling' and enforce a low-spin state without breaking the global cubic symmetry. However, a word of caution here is that this rests on the assumption that electron–electron interactions *per se* always favor the high-spin electronic configuration. It is therefore intriguing that recent theoretical calculations show that Hund's first rule can be violated without the necessity of the dynamic Jahn–Teller effect [64]. For an average of three electrons per molecule (C_{60}^{3-}), an effective attraction (pair-binding) occurs making it favorable to place four electrons on one molecule (C_{60}^{4-}) and two on a second (C_{60}^{2-}). This implies that a dominantly electronic mechanism of superconductivity in the fullerides is not precluded by the currently available experimental findings and remains a key area of future investigation in this field.

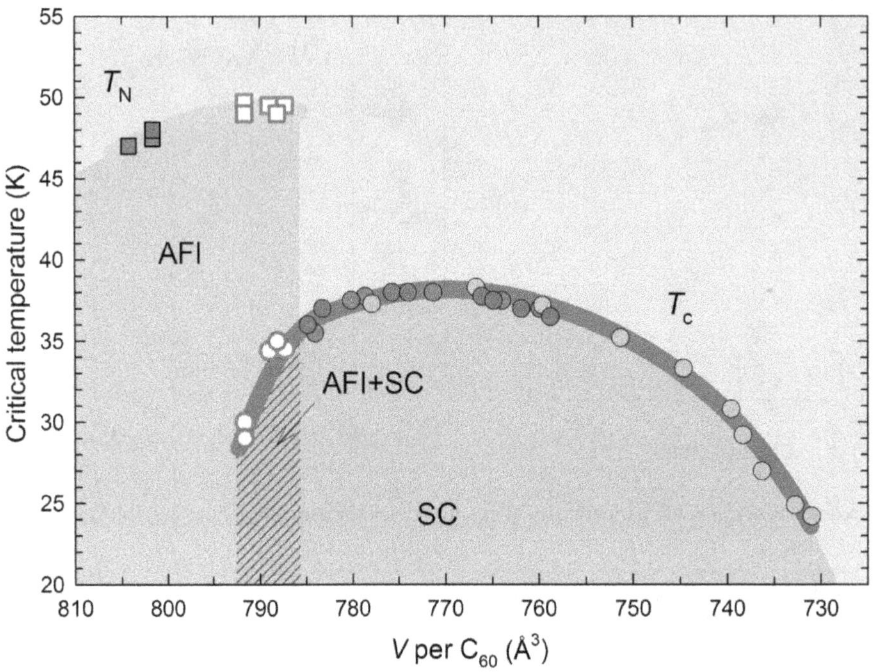

FIGURE D3.13 Electronic phase diagram of A15 Cs_3C_{60}.

D3.6 Traversing the Superconductivity Dome – the Emergence of the Jahn–Teller Metallic State at Optimal T_c

According to the currently prevailing interpretation, both Cs_3C_{60} polymorphs can be classified as magnetic Mott–Jahn–Teller (MJT) insulators with the on-molecule dynamic distortion creating the $S = ½$ ground state, which produces the magnetism from which the superconductivity emerges. Contraction of the MJT insulator through chemical pressurization in the ternary solid solution fcc compositions, $Rb_xCs_{3-x}C_{60}$ ($0 \le x \le 3$) yields a metallic state which becomes superconducting on cooling. Infrared spectroscopy has revealed that the insulator-to-metal crossover is not immediately accompanied by the suppression of the molecular Jahn–Teller distortions [61]. The metallic state which emerges following the destruction of the Mott insulator is unconventional – correlations sufficiently slow carrier hopping and the intramolecular JT effect co-exists with metallicity. This inhomogeneous JT metallic state of matter shows both molecular (dynamically JT-distorted C_{60}^{3-} ions) and free-carrier (electronic continuum) features. The JT metal exhibits a strongly enhanced spin susceptibility relative to that of a conventional Fermi liquid, characteristic of the importance of strong electron correlations. As the fulleride lattice contracts further, there is a crossover from the JT metal to a conventional Fermi liquid state as we move further away from the Mott boundary toward the underexpanded regime with the molecular distortion arising from JT effect disappearing and the electron mean-free path extending to more than a few intermolecular distances [61]. Remarkably, the JT to conventional metal crossover occurs exactly where the maximum T_c is found for optimally expanded fullerides (Figure D3.14).

Therefore, the crossover of the electronic states across the bandwidth-controlled phase diagram from a MJT insulator to a JT metal and then to a Fermi liquid state directly affects the superconducting states which lie underneath. NMR spectroscopy reveals the opening of a single isotropic BCS-like (s-wave) superconducting gap, Δ_0 below T_c across the entire electronic phase diagram from over- through optimally- to underexpanded (both chemically- and physically pressurized) A_3C_{60} fullerides [61, 64]. The normalized gap ratios, ($2\Delta_0/kT_c$) for underexpanded conventional metals far away from the Mott insulator boundary take values characteristic of weak-coupling BCS superconductors (~3.52). However, as the lattice expansion reaches optimal values and beyond, the gap ratio begins to increase monotonically, approaching and then exceeding a value of 5 [61, 64, 65] for superconductors emerging from JT metals upon cooling. The superconducting gap does not correlate with T_c in the overexpanded regime and in contrast to the dome-shaped dependence of T_c on packing density, V, the gap increases monotonically with interfullerene separation (Figure D3.15). Notably, the maximum T_c occurs at the crossover between the two types of gap behavior.

The dichotomy in the dependence of the superconducting gap and transition on packing density, V is mirrored by the upper critical magnetic field, H_{c2}, which is determined by both the strength of the pairing potential and the coherence length.

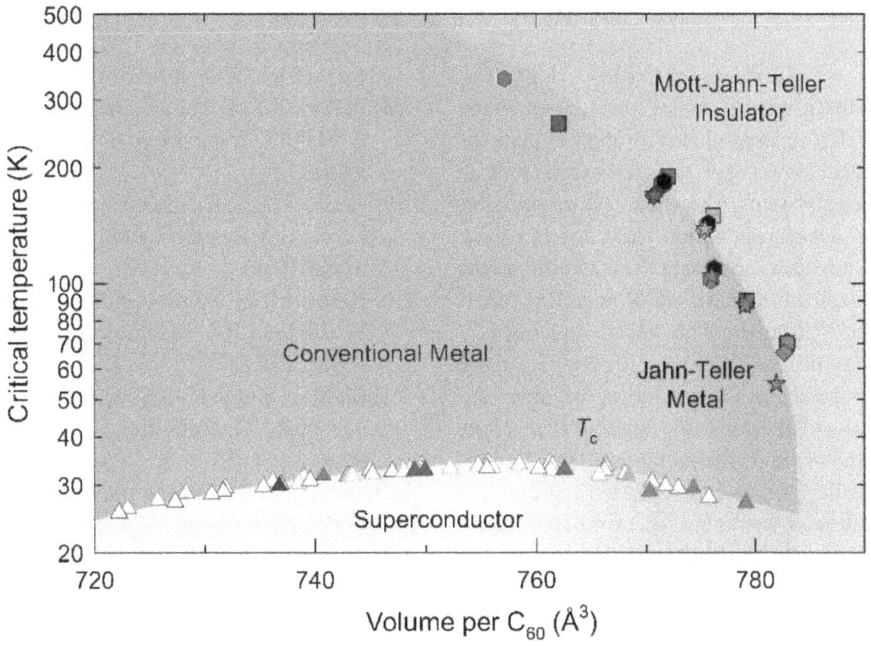

FIGURE D3.14 Global electronic phase diagram of fcc-structured A_3C_{60} fullerides.

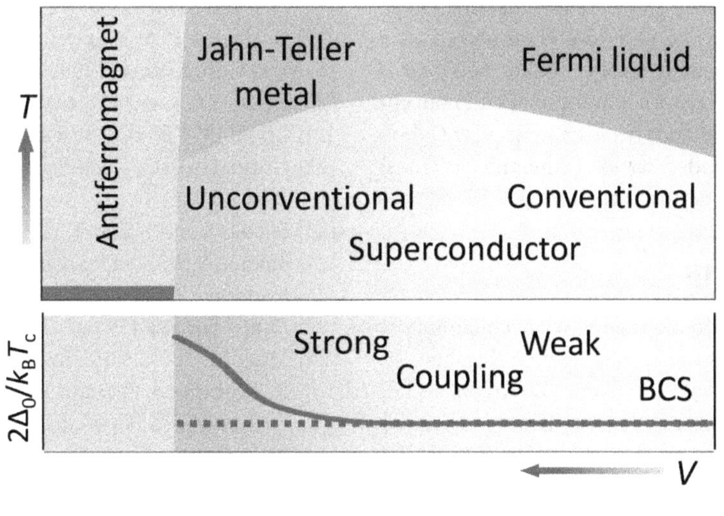

FIGURE D3.15 Contrasting behavior of the dependence of the superconducting gap and T_c on fulleride packing density.

H_{c2} is found to track the response of the gap and reach values in excess of 90 T close to the Mott boundary [66] – these are the highest known for any 3D superconductor.

D3.7 Summary

The current understanding of fullerene intercalation compounds and their superconductivity has been reviewed. Intercalation compounds of C_{60} have been classified into two groups: one is alkali metal-doped systems, which provide partially filled t_{1u} bands, and the other is alkaline-earth metal-doped systems, which provide partially filled t_{1g} bands.

Superconductivity in the former group occurs exclusively in the fcc structure and exclusively at the half-filled state. Once these conditions are violated, T_c rapidly decreases and finally disappears. The electronic ground state of the non-superconducting compounds is either a magnetic or a nonmagnetic insulator depending on how the condition is violated. In this sense, superconductivity in the alkali-doped systems is rather fragile. Nonetheless, once the condition is fulfilled, critical temperatures and magnetic fields can be very high. These various instabilities are caused by competition or cooperation between the electron–phonon interaction and electron–electron interactions. Also, the symmetry of the molecule and crystal structure seems quite important for this group of materials.

Superconductivity in the alkaline-earth metal-doped C_{60}, on the other hand, is insensitive to crystal structure and band filling. It looks likely that superconductivity appears once the t_{1g} state is partially filled. However, T_c does not exceed 10 K, in sharp contrast to the t_{1u} systems. These features are possibly explained by the broadened conduction band due to orbital mixing. The larger bandwidth suppresses the instability to the insulating and stabilizes the metallic states and thus superconductivity, however, $N(E_F)$ is therefore lower, reducing T_c.

The recent discovery of Cs_3C_{60} and the $Rb_xCs_{3-x}C_{60}$ alloys has dramatically advanced the understanding of superconductivity of the trivalent fullerides. The highest T_c reaching 38 K is achieved at the verge of the metal–insulator boundary, displaying a peculiar dome-shaped phase diagram. The commonalities of fullerides with cuprate, iron-based superconductors, organic superconductors and heavy fermion superconductors have proven that fullerenes are members of the family of strongly correlated superconductors. In particular, the highest T_c emerges from an anomalous, highly inhomogeneous metallic state in which itinerant metallic electrons co-exist with localized electrons, which produce on-molecule Jahn–Teller distortions – the Jahn–Teller metallic state. Therefore, the optimal T_c in the fullerides is associated with a strongly coupled (or extremely stable) Cooper pair and is found at the boundary between unconventional and conventional behavior, where the balance between molecular (Jahn–Teller distortion) and extended lattice (itinerant electrons) features of the electronic structure is optimized.

Acknowledgements

The authors are indebted to their long-term collaborators and particularly to H. Shimoda, T. Takenobu, Y. Kasahara, Y. Takabayashi, M. J. Rosseinsky, A. Y. Ganin, D. Arcon, and E. Tosatti. This work was supported in part by Grants-in-Aid for Specially Promoted Research (No.25000003), JSPS KAKENHI 15H05886, the Mitsubishi Foundation, the "World Premier International (WPI) Research Center Initiative for Atoms, Molecules and Materials," Ministry of Education, Culture, Sports, Science, and Technology of Japan, and SICORP-LEMSUPER FP7-NMP-2011-EU-Japan project (No.283214).

References

[1] Osawa E 1970 *Kagaku* **25** 854 (in Japanese)
[2] Kroto H W, Heath J R, O'Brien S C, Curl R F and Smalley R E 1985 *Nature* **354** 56
[3] Krätchmer W, Lamb L C, Fostiropoulos K and Huffman D R 1990 *Nature* **347** 354
[4] Haddon R C, *et al* 1991 *Nature* **350** 320
[5] Hebard A F, Rosseinsky M J, Haddon R C, Murphy D W, Glarum S H, Palstra T T M, Ramirez A P and Kortan A R 1991 *Nature* **350** 600
[6] Rosseinsky M J, *et al* 1991 *Phys. Rev. Lett.* **66** 2830
[7] Tanigaki K, Ebbesen T W, Saito S, Mizuki J, Tsai J S, Kubo Y and Kuroshima S 1991 *Nature* **352** 222
[8] Palstra T T M, Zhou O, Iwasa Y, Sulewski P E, Fleming R M and Zegarski B R 1994 *Solid State Commun.* **93** 327
[9] Ganin A Y, Takabayashi Y, Khimyak Y Z, Margadonna S, Tamai A, Rosseinsky M J and Prassides K 2008 *Nature Mater.* **7**, 367
[10] Fleming R M, Ramirez A P, Rosseinsky M J, Murphy D W, Haddon R C, Zahurak S M and Makhija A V 1991 *Nature* **352** 787
[11] Zhou O, Vaughan G B M, Zhu Q, Fischer J E, Heiney P A, Coustel N, McCauley J P Jr and Smith A B III 1992 *Science* **255** 833
[12] Diederichs J, Schilling J S, Herwig K W and Yelon W B 1997 *J. Phys. Chem. Solids* **58** 123
[13] Ramirez A P 1994 *Superconductivity Review* **1** 1
[14] Gerfant M P 1994 *Superconductivity Review* **1** 103
[15] Gunnarsson O 1994 *Rev. Mod. Phys.* **69** 575
[16] Prassides K 1997 *Curr. Opin. Solid State Mater. Sci.* **2** 433
[17] Rosseinsky M J 1998 *Chem. Mater.* **10** 2665
[18] Lof R W, Van Veenendaal M A, Koopmans B, Jonkman H T and Sawatzky G A 1992 *Phys. Rev. Lett.* **68** 3924
[19] Bruhwiler P A, Maxwell A J, Nilsson A, Martensson N and Gunnarsson O 1992 *Phys. Rev. B* **48** 18296
[20] Erwin S C and Pickett W E 1991 *Science* **254** 842
[21] Lu J P 1994 *Phys. Rev. B* **49** 5687
[22] Gunnarsson O, Koch E and Martin R M 1996 *Phys. Rev. B* **54** R11026
[23] Suzuki S and Nakao K 1995 *Phys. Rev. B* **52** 14206
[24] Takada Y 1992 *J. Phys. Soc. Japan.* **65** 1454
[25] Haddon R C, Brus L E and Raghavachari K, 1986 *Chem. Phys. Lett.* **125** 459
[26] Murphy D W, Rosseinsky M J, Fleming R M, Tycko R, Ramirez A P, Haddon R C, Siegrist T, Dabbagh G, Tully J C and Walstedt R E 1992 *J. Phys. Chem. Solids* **53** 1321
[27] Zhou O, Fischer J E, Coustel N, Kycia S, Zhu Q, McGhie A R, Romanow W J, McCauley J P Jr, Smith A B III and Cox D E 1991 *Nature* **351** 462
[28] Kortan A R, Kopylov N, Glarum S, Gyorgy E M, Ramirez A P, Fleming R M, Zhou O, Thiel F A, Trevor P L and Haddon R C 1992 *Nature* **360** 566
[29] Fleming R M, Rosseinsky M J, Murphy D W, Ramirez A P, Haddon R C, Siegrist T, Tycko R, Dabbagh G and Hampton C 1991 *Nature* **352** 701
[30] Baenitz M, Heinze M, Lüders K, Werner H, Schlögl R, Weiden M, Sparn G and Steglich F 1995 *Solid State Commun.* **96** 539
[31] Brown C M, Taga S, Gogia B, Kordatos K, Margadonna S, Prassides K, Iwasa Y, Tanigaki K, Fitch A N and Pattison P 1999 *Phys. Rev. Lett.* **83** 2258
[32] Kortan A R, Kopylov N, Fleming R M, Zhou O, Thiel F A and Haddon R C 1993 *Phys. Rev. B* **47** 13070

[33] Kortan A R, Kopylov N, Özdas N, Ramirez A P, Fleming R M and Haddon R C 1994 *Chem. Phys. Lett.* **223** 501
[34] Eklund P C, Zhou P, Wang K A, Dresselhaus G and Dresselhaus M S 1992 *J. Phys. Chem. Solids* **53** 1992
[35] Chen X H, Taga S and Iwasa Y 1999 *Phys. Rev. B* **60** 4351
[36] Saito S and Oshiyama A 1993 *Phys. Rev. Lett.* **71** 121
[37] Erwin S C and Pederson M R 1993 *Phys. Rev. B* **47** 14657
[38] Yildirim T, Barbedette L, Fischer J E, Lin C L, Robert J, Petit P and Palstra T T M 1996 *Phys. Rev. Lett.* **77** 167
[39] Kortan A R, Kopylov N, Glarum S, Gyorgy E M, Ramirez A P, Fleming M, Thiel F A, Trevor P L and Haddon R C 1992 *Nature* **355** 529
[40] Iwasa Y, Hayashi H, Furudate T and Mitani T 1996 *Phys. Rev. B* **54** 14960
[41] Iwasa Y, Kawaguchi M, Iwasaki H, Mitani T, Wada T and Hasegawa T 1998 *Phys. Rev. B* **57** 13395
[42] Zhou O, Fleming R M, Murphy D W, Rosseinsky M J, Ramirez A P, van Dover R B and Haddon R C 1993 *Nature* **362** 433
[43] Rosseinsky M J, Murphy D W, Fleming R M and Zhou O 1993 *Nature* **364** 425
[44] Shimoda H, Iwasa Y, Miyamoto Y, Maniwa Y and Mitani T 1996 *Phys. Rev. B* **54** R15653
[45] Zhou O, Palstra T T M, Iwasa Y, Fleming R M, Hebard A F and Sluewski P E 1995 *Phys. Rev. B* **52** 483
[46] Iwasa Y, Shimoda H, Palstra T T M, Maniwa Y, Zhou O and Mitani T 1996 *Phys. Rev. B* **53** R8836
[47] Allen K M, Heyes S J and Rosseinsky M J 1996 *J. Mater. Chem.* **6** 3956
[48] Prassides K, Margadonna S, Arcon D, Lappas A, Shimoda H and Iwasa Y 1999 *J. Am. Chem. Soc.* **121** 11227
[49] Tou H, Maniwa Y, Shimoda H, Iwasa Y and Mitani T 2000 *Phys. Rev. Lett.* **B62** R775
[50] Kiefl R F, *et al* 1992 *Phys. Rev. Lett.* **69** 2005
[51] Iwasa Y and Kaneyasu T 1995 *Phys. Rev. B* **51** 3678
[52] Erwin S C 1992 *Buckminsterfullerenes*, eds W E Billups and M A Ciufolini (New York: VCH Publishers)
[53] Knupfer M and Fink J 1997 *Phys. Rev. Lett.* **79** 2714
[54] Ganin A Y, Takabayashi Y, Jeglič P, Arčon D, Potočnik A, Baker P J, Ohishi Y, McDonald M T, Tzirakis M D, McLennan A, Darling G R, Takata M, Rosseinsky M J and Prassides K 2010 *Nature* **466**, 221
[55] Ihara Y, Alloul H, Wzietek P, Pontiroli D, Mazzani M and Riccò M 2010 *Phys. Rev. Lett.* **104**, 256402
[56] Darling G R, Ganin A Y, Rosseinsky M J, Takabayashi Y and Prassides K 2008 *Phys. Rev. Lett.* **101**, 136404
[57] Takabayashi Y, Ganin A Y, Jeglič P, Arčon D, Takano T, Iwasa Y, Ohishi Y, Takata M, Takeshita, N, Prassides K and Rosseinsky M J 2009 *Science* **323**, 1585
[58] Han J E, Gunnarsson O and Crespi V H 2003 *Phys. Rev. Lett.* **90**, 167006
[59] Capone M, Fabrizio M, Castellani C and Tosatti E 2009 *Rev. Mod. Phys.* **81**, 943
[60] Nomura Y, Sakai S, Capone M and Arita R. 2015 *Sci. Adv.* **1**, e1500568
[61] Zadik R H, Takabayashi Y, Klupp G, Colman R H, Ganin A Y, Potočnik A, Jeglič P, Arčon D, Matus P, Kamarás K, Kasahara Y, Iwasa Y, Fitch A N, Ohishi Y, Garbarino G, Kato K, Rosseinsky M J and Prassides K 2015 *Sci. Adv.* **1**, e150059
[62] Klupp G, Matus P, Kamarás K, Ganin A Y, McLennan A, Rosseinsky M J, Takabayashi Y, McDonald T M and Prassides K. 2012 *Nature Commun.* **3**, 912
[63] Potočnik A, Ganin A Y, Takabayashi Y, McDonald M T, Heinmaa I, Jeglič P, Stern R, Rosseinsky M J, Prassides K and Arčon D. 2014 *Chem. Sci.* **5**, 3008
[64] Potočnik A, Krajnc A, Jeglič P, Takabayashi Y, Ganin A Y, Prassides K, Rosseinsky M J and Arčon D. 2014 *Sci. Rep.* **4**, 4265
[65] Wzietek P, Mito T, Alloul H, Pontiroli D, Aramini M and Riccò M. 2014 *Phys. Rev. Lett.* **112**, 066401
[66] Kasahara Y, Takeuchi Y, Zadik R H, Takabayashi Y, Colman R H, McDonald R D, Rosseinsky M J, Prassides K and Iwasa Y 2017 *Nature Commun.* **8**, 14467

D4

Future High-T_c Superconductors

Ching-Wu Chu, Liangzi Deng, and Bing Lv

D4.1 Introduction

The search for novel superconducting materials has always been an important integral part of superconductivity research for science and technology. Prior to 1986, during the low-temperature superconductivity (LTS) era, it has helped raise the superconducting transition temperature (T_c), broaden the material base, unravel the mystery of superconductivity, and demonstrate the viability of superconductivity technology. The development of the Bardeen–Cooper–Schrieffer (BCS) theory on LTS was greatly assisted by the then-available data on a wide range of high-quality compounds. Today's powerful magnetic resonance imaging technology for medical diagnoses and the omnipotent accelerator technology for particle physics research would not have been possible without superconductors, although having a low T_c. In the process, many new non-superconducting compounds were also discovered, as were the many new associated physical phenomena, resulting in the development of new physics and theories. Itinerant ferromagnetism in $ZrZn_2$ [1], the charge-density waves in the layered transition metal dichalcogenides [2], and the recently discovered Majorona fermion in insulator-Nb structures [3] are just a few of the examples. The discovery of the 30-K high-temperature superconducting (HTSg) Ba-doped La_2CuO_4 compound in 1986 [4] and the subsequent discovery of the first liquid nitrogen superconductor, in the Y-Ba-Cu-O system with T_c above 93 K in 1987 [5], did not end the search for novel superconductors nor did it lessen the importance of such a search. This is borne out by the discovery of many new HTSg compounds in the ensuing three decades [6, 7] leading to the rapid rise of T_c (Figure D4.1); the discoveries of several unconventional superconducting systems [7]; the development of powerful tools to prepare the high-temperature superconductors (HTSrs) and to probe the origin of high-temperature superconductivity (HTSy) [8]; the revelation of the unusual HTSg properties [9]; the proposition of various theoretical models [10]; and the development of HTSg prototype devices [11]. The great resurgence of interest in the scientifically significant and technologically important colossal magneto-resistance compounds [12], as well as the topological insulators and superconductors, which are scientifically intriguing and hold technological potential [13], are yet more examples of the collateral effects of HTSg material study. In spite of this impressive progress, no commonly accepted microscopic theory of HTSy exists and viable commercialization of HTSrs remains beyond our reach.

We believe that the impediment can be overcome to a large extent by the attainment of HTSrs with prescribed properties to be discussed later. They consist of existing compounds with modified and improved characteristics, as well as new ones yet to be discovered. Such future HTSrs will help provide answers to the existing scientific questions and expedite the commercialization of HTSg devices. As higher T_cs above the present record of 134 K at ambient [14] and 164 K under pressure [15] are attained in the stable cuprates and other stable compound systems, and the reported T_cs above 200 K in unstable superhydrides—203 K in the H-S system under 155 GPa [16], 260 K in LaH_{10} under 190 GPa [17], and up to 287 K in C-H-S under 267 GPa [18]—are stabilized and reproduced, new challenges for scientists and new promises for technologists will undoubtedly arise. The recent progress in hydrides mentioned above has heralded a new era of room-temperature superconductivity (RTS) with immense technological promise. Indeed, RTS has lifted the temperature barrier for the ubiquitous application of superconductivity. Unfortunately, formidable pressure is required to attain such high T_{cs}. The most effective relief to this impasse is to remove the pressure needed while retaining the pressure-induced T_c without pressure. Recently our group at Houston indeed showed such a possibility in the pure and doped HTSr FeSe by retaining, at ambient via pressure-quenching (PQ), its T_c up to 37 K (quadrupling that of pristine FeSe) and other pressure-induced phases, offering a possible relief to such an impasse. We strongly believe that the PQ technique developed can be adapted to the RTS hydrides and other materials of value with minimal effort. In this chapter, we shall summarize the material issues relevant to the present scientific and technological challenges, describe the prescribed

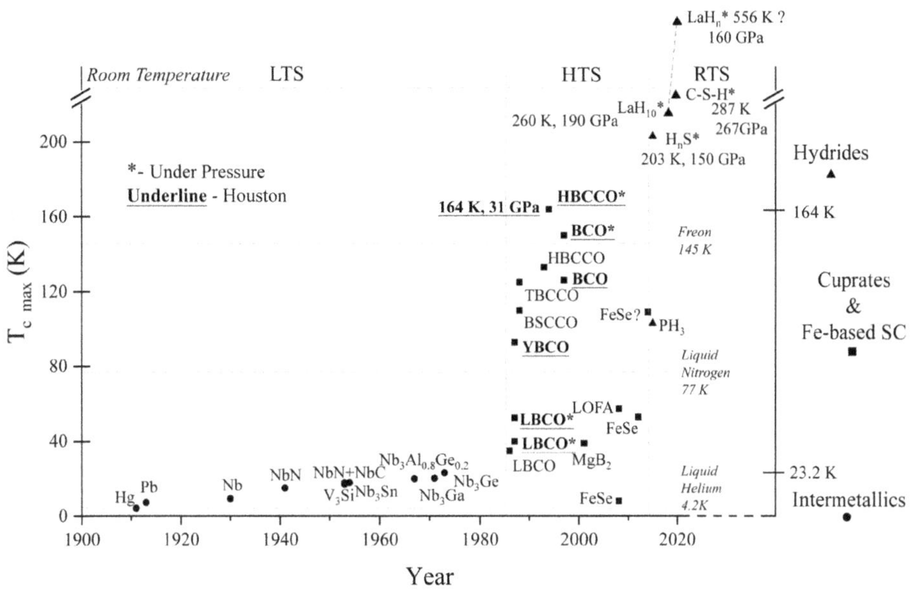

FIGURE D4.1 The evolution of T_c with time, i.e. from low-temperature superconductivity (LTS) to high-temperature superconductivity (HTS) and to very high-temperature superconductivity (VHTS) or even room temperature superconductivity (RTS).

properties of the future HTSrs, and propose possible ways to attain such future HTSrs.

In general, any means proposed by a practicing experimentalist to achieve anything with a desired outcome should be taken with a grain of salt. This is because if what is proposed is so promising for producing the great result expected, it would have been done by the proposer, without giving others a chance to try. However, the field of HTSy is so vast and encompassing, no single person or group has the energy and resources to blanket-try all of the methods that are conceived even just by one person or one group. We would also like to note that all approaches mentioned below do include some that have been explored by us but none that have been tested not to work. The only limitation of the possible approaches to future HTSrs lies in one's imagination and understanding of the complex HTSg materials and their relation to science and technology, as well as one's ingenuity to obtain the necessary data.

D4.2 Materials Issues

D4.2.1 Scientific Consideration

D4.2.1.1 Cuprates

As always in the study of solids, the successful development of theories depends critically on the attainment of the intrinsic properties of solids. While making a sample that displays a high T_c after its discovery is relatively effortless, preparing a HTSg sample in a state such that the intrinsic properties of the compound can be obtained unambiguously is a challenge. This is because HTSg cuprates and the newly discovered Fe-pnictides and -chalcogenides are physically intricate, chemically complex, and chemically unstable, not to mention the existence of many almost degenerate ground states in these HTSg compounds. Although impressive progress has been made in synthesizing compounds of high quality and in obtaining data good enough to address some of the scientific issues, other issues remain to be resolved and some of these are rather fundamental. To differentiate the central from the trivial observations becomes even more important as more and more unusual features of HTSrs are uncovered due to the rapid advancement of the characterization tools.

All known bulk HTSrs with a T_c above 77 K, the liquid nitrogen boiling point, are perovskite-like cuprates except for the recently reported unit-cell FeSe ultrathin film on a STO substrate [19]. However, the Fe-chalcogenides and their isostructural Fe-pnictide families have formed a nice platform to help reveal the role of magnetism in HTSy because of the presence of a large amount of the magnetic Fe-element and their relatively simple structure. Magnetism has been proposed by many to be a crucial ingredient in the occurrence of HTSy in the cuprates [10]. This chapter will discuss both systems but focusing mostly on cuprates. The cuprates can be represented by the generic formula $A_mE_2R_{n-1}Cu_nO_{2n+m+2}$ and designated as A-$m2(n$-$1)n$, where A, E, and R are cations often with A = Y, rare-earth element, Bi, Tl, or Hg; E = Ba, Sr, or Cu; and R = Ca or a rare-earth element. For some cases, R can be replaced by a (RO)-layer or by a more complex oxide slab. When A is absent, they are designated as $02(n$-$1)n$. These HTSg cuprates possess a layered structure (Figure D4.2) [20] consisting of n CuO_2-layers per unit formula separated by n-1 R-layers, denoted by $\{(CuO_2)[R(CuO_2)]_{n-1}\}$, known as the active block (AB), and m AO-layers sandwiched between two EO-layers, denoted by $[(EO)(AO)_m(EO)]$, known as the charge reservoir block (CRB).

Superconductivity is induced in these compounds from their insulating state by chemical, pressure [21], photon [22], or electric field [23] doping. The chemical approach consists of

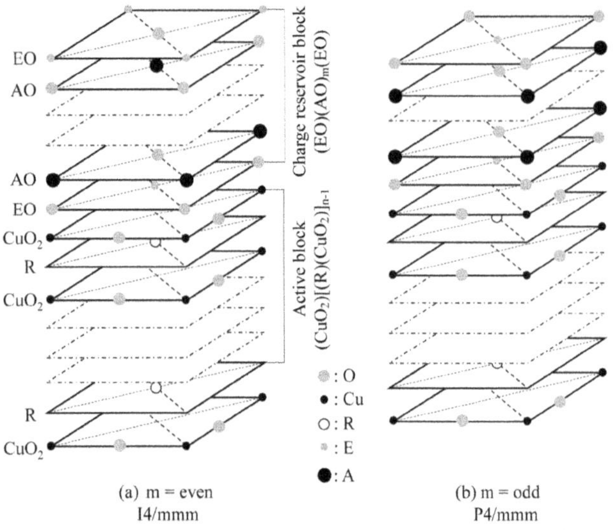

FIGURE D4.2 The schematic layered structure of cuprate HTSrs, $A_mE_2R_{n-1}Cu_nO_{2n+m+2}$ [A-$m2(n$-1)n or -02(n-1)n when m = 0], for m = even (a) and odd (b).

anion doping, cation doping, or both. Such chemical dopings are commonly carried out in the CRB. However, cation doping can also be performed in the AB by partial replacement of R by other cations of different valences. As a result, charge carriers are added to or removed from the AB without introducing defects to the CuO_2-layers in the CRB, which plays a dominant role in HTSy, similar to modulation doping in semiconductor superlattice systems. Recently, it has been demonstrated that an externally applied electric field [24, 25] can also help extend the doping range of a compound either reversibly or irreversibly upon the removal of the field. A generic phase diagram has been constructed based on the results of extensive studies on HTSrs of different dopings [26], as shown in Figure D4.3.

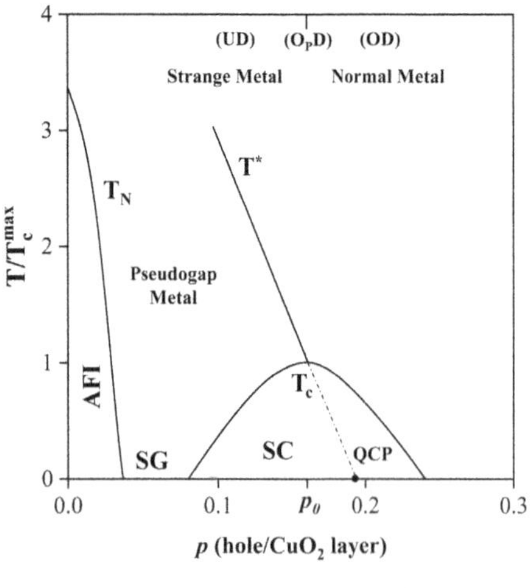

FIGURE D4.3 The generic phase diagram of cuprate HTSrs: UD - underdoped; OPD - optimally doped; OD - overdoped; AFI – antiferromagnetic insulating; SG - spin glass; SC – superconducting; and QCP - quantum critical point.

The normalized transition temperature (T_c/T_c^{max}) of a compound varies parabolically with the hole concentration (p) in a universal manner, i.e. $(T_c/T_c^{max}) = [1 - 82.6(p - p_o)^2]$, where T_c^{max} is the maximum T_c of the compound when optimally doped at $p = p_o \sim 0.16$ hole/CuO_2-layer. The compound is underdoped when $p < p_o$ and overdoped when $p > p_o$. The underdoped compound but not the overdoped one exhibits anomalies in a wide range of properties [27], which include dc electrical, dc magnetic, optical, nuclear magnetic resonance, thermoelectric power, specific heat, photoemission, neutron scattering, and tunneling, at a characteristic temperature $T_s > T_c$. They suggest the appearance of a so-called pseudogap. The underdoped compound is therefore called "strange" metal whereas the overdoped compound is called "normal" metal. HTSrs also show a rather complex behavior between the upper critical field line $H_{c2}(T)$ and the lower critical field $H_{c1}(T)$, represented by the melting line H_m(T) or the irreversibility line $H_i(T)$ in the magnetic phase diagram [28], as displayed in Figure D4.4. Almost all proposed models of HTSy have been constructed mainly based on the results summarized in Figures D4.2, D4.3, and D4.4 for HTSs in general and cuprates in particular. Unfortunately, the uncertain state of the samples raises questions concerning the interpretation of some of the results, and it is not known whether a cuprate is necessary for HTSy. The recent report of 109 K in unit-cell FeSe film on a STO substrate [19(d)], if reproduced and confirmed, will demonstrate that HTSy need not be synonymous with cuprates.

As mentioned earlier, the HTSg cuprates display a layered structure and, thus, highly anisotropic properties. The electrical resistivity perpendicular to the CuO_2-plane can be as large as ~ 10^6 times that along the plane [29], showing the need for structurally perfect single crystals for some critical experiments. The oxygen diffusion coefficient along the CuO_2-layer can therefore be ~ 10^6 times as large as that perpendicular to the plane [30], suggesting that it is extremely difficult to achieve single-crystalline samples of size with a

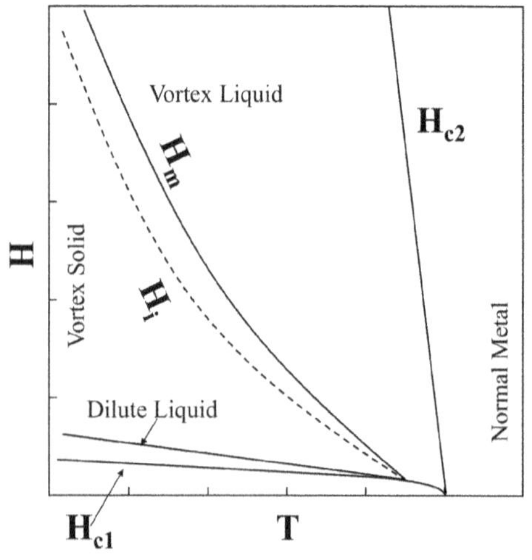

FIGURE D4.4 The schematic magnetic phase diagram of cuprate HTSrs.

uniform oxygen distribution, especially upon O-removal or O-addition after the crystal is formed. Therefore, HTSg samples may be more-often-than-not inhomogeneous electrically and magnetically, since the electric and magnetic properties of the compounds depend sensitively on the oxygen content or doping. Further complications are the observations of phase separation [31] and a stripe phase [32] in the underdoped HTSg cuprates of both static and dynamic nature, not to mention the appearance of the charge-density waves state [33], all suggesting a possibly intrinsic inhomogeneous nature of the HTSg samples examined. Against the above issues of HTSg materials, the great majority of the HTSy models proposed to date have been built on a homogeneous 2D electronic system of the CuO_2-layers in the HTSrs.

While many of the experiments can be made on very small samples due to the impressive advancement in the characterization techniques, some require large defect-free samples. For example, neutron studies require a large single crystal to provide enough signal strength for the determination of the charge and spin excitation energy spectra, which are crucial for testing and developing the theory. Unfortunately, due to the complex chemistry, high-quality single crystals of large sizes have only been prepared for limited HTSg cuprate systems [34], such as $La_{1-x}Sr_xCuO_4$ and $YBa_2Cu_3O_7$. Fortunately, recent developments in angle-resolved photoemission spectroscopy (ARPES) using laser [35] or synchrotron sources [8] render large samples unnecessary for the determination of the electronic structures of solids.

D4.2.1.2 Fe-Based Superconductors

The discovery [36] of Fe-based superconductors up to 56 K since 2008 has generated abundant excitement for theoretical and practical reasons. The presence of Fe, which was previously believed to be antithetical to superconductivity, let alone HTSy, can provide a unique opportunity for examining the role of magnetism in HTSy. Furthermore, the existence of a large number of other pnictides and chalcogenides of the same structural type may offer a new path to even higher T_c superconductivity. Many Fe-based superconductors have been discovered in the ensuing years, which mainly include the 1111 phase (*R*FeAsO, where *R* = rare earth) [36]), the 122 phase ($AeFe_2As_2$, AFe_2As_2, and $A_{1-x}Fe_{2-y}Se_2$ [37], where *Ae* = alkaline earth and *A* = alkaline metal), the 111 phase (*A*FeAs, where *A* = alkaline metal) [38]), and the 11 phase (FeSe) [39]. The commonly existing single layer of tetragonally coordinated $(FeAs)_2$ in these compounds is generally believed to be the charge carrier block, whereas the *R*O-, *A*-, or *Ae* layers are the charge reservoir block (CRB). By modifying the CRB through intercalation and intergrowth, more superconducting compounds were discovered. But different from cuprates, the active charge layers remain as the single layer of $(FeAs)_2$-slab, which may hinder the further T_c increase in Fe-based superconductors by building more complex structures as was done in cuprates, except for the interface-enhanced scenario observed in the FeSe/STO system [19].

Despite the fact that the parent compounds of Fe-based superconductors are found to be poor metals with multiband nature and spin density wave/spin fluctuations, rather than Mott insulators as in the cuprates, the phase diagram of the Fe-based superconductors is strikingly similar to those of several other classes of unconventional superconductors, including the cuprates, organics, and heavy-fermion superconductors. Most Fe-based superconductors are readily available in single-crystalline form, which helps detailed characterizations [40] through e.g. ARPES, scanning tunneling and nuclear magnetic resonance microscopy, neutron, muon, etc. to elucidate the complicated band structure features and understanding of the pairing mechanism. Although the exact pairing mechanism for Fe-based superconductors remains unclear ($s^{\pm}$ or s^{++}, spin and/or orbital fluctuation-mediated pairing) at this moment, it is apparently quite different from the d-wave pairing accepted in cuprates.

Another distinct feature setting Fe-based superconductors apart from cuprate superconductors is the ability to substitute or dope directly into the active pairing layer [41]. Taking Ba122 as an example, superconductivity can be induced not only through charge reservoir layer doping (Ba with K, Rb, Cs, and La), but also through active layer doping (at the Fe site with various transition metal doping such as Co, Ni, Rh, Pt, Pd, or at As site with P). Pressure is also found to be quite useful to induce superconductivity in the parent compounds, which is different from the cuprates, where no superconductivity so far has been induced through applying pressure to parent compounds. The superconductivity is found to be sensitive to applied pressure media or strains caused by nonhydrostatic pressure as well [42].

D4.2.1.3 Hydrides

The metallization of the hydrogen-rich molecular solids under ultrahigh pressure is known to be the prerequisite for superconductivity. The rapid development of the high-pressure diamond anvil cell by David Mao and Peter Bell since the 1970s has provided the means to attain the required pressure and the calibrated ultrahigh pressure scale. As a result, Mikhail Eremets' announcement in 2015 of superconductivity up to 203 K in H_3S at ~ 155 GPa heralded in the new era of very-high-temperature superconductivity in hydrogen-rich molecular solids. In 2017, Hemley and colleagues examined and predicted the structural stability of high-pressure phases of lanthanum hydrides with strong electron-phonon interaction for a T_c of 274-286 K for LaH_{10} at 210 GPa. In 2019, Hemley and colleagues successfully synthesized LaH10 inside the diamond anvil cell through laser heating and achieved a T_c of 260 K under 180-200 GPa, very close to room temperature, opening up the era of room-temperature superconductivity. By continued thermal annealing, they reported a trace of superconductivity up to 556 K in the lanthanide "super hydrides" at 160 GPa, heralding in a new era of "hot hydride superconductivity" as coined in their recent 2020 article. Later, Dias *et al.* reported the 288-K superconductivity at 267 GPa in a C-S-H compound formed under pressure by laser heating CH_4 and H_2S in a diamond anvil cell. The era of the room-temperature superconductor is thus born and it holds immense promise for technologists while offering great challenges for scientists.

To fully realize the immense technological potential and to meet the great scientific challenges of the room-temperature superconductivity of the hydrogen-rich molecular compounds, the exact nature of the ultrahigh-pressure-induced phase transitions must be understood and possible ways to retain this ultrahigh-pressure phase at ambient need to be explored. The primary challenges of the present experiments, in sample synthesis, handling, and characterization, are extraordinary, mainly arising from the very small sample size on the μm scale and the ultrahigh-pressure environment above 100 GPa. As a result, there are only a few teams in laboratories around the world who can carry out these types of experiments with rigor. Reports of superconductivity in hydrogen-rich molecular solids to date have been based on the agreement of experiment with theory and the detection of all or most of the following: a drastic drop in resistance to zero, suppression of the transition by magnetic field, and a diamagnetic shift in magnetic susceptibility and the isotope effect, all of which are expected of a standard superconductor. Given the significance of the reports and the constraints in the experimental conditions, the following points may be worthy of further study [43]:

- The existence of a pressure-induced insulator-metal (I-M) transition at a temperature T_m above T_c – this is important for exacting information about the normal-state behavior of the superconductor.
- The achievement of a real "zero-resistance" state – it is extremely challenging to determine the "zero-resistance" or more so the "zero-resistivity" state due to the small size of the sample and the large change of resistance at the transition.
- The field effect on the transition – a similar effect has been observed for the not-superconducting I-M transition.
- The isotope effect on the transition – a similar effect has been observed for the not- superconducting I-M transition.
- The diamagnetic shift in ac magnetic susceptibility – a similar effect has been detected in a temperature region where resistance changes greatly due to the eddy current.
- Only the detection of the true Meissner effect (in the field-cooled mode) can clarify the above confusion, although the possible inherent defects in the sample under pressure may make such a test difficult. However, the sharpness of the transition may imply that the defects in the sample are small.
- The very sharp transition and the almost downward shift of the transition in the presence of field suggest that the absence of flux flow is very puzzling for a type-II superconductor with such high a T_c.
- The absence of a systematic experiment on the same individual sample for different types of measurements makes difficult the judgement of the reproducibility of the experiment – this is especially critical in determining the isotope effect.
- The exact role of hydrogen in the samples investigated appears not to be clear – for instance, the role of B in the B-rich superconducting ZrB_{12} is rather limited.
- The retention at ambient without pressure of the ultra-high-pressure-induced room-temperature superconducting phase in these hydrogen-rich molecular solids should be the most exciting and rewarding endeavor in superconductivity science and technology research and development. Recent preliminary work done by our group at Houston on several superconducting elements and compounds has demonstrated such a possibility.

D4.2.2 Technological Consideration

The availability of high-performance HTSg materials and devices that can be prepared in large quantities at low cost is extremely important to the successful commercialization of HTSrs. While the intrinsic properties of the HTSg materials are crucial for scientific pursuit, the performance of the HTSg materials is paramount for applications and is also dictated by additional factors, such as processing and material modification. As in scientific studies, the unusual physical and chemical complexity of HTSg materials pose great challenges to meeting the stringent material and cost requirements for commercially viable devices.

In a cryogenic environment, the higher the operating temperature is, the greater the efficiency of the device and the lower its operational cost will be. The discovery of HTSg compounds has made applications more practical [44]; i.e. one may use either the inexpensive liquid nitrogen as a coolant or a single-stage cryocooler to achieve the superconducting state of the device for its operation. Materials with a T_c at ambient pressure higher than the present record are always desirable, so more readily available cooling techniques or coolants such as air conditioning or dry ice can be used. Ultimately, one would like to be able to operate a superconducting device at room temperature without coolant. Equally important is that the operating temperature for a large current superconducting device usually is substantially lower than its T_c. This is because the critical current density of a superconductor increases as the operating temperature decreases, e.g. to about 90% of its maximum value at about 70% of its T_c. Therefore, a superconductor with a $T_c \geq 110$ K is required in order to be able to take 90% of the full advantage of a large current superconducting device when it is operated at the liquid nitrogen boiling point of 77 K.

The coherence length (ξ) of a superconductor is a measure of the extent to which the superconducting Cooper pair electrons interact. It is proportional to v_F/T_c, where v_F is the Fermi velocity and increases with carrier concentration (p).

For HTSrs, p and T_c are about 100 times smaller and about 10 times higher than those of the LTSrs [45], respectively. ξ of the HTSrs is therefore much smaller than that of the LTSrs, rendering the HTSrs much less forgiving to retain their superior superconducting characteristics throughout the material. In other words, superconductivity cannot extend over defects of dimensions greater than $\xi \sim 10$ Å in these compounds. A chemically pure material is therefore needed. Unfortunately, impurity phases often appear in the materials due to the complex chemistry in the formation of most HTSrs. Some tedious and cumbersome processing steps have been developed to minimize impurities and defects, adding to the material processing cost. One way to alleviate the processing burden, perhaps, is to lower its operating temperature, i.e. to optimize its performance by adjusting the material and operating environment parameters. Another way is to look for new materials with specific favorable processing features. For instance, Fe(Se,Te) has been demonstrated [46] to show attractive large critical current density in a strong magnetic field. Although the T_c is relatively low, the ease of fabrication due to its small anisotropy and small intergrain scattering makes the compound a potential candidate for high-field magnets operated below 4 K.

Since HTSg cuprates have a layered structure, the superconducting critical current density along the CuO_2-plane is much greater than that perpendicular to the plane [47]. The HTSg materials for devices have to be fabricated to atomic scale perfection. In other words, all HTSg devices, large or small, must be made from materials with all of the CuO_2-layers aligned perfectly, i.e. in single-crystalline form, throughout the devices in order to achieve their best performance. Various film techniques [48] and the so-called melt-texturing methods [49] have been developed to fabricate HTSg materials in thin film, thick film, wire, and bulk forms with the needed atomic alignment. An overdoped HTSr layer has been reported [50] to be able to provide relief to such an impasse of atomic misalignment. If proven to be effective, long lengths of polycrystalline HTSg wires will become a reality, reducing the processing cost immensely. HTSg materials are intrinsically brittle. Except in their bulk form, they require a substrate to provide the physical support needed and sometimes also to act as the template for the epitaxial growth of thin or thick films. All processing techniques are tedious and time-consuming because of the complicated chemical phase diagrams of HTSg compound formation [51]. Methods for melt-texturing wires and bulk, and sometimes thick films, are carried out at high temperatures, i.e. near and above their peritectic temperature of ~ 1000°C, at which point the compound decomposes, becoming a mixture of liquid and solid phases. The decomposed HTSrs are extremely corrosive at this high processing temperature. Often only pure and slightly doped Ag and Au have been found to be compatible with the HTSg cuprates at the processing temperature. Ag and Au have the added benefit of being permeable to O_2. Presently, pure and slightly doped Ag are therefore the most used sheath materials for HTSg wires or tapes, which are prepared by the well-developed powder-in-tube [52] or chemical vapor deposition [53] techniques. Ni substrates coated with a buffer layer or processed with preferred grain orientation have also been used with some success [54].

HTSg cuprates are known to be type II superconductors, where magnetic fluxoids exist in their superconducting state when the external field exceeds the lower critical field. In the presence of an electrical current, the fluxoid moves under the influence of the Lorentz force due to the interaction of the fluxoid with the current, resulting in energy dissipation. A pinning force is needed to prohibit the movement of the magnetic fluxoid and thus to prevent the energy dissipation. Defects of a dimension ~ ξ are introduced to provide the pinning force. Because of the small ξ of the HTSrs, the conventional metallurgical method of impurity phase precipitation, used to create the pinning centers in LTSrs, does not work well for HTSrs. The weak interlayer coupling reduces the effectiveness of point defects to pin the fluxoids. Columnar defects induced by heavy ion bombardment or local nuclear fission have been shown to enhance the pinning greatly [55]. Unfortunately, residual radioactivity and cost are serious obstacles to their practical employment.

The commercial viability of a device is intimately related to the lifetime of the device. For instance, the lifetime of a power transmission line must exceed three decades before a utility company will even consider it for possible deployment. The lifetime of a device depends on the chemical stability and physical integrity of the device material in the operating environment. HTSg materials are known to be chemically unstable and physically fragile. The HTSg cuprates degrade in air, stemming from the formation of carbonates of some constituents in the compounds when exposed to air. The degradation usually initiates at defects in the sample caused by the mismatch between layers of the compound. Passivating the HTSr surface and coating the HTSr with a protective layer have been two methods proposed to prohibit the degradation of the HTSg material. Atomic size matching has also been adopted [56] to reduce the internal stress and thus to improve the chemical stability of $YBa_2Cu_3O_7$. Since HTSy has been in existence for about 30 years, it has yet to be determined if HTSg materials can last longer than 30 years.

D4.3 Future HTSrs

According to the above discussion, future HTSrs should possess all or some of the following desirable characteristics to enhance our understanding of HTSy science and to expedite the commercialization of HTSg devices:

- *higher T_c – to facilitate the testing of the limit of some existing theoretical models and to make applications more practical*

The previous ceiling in the theoretical T_c of 30 K was removed following the discovery of HTSy. Although the prediction of T_c is beyond the reach of current theories, some models do suggest that the T_c achieved to date is far too low [57]. It has been reasoned [27] that the partially filled 3d-shell of Cu in the HTSg cuprates may place them in the class of transition metal oxides. Transition metal oxides display various electronic phase transitions at temperatures exceeding 300 K, driven by the strong electron–electron correlations and strong electron–phonon interactions. If HTSy arises from similar interactions as suggested by these models, a T_c much higher than the present record appears not to be impossible. To obtain a T_c in cuprate and related materials much higher than the present record will therefore allow one to determine the validity and test the limit of current models.

A higher T_c enables one to operate a HTSg device at a higher temperature and thus more efficiently. A HTSg device with a higher T_c also has a greater safety margin to avoid accidental quenching than its lower T_c counterpart when operated at the same temperature, due to its higher heat capacity [58]. The ultimate goal is to attain a T_c above room temperature so that no cryogen or cryocooler is needed. When this is accomplished, HTSrs may be put on an equal footing with semiconductors in terms of some consumer electronic and other large current applications.

While higher T_c is the major goal in HTSy research, it should be mentioned that the rigidity of a superconducting state to support the superconducting current is inversely related to the T_c [59]. Enhancing the T_c while maintaining the critical current-carrying capacity of a HTSr by reducing the anisotropy and introducing flux pinning of the material is important for a practical HTSr.

- *reduced anisotropy – to provide a material system to test the current models based on 2D systems and to simplify the preparation of quality materials and high-performance devices*

Cuprate HTSrs have a layered structure and display a large anisotropy in many physical properties. The significance of the CuO_2-layers in these cuprates cannot be overstated in the current theoretical models of HTSy [60]. They have two salient features, namely, the 2D-nature and the unique square-planar atomic arrangement of Cu and O. It is known that in a strictly 2D system, fluctuations will prevent any long-range order, and thus a thermodynamic phase transition, from taking place. Interlayer coupling between the CuO_2-layers or interblock coupling between the ABs has to play a role in facilitating the occurrence of the bulk superconducting transition observed in cuprates. The question is to what extent the coupling affects HTSy. Studying HTSg compounds with different anisotropies under various pressures will help answer this question.

The large anisotropy of the existing HTSg materials puts serious constraints on material processing and device fabrication, since HTSg materials must be atomically aligned or in single-crystalline form for devices. A material with a reduced anisotropy will relax such a stringent requirement of atomic alignment for devices and thus greatly simplify the material processing and device fabrication. MgB_2 [61] and Fe(Se,Te) [46] appear to be two such examples, although with low T_cs. An additional benefit of such a material with a reduced anisotropy will be the improvement in flux pinning [46] and hence the device performance.

- *non-cuprates or non-oxides – to determine if cuprate (or Fe-pnictide or -chalcogenide) is synonymous with HTSy and to make the development of HTSr/semiconductor hybrid devices easier*

Presently, all HTSrs with a T_c above the record 23 K for the conventional LTSrs are limited to layered cuprates, Fe-pnictides and -chalcogenides, MgB_2, and fullerites [7, 61]. The great majority of the models proposed for HTSrs above 77 K has focused on the charge and spin dynamics associated with the Cu- and O-ions in the 2D CuO_2-layers of the cuprates [62] and the specific band structures characteristic of the unit-cell layer FeSe on a STO substrate [63]. However, the fundamental question, whether properly doped CuO_2-layers or unit (FeSe)-layers on STO are necessary or sufficient for HTSy to occur, remains open. An answer will be found by the discovery of HTSy in a non-cuprate layered compound, a 3D oxide, a 3D chalcogenide, or, better yet, a non-oxide or a non-chalcogenide/STO film or bulk, enabling the development of a comprehensive microscopic theory of HTSy. The recently reported unit-cell FeSe/STO with a T_c above 45 K [19] and the field effect superconductivity (FES) up to 47 K in non-superconducting ZrNCl [23], MoS_2 [24], and FeSe [25] may provide an avenue to HTSr/semiconductor hybrid three-terminal devices. As mentioned earlier, an applied electric field may induce superconductivity irreversibly or reversibly upon the removal of the field. We call the former electro-chemical doping (in most reported cases) and the latter, possible electro-physical doping. For the success of electro-physical doping (but not electro-chemical), a long-sought-after three-terminal FES (field-effect-superconducting) device, the equivalent of the FET (field-effect-transistor), may thus be made. However, given the nature of electro-doping, where an electrolyte, be it solid or liquid, is always used, to reduce the switching time may be a formidable challenge.

While HTSg devices without a non-superconducting substitute, such as the superconducting quantum interference device, do exist, most conceived or developed so far are complementary to or replacements of non-superconducting ones with better performance [64]. The development of HTSr/semiconductor hybrid devices will accelerate the impact of HTSy, and many such devices have been proposed [65]. This is because, in the last few decades, semiconductor technology has continued to advance in sophistication and to demonstrate its profound impact on our modern technological world with no peer. Unfortunately, with the exception of unit-cell FeSe/STO, a major hurdle to developing HTSr/semiconductor hybrid devices exists due to the mismatch between the preparation environments associated with the distinct characteristics of these two classes of materials, i.e. oxidizing for HTSrs but reducing for semiconductors and a relatively low processing temperature for HTSrs but high for semiconductors. The discovery of non-cuprate and non-oxide HTSrs, in addition to unit cell FeSe/STO film, will alleviate the impasse. The pursuit of oxide-HTS/oxide-semiconductor hybrid three-terminal devices is definitely worthwhile.

- *chemical homogeneity and stability and structural perfection – to make the easy acquisition and correct interpretation of experimental data possible and to improve the performance and lifetime of the devices*

HTSg materials have a complex chemistry and a layered structure. As a result, impurity phases, albeit minute, are often present, and intralayer breakage, interlayer linkage, and nonuniform oxygen distribution frequently occur. The impurity and the oxygen-deficient phases often generate magnetic signals, which can be confused with the intrinsic magnetism assigned to the CuO_2-layers as the origin of HTSy by many models [60]. The cross-layer linkage in a sample with an imperfect structure can confound the intralayer properties with the very different interlayer properties. The attainment of the intrinsic properties of the HTSg materials is indispensable for the development of a comprehensive microscopic HTSy theory. This requires pure and uniform samples with perfect structures.

The presence of impurity and structurally defected phases of dimensions greater than ξ limits the current-carrying capacity and thus degrades the performance of the HTSg material. Since many of the impurity phases easily form carbonates and/or hydroxides and thus act as the degradation centers of the HTSg material, they shorten the lifetime of the HTSg devices. The loss of oxygen and the relatively high chemical reactivity of some HTSg compounds, especially in their imperfect form, can further reduce the lifetime and performance of the HTSg devices. Therefore, chemically stable and structurally perfect HTSg materials are needed to manufacture commercially viable HTSg devices. A protective thin coating will help.

The above challenges may be reduced for the Fe-based superconductors in bulk with a T_c much lower than 77 K or in unit-cell thin film covered by a protective layer. Unfortunately, the preparation of the unit-cell FeSe/STO ultrathin film requires the deployment of the sophisticated molecular beam epitaxy (MBE) technique, making it less appealing.

- *high flux pinning force – to enhance the superconducting current-carrying capacity and the power-handling ability of HTSg devices*

Since cuprates are type II superconductors, magnetic flux lines exist in HTSrs in their superconducting state in a field above their lower critical fields. Any movement of these flux lines in the presence of an electric current will result in an energy loss. A strong pinning force has to be incorporated into the HTSg materials, e.g. by irradiation [55], to prevent the magnetic flux lines from rattling or moving, and thus to enhance the current-carrying capability and power-handling ability with no energy loss.

- *other factors – to reduce costs of material processing and device manufacturing, which is paramount for commercialization*

Several prototype HTSg devices have been constructed and have shown properties superior to those of their non-superconducting counterparts [11]. Unfortunately, a viable commercial product is yet to be identified. This stems from the high costs associated with HTSg material processing and device manufacturing. Either modified or novel HTSg materials will help reduce the costs of the devices if they allow the use of simplified material processing steps and if they are more forgiving in regard to their device manufacturing and operating environments. The 40-K superconductor MgB_2 [60] or the 15-K $FeSe_{0.5}Te_{0.5}$ [46] with a small intergrain effect on its transport current are two such examples.

D4.4 Possible Approaches to Future HTSrs

Future HTSrs, known HTSrs that will be modified as well as new HTSrs that will be discovered, should be materials with prescribed properties that will help advance HTSy science and expedite HTSy commercialization. We shall group the possible approaches in terms of the specific desired characteristics of the future materials.

- *higher T_c – by modifying the layered cuprates and discovering new cuprate and non-cuprate compounds*

Experiments show that a superconductor in a material system of either the intermetallic LTSrs or the non-intermetallic HTSrs becomes unstable as its T_c rises [66]. The instabilities are electron- and/or phonon-driven. In addition to the superconducting transition, the strong electron–electron correlation and/or the electron–phonon interaction generate a diversity of transitions such as charge-density waves, spin-density waves, magnetism, charge-order, metal-insulator, structure, or even the collapse of a crystal structure. In other words, the superconducting state must always compete against many other possible ground states, and the competition becomes more severe as the T_c gets higher. For instance, a large electron–phonon interaction enhances the T_c of LTSrs based on the BCS theory, and a strong electron–electron correlation promotes the T_c of HTSrs according to some current HTSy models. However, at the same time, a strong electron–phonon interaction can trigger the formation of charge-density waves or a structural transition, and a large electron–electron correlation can promote the formation of spin-density waves or a magnetic transition. Therefore, the best approach to achieve a higher T_c is to optimize the overall interactions for superconductivity instead of maximizing only certain interactions. One may apply the same rule to take advantage of the earlier suggestion that a much higher

T_c may not be impossible in view of the many high transition temperatures in various transition metal oxides to their respective ground states [20, 27].

It appears that the T_c of the layered cuprates can be further enhanced if the instabilities associated with the Coulomb repulsion are overcome. Immediately after the discovery of the 93-K YBCO, it was proposed that T_c could be enhanced by increasing the number of CuO_2-layers per unit cell (*n*), based on a simple density-of-states consideration [5]. This is indeed the case for $n \leq 3$ or 4. However, T_c decreases for $n \geq 3$ or 4. This observation has been attributed [67] to the non-uniform charge distribution from one CuO_2-layer to the other in a unit cell, similar to the C-layers in the intercalated graphite. In other words, there are more holes in the outer layers than in the inner ones and, hence, the inner layers cannot effectively participate in the superconducting process. The suggestion can be understood in terms of the Coulomb repulsion between the excessive holes accumulated in the closely packed CuO_2-layers in the unit cell [66]. This is consistent with observations [68] of increasing difficulties in forming compounds or in optimally doping the compounds with $n > 3$ or 4. In fact, the cuprates with $n > 3$ reported are always found to be underdoped, i.e. the average $p < 0.16$, and their T_cs may yet be maximized by proper doping. To overcome the structural and doping instabilities, one has to develop strategies to dope the CuO_2-layers more uniformly from layer to layer as *n* becomes greater than 3 without degrading the CuO_2-layer integrity, while increasing the doping to the average optimal level. Two possible strategies are proposed, i.e. to dope by partial substitution of the *R*-layer inside the AB and to decrease the coupling between the CuO_2-layers. Preliminary data did show a T_c-enhancement of a few degrees by the former strategy. Later data indicated [69] that the doping effect on HTSy depends on the doping site chosen and therefore one more factor must be taken into consideration. High pressure also helps suppress structural instabilities in cuprates, as evidenced by the successful synthesis of many compounds with $n > 3$ [68, 70]. Various modes of doping besides chemical may prove to be fruitful.

All HTSg cuprates display low carrier density and their insulating parent can easily be turned into a superconductor with slight chemical doping. Therefore, soon after the discovery of YBCO, we attempted to induce superconductivity in the parent compound by an electric field *E* using the thin film capacitor arrangement but failed due to the breakdown of the defected thin STO and PZT film with a dielectric constant ϵ before *E* is large enough to generate the required surface charge density $\sigma = \epsilon E$. As discussed earlier, recent advancement in field-induced doping has extended the doping range where conventional chemical doping mentioned earlier fails by overcoming the electric field breakdown. For instance, superconductivity has been induced or enhanced in ZrNCl [23], MoS_2 [24], FeSe [25], and others by an external electric field when a liquid or solid electrolyte is used. The electric field–induced effects may be grouped into two categories depending on the electrolyte used: one in which the effect is reversible and disappears upon the removal of the applied field and the other in which the effect is irreversible and remains on the removal of the field. The former may be called electro-physical doping, and the latter electro-chemical doping. Electro-physical doping may involve only the motion of electrons and thus the process would be fast and good for switch devices, while electro-chemical doping involves motion of the ions and thus the process would have to be slow and not suitable for switches, but good for ordinary chemical doping with extended range. Electro-doping may thus avail us an extra handle to investigate the doping process in the parent compounds. We believe that electric field may remove the constraints that govern the universal quadratic rule of doping on T_c [26], as will pressure, as discussed below. The added challenge in these modes of doping is the small size of the samples used to make them effective, restricting them to small current devices but difficult for large current devices.

Pressure has played a crucial role in the development of HTSy due to its simplicity by varying the basic parameter of interatomic distance without introducing complications associated with altering the chemistry of the compound (assuming that there is no pressure-induced phase transition). It was high pressure that led to our discovery of YBCO and related HTSrs and to our attainment of the record high T_c of 164 K under 31 GPa in the stable Hg1223 (vs. the 287 K under 267 GPa in the unstable C-H-S system). Pressure on a compound can change the carrier concentration and shift the Fermi level and sometimes even alter its Fermi surface topology, leading to a Lifshitz transition. A thorough review of pressure effects on the T_c of cuprates by us [15, 71, 72] and on Fe-pnictides and -chalcogenides by us [42(c), 72] and others [42(e)] shows that a new record $T_c > 164$ K or even ~ 300 K is not just possible but probable, particularly in cuprates, where the constraint on the universal inverse quadratic (T_c/T_{cmax}) carrier per Cu relation may be lifted by pressure [68], especially when an electronic transition in the CuO_2-layers of the AB is induced [71, 74].

Cuprates with simpler CRBs, i.e. smaller *m*, and/or Ba-containing CRBs (vs. Sr-containing) appear to possess a higher T_c. This may be associated with the closer interblock coupling between the ABs and the higher atomic polarizability of Ba. Following this observation, a new interstitially doped $Ba_2Ca_{n-1}Cu_nO_z$ was successfully synthesized under high pressure and was found to display a T_c up to 126 K [75]. Unfortunately, these compounds appear to be less stable at ambient.

The search for materials conducive to novel superconducting mechanisms has long been considered the route to superconductors of higher T_c. In fact, it was the search for polaronic superconductors in perovskite oxides that led to the discovery of the first cuprate HTSrs in 1986 [4], although the superconducting mechanism in cuprates is yet to be determined. Many exotic non-phononic mechanisms have been suggested for both LTSrs and HTSrs. Unfortunately, little information has been reported concerning the material systems for novel mechanisms, the detection of such mechanisms, and the differentiation of such mechanisms from the conventional one. Since the necessary and sufficient conditions for the existence of superconductivity are zero resistivity and zero magnetic induction below T_c, neither is sensitive to the mechanism responsible for its existence. The appearance of an unusually high T_c has often been taken as evidence for the occurrence of a novel superconducting mechanism. Unfortunately, the history of superconductivity shows that this method is often short-lived, valid only until a modification is made on conventional theories.

One possible exception may be interfacial superconductivity, through which a high T_c may be obtained through the exchange of excitons of much higher characteristic energy between electrons, first proposed in 1964 by Ginzburg [76]. However, the first rigorous model analysis based on a metal/semiconductor interface did not take place until 1973 by Allender, Bray, and Bardeen [77]. Many experiments have been carried out to detect such interfacial enhancement or induction effect on superconductivity in the ensuing years. A nice comprehensive review article has appeared in 2011 [78], summarizing all efforts made until then. Unfortunately, clear and unambiguous evidence is still lacking. Recently, we succeeded in mixing two non-superconducting phases of $CaFe_2As_2$ reversibly through low-temperature annealing over different time periods and detected superconductivity at 25 K when the mixture of these two phases appears [79]. Careful XRD study and simulation confirm the existence of such mixtures. Further transmission electron microscopy investigation shows that the mixtures appear as nano-dispersions of dimensions between 8 and 20 nm [80]. The observations demonstrate clearly interfacial induction effect of superconductivity at high temperature, providing a new avenue to high T_c. Whether the effect is of electron–electron or electron–phonon nature is yet to be determined.

To exploit the path of novel superconducting mechanisms to a higher T_c, we may examine the following:

- unusual material systems such as immiscible alloys of metals and semiconductors, artificially made superlattices of different types of materials that do not mix, including oxides, artificially made heterogeneous material systems, inorganic/organic "composites," and organic compounds;
- low-T_c compounds that are supposed to have a high T_c;
- superconductors that are not supposed to be superconducting;
- compounds very close to the metal-insulator phase boundary;
- compounds with low dimensionality;
- compounds whose universal quadratic T_c – carrier dependence can be broken by electro-physical or electro-chemical doping;
- compounds whose inverse quadratic T_c – carrier dependence can be lifted by pressure;
- metastable compounds synthesized through novel chemical routes at ambient conditions or physical routes under extreme conditions such as high pressure and ultrafast quenching; and
- metastable phases of compounds that exist only under high pressure or a large pressure gradient coupled with high pressure.
- pressure-induced metastable superconducting phases with very high T_cs stabilized at ambient via pressure quenching at low temperature [CW Chu et al., submitted for publication]
- *broader material base –*
  by doping non-cuprate perovskite oxides and examining non-oxides

In the absence of any theoretical guidance, only a few general examples are chosen and discussed here. There exists a striking similarity between the cuprate HTSrs and perovskite (particularly layered non-cuprate) oxides in terms of their basic building blocks, such as the oxygen octahedra, pyramids, and planar squares. Soft-phonon modes associated with these building blocks are known to be the main source for ferroelectricity in perovskite oxides [81]. In the scheme of electron–phonon mechanism, soft-phonon modes favor high T_c. This seems to be consistent with the highest T_c observed at the metal-insulator phase boundary of the doped WO_3 [82] (but not for the cuprates). A diagrammatic analysis of the average valence electron numbers and the differential spectroscopic electronegativities of a wide range of compounds indicates [83] that HTSrs and ferroelectrics fall into two islands with a small overlapping region. This suggests that properly doping a ferroelectric, either electro-physically or electro-chemically, in addition to the conventional chemical doping, in this small overlapping region may result in HTSy. It will also be interesting to dope the layered ferroelectrics under high pressure or fast quenching condition, since conventional chemical doping under ambient has yet to be achieved without causing a structural collapse. It should be noted that superconductivity has been induced [82, 84] by electron doping the ferroelectric WO_3 and the incipient ferroelectric $SrTiO_3$ with their T_c below 6 K, and hole doping the spinel $LiTi_2O_4$ with a T_c up to 13 K [85]. Careful examination of samples near the metal-insulator phase boundary may be fruitful. However, no ferroelectricity has been detected in the insulating parents of the current cuprate HTSrs to date, except in a defected $(La,Ba)_2CuO_4$ [86]. Defect-induced electronic states may be an effective way to dope this class of stable materials [87].

HTSg compounds have been shown to be highly correlated electron systems. Such a system often displays many degenerate ground states. For example, known high-T_c superconductors exhibit multiple interactions, and magnetic interaction has been considered by many to play an important role. The question arises whether ferroelectric interaction can play a beneficial role for superconductivity as depicted in Figure D4.5. Given the high transition for ferroelectric or antiferroelectric ordering temperature, a beneficial effect on superconductivity similar to that of magnetism may not be impossible.

Multiferroics exhibit multiple competing interactions and display concurrently various ground states, e.g. the coexistence of magnetism and ferroelectricity. Studies by us and others show that ferroelectricity in multiferroics can be induced by magnetism, pressure, or an external magnetic field [88]. It has also been shown recently that superconductivity can coexist with ferromagnetism and that magnetic field can

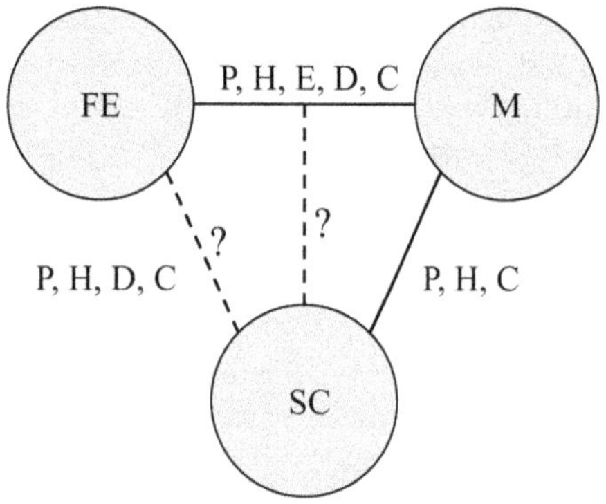

FIGURE D4.5 The multiple interactions in a highly correlated electron system, such as superconductivity (SC), magnetism (M), and ferroelectricity (FE). P – pressure; H – magnetic field; E – electrical field; D – defect; and C – chemical doping..

induce superconductivity [89]. This raises the prospect that superconductivity may evolve from ferroelectricity directly or indirectly through magnetism. Since multiferroics belong to the class of highly correlated electron material systems that possess transition temperatures up to or above room temperature, it is conjectured that it may not be impossible to achieve superconductivity with a very high T_c by optimizing the various interactions present. The first order of business is to metallize the multiferroics. The conventional chemical doping approach fails because of the high stability of the chemical bonds in the compound. We therefore have decided that the most effective way is through high pressure, defect induction, electro-doping, or a combination of the above. Unfortunately, no detection of superconductivity has been achieved by us to date. We have attributed the failure in part to the Mn element that often appears in the multiferroics, since Mn is a killer of superconductivity (to the best of our knowledge there exists only two low-T_c superconductors recently discovered [90]). We are now looking at multiferroics without Mn. Preliminary results appear encouraging.

Double perovskites $A'A''B'B''O_6$ that show three distinct B-cation arrangements [91] may offer another avenue to high T_c. They are arranged in random, rock salt, or layered form, depending on the ionic size of the cations and the charge difference between A' and A or between B' and B. They form a large class of compounds with over 300 members. Unusual triple-pairing superconductivity with a spin $S=1$ has been proposed [92] for some compounds of this class of materials. Some of them consist of CuO_2-layers, such as La_2CuSnO_6, and others contain CuO_6-octahedra but in different configurations, such as $La_4Cu_3MoO_{12}$. Cation doping and/or high-pressure tuning may induce superconductivity in these compounds and allow for the further study of the role of CuO_2-layers and CuO_6-octahedra in HTSy. Another interesting compound is Sr_2YRuO_6, where a small superconducting signal up to 60 K has been reported [93] when Ru is partially replaced by Cu. The observation will be extremely important when the complete absence of any superconducting $YSr_2(Ru_xCu_{3-x})O_{3+\delta}$ phase in the samples investigated is ascertained. Additionally, the possible appearance of various magnetic orderings in different parts of the unit cell of this and related compounds may offer a unique opportunity for the study of the interplay between magnetism and superconductivity and for the test of the dimensionality effect on HTSy. The studies [94] on the so-called superconducting ferromagnets, $RuSr_2GdCu_2O_8$ and $RuSr_2(Gd,Ce)_2Cu_2O_{10}$, which undergo a magnetic transition prior to a superconducting transition on cooling, serve as an example.

- *enhanced pinning and reduced anisotropy – by modifying the CRB of the cuprates, by exploiting the small anisotropy in Fe-based pnictides, and by searching for novel non-layered HTSrs*

The weak flux pinning of the cuprate HTSrs is believed to stem from the short coherence length and the weak coupling between the ABs of the cuprates due to the large anisotropy associated with their layered structure. However, it has been observed that the irreversibility line (H_i) or the melting line (H_m), which is a measure of the flux pinning strength, can be different for cuprate HTSrs with the same AB but different CRBs. For instance, compounds with a CRB of a smaller m [95] (e.g. Cu-1212 or $YBa_2Cu_3O_7$ vs. Bi-2212 or Tl-1212 vs. Tl-2212) or a more metallic CRB [96] (e.g. rare-earth-doped Hg-1223 vs. undoped Hg-1223) display a greater flux pinning and a higher H_i or H_m. A CRB with a smaller m or a more metallic CRB provides a better communication between the ABs and hence a reduced anisotropy of the cuprates. Therefore, enhanced flux pinning and reduced anisotropy may be achieved in cuprates by reducing the thickness and improving the metallic characteristic of their CRBs via crystal modification or doping. These may also be accomplished by discovering less 2D-like or 3D HTSg compounds, as described earlier.

The stronger interlayer coupling in the Fe-based pnictides and chalcogenides [97] offers greater flux pinning to obtain high critical current-carrying capability. The challenge is to raise their T_c. Although the T_c of this family of superconductors in general is low, the recent report of a T_c of 109 K in FeSe [19(d)], if reproduced and confirmed, suggests that the challenge may be surmountable.

- *higher chemical stability and improved homogeneity – by doping and micromaterial engineering the cuprate HTSrs*

There exists a diverse degree of chemical instabilities in cuprate HTSrs, ranging from the inability to form the compound to the loss of oxygen, and the separation of phases to the degradation of the compound once formed. These instabilities arise from the inherent defected structures of the compounds with respect to the perfect perovskite structure, partially due to the size mismatch of the atoms [56] and the strong Coulomb repulsion [68] associated with the excess charge accumulation in the ABs. The ease of the alkaline earth elements in cuprates to form stable carbonates in air adds to the complications. This is particularly serious when the material is not made in its perfect single-crystalline form. It has been demonstrated that many metastable HTSg compounds, such as those with $n > 3$, (Cu,C)-12(n-1)n [98], 02(n-1)n [74], and $YSr_2Cu_3O_{7.5}$ [99], can be synthesized successfully under high pressures by tuning the interatomic distances between various elements. By matching the interatomic distances in the various layers of the compound through doping at ambient, one succeeds in reducing the stress introduced by the size mismatch, thus achieving great chemical stability in $Y_{0.6}Ca_{0.4}Ba_{1.6}La_{0.4}Cu_3O_{7-\delta}$ [56]. It has also been shown that a proper precursor or chemical route chosen can affect the formation and the purity of the final product [75]. For example, the use of $Ba_2Cu_3O_{5+\delta}$ and CaO as the precursor leads to the synthesis of the 0223 instead of the (Cu,C)-1223 when BaO, CuO, and CaO are used, even after precautions are taken to minimize possible C-contamination from the air. Nitrate precursors produce carbonate-free cuprates more readily than carbonate or simple oxide precursors. In general, cuprates with Sr in the CRB exhibit a greater O-affinity than those with Ba given that everything else is the same, for reasons not yet known, although the greater atomic polarizability of Ba has been suggested [75]; and diffusion is greater for oxygen than that for cations in solids. Therefore, cuprates with a greater chemical stability and physical integrity may be obtained by preparing cuprates free of impurities in a perfect crystalline form via a proper chemical route and doping; and cuprates with greater homogeneity may be attained by adopting cation doping when possible. It should be noted that, while a cuprate with a Sr-containing CRB usually displays a greater O-affinity, its T_c tends to be lower than its counterpart with a Ba-containing CRB; and while cation doping minimizes anion diffusion in the sample and hence improves the sample homogeneity, different dopings may not result in the same maximum T_c in a cuprate due to differences between doping sites.

- *lower processing cost – by modifying the cuprate phase diagram and/or lowering the operating temperature*

The present high cost of HTSg devices originates mainly from material processing and device fabrication. Because of the complex chemistry of formation and the stringent quality requirement of the cuprate HTSg material, processing of HTSg material requires multiple elaborate steps, usually in a well-controlled environment and on a highly demanding substrate [48, 49]. The process is tedious and time-consuming, and the substrate is costly. Layered cuprates have to be prepared in their atomically perfect forms for scientific study and device application. Melt-texturing at high temperature and epitaxial growth in a vacuum environment are the two techniques most commonly employed to date: the former limits the substrate material only to the expensive pure or doped Ag and Au and the latter is commercially unattractive for large-current HTSg device fabrication. It has been shown that the HTSr/Ag interface lowers the texturing temperature [100], and a less oxidizing or reducing atmosphere suppresses the formation temperature of cuprates [101]. Therefore, by modifying the phase diagram of cuprate phase formation via doping and controlling the processing atmosphere, one can lower the compound formation and texturing temperatures to simplify the material processing and device fabrication. It has been demonstrated [50] that a Ca-overdoped layer of YBCO can improve the superconducting current transporting across the grain boundaries in the misoriented YBCO layer underneath it. This suggests that the stringent requirement on atomic alignment can perhaps be relaxed. It is also known that the intergrain coupling grows and the critical current increases at lower temperature. To reduce the HTSr processing cost by lowering the operating temperature to, e.g., 0.7 T_c may provide a meaningful relief without seriously sacrificing the performance of the HTSg device.

Acknowledgements

The work in Houston is supported in part by the U.S. Air Force Office of Scientific Research (AFOSR) Grants FA9550-15-1-0236 and FA9550-20-1-0068, the T. L. L. Temple Foundation, the John J. and Rebecca Moores Endowment, and the State of Texas through the Texas Center for Superconductivity at the University of Houston. B. Lv also acknowledges support from AFOSR Grant FA9550-19-1-0037.

References

[1] Matthias BT, Bozorth RM (1958) Ferromagnetism of a zirconium-zinc compound. Phys. Rev. 109:604–605.

[2] Wilson JA, Di Salvo FJ, Mahajan S (1974) Charge-density waves in metallic, layered, transition-metal dichalcogenides. Phys. Rev. Lett. 32:882–885.

[3] He QL et al. (2017) Chiral Majorana fermion modes in a quantum anomalous Hall insulator–superconductor structure. Science 357:294–299.

[4] Bednorz JG, Müller KA (1986) Possible high T_c superconductivity in the Ba–La–Cu–O system. Z. Phys. B 64:189–193.

[5] Wu MK et al. (1987) Superconductivity at 93 K in a new mixed-phase Y-Ba-Cu-O compound system at ambient pressure. Phys. Rev. Lett. 58:908–910.

[6] Chu CW (2011) Materials. In: 100 Years of Superconductivity (Rogalla H and Kes PH, eds), pp. 233–311. New York: CRC Press.

[7] Hirsch JE, Maple MB, Marsiglio F (2015) Chapter 2: Possibly unconventional superconductors and Chapter 3: Unconventional superconductors. Physica C 514:152–434.

[8] Damascelli A, Hussain Z, Shen ZX (2003) Angle-resolved photoemission studies of the cuprate superconductors. Rev. Mod. Phys. 75:473–541.

[9] Timusk T, Statt B (1999) The pseudogap in high-temperature superconductors: an experimental survey. Rep. Prog. Phys. 62:61–122; Doiron-Leyraud N et al. (2007)

Quantum oscillations and the Fermi surface in an underdoped high-T_c superconductor. Nature 447:565–568; Fernandes RM, Chubukov AV, Schmalian J (2014) What drives nematic order in iron-based superconductors? Nat. Phys. 10:97–104.

[10] Zannen J (2011) Theory. In: 100 Years of Superconductivity (Rogalla H and Kes PH, eds), pp. 51–145. New York: CRC Press; Lee PA (2008) From high temperature superconductivity to quantum spin liquid: progress in strong correlation physics. Rep. Prog. Phys. 71:012501.

[11] (2011) Chapters 5-12. In: 100 Years of Superconductivity (Rogalla H and Kes PH, eds), pp. 311–830. New York: CRC Press.

[12] Ramirez AP (1997) Colossal magnetoresistance. J. Phys. Condes. Matter 9:8171–8199.

[13] Hasan MZ, Kane CL (2010) Colloquium: topological insulators. Rev. Mod. Phys. 82:3045–3067; Qi XL, Zhang SC (2011) Topological insulators and superconductors. Rev. Mod. Phys. 83:1057–1110.

[14] Schilling A, Cantoni M, Guo JD, Ott HR (1993) Superconductivity above 130 K in the Hg-Ba-Ca-Cu-O system. Nature 363:56–58.

[15] Chu CW, Gao L, Chen F, Huang ZJ, Meng RL, Xue YY (1993) Superconductivity above 150 K in $HgBa_2Ca_2Cu_3O_{8+\delta}$ under pressure. Nature 365:323–325; Gao L et al. (1994) Superconductivity up to 164 K in $HgBa_2Ca_{m-1}Cu_mO_{2m+2+\delta}$ (m = 1, 2 and 3) under quasihydrostatic pressures. Phys. Rev. B. Rapid Comm. 50:4260–4263.

[16] Drozdov AP, Eremets MI, Troyan IA, Ksenofontov V, Shylin SI (2015) Conventional superconductivity at 203 kelvin at high pressures in the sulfur hydride system. Nature 525:73–76.

[17] Somayazulu, M et al. (2019) Evidence for superconductivity above 260 K in lanthanum superhydride at megabar pressures. Phys. Rev. Lett. 122:027001.

[18] Snider E et al. (2020) Room-temperature superconductivity in a carbonaceous sulfur hydride. Nature 586:373–377.

[19] (a) Wang QY et al. (2012) Interface-induced high-temperature superconductivity in single unit-Cell FeSe films on $SrTiO_3$. Chin. Phys. Lett. 29:037402; (b) He SL et al. (2013) Phase diagram and electronic indication of high-temperature superconductivity at 65 K in single-layer FeSe films. Nat. Mater. 12: 605-610; (c) Deng LZ et al. (2014) Meissner and mesoscopic superconducting states in 1–4 unit-cell FeSe films. Phys. Rev. B 90:214513; (d) Ge JF et al. (2015) Superconductivity above 100 K in single-layer FeSe films on doped $SrTiO_3$. Nat. Mater. 14:285–289.

[20] Chu CW (2008) A possible path to RTS. AAPPS Bulletin 18:9–21.

[21] Chu CW, Huang ZJ, Meng RL, Gao L, Hor PH (1988) High-pressure study on 60- and 90-K $EuBa_2Cu_3O_{7-\delta}$. Phys. Rev. B 37:9730–9733.

[22] Kudinov VI, Kirilyuk AI, Kreines NM, Laiho R, Lahderanta E (1990) Photoinduced superconductivity in YBaCuO films. Phys. Lett. A 151:358–363.

[23] Ye JT et al. (2010) Liquid-gated interface superconductivity on an atomically flat film. Nat. Mater. 9:125–128.

[24] Ye JT, Zhang YJ, Akashi R, Bahramy MS, Arita R, Iwasa Y (2012) Superconducting dome in a gate-tuned band insulator. Science 338:1193–1196.

[25] Lei B et al. (2017) Tuning phase transitions in FeSe thin flakes by field-effect transistor with solid ion conductor as the gate dielectric. Phys. Rev. B 95:020503(R).

[26] Presland MR, Tallon JL, Buckley RG, Liu RS, Flower NE (1991) General trends in oxygen stoichiometry effect on T_c in Bi and Tl superconductors. Physica C 176:95–105.

[27] Rice TM (1997) Reviews, prospects and concluding remarks. High-T_c superconductivity - Where next? Physica C 282–287:xix; Uchida S (1997) Spin gap effects on the c-axis and in-plane charge dynamic of high-T_c cuprates. Physica C 282-287:12–17.

[28] Blatter G (1997) Vortex matter. Physica C 282-287:19–26.

[29] Martin S, Fiory AT, Fleming RM, Espinosa GP, Cooper AS (1989) Anisotropic critical current density in superconducting $Bi_2Sr_2CaCu_2O_8$. Appl. Phys. Lett. 54:72–74.

[30] Routbort JL, Rothman SJ (1994) Oxygen diffusion in cuprate superconductors. J. Appl. Phys. 76:5615–5628.

[31] Tranquada JM, Axe JD, Ichikawa N, Moodenbaugh AR, Nakamura Y, Uchida S (1997) Coexistence of and competition between superconductivity and charge-stripe order in $La_{1.6-x}Nd_{0.4}Sr_xCuO_4$. Phys. Rev. Lett. 78: 338–341; Emery V, Kivelsen SA (1995) Importance of phase separations in superconductors with small superfluid density. Nature 374:434–437.

[32] Tranquada JM, Sternlieb BJ, Axe JD, Nakamura Y, Uchida S (1995) Evidence for stripe correlations of spins and holes in copper oxide superconductors. Nature 375:561–563; Kivelson SA et al. (2003) How to detect fluctuating stripes in the high-temperature superconductors. Rev. Mod. Phys. 75:1201–1241.

[33] Chang J et al. (2012) Direct observation of competition between superconductivity and charge density wave order in $YBa_2Cu_3O_{6.67}$. Nat. Phys. 8:871–876.

[34] Schneemeyer LF (1993) Growth of single crystals of various high-T_c superconductors. In: Processing and Properties of High-T_c Superconductors (Jin SH, ed), pp. 45–86. Singapore: World Scientific.

[35] Liu GD et al. (2008) Development of a vacuum ultraviolet laser-based angle-resolved photoemission system with a super-high energy resolution better than 1 meV. Rev. Sci. Instrum. 79:023105.

[36] Kamihara Y, Watanabe T, Hirano M, Hosono H (2008) Iron-based layered superconductor $La[O_{1-x}F_x]FeAs$ (x = 0.05–0.12) with T_c = 26 K. J. Am. Chem. Soc. 130: 3296–3297.

[37] Rotter M, Tegel M, Johrendt D (2008) Superconductivity at 38 K in the iron arsenide $(Ba_{1-x}K_x)Fe_2As_2$. Phys. Rev. Lett. 101:107006; Sasmal K et al. (2008) Superconducting Fe-based compounds $(A_{1-x}Sr_x)Fe_2As_2$ with A = K and Cs with transition temperatures up to 37 K. Phys. Rev.

Lett. 101:107007; Guo JG et al. (2010) Superconductivity in the iron selenide $KxFe_2Se_2$ ($0 \leq x \leq 1.0$). Phys. Rev. B 82:180520(R).
[38] Tapp JH et al. (2008) LiFeAs: an intrinsic FeAs-based superconductor with $T_c = 18$ K. Phys. Rev. B 78:060505(R).
[39] Hsu FC et al. (2008) Superconductivity in the PbO-type structure α-FeSe. Proc. Natl. Acad. Sci. USA 105:14262–14264.
[40] Ishida K, Nakai Y, Hosono H (2009) To what extent iron-pnictide new superconductors have been clarified: a progress report. J. Phys. Soc. Jpn. 78:062001; Hirschfeld PJ, Korshunov MM, Mazin II (2011) Gap symmetry and structure of Fe-based superconductors. Rep. Prog. Phys. 74:124508; Hoffman JE (2011) Spectroscopic scanning tunneling microscopy insights into Fe-based superconductors. Rep. Prog. Phys. 74:124513; Dagotto E (2013) Colloquium: the unexpected properties of alkali metal iron selenide superconductors. Rev. Mod. Phys. 85:849–867; Johnston DC (2010) The puzzle of high temperature superconductivity in layered iron pnictides and chalcogenides. Adv. Phys. 59:803–1061; Stewart GR (2011) Superconductivity in iron compounds. Rev. Mod. Phys. 83:1589–1652; Chubukov AV, Khodas M, Fernandes RM (2016) Magnetism, superconductivity, and spontaneous orbital order in iron-based superconductors: which comes first and why? Phys. Rev. X 6:041045; Bang Y, Stewart GR (2017) Superconducting properties of the $s^{\pm}$-wave state: Fe-based superconductors. J. Phys. Condens. Matter 29:123003.
[41] Ishida K, Nakai Y, Hosono H (2009) To what extent iron-pnictide new superconductors have been clarified: a progress report. J. Phys. Soc. Jpn. 78:062001; Stewart GR (2011) Superconductivity in iron compounds. Rev. Mod. Phys. 83:1589–1652; Hosono H, Kuroki K (2015) Iron-based superconductors: current status of materials and pairing mechanism. Physica C 514:399–422; Luo X, Chen XH (2015) Crystal structure and phase diagrams of iron-based superconductors. Sci. China Mater. 58:77–89.
[42] (a) Torikachvili MS, Bud'ko SL, Ni N, Canfield PC (2008) Pressure induced superconductivity in $CaFe_2As_2$. Phys. Rev. Lett. 101, 057006; (b) Yu W et al. (2009) Absence of superconductivity in single-phase $CaFe_2As_2$ under hydrostatic pressure. Phys. Rev. B 79, 020511; (c) Chu CW, Lorenz B (2009) High pressure studies on Fe-pnictide superconductors. Physica C 469:385–395; (d) Sefat SA (2011) Pressure effects on two superconducting iron-based families. Rep. Prog. Phys. 74:124502; (e) Sun LL et al. (2012) Re-emerging superconductivity at 48 Kelvin in iron chalcogenides. Nature 483:67–69.
[43] Chu CW (2021) Room-temperature superconductivity – What more needs to be studied further! IEEE CSC & ESAS Superconductivity News Forum 14(49):STH62; and references therein.
[44] Sheahen TP (1994) Refrigeration. In: Introduction to High Temperature Superconductivity (Sheahen TP, ed), pp. 37–64. New York: Plenum Press.
[45] Ginsberg DM (1989) Introduction, History and Overview of HTS Physical Properties of High Temperature Superconductors. pp. 1–38. Singapore: World Scientific.
[46] Si W et al. (2013) High current superconductivity in $FeSe_{0.5}Te_{0.5}$-coated conductors at 30 tesla. Nat. Commun. 4:1347.
[47] Kleiner R, Müller P (1994) Intrinsic Josephson effects in high-T_c superconductors. Phys. Rev. B 49:1327–1341.
[48] Chu CW (1997) Superconductivity, high-temperature. In: Encyclopedia of Applied Physics, Vol 20. pp. 213–247. New York: VCH Publishers; Humphreys RG et al. (1990) Physical vapor deposition techniques for the growth of $YBa_2Cu_3O_7$ thin films. Super. Sci. Tech. 3:38–52.
[49] Chu CW (1997) Superconductivity, high-temperature. In: Encyclopedia of Applied Physics, Vol 20. pp. 213–247. New York: VCH Publishers; Salama K, Selvamanickam V, Lee DF (1993) Melt processing and properties of YBCO. In: Processing and Properties of High-T_c Superconductors (Jin SH, ed), pp. 155–211. Singapore: World Scientific; Murakami M (1993) Melt process, flux pinning and levitation ibid., pp. 213–270.
[50] Hammerl G et al. (2000) Enhanced supercurrent density in polycrystalline $YBa_2Cu_3O_{7-\delta}$ at 77 K from calcium doping of grain boundaries. Nature 407:162–164.
[51] Scheel HJ, Licaci F (1987) Crystal growth of $YBa_2Cu_3O_{7-x}$. J. Crystal Growth 85:607–614.
[52] Sato K et al. (1991) High-J_c silver-sheathed Bi-based superconducting wires. IEEE Trans. Magn. 27:1231–1238.
[53] Selvamanickam V et al. (2001) High-current Y-Ba-Cu-O coated conductor using metal organic chemical-vapor deposition and ion-beam-assisted deposition. IEEE Trans. Appl. Supercond. 11:3379–3381.
[54] Wu XD et al. (1995) Properties of $YBa_2Cu_3O_{7-\delta}$ thick films on flexible buffered metallic substrates. Appl. Phys. Lett. 67:2397–2399; Meng RL et al. (1996) Processing of highly oriented $(Hg_{1-x}Re_x)Ba_2Ca_2Cu_3O_{8+\delta}$ tape with $x \sim 0.1$. Appl. Phys. Lett. 68:3177–3179.
[55] Krusin-Elbaum L et al. (1998) Anisotropic rescaling of a splayed pinning landscape in Hg cuprates: strong vortex pinning and recovery of variable range hopping. Phys. Rev. Lett. 81:3948–3951 and references therein.
[56] Zhou JP et al. (1995) Improved corrosion resistance of cation substituted $YBa_2Cu_3O_{7-\delta}$. Appl. Phys. Lett. 66:2900–2902.
[57] Lee PA, Read N (1987) Why is T_c of the oxide superconductors so low? Phys. Rev. Lett. 58:2691–2694.
[58] Collings EW (1989) Conductor design with high-T_c ceramics: a review. In: Advances in Superconductivity, Vol 2 (Ishiguro T and Kajimura K, eds), pp. 325–333. Tokyo: Springer-Verlag.
[59] Beasley MR (2011) Will higher T_c superconductors be useful? Fundamental issues from the real world. MRS Bulletin 36:597–600.
[60] Pines D (1996) Spin fluctuations, magnetotransport and $d_{x^2-y^2}$ pairing in the cuprate superconductors.

In: Proceedings of the 10th Anniversary HTS Workshop (Batlogg B et al., eds), pp. 471–476. Singapore: World Scientific; Scalapino DJ (1996) Pairing mechanism in the two dimensional Hubbard model ibid., pp. 477–480; and references therein.

[61] Nagamatsu J, Nakagawa N, Muranaka T, Zenitani Y, Akimitsu J (2001) Superconductivity at 39 K in magnesium diboride. Nature 410:63–64.

[62] Anderson PW (1997) The Theory of Superconductivity in the High-T_c Cuprate superconductors. Princeton, NJ: Princeton University Press.

[63] Lee DH (2015) What makes the T_c of FeSe/$SrTiO_3$ so high? Chin. Phys. B 24:117405.

[64] Chu CW (1997) Superconductivity, high-temperature. In: Encyclopedia of Applied Physics, Vol 20. pp. 213–247. New York: VCH Publishers.

[65] Tanaka S (1997) Reviews, prospects and concluding remarks. Materials needs for applications Physica C 282–287:xxxi–xxxix; Tanaka S (2000) Status and future perspectives of applications of high temperature superconductors. Physica C 341-348:31–35.

[66] Chu CW (1995) Possible approaches to the discovery of superconductors with higher T_c's. In: Chen Ning Yang: A Great Physicist of the Twentieth Century (Liu CS and Yan ST, eds), pp. 73–86. Cambridge, MA: International Press.

[67] Di Stasio M, Müller KA, Pietronero L (1990) Nonhomogeneous charge distribution in layered high T_c superconductors. Phys. Rev. Lett. 64:2827–2830.

[68] Lin QM (1996) Instabilities and high temperature superconductivity in $HgBa_2Ca_{n-1}Cu_nO_{2n+2+\delta}$. Ph.D. thesis, University of Houston.

[69] Merz M et al. (1998) Site-specific x-ray absorption spectroscopy of $Y_{1-x}Ca_xBa_2Cu_3O_{7-\delta}$: overdoping and role of apical oxygen for high temperature superconductivity. Phys. Rev. Lett. 80:5192–5195.

[70] Scott BA, Snard EY, Tsuei CC, Mitzi DB, McGuire TR, Chen BH (1994) Layered dependence of the superconducting transition temperature of $HgBa_2Ca_{n-1}Cu_nO_{2n+2+\delta}$. Physica C 230:239–245.

[71] Deng LZ et al. (2019) Higher superconducting transition temperature by breaking the universal pressure relation. Proc. Natl. Acad. Sci. USA 116:2004–2008.

[72] Muramatsu T, Pham D, Chu CW (2011) A possible pressure-induced superconducting-semiconducting transition in nearly optimally doped single crystalline $YBa_2Cu_3O_{7-\delta}$. Appl. Phys. Lett. 99:052508.

[73] Lorenz B, Chu CW (2005) High pressure effects on superconductivity. In: Frontiers in Superconducting Materials (Narlikar AV, ed), pp. 459–497. Berlin: Springer-Verlag.

[74] Chen XJ, Struzhkin VV, Yu Y, Goncharov AF, Lin CT, Mao HK (2010) Enhancement of superconductivity by pressure-driven competition in electronic order. Nature 466:950–953.

[75] Chu CW et al. (1997) Superconductivity up to 126 Kelvin in interstitially doped $Ba_2Ca_{n-1}Cu_nO_x$ [02(n-1)n-Ba]. Science 277:1081–1083.

[76] Ginzburg VL (1964) On surface superconductivity. Phys. Lett. 13:101–102.

[77] Allender D, Bray J, Bardeen J (1973) Model for an exciton mechanism of superconductivity. Phys. Rev. B 7:1020–1029.

[78] Pereiro J, Petrovic A, Panagopoulos C, Božović I (2011) Interface superconductivity: history, development and prospects. Phys. Express 1:208–241.

[79] Zhao K, Lv B, Deng LZ, Huyan SY, Xue YY, Chu CW (2016) Interface-induced superconductivity at ~ 25 K at ambient pressure in undoped $CaFe_2As_2$ single crystals. Proc. Natl. Acad. Sci. USA 113:12968–12973.

[80] Huyan SY et al. (2019) Low-temperature microstructural studies on superconducting $CaFe_2As_2$. Sci. Rep. 9:6393.

[81] Maradudia AA (1967) Ferroelectricity and lattice anharmonicity. In: Ferroelectricity (Weller EF, ed), pp. 72–100. Amsterdam: Elsevier.

[82] Shanks HR (1994) Enhancement of the superconducting transition temperature near a phase instability in Na_xWO_3. Solid State Commun. 15:753–756.

[83] Villars P, Phillips JC (1988) Quantum structural diagrams and high-T_c superconductivity. Phys. Rev. B 37:2345–2348.

[84] Schooley JF, Hosla WR, Cohen ML (1964) Superconductivity in semiconducting $SrTiO_3$. Phys. Rev. Lett. 12:474–475.

[85] Johnston DC, Prakash H, Zachariasen WH, Viswanathan R (1973) High temperature superconductivity in the Li-Ti-O ternary system. Mat. Res. Bull. 8:777–784.

[86] Viskadourakis Z, Radulov I, Petrović AP, Mukherjee S, Andersen BM, Jelbert G (2012) Low-temperature ferroelectric phase and magnetoelectric coupling in underdoped La_2CuO_{4+x}. Phys. Rev. B 85, 214502.

[87] Zhu XY et al. (2013) Disorder-induced bulk superconductivity in $ZrTe_3$ single crystals via growth control. Phys. Rev. B 87:024508.

[88] Lorenz B, Litvinchuk AP, Gospodinov MM, Chu CW (2004) Field-induced reentrant novel phase and a ferroelectric-magnetic order coupling in $HoMnO_3$. Phys. Rev. Lett. 92:087204; Dela Cruz CR, Lorenz B, Sun YY, Wang Y, Park S, Cheong SW (2007) Pressure-induced enhancement of ferroelectricity in multiferroic RMn_2O_5 (R=Tb,Dy,Ho). Phys. Rev. B 76:174106.

[89] Lorenz B, Chu CW (2005) Superconducting ferromagnets: ferromagnetic domains in the superconducting state. Nat. Mater. 4:516–517; Dikin DA, Mehta M, Bark CW, Folkman CM, Eom CB, Chandrasekhar V. (2011) Coexistence of superconductivity and ferromagnetism in two dimensions. Phys. Rev. Lett. 107:056802.

[90] Cheng JG et al. (2015) Pressure induced superconductivity on the border of magnetic order in MnP. Phys. Rev.

Lett. 114:117001; Hung TL et al. (2020) Pressure induced superconductivity in MnSe. arXiv:2011.01510 [cond-mat. supr-con] (Nat. Commun., submitted).

[91] Anderson MT, Greenwood KB, Taylor GA, Poeppelmeier KR (1993) B-cation arrangement in double perovskites Prog. Solid State Chem. 22:197–233.

[92] Pickett WE (1996) Single spin superconductivity. Phys. Rev. Lett. 77:3185–3188.

[93] Wu MK, Chen DY, Chien FZ, Sheen SR, Ling DC, Tai CY (1996) Anomalous magnetic and superconducting properties in a Ru-based double perovskite. Z. Phys. B 102:37–41.

[94] Chen DY et al. (1997) Superconductivity in Ru-based double perovskite–the possible existence of a new superconducting pairing state. Physica C 282-287:73–76.

[95] Huang ZJ, Xue YY, Meng RL, Chu CW (1994) Irreversibility line of the $HgBa_2CaCu_2O_{6+\delta}$ high temperature superconductors. Phys. Rev. B 49:4218–4221.

[96] Hahakura S, Shimoyama J, Shiino O, Hasegawa T, Kitazawa K, Kishio V (1994) Chemical stabilization of Hg-based superconductors. Physica C 235-240:915–916.

[97] Tarantini C et al. (2012) Artificial and self-assembled vortex-pinning centers in superconducting $Ba(Fe_{1-x}Co_x)_2As_2$ thin films as a route to obtaining very high critical-current densities. Phys. Rev. B 86:214504.

[98] Ihara H et al. (1994) New high-T_c superconductor family of Cu-based $Cu_{1-x}Ba_2Ca_{n-1}Cu_nO_{2n+4-\delta}$ with $T_c > 116$ K. Jpn. J. Appl. Phys. 33:L503–L506; Kawashima T, Matsui Y, Takayama-Muromachi E (1994) New oxycarbonate superconductors $(Cu_{0.5}C_{0.5})Ba_2Ca_{n-1}Cu_nO_{2n+3}$ (n=3,4) prepared at high pressure. Physica C 224:69–74; A new series of oxycarbonate superconductors $(Cu_{0.5}C_{0.5})_2Ba_3Ca_{n-1}Cu_nO_{2n+5}$ (n=4,5) prepared at high pressure, ibid 227:95–101.

[99] Okai B (1990) High-pressure synthesis of superconducting $YSr_2Cu_3O_7$. Jpn. J. Appl. Phys. 29:L2180–L2182.

[100] Larbalestier DC et al. (1994) Position-sensitive measurements of the local critical current density in Ag sheathed high-temperature superconductor $(BiPb)_2Sr_2Ca_2Cu_3O_y$ tapes. Physica C 221:299–303.

[101] Forster KM, Formica JP, Milonopoulou V, Kulik J, Richardson JT and Luss D (1992) in situ kinetic study of YBCO formation. In: Proceedings of the 10th Anniversary HTS Workshop (Batlogg B et al., eds), pp. 411–416. Singapore: World Scientific.

D5

Fe-Based Chalcogenide Superconductors

Ming-Jye Wang, Phillip M. Wu, and Maw-Kuen Wu

D5.1 Introduction

Fe-chalcogenide superconductors are the family of materials that contain iron and chalcogens. Since 2008, interest in this family of materials has grown dramatically due to the high transition temperature T_c, which rivals those of some cuprate superconductors. The first Fe chalcogenide discovered, FeSe, has tetragonal P4/nmm symmetry at room temperature and is composed of a stack of edge-sharing FeSe4-tetrahedra layer by layer, which makes it the simplest crystal structure of all the Fe-based compounds (Hsu *et al.*, 2008). Despite this simplicity, T_c of FeSe increases dramatically from 8.5 K to 30–40 K when the system undergoes either external pressure (Mizuguchi *et al.*, 2008) or internal pressure via chemical substitutions (Yeh *et al.*, 2008) and intercalations (Guo *et al.*, 2010). The materials discovered could be categorized into four major groups as tabulated in Table D5.1, and their crystal structures are shown in Figure D5.1.

More recently, thin film of even one single layer (Wang *et al.*, 2012a) shows evidence for even higher T_c's approaching liquid nitrogen temperatures, and nanostructured materials (Chen *et al.*, 2014) provide rich ground for detailed study of the mechanism responsible for superconductivity. Many new exciting results that are of both fundamental and technological interest have emerged in a relatively short amount of time. There exist several review articles that cover wide range of properties specifically on Fe chalcogenides (Wu *et al.*, 2009, Stewart, 2012, Deguchi *et al.*, 2013, Wu *et al.*, 2013, Dagotto, 2013, Chang *et al.*, 2015, Wu *et al.*, 2015). Interesting readers are encouraged to look into these review articles.

After a brief introduction, we describe the various members of this family of superconductors and then followed with details of their synthesis methods. In Section D5.4, we summarize the experimental results that are relevant to better understand the origin of superconductivity in this family of materials. The topics covered include the structural distortion at low temperature, the observation of gap-opening above the structural transition, the presence of Fe-vacancy in Fe chalcogenides and its relation with the phase diagram, the structure under high pressure, and the symmetry of superconducting gap. Section D5.5 presents a picture emerges from the experimental facts on the possible origin of superconductivity. Importantly for applications, high critical fields H_{C2} have been reported for nearly all Fe-based superconductors, making these materials strong candidates for superconducting wires. Thus, we present the current efforts in superconducting wires and devices fabrication briefly in Section D5.6. Finally, we provide some conclusions and future directions for this fascinating family of superconductors.

D5.2 Material Systems

D5.2.1 FeSe, FeSeTe, and FeSeS

Superconducting FeSe has tetragonal P4/nmm symmetry (PbO crystal structure) at room temperature and is one of the stable phases of the Fe–Se binary compounds. Numerous works and reviews provide in-depth details regarding crystal synthesis (Wu *et al.*, 2009). Systematic Te substitution to layered PbO type FeSe has been carried out (Deguchi *et al.*, 2013). The larger Te atom replaces Se, resulting in the room temperature crystal structure changing from tetragonal to monoclinic. At complete substitution, FeTe maintains the tetragonal structure but is not superconducting. Crystal structure evolution is correlated to changes in Tc, with a maximum $T_c = 15$ K observed at 50% Te substitution.

$FeSe_{1-x}S_x$ (x=0 to 0.5) can be prepared with maximum onset $T_{c,onset}$ = 15.5 K at x=0.2 (Deguchi *et al.*, 2013). X-ray diffraction (XRD) patterns show both a and c lattice parameters decreased in this compound compared to FeSe. Sulfur has a smaller ionic radius than Se, and the effect of substitution is similar to that of applying external pressure, hence giving rise to the larger onset transition temperatures. However, superconductivity in tetragonal FeS was recently reported with T_c at about 5 K (Lai *et al.*, 2015). The material was synthesized by the hydrothermal reaction of iron powder with sulfide solution, and the obtained samples were highly crystalline and less air-sensitive.

TABLE D5.1 Fe-Chalcogenide Superconductors

Superconductor	T_c, zero (ambient pressure)	Lattice Constant a, c (nm)	$\xi_{ab}(0)$, $\xi_c(0)$ (nm)	$\lambda_{ab}(0)$, $\lambda_c(0)$ (nm)
$FeSe_{1-x}$	8 K	0.3774, 0.5525	$\xi_{ab}(0)=4.5$	$\lambda_{ab}(0)=445$
$FeSe_{1-x}Te_x$	8–15 K	0.3791, 0.5957 (x=0.5)	$\xi_{ab}(0)=1.5$ $\xi_c(0)=0.4$	$\lambda_{ab}(0)=430$ $\lambda_c(0)=1600$
$A_xFe_{2-y}Se_2$ (A=alkali metals, and alkaline earths, (Tl,K), (Tl,Rb))	27–32 K	0.3914, 1.4037 (A=K)	$\xi_{ab}(0)=1.88$ $\xi_c(0)=3.28$	
$A_x(NH_3)_yFe_2Se_2$ (A=Li,Na,K,Ba,Sr,Ca,Yb,Eu)	30~46 K	0.3785, 1.7432 (A=Na, x=1)	note: β-FeSe+A+liquid ammonia (LA) sealed in the autoclave	
$(LiFeO_2)Fe_2Se_2$	43 K	0.3793, 0.9285		

D5.2.2 FeTeS

Superconductivity is recovered when sulfur is substituted for Te in $FeTe_{1-x}S_x$, with an onset $T_c \sim 10$ K (zero resistance $T_c = 7.8$ K) for $x = 0.2$ (Deguchi *et al.*, 2013). At room temperature, $FeTe_{1-x}S_x$ is tetragonal with P4/nmm symmetry. FeTe undergoes a structural phase transition around 70 K. Sulfur substitution of Te suppresses the structural phase transition: at $x = 0.1$ sulfur doping the transition shifts below 50 K, and disappears for $x = 0.2$. The upper critical field $H_{C2}(0) = 70$ T is quite large, making the material a suitable candidate for wire applications.

D5.2.3 Transition Metal Substitutions $Fe_{1-x}(Ti, V, Cr, Mn, Co, Ni, Cu)_xSe_{0.85}$

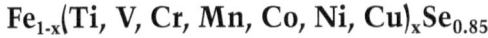

Several groups performed systematic transition metal substitutions to the Fe site (Wu *et al.*, 2009) in FeSe. Ti-, V-, and Cr-substituted samples were found to be no longer superconducting. Low level of Co (up to 2%) and Ni (to 5%)-doping retains superconductivity in FeSe (Deguchi *et al.*, 2013). Mn-substituted samples (up to 10%) showed increased normal-state resistivity with the superconductivity essentially the same as in FeSe. Detailed powder XRD shows that the tetragonal to orthorhombic transitions still takes place in $Fe_{0.9}Mn_{0.1}Se_{0.85}$. Superconductivity is preserved for Cu substitutions $Fe_{1-x}Cu_xSe_{0.85}$ for x up to 0.02, however, above 3% doping, superconductivity is completely suppressed. Detailed low-temperature structure studies indicate $Fe_{0.9}Cu_{0.1}Se_{0.85}$ remains in tetragonal symmetry without deformation down to low temperatures (Huang *et al.*, 2010). These results suggested strong correlation between the occurrence of superconductivity in FeSe and the low-temperature structure distortion from tetragonal to orthorhombic symmetry.

D5.2.4 Non-Transition Metal Substitutions $Fe_{1-x}(Al, Ga, Sm, Ba)_xSe_{0.85}$

Partial substitutions of Fe by nonmagnetic 3+ ions from group III were investigated, as well as substitutions of nonmagnetic

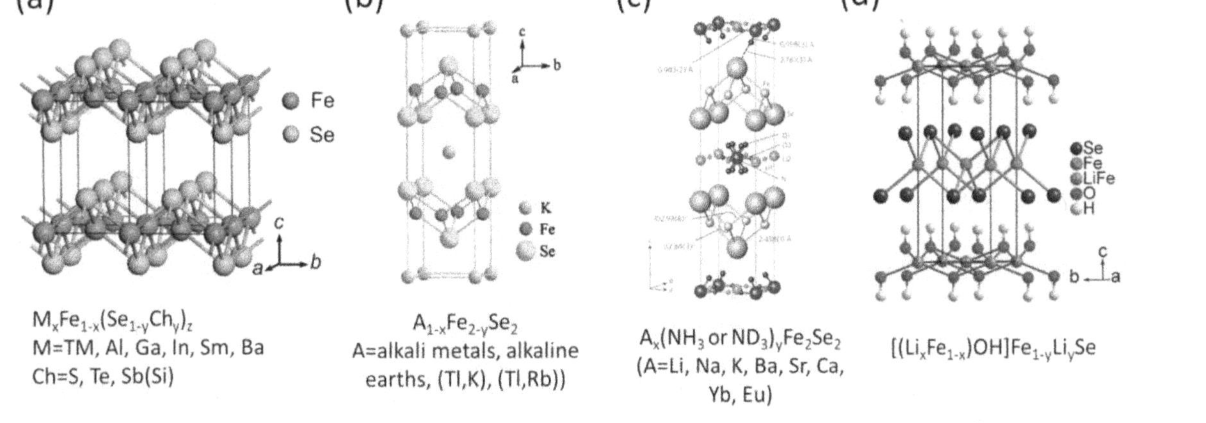

FIGURE D5.1 The crystal structure of Fe-chalcogenide superconductor. (a) $(M_xFe_{1-x})(Se_{1-y}Ch_y)_z$ 11-type (Hsu *et al.*, 2008), M=TM, Al, Ga, In, Sm, and Ba, and Ch=S, Te, Sb, Si. (b) $A_xFe_{2-y}Se_2$ type (Guo *et al.*, 2010a), A=alkali metals, alkaline earths, (Tl,K), and (Tl,Rb). (c) $A_x(NH_3$ or $ND_3)_yFe_2Se_2$, A=Li, Na, K, Ba, Sr, Ca, Yb, and Eu (Burrard-Lucas *et al.*, 2013). (d) $[(Li_xFe_{1-x})OH]Fe_{1-y}Li_ySe$ (Pachmayr *et al.*, 2015).

Ba2+ and magnetic rare-earth Sm (Wu *et al.*, 2009). XRD results showed that $Fe_{1-x}(Al,Ga,Sm)_xSe_{0.85}$ were tetragonal PbO structure at room temperature with only trace amounts of impurity phase for low doping. Onset of superconductivity at ~ 8.5 K was seen in resistivity versus temperature in $Fe_{1-x}Al_xSe_{0.85}$ for $x = 0.1$ to 0.25. Similarly, onset T_Cs ~ 6.8 K was found in Ga-substituted samples with concentrations also at $x = 0.1$ to 0.25. In contrast to these, 10% Sm doping raised the onset T_c to about 10.6 K, but dropped at 25% doping. 10% Ba substitution was attempted, but the samples were not crystalline.

D5.2.5 Alkali-Doped $A_{1-x}Fe_{2-y}Se_2$ (A = K, Cs, Rb, Tl)

Intercalating alkali metals between FeSe layers form $A_{1-x}Fe_{2-y}Se_2$, A = K, Cs, Rb, Tl, with high superconducting T_Cs (Guo *et al.*, 2010b, Chang *et al.*, 2015). Its superconducting transition has been enhanced to above 30 K. One of the most studied is the potassium intercalated system, with $ThCr_2Si_2$ type tetragonal lattice and I4/mmm space group. Subsequent local probe studies such as SEM, TEM, and STM showed that strong phase separation exists in these crystals (Dogotto 2013). Recent detailed micro-XRD also confirms this result (Ricci *et al.*, 2015).

D5.2.6 Organic Intercalations

Krzton-Maziopa *et al.* (Krzton-Maziopa *et al.*, 2012) reported a new synthesizing method to intercalate alkaline metal to FeSe, with the general formula $A_x(C_5H_5N)_yFe_{2-z}Se_2$ (A = Li, Na, K, Rb) that shows superconducting onset T_c ~ 45 K. Post-annealing of intercalated material $[Li_x(C_5H_5N)_yFe_{2-z}Se_2]$ at elevated temperatures drastically enlarges the c-parameter of the unit cell (~44%) and increases the superconducting shielding fraction to nearly 100%. It is noted that samples prepared this way typically exhibit two phases based on resistivity measurements. Several studies have reported enhanced T_Cs from intercalating molecular spacers between the FeSe layers.

D5.3 Synthesis Methods

D5.3.1 Crystal Growth

D5.3.1.1 FeSe Crystals

Because of the nonstoichiometry and intrinsically Se-deficient nature, the superconductivity in β-FeSe is very sensitive to composition and disorder (Wu *et al.*, 2009). McQueen *et al.* have successfully demonstrated a well-controlled thermal process for the fabrication of bulk polycrystalline samples (McQueen *et al.*, 2009). Methods using KCl or KBr as flux have been reported to grow large size PbO-type β-FeSe single crystals (Wu *et al.*, 2009). There were reports on the growth of β-$FeSe_{1-x}$ crystals using a physical vapor transport (PVT) approach. The cubic-anvil high-pressure technique was also performed to grow FeSe crystals.

FeSe crystals grown by one of the techniques mentioned above always suffer an inevitable intergrowth of δ-FeSe and a nonuniform compositional distribution and therefore display a wide superconducting transition. Annealing the crystals at 400°C *in situ* during the cooling program appears to be beneficial to the properties of the crystals (Wu *et al.*, 2009).

D5.3.1.2 $FeSe_{1-x}Te_x$ Crystal Growth

Crystals of $FeSe_{1-x}Te_x$ can be easily grown from the high-temperature melt using the Bridgeman method. $FeSe_{1-x}Te_x$ single crystals can also be grown by optical zone-melting technique (Wu *et al.*, 2009). The samples were zone-melted using a focused halogen light source and moved at a translation rate of 1–2 mm/h to control the crystallization rate.

D5.3.1.3 $K_xFe_2Se_2$ Crystal Growth

$K_{0.86}Fe_2Se_{1.82}$ single crystals with $T_{C,onset}$ ~31 K are easily obtained by melting the proper stoichiometry materials (Liu *et al.*, 2012). The superconducting fraction can be close to 100% for these crystals if the preparation was optimized. $K_xFe_2Se_2$ crystals are easy to cleave, and thin crystals with thickness less than 100 μm can be obtained. However, a problem in these crystals is that they inevitably exhibits multiphase features.

D5.3.2 Thin Film of β-$FeSe_{1-x}Te_x$ Superconductors

D5.3.2.1 Preparation of Superconducting β-$FeSe_{1-x}Te_x$ Films

Superconducting β-$FeSe_{1-x}Te_x$ films can be prepared by pulsed laser deposition (PLD) (Wang *et al.*, 2009), metal-organic chemical vapor deposition (MOCVD)(Li *et al.*, 2011), molecular beam epitaxy (MBE)(Agatsuma *et al.*, 2010), and electrochemical synthesis (Demura *et al.*, 2012). The requirement on base pressure of deposition chamber is moderate, typically in the order of 10^{-4} Pa. The $FeSe_{1-x}Te_x$ phase can be grown in a wide temperature range (200°C–600°C) on various substrates such as MgO, $LaAlO_3$, $SrTiO_3$, Si, SiO_x/Si, GaAs, R-Al_2O_3, CaF_2, and *LSAT* (lanthanum–strontium aluminum tantalite).

The FeSe films deposited at low temperature (near 300°C) have a (001) preferred orientation out of substrate plane, even grown on (100)-Si (44% lattice mismatch) and amorphous SiO_x substrates (Wu *et al.*, 2013). The orientation of FeSe films grown at high temperature (>500°C) depends on the mismatch of FeSe and substrate. The orientation changes to (101) direction if the film is grown on a substrate with a relative large mismatch (Wang *et al.*, 2009). Most reported Te-substituted films were grown along (001) orientation on various substrates. Similar to bulk samples, the issue of nonuniform Te distributions was reported in low Te-substituted films in early literature (Wu *et al.*, 2009). This phase separation issue in low Te-substitute film was resolved successfully by using CaF_2 substrates (Imai *et al.*, 2015).

D5.3.2.2 Superconducting Monolayer FeSe Films

The monolayer FeSe films grown by molecular beam epitaxy (MBE) method on Se-treated $SrTiO_3$ substrates demonstrate an unexpected high superconducting transition temperature (Wang *et al.*, 2012a). These monolayer films show a superconducting transition temperature above 50 K by an *ex situ* resistive measurement and a superconducting gap of 20 mV at low temperature (*in situ* scanning tunneling microscopy [STM] measurement), which implies a T_c above 77 K. A superconducting transition around 40 K ($T_{c,onset}$) was also reported in two monolayer FeSe films (Sun *et al.*, 2014).

Most of the high-T_c reports on monolayer FeSe are based on the STM (Wang *et al.*, 2012a) and ARPES (He *et al.*, 2013) results. T_c higher than 50 K has yet to be unambiguously demonstrated in transport or susceptibility measurement. Recently, Ge *et al.* (Ge *et al.*, 2015) reported a novel *in situ* four-point-probe technique to measure the transport properties of FeSe monolayer with magnetic field up to 11 T without moving the sample out of ultrahigh vacuum chamber. They report resistance-versus-temperature plots extracted from I–V measurements of the four separate probes, as shown in Figure D5.2. The data show a resistive transition at temperature higher than 100 K. While these results have yet to be verified, it is certainly intriguing and suggests there is more to come from this exciting system.

D5.3.3 Synthesis of Fe-Chalcogenide Nanomaterials

There are several interesting approaches to prepare nanostructured FeSe and related compounds. A rapid, solvent-less reaction under autogenic pressure at elevated temperature (RAPET) process was used to synthesize superconducting Fe–Se nanoparticles (Chang *et al.*, 2015). Detailed magnetization measurements on these nanoparticles show that in addition to the superconducting transition at ~10 K, an anomaly in magnetic susceptibility suggesting the presence of superconducting transition at ~40 K.

Highly crystalline FeSe nanowires (NWs) can be obtained by annealing thin film grown by pulsed laser deposition. High-resolution transmission electron microscope (HRTEM) images of these NWs show excellent crystalline tetragonal structure with the growth along [100] direction. The EDS results show uniform composition with stoichiometry Fe_4Se_5 in most FeSe NWs (Chen *et al.*, 2014). HRTEM and X-ray diffraction confirm that these NWs are tetragonal with ordered Fe-vacancy. Transport measurements of FeSe(Te) nanowires show onset superconducting transition at 10–14 K (Chang *et al.*, 2014).

D5.4 Structure–Property Correlations of Superconducting Fe Chalcogenides

There exist in Fe-chalcogenide superconductors competing magnetic, nematic, and superconducting phases, as well as other effects, such as orbital ordering and structural phase transition. The properties found below the structural phase transition and the corresponding fluctuations might be related to the pairing mechanism of superconductivity (Dagotto, 2013). Possible candidates for the many-body state responsible for the structural phase transition are the electronic nematicity (Fradkin *et al.*, 2010) or the orbital order (Lee *et al.*, 2012). Here we present several observations that address the importance of these competing orders.

D5.4.1 Low-Temperature Structure Distortion and Superconductivity

Detailed X-ray diffraction refinement of β-FeSe shows that a structural transformation from tetragonal (P4/nmm) to orthorhombic symmetry at low temperature, as shown in Figure D5.3, and below 90 K, its space symmetry group is

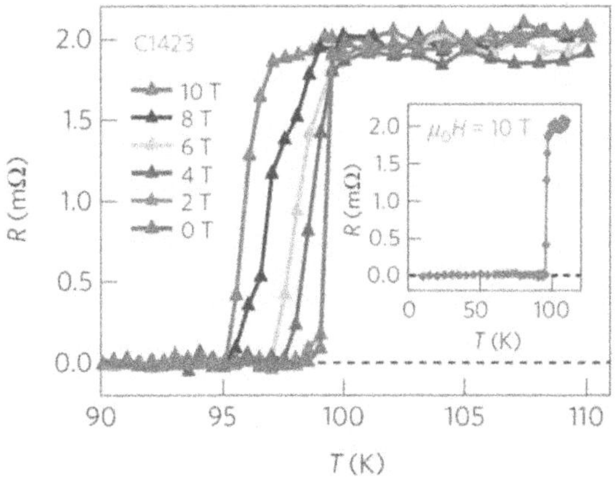

FIGURE D5.2 Superconducting transition of monolayer FeSe film under magnetic field. The measurement was carried out by an in situ four-point-probe technique. (Ge *et al.*, 2015.)

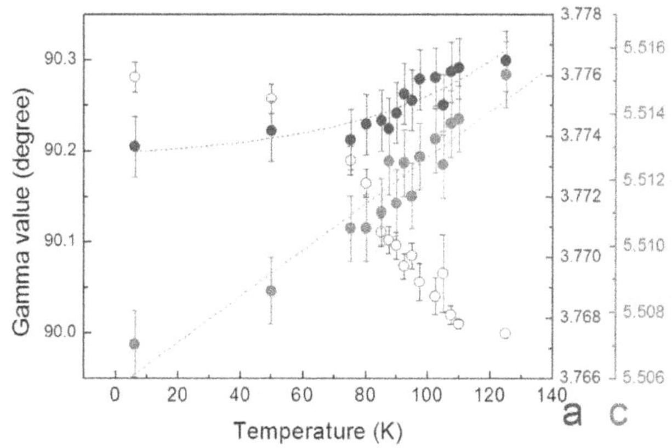

FIGURE D5.3 The crystal lattice parameters of FeSe as a function of temperature showing a distortion occurs at about 100 K.

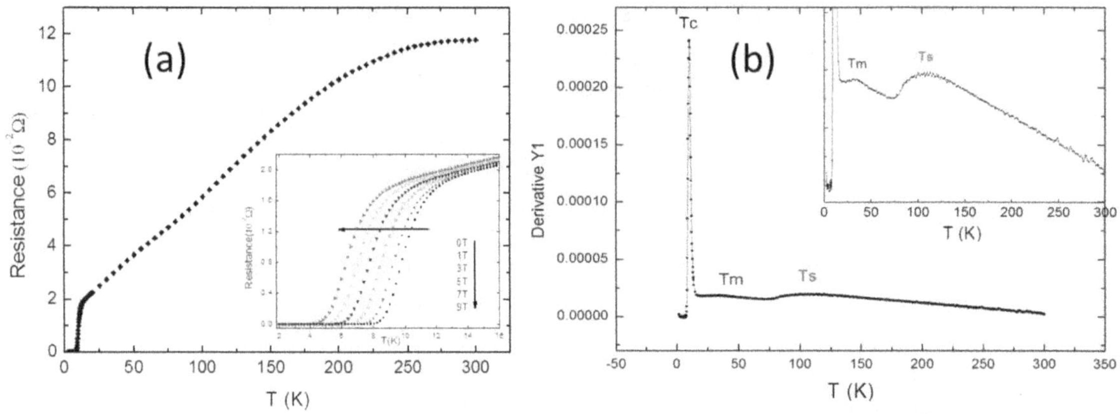

FIGURE D5.4 (a) Temperature dependence of resistivity of β-$FeSe_{1-x}$ single crystals, inset is the magnetic field dependence of the resistive transition; (b) derivative dR/dT of the resistivity of β-$FeSe_{1-x}$ single crystal as a function of the temperature. The inset is a plot with different scale to reveal more clearly the transitions. (Wu *et al.*, 2013).

Cmma. This structural deformation correlates closely with the occurrence of superconductivity at low temperature. Similar low-temperature structural phase transition was also observed in Te-, S-doped, and Mn-doped FeSe superconductors (Huang *et al.*, 2010).

Figure D5.4(a) is the temperature-dependent resistivity of FeSe single crystal. More detailed analysis of the data, as shown in the inset of Figure D5.4(b), which displays the temperature derivative of the resistivity as a function of temperature, shows an anomaly with onset at about 100 K (Ts). This is a clear signature related to the low-temperature structural distortion observed. A second feature is the observation of a small resistivity drop at about 40 K. This anomaly suggests that the 40-K superconducting phase, which is observable in β-$FeSe_{1-x}$ under high pressure, may exist at ambient pressure.

Temperature dependence of magnetoresistance (MR), resistivity (ρ), and Hall coefficient (R_H) of $FeSe_{1-x}Te_x$ with different Te contents has been reported (Wu *et al.*, 2013). The data imply a significant increase of impurity scattering rate of the carriers in highly Te-substituted samples. In addition, the magnitude of MR increases around 100 K, where the structural distortion occurs, for both pure FeSe and highly Te-substituted samples. The low-temperature thermoelectric power of β-$FeSe_{1-x}$ also exhibits a maximum at about 100 K (Wu *et al.*, 2015).

Symmetry analysis of FeSe shows four Raman active modes in this phase [$G_{Raman}=A_{1g}(Se)+B_{1g}(Fe)+2E_g(Se, Fe)$]. There were reports (Wu *et al.*, 2015) showing the frequency of E_g mode saturates below 90 K, which is consistent with the observation of the structural distortion at that temperature. In addition, optical spectroscopy with ultrafast laser excitation pulses also shows propagation of coherent acoustic phonons in FeSe and $FeSe_{1-x}Te_x$ indicating the softening of acoustic phonons near T_S (Wen *et al.*, 2012).

An interesting feature noted is the magnetic susceptibility of $FeSe_{1-x}$ crystal, which shows orientation dependence. There is a strong difference in low-temperature dependence of the susceptibilities with the applied field in parallel or in perpendicular to the (101) direction of the crystal, suggesting the charges along (101) may behave more like free electrons, whereas those perpendiculars to (101) are more localized (Wu *et al.*, 2013). Additionally, the ZFC and FC curves separate at ~230 K, suggesting an additional anomaly occurs at this temperature, which may correlate with the sign change observed in the Seebeck coefficient (Wu *et al.*, 2013).

D5.4.2 The Observation of Gap-Opening at Temperatures above the Structural Transition

With quasi-backscattering configuration, the Raman spectrum exhibits two distant peaks at 182 cm^{-1} for A_{1g}(Se) mode and 206 cm^{-1} for B_{1g}(Fe) mode. From the temperature-dependent frequency shifts of the phonon modes, a large (~6.5%) hardening of the B_{1g}(Fe) mode was observed and attributed to the suppression of local fluctuations of the iron spin state with a gradual decrease of the iron paramagnetic moment (Gnezdilov *et al.*, 2013). The temperature-dependent Raman spectra of the FeSe revealed quasielastic light scattering with electronic excitations develop below ~140 K, which is above the structural distortion temperature. This observation was attributed to the opening of an energy gap between low ($S = 0$) and higher spin states, which prevents magnetic order in FeSe. This feature may correlate with the Hall coefficient result, which shows a sharp rise at a temperature close to 140 K.

Two high-frequency phonon modes at 1423 and 1519 cm^{-1}, respectively, were examined in detail. The higher energy mode (1519 cm^{-1}) became highly asymmetric at temperatures below 140 K, and can be fit to two separate peaks. The partial density of states calculation of FeSe shows the zx/yz bands are mostly around 300 meV, which corresponds with the Raman mode near 1519 cm^{-1}. The result provides direct evidence that

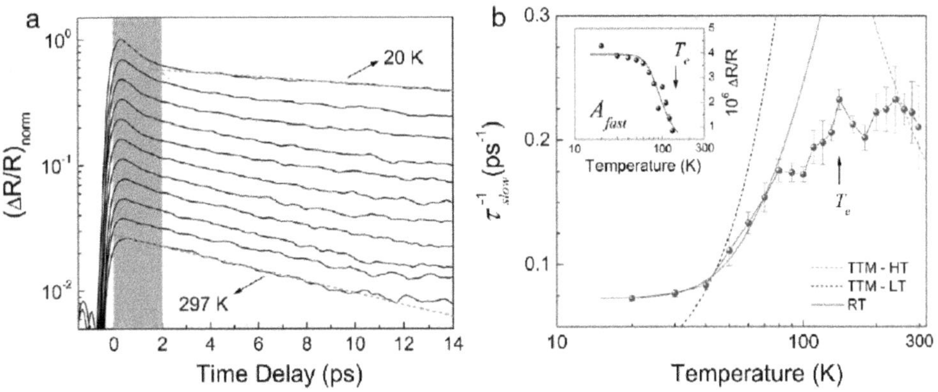

FIGURE D5.5 (a) Picosecond response of transient optical reflectivity of FeSe measured at different temperatures (solid lines from top to bottom: 20, 40, 60, 80, 120, 140, 180, 220, 260, 297 K). The data are normalized and vertically shifted for clarify. The dotted lines depict relaxation processes. (b) Temperature-dependent inverse relaxation time of the picosecond relaxation and amplitude of the sub-ps relaxation process (inset). (Wen *et al.*, 2012.)

at around 140 K d_{zx}/d_{yz}, orbitals split into two energy level $d_{zx/yz}$ and $d_{x^2-y^2}$ (Wu *et al.*, 2015).

Beyond the steady-state material properties, quasiparticle dynamics in $FeSe_{1-x}Te_x$, Figure D5.5(a), shows the optical reflectivity changes of superconducting FeSe in the picosecond timescale after a pulsed laser excitation (Wen *et al.*, 2012). The high-temperature traces can be well described by a single relaxation, whereas an additional sub-ps relaxation was observed at low temperatures. The picosecond relaxation (*slow* process) is attributed to carrier-phonon (*c-p*) thermalization that is ubiquitous at all temperatures and reflects the *c-p* coupling strength. The sub-ps relaxation (*fast* process) is absent at high temperatures and emerges below 130–140 K, reflecting the appearance of new electronic structure near the Fermi level. The amplitude of the fast relaxation process [Inset of Figure D5.5(b)] is found to gradually increase with decreasing *T* and saturate at ~70 K. This temperature dependency allows for attributing this process to the relaxation of gap-like quasiparticles. With a very crude approximation, the effective gap size was estimated to be ~36 meV. This high-temperature energy gap was observed above the structural phase transition and can be explained in terms of the short-range orbital and/or charge orders.

Earlier, Raman studies of iron-based superconductors identified the modes above 800 cm^{-1} as the electronic Raman scattering involving the *d*-orbitals of iron (Kumar A. *et al.*, 2010; Kumar P. *et al.*, 2010; Sanjuán ML *et al.*, 1992; Zhao SC *et al.*, 2009). More detailed studies (Hsiung HI *et al.*, 2018) reveal the presence of a high-frequency mode 1423 cm^{-1}, (labeled as S10 in the figure), which shows no obvious changes in frequency and FWHM (full width half maximum) as temperature decreases. The other mode near 1519 cm^{-1}, labeled as S11, can be fitted well to only one Lorentzian function at high temperature. However, when the temperature is below 130 K, this mode becomes highly asymmetric and fits well to a sum of two Lorentzian functions, labeled as S12 and S13, as shown in Figure D5.6(a). The temperature dependence of frequency of these three modes, S11, S12, and S13, are shown in Figure D5.6(b). The frequency of S13 shows a red shift, while the S12 mode shows no obvious change in frequency.

The energy diagram of the d-orbitals is illustrated schematically in Figure D5.6(c). The calculation shows that the d_{zx}/d_{yz} bands are mostly around 300 meV, which corresponds well with the Raman mode S11 observed. The detailed spectroscopy shows the splitting in d_{zx}/d_{yz}, as shown in Figure D5.6(b), correlates well with the temperature evolution of frequencies of S11, S12, and S13. The results indicate that when the temperature is lower than 130 K, S11 mode splits into two modes,

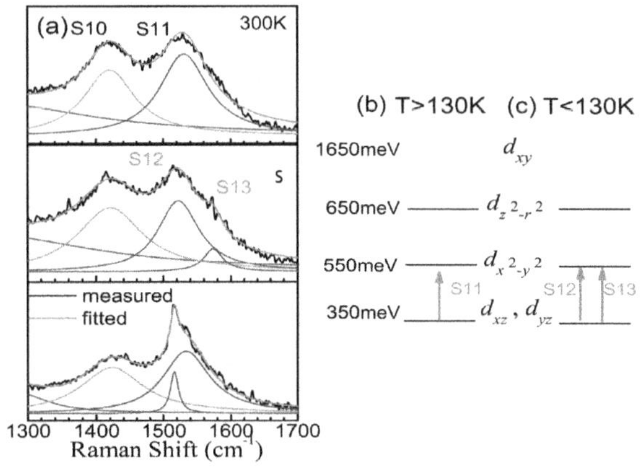

FIGURE D5.6 (a) Temperature evolution of high-frequency (1300–1700 cm^{-1}) modes. The thick line is the total fit to the experimental data, and the other lines represent the individual modes S10, S11, S12, and S13, respectively. (b) The crystal field energy level diagram for iron d-orbitals of FeSe with degeneracy of d_{zx} and d_{yz} orbitals. (c) The crystal field energy level diagram shows that below 130 K the d_{zx} and d_{yz} orbitals are lifted. (The energy values are not to scale). (From Hsiung I, 2018.)

S12 and S13. As shown in the schematics in Figure D5.3(d), the energy difference of these two modes, which can be associated with the energy difference between d_{zx} and d_{yz} orbitals when the degeneracy is lifted, decreases as temperature decreases. It implies that the splitting of d_{zx} and d_{yz} orbitals occurs around 130 K, and they become closer to achieve balance as temperature decreases. Subsequently, the anisotropy of the d_{zx} and d_{yz} orbitals would lead to the structural phase transition from tetragonal to orthorhombic (Lv W. *et al.*, 2009; Kumar A. *et al.*, 2010; Kumar P. *et al.*, 2010). It is noted that this temperature at which the *d*-orbitals splitting occurs is consistent with the temperature of energy gap-opening observed by transient optical spectroscopy discussed earlier (Wen YC *et al.*, 2012).

Terahertz (THz) spectroscopy was applied to study the superconducting $Rb_{1-x}Fe_{2-y}Se_2$ (Wang *et al.*, 2014). The temperature dependence of the optical conductivity and dielectric constant indicate a metal-to-insulator-type, orbital-selective transition at 90 K (T_{met}). At $T_{gap} = 61$ K, a gap-like suppression of the optical conductivity was observed and was followed by the occurrence of the superconducting transition at $T_c = 32$ K. This hierarchy of temperature $T_c < T_{gap} < T_{met}$ implies that the quasiparticles in the d_{xy} band are more strongly correlated.

Recently, ARPES studies suggested an orbital selective Mott transition to occur in the normal state of iron-selenide superconductor (Yi *et al.*, 2013), where some of the five d-orbitals independently undergo metal-to-insulator-like transitions. ARPES results on the single crystal FeSe indicated a strongly orbital-dependent renormalization (Maletz *et al.*, 2014). A comparable renormalization was found for the d_{xz}/d_{yz}, while the d_{xy} shows a three times larger renormalization. In this aspect, the d_{xy} band shows the most peculiar behavior: it is not subject to k_z dispersion, shows a stronger renormalization than the other two bands, and is shifted to higher binding energies, thus not taking part in the formation of the Fermi surface. It is noted that an energy gapping at ~125 K was reported by ARPES in $FeSe/SrTiO_3$ thin films (Tan *et al.*, 2013). On the other hand, in the high-energy region, the Raman spectra of $A_xFe_{2-y}Se_2$ exhibit a broad, asymmetric peak around 1600 cm^{-1}, which was identified as a two-magnon process involving optical magnons (Zhang *et al.*, 2012a). The intensity of the two-magnon peak falls sharply on entering the superconducting phase, suggesting a complete mutual proximity effect occurring within a microscopic structure based on nanoscopic phase separation.

D5.4.3 Fe-Vacancy in Fe Chalcogenides

The exact crystal structure of the superconducting alkaline metal intercalated A-FeSe with a $T_C = 30$ K remains controversial. Detailed studies revealed that the $A_xFe_{2-y}Se_2$ system shows strong phase separation in superconducting samples including crystals. A charge balanced compound $A_2Fe_4Se_5$ with a formal oxidation state close to +2 for Fe was realized. This $A_2Fe_4Se_5$ compound exhibits an ordered Fe-vacancy that could be described by a √5×√5×1 supercell (a tetragonal *I4/m* unit cell) below an ordering temperature $T_S = 500$–578 K depending on the intercalated metal A (Bao *et al.*, 2011, Dagotto, 2013). Neutron scattering data suggest a block antiferromagnetic order developed slightly below T_S ($T_N = 471$–559 K) with an ordered magnetic moment ~3.3 μ_B/Fe at 10 K (Bao *et al.*, 2011).

Raman studies on $K_{0.8}Fe_{1.6}Se_2$, $Tl_{0.5}K_{0.3}Fe_{1.6}Se_2$, and $Tl_{0.5}Rb_{0.3}Fe_{1.6}Se_2$ have been carried out together with the first-principles calculations (Zhang *et al.*, 2012b). The authors show that the abundant phonon modes in $A_xFe_{2-y}Se_2$ are the consequence of the superstructure with ordered iron vacancies. The first STM study of the alkaline metal Fe–Se was on thin films of $K_xFe_{2-y}Se_2$ grown by MBE (Li *et al.*, 2011b) showed that the samples consist of rich mixture of various Fe-vacancy phases, with √5×√5, √2×√2 or even √2×√5 super-structure.

Figure D5.7 displays the SEM and high-resolution transmission electron microscope (HRTEM) images of the β-FeSe

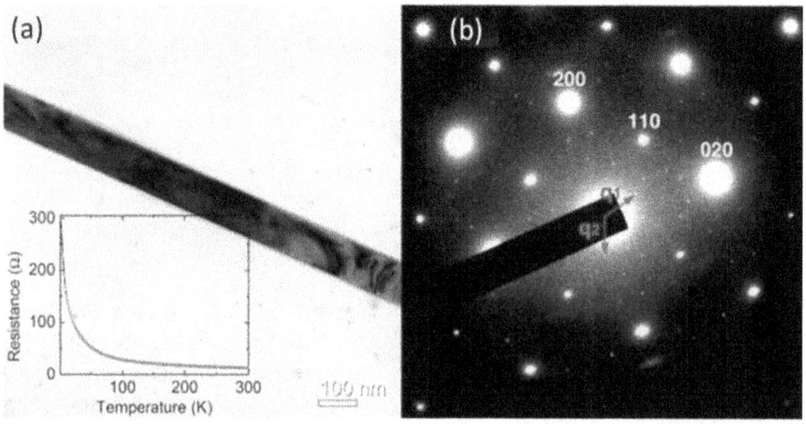

FIGURE D5.7 (a) TEM image of a FeSe nanowire. (Inset) Temperature-dependent transport property of the same nanowire. (b) The SAED pattern of the nanowire, revealing a tetragonal lattice along the [001] zone-axis direction. Superstructure wave vectors $q_1 = (1/5, 3/5, 0)$ and $q_2 = (3/5, 1/5, 0)$ are indicated by arrows. (Chen *et al.*, 2014.)

nanowires (NWs), showing excellent crystalline tetragonal structure (Chen *et al.*, 2014). From energy-dispersive spectroscopy (EDS) results, the NWs are uniform in composition with Fe/Se ratio of 4/5. In addition, the results of HRTEM and X-ray diffraction confirm the presence of $\sqrt{5}\times\sqrt{5}$ Fe-vacancy order as that observed in $A_2Fe_4Se_5$. This observation resolved the puzzle of the absence of superconductivity in the FeSe nanowires, which instead were found to be insulators. After careful examination, the presence of $\sqrt{5}\times\sqrt{5}\times1$ Fe-vacancy order also exists in Fe_4Se_5 nanoparticles and nanosheets, and in samples from $K_2Fe_4Se_5$ crystals after extracting K by iodine (Chen *et al.*, 2014). It is noted that the scanning tunneling microscopy (STM) topography and spectroscopy results on FeSe film grown on bilayer graphene by MBE techniques (Song *et al.*, 2011) show $\sqrt{5}\times\sqrt{5}$ reconstruction, which indicates the presence of the ordered Fe-vacancy on sample surface.

D5.4.4 Phase Diagram of Fe Chalcogenides

In an attempt to gain more insight into the exact stoichiometry and structure of the FeSe superconductors, it was found that the synthesized nanoparticles become superconducting with an onset transition temperature at ~20 K (Chang *et al.*, 2012). This result provides an opportunity to better understand the physics behind the observation of the sensitive stoichiometry (Hsu *et al.*, 2008, McQueen *et al.*, 2009), the dramatic T_C enhancement in FeSe by applying pressure, and the exact phase diagram of the FeSe system.

As mentioned in previous section, the presence of $\sqrt{5}\times\sqrt{5}$ Fe-vacancy was observed in FeSe nanowires, the as-grown Fe_4Se_5 nanoparticles and nanosheets. Figure D5.8(a) shows the temperature dependence of resistance of the nanosheet. The data fit well to the variable-range-hopping model with $T^{-1/3}$ power, suggesting it is a Mott insulator. Magnetic susceptibility measurements of the nanosheets did not detect superconducting signal. However, superconductivity emerges after annealing the samples at 700°C for a few hours, as shown in Figure D5.8(b).

Besides the $\sqrt{5}\times\sqrt{5}\times1$ Fe-vacancy order, several different types of Fe-vacancy order in tetragonal β-$Fe_{1-x}Se$ were observed. Detailed structure calculations and simulations suggested the presence of β-Fe_3Se_4 ($x = 0.2$) that exhibits $\sqrt{2}\times\sqrt{2}$ superstructure with d_{100} shift every other (001) plane (the $FeSe_4$ tetrahedron layer). The third type of Fe-vacancy order, β-Fe_9Se_{10} ($x = 0.1$), yielded a tetragonal lattice with a twinned Fe-vacancy order of $\sqrt{10}\times\sqrt{10}$ with $\frac{1}{2}d_{310}$ shift every other (001) plane yielding.

The above results reveal that the PbO-type tetragonal β-$Fe_{1-x}Se$ compound exists in a rich composition range (McQueen *et al.*, 2009), to a region with great deficiency of iron. It is known that superconductivity of the molecular beam epitaxy (MBE)-grown monolayer FeSe films could be completely destroyed by Se doping (Song *et al.*, 2011). The films were semiconducting where the extra Se dopants were ordered into $\sqrt{5}\times\sqrt{5}$ superstructure at high doping concentration. Such an observation is consistent with the presence of Fe-vacancies in a Se-rich growth environment and ordering into superstructures in these ultra-thin FeSe films.

Based on the above observations, a temperature-doping phase diagram for the Fe–Se superconducting system was proposed (Chen *et al.*, 2014), as illustrated in Figure D5.9, which is very similar to the phase diagram of the superconducting La_2CuO_{4+y}. The magnetic and insulating/semiconducting phases of β-$Fe_{1-x}Se$ with Fe-vacancy orders act as the parent phase of FeSe superconductor, instead of the previously argued β-$Fe_{1+\delta}Te$, which has different magnetic and electronic features compared to β-$Fe_{1+\delta}Se$. The superconducting β-$Fe_{1.01}Se$ with $T_c = 8.5$ K first discovered could result from the β-Fe_4Se_5 nanosheets by annealing at 700°C in vacuum and then quenching to room temperature.

A recent detailed magneto-transport study on the Fe-deficient $Fe_{4+\delta}Se_5$ nanowire (Yeh *et al.*, 2020)

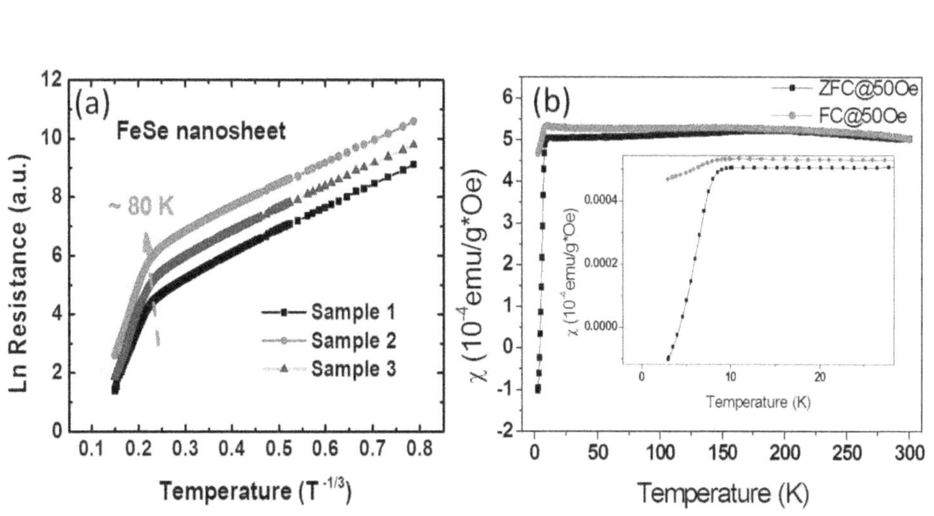

FIGURE D5.8 (a) Temperature dependence of resistance for FeSe nanosheets plots in terms of the variable-range-hopping model with $T^{-1/3}$ power. (b) The magnetic susceptibility of the same nanosheets after annealing at 700°C, showing superconductivity at ~10 K. (Wu *et al.*, 2015.)

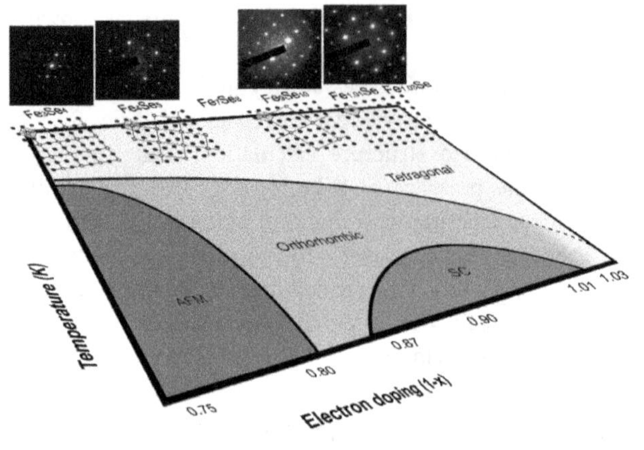

FIGURE D5.9 Schematic of the proposed temperature-doping phase diagram of Fe–Se superconducting system, showing regions for antiferromagnetism (AFM) and superconductivity. (Chen *et al.*, 2014.)

unambiguously presented an important observation: the $Fe_{4+\delta}Se_5$ with Fe-vacancy order is a Mott insulator showing a Verwey-like transition. In this work, it first observed a first-order metal–insulator (MI) transition with an onset transition temperature at $T \approx 28$ K at zero magnetic fields. The authors used an alternative-current excitation to investigate the underlying nature of the transition. The data show, as shown in Figure D5.10, similar to that observed in Fe_3O_4 nanowire, a strong frequency dependence of the metal–insulator resistive transition, and the spin–orbit coupling is crucial for the transition observed. Detailed magneto-transport measurements were also used to better understand the spin-relevant correlation with the phase transition. The observation of weak magnetic field in the nanowires below 17 K provides additional evidence that the low-temperature state is an antiferromagnetic charge order state.

Another study on the crystal structure and properties of $Fe_{4+x}Se_5$ compounds using rapid-thermal-annealing process, by properly controlling the temperature and time, to investigate the evolution from insulating state to superconducting state has been performed. The materials used in this study were single-crystal-like nanoparticles. The findings further unambiguously demonstrated that the $Fe_{(1-x)}Se$ with Fe-vacancy order is the parent compound of FeSe superconductors (Yeh, 2020), as shown in Figure D5.11. Detailed XPS measurements on samples after RTA treatment show a direct correlation between the percentage of Fe^{3+} state and superconducting transition temperature T_c.

In the alkaline metal–doped FeSe, except the stable $A_5Fe_4Se_5$ non-superconducting matrix in the superconducting crystals, generally a separated phase with an expanded *c*-axis and a composition close to $A_xFe_2Se_2$ (Dagotto, 2013) was found and assigned to be the superconducting phase with $T_C \sim$ 29–32 K. There were also reports attribute the superconductivity to originate from a parent phase of semiconducting antiferromagnetic $A_2Fe_3Se_4$ with rhombus ($\sqrt{2}\times2\sqrt{2}$) Fe-vacancy order, of $A_2Fe_7Se_8$ parallelogram structure or of $A_3Fe_4Se_6$ (Wu *et al.*, 2015). Recent results on the $K_2Fe_{4+x}Se_5$ samples, which are chemically homogeneous and stoichiometric (Wang *et al.*, 2015a), show the possibility to change the sample from non-superconducting state, which is a Mott insulator with strong Fe-vacancy order, to a superconducting state, in which Fe-vacancy order is completely suppressed. These results are consistent with the phase diagram for the FeSe system proposed.

A series of ARPES measurements on FeSe films with different thickness (Tan *et al.*, 2013) found significant difference in the Fermi surfaces of the monolayer film to the multilayer

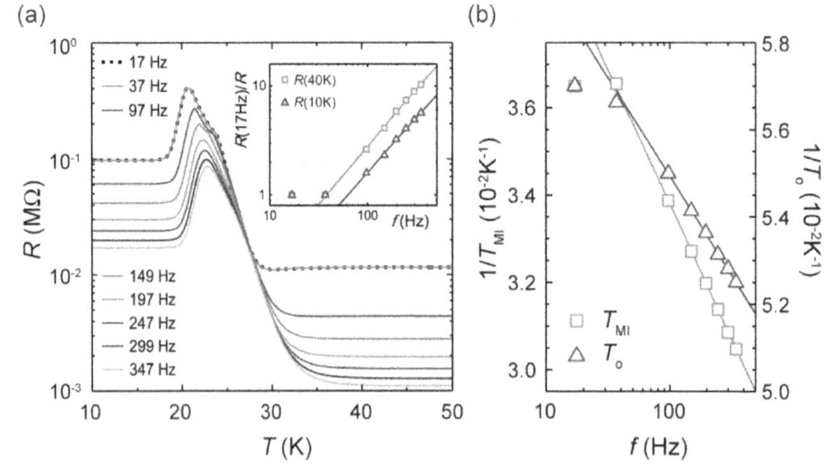

FIGURE D5.10 The AC-excitation frequency (f) dependence on the transition. (a) $R(T)$ curves measured in warming at different f. The curves of 17 Hz and 37 Hz are almost overlapped. The inset shows the f dependence of conductance $1/R$(10 K) and $1/R$(40 K), normalized by the values at $f = 17$ Hz. The solid lines show $1/R \propto f^{\alpha}$, where α is 1.06 for $1/R$(40 K) and 1.02 for $1/R$(10 K). (b) $1/T_a$ versus f, where T_a is T_{MI} or T_o. The solid lines depict $1/T_a = (-k_B/E_a)\ln(f) + constant$, an alternative expression of the Arrhenius law. E_a is 32.3 meV for T_{MI} and 44.0 meV for T_o. (Yeh, *et al.*, 2020.)

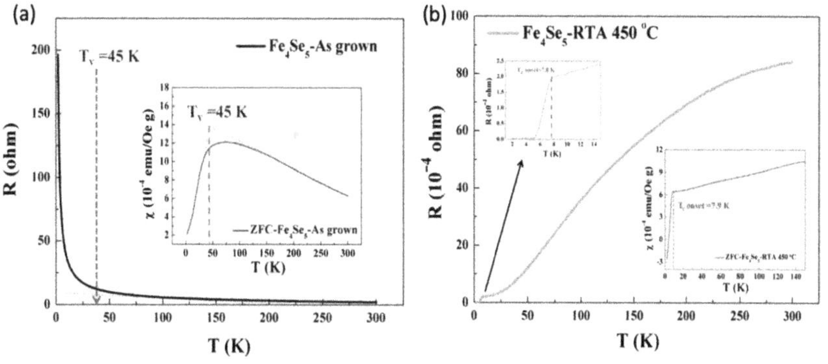

FIGURE D5.11 (a) The temperature dependence of resistance and magnetic susceptibility of the as-grown Fe_4Se_5 polycrystal. The data reveal the metal–insulator transition at low temperature. (b) The same sample heat treated by RTA process becomes superconducting with T_c ~ 8K, as shown in both magnetic and resistive measurements.

film. A phase diagram for FeSe as a function of lattice constant was proposed, Figure D5.12a, that the competition between SDW and superconductivity, with the "lattice constant" as the tuning parameter, is the underlying reason why superconductivity only observed in monolayer film. A support to this proposal was the success to enhance T_c of the monolayer FeSe, with Nb:STO as buffer, epitaxially grown on $KTaO_3$, which has relatively larger lattice parameter.

Another independent work proposed a different phase diagram based on the carrier concentration tuning by a series of annealing procedures on the monolayer FeSe films (Figure D5.12b) (He *et al.*, 2013). They identified three evolution phases: the initial non-superconducting *N* phase, the intermediate mixture of *N* phase and superconducting *S* phase, and the final *S* phase. Neither the *N* phase nor the *S* phase shows bands crossing the Fermi level in the center of BZ. On the other hand, the *S* phase shows two degenerate electron pockets at the BZ corner. Under an optimized annealing condition, the highest T_c is enhanced to ~65 K, giving $2\Delta/k_BT_c$ to the order of 6–7.

The ARPES results seem all to imply that the substrate–film interface plays a crucial role, e.g., the lattice mismatch-induced strain, the charge carrier transfer between film and substrate. However, other effects such as the excess selenium or oxygen vacancies forming during the annealing and the rearrangement of the substrate–film interface need to be considered.

D5.4.5 Structure under High Pressures

Among all Fe-based superconductors, FeSe has the most dramatic response to applied pressure, with T_C increasing from $T_{C,onset}$ = 12 K at ambient pressure to 37 K at 4–6 GPa (Mizuguchi *et al.*, 2008, Medvedev *et al.*, 2009) and T_c drops above 6 GPa. It is noted that FeSe undergoes an orthorhombic to hexagonal transition at ~6 GPa. Nuclear magnetic resonance studies were carried out to better understand the electronic and magnetic changes involved in the higher T_C. The results showed that the antiferromagnetic fluctuations are also enhanced under pressure (Imai *et al.*, 2009), which

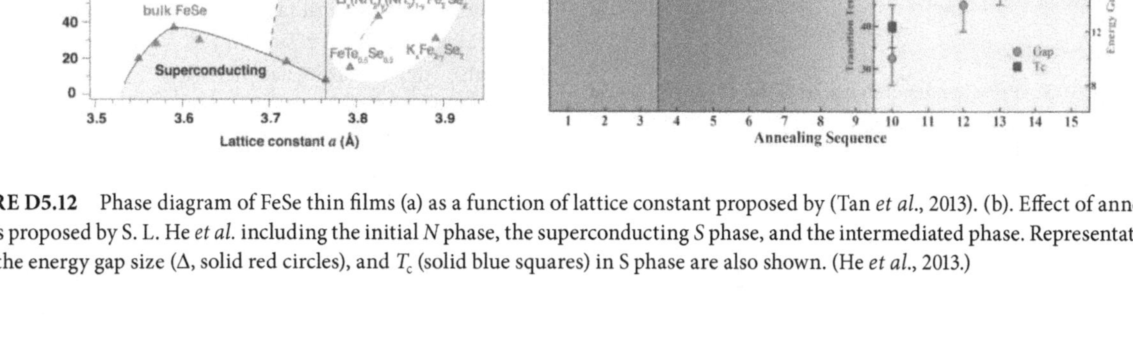

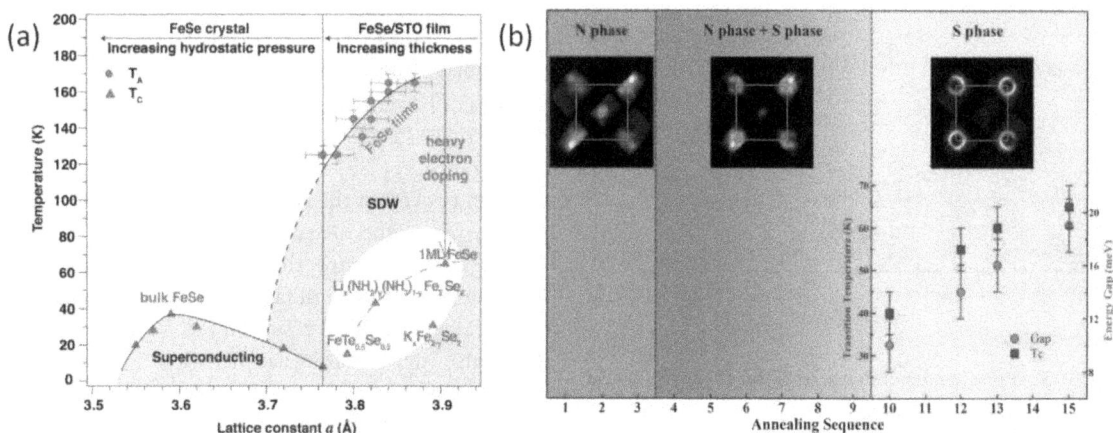

FIGURE D5.12 Phase diagram of FeSe thin films (a) as a function of lattice constant proposed by (Tan *et al.*, 2013). (b). Effect of annealing process proposed by S. L. He *et al.* including the initial *N* phase, the superconducting *S* phase, and the intermediated phase. Representative FS maps, the energy gap size (Δ, solid red circles), and T_c (solid blue squares) in S phase are also shown. (He *et al.*, 2013.)

further supports the picture that superconductivity in FeSe is correlated with magnetic fluctuations.

Under applied external pressure, no structural transition up to ~12 GPa for $A_xFe_{1-y}Se_2$ (A = Cs, Rb, K) samples was observed (Svitlyk *et al.*, 2011). T_C was found to decrease monotonically with increasing pressure for $K_{0.8}Fe_{1-x}Se_2$ and $Tl_{0.6}Rb_{0.4}Fe_{1.67}Se_2$ and was completely suppressed at a critical pressure of ~10 GPa (Sun et al., 2012). Surprisingly, above 10 GPa, a new superconducting phase with higher T_C (~48 K) appeared and again was suppressed above 13.2 GPa (Sun *et al.*, 2012).

It was found that $K_{0.8}Fe_{1-x}Se_2$ superconductors transformed from an antiferromagnetic state with *I4/m* symmetry to a paramagnetic state with *I4/mmm* symmetry in the pressure range of 9.2–10.3 GPa, at which superconductivity tends to disappear, shown in Figure D5.13. On the other hand, the non-superconducting $K_2Fe_4Se_5$ insulator was found to transform into an intermediate metallic state and coexisted with the insulating state over a significant range of pressure up to ~10 GPa (Gao *et al.*, 2014). The Fe-vacancy order was fully suppressed at 11 GPa (Ying *et al.*, 2013), similar to the pressures where the second superconducting phase in $K_{0.8}Fe_{1-x}Se_2$ samples re-emerged.

D5.4.6 Symmetry of Superconducting Gap

Several STM studies addressed the superconducting pairing symmetry. Hanaguri *et al.* first argued for $s_\pm$ order parameter in Fe(Se,Te), and the gapped spectra suggested no nodes in the order parameter, though there was indication for gap anisotropy (Hanaguri *et al.*, 2010). In contrast, Fridman *et al.* observed a V-shaped gap in FeTeSe, suggesting the presence of

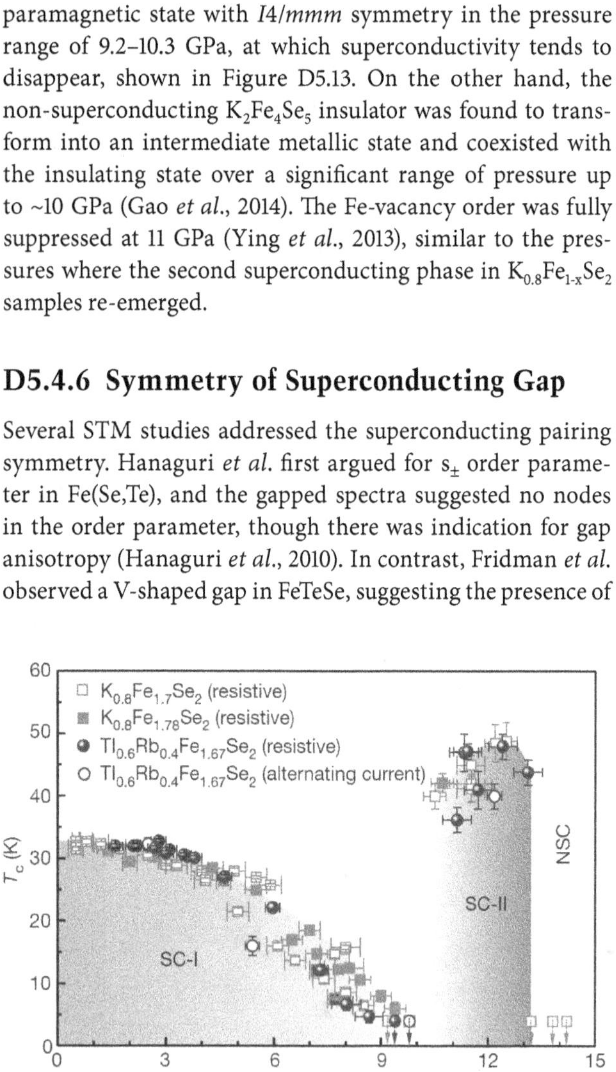

FIGURE D5.13 Pressure dependence of T_C for three $K_{0.8}Fe_{1.7}Se_2$, $K_{0.8}Fe_{1.78}Se_2$, and $Tl_{0.6}Rb_{0.4}Fe_{1.67}Se_2$. The symbols represent the pressure–temperature conditions for which T_C values were observed from the resistive and alternating current susceptibility measurements; symbols with downward arrows represent the absence of superconductivity to the lowest temperature (4 K). (Sun *et al.*, 2012.)

nodes (Fridman *et al.*, 2011). An early specific heat measurement under high magnetic field also suggested a nodal energy gap in FeSe crystal (Wu *et al.*, 2009). More recently, evidences for nodal order parameter were found in MBE-grown FeSe (Song *et al.*, 2011). This discrepancy in the gap features was attributed to the difference in interlayer coupling.

For the superconducting gap in $A_xFe_{2-y}Se_2$, a nodeless, isotropic superconducting gap of 10–15 meV is observed at the electron-like pockets at the BZ corner (Dagotto, 2013). Similar isotropic superconducting gap (~6.2 meV) at the Z-centered electron pocket is observed in $(Tl,K)Fe_{1.78}Se_2$ (Wang *et al.*, 2011a). Whether the superconducting gap should be nodeless is still debatable. ARPES results generally report isotropic behavior of the superconducting gaps, however, some bulk-sensitive experiments indicate gap anisotropy. The complexity of the bulk- and surface-related behavior maybe the cause to the controversial views of the paring symmetry. Another essential issue is that $A_xFe_{2-y}Se_2$ intrinsically exhibits mesoscopic phase separation including superconducting, semiconducting, or even AFM insulating phase (Chen *et al.*, 2011), which shows vacancy order.

ARPES (Moreschini *et al.*, 2014) was used to investigate the broken translational symmetry in the superconductor $Fe_{1+y}Te_{1-x}Se_x$. The spectral weight distribution follows the one-Fe periodicity rather than the conventional BZ defined by the two-Fe unit cell. The results also pointed out that the form of the perturbing potential and the symmetries of the Fe d-orbitals lead to differences in the orbital character of the bands even at the nominally equivalent locations in the reciprocal space. Such parity switching is unusual and could be unique to the family- of iron-based superconductors.

Several ARPES measurements have also been performed on $A_xFe_{2-y}Se_2$. Electron-like Fermi pockets are observed at the corner of the BZ (M), and the hole-like bands are absent at the zone center (Qian *et al.*, 2011). However, there is a small electron-like (κ band) feature enclosing at Z (k_z=π). Regarding the missing hole pocket, it was reported that the top of the hole band at Γ sinks to the ~90 meV below the Fermi level (Qian *et al.*, 2011). With the hole pockets missing at the zone center, the idea of Fermi nesting may no longer be the key for superconductivity in this system. It was suggested that the Fermi nesting in $A_xFe_{2-y}Se_2$ might still happen between the circular electron pockets at the BZ corner through interband scattering. Owing to the inequivalent Se potential with respect to the wave vector connecting the two-electron pockets, such interband scattering will favor the opposite signs of the pairing parameter. ARPES result on (Tl, K)$Fe_{1.78}Se_2$ (Wang *et al.*, 2011a) also supports a similar picture.

There were intensive investigation on the superconducting gap of FeSe, however, due to the complexity of multiorbital nature, the experimental results are controversial whether it is nodeless (Dong J *et al.*, 2009; Bourgeois-Hope P. *et al.*, 2016; Lin J. *et al.*, 2011; Jiao L *et al*, 2017; Abdel-Haez *et al*, 2013; Khasanov R. *et al.*, 2010; Sprau P. *et al.*, 2017; Song C *et al.*, 2011;

Kasahara S *et al.*, 2014 16–24) or nodal (Hashimoto T *et al.*, 2018; Suzuki Y. *et al.*, 2015; Kreisel A. *et al.*, 2017 25–27). It also remains highly controversial on the orbital nature of the low-energy excitations that are relevant to superconductivity (Sprau P. *et al.*, 2017; Song C *et al.*, 2011; Kasahara S *et al.*, 2014; Kreisel A. *et al.*, 2017; Tanatar M. *et al.*, 2016; Boehmer A. *et al.*, 2018; Back S. *et al.*, 2015; Boehmer A. *et al.*, 2015).

Recently, an unexpected new result (Liu D. *et al.*, 2018) on the electronic structure and superconducting gap of the bulk FeSe superconductor (*Tc* 8.0 K) in the nematic state based on high-resolution laser-based ARPES measurements was reported. The authors observed a highly anisotropic Fermi surface around the Brillouin zone (BZ) center with the aspect ratio of ~3 between the long axis (along *ky*) and short axis (along *kx*). The superconducting gap along the Fermi pocket is extremely anisotropic, varying between ~3 meV along the short axis of the Fermi surface and zero along the long axis within the experimental precision (±0.2 meV). Detailed band structure analysis, combined with band structure calculations, indicates that the Fermi surface is dominated by a single d_{xz} orbital. Moreover, the authors find that the superconducting gap size shows an anticorrelation with the d_{xz} spectral weight near the Fermi level.

The laser ARPES system allows the authors to collect data points, as shown in Figure D5.14, of the superconducting gap along the entire Fermi surface measured with superhigh resolution so that detailed quantitative fitting with various possible gap functions can be carried out. The twofold symmetry of the superconducting gap has excluded the possibility of a pure s-wave which would have fourfold symmetry. The measured gap is fitted by the *p*-wave form $\Delta p = |\Delta_0 \cos(\theta)|$ and the fitted curve is shown as a lighter line in Figure D5.14(d). An alternative gap form is *s+d* type, which can also assume twofold symmetry (Xu H. *et al.*, 2016; Johnston D., 2010). In Figure D5.14(d), the data are also fitted with two types of *s+d* form. The first was to try $\Delta s, d = |\Delta_0 + \Delta_1 \cos(2\theta)|$, which contains a simple s-wave form Δ_0 and a *d*-wave form $\Delta_1 \cos(2\theta)$. The data do not fit well with this relation as shown the darker line in

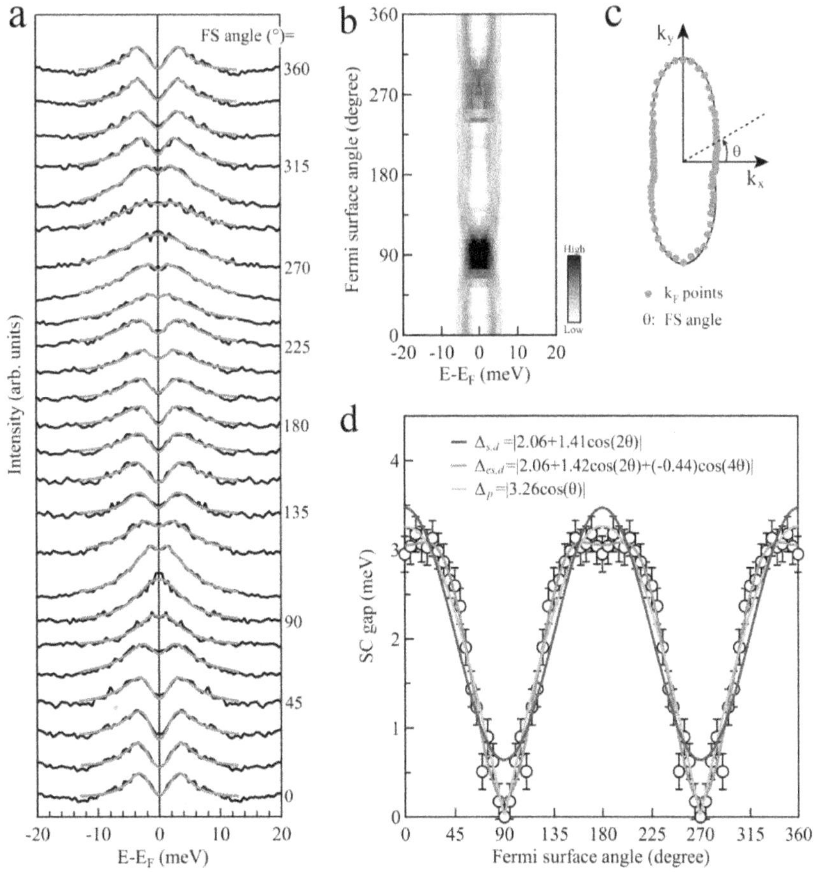

FIGURE D5.14 Momentum dependence of the superconducting gap for FeSe measured at 1.6 K. (a) Symmetrized EDCs at different Fermi momenta along the Fermi surface measured at 1.6 K. These curves are fitted with a phenomenological gap formula (red curves). The same data are plotted in (b) as a false color image where the variation of the superconducting peak position can be directly visualized. The location of the Fermi momentum is defined by the Fermi surface angle θ as shown in (c). (d) Momentum dependence of the superconducting gap derived from fitting the symmetrized EDCs in (a). To enhance the data statistics and keep the twofold symmetry, the gap value is obtained by averaging over the four quadrants. The measured gap (empty circles) is fitted by *s+d* (darker curve), extended -*s+d* (intermediate-shaded curve), and *p* (lighter curve) wave pairing gap forms. (From Liu D. 2018.)

Figure D5.14(d), particularly near the minimal and maximal gap regions. A second fit used $\Delta es,d = |\Delta_0 + \Delta_1 \cos(2\theta) + \Delta_2 \cos(4\theta)|$, which contains an anisotropic s-wave $\Delta_0+\Delta_2 \cos(4\theta)$ and a d-wave form $\Delta_1 \cos(2\theta)$. The fitted curve is marked as an intermediate-shaded line in Figure D5.14(d). We note that within our experimental precision (~0.2 meV), we cannot differentiate between the cases of zero node, two nodes, and four nodes along the Fermi surface for the $es + d$ gap form.

Additional support to p-wave paring in FeSe is that no drop of the Knight shift across T_c is observed in NMR measurements of bulk FeSe, which is also compatible with a possible triplet pairing (Watson M. *et al.*, 2016; Okazaki K. *et al.*, 2012). Other relevant observations include the coexisting N′eel and stripe spin fluctuations from neutron scattering (Lei B. *et al.*, 2016; Boehmer A., 2013 51,52), and the observation of charge ordering in FeSe suggests the presence of an additional magnetic fluctuation with a rather small wave vector (Zhang Y. *et al.*, 2017) that is related to intra-pocket scattering around the Γ point (Tam Y. *et al.*, 2015). The measured gap function can be fitted to a simple p-wave gap function is intriguing. This result seems to support the recent intensive discussions regarding whether FeSe and related compound connote topological superconductors. We shall present this subject in the next section.

D5.4.7 A Platform for Detecting the Majorana Zero Mode

One of the most exciting recent developments in the study of Fe-based superconductors is that they can be associated as connate topological superconductors (Shi X *et al.*, 2017 9), which can be viewed as an internal hybrid system that has conventional superconductivity in the bulk but topological superconductivity on the surface (Shi X *et al.*, 2017; Xu G. *et al.*, 2016 9, 10). Hao and Hu first noted (Hao & Hu, 2014) a tiny band gap around the M point in the band structure of monolayer FeSe/SrTiO3 (FeSe/STO) (Wang Q. *et al.*, 2016 39) from ARPES measurement (He S. *et al.*, 2013; Tan S. *et al.*, 2013; Liu D. *et al.*, 2012; Lee JJ *et al.*, 2014; Miyata Y. *et al.*, 2015; Peng R. *et al.*, 2014), and soon proposed the first theoretical study of non-trivial band topology for the single-layer FeSe/STO, in which a band inversion can take place at the M points to create non-trivial topology. It was further found that the band inversion can easily take place at the Γ point if the anion height from the Fe layers is high enough. For FeSe, the height can be increased by substituting Se with Te (Wu X. *et al.*, 2015; Wang Z. *et al.*, 2015). For iron pnictides, the As height is predicted to be high enough in the 111 series, LiFeAs, to host the non-trivial topology (Zhang P. *et al.*, 2019 14).

Besides these intrinsic topological properties from the Fe d-orbitals, non-trivial topology can also stem from bands outside the Fe layers. For example, the As p-orbitals in the As layers of 122 CaFeAs2 are shown to be described by a model similar to the Kane–Mele model in graphene (Wu X. *et al.*, 2015). Most recently, because of the improvement of sample quality and experimental resolution, there has been increasing experimental evidence for topological properties in iron-based superconductors (Zhang P. *et al.*, 2018; Wang D. *et al.*, 2018; Liu O. *et al.*, arXiv 1807,0127816–18). The theoretically predicted band inversions, together with the topologically protected surface states, have been directly observed. Majorana-like modes have been observed in several iron-chalcogenide materials (Wang D. *et al.*, 2018; Liu O. *et al.*, arXiv 1807,0127817,18). All of this progress has made iron-based superconductors a new research frontier for topological superconductivity.

The topological insulator phase emerges around the zone corner in which a topological bulk gap is induced by a combination of the effects of SOC and substrate stress. In 2015, evidence of p–d-orbital inversion (Wang ZJ *et al.*, 2015) by angle-resolved photoemission spectroscopy (ARPES) and evidence of iron impurity zero-bias conductance peaks (ZBCPs) by scanning tunneling spectroscopy (STS) (Yin JX *et al.*, 2015) were confirmed in bulk crystal $FeTe_{0.55}\ Se_{0.45}$. These observations clearly push forward the idea of achieving topological insulating band structure by chemical substitution in a bulk Fe-based material. The definitive evidence came from high-resolution spin-resolved ARPES in 2018 that observed a spin–momentum locking pattern of the Dirac surface band, a hallmark of a topological insulator, in $FeTe_{0.55}\ Se_{0.45}$ (Zhang P. *et al.*, 2018). More remarkably, this Dirac surface band opens a reasonably sized superconducting gap below the bulk T_c, which is identical to the interfacial state in the Fu–Kane model (Fu and Kane, 2008), but with the advantage of much larger Δ/E_F in $FeTe_{0.55}Se_{0.45}$. This advantage, which would separate the non-topological bound states from the MBS in a superconducting vortex core, immediately motivated a high-resolution STM study on $FeTe_{0.55}Se_{0.45}$ (Wang DF *et al.*, 2018). Indeed, a spatial non-split ZBCP was observed in the vortex cores of $FeTe_{0.55}Se_{0.45}$ (Wang DF *et al.*, 2018), which is a fingerprint of the Majorana bound state (MBS) (Figure D5.15). The behaviors of the observed ZBCPs were studied carefully in $FeTe_{0.55}\ Se_{0.45}$ (Wang DF *et al.*, 2018) under different temperatures, magnetic fields, and tunneling barriers. Its full width at half maximum approaches the STM resolution limit, indicating a single peak. The intensity line profile of the ZBCPs was reproduced very well by a theoretical Majorana wave function with the parameters extracted from the experimental values of the Dirac surface state. An additional suppression mechanism of the ZBCPs is also identified by temperature-dependent measurements. Although an unambiguous demonstration of the Majorana behavior needs more evidence, those results drew a self-consistent picture in which the observed ZBCPs are the MBSs associated with the surface Dirac electrons. Therefore, the unique features of the FeSe-based superconducting platform for Majorana research have attracted much attention from the community.

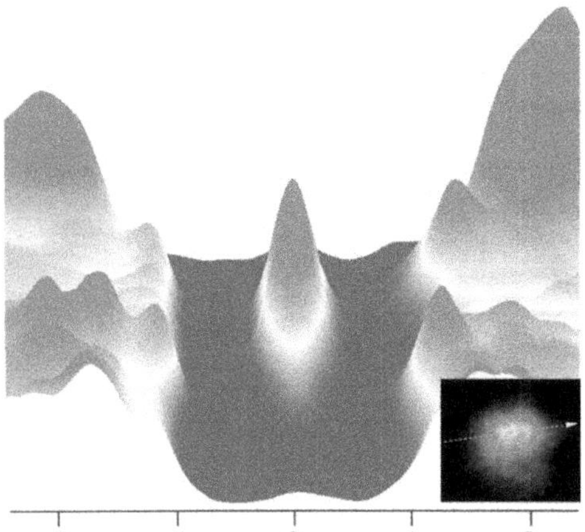

FIGURE D5.15 A robust Majorana bound state observed in FeSC. A 3D display of a line-cut intensity plot of Majorana bound states on the vortex core (inset) of FeTe0.55Se0.45. It is visualized that the spatial non-split zero bias peaks around the vortex core. (From Kong LY & Ding H, 2019.)

There are many important unsolved issues remaining for future studies, such as MBS-induced spin-selective Andreev reflection, quantized conductance of $2e^2/h$, the Majorana interference pattern, and non-Abelian statistics demonstrated in a Majorana braiding process. Another puzzle that needs to be resolved is the absence of MBS on many vortices in $FeTe_{0.55}$ $Se_{0.45}$ (Wang DF *et al.*, 2018; Kong LY and Ding H, 2019). Although the MBSs in the vortex of Fe(Te, Se) are purer and the large quasiparticle gap may protect the fermion parity information carried by a pair of MBSs robust against perturbations, which is a good feature required by fault-tolerant quantum computing, a realistic method for experimentally realizing braiding for MBSs in vortices has yet to be proposed. The scenario of MBSs in a vortex was recently put forward to explain the robust ZBCPs observed on iron-impurity sites of $FeTe_{0.55}Se_{0.45}$, suggesting that an anomalous vortex core is induced by a single impurity without an extrinsic magnetic field (Jiang K *et al.*, 2019). The successes with $FeTe_{0.55}Se_{0.45}$ have also prompted researchers to pursue better platforms with higher T_c, such as Fe(Te, Se) monolayer (Shi X *et al.*, 2017), Li(Fe, Co)As (Zhang P. *et al.*, 2019), and (Li, Fe) OHFeSe (Liu O *et al.*, 2018). All of this progress has made iron-based superconductors a new research frontier for topological superconductivity.

D5.5. Possible Origin of Superconductivity for Fe Chalcogenide

There were enormous theoretical efforts to investigate the interplay between the structural phase transition, magnetism, and superconductivity of iron chalcogenides. The structural phase transition, which leads to anisotropy in several physical properties at low temperature, suggests it originates from the strong electron correlation. The corresponding fluctuations might also be related to the pairing mechanism of superconductivity. Therefore, a thorough understanding of the electronic state of matters giving rise to the structural phase transition is essential to better understand the iron-based superconductors.

One natural candidate is the electronic nematicity (Fradkin *et al.*, 2010). The electronic nematicity is the state breaking C_4 rotational symmetry due to a *d*-wave interaction. For iron-based superconductors, the measurable "order parameters" include the difference in lattice constant along *x* and *y* directions, the difference in occupation number in *xz* and *yz* orbitals ($n_{nem} = n_{xz}$-n_{yz}), and the magnetic torque measurement that is related to the difference in the zero-frequency spin susceptibility along (q_x,0) and (0,q_y) direction [$\chi_{nem} = \chi(q_x,0) - \chi(0,q_y)$]. While the first order parameter is related to the lattice degrees of freedom, the latter two are both of electronic origins but with very different microscopic mechanisms. The main difficulty in resolving this debate experimentally lies in the fact that whenever one of these order parameters is non-zero, all others become non-zero simultaneously, for all of them break exactly the same C_4 rotational symmetry (Lee *et al.*, 2012). Therefore, typical thermodynamic properties used to characterize symmetry-breaking phases are blind to the orbital and the spin scenarios.

For the pairing symmetry of the superconductivity, the orbital fluctuation near zero wave-vector favors $s_\pm$ pairing symmetry. For the spin scenario, since the superconductivity is mediated by the magnetic fluctuations at (π, 0) and (0, π), the pairing symmetry would favor a sign change between different Fermi surface pockets. As a result, $s_\pm$ symmetry for systems with both electron and hole pockets and d-wave gap symmetry for systems without hole pockets are the most robust consequences in the spin scenario. Proposals for some pairing states that break time-reversal symmetry have been also proposed for some materials in situations where two competing pairing states exist. As for the normal state, one unique feature of the spin scenario is that a meta-nematic state could exist between the nematic and the stripe-like antiferromagnetic states (Dagotto, 2013. Experimental evidences suggest that Fe chalcogenide superconductors might lean toward the orbital scenario.

On the other hand, it is important to pinpoint, from the material point of view, what is the material origin for the superconductivity in Fe-chalcogenide materials, A recent novel synthesis approach developed made possible the preparation of single-phase $K_{2-x}Fe_{4+y}Se_5$ samples with homogenous crystal phase and exact stoichiometry (Wu *et al.*, 2015). And the detailed studies of the correlation between its structure and superconductivity (Wang *et al.*, 2019) provide deeper insight into the understanding of this issue. Figure D5.16 displays the data of the $K_{2-x}Fe_{4+y}Se_5$ samples prepared by under different conditions. Detailed X-ray and TEM works on a low temperature (LT-process at 300°C) annealed $K_2Fe_4Se_5$ samples show √5×√5 Fe-vacancy

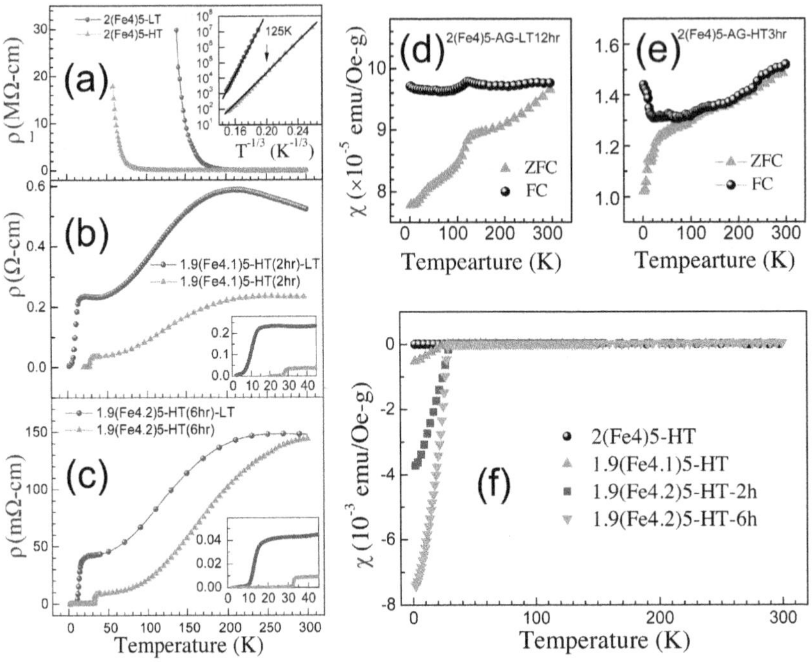

FIGURE D5.16 (a)–(c) The temperature dependence of $K_{2-x}Fe_{4+y}Se_5$ prepared by LT and HT processes. (d), (e) The magnetic susceptibility of LT and HT treated $K_2Fe_4Se_5$ samples. (f) The magnetic susceptibility of HT treated $K_{2-x}Fe_{4+y}Se_5$ samples with various extra irons. (Replotted figures of Wang *et al.*, 2015.)

order in these samples. Magnetic measurements exhibit a transition at 125 K. Resistivity measurements confirm that this sample is a Mott insulator with $ln(\rho)\sim T^{(-1/3)}$, and the resistivity becomes extremely high at the temperature close to the 125 K transition observed.

The high temperature processed $K_2Fe_4Se_5$ (HT-process at 750C and then quenched to room temperature) reveals no magnetic feature in between 50 and 300 K, but shows two superconducting-like transitions at 29 and 11 K. The corresponding XRD pattern shows the √5×√5 Fe-vacancy order is suppressed. The sample resistivity is reduced though still behaves like semiconductor. The result suggests the sample contains low density of disconnected superconducting puddles.

Adding extra iron and preparing the samples with high-temperature heat-treatment then quenched to room temperature is found to effectively suppress the √5×√5 Fe-vacancy order. The obtained stoichiometric single-phase sample shows substantial increase in superconducting volume. The resistivity of the sample with highest superconducting volume is essentially metallic, with a sharp T_c at 31 K.

The detailed structural studies (Wang *et al.*, 2019) reveal more subtle features for better understanding the origin to superconductivity in these materials. Figure D5.17 shows the detailed X-ray diffractions at different temperatures up to 750°C for the $K_2Fe_4Se_5$ (2(4)5) and excess-Fe $K_{1.9}Fe_{4.2}Se_5$ (1.9(4.2)5) samples under both warming and cooling cycles, which exhibit no difference in the diffraction patterns. The significant shifts in diffraction peaks (008) and (051) at 275C in warming, as shown in Figure D5.15(a), are the indication of the transition from the Fe-vacancy order to disorder state. The superstructural peaks related to the vacancy order state completely disappear above the ordering temperature. The authors obtained the best fit to the diffraction pattern with *I4*/*m* symmetry by considering that the high-temperature phase Fe could occupy both 16*i* (fully occupied) and 4*d* site (originally empty site). This result indicates the crystal structure of 245 could be well described by single phase with I4/m symmetry even above 285C (Bao W., 2015).

More carefully examined the diffraction patterns, as shown in Figures D5.17(a) and D5.17(b), the data for 2(4.2)5 sample are found to best fit with two different sets of parameters, one with Fe-vacancy ordered and the other disordered Fe-vacancy, under the same *I4*/*m* symmetry. Figure D5.17(c) and D5.17(d) are the lattice parameters of 245 and 2(4.2)5 samples, respectively, below (both fit with *I4*/*m* symmetry) and above the Fe-vacancy order temperature. There are two phases coexisting in excess-Fe sample shown in Figure D5.18(a). However, from Figure D5.18(b), those additional features can be associated with the planes (002), (130), (132), (134), (136) …, based on *I4*/*m* symmetry, which are all crossing the Fe-4*d* site. This result indicates that the presence of excess Fe atoms shall begin to fill the original empty 4*d* sites. This observation suggests that the added excess-Fe atoms play a critical role in maintaining the crystal lattice (high-temperature phase) with the *I4*/*m* symmetry as that of the vacancy-ordered state. Above the vacancy order temperature, the lattice remains with the *I4*/*m* symmetry but with disorder occupation of the Fe atom to all possible Fe site so that the additional features disappear.

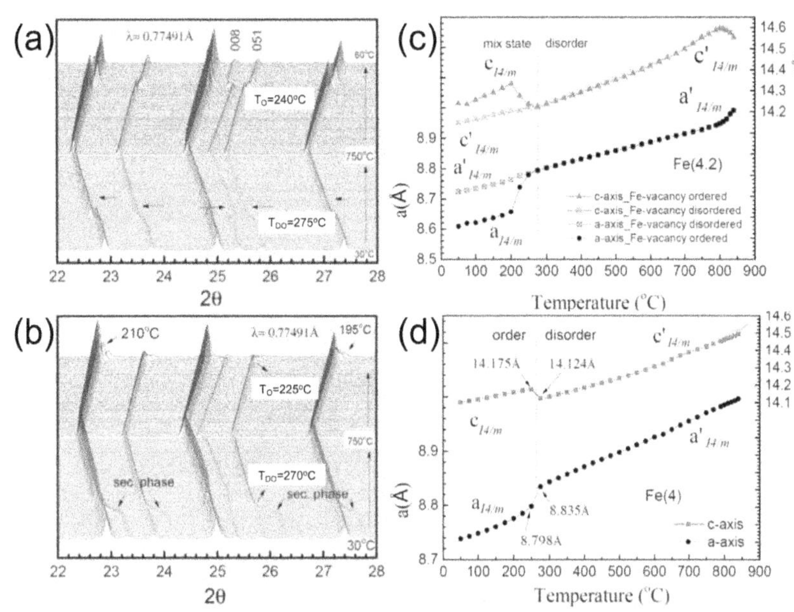

FIGURE D5.17 Evolution of the temperature dependence x-ray powder diffraction pattern of (a) $K_2Fe_4Se_5$ (245) (b) $K_{1.9}Fe_{4.2}Se_5$ (2(4.2)5). The lighter shaded data indicate increasing temperature; the darker shaded data indicate decreasing temperature. Refined lattice parameters of $K_2Fe_4Se_5$ and $K_{1.9}Fe_{4.2}Se_5$ samples are displayed in figure (c) and (d) respectively. All data are refined using the *I4/m* unit cell. A cell with compressed *c*-axis and expanded *ab*-plane is observed above vacancy disorder temperature in 245 sample; two cells with different lattice parameters, one for vacancy-ordered cell, the other is vacancy disordered, are found to best fit the data for 2(4.2)5 sample below Fe-vacancy disorder temperature. (From Wang *et al.*, 2019.)

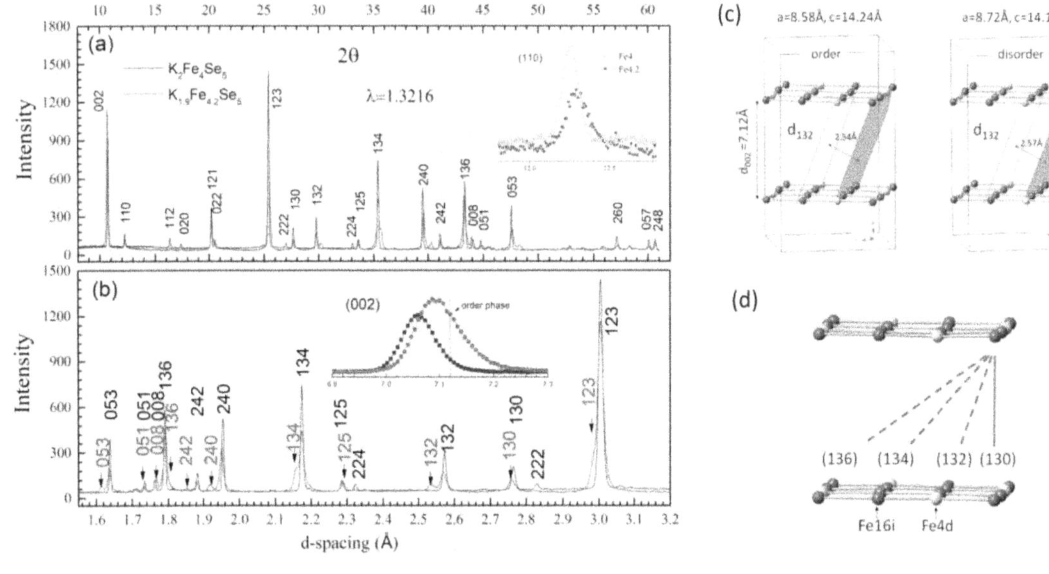

FIGURE D5.18 Synchrotron powder diffraction patterns of the polycrystalline $K_{2-x}Fe_{4+y}Se_5$ samples. (a) Superconducting 1.9(4.2)5 and non-superconducting parent 245 x-ray diffraction pattern in diffraction angles. The superlattice peaks, e.g., (222) and (224), are almost suppressed in the superconducting 1.9(4.2)5 sample. Inset shows the (110) diffraction profile. (b) The same diffraction patterns displayed with lattice *d*-spacing show that there are two similar phases in 1.9(4.2)5 sample. One is similar to 245 parent phase but without (222) superlattice, and the other with slightly smaller *d*-spacing analyze. Inset shows the (002) profile. (c) The two similar phases with difference d-spacing for (132) plane. (d) Schematic of (130), (132), (134), and (136) plane. The iron-vacancy site 4*d* is marked by white symbol, and the 16*i* site is marked using khaki color. (From Wang *et al.*, 2019.)

The experimental observations unambiguously demonstrate that the random occupation of Fe atom in the lattice is key for superconductivity. These results also support the viewpoint that $A_xFe_{2-y}Se_2$ tends toward mesoscopic phase separation into superconducting, semiconducting, or AFM insulating phase proposed (Chen *et al.*, 2011) based on the $K_xFe_{2-y}Se_2$ ARPES study, and the results by Ricci *et al.* using nano-focus x-ray diffraction (Ricci *et al.*, 2015). It certainly will be essential to study in detail how the Fermi surface evolves from the non-superconducting parent state with Fe-vacancy order to the disordered superconducting state in the well-characterized $K_{2-x}Fe_{4+y}Se_5$ samples.

Based on the existing observations, we can draw the following picture regarding the origin of superconductivity of the Fe chalcogenides: There exists a Fe-vacancy–ordered non-superconducting parent phase, which is a Mott insulator and exhibits a magnetic (or Verwey-like) transition at low temperature. With additional Fe filling the vacancies and processed in high temperature to make the vacancies disorder, superconductivity emerges. For a superconducting sample, several relevant features are observed: the magnetic transition observed in the parent compound disappears, while a *d*-orbital splitting accompanying with a gap-opening is observed and appears at about the same temperature, and at a lower temperature a structural distortion appears before the onset of superconductivity. Recent detailed x-ray studies reveal that superconductivity in this system is closely associated with the presence of nanoscale phase separation in the material (Ricci *et al.*, 2015, Wang *et al.*, 2015). Strong evidences suggest that superconductivity originates from the disorder of Fe-vacancy in the material. The similarities for this system to the cuprates are striking. Thus, the long pending question whether magnetic and superconducting state are competing or cooperating for cuprate superconductors may also apply to the Fe-chalcogenide superconductors. It is also important to look into whether the Fe-based- and the cuprate superconductors share the same origin for superconductivity.

D5.6 Potential Applications of Fe-Chalcogenide Superconductors

The critical current density (J_C) of $FeSe_{1-x}Te_x$ film is higher than 1 MA/cm^2 (Si *et al.*, 2013). Using an *in situ* measurement technique, J_C of 1.3×10^7 A/cm^2 is reported in monolayer FeSe films (Ge *et al.*, 2015). Furthermore, J_C up to 10^5 A/cm^2 in $FeSe_{0.5}Te_{0.5}$-coated conductors at 30 T was achieved, which demonstrates a great potential on high magnetic field magnet application (Si *et al.*, 2013). Josephson junctions made of $FeSe_{0.5}Te_{0.5}$ nano-bridges were demonstrated by focus ion beam (FIB) (Wu *et al.*, 2013). Their I-V characteristics follow RCSJ model quite well with Jc of hundreds microamperes. Further applications such as SQUID are feasible based on this kind of Josephson junction.

The superconducting $FeTe_{1-x}Se_x$, with its advantage of the low anisotropy, has been made into wire form using the powder-in-tube method with Fe sheath and FeSe powder. Good connections between the Fe sheath and $FeTe_{1-x}Se_x$ superconducting phase were obtained at the edge of the cross-section. The estimated critical current density for single core wire is up to 600 A/cm^2 at 4 K without any magnetic field (Ding *et al.*, 2012). The J_C of wire can be improved to over 10^3 A/cm^2 by multi-core form (Ozaki *et al.*, 2012). However, the J_C of wire is degraded by one to two orders of magnitude at presence of external field. There are obviously plenty of rooms for the development of Fe-chalcogenide superconducting wire.

Acknowledgements

The authors thank all members in the Superconductor Laboratory in the Institute of Physics, Academia Sinica, in particular, Drs. T.K. Chen, A.Y. Fang, F.C. Hsu, J.Y. Luo, T.W. Huang, C.H. Wang, Y.C. Wen, K.W. Yeh, and C.C. Chang for their long-term support and contributions. We also acknowledge the financial support from Academia Sinica and the Ministry of Science and Technology of Taiwan.

References

Abdel-Haez M, et al., (2013) Temperature dependence of lower critical field H_{c1}(T) shows nodeless superconductivity in FeSe. Phys. Rev. B 88: 174512.

Agatsuma S, Yamagishi T, Takeda S, Naito M (2010) MBE growth of FeSe and $Sr_{1-x}K_xFe_2As_2$. Physica C: Superconductivity 470: 1468–1472.

Aprau P, et al., (2017) Discovery of orbital-selective Cooper pairing in FeSe. Science 357: 75.

Back S, et al., (2015) Orbital-driven nematicity in FeSe. Nat. Mater. 14, 210.

Bao W, et al., (2011) A novel large moment antiferromagnetic order in K 0.8 Fe 1.6 Se 2 superconductor. Chinese Physics Letters 28: 086104.

Boehmer A, et al., (2013) Lack of coupling between superconductivity and orthorhombic distortion in stoichiometric single-crystalline FeSe. Phys. Rev. B 87: 180505(R).

Boehmer A, et al., (2015) Origin of the tetragonal-to-orthorhombic phase transition in FeSe: a combined thermodynamic and NMR study of nematicity. Phys. Rev. Lett. 114: 027001.

Boehmer A, et al., (2018) Nematicity, magnetism and superconductivity in FeSe. J. Phys.: Condens. Matter 30: 023001.

Bourgeois-Hope P, et al., (2016) Thermal conductivity of the iron-based superconductor FeSe: Node-less gap with a strong two-band character. Phys. Rev. Lett. 117: 097003.

Burrard-Lucas M, et al., (2013) Enhancement of the superconducting transition temperature of FeSe by intercalation of a molecular spacer layer. Nat. Mater. 12: 15–19.

Chang CC, et al., (2012) Superconductivity in PbO-type tetragonal FeSe nanoparticles. Solid State Commun. 152: 649–652.

Chang CC, et al., (2015) Superconductivity in Fe-chalcogenides. Physica C-Superconductivity and Its Applications 514: 423–434.

Chang HH, et al., (2014) Growth and characterization of superconducting b-FeSe tupe iron chalcogenide nanowires. Supercond. Sci. Technol. 27: 025015.

Chen F, et al., (2011) Electronic identification of the parental phases and mesoscopic phase separation of KxFe2–ySe2 superconductors. Phys. Rev. X 1: 021020.

Chen TK, et al., (2014) Fe-vacancy order and superconductivity in tetragonal beta-Fel-xSe. Proc. Natl. Acad. Sci. U S A 111: 63–68.

Dagotto E (2013) Colloquium: the unexpected properties of alkali metal iron selenide superconductors. Rev. Mod. Phys. 85: 849–867.

Dai P 2015 Antiferromagnetic order and spin dynamics in iron-based superconductors. Rev. Mod. Phys. 87: 855.

Deguchi K, Takano Y, Mizuguchi Y (2013) Physics and chemistry of layered chalcogenide superconductors. Sci. Technol. Adv. Mater. z13: 054303

Demura S, et al., (2012) Electrochemical synthesis of iron-based superconductor FeSe films. J. Phys. Soc. Jpn. 81: 043702.

Ding QP, et al., (2012) Magneto-optical imaging and transport properties of FeSe superconducting tapes prepared by the diffusion method. Supercond. Sci. Technol. 25: 025003.

Dong J, et al., (2009) Multigap nodeless superconductivity in FeSex: evidence from quasiparticle heat transport. Phys. Rev. B 80: 024518.

Fradkin E, et al., (2010) Nematic Fermi fluids in condensed matter physics. Annual Review of Condensed Matter Physics 1: 153–178.

Fridman I, Yeh KW, Wu MK, Wei JYT (2011) STM spectroscopy on superconducting FeSel–xTex single crystals at 300mK. J. Phys. Chem. Solids 72: 483.

Fu L and Kane CL (2008) Superconducting proximity effect and Majorana fermions at the surface of a topological insulator. Phys. Rev. Lett. 100: 096407.

Gao PW, et al., (2014) Role of the 245 phase in alkaline iron selenide superconductors revealed by high-pressure studies. Phys. Rev. B 89: 094514.

Ge JF, et al., (2015) Superconductivity above 100 K in single-layer FeSe films on doped $SrTiO_3$. Nat. Mater. 14: 285–289.

Gnezdilov V, et al., (2013) Interplay between lattice and spin states degree of freedom in the FeSe superconductor: dynamic spin state instabilities. Phys. Rev. B 87: 144508.

Guo J, et al., (2010) Superconductivity in the iron selenideKxF e_2Se_2(0≤x≤1.0). Phys. Rev. B 82: 180520(R).

Hanaguri T, Niitaka S, Kuroki K, Takagi H (2010) Unconventional s-wave superconductivity in Fe(Se,Te). Science 328: 474–476.

Hao N and Hu J (2014) Topological phases in the single-layer FeSe. Phys Rev X 4: 031053.

Hashimoto T, et al., (2018) Superconducting gap anisotropy sensitive to nematic domains in FeSe. Nat. Commun. 9: 282.

He S, et al., (2013) Phase diagram and electronic indication of high-temperature superconductivity at 65 K in single-layer FeSe films. Nat. Mater. 12: 605–610.

Hsiung HI, et al., (2018) Observation of iron d-orbitals modifications in superconducting FeSe by Raman Spectra Study Physica C-superconductivity and its applications 552: 61–63.

Hsu FC, et al., (2008) Superconductivity in the PbO-type structure alpha-FeSe. Proc. Natl. Acad. Sci. U S A **105**: 14262–14264.

Hu J, et al., (2012) Unified minimum effective model of magnetic properties of iron-based superconductors. Phys. Rev. B 85: 144403.

Huang TW, et al., (2010) Doping-driven structural phase transition and loss of superconductivity in *M*xFel-xSe delta (*M*=Mn, Cu). Phys. Rev. B 82: 6.

Imai T, Ahilan K, Ning FL, McQueen TM, Cava RJ (2009) Why does undoped FeSe become a high-T_c superconductor under pressure? Phys. Rev. Lett. 102.

Imai Y, Sawada Y, Nabeshima F, Maeda A (2015) Suppression of phase separation and giant enhancement of superconducting transition temperature in FeSe(1-x)Te(x) thin films. Proc. Natl Acad. Sci. U S A 112: 1937–1940.

Jiang K, et al., (2019) Quantum anomalous vortex and Majorana zero mode in iron-based superconductor Fe(Te,Se). Phys. Rev. X 9: 011033.

Jiao L, et al., (2017) Superconducting gap structure of FeSe. Sci. Rep. 7: 44024.

Johnston D. (2010) The puzzle of high temperature superconductivity in layered iron pnictides and chalcogenides. Adv. Phys. 59: 803.

Kasahara S, et al., (2014) Field-induced superconducting phase of FeSe in the BCS-BEC cross-over. Proc. Natl. Acad. Sci. USA. 111: 16309.

Khasanov R., et al., (2010) Evolution of two-gap behavior of the superconductor $FeSe_{1-x}$. Phys. Rev. Lett. 104: 087004.

Kong L.Y. and Ding H. (2019) Majorana gets an iron twist. National Science Review 6: 196.

Kreisel A, et al., (2017) Orbital selective pairing and gap structures of iron-based superconductors. Phys. Rev. B 95: 174504.

Krzton-Maziopa A, et al., (2012) Synthesis of a new alkali metal-organic solvent intercalated iron selenide superconductor with T_c approximate to 45 K. J. Phys.: Condens. Mat. 24.

Kumar A, et al., (2010) First-principles analysis of electron correlation, spin ordering and phonons in the normal state of $FeSe_{1-x}$. J. Phys.: Condens. Mat. 22: 385701.

Kumar P, et al., (2010) Anomalous Raman scattering from phonons and electrons of superconducting. Solid State Communications 150: 557–560.

Lai X, et al., (2015) Observation of superconductivity in tetragonal FeS. J. Am. Chem. Soc. 137: 10148–10151.

Lee JJ, et al., (2014) Interfacial mode coupling as the origin of the enhancement of T_c in FeSe films on $SrTiO_3$. Nature 515: 245–248.

Lee WC, Lv W, Tranquada JM, Phillips PW (2012) Impact of dynamic orbital correlations on magnetic excitations in the normal state of iron-based superconductors. Phys. Rev. B 86: 094516.

Lei B, et al., (2016) Evolution of high-temperature superconductivity from low-T_c phase tuned by carrier concentration in FeSe thin akes. Phys. Rev. Lett. 116: 077002.

Li L, et al., (2011) Superconductivity and magnetism in FeSe thin films grown by metal–organic chemical vapor deposition. Supercond. Sci. Technol. 24: 015010.

Lin J, et al., (2011) Coexistence of isotropic and extended s-wave order parameters in FeSe as revealed by low-temperature specific heat. Phys. Rev. B 84: 220507.

Liu D, et al. (2012) Electronic origin of high-temperature superconductivity in single-layer FeSe superconductor. Nat. Commun. 3: 931.

Liu D, et al. (2018) Orbital origin of extremely anisotropic superconducting gap in nematic phase of FeSe superconductor Phys. Rev. X 8: 031033.

Lu XF, et al., (2014) Superconductivity in LiFeO2Fe2Se2with anti-PbO-type spacer layers. Phys. Rev. B 89: 020507(R).

Lv W, et al., (2009) Orbital ordering induces structural phase transition and the resistivity anomaly in iron pnictides. Phys. Rev. B 80: 224506.

Maletz J, et al., (2014) Unusual band renormalization in the simplest iron-based superconductorFeSe_{1-x}. Phys. Rev. B 89: 220506(R).

McQueen TM, et al., (2009) Tetragonal-to-orthorhombic structural phase transition at 90 K in the superconductor Fe1.01Se. Phys. Rev. Lett. 103.

Medvedev S, et al., (2009) Electronic and magnetic phase diagram of beta-Fe1.01Se with superconductivity at 36.7 K under pressure. Nat. Mater. 8: 630–633.

Miyata Y, et al., (2015) High-temperature superconductivity in potassium-coated multilayer FeSe thin films. Nat. Mater. 14: 775.

Mizuguchi Y, et al., (2008) Superconductivity at 27 K in tetragonal FeSe under high pressure. Appl. Phys. Lett. 93: 152505.

Moreschini L, et al., (2014) Consequences of broken translational symmetry inFeSexTe_{1-x}. Phys. Rev. Lett. 112.

Okazaki K, et al., (2012) Evidence for a cos(4') modulation of the superconducting energy gap of optimally doped FeTe0:6Se0:4 single crystals using laser angle-resolved photoemission spectroscopy. Phys. Rev. Lett. 109: 237011.

Ozaki T, Deguchi K, Mizuguchi Y, Kawasaki Y, Tanaka T, Yamaguchi T, Kumakura H, Takano Y (2012) Fabrication of binary FeSe superconducting wires by diffusion process. Journal of Applied Physics 111: 112620.

Pachmayr U, et al., D (2015) Coexistence of 3d-Ferromagnetism and Superconductivity in (Li1-xFex)OH (Fe1-yLiy) Se. Angewandte Chemie-International Edition 54: 293–297.

Peng R et al., (2014) Tuning the band structure and superconductivity in single-layer FeSe by interface engineering. Nat. Commun. 5: 5044.

Qian T, et al., (2011) Absence of a Holelike Fermi surface for the iron-based K0.8Fe1.7Se2 superconductor revealed by angle-resolved photoemission spectroscopy. Phys. Rev. Lett. 106: 187001.

Ricci A, et al., (2015) Direct observation of nanoscale interface phase in the superconducting chalcogenideKxFe_2–y$Se_2$with intrinsic phase separation. Phys. Rev. B 91: 020503(R).

Sanjuán ML, et al., (1992) Electronic Raman study of Fe^{+2} in $FePX_3$ (X = S, Se) layer compounds, Phys. Rev. B 46: 11501–11506.

Shi X, et al., (2017) $FeTe_{1-x}Se_x$ monolayer films: towards the realization of high-temperature connate topological superconductivity. Sci. Bull. 62: 503–507.

Si W, et al., (2013) High current superconductivity in FeSe0.5Te0.5-coated conductors at 30 tesla. Nat. Commun. 4: 1347.

Song C, et al., (2011) Direct observation of nodes and twofold symmetry in FeSe superconductor. Science 332: 1410.

Song CL, et al., (2011) Molecular-beam epitaxy and robust superconductivity of stoichiometric FeSe crystalline films on bilayer graphene. Phys. Rev. B 84: 020503(R).

Sun L, et al., (2012) Re-emerging superconductivity at 48 kelvin in iron chalcogenides. Nature 483: 67–69.

Sun, Y., Zhang, W., Xing, Y. *et al*, (2014) High temperature superconducting FeSe films on $SrTiO_3$ substrates. Sci. Rep. **4**, 6040. https://doi.org/10.1038/srep06040

Suzuki Y, et al., (2015) Momentum-dependent sign inversion of orbital order in superconducting FeSe. Phys. Rev. B 92: 205117.

Svitlyk V, et al., (2011) Temperature and pressure evolution of the crystal structure of A(x)(Fe(1-y)Se)2 (A = Cs, Rb, K) studied by synchrotron powder diffraction. Inorg. Chem. 50: 10703–10708.

Tam Y, et al., (2015) Itinerancy-enhanced quantum fluctuation of magnetic moments in iron-based superconductors. Phys. Rev. Lett. 115: 117001.

Tan S, et al., (2013) Interface-induced superconductivity and strain-dependent spin density waves in FeSe/SrTiO3 thin films. Nat. Mater. 12: 634–640.

Tanatar M, et al., (2016) Origin of the resistivity anisotropy in the nematic phase of FeSe. Phys. Rev. Lett. 117: 127001.

Wang CH, et al., (2015) Disordered Fe vacancies and superconductivity in potassium-intercalated iron selenide ($K_{2-x}Fe_{4+y}Se_5$). Europhys. Lett. 111: 27004.

Wang CH, et al., (2019) Role of the extra Fe in $K_{2-x}Fe_{4+y}Se_5$ superconductors. Proc. Natl. Acad. Sci. U S A 116(4): 1104–1109.

Wang D, Kong L, Fan P, et al., (2018) Evidence for Majorana bound states in an iron-based superconductor. Science 362(6412): 333–335.

Wang DF, et al., (2018) Science 362: 333–335.

Wang M, et al., (2012) Spin waves and magnetic exchange interactions in insulating $Rb_{0.89}Fe_{1.58}Se_2$. Nat. Commun. 2: 580.

Wang MJ, et al., (2009) Crystal orientation and thickness dependence of the superconducting transition temperature of tetragonal FeSe1–x thin films. Phys. Rev. Lett. 103: 117002.

Wang Q, et al., (2016) Strong interplay between stripe spin fluctuations, nematicity and superconductivity in FeSe. Nat. Mater. 15: 159.

Wang QY, et al., (2012) Interface-induced high-temperature superconductivity in single unit-cell FeSe films on $SrTiO_3$. Chinese Phys. Lett. 29: 037402.

Wang XP, et al., (2011) Strong nodeless pairing on separate electron Fermi surface sheets in (Tl, K) Fe1.78Se2 probed by ARPES. Europhys. Lett. 93: 57001.

Wang Z, et al., (2014) Orbital-selective metal-insulator transition and gap formation above TC in superconducting Rb(1-x)Fe(2-y)Se2. Nat. Commun. 5: 3202.

Wang Z, et al., (2015) Archimedean solidlike superconducting framework in phase-separated K0.8Fe1.6+xSe2 (0≤x≤0.15). Phys. Rev. B 91: 064513.

Wang ZJ, et al., (2015) Topological nature of the FeSe0.5Te0.5 superconductor. Phys. Rev. B 92: 115119.

Watson M, et al., (2016) Evidence for unidirectional nematic bond ordering in FeSe. Phys. Rev. B 94: 201107(R).

Wen YC, et al., (2012) Gap opening and orbital modification of superconducting FeSe above the structural distortion. Phys. Rev. Lett. 108: 267002.

Wu CH, et al., (2013) Transport properties in FeSe0.5Te0.5 nanobridges. Appl. Phys. Lett. 102: 222602.

Wu, MK, et al., (2009) The development of the superconducting PbO-type beta-FeSe and related compounds. Physica C-Superconductivity and Its Applications 469: 340–349.

Wu MK, et al., (2015) An overview of the Fe-chalcogenide superconductors. J. Phys. D: Appl. Phys. 48: 323001.

Wu MK, Wang MJ, Yeh KW, (2013) Recent advances in beta-FeSe1-x and related superconductors. Science and Technology of Advanced Materials 14: 014402.

Wu X, Qin S, Liang Y, et al., (2015) CaFeAs2: a staggered intercalation of quantum spin Hall and high-temperature superconductivity. Phys. Rev. B 2015; 91: 081111.

Wu X, Qin S, and Liang Y, et al., (2016) Topological characters in Fe(Te1 – xSex) thin films. Phys. Rev. B 93: 115129.

Xu G, Lian B and Tang P, et al., (2016) Topological superconductivity on the surface of Fe-based superconductors. Phys. Rev. Lett. 117: 047001.

Xu H. et al., (2016) Highly anisotropic and twofold symmetric superconducting gap in nematically ordered FeSe0:93S0:07. Phys. Rev. Lett. 117: 157003.

Xu M, Ge QQ, Peng R, et al., (2012) Evidence for an s-wave superconducting gap in KxFe2-ySe2 from angle-resolved photoemission. Phys. Rev. B 85, 220504(R).

Ye F, et al., (2011) Common crystalline and magnetic structure of superconducting A2Fe4Se5 (*A*=K,Rb,Cs,Tl) single crystals measured using neutron diffraction. Phys. Rev. Lett. 107: 137003.

Yeh K. W., et al., (2008) Tellurium substitution effect on superconductivity of the alpha-phase iron selenide. Europhys. Lett. 84: 4.

Yeh K.Y. (2020) Ph.D. dissertation, National Tsing Hua University, Hsinchu, Taiwan.

Yeh K. Y, et al., (2020) Magneto-transport studies of Fe-vacancy ordered $Fe_{4+\delta}Se_5$ nanowires. Proc. Natl. Acad. Sci. U S A 117(23): 12606–12610.

Yi M, et al., (2013) Observation of temperature-induced crossover to an orbital-selective Mott phase inAxFe2–ySe2(*A*=K, Rb) superconductors. Phys. Rev. Lett. 110: 067003.

Ying T, et al., (2013) Superconducting phases in potassium-intercalated iron selenides. J. Am. Chem. Soc. 135: 2951–2954.

Zhang AM, et al., (2012a) Effect of iron content and potassium substitution inA0.8Fe1.6Se2(*A*=K, Rb, Tl) superconductors: a Raman scattering investigation. Phys. Rev. B 86.

Zhang AM, et al., (2012b) Two-magnon Raman scattering inA0.8Fe1.6Se2systems (*A*=K, Rb, Cs, and Tl): competition between superconductivity and antiferromagnetic order. Phys. Rev. B 85.

Zhang P, et al., (2018) Observation of topological superconductivity on the surface of an iron-based superconductor. Science 360: 182–186.

Zhang P, et al., (2019) Multiple topological states in iron-based superconductors Nat. Phys. 15: 41–47.

Zhang Y, et al., (2017) Electronic evidence of temperature induced Lifshitz transition and topological nature in ZrTe5. Nat. Commun. 8: 15512.

Zhao S. C., et al., (2009) Raman spectra in iron-based quaternary $CeO_{1-x}F_xFeAs$ and $LaO_{1-x}F_xFeAs$. Supercond. Sci. Technol. 22: 015017.

D6

Interface Superconductivity

Jörg Schmalian

D6.1 Introduction

The realization and characterization of interface superconductivity, i.e. of two-dimensional (2D) superconducting phases in various different systems, constitute a topic of significant current interest [1–10]. In addition to the fundamental interest in superconducting transitions in reduced dimensions, this is motivated by the high degree of gate tunability of the electronic properties [2, 3, 5, 7–10], allowing to gradually change the carrier concentration. In addition, the promising role played by 2D superconductors in the search for topological Majorana modes and related applications [11] has attracted significant attention.

Superconductivity is most-widely studied in bulk systems. Some of the most widely investigated and interesting systems, such as the cuprates, organic charge transfer salts, and some of the iron-based materials are structurally and electronically highly anisotropic. Still, they do form three-dimensional superconductors. Systems where superconductivity is confined to a single two-dimensional interface behave rather distinct from such layered system. In this paragraph we give several interesting examples of interface superconductivity and discuss some of the peculiarities that occur in genuinely two-dimensional systems. We focus on aspects that are related to the nature of superconducting fluctuations and the fact that in generic interfaces the three-dimensional inversion symmetry is broken.

D6.2 Materials

Prominent materials with interface superconductivity are the $LaAlO_3/SrTiO_3$ heterostructures that display very rich electronic behavior [12, 13], single-layer FeSe on the [001] surface of $SrTiO_3$ with significantly enhanced transition temperatures compared to its bulk value [4], and twisted bilayer graphene that emerges in moiré structures for non-commensurate values of the electron filling [8]. The underlying electronic structures of these materials are very different, not pointing toward a straightforward unifying description. However, as we will see below, generic insights can be obtained from an analysis of the modified point symmetry of these rather distinct materials. The crucial states near the Fermi energy of the oxide interface of the $LaAlO_3/SrTiO_3$ heterostructures are made up of titanium t_{2g} orbitals. Depending on the termination of the $SrTiO_3$ substrate one finds interfaces with two-, three-, or fourfold rotation symmetry. In distinction, flat bands in twisted bilayer graphene emerge from resonant scattering between two almost overlapping Dirac cones of the two graphene sheets [14]. As a result, a honeycomb lattice with a threefold rotation symmetry emerges with a much enlarged lattice constant and with additional valley degree of freedom. Another 2D system with threefold symmetry, is MoS_2 that was shown to be superconducting via liquid ion gating [5, 7]. These engineered systems behave in many ways distinct from corresponding bulk counter parts. In some cases such as twisted bilayer graphene, no such bulk system exists. In parallel, there is also an interesting development to investigate superconductivity in single or few atomic layer versions of correlated materials such as the cuprates [15], Sr_2RuO_4 [16], UPt_3 [17], or CeCoIn5 [18]. This has the additional potential to obtain clearer insights into the electronic properties of the related bulk material.

D6.3 Statistical Mechanics of Two-Dimensional Superconductors

A well-established aspect of two-dimensional superconductivity is the distinct nature of superconducting fluctuations. The Hohenberg theorem implies that a two-dimensional superconductor will not give rise to a finite expectation value of the order parameter [19]. The original proof of the theorem was performed for neutral superfluids, where long wavelength fluctuations of the order parameter phase θ suppress a finite order parameter $\Psi=|\Psi|e^{i\theta}$. If one considers charged superfluids with long-range longitudinal Coulomb interaction, a Higgs mechanism changes the nature of the excitation spectrum [20, 21]. However, the key conclusion of the Hohenberg theorem, namely the absence of spontaneous symmetry breaking, remains unchanged [22-24].

The most dramatic phenomenon in two-dimensional superconductivity is, however, the fact that they have, despite the absence of a finite order parameter expectation value, a finite superfluid stiffness. The stiffness $\rho_s(T)$ is defined as the renormalized coefficient of a long-wavelength phase twist with energy

$$F_{\text{twist}} = \frac{1}{2}\rho_s(T)\int d^2r(\nabla\theta(r))^2. \quad \text{(D6.1)}$$

A low-temperature phase with bound vortex–antivortex pairs and finite stiffness is separated from a high-T phase with unpaired vortex excitations at the Berezinskii–Kosterlitz–Thouless (BKT) transition temperature T_{BKT}. At this transition, the stiffness in units of T_{BKT} takes a universal value [25]

$$\rho_s(T_{\text{BKT}}) = \frac{2}{\pi}k_B T_{\text{BKT}}. \quad \text{(D6.2)}$$

The applicability of the BKT description of superconductors is based on the assumption that one can ignore fluctuations of the electromagnetic field. For sufficiently long-length scales this assumption is not applicable. The electromagnetic coupling between vortices changes and no BKT transition takes place. The characteristic length scale when the vortex interaction is affected by electromagnetic coupling is the Pearl length $\Lambda = 2\lambda^2/d$ with superconducting penetration depth λ and film thickness d [26]. For thin films $d \ll \lambda$ in many cases of interest and the BKT formalism can be applied, except for a very narrow regime right at the putative BKT transition where the superconducting correlation length becomes comparable to the system size. For a lucid discussion of these effects and the new physics that emerges in samples of size smaller than the Pearl length, see Ref. [27]. If one considers sufficiently large samples and systems with a coherence length shorter than λ, the BKT description should suffice. Above T_{BKT} a number of unconventional transport properties can be linked to the vortex unbinding dynamics, most notably a nonlinear current-voltage relation $V \propto I^3$ that is a consequence of the fact that the number of vortices grows with the current according to I^2.

For superconductors where BCS theory offers a valid description in the absence of two-dimensional critical fluctuations, Equation (D6.2) should not be read as an equation that determines T_{BKT}. The latter is rather close to the value of the mean-field BCS theory. Instead Equation (D6.2) determines the smallest value of the stiffness that a two-dimensional superconductor can sustain. However, recently Equation (D6.2) was employed to draw rigorous conclusions about the value of the transition temperature of a two-dimensional superconductor. Combining Equation (D6.2) with an optical sum rule, an exact result can be obtained for the BKT transition temperature of a 2D superconductor with Galilean-invariant electronic spectrum [28]:

$$k_B T_{\text{BKT}} \leq \frac{1}{8}E_F. \quad \text{(D6.3)}$$

E_F is the Fermi energy. In the BCS regime the transition temperature is small compared to E_F, and the above inequality is not really restrictive. However, systems near the crossover to Bose–Einstein condensation of Cooper pairs turn out to be surprisingly close to the above bound [28]. Thus, for 2D materials with low carrier concentrations there is a fundamental limitation for the maximally allowed transition temperature.

In order to further illustrate the power of statistical mechanics arguments in two-dimensional systems, we consider a triplet superconductor with weak spin–orbit interaction. For the moment we ignore the spin–orbit coupling all together. The system forms triplet pairs with pair wave function

$$\Psi(\mathbf{r},\boldsymbol{k},\alpha,\beta) = \sum_{\mu=1}^{3}\psi_\mu(\mathbf{r})\chi^\mu_{\alpha\beta}(\mathbf{k}).. \quad \text{(D6.4)}$$

Here $\Psi = (\psi_1,\psi_2,\psi_3)^T$ is the three-component complex superconducting order parameter of the system. $\chi^\mu_{\alpha\beta}(\mathbf{k}) = \chi(\mathbf{k})(i\sigma^y\sigma^\mu)_{\alpha\beta}$ describes the pairing wave function in coordinate and spin space. $\boldsymbol{k}$ is the two-dimensional crystal momentum. The symmetry group of the problem is $g_0 \times SU(2) \times U(1)$. Here g_0 is the point group of the problem, $SU(2)$ is the rotation invariance in spin space, and $U(1)$ corresponds to invariance under a global change of the phase of the fermionic fields. This yields the most general Ginzburg–Landau expansion:

$$S = \int d^2r(\Psi^\dagger \cdot \Psi + \nabla\Psi^\dagger \cdot \nabla\Psi) + S^{(4)}, \quad \text{(D6.5)}$$

where the only allowed non-linear terms are:

$$S^{(4)} = u_0\int d^2r(\Psi^\dagger \cdot \Psi) + u_1\int d^2r(\Psi^\dagger \times \Psi)\cdot(\Psi^\dagger \times \Psi). \quad \text{(D6.6)}$$

In a two-dimensional system, the order parameter vanishes due to the Hohenberg–Mermin–Wagner theorem for any finite temperature $\langle\Psi\rangle = 0$. It also does not establish algebraic order with finite stiffness, due to the strong fluctuations in spin space. The same is true for the composite order parameter $\langle\Psi^\dagger \times \Psi.\rangle$ For $u_1 > 0$, it is, however, possible to develop algebraic order of $\varphi = \Psi\cdot\Psi$. φ corresponds to a singlet in spin space, yet it breaks the $U(1)$ symmetry in the sense of a charge-4e superconductor. The inevitable state of order is a superconductor made up of bound states of triplets. It will undergo a genuine BKT transition in the sense discussed in the previous paragraph and with a transition temperature that is of the same order as the mean-field transition temperature of the underlying triplet state. If one determines the coupling constant u_1 from a microscopic theory within a weak coupling approach of a single band system, one does find that it is positive, as required for the charge-4e pairing state. Allowing for a finite value of the spin–orbit coupling, the charge-4e pairing state remains stable in a finite but narrow regime of parameters and for intermediate temperatures. Such states have been discussed in the context of ^{3}He films [29] and spin-1 bose condensates [30, 31]. Phase transitions of composite order parameters such as φ have also been discussed in the context of so-called vestigial order where symmetry-breaking superconducting fluctuations condense [32–34].

A final comment on these considerations of the statistical mechanics of two-dimensional superconductors follows from the analysis of Ref. [35]. If one includes longitudinal and transverse fluctuations of the electromagnetic field, the corresponding quantum Ginzburg–Landau theory becomes a genuine gauge theory. This implies that the usual order parameter is not a gauge-invariant object and should not have physical meaning in the sense of an observable. In this regime, one can demonstrate that the proper description of a superconductor is in terms of a topological quantum field theory [35]. A superconductor, even a *s*-wave pairing state as it occurs in conventional materials, is in fact a state of topological order. This behavior is closely connected to the fact that Bogoliubov quasiparticles of charged superconductors are truly fractionalized degrees of freedom [36]. This topological order is in many ways a more fundamental topological state than what is often discussed in the context of topological superconductivity with Majorana edge states [11]. The topological quantum field theory allows to link the degeneracy of the many-body ground state to the genus of the manifold on which the superconductor exists [35] and offers a fundamental explanation for the known degeneracy of the spectrum of superconducting rings with external flux [37]. Despite these subtle phenomena it is reassuring that one can still employ the Ginzburg–Landau formalism to draw conclusions about observable quantities. Non-local, yet gauge-invariant observables of a superconductor can, under a specific gauge choice, be directly related to the results that follow the usual analysis of the Ginzburg–Landau theory, see Ref. [35]. Thus, if properly interpreted, a Ginzburg–Landau formalism of two-dimensional superconductors continues to be possible despite the exotic topological nature of the superconducting state.

D6.4 Broken Inversion Symmetry and Cooper Pair Selection Rules

A rather generic aspect of interface superconductivity is that the three-dimensional inversion symmetry is broken. This happens when there is no center of inversion in the two-dimensional material, even if it was in vacuum or, more generically, when the materials above and below the interface are distinct, see Figure D6.1. An immediate implication of the lack of inversion symmetry is that the Kramers degeneracy of electronic bands is lifted even if time-reversal symmetry remains intact. Depending on the specific symmetry of the system and the relevant electronic orbitals involved this gives rise to Rashba [38] or Dresselhaus [39] coupling terms in the band structure. For a detailed discussion of such terms for the electronic structure of the 3*d* orbitals in $LaAlO_3/SrTiO_3$ heterostructures, see Ref. [13, 40].

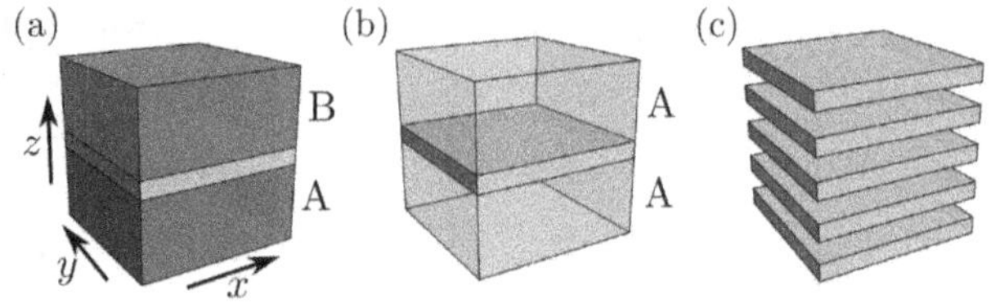

FIGURE D6.1 From a symmetry point of view, 2D systems can be grouped into those realized in an asymmetric (a) or symmetric (b) environment and should be distinguished from layered materials consisting of weakly coupled sheets as shown in (c); after Ref. [43].

Another important implication of the broken inversion symmetry for interfaces is due to the fact that the superconducting order parameter behaves precisely like a two-particle wave function. Let us consider the pair wave function $\Psi(\boldsymbol{r}, \boldsymbol{\rho}, \alpha, \beta)$ as function of the center of gravity coordinate $\boldsymbol{r}$ and the relative coordinate $\boldsymbol{\rho}$ as well as the spins α and ß of the two electrons of the pair, see also Equation (D6.4). We can now expand this wave function according to

$$\Psi(\mathbf{r},\boldsymbol{\rho},\alpha,\beta) = \Psi_s(\mathbf{r},\boldsymbol{\rho})(i\sigma^y)_{\alpha\beta} + \Psi_t(\mathbf{r},\boldsymbol{\rho})\cdot(i\sigma^y\sigma)_{\alpha\beta}, \quad \text{(D6.7)}$$

where the singlet and triplet part obey under particle exchange

$$\begin{aligned} \Psi_s(\mathbf{r},\boldsymbol{\rho}) &= \Psi_s(\mathbf{r},-\boldsymbol{\rho}), \\ \Psi_t(\mathbf{r},\boldsymbol{\rho}) &= -\Psi_t(\mathbf{r},-\boldsymbol{\rho}). \end{aligned} \quad \text{(D6.8)}$$

These relations are a consequence of the Fermi statistics of the electrons. Let us consider a system with inversion symmetry, where the wave function is either even or odd under $\boldsymbol{\rho} \mapsto -\boldsymbol{\rho}$. The spin is a pseudo-vector, i.e. it does not change under parity. Thus, it must hold that

$$\Psi(\mathbf{r},\boldsymbol{\rho},\alpha,\beta) = \pm\Psi(\mathbf{r},-\boldsymbol{\rho},\alpha,\beta). \quad \text{(D6.9)}$$

With inversion symmetry a superconductor is then either made up of singlet or of triplet Cooper pairs, at least in the vicinity of the transition temperature. For a combination of singlet and triplet pairing, the total wave function would have no well-defined parity eigenvalue. The above proof is also valid if one includes spin–orbit interaction. In this case, the labels α and β refer to the pseudo-spin quantum numbers that form Kramers pairs. For systems without inversion symmetry, such as interfaces, both pairing states always exist simultaneously, even if in specific instances one may dominate over the other. For a discussion of the magnetic properties of such a mixed singlet–triplet state in an interface with Rashba spin–orbit coupling, see Ref. [41]. Mixed singlet–triplet pairing should, for example, arise in electrical contact areas and influence proximity effects and spin injection phenomena. To determine the nature of the triplet component of a superconductor that would be singlet in bulk, one can utilize symmetry arguments, see Ref. [42].

General conclusion can be drawn about the nature of superconducting pairing states in interfaces, whenever the mentioned Rashba splitting of the normal-state bands is larger than the magnitude of the superconducting gap. Let Θ be the time-reversal operator. If the normal state is time-reversal symmetric, we know that $\Theta\phi_{\mathbf{k},a}$ is an eigenfunction of the normal-state Bloch Hamiltonian $h_{-\mathbf{k}}$, where $\phi_{\mathbf{k},a}$ is an eigenfunction of $h_{\mathbf{k}}$. α labels spin and orbital degrees of freedom as well as sub-band flavors that might occur in interface problems. Because of the fact that the bands without inversion symmetry are generically non-degenerate, it follows that $\phi_{\mathbf{k},a}$ equals $\Theta\phi_{-\mathbf{k},a}$ up to a phase factor. Following Ref. [43] one can determine the transformation properties of the superconducting gap $\Delta_{\mathbf{k},a}$ as function of the in-plane momentum $\boldsymbol{k}$:

$$\Delta_{\mathbf{k},a} = \pm\Delta_{-\mathbf{k},a} \tag{D6.10}$$

if $\Theta^2 = \mp 1$. If one operates in a very large magnetic field that polarizes the electron spin, yielding effectively spinless fermions, the time-reversal operator obeys $\Theta^2 = 1$. Otherwise one obtains the behavior $\Theta^2 = -1$ of usual spin-$\frac{1}{2}$ fermions, the case we will consider next. Since the transformation $\boldsymbol{k} \rightarrow -\boldsymbol{k}$ in two dimensions can be realized by a twofold rotation about an axis perpendicular to the interface, the point group symmetry of the interface gives rise to significant restrictions for allowed pairing states. As shown in Ref. [43], it follows that one can obtain *selection rules* for Cooper pairing in interfaces. First, in interfaces, time-reversal symmetry can only be broken in systems with threefold rotation symmetry. For example, no time-reversal symmetry breaking of the [001] interface of a tetragonal system allows while it is possible on top of a [111] surface of a cubic system. One can even demonstrate a second rather strong rule: if a superconductor transforms according to a complex or multi-dimensional irreducible representation of the normal-state point group, it must break time-reversal symmetry. If superconductivity in MoS_2 occurs with a symmetry breaking that corresponds to a higher-dimensional irreducible representation, time-reversal symmetry must be broken at T_{BKT}. These examples demonstrate that the nature of allowed unconventional pairing states in interfaces behaves rather distinct from those of inversion symmetric three-dimensional superconductors.

To summarize, superconductivity in interfaces is a thriving area of research with a range of distinct materials that can be tuned much easier than their three-dimensional counterparts. The most impressive example to date is clearly twisted bilayer graphene, where the carrier concentration can be changed by several electrons per unit cell. The novel physics associated with enhanced critical fluctuations, strong electronic correlations caused by notoriously low carrier densities, and symmetry dictated selection rules for pairing symmetries promise to give rise to an exciting new phenomenon in these low-dimensional structures.

References

[1] N. Reyren, S. Thiel, A. D. Caviglia, L. F. Kourkoutis, G. Hammerl, C. Richter, C. W. Schneider, T. Kopp, A. S. Rüetschi, D. Jaccard, M. Gabay, D. A. Muller, J.M. Triscone, and J. Mannhart, Science 317, 1196 (2007).

[2] A. D. Caviglia, S. Gariglio, N. Reyren, D. Jaccard, T. Schneider, M. Gabay, S. Thiel, G. Hammerl, J. Mannhart, and J. M. Triscone, Nature 456, 624 (2008).

[3] K. Ueno, S. Nakamura, H. Shimotani, H. T. Yuan, N. Kimura, T. Nojima, H. Aoki, Y. Iwasa, and M. Kawasaki, Nat. Nano 6, 408 (2011).

[4] W. Qing-Yan, L. Zhi, Z. Wen-Hao, Z. Zuo-Cheng, Z. Jin-Song, L. Wei, D. Hao, O. Yun-Bo, D. Peng, C. Kai, W. Jing, S. Can-Li, H. Ke, J. Jin-Feng, J. Shuai-Hua, W. Ya-Yu, W. Li-Li, C. Xi, M. Xu-Cun, and X. Qi-Kun, Chinese Phys. Lett. 29, 037402 (2012).

[5] J. T. Ye, Y. J. Zhang, R. Akashi, M. S. Bahramy, R. Arita, and Y. Iwasa, Science 338, 1193 (2012).

[6] Y. L. Han, S. C. Shen, J. You, H. O. Li, Z. Z. Luo, C. J. Li, G. L. Qu, C. M. Xiong, R. F. Dou, L. He, D. Naugle, G. P. Guo, and J. C. Nie, Appl. Phys. Lett. 105, 192603 (2014).

[7] Y. Saito, Y. et al., Nat. Phys. 12, 144 (2016).

[8] Y. Cao, V. Fatemi, S. Fang, K. Watanabe, T. Taniguchi, E. Kaxiras, and P. Jarillo-Herrero, Nature 556, 43 (2018).

[9] M. Yankowitz, S. Chen, H. Polshyn, Y. Zhang, K.Watanabe, T. Taniguchi, D. Graf, A. F. Young, and C. R. Dean, Science 363, 1059 (2019).

[10] X. Lu, P. Stepanov, W. Yang, M. Xie, M. A. Aamir, I. Das, C. Urgell, K. Watanabe, T. Taniguchi, G. Zhang, A. Bachtold, A. H. MacDonald, and D. K. Efetov, arXiv e-prints, arXiv:1903.06513 (2019).

[11] B. A. Bernevig, Topological Insulators and Topological Superconductors (Princeton University Press, 2013).

[12] J. Mannhart and D. G. Schlom, Science 327, 1607 (2010).

[13] S. Gariglio, M. Scheurer, J. Schmalian, A. M. R. V. L. Monteiro, S. Goswami, and A. Caviglia, Chapter 7 in Small Superconductors, ed. A.V. Narlikar, Clarendon Press, Oxford (2016).

[14] R. Bistritzer and A. H. MacDonald, Proc. Natl. Acad. Sci. 108, 12233 (2011).

[15] G. Logvenov, A. Gozar, and I. Bozovic, Science 326, 699 Oct (2009).

[16] Y. Krockenberger, M. Uchida, K. S. Takahashi, M. Nakamura, M. Kawasaki, and Y. Tokura, Appl. Phys. Lett. 97, 082502 (2010).

[17] M. Huth, S. Reber, C. Heske, P. Schicketanz, J. Hessert, P. Gegenwart, and H. Adrian, J. Phys. Condens. Matter 8, 8777 (1996).

[18] Y. Mizukami, H. Shishido, T. Shibauchi, M. Shimozawa, S. Yasumoto, D. Watanabe, M. Yamashita, H. Ikeda, T. Terashima, H. Kontani, and Y. Matsuda, Nat. Phys. 7, 849 (2011).

[19] P. C. Hohenberg, Phys. Rev. 158, 383 (1967).

[20] P. W. Anderson, Phys. Rev. 110, 827 (1958).
[21] P. B. Littlewood and C. M. Varma, Phys. Rev. B 26, 4883 (1982).
[22] S. Fischer, M. Hecker, M. Hoyer, and J. Schmalian, Phys. Rev. B 97, 054510 (2018).
[23] V. L. Berezinskii, Sov. Phys. JETP, 32, 493 (1971); Sov. Phys. JETP, 34, 610 (1972).
[24] J. M. Kosterlitz and J. D. Thouless, J. Phys. C (Solid State Phys.), 6, 1181 (1973).
[25] D. R. Nelson and J. M. Kosterlitz, Phys. Rev. Lett. 39, 1201 (1977).
[26] J. Pearl, Appl. Phys. Lett. 5, 65 (1964).
[27] V. G. Kogan, Phys. Rev. B 75, 064514 (2007).
[28] T. Hazra, N. Verma, and M. Randeria Phys. Rev. X 9, 031049 (2019).
[29] S. E. Korshunov, Zh. Eksp. Teor. Fiz. 89, 532 (1985).
[30] A. J. A. James and A. Lamacraft, Phys. Rev. Lett. 106, 140402 (2011).
[31] S. Mukerjee, C. Xu, and J. E. Moore, Phys. Rev. Lett. 97, 120406 (2006).
[32] E. Berg, E. Fradkin, and S. A. Kivelson, Charge-4e superconductivity from pair-density-wave order in certain high-temperature superconductors, Nat. Phys. 5, 830 (2009).
[33] R. M. Fernandes, P. P. Orth, and J. Schmalian, Ann. Rev. Condens. Matter Phys. 10, 133 (2019).
[34] M. Hecker and J. Schmalian, NPJ Quantum Mater. doi 10.1038/s41535-018-0098-z (2018).
[35] T. H. Hansson, V. Oganesyan, and S. L. Sondhi, Ann. Phys. 313, 497 (2004).
[36] S. A. Kivelson, D. S. Rokhsar, Phys. Rev. B 41, 11693 (1990).
[37] J. R. Schrieffer, Theory of Superconductivity, Adison Wesley, New York, (1971).
[38] E. I. Rashba, Sov. Phys. Solid State 2, 1109 (1960).
[39] G. Dresselhaus, Phys. Rev. 100, 580 (1955).
[40] M. Scheurer and J. Schmalian, Nat. Commun. 6, 6005 doi: 10.1038/ncomms7005 (2015).
[41] L. P. Gor'kov and E. I. Rashba, Phys. Rev. Lett. 87, 037004 (2001).
[42] J. Schmalian and W. Hübner, Phys. Rev. B 53, 11860 (1996).
[43] M. S. Scheurer, D. F. Agterberg, and J. Schmalian, NPJ Quantum Mater. 2, 9 (2017).

D7

Topological Superconductivity

Panagiotis Kotetes

D7.1 Introduction

From the very beginning of its discovery, the phenomenon of superconductivity has been a source of inspiration for numerous breakthrough concepts and applications. One of the most recent and exciting developments in the field relates to the possibility of spontaneously formed or artificially engineered topological superconductors (TSCs). TSCs constitute systems which harbor exotic charge-neutral excitations termed Majorana fermions (MFs) [1–3]. These are closely related to the particles that Ettore Majorana put forward in 1937 [4] as self-conjugate solutions of the Dirac equation, with the neutrino being considered as the most prominent candidate for a Majorana particle. The crucial difference is that Majorana particles are fundamental and indivisible, while Majorana excitations encountered in TSCs define quasiparticles emerging only in the presence of electron-electron interactions.

Out of all the possible types of MF quasiparticles, special attention has so far been paid to the zero energy and simultaneously spatially localized ones. These are usually termed Majorana zero modes (MZMs), and can be trapped at 0D defects of TSCs [1–3, 5]. The charge neutrality of MZMs endows them with additional nonstandard properties, which are mainly reflected in their non-Abelian exchange statistics [2]. Besides the paramount significance of realizing anyonic statistics in low-dimensional condensed matter systems, the possibility of being in a position to generate and manipulate MZMs additionally opens perspectives for implementing fault-tolerant quantum computing [6, 7].

TSCs first appeared in the context of spin-triplet p-wave superconductors [5, 8] in connection to Sr_2RuO_4 [9] and noncentrosymmetric superconductors [10, 11]. However, the lack of an unambiguous experimental verification of the topological nature of these materials blocked the development of this direction. This hindrance was circumvented by the proposal of artificial TSCs in diverse hybrid platforms [12–26] with, usually, a conventional superconductor being an integral component. Notably, strong spectroscopic evidences for MZMs have been experimentally captured in numerous hybrid systems [27–38]. More recently, theoretical predictions [39] guided a number of experimental efforts to confirm the appearance of MZMs also in iron-based materials. In particular, FeTeSe compounds harbor helical surface states [40], which ensure the appearance of MZMs in the presence of superconducting vortices [41–43].

In the following sections we give an account of the properties of MF quasiparticles and topological superconductivity. Specifically, Section D7.2 discusses fundamental properties of MFs. Section D7.3 focuses on the classification of TSCs, while Section D7.4 categorizes the various types of MF quasiparticles that become accessible in these systems. Section D7.5 describes some of the most prominent candidate TSC systems which exclusively rely on spin-singlet pairing. In Section D7.6, we briefly present breakthrough experiments, in which, fingerprints of MZMs have been detected by means of tunneling spectroscopy. Finally, Section D7.7 enlists yet unresolved problems and discusses further directions and perspectives.

D7.2 Fundamentals of Majorana Fermions

In the following paragraphs we discuss a number of characteristic properties of MFs, with a focus on Majorana quasiparticles emerging in condensed matter systems, and in particular MZMs.

D7.2.1 Electrons Viewed as Majorana Fermion Composites

Identifying materials and synthesizing systems which harbor MZMs appear pivotal for leveling up the field of fault-tolerant, also termed topological, quantum computing. An appealing concept that has so far been successfully employed to guide this search is to mathematically describe an electron as a composite particle comprised by two MFs. This legitimate representation of electrons fully complies with their indivisible and fundamental nature, and mainly serves as a convenient approach to understand a number of phenomena taking place in TSCs. Within this picture, MFs in topologically trivial superconductors are kept well-hidden inside the electrons and, thus, are neither accessible nor detectable degrees of freedom. This is because, in these systems, MFs are still bound to come in pairs. Each MF pair behaves as an electron and unavoidably couples to local charge probes. In contrast, TSCs are exactly those materials which allow us to access the "internal" structure of an electron and probe its two MF constituent parts since, in this case, the MFs making up a pair become spatially split and, thus, unpaired MFs can be isolated.

Using the formalism of second quantization, we now express the above concept mathematically. For this purpose, we introduce the creation $c_\alpha^\dagger(\boldsymbol{r})$ and annihilation $c_\alpha(\boldsymbol{r})$ operators of an electron, which satisfy the anticommutation relations $\{c_\alpha(\boldsymbol{r}), c_\beta^\dagger(\boldsymbol{r}')\} = \delta_{\alpha\beta}\delta(\boldsymbol{r}-\boldsymbol{r}')$ and $\{c_\alpha(\boldsymbol{r}), c_\beta(\boldsymbol{r})\} = 0$. The index α denotes the spin projection $\uparrow$ and $\downarrow$, $\boldsymbol{r}$ corresponds to the spatial coordinate, and δ stands for the Dirac and Kronecker delta functions. One then writes:

$$c_\alpha(\boldsymbol{r}) = \frac{\gamma_\alpha(\boldsymbol{r}) + i\tilde{\gamma}_\alpha(\boldsymbol{r})}{\sqrt{2}} = \begin{matrix}\gamma_\alpha(\boldsymbol{r}) \\ \tilde{\gamma}_\alpha(\boldsymbol{r})\end{matrix} \tag{D7.1}$$

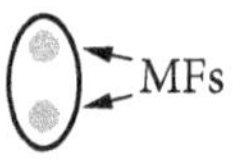

where we introduced the self-conjugate (or simply real) MF operators $\gamma_\alpha(\boldsymbol{r})$ and $\tilde{\gamma}_\alpha(\boldsymbol{r})$ satisfying $\gamma_\alpha(\boldsymbol{r}) = \gamma_\alpha^\dagger(\boldsymbol{r})$ and $\tilde{\gamma}_\alpha(\boldsymbol{r}) = \tilde{\gamma}_\alpha^\dagger(\boldsymbol{r})$. The MF operators additionally fulfill:

$$\begin{aligned}&\{\gamma_\alpha(\boldsymbol{r}), \gamma_\beta(\boldsymbol{r}')\} = \{\tilde{\gamma}_\alpha(\boldsymbol{r}), \tilde{\gamma}_\beta(\boldsymbol{r}')\} = \delta_{\alpha\beta}\delta(\boldsymbol{r}-\boldsymbol{r}') \quad and \\ &\{\gamma_\alpha(\boldsymbol{r}), \tilde{\gamma}_\beta(\boldsymbol{r}')\} = 0, \quad \forall \alpha,\beta = \uparrow,\downarrow.\end{aligned} \tag{D7.2}$$

Remarkably, the above equation leads to the nonstandard relations $2\gamma_{\uparrow,\downarrow}^2(\boldsymbol{r}) = 2\tilde{\gamma}_{\uparrow,\downarrow}^2(\boldsymbol{r}) = \delta(\boldsymbol{0})$, which have to be contrasted with the ones obeyed by usual fermionic operators, i.e., $c_{\uparrow,\downarrow}^2(\boldsymbol{r}) = 0$. This extraordinary property of MFs implies that the action of only a single self-conjugate operator γ (or $\tilde{\gamma}$) on the ground state cannot create or annihilate an electron.

D7.2.2 Unpaired Majorana Fermions in the Kitaev Chain Model

Alexei Kitaev [3] employed the above representation to explicitly demonstrate that there exist systems, in which, MFs that were originally forming local pairs become spatially separated and even completely unpaired. The paradigm of such systems is the so-called Kitaev chain model, which describes spinless electrons that form Cooper pairs. These Cooper pairs necessarily consist of electrons residing on different (here neighboring) lattice sites. In terms of second quantization, the Kitaev chain model reads:

$$\begin{aligned}H = &-\mu\sum_{n=1}^{N} c_n^\dagger c_n - t\sum_{n=1}^{N-1}\left(c_n^\dagger c_{n+1} + c_{n+1}^\dagger c_n\right) \\ &-\frac{\Delta_p}{2}\sum_{n=1}^{N-1}\left(c_n^\dagger c_{n+1}^\dagger - c_{n+1}^\dagger c_n^\dagger + c_{n+1}c_n - c_n c_{n+1}\right),\end{aligned} \tag{D7.3}$$

and describes a 1D lattice model for a p-wave superconductor with an order parameter of an energy scale denoted Δ_p. The index n labels the sites of the chain with the lattice constant set to unity. The electrons feel a chemical potential μ, and they are governed by a nearest-neighbor hopping of a matrix element t. While the above model is often seen as a toy model, it may be actually relevant for the so-called ferromagnetic superconductors, e.g. UGe_2, UCoGe and others [44], where superconductivity has been already experimentally observed deep inside the spin-polarized phase.

A topologically nontrivial superconducting phase is obtained for $|\mu| < 2|t|$, with a single MZM appearing on each edge of the chain. While we provide further details regarding the topological phase diagram later, here we consider the sweet-spot case that was identified by Kitaev, i.e. $\mu = 0$ and $|\Delta_p| = |t|$, where a single MZM becomes unpaired at the first ($n = 1$) and last ($n = N$) sites of the chain. Using the MF composite picture of electrons, the Hamiltonian obtains a transparent form for $\mu = 0$ and $\Delta_p = +t$ (the choice $\Delta_p = -t$ is discussed later), i.e.:

$$\begin{aligned}H &= -t\sum_{n=1}^{N-1}\left(c_n^\dagger c_{n+1} - c_n c_{n+1}^\dagger + c_n^\dagger c_{n+1}^\dagger - c_n c_{n+1}\right) \\ &= 2ti\sum_{n=1}^{N-1}\frac{c_n - c_n^\dagger}{\sqrt{2}i}\frac{c_{n+1} + c_{n+1}^\dagger}{\sqrt{2}} = 2ti\sum_{n=1}^{N-1}\tilde{\gamma}_n\gamma_{n+1}.\end{aligned} \tag{D7.4}$$

We directly notice that the MF operators γ_1 and $\tilde{\gamma}_N$ are not present in the above Hamiltonian, i.e., they are unpaired. This is also illustrated in Figure D7.1. These two operators are defined exactly at the end points of the chain and one can employ them to construct the nonlocal zero-energy electronic operator $d_0 = (\gamma_1 + i\tilde{\gamma}_N)/\sqrt{2}$, satisfying $d_0^2 = 0$ and $\{d_0, d_0^\dagger\} = 1$. We remark that the alternative choice $\Delta_p = -t$ renders instead the MF operators $\tilde{\gamma}_1$ and γ_N unpaired.

When the parameters are detuned from the above special values, but with $|\mu| < 2|t|$ still being fulfilled, the wave functions of the arising MZM excitations have spatial supports which are not only concentrated at the first or last sites, but also extend to other sites. Hence, in the most general and realistic case, the MZM wave functions have a nonzero overlap and the two MZMs hybridize into a nonzero-energy electron d_ε. This, at the same time implies that the MZMs can be strictly defined only for an infinite chain, while for finite-sized systems their exotic properties remain accessible only as long as the length of the system L is much larger than the spatial extent ξ_M of the MZM wave functions. Consequently, for MZM-based applications it is desirable to achieve an as small as possible value for the ratio ξ_M / L, since this also controls the resulting energy splitting of the d_ε electron. This energy splitting sets in turn a characteristic frequency and a time scale for performing operations with MZMs. The optimal operational frequency is required to be much smaller than the frequency-equivalent of the bulk energy gap of the system

(a) Topologically-Trivial Phase

$$H = \left(-\mu \sum_{n=1}^{N} c_n^\dagger c_n \right) - \sum_{n=1}^{N-1} \left(t c_n^\dagger c_{n+1} + \Delta_p c_n^\dagger c_{n+1}^\dagger + \text{h.c.} \right)$$

$$\Delta_p = t = 0 \quad \Rightarrow \quad H = \mu \sum_{n=1}^{N} i \widetilde{\gamma}_n \gamma_n$$

γ_n

$\widetilde{\gamma}_n$

(b) Topologically-Nontrivial Phase

$$H = -\mu \sum_{n=1}^{N} c_n^\dagger c_n - \sum_{n=1}^{N-1} \left(t c_n^\dagger c_{n+1} + \Delta_p c_n^\dagger c_{n+1}^\dagger + \text{h.c.} \right)$$

$$\mu = 0 \ \& \ \Delta_p = t \quad \Rightarrow \quad H = 2t \sum_{n=1}^{N-1} i \widetilde{\gamma}_n \gamma_{n+1}$$

γ_n

$\widetilde{\gamma}_n$

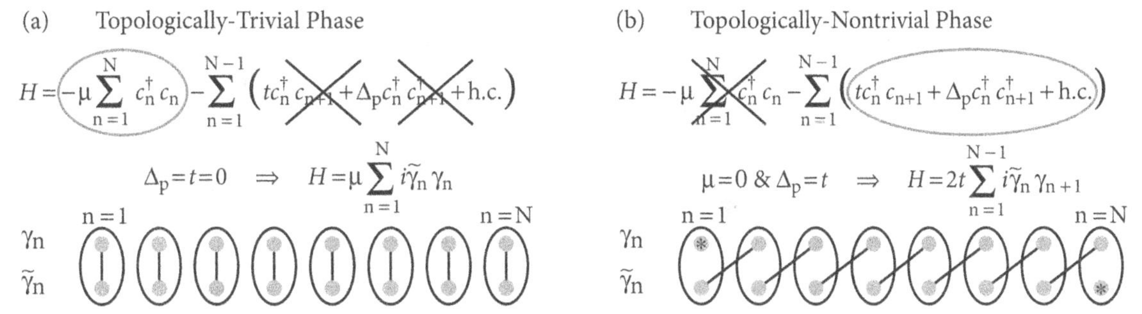

FIGURE D7.1 Accessible phases of the Kitaev chain model for $\Delta_p = t$. (a) Topologically trivial phase. A special case with $\Delta_p = t = 0$ and $\mu \neq 0$ is depicted. All MFs pair up in a local fashion. (b) Topologically nontrivial phase. For the choice $\Delta_p = t \neq 0$ and $\mu = 0$, we obtain two unpaired edge MFs (marked with $*$), which define a nonlocal electronic degree of freedom.

since, otherwise, quasiparticles become excited and poison the device. The presence of the latter is undesired, since these introduce decoherence and dephasing in the MZM-defined states and quantum bits (qubits).

D7.2.3 Many-Particle Ground-State Degeneracies and Braiding

The presence of two non-overlapping edge MZMs allows us to construct a zero-energy electron which is nonlocal, i.e., possessing a wave function with an equal support near the two edges, as determined by the profile of the respective MZM wave functions. See Figure D7.2(a). What is most important, however, is that the presence of this zero-energy electron leads to a twofold degeneracy of the many-body ground state [7], since occupying this state or leaving it empty does not modify the energy of the system. Consequently, the state in which the electron is not occupied, which is here denoted $|0>$ and satisfies $d_0 |0> = 0$, is degenerate with the occupied state $|1> = d_0^\dagger |0>$. Notably, the particular many-body degeneracy features a topological protection, which is inherited from the respective protection of the MZMs.

The arising degenerate two-state subspace can be employed to define a qubit, which constitutes the fundamental building block for MZM-based quantum computing [7]. The two states are distinguished by their different fermion parity, i.e., the number of electrons modulo 2. Assuming that all the remaining electrons take part in the formation of Cooper pairs, the states $|1>$ and $|0>$ are superpositions of states with an odd and even number of electrons, respectively. This implies that for a superconductor with a fixed fermion parity, the two states have zero overlap. As a consequence, the only accessible qubit operations consist of phase gates [7]. When a phase gate is effected, the two ground states generally pick up different phases, thus allowing for Abelian and non-Abelian operations. In the former (latter) case, the two states pick up the same (an opposite) phase. Notably, only non-Abelian phase gates are exploitable for MZM-based quantum information processing. In general, an arbitrary phase gate is obtainable by first hybridizing the two MZMs, then letting the system evolve for a given time depending on the phase gate we wish to implement, and finally uncoupling the MZMs [6]. However, such a protocol does not take advantage of the exotic properties of MZMs and, thus, is unprotected against noise and decoherence. Harnessing instead the non-Abelian exchange statistics of two uncoupled MZMs allows for a single topologically protected phase gate, which yields the result $(|0>, |1>) \mapsto \left(e^{i\pi/4} |0>, e^{-i\pi/4} |1> \right)$. This operation is termed braiding [2, 6, 7], and is effected by exchanging

(a) $d_0 = \dfrac{\gamma_1 + i\gamma_2}{\sqrt{2}}$

γ_1

γ_2

Zero-Energy Nonlocal Fermion d_0

(b) $\gamma_1 \mapsto -\gamma_2$

$\gamma_2 \mapsto +\gamma_1$

γ_1

Braiding

γ_2

(c) $\Delta_p e^{i\varphi \ell}$ T $\Delta_p e^{i\varphi r}$

$n = 1$ $n = N$ $n = 1$ $n = N$

$c_{N,\ell} \approx \dfrac{i\widetilde{\gamma}_{N,\ell}}{\sqrt{2}}$ Weak Link $c_{1,r} \approx \dfrac{\gamma_{1,r}}{\sqrt{2}}$

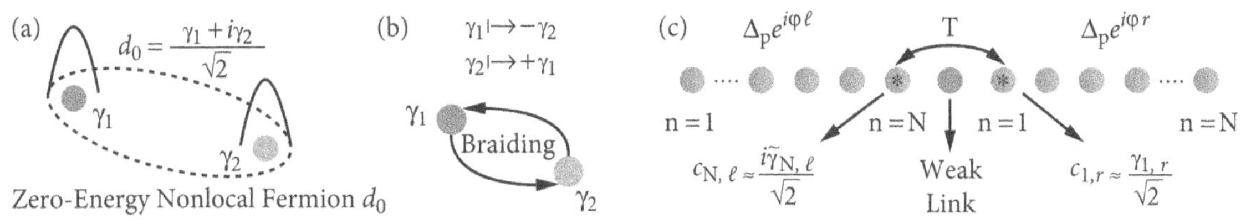

FIGURE D7.2 (a) Nonlocal charged fermion built up from two spatially separated zero-energy Majorana fermions. (b) Exchanging Majorana fermions in 2D real space leads to a topologically protected single-qubit operation termed braiding. In qubit space, this implements a phase gate where the qubit states pick up a relative phase of $\pi/2$. (c) Electron tunneling mediated by two unpaired Majorana fermions ($*$) which are located near the edges of two topological superconductors which feel a different superconducting phase. At low energies, only the MF contribution enters the Josephson current, which is 4π-periodic in the phase difference $\Delta\varphi = \varphi_\ell - \varphi_r$. This is distinct to the 2π-periodicity encountered in topologically trivial superconductors.

two MZMs in the 2D space, as it is depicted in Figure D7.2(b). This manipulation needs to be adiabatic in order to guarantee that no decoherence-introducing quasiparticles become excited, and that the accumulated phase is independent of the exchange path. Notably, the path independence of the phase gate reflects its topological character.

The fact that only a single operation can be implemented using two MZMs in a path-independent and thus topologically protected manner implies that MZM-based quantum computation is not universal. This can be alternatively attributed to the Ising anyon character of MZMs, while more complex types of anyons, e.g., Fibonacci [6], do not lead to such universality limitations. Nevertheless, if one insists to employ MZMs to define a fully functional qubit, then a minimum of four MZMs is required. Four MZMs can define two standard zero-energy fermionic degrees of freedom, which span a four-state degenerate space $\{|0,0>,|1,1>\}$ and $\{|0,1>,|1,0>\}$. A logical qubit, i.e., one that supports all single-qubit operations, can be defined using either the former or latter subspace of fixed fermion parity.

D7.2.4 4π-Periodic Josephson Effect

One of the hallmarks of superconductivity is the Josephson effect, i.e., the appearance of a net current flow through a link coupling two superconductors whose order parameters feel a phase difference $\Delta\varphi$. For two conventional superconductors linked by an insulating material, the Josephson current is a result of the tunneling of Cooper pairs across the interface. Instead, when the link is a quantum dot or a quantum point contact, the current is mediated by the so-called Andreev bound states (ABSs) which are fermions localized near the interface. In either case, the current is a 2π-periodic function of $\Delta\varphi$. Remarkably, TSCs deviate from the above behavior due to the MZMs that they induce near the junction. These MZMs allow for a Josephson transport which at low energies is mediated by $1e$ tunneling. This is effected by the d_ε fermion defined using the two MZMs near the junction. This fermion has $1e$ charge, and its energy ε lies inside the bulk energy gap. The apparent difference arising between the conventional and TSC cases is further reflected in the 4π-periodicity [3, 45] of the energy dispersion $\varepsilon(\Delta\varphi)$ of the d_ε ABS.

We explicitly demonstrate the 4π-periodicity of $\varepsilon(\Delta\varphi)$ by considering two TSCs of the p-wave type discussed earlier, each one consisting of N sites. See also Figure D7.2(c). These are dictated by the pairing terms $\Delta_p e^{i\varphi_s} c^\dagger_{n,s} c^\dagger_{n+1,s} + h.c.$, where $s = \ell, r$ labels the left and right superconductor. Here, the electronic coupling between the two superconductors is achieved by means of electron tunneling near the coupled edges of the two chains. A weak link mediates a tunnel coupling between the two TSCs, which is expressed by the Hamiltonian $H_T = \sum_{n,m} T_{mn}(c^\dagger_{m,r} c_{n,\ell} + h.c.)$ with T_{mn} being a monotonically decaying function with increasing distance between the sites labelled by m and n. Its exact form depends on the details of the interface and is not important for the present discussion. Since the superconducting phase is assumed to be constant within a given superconducting segment, it can be removed from each pairing term by a suitable gauge transformation and get transferred to the tunnel coupling term, so that the latter reads $H_T = \sum_{n,m} T_{mn}(e^{i\Delta\varphi/2} c^\dagger_{m,r} c_{n,\ell} + h.c.)$. To proceed, we consider without loss of generality the sweet-spot case of the Kitaev chain model, where the MZMs live only at the end sites. Assuming otherwise identical superconductors and restricting to the low-energy limit, it is eligible to replace the electronic operators by the unpaired MF operators, i.e., $c_{N,\ell} \approx i\tilde{\gamma}_{N,\ell}/\sqrt{2}$ and $c_{1,r} \approx \gamma_{1,r}/\sqrt{2}$, thus providing $H_T = T\cos(\Delta\varphi/2) i\gamma_{1,r}\tilde{\gamma}_{N,\ell} \equiv \varepsilon(\Delta\varphi)(2d^\dagger_\varepsilon d_\varepsilon - 1)/2$, with T the resulting strength of the coupling, $\varepsilon(\Delta\varphi) = T\cos(\Delta\varphi/2)$, and $d_\varepsilon = (\gamma_{1,r} + i\tilde{\gamma}_{N,\ell})/\sqrt{2}$. The Josephson current reads:

$$J_s = -\frac{2e}{\hbar}\left\langle \frac{dH_T}{d\Delta\varphi} \right\rangle = -\frac{e}{\hbar}\frac{d\varepsilon}{d\Delta\varphi} P_\varepsilon, \quad \text{(D7.5)}$$

where we introduced the parity of the electronic state $P_\varepsilon = 2\langle d^\dagger_\varepsilon d_\varepsilon \rangle - 1$. The fermion parity takes the value +1(−1) depending on whether the state is occupied (empty). If the fermion parity is fixed for the Josephson setup, the resulting current is 4π-periodic, i.e., $J_s \sim \sin(\Delta\varphi/2)$. Thus, the current is extremal for $\Delta\phi = \pi$, which is the phase difference value for which the two MZMs become uncoupled, thus leading to a protected crossing for the electron and hole branches $\pm\varepsilon(\Delta\varphi)$ [45]. In contrast, in topologically trivial superconductors, the Josephson current originating from an ABS d_ε is 2π-periodic, since in that case $\langle d^\dagger_\varepsilon d_\varepsilon \rangle \equiv n_F(\varepsilon)$, i.e., the occupation is determined by the energy of the ABS with $n_F(\varepsilon)$ the Fermi–Dirac distribution. This periodicity difference is in principle experimentally detectable.

D7.3 Classification of Topological Superconductors

Kitaev's work [3] played a catalytic role in boosting the pursuit of topological superconductors, as it was particularly successful in providing a transparent analysis of the arising MZMs, as well as in bringing forward a suitable topological invariant predicting their appearance. Previous seminal works had focused on the emergence of MZMs in vortices appearing in chiral $p_x \pm ip_y$ superconductors [1, 2]. These works, along with the discovery of topological insulators [46], as well as the proposals for artificial topological superconductivity by Fu and Kane [12], and later on by others [13–26], sparked the systematic study of topological phases of matter. These advances led to parallel developments in the classification of TSCs [47], a direction which had already been initiated by the groundbreaking work of Altland and Zirnbauer [48]. In the following, we briefly present the general formalism and symmetry classification of TSCs.

D7.3.1 Real Space Bogoliubov–de Gennes Hamiltonian

To discuss the emergence of MZMs in a general setting, we consider the following mean-field Hamiltonian [8]

$$H=\sum_{n,m=1}^{N}\sum_{\alpha,\beta=\uparrow,\downarrow}\left[c_{n,\alpha}^{\dagger}h_{nm}^{\alpha\beta}c_{m,\beta}+\frac{1}{2}c_{n,\alpha}^{\dagger}\Delta_{nm}^{\alpha\beta}c_{m,\beta}^{\dagger}+h.c.\right], \quad \text{(D7.6)}$$

where n,m label the position vectors $\mathbf{R}_{n,m}$ of the underlying lattice, and $\hat{\Delta}_{nm}=(\psi_{nm}1_\sigma+\mathbf{d}_{nm}\cdot\boldsymbol{\sigma})i\sigma_y$ defines the most general pairing term in lattice and spin spaces. 1_σ and $\boldsymbol{\sigma}$ denote the unit and Pauli matrices in spin space. The term $\hat{\Delta}_{nm}$ consists of the spin-singlet and -triplet parts ψ_{nm} and $\mathbf{d}_{nm}$, respectively, which satisfy $\psi_{nm}=\psi_{mn}$ and $\mathbf{d}_{nm}=-\mathbf{d}_{mn}$. These constraints result from the antisymmetry of the pairing term under the exchange of the quantum numbers of the electrons comprising a Cooper pair. To proceed, we rewrite the Hamiltonian as:

$$H=\frac{1}{2}\sum_{n,m=1}^{N}\left(c_n^{T\dagger}c_n^{T}\right)\hat{H}_{nm}\begin{pmatrix}c_m\\ c_m^{\dagger}\end{pmatrix}+Constant$$

$$where\ \hat{H}_{nm}=\begin{bmatrix}\hat{h}_{nm} & \hat{\Delta}_{nm}\\ -\hat{\Delta}_{nm}^{*} & -\hat{h}_{nm}^{*}\end{bmatrix}, \quad \text{(D7.7)}$$

with $\hat{h}_{nm}^{*}=\hat{h}_{mn}^{T}$, where ^{T} denotes matrix transposition in spin space. For compactness, we introduced the spinor $\mathbf{c}_n^{\dagger}=(c_{n,\uparrow}^{\dagger},c_{n,\downarrow}^{\dagger})$. The above Hamiltonian can be diagonalized by expanding the electronic operators as:

$$\begin{pmatrix}\mathbf{c}_n\\ \mathbf{c}_n^{\dagger}\end{pmatrix}=\sum_{\nu=1}^{N}\begin{pmatrix}\hat{U}_{n,\nu} & \hat{V}_{n,\nu}^{*}\\ \hat{V}_{n,\nu} & \hat{U}_{n,\nu}^{*}\end{pmatrix}\begin{pmatrix}\mathbf{d}_\nu\\ \mathbf{d}_\nu^{\dagger}\end{pmatrix}+\sum_{\nu=0}^{N_M}\begin{pmatrix}\mathbf{u}_{n,\nu}\\ \mathbf{u}_{n,\nu}^{*}\end{pmatrix}\gamma_\nu \Rightarrow$$

$$H=\frac{1}{2}\sum_{\nu=1}^{N}\sum_{\alpha=\uparrow,\downarrow}E_{\nu,\alpha}\,(2d_{\nu,\alpha}^{\dagger}d_{\nu,\alpha}-1), \quad \text{(D7.8)}$$

where the Bogoliubov eigenoperators $\mathbf{d}_\nu^{\dagger}=(d_{\nu,\uparrow}^{\dagger},d_{\nu,\downarrow}^{\dagger})$ satisfy standard anticommutation relations and correspond to the eigenergies $E_{\nu,\alpha}$. In contrast, γ_ν define MF operators of a number of N_M MZMs, and fulfill the relation $2\gamma_\nu^2=1$. Note that in the general case discussed here, the MF operators are labelled by a quantum number ν instead of the coordinate vector $\mathbf{R}_n$. As a result, the spatial distribution of their wave functions is set by $\mathbf{u}_{n,\nu}$. The topological protection of MFs follows from the charge-conjugation symmetry $\Xi^{\dagger}\hat{H}_{nm}\Xi=-\hat{H}_{nm}$ which dictates the so-called Bogoliubov–de Gennes (BdG) Hamiltonian $\hat{H}_{nm}$. Here $\Xi^2=1$, while the representation of this symmetry transformation in the given basis is $\Xi=\tau_x K$, where K denotes complex conjugation [47, 48]. This charge-conjugation symmetry further implies that the eigenvector corresponding to the eigenenergy $E_{\nu,\alpha}$ is obtained by acting with Ξ on the eigenvector corresponding to $-E_{\nu,\alpha}$, and vice versa. Remarkably, the antiunitary symmetry Ξ results from the antisymmetric behavior of the pairing term reflected in the relation $\hat{\Delta}_{mn}^{T}=-\hat{\Delta}_{nm}$ and, thus, it is not removable. We remark that the here-exposed formalism assumes boundary conditions which are suitable for studying MZMs, i.e., it implies the presence of 0D defects, which fully break translational invariance. If we are instead interested in the emergence of dispersive Majorana boundary modes, then periodic boundary conditions need to be partly employed. In this case, the dispersive Majorana mode operators satisfy $\gamma_{\nu,\mathbf{k}_c}=\gamma_{\nu,-\mathbf{k}_c}^{\dagger}$, where the wave vector $\mathbf{k}_c$ consists of the components of $\mathbf{k}$ which remain conserved in the presence of the boundary.

D7.3.2 Bulk Bogoliubov–de Gennes Hamiltonian and Symmetry Classification

When the various Hamiltonian terms in Equation (D7.7) depend only on the difference of the position vectors, it is preferrable to employ a $\mathbf{k}$-space description. Here, $\mathbf{k}$ defines the wave vector, which for a continuum (crystalline) d-dimensional model is defined in $\mathbb{R}^d$ space ($\mathbb{T}^d$ Brillouin zone). The $\mathbf{k}$-space analog of the BdG Hamiltonian in Equation (D7.7) reads:

$$H=\frac{1}{2}\sum_{\mathbf{k}}\left(\mathbf{c}_{\mathbf{k}}^{T\dagger}\mathbf{c}_{-\mathbf{k}}^{T}\right)\hat{H}_{\mathbf{k}}\begin{pmatrix}\mathbf{c}_{\mathbf{k}}\\ \mathbf{c}_{-\mathbf{k}}^{T}\end{pmatrix}+Constant$$

$$where\ \hat{H}_{\mathbf{k}}=\begin{pmatrix}\hat{h}_{\mathbf{k}} & \hat{\Delta}_{\mathbf{k}}\\ -\hat{\Delta}_{-\mathbf{k}}^{*} & -\hat{h}_{-\mathbf{k}}^{*}\end{pmatrix} \quad \text{(D7.9)}$$

and allows us to construct a bulk topological invariant to predict the emergence of MFs. Since $\hat{\Delta}_{-\mathbf{k}}^{T}=-\hat{\Delta}_{\mathbf{k}}$, also the $\mathbf{k}$-space BdG Hamiltonian possesses the charge-conjugation symmetry Ξ of its real space counterpart, that is now defined as $\Xi^{\dagger}\hat{H}_{\mathbf{k}}\Xi=-\hat{H}_{-\mathbf{k}}$. In analogy to the real space description, the presence of this symmetry sets constraints on the bulk energy spectrum. Specifically, for every solution $\phi_{\mathbf{k}}$ corresponding to energy $E_{\mathbf{k}}$, i.e., $\hat{H}_{\mathbf{k}}\phi_{\mathbf{k}}=E_{\mathbf{k}}\phi_{\mathbf{k}}$, there exists a charge-conjugate solution $\Xi\phi_{-\mathbf{k}}=\tau_x\phi_{-\mathbf{k}}^{*}$ corresponding to energy $-E_{-\mathbf{k}}$.

A generic BdG Hamiltonian which is characterized by the above built-in charge-conjugation symmetry Ξ, falls into one of the BDI, DIII and D symmetry classes [47]. Notably, this classification assumes that apart from the charge-conjugation symmetry, the BdG Hamiltonian may also have a chiral and simultaneously a generalized time-reversal symmetry, effected by the operators Π and Θ, respectively. Depending on the dimensionality of the system and the class, various topologically nontrivial phases become accessible. When $\Theta^2=+1$ ($\Theta^2=-1$) the system belongs to the BDI (DIII) symmetry class and supports real (Kramers degenerate) solutions. In more detail, Figure D7.3 depicts the accessible so-called strong topological invariants in $d=1,2,3$ which can be of the $\mathbb{Z}$ or $\mathbb{Z}_2$ type [47].

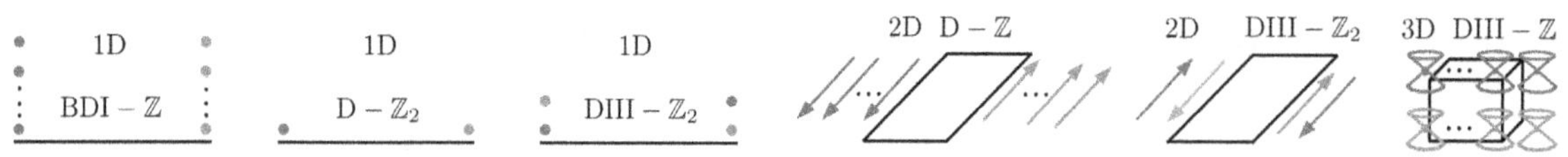

FIGURE D7.3 Accessible types of topological superconductivity and Majorana excitations in $d=1,2,3$ spatial dimensions. We consider all the possible classes in which the Bogoliubov–de Gennes Hamiltonian is dictated by a charge-conjugation symmetry Ξ, with $\Xi^2=1$. These consist of the BDI, D and DIII classes. In each case, we provide the respective $\mathbb{Z}$ or $\mathbb{Z}_2$ strong topological invariant. (i) Dots, (ii) arrows and (iii) cones, schematically depict (i) MZMs, (ii) dispersive chiral and helical MF edge modes and (iii) helical MF surface modes.

D7.4 Classification of Majorana Fermion Quasiparticles

In this section, we present the types of MFs which become accessible in the BDI, DIII and D symmetry classes in $d=1,2,3$ spatial dimensions. See Figure D7.3. Since systems of the same topological class demonstrate the same topological behavior, we here restrict our analysis to representative spinless and spinful models of p-wave superconductors.

D7.4.1 Majorana Zero Modes in 1D Models

- **Spinless Kitaev chain model.** Here we consider both the lattice and continuum versions, which in k-space are described by a BdG Hamiltonian of the form $\hat{H}(k)=\mathbf{g}(k)\cdot\boldsymbol{\tau}$ with:

$$\mathbf{g}(k)=(0,\Delta_p\sin k,-2t\cos k-\mu)\quad\text{and}\quad\mathbf{g}(k)=(0,\Delta_p k,\varepsilon(k)) \qquad \text{(D7.10)}$$

where $\boldsymbol{\tau}$ (1_τ) define(s) Pauli matrices (the unit matrix) in electron-hole Nambu space, and $\varepsilon(k)=(\hbar k)^2/(2m)-\mu$. The symmetry class of these models is BDI since, in addition to the built-in charge-conjugation symmetry effected by Ξ, the BdG Hamiltonian exhibits a chiral symmetry with generator $\Pi=\tau_x$ and a generalized time-reversal symmetry with $\Theta=1_\tau K$. Systems of this class support a $\mathbb{Z}$ topological invariant in this dimensionality which, when it is nonzero, its absolute value yields the number of MZMs appearing on each edge of the 1D TSC.

Given the simple and illustrative form of the Hamiltonian, we define the ensuing topological invariant as the winding number of the two-component unit vector $\hat{\mathbf{g}}(k)=\mathbf{g}(k)/|\mathbf{g}(k)|$ in k space:

$$w=-\frac{1}{2\pi}\int dk\left(\hat{\mathbf{g}}(k)\times\frac{\partial\hat{\mathbf{g}}(k)}{\partial k}\right)_x, \qquad \text{(D7.11)}$$

while for more complex BDI Hamiltonians there exists a generic topological-invariant construction, cf. Ref. [47]. Straightforward calculations yield $w=0$ for $|\mu|>2|t|$ ($\mu<0$) for the lattice (continuum) model, and $|w|=1$ otherwise. The critical boundaries in parameter space separating the various phases are given by the parameter values for which the winding number becomes ill defined. These are determined by $|\mathbf{g}(k)|=0$, i.e., the gap closings of the bulk energy spectrum. Notably, this excludes gap closings which modify the symmetry class, e.g., $\Delta_p=0$. In the lattice model one finds two gap closings for $\Delta_p\neq 0$, occuring at $k=0$ and $k=\pi$, for $\mu=-2t$ and $\mu=2t$, respectively. In contrast, the continuum model features only one gap closing at $k=0$ for $\mu=0$. Note that a gap closing is not possible for $k=\pm\infty$, which here correspond to $k=\pi\equiv-\pi$. Thus, both lattice and continuum models are defined in a compact space, which becomes homotopically equivalent to a circle, i.e., a $\mathbb{S}^1$ sphere. In fact, the winding number is the linking number effecting the homotopy mapping between k (base) space and the Hamiltonian (target) space [5]. This is because the Hamiltonian is parametrized by the two-component unit vector $\hat{\mathbf{g}}(k)$, which also takes values on a circle $\mathbb{S}^1$, since $|\hat{\mathbf{g}}(k)|=1$. Consequently, the winding number is the $\mathbb{Z}$ number performing the mapping $\mathbb{S}^1\mapsto\mathbb{S}^1$, and counts how many full circles are covered in target space for a single full circle covered in base space. For an illustration see Figures D7.4(a–c).

The so-called bulk-boundary correspondence [47] implies that the predictions of the topological invariant are confirmable by considering open boundary conditions. In this case, the wave vector k is no longer a good quantum number, but instead is a function of energy. The latter remains conserved despite the violation of the translation invariance. Remarkably, the zero-energy solutions can be labelled by the eigenvalues $\tau=\pm1$ of the chiral-symmetry operator $\Pi=\tau_x$, i.e. $\phi_{E=0,\tau}$. The respective wave vector is generally complex, thus implying that the solutions decay or increase exponentially in position space. Depending on the location of the boundary, i.e., whether it is on the left or right edge, the MZM wave function has to be accordingly normalizable. Thus, we find that the MZM solution on a given edge corresponds to an eigensolution of a given chirality and appears only for $\mu>0$, i.e. in agreement with the outcome of the topological invariant. Even more, the two edges always support MZMs with wave functions of opposite chiralities. Thus, the degrees of freedom of the BdG spinor, in which both chiralities are present, become fractionalized near the edge due to the fixed chirality of the MFs.

- **Extended spinless Kitaev chain models.** In this paragraph, we study two extensions of the Kitaev chain model. These concern the emergence of multiple MZMs per edge, and the violation of chiral symmetry.
  a. **Higher winding numbers.** As found above, the models of Equation (D7.10) support a single MZM per edge. However, the lattice model allows for

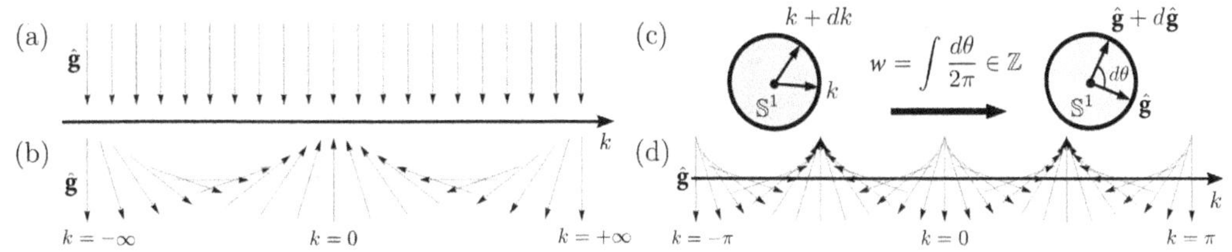

FIGURE D7.4 In the topologically trivial phase, the $\hat{\mathbf{g}}(k)$ posseses a zero winding in k-space and, thus, it can be smoothly deformed into a configuration with a constant orientation as shown in (a). (b) Example of a topologically nontrivial configuration of the unit $\hat{\mathbf{g}}(k)$ vector with a single twist that leads to $|w|=1$. In both (a) and (b) cases, $\hat{\mathbf{g}}(k)$ has the same orientation at $\pm\infty$. This allows the compactification of the 1D k-space from $\mathbb{R}^1$ to $\mathbb{S}^1$. (c) The two $\mathbb{S}^1$ spheres are homotopically mapped via the winding number w, which counts how many times an angle of 2π is covered in $\hat{\mathbf{g}}$ space, when an angle of 2π is covered in the compactified k-space. (d) A topologically nontrivial configuration of $\hat{\mathbf{g}}(k)$ with a double twist from $-\pi$ to π. The double winding results into $|w|=2$ and implies the appearance of two MZMs per edge for an infinite system with open boundary conditions. The presence of more than one MZMs per edge is allowed by the here-assumed chiral symmetry of the system.

a richer topological phase diagram. This is possible by including hopping and pairing terms with a range longer than that of nearest neighbors. As an example, let us first consider the situation in which the lattice Hamiltonian has the same form as in Equation (D7.10), but now with Δ_p and t corresponding to next-nearest-neighbor terms. In this case, we obtain two interpenetrating chains, both residing in the topologically nontrivial phase for $|\mu|<2|t|$. As a result, each one of them hosts one MZM per edge, leading to the following unpaired MF operators: γ_1, γ_2, $\tilde{\gamma}_{N-1}$ and $\tilde{\gamma}_N$. The appearance of multiple MZMs per edge is possible by virtue of chiral symmetry, which allows for a $\mathbb{Z}$ topological invariant. By transferring to k-space, the BdG Hamiltonian has the usual $\hat{H}(k)=\mathbf{g}(k)\cdot\boldsymbol{\tau}$ form, but with $\mathbf{g}(k)=(0,\Delta_p\sin(2k),-2t\cos(2k)-\mu)$. The factor of 2 in front of k leads to $|w|=2$. This result is easy to grasp by depicting the winding of the $\mathbf{g}$-vector in k-space. For an illustration, see Figure D7.4(d).

b. **Violation of chiral symmetry and Majorana number.** The violation of chiral symmetry effects the symmetry-class transition BDI→D, with the latter class supporting a $\mathbb{Z}_2$ strong topological invariant, also termed as Majorana number M [3]. As a result, phases with an even number of MZMs per edge become trivial, since the MZMs on a given edge can hybridize into nonzero energy ABSs. Thus, only phases with an odd number of MZMs remain topologically nontrivial, since there is always a single MZM remaining uncoupled. As pointed out first by Kitaev [3], the presence of Ξ symmetry implies that we can define a matrix $\hat{B}(k)=\tau_x\hat{H}(k)$ which is skew-symmetric at the inversion-symmetric points $k_I=0,\pi$ $(k=0,+\infty\equiv-\infty)$ for the lattice (continuum) model. In turn, the skew-symmetric character of $\hat{B}$ allows us to introduce its Pfaffian $Pf(\hat{B})$, which squares to its determinant, i.e., $det(\hat{B})=[Pf(\hat{B})]^2$. The determinant and the Pfaffian share the same information regarding the bulk energy spectrum, including the occurence of the gap closings which determine the topological phase diagram. Since the latter can only occur at k_I points, the Majorana number is defined as $M=sgn\{Pf[\hat{B}(0)]Pf[\hat{B}(\pi)]\}$. By employing M, we confirm that the Kitaev chain model supports a single MZM per edge for $|\mu|<2|t|$, in which regime $M=-1$.

- **Majorana Kramers pairs in 1D.** While DIII BdG Hamiltonians also possess chiral- and time-reversal symmetries, their crucial difference is that the ensuing time-reversal operator satisfies $\Theta^2=-1$, which imposes a Kramers degeneracy. The simplest example of such a system is obtained by considering two identical Kitaev chain models, one per spin. Under this condition, the two blocks harbor a a single MZM per edge simultaneously, and one ends up with a MZM pair per edge. However, such a MZM pair is protected by the presence of spin-rotational symmetry. This is reflected in the block diagonal form of the Hamiltonian, which also implies that the correct symmetry class is BDI⊕BDI. Breaking rotational symmetry, while simultaneously preserving Θ, is required to effect the transition to the more general DIII case. This can take place by, for instance, introducing odd-under-inversion spin-orbit coupling (SOC) terms which mix the two spin components. This type of systems are classified by a $\mathbb{Z}_2$ invariant which can be defined in analogy to the one introduced earlier for class D. In fact, by writing $\Theta=\hat{U}_\Theta K$, one can further define the matrix $\hat{W}(k)=\hat{U}_\Theta\hat{H}(k)$, which is skew-symmetric at k_I points.

D7.4.2 Majorana Dispersive Modes in 2D and 3D Models

In the following we discuss dispersive MF excitations appearing in 2D and 3D TSCs. We restrict ourselves to TSCs dictated by a fully gapped bulk energy spectrum and a strong topological invariant.

- **Chiral Majorana edge modes.** This type of MF excitations was the first to be discussed in the condensed matter context. This happened after experimental findings claiming the presence of chiral $p_x \pm ip_y$ superconductivity in Sr_2RuO_4 came to light [9]. Notably, chiral superconductors can also harbor MZMs in vortex cores [1]. However, the underlying mechanism is different, and their emergence cannot be predicted by the bulk band topology alone. Importantly, in this case, the defect's properties need to be also accounted for [5, 47].

A chiral $p_x + ip_y$ superconductor is defined by $\mathbf{g}(\mathbf{k}) = (\Delta_p k_y, \Delta_p k_x, \varepsilon(\mathbf{k}))$, with $\varepsilon(\mathbf{k}) = (\hbar\mathbf{k})^2/(2m) - \mu$. The resulting BdG Hamiltonian belongs to symmetry class D, and supports a $\mathbb{Z}$ topological invariant. In analogy to the winding number, here, the topological invariant is given by the first Chern number C_1 of $\hat{\mathbf{g}}(\mathbf{k}) = \mathbf{g}(\mathbf{k})/|\mathbf{g}(\mathbf{k})|$:

$$C_1 = \int \frac{dk_x dk_y}{4\pi} \hat{\mathbf{g}}(\mathbf{k}) \cdot \left[\frac{\partial \hat{\mathbf{g}}(\mathbf{k})}{\partial k_x} \times \frac{\partial \hat{\mathbf{g}}(\mathbf{k})}{\partial k_y} \right]. \tag{D7.12}$$

In the present situation, the topological invariant effects a homotopy mapping from the $\mathbf{k}$ base space to the target space defined by $|\hat{\mathbf{g}}| = 1$. By virtue of the limiting behavior of $\hat{\mathbf{g}}(\mathbf{k})$ for $|\mathbf{k}| \to \infty$, the $\mathbb{R}^2$ $\mathbf{k}$-space is compactifiable to a sphere $\mathbb{S}^2$, while the space spanned by the three-component unit vector $\hat{\mathbf{g}}$ is also $\mathbb{S}^2$. The first Chern number performs the mapping $\mathbb{S}^2 \mapsto \mathbb{S}^2$, and yields the solid angle $4\pi C_1$ covered in $\hat{\mathbf{g}}$-space when a full sphere corresponding to a solid angle of 4π is covered in $\mathbf{k}$-space [5].

A direct calculation of C_1 for the above model yields $C_1 = 1$ for $\mu > 0$ and zero otherwise. When the Chern number is nonzero, the system resides in the topologically nontrivial phase with a single chiral Majorana mode per edge when open boundary conditions are imposed in any direction, say in the x direction for the following. The chiral Majorana modes on opposite edges propagate in opposite directions, similar to the chiral edge modes encountered in quantum Hall effect systems [1]. The difference is that the former are electrically neutral while the latter charged. To obtain the dispersions for the chiral edge modes in the low-energy regime and thus small k_y, we rely on the MZMs obtained for the Kitaev chain model. We thus project the Hamiltonian onto the basis of the MZM wave functions $\phi_{E=0,\tau} \equiv \phi_{k_y=0,\tau}$ obtained from the 1D model, since $E_{k_y=0} = 0$ due to the ensuing charge-conjugation symmetry of the BdG Hamiltonian. This projection immediately yields the two chiral energy dispersions for small k_y, which read $E_{\text{left}}(k_y) = |\Delta_p| k_y$ and $E_{\text{right}}(k_y) = -|\Delta_p| k_y$, with the sign of the dispersion being fixed by the chirality eigenvalue of the respective MZMs appearing for the corresponding 1D model, which is here retrieved by setting $k_y = 0$. The corresponding MF operators γ_{k_y} satisfy $\gamma_{k_y} = \gamma^{\dagger}_{-k_y}$, as a result of the charge neutral character of the chiral MF modes. Hence, the solutions for $\pm k_y$ are not independent.

- **Helical Majorana edge modes.** Constructing a 2D Hamiltonian belonging to the DIII symmetry class is possible along the same lines as in 1D. To define a time-reversal invariant TSC, we consider that the Hamiltonian for each spin sector is of the chiral p-wave type but with opposite Chern numbers. Such a construction leads to a symmetry class D⊕D. A transition D→DIII occurs upon introducing, for instance, odd-under-inversion SOC terms. Such a type of SOC is naturally present in noncentrosymmetric materials [10], and generally leads to the coexistence of spin-singlet and -triplet pairing [10, 11]. In fact, topologically nontrivial phases become accessible when the triplet pairing is dominant [11], and are characterized by a $\mathbb{Z}_2$ topological invariant [47].
- **Helical Majorana surface modes.** In 3D, the only class that supports a strong topological invariant is DIII. In this case, a $\mathbb{Z}$ number of MFs appear on every 2D termination surface of the 3D system. As an example, consider the spinful BdG Hamiltonian $\hat{H}(\mathbf{k}) = \varepsilon(\mathbf{k})\tau_z + \Delta_p \tau_y \mathbf{k} \cdot \boldsymbol{\sigma}$, defined for the spinor $\psi^{\dagger}_{\mathbf{k}} = \left(c^{\dagger}_{\mathbf{k},\uparrow}, c^{\dagger}_{\mathbf{k},\downarrow}, c_{-\mathbf{k},\downarrow}, -c_{-\mathbf{k},\uparrow}\right)$, with $\boldsymbol{\sigma}$ ($\boldsymbol{\tau}$) acting in spin (Nambu) space. Assuming open boundary conditions in the z direction, the Hamiltonian for the helical Majorana surface modes is $\sim k_x\sigma_x + k_y\sigma_y$, and the MF operators satisfy $\gamma_{k_x,k_y} \equiv \gamma^{\dagger}_{-k_x,-k_y}$.

D7.4.3 Majorana Zero Modes Trapped at Vortex Defects

The mechanisms stabilizing MZMs at the edges of a Kitaev chain and at vortices are different. In the former case, the appearance of MZMs can be solely understood in terms of the bulk topological properties. However, this does not hold for MZMs appearing in 0D defects in 2D hosts, since the properties of the defects themselves are crucial. For details see Refs. [5, 47]. MZMs are generally trapped at vortices appearing in both intrinsic and artificial chiral p-wave condensates, such as, p-wave superfluids [5], axion strings [49], superconductor hybrid systems [12] and cold atoms [50]. In fact, in artificial chiral p-wave platforms, the required vortices do not have to be induced necessarily in the condensate's order parameter. Such a possibility was discussed in Ref. [50], where it was shown theoretically that MZMs become accessible in cold atom condensates when considering vortex defects in SOC-like fields.

In the following, we focus on how MZMs emerge in superconducting vortices of spinless chiral p-wave superconductors. We employ the real space version of the chiral p-wave superconductor Hamiltonian of Section D7.4.2, that reads:

$$\hat{H}(\hat{\boldsymbol{p}}, \boldsymbol{r}) = \left(\frac{\hat{\boldsymbol{p}}^2}{2m} - \mu\right)\tau_z + \left(\frac{1}{2}\left\{|\Delta_p(\boldsymbol{r})| e^{i\varphi(\boldsymbol{r})}, -i(\hat{p}_x + i\hat{p}_y)\right\}\frac{\tau_x + i\tau_y}{2} + h.c.\right). \tag{D7.13}$$

For a superconducting vortex defect, the superconducting order parameter $\Delta_p(\boldsymbol{r}) = |\Delta_p(\boldsymbol{r})| e^{i\varphi(\boldsymbol{r})}$ is spatially dependent, and for a vortex of a -1 unit of vorticity we can write

$\Delta_p(\boldsymbol{r}) = \Delta_p(\rho,\theta) = |\Delta_p(\rho)| e^{-i\theta}$, where we introduced the cylindrical coordinates $\rho = \sqrt{x^2+y^2}$ and $\tan\theta = y/x$. The radial dependence of the modulus is here assumed to be of the following form $|\Delta_p(\rho)| = \Delta_p \Theta(\rho - R)$, i.e., we consider an extended vortex, where the order parameter is exactly zero for $\rho < R$ and recovers its bulk value for $\rho \geq R$. The BdG Hamiltonian in cylindrical coordinates reads:

$$\hat{H}(\hat{p}_\rho, \hat{p}_\theta, \rho, \theta) = -\frac{\hbar^2}{2m}\left(\frac{\partial^2}{\partial\rho^2} + \frac{1}{\rho}\frac{\partial}{\partial\rho} + \frac{1}{\rho^2}\frac{\partial^2}{\partial\theta^2} + k_F^2\right)\tau_z + \frac{\{|\Delta_p(\rho)|, \hat{p}_\rho\}}{2}\tau_y + \frac{\hbar}{i\rho}\frac{|\Delta_p(\rho)|}{2}\tau_y + |\Delta_p(\rho)|\hat{p}_\theta\tau_x. \tag{D7.14}$$

One finds that the defect's vorticity exactly cancels out the phase factor $e^{i\theta}$ introduced by the p-wave pairing. Since the Hamiltonian of Equation (D7.14) depends only on derivatives of the variable θ, it can be diagonalized using the eigenfunctions $\sim e^{in\theta}$ of the orbital angular momentum operator $\hat{L}_z = \frac{\hbar}{i}\partial/\partial\theta$. After taking into account that a MZM is obtained only for $n = 0$, we find that the spinor part χ of the MZM eigenvector $\Phi_M(\rho) = \Phi_M(\rho)\chi$ is an eigenstate of the emergent chiral-symmetry operator τ_x. For $\rho < R$ the spatial part reads $\Phi_M(\rho) = J_0(k_F\rho)$, while for $\rho \geq R$ it is decomposable into two parts, i.e., $\Phi_M(\rho) = F(\rho)e^{-(\rho-R)/\xi_M}$. The decay factor stems from the pairing gap where $\xi_M = \hbar/(m\Delta_p)$ and J_0 denotes the zeroth order Bessel function of the first kind. The requirement of a normalizable wave function for $\rho \to \infty$ implies that a MZM is accessible only for the $\tau_x = 1$ eigenstate. When the Fermi energy is much larger than the p-wave order parameter, the $F(\rho)$ factor shows oscillations with a wave length set by the inverse Fermi wave vector k_F. Within this so-called quasiclassical approximation we find $F(\rho) \simeq J_0(k_F\rho)$. Notably, $F(\rho)$ remains nonzero also for a point-like vortex where $R \to 0$, since $J_0(0) = 1$.

D7.5 Topological Spin-Singlet Superconductors: Devices and Materials

TSCs are also accessible when only spin-singlet pairing is considered. However, this becomes possible only as long as suitable SOC or/and magnetic field terms are present. Below, we give an account of prominent quantum devices and materials, in which, strong and systematic experimental evidence for MZM fingerprints has been already provided.

- **Topological insulator–superconductor hybrids.** The nonfully confirmed topological superconductivity in ruthenates [9] and heavy fermion compounds [8, 10] hindered the development of the field. Nonetheless, the interest in topological superconductivity resurfaced after the discovery of topological insulators [46], which support electron-like helical boundary modes. These feature spin-momentum locking and are described by 1D (2D) Dirac-like Hamiltonians, e.g., of the form: $\hat{H}(k_x) = k_x\sigma_x$ $(\hat{H}(\mathbf{k}) = k_x\sigma_x + k_y\sigma_y)$. In 2008, Fu and Kane [12] put forward a number of blueprints for engineering topological superconductivity, all relying on the proximity of these protected helical boundary modes to a conventional superconductor. They showed that MZMs appear in (i) 2D surfaces or trijunctions, where a continuous or a discrete vortex is induced in the superconductor enabling the proximity effect and (ii) Josephson junctions with 1D topologically-protected helical channels as junction links. These ideas of Fu and Kane not only motivated a number of experiments where MF fingerprints have been already observed, but more importantly, they fuelled further theoretical developments in the field of engineered topological superconductivity. These subsequent efforts primarily focused on predicting alternative mechanisms to obtain the spin-momentum coupling found in the helical surface states of topological insulators.
- **Semiconductor–superconductor hybrids.** A number of pioneering theoretical works [13–15] brought III–V semiconductors forward as prominent topologically trivial systems featuring spin-momentum locking. It is well established that low-dimensional electron gases in confined quantum wells and nanowires of semiconductors, such as InAs and InSb, exhibit substantial Rashba SOC. In fact, in the vicinity of the high-symmetry points of the Brillouin zone, the Hamiltonian describing the electron gas may resemble that of the helical boundary states in topological insulators, with a crucial difference. The number of Dirac points appearing in an electron gas come in pairs, i.e., there exists an even number of bulk helical branches, which is in full accordance with the Nielsen–Ninomiya fermion-doubling theorem. In contrast, topological insulators harbor only a single topologically protected helical branch per edge or surface. Thus, engineering a p-wave superconductor that harbors a single MZM in vortices or other 0D defects, requires to isolate a single helical electron branch in an electron gas.

A significant boost in this research direction took place after the proposal [13] to employ an external magnetic field to isolate a single helical branch of an electron gas. Notably, a similar mechanism had been considered earlier in connection with noncentrosymmetric superconductors dictated only by spin-singlet pairing [11]. These ideas further inspired two theory groups [14, 15] to investigate the topological properties of a single-channel quantum nanowire in the presence of (i) Rashba SOC, (ii) an applied magnetic field and (iii) a proximity-induced superconducting gap Δ. See related Figures D7.5(a–c). They showed that such a system becomes equivalent to a spinless Kitaev chain model for sufficiently

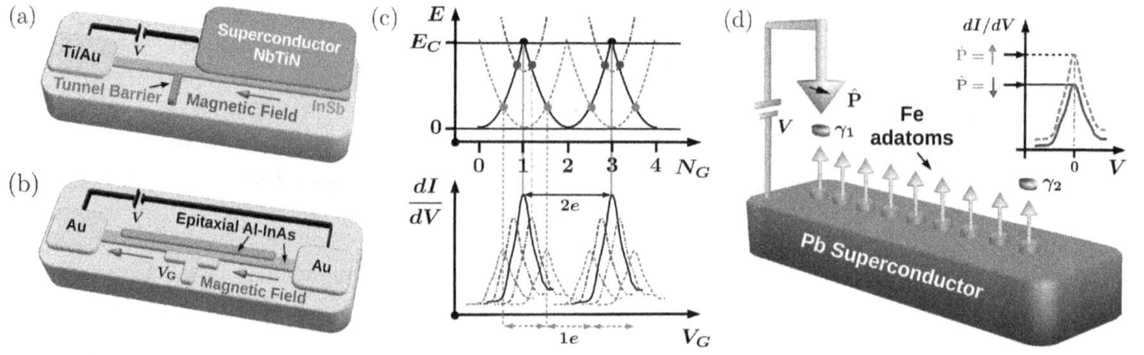

FIGURE D7.5 (a) Semiconductor nanowire (InSb) – Superconductor (NbTiN) heterostructure. A nanowire edge is probed by means of a voltage-biased normal Ti/Au lead, in the additional presence of a tunnel barrier. The resulting differential conductance dI/dV shows a zero-bias peak which is attributed to an edge MZM. (b) Semiconductor nanowire (InAs) – Superconductor (Al) epitaxial hybrid platform in the Coulomb-blockade regime. The differential conductance for a given bias voltage V is measured upon varying the gate voltage V_G and, thus, the number of excess electrons N_G on the island. (c) At zero external magnetic field, i.e., in the topologically trivial regime, coherent transport is mediated by Cooper pairs and occurs when the energy of the island E becomes equal to the charging energy E_C. In contrast, in the topologically nontrivial regime, coherent transport occurs due to states differing by a $1e$ charge and is attributed to the nonlocal electron originating from the two-edge MZMs. (d) Topological ferromagnetic chain consisting of Fe magnetic adatoms deposited on Pb, which is dictated by a substantial Rashba spin-orbit coupling. Probing the chain using spin-resolved scanning tunneling microscopy (STM) allows detecting spin-selective phenomena associated with the MZM wave functions. When the STM lead probes a single MZM, say γ_1, the resulting differential conductance depends on the degree (P_s) and direction ($\hat{\mathbf{P}}$) of the spin-polarization of the tip.

high magnetic fields. Notably, this is possible only for an orientation of the field which is primarily orthogonal to the orientation of the SOC term. The critical value of the respective Zeeman energy E_Z that has to be exceeded for the hybrid system to enter the topologically nontrivial phase is given by $E_Z^{crit} = \sqrt{\Delta^2 + \mu^2}$, where μ defines the chemical potential of the nanowire. Remarkably, while the SOC strength does not enter the topological criterion, its value is crucial, as it determines the spatial extent of the MZM wave functions. Thus, its value should be such so to minimize the overlap of the MZMs on opposite edges. Employing semiconductors for crafting topological superconductivity appears particularly advantageous for experiments, since their chemical potential is gate-tunable, and their effective Landé factor g satisfies $|g| \gg 2$. The latter allows for the topological transition to occur at weak fields, i.e., typically smaller than 1T.

- **Topological magnetic chains.** An alternative route to engineer topological nanowires without relying on the quantum confinement of an electron gas becomes possible in magnetically doped conventional superconductors. See Figure D7.5(d). When a point-like magnetic adatom is added to a conventional superconductor, so-called Yu–Shiba–Rusinov (YSR) bound states appear in its vicinity. In the classical limit, the magnetic moments act as local magnetic fields, and the spectral weight of the YSR states is concentrated only inside the superconductor. The energy of a YSR state for negligible SOC is given by $\varepsilon = \Delta[1-(\pi\nu_F JS)^2]/[1+(\pi\nu_F JS)^2]$, with Δ denoting the superconducting gap of the host, ν_F is the host's normal-phase density of states (DOS) at the Fermi level, J is the exchange energy dictating the coupling between the magnetic moment and the host's electrons, and S defines the adatom's spin moment. In the so-called deep YSR limit, the energy scale of the YSR states satisfies $\varepsilon \ll \Delta$. In this regime, the YSR states govern the system's low-energy properties and topological behavior.

Current scanning tunnel microscopy techniques allow for the manipulation of the position of the magnetic adatoms [37], therefore opening perspectives for engineering artificial chains and lattices originating from coupled YSR states, whose locations track those of the magnetic adatoms. Integrating out the continuum degrees of freedom yields an effective YSR nanowire lattice model which resembles that of semiconductor nanowires with proximity-induced superconductivity. However, there exists a number of crucial differences. First, the hopping and spin-singlet pairing terms in the YSR nanowire model are generally long ranged [20], a property that allows for multiple MZMs per chain edge when chiral symmetry is present [25]. Second, when the superconductor is a thin film, structural inversion asymmetry becomes important and a Rashba-type of SOC is present. The latter leads to additional spin-triplet p-wave pairing terms for the YSR Hamiltonian [23–25]. Of course, on top of the above-mentioned terms, the YSR states feel the exchange field stemming from the magnetic adatoms.

The type of magnetic configuration which becomes stabilized in the adatom chain depends on the adatom-lattice constant, the strength of the exchange coupling of the adatoms to the superconductor's electrons, the strength of the SOC and the possible influence of crystal fields [25]. When the latter are dominant, ferro- and antiferromagnetic grounds states are favored. Instead, if these are negligible, a magnetic spiral profile is prominent, with a wave vector determined by the Fermi-surface characteristics of the host [21]. In general, a complex phase diagram exhibiting an interplay of collinear and spiral phases appears [25]. When the ground state is ferromagnetic, a topologically nontrivial phase can be accessed only in the presence of Rashba SOC. While the presence of Rashba SOC is a sufficient requirement also in the antiferromagnetic case, the even-under-inversion SOC induced by the staggered magnetization opens additional perspectives. Specifically, it has been shown that the combined application of external Zeeman fields and supercurrents can induce a topologically nontrivial phase in antiferromagnetic chains [22]. On the other hand, the onset of magnetic spiral order simultaneously induces the nontrivial topology [20, 21], since such a field is equivalent to Rashba SOC in the presence of a perpendicular ferromagnetic moment, cf. Refs. [17] and [21] and additional works mentioned therein.

- **Iron-based superconductors.** We conclude with a class of spin-singlet superconducting materials which harbor MZMs in vortices of the pairing order parameter. In its nonsuperconducting phase, the $FeTe_{0.55}Se_{0.45}$ compound is a topological insulator and harbors helical surface states [39, 40]. The SOC responsible for the emergence of these protected surface states stems from the multi-orbital character of the electrons in these materials, and specifically from $d-p$ orbital mixing. As it follows from the Fu–Kane proposal [12], such systems are expected to harbor MZMs in vortices emerging in the superconducting phase [39, 40, 42]. Remarkably, MZMs have been indeed experimentally observed in superconducting vortices [41, 43], which in these systems become stabilized either by the presence of magnetic dopants [41] or an external magnetic field [43].

D7.6 Spectroscopic Signatures of Majorana Zero Modes in Experiments

So far, one of the most popular strategies to search for MZMs has relied on quantum transport measurements. In such experiments, one couples the TSC to either a normal or a superconducting lead and records the current flowing through it for a given voltage bias V. The presence of MZMs leads to a peak in the differential tunneling conductance dI/dV at zero bias when this is measured using a normal lead [51]. Moreover, for a spin-polarized lead the tunneling conductance generally shows spin selectivity [52], i.e., it depends on the orientation $\hat{\mathbf{P}}=\mathbf{P}/|\mathbf{P}|$ of the spin-polarization field $\mathbf{P}$ characterizing the lead. Below, we restrict to normal probe leads. For a voltage bias which yields an energy scale which is much smaller than the bulk energy gap of the TSC, the tunneling current is solely governed by the MZM contribution. This holds as long as other current sources [53] can be excluded. Assuming that the lead couples to only a single MZM, the zero-temperature differential tunneling conductance in the steady state reads [51, 52]:

$$\frac{dI}{dV}=\frac{2e^2}{h}\frac{\Gamma_{\mathrm{P}}^2}{(eV)^2+\Gamma_{\mathrm{P}}^2} \qquad \textit{with} \qquad \Gamma_{\mathrm{P}}=\Gamma\mathbf{u}_{n_{\mathrm{lead}}}^{\dagger}\frac{1+\mathrm{P_s}\hat{\mathbf{P}}\cdot\boldsymbol{\sigma}}{2|\mathbf{u}_{n_{\mathrm{lead}}}|^2}\mathbf{u}_{n_{\mathrm{lead}}} \tag{D7.15}$$

where $\mathbf{u}_{n_{\mathrm{lead}}}$ defines the electron part of the corresponding MZM eigenvector, evaluated at the position $\mathbf{R}_{n_{\mathrm{lead}}}$ of the TSC which is probed by the lead. Moreover, within the so-called wideband approximation [51], Γ defines the linewidth broadening experienced by the MZM when its wave function and the lead see the same spin polarization, with $\mathrm{P_s}\hat{\mathbf{P}}\cdot\boldsymbol{\sigma}\mathbf{u}_{n_{\mathrm{lead}}}=\mathbf{u}_{n_{\mathrm{lead}}}$. $\mathrm{P_s}$ denotes the spin-polarization degree of the lead, which is expressed as the difference between the normalized DOS $\nu_{\uparrow,\downarrow}$ of the up and down spin bands of the lead, i.e., $\mathrm{P_s}=\nu_{\uparrow}-\nu_{\downarrow}$ ($\nu_{\uparrow}+\nu_{\downarrow}=1$).

The first experimental claim regarding the discovery of MZMs came from the Delft group [27]. This relied on the observation of a zero-bias peak in the conductance measured by means of a tunnel barrier in the NbTiN–InSb nanowire hybrid platform depicted in Figure D7.5(a). When such a system resides deep in the topologically nontrivial regime, the applied magnetic field is sufficiently strong to assume that the nanowire electrons are spin polarized. When the conductance is measured using a nonmagnetic lead, and under the assumption that only a single MZM is probed, a peak at zero energy appears in the topologically nontrivial phase with a height of $2e^2/h$. This quantized conductance reflects the topologically nontrivial character of the hybrid system and results from the perfect local Andreev reflection mediated by the MZM [51], which fully converts an incident electron into a hole, while a Cooper pair is at the same time absorbed by the topological system. In the 2012 Delft experiment, however, the zero-bias peak height was not quantized. Apart from possible MZM-unrelated scenarios leading to such a peak, this deviation was also attributed to temperature broadening or finite-size effects. See, for instance, the works mentioned in Ref. [53].

Very soon after the observations by the Delft team, further experimental efforts [28] backed the emergence of MZMs in nanowire hybrids. Later on, tremendous progress was made in both the fabrication and design of semiconductor–superconductor devices [54], which apart from enabling the development of novel epitaxial hybrids [32–34], it eventually led to the observation of a quantized zero-bias peak in InAs [34] and InSb [36] platforms, thus, solidifying the evidence for MZMs in these systems. In more detail, zero-bias peaks were observed by the Copenhagen group in epitaxial Al–InAs hybrids, both in nanowires coupled to a quantum dot [33], as well as in 1D channels defined in 2D electron gases [34]. The latter setup additionally motivated the theoretical proposal of flux-controllable 2D devices [26], in which, fingerprints of

MZMs were afterward observed [38]. Even more, conductance measurements were reported in Ref. [32] using mesoscopic islands of epitaxial Al–InAs nanowires in the Coulomb-blockade regime. Such devices allow measuring the fermion parity of the ground state, i.e., whether it contains an even or odd number of electrons. In the absence of energetically low-lying quasiparticles of an extrinsic origin, the charge of a TSC in the trivial regime can only change by $2e$, since only Cooper pairs can be added or subtracted. Instead, in the topologically nontrivial regime, the two MZMs expected to be located at the opposite edges of the device give rise to a $1e$-charged nonlocal fermion which, in turn, allows for charge variations by $1e$ [55]. Hence, the appearance of two MZMs in a TSC island can be experimentally accessed by inspecting the periodicity of the tunneling conductance as a function of the voltage gating the island. The experiment of Ref. [32], details of which are given in Figures D7.5(b,c), has recorded such a change in the periodicity upon the increase of the applied magnetic field, thus, strongly suggesting the presence of MZMs.

In parallel to the experiments in nanowires, several groups performed scanning tunneling microscopy (STM) measurements on topological magnetic chains, cf. Refs. [29, 30, 35, 37]. In 2014, the Princeton group was the first to observe a zero-bias peak in Fe adatom chains on Pb, which was though nonquantized. In contrast to the approach followed in nanowire hybrids, the employment of a STM tip allows the local measurement of the tunneling conductance. In fact, this enabled the experimentalists to confirm that the zero-bias peak originated from edge states which they associated with MZMs. Subsequent efforts provided additional evidence hinting toward the emergence of MZMs in magnetic chains [30]. Later on, two groups [35, 37] carried out high-resolution spin-polarized STM experiments on topological magnetic chains. This gave them the opportunity to access the spin content of the bound states probed. See Figure D7.5(d). According to Equation (D7.15), when the STM tip is fully spin-polarized, there exists a direction of $\hat{\mathbf{P}}$ for which $\hat{\mathbf{P}} \cdot \boldsymbol{\sigma} \mathbf{u}_{n_{\text{lead}}} = -P_s \mathbf{u}_{n_{\text{lead}}}$ and, hence, the conductance is zero [52]. However, in real experiments it is extremely difficult to have a completely spin-polarized tip, as well as to fully control the direction of the spin polarization. As a result, the complete suppression of the current flow is not expected. Nevertheless, the resolution of the experiments in Refs. [35, 37] was sufficient to track spin-selective transport phenomena, which were attributed to MZMs.

STM has also proven to be a successful tool to detect MZMs trapped at the vortex cores of topological insulators in proximity to conventional superconductors and, more recently, in iron-based superconductors. Specifically, the authors of Ref. [31] employed spin-resolved STM to observe the change in the conductance upon flipping the polarization of the STM tip. See Figure D7.5(d). Indeed, they obtained spin-resolved results which are consistent with the presence of MZMs. In the case of iron-based materials, and in particular, the family of FeTeSe compounds, there has been a cascade of experimental results providing strong evidence of MZMs trapped in the cores of superconducting vortices with [43] and without [41] the application of a magnetic field. The latter observations have been so far understood in terms of MZMs pinned by iron-adatom dopants [42]. Experiments in FeTeSe [43] have also shown good agreement with the theoretical predictions regarding the low-energy level hierarchy of the bound states trapped at vortex cores.

D7.7 Open Problems and Perspectives

The advances presented in the previous paragraphs reflect only a small fraction of the community's ongoing efforts to pin down the presence of MZMs. For a broader coverage of the research activity in the field, the reader can further refer to the reviews by Beenakker (Annu. Rev. Condens. Matter Phys. 2013), Alicea (Rep. Prog. Phys. 2012), Leijnse and Flensberg (Semicond. Sci. Technol. 2012), Elliott and Franz (RMP 2015), Aguado (Riv. Nuovo Cimento 2017), Sato and Ando (Rep. Prog. Phys. 2017), Lutchyn *et al.* (Nat. Rev. Mater. 2018) and Pawlak *et al.* (Progress in Particle and Nuclear Physics 2019). The picture that surfaces from the existing experiments is that, while strong evidences for the emergence of MZMs exist, these mainly stem from conductance measurements. Another downside is that fabrication issues and design factors in quasi-1D devices impose restrictions that only allow the witnessing of a single MZM out of the anticipated pair of MZMs on opposite edges. Recent experiments aimed at answering these timely questions. Wiedenmann *et al.* and Laroche *et al.* (Nat. Commun. 2016 and 2019) focused on the Josephson effect in HgTe and InAs platforms, and detected a 4π-periodicity which they attributed to MZMs. Moreover, Ménard *et al.* (Nat. Commun. 2019) have recently claimed the observation of a pair of MZMs in 2D platforms with magnetic adatoms. Despite being very promising, the precise mechanism stabilizing the MZM pair is not yet fully understood. Hence, a remaining pressing challenge of future experiments is to provide compelling evidence for the simultaneous presence and nonlocality of pairs of MZMs (cf. H. Zhang *et al.* Nat. Commun. 2019).

The possible feasibility of multi-MZM manipulations also promises to unlock topological quantum computing, and novel transport phenomena including the so-called topological Kondo effect (Béri and Cooper, PRL 2012). So far, a number of advantageous quantum computing protocols have been proposed, cf. van Heck *et al.* (New J. Phys. 2012), Aasen *et al.* (PRX 2016), and Karzig *et al.* (PRB 2017), which circumvent the need for moving MZMs in real space. Further theoretical progress is also required for the faithful description of realistic semiconductor hybrid devices. This requires considering that in typical experiments more than one confinement channels become relevant. Multichannel situations further imply that the orbital magnetic field effects become non-negligible (Nijholt and Akhmerov, PRB 2016). In fact, there even exist situations in which magnetic fluxes alone can drive the topological phase transition (cf. Kotetes PRB 2015, Vaitiekėnas *et al.* and Lutchyn *et al.* preprints 2018). Even more, the characteristics of the superconductor–semiconductor hybridization (cf. Potter and Lee, PRB 2011; Vuik *et al.* New J. Phys. 2016; Mikkelsen *et al.* and Antipov *et al.* PRX 2018; Woods *et al.* PRB 2018) need to be understood in further depth.

Our discussion has so far mainly focused on MZMs, since these constitute the most extensively studied MF excitations, and hold promise for topological quantum computing. At the same time, the experimental observations regarding the emergence of Majorana edge and surface modes remain to a large extent inconclusive. This is because most of the well-established candidate systems constitute correlated materials, e.g. heavy fermions and ruthenates, which exhibit a complex phase diagram. In fact, experimental claims for topological superconductivity in these materials are usually incompatible with other observations. More recent experiments (Sasaki *et al.*, PRL 2011) strongly hinted that $Cu_xBi_2Se_3$ constitutes a nematic time-reversal invariant TSC (Fu and Berg, PRL 2010). Promising signatures of chiral (and gapped helical) Majorana edge modes have been also provided in a 2D magnetic adatom - superconductor hybrid platform (Ménard *et al.*, Nat. Commun. 2017). Evidence for chiral TSC has been also advocated in Chern insulator - superconductor heterostructures by He *et al.* (Science 2017). However, very recent findings by Kayyalha *et al.* (Science 2020) have challenged these claims. Nonetheless, this route for engineering TSCs (Qi *et al.*, PRB 2010) still appears promising, especially after the proposal that chiral Majorana edge modes are employable for quantum computing (Lian *et al.* PNAS 2018). Detecting chiral Majorana edge modes in future experiments may rely on their conversion into charged chiral edge modes (cf. Fu and Kane, and Akhmerov *et al.*, PRL 2009), thermal responses (Ryu *et al.* 2012), and/or a fractionally quantized conductance as in He *et al.* (Science 2017).

We conclude by re-emphasizing that MFs possess exotic properties which can be harnessed for cutting edge technologies and applications. More importantly, MZMs belong to the simplest type of anyons that can emerge in interacting electronic systems. Another exotic type of quasiparticle is the Fibonacci anyon [6] which enables universal topological quantum computing. Blueprints for designer Fibonacci-anyon platforms have also appeared, amongst which, one finds proposals for fractional quantum Hall - superconductor hybrids (cf. Mong *et al.* and Vaezi PRX 2014). Experiments paving the way toward the realization of these ideas have already been reported (e.g., Amet *et al.*, Science 2016).

Acknowledgements

The author wishes to thank Alex Braginski and Peter Littlewood for giving him the opportunity to contribute to this handbook, which came after the kind recommendation of Gerd Schön. Peter Littlewood, Brian Møller Andersen and Tilen Čadež are also gratefully acknowledged for their suggestions on the manuscript. Finally, note that parts of the material presented here have already appeared (will appear) in some form in the already published (upcoming) book "Topological Insulators" ("Topological Superconductors") by the author and Morgan & Claypool publishers.

References

[1] N. Read and D. Green, Phys. Rev. B **61**, 10267 (2000); G. E. Volovik, JETP Lett. **70**, 609 (1999).
[2] D. A. Ivanov, Phys. Rev. Lett. **86**, 268 (2001).
[3] A. Y. Kitaev, Phys. Usp. **44**, 131 (2001).
[4] E. Majorana, Nuovo Cim. **14**, 171 (1937).
[5] G. E. Volovik, *The Universe in a Helium Droplet*, (Oxford University Press, 2003).
[6] C. Nayak *et al.*, Rev. Mod. Phys. **80**, 1083 (2008).
[7] J. Alicea *et al.*, Nat. Phys. **7**, 412 (2011).
[8] M. Sigrist and K. Ueda, Rev. Mod. Phys. **63**, 239 (1991).
[9] Y. Maeno *et al.*, Nature (London) **372**, 532 (1994); T. M. Rice and M. Sigrist, J. Phys. Condens. Matter **7**, L643 (1995); G. M. Luke *et al.*, Nature **394**, 558 (1998).
[10] E. Bauer *et al.*, Phys. Rev. Lett. **92**, 027003 (2004); P. Frigeri *et al.*, Phys. Rev. Lett. **92**, 097001 (2004).
[11] S. Fujimoto, Phys. Rev. B **77**, 220501(R) (2008); M. Sato and S. Fujimoto, Phys. Rev. B **79**, 094504 (2009).
[12] L. Fu and C. L. Kane, Phys. Rev. Lett. **100**, 096407 (2008).
[13] J. D. Sau *et al.*, Phys. Rev. Lett. **104**, 040502 (2010).
[14] R. M. Lutchyn, J. D. Sau, and S. Das Sarma, Phys. Rev. Lett. **105**, 077001 (2010).
[15] Y. Oreg, G. Refael, and F. von Oppen, Phys. Rev. Lett. **105**, 177002 (2010).
[16] T. P. Choy *et al.*, Phys. Rev. B **84**, 195442 (2011).
[17] M. Kjaergaard, K. Wölms, and K. Flensberg, Phys. Rev. B **85**, 020503(R) (2012).
[18] S. Nadj-Perge *et al.*, Phys. Rev. B **88**, 020407(R) (2013).
[19] S. Nakosai, Y. Tanaka, and N. Nagaosa, Phys. Rev. B **88**, 180503(R) (2013).
[20] F. Pientka, L. I. Glazman, and F. von Oppen, Phys. Rev. B **88**, 155420 (2013).
[21] B. Braunecker and P. Simon, Phys. Rev. Lett. **111**, 147202 (2013); J. Klinovaja *et al.*, Phys. Rev. Lett. **111**, 186805 (2013); M. M. Vazifeh and M. Franz, Phys. Rev. Lett. **111**, 206802 (2013).
[22] A. Heimes, P. Kotetes, and G. Schön, Phys. Rev. B **90**, 060507(R) (2014).
[23] P. M. R. Brydon *et al.*, Phys. Rev. B **91**, 064505 (2015).
[24] J. Li *et al.*, Phys. Rev. B **90**, 235433 (2014).
[25] A. Heimes, D. Mendler, and P. Kotetes, New J. Phys. **17**, 023051 (2015).
[26] M. Hell, M. Leijnse, and K. Flensberg, Phys. Rev. Lett. **118**, 107701 (2017); F. Pientka *et al.*, Phys. Rev. X **7**, 021032 (2017).
[27] V. Mourik *et al.*, Science **336**, 1003 (2012).
[28] M. T. Deng *et al.*, Nano Lett. **12**, 6414 (2012); A. Das *et al.*, Nat. Phys. **8**, 887 (2012).

[29] S. Nadj-Perge *et al.*, Science **346**, 602 (2014).
[30] M. Ruby *et al.*, Phys. Rev. Lett. **115**, 197204 (2015); R. Pawlak *et al.*, Npj Quantum Inf. **2**, 16035 (2016).
[31] H. H. Sun *et al.*, Phys. Rev. Lett. **116**, 257003 (2016).
[32] S. M. Albrecht *et al.*, Nature **531**, 206 (2016).
[33] M. T. Deng *et al.*, Science **354**, 1557 (2016).
[34] F. Nichele *et al.*, Phys. Rev. Lett. **119**, 136803 (2017).
[35] S. Jeon *et al.*, Science **358**, 772 (2017).
[36] H. Zhang *et al.*, Nature **556**, 74 (2018).
[37] H. Kim *et al.*, Sci. Adv. **4**, eaar5251 (2018).
[38] A. Fornieri *et al.*, Nature **569**, 89 (2019); H. Ren *et al.*, Nature **569**, 93 (2019).
[39] Z. Wang *et al.*, Phys. Rev. B **92**, 115119 (2015); G. Xu *et al.*, Phys. Rev. Lett. **117**, 047001 (2016).
[40] Z. F. Wang *et al.*, Nat. Mat. **15**, 968 (2016); P. Zhang *et al.*, Science **360**, 182 (2018).
[41] J.-X. Yin *et al.*, Nat. Phys. **11**, 543 (2015).
[42] K. Jiang, X. Dai, and Z. Wang, Phys. Rev. X **9**, 011033 (2019).
[43] D. Wang *et al.*, Science **362**, 333 (2018); L. Kong *et al.*, Nat. Phys. **15**, 1181 (2019).
[44] D. Aoki, K. Ishida, and J. Flouquet, J. Phys. Soc. Jpn. **88**, 022001 (2019).
[45] L. Fu and C. L. Kane, Phys. Rev. B **79**, 161408(R) (2009).
[46] M. Z. Hasan and C. L. Kane, Rev. Mod. Phys. **82**, 3045 (2010); X. L. Qi and S. C. Zhang, *ibid* **83**, 1057 (2011).
[47] A. P. Schnyder *et al.*, Phys. Rev. B **78**, 195125 (2008); A. Kitaev, AIP Conf. Proc. **1134**, 22 (2009); S. Ryu *et al.*, New J. Phys. **12**, 065010 (2010); J. C. Y. Teo and C. L. Kane, Phys. Rev. B **82**, 115120 (2010).
[48] A. Altland and M. R. Zirnbauer, Phys. Rev. B **55**, 1142 (1997).
[49] M. Sato, Phys. Lett. B **575**, 126 (2003).
[50] M. Sato, Y. Takahashi, and S. Fujimoto, Phys. Rev. Lett. **103**, 020401 (2009).
[51] K. T. Law, P. A. Lee, and T. K. Ng, Phys. Rev. Lett. **103**, 237001 (2009); K. Flensberg, Phys. Rev. B **82**, 180516 (2010).
[52] J. J. He *et al.*, Phys. Rev. Lett. **112**, 037001 (2014); P. Kotetes *et al.*, Physica E **74**, 614 (2015).
[53] E. J. H. Lee *et al.*, Phys. Rev. Lett. **109**, 186802 (2012); D. I. Pikulin *et al.*, New J. Phys. **14** 125011 (2012); G. Kells, D. Meidan, and P. W. Brouwer, Phys. Rev. B **86**, 100503(R) (2012); C. X. Liu *et al.*, Phys. Rev. B **96**, 075161 (2017).
[54] P. Krogstrup *et al.*, Nat. Mater. **14**, 400 (2015); W. Chang et al., Nat. Nanotechnol. **10**, 232 (2015); J. Shabani *et al.* Phys. Rev. B **93**, 155402 (2016); Ö. Gül et al., Nano Lett. **17**, 2690 (2017); S. Gazibegovic *et al.*, Nature **548**, 434 (2017).
[55] L. Fu, Phys. Rev. Lett. **104**, 056402 (2010).

Glossary

A-15 compound: a generic term for intermetallic compounds with a composition of A_3B, where B atoms form a body-centered cubic lattice, and A atoms form a one-dimensional chain in x, y, and z directions in the cubic lattice. Superconducting materials such as Nb_3Sn, V_3Ga, and Nb_3Al have this crystal structure.

AC or ac: alternating current (generally varying in time).

AC or ac loss: energy loss that occurs when a superconductor is exposed to a time-varying (alternating) current and/or magnetic field. According to its origin, this may be classified as "magnetization loss" (due to a time-varying magnetic field) or "transport loss" (due to a time-varying transport current). For certain types of superconducting material, e.g., multifilamentary wire and striated/non-striated coated conductor, additional ac loss components can arise, such as "coupling loss" (due to electromagnetic coupling of filaments) and "eddy current loss" (due to induced eddy currents in any normal conducting material).

ADC: analog-to-digital converter.

AFM: atomic force microscopy.

Adiabatic quantum computing and quantum annealing: family of metaheuristics for quantum algorithms that relies on slow global controls.

adiabatic stabilization: a superconducting magnet design concept which avoids the superconducting-to-normal instability. The "adiabatic" term reflects the condition in which stability is ensured even without relying on heat removal.

ALD: atomic layer deposition.

AM: additive manufacturing.

Andreev reflection (AR): a type of particle scattering which occurs at interfaces between a superconductor (S) and a normal-state material (N). It is a charge-transfer process by which normal current in N is converted to supercurrent in S.

Anharmonicity: difference of the qubit transition frequency between the ground and the first excited state to that between the first and the second excited state.

AQFP: adiabatic quantum flux parametron.

ARPES: angle-resolved photoemission spectroscopy.

aspect ratio (of a conductor): ratio of the longer to the shorter transverse dimensions of a conductor.

asynchronous machine: ac motor or generator, such as an induction motor, operating over a range of frequencies up to the synchronous speed, determined by the frequency of the three-phase supply to the stator and the number of pole pairs.

axial stress: stress on the windings of a coil in the direction of a magnetic field or during thermal cooldown. Compressive stress has a negative sign.

B-1 compound: a generic term for transient metallic carbides, nitrides and oxides with a NaCl-type crystal structure. Superconducting materials such as NbN, NbC, and MoN have this crystal structure.

Bi-2212: bismuth strontium calcium copper oxide superconducting compound with chemical formula $Bi_2Sr_2CaCu_2O_{8+x}$ and T_c of ~96 K. Included in the HTS class of superconducting materials.

Bi-2223: Bismuth strontium calcium copper oxide superconducting compound with chemical formula $Bi_2Sr_2Ca_2Cu_3O_{10+x}$ and T_c of ~108 K. Included in the HTS class of superconducting materials.

BCS theory: a microscopic theory of superconductivity by John Bardeen, Leon Cooper, and John Robert Schrieffer based on the formation of Cooper pairs of electrons in the vicinity of the Fermi level that couple via lattice distortions (i.e., phonons).

Bean (critical state) model: a macroscopic model for the magnetization behavior in type II superconductors that assumes that the critical current density is constant with respect to the magnetic field. Sometimes also referred to as Bean–London model.

biaxially textured: a polycrystalline material with grains sharing the same (within a scatter of a few degrees) orientation of at least two crystallographic axes. Typically refers to the grains in a film or tape with closely aligned crystallographic a and b axes lying in the plane of the film.

binary phase diagram: thermodynamic equilibrium diagram showing the phases possible for two elements A and B as a function of temperature.

bi-SQUID: a two-loop SQUID containing three overdamped Josephson junctions. This device was developed to linearize its voltage response to the applied magnetic signal.

Bloch sphere: geometric representation of the quantum state of a two-level system through the expectation values of the three Pauli matrices.

boil-off: relates to the rate of a cryogen evaporation, such as liquid helium or nitrogen. It is expressed in liters, centiliters, or milliliters per hour.

bolometer: thermal detector for small heat fluxes, frequently works on the transition edge (T_c) of a superconductor.

braid: a narrow tubular or flat structure produced by intertwining strands of materials to form a definite pattern.

Brayton cryocooler: a recuperative type (steady pressures and flow) of cryocooler operating with the Brayton cycle, which uses expansion work at the cold end to provide refrigeration.

bronze process: a method used to produce A15 superconductors of the formula A_3B, usually Nb_3Sn, in a Cu-based matrix, where components of element A (typically rods) are embedded in a ductile "bronze" alloy of Cu and element B. In this form, the composite can be mechanically deformed into the desired final geometry before heat treating to form the brittle A15 in the A components with the B element being supplied by the bronze.

BSCCO: family of bismuth strontium calcium copper oxide superconductors with general formula $Bi_2Sr_2Ca_{n-1}Cu_nO_{2n+4+x}$, where n ranges from 1 to 3. Included in the HTS class of superconductors with T_c values ranging from ~33 K to ~108 K.

BSE: backscattered electrons.

buffer layer: intermediary layer usually used to control the interfaces between one material and another either to reduce reactivity or improve crystallinity.

cable (concentric lay conductor): a conductor constructed by assembling multiple wires or tapes. A cable may have a central core surrounded by one or more layers of helically laid wires, or may be obtained as a multistage assembly of wires and sub-cables, each helically wound with a twisted structure. Depending on the layout and on the final shape obtained by rolling, drawing, or by other means, there exist several types of cables including compact round, flat, conventional concentric, equilay, parallel core, rope-lay, unidirectional and unilay.

cable-in-conduit conductor: a composite conductor consisting of a cable inside a metal conduit providing mechanical stiffness under stress. The conduit provides mechanical stiffness and allows forced-flow cooling of the cable to improve the thermal stability of the conductor.

calcination: the reaction of precursor compounds (or elemental materials) at elevated temperature to form a single- or multiphase compound.

capacitively shunted flux qubit (CSFQ): superconducting qubit design consisting of two large and one small Josephson junctions arranged in a loop with a capacitive shunt. Level spectrum has a positive anharmonicity.

CC: coated conductor.

CCD: charge coupled device/detector, usually in the form of an array capable of spectroscopy or imaging.

CFHX: counter flow heat exchanger.

chain layer: The cases of $YBa_2Cu_3O_{7-\delta}$, $Y_2Ba_4Cu_7O_{15-\delta}$, and $YBa_2Cu_4O_8$ refer to the $CuO_{1-\delta}$ layer in the structure which, when fully occupied (δ=0), extends as linear structures in the crystallographic *b*-direction. In Y123 and Y247, the variable oxygen content of the chains allows the doping state to be varied. In Y124, the chains are always fully occupied and appear as double chains or Cu_2O_2 ladders.

charge reservoir: usually refers to the non-conducting layers that supply charge to the CuO_2 planes in cuprate superconductors.

Charging energy: the energy associated with placing one electron charge on a conducting island.

Chevrel phase: a generic term for compounds with a typical composition of MMo_6X_8, where M refers to metallic elements such as Pb, Sn, Cu, and La, and X corresponds to chalcogen elements such as S, Se, and Te.

CIP: cold isostatic press.

Circuit Quantum Electrodynamics (also circuit QED or cQED): family of physical systems in which nonclassical quantum states of the electromagnetic field are used. Circuit QED requires the use of a nonlinear quantum system such as a Josephson junction-based circuit.

CLJ: current-limiting junction.

Claude cycle: a refrigeration cycle of the recuperative type (steady pressures and flow) used by most large liquefaction systems in which expansion work is used to precool a final Joule–Thomson expansion stage where liquefaction occurs.

CMP: chemical mechanical polishing, planarization process used in thin-film multilayer technology.

coated conductor: typically refers to quasi-epitaxial superconducting thin-film coated onto a flexible tape substrate usually comprising a metal base with buffer layers between the metal and the superconductor that protect the superconductor from reactive contamination and allows lattice matching for epitaxy.

co-evaporation: deposition approach utilizing multiple evaporation sources to control deposition stoichiometry for multicomponent systems.

coherence length, ξ: (1) a temperature-independent quantity, defined in BCS theory, describing the spatial extent of Cooper pairs. (2) A temperature-dependent quantity defined in Ginzburg–Landau theory describing the spatial variation of the order parameter.

coldhead: see *cryocooler.*

compact round conductor: a conductor constructed with a central core surrounded by one or more layers of helically laid wires and formed into final shape by rolling, drawing, or by other means.

compact stranded conductor: a conductor composed of helically laid monolithic or stranded wires and formed into a final plate-like shape by rolling, drawing, or other by means.

compensation: cancellation of gross field inhomogeneities using either passive or active shimming.

compliance: the thermoacoustic term relating the mass storage term in an oscillating gas flow to the associated pressure change. The term is analogous to the capacitance in an AC electric circuit.

composite conductor: a conductor consisting of two or more types of material, each type being plain, clad, or coated, and assembled together to operate mechanically and electrically as a single conductor.

composite diffusion process: a fabrication process for composite conductors, where the composite is cold-worked into a final shape with or without intermediate anneals. Solid state diffusion within the components of the composite is used to form the desired superconducting phase or an appropriate microstructure containing superconducting and normal conducting phases.

composite superconductor: a conductor incorporating superconducting and normal material. There exist several types of composite superconductors including filamentary, coreless, tape, tubular, and hollow conductors.

condensation energy, superconducting, δG: energy that has to be removed to convert a superconductor from the normal to the superconducting state.

conduction cooled: instead of using liquid cryogens such as liquid helium or nitrogen to cool down the material, cooling is achieved by using a thermal link between the material and the cooling source such as a coldhead or a cooling pot.

conventional concentric conductor: conductor constructed with a central core surrounded by one or more layers of helically laid round wires. The direction of lay is reversed in successive layers, generally with increasing length of lay.

cooling: a method to lower the temperature of a magnet or material.

Cooper pair: the unit of charge carrier in a BCS-type superconductor. Electrons in a Cooper pair are of equal and opposite wave number/vector (momentum) and spin. Cooper pairs are not scattered by the lattice and may move within a superconductor without loss of energy.

Cooper pair box: an early qubit design consisting of a superconducting island connected to ground through high-resistance tunnel junctions.

coplanar waveguide: a microwave transmission line typically consisting of a planar electrode structure on a substrate with a center signal conductor and two side ground plane conductors each separated by a gap from the center conductor.

coreless conductor: a conductor constructed with one or more layers of helically laid wires and formed into final shape by rolling, drawing, or by other means.

coupling current: a current flowing, in an alternating or pulsed magnetic field, between superconducting filaments or strands separated by normal conducting materials.

coupling loss: an ac loss caused by Ohmic dissipation associated with a coupling current due to electromagnetic coupling of individual superconducting filaments and a normal-metal conducting path, e.g., the normal-metal matrix in a multifilamentary wire. The coupling loss is sometimes referred to as eddy current loss (see also *eddy current loss*).

coupling time constant: a time interval for which a coupling current caused by an alternating or pulsed magnetic field decays. The coupling time constant is proportional to the square of the twist pitch and inversely proportional to the matrix resistivity in the direction perpendicular to the filament axis.

CRB: charge reservoir block. See also *charge reservoir.*

critical current, I_c: the maximum current a superconductor can carry while being in the superconducting state.

critical field, lower, H_{c1} or B_{c1}: the magnetic field strength above which magnetic flux vortices begin to penetrate the bulk of a type II superconductor in the absence of demagnetizing effects. Type II superconductors are in a Meissner state below this field.

critical field, thermodynamic, H_c or B_c: the maximum magnetic field below which a type I superconductor exhibits superconductivity at zero current and temperature. In general, the condensation energy is equal to $\mu_0 H_c^2/2$.

critical field, upper, Hc_2 or B_{c2}: the magnetic field strength above which the mixed state in a type II superconductor is destroyed and the material reverts to the normal conducting state.

critical state: regions within a superconductor in the critical state that carry the critical current density and which are penetrated by magnetic field.

cryocooler: refrigerator designed to reach cryogenic temperatures (typically below 150 K).

cryogen-free system: a cryostat or system that uses no cryogens and produces cooling with the aid of a cryocooler.

cryostat: a vessel to maintain a material or a device at low (cryogenic) temperature.

crystal pulling: the growth of large crystals through the enlargement of a seed crystal in contact with a melt. The seed is usually slowly removed from the melt while maintaining contact with the meniscus.

CSD: chemical solution deposition.

cuprate superconductors: family of superconductors whose structure is based on layers of copper oxides. Most notably the first of the HTS superconductors to be discovered, lanthanum barium copper oxide, and the most commonly used, yttrium barium copper oxide ($YBa_2Cu_3O_{7-x}$).

current, critical, I_c: the maximum electrical current below which a superconductor exhibits superconductivity at a given temperature and magnetic field. The critical current is usually defined via a value of resistivity or an electric field.

current, persistent: current flowing through a closed, superconducting loop circuit which does not measurably decay with time.

current density J: current per unit area.

current density, critical, J_c: the critical current of a superconductor divided by its cross-sectional area.

current density, engineering, J_e: the critical current flowing in a composite conductor divided by its cross-sectional area.

current leads: connections between the superconducting joints of the coils and the external power supplies. However, this often refers to the leads between the superconducting joints of the magnet and the top of the cryostat which can be made using a combination of HTS and LTS.

current-phase relation: describes the relation between the current and the phase in the static (dc) Josephson effect in various types of superconducting junctions.

CVD: chemical vapor deposition. The use of molecular precursors over a hot substrate that decomposes to give the desired compound material.

DAC: digital-to-analog converter.

Dayem bridge: a sufficiently narrow strip patterned between two electrodes in a single-layer superconductor thin film that can exhibit Josephson-junction-like behavior.

DC or dc: direct (i.e., constant in time) electric current.

Debye-Waller coefficient: refers to the reduction in intensity of an X-ray diffraction peak caused by thermally induced lattice vibrations (phonons).

decay: has several associations. (1) see *drift:* the decay of a magnet in relation to its stability reflects the rate of change in the magnetic field as a function of time. This normally applies to magnets operated in a persistent mode. It is given in units of parts per million or proton hertz per hour. For NMR magnets, the specification is −0.01 ppm/h. (2) The magnetic field decreases in intensity with distance.

decoherence: process that converts nonclassical quantum states into their classical limiting case.

demagnetization factor (demagnetizing factor), N or D_m: the ratio of the average demagnetizing field to the average magnetization in a magnetic or superconducting material of finite size. Defined for an ellipsoid by H¼ $H_0 - NM$, where H is magnetic field strength, H_0 is applied field, and M is the magnetization. Also used qualitatively for non-ellipsoid samples. Samples with large demagnetizing factors cause field distortions and concentrations near the sample edges.

dephasing time, $T\varphi$, the characteristic timescale for the loss of coherence in a qubit superposition state.

derivative removal by adiabatic gate (DRAG): family of strategies for envelope shaping of microwave pulses that control superconducting quantum bits. DRAG is designed to minimize the occupation of non-computational states.

Dewar: vessel used for containing cryogenic liquids that extends storage time by the insulation properties of a partial vacuum between the walls of the container.

diamagnetic material: a material with negative susceptibility.

diamagnetism, perfect: magnetic susceptibility equal to −1 and magnetic permeability equal to 0, exhibited by a type I superconductor below the critical magnetic field, H_c, and a type II superconductor below the lower critical field, H_{c1}. The bulk of a superconductor exhibiting perfect diamagnetism is shielded from magnetic fields.

diamagnetism: magnetism where the field inside a substance is less than the applied field. In this case, the magnetic susceptibility, χ, is negative and the permeability, μ, lies between 0 and 1.

differential resistance: a resistance defined for any point of I-V curve of a device as derivative $R_d = dV / dI$.

dip coating: process methodology whereby a sample is lowered into one or more precursor solutions or slurries after which unwanted materials are burned out, and the remainder is thermally processed to the form desired material.

dipole coil: an electromagnetic coil that provides a uniform interior field with one axis (in particle accelerators this dipole property is used to bend the particle beam).

DMP: decision-making pair.

double pancake coil: a pair of pancake coils so connected to have their conductor ends appears at the outer circumference of the coil. A magnet is constructed by stacking double pancake coils connected in series.

drift: the rate of change in the magnetic field as a function of time. This normally applies to magnets operated in a persistent mode. It is given in units of parts per million or proton hertz per hour. For NMR magnets, the specification is −0.01 ppm/h.

DSV: diameter spherical volume. Defines a homogeneity specification in a sphere defined by a certain diameter.
DOS: density of states.
DTA: differential thermal analysis, a technique to provide data on temperature-related transformations in a material by comparing the temperature in the sample with an inert reference over an identical thermal cycle.
dynamic stabilization: the prevention of flux jumps by slowing down the rate of flux motion or by increasing the rate of cooling, usually achieved by embedding the material in a high conductivity material, such as copper.

EBIC: electron beam induced current.
EBIV: electron beam induced voltage.
EBSD: electron backscattered diffraction.
ED: electron diffraction.
eddy current loss: an ac loss resulting in an Ohmic dissipation associated with either induced eddy currents in any normal conducting material within the superconducting wire architecture or with a coupling current (see also *coupling loss*).
EDM: Electrical discharge machining.
EDS: electron dispersive X-ray spectroscopy. A technique available in electron microscopes that utilizes the X-rays generated by the electron beam to identify and quantify elements in a sample. In conjunction with a rastered electron beam, it can provide elemental maps.
EDX: see *EDS*.
EELS: electron energy loss spectroscopy. In a transmission electron microscope, measurement of the energy distribution of the transmitted beam can be used to provide a variety of information including light element composition, sample thickness, and oxidation state. In conjunction with a rastered electron beam, it can provide energy-filtered maps.
electromagnet: a type of magnet generated using an electrical current.
electron beam lithography (e-beam lithography): a maskless lithography used for patterning of thin films in conjunction with an etching process. A rastered electron beam is used for the exposure of a photoresist by direct writing.
energization: process of ramping or putting current in a magnet.
energy gap, superconducting, Δ: one half of the minimum value of energy required to break a Cooper pair. The energy gap is temperature dependent and is usually cited at absolute zero in BCS theory.
environmental protection: the course of action required to protect the local and extended environment from the proposed laboratory or industrial activities, a portion of which may be specified by local or national codes.
epitaxial: refers to the deposition of a crystalline film on a (quasi-)crystalline substrate. The coated film is called an epitaxial film or epitaxial layer if the crystalline substrate induces a matching of the lattice parameters and orientation of the film.
equilay conductor: conductor constructed with a central core surrounded by more than one layer of helically laid wires, with all layers having a common length of lay, and the direction of lay being reversed in successive layers.
error syndromes: results of measurements in a quantum computer that reveal information about errors without direct information about the state of a quantum register.
ESEM: environmental scanning electron microscope. A type of SEM that uses a combination of differential vacuum pumping, electron beam transfer, and specialized detectors to allow examination of organic or insulator samples in a gaseous environment, avoiding the need for coating with a conducting element or for wet sample drying techniques that would alter their condition.
ERSFQ: energy-efficient rapid single-flux quantum.
eSFQ: energy-efficient single-flux quantum.
ESA: electrically small antenna—an antenna where the maximum physical dimensions are small compared to the wavelength, λ, of interest, usually less than λ/10.
eutectic alloy: an alloy having the composition of its eutectic point.
eutectic point: The composition of a liquid phase in equilibrium with two solid phases.
eutectic reaction: the formation of a mixture of two solid phases from a liquid at a fixed temperature (the eutectic temperature).
external diffusion process: a fabrication process typically for Nb_3Sn and V_3Ga composite conductors, where a Cu jacket with axially drilled holes containing Nb or V rods is cold-worked into a final shape, coated on its surface with Sn or Ga and heated to diffuse the Sn or Ga initially into the matrix and then to form a Nb_3Sn or V_3Ga layer at the interface between the Cu matrix and the Nb or V cores.

fault current limiter: a device in series with a current supply line on a power distribution system that increases its resistance rapidly when the current exceeds the rated value. Superconductors are appropriate for this application since they enter the flux flow and, subsequently, the normal state when the current density exceeds its critical value.
fault-tolerant quantum computation: method of quantum computing involving quantum error correction that allows to perform quantum algorithms at an error rate much lower than those of the constituent quantum devices.

FC: field cooled; a superconductor is in the field-cooled condition if it has been cooled below its critical temperature while immersed in a magnetic field.

ferromagnet: a material containing unpaired electron spins that are spontaneously aligned parallel to each other in the absence of an applied magnetic field. Often there are ferromagnetic domains of parallel spins that can be aligned by the application of an external magnetic field.

FIB: focused ion beam; a nanoscale patterning technique for thin-film materials is ablated by a focused ion beam. Also used in SEMs to provide nanomachining and tomographic sectioning capability. For SEM nanomachining, it is typically combined with a gas injection system to deposit material using the electron beam.

filament (elementary filament): a thin, elongated fiber of superconducting material contained in a composite conductor.

filamentary (multifilamentary) conductor: a composite superconductor consisting of more than one superconducting filament embedded in a non-superconducting matrix.

film boiling: a phenomenon in which the surface of a material being cooled is completely covered with a film of evaporated coolant.

fluence, *F*: total number of particles per unit area to which a material is exposed. $F = \Phi \cdot t$, where Φ is the flux, and t is the time.

flux (flux density), Φ: number of particles per unit area and time impinging on a material. $\Phi = n'v$, where n' is the density (per volume) of incident particles, and v is their mean velocity. In a superconducting context, the flux density is the flux per unit area normal to the magnetic field induction.

fluxonium: superconducting qubit design incorporating a high-impedance superinductor with enhanced qubit lifetimes for certain bias points.

fishtail: a feature in magnetization–field (M–H) plot where the magnetization passes through a second peak with increasing field, often at a so-called matching field where the average spatial distribution of pinning defects matches the separation of vortices in the vortex lattice. Often called a "second-peak" effect.

flux creep: a phenomenon in which fluxoids pinned in a superconductor move due to thermal activation.

flux flow: a phenomenon in which fluxoids in a superconductor move when the Lorentz force exceeds the pinning force.

flux growth: the use of a non-reactive additive to facilitate the dissolution and subsequent growth of crystals of the desired material. It may be a minor constituent or used like a solvent.

flux flow resistivity: electrical resistance arising from the flow of magnetic fluxoids in a type II superconductor.

flux jump: the collective, discontinuous motion of fluxoids in a superconductor, produced by mechanical, thermal, magnetic, or electrical perturbation in the material.

flux-jump stabilization: a superconducting magnet design concept to avoid instability caused by internal magnetic disturbances. Achieved by subdividing the superconducting material into fine filaments small enough so that the energy liberated after an internal magnetic disturbance is sufficiently small to avoid a flux jump.

flux-line lattice: an array of magnetic flux lines in a type II superconductor. The order of the lattice depends on the degree of flux pinning, with a perfectly ordered array in a superconductor that contains no pinning centers.

flux pinning: the trapping of fluxoids at defects in a type II superconducting material.

flux quantum, Φ_0: the quantum of magnetic flux has a fixed value that combines the Planck constant h and the electron charge: $\Phi_0 = h/(2e) \approx 2.067833848\times10^{-15}$ Wb.

fluxoid (fluxon): a quantized line of magnetic flux within a type II superconductor containing one flux quantum, Φ_0. Behaves as a Faraday line of magnetic force.

forced cooling: a cooling method for superconducting magnets or conductors by the forced flow of liquid helium through cooling channels.

Four-terminal (or contact) measurement: a technique to measure the resistivity of metals and superconductors which avoids the effects of contact resistance. Two terminals are used to inject and remove the current from the sample and two are used to monitor the voltage drop across the sample. Typically, large area, low-resistance current contacts are used to minimize heating effects, and a high-impedance voltmeter is used to monitor the voltage.

fringe field: commonly known as static stray field, is a field gradient external to a magnet that is due to the returning flux produced by the magnet. The stray field is inversely related to the third power of the distance center of the magnet and for high field magnets, the volume over which the stray field has influence can be extensive and must be considered when designing magnet facilities with regard to safety and the impact on items that are sensitive to magnetic fields. The dynamic fringe field is concerned with how the static fringe field changes during a quench.

full stabilization: a design concept in which the amount of high conductivity material included in a composite superconductor and the level of cooling provided are such that should the superconducting component quench, thus diverting all the current into the normal material, the temperature of the composite will remain below the critical temperature of the superconducting component. In these circumstances, the temperature of the superconducting component will always recover to its original level and the current will then transfer back.

fullerene: the term for a series of 3D closed structures primarily containing carbon, of which the canonical example is C60 or Buckminsterfullerene.

gate: elementary operation in a quantum computer program (in the context of a quantum gate or quantum logic gate).

GB: see *grain boundary.*

generation temperature: this relates to the temperature at which current in a composite superconductor begins to flow in the matrix. It delineates the normal and the superconducting states.

GeV: energy unit, 1 GeV = 1.6×10^{-10} J.

Gifford–McMahon cryocooler: A regenerative type (oscillating pressure and flows) of cryocooler that uses modified commercial air-conditioner compressors to provide a steady low- and a high-pressure source of helium, which is then converted by valves to an oscillating pressure for the cold finger made up of a displacer and regenerator.

Ginzburg–Landau (GL) parameter, κ: the ratio of penetration depth, λ, to coherence length, ξ: $\kappa = \lambda/\xi$. A superconductor is type I if κ is less than $1/\sqrt{2}$ and type II if it is greater than $1/\sqrt{2}$.

Ginzburg–Landau (GL) theory: the first quantum phenomenological theory of superconductivity that describes the properties of both type I and type II superconductors just below the critical temperature.

grazing-incidence small-angle neutron scattering (GISANS): a neutron scattering technique where the incident beam intersects the sample at a low angle, typically employed for analyzing the structure of the sample surface.

gradient (in the context of MRI magnets): the variation of field with distance generated by gradient coils during MRI scanning.

grading: the use of conductors or cables with different properties or dimensions at different locations in a magnet to optimize the use of conductor for the designed local field primarily to reduce conductor cost.

grain boundary (GB): the interface separating two grains (crystallites) in a bulk or thin-film material.

grain boundary resistance: the impedance to thermal conduction or electronic carrier transport across a grain boundary. Normally, it is a function of the grain boundary angle, impurity content, and defects in the boundary.

granularity, magnetic: an inhomogeneous field profile typically related to the microstructure of the superconductor.

gradiometer: difference circuit of SQUID pickup loops to reduce noise pickup.

Hastelloy C276: a tungsten containing nickel-chromium-molybdenum alloy frequently used as a substrate for superconducting tape. It is resistant to corrosion in a variety of chemical processing conditions.

H_{c2} anisotropy: anisotropy of the upper critical field in an anisotropic superconductor. It is defined by $H_{c2}^{//ab}/H_{c2}^{//c}$, the ratio between the H_{c2} values in the main crystallographic directions.

heat capacity: the heat required to raise the temperature of a specified object by 1 K. The specific heat capacity of a substance is the heat capacity of an object of that substance divided by the mass of the object. The heat capacity of an object can depend strongly on the starting temperature of the object.

heavy fermion superconductor: superconductors that contain rare earth (predominantly Ce) or actinide (predominantly U) ions and in which some of the electron bands have very large effective masses, typically a few hundred times larger than the free electron mass.

heavy ion irradiation: heavy ions, such as Pb, being produced in an accelerator with energies in the GeV range, are able to create "columnar tracks" in the cuprates, which are useful tools for studying flux pinning in these materials.

helium, supercritical: ^{4}He above its critical temperature, T_c (5.22 K), and critical pressure, P_c (0.227 MPa).

Helmholtz coils: a split pair of specially constructed coils each with a common radius and number of turns, separated by a distance equal to the coil radius. Current passed through the coils produces an extremely uniform magnetic field in the space between the coils.

Helmholtz pair: a matched pair of Helmholtz coils.

high electron mobility transistor (HEMT): a low-noise cryogenic semiconductor amplifier in the microwave range that is commonly used for superconducting qubit readout.

HgBCCO: family of bismuth strontium calcium copper oxide superconductors with general formula $HgB_2Ca_{n-1}Cu_nO_{2n+2+x}$, where n ranges from 1 to 3 and T_c values from 94 to 134 K. Mercurocuprates are included in the HTS and cuprate classes of superconductors.

HIP: hot isostatic press(ing). A heat treatment technique that applies a high isostatic inert gas pressure (> 50 MPa) to consolidate a metal or ceramic at high temperature.

HMIS: hazardous materials information system.

HOMO: highest occupied molecular orbital.

Homologous: in relation to HTS compounds or structures refers to a sequence of structures with the same blocking layers but increasing numbers of active planes such as CuO_2 layers. This is exemplified by $HgBa_2Ca_{n-1}CunO_{2+2n}$, where n = 1,2,3,4,5...

HOT: high operating temperature.

HPCVD: hybrid physical–chemical vapor deposition. A modified CVD process, see *CVD*.

HRTEM (or HREM): high-resolution transmission electron microscope, a microscopy technique with sufficient spatial resolution to resolve the atomic structure of materials.

HTS: high-temperature superconductor. Not strictly defined but typically refers to classes of superconducting compound that have their highest T_c above 77 K (the boiling point of liquid nitrogen).

homogeneity: refers to the uniformity of a magnetic field within a defined volume. There are two types of homogeneity. Spatial homogeneity describes the variation of the field within a given volume and is expressed in ppm, parts per million. Spatial homogeneity is optimized during the magnet design. Temporal homogeneity refers to the stability or change of the field as a function of time. Temporal homogeneity is primarily affected by the decay of joints. Both homogeneities are important and relevant during scanning of patients.

homopolar motor: a direct current electric motor with two magnetic poles producing static working flux. The rotating conductor of the homopolar (another name unipolar) motor cuts flux lines unidirectionally and a constant EMF is generated without any commutation inherent to other DC machines. Slip rings, however, are an essential part of the homopolar motor.

hoop stress: stress arising due to the energization of a circular coil due to the Lorentz force (positive tension) or thermal cooldown (negative tension).

hybrid magnet: a magnet consisting of normal conducting and superconducting magnets. Also, a magnet incorporating HTS and LTS coils.

hysteresis loss (pinning loss): ac loss associated with the unpinning of flux lines in a superconductor in an alternating magnetic field. Equivalent to the work done in cycling the magnetic field applied to a superconductor below its irreversibility field. Also applies to magnetic hysteresis.

hysteresis loss (pinning loss): ac loss associated with the unpinning of flux lines in a superconductor in an alternating magnetic field. Equivalent to the work done in cycling the magnetic field applied to a superconductor below its irreversibility field. Also applies to magnetic hysteresis.

IBAD: ion beam assisted deposition.

impregnated coil: a coil impregnated with appropriate resin to improve its mechanical stability and electrical insulation within a magnet structure.

in situ A15 process: a fabrication process, typically for Nb_3Sn and V_3Ga conductors, where a Cu–Nb or Cu–V alloy ingot containing Nb or V fine dendrites dispersed in a Cu matrix is cold-worked into a final shape. The arrangement is then coated with Sn or Ga and heated initially to disperse Sn or Ga throughout the matrix and then to form a Nb_3Sn or V_3Ga layer at the interfaces between the Cu matrix and the discrete, elongated Nb or V dendrites.

inertance: the thermoacoustic term relating the inertia of an oscillating gas to the resulting pressure difference. The thermoacoustic term is analogous to the inductance in an AC electric circuit.

insulation, electrical: material covering electrical wires so that they do not short to neighboring conducting materials.

integrated stresses: this refers to the summation of stresses over (1) a range of temperature during cooldown or (2) individual turns in coils to take into account the different levels of stresses which are high on the inside turns and lesser on the outside turns.

intergranular currents: currents flowing across grain boundaries which sum up to the macroscopic (transport) current in polycrystalline materials.

intragranular currents: currents flowing within individual grains not crossing grain boundaries.

interfacial energy: the contribution to the Gibbs energy of a material, due to interfaces between constituent phases.

intergrowth: the inclusion of a second, structurally related, phase in a single crystal region of a material. Such growths, e.g., the presence of Bi-2212 phase in a Bi-2223 grain can be completely epitaxial or they can create structural disorder.

intermediate state: the coexistence of superconducting and normal conducting regions in the bulk of a type I superconductor with finite demagnetizing factor placed in a magnetic field below the critical value.

intermetallic compound: a chemical compound containing two or more elements that has a reasonably well-defined composition and some metallic properties, especially high electrical conductivity, e.g., Nb_3Sn.

internal tin process: a fabrication process for Nb_3Sn composite wires, where a combination of Nb (typically alloyed with Ta or Ti) rods and Sn (or Sn alloyed) rods are assembled in a Cu matrix so that, when cold-drawn into a wire and then heat-treated, the Sn diffuses through the matrix to react with the Nb to form Nb_3Sn.

inverse problem: estimation of model-based description of source configuration from measured data.

IR: infrared radiation. Refers to light wavelengths longer than those typically visible by the human eye (> 700 nm) and up to 1 mm.

iron yoke: This is used in specialty magnets where the magnetized iron gives an additional contribution to the field generated by the coil although the iron tends to saturate at high fields. Yoke magnets are usually heavy. Iron yokes are also used in accelerator magnets to provide magnetic shielding.

irreversibility field: the field at which the properties of a superconductor become reversible. The effects of flux pinning are negligible above this field.

irreversibility line: a graph of the irreversibility field as a function of temperature that separates regions in which a superconductor behaves reversibly and irreversibly. In general, the use of superconductors is limited to operation below the irreversibility line.

irreversible magnetization: the component of magnetization, arising from bulk currents, that varies irreversibly with applied magnetic field. Irreversibility is a measure of flux pinning.

isothermal line: line in a phase space connecting points with the same temperature.

isotope effect: the inverse relation between the critical temperature and the square root of the atomic weight of the component elements in a superconductor.

JCPDS-ICDD: data base of powder diffraction patterns maintained by the Joint Committee on Powder Diffraction Standards.

jelly roll process: a fabrication process typically used for Nb_3Sn conductors, where a foil of Cu–Sn bronze and a foil of Nb are lapped and rolled spirally into a cylinder, cold-worked to a final shape and heated to form Nb_3Sn at the interfaces between the Cu–Sn and the Nb. In the modified jelly roll (MJR) process, the Nb sheet is periodically slit so that it expands into filaments on wire drawing.

joint, persistent: when lengths of superconductor are joined together it is important that the resistance in the connection region is very low; persistent jointing has a sufficiently low resistance that it allows a magnet to be operated in persistent (closed loop) mode so that it can achieve high field stability of routinely better than 0.01 ppm/h.

Josephson arbitrary waveform synthesizer (JAWS): pulse-driven Josephson voltage standard for ac voltage based on arrays of overdamped Josephson junctions.

Josephson effect, dynamic (ac): occurs when the applied dc current I exceeds critical current I_c of the Josephson junction (weak link between two superconducting electrodes) and therefore is carried by sum of the superconducting and normal current components: $I=I_S+I_N$, which have to oscillate in anti-phase since the normal component provides voltage drop $V = I_N R_N$ (here R_N is normal resistance of the junction) forcing Josephson phase ϕ to change in accordance with the Josephson effect stating relation between the voltage drop and the time derivative of the phase as follows: $eV = \hbar d\varphi / dt$. Thus, the superconducting current component $I_s(\varphi)$ [in many cases, this is $I_c sin(\varphi)$], the normal current component I_N, and the voltage drop V oscillate with angular frequency $\Omega = e\underline{V} / \hbar$, where $\underline{V}$ is dc component of the voltage drop.

Josephson effect, static (dc): occurs when the applied dc current $I < I_c$ is carried via Josephson junction (weak link between two superconducting electrodes) by the only superconducting (Cooper pair) current component I_S. This component is a 2π-periodic function of the phase difference ϕ (called as Josephson phase) of the wave functions of the Cooper pair condensates in the electrodes; in the simplest and mostly often case, this is sinusoidal dependence $I_S = I_c \cdot \sin(\phi)$, where I_c is the maximum superconducting current, i.e., critical current of the junction.

Josephson effects: effects occurring when two superconducting electrodes are weakly connected such that their wave functions of Cooper pair condensate overlap only partly, with Cooper pair (superconducting) and quasiparticle (normal) current components flowing between these electrodes. The effects determine relations between the flowing superconducting current and the phase difference ϕ of the wave functions, as well as between the time derivative of the phase difference and the voltage across the weak link.

Josephson energy: the characteristic energy scale associated with a Josephson junction.

Josephson junction: a two-terminal device in which two superconducting electrodes are only weakly electrically coupled and Josephson effects occur.

Josephson junction, biepitaxial: planar high-T_c grain boundary junction, in which the grain boundary (GB) is created by first depositing on single crystal substrate a patterned very thin seed level growing with in-plane crystal axes at high angle to those of the substrate.

Josephson junction, grain boundary: high-T_c junction in which coupling occurs via high-angle grain boundary (GB) in epitaxial film deposited on single crystal substrate with such boundary.

Josephson junction, intrinsic: Josephson junction, stacks of which intrinsically exist between conducting Cu–O planes within extremely anisotropic single crystals of BSCCO or another such high-T_c compound. Transport current occurs in c-axis direction.

Josephson junction, locally damaged: single thin-film layer high-T_c junction, in which the weak link between planar electrodes is created by locally damaging the crystal lattice, for example, by focused ion beam (FIB) writing.

Josephson junction, ramp: planar junction over a ramp between electrodes on two substrate surface levels, where the ramp material assures weak coupling between electrodes; presently high-T_c device, historically also low-T_c device.

Josephson junction, step edge: high-T_c junction in which coupling occurs via one or two high-angle grain boundary(ies) in epitaxial film deposited on single crystal substrate across a step having one or two sharp edges; a grain boundary forms only on a sharp edge.

Josephson junction, tunnel: Josephson junction in which weak coupling between electrodes occurs by Cooper pair and quasiparticle tunneling through a thin insulating layer.

Josephson junction capacitance: capacitance *C* between junction electrodes.

Josephson junction critical current: current between junction electrodes, I_0 or I_c, above which voltage appears across the junction.

Josephson junction (normal) resistance: resistance R_N or R_n that appears when current through the junction exceeds the critical value I_0 or I_c.

Josephson junction, SIS, SNS, SINIS, SIFS, and the like: acronyms indicating the layer sequence in thin-film Josephson junctions (S—superconductor, I—insulator, N—normal metal, F—ferromagnet).

Josephson voltage standard: quantum standard enabling the reference of the unit of voltage, the volt, to physical constants. Josephson voltage standards are based on Josephson junction arrays consisting of tens of thousands of Josephson junctions fabricated in thin-film technology; different types: dc Josephson voltage standard, programmable Josephson voltage standard for ac voltage, pulse-driven Josephson voltage standard for AC voltage.

Josephson photomultiplier (JPM): a device based on a Josephson junction that can convert a microwave photon into a measurable electrical signal.

Josephson parametric amplifier (JPA): JPAs and traveling wave parametric amplifiers (TWPAs) that incorporate a pumped Josephson nonlinearity to achieve parametric gain; capable of achieving added noise levels approaching the quantum limit.

Joule-Thomson cryocooler: a recuperative type (steady pressures and flow) of cryocooler in which cooling occurs at the cold end during irreversible expansion of a non-ideal working fluid through a resistive flow impedance.

JT: Joule–Thomson, referring to the Joule–Thomson effect, cooling of a fluid by isenthalpic expansion (constant enthalpy).

keystone stranded conductor: a compact stranded conductor of trapezoidal geometry.

Kikuchi pattern: a diffraction pattern formed by inelastically scattered electrons in TEMs and SEMs (see *EBSD*) that can be used to determine crystallographic orientations. Rapid automated indexing of these patterns can produce high-resolution crystallographic orientation maps.

Kim (critical state) model: a critical state model, proposed by Kim, Hempstead, and Strnad for the magnetization process in type II superconductors. The model assumes that the critical current density is inversely proportional to the internal field.

Kramer's law: the scaling law for pinning force density proposed by E. J. Kramer. It describes the pinning behavior of Nb_3Sn at higher fields.

lattice defect: local disorder in a crystallite. Can be a point or extended defect. Impurities are often considered lattice defects.

Laves phase compound: a generic term for intermetallic compounds with a composition of AB_2. HfV_2, and ZrV_2 have this type of crystal structure.

layered perovskite compound: perovskite is a generic term for oxide compounds of composition ABO_3, where A and B refer to metallic elements, and O is oxygen. The oxygen ions form an octahedral structure. YBCO, BSCCO, and TBCCO belong to this general crystal structure.

LEED: low-energy electron diffraction.

lithography: the process of imprinting patterns on a material to define its shape or to define that of a subsequent material that will be deposited on the first. Employed for image definition as well as for the development of complex multilayer structures.

liquid metal current collector, LMCC: a device which uses liquid conductive medium as the mean of transferring electric currents between moving and stationary parts of electric machines. Its function is similar to that of slip rings, but much higher surface speeds and current densities are achievable compared to conventional solid brushes. Low melting point alloys or metals are generally used.

LNA: low-noise amplifier.

London equations: equations describing the magnetic properties of superconductors derived from the phenomenological theory of F and H London.

London equations: equations describing the magnetic and electrical properties of superconductors derived from the phenomenological theory of Fritz and Heinz London.

Lorentz force, F: the electromagnetic force experienced by electrons moving in a magnetic field, with a direction perpendicular to both the electron motion and the magnetic field. In the case of a superconductor, the Lorentz force acts on fluxoids in a direction perpendicular to both the transport current and the applied magnetic field.

low-angle grain boundary: a boundary typically of less than 15° between adjacent grains in a bulk or thin film superconductor.

LSCO: La–Sr–Cu–O cuprate compound.

LTS: low-temperature superconductor. Not strictly defined but typically refers to the alloy and intermetallic superconductor classes discovered before the advent of HTS in 1986, with a highest T_c of 23 K for Nb_3Ge. Superconductor classes with highest T_c values between 23 and 77 K are sometimes referred to as intermediate-temperature superconductors.

LUMO: lowest unoccupied molecular orbital.

macroscopic quantum tunneling (MQT): MQT refers to tunneling out of a metastable potential well in a Josephson junction and is measurable through switching current distributions.

magnetic pattern (diffraction pattern, Fraunhofer's pattern): Josephson effect dependence of the critical current on an externally applied magnetic field.

magnetostriction: the length changes of a material due to an applied magnetic field.

Maddock's stability criterion: a modification of Steckly's stability criterion that takes into account thermal conduction along the superconductor length, resulting in a reduction in the amount of normal conducting metal surrounding the superconductor.

magnetic field strength, *H* (or μ_0H): the intensity of a magnetic field at a point, measured in ampere per meter, A/m in the SI system of units.

magnetic flux, Φ: the product of the magnetic induction, *B*, through a surface and the area of the surface, *A*.

magnetic induction, B: magnetic flux density or magnetic flux per unit area. In a material, *B* is the average of the magnetic field on a microscopic scale.

magnetic moment: the moment of a current loop of area δS carrying a current *i* is $m = i\delta S$. The magnetic moment of a body with no current across its boundaries is $m_o = \Sigma\, i\delta S$. This can also be written $m_o = 1/2\int r \times j\mathrm{d}V$, where *r* is the radius vector from an arbitrary origin, and *j* is the current density.

magnetic pattern: the Josephson effect dependence of the critical current on an externally applied magnetic field.

magnetic separation: the removal of small magnetic particles from a non-magnetic slurry under the influence of a gradient in magnetic field. Also used to separate materials of different magnetic susceptibilities.

Majorana fermions: quantum particles that are their own antiparticles.

magnetic shielding: the shielding of magnetic field by high permeability or superconducting materials.

magnetization, *M*: the magnetic dipole moment per unit volume.

magnetization ac loss: an ac loss due to the purely magnetic hysteresis behavior of a hard superconductor carrying no current, which arises because of magnetic flux pinning combined with the movement of vortices when exposed to a time-varying magnetic field.

Matching field: the field strength where the average spatial distribution of pinning defects matches the separation of vortices in the vortex lattice. This gives rise to a "fishtail" or "second-peak" effect in a magnetization–field (*M–H*) plot, where the magnetization passes through a second peak with increasing field, located at the matching field.

matrix (of composite superconductor): normal conductor in which superconducting wire filaments are embedded for the purpose of stabilization.

MBE: molecular beam epitaxy.

Meissner effect: the expulsion of magnetic flux from a superconductor as the temperature is lowered through the critical value.

melt processing: fabrication of refractory bulk or thick-film superconductors from a partially or fully molten state.

MEMS: microelectromechanical systems.

MeV: energy unit, 1 MeV = 1.6×10^{-13} J.

MG: melt grown.

microcooler: cryocooler of maximum dimensions less than ~100 mm.

microstrip: a conducting strip separated from a conducting ground plane by a dielectric layer or substrate. Used for microwave signal transmission in planar circuits.

minimum propagation zone (MPZ) theory: a superconducting magnet stabilization theory which defines the magnet cooling conditions to prevent the propagation of the normal zone, and hence quenching, by estimating the magnitude of localized disturbances in the superconductor.

MIS: metal–insulator superconductor.

mixed matrix (of composite superconductor): matrix composed of more than one component.

mixed state: the state of coexistence of superconducting and normal conducting regions in the form of a vortex lattice in a type II superconductor placed in a magnetic field of magnitude between the lower and the upper critical fields.

MJJ: magnetic Josephson junction.

MMC: metallic magnetic calorimeter, a cryogenic radiation detector.

MMM: mechanical micro-machining.

MOCVD: metal-organic chemical vapor deposition.

MOD: metal-organic deposition.

monolithic conductor: a composite conductor containing superconductor and stabilizer material, and possibly reinforcement and insulating materials, contiguously assembled with one another to form a solid structure that allows no relative motions of the components.

MOI: magneto-optical imaging, the visualization of magnetic flux using a polarized light microscope combined with magneto-optical Faraday rotation indicating films.

MQE: macroscopic quantum effect(s).

MQE: minimum quench energy. The disturbance below which a wire or coil remains stable and thus will not quench.

MPMG: melt process melt growth.

MRI: magnetic resonance imaging.

MSDS: material safety data sheet consisting of extensive documentation of the safety and environmental hazards associated with a material.

MSR: magnetically shielded room—man sized magnetic shielding to walk in.

MTG: melt-textured growth.

MTTF: mean time to failure.

MuMETAL: a high nickel content alloy with a high relative magnetic permeability that is used for magnetic shielding.

mutual inductance: the property of a coil or circuit to oppose change in the current of another coil to which they are coupled.

μSR: muon spin rotation (for measurements in a transverse field) or muon spin relaxation (for measurements in zero or longitudinal field).

NbN: niobium nitrate. The cubic δ-phase of NbN can have a T_c as high as 17.3 K.

NbTi: a commonly used abbreviation for alloys of Nb–Ti that have chemical compositions close to Nb-47wt%Ti. This is the most widely used superconductor.

neutron diffraction: structural determination for a crystal or polycrystalline material employing a beam of neutrons, especially useful for light elements, and for magnetically ordered materials.

neutron irradiation: the exposure of superconductors to neutron irradiation is usually done in a fission reactor, but only "fast neutrons", i.e., those with energies greater than 0.1 MeV, are able to modify the defect structure of the material on the scale of the coherence length ξ and are therefore useful for the study of flux pinning. Alternative neutron sources are based on the D–T reaction and produce neutrons with an energy of 14 MeV but reach only comparatively low fluences.

***n*-value:** a measure of the quality of superconducting wires. It relates to the rise of the short sample critical current transition.

noisy intermediate-scale quantum technology (NISQ): approach to quantum technology in which technological primitives are used without further error correction, hence involving physical noise.

normal (conducting) state: the thermodynamic state in which a superconducting material no longer exhibits any superconducting characteristics.

normal zone: a region in a conductor or winding in which the superconductor has transformed to the normal state.

normal zone propagating velocity: the velocity at which the envelope of a normal zone advances along a conductor or through a winding during a quench.

nSQUID: negative-inductance superconducting quantum interference device.

nucleate boiling: a phenomenon in which the rate of cooling becomes so large as to cause the formation and detachment of bubbles of vapor at a cooled surface. Nucleate boiling gives way to film boiling as the rate of cooling is further increased.

occupational medicine: is concerned with the protection of the health of people in the workplace, the prevention of occupational injuries and disease and related environmental issues.

occupational safety: the study and implementation of controls specific to workplace safety.

operating current: current for the intended designed magnetic field induction.

operating temperature: temperature for the intended designed magnetic field induction.

OPIT: oxide powder in tube.

optimal (or optimum) doping, p_o: the carrier (usually hole) density $p = p_o$ that optimizes T_c in HTS cuprates; it can be achieved by varying oxygen contents or by cation substitution; the states with $p < p_o$ are called *underdoped* and with $p > p_o$, *overdoped*.

order parameter: thermodynamic quantity defined in GL theory. The order parameter is a measure of the density of Cooper pairs and is roughly proportional to the superconducting energy gap.

organic superconductor: an organic material which exhibits superconductivity under appropriate conditions.

overdoping (overdoped): carrier concentration which exceeds optimal doping, and T_c is thereby reduced below its maximum value.

overpressure processing: a modification of HIP for processing oxide-powder-in-tube (OPIT) conductors with a total pressure ≤ 20 MPa and in a flowing atmosphere to better control oxygen partial pressure.

oxide superconductor: an oxide which exhibits superconductivity under appropriate conditions.

pancake coil: a flat, 2D coil wound from a conductor unit length (i.e., superconducting tape, cable, or cable-in-conduit conductor).

parallel core conductor: conductor constructed with a central core of parallel-laid wires surrounded by one layer of helically laid wires.

Paschen's law: provides the breakdown voltage between electrodes as a function of pressure and separation distance for a given medium.

passive shimming: the magnet is shimmed using ferromagnetic materials such as iron or permanent magnet plates. In contrast, active shimming is achieved using a coil and requires a power supply.

PCB: printed circuit board.

peak effect: the occurrence of a peak in a plot of magnetic moment or critical current against magnetic field for a type II superconductor, usually at a field significantly greater than the lower critical field.

penetration depth, λ: the penetration depth of magnetic field into the surface of a superconductor in the Meissner state.

peritectic reaction: the formation of a single solid phase from a liquid and a different solid phase at a fixed temperature (the peritectic temperature).

permeability: the ratio of B to H in a linear material.

persistence: see *drift*.

persistent current switch: a short across the terminals of a superconducting magnet, which can be made either superconducting or normal, that allows the magnet to operate in persistent mode.

persistent mode: this is a state when the current in a magnet circulates in the loop formed by the magnet and the switch. The current power supply is removed. This is the case for most MRI and NMR magnets where a temporal stability of the magnetic field is required.

phonon: a lattice vibration such as that used to couple electrons in a Cooper pair.

pinning center: a defect in a superconductor at which fluxoids are pinned. Defects that act as pinning centers include various lattice defects, precipitates, grain boundaries, twins, etc.

pinning force density, F_p: pinning force per unit volume of pinning centers preventing the movement of fluxoids in a type II superconductor.

PIT: powder in tube. A processing technique for the manufacture of wires and tapes. For the specific case of using oxide powders is called OPIT.

PLD: pulsed lased deposition.

PLM: polarized light microscope.

pO_2: partial oxygen pressure.

poloidal magnet: a magnet used in a Tokamak fusion reactor to generate a field perpendicular to the toroidal field. The axis of the coil is coincident with that of the torus. In a Tokomak, the poloidal field is used to shape the plasma and maintain stability of confinement.

pool cooling: a cooling method for superconducting magnets or conductors by direct immersion in liquid helium.

potting: impregnation of a coil with a filling or a bonding material.

powder metallurgy process: a fabrication process that uses precursor powders as a starting point in the fabrication of a component and usually involves one or more subsequent mechanical processing steps.

precursor compound/powder: the starting or raw material/s from which a given material composition is made.

prepreg: stands for a pre-impregnated material composite which consists of a cloth or felt material impregnated with partially cured epoxy resin. It is used as an interleave between layers. The coil is then cured in an autoclave.

proximity effect (PE): penetration of superconducting electrons into normal conducting regions adjacent to superconducting regions. Describes phenomena that occur when a superconductor (S) is placed in contact with a "normal" (N) conductor.

pseudogap: a gap, or partial gap, in the normal-state electronic density of states near the Fermi level. A pseudogap occurs in underdoped cuprate superconductors.

pulse tube cryocooler: a regenerative type (oscillating pressures and flows) of cryocooler that has no displacer at the cold end but uses a resistive impedance or an inertance tube at the warm end to provide a necessary phase shift between flow and pressure oscillations. The pulse tube is an empty tube with sufficient volume to provide a thermal buffer between the high- and low-temperature ends.

PVD: physical vapor deposition.

QCP: quantum critical point.

QP(s): quasiparticle(s), unpaired or normal electrons in a superconductor.

quadrupole coil: a coil consisting of four saddle coils arranged in a quadraxis symmetry to generate a magnetic field with quadrupolar components.

qubit: quantum bit, a unit of quantum information in quantum computing (QC).

qubit, superconducting: a qubit implemented using superconducting electronic circuits.

quantum critical point: location of a phase transition occurring at $T = 0$ K and therefore mediated only by quantum fluctuations, not by thermal fluctuations.

quench: an abrupt and uncontrolled loss of superconductivity.

quench protection: means utilized to minimize the impact of quench, i.e., enable the magnet to quench safely by diverting current away from the quenching coils.

quench stresses: stresses developed during a quench as per coil.

quench temperature: temperature developed during a quench as per coil.

quench voltage: voltage developed during a quench as per coil.

quenching: rapid cooling of a material from elevated temperature.

Rabi frequency: frequency of oscillation of the population of quantum states under resonant drive. The Rabi frequency is proportional to the amplitude of the driving field.

RABiTS: rolling-assisted biaxially textured substrate.

racetrack coil: a form of coil in accelerator magnets to guide a beam, so called because it follows the shape of a racetrack with two parallel straight sections join by semicircular curves at each end.

radial stress: stress on the outer edge of a material or a coil. Radial stress is a consequence of hoop stress.

RAM: random access memory.

Raman scattering: a light scattering technique in which small frequency or wavelength shifts are measured.

rapid quench: a design concept in which the normal zone propagation velocity perpendicular to the conductor is enhanced artificially to protect the winding from burnout. In this process, the stored energy is dissipated more uniformly throughout the winding, which produces a more rapid growth of the normal zone, and hence a more rapid quench.

react and wind method: a processing technique for the fabrication of a superconducting coil. In this, a conductor containing component elements of a required superconductor is wound into a coil after having been heat-treated to form the superconductor.

recuperative cycle: a refrigeration (or engine) cycle that uses a steady flow and steady low- and high-pressures in the system.

recuperator: a type of heat exchanger with separate flow channels for the low- and high-pressure flow streams and used in recuperative cycle cryocoolers.

regenerative cycle: a refrigeration (or engine) cycle that uses oscillating pressure and flow in the system and a means to provide a desired phase between the two. Helium gas is usually the working fluid.

regenerator: a type of heat exchanger used in regenerative cycle cryocoolers that has one flow channel filled with either stacked screens or packed spheres with sufficient heat capacity to absorb and give up heat during a half-cycle from the oscillating working fluid (usually helium gas).

relaxation time, T_1: the characteristic timescale for a qubit excited state to decay to the ground state.

residual resistivity ratio (RRR): the ratio of the normal-state electrical resistivity at 293 K to that at 4.2 K, however, for superconductors to be in the normal state, the lower temperature must be above the transition temperature so the low-temperature resistivity is measured just above the critical temperature, *Tc*.

residual resistivity ratio (RRR): the ratio of the electrical resistivity at room temperature, 273 K, to that at sufficiently low temperature such that the resistivity is dominated by impurities rather than by phonon scattering. This lower temperature is taken usually just above the critical temperature, *Tc*.

resistive magnet: a magnet made using resistive wire or disc.

resistivity, residual: finite electrical resistivity that remains in a normal metallic conductor at absolute zero temperature.

reversible magnetization: the component of magnetization that varies reversibly with applied magnetic field. The reversible magnetization is independent of flux pinning.

RHEED: reflection high-energy electron diffraction.

Rietveld refinement: a mathematical technique for improving the accuracy of the analysis of powder diffraction data from crystalline materials.

rope-lay conductor: conductor constructed of a bunch-stranded or a concentric-stranded member or members as a central core, around which are laid one or more helical layers of such members.

RQL: reciprocal quantum logic.

RSFQ circuit: superconducting digital or mixed-signal circuit based on rapid single flux quantum processing (see also *SFQ circuit*).

ruthenate: a ruthenium oxide compound, typically a perovskite, containing a combination of other cations often including Sr, Y, Ca, Mg, Co. A common example of a ruthenate is Sr_2RuO_4, which is a spin-triplet superconductor.

saddle coil: a coil wound in the shape of a saddle to generate a magnetic field in a direction perpendicular to the coil axis.

SANS: small-angle neutron scattering. A neutron scattering technique where an incident neutron beam transmits through the bulk sample volume. Neutrons scattered through their interaction with the sample are detected at low momentum transfers (small scattering angles typically less than five degrees). Experimental technique often used for studying and imaging of the magnetic flux-line lattice in the bulk of type II superconductors. See also *GISANS*.

sausaging: the development of significant filament or wire diameter variations along the length of a wire during its production. Typically found in composites that contain materials with very different mechanical properties.

scaling law for pinning force density: an empirical law describing the dependence of the pinning force density, F_p, on magnetic field, temperature, and strain.

scintillation counter: a device for detecting and measuring radiation by means of counting luminescent events produced by radiation when it strikes a radiation-sensitive substance such as a phosphor. Phosphors used in scintillation counters include zinc sulfide, sodium iodide, various liquids, and organic phosphors.

screening: shielding the magnet from any adverse effects of external disturbances from moving heavy objects such as lifts, trams, trains, cars, and trucks.

SDW: spin density wave.

second-peak effect: often referred to as a "fishtail effect". It is a feature in a magnetization–field (*M–H*) plot, where the magnetization passes through a second peak with increasing field, often at a so-called matching field where the average spatial distribution of pinning defects match the separation of vortices in the vortex lattice.

segregation: refers to the separation of elements or phases in a material as driven by overall and local thermodynamic considerations. Typical examples are the phase segregation of Y-211 in Y-123 or grain boundary segregation of elements often found in heat-treated materials.

self-field: magnetic field generated by the sample. The largest contribution usually stems from critical currents.

SEM: scanning electron microscope.

SFDR: spurious-free dynamic range.

SFES: superconducting flywheel energy storage.

SFQ circuit: superconducting digital or mixed-signal circuit based on single flux quantum (SFQ) processing.

shadow evaporation: a technique for fabricating small-area tunnel junctions involving evaporation of aluminum from two different angles through a nanoscale shadow mask.

shielding factor: the factor by which the magnetic field is attenuated due to magnetic shielding. The shielding factor is defined as the ratio of the magnetic field without shielding to the magnetic field inside the magnetic shielding. Usually, shielding factors are given for each spatial direction.

shim coil: a coil to compensate an inhomogeneity in a magnetic field generated by a main coil.

SHPM: scanning Hall probe microscope.

sintering: the process of densification of inherently porous material including compacted powders by heating to a temperature typically of between 90 and 95% of the melting point of the material.

sitting stresses: a result of iron in the floor which exerts a force on the magnet coils.

skin effect: an effect in which the intensity of an applied ac electric or magnetic field falls off exponentially with depth from the surface of a substance. It is characterized by a skin depth $d = (2/\mu_0\sigma\omega)^{1/2}$, where σ is the electrical conductivity, and ω is the angular frequency.

SMES: superconducting magnetic energy storage.

SMG: seeded melt growth.

SNSPD: superconducting nanostripe/nanowire single-photon detector.

sol–gel: a sol is a fluid colloidal solution usually formed from inorganic or organometallic precursors. Tends to provide intimate contact between the reagents upon gelation and requires subsequent thermal processing to form the desired materials.

solenoid coil: a coil wound helically around an axis using a wire, strand, or cable to generate an approximately uniform magnetic field at its center.

sorption compressor: a compressor operating on the basis of cyclic adsorption and desorption of a fluid on a sorbent such as activated carbon.

specific heat capacity: the heat required to raise the temperature of unit mass (or volume) of a substance by 1 K.

split pair coil: a pair of solenoid coils with a common axis but separated by a distance.

sputtering: application of energetic ions from the ambient atmosphere in a vacuum system as typically generated by dc or rf fields to remove material from a target for deposition of thin films.

SQA: superconducting quantum array. A uniform periodic structure composed of identical superconducting (SQUID-like) cells with a linear voltage response to applied magnetic signal.

SQUID: superconducting quantum interference device. A device used to detect extremely weak magnetic fields using the Josephson effect and the effect of macroscopic quantum interference in superconducting loop containing one Josephson junction (ac SQUID) or two Josephson junctions (dc SQUID). A combination of the two SQUIDs in one device (a two-loop device containing three junctions) named bi-SQUID can be used to obtain highly linear voltage response.

stabilization: a design concept in which quenching of the superconducting state is prevented.

stabilizer: a metal, but not necessarily the matrix, in electrical contact with a superconductor, to act as an electric shunt in the event that the superconductor reverts to the normal state.

Steckly's stability criterion: a superconducting magnet stability criterion which states that a superconducting magnet is in a stable condition if the heat evolved in the normal superconducting metal on quench is less than that removed from the surface of the metal by the coolant. Thermal conduction is taken into consideration essentially only for the direction perpendicular to the superconductor.

Stirling cryocooler: a regenerative type (oscillating pressure and flows) of cryocooler that uses an oscillating piston to produce an oscillating pressure in the entire system and an oscillating displacer in the cold finger to provide the proper phase between pressure and flow at the cold end to produce cooling.

STJ: superconducting tunnel junction.

STM: scanning tunneling microscope.

stranded conductor: a conductor composed of a group of wires, usually twisted together, or of any combination of such groups of wires.

strain: change in dimension of a material divided by its starting length, usually expressed as a percentage (on a scale of 1 to 100) or in microstrain, $\mu\varepsilon$, (on a scale of 1 to 1,000,000). Strain is measured using a variety of methods. Strain can be positive when the material is expanding (under tension) or negative when the material is contracting (compression).

stray field: see *fringe field*.

stress: the ratio of load divided by cross-sectional area of the contact area. It is a result of strain but one cannot directly measure stress. Stress is a three-dimensional property and is usually expressed in the form of a tensor.

stress effect/strain effect: a change of superconducting properties due to a mechanical or electromagnetic stress/strain applied to a superconductor.

superconducting magnetic bearing: low loss bearing based on the stable repulsive force that occurs between a permanent magnet and a superconductor.

superconducting transition: the combination of values of temperature, *T*, current density, *J*, and magnetic field, *H*, at which a transition from the superconducting state to the normal state takes place.

superconductivity: a property of many elements, alloys and compounds in which their electrical resistivity to the flow of dc electrical current becomes zero.

superfluid density: the volume density, n_s, of Cooper pairs in a superconductor. The superfluid density characterizes the rigidity of the superconducting state to fluctuations in phase. Because $\lambda^{-2} \sim n_s/m^*$, where λ is the London penetration depth, and m^* is the effective electron mass, then λ^{-2} is often loosely referred to as the superfluid density.

superinductor: a compact superconducting circuit capable of achieving low loss, high impedances at microwave frequencies relevant for superconducting qubits.

surface code: a scheme for quantum error correction and fault-tolerant computation that uses topological properties on a surface.

surface diffusion process: a fabrication process for compound superconductors such as V3Ga and Nb3Sn, where a tape of V or Nb is first dipped into a Ga or Sn bath, and heat-treated at high temperature to form V_3Ga or Nb_3Sn. A second option is to hold the bath at the reaction temperature so that the V_3Ga or Nb_3Sn are formed during the immersion process.

surface pinning: pinning of magnetic flux at the surface of a superconductor. The interface between superconducting filaments and the matrix in a multifilamentary conductor provides an effective source of surface pinning.

surface impedance: the ratio of the electric field, *E*, to the magnetic field, *H*, (measured in ohms) at the surface of a conductor at microwave frequencies.

surface superconductivity: superconductivity within a region approximately equal to the coherence length of the surface of a type II superconductor that persists beyond the upper critical field, H_{c2}, up to a magnetic field termed H_{c3} $(= 1.69H_{c2})$.

susceptibility, effective: the ratio of the total magnetic moment per unit volume of a sample to the applied magnetic field, H_0.

susceptibility, magnetic, χ: the ratio of magnetization *M* to the field *H*. $\chi = M/H$.

SWaP: size, weight, and power.

switch: see *persistent current switch*.

switching current distribution (SCD): in underdamped Josephson junctions, switching from the superconducting to the finite voltage state is a stochastic process resulting in a distribution of current values at which switching from superconducting to normal state occurs.

synchronous machine: ac motor or generator operating at a constant speed of rotation, determined by the frequency of the three-phase supply to the stator and the number of pole pairs.

tape conductor: a conductor constructed in the form of flat ribbon or strip.

TEM: transmission electron microscope.

temperature absolute: (1) temperature measured in Kelvin (K) on the thermodynamic scale; (2) temperature measured from absolute zero (–273.16°C). The numerical values are the same for the Kelvin and the ideal gas temperature scales.

temperature, critical, T_c: the maximum temperature below which a superconductor exhibits superconductivity at zero magnetic field and current.

transition-edge sensor (TES): a cryogenic detector that takes advantage of the strong temperature dependence of the resistance of a superconducting film at its superconducting transition to measure the increase in temperature of a particle or flux of particles, especially photons.

thermal expansion: the length change of a material due to a temperature change.

thermal conductance: the amount of heat that passes, in unit time, through some link between two heat reservoirs that differ by 1 K. The thermal conductivity of a substance is the thermal conductance of a link of that substance multiplied by the length of the link and divided by the cross-sectional area. Thermal conductance can be strongly temperature dependent.

thermal conductivity: the ability of a material to conduct and transfer heat.

thermocouple: a device to measure temperature based on the potential difference that occurs between different metals in contact under controlled conditions.

Thouless energy: a characteristic energy scale of diffusive disordered conductors.

three-component conductor: a composite conductor composed of a superconductor and two different matrices. An example of this type of conductor is an NbTi-based conductor with copper and Cupronickel matrices, the former for thermal stabilization and the latter for the reduction of coupling losses.

three-stage extrusion process: a fabrication process for NbTi conductors where a composite produced by a two-stage extrusion process is again divided into lengths which are again put together side by side in a can and then extruded for a third time.

TlBCCO: Tl–Ba–Ca–Cu–O.

toroidal magnet: a closed winding constructed such that the planes of individual equispaced turns or winding sections lie along the radii of a cylinder whose axis is outside the turns or sections. A line joining the centers of the turns or sections forms a closed circle with its center on the axis of the cylinder. The solid shape thus formed is a toroidal or torus. The best-known examples are the windings used to generate the steady field required for plasma confinement in a Tokamak fusion reactor.

training effect: an effect where a magnet quenches below its designed current or magnetic field but repeated energizations progressively increase the current or magnetic field.

transmission line: a linear conductor for the transmission of either electrical power (dc or ac) at low frequencies or an electrical signal at microwave and rf frequencies.

transport current loss: an ac loss due to the combined effect of a dc or ac transport current with an ac magnetic field.

transposed conductor: a composite conductor in which filaments or strands are plaited together to occupy different distances from the conductor axis in a regular arrangement along its length.

transposition length: the region in a transposed conductor in which a filament or strand returns to its original relative position.

trapped (magnetic) flux: the magnetic flux retained in a superconductor when the applied magnetic field is reduced to zero.

tube process: a fabrication process typically for Nb_3Sn composite conductors, where a Cu tube containing Sn bars is inserted into a Nb tube which is then further inserted in a Cu tube to form a composite. Such composites are inserted into a larger Cu tube, cold-worked into a final shape and heated to diffuse the Sn into the matrix and then to form a Nb_3Sn layer at the interfaces between the Cu and Nb tubes.

tunneling: the transition of a charge carrier from an energy state in one metal to an unoccupied state in a second metal separated from the first by a thin insulating sheet.

twin planes: planes defined by shared lattice points between adjacent crystallographic twins.

twins: partial symmetrical displacement of crystallographic planes between adjacent regions of a crystalline material which give rise to a periodic structure of parallel lines when viewed in a microscope.

twist: the number of turns per unit length of a filament or strand about a conductor axis.

twisted conductor: a composite conductor in which the filaments or strands form helices about the conductor axis.

twist pitch: the length over which a filament or strand returns to its original relative position in a twisted conductor.

two-fluid model: the separation of charge carriers into normal electrons and superelectrons in a superconductor below its transition temperature. The relative ratio of these types of carriers varies with temperature.

TLS (two-level system): defects commonly found in amorphous dielectric layers capable of absorbing microwave energy at low temperatures, thus impacting superconducting qubit performance.

transmon: superconducting qubit design consisting of a Josephson element shunted by a large capacitance with minimal sensitivity to charge noise.

type I superconductor: a superconductor in which superconductivity with perfect diamagnetism appears below the critical magnetic field, H_c, but disappears above H_c.

type II superconductor: a superconductor in which superconductivity appears with perfect diamagnetism up to the lower critical magnetic field, H_{c1}, persists in a mixed state for the magnetic field range between H_{c1} and the upper critical field, H_{c2}, and disappears above H_{c2}. Surface superconductivity within a region approximately equal to the coherence length of the superconductor persists beyond the upper critical field, H_{c2}, up to a magnetic field termed H_{c3} ($= 1.69H_{c2}$).

underdoping (underdoped): carrier concentration which is less than optimal doping and T_c is thereby less than its maximum value.

unidirectional conductor: a conductor constructed with a central core surrounded by more than one layer of helically laid wires, with all layers having a common lay direction, with an increase in length of lay for each successive layer.

unilay conductor: conductor constructed with a central core surrounded by more than one layer of helically laid wires, with all layers having a common length and direction of lay.

UV: ultraviolet. Refers to light wavelengths between approximately 10 and 400 nm.

vacuum impregnation: the impregnation of a coil or component with embedding medium such as epoxy resin under vacuum to protect it from mechanical or electromagnetic forces. In this process, the medium infiltrates the smallest recesses of the winding or open pores so that no air bubbles remain.

VAP(s): vortex–antivortex pair(s).

variational quantum eigensolver (VQE): family of quantum algorithms that involve variational calculus on a classical computer and a cost function prepared on a quantum computer.

Vicker's hardness: a measure of the resistance of the surface of a material to plastic deformation. Given by the ratio of the applied force to the projected cross-section of the area indented into the sample surface.

VLSI: very large-scale integration (of electronic devices/components).

voltage breakdown: the minimum voltage which causes electrical conduction across a gap or an insulator through a spark or acing.

voltage isolation: voltage required to ensure that no arcing occurs between the coils and any conducting surface, and this can be achieved by either distance or insulation.

von Mises stresses: a combination of the components of the stress tensor, with the hydraulic components subtracted. It is a measure of the maximum stress to which a component is subjected.

vortex lattice melting: the transition of an ordered array of fluxons (a vortex lattice) to a state of lower order under the influence of temperature or magnetic field.

VSM: vibrating sample magnetometer.

washboard potential: in the RSJ model of a Josephson junction, the relation between the potential and the superconducting phase difference between the two electrodes.

WDX: wavelength dispersive X-ray analysis (also WDS).

WHH relation: relation describing the temperature dependence of *Hc2* as described by Werthamer, Helfand, and Hohenberg.

wind and react method: a processing technique for the fabrication of a superconducting coil, where a conductor containing component elements of a required superconductor is wound into a coil and subsequently heat-treated to form the superconductor.

Y-123: Yttrium barium copper oxide compound with general formula $YBa_2Cu_3O_{7-x}$.

YBCO: Yttrium barium copper oxide. Various compositions exist with different ratios of Y:Ba:Cu including Y-123,Y-124, and Y-247, however, YBCO is commonly used as shorthand for Y-123, which has a general formula of $YBa_2Cu_3O_{7-x.}$

Young's modulus: the ratio of mechanical stress to strain for a material undergoing linear, elastic deformation.

zero boil-off: this applies to a magnet which does not lose any cryogen with time or during operation, notably helium. This is because a cryocooler recondenses any boiled-off helium back into the magnet environment.

ZFC: zero field cooling: The procedure in which a superconductor is cooled below its critical temperature when there is no field applied to it.

(July 1st, 2020)

Reprinted, with modifications and additions, and rearranged in alphabetical order, with permission from Cryogenics, Vol 35, Wada H *et al.* Terminology for Superconducting Materials, 1995, Elsevier Science Ltd, Oxford, England.

Index

Note: Locators in *italics* represent figures and **bold** indicate tables in the text.

E

F

I

J

K

L

M

www.ingramcontent.com/pod-product-compliance
Lightning Source LLC
Chambersburg PA
CBHW080129220326
41598CB00032B/5006
9780367687670